Atomic numb

Symbol

At

	IB	IIB	IIIA	IVA	VA	VIA	VIIA	VIIIA
								2 **He** Helium 4.003
			5 **B** Boron 10.81	6 **C** Carbon 12.01	7 **N** Nitrogen 14.01	8 **O** Oxygen 16.00	9 **F** Fluorine 19.00	10 **Ne** Neon 20.18
			13 **Al** Aluminum 26.98	14 **Si** Silicon 28.09	15 **P** Phosphorus 30.97	16 **S** Sulfur 32.06	17 **Cl** Chlorine 35.45	18 **Ar** Argon 39.95
28 **Ni** Nickel 58.71	29 **Cu** Copper 63.55	30 **Zn** Zinc 65.38	31 **Ga** Gallium 69.72	32 **Ge** Germanium 72.59	33 **As** Arsenic 74.92	34 **Se** Selenium 78.96	35 **Br** Bromine 79.90	36 **Kr** Krypton 83.80
46 **Pd** Palladium 106.4	47 **Ag** Silver 107.9	48 **Cd** Cadmium 112.4	49 **In** Indium 114.8	50 **Sn** Tin 118.7	51 **Sb** Antimony 121.8	52 **Te** Tellurium 127.6	53 **I** Iodine 126.9	54 **Xe** Xenon 131.3
78 **Pt** Platinum 195.1	79 **Au** Gold 197.0	80 **Hg** Mercury 200.6	81 **Tl** Thallium 204.4	82 **Pb** Lead 207.2	83 **Bi** Bismuth 209.0	84 **Po** Polonium (210)*	85 **At** Astatine (210)*	86 **Rn** Radon (222)*

63 **Eu** Europium 152.0	64 **Gd** Gadolinium 157.2	65 **Tb** Terbium 158.9	66 **Dy** Dysprosium 162.5	67 **Ho** Holmium 164.9	68 **Er** Erbium 167.3	69 **Tm** Thulium 168.9	70 **Yb** Ytterbium 173.0	71 **Lu** Lutetium 175.0
95 **Am** Americium (243)*	96 **Cm** Curium (247)*	97 **Bk** Berkelium (247)*	98 **Cf** Californium (251)*	99 **Es** Einsteinium (254)*	100 **Fm** Fermium (257)*	101 **Md** Mendelevium (258)*	102 **No** Nobelium (259)*	103 **Lr** Lawrencium (260)*

CHEMISTRY

GENERAL, ORGANIC, BIOLOGICAL

CHEMISTRY
GENERAL, ORGANIC, BIOLOGICAL

Jacqueline I. Kroschwitz
Formerly of Kean College of New Jersey

Melvin Winokur
Bloomfield College

McGraw-Hill Book Company

New York St. Louis San Francisco Auckland Bogotá
Hamburg London Madrid Mexico
Montreal New Delhi Panama Paris São Paulo
Singapore Sydney Tokyo Toronto

*This book was set in Times Roman by General Graphic Services.
The editors were Stephen Zlotnick, Sibyl Golden, and Susan Hazlett;
the designer was Joan E. O'Connor;
the production supervisor was Phil Galea.
New drawings were done by J & R Services, Inc.
The cover art was executed by George Kanelous.
Von Hoffmann Press, Inc., was printer and binder.*

CHEMISTRY: General, Organic, Biological

3 4 5 6 7 8 9 0 DOCDOC 8 9 8

ISBN 0-07-035535-5

Library of Congress Cataloging in Publication Data

Kroschwitz, Jacqueline I.
Chemistry: general, organic, biological

Includes index.
1. Chemistry. I. Winokur, Melvin. II. Title.
QD31.2.K77 1985 540 84-12224
ISBN 0-07-035535-5

In loving memory
Irene Mizerski Kroschwitz
(1917–1983)

CONTENTS

PREFACE

Aware that we were writing for a captive audience, we have striven to satisfy our desire to teach essential chemical principles while maintaining student interest and answering their questions "Why must I take chemistry?" and "How does it relate to my career goals?" This textbook is designed for students preparing for careers in the allied health professions such as nursing, x-ray technology, inhalation therapy, and medical technology and for students of nutrition, agriculture, and forestry.

We believe we have achieved our goal of writing a text that covers all of the appropriate topics in a manner that is clear and interesting for our audience. As we wrote each chapter we never lost sight of what our students *need to know* about chemistry; we also tried to provide interesting applications that they might *like to know*. Consequently most chapters contain material dealing with specific topics in health, medicine, food, nutrition, agriculture, or household chemistry. These applications are integrated into the general flow of the text so students can see how they relate to basic chemical principles, and are set in color type for easy identification. Connections are drawn between chemical principles and physiological phenomena or other processes of everyday interest. This enhances the likelihood of understanding and maintains student interest. In further efforts to make the textbook "student friendly" we use a conversational tone, abundant illustrations, some humor, color plates, and we encourage students to visualize microscopic phenomena.

However, we have not sacrificed an appropriate level of chemical rigor in basic areas such as molecular shape, intermolecular forces, equilibrium, solutions, acid-base chemistry, enzymes, and energy transfer in metabolism. We have emphasized unifying concepts such as the tendency to minimum energy, the importance of weak interactions—particularly hydrogen bonding, and the relationship of structure-shape to physiological function. Consequently, we hope our students finish the course surprised to know how much pure chemistry they learned as they pursued an understanding of the principles of interest and relevance to them.

Most students enter this course with little or no background in chemistry or mathematics. Since many are phobic toward mathematical concepts, we have taken nothing for granted whether it be prior scientific knowledge, computational skill, or prior experience in deduction. We offer complete, logical explanations which include all necessary steps in the deductive reasoning process, rather than assume the student already has the ability to make deductive leaps. This approach gives students the opportunity to glimpse the complete development of scientific thought, which is usually not their natural way of thinking, and to develop their own reasoning skills as they explore increasingly complex phenomena.

Math anxiety is somewhat relieved by the use of the Unit Conversion Method whenever possible and appropriate. When other arithmetic operations, such as logarithm manipulations in pH calculations, are needed, they are explained and illustrated and not assumed as prior knowledge. Discussions of moles, stoichiometry, and molarity are more complete but not more sophisticated than in most texts of this kind. Sample exercises are worked out in careful stepwise detail. No knowledge of algebra is necessary to understand this text.

Summarizing the convenient features and learning aids in the text:

Color Highlighted Sections discussing practical applications

Color Photographic Inserts depicting chemical phenomena and applications most easily visualized in color

Sample Exercises carefully worked out in stepwise detail

Problems interspersed within the chapter for students' self-testing of concepts explained

Numerous Problems at varying levels of difficulty at the end of each chapter

Answers to all in-chapter problems and to alternate end-of-chapter problems

Stepwise Procedural Rules or guidelines for significant manipulations; for example, see the procedure for naming alkanes in Chapter 12

Numerous Tables to organize and summarize information

Illustrations as an integral part of explanations

Cross-referencing to encourage review and inform students of "coming attractions"

Chapter Accomplishments in sufficient detail so that students know what performance is expected

Fully-defined Key Words that are boldface in the text and italicized in the index for ready location of definitions

The order of the major topics of this textbook is basically traditional—(1) general chemistry (Chapters 1–11), (2) organic chemistry, (3) biological chemistry. However, to achieve the integration necessary to appreciate physiological processes, organic compounds are introduced early along with covalent bonding, and used in illustrations of molecular shape, intermolecular forces, and colligative properties. Carbohydrates are discussed in Chapter 15 immediately after alcohols, aldehydes, and ketones in Chapter 14; this establishes immediate relevancy to the student studying the alcohol and carbonyl functional groups and reinforces the similar chemistry of these groups. Lipids and proteins follow the discussion of esters and amides. We have used this organization in our lectures for years and have found that students vastly prefer the interweaving.

For those who prefer a strictly traditional order, Chapters 12, 13, 14, 16, and 17 cover the organic functional groups; Chapters 15, 18, 19, and 20 cover biomolecules. Chapter 21 is devoted to enzymes, and Chapters 22–25 cover metabolism. That is, only carbohydrates are interspersed in the organic section and coverage of them could be delayed without loss of continuity.

A complete set of supplements designed to help you and your students accompanies this text:

Instructor's Manual/Test Bank summarizes chapters, provides all answers to problems, and contains approximately 750 short-answer and multiple-choice examination questions.

Study Guide (by Erwin Boschmann and Charles Baker) contains chapter reviews, new term lists, chapter outlines, worked-out study problems, and self-tests.

Chemistry in Action: A Laboratory Manual for General, Organic, and Biological Chemistry (by Erwin Boschmann and Norman Wells) consists of 41 class-tested experiments, including background information, step-by-step procedures, and tear-out report sheets.

Instructor's Manual to Accompany Chemistry in Action explains how each experiment ties in with the text and contains further information about materials needed and helpful hints about procedures.

MicroExaminer, a computerized version of the examination questions in the Instructor's Manual/Test Bank for use on microcomputers.

Transparencies of line drawings and tables in the text.

We earnestly invite your comments and suggestions toward the improvement of this textbook as a learning device.

ACKNOWLEDGMENTS

We would like to acknowledge and thank the many individuals who helped and encouraged us as we developed this text. First, there are the nursing students and pre-physical therapy majors at Bloomfield College and Kean College of New Jersey. Our colleagues there also offered encouragement and we are especially grateful to Bryan Lees, who shot our photographs, and Joanne Petrin, who reviewed our problems with special care. Many typists labored over the manuscript. We especially thank Robin Wallach. Thanks also to Virginia Scardelli, Mary Ann Van Esselstine, Merry Gordon, and Peggy Carvill.

We are indebted to Erwin Boschmann of Indiana University-Purdue University at Indianapolis for his careful review of our manuscript and for providing (with coauthors) an excellent laboratory manual and study guide as accompaniments for this text. We also thank the following reviewers for their thoughtful suggestions: David L. Adams, North Shore Community College; Robert W. Batch, Canada College; Thomas Berke, Brookdale Community College; Thomas B. Brill, University of Delaware; David S. Byrd, Northeast Louisiana University; Andrew C. Dachauer, University of San Francisco; Judith E. Durham, Saint Louis University; Stanley Grenda, University of Nevada—Las Vegas; John Griswold, Cedar Crest College; Lidija Kampa, Kean College; Loretta Koechel, Molloy College; William L. Leoschke, Valparaiso University; Howard W. Lyon, University of Northern Iowa; Nancy Sandelin Paisley, Montclair State College; Irvin Rothberg, Rutgers University; William M. Scovell, Bowling Green State University; David Tuleen, Vanderbilt University; James E. Vaux, Jr., University of Pittsburgh; Wanda P. Wehner, University of Northern Iowa; and Jim Zeller, Louisiana School for Mathematics, Science, and Arts.

Special thanks are extended to Sibyl Golden, our editor at McGraw-Hill, who offered so many valuable suggestions for improvement of the text and illustrations.

Finally we acknowledge that we are exhausted, but very pleased with the fruits of our labor.

Jacqueline I. Kroschwitz
Melvin Winokur

CHEMISTRY
GENERAL, ORGANIC, BIOLOGICAL

CHAPTER ONE

MATTER

1.1 INTRODUCTION

It may sound like the boast of enthusiastic chemists, but it is impossible to appreciate fully any area of human activity dealing with the material world without some understanding of chemistry. This is because **chemistry** is the study of matter, which is everything you see around you, this book, your hand, air, water, a tree. Two characteristics define **matter:** matter occupies space and has mass.

This pervasiveness of matter makes its study an essential part of your curriculum whether that be agriculture, biology, engineering, geology, nursing, nutrition, or any medical field. We all must understand the fundamental character of the matter with which we work and have some knowledge of its interaction with other matter. How do different soils affect crop yields? Why must we include protein in our diet? What is the antidote for a poison? Why do we look like our relatives? These are all ultimately questions of chemistry.

In our increasingly complex technical world it is difficult to function as a responsible citizen without some chemical knowledge. Many environmental and quality-of-life issues cannot be fully understood in the absence of knowledge about underlying scientific principles. Thus the study of matter has become almost as important as a study of the "three R's."

1.2 CLASSIFICATION OF MATTER

As you begin this text, your view of matter resembles that of the earliest chemists who set out to study matter. That is, you probably perceive the world around you as boasting an unlimited number of different forms of matter. As it turns out, we can actually classify matter in a surprisingly small number of categories. Discovering useful classifications of matter will be our first task in this text, as it was for the earliest investigators of matter.

As citizens of the twentieth century, you have a decided advantage over the early investigators because you have absorbed some of the compiled scientific knowledge of the last 200 years in your everyday experience. For example, consider the following list of samples of matter and think about how you might group the samples into three categories based on characteristics of the samples that you have observed.

Samples of Matter (See Figure 1.1)

Steam	Aspirin
Blood	Ice
Sand	Oxygen
Water	Carbon dioxide
Mercury	Alcohol

General
OXYGEN
RITE AID
DISTILLED WATER
MERCURY
rubbing alcohol
CVS
aspirin tablets
SAND

FIG. 1.1
Matter is everything you see around you. Here are some familiar examples classified in Section 1.2. Most samples are labeled; the fire extinguisher contains carbon dioxide. Steam is not shown because of its elusiveness. *(Photograph by Bryan Lees.)*

A useful classification system that you may have chosen involves identifying these samples as *solids, liquids,* or *gases:*

Solids	Liquids	Gases
Sand	Blood	Steam
Aspirin	Water	Oxygen
Ice	Mercury	Carbon dioxide
	Alcohol	

This is the classification of matter according to *physical state*. Because you have long been aware of the characteristics of the three physical states, you now recognize solids, liquids, and gases without realizing that you are examining the definiteness of the shape and volume of the sample when you make your classification. We have discussed physical states here as a simple example of how matter can be classified into a small number of categories through observation of simple distinguishing properties, but we will save a further examination of them for Chapter 8.

1.3 PURE SUBSTANCES VERSUS MIXTURES

In chemistry, classification of matter into two broad categories, **pure substances** and **mixtures,** is more useful. All matter can be separated into these categories, based on the *constancy of composition* of the sample, which is reflected in the properties of the sample. **Pure substances** have a constant composition, or makeup, and this is demonstrated by the fact that they always have the same properties regardless of their origin. For example, copper metal has the same properties whether it is found in ore in the United States or in South America or is reclaimed from copper tubing in an old building. Other examples of pure substances are any pure metal, distilled water, table salt (uniodized), and refined sugar.

Mixtures, on the other hand, display varying properties as the proportions of the components of the mixture vary. For example, a cup of coffee is a mixture. The properties of a cup of coffee are described as strong or weak, bitter or mellow, depending on how much material has been removed from various coffee beans by water. Similarly, a piece of steak is a complex mixture. It can be juicy and tasty or dry and tasteless, depending on its particular composition. Paint is a mixture. There is no definite composition, and hence no definite properties, corresponding to paint. It can contain water or no water; it can contain yellow pigment or blue pigment or other pigments in varying amounts. These different compositions result in different solubility and color properties.

Problem 1.1 Are the following mixtures or pure substances?

a Vegetable soup
b Blood
c Pure iron metal
d A rusty nail

A second characteristic of mixtures is that their components may be separated by physical means. For example, a mixture of sand and water can be separated by filtration. This test of the separability of components is a good experiment that distinguishes between mixtures and pure substances. In Problem 1.1 you were asked to make the classification without doing any experimental testing, and this is possible because in each of the chosen examples the constancy of composition is discernible through observation or through your previous everyday experiences.

Early investigators were able to identify two categories of pure substances, **elements** and **compounds.** They made their classification by observations that some pure substances (compounds) could be changed into distinctly different materials by certain experiments, whereas other pure substances (elements) defied attempts to alter them. Table 1.1 lists the results of two experimental tests (heating and passage of an electric current) on pure substances and shows how we use the results of these tests to classify materials as elements or compounds.

Notice that there are three possible outcomes for each experiment. One possible outcome is no change. A second possible outcome is a change in physical state. This frequently happens when the sample is heated; for example, water boils, producing steam, and tin melts. This kind of change is a physical change and results in no new substance formation. Physical changes are reversible; for example, when steam is cooled, liquid water re-forms, and when the melted tin is cooled, it resolidifies. The third possible outcome is the appearance of a new substance, as observed, for example, for mercuric oxide. This is a chemical change, and a chemical change by heating in the absence of air or other substances indicates that the pure substance tested is a compound.

Not all substances can be correctly classified by these two simple experiments of heating and passage of an electric current. However, experimentation can subdivide all pure substances into elements and compounds, and these simple tests give you an idea of how it's done.

1.4 PHYSICAL VERSUS CHEMICAL CHANGE

Let us explore the distinctions between changes called physical and those called chemical, since these distinctions help in distinguishing elements from compounds.

Changes in state are examples of physical changes. A **physical change**

TABLE 1.1 Effects of Experiments on a Selection of Pure Substances

Pure Substance	Description	Effect of Heating (in a Vacuum)*	Effect of an Electrical Current	Conclusion and Classification
Mercuric oxide	Red powder	Turns black and a gas is given off.	No further testing necessary.	New materials are formed. Mercuric oxide is a *compound.*
Malonic acid	White solid	A gas is given off and a white solid with different properties is formed. Cooling does not restore the original.	No further testing necessary.	New materials are formed. Malonic acid is a *compound.*
Water	Clear liquid	Heating produces steam; cooling restores the water.	Bubbles of gas with properties other than those of steam are released.	New materials are formed. Water is a *compound.*
Tin	Shiny solid	Can be melted and resolidified unchanged.	Tin conducts electricity, but it is not changed by it.	Tin remains unaltered in the experiments. Tin is an *element.*
Iodine	Purple solid	Sublimes† and the solid re-forms upon cooling.	Does not conduct.	No change. Iodine is an *element.*
Helium	Colorless gas	Remains the same.	Unchanged.	No change. Helium is an *element.*
Table salt	White solid	Melts at a very high temperature and resolidifies unchanged.	Passing an electric current through melted table salt produces a shiny metal and a yellowish green gas.	New materials are formed. Table salt is a *compound.*
Mercury	Silvery liquid	Vaporizes at high temperature, but the liquid re-forms upon cooling.	Mercury conducts electricity, but it is not changed by it.	No change. Mercury is an *element.*

* A vacuum is an environment in which essentially no other matter except the sample is present. It is necessary to exclude other matter so that it does not interfere with our observations of the sample we are studying.

† Sublimation is the conversion of a solid into a gas.

causes no change in the basic nature or composition of pure substances in a sample of matter. Thus the freezing of water is a physical change. No change in the composition of water occurs in this process; no new kind of matter is produced. The basic nature of the matter remains the same; only the state changes; i.e., liquid water becomes solid water.

Another common example of physical change is a change in size or shape of matter. For example, tearing aluminum foil into pieces is a physical change; the chemical composition is still aluminum, but we have several smaller pieces rather than one larger one.

Mixing together two pure substances, such as iron filings and salt, *without* changing the composition of either, is also a physical change. Likewise, separating the components of a mixture is a physical change as long as the composition of the individual components remains unchanged. In Section 1.3 we mentioned separating sand and water by filtration, a physical means of separation. In this case the composition of the *mixture* is altered, but the compositions of the pure-substance components (water and sand) are *unchanged. Physical changes* **never** *affect the composition of pure substances.* No new pure substance is produced during a physical change.

In contrast, **chemical changes** produce a change in the basic nature or composition of pure substances in a sample of matter. Old substances are converted into new ones. Burning coal is an example of a process that produces chemical change. The solid black coal burns, giving off heat and light. Colorless carbon dioxide gas forms. This new material has a composition and a set of properties that differ from those of coal. Note that the new substance also has a different physical state from that of the old material. However, we do not call this a physical change, because the solid coal is not becoming liquid coal or gaseous coal. Rather, the different physical state arises because the new material that has been formed happens to exist in a physical state that is different from that of the old material, which has disappeared. That is, the new material has a different set of physical properties.

Another example of a chemical change is the electrolysis of table salt (described in Table 1.1), wherein the familiar white crystals are changed into shiny sodium metal and yellowish green chlorine gas (see Figure 1.2). *Chemical changes* **always** *affect the composition of pure substances.* A new pure substance is always produced during a chemical change.

Problem 1.2 Are the following physical or chemical changes?

a Food is digested.
b A ham is sliced.
c Perspiration evaporates.
d Hydrogen peroxide, an antiseptic and hair bleach, decomposes to form water and oxygen.

We have used the term *property* (or *characteristic*) in discussing classification. Properties are used to classify things into groups or categories, which is another way of saying that they serve to identify things. Pure substances can be identified on the basis of their chemical and physical properties.

Chemical properties are observed when a substance undergoes a chemical change, i.e, a change in composition. Flammability is a chemical property: some substances burn (a chemical change); others do not. Therefore, this property distinguishes some substances from others. Iron does not burn, but it does have the chemical property of rusting. This property uniquely identifies iron. Stainless steel, on the other hand, does not rust. Food has the chemical property of being digestible. The explosive power of TNT is a chemical property of that substance. Whenever a chemical change is described, a chemical property of the material undergoing the change is also being described.

Substances can be identified by their chemical properties, but more often we identify substances by their physical properties. **Physical properties** can be observed and measured without *chemically* changing the composition of the substance observed. Size, color, taste, odor, and physical state are some physical properties. The ability to conduct heat or electricity and flexibility or brittleness are other physical properties.

Pure substances have characteristic temperatures at which they change physical state, that is, melt or freeze and boil or condense. Table 1.2 lists the melting and boiling points of some common pure substances. Knowledge about the temperature at which physical states change enables one to predict the physical state of a pure substance at any temperature.

Note how the experimentally determined physical properties in Table 1.2

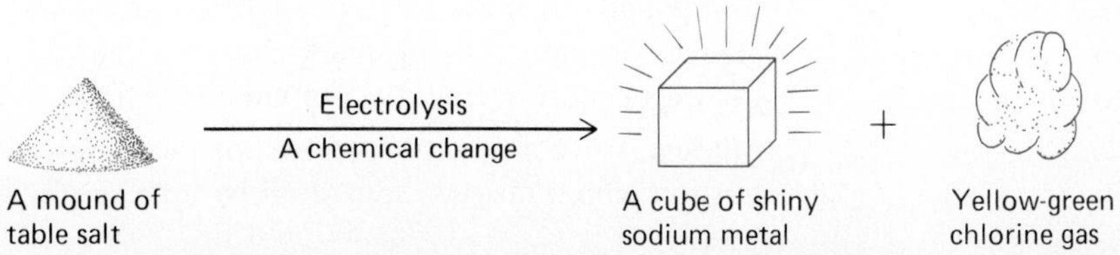
Electrolysis
A chemical change
+
A mound of table salt
A cube of shiny sodium metal
Yellow-green chlorine gas

FIG. 1.2
Chemical changes produce new materials with different properties.

TABLE 1.2 Some Physical Properties of a Few Common Pure Substances

Substance	Familiarity	Color	Odor*	Taste*	Melting Point, °C	Boiling Point, °C
Acetic acid	Vinegar is a 5% solution of acetic acid	Colorless	Pungent	Vinegary	16.6	118.1
Carbon tetrachloride	Dry-cleaning solvent	Colorless	"Cleaning fluid"	?†	−23.0	76.5
Chlorine	Swimming pool disinfectant	Yellow-green	Irritating	?	−101.0	−34.6
Iron	Nails, etc.	Gray	Odorless	"Irony"‡	1530	3000
Methane	Natural gas	Colorless	Odorless§	?	−182.5	−161.5
Oxygen	Take a deep breath!	Colorless	Odorless	Tasteless	−218.8	−183.0
Sodium chloride	Table salt	White	Odorless	"Salty"	801	1473
Sucrose	Table sugar	White	Odorless	Sweet	186 decomposes	Decomposes

* Notice how odor and taste are often described in terms of our familiarity with the substance.
† Because carbon tetrachloride is a known carcinogen (cancer-producing agent), it is unlikely that the taste will be determined.
‡ The characteristic taste of blood is saltiness plus the taste of iron.
§ A substance is added to natural gas to give it a detectable odor for reasons of safety. The methane itself has no odor.

enable you to distinguish between look-alikes. For example, salt and sugar are both white solids at room temperature and difficult to distinguish by observation alone. However, the simple experiment of heating the materials readily identifies them. Sugar melts easily (186°C), but salt does not melt in a match or bunsen burner flame. If you heat too vigorously, your identification will be made on the basis of a chemical property; i.e., sugar burns and salt does not. This simple example illustrates how experimentation can offer more information than observation alone.

Obviously, to distinguish between sugar and salt, we could taste them, the way we would in our kitchen. However, in the chemistry laboratory it is generally wise to refrain from tasting.

The properties of mixtures are not usually recorded because the properties vary as the proportion of ingredients in the mixture varies. For example, three lumps of sugar give a sweeter taste to the same volume of water than two lumps of sugar. Also, the melting point of the solution decreases and the boiling point increases as one, two, or three lumps of sugar are added.

Problem 1.3 Chloroform looks like and smells somewhat like carbon tetrachloride. Its boiling point is 61°C. How might you identify two unlabeled liquids if you already know that one is chloroform and the other carbon tetrachloride?

1.5 ELEMENTS, COMPOUNDS, AND MIXTURES

So far we have classified matter as pure substances or mixtures and have further classified pure substances as elements or compounds. This is the most useful classification of matter. Now let us distinguish among the three categories by the following definitions:

Element A pure substance which cannot be broken down into simpler substances by ordinary chemical changes.

Compound A pure substance which can be broken down into simpler substances (elements) by ordinary chemical changes. Compounds are composed of two or more elements combined in a fixed proportion.

Mixture A combination of two or more pure substances in *no* fixed proportion which may be separated by a mere physical change.

Elements

Elements are the simplest substances, the building blocks from which all other matter is made up by various combinations. Whereas there are unlimited numbers of compounds and mixtures, there are only 106 known elements. Look inside the front cover of this text and you will see the names of all the elements arranged in a so-called periodic table, which we will discuss in Chapter 4. In this text we will rarely deal with more than 35 of the elements. Table 1.3 lists the names of the elements with which you need to be familiar. Notice in Table 1.3 that each element has a symbol (see Section 1.6) and that elements are classified as either **metals** or **nonmetals.**

The classification of elements as metals or nonmetals arose when chemists first began studying the properties of elements. They observed two distinct sets of elements, each displaying certain physical properties. Within one set the elements are shiny, conduct electricity, and bend without breaking. These substances are called *metals*. You are already familiar with metals from everyday life, and you can probably describe the properties of such metals as aluminum, copper, or gold. *Nonmetals,* on the other hand, have the opposite characteristics. They lack luster, do not conduct electricity, and are brittle.

As you become familiar with the elements, it is a good idea to distinguish metals from nonmetals from the very start. This classification into two types of elements which have opposite properties is nearly as important a concept in chemistry as the concept of male and female is in biology.

It is not necessary to memorize which elements are metals and which are nonmetals, because the periodic table contains this information. Look at the

TABLE 1.3 Some Common Elements Classified as Metals and Nonmetals

METALS		NONMETALS	
Element	Symbol	Element	Symbol
Aluminum	Al	Boron	B
Barium	Ba	Bromine	Br
Calcium	Ca	Carbon	C
Chromium	Cr	Chlorine	Cl
Cobalt	Co	Fluorine	F
Copper	Cu	Helium	He
Gold	Au	Hydrogen	H
Iron	Fe	Iodine	I
Lead	Pb	Neon	Ne
Lithium	Li	Nitrogen	N
Magnesium	Mg	Oxygen	O
Manganese	Mn	Phosphorus	P
Mercury	Hg	Radon	Rn
Nickel	Ni	Selenium	Se
Palladium	Pd	Silicon	Si
Platinum	Pt	Sulfur	S
Potassium	K		
Radium	Ra		
Silver	Ag		
Sodium	Na		
Strontium	Sr		
Tin	Sn		
Uranium	U		
Vanadium	V		
Zinc	Zn		

TABLE 1.4 Elements in the Body and Their Deficiency Symptom

Element	Deficiency Symptom
MAJOR METALLIC	
Calcium	Children: Improper bone growth, rickets Adults: Bone structure breakdown, osteoporosis
Potassium	Heart arrhythmias, muscle weakness
Sodium	Nausea, muscle weakness, cramps
Magnesium	Neuromuscular irritability
Iron	Anemia
Zinc	Slow wound healing, growth retardation, delayed sexual maturity
TRACE METALLIC	
Copper	Anemia, low white blood cell count
Manganese	Deficiency unknown in humans
Molybdenum	Deficiency unknown in humans
Chromium	Impaired glucose tolerance
Cobalt	Anemia
MAJOR NONMETALLIC	
Chlorine	Improper body fluid balance, vomiting
TRACE NONMETALLIC	
Fluorine	Increased dental caries
Iodine	Goiter, cretinism
Selenium	Possible increased heart disease

periodic table and note the stepped diagonal line toward the right side. This line divides metals from nonmetals. The only exception is hydrogen, which is usually treated as a nonmetal despite its being to the left of the line.

Most of the human body's mass is composed of the nonmetallic elements carbon, hydrogen, oxygen, nitrogen, phosphorus, and sulfur. These elements are not found free in the body; rather they are combined in intricate ways in various compounds. Only the element oxygen is found in the elemental state in the body. Other nonmetals (chlorine, fluorine, iodine, and selenium) are required in the body in lesser amounts.

Many of the metallic elements are also required for good health, and their absence or insufficiency in the body leads to a deficiency symptom. Metals are always found in their combined form in the body, never free. Table 1.4 lists some elements and the symptoms experienced if the body is deficient in that element. The elements classed as major appear in measurable amounts in the body, whereas trace elements are present in only minute amounts.

Compounds

Compounds are formed when two or more elements are chemically combined in a fixed ratio. *Chemical combination* means that a chemical change has occurred, that the combining elements disappear and a new substance with entirely different properties, the compound, appears. Because the elements are chemically combined to form the compound, a chemical change is necessary to separate them. Compounds are distinguished most readily from elements by experiments such as those described in Table 1.1.

The fixed-ratio notion is similar to the idea that to make a certain cake turn out the same way every time, you always follow the same recipe, with the specified amounts of each ingredient. This *constant* fixed ratio of elements in compounds gives compounds the *constancy of composition* that is a characteristic of pure substances, as described in Section 1.3. Composition is constant in elements because there is only one component; composition is constant in compounds because the ratio of component elements is fixed. This property of a compound that its component elements always appear in a fixed ratio by weight was first noticed and formalized as the **Law of Definite Composition** (or Proportion) in 1797.

Water is a common example of a compound. The elements hydrogen, a flammable gas, and oxygen, the gas needed for life, combine in the fixed proportion 1:8 by weight to give the compound water, which has the properties with which we are familiar. Only a chemical change such as that produced by an electric current can separate water into its components, as described in Table 1.1. Table salt (sodium chloride) is another compound. It is the chemical combination of the elements sodium and chlorine in the ratio 23:35.5 by weight. We looked at the very different properties of these elements and this compound in Section 1.4 and Figure 1.2. In Section 6.13 you will see how the fixed combining ratios of elements in compounds are determined.

Mixtures

Mixtures are physical combinations of two or more elements, or of elements and compounds, or of two or more compounds. Because the combination is merely a physical mixing, the components of a mixture can be separated by physical means. For example, consider a mixture of iron filings (an element) and salt (a compound). We know this combination is a mixture because we can separate the components by a physical process. We can use a magnet to attract the iron away from the salt. Or we can place the mixture in water (which dissolves the salt) and filter, thereby separating the iron filings. The salt (saline) solution is also a mixture. The mixed compounds, salt and water, can be separated by the physical change of boiling the water away and thereby leaving the salt behind (see Figure 1.3).

In addition to physical separability, the other property of mixtures is the lack of a fixed proportion of components; i.e., the composition of a mixture is *not constant*. Consider the preceding mixture. We could have a 50:50 mix of iron filings and salt, or we could have more iron and less salt, or much salt and little iron. Similarly, in making a salt solution, we could dissolve just a pinch of salt in a large amount of water or we could dissolve several tablespoons. This is what is meant by no fixed proportion, or nonconstant composition. The lack of constant composition of a mixture shows up in its physical properties.

FIG. 1.3
Separation of mixtures by physical means. (*a*) After a magnet is applied to the iron filing–salt mixture, some of the iron has been separated from the mixture because it is attracted to the magnet while the salt is not. (*b*) Water can be separated from salt in a salt solution by boiling the water away into the air, thus leaving the solid salt behind in the beaker.

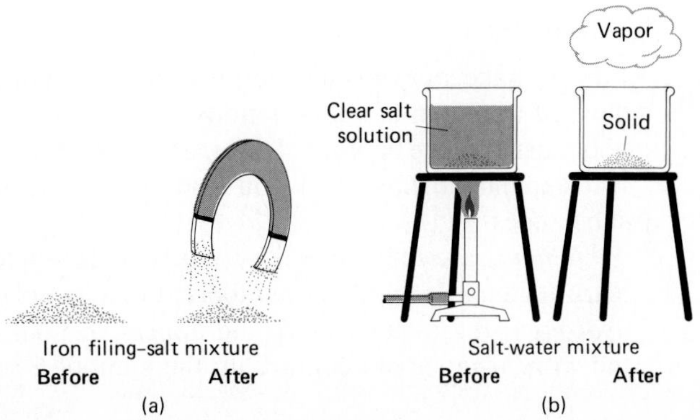
Vapor
Clear salt solution
Solid
Iron filing–salt mixture
Before
After
(a)
Salt-water mixture
Before
After
(b)

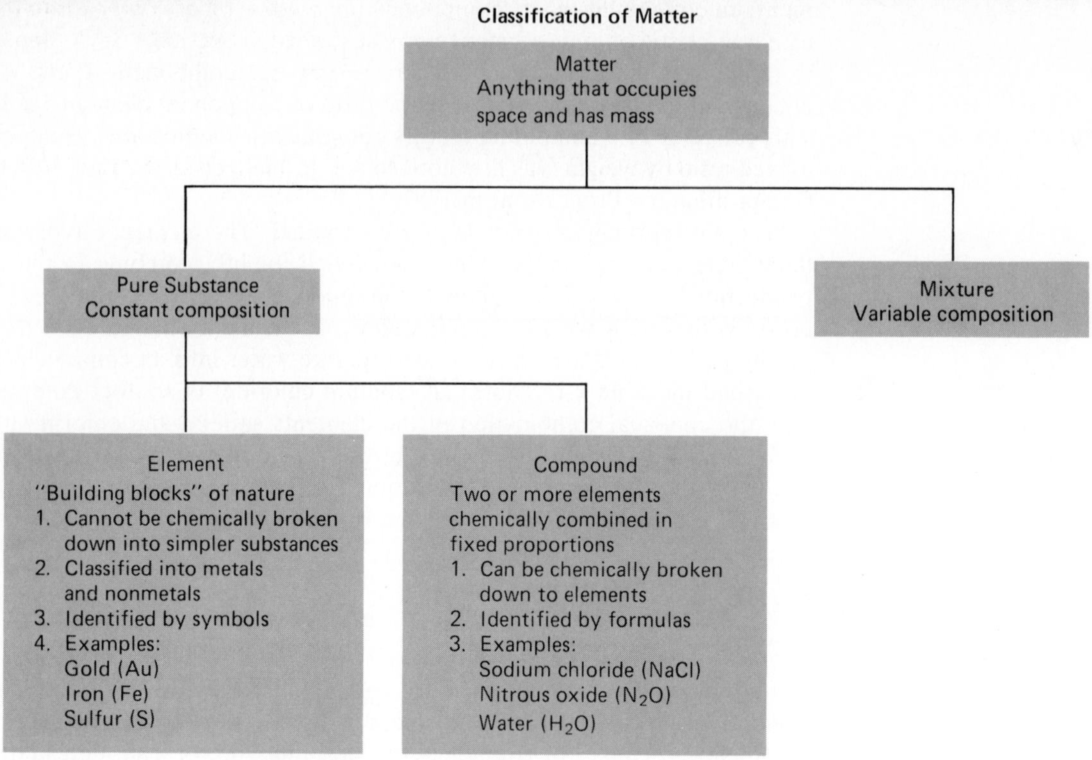
Classification of Matter
Matter
Anything that occupies space and has mass
Pure Substance
Constant composition
Mixture
Variable composition
Element
"Building blocks" of nature
1. Cannot be chemically broken down into simpler substances
2. Classified into metals and nonmetals
3. Identified by symbols
4. Examples:
Gold (Au)
Iron (Fe)
Sulfur (S)
Compound
Two or more elements chemically combined in fixed proportions
1. Can be chemically broken down to elements
2. Identified by formulas
3. Examples:
Sodium chloride (NaCl)
Nitrous oxide (N_2O)
Water (H_2O)

FIG. 1.4
All matter is divided into two large categories: pure substances and mixtures. Pure substances are further divided into elements and compounds. The properties of each subdivision are described briefly.

Problem 1.4 Classify the following as elements, compounds, or mixtures:

a Beer
b Chalk, a 40:12:48 (by weight) combination of calcium, carbon, and oxygen
c Carbon
d Pure aspirin a 27:2:16 (by weight) combination of carbon, hydrogen, and oxygen

Figure 1.4 summarizes the overall scheme most useful for classifying matter. All matter can be classified as either mixtures or pure substances. Furthermore, pure substances can be divided into elements or compounds.

Problem 1.5 Which of the materials listed in Problem 1.4 are pure substances?

1.6 CHEMICAL REPRESENTATION

For the sake of convenience and precise communications, chemists have devised a system of abbreviations for elements and compounds with which you must become familiar. To some extent your success in chemistry courses will depend on how well you understand the symbolic representation of substances.

Elements are represented by *symbols*. It would be very helpful to you to learn to think in terms of symbols. For example, when you hear the word *hydrogen*, the symbol H should appear in your mind's eye. The sound of the word *iron* should conjure up the symbol Fe along with (or rather than) the letters *i r o n*. Symbols for elements always consist of either one capital letter or one capital followed by one small letter. These are the only correct forms. Carbon is always C, never c: calcium is always Ca, never CA, ca, or cA. Sticking to these correct forms for symbols is essential if you are to write correct representations of compounds. You should memorize the symbols for all the elements in Table 1.3.

Compounds are represented by *formulas* which indicate which elements have chemically combined to form the compound. For example, the compound table salt has the formula NaCl, which immediately tells you sodium and chlorine have combined chemically. Similarly, the formula for water, H_2O, tells you water is a chemical combination of hydrogen and oxygen. In Chapter 5 you will see how these formulas also tell you the fixed proportion in which these chemical combinations occur and you will learn what the numbers in the formulas mean. For now you should realize that the formula tells you the elements that have chemically combined. Also recognize that a correctly written formula employs correct symbols for the elements and that the numbers appear as *subscripts* (i.e., they appear below the line on which the symbols are written).

1.7 ENERGY

As you learned in Section 1.1, matter is defined as anything that occupies space and has mass. In addition to mass, all matter has energy. **Energy** is the ability to do work, or we might say the ability to cause some change. The change may be in such things as the position of an object or its temperature.

There are different types or forms of energy: heat energy, electric energy, solar energy, atomic energy, to name a few. All forms of energy can be classified as either *kinetic* or *potential* energy. **Kinetic energy** is the energy of matter in motion. A moving car possesses kinetic energy which lets it do the work of carrying people or pulling a trailer. Notice that this work effects a change in position.

In chemistry, we talk more about **potential energy,** which is the energy matter possesses because of its position or composition. A rock at the top of a hill has potential energy because of its position. If it begins to roll down the hill, the potential energy is converted into kinetic energy. You contain a great deal of potential energy. Move your arm. You just changed some of your potential energy into kinetic energy. Your potential energy comes from the composition of the matter in your body. All matter contains **chemical energy,** the potential energy which comes from the way the matter is put together. Gasoline contains chemical energy, which is released when it burns. Note that chemical energy and all potential energy are *stored* energy. Stored potential energy is used by changing it to a form of kinetic energy.

The energy changes that occur within an automobile cylinder offer good examples of what is meant by a change in the form of energy. The gasoline-air mixture, rich in chemical energy, is introduced into the cylinder. Electric energy produces the spark that ignites the gasoline, converting the chemical energy of gasoline into heat energy and the mechanical energy of the moving piston. Thus the potential energy of gasoline is converted into the kinetic energy of the moving piston and moving car.

Although energy can change form, the total amount of energy in the universe is a constant. This is one way of stating the **Law of Conservation of Energy.** Another way to state this law is to say that energy is neither created nor destroyed in any change in matter. Both statements tell us that we can never make new energy and we can never destroy what we have.

What is meant, then, when we hear on the news that we must save energy? The answer is that saving energy means keeping it in a useful, potential (stored) form. The message is that we must save fuels, rich sources of chemical energy. If we burn all our fuel, the universe will still contain the same amount of energy, but it will not be in a form that is useful to us.

CHAPTER ACCOMPLISHMENTS

After completing this chapter you should be able to define **key words** and do the following:

1.1 INTRODUCTION

1.2 CLASSIFICATION OF MATTER

1. Classify common samples of matter according to physical state.

1.3 PURE SUBSTANCES VERSUS MIXTURES

2. Distinguish between the characteristics of pure substances and mixtures.
3. Classify common samples of matter as pure substances or mixtures.
4. Given experimental data, classify pure substances as elements or compounds.

1.4 PHYSICAL VERSUS CHEMICAL CHANGE

5. Distinguish between physical and chemical changes.
6. Distinguish between physical and chemical properties.
7. Given a list of samples and their melting and boiling points, classify the substances according to their physical state at a given temperature.
8. Given a description of physical and chemical properties, distinguish between samples of matter.

1.5 ELEMENTS, COMPOUNDS, AND MIXTURES

9. Classify common samples of matter as elements, compounds, or mixtures.
10. Given a periodic table, classify elements as metals or nonmetals.

1.6 CHEMICAL REPRESENTATION

11. Given the name, give the symbol of each element in Table 1.3.
12. Given the symbol, give the name of each element in Table 1.3.
13. Recognize a correctly written formula.

1.7 ENERGY

14. Distinguish between potential and kinetic energy.
15. Given a description of an energy transformation, identify the energy forms as kinetic or potential.
16. State the Law of Conservation of Energy.

PROBLEMS

1.6 What does the study of chemistry involve?

1.7 What is matter?

1.8 How is matter classified according to physical states?

1.9 How is matter classified according to constancy of composition?

1.10 How is matter classified according to whether its components are separable by physical means?

1.11 Classify each of the following as a pure substance or a mixture:

a Ice cream
b Table sugar
c Bread
d Dry ice (frozen CO_2)
e Snow
f Cola
g A dime
h Gold
i Ink

1.12 A pure blue powder, when heated in a vacuum, releases a greenish gas and leaves behind a white solid. Is the original blue powder a compound or an element?

1.13 A shiny, metalliclike substance conducts an electric current without a change in its properties. The substance is heated until it liquefies, and then an electric current is passed through the liquid again without a change in properties. Is the substance likely to be an element or a compound?

1.14 Classify each of the following as a chemical or physical property:

a Aluminum melts at 660°C.
b Aluminum forms a white, powdery oxide coating in air.

c Napthalene sublimes at room temperature.
d Mercuric oxide forms mercury and oxygen when heated.
e Butter turns rancid when left unrefrigerated in the open air.
f Water evaporates from an open container.
g Sugar dissolves in water.
h Oil burns, giving off carbon dioxide, water, and heat.

1.15 Classify each of the following as a physical or chemical change:
a Photosynthesis (CO_2 and H_2O are converted to carbohydrates in plants).
b Antifreeze boils out of a radiator.
c A dish of cherries jubilee is flamed with brandy.
d A firefly emits light.
e A nail is magnetized.
f A nail rusts.
g Leaves turn color in autumn.
h Food spoils.

1.16 The electrolysis of water produces bubbles of hydrogen and oxygen gas. Boiling water produces steam bubbles. How does the composition of the bubbles tell us that one process is chemical and the other is physical?

1.17 Pure substances can be classified into how many categories? Identify them.

1.18 Classify the pure substances in Problem 1.11 as elements or compounds.

1.19 Are there more elements or more compounds in the world? Explain.

1.20 Classify the following as elements, compounds, or mixtures:
a Methane (1 part H to 3 parts C by weight)
b Carbon particles in a hydrogen gas atmosphere
c Pizza
d Milk shake
e Zinc
f Laughing gas (14 parts N to 8 parts O by weight)

1.21 Name some properties that are useful in identifying specimens of pure substances.

1.22 What properties distinguish water from other colorless liquids such as alcohol, benzene, and acetone?

1.23 What name do we give to the ability of matter to do work?

1.24 What are the two major classifications of energy?

1.25 For each of the following, state whether the energy described is potential or kinetic:
a A book is poised at the edge of a table.
b The book is falling.
c You are walking.
d Food.
e A stretched rubber band.

1.26 Name the elements whose symbols are Cu, Hg, Pb, Sn, S, and P.

1.27 Give symbols for the elements potassium, sodium, chlorine, iron, hydrogen, silver, and gold.

1.28 Classify the elements in Problems 1.26 and 1.27 as metals or nonmetals.

1.29 Tell whether the following symbols and formulas are written correctly or not. If not, tell what is wrong.
a CH_4
b AL
c I^2R
d Zn
e h

1.30 Classify the following as elements, compounds, or mixtures:
a KNO_3
b Ca and C
c B
d CaC_2
e CO and NO

1.31 Match each of the items on the left with as many descriptions on the right as apply:

a C and P	*1* A pure compound
b $C_6H_{12}O_6$ (blood sugar)	*2* A mixture of compounds
c C_2H_5OH and H_2O (alcohol in water)	*3* An element
d A chocolate chip cookie	*4* A mixture of elements
e Oxygen dissolved in water	*5* A mixture of an element and a compound

CHAPTER TWO

MEASUREMENT

2.1 INTRODUCTION

It is not possible to function successfully in everyday life, let alone in a technical, scientific, or medical capacity, without some knowledge of weights and measures. How could we understand different drug dosages, ascertain body temperature or blood pressure, or even buy proper amounts of meat and potatoes for supper without understanding measurement?

Because the understanding of scientific concepts is often based on measurements, it is very important that we consider exactly what the process of measurement is. When you *measure* something, you are *comparing* it with some *reference standard.* For example, when you step on a bathroom scale calibrated or marked off in pounds, you are comparing your weight with the *reference standard,* the pound. The scale tells you how heavy you are *relative* to 1 lb. All systems of measurement involve this idea of comparison with a reference standard. All measurements are relative to a standard. This is an important concept which comes up repeatedly in chemistry. It is a good idea to think it through for some familiar measurements before tackling unfamiliar ones.

2.2 THE METRIC SYSTEM

The reference standards scientists use are those of the metric system, a highly logical, easy-to-use system based on multiples of 10. This system was originally developed during the French Revolution and at that time was adopted throughout the world except in English-speaking countries. However, the British Commonwealth has converted to the metric system, and as you know, the United States is finally changing also—new road signs in kilometers and food package weights in grams are just the beginning.

Your children will grow up with the metric system and will be able to skip this section of the text. Those of us who have been brought up with the English system must now learn the metric system. Around 1960, scientists slightly revised and updated the metric system and called it **SI** after the French name, *S*ystéme *I*nternational D'Unités (International System of Units).

Length, the Distance between Two Points

For **length,** the distance between two points, the reference standard is represented by a meter stick (1 meter, abbreviated m). The meter is defined as the distance light will travel in $\frac{1}{299{,}792{,}458}$ of a second. This may seem inconvenient and unwieldy, but the definition was chosen because this distance is unchangeable and readily measured with great accuracy by modern instruments. Meter sticks are made the length of this distance as accurately as possible; fortunately they are tangible. A yardstick (36 inches, abbreviated in) is somewhat shorter than a meter stick (39.37 in).

The meter is the base unit, and all length measurements could be expressed in meters. However, it is more convenient to have some larger and smaller units (just as in the English system we have inches, feet, yards, and miles). The beauty of the metric (SI) system is that all the units are related by multiples (or divisions) of 10, so the names of the units are constructed from the base unit name preceded by a prefix which tells which multiple of 10 is involved.

Prefix	**Base unit**
Tells the multiple of 10	Unit

For example,

kilo-	**meter**
Multiple of 10 is 1000	Base unit of length

Prefixes, their mathematical meanings, and their abbreviations appear in Table 2.1.

Mass, the Amount of Matter in an Object

The metric reference standard for mass is a metal cylinder made of a platinum-iridium alloy stored in a vault near Paris, France. The **mass** of this block of metal is defined as 1 kilogram (abbreviated kg). In the English system this is about 2.2 lb. The base unit of mass is the gram (abbreviated g), and from the foregoing discussion about prefixes you can see that 1 kg must equal 1000 g. Other commonly used relationships among the units of mass are

$$1 \text{ mg (milligram)} = 0.001 \text{ g}$$

$$1 \ \mu\text{g (microgram)} = 0.000001 \text{ g}$$

Scientists prefer to measure *mass,* the *fixed* amount of matter in an object, rather than *weight,* the amount of gravitational attraction pulling on an object. This is so because *mass is always the same* no matter where it is measured, whereas *weight varies.* The measured value for weight depends on the dis-

TABLE 2.1 Commonly Encountered Prefixes of the Metric System and Their Mathematical Meanings

Prefix	Abbreviation	Multiply Base Unit by: Common Number	Multiply Base Unit by: Exponential Number	Examples (Prefix + Length Base Unit)
Kilo-	k	1000	10^3	1 km (kilometer) = 1000 m
Deci-	d	1/10, or 0.1	10^{-1}	1 dm (decimeter) = 0.1 m
Centi-	c	1/100, or 0.01	10^{-2}	1 cm (centimeter) = 0.01 m
Milli-	m	1/1000, or 0.001	10^{-3}	1 mm (millimeter) = 0.001 m
Micro-	μ	1/1,000,000, or 0.000001	10^{-6}	1 μm (micrometer) = 0.000001 m

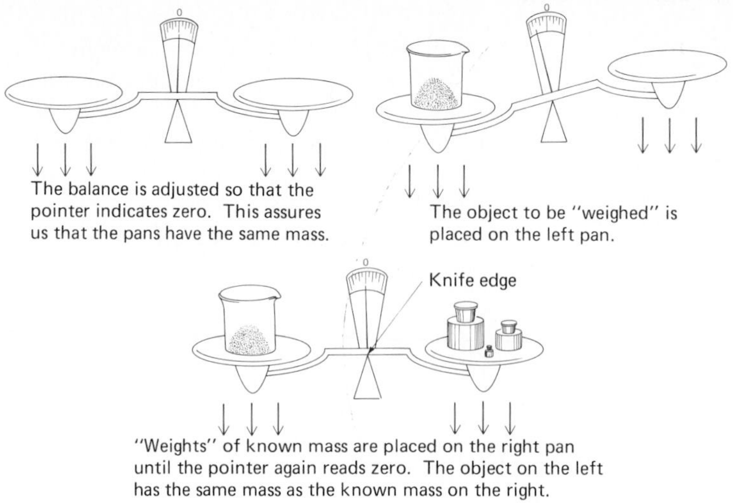
0
The balance is adjusted so that the pointer indicates zero. This assures us that the pans have the same mass.
The object to be "weighed" is placed on the left pan.
Knife edge
"Weights" of known mass are placed on the right pan until the pointer again reads zero. The object on the left has the same mass as the known mass on the right.

FIG. 2.1 The use of the two-pan balance.

Notice that gravity (↓ ↓ ↓) is pulling on *both* pans at all times. Because it affects both pans equally we do not observe the effect of gravity and can be assured that we are truly measuring mass by comparison to some reference standard.

tance the object being weighed is from the center of the earth. The weight of an object measured in Death Valley (low altitude) is slightly greater than the weight measured on top of Mount Everest (high altitude). The mass is identical in both locations.

Mass is measured by using a double-pan balance, as shown in Figure 2.1. Because gravity works on each pan equally at any altitude or on any planet, we are assured that we are actually measuring mass by comparing the unknown object with the known masses. Modern single-pan lab balances compensate for gravity in a less obvious way than does the double-pan balance in Figure 2.1, but the idea is the same. The crucial feature of a device for measuring mass is the knife-edge on which the weighing pan is balanced.

Although mass and weight are not the same thing, it is common practice to use the words interchangeably because we use the verb *weigh* to describe the measurement of either mass or weight. Because chemists always use balances to weigh things, chemists always measure mass whether it is called mass or incorrectly called weight.

Volume, the Amount of Space Occupied by a Three-Dimensional Object

The reference standard for **volume,** the amount of space occupied by a three-dimensional object, in the metric system is the liter, which is just slightly larger than a quart.[1] The *liter* is the volume occupied by a perfect cube, 10 cm on each edge. Liter is abbreviated L. Picture the cube as an empty box. We can calculate the volume of any rectangular solid in the unit cubic centimeter (cm^3) by multiplying the length times the width times the height. So, in this case,

$$\text{Volume} = 10\ \text{cm} \times 10\ \text{cm} \times 10\ \text{cm} = 1000\ \text{cm}^3$$

[1]The metric system is the precursor of the SI system. The actual SI reference standard for volume is the cubic meter, but this unit is impractical in many settings, for example, in medical laboratories. Therefore, the older unit, liter, is maintained and discussed in this text.

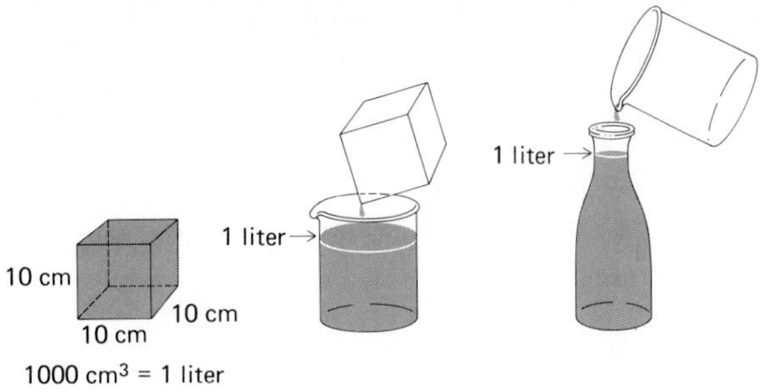
10 cm
10 cm
10 cm
1000 cm3 = 1 liter
1 liter
1 liter

FIG. 2.2
Although the shape may change as the shape of the container changes, 1 L of water always occupies the the same volume (1 L = 1000 cm^3).

Therefore, 1 L = 1000 cm^3. An alternative abbreviation for cubic centimeter often used in medicine is cc.

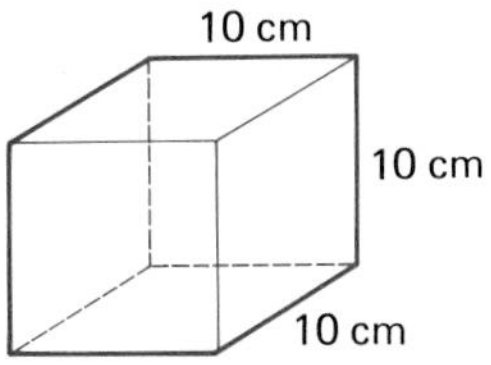
10 cm
10 cm
10 cm

Now imagine filling up the box with water. The volume of the water is 1000 cm^3, or 1 L (see Figure 2.2). If we pour the 1 L of water into different containers, it is still 1 L, despite the fact that the shape changes. Volume is an amount of space occupied and does not depend on shape.

We are most often concerned with measuring the volumes of liquids. As shown in Figure 2.3, chemists have invented several kinds of equipment to do this conveniently. Notice the hypodermic syringe, which is also used medically. These devices are all calibrated in milliliters. As the prefix tells you, 1 mL is a tiny part, namely, one thousandth of a liter (1 mL = 0.001 L), or 1000 mL = 1 L. Liters and milliliters (or cubic centimeters) are the most common volume units used by scientists and health professionals. Deciliters (dL) are also encountered in clinical reports, such as the one shown in Figure 2.4. The wine industry uses centiliters (cL). For example, the volume of wine in a standard bottle is typically 75 cL.

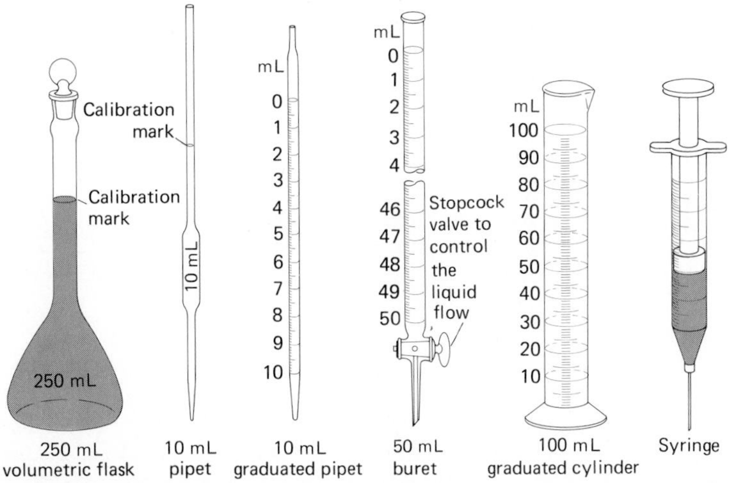
Calibration mark
Calibration mark
250 mL
10 mL
mL
0 1 2 3 4 5 6 7 8 9 10
mL
0 1 2 3 4
46 47 48 49 50
Stopcock valve to control the liquid flow
mL
100 90 80 70 60 50 40 30 20 10
250 mL volumetric flask
10 mL pipet
10 mL graduated pipet
50 mL buret
100 mL graduated cylinder
Syringe

FIG. 2.3
Various types of common laboratory equipment used to measure the volume of liquid samples.

METPATH

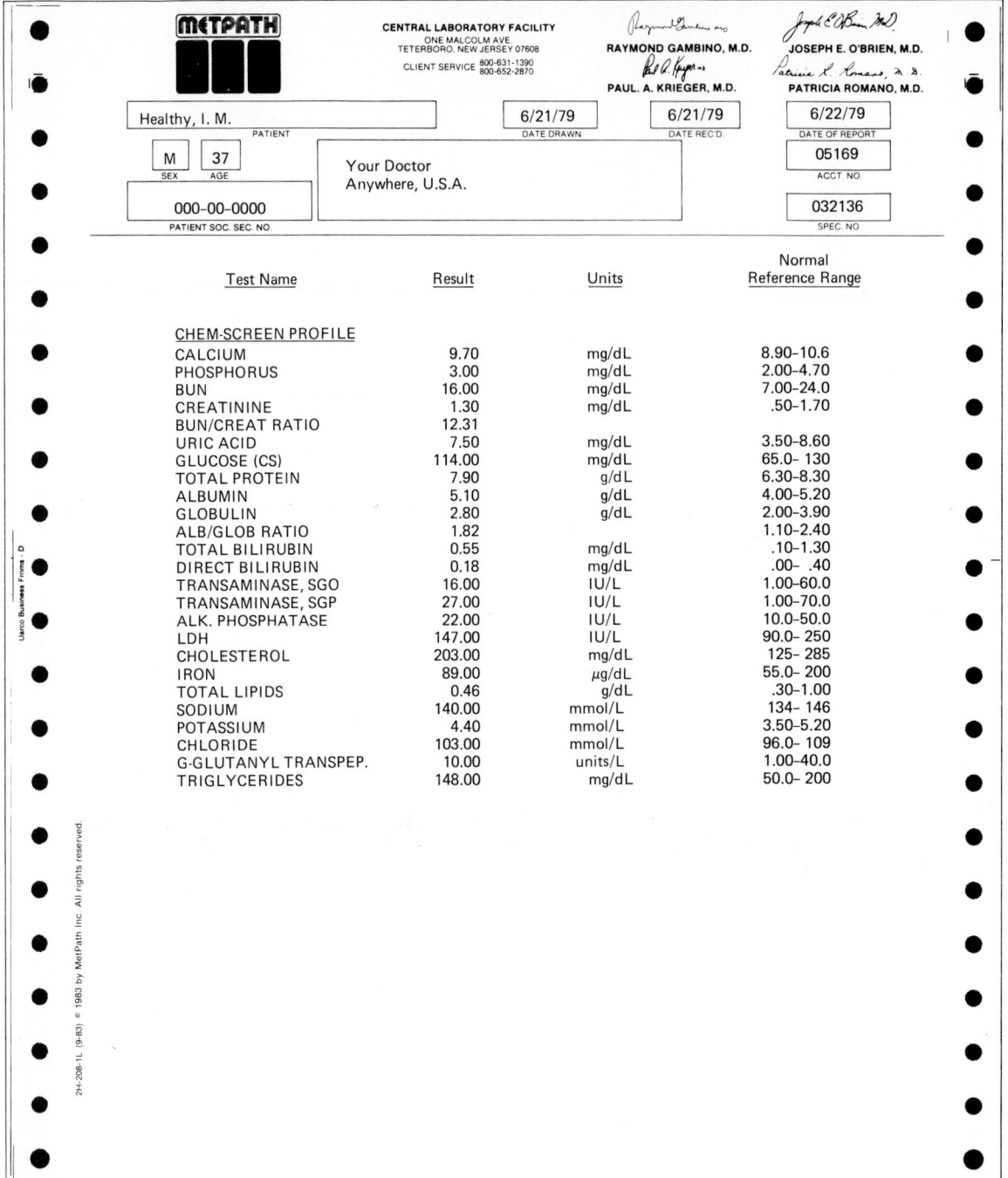

CENTRAL LABORATORY FACILITY
ONE MALCOLM AVE.
TETERBORO, NEW JERSEY 07608
CLIENT SERVICE 800-631-1390 800-652-2870

RAYMOND GAMBINO, M.D. — JOSEPH E. O'BRIEN, M.D.
PAUL. A. KRIEGER, M.D. — PATRICIA ROMANO, M.D.

PATIENT	Healthy, I. M.
DATE DRAWN	6/21/79
DATE REC'D	6/21/79
DATE OF REPORT	6/22/79
SEX	M
AGE	37
PATIENT SOC. SEC. NO.	000-00-0000
ACCT. NO.	05169
SPEC. NO.	032136

Your Doctor
Anywhere, U.S.A.

Test Name	Result	Units	Normal Reference Range
CHEM-SCREEN PROFILE			
CALCIUM	9.70	mg/dL	8.90-10.6
PHOSPHORUS	3.00	mg/dL	2.00-4.70
BUN	16.00	mg/dL	7.00-24.0
CREATININE	1.30	mg/dL	.50-1.70
BUN/CREAT RATIO	12.31		
URIC ACID	7.50	mg/dL	3.50-8.60
GLUCOSE (CS)	114.00	mg/dL	65.0- 130
TOTAL PROTEIN	7.90	g/dL	6.30-8.30
ALBUMIN	5.10	g/dL	4.00-5.20
GLOBULIN	2.80	g/dL	2.00-3.90
ALB/GLOB RATIO	1.82		1.10-2.40
TOTAL BILIRUBIN	0.55	mg/dL	.10-1.30
DIRECT BILIRUBIN	0.18	mg/dL	.00- .40
TRANSAMINASE, SGO	16.00	IU/L	1.00-60.0
TRANSAMINASE, SGP	27.00	IU/L	1.00-70.0
ALK. PHOSPHATASE	22.00	IU/L	10.0-50.0
LDH	147.00	IU/L	90.0- 250
CHOLESTEROL	203.00	mg/dL	125- 285
IRON	89.00	μg/dL	55.0- 200
TOTAL LIPIDS	0.46	g/dL	.30-1.00
SODIUM	140.00	mmol/L	134- 146
POTASSIUM	4.40	mmol/L	3.50-5.20
CHLORIDE	103.00	mmol/L	96.0- 109
G-GLUTANYL TRANSPEP.	10.00	units/L	1.00-40.0
TRIGLYCERIDES	148.00	mg/dL	50.0- 200

Uarco Business Forms - D

FIG. 2.4
Clinical Report Sheet. Notice the use of the metric units: g, mg, μg, L, and dL. IU stands for International Units; they are employed in measuring enzyme concentrations as you will see in Chapter 21. Chemists use moles (mol) and mmol as you will learn in Chapter 7. *(Report form and data courtesy of Metpath Laboratories.)*

Because 1 L = 1000 cm^3 and 1 L = 1000 mL, it must be true that 1000 cm^3 = 1000 mL. Dividing each side of the equation by 1000, it becomes apparent that 1 cm^3 = 1 mL. Thus the units abbreviated cm^3, cc, and mL all specify the same volume. The liquid contents of a cube 1 cm on a side come up to the 1-mL mark of a graduated cylinder or the 1-cc mark on a hypodermic syringe.

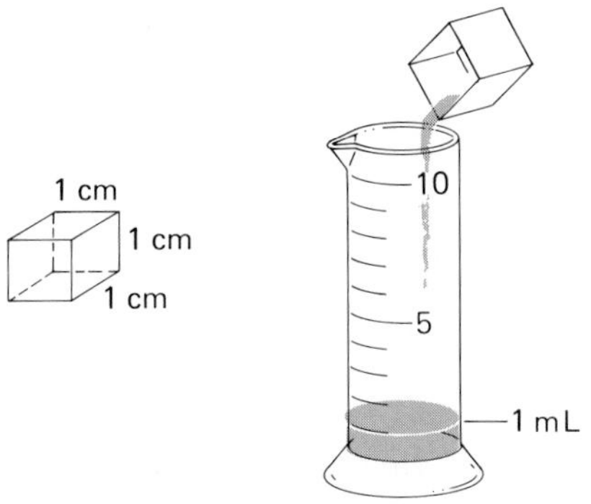
1 cm
1 cm
1 cm
10
5
1 mL

2.3 APOTHECARIES' MEASURES

You are probably familiar with the common English units of measure such as the foot, pound, and quart. Some less familiar older English units of the so-called apothecaries' system are still used by physicians and pharmacists in drug prescriptions. Of these, we most often encounter the grain (gr). A grain is a small unit, approximately one-fifteenth of a gram (1 g = 15.43 gr).

In measuring liquid volumes, larger pharmaceutical units are the familiar quart, pint, and liquid ounce (1 qt = 2 pt = 32 liquid oz). The liquid ounce is subdivided into the smaller units of fluidrams and minims (1 liquid oz = 8 fluidrams = 480 minims).

2.4 UNIT CONVERSION

If you measure the distance between the two vertical lines marked off below, the measurement you report might be 12.7 cm, 127 mm, or 5 in, depending on the ruler used and the scale used. Notice that the number reported is meaningless without reporting the proper unit.

Measure this distance:

It is frequently necessary to be able to convert the units of a measurement to different units. The method we recommend to perform these calculations is the *Unit Conversion Method*. It is well worth your time to master this method because later you can apply it to a variety of chemical problems.

Mathematical Background

The Unit Conversion Method depends on two mathematical facts: (1) any equality can be used to write a fraction equal to 1, and (2) like quantities in the numerators and denominators of fractions can be "canceled out."

A quantity divided by itself is equal to 1. For example, clearly 8 ft/8 ft = 1. The equality 1 m = 100 cm tells us that the 1 m and 100 cm represent exactly the same distance. Therefore, dividing 1 m by 100 cm is the same as dividing 1 m by itself, and therefore the fraction 1 m/100 cm = 1. Similarly, 100 cm/1 m = 1. *Any equality can be made into a fraction equal to 1; we call that fraction a* **conversion factor.** The following fraction which has a meter stick in both the numerator (100 cm) and denominator (1 m) demonstrates this visually.

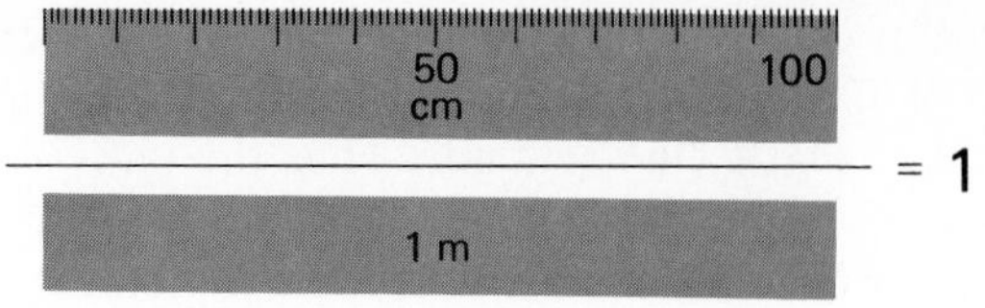
50
cm
100
= 1
1 m

SAMPLE EXERCISE 2.1 Write conversion factors, i.e., fractions equal to 1, from the following equalities:

a 1 kg = 1000 g
b 1000 cm^3 = 1000 mL

Solution

a Both 1 kg/1000 g = 1 and 1000 g/1 kg = 1 because both 1 kg and 1000 g represent an identical mass.
b Both 1000 cm^3/1000 mL = 1 and 1000 mL/1000 cm^3 = 1 because both 1000 cm^3 and 1000 mL represent an identical volume. Notice also that 1000 can be canceled out, so that 1000 cm^3/1000 mL = 1 cm^3/1 mL.

Useful equalities relating the metric and English systems appear in Table 2.2.

Problem 2.1 Write conversion factors for the following equalities:

a 1 L = 1000 mL
b 1 kg = 2.20 lb

Multiplication by 1

As you know, multiplication by 1 does not change the quantity that is being multiplied, for example 8 ft × 1 = 8 ft. However, units can be changed when multiplying by conversion factors. For example, if we multiply 8 kg by the factor equal to 1, that is, 1000 g/1 kg, we get 8000 g:

$$8\ \cancel{\text{kg}} \times \frac{1000\ \text{g}}{1\ \cancel{\text{kg}}} = 8000\ \text{g}$$

We can be certain that 8000 g is the same quantity as 8 kg because the multiplication is by 1, but we have done a *unit conversion,* i.e., changed the units from kilograms to grams.

For this simple example, of course, we could have used the following reasoning: if 1 kg is 1000 g, then 8 kg must be 8000 g. Frequently, however, unit conversions are more complex, and so we recommend the general method of problem solving that follows.

The Unit Conversion Method

The steps to be followed in reading a problem and setting up a calculation by the Unit Conversion Method are given here and applied in Sample Exercise 2.2:

TABLE 2.2 Some Equalities between the Metric and English Systems

Length	Mass	Volume
1 in = 2.54 cm	2.20 lb = 1 kg	1.06 qt = 1 L
39.4 in = 1 m	1 lb = 454 g	1 liquid oz = 29.6 mL
1 mi = 1.61 km	1 oz = 28.4 g	
	15.4 gr = 1 g	

1 Identify the **given quantity** and **unit** and write them down.
2 Identify the **new quantity** to be determined and write down the **new units** it is to have.
3 Determine the *conversion factor(s)* that will change the given into the new quantity and unit. The factor will have given units in the denominator and new units in the numerator.
4 Set up the calculation according to the following format:

$$\textbf{Given quantity and unit} \times \textbf{conversion factor(s)} = \textbf{new quantity and unit}$$

$$\_\_\_\_\_\_ \ \text{given unit} \times \frac{\_\_\_\_\_\_ \ \text{new unit}}{\_\_\_\_\_\_ \ \text{given unit}} = \_\_\_\_\_\_ \ \text{new unit}$$

SAMPLE EXERCISE 2.2

How many meters are there in 76 cm?

Solution

Follow the steps just outlined:

STEP 1 The given (old) quantity is 76 and the given unit is *centimeters.*

STEP 2 The new quantity to be determined is numbers of meters. Thus the new unit is the *meter.*

STEP 3 The equality relating meters and centimeters can be found in Table 2.1 as 0.01 m = 1 cm, or rewritten as 1 m = 100 cm. Thus the conversion factor to be used must be either 1 m/100 cm or 100 cm/1 m. Choose the one that has the given units in the denominator, which in this case is 1 m/100 cm. This conversion factor is chosen so that the given units will cancel out.

STEP 4 Set up in the proper format:

Given quantity and unit × conversion factor = new quantity and unit

$$76 \ \text{cm} \times \frac{1 \ \text{m}}{100 \ \text{cm}} = \_\_\_\_\_\_ \ \text{m}$$

Notice that the given units cancel to give the desired new unit, meters. The setup of the fractions tells you to divide 76 by 100, that is,

$$76 \ \text{cm} \times \frac{1 \ \text{m}}{100 \ \text{cm}} = \frac{76 \ \text{m}}{100} = 0.76 \ \text{m}$$

Multistep Conversions

This method can be employed for more complex conversions in which equalities may not be given directly in the tables. The idea in such a case is to examine the table and assemble all needed equalities in order to establish the one you want.

SAMPLE EXERCISE 2.3

How many milligrams are there in 0.53 kg?

Solution Follow the preceding steps.

STEP 1 The given quantity is 0.53 and the given unit is *kilograms.*

STEP 2 The new unit to be determined is *milligrams.*

STEP 3 Applying the prefixes of Table 2.1 to the base unit, gram, does not offer a direct equality between kilograms and milligrams. However, it does relate kilograms to grams and grams to milligrams, (1 kg = 1000 g and 1 g = 1000 mg). Therefore, if we use the conversion factors corresponding to these equalities, we should be able to convert kilograms to grams to milligrams. The factors are always in the form new unit/given (old) unit, so in this case we use 1000 g/1 kg and 1000 mg/1 g.

STEP 4 The setup is then

Given quantity and unit × conversion factors = new quantity and unit

$$0.53\ \not{\text{kg}} \times \frac{1000\ \not{\text{g}}}{1\ \not{\text{kg}}} \times \frac{1000\ \text{mg}}{1\ \not{\text{g}}} = 530{,}000\ \text{mg}$$

The units cancel to give milligrams, and the setup tells you to multiply 0.53 × 1000 × 1000.

Problem 2.2 How many milliliters are in 0.034 L?

With the help of Table 2.2, you can also do conversions between the metric and English systems. For example, you can now prove that the three measurements reported at the very beginning of Section 2.4 are all correct. The three values given for the distance between the two lines were 12.7 cm, 127 mm, and 5 in.

$$12.7\ \not{\text{cm}} \times \frac{1\ \not{\text{m}}}{100\ \not{\text{cm}}} \times \frac{1000\ \text{mm}}{1\ \not{\text{m}}} = \frac{12.7 \times 1000\ \text{mm}}{100} = 127\ \text{mm}$$

$$12.7\ \not{\text{cm}} \times \frac{1\ \text{in}}{2.54\ \not{\text{cm}}} = \frac{12.7 \times 1\ \text{in}}{2.54} = 5.00\ \text{in}$$

Problem 2.3 How many pounds are in 273 g?

The Bottom Line

The Unit Conversion Method is a powerful tool. Remember,

1 Whenever you have an equality, you can construct a fraction equal to 1.
2 Fractions equal to 1 can be used as conversion factors to convert units.

We will continue to use the Unit Conversion Method throughout this book.

2.5 DENSITY

Neither the mass nor the volume of a substance is a characteristic property because both mass and volume vary with the size of the sample. For example, consider two samples of gold, one a small gold nugget and the other a gold bar from Fort Knox (see Figure 2.5). Clearly the gold bar has a larger volume (occupies more space) and weighs more (has a larger mass), but you will see that the ratio of the mass to the volume, i.e., mass/volume, is a constant, characteristic property. This ratio is called the **density.**

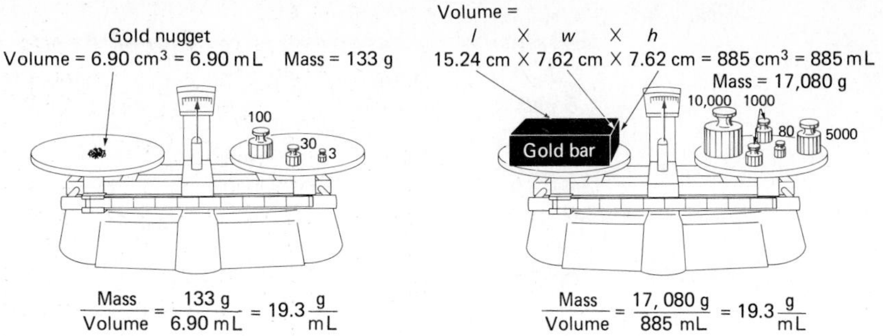
Gold nugget
Volume = 6.90 cm³ = 6.90 mL
Mass = 133 g
100
30
3
Mass / Volume = 133 g / 6.90 mL = 19.3 g/mL
Volume =
l X w X h
15.24 cm X 7.62 cm X 7.62 cm = 885 cm³ = 885 mL
Mass = 17,080 g
10,000
1000
80
5000
Gold bar
Mass / Volume = 17,080 g / 885 mL = 19.3 g/mL

FIG. 2.5 The masses and volumes of the two gold samples are very different, but the mass/volume ratio is a constant characteristic property. This property is called **density.**

Density is determined in the laboratory by measuring both the mass and the volume of a sample and dividing the mass by the volume.

$$\text{Density} = \frac{\text{mass}}{\text{volume}}$$

Because the mass is usually measured in grams and the volume in milliliters, density has the unit grams per milliliter (abbreviated g/mL).

If you were to measure the masses and volumes of the two gold samples shown in Figure 2.5, you would find very different values, as you would expect.

	Gold Nugget	Gold Bar
Mass	133 g	17,080 g
Volume	$6.90\ cm^3 = 6.90$ mL	$885\ cm^3 = 885$ mL

But consider the mass/volume ratios for the two samples:

$$\text{Density} = \frac{\text{mass}}{\text{volume}} = \frac{133\ \text{g}}{6.90\ \text{mL}} = 19.3\ \frac{\text{g}}{\text{mL}} \qquad \frac{17{,}080\ \text{g}}{885\ \text{mL}} = 19.3\ \frac{\text{g}}{\text{mL}}$$

We see that the density is a constant. A density of 19.3 g/mL is a characteristic property of any sample of gold of any size.

Problem 2.4 What is the density of dry-cleaning solvent if 15 mL weighs 24 g?

Density varies slightly with temperature. Therefore, when densities are reported, the temperature at which the measurement was done is usually noted. Table 2.3 lists the densities of some common materials.

When working in the laboratory, you will find that it is particularly useful to know the densities of liquids. When you wish to know the mass of a liquid, it is generally easier to measure the volume and use the known density to calculate the mass. Density relates mass and volume and allows the calculation of one from the other either by algebraic manipulation of the equation density = mass/volume or by regarding density as a conversion factor relating mass and volume units.

Density can be regarded as a conversion factor because the mass and volume cited represent the same sample. For example, consider the gold nugget in Figure 2.5. Because 133 g of gold and 6.90 mL of gold represent the same sample, the fraction 133 g/6.90 mL, or its reduced form 19.3 g/mL, is a fraction equivalent to 1 and can be used as a conversion factor. This is true of all densities.

TABLE 2.3 Densities of Common Materials

Solids	Density, g/mL, 20°C	Liquids	Density, g/mL, 20°C	Gases	Density, g/L, 0°C*
Gold	19.3	Water	1.00	Air	1.29
Lead	11.3	Gasoline	0.67	Oxygen	1.43
Salt	2.16	Milk	1.03	Hydrogen	0.090
Paper	0.70	Seawater	1.03	Helium	0.178
Balsawood	0.20	Blood	1.06	Carbon dioxide	1.96
Redwood	0.44	Mercury	13.6		
Rubber	1.1	Olive oil	0.92		
Ice	0.92	Alcohol	0.79		
		Vinegar	1.01		
		Ether	0.70		
		Carbon tetrachloride	1.59		

* Notice that for gases the units are grams per liter and the temperature is 0°C. This is because gases are much less dense than liquids or solids and change more dramatically with temperature. Density increases as temperature decreases.

SAMPLE EXERCISE 2.4

What is the mass of 3.0 mL of ether?

Solution Approach this by the Unit Conversion Method.

STEP 1 The given is 3.0 mL.

STEP 2 The new quantity to be determined is the mass, i.e., the number of *grams* of sample.

STEP 3 Density relates grams and milliliters. The density of ether from Table 2.3 is equal to 0.70 g/mL.

STEP 4 Set up in the proper format:

Given quantity and unit × conversion factor = new quantity and unit

$$3.0\ \cancel{\text{mL}} \times \frac{0.70\ \text{g}}{1.0\ \cancel{\text{mL}}} = \frac{3.0 \times 0.70\ \text{g}}{1.0} = 2.1\ \text{g}$$

SAMPLE EXERCISE 2.5

What is the volume of 27.2 g of mercury?

Solution STEP 1 The given is 27.2 g.

STEP 2 The new quantity is the volume, i.e., the number of *milliliters.*

STEP 3 Density relates grams and milliliters. The density of mercury is 13.6 g/mL, which can also be written as 1.00 mL/13.6 g.

STEP 4 The setup is

Given quantity and unit × conversion factor = new quantity and unit

$$27.2\ \cancel{g} \times \frac{1.00\ mL}{13.6\ \cancel{g}} = 2.00\ mL$$

Problem 2.5 What volume does 48 g of carbon tetrachloride occupy?

2.6 SPECIFIC GRAVITY

In clinical chemistry and medical laboratories the property of specific gravity, rather than density, is used. The two properties are essentially identical at room temperature. **Specific gravity** is defined as the ratio of the density of a sample to the density of water, i.e.,

$$\text{Specific gravity} = \frac{\text{density of a sample}}{\text{density of water}}$$

Because the density of water is 1.00 g/mL (from 0 to 30°C), we see that the numerical values of the specific gravity and density of a sample are the same. However, specific gravity is unitless because the units cancel out. For example, the density of ethyl alcohol is 0.79 g/mL and its specific gravity is 0.79.

$$\text{Specific gravity of ethyl alcohol} = \frac{0.79\ \cancel{g/mL}}{1.00\ \cancel{g/mL}} = 0.79$$

Densities or specific gravities can always be measured by determining both the mass and volume of a sample. Another method of determination is to use a hydrometer (see Figure 2.6). A hydrometer floats in a liquid to a depth dependent on the specific gravity (density) of the liquid; the denser the liquid, the higher the hydrometer floats. Hydrometers are frequently used to determine the densities of the liquid in car radiators or car batteries.

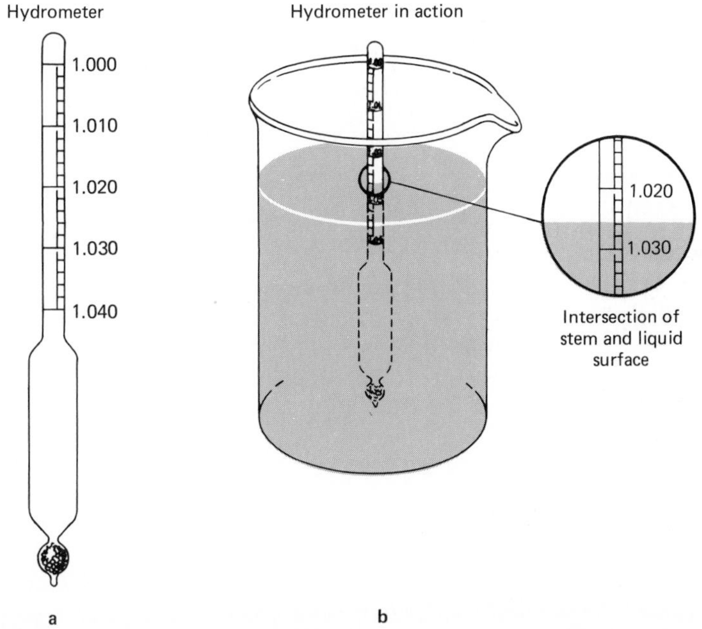
Hydrometer
1.000
1.010
1.020
1.030
1.040
a
Hydrometer in action
1.020
1.030
Intersection of stem and liquid surface
b

FIG. 2.6 (*a*) The hydrometer shown is calibrated to measure specific gravities of liquids having values greater than 1. (*b*) The hydrometer floats in the liquid with its stem submerged to some depth depending on the specific gravity of the liquid. The specific gravity of the liquid is the scale reading at the intersection of the stem and the liquid surface, here about 1.026. A hydrometer specially calibrated to measure specific gravities of typical urine samples (1.018–1.025) is called a urinometer.

The densities reflect the composition of the liquids and indicate whether more antifreeze is required in the radiator, or how reliable the battery is.

A hydrometer especially calibrated to determine the specific gravity of urine samples is called a urinometer. Healthy "morning" urine has a specific gravity in the range 1.018–1.025. Values outside the range indicate some kidney malfunction.

2.7 TEMPERATURE

Temperature is a measure of the hotness or coldness of an object. There is a difference between temperature and heat which will unfold in this section and the next. Because the United States is "going metric," weather reports now commonly give two values for the daily temperature. One is the familiar Fahrenheit temperature (°F); the other is the metric system equivalent, i.e., the Celsius temperature (°C). By choosing two fixed points, namely, the normal boiling point (bp) and freezing point (fp) of water, we can determine a mathematical relationship between degrees Fahrenheit and degrees Celsius. If we use a Fahrenheit thermometer to measure these fixed points, the values obtained are

Boiling point = 212°F
Melting point = 32°F
180° separate the two points.

If we use a Celsius thermometer, the values are

Boiling point = 100°C
Melting point = 0°C
100° separate the two points.

From these measurements we can construct an equation relating degrees Fahrenheit and Celsius. The equation reflects the difference in degree size on the two scales, i.e., the degree ratio 1.8 (180/100), and it also takes into account the displacement of the two scales, i.e., the freezing point at 0 for °C and at 32 for °F. One form of the equation is

$$°F = 1.8°C + 32 \qquad (2.1)$$

This form is convenient to use when you are given a Celsius temperature and asked to calculate a Fahrenheit temperature because the unknown (°F) is isolated.

SAMPLE EXERCISE 2.6

A scientist measures the temperature in her lab as 25°C. What is the temperature in degrees Fahrenheit?

Solution

We are *given* the number of degrees Celsius. Therefore, we substitute this number in Equation (2.1):

$$°F = 1.8°C + 32 \qquad (2.1)$$

Substituting, we have

$$°F = 1.8(25) + 32$$

To complete this calculation, we must multiply before we add because the plus sign is *outside* the parentheses. Thus,

°F = 45 + 32 = 77°

Equation (2.1) can be rearranged to a form that is more convenient for calculating degrees Celsius, given degrees Fahrenheit. Let us do the rearrangements step by step. Subtract 32 from each side:

$$\begin{array}{l} °F \quad = 1.8°C + 32 \\ -32 \qquad\qquad\quad -32 \\ \hline °F - 32 = 1.8°C \end{array} \qquad (2.1)$$

Divide each side by 1.8:

$$\frac{(°F - 32)}{1.8} = \frac{\cancel{1.8}°C}{\cancel{1.8}} = °C$$

Rewrite the equation putting degrees Celsius on the left:

$$°C = \frac{(°F - 32)}{1.8}$$

Then if you are given a Fahrenheit temperature, use this equation to calculate the temperature in degrees Celsius. After substituting the given °F value, be sure to subtract before you divide because the minus sign is *inside* the parentheses.

The general method of problem solving for temperature coversion is:

1 Identify the kind of degrees given.
2 Select a correct equation relating degrees Celsius and degrees Fahrenheit, and substitute for the degrees given.
3 Complete the calculation, being careful to multiply or add, subtract or divide in the proper order.

Notice that temperature conversion cannot be done by the Unit Conversion Method because the relationship between degrees Celsius and degrees Fahrenheit involves addition or substraction as well as multiplication or division. You must always add or subtract 32 at some point in the calculation.

Problem 2.6 Americans have been asked to save energy by keeping thermostats set at 68°F in winter. What is this temperature in degrees Celsius?

2.8 UNITS OF ENERGY

Heat is a form of energy. Temperature is an indicator of the tendency of heat energy to be transferred. Heat energy flows from objects of higher temperature to objects of lower temperature.

We saw in Section 1.7 that there are different forms of energy. Historically for chemists, the energy changes that were most useful and most easily measured experimentally were those involving heat energy. For this reason chemists traditionally use *calories* (abbreviated cal), units of heat energy, for all energy forms.[1]

One **calorie** is the amount of heat necessary to raise the temperature of one gram of water by one degree Celsius.[2]

[1]This practice is slowly being abandoned. The SI unit of energy is the joule.
[2]To be absolutely accurate, the calorie is defined as the heat required to raise one gram of water from 14.5 to 15.5°C.

TABLE 2.4 Specific Heats of Some Common Substances

Substance	Specific Heat, cal/(g·°C)
Aluminum	0.22
Copper	0.093
Ethyl alcohol	0.51
Iron	0.11
Lead	0.31
Olive oil	0.50
Silver	0.056
Table salt	0.21
Water	1.0

Notice that temperature does *not* measure heat energy. The amount of heat energy contained in a body depends on:

1 Nature of the substance
2 Mass of the substance
3 Temperature

These three items are all reflected in the definition of a calorie:

1 cal = amount of heat to raise 1 g of water by 1°C

↑ 1 g: Mass ↑ water: Nature ↑ 1°C: Temperature

For every pure substance we can measure a physical property called the **specific heat** of that substance. The specific heat tells us the amount of heat necessary to raise the temperature of one gram of material by one degree Celsius. Table 2.4 lists specific heats for some common materials. Notice that the units of specific heat are cal/(g·°C). The specific heat of water is 1.0 cal/(g·°C); that is, it takes 1 cal to raise 1 g of water by 1°C. For iron the specific heat is 0.11 cal/(g·°C); that is, it takes only 0.11 cal to raise 1 g of iron by 1°C.

SAMPLE EXERCISE 2.7

How much heat energy must be applied to raise the temperature of 10 g of water by 3°C?

Solution It requires 1 cal to raise the temperature of 1 g of water by 1°C. Therefore, 10 cal must be applied to raise the temperature of 10 g of water by 1°C, but we want the temperature to go up 3°C; therefore we need 3 times the 10 cal, or 30 cal.

We can also do this problem by the Unit Conversion Method. Follow the usual steps:

STEP 1 The given quantities and units are 10 g of water and 3°C change in temperature.

STEP 2 The new unit to be determined is calories, the unit of heat energy.

STEP 3 The conversion factor relating calories, grams, and degrees Celsius is the specific heat of the given substance, in this case water.

STEP 4 The setup is therefore

Given quantities and units	× conversion factor	= new quantity and unit
10 g of water	$\times 3°C \times \frac{1\ cal}{g \cdot °C}$	= 30 cal

In general the heat energy change for a given mass of substance m undergoing a temperature change t_1 to t_2 is given by the equation

$$m \times (t_2 - t_1) \times \text{specific heat} = \text{heat energy change}$$

$$\text{Grams} \times °C \quad \times \frac{\text{calories}}{g \cdot °C} = \text{calories}$$

Problem 2.7 How much heat energy must be applied to raise the temperature of 7 g of iron by 4°C?

2.9 FOOD CALORIES

The prefix *kilo-* is used with units of energy in the usual manner, that is, 1 kilocalorie = 1000 cal. When nutritionists speak of Calories (with a capital C, abbreviated Cal) as applied to foods, they mean kilocalories (abbreviated kcal). Because energy can change form, we can use calories and kilocalories as the units for any form of energy, even though they have been defined only in terms of heat energy.

The energy content (caloric content) in food is measured by totally burning the food and measuring the heat given off by the burning food. This amount of heat energy is exactly equivalent to the amount of energy the food can potentially provide to the body for moving muscles, conducting electrical impulses along nerves, maintaining body temperature, etc. However, not all of the energy in food is completely available to the body because of inefficiencies in digestion and absorption. For example, although fat contains 9.45 kcal/g, fat contributes only 9.0 kcal/g to the body.

Of the three major foodstuffs, fats provide the most energy per gram, approximately 9 kcal/g, or in nutritionists' notation, 9 Cal/g. Carbohydrates and proteins each provide about 4 Cal/g. Later chapters in this book are devoted to these foodstuffs and their metabolism: Chapters 15 and 23 treat carbohydrates, Chapters 18 and 24 discuss fats, and Chapters 19 and 25 take up proteins. Chapter 26 discusses nutritional aspects of the various foodstuffs. Table 2.5 shows the makeup and caloric content of some common foods.

Table 2.6 gives some approximate values for caloric use during certain

TABLE 2.5 Caloric Value of Some Common Foods

	Approximate Makeup (Percentage Composition)						
Food	**Carbohydrate**	**Fat**	**Protein**	**Water**	**Cal/g**	**Typical Serving Size**	**Caloric Content of Serving**
Apples (raw)	14	0.5	0.4	85	0.6	1 apple (~175 g)	105
Bread (white enriched)	52	3	9	36	2.8	1 slice	68
Cheese (cheddar)	4	37	28	31	4.7	1 oz	133
Chocolate	51	35	8	6	5.5	1 oz	156
Eggs	0.7	10	13	76	1.6	1 egg	80
Hamburger	0	30	22	48	3.6	$\frac{1}{4}$ lb	409
Milk	5	4	3.3	88	0.7	1 glass	160
Peanuts	22	39	26	13	5.6	1 oz	159
Sugar	100	—	—	—	4.0	2 tsp	25

TABLE 2.6 Approximate Energy Expenditure by a 150-lb Person in Various Activities

Activity	Calories Expended per Hour
Lying down or sleeping	80
Sitting	100
Driving an automobile	120
Standing	140
Domestic work	180
Walking, $2\frac{1}{2}$ mi/h	210
Bicycling, $5\frac{1}{2}$ mi/h	210
Gardening	220
Golf, lawn mowing with power mower	250
Bowling	270
Walking, $3\frac{3}{4}$ mi/h	300
Swimming, $\frac{1}{4}$ mi/h	300
Square dancing, volleyball, roller skating	350
Wood chopping or sawing	400
Tennis	420
Skiing, 10 mi/h	600
Squash and handball	600
Bicycling, 13 mi/h	660
Running, 10 mi/h	900

Source: Based on material prepared by Robert E. Johnson, M.D., Ph.D., and colleagues, University of Illinois, House and Garden Bulletin No. 232, Depts. of Agriculture and Health and Human Services.

activities. Most people's daily activities require 2000 to 3000 Cal of energy. If they take in more Calories in food than they require, they will gain weight. Taking in less Calories leads to loss of weight. Since 1 lb of body weight is equivalent to 3500 Cal, the difference between intake and outgoing energy must be 3500 Cal in order to lose 1 lb.

2.10 SIGNIFICANT FIGURES

Whenever you measure something, there is *always* some *uncertainty* associated with the measurement. The uncertainty comes about because of the nature of the object measured and the limitations of the measuring device. For example, consider the attempts at measuring the length of a common ball-point pen shown in Figure 2.7. There is an uncertainty because the shape of the pen makes it difficult to line up the ends for measurement accurately. This is an unavoidable uncertainty, but we can see that the measuring device also determines accuracy. In Figure 2.7*a*, the use of a meter stick calibrated only in decimeters allows us to say with certainty that the pen measures between 0.1 and 0.2 m. We can *estimate* that it is about 0.14 m.

In Figure 2.7*b*, using a meter stick calibrated in centimeters allows us to say with certainty that the pen measures between 14 cm (0.14 m) and 15 cm (0.15 m). We can *estimate* that it is about 14.5 cm (0.145 m).

In the first measurement (Figure 2.7*a*), we say that the measurement is good to 2 significant figures, one we know for certain and one we guessed.

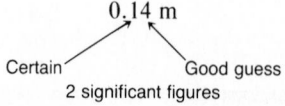

0.14 m

Certain — Good guess

2 significant figures

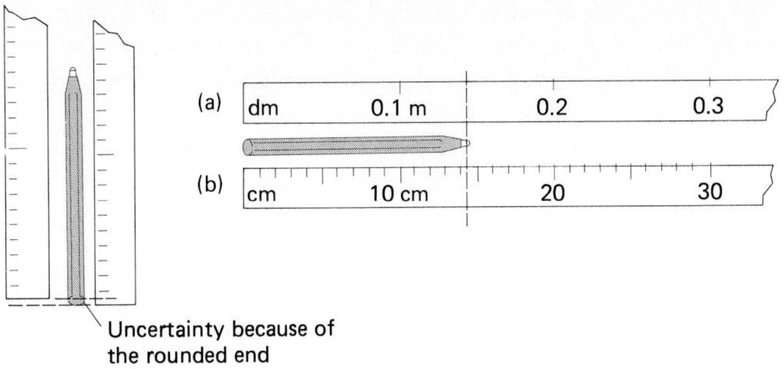
(a)
dm
0.1 m
0.2
0.3
(b)
cm
10 cm
20
30
Uncertainty because of
the rounded end

FIG. 2.7
Uncertainty in measurement. The nature of the measured object introduces uncertainty. The measuring device has limitations. (*a*) With the metric ruler, one can say with certainty that the pen measures between 0.1 and 0.2 m. We can estimate that it is 0.14 m. (*b*) With the centimeter ruler, one can say with certainty that the pen measures between 14 and 15 cm (between 0.14 and 0.15 m). We can estimate that it is 14.5 cm, or 0.145 m.

In the second measurement, using a more sensitive measuring device, we say that the measurement is good to 3 significant figures.

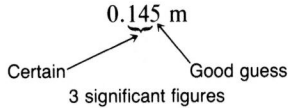

0.145 m

Certain — Good guess

3 significant figures

For any measurement, the number of **significant figures** *that can be reported is the number of figures that can be read accurately from the measuring device plus one more figure that must be estimated.*

SAMPLE EXERCISE 2.8

How many significant figures can be reported in each of the following measurements?

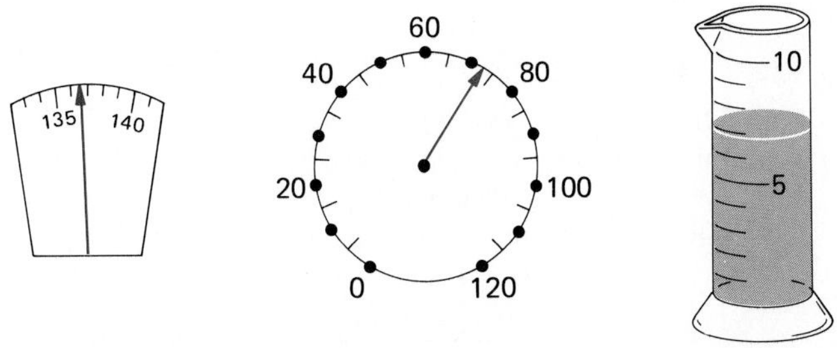
135
140
60
40
80
20
100
0
120
10
5

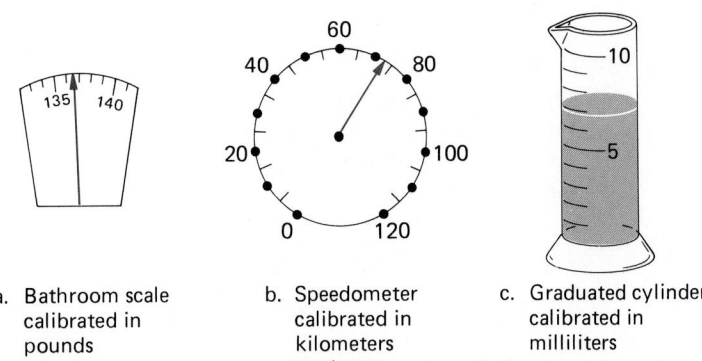

a. Bathroom scale calibrated in pounds

b. Speedometer calibrated in kilometers per hour

c. Graduated cylinder calibrated in milliliters

Solution *a* We can tell with accuracy that the weight is between 136 and 137 lb. Then we can estimate that the pointer is midway between 136 and 137 and report 4 significant figures.

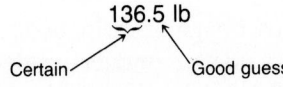

136.5 lb

Certain — Good guess

b The speed is definitely between 70 and 75 km/h. Because there are no calibration lines between 70 and 75, we must estimate the nearest kilometer. Our guess is 72 km/h. Therefore, we are allowed 2 significant figures: the 7 is certain, and the 2 is a good guess.

c The volume is between 6 and 7 mL. We estimate that it is 6.8 mL. Two significant figures are allowed.

When we actually do or see the measurement, as in Sample Exercise 2.8, it is easy to determine how many significant figures are allowed and are represented by the reported number. More often you are given a measured number and asked to tell how many significant figures it contains. To do this requires a set of rules.

Rules for Counting Significant Figures in a Given Number

1 All *nonzero* figures are significant.
2 All *zeros between nonzero* figures are significant.
3 When a decimal point is shown, *zeros to the right of nonzero* figures are significant. (When a decimal point is not shown, the number is ambiguous and the number of significant figures cannot be determined.)
4 *Zeros to the left* of the first nonzero figure are *not* significant.

An example demonstrating all the rules is

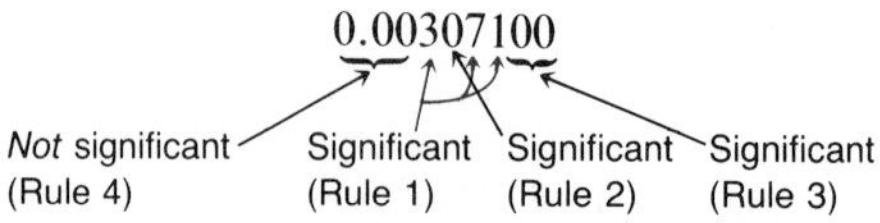
0.00307100
Not significant (Rule 4)
Significant (Rule 1)
Significant (Rule 2)
Significant (Rule 3)

This example contains a total of 6 significant figures.

SAMPLE EXERCISE 2.9 How many significant figures are shown in each of the following measurements?

a 4.065 m
b 0.32 g
c 57.98 mL
d 20.00 s
e 0.00040 km
f 604.0820 kg

Solution

a 4 significant figures (Rules 1 and 2)
b 2 significant figures (Rules 1 and 4)
c 4 significant figures (Rule 1)
d 4 significant figures (Rules 1 and 3)
e 2 significant figures (Rules 1, 3, and 4)

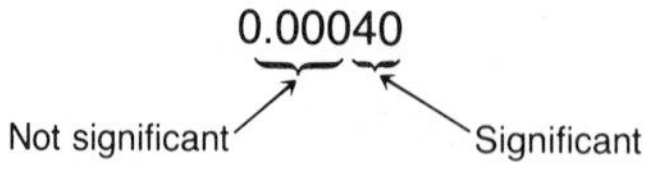
0.00040
Not significant
Significant

f 7 significant figures (Rules 1, 2, and 3)

Problem 2.8 Tell the number of significant figures in each of the following:

a 750.3 cm
b 0.06020 g
c 30.0 mL
d 0.0001 kg

2.11 ROUNDING OFF

In doing calculations with significant figures, you will frequently need to "round off" answers to smaller numbers of digits. Use the following rules:

1 If the first figure to be dropped is less than 5, the preceding digit is not changed. Therefore, 1.234 is rounded to 1.23 (3 significant figures) or 1.2 (2 significant figures) or 1 (1 significant figure).
2 If the first figure to be dropped is 5 or more, the preceding digit is increased by 1. Therefore, 1.756 is rounded to 1.76 (3 significant figures) or 1.8 (2 significant figures) or 2 (1 significant figure).

2.12 SIGNIFICANT FIGURES IN CALCULATIONS

Multiplication and Division

If you multiply the measured numbers 2.3 × 5.69 on a calculator, it will read out 13.087. But this is not a proper answer to the multiplication because *in multiplying or dividing measurements, the answer should contain no more significant figures than the measurement with the fewest number of significant figures*. Therefore, in multiplying 2.3 (2 significant figures) × 5.69 (3 significant figures), we are allowed only 2 significant figures in the answer. So we must round off to 2 significant figures:

$$2.3 \times 5.69 = 13.087 = 13$$

SAMPLE EXERCISE 2.10

The mass and volume of a sample of blood are measured and found to be 12.193 g and 11.5 mL, respectively. Determine the density and express the answer in the proper number of significant figures.

Solution

$$\text{Density} = \frac{\text{mass}}{\text{volume}}$$

Using a *calculator,*

$$\text{Density} = \frac{12.193\ \text{g}}{11.5\ \text{mL}} = 1.0602609\ \text{g/mL}$$

But because we know the volume to only 3 significant figures, we are allowed only 3 significant figures in the final answer. Therefore, we round off and report the density as 1.06 g/mL.

Even when there is a series of multiplications and divisions, the measurement with the smallest number of significant figures governs the number of significant figures in the answer. Consider doing the following operations with a calculator:

$$\frac{4 \times 2.546 \times 3.12}{7.0012} = 4.5383762$$

Calculator answer

Because one of the multipliers (4) has only 1 significant figure, the correct answer to report must have only 1 significant figure. In this case, the answer would be 5.

Problem 2.9 Assuming that the following numbers are measurements, perform the calculation and report the answer to the correct number of significant figures:

$$\frac{5.42 \times 21.003}{4.6} = ?$$

The idea of significant figures applies only to *measured* numbers. Some numbers are *defined* or *exact* numbers and when used in a calculation do not affect the numbers of significant figures allowed in the answer. For example, 1 dozen eggs is defined as 12 eggs. To determine the number of eggs in 11 dozen, we multiply 11 dozen × 12 eggs/1 dozen = 132 eggs. We do not round off here because we are dealing with exact numbers, *not* measured numbers.

All the equalities given in Table 2.1 are exact or defined. For example, 1 m is defined as exactly 100 cm, just as 1 dozen eggs is defined as 12 eggs, so the use of conversion factors involving meters and centimeters will not affect the number of significant figures allowed. The equalities in Table 2.2 are inexact because they are measured and must be considered in determining significant figures allowed. They are expressed in such a way that the 1 is exact, but the other number is good only to the number of significant figures shown. For example, for 1 lb = 454 g, the 1 is exact or defined and will not affect a calculation, but the 454 is measured and will limit a calculated answer to 3 significant figures.

To summarize, *only* **measured** *numbers* and **inexact** *equalities* limit the number of significant figures in a calculated answer. In practice, tables of equalities are constructed so that they do not limit the number of significant figures allowed any more than the measured number does.

Addition and Subtraction

Addition and subtraction of measured numbers follow a different rule. In adding and subtracting measurements, the final answer must not contain any more digits to the right of the decimal point than does the measured number with the *least number of digits to the right of the decimal point.*

For example, in adding 937.3 g + 15.224 g + 71.04 g, the number 937.3 governs the significant figures in the answer—there may be only *one digit* to the right of the decimal point.

$$\begin{array}{r} 937.3\ \ \ \text{g} \\ 15.224\ \text{g} \\ \underline{71.04\ \ \text{g}} \\ 1023.564\ \text{g} \end{array} = 1023.6\ \text{g}$$

Because one mass is known only to the nearest tenth, the answer must be rounded off to tenths.

SAMPLE EXERCISE 2.11

Report the result of the following subtraction to the correct number of significant figures:

2.572 m − 0.41 m = ?

Solution

$$\begin{array}{r} 2.572\ \text{m} \\ \underline{-0.41\ \ \ \text{m}} \\ 2.162\ \text{m} \end{array}$$

← **This is the limiting number because there are fewer digits to the right of the decimal point.** (refers to 0.41 m)

The answer must be rounded off to 2.16 m.

Problem 2.10 Add up the following weights and report the sum to the proper number of significant figures:

2.2332 g + 1004.12 g + 26.557 g = ?

Comment on the Use of Significant Figures in This Book

The solutions to all the sample exercises in this book and the solutions to the problems in Appendix 1 are reported to the correct number of significant figures. You should always round off all your final answers to the proper number of significant figures.

2.13 SCIENTIFIC NOTATION

Thus far all calculations have been done using numbers expressed in the usual manner to which you are accustomed. In scientific work it is often necessary to use extremely large and very small numbers. For example, the mass of an electron is 0.0000000000000000000000000009107 g, and in Chapter 7 you will encounter Avogadro's number, 602,300,000,000,000,000,000,000. We need a more convenient way to express such numbers. A way that eliminates the need to write all the zeros directly is desirable if for no other reason than to save ink and elbow grease. The method that is used is known as scientific notation. Let us review some of its basic principles.

Exponents and Bases

You are probably already familiar with the idea of raising a number to a *power*. For example, 10^2 is read as 10 to the second power, and it is equal to 10×10, or 100. The 10 is known as the *base* and the 2 is called the *exponent*.

10^2 ← Exponent (the 2); Base (the 10)

The exponent tells us the number of times the base is a factor in the multiplication. Review the following examples:

Exponent

$$10^6 = \underbrace{10 \times 10 \times 10 \times 10 \times 10 \times 10}_{\text{Six factors of ten}} = 1{,}000{,}000$$

Base

$$10^{10} = 10 \times 10 \times 10 \times 10 \times 10 \times 10 \times 10 \times 10 \times 10 \times 10$$

$$= 10{,}000{,}000{,}000$$

$$10^1 = 10$$

A base can also have a negative exponent, for example, 10^{-1}, 10^{-3}. We can always write a base with a negative exponent as the reciprocal, with the sign of the exponent changed to positive. The reciprocal of a number A is defined as $1/A$.

Negative exponent   Reciprocal of original

$$10^{-1} = \frac{1}{10^1} = 0.1$$

Same exponent with a positive sign

$$10^{-3} = \frac{1}{10^3} = \frac{1}{1000} = 0.001$$

Since $10^1 = 10$, and $10^{-1} = 0.1$, it is reasonable that $10^0 = 1$.

Writing Numbers in Scientific Notation

Any ordinary decimal number can be expressed as a decimal number between 1 and 10 multiplied by some power of 10. For example, the number 392,000 written in scientific notation looks like this:

$$\underbrace{3.92}_{\text{Decimal number between 1 and 10}} \times \underbrace{10^5}_{\text{Power of 10}}$$

The ordinary decimal number 0.00432 written in scientific notation looks like this:

$$\underbrace{4.32}_{\text{Decimal number between 1 and 10}} \times \underbrace{10^{-3}}_{\text{Power of 10}}$$

The rules for converting ordinary numbers to scientific notation are subdivided into three cases depending on the size of the number.

General Rules for Converting Ordinary Decimal Numbers into Scientific Notation

CASE 1 If the number is equal to or greater than 10:

1 Move the decimal point to the left, counting the number of places you must move it until you have a decimal number between 1 and 10.
2 Multiply this decimal number by 10 raised to a *positive* power equal to the number of places you moved the decimal point.

SAMPLE EXERCISE 2.12

Write 138.34 in scientific notation.

Solution We recognize that the 138.34 is greater than 10 (Case 1). Therefore, we proceed as follows:

STEP 1 We move the decimal point to the left until we obtain a number between 1 and 10: 138.34 becomes 1.3834. This requires a movement of two places.

STEP 2 Now we write down the new number and multiply it by 10^2 because the decimal was moved two places. The sign is positive because the number is greater than 10.

ANSWER 1.3834×10^2

CASE 2 If the number is less than 1:

1 Move the decimal point to the right, counting the number of places you must move it until you have a decimal number between 1 and 10.
2 Multiply this decimal number by 10 raised to a *negative* power equal to the number of places you moved the decimal point.

SAMPLE EXERCISE 2.13 Write 0.000108 in scientific notation.

Solution We recognize that 0.000108 is less than 1 (Case 2). Therefore, we proceed as follows:

STEP 1 We move the decimal point to the right until we obtain a number between 1 and 10: 0.000108 becomes 1.08. This requires a movement of four places.

STEP 2 We write down the new number and multiply it by 10^{-4} because the decimal was moved four places. The sign is negative because the number is less than 1.

ANSWER $0.000108 = 1.08 \times 10^{-4}$

CASE 3 If the number is equal to or greater than 1.0 and less than 10.0, simply write down the number and multiply it by 10^0 because no (zero) movement of the decimal point is required. For example, the decimal number 3.85 written in scientific notation is 3.85×10^0, or simply 3.85.

Problem 2.11 Write the following in scientific notation:

a 18.9	*c* 5,342,000	*e* 773
b 0.013	*d* 0.00000524	*f* 7.2

Converting Scientific Notation to Decimal Numbers

To convert a number in scientific notation, such as 4.923×10^2, into an ordinary decimal number, start by examining the exponent. If the exponent is positive, move the decimal to the right a number of places equal to the value of the exponent. You may have to fill in zeros. Remember that positive exponents are associated with numbers greater than 10.

If the exponent is negative, move the decimal to the left a number of places equal to the value of the exponent. Again, you may have to fill in zeros. Remember, negative exponents are associated with numbers that are less than 1.

SAMPLE EXERCISE 2.14 Write 4.923×10^2 as an ordinary decimal number.

Solution The exponent is positive 2, so we move the decimal point two places to the right.

$4.923 \times 10^2 = 492.3$

SAMPLE EXERCISE 2.15 Write 9.23×10^{-5} as an ordinary decimal number.

Solution The exponent is negative 5, so we move the decimal point five places to the left, filling in zeros.

.00009.23 × 10^{-5} = 0.0000923

Problem 2.12 Write the following as decimal numbers:

a 1.19×10^4
b 8.73×10^{-2}
c 5.93×10^8
d 2.19×10^{-6}

2.14 CALCULATIONS IN SCIENTIFIC NOTATION

You will do calculations using scientific notation less often than with ordinary decimal numbers. We include here a basic statement of the rules for multiplication, division, addition, and subtraction, and some examples.

To multiply two numbers written in scientific notation, multiply the two decimal numbers together in the usual manner and then *add* the two exponents, being careful to treat the exponents as signed numbers. The sum is the correct power of 10 in the product.

$$(a \times 10^x)(b \times 10^y) = ab \times 10^{x+y}$$

For example,

$$(2 \times 10^4)\ (3 \times 10^7) = 6 \times 10^{4+7}$$
$$= 6 \times 10^{11}$$

To divide two numbers written in scientific notation, divide one decimal number by the other in the usual manner and then *subtract* the exponent in the denominator (divisor) from the exponent in the numerator (dividend), being careful to treat this as the subtraction of signed numbers. The difference is the correct power of 10 in the quotient.

$$\frac{a \times 10^x}{b \times 10^y} = \frac{a}{b} \times 10^{x-y}$$

For example,

$$\frac{8 \times 10^4}{4 \times 10^7} = 2 \times 10^{4-7} = 2 \times 10^{-3}$$

You will very rarely need to be able to add or subtract numbers written in scientific notation. If the need does arise, an easy way to perform the addition or subtraction is to convert to ordinary numbers and then add or subtract in the usual manner. For example, to add $(3.71 \times 10^2) + (4.55 \times 10^3)$, convert to 371 + 4550 = 4921. The answer can then be put back into scientific notation as 4.921×10^3.

SAMPLE EXERCISE 2.16 How many significant figures are represented by each of the following measurements?

a 3.24×10^{-8} cm
b 5.0×10^3 g
c 7.0300×10^2 mL
d 4.705×10^{-4} km

Solution To determine the number of significant figures in a number written in scientific notation, apply the rules to the number preceding the times sign. You will find that if the number is in the correct notation form, then all digits are significant.

a 3
b 2
c 5
d 4

CHAPTER ACCOMPLISHMENTS

After completing this chapter you should be able to define **key words** and do the following:

2.1 INTRODUCTION
1. State the role of a reference standard in measurement.

2.2 THE METRIC SYSTEM
2. State the metric base units used for length, mass, and volume measurements.
3. State the meanings of the common metric prefixes (Table 2.1).
4. State the equalities between common metric units (Table 2.1).
5. Distinguish between mass and weight.
6. State the advantage of measuring mass rather than weight.

2.3 APOTHECARIES' MEASURES
7. Recognize apothecaries' measures commonly used in drug prescriptions.

2.4 UNIT CONVERSION
8. Construct conversion factors from equalities.
9. Beginning with a given quantity and unit, use conversion factors to calculate some new quantity and unit.

2.5 DENSITY
10. State the usual units in which density is expressed.
11. Given the mass and volume of a sample, calculate the sample's density.
12. Recognize density as a conversion factor.
13. Given the density, convert the given volume of a sample to the mass of the sample, or convert the given mass of a sample to the volume of the sample.

2.6 SPECIFIC GRAVITY
14. Given a density, state the specific gravity.

2.7 TEMPERATURE
15. Given a temperature in degrees Fahrenheit or degrees Celsius, convert from the given scale to the other scale.

2.8 UNITS OF ENERGY
16. Distinguish between heat and temperature.
17. State the units of energy most commonly used by chemists.

2.9 FOOD CALORIES

2.10 SIGNIFICANT FIGURES
18. Given a measuring device, state the number of significant figures that can be reported in a measurement.
19. Given a measured number, state the number of significant figures it has.

2.11 ROUNDING OFF
20. Round off numbers to some indicated number of significant figures.

2.12 SIGNIFICANT FIGURES IN CALCULATIONS
21. Express the result of the multiplication or division of measured numbers to the correct number of significant figures.
22. Express the result of the addition or subtraction of measured numbers to the correct number of significant figures.

2.13 SCIENTIFIC NOTATION
23. Indicate the base and exponent of an exponential number.
24. Write any decimal number in scientific notation.
25. Convert a number in scientific notation into an ordinary decimal number.

2.14 CALCULATIONS IN SCIENTIFIC NOTATION
26. Multiply or divide two numbers in scientific notation.

PROBLEMS

2.13 Give five common prefixes of the metric system and their mathematical meanings.

2.14 State the equalities between the following metric units:
- *a* Grams and kilograms
- *b* Liters and milliliters
- *c* Meters and centimeters
- *d* Micrometers and meters
- *e* Milliliters and cubic centimeters

2.15 For each equality that you wrote in Problem 2.14, write two conversion factors.

2.16 Using the Unit Conversion Method, carry out the following one-step conversions:

Within the metric system	*Between metric and English*
a 13.2 m to kilometers	*a* 22.0 kg to pounds
b 3.1 g to milligrams	*b* 322.0 km to miles
c 12.0 cm to meters	*c* 17.3 cm to inches
d 541.0 mL to liters	*d* 4.2 lb to grams
e 0.04 kg to grams	*e* 3.00 L to quarts
f 6.7 m to millimeters	*f* 10.0 m to inches
g 895 cm^3 to milliliters	*g* 145.0 lb to kilograms
h 0.76 L to milliliters	*h* 7.5 qt to liters

2.17 Carry out the following two-step conversions:
- *a* 0.06 km to centimeters
- *b* 75 mm to centimeters
- *c* 4330. mg to kilograms
- *d* 2.00 ft to meters
- *e* 674 mg to pounds
- *f* 0.43 qt to milliliters

2.18 Carry out the following multistep conversions:
- *a* 4.00 mi to centimeters
- *b* 750. mL to gallons
- *c* 3.1 oz to grams (16 oz = 1 lb)

2.19 Do the following conversions:
- *a* 35 cm^3 to milliliters
- *b* 2.00 lb of sugar to grams of sugar
- *c* 2.00 qt of milk to milliliters of milk
- *d* $\frac{1}{4}$ lb of liverwurst to grams of liverwurst
- *e* 3.00 yd to millimeters
- *f* 17 m to feet

2.20 A highway sign tells you that you are 267 km away from home. How many miles away are you?

2.21 How long is a standard football field (100 yd) in meters?

2.22 Men's shirt sizes are based on the circumference of men's necks measured in inches. Size 17 means the neckband measures 17 in. What is the metric size (in centimeters) corresponding to size 17?

2.23 A swimmer swims 70 lengths of a pool each day. The pool is 25 m long. What distance does she swim each day (*a*) in meters, (*b*) in miles?

2.24 In France gasoline is sold by the liter. If you wanted 10. gal of gasoline, how many liters should you ask for (1 gal = 4 qt)?

2.25 The national speed limit is 55 mi/h. What is this speed limit in kilometers per hour?

2.26 If your car gets 21 mi/gal, how many kilometers per liter does it get?

2.27 The sun is 92 million mi from the earth.
- *a* How far away is this in centimeters? (Use scientific notation.)
- *b* Given that light travels 3.00×10^{10} cm/s, how many seconds does it take for light to travel from the sun to earth?

2.28 A fish tank is 20. in long, 20. in deep, and 10. in high.
- *a* How many liters of water are required to fill it?
- *b* How many gallons are needed?

2.29 A typical aspirin tablet contains 5 gr of the analgesic compound. The rest of the tablet is starch. How many *grams* of aspirin are in two aspirin tablets?

2.30 Calculate the density of a liquid, 24.90 mL of which weighs 25.86 g.

2.31 What is the specific gravity of the liquid described in Problem 2.30?

2.32 A piece of metal weighs 63.92 g and occupies a volume of 5.69 mL.
- *a* What is the metal's density?
- *b* Consult Table 2.3 and identify the metal.

2.33 Use Table 2.3 to calculate the mass in grams of each of the following samples:
- *a* 20.0 mL of gasoline
- *b* 470.0 mL of vinegar
- *c* 6.25 mL of olive oil

2.34 Use Table 2.3 to calculate the volume of each of the following samples:
- *a* 80. g of alcohol
- *b* 49.6 g of gold
- *c* 272 g of mercury

2.35 How much does 1 pt of blood weigh? (See Table 2.3; 1 qt = 2 pt.)

2.36 How could you use a urinometer to determine the *mass* of 250 mL of urine?

2.37 Consult Table 2.3 and calculate the mass in pounds of the following:
- *a* 1 qt of milk
- *b* 1 qt of mercury

2.38 Do the following temperature conversions:
- *a* 149°F to degrees Celsius
- *b* 0°F to degrees Celsius
- *c* 20°C to degrees Fahrenheit
- *d* −40°C to degrees Fahrenheit
- *e* 11.43°F to degrees Celsius

2.39 Which is the higher temperature, −15°C or +4°F?

2.40 A typical antifreeze mixture protects your radiator down to −20°F. What is this in degrees Celsius?

2.41 Which is colder, −90°C or −125°F?

2.42 How many calories are there in 54.3 kcal?

2.43 How much heat must be applied to raise the temperature of 55 g of olive oil by 100°C? (See Table 2.4.)

2.44 How much heat must be removed to cool 76 g of lead from 95 to 16°C? (See Table 2.4.)

2.45 When a diamond weighing 2.0 g absorbs 12 cal, the temperature changes from 25 to 76°C. What is the specific heat of diamond?

2.46 How many significant figures are in each of the following measured numbers?

a 0.006
b 11.1
c 0.2001
d 1.4×10^{-6}
e 400.0
f 786,800.9
g 3.00×10^{9}
h 71.050
i 0.0172
j 0.0610

2.47 A hospital patient has an oral temperature of 39.5°C and weighs 190 lb. He is to receive a drug, the dosage of which is 50 mg/kg of body weight. The drug is dissolved in water (25 mg/mL).

a What is his temperature in degrees Fahrenheit?
b What is his metric weight?
c What mass of pure drug should he receive?
d How many milliliters of the drug solution should he drink?

2.48 Round off each of the following to 3 significant figures:

a 235.674 g
b 10.528 mL
c 7.2457 g/mL
d 0.000328730 kg

2.49 Do the following multiplications or divisions and express your answers to the proper number of significant figures:

a 5.01×2.0
b 0.03×0.578
c $\dfrac{0.04}{17.5}$
d $\dfrac{286.3}{2.00 \times 10^{1}}$
e 0.00571×78.11

2.50 Do the following additions or subtractions and express your answers to the proper number of significant figures:

a 5.79 + 6.2543 + 0.6
b 25.856 − 1.50
c 1.354 + 65.610 + 3.7504
d 0.4531 − 1.480

2.51 Do the following calculations and report the answers to the proper number of significant figures:

a Given 1 lb = 453.6 g, convert 2973 g to pounds.
b Given 1 lb = 454 g, convert 2973 g to pounds.

2.52 Calculate the density of chloroform to the proper number of significant figures, given the data that for a certain sample, mass = 29.810 g and volume = 20. mL.

2.53 What is the total volume accumulated in a beaker by successive additions of liquid in the amounts 21.05 mL, 4.10 mL, 3.2 mL, and 0.575 L? Report the answer to the proper number of significant figures.

2.54 Write the following numbers in scientific notation:

a 34.9
b 0.092
c 932,000,000
d 1032
e 98.04
f 0.00000491

2.55 Calculate the following and express your answer in scientific notation:

a $(3.0 \times 10^{11})(3.0 \times 10^{3}) =$
b $(1.32 \times 10^{4})(1.11 \times 10^{-9}) =$
c $\dfrac{9.3 \times 10^{5}}{3.1 \times 10^{2}} =$
d $\dfrac{9.5 \times 10^{5}}{1.9 \times 10^{-7}} =$

CHAPTER THREE

ATOMIC STRUCTURE AND NUCLEAR CHEMISTRY

3.1 INTRODUCTION

Thus far the discussion of matter has been in *macroscopic* terms. That is, we have explained the classification of matter by examining large (*macro-* means "large") "chunks" of matter which we can see and hold. For a true and clearer understanding of matter, it must be explored in *microscopic* (*micro-* means "small") terms. We should really say *sub*microscopic terms, because the smallest particles of matter cannot be seen even by the most sensitive electron microscopes. As soon as you learn the basic makeup of **atoms,** the tiniest particles of matter, you will be able to understand such topics of everyday interest as radioactivity, nuclear medicine, and nuclear power plants.

3.2 ATOMIC THEORY

We assume that you have already heard that all matter is made up of tiny particles called atoms because of the publicity atomic theory receives through such things as the atomic bomb and atomic energy. Philosophically there are only two ways matter could be constructed: (1) it might be *continuous,* i.e., divisible into ever smaller parts so that no smallest particle exists; or (2) it might be *discrete,* i.e., made up of characteristic small particles which if further divided no longer have the properties of the matter which we are examining. Experiment indicates that matter is discrete, i.e., made up of atoms.

The concept of the atom is an old one. The word was coined by the Greek Democritus, who first suggested the atomic idea in 400 B.C. Scientists and philosophers generally did not believe in atoms until the early nineteenth century. At that time an English schoolteacher, John Dalton, stated the Atomic Theory of Matter in a clear manner which explained many known scientific laws and experiments and also accounted for the classification of matter into elements, compounds, and mixtures. Following are some of the key statements of Dalton's theory:

1 All matter is made up of atoms which are indestructible by ordinary means.
2 All atoms of a given element are identical in chemical properties.
3 Atoms of different elements have different chemical properties.
4 Atoms of different elements can chemically combine in simple, whole-number ratios to form compounds.

Now we can define elements, compounds, and mixtures in terms of their atomic makeup.

Element Matter which is composed of one kind of atom

Compound Matter which is composed of different kinds of atoms chemically combined in simple, whole-number ratios

Mixture Matter which is a physical combination of the particles of elements or compounds

Compare these definitions with those given in Section 1.5.

Problem 3.1 Explain how the new atomic definition of compound relates to the former definition given in Section 1.5.

Dalton pictured atoms as tiny spherical particles. We will find that this is not a completely accurate picture, but it provides a useful mental image for understanding the composition of matter. Figure 3.1 attempts to convey the extreme tininess of the atom.

3.3 INSIDE THE ATOM

Atoms are not really hard little balls as Dalton imagined them. Actually, atoms are mostly empty space because of the arrangement of the *subatomic* particles of which they are made. That is, by the end of the nineteenth century it had become clear that atoms were not the smallest particles. Rather, atoms can be subdivided into smaller *subatomic* particles. By today, atomic physicists have identified hundreds of indescribably small and often short-lived particles. Luckily, to understand chemistry one need be aware of only the three major types of subatomic particles: protons, electrons, and neutrons. Atoms of different elements differ in the *numbers* of these particles contained. Figure 3.2 shows the arrangement of these particles and explains the statement that atoms are mostly empty space.

Characteristics of Subatomic Particles

The important characteristics of protons, electrons, and neutrons are their relative masses and electrical charges. These are summarized in Table 3.1.

Protons and neutrons have approximately the same mass. Electrons are much smaller and lighter; the electron mass is only roughly 1/1840 the mass of a proton. To express these mass relationships conveniently, we assign the proton a mass of 1; then the neutron mass must be 1, and the electron mass must be 1/1840. The mass of the electron is so much less than that of the proton or neutron, we can ignore it. Because electrons contribute practically no mass to an atom, we assign the electron a relative mass of zero.

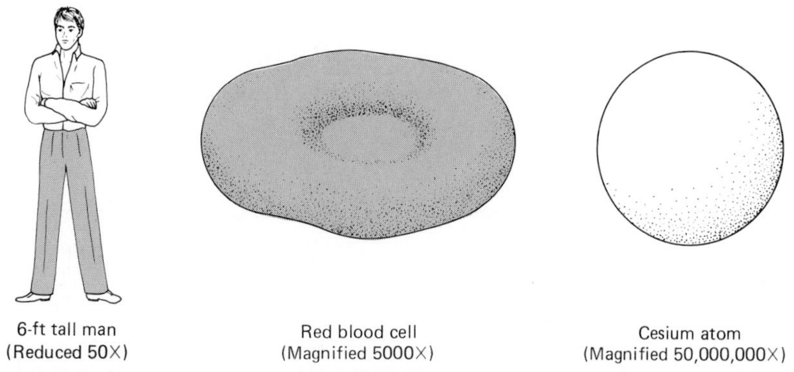
6-ft tall man
(Reduced 50X)
Red blood cell
(Magnified 5000X)
Cesium atom
(Magnified 50,000,000X)

FIG. 3.1 We can draw objects in any convenient size for illustration, and so here a man, a cell, and an atom are shown all seemingly of the same approximate dimensions on the page. However, the magnification or reduction compared with the real objects shows the dramatic difference in size. Cesium atoms are the largest atoms.

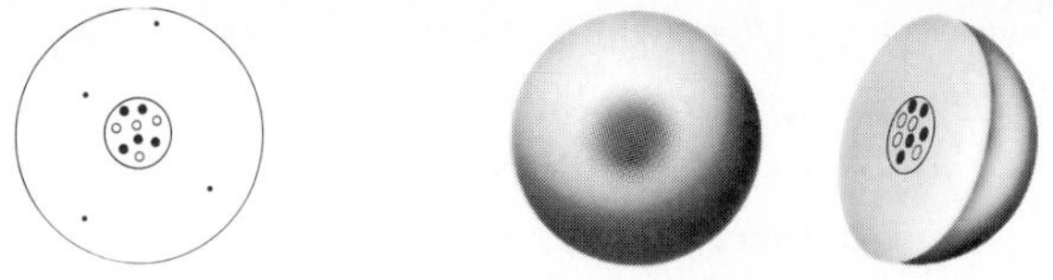

FIG. 3.2 The arrangement of subatomic particles in the atom.

(a) A cross section of an atom shows the protons (○) and neutrons (●) clustered in the nucleus and the electrons outside the nucleus and fairly far apart.

(b) A better representation of the atom is a fuzzy cloud for the electrons to show their constant motion and to show that atoms are three-dimensional. Cutting the cloud in half reveals the nucleus in the center.

In Section 3.8 we will examine the units scientists use for these relative masses.

You are probably familiar with the positive (+) and negative (−) character of electricity. Look at any battery; there is a positive pole and a negative pole. If you study physics, you will find that electricity is a flow of electrons and that electrons have a negative (−) charge. Atoms are electrically neutral (no net charge). This is so because for every electron with a −1 charge, there is a proton which has a +1 charge. Just as adding signed numbers of equal magnitude and opposite signs results in zero, equal numbers of positive protons and negative electrons result in zero charge (electric neutrality). Neutrons are so named because they are electrically neutral (zero charge).

Arrangement of Subatomic Particles

The arrangement of protons, electrons, and neutrons was shown in Figure 3.2. The mass of an atom is concentrated in the center because the protons and neutrons are clustered there. The center of the atom occupied by protons and neutrons is called the **nucleus.** The nucleus is described as being dense, or compact, because most of the mass of the atom is contained in a small volume. In addition, the nucleus bears a positive charge because of the positive protons.

Sprinkled around the dense, positive nucleus are the "weightless" negative electrons. In Chapter 4 you will learn more details about the arrangement of electrons. For now, it is sufficient to know the following:

1 Electrons are outside the nucleus.
2 Electrons are constantly moving.
3 Because of the spaces between electrons and between the electrons and the nucleus, atoms are mostly empty space.
4 Because electrons bear negative charges and are constantly moving, they appear to be a negatively charged cloud with fuzzy borders.

Look out your window at the clouds. Pick out a nice round or egg-shaped cloud. Imagine that the cloud is negatively charged and that buried deep within the cloud is a tiny positively charged pebble. You have just developed a mental picture of an atom.

TABLE 3.1 Characteristics of Subatomic Particles

Particle	Symbol	Relative Charge	Relative Mass
Proton	p^+	+1	1
Electron	e^-	−1	0 (1/1840)
Neutron	n^0	0	1

3.4 ATOMIC NUMBER

The number of protons in an atom is called the **atomic number**. Atoms and elements are identified by their atomic number. Look at the periodic table inside the front cover. The numbers above the symbols are the atomic numbers, that is, the number of protons in atoms of that element. Any atom with six protons must be carbon. An atom with just one proton must be hydrogen. For an element to be gold, there must be 79 protons in the nucleus of each atom.

Problem 3.2
a What are the atomic numbers of Al, Zn, and As?
b How many protons are there in Al, Zn, and As?

Problem 3.3 Identify the elements for which the following information is given:
a Atomic number is 35.
b Number of protons is 12.

Atoms are electrically neutral if for every proton (+1) there is an electron (−1). Therefore, the atomic number also tells you the number of electrons in a *neutral* atom. Carbon atoms with atomic number 6 have six protons and six electrons.

Problem 3.4 How many electrons are there in neutral atoms of Li, P, and Hg?

Problem 3.5
a What are the atomic numbers of elements with 11, 5, and 26 electrons in their neutral atoms?
b What are the names of these atoms?

3.5 CHARGED ATOMS

Eventually we will see that atoms often gain or lose electrons. When this happens, charged atoms called **ions** result because the number of protons and the number of electrons are no longer equal. It is important to realize that *atoms* and *ions* of an element are very different in physical, chemical, and biological properties.

For example, chlorine atoms (atomic number 17) often gain one electron and become so-called chloride ions. Pictorially:

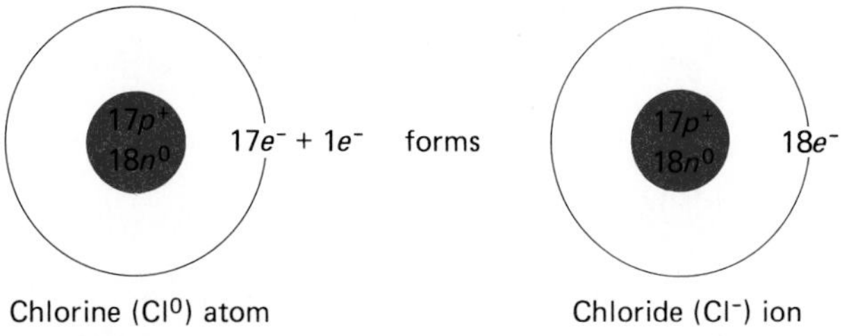

17p⁺
18n⁰
17e⁻ + 1e⁻
forms
17p⁺
18n⁰
18e⁻
Chlorine (Cl⁰) atom
Chloride (Cl⁻) ion

Whereas chlorine atoms are electrically neutral because the +17 charge from 17 protons is just cancelled by the −17 charge from electrons, chloride ions have a net charge of −1. This is because of the one extra electron in the chloride ion (−18 + 17 = −1). *The properties of chlorine atoms and chloride ions are very different.* In particular, chloride ions are very common and have certain essential bodily functions (see Section 19.12). Chlorine atoms, on the other hand, are "rare birds" as we'll see in Section 6.3 and would very likely disrupt body chemistry rather than perform the essential function of chloride ions.

3.6 ISOTOPES

Whereas the number of protons (atomic number) identifies atoms of a particular element, the atoms of the same element may contain a different number of neutrons. For example, *all* carbon atoms have six protons and most carbon atoms have six neutrons. But about 1 percent of carbon atoms contain seven neutrons, and there are some carbon atoms with eight neutrons. The six protons in the atom make that atom the element carbon; the number of neutrons may vary. Atoms with the same number of protons but different numbers of neutrons are called **isotopes**.

There are three isotopes of hydrogen (Figure 3.3). All isotopes of hydrogen have one proton in the nucleus. Most isotopes of hydrogen have no neutrons, some have one neutron, and a few have two neutrons.

Isotope is really just another word for *atom*. The two words can often be used interchangeably. Notice: "All *atoms* of nitrogen contain seven protons. All *isotopes* of nitrogen contain seven protons." Keep this idea in mind as further discussions of isotopes arise.

Problem 3.6 How many protons are in the nucleus of all isotopes of chlorine?

3.7 MASS NUMBER

The sum of the number of protons and the number of neutrons in an atom is called the **mass number.** Remember that the mass of the atom is almost totally accounted for by the protons and neutrons because the mass of an electron is effectively zero.

Knowing the atomic number and the mass number for an isotope means that you can figure out the number of protons, electrons, and neutrons in that atom. For a neutral atom,

ATOMIC NUMBER = number of protons = number of electrons

Mass number = number of protons + number of neutrons

or we can substitute atomic number for number of protons. Then

Mass number = atomic number + number of neutrons

Now subtract **atomic number** from each side of the equation, and you get

MASS NUMBER − ATOMIC NUMBER = number of neutrons

Problem 3.7 How many protons, neutrons, and electrons are there in neutral atoms which have:

a Atomic number 17, mass number 37
b Atomic number 12, mass number 24
c Atomic number 19, mass number 39

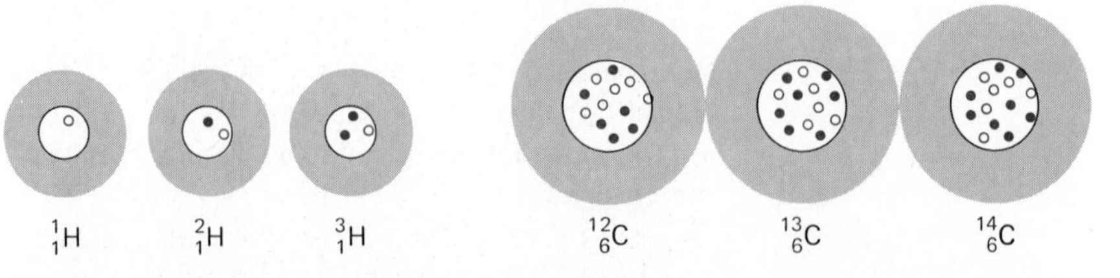

^1_1H
^2_1H
^3_1H
$^{12}_6C$
$^{13}_6C$
$^{14}_6C$

FIG. 3.3 Isotopes of an element differ in the numbers of neutrons they possess. This also means that the mass numbers will vary.

(a) Three isotopes of hydrogen with 0, 1, and 2 neutrons (•) respectively. Note in each case the atom has 1 proton (○); this makes it hydrogen.

(b) Three isotopes of carbon with 6, 7, and 8 neutrons (•) respectively. Note in each case the atom has 6 protons (○); this makes it carbon.

Atomic number and mass number data are conveniently given in shorthand symbols. For example, the symbol that represents neutral atoms of carbon with six protons and six neutrons is $^{12}_{6}C$. The atomic number is written as a *sub*script to the left of the symbol for carbon. The mass number is written as a *super*script to the left of the symbol. This method can be represented in general by the following illustration, in which Sy stands for any chemical symbol:

$$^{\text{mass number}}_{\text{atomic number}}\text{Sy}$$

The symbol $^{12}_{6}C$ can also be read as carbon-12. Carbon-13 is $^{13}_{6}C$. Remember, all atoms of carbon have the same atomic number. But different isotopes of carbon will have different mass numbers because the number of neutrons varies. We can also define **isotopes** as atoms with the same atomic number but different mass numbers.

Often a particular isotope of an element is indicated only by its name or symbol and mass number, e.g., carbon-12, ^{12}C, or uranium-238, ^{238}U. It is not necessary to include the atomic number since the name or symbol is sufficient to identify the element with a unique atomic number.

SAMPLE EXERCISE 3.1

Determine the number of protons (p^+), electrons (e^-). and neutrons (n^0) in the neutral atoms indicated. Also, identify the atoms.

a $^{65}_{30}X$ *b* $^{27}_{13}X$ *c* $^{28}_{13}X$ *d* $^{28}_{14}X$ *e* $^{64}_{29}X$

Solution

$^{\text{mass number}}_{\text{atomic number}}\text{Sy}$	Atomic Number = p^+ = e^-	Mass Number − Atomic Number = n^0
a $^{65}_{30}X$	$30 = p^+ = e^-$	$65 - 30 = n^0$ $35 = n^0$
b $^{27}_{13}X$	$13 = p^+ = e^-$	$27 - 13 = n^0$ $14 = n^0$
c $^{28}_{13}X$	$13 = p^+ = e^-$	$28 - 13 = n^0$ $15 = n^0$
d $^{28}_{14}X$	$14 = p^+ = e^-$	$28 - 14 = n^0$ $14 = n^0$
e $^{64}_{29}X$	$29 = p^+ = e^-$	$64 - 29 = n^0$ $35 = n^0$

The atomic number identifies the atoms. Consult the periodic table to see that the atoms are (*a*) Zn, (*b*) Al, (*c*) Al, (*d*) Si, and (*e*) Cu.

SAMPLE EXERCISE 3.2

Refer to Sample Exercise 3.1. Which of the atoms are isotopes of the same element?

Solution $^{27}_{13}X$ and $^{28}_{13}X$ are isotopes of Al. They are atoms with the same number of protons and different numbers of neutrons. They are atoms with the same atomic number and different mass numbers.

Problem 3.8 Consider the neutral atoms $^{24}_{11}Q$, $^{24}_{12}Q$, $^{79}_{35}Q$, $^{25}_{12}Q$, and $^{81}_{35}Q$.

a Indicate the numbers of protons, neutrons, and electrons in each.
b Identify each.
c Pick out any sets of isotopes.

3.8 RELATIVE ATOMIC MASS

A single atom is much too tiny to be weighed. We cannot directly determine the mass of an individual atom. Fortunately, scientists realized a long time ago that it is not necessary to weigh individual atoms. What we need to know are the masses of atoms in one element *relative* to atoms in other elements. Knowing which atoms are heavier or lighter than others and by what factor enables us to set up all needed weight relationships for elements and compounds.

In Section 2.1 we discussed the idea that all measurements are made relative to some standard. (Refer also to Section 2.2, which discussed the various standards in the metric system.) The standard chosen for the atomic-weight scale is the most abundant isotope of carbon, $^{12}_{6}C$ (carbon-12). In 1961 carbon-12 was assigned a mass of 12 *a*tomic *m*ass *u*nits (amu). Then it was experimentally determined how much heavier or lighter atoms of other elements were. From these experiments, the atomic-weight scale was established.

For example, it was found that helium atoms are one-third as heavy as carbon-12 atoms, and titanium atoms are about 4 times as heavy. This means that the mass of a helium atom is

$$\frac{1}{3} \times 12 \text{ amu} = 4 \text{ amu}$$

and the mass of a titanium atom is

$$4 \times 12 \text{ amu} = 48 \text{ amu}$$

These multiplying factors can be found for all atoms, and thus relative atomic masses can be assigned to all atoms.

Problem 3.9 Atoms of element Q are twice as heavy as ^{12}C atoms. What is the atomic mass of these atoms?

On the carbon-12 scale, the protons and neutrons weigh approximately 1 amu. Because this is true, the mass number of an atom is approximately equal to its atomic mass in atomic mass units. The approximation is usually good to at least the first decimal place.

3.9 AVERAGE ATOMIC WEIGHT

Look at the square devoted to the element carbon in the periodic table. Above the symbol, the number 6 is the atomic number. Below the symbol is the **average atomic weight,** 12.01. The average atomic weight reflects the fact that elements are made up of atoms with different masses, i.e., isotopes. In the case of carbon, most carbon atoms are carbon-12, mass = 12 amu. But about 1 percent of carbon atoms are carbon-13, mass = 13 amu, and there are a very small number of carbon-14 atoms, mass = 14 amu. The weighted average turns out to be 12.01 amu. The weighted average takes into account the amount of each isotope as well as the weight of each isotope. Notice the other atomic weights (the numbers below the symbols) in the periodic table. They are not whole numbers because they are weighted averages of the masses of all isotopes of that element.

In order to determine an average atomic weight, one needs to know (1) which isotopes exist for a given element and (2) the percentage abundance of each isotope.

For example, the element magnesium is made up of three isotopes: 78.7 percent ^{24}Mg, 10.1 percent ^{25}Mg, and 11.2 percent ^{26}Mg. The weight contributed by each isotope is its fractional abundance (i.e., the percentage in fractional or decimal form) multiplied by the isotopic mass. Therefore, the weight contributions for magnesium are:

	Fractional Abundance	×	Isotopic Mass	=	Weight Contribution
^{24}Mg contribution:	78.7/100	×	24.0	=	18.9
^{25}Mg contribution:	10.1/100	×	25.0	=	2.53
^{26}Mg contribution:	11.2/100	×	26.0	=	2.91
					24.34

The average atomic weight is the sum of the contributions. In this example, the average weight of Mg is 24.3 amu.

3.10 ELECTRONIC VERSUS NUCLEAR CHANGES

As you will see shortly in Chapter 4, the chemistry that we ordinarily encounter in everyday life involves electrons. The numbers and/or arrangements of electrons in atoms change as chemical reactions occur, but the nucleus containing protons and neutrons remains unaltered. However, the nuclei of some isotopes of many elements do undergo changes in their numbers of protons and/or neutrons during a process called radioactive decay or **radioactivity.** Such isotopes are called **radioisotopes.**

For example, carbon-14 is radioactive and its nucleus changes in such a way that it acquires an extra proton. This nuclear change converts the carbon atom (with six protons) to a nitrogen atom (with seven protons). The most abundant isotope of carbon, ^{12}C, has no tendency to undergo such a nuclear change. Carbon is typical of the common elements. The most abundant isotopes of the common elements are not radioactive, but usually a small amount of some radioisotope (generally of higher mass) does exist.

On the other hand, all isotopes of the element uranium are radioactive. This behavior is typical of the heavier elements (those of larger atomic numbers) such as uranium. The most abundant isotope of uranium is ^{238}U; the isotope of uranium that is most useful as a nuclear fuel is ^{235}U.

The rest of this chapter is devoted to the consequences of nuclear changes. However, you should bear in mind that this is a special branch of chemistry, **nuclear chemistry.** The chemistry of everyday life, outside the x-ray lab or nuclear power plant, involves electrons and electrons only.

3.11 RADIOACTIVITY

A great deal of radiant energy is emitted by the nucleus during nuclear change. This radiant energy is like the energy of a visible light beam, but it is much more powerful than visible light. Nuclear radiation was originally discovered accidentally because of its similarity to light rays. Like visible light, the radiant energy from nuclei can expose photographic film.

In 1896 in France, Henri Becquerel happened to place a sample of a uranium compound on top of a photographic plate wrapped in opaque black paper in a dark drawer. Several days later he developed the photographic plate and found the image of the outline of the uranium crystal on the plate. He concluded that it was the uranium sample that gave off radiant energy since the photographic plate had been subjected to no other source of radiation.

Marie (Sklodowska) and Pierre Curie took up the study of the mysterious emissions from uranium, and it was they who coined the term *radioactive* to describe elements which spontaneously give off radiation. In the course of their studies of radioactivity, they discovered two previously unknown elements, polonium (Po) (named for Marie's homeland, Poland) and radium

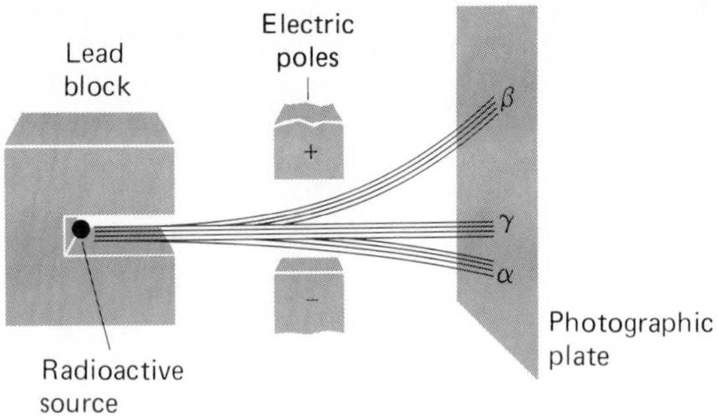
Electric
poles
Lead
block
+
−
β
γ
α
Photographic
plate
Radioactive
source

FIG. 3.4
The existence of three types of radiation can be demonstrated by passing a beam from a radioactive source between electrical plates. The radiation that is unaffected is called gamma (γ). That bent toward the positive plate is called beta (β). That bent toward the negative plate is called alpha (α). The fact that β-particles are deflected more than α-particles indicates that β-particles must be lighter than α-particles.

(Ra). In 1903 the Curies shared the Nobel Prize in physics with Henri Becquerel for their discoveries.

Meanwhile in England, Ernest Rutherford was also studying radiation, and he was able to establish that there were two types of radiation, which he called alpha (α) and beta (β). (Alpha and beta are the first two letters of the Greek alphabet.) His experiments also showed that these two types of nuclear radiation involve the emission of particles from the nucleus.

Another scientist demonstrated the existence of a third type of radiation which does not involve particles. This is γ-radiation, named for the third letter of the Greek alphabet, gamma. An experimental setup that identifies the three types of radiation is shown in Figure 3.4. When a beam of nuclear radiation is passed between electric or magnetic poles, the beam splits into three separate beams. One beam, α, bends toward the negative plate, another, β, bends toward the positive plate, and the third, γ, is unaffected by the electric poles and passes along the same line as the original emission.

3.12 PROPERTIES OF α-, β-, AND γ-RADIATION

The experiment shown in Figure 3.4 clearly shows that α-radiation is positive because it is attracted toward the negative plate, β-radiation is negative because it is attracted to the positive plate, and γ-radiation is electrically neutral since it is attracted to neither plate. The magnitudes of these charges were also determined by experiment and found to be α, $+2$; β, -1; and γ, 0, using the same relative scale as for the subatomic particles. These and other properties are summarized in Table 3.2.

Alpha (α) Radiation

Alpha radiation involves the emission of particles from a radioactive nucleus. From a determination of the relative charge ($+2$) and relative mass (4 amu) of α-particles, scientists were able to conclude that α-particles are identical to helium nuclei. Identity is symbolized $\equiv$.

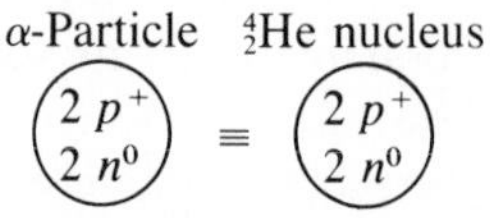
α-Particle
⁴₂He nucleus
2 p⁺
2 n⁰
≡
2 p⁺
2 n⁰

TABLE 3.2 Characteristics of the Three Types of Nuclear Radiation

Type of Radiation	Symbol	Composition	Charge	Relative Mass	Velocity*	Penetrating Power
Alpha	α, ${}^4_2$He	Identical to the He nucleus	+2	4	Variable, 10% c	Low
Beta	β, ${}^{\ 0}_{-1}e$	Identical to the electron	−1	1/1840	Variable, as fast as 90% c	Moderate
Gamma	γ, ${}^0_0\gamma$	High-energy radiation	0	0	c	High

* c is the standard symbol for the speed of light; $c = 3 \times 10^{10}$ cm/s.

Because α-particles are the heaviest of the three radiations, they are the slowest-moving and least-penetrating of the three types. They are readily stopped by a few pieces of paper or a layer of human skin. Therefore, α-radiation is usually not hazardous to living organisms unless swallowed or inhaled. Figure 3.5 shows the relative penetrating power of the three radiation forms.

Beta (β) Radiation

Beta radiation is also particulate. The mass and charge of the β-particle indicate that it is identical to an electron:

β-Particle $\equiv e^-$

Because β-particles are smaller and faster-moving, they have about 100 times the penetrating power of α-particles. An aluminum plate, a block of wood, or heavy protective clothing is necessary to stop β-radiation. Although most β-radiation is not sufficiently energetic to reach the internal organs of the body, it goes deep enough within the outer layers of skin to cause damage (similar to severe sunburn) and represents a special hazard to your eyes.

FIG. 3.5
The penetrating powers of α-, β-, and γ-radiation vary greatly. Alpha particles are stopped by paper alone. Beta particles are stopped by a wood block or aluminum plate. Gamma rays are very penetrating. Notice that a small amount of γ-radiation is able to get through even a lead block approximately 5 cm thick.

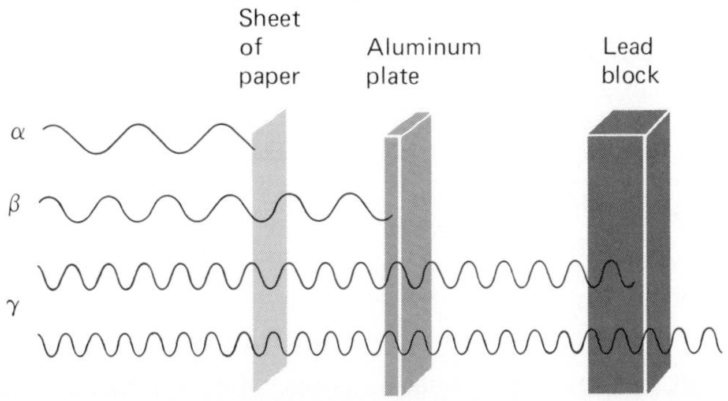
Sheet of paper
Aluminum plate
Lead block
α
β
γ

Gamma (γ) Radiation

Gamma radiation is pure energy; γ-rays are not particles. This high-energy radiation, which is similar to x-rays, travels at the speed of light and can be stopped only by a block of several layers of lead. Gamma radiation easily penetrates the skin and can cause severe internal damage. We assume you have an idea of the properties of x-rays and know about their ability to travel through space (or the body) from your contact with them in medical and dental diagnosis. Your notions are sufficient for an understanding of the concepts being presented.

3.13 IONIZING RADIATION

Nuclear radiation, like x-radiation, is also referred to as *ionizing radiation*. This is so because the interaction of α-, β-, and γ-radiation and x-rays often causes ionization of an atom or molecule (group of atoms linked together). The radiation collides with the atom and "kicks out" an electron or two. For example, visualize the effect on an argon atom:

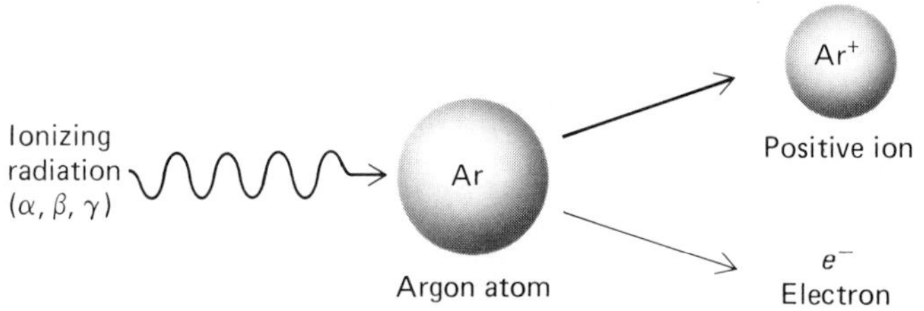
Ionizing radiation (α, β, γ)
Ar
Argon atom
Ar^+
Positive ion
e^-
Electron

Recall from Section 3.5 that ions and atoms of an element have very different properties.

Within the cells of the body, ionization effects can be devastating. Smooth-working body chemistry depends on the presence of particular substances participating in certain reactions. Alteration of these substances through ionization interferes with necessary cell reactions and leads to other undesirable reactions. The result is a total disruption of cell activity.

The ionizing power of nuclear radiation is inversely related to its penetrating power; that is, γ-radiation is most penetrating but produces the fewest ions per unit volume. On the other hand, the least penetrating radiation, α, produces the largest number of ions, and β-radiation is moderate in penetrating power and intermediate in ionizing ability. Many of the units which measure the power or biological effectiveness of a radioactive source relate to ionizing ability. See Table 3.3 for the definition of units which are used to measure radiation.

Many of the devices for detecting and measuring radioactive emissions depend on the fact that nuclear radiation produces ions. For example, probably the best known device for detecting and measuring radiation is the Geiger-Müller tube, or Geiger counter. This instrument is a tube filled with gas such as argon (see Figure 3.6). Entering radiation ionizes the gas, and an electric current is produced as electrons go to the positive electrode and argon (+) ions go to the negative electrode. The amount of current is proportional to the amount of radiation. Frequently, the electric current is used to produce audible clicks, the speed and intensity of which increase with increasing radiation.

Radiation intensity decreases rapidly with increasing distance from the source of radiation. Doubling your distance from a radiation source ensures that the radiation felt will be one-fourth of the original value; at triple the distance, the radiation is one-ninth the original value. Quantitatively this is

TABLE 3.3 Units of Radiation Measurement

Unit	What It Measures	Description	Example
Curie (Ci)	Activity	An amount of radioactive material that undergoes 3.7×10^{10} disintegrations per second.	^{60}Co γ-ray sources delivered to hospitals for radiation therapy are typically in the milliCurie range.
Roentgen (R)	Intensity or exposure	An amount of x- or γ-radiation that produces 1 electrostatic unit of charge (~ 20 billion ion pairs) in 1 mL of dry air at 0°C and 1 atm pressure.	The output from an x-ray machine is described in terms of roentgens.
Radiation absorbed dose (rad)	Absorbed dose	An amount of radiation which when absorbed by tissue delivers 100 ergs per gram of tissue (1 erg = 2.4×10^{-8} cal). For x- and γ-rays 1 R delivers 1 rad.	Radiation therapists quote dosages in rads.
Relative biological effectiveness (rbe)	Biological effectiveness	A quality factor which accounts for the fact that the same amount of exposure to different forms of radiation produces different amounts of biological damage.	X-, γ-, and β-rays all have approximately the same effectiveness and are given an rbe = 1. α-rays are much more damaging and are rated at 10 rbe.
Roentgen equivalent for man (rem)	Dose equivalent weighted for different kinds of radiation	An amount of radiation which produces the same damage to tissue as 1 R of x-radiation. Dose in rems = dose in rads × rbe. Dosages in rems are additive; thus cumulative effects are expressed by this unit.	Preferred dosage statement because it compensates for the differences in effectiveness among the forms of radiation.

FIG. 3.6
Geiger-Müller counter. Ionizing radiation enters the Geiger tube through the thin window at the left of the apparatus. As the radiation passes through the gas inside the tube, it ionizes argon atoms along its path by kicking out electrons. The Ar^+ ions are attracted to the negatively charged walls. The electrons are attracted to the central rod, which is positively charged. This produces an electrical current in the circuit which is amplified and recorded or heard as a click.

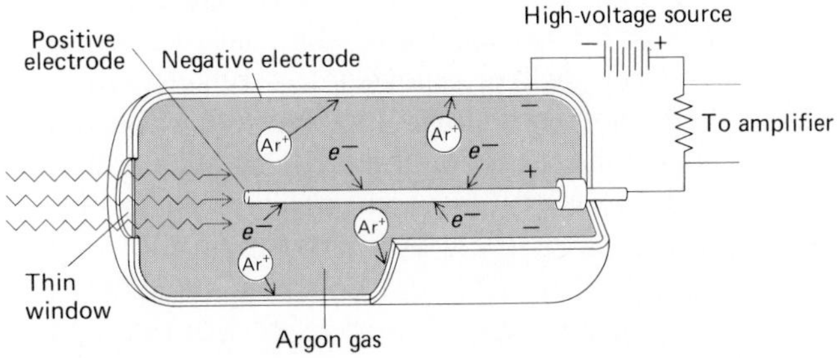
High-voltage source
Positive electrode
Negative electrode
To amplifier
Ar+
e−
Thin window
Argon gas

known as the inverse square law: intensity I is inversely proportional to the square of the distance from the source.

$$I \propto \frac{1}{d^2}$$

This diminishment of radiation with distance offers the simplest protection against radiation. Distance and shielding are the principal safety measures practiced around the radiation laboratory and clinic.

3.14 PHYSIOLOGICAL EFFECTS OF RADIATION

Ionizing radiation is most damaging to the *nuclei* of living cells. The cell nucleus contains the "blueprints" for producing more identical cells. Because the cell nucleus directs division and replication, cells that are dividing most rapidly are the first to show the effects of radiation. Genetic damage frequently occurs because the principal genetic material resides in the cell nucleus (see Section 20.2). The cells most susceptible to radiation damage are those in the lymphatic system, bone marrow, intestinal tract, reproductive organs, and lens of the eye.

The body can tolerate exposure to small amounts of ionizing radiation without apparent symptoms. Thus, background radiation (from the soil and outer space) and medical x-rays produce no noticeable harm. However, small doses may be cumulative, and it is important for persons who work near sources of radiation to monitor their exposure. A common device employed for this purpose is the film badge which makes use of the fact that photographic film detects radiation. Film badges are developed periodically to determine the extent of exposure.

One source of background radiation in homes is the gas radon. Specialists fear it may be appearing in many homes in amounts greater than previously believed. The source of the radon is soil or rocks on which a house is built or the actual building materials.

Rocks, especially granite, contain minute amounts of radium-226, a solid which, as such, stays in the rock. However, ^{226}Ra emits alpha particles and is thereby converted to radon. (You will see this conversion in Sample Exercise 3.3.) The radioactive gas radon goes into the air. Outdoors the radon atoms are quickly dispersed and are harmless. Indoors, on the other hand, they may accumulate.

In some homes, background radiation from radon may provide more unwanted exposure to radiation in the course of a year than the medical x-rays a person may have. Increased incidence of lung cancer from this exposure is especially likely, because radon is a gas, and a source of very damaging alpha radiation can therefore be inhaled into the lungs.

Even in the absence of noticeable symptoms of illness, excessive low-level exposure could lead to sterility or birth defects because reproductive cells and fetal tissue are especially sensitive to radiation. The first clinical symptom for higher levels of exposure is a drop in white blood cell count. White blood cells have short life spans; therefore damage to tissue producing these cells shows up rapidly. A reduction in white blood cell count increases susceptibility to infection because a person's natural resistance is lowered. Red blood cells are also affected, and anemia may result.

Higher doses of radiation cause symptoms such as nausea, vomiting, and diarrhea because of damage to cells in the intestinal tract.

The highest doses may produce burns to the skin, clouding of the eye lens (cataracts), and frequently death because of damage to so many essential

TABLE 3.4 Effects on Human Beings of Short-Term Whole-Body Radiation Exposure

Dose, rems	Effects
0–25	No detectable clinical effects.
25–100	Slight short-term reduction in number of some blood cells. Disabling sickness not common.
100–200	Nausea and fatigue, vomiting if dose is greater than 125 rems, longer-term reduction in number of some blood cells.
200–300	Nausea and vomiting first day of exposure, up to a 2-week latent period followed by appetite loss, general malaise, sore throat, pallor, diarrhea, and moderate emaciation. Recovery in about 3 months unless complicated by infection or injury.
300–600	Nausea, vomiting, and diarrhea in first few hours. Up to a 1-week latent period followed by loss of appetite, fever, and general malaise in the second week, followed by hemorrhage, inflammation of mouth and throat, diarrhea, and emaciation. Some deaths in 2 to 6 weeks. Eventual death for 50% if exposure is above 450 rems; others recover in about 6 months.
600 or more	Nausea, vomiting, and diarrhea in first few hours. Rapid emaciation and death as early as second week. Eventual death of nearly 100%.

From *Medical Aspects of Radiation Accidents,* E. L. Saenger, ed. (Washington, D.C.: United States Atomic Energy Commission, 1963), p. 9.

body functions. If a person does survive massive exposure to radiation, the likelihood of developing cancer, particularly leukemia or blood cancer, is greatly increased. Table 3.4 lists some of the effects on human beings of short-term radiation exposure.

3.15 NUCLEAR DECAY

Let's get back to the fact that the emission of radiation accompanies a change in the nucleus of the emitting radioactive atom. The change in the nucleus produces an atom of a new element. For example, in Section 3.10, the conversion of $^{14}_{6}C$ to $^{14}_{7}N$ was described. The change from one element to another is called **transmutation.** Because radioactive elements disappear as they emit radiation and are transmuted into other elements, they are said to disintegrate or *decay*. Because we know the composition of α- and β-particles, we can predict the identity of the new element formed from the emitting element as it decays. Using symbols, we can record this conversion, or nuclear reaction, by a nuclear equation of the form:

$$\text{Emitting element} \longrightarrow \text{emitted particle} + \text{new element}$$

α-Emission

Uranium-238 is an α-emitter. This means $^{238}_{92}U$ nuclei lose α-particles, which we know to be composed of two protons and two neutrons, i.e., the helium nucleus, $^{4}_{2}He$. The nuclear equation for this event begins with the uranium-238 emitter and shows the emitted α-particle represented by its nuclear symbol, $^{4}_{2}He$.

$$^{238}_{92}U \longrightarrow {}^{4}_{2}He + ?$$

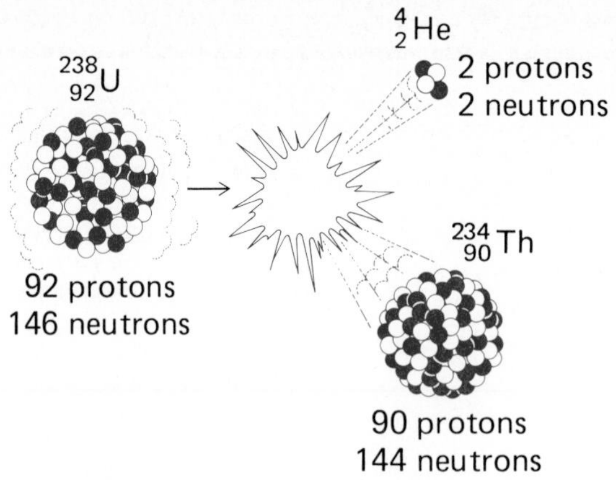

$^{4}_{2}He$
2 protons
2 neutrons
$^{238}_{92}U$
92 protons
146 neutrons
$^{234}_{90}Th$
90 protons
144 neutrons

$$^{238}_{92}U \longrightarrow\ ^{4}_{2}He + {}^{234}_{90}Th$$

FIG. 3.7 Alpha emission from uranium-238. Two protons and two neutrons leave the uranium-238 nucleus, leaving behind 90 protons and 144 neutrons. A nucleus with 90 protons is the element thorium.

Because of the loss of two protons in the α-particle, after emission the original nucleus contains only 90 protons $(92-2)$. Because the number of protons characterizes an atom, this nucleus must be the element thorium, which we determine by looking up atomic number 90 in the periodic table. The mass number of the Th atom is 4 less than the original U atom because a total of four particles, each of mass 1, have been lost. The nuclear equation is completed by showing the symbol for the thorium isotope.

$$^{238}_{92}U \longrightarrow\ ^{4}_{2}He + {}^{234}_{90}Th$$

In this, as in all nuclear equations, there is a balance (that is, their sums are equal) of both atomic numbers $(92 = 2 + 90)$ and mass numbers $(238 = 4 + 234)$ on the two sides of the arrow, which is like an equals sign. Nothing is lost; all protons and neutrons can be accounted for. Figure 3.7 illustrates the reaction and equation.

Another α-emitter is $^{218}_{84}Po$. The new element formed upon α-emission is an isotope of lead.

$$^{218}_{84}Po \longrightarrow \overbrace{\underbrace{^{4}_{2}He + {}^{214}_{82}Pb}_{84}}^{218}$$

SAMPLE EXERCISE 3.3

Write the nuclear equation for the change that occurs in radium-226 when it emits an α-particle. Radium-226 was the first radioisotope used to treat cancer.

Solution

STEP 1 Write the symbol of the emitter, including atomic number and mass number, on the left-hand side of the equation. In this case, the mass number is given, and you must find radium's atomic number in the periodic table.

$$^{226}_{88}Ra \longrightarrow$$

STEP 2 Write the symbol for the α-particle which shows that it is a helium nucleus on the right side of the equation:

$$^{226}_{88}Ra \longrightarrow\ ^{4}_{2}He + ?$$

STEP 3 Complete the equation by writing a symbol for an isotope that has an atomic number 2 less than the original isotope (88 − 2 = 86), and a mass number 4 less than the original (226 − 4 = 222). Use the periodic table to ascertain that atomic number 86 identifies radon, Rn.

$$^{226}_{88}Ra \longrightarrow {}^{4}_{2}He + {}^{222}_{86}Rn$$

STEP 4 Check the equation to see that the mass numbers and atomic numbers are balanced, that is, that the totals on each side of the arrow are equal.

$$226 = 4 + 222$$

$$88 = 2 + 86$$

Problem 3.10 Write a nuclear equation for α-emission from $^{239}_{94}Pu$.

β-Emission

Because you know that β-particles are electrons, you may be wondering how an electron emerges from the nucleus which contains only protons and neutrons. The answer is that, in effect, a neutron is transformed into a proton which stays in the nucleus and an electron which leaves the nucleus. The overall result is that the number of protons in the nucleus increases by 1 (atomic number increases by 1), the number of neutrons decreases by 1, and the mass number remains the same, since the sum of the number of protons and neutrons is not altered. For example, $^{90}_{38}Sr$, a product of radioactive fallout from nuclear testing, is a β-emitter. One of its 52 neutrons is transformed into a proton, and an electron is emitted. Thus the original nucleus contained 52 neutrons and 38 protons, and the new nucleus contains 51 neutrons and 39 protons. The element with atomic number 39 is yttrium. The mass number of the new element is 90 because 51 neutrons + 39 protons has that mass. This reaction is illustrated in Figure 3.8 and can be summarized by the nuclear equation

$$^{90}_{38}Sr \longrightarrow {}^{0}_{-1}e + {}^{90}_{39}Y$$

The use of a symbol $^{0}_{-1}e$ for the β-particle enables us to check that the

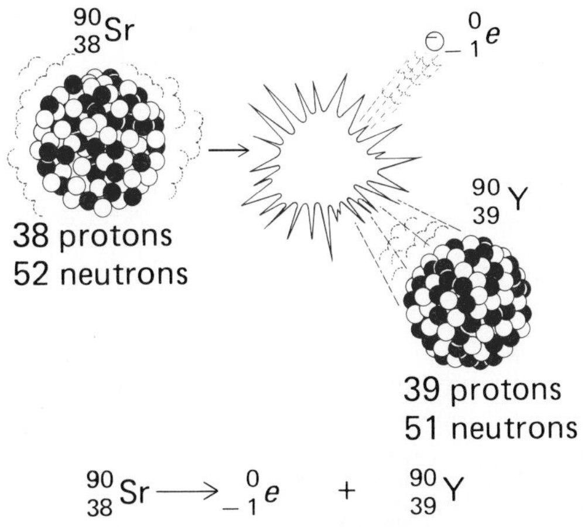

$^{90}_{38}Sr$
$^{0}_{-1}e$
$^{90}_{39}Y$
38 protons
52 neutrons
39 protons
51 neutrons
$^{90}_{38}Sr \longrightarrow {}^{0}_{-1}e + {}^{90}_{39}Y$

FIG. 3.8
Beta emission from strontium. One of strontium's neutrons decomposes to an electron and a proton. The electron leaves the nucleus as a β-particle. The nucleus now contains one more proton (39) and one less neutron (51). A nucleus with 39 protons is the element yttrium.

nuclear equation is balanced.

$$90 = 0 + 90$$
$$38 = -1 + 39$$

Another β-emitter is $^{210}_{83}Bi$. The new nucleus formed has 84 protons and the same mass number.

$$^{210}_{83}Bi \longrightarrow {}^{0}_{-1}e + {}^{210}_{84}Po$$

SAMPLE EXERCISE 3.4 Write the nuclear equation for the change that occurs in cobalt-60, the usual medicinal source of γ-radiation, when it emits a β-particle.

Solution STEP 1 Write the symbol of the emitter, including atomic number and mass number, on the left-hand side of the equation. From the periodic table, the atomic number of cobalt is 27. The mass number 60 is given.

$$^{60}_{27}Co \longrightarrow$$

STEP 2 Write the symbol for the β-particle which shows that it is an electron on the right side of the equation.

$$^{60}_{27}Co \longrightarrow {}^{0}_{-1}e + ?$$

STEP 3 Complete the equation by writing a symbol for an isotope that has an atomic number 1 greater than the original isotope ($27 + 1 = 28$) and the same mass number. Use the periodic table to ascertain that atomic number 28 identifies nickel, Ni.

$$^{60}_{27}Co \longrightarrow {}^{0}_{-1}e + {}^{60}_{28}Ni$$

STEP 4 Check the equation by seeing whether the mass numbers and atomic numbers are balanced.

$$60 = 0 + 60$$
$$27 = -1 + 28$$

Problem 3.11 Write a nuclear equation for β-emission from $^{32}_{15}P$.

γ-Emission

Virtually all α- and β-emissions are accompanied by γ-emission. Because γ-radiation produces no change in nuclear contents, it is usually ignored in writing nuclear equations. For example, γ-emission from cobalt-60 is used in the treatment of cancerous tumors. However, in Sample Exercise 3.4, only β-emission was mentioned. We can include the fact that γ-emission accompanies β-emission in the nuclear equation. Notice that the symbol $^{0}_{0}\gamma$ does not alter the balanced equation.

$$^{60}_{27}Co \longrightarrow {}^{0}_{-1}e + {}^{60}_{28}Ni + {}^{0}_{0}\gamma$$

3.16 HALF-LIFE, $t_{1/2}$

The rate at which radioactive elements decay varies greatly. Rate of radioactive decay is usually measured in terms of an element's **half-life,** symbolized $t_{1/2}$, *the time required for the decay of one-half of a radioactive sample*. For example, the half-life of ^{51}Cr, a radioisotope used to study blood volumes, is 28 days. This means that if you have 100 mg of ^{51}Cr today, then 28 days from today you would have only 50 mg because one-half the sample (50 mg) would have decayed. During the following 28 days, half the 50 mg would decay, so that only 25 mg would remain, and the process would continue.

$$100 \text{ mg } {}^{51}\text{Cr} \xrightarrow{\text{28 days}} 50 \text{ mg } {}^{51}\text{Cr} \xrightarrow{\text{28 days}} 25 \text{ mg } {}^{51}\text{Cr} \xrightarrow{\text{28 days}} 12.5 \text{ mg } {}^{51}\text{Cr} \xrightarrow{\text{28 days}} \cdots$$

The listing of half-lives of other common radioisotopes in Table 3.5 points out the wide variation in half-life, from a few hours to millions of years.

For medical diagnostic applications, radioisotopes with shorter half-lives are desirable. Such isotopes can be introduced into the body and used to advantage; they will continue to decay rapidly so that they essentially disappear and pose no long-term exposure threat to the patient. On the other hand, long-lived radioisotopes, such as ^{238}U, would not be suitable for internal medical applications and they present storage problems. Radioactive waste from nuclear power plants is mostly long-lived uranium isotopes which must be shielded from the environment for thousands of centuries before the uranium decays sufficiently to be harmless.

Calculations relating amounts of radioactive material and $t_{1/2}$ can be done

TABLE 3.5 Half-lives and Uses of Selected Radioisotopes

Name	Symbol	Half-life	Radiation Emitted	Usefulness
Carbon-14	$^{14}_{6}C$	5720 years	β	Radioactive dating and labeling
Sodium-24	$^{24}_{11}Na$	15 h	β, γ	Blood-circulation studies
Phosphorus-32	$^{32}_{15}P$	14 days	β	Fertilizer uptake monitor
Sulfur-35	$^{35}_{16}S$	88 days	β	Protein label for metabolic studies
Potassium-42	$^{42}_{19}K$	12 h	β	Plant and animal nutrition studies
Calcium-45	$^{45}_{20}Ca$	165 days	β	Animal nutrition studies
Iron-59	$^{59}_{26}Fe$	45 days	β	Red blood cell studies
Cobalt-60	$^{60}_{27}Co$	5.3 years	β, γ	Radiation therapy
Yttrium-90	$^{90}_{39}Y$	64 h	β	Pituitary implant radiation therapy
Technetium-99	$^{99}_{43}Tc$	6 h	γ	Brain scans
Iodine-123	$^{123}_{53}I$	13 h	γ	Thyroid radiation therapy
Iodine-131	$^{131}_{53}I$	8 days	β, γ	Thyroid activity studies
Uranium-235	$^{235}_{92}U$	710 million years	α, γ	Nuclear reactors

by using the step-by-step format shown earlier for the decay of ^{51}Cr. Alternatively, you can use the following equation which relates the quantity of material remaining to the number n of half-lives elapsed.

$$\text{Quantity remaining after } n \text{ half-lives} = \text{original quantity} \times (\tfrac{1}{2})^n \qquad (3.1)$$

For example, if you begin with 8.0 g of ^{131}I, which has $t_{1/2} = 8$ days, and want to know how much of this sample is left after 32 days, determine the number of half-lives elapsed and use Equation (3.1). Note that 32 days represents four half-lives ($n = 4$). Therefore,

$$\text{Quantity remaining after 4 half-lives} = \underset{\text{Original sample}}{8.0 \text{ g}} \times (\tfrac{1}{2})^4$$

$$= 8.0 \text{ g} \times \frac{1}{2 \times 2 \times 2 \times 2} \qquad (3.1)$$

$$= 8.0 \text{ g} \times \frac{1}{16}$$

$$= 0.5 \text{ g}$$

SAMPLE EXERCISE 3.5 The half-life of ^{24}Na is 15 h. If you have a 240-mg sample of this radioisotope at noon on a Monday, how many milligrams will remain at 3 P.M. on Thursday?

Solution In order to use Equation (3.1), it is necessary to determine the total time elapsed and from that the number n of half-lives that have elapsed.

$$\text{Monday noon} \xrightarrow{24\text{ h}} \text{Tuesday noon} \xrightarrow{24\text{ h}} \text{Wednesday noon} \xrightarrow{24\text{ h}}$$
$$\text{Thursday noon} \xrightarrow{3\text{ h}} \text{Thursday 3 P.M.}$$

Thus, 75 h has elapsed. Because the half-life is 15 h, five half-lives elapse:

$$75 \not{h} \times \frac{1 \text{ half-life}}{15 \not{h}} = 5.0 \text{ half-lives}$$

Now use Equation (3.1).

$$\text{Quantity remaining} = 240 \text{ mg} \times \frac{1}{2 \times 2 \times 2 \times 2 \times 2}$$

$$= 240 \text{ mg} \times \frac{1}{32}$$

$$= 7.5 \text{ mg}$$

Problem 3.12 The half-life of ^{214}Bi is 20 min. How much of an original 16-g sample remains after 2 h 20 min?

The half-life concept can be used to determine the age of objects which contain radioisotopes. The age of museum relics can often be determined by carbon-14 dating. This procedure is based on the fact that all living things—plants and animals alike—are composed principally of the element carbon. Furthermore, whereas most carbon atoms are the nonradioactive $^{12}_{6}C$ isotopes, a small percentage of carbon atoms are radioactive $^{14}_{6}C$ isotopes.

While a plant is alive, the ratio of ^{14}C to ^{12}C isotopes is a constant because although ^{14}C is continually decaying in the plant, it is being replaced by ^{14}C from carbon dioxide containing ^{14}C ($^{14}CO_2$) in the atmosphere. Similarly, the ^{14}C:^{12}C ratio is constant in animals because they continually eat fresh plants and inhale some $^{14}CO_2$. When a plant or animal dies, the ^{14}C continues to decay but is no longer replaced by new ^{14}C, and the ^{14}C:^{12}C ratio diminishes.

The Dead Sea Scrolls were dated by examining the ^{14}C:^{12}C ratio in their paper (paper is made from trees). It was found that the ratio in these scrolls was only 79.5 percent of the ratio in a living plant. From this and the known $t_{1/2}$ of carbon-14 (5720 years), archaeologists calculated that the scrolls are approximately 1900 years old.

Similar dating procedures are applied with other radioisotopes. For example, rocks can be dated from their ^{238}U or ^{87}Rb content.

3.17 MEDICAL USES OF RADIOISOTOPES

Radioisotopes have been found to be very useful in chemistry, biology, industry, and especially medicine because of their following properties:

1 They are easily detected in even minute amounts.
2 The chemical reactivity of radioisotopes (that is, electron activity as described in Section 3.10) is identical to that of nonradioactive isotopes.
3 Radiation damages cells, particularly those which divide rapidly.

The first two properties are used in medical diagnosis. Some medicinal treatments employ the latter two properties. Other properties desirable in radioisotopes used medicinally are a short half-life, so that exposure is not long-term, and high energy, so that they are intense and easily detected.

Radioactive Labeling

Radioisotopes react and combine with other elements just like nonradioactive elements. For example, the compound sodium iodide (NaI), which can be used to study thyroid activity, is ordinarily composed of nonradioactive ^{23}Na and ^{127}I. However, ^{23}Na can equally readily combine with radioactive ^{131}I. This sodium iodide ($^{23}Na^{131}I$) is radioactively *labeled*. If it is introduced into the body, such a labeled compound can be readily detected and followed, that is, *traced* throughout the body. Tracing is the basis of medical diagnostic techniques.

Medical Diagnosis

The rate at which a thyroid gland absorbs ^{131}I from a $Na^{131}I$ solution a patient has drunk can be readily monitored by a Geiger counter. This rate of absorption indicates whether the thyroid is working properly or is underactive or overactive.

A particularly useful radioisotope with several applications is $^{99}_{43}Tc$. For example, labeled sodium pertechnitate ($NaTcO_4$), a combination of Na^+ ions and TcO_4^- ions, is often used for brain studies. Ordinarily, an ion such as TcO_4^- cannot pass the blood-brain barrier, the body's mechanism for protecting brain cells. However, certain tumors or other abnormalties seem to ignore this barrier and hence radioactive TcO_4^- can enter brain tissue. The resulting accumulation of ^{99}Tc in brain tissue indicates to doctors that brain abnormalities are present.

^{99}Tc combined in other forms has an affinity for tissue types other than brain tissue. $Tc_2P_2O_7$, technetium pyrophosphate, selectively collects in bone tissue; a ^{99}Tc-sulfur combination is taken up preferentially by cells of the liver, spleen, and bone marrow. Studies of the lung and kidneys are also possible using technetium radiopharmaceuticals.

Other radioisotopes are also used for medical diagnosis. Iron-59 is used to study the formation of red blood cells because the compound hemoglobin in red blood cells contains iron. Sodium-24 is commonly used to study blood circulation. A ^{24}Na sample is injected into the bloodstream, and its course throughout the body is followed by a Geiger tube.

Medical Treatment

Radioisotopes have been used for medical treatment almost from the very first moment of the discovery of radium. Malignant tissues are irradiated with γ-radiation, usually from a cobalt-60 source. Ionizing radiation is more damaging to fast-growing cancer cells than it is to normal tissues, so it is possible to kill the cancer cells while leaving normal cells relatively unharmed. The rays are directed as much as possible toward the tumor, while normal tissue is shielded.

Another treatment technique is to use radioisotopes that show selectivity for certain kinds of tissues. Thus ^{131}I can be used to treat cancerous thyroid glands because iodine is absorbed only by thyroid tissue and accumulates there. ^{123}I may also be used. It has two advantages over ^{131}I: its half-life is shorter (13 h) and it emits only γ-radiation.

Radioactive iodine is also sometimes used to destroy healthy (nonmalignant) thyroid tissue. In Graves' disease, the thyroid is dangerously overactive. One treatment involves ingestion of a pill containing ^{131}I. The radioactive iodine accumulates in the thyroid gland, emits radiation, and destroys some thyroid tissue, thus diminishing the activity of the gland.

Leukemia can be treated by phosphorus-32 which becomes part of bone structure. Bones normally incorporate phosphorus from dietary sources into their makeup. ^{32}P in bone tissue emits beta radiation; the excess white blood cells in the leukemia patient's bone marrow are thus exposed to radiation and many are killed.

3.18 NEW ELEMENTS THROUGH BOMBARDMENT

Radioactive decay is a natural process that occurs spontaneously. However, it is also possible to induce nuclear reactions by bombarding stable nuclei with nuclear-sized "bullets." For example, if a high-speed neutron is shot at the stable isotope of aluminum $^{27}_{13}$Al, a radioactive aluminum isotope is produced:

$$^{27}_{13}\text{Al} + {}^{1}_{0}n \longrightarrow {}^{28}_{13}\text{Al} \qquad \textbf{neutron bombardment}$$

In this case the force of the collision has simply caused neutron capture: that is, the neutron is taken into the aluminum nucleus. The $^{28}_{13}$Al is a β-emitter and decays:

$$^{28}_{13}\text{Al} \longrightarrow {}^{0}_{-1}e + {}^{28}_{14}\text{Si} \qquad \textbf{decay}$$

Neutron bombardment also occurs naturally and produces other results besides simple capture. Carbon-14 exists in the atmosphere because of high-speed collisions between neutrons from high-energy cosmic rays in the upper atmosphere and the common isotope of nitrogen:

$$^{14}_{7}\text{N} + {}^{1}_{0}n \longrightarrow {}^{14}_{6}\text{C} + {}^{1}_{1}\text{H}$$

In this case the bombardment knocks a proton (^1_1H) out of the nucleus as the neutron enters.

Subatomic particles other than neutrons can be involved in bombardment. For example, bombardment of $^{14}_7N$ by an α-particle produces a nonradioactive isotope of oxygen:

$$^{14}_{7}N + {}^{4}_{2}He \longrightarrow {}^{17}_{8}O + {}^{1}_{1}H$$

Notice that these nuclear equations are balanced; that is, the sum of the mass numbers and the sum of the atomic numbers are identical on both sides of the equation.

Perhaps the most exciting application of bombardment reactions is in the preparation of non-naturally-occurring elements. Indeed, most of the radioisotopes routinely used in medicine do not occur naturally and are made artificially. The name *technetium,* given to the first element made artificially, is derived from a Greek word meaning "artificial."

Elements with atomic numbers greater than 92 are called *transuranium elements;* these elements do not occur naturally, but rather have been synthesized by bombardment reactions. For example, $^{241}_{94}Pu$ can be made by α-bombardment of $^{238}_{92}U$:

$$^{238}_{92}U + {}^{4}_{2}He \longrightarrow {}^{241}_{94}Pu + {}^{1}_{0}n$$

Other bombardment reactions lead to other transuranium elements. Also, the decay of these elements, all of which are radioactive, often leads to new elements. For example, the β-decay of $^{241}_{94}Pu$ produces americium-241:

$$^{241}_{94}Pu \xrightarrow{\text{decay}} {}^{241}_{95}Am + {}^{0}_{-1}e$$

Americium is used commercially in smoke detectors.

Problem 3.13 What radioisotope is produced when $^{242}_{96}Cm$ is bombarded by an α-particle and a neutron is ejected, that is,

$$^{242}_{96}Cm + {}^{4}_{2}He \longrightarrow {}^{1}_{0}n + ?$$

3.19 NEUTRON BOMBARDMENT IN RADIATION THERAPY

The medical use of neutron bombardment reactions offers certain advantages over conventional irradiation techniques. Consider the use of this technique in the treatment of *glioblastoma,* a highly malignant, fast-growing brain cancer. Neutron bombardment of boron-10 incorporated in glioblastoma tissue produces α-radiation in the malignant tissue and may cause regression of the tumor.

$$\underset{\text{Neutron}}{^{1}_{0}n} + \underset{\text{In a borate salt}}{^{10}_{5}B} \longrightarrow {}^{7}_{3}Li + \underset{\alpha\text{-Particle}}{^{4}_{2}He}$$

Conventional γ-radiation can be directed to brain tumors, but there is always some damage to surrounding tissues. The bombardment technique allows pinpoint focusing of the neutron beam on the tumor. Radiation is nearly confined to the tumor because the tumor selectively absorbs the boron-10. The blood-brain barrier tends to retard entry of the boron (as borate, a polyatomic ion; see Section 5.8) into normal brain tissue.

The alpha particles produced in the bombardment reaction have a relative biological effectiveness (rbe) 10 times that of γ-radiation; therefore, α-radia-

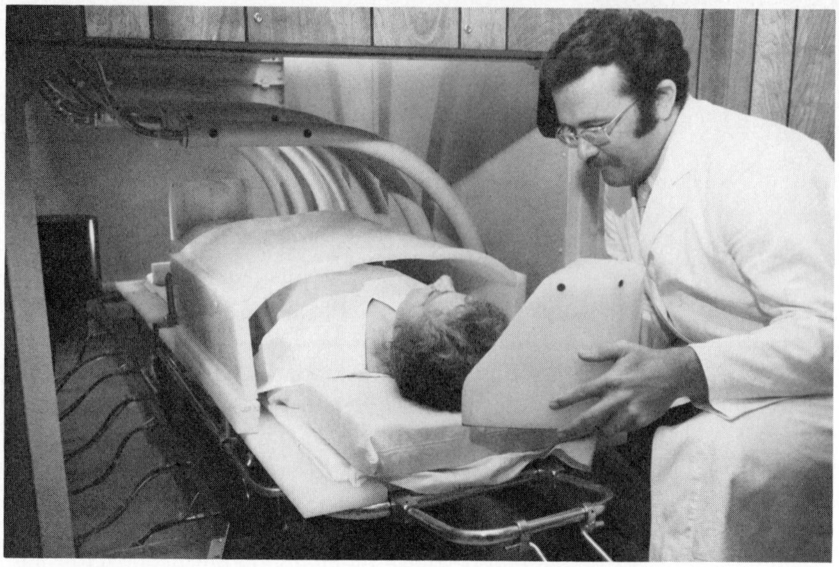

FIG. 3.9
This facility allows for Total Body Neutron Activation (TBNA). The neutron sources are located in pipes above and below the patient; the upper pipes are shown. The sources are elevated in the pipe containers for activation and lowered into shielded containers underground when not in use. This facility is located in the Medical Research Center of Brookhaven National Laboratory, which we thank for this photograph.

tion is more damaging to cancerous tissue per given dose. Alpha radiation will not penetrate skin. Consequently only special techniques allow the use of α-radiation internally. In the glioblastoma treatment, boron-10 is injected into the bloodstream and selectively absorbed by tumor cells. A neutron beam directed to the cancerous tissue produces α-radiation in the tumor when neutrons strike boron-10 isotopes.

Neutron sources at medical facilities are rare, but more are being built. The bombardment reaction will increase the treatment repertoire of the radiation therapist. Figure 3.9 shows a neutron facility.

3.20 NUCLEAR ENERGY

Radioisotopes and nuclear reactions have many other practical applications besides medical uses. For example, they provide an important energy source. The nuclear power plants in operation today all employ *nuclear fission* reactions. Nuclear fission produces a great deal of heat energy, which is used to generate steam to drive turbines and produce electricity in much the same way that heat energy from fossil fuels is used.

Nuclear fission is the splitting of an atom upon bombardment. For example,

$$^{235}_{92}\text{U} + {}^{1}_{0}n \longrightarrow {}^{141}_{56}\text{Ba} + {}^{92}_{36}\text{Kr} + 3\,{}^{1}_{0}n$$

Because more neutrons (three) are produced than are required for the fission to occur, a self-sustaining *chain reaction* will occur as long as there is a sufficient mass of ^{235}U present for the new neutrons to interact with. See Figure 3.10. This sufficient mass is called the *critical mass*. The chain reaction is self-sustaining because the product neutrons can react with the ^{235}U. After the first fission, no additional energy need be added. Indeed, a great deal of energy is released.

A major disadvantage of the use of nuclear fission in power plants is the fact that many of the products of the fission are themselves radioactive. For example, the ^{141}Ba and ^{92}Kr shown in the fission equation are both radioactive and decay to other radioactive species. This presents a problem for radioactive-waste disposal; the products as well as the reactants are radioactive.

Much more energy can be generated without the production of radioactive waste through nuclear *fusion,* which is the *combination* of two small nuclei into one larger nucleus. For example, the energy of the sun comes about

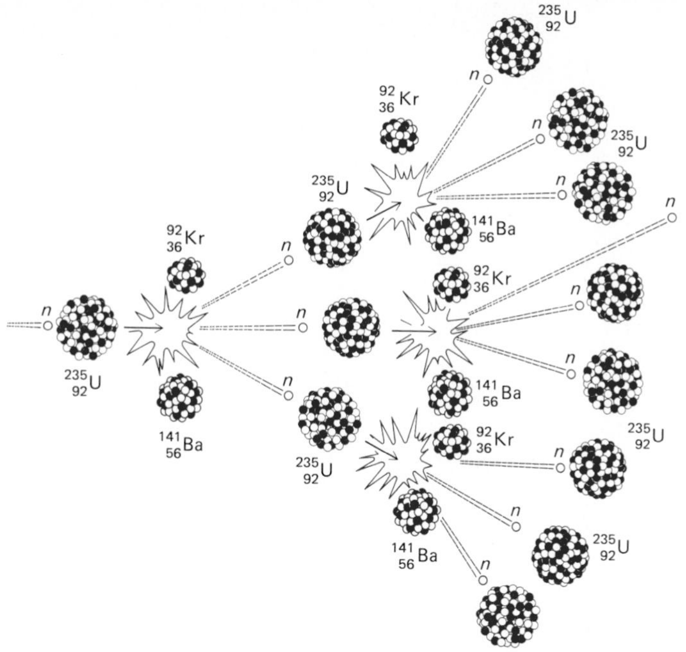
n
$^{235}_{92}U$
$^{92}_{36}Kr$
$^{141}_{56}Ba$

FIG. 3.10
The fission of one U-235 nucleus by one neutron can set off a chain reaction if there are sufficient numbers of the U-235 nuclei present (critical mass) to be split by the neutrons generated by the original fission.

through the fusion reaction

$$^{2}_{1}H + ^{3}_{1}H \longrightarrow ^{4}_{2}He + ^{1}_{0}n$$

Fusion is preferable to fission as an energy source because (1) fusion reactions produce more energy per amount of starting material than do fission reactions, (2) the starting materials for fusion are more abundant than fissionable isotopes, and (3) the products of fusion are safe, i.e., not radioactive. However, there are extreme technological problems to be overcome before nuclear fusion can be a practical means of energy production. Most notably, fusion reactions require extremely high temperatures (1–5 million°C) for initiation, and no known substance can withstand these temperatures and act as a container for the reaction. However, research continues, and it is hoped that fusion will meet the energy needs of the twenty-first century.

The harnessing of nuclear energy to generate electric power is probably the most significant industrial use of radioisotopes. Tagging or labeling techniques, i.e., the replacement of minute numbers of inactive atoms by radioisotopes, are used for wear and corrosion tests. Smoke detection devices contain small amounts of radioactive materials; the detection is based on the interaction of smoke particles and radiation which produces an electric current and sounds the alarm.

CHAPTER ACCOMPLISHMENTS

After completing this chapter you should be able to define **key words** and do the following:

3.1 INTRODUCTION
3.2 ATOMIC THEORY
1. Relate the atomic definitions of elements, compounds, and mixtures to the macroscopic definitions.

3.3 INSIDE THE ATOM

2. State the names and symbols of the three subatomic particles.
3. State the relative charges and masses of the subatomic particles.
4. Form a mental picture of atoms, showing the arrangement of the subatomic particles.

3.4 ATOMIC NUMBER

5. Given the atomic number, determine the number of protons and electrons in a neutral atom, or given the number of protons or electrons in a neutral atom, write the atomic number.

3.5 CHARGED ATOMS

6. Given the number of protons and electrons in a species, recognize the species as a neutral atom or an ion.

3.6 ISOTOPES

7. Given the number of protons and neutrons in two atoms, indicate whether the atoms are isotopes.

3.7 MASS NUMBER

8. Given the atomic and mass numbers, determine the number of subatomic particles in an atom.
9. Given the number of protons and neutrons in an atom and a periodic table, write a correct symbol showing the atomic and mass numbers.
10. Given the atomic and mass numbers of a set of atoms, indicate which atoms are isotopes.

3.10 ELECTRONIC VERSUS NUCLEAR CHANGES

11. Distinguish between the chemistry of everyday life and nuclear chemistry.

3.11 RADIOACTIVITY

12. Name the three types of nuclear radiation.

3.12 PROPERTIES OF α-, β-, AND γ-RADIATION

13. Write symbols which indicate the composition of α-, β-, and γ-radiation.
14. State the charge and relative mass of each of the three types of radiation.
15. State the relative penetrating powers of α-, β-, and γ-radiation.

3.13 IONIZING RADIATION

16. State the effect on matter of interaction with α-, β-, and γ-radiation.
17. State the relative ionizing abilities of α-, β-, and γ-radiation.

3.14 PHYSIOLOGICAL EFFECTS OF RADIATION

18. Describe some of the physiological effects of exposure to nuclear radiation.

3.15 NUCLEAR DECAY

19. Write a nuclear equation representing the emission of an α-particle from a specified radioisotope.
20. Write a nuclear equation representing the emission of a β-particle from a specified radioisotope.

3.16 HALF-LIFE, $t_{1/2}$

21. Given the amount of a radioactive sample and the half-life of the radioisotope, calculate the amount of sample remaining after some specified time interval.

3.17 MEDICAL USES OF RADIOISOTOPES

22. State at least two attractive properties of radioisotopes as medical diagnostic and treatment aids.
23. Describe at least one use of radioisotopes in medical diagnosis.
24. Describe at least one use of radioisotopes in medical treatment.

3.18 NEW ELEMENTS THROUGH BOMBARDMENT

3.19 NEUTRON BOMBARDMENT IN RADIATION THERAPY

3.20 NUCLEAR ENERGY

25. State at least one advantage of nuclear fusion over fission as an energy source.
26. State the technological problem that prohibits the practical use of fusion today.

PROBLEMS

3.14 In what way do atoms of different elements differ?

3.15 In what ways are atoms of different elements similar?

3.16 How do elements and compounds differ in terms of the Atomic Theory of Matter?

3.17 Draw a picture of an atom the way you visualize it.

3.18 Name the three subatomic particles.

3.19 Which subatomic particle has zero mass?

3.20 Which subatomic particle has zero charge?

3.21 Compare the relative masses of the three subatomic particles.

3.22 Compare the relative charges of the three subatomic particles.

3.23 Which subatomic particles account for the charge on the nucleus?

3.24 Complete the following table for the neutral atoms indicated:

Element Name	Atomic Number	Mass Number	Number of Protons	Number of Neutrons	Number of Electrons
	32	72			
	13			13	
			19	20	
		51			23

3.25 Tell the numbers of protons, neutrons, and electrons in neutral atoms which have the following mass numbers and number of neutrons:

	Mass Number	Number of Neutrons
a	56	30
b	31	16
c	11	6

3.26 With the help of the periodic table, tell the number of protons, neutrons, and electrons in the following isotopes. Nuclear fallout (or radioactive residue in the atmosphere) from the testing of nuclear bombs contains these radioisotopes, and consequently they appear in higher concentrations in milk and other food products in contaminated areas.
a Strontium-90
b Cesium-137

3.27 Complete the following table:

Symbol	Number of Protons	Number of Neutrons	Number of Electrons
$^{27}_{13}Al$			
$^{79}_{35}Br$			
$^{40}_{20}Ca$			
$^{84}_{36}Kr$			

3.28 Write symbols for atoms with the following number of protons and neutrons:

	Number of Protons	Number of Neutrons
a	9	10
b	31	39
c	25	30

3.29 Which of the following atoms are isotopes?
a $^{24}_{12}X$ *d* $^{6}_{3}X$
b $^{12}_{6}X$ *e* $^{24}_{11}X$
c $^{52}_{24}X$ *f* $^{23}_{11}X$

3.30 Consult the periodic table and identify each of the elements represented in Problem 3.29.

3.31 *a* An atom contains seven protons, eight neutrons, and seven electrons. Write the symbol for this atom.
b Write the symbol for a possible isotope of this atom.

3.32 Two isotopes of the same element are alike in what way? In what way do they differ?

3.33 Calculate the charges on the following ions:

	Ion	Number of Protons	Number of Electrons	Charge
a	Mg	12	10	
b	F	9	10	
c	S	16	18	
d	Au	79	76	

3.34 Complete the table below.

Isotope	Symbol	Atomic Number	Mass Number	Number of Protons	Number of Neutrons	Number of Electrons
Sodium-23						11
	$^{56}_{26}Fe$					
		27			32	
			15	7		
	^{37}Cl					
Magnesium-24						

3.35 Three isotopes of hydrogen occur naturally: protium (mass number = 1) deuterium (mass number = 2), and tritium (mass number = 3). Tritium is radioactive.
a Write symbols for these three isotopes that distinguish among them.
b Which isotope could agriculture researchers use to study water (H_2O) uptake by plants?

3.36 What is the difference between reactions done in an ordinary chemistry laboratory and nuclear reactions?

3.37 What is radioactivity?

3.38 How many types of nuclear radiation are there? Name them.

3.39 Describe the mass and charge characteristics of α-, β-, and γ-radiation.

3.40 Write the proper nuclear symbols for α- and β-particles.

3.41 If only a heavy-cloth curtain stood between you and a radioactive source, would you be endangered more by α- or γ-radiation? Explain.

3.42 What is the effect of nuclear radiation on matter?

3.43 Explain how radiation causes a Geiger counter to click.

3.44 What part of a living cell is most susceptible to radiation damage?

3.45 What is the first clinical symptom of exposure to higher levels of radiation?

3.46 Besides immediate clinical symptoms, what are the possible long-term effects of radiation exposure?

3.47 Give an example of transmutation.

3.48 Why does α-emission lead to transmutation?

3.49 Why does β-emission lead to transmutation?

3.50 Write nuclear equations for α-emission from:
a $^{219}_{86}Rn$ *b* $^{227}_{89}Ac$ *c* $^{234}_{92}U$

3.51 Write nuclear equations for β-emission from:
a $^{42}_{19}K$ *b* $^{131}_{53}I$ *c* $^{14}_{6}C$

3.52 The habits of insect pests can be studied by tagging them with radioactive labels. Three such studies have involved the following insects and labels:

Insect	Label
Wireworm beetles	$^{226}_{88}Ra$ (α-emitter)
Mosquitoes	$^{89}_{38}Sr$ (β-emitter)
Cockroaches	$^{95}_{40}Zr$ (β-emitter)

Write nuclear equations for the emissions from the labels.

3.53 Based on the following partial equations, decide whether the emitting isotopes are α- or β-emitters:
a $^{3}_{1}H \longrightarrow ^{3}_{2}He + ?$ *c* $^{211}_{83}Bi \longrightarrow ^{207}_{81}Tl + ?$
b $^{218}_{84}Po \longrightarrow ^{214}_{82}Pb + ?$ *d* $^{35}_{16}S \longrightarrow ^{35}_{17}Cl + ?$

3.54 ^{82}Br is a β-emitter which has been used to study abscesses because of selective absorption of this element by abscesses. What is the decay product of ^{82}Br?

3.55 Compare ^{24}Na and ^{14}C in Table 3.5. Why is ^{24}Na a better choice for use in blood-circulation studies than ^{14}C?

3.56 In an experiment, 1.6 g of ^{90}Sr ($t_{1/2}$ = 28 years) was buried in 1952. How much of this sample will be left in 2036?

3.57 A nuclear chemist needs 12 mg of $^{68}_{29}Cu$ ($t_{1/2}$ = 32 s) to do a particular experiment. At 10 A.M. a colleague brings him 750 mg of ^{68}Cu which he has just synthesized. The chemist is distracted and does not begin to work until 10:10 A.M. Is there still sufficient ^{68}Cu for this experiment?

3.58 A hospital has a 24-g supply of ^{131}I ($t_{1/2}$ = 8 days) on January 1. Even if none is used by the staff, by what date will the sample have diminished to only 3 g?

3.59 ^{32}P ($t_{1/2}$ = 14.3 days) can be used to locate brain tumors because when injected into the body it is preferentially absorbed by diseased brain tissue. ^{99}Tc ($t_{1/2}$ = 6 h) works just as well for this purpose. Why is the newer isotope (^{99}Tc) used more often?

3.60 The element selenium occurs in crops in certain regions and is toxic to animals. The radioisotope ^{75}Se can be used to label plants, and the uptake of selenium by animals can then be measured readily. The half-life of ^{75}Se is 120 days. If an animal ingested 0.1 mg on January 1, 1985, in what year will there be less than 1 μg (1×10^{-6} g) in the animal (assuming none is excreted)?

3.61 Identify the following as fission or fusion reactions:

a $^{235}_{92}U + ^{1}_{0}n \longrightarrow ^{94}_{38}Sr + ^{139}_{54}Xe + 3\,^{1}_{0}n$

b $^{3}_{1}H + ^{1}_{1}H \longrightarrow ^{4}_{2}He$

c $^{13}_{6}C + ^{1}_{0}n \longrightarrow ^{10}_{4}Be + ^{4}_{2}He$

CHAPTER FOUR

ELECTRON STRUCTURE OF THE ATOM

4.1 INTRODUCTION

Now that we have explored the nuclei of atoms and the dramatic consequences of nuclear changes, henceforth we will largely ignore the subatomic particles of the nucleus. This is so because most chemistry we encounter in our lives involves only the electrons.

Whereas we have been satisfied to accept the neutrons and protons as simply "bunched" together in the nucleus, we must know more of the details of how the electrons are arranged about the nucleus, in order to understand why compounds form and chemical reactions occur. This arrangement of electrons is called **electron structure.**

As soon as you know about electron structure, you will be able to understand how elements combine to form compounds. Compound formation will be discussed in Chapters 5 and 6, and the discussions rely heavily on principles in this chapter. The theoretical principles behind metal ion analysis in blood and in the environment also depend on electron structure and will unfold in this chapter.

4.2 ENERGY REVISITED

In Section 1.7 **energy** was defined as the *ability to do work,* and we discussed the idea that energy can be witnessed only when work is done and some *energy change* occurs. It was also pointed out that all matter has energy stored within it. The amount of stored energy depends on how the sample of matter is constructed. One important aspect of the construction is the arrangement of the electrons in atoms.

Different amounts of energy are associated with different electron arrangements. As electron arrangements change, we can observe changes in the energy content of samples of matter. It is impossible to consider electron arrangements without considering the energies of such arrangements.

It would probably be a good idea to reread Section 1.7 before continuing to read this chapter.

4.3 CONCEPT OF MINIMUM ENERGY

Scientists talk about **energy states**. It is hoped that you approach your studies in an *excited* energy state, because this is a high-energy condition. Matter in a high-energy state has a great deal of energy. If a sample of matter loses energy, then it will be in a lower-energy state. So, modification of the term **state** is used to describe how much energy something has.

It is a principle of nature that all things try to reach a minimum- (lowest-) energy state. For example, lay your pencil at the edge of your desk. You know that there is a natural tendency for that pencil to roll off and fall to the floor. In terms of energy states, the pencil on the desk is in a high-potential-energy state. When the pencil hits the floor, it has reached a minimum-energy state for the room and in so doing has lost the higher potential energy it possessed when it hovered at the edge of the desk. It still has some smaller amount of potential energy—for example, if you were to cut a hole in the floor, the pencil would fall through. In order for the pencil to return to the higher-energy state that it occupied while it lingered at the edge of the desk, energy must be put in. If you were to scoop the pencil up, you would be transferring the energy from yourself to the pencil. When the pencil is back on the desk, it has the same tendency to fall again to the lower state.

This natural tendency toward minimum energy is true for the chemical energy of matter also. Matter is constructed so that it is in a low-energy state. Though it can be excited into a higher-energy state temporarily through the addition of outside energy, it exhibits a natural tendency to return to the low-energy state (called the **ground state**). Because matter will stay in the ground state if it is left alone, its condition in this state is labeled a **stable** condition. Something that is stable or has **stability** is in a low-potential-energy state. Higher-energy states are less stable. This idea that **high energy corresponds to low stability and low energy corresponds to high stability** is very important but can be confusing. You will find it well worth the effort to clarify these terms for yourself now to save confusion later. Think about the pencil at the edge of the desk. The pencil is **unstable** (exhibits low stability) because it has high potential energy. It will fall to the ground state, or lowest-energy state, and become more stable.

SAMPLE EXERCISE 4.1

When gasoline burns, the products of the combustion are carbon dioxide (CO_2) and water (H_2O), which are more stable than gasoline. Describe the relative energy contents of gasoline, CO_2, and H_2O.

Solution

We are told that gasoline is less stable than CO_2 and H_2O; therefore, it must possess *more energy* than CO_2 and H_2O.

Problem 4.1

Food contains more energy than the products of metabolism into which it is converted in the body. Is food more or less stable than its metabolic products?

Though it was not mentioned in Chapter 3, the reason that some nuclei are radioactive and decay is that those nuclei are unstable, that is, of high energy. They decay to lower-energy, more stable states. Nuclear energy is the energy released in going from the higher- to lower-energy states. These energy changes for nuclei are quite large.

4.4 MINIMUM ENERGY IN THE ATOM

In terms of energy, electrons in atoms obey the same rules as all other matter in the universe; they prefer to be arranged in a condition (state) of minimum (lowest) energy and hence maximum (highest) stability.

Electrostatics

Let us divert our attention from atoms for a moment to discuss some aspects of electrostatics that will be useful in discovering the minimum-energy electron structure (arrangement) of the atom. **Electrostatics** deals with interactions between charged particles. In terms of "male-female electricity" the concept of "opposites attract and likes repel" is well established, although there are exceptions. In electrostatics, *without exception,* **objects with opposite electrical charge (one + and one −) attract each other; objects with**

identical charge (both + or both −) repel each other. The attraction or repulsion increases as the objects come closer together.

In terms of energy states, oppositely charged bodies are in a low-energy stable state when they remain close to each other; work must be performed and energy thereby increased in order for them to move farther apart. The low-energy stable condition for identically charged bodies occurs when they are far apart. Work must be performed and energy increased in order to push them closer together.

The low-energy arrangement of electrons in an atom is thus based on the ideas that:

1 **The close approach of a negative electron to the positive nucleus is a low-energy condition.**
2 **The far separation of two negative electrons is a low-energy condition.**

To obtain the minimum-energy condition for the atom as a whole, these two conditions must be met simultaneously. That is, electrons prefer to be close to the nucleus, but they must avoid other electrons. These simple ideas are the entire foundation for the following discussion of the electron arrangements in atoms.

Problem 4.2 In which case would the overall energy of the system be lower?

a Two electrons inside a bread box
b Two electrons inside a thimble

Hydrogen, the Smallest Atom

Consider the H atom with its one (+) proton in the nucleus and one (−) electron outside the nucleus. From the foregoing discussions you might immediately conclude that the (−) electron must "fall into" the nucleus, i.e., draw as close as possible to the (+) proton. But we must accept a puzzling but firmly established fact: although the electron may get very close to the nucleus, it never "falls in." This contradiction of classical electrostatics puzzled scientists early in this century. The person who offered the first solution to the problem was Niels Bohr, the Danish physicist who eventually won the Nobel Prize for this and other brilliant work.

Bohr said that electrons could have only certain specific energies and therefore could occupy only certain specific orbits around the nucleus, just as planets orbit around the sun. Because of attractions between matter, pathways (orbits) at specific distances are associated with specific energies. The Bohr atom is shown in Figure 4.1*a*. The lowest-energy state for the hydrogen atom occurs when the electron is in orbit number 1, the one closest to the nucleus. If the electron were in orbits farther from the nucleus, then

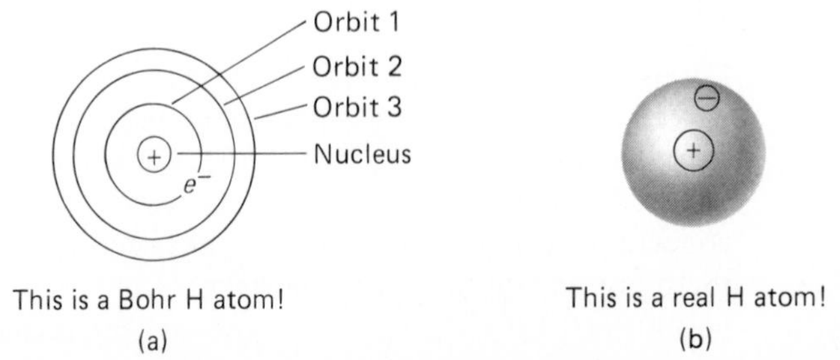
Orbit 1
Orbit 2
Orbit 3
Nucleus
+
e⁻
This is a Bohr H atom!
(a)
−
+
This is a real H atom!
(b)

FIG. 4.1 (*a*) The Bohr atom restricts electrons to definite orbits. The figure shows hydrogen's one electron in the lowest-energy orbit closest to the nucleus. (*b*) In the real atom the electron moves about the nucleus within a sphere which has the appearance of a negatively charged cloud.

the atom would exhibit higher energy and be in an excited state. The electron can never be in the regions between the orbits.

Bohr's model of the atom is not entirely correct. The Bohr model did not agree with experimental observation for atoms containing more than one electron and hence required further modification. However, his idea that electrons possess only certain definite energies associated with certain regions in space is correct and is very important. This is the idea of *quantization*. Electron energies in atoms are quantized; that is, there are specific energy states that electrons may have and they never have intermediate energies. Bohr's introduction of quantization eventually led to the development of *quantum mechanics,* which fully describes electron energies and arrangements mathematically.

Whereas the mathematical description is the only precise one, most people find it easier and more useful to develop a mental picture of the atom. Happily, it is possible to do so using the ideas we have already discussed. To recap, (1) the close approach of a negative electron to the positive nucleus is a low-energy condition, and (2) the far separation of two negative electrons is also a low-energy condition. The only exception to these two principles that we need to remember is that electrons do not enter the nucleus.[1] If we picture electrons as continuously moving rapidly, then a snapshot would capture them as a fuzzy cloud. The cloud is negatively charged because electrons bear negative charges. Figure 4.1*b* shows the picture of a hydrogen atom implied by these ideas. The one negative electron has a definite lowest-energy state which confines it to movement within a region close to the positive nucleus.

The fuzzy cloud, or more scientifically, the volume, in which the electron is likely to be found is called an **orbital.** An **orbital** in general is defined as a volume in space in which an electron or a pair of electrons is found. As you continue reading you will find that no more than two electrons can occupy an orbital. There is a specific energy associated with every orbital.

Helium Atoms

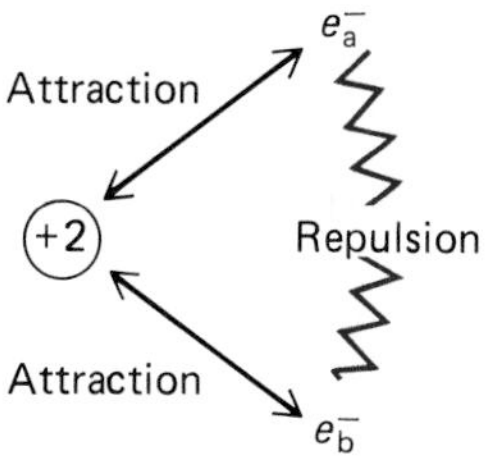
e_a^-
Attraction
+2
Repulsion
Attraction
e_b^-

Let us go on to the helium atom, with its two-proton (+2) nucleus and its two electrons (−). There are attractive forces between the nucleus and each of the electrons, and the attractive forces will attempt to keep each electron as close to the nucleus as possible. However, there is an additional force in this system of charges: the repulsive force operating between the two like-charged (−) electrons. To answer the question of how the electrons are arranged in the helium atom, one must ask: What is the arrangement of minimum energy of this three-component system which includes the two negative electrons and the positive nucleus?

It turns out that the volume of space within which the electron of the hydrogen atom is found is large enough to accommodate another electron. Placing the two electrons in a spherical volume, or orbital, surrounding the nucleus satisfies the two principles which lead to a minimum-energy condition, namely, proximity to the nucleus and the ability to avoid one another (see Figures 4.1 and 4.2). This sphere within which the two electrons of lowest energy in any atom reside is called the **1*s* orbital.**

[1]You may be wondering how so many positive particles (the protons) can cluster together in the nucleus. This question is one of the major puzzles that physicists are trying to unravel. Fortunately for us, we need to know only that protons do cluster, and we can regard the nucleus as one clump of positive charge.

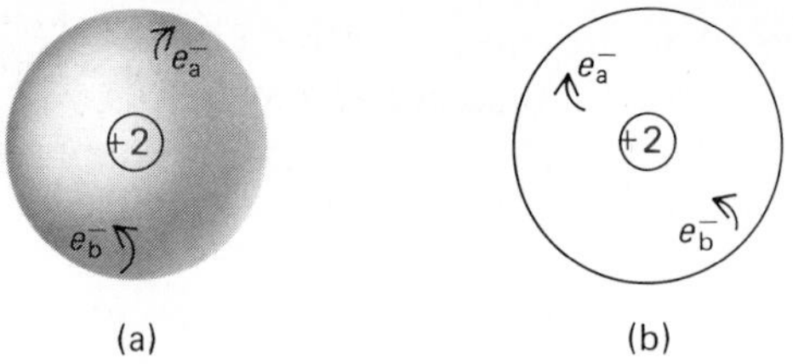

e_a^-
+2
e_b^-
(a)
e_a^-
+2
e_b^-
(b)

FIG. 4.2
The 1*s* orbital for He. (*a*) Spherical volume occupied by the two electrons. (*b*) Cross section of the orbital. The arrows denote the property of spin, which is discussed in Section 4.4.

Li and Be

What is the minimum-energy picture for an atom containing three electrons? If all three electrons are put in the 1*s* spherical volume (orbital), calculations show that the repulsion between electrons in that small volume is greater than the attractive energy from the pull of the nucleus on the electrons. Minimum energy can be obtained only when the *third* electron is located in a spherical volume around the nucleus that is larger than the volume of the 1*s* orbital. This new spherical volume is called the 2*s* orbital. You can envision the relationship between the 1*s* and 2*s* orbitals as a softball (1*s*) within a basketball (2*s*). The 2*s* orbital, like the 1*s*, can contain a maximum of two electrons. (See Figure 4.3.)

SAMPLE EXERCISE 4.2

How are the four electrons of Be arranged?

Solution Two electrons are in the 1*s* and two are in the 2*s*. This maintains the electrons close to the nucleus, but does not exceed the limit of two electrons per orbital.

Orbitals Have Various Shapes

What is the minimum-energy picture for an atom containing five electrons? The fifth electron cannot be in the 2*s* orbital because of repulsive forces with the two electrons already in that orbital. However, to minimize energy, as always it is desirable to keep electrons close to the nucleus rather than move them significantly farther away than the 2*s* orbital. Orbitals of a different shape accomplish the juggling act of keeping the fifth electron close to the nucleus, but at the same time away from other electrons. These new orbitals are called 2*p* and are shaped like a long balloon pinched in the middle or an iceskater's figure eights (see Figure 4.4). We can think of electrons in these 2*p* orbitals as being only slightly farther from the nucleus, on the average, than they would be if they were in the 2*s*. However, repulsion between the 2*s* electrons and them is minimized by the fact that they are located in different three-dimensional volumes.

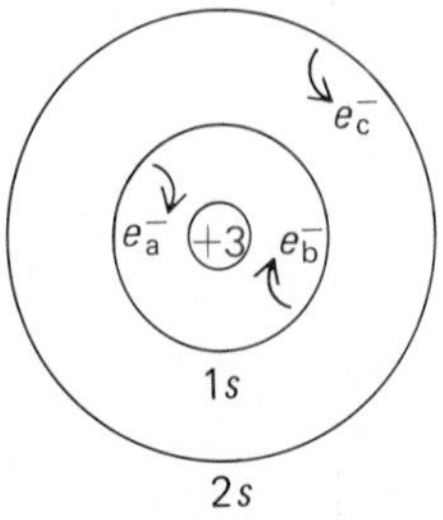

e_c^-
e_a^-
+3
e_b^-
1s
2s

FIG. 4.3
Cross section of the Li atom. The third electron occupies the 2*s* orbital to minimize electron repulsion. The arrows denote the property of spin, which is discussed in Section 4.4.

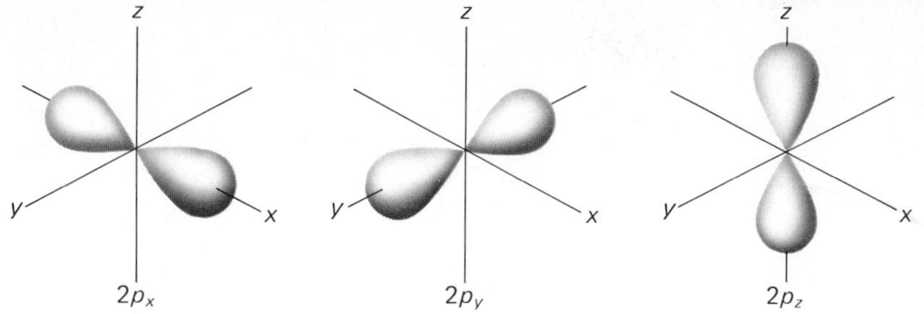
z
y
x
$2p_x$
z
y
x
$2p_y$
z
y
x
$2p_z$

FIG. 4.4
The three 2*p* orbitals directed along the three mutually perpendicular cartesian axes.

The three dimensions of space are represented mathematically by three mutually perpendicular axes, the *x*, *y*, and *z* (Figure 4.4). There are three 2*p* orbitals: $2p_x$, $2p_y$, and $2p_z$, where the subscript tells us along which axis in space the orbital is oriented. Each directed 2*p* orbital can hold a maximum of two electrons, just as the 1*s* and 2*s* orbitals each hold a maximum of two electrons.

Orbitals and Energy Levels

Perhaps you have begun to see the idea of energy levels surrounding the nucleus. By far the lowest-energy level, because it allows electrons the closest proximity to the nucleus, is the first energy level. The volume corresponding to the first energy level is only large enough to accomodate one orbital, the 1*s*. Because this is so, only two electrons in any atom can have this lowest-energy state.

Next in energy is a larger level, in which electrons can travel in a spherical volume, the 2*s* orbital, or in figure eights, the 2*p* orbitals. Because this second energy level contains two types of orbitals of slightly different energies, it is said to consist of sublevels, with the lower-energy 2*s* sublevel consisting of one 2*s* orbital, and the slightly higher-energy 2*p* sublevel consisting of the three perpendicular 2*p* orbitals (see Figure 4.5). You will see shortly that a third level contains three sublevels, the 3*s*, 3*p*, and 3*d*. Figure 4.6 summarizes the maximum numbers of electrons that can occupy the main energy levels.

With the information given so far about 1*s*, 2*s*, and 2*p* orbitals, it is possible to predict the proper location for the 10 electrons of lowest energy in any atom. First, two electrons occupy the 1*s* sublevel, or orbital; then electrons three and four fit into the 2*s* sublevel or orbital; then electrons five through 10 occupy the 2*p* sublevel. The 2*p* sublevel can hold six electrons, with two in each of the three 2*p* orbitals. Try to visualize the arrangement of the 10 electrons in a neon atom by looking at Figure 4.5*b*. Two electrons are in the 1*s* orbital, two in the 2*s*, and two each in the $2p_x$, $2p_y$, and $2p_z$ orbitals.

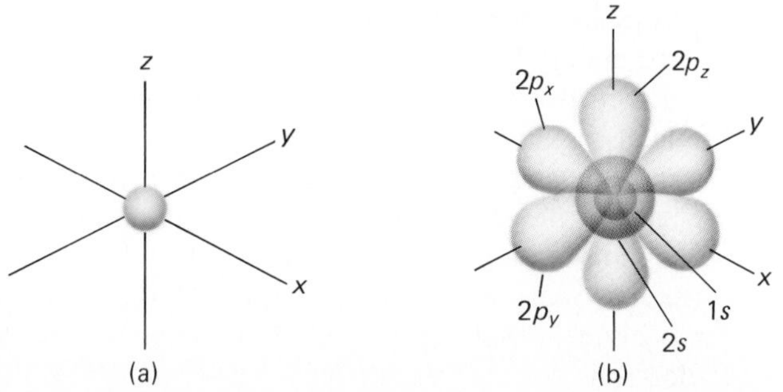
z
y
x
(a)
z
$2p_x$
$2p_z$
y
x
$2p_y$
1s
2s
(b)

FIG. 4.5
(*a*) The volume corresponding to the first energy level is only large enough to accommodate one orbital, the 1*s*. (*b*) The volume corresponding to the second energy level contains the 2*s* orbital and the three perpendicular 2*p* orbitals. Notice the 1*s* "buried" inside the 2*s*.

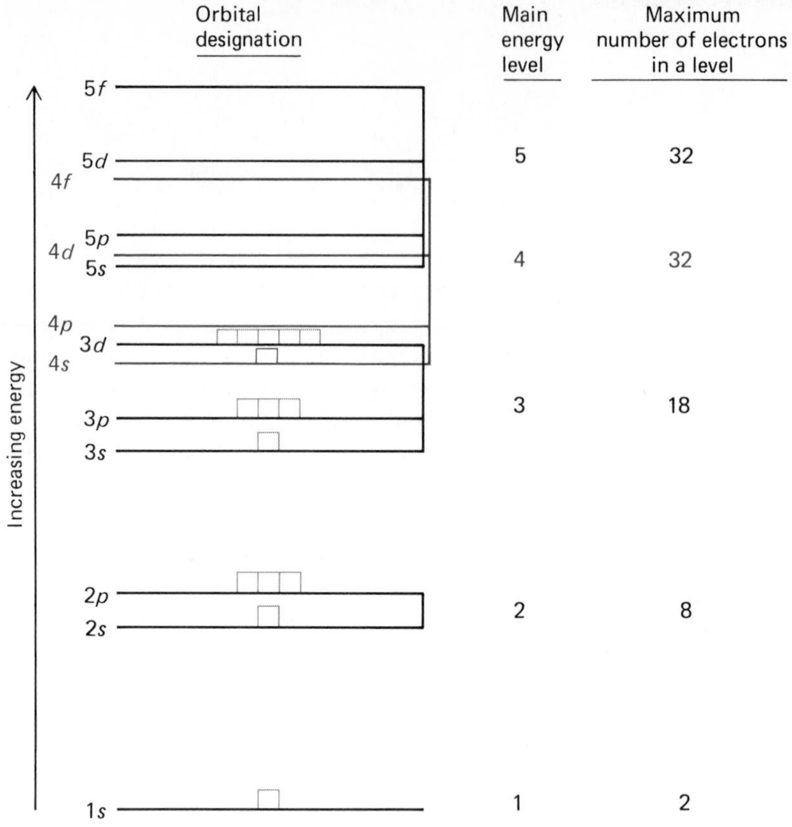
Orbital designation
Main energy level
Maximum number of electrons in a level
Increasing energy
5f
5d
4f
5p
4d
5s
4p
3d
4s
3p
3s
2p
2s
1s
5
32
4
32
3
18
2
8
1
2

FIG. 4.6
Electrons are in orbitals (□) within *s*, *p*, *d*, or *f* sublevels within numbered main levels. The energy of the main levels begins to overlap at level 4. The spacings between the levels vary for different elements.

The Third Energy Level

Electrons 11 and 12 in any atom are found to occupy a 3*s* *orbital,* another spherical orbital similar in shape to the 1*s* and 2*s*, but farther away form the nucleus. Electrons 13 to 18 are placed in a 3*p* sublevel (consisting of $3p_x$, $3p_y$, and $3p_z$ orbitals). The 3*p* orbitals have the same shape as the 2*p*, but on the average they are farther away from the nucleus. Within the third energy level there is another set of orbitals with unique shapes, called 3*d* orbitals. However, after electron 18 completes the filling of the 3*p* orbitals, electrons 19 and 20 go into the 4*s* orbital. The 4*s* orbital penetrates closer to the nucleus than the 3*d*, making the overall energy of an electron in a 4*s* orbital lower than one in a 3*d* orbital. Consequently, the total energy of an atomic system is found to be at a minimum when electrons 19 and 20 are contained in a 4*s* orbital. Another way of saying this is that the third and fourth energy levels overlap, as shown in Figure 4.6. It is this overlap that allows two electrons to enter the fourth energy level (4*s* sublevel) before the third energy level is filled.

Electrons 21 through 30 occupy the five 3*d* orbitals. Once again we find that a maximum of two electrons can fit into each 3*d* orbital. The most important and most common elements encountered, especially in body chemistry, are the simpler ones with atomic numbers less than 20. These simpler elements do not have occupied *d* orbitals, and consequently the shapes of these orbitals need not concern you. You need know only that there are five *d* orbitals in the *d* sublevel and that each orbital can hold a maximum of two electrons.

Spin

Scientists have found that electrons spin in either a clockwise or a counterclockwise direction. Different magnetic properties are associated with spins in opposite directions. Because an understanding of these different properties requires a knowledge of physics beyond that which most of you have acquired, we ask you to accept this fact without further discussion. The important implication of different electron spins is that **when two electrons are placed in one orbital, energy will be lower when the two electrons have opposite spins than when their spins are the same.**

Return to Figures 4.2 and 4.3 and notice that each electron has been assigned a spin which is indicated by a curved arrow denoting either clockwise or counterclockwise motion. This requirement of opposite spins to assure low energy also accounts for the existence of a maximum of two electrons per orbital. If more than two electrons are placed in an orbital, then two electrons would necessarily have the same spin.

Summary of Electron Arrangements

Let us try to summarize briefly the picture of electrons in atoms that you should have developed by now:

For negatively charged electrons, low energy is linked with being close to a positive nucleus. However, at the same time, electrons must avoid one another. Thus, they are located within different three-dimensional patterns (orbitals) as close to the nucleus as their need to avoid other electrons allows. Each orbital can contain two electrons with opposite spins.

4.5 ELECTRON CONFIGURATION NOTATION

Chemists employ a system of shorthand notation to show the **electron configuration** of an atom, i.e., how the electrons within a given atom are arranged in orbitals within energy sublevels. For example, the fact that hydrogen's one electron occupies the lowest-energy 1*s* orbital is abbreviated $1s^1$. The superscript 1 refers to the one electron in the 1*s* orbital. Helium has two electrons in the 1*s*, and the superscript 2 is used:

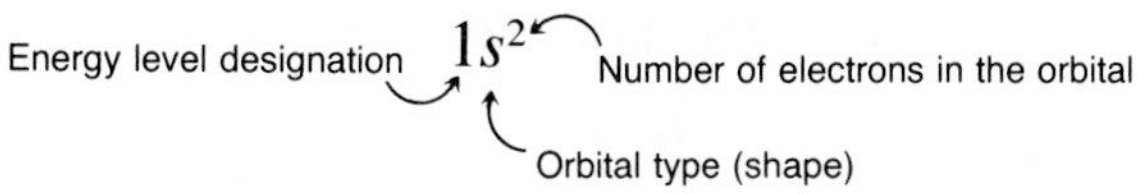
Energy level designation
1s²
Number of electrons in the orbital
Orbital type (shape)

Now consider fluorine, which has nine electrons to be arranged. The lowest-energy arrangement calls for two electrons in the 1*s* orbital, $1s^2$, and two electrons in the 2*s* orbital, $2s^2$, and the last five electrons distributed among the $2p_x$, $2p_y$, and $2p_z$ orbitals (abbreviated $2p^5$). The full fluorine configuration is abbreviated $1s^2 2s^2 2p^5$.

To write an electron configuration for an atom, begin by determining the atom's atomic number. Because in a neutral atom the number of protons always equals the number of electrons, the atomic number tells you the number of electrons in a neutral atom as well as the number of protons. Place the electrons in orbitals of increasing energy (Figure 4.6) until the total number of electrons is accounted for. In this notation system, *p* designates the set of three *p* orbitals (p_x, p_y, p_z), or more rigorously, the sublevel of energy which distinguished *s* orbitals from *p* orbitals. Consequently the superscript of *p* may be as large as 6 if all three *p* orbitals are occupied. Similarly, the superscript for *d* may be as high as 10 because *d* designates the set of five *d* orbitals. Table 4.1 shows the lowest-energy arrangement of electrons for the first 18 elements.

TABLE 4.1 Electron Configuration of First 18 Elements in the Periodic Table

Element	Atomic Number	Notation
Hydrogen	1	$1s^1$
Helium	2	$1s^2$
Lithium	3	$1s^22s^1$
Beryllium	4	$1s^22s^2$
Boron	5	$1s^22s^22p^1$
Carbon	6	$1s^22s^22p^2$
Nitrogen	7	$1s^22s^22p^3$
Oxygen	8	$1s^22s^22p^4$
Fluorine	9	$1s^22s^22p^5$
Neon	10	$1s^22s^22p^6$
Sodium	11	$1s^22s^22p^63s^1$
Magnesium	12	$1s^22s^22p^63s^2$
Aluminum	13	$1s^22s^22p^63s^23p^1$
Silicon	14	$1s^22s^22p^63s^23p^2$
Phosphorus	15	$1s^22s^22p^63s^23p^3$
Sulfur	16	$1s^22s^22p^63s^23p^4$
Chlorine	17	$1s^22s^22p^63s^23p^5$
Argon	18	$1s^22s^22p^63s^23p^6$

SAMPLE EXERCISE 4.3

Write out the electron configuration of sodium.

Solution The atomic number of sodium is 11, which indicates that there are 11 electrons in the sodium atom. Place these electrons in orbitals in order of increasing energy, in accord with the maximum number that can fit into each sublevel: two electrons in *s;* six electrons in *p;* and 10 electrons in *d* when necessary.

Na $1s^22s^22p^63s^1$

The total of the superscripts should equal the number of electrons in the atom, in this case 11.

Problem 4.3 Write out the electron configuration for silicon (without looking at Table 4.1).

Remembering the orbitals or sublevels in order of increasing energy is important; fortunately there is a device, shown in Figure 4.7, for easily determining the energy order.

SAMPLE EXERCISE 4.4

Write out the electron configuration for rubidium (Rb).

Solution The atomic number of rubidium is 37, which indicates that there are 37 electrons in the rubidium atom. Place electrons into the sublevels in order of increasing energy until 37 electrons have been accounted for:

Rb $1s^22s^22p^63s^23p^64s^23d^{10}4p^65s^1$

Problem 4.4 Write out the electron configuration for cesium.

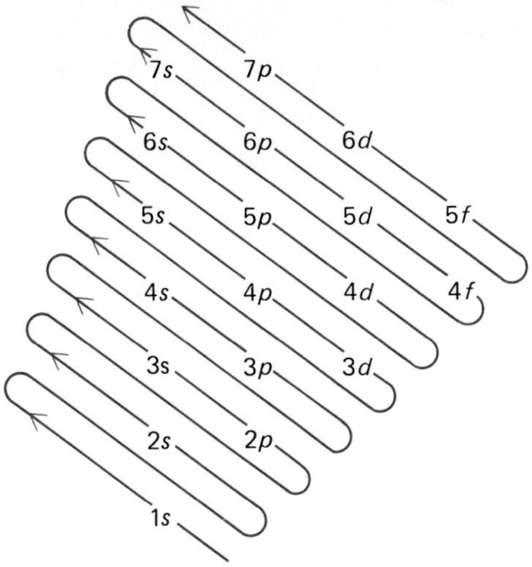
7s
7p
6s
6p
6d
5s
5p
5d
5f
4s
4p
4d
4f
3s
3p
3d
2s
2p
1s

FIG. 4.7
This is a shortcut to learning the order in which orbitals fill. Write the orbitals in columns in numerical order as shown. Draw diagonal arrows beginning at the bottom of each column. Follow the arrows from head to tail in order. This gives the same order of increasing energy as shown in Figure 4.6.

Aufbau Principle

The orderly placement of electrons, first in lower-energy orbitals and then in higher-energy orbitals, is known as the *Aufbau principle. Aufbau* means "buildup" in German. Another way of showing the electron configuration is by using boxes for orbitals and arrows for electrons. Consider Na, atomic number 11, as a first example.

Na $1s^2$ $2s^2$ $2p^6$ $3s^1$

[↑↓] [↑↓] [↑↓|↑↓|↑↓] [↑]

Increasing energy as distance from (+) nucleus increases →

The arrows pointing in opposite directions show that two electrons in an orbital have opposite spins. The separation of boxes indicates differences in energy. The drawing of three 2*p* boxes together indicates that the three orbitals all have the same energy; i.e., the 2*p* sublevel contains three orbitals all of the same energy. Consider Cl, atomic number 17, as another example.

Cl $1s^2$ $2s^2$ $2p^6$ $3s^2$ $3p^5$

[↑↓] [↑↓] [↑↓|↑↓|↑↓] [↑↓] [↑↓|↑↓|↑]

N, atomic number 7, presents us with another consideration.

N $1s^2$ $2s^2$ $2p^3$

[↑↓] [↑↓] [↑|↑|↑]

Notice that the electrons fill the 2*p* orbitals singly with the same spin before any pairing takes place. This is completely consistent with our idea that low energy is associated with maximum separation of electrons, but at the minimum distance from the nucleus. This lowest-energy arrangement for the three electrons in the *p* orbitals of nitrogen is shown in Figure 4.8.

This principle that electrons enter orbitals of equal energy singly with the same spin before they become paired is called *Hund's rule*.

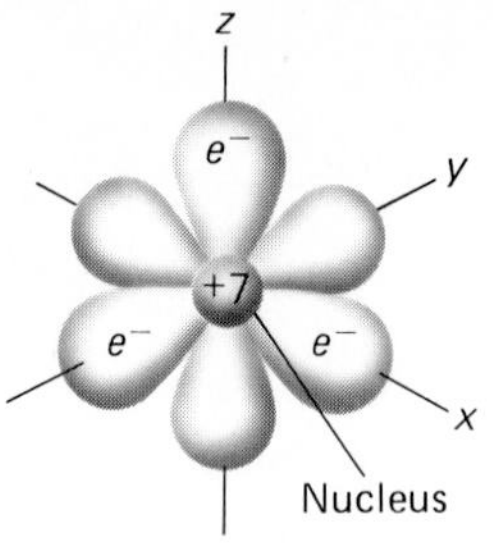
z
y
x
e⁻
e⁻
e⁻
+7
Nucleus

FIG. 4.8
The three electrons are all the same distance from the nucleus and thus feel the same attractive force, but are far apart from one another.

At this point you should be able to make a diagram complete with boxes and arrows for the electron configuration of any element in a vertical column of the periodic table labeled A. Remember:

1 The atomic number tells you the total number of electrons in the atom.
2 Lower-energy sublevels are completely filled with electrons before electrons enter higher-energy sublevels.
3 Use Hund's rule when placing electrons in *p* and *d* orbitals.

SAMPLE EXERCISE 4.5

How many unpaired electrons are there in oxygen?

Solution The distribution of electrons in oxygen is shown in Table 4.1 to be $1s^22s^22p^4$. Showing the boxes-and-arrows diagram and following Hund's rule, we see:

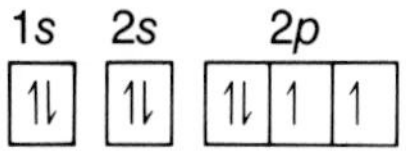
1s
2s
2p

Thus we find *two* unpaired electrons in the 2*p* sublevel.

Problem 4.5 How many unpaired electrons are there in phosphorus?

4.6 EVIDENCE AND USES OF THE ELECTRON ENERGY LEVELS

You have now seen that the arrangement of electrons in an atom is based on the buildup of electrons in discrete energy levels, with low-energy levels close to the nucleus and higher-energy levels as we move farther from the nucleus. In the undisturbed ground state, the electrons occupy the lowest-energy levels consistent with the ideas presented in Section 4.4.

However, energy can be put into an atom. For example, if we heat atoms in a flame, we are adding heat energy. As a result of this added energy, one or more electrons may be raised to a higher-energy level and the atom is then said to be in an excited state. In time the electrons will fall back to the lower levels of the ground state when the excess energy is given off. Figure 4.9 summarizes the excitement of an He atom by heat energy and its relaxation when light energy is emitted.

Light is a form of radiant energy or radiation. You know it is energy because it is capable of doing work or effecting a change. For example, light provides you with a suntan or sunburn, or it can develop a photographic film. Visible light from the sun or a light bulb is usually white light. A prism can divide white light into a rainbow. Scientists call the array of colors in the rainbow a *spectrum.* (Raindrops in the atmosphere sometimes act as prisms in the sky and produce a rainbow.) What the prism and rainbow (spectrum) tell us is

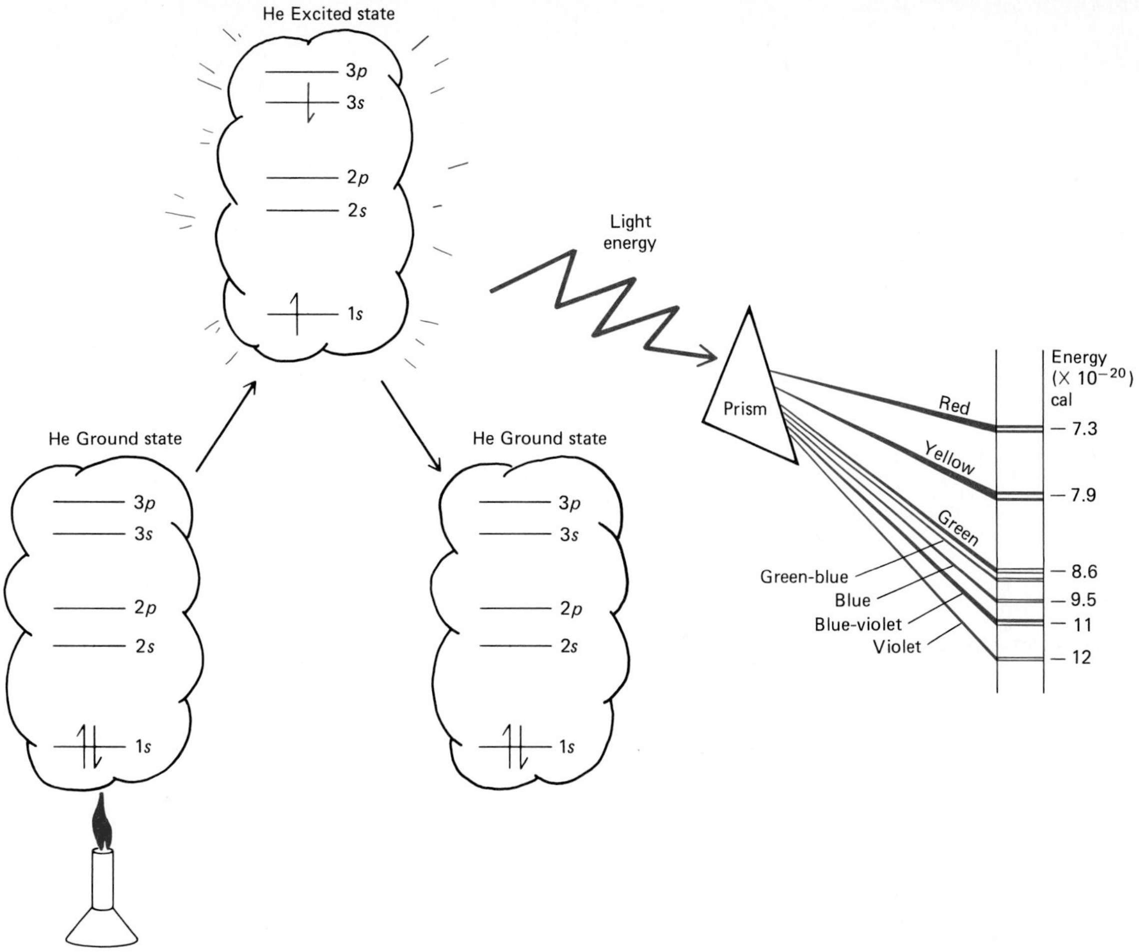
He Excited state
3p
3s
2p
2s
1s
Light
energy
Prism
Red
Yellow
Green
Green-blue
Blue
Blue-violet
Violet
Energy
(X 10−20)
cal
− 7.3
− 7.9
− 8.6
− 9.5
− 11
− 12
He Ground state
3p
3s
2p
2s
1s
He Ground state
3p
3s
2p
2s
1s

FIG. 4.9
Heat energy from the flame promotes electrons to higher-energy levels. In this representative case we see an electron promoted from the 1*s* to the 3*s*. As electrons fall back to the ground state, excess energy is emitted as light energy. Each line in the spectrum corresponds to some energy level transition.

that white light is a mixture of light rays of different colors. Each color of visible light has associated with it a distinct energy, the shades of violet corresponding to the higher energies and red to the lower energies. The actual definition of a *spectrum* is that it is an array of energies.

Excited atoms do not emit white light, i.e., the total spectrum of energies; rather, they emit a spectrum of energies unique to that atom. This is so because the energy levels of each element (sodium, potassium, lead, etc.) are uniquely spaced and electrons can occupy only these states; therefore, a particular excited element will emit its own unique spectrum of energies as electrons fall back to the ground state. This in turn means that there will be a unique spectrum of discrete colors in the visible light that is emitted. This emitted radiation can be precisely analyzed with an instrument known as a spectroscope. The presence or absence of a particular element can be confirmed by analyzing the emitted light of its excited state in a spectroscope. See Figure 4.10.

The concentration of sodium ion in blood plasma and in various foodstuffs (after being made water-soluble) can be readily determined by flame emission spectroscopy. A solution containing the sodium ion is sprayed into a high-

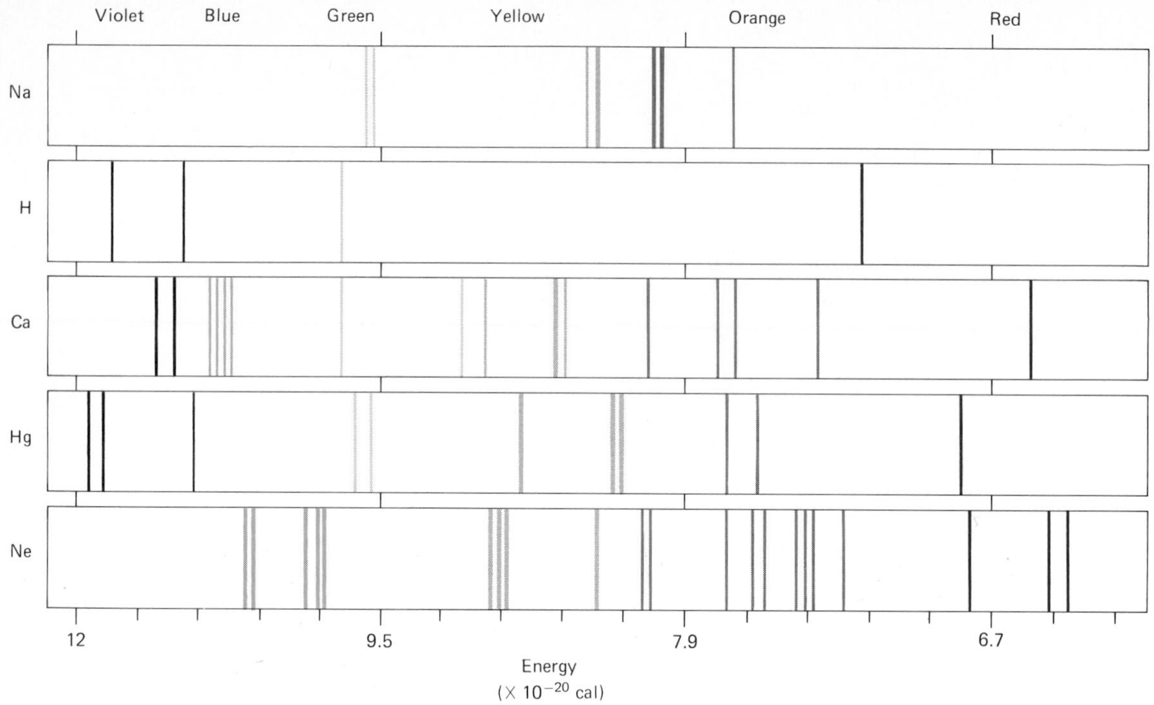

Violet
Blue
Green
Yellow
Orange
Red
Na
H
Ca
Hg
Ne
12
9.5
7.9
6.7
Energy
(× 10^{-20} cal)

FIG. 4.10
The emitted light spectrum is different and unique for every element.

intensity flame, which excites the sodium. The ion then emits its light spectrum. The emitted light passes through a filter which is selective for the yellow light in the sodium spectrum. The intensity of the light passing through the filter is proportional to the concentration of the sodium ion in the original solution.

The concentrations of many other ions in physiological fluids can be measured by a procedure known as atomic absorption spectroscopy. Here a selected amount of energy, which can be varied for different elements, is pumped into a sample; electrons are promoted to higher levels by this energy. The amount of energy absorbed by the sample can be quantitatively measured and related to the concentration of the metallic ion in question.

Using these two analytical methods, medical and food testing laboratories can quickly and precisely obtain concentrations of almost every metal ion in physiological fluids, foodstuffs, and other samples of biomass. Forensic specialists and ecologists also employ these techniques extensively.

4.7 PERIODIC TABLE

Now that we have considered electron structure, we can delve more deeply into the wealth of information which the periodic table contains. You have already seen how to distinguish metals from nonmetals in the periodic table (Section 1.5), but no explanation of the arrangement of the elements in the periodic table has been offered as yet. Figure 4.11 is a modern periodic table. The elements in the modern periodic table are arranged in order of increasing atomic number, the number above the symbol. The average atomic weight appears below the symbol.

Historically the periodic table was first constructed by arranging the elements in order of increasing atomic weights. This was so because the original construction was done independently in the 1860s by two scientists, the Russian Dimitri Mendeleev and the German Lothar Meyer, and at that time

Elements essential for health

Group | Metals | Nonmetals

Periods	IA	IIA	IIIB	IVB	VB	VIB	VIIB	VIIIB	VIIIB	VIIIB	IB	IIB	IIIA	IVA	VA	VIA	VIIA	VIIIA
1	1 **H** 1.008																	2 **He** 4.003
2	3 **Li** 6.941	4 **Be** 9.012											5 **B** 10.81	6 **C** 12.01	7 **N** 14.01	8 **O** 16.00	9 **F** 19.00	10 **Ne** 20.18
3	11 **Na** 22.99	12 **Mg** 24.31											13 **Al** 26.98	14 **Si** 28.09	15 **P** 30.97	16 **S** 32.06	17 **Cl** 35.45	18 **Ar** 39.95
4	19 **K** 39.10	20 **Ca** 40.08	21 **Sc** 44.96	22 **Ti** 47.90	23 **V** 50.94	24 **Cr** 52.00	25 **Mn** 54.94	26 **Fe** 55.85	27 **Co** 58.93	28 **Ni** 58.71	29 **Cu** 63.55	30 **Zn** 65.38	31 **Ga** 69.72	32 **Ge** 72.59	33 **As** 74.92	34 **Se** 78.96	35 **Br** 79.90	36 **Kr** 83.80
5	37 **Rb** 85.47	38 **Sr** 87.62	39 **Y** 88.91	40 **Zr** 91.22	41 **Nb** 92.91	42 **Mo** 95.94	43 **Tc** 98.91	44 **Ru** 101.1	45 **Rh** 102.9	46 **Pd** 106.4	47 **Ag** 107.9	48 **Cd** 112.4	49 **In** 114.8	50 **Sn** 118.7	51 **Sb** 121.8	52 **Te** 127.6	53 **I** 126.9	54 **Xe** 131.3
6	55 **Cs** 132.9	56 **Ba** 137.3	f block	72 **Hf** 178.5	73 **Ta** 180.9	74 **W** 183.9	75 **Re** 186.2	76 **Os** 190.2	77 **Ir** 192.2	78 **Pt** 195.1	79 **Au** 197.0	80 **Hg** 200.6	81 **Tl** 204.4	82 **Pb** 207.2	83 **Bi** 209.0	84 **Po** (210)	85 **At** (210)	86 **Rn** (222)
7	87 **Fr** (223)	88 **Ra** 226.0																

FIG. 4.11
The stepped color line divides the periodic table into metals on the left and nonmetals on the right. The one exception is *hydrogen*. Notice that there are many more metals than nonmetals. The elements shaded in color are found in the body and are essential for health. This periodic table does not include all the known elements, just the ones you are likely to encounter in this course. A complete periodic table is printed inside the front cover of this book. There are other systems for labeling A and B Groups in the periodic table. The one we have used here is most useful for distinguishing between the characteristic (main group representative) and transition (nonrepresentative) elements and organizing your studies thereby.

atomic weights could be determined experimentally, but the concept of atomic number had not been developed. Refinement of atomic theory and the discovery of the subatomic particles did not begin until the 1890s, and atomic numbers were first assigned by Moseley around 1914.

Mendeleev and Meyer noticed that if they arranged the elements in order of increasing atomic weight, certain chemical properties repeated themselves *periodically*. For example, at regular intervals the combining ratio of elements with oxygen is repeated, as shown by the following partial listing. The color subscripts are atomic numbers. The black subscripts in the formulas indicate the combining ratios. For example, in the first column the combining ratio is two metal atoms to one oxygen atom.

IA	IIA	IIIA	IVA	VA	VIA	VIIA	VIIIA
${}_{3}Li_2O$	${}_{4}BeO$		${}_{6}CO_2$				${}_{10}Ne$
${}_{11}Na_2O$	${}_{12}MgO$	${}_{13}Al_2O_3$	${}_{14}SiO_2$	${}_{15}P_2O_5$	${}_{16}SO_3$	${}_{17}Cl_2O_7$	${}_{18}Ar$
${}_{19}K_2O$	${}_{20}CaO$						
2:1	1:1	2:3	1:2	2:5	1:3	2:7	No combination

Mendeleev and Meyer arranged the elements in order of increasing atomic weight reading horizontally from left to right and so that the elements that reacted similarly were in the same vertical columns, which they called **groups.** The horizontal rows are called **periods.** Not all the elements had been discovered at this time, but Mendeleev predicted the existence of undiscovered elements based on the observation of periodicity.

The usefulness of the periodic table is in the knowledge that elements in the same **group** or **family** have similar properties. Therefore, if you know the chemical reactivity of one member of a group, you know the reactivity of all members.

Notice that the *groups* are designated by roman numerals and a capital letter A or B. We will consider mostly elements in Group A, because they are more prevalent and their behavior is more predictable based on theory and their position in the periodic table. The names of the A Groups appear in Table 4.2.

4.8 PERIODIC GROUPS

Let us look at some of the characteristics of the A Groups.

Alkali Metals

Among the alkali metals, the most abundant and most physiologically important are sodium and potassium. These elements occur in the body as ions (Na^+ and K^+); they play important roles in nerve transmission and are

TABLE 4.2 Names of the A Groups

Group	Name
IA	Alkali metals
IIA	Alkaline earth metals
IIIA	Boron family
IVA	Carbon family
VA	Nitrogen family
VIA	Oxygen family (chalcogens)
VIIA	Halogens
VIIIA	Inert (noble) gases

principally responsible for the maintenance of proper fluid balance between the cells and tissues of the body (see Section 9.16). Lithium does not naturally appear in the body, but in the compound lithium carbonate (Li_2CO_3) it is useful in the treatment of manic depression.

Alkaline Earth Metals

Calcium, an alkaline earth metal, is the most abundant metal in the body, occurring principally in bones and teeth. It is the chemical similarity of its "group mate" strontium that poses the danger of radioactive strontium-90. Strontium-90 ions can replace calcium ions in bones and emit β-radiation within the body for years. Magnesium is essential to enzyme activity in the body and is part of the structure of chlorophyll in plants. The extreme insolubility of barium salts is used to good advantage in certain medical diagnostic techniques (see Section 9.3).

The Carbon Family

Carbon is far and away the most important member of its family. All biological compounds contain carbon, and there are so many carbon compounds that an entire branch of chemistry, organic chemistry, is devoted to this element. We will begin to examine organic chemistry in Chapter 12.

The Nitrogen Family

Air is principally (80 percent) nitrogen gas. In combined form nitrogen is essential to life as a significant component of proteins. Phosphorus appears in proteins and also in lipids. The chemical similarity of arsenic to phosphorus allows arsenic to enter biochemical systems as a substitute for phosphorus atoms. However, arsenic does not behave in a biologically identical fashion to phosphorus and can be quite toxic.

Chalcogens

Oxygen is the only element which can be used by the body in its elemental form. All other elements enter the body in compounds or chemically combined forms. Sulfur is found in proteins and in compounds of metabolic importance.

Halogens

All of the halogens except bromine have a place in body chemistry. They occur as negatively charged ions (F^-, Cl^-, and I^-). The names of the ions end in *-ide* rather than *-ine* (see Section 5.10). Fluoride is essential to dental health, chloride is part of the fluid-balance control system (see Section 9.16), and iodide is necessary for the proper activity of the thyroid gland. In their elemental form the halogens exist as diatomic molecules. A molecule is several atoms hooked together. The prefix *di-* always means "two." Therefore, a diatomic molecule is two atoms hooked together. Other common elements that exist as diatomic molecules are hydrogen, nitrogen, and oxygen. Because of this diatomic elemental makeup, these elements are represented symbolically: H_2, N_2, O_2, F_2, Cl_2, Br_2, and I_2.

Inert Gases

A lack of chemical reactivity characterizes these elements. You will see in Section 5.2 that their inertness suggested a theory of chemical bonding.

4.9 PERIODICITY EXPLAINED

The periodic recurrence of similar properties among elements is an experimental fact. But why is this behavior displayed? The answer lies in electron configurations. Let us write the electron configuration of lithium, sodium, potassium, and rubidium, elements from Group IA of the periodic table, and see.

Atomic Number	Element	Electron Configuration
3	Li	$1s^2 2s^1$
11	Na	$1s^2 2s^2 2p^6 3s^1$
19	K	$1s^2 2s^2 2p^6 3s^2 3p^6 4s^1$
37	Rb	$1s^2 2s^2 2p^6 3s^2 3p^6 4s^2 3d^{10} 4p^6 5s^1$

These four elements exhibit very similar chemical and physical properties. Can you detect the similarity in their electron configuration?

In each of these elements there is one electron in the outermost energy level. The outermost levels are those designated 2, 3, 4, and 5 for Li, Na, K, and Rb, respectively. Could it be that the properties of an element are determined by the number of electrons in the outermost energy level? Let us examine the electron configurations of some elements in Group VIIA of the periodic table. Notice that we have not arranged the sublevels in the order in which they fill, but rather have grouped sublevels together within the major energy level in order of increasing distance from the nucleus.

Atomic Number	Element	Electron Configuration
9	F	$1s^2 2s^2 2p^5$
17	Cl	$1s^2 2s^2 2p^6 3s^2 3p^5$
35	Br	$1s^2 2s^2 2p^6 3s^2 3p^6 3d^{10} 4s^2 4p^5$
53	I	$1s^2 2s^2 2p^6 3s^2 3p^6 3d^{10} 4s^2 4p^6 4d^{10} 5s^2 5p^5$

If chemical tests are performed on these elements, it is found that fluorine, chlorine, bromine, and iodine all have very similar chemical properties. Each of these elements has seven electrons in the outermost level. The outermost levels are those designated 2, 3, 4, and 5 for F, Cl, Br, and I, respectively. This reinforces the idea that the chemistry of an element is determined by the number of electrons in the outermost level. Note also that the group number in the periodic table is simply the number of electrons in the outermost level for elements within that group. Electrons in the outermost level are called **valence electrons.**

In general, for the A Group elements, all the elements within a group have the same number of valence electrons. The number of valence electrons is the group number. The properties of the elements repeat themselves at certain fixed intervals because a given number of valence electrons is repeated at certain fixed intervals.

SAMPLE EXERCISE 4.6

How many electrons are there in the outermost level of Group V elements?

Solution

The group number tells you there are five. Check by examining the electron configuration of any member of the group, for example, N: $1s^22s^22p^3$. There are five electrons in the outermost second level.

Problem 4.6 How many electrons are there in the outermost level of Group VIIIA elements?

In Chapter 5 you will discover the key role valence electrons play in the bonding properties of atoms. Indeed, you will see that only the valence electrons are involved in bonding and chemical reactions.

4.10 LEWIS DOT STRUCTURES

Because of the importance of the valence electrons, chemical symbols which represent them are often used. These symbols are called **Lewis electron dot structures** because the theoretical chemist G. N. Lewis developed them.

A Lewis structure for an element consists of the element's symbol and surrounding dots to represent the number of valence electrons. To write a Lewis dot structure, write down the symbol of the element and surround the symbol with a number of dots corresponding to the number of valence electrons. Do not put more than two dots around any edge of the symbol. Whether the dots are paired, for example, :B, or spread out, for example, ·B·, is not important. The best method depends on the ultimate use to be made of the symbol. Remember that for an A Group element the group number is equal to the number of valence electrons.

H· ·Ö: K· :Ḟ̣: ·Ȧl

Lewis structures will be a very helpful notation in your study of chemical bonding, particularly covalent bonding, in Chapter 6.

Problem 4.7 Write Lewis dot structures for the following elements: Na, N, Br, C, Ne, and Mg.

4.11 PERIODIC TRENDS

The arrangement of electrons with which you are now familiar not only explains the existence of chemical groups, it also accounts for other observable trends in properties. Look at the special periodic table in Figure 4.12. This periodic table tells the atomic size and ionization energy for every representative group element. The top and side of this periodic table are marked with the trends in these properties among the elements.

Ionization energy is defined as the amount of energy needed to remove one electron from each of the atoms in a gaseous sample, the mass of which in grams is numerically equal to the atomic weight of the element. For example, the ionization energy of sodium is 118 kcal for 23 g. Notice in Figure 4.12 that the ionization energy increases as one proceeds across a period from left to right; that is, it becomes more difficult to remove an electron as one moves across the period from metals to nonmetals. In addition, the atomic size of the elements shrinks from left to right across a period.

Both effects are based on the same feature of electron structure: the nuclear charge (atomic number) increases from left to right, but the added electrons are being placed in the same outermost or valence level. For example, the buildup of electrons in period 3 can be represented as follows:

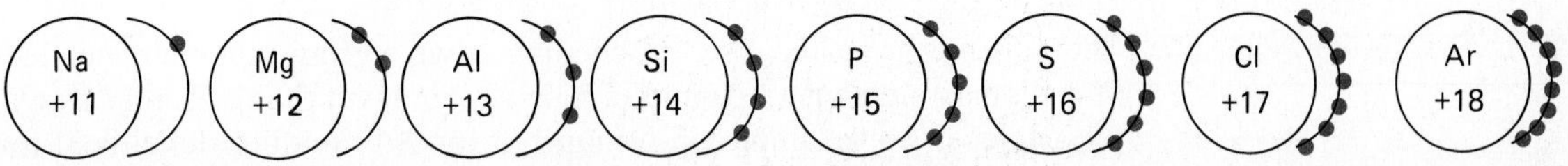

Na
+11
Mg
+12
Al
+13
Si
+14
P
+15
S
+16
Cl
+17
Ar
+18

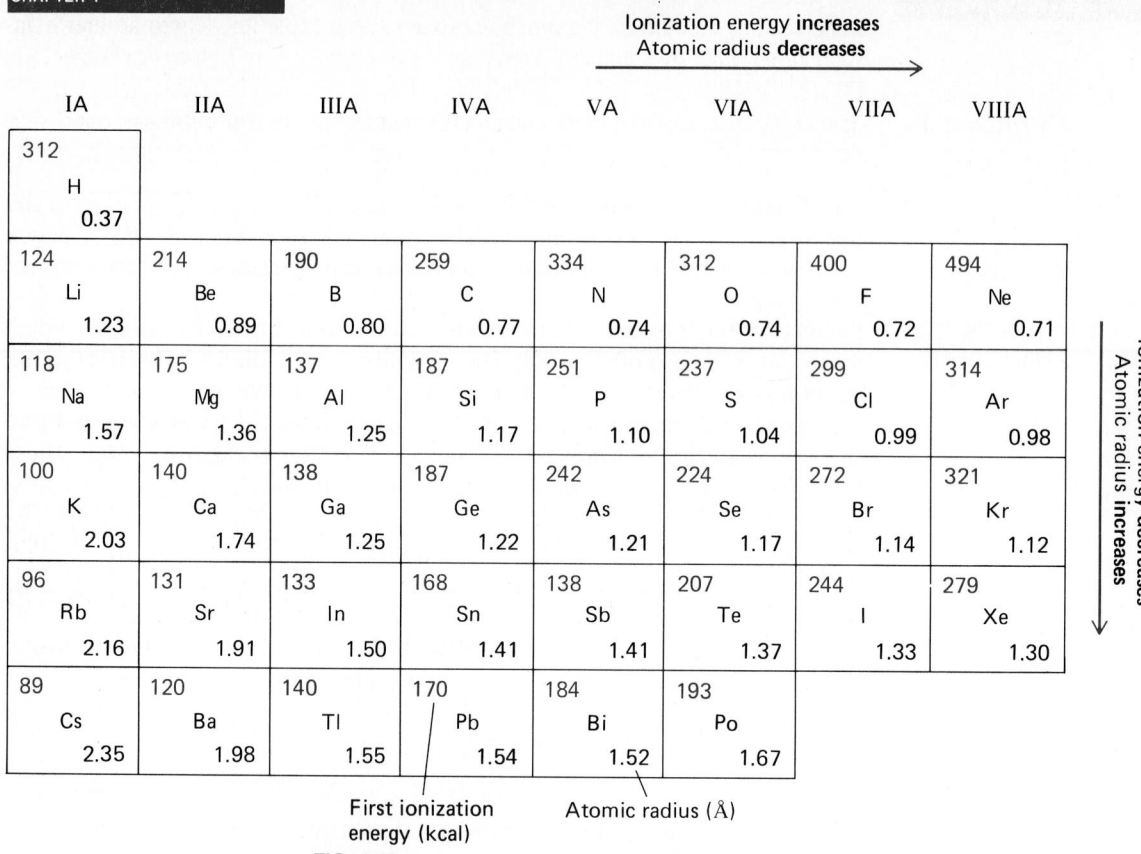

Ionization energy **increases**
Atomic radius **decreases** →

Ionization energy **decreases**
Atomic radius **increases** ↓

IA	IIA	IIIA	IVA	VA	VIA	VIIA	VIIIA
312 H 0.37							
124 Li 1.23	214 Be 0.89	190 B 0.80	259 C 0.77	334 N 0.74	312 O 0.74	400 F 0.72	494 Ne 0.71
118 Na 1.57	175 Mg 1.36	137 Al 1.25	187 Si 1.17	251 P 1.10	237 S 1.04	299 Cl 0.99	314 Ar 0.98
100 K 2.03	140 Ca 1.74	138 Ga 1.25	187 Ge 1.22	242 As 1.21	224 Se 1.17	272 Br 1.14	321 Kr 1.12
96 Rb 2.16	131 Sr 1.91	133 In 1.50	168 Sn 1.41	138 Sb 1.41	207 Te 1.37	244 I 1.33	279 Xe 1.30
89 Cs 2.35	120 Ba 1.98	140 Tl 1.55	170 Pb 1.54	184 Bi 1.52	193 Po 1.67		

First ionization energy (kcal) — Atomic radius (Å)

FIG. 4.12
First ionization energies and atomic radii for the representative elements. (Å is the abbreviation for angstrom, a unit of length equivalent to 1×10^{-8} cm.)

Because the valence electrons are all in the same main level, the distance between the positively charged nucleus and the valence electrons is not significantly increased going across a period. But the positive charge on the nucleus becomes larger, and the bigger the charge, the stronger the attractive pull of the nucleus on the electrons. The size of the atoms decreases because the nucleus pulls electrons closer. Ionization energy increases because more energy must be applied to remove electrons from the stronger nuclear pull. Summarizing horizontal trends, we find:

Decreasing atomic size
→
Increasing ionization energy

Let us now examine vertical trends, i.e., trends within a group. Figure 4.12 shows you that atomic size increases and ionization energy decreases moving down a group.

Increasing atomic size ↓ Decreasing ionization energy

These trends arise because, proceeding down a group, the valence electrons of each new element are found in a new main level that is farther away from the nucleus. For example, in Group IA, for Na the outer level is 3; for K it is 4; for Rb, 5; and for Cs, 6. The atomic size increases because electrons are found at larger distances from the nucleus. Because it is easier to remove electrons from an atom when they are far away from the nucleus and therefore feel the nuclear pull less strongly, ionization energy decreases as atoms become larger.

SAMPLE EXERCISE 4.7

Using only the periodic table in Figure 4.11 or on the inside cover of this text, predict which atom in each of the following sets is larger and which has the higher ionization energy:

a As and N *b* Ca and Br *c* Cl and Rb

Solution Refer to the preceding summarization of horizontal (⟶) and vertical (↓) trends.

a These elements are in the same group. Arsenic is larger because it is below nitrogen in the periodic table. The ionization energy of arsenic is smaller for the same reason.

b These elements are in the same period. Bromine is smaller than calcium because it is farther to the right. Bromine has a larger ionization energy for the same reason.

c Rubidium is larger because it is both below and to the left of chlorine. The ionization energy of rubidium is smaller for the same reason.

Problem 4.8 *a* Which of the A group elements have the lowest ionization energy?
b Are these metals or nonmetals?

Problem 4.9 In general, across a period, which are the larger atoms, metals or nonmetals?

Problem 4.10 Consulting *only* the periodic table in Figure 4.11,
a Which atom do you predict has the highest ionization energy?
b Which do you predict has the lowest ionization energy?

This discussion of periodic trends has been restricted to the A Groups in the periodic table. The elements in these groups are known as the **representative,** or **main group,** elements. For these A Group elements, all the sublevels in the levels below the outermost level are completely filled.

For the **nonrepresentative B Group elements,** the inner levels are not always completely filled. For example, the electron configuration of vanadium (atomic number 23) is V: $1s^22s^22p^63s^23p^63d^34s^2$. The third energy level is not filled even though there are two electrons in the 4*s* orbital. The consequence of this is that these elements do not exhibit the dramatic periodic trends seen for A Group elements. For example, there is little variation in ionization energy among the elements with atomic numbers 21 to 30 (Sc to Zn). The inner 3*d* sublevel is filled as you go from $_{21}Sc$ to $_{30}Zn$, but there is no change in outer electron configuration and thus essentially no change in ionization energy. We see a similar pattern as the 4*d* orbitals fill in $_{39}Y$ through $_{48}Cd$.

The B Group elements, which are all metals, are also called **transition elements** or transition metals. Many of them are quite important in body chemistry. For example, iron occurs in hemoglobin of red blood cells and is essential to the oxygen-carrying capacity of the blood. Other transition metals which occur in essential enzymes are chromium, cobalt, copper, manganese, and zinc.

As you progress in your study of chemistry you will find numerous other uses for this unique compilation known as the periodic table, and it should be as helpful to you in studying chemical properties as your electronic calculator is in doing numerical calculations.

CHAPTER ACCOMPLISHMENTS

After completing this chapter you should be able to define **key words** and do the following:

4.1 INTRODUCTION
4.2 ENERGY REVISITED
4.3 CONCEPT OF MINIMUM ENERGY
1. Explain the relationship between minimum energy and maximum stability.

4.4 MINIMUM ENERGY IN THE ATOM
2. Explain the electrostatic factors that determine how electrons are arranged in atoms.
3. State the relationships between orbitals, sublevels, and main energy levels.
4. State the spin relationship between electrons in the same orbital.

4.5 ELECTRON CONFIGURATION NOTATION
5. List the order in which atomic orbitals are filled (Aufbau principle).
6. Write the electron configuration of any A Group element.
7. State and apply Hund's rule for writing electron configuration and determining the number of unpaired electrons in an atom.

4.6 EVIDENCE AND USES OF THE ELECTRON ENERGY LEVELS
8. Describe the result of putting energy into an atom.
9. Explain why the light spectrum emitted by excited atoms is unique for each element.

4.7 PERIODIC TABLE
10. Give the general names applied to the vertical columns and horizontal rows of the periodic table.
11. Explain the practical significance when elements appear in the same *group* in the periodic table.

4.8 PERIODIC GROUPS
12. Given a periodic table, indicate the name of the family for an A Group element.

4.9 PERIODICITY EXPLAINED
13. Explain the relationship between electron arrangement and the periodic group concept.
14. Use the periodic table to obtain the number of valence electrons for any A Group element.

4.10 LEWIS DOT STRUCTURES
15. Write the Lewis electron dot structure for any element, given the number of valence electrons.

4.11 PERIODIC TRENDS
16. Explain the trends in ionization energy and atomic size across a period and within a group of the periodic table.
17. Given a periodic table, recognize the *transition metals.*
18. State the distinguishing difference in electron arrangement between the A and B Group elements.

PROBLEMS

4.11 Explain in terms of an energy change why:
a Apples fall from a tree to the ground.
b Water runs down, not up, a hill.
c Electrons try to remain close to the nucleus.

4.12 *a* What forces must be overcome to remove an electron from an atom?
b What force would be relieved by removing an electron from a neutral atom?
c Is energy required or is energy released in removing an electron from a neutral atom?
d Is the type of force in part *a* greater or lesser than that in part *b*? Explain how you arrived at your answer.

4.13 Explain why there is only one *s*-type orbital, but three *p*-type orbitals at any given level (above the first main level).

4.14 For each of the following atoms, tell how the electrons are arranged:
a Sulfur (16 electrons)
b Phosphorus (15 electrons)
c Magnesium (12 electrons)

4.15 Using only the periodic table on the inside front cover, write out the electron configurations of:
a H *d* Na
b C *e* Si
c O

4.16 Using only the periodic table on the inside front cover, write the electron configurations of:
a Potassium *d* Tin
b Bromine *e* Radium
c Barium

4.17 Give the symbols of the elements with the following electron configurations:
a $1s^22s^22p^1$
b $1s^22s^22p^63s^23p^64s^1$
c $1s^22s^22p^63s^23p^64s^23d^7$
d $1s^22s^22p^63s^23p^64s^23d^{10}4p^1$
e $1s^22s^22p^63s^23p^64s^23d^{10}4p^6$

4.18 Correct any errors in the following electron configurations:
a $1s^22s^32p^1$ *c* $1s^32s^22p^63s^23p^54s^1$
b $1s^22s^22p^63s^23d^8$ *d* $1s^22s^22p^63s^23p^63d^7$

4.19 Give the number of electrons in the main energy levels for each of the first 20 elements.

4.20 Give the symbols of all the sublevels within the third main energy level.

4.21 Use the boxes-and-arrows notation to show the electron configuration of:
a Li *b* C *c* Ne *d* Ca *e* Br *f* Fe

4.22 How many unpaired electrons are there in each of the atoms in Problem 4.21?

4.23 Fill in the word *emitted* or *absorbed:*
In order for an electron to be promoted to an excited state, energy must be ______. When an electron falls back to the ground state, energy is ______.

4.24 Explain in your own words the difference between absorption and emission spectra.

4.25 When we see a yellow flame, what are we seeing on an atomic level?

4.26 After being heated by a bunsen burner, a crucible may glow cherry red. What is the red color due to?

4.27 What similarity exists in the electron configurations of elements in Group VIA? Group VIIA?

4.28 The atomic numbers of the elements within a certain set are 6, 14, 32, 50, and 82. What similarity exists in the electron configuration of these elements?

4.29 The atomic numbers of the elements within a certain set are 4, 5, 18, 20, 34, and 38. Which elements will have similar chemical properties?

4.30 Write out the electron configuration of:
a Strontium
b Iodine
c Bismuth
d A synthetic element with atomic number 104

4.31 *a* What is the maximum number of electrons in the first main energy level?
b The second main energy level?
c The third main energy level?

4.32 How many valence electrons do each of the following elements have?
a Phosphorus *c* Xenon
b Bromine *d* Cesium

4.33 Indicate the number of valence electrons found in each element of the following groups:
a Group VIIA *c* Group IVA
b Group IA *d* Group VIA

4.34 Consult a periodic table and name three alkaline earth metals.

4.35 Consult a periodic table and name the halogens.

4.36 Give an example of each of the following:
a An alkaline earth metal *c* A chalcogen
b An alkali metal *d* An inert gas

4.37 Which elements exist as diatomic molecules?

4.38 *a* Which groups in the periodic table are the representative elements?
b Give the symbol of a nonrepresentative element.

4.39 Define the term *transition metal*.

4.40 What is the main difference in electron configuration between a representative and a transition element?

4.41 Write Lewis dot structures for the elements K, P, Cl, Si, Kr, and Ca.

4.42 *a* Explain why atomic size decreases proceeding across a period from left to right.
b Why does the ionization energy increase as the atomic size decreases?

4.43 *a* Where in the periodic table are elements with low ionization energies found?
b Where in the periodic table does one find elements with high ionization energies? What types of elements are these?
c Which group of elements in the periodic table has the highest ionization energies? What does this say about the stability of the electron arrangement for this group?

4.44 Arrange each set of elements in order of increasing ionization energy:
a Mg, Sr, Ba, Be
b Si, Mg, Na, P, Cl, S
c Se, Br, K

4.45 Circle the correct choice in each set:
a Highest ionization energy: Na, K, C, Si
b Smallest size: C, Li, Ge, F
c Largest size: K, Cs, Se
d Smallest ionization energy: K, Cs, Se

4.46 Which elements tend to have higher ionization energies, metals or nonmetals?

4.47 State at least five characteristics of atoms of an element that you can glean from a periodic table.

CHAPTER FIVE

IONIC COMPOUNDS

5.1 INTRODUCTION

Elements are rarely found in uncombined form. Rather, they are combined with other elements to form compounds. There are two major types of compounds: (1) **ionic compounds,** in which the ionic bond holds the elements together, and (2) **molecular compounds,** in which the covalent bond is the holding force. Note that a chemical **bond** is the force that holds elements together in compounds.

In this chapter we will explore the nature of the ionic bond and ionic compounds. The concept of ion formation was established much earlier in Section 3.5. An ion is a charged atom; the charge develops as a result of a neutral atom's gaining or losing negatively charged electrons.

Ionic compounds play quite important roles in the body, especially with regard to electrolyte and fluid balance. *Electrolyte* is the name given to a solution that contains ions. Section 5.13 discusses metal ion levels in blood.

5.2 HOW CAN ATOMS ACHIEVE LOWER-ENERGY STATES?

In Chapter 4 you learned that there is a natural tendency for matter to attain a lower-energy state. Therefore, the electrons in an atom are arranged to give the most stable (lowest-energy) electron configuration for that atom. An atom left to itself does the best it can, given the number of electrons that it has, to arrange those electrons in such a way as to produce the lowest possible energy state for that atom. However, the atoms of most elements interact with atoms of other elements; this enables them to form electron arrangements of lower energy than they could achieve as individuals, and this interaction leads to compound formation. Compound formation can be explained by an extension of the same principles of low-energy arrangements that you have seen for elements.

There is a family of elements in the periodic table, Group VIIIA, the **inert** or **noble gases,** that has essentially no tendency to interact with other elements to form compounds.[1] The electron configurations of the noble gases are apparently the most stable (possessing the lowest energy) of all the elements, since these gases have so little tendency to undergo change in their electron configuration. They are stable just as they are. Because the noble gases are a periodic *group,* we know that their electron configurations are similar.

[1]Since 1962 several compounds of the noble gases xenon, radon, and krypton have been prepared. However, these compounds are exceptions to the rule. In this book, we will regard the noble gases as inert.

Element	Symbol	Electron Configuration
Helium	He	$1s^2$
Neon	Ne	$1s^22s^22p^6$
Argon	Ar	$1s^22s^22p^63s^23p^6$
Krypton	Kr	$1s^22s^22p^63s^23p^63d^{10}4s^24p^6$
Xenon	Xe	$1s^22s^22p^63s^23p^63d^{10}4s^24p^64d^{10}5s^25p^6$
Radon	Rn	$1s^22s^22p^63s^23p^63d^{10}4s^24p^64d^{10}4f^{14}5s^25p^65d^{10}6s^26p^6$

The consistent lowest-energy pattern in these configurations is that the outermost *s* and *p orbitals* are *completely filled* with paired electrons. For helium this corresponds to two valence (outer-level) electrons. For all the other noble gases, completely filled outermost *s* and *p* orbitals require *eight* valence electrons. Eight valence electrons are often referred to as an **octet.**

Most elements react to form compounds through a process whereby their atoms gain, lose, or share valence electrons in order to achieve the highly stable (octet) electron configuration of a noble gas element. Atoms can gain, lose, or share electrons and form compounds because by so doing they acquire lower-energy electron arrangements.

By gaining, losing, or sharing electrons, an element will acquire the electron configuration of the noble gas to which it is closest in the periodic table. For H, Li, and Be this means they strive to attain the He valence electron configuration of $1s^2$, *two* valence electrons, a **duet.** Other elements strive to achieve eight valence electrons. This is often called the "rule of eight" or the **"octet rule."**

Elements do not always form a noble-gas electron configuration, but that is far and away the general rule. Chapter 6 will discuss the attainment of noble-gas octets by the process of electron sharing. In this chapter, electron loss and gain to form ions and ionic compounds will be explored.

5.3 METALS LOSE ELECTRONS

Your study of the trends in ionization energy in Section 4.11 revealed that electrons can be removed from metals more readily than from nonmetals. The number of electrons that a metal readily loses is the number that will give the metal ion the stable electron configuration of a noble gas. For example, sodium atoms have 11 electrons and lose one electron to attain the same configuration as neon, which has 10 electrons.

$$1s^22s^22p^63s^1 \xrightarrow[\text{from the 3}s\text{ orbital}]{\text{remove the electron}} 1s^22s^22p^6$$

Na atom configuration — Na^+ ion configuration = Ne atom configuration

Predicting the number of electrons that a metal will lose is done easily for an A Group metal. A Group metals lose a number of electrons equal to the group number because loss of this number will yield a metal ion with a noble-gas configuration. Table 5.1 offers some examples.

Electron loss always produces a positively charged metal ion. This is so because upon electron loss from an atom, the positively charged protons in the nucleus outnumber the negatively charged electrons. A positive ion is called a **cation.** The A Group number also tells you the magnitude of the positive charge on a metal cation. A loss of one electron produces an ion with a +1 charge, as exemplified by sodium:

TABLE 5.1 Noble-Gas Configurations Attained by Electron Loss by Metals

Metal	Group Number	Metal Configuration	Noble-Gas Configuration Achieved	Number of Electrons Lost*
Li	IA	$1s^22s^1$	He: $1s^2$	1
Mg	IIA	$1s^22s^22p^63s^2$	Ne: $1s^22s^22p^6$	2
Al	IIIA	$1s^22s^22p^63s^23p^1$	Ne: $1s^22s^22p^6$	3

* Notice how the number of electrons lost always equals the group number.

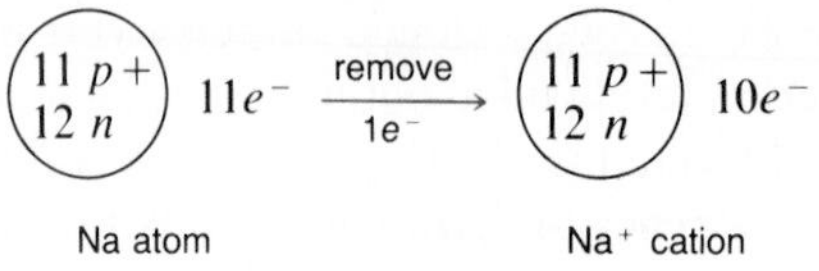
11 p+
12 n
11e⁻
remove
1e⁻
11 p+
12 n
10e⁻
Na atom
Na⁺ cation

Sodium is in Group IA.

SAMPLE EXERCISE 5.1

How many electrons must Ca lose to achieve a noble-gas configuration? Which noble-gas configuration will the calcium ions have?

Solution Calcium is in Group IIA, which tells us that it will lose two electrons and become Ca^{2+}. A loss of two electrons results in the argon configuration.

$$\underset{1s^22s^22p^63s^23p^64s^2}{\text{Ca }(20e^-)} \xrightarrow[2e^-]{\text{remove}} \underset{1s^22s^22p^63s^23p^6}{\text{Ar }(18e^-)\text{ and }Ca^{2+}\ (18e^-)}$$

5.4 NONMETALS GAIN ELECTRONS

Nonmetals do not readily lose electrons. Remember that ionization energy increases going from left to right across the periodic table, so that more energy is required to remove electrons from nonmetals than from metals. Also, nonmetals can more easily achieve noble-gas configurations by gaining electrons than by losing them. For example, consider chlorine, which has 17 electrons:

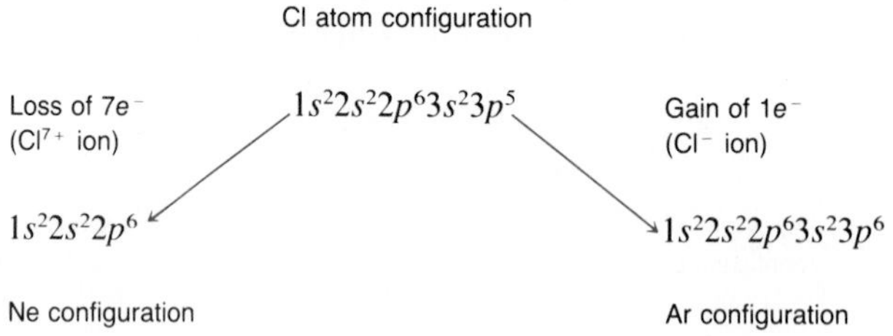
Cl atom configuration
Loss of 7e⁻
(Cl⁷⁺ ion)
1s²2s²2p⁶3s²3p⁵
Gain of 1e⁻
(Cl⁻ ion)
1s²2s²2p⁶
1s²2s²2p⁶3s²3p⁶
Ne configuration
Ar configuration

It is much easier to *gain one* electron than to *lose seven*. **Electron affinity** is the name given to the energy change that occurs when an electron is added to a gaseous atom or ion.

Nonmetals gain electrons to form anions with a noble-gas configuration, that is, to gain an octet (eight electrons). For an A Group nonmetal the number of electrons gained is equal to the difference between eight and the group number (8 − group number). Chlorine is in Group VIIA, so it gains $8 - 7 = 1$ electron. Table 5.2 offers other examples.

Electron gain always produces negatively charged nonmetal ions. A negative ion is called an **anion.** This is so, because the gain causes the electrons

TABLE 5.2 Noble-Gas Configurations Attained by Electron Gain by Nonmetals

Nonmetal	Group Number	8 − Group Number	Nonmetal Configuration	Noble-Gas Configuration Achieved	Number of Electrons Gained*
F	VIIA	8 − 7 = 1	$1s^22s^22p^5$	Ne: $1s^22s^22p^6$	1
O	VIA	8 − 6 = 2	$1s^22s^22p^4$	Ne: $1s^22s^22p^6$	2
P	VA	8 − 5 = 3	$1s^22s^22p^63s^23p^3$	Ar: $1s^22s^22p^63s^23p^6$	3

* Notice how the number of electrons gained always equals 8 − group number.

to outnumber the protons in the nucleus. The difference between eight and the nonmetal anion's group number also tells you the magnitude of the negative charge on a nonmetal anion. A gain of one electron produces an ion with a −1 charge, as exemplified by chlorine:

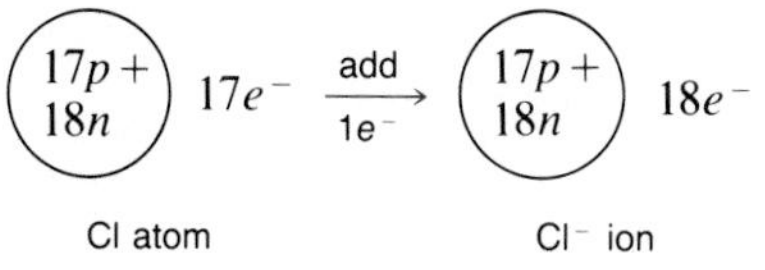
17p+
18n
17e⁻
add
1e⁻
17p+
18n
18e⁻
Cl atom
Cl⁻ ion

SAMPLE EXERCISE 5.2

How many electrons must N gain to achieve a noble-gas configuration? Which noble-gas configuration will the ions have?

Solution N is in Group VA, which tells us that it will gain 8 − 5 = 3 electrons and become N^{3-}. A gain of three electrons results in the Ne configuration.

$$\underset{1s^22s^22p^3}{\text{N }(7e^-)} \xrightarrow[3e^-]{\text{gain}} \text{Ne }(10e^-) \text{ and } \underset{1s^22s^22p^6}{N^{3-}\ (10e^-)}$$

Problem 5.1

a How many electrons must aluminum lose to achieve a noble-gas configuration? Which noble-gas configuration will the aluminum ions have?

b How many electrons must sulfur gain to achieve a noble-gas configuration? Which noble-gas configuration will the sulfur ions have?

c Identify the ions in parts *a* and *b* as cations or anions and give the magnitude of the ionic charge.

You will find it necessary to know the ionic charges which common metals and nonmetals acquire. Table 5.3 shows that with a periodic table in hand there is no need to memorize the charges of the A Group ions. Simply apply

TABLE 5.3 Common Ions of the Main Group Elements*

IA	IIA		IIIA	IVA	VA	VIA	VIIA
H^+							
Li^+	Be^{2+}				N^{3-}	O^{2-}	F^-
Na^+	Mg^{2+}	Transition	Al^{3+}		P^{3-}	S^{2-}	Cl^-
K^+	Ca^{2+}	elements	Ga^{3+}				Br^-
Rb^+	Sr^{2+}		In^{3+}	(Sn^{4+})			I^-
Cs^+	Ba^{2+}		Tl^{3+}	(Pb^{4+})			

* Atoms such as carbon which do not commonly form ions are omitted from the table. Tin and lead are shown in parentheses because they more commonly form +2 ions (see Table 5.4).

the relationships among group number, number of electrons gained or lost, and ionic charge acquired.

5.5 ELECTRON TRANSFER

Because metals tend to achieve a noble-gas configuration by losing electrons and nonmetals tend to do this by gaining electrons, there is a perfect setup for electron transfer. Ionic compounds form through **electron transfer**—metals give electrons to nonmetals, thus forming ions. Ionic compounds are so named because they are combinations of ions. In the formation of any ionic compound, the total number of electrons lost by the metal *must* equal the total number gained by the nonmetal so that electrical neutrality is maintained.

When the compound sodium chloride forms, each sodium atom (Group IA) gives up *one* electron to one chlorine atom (Group VIIA).

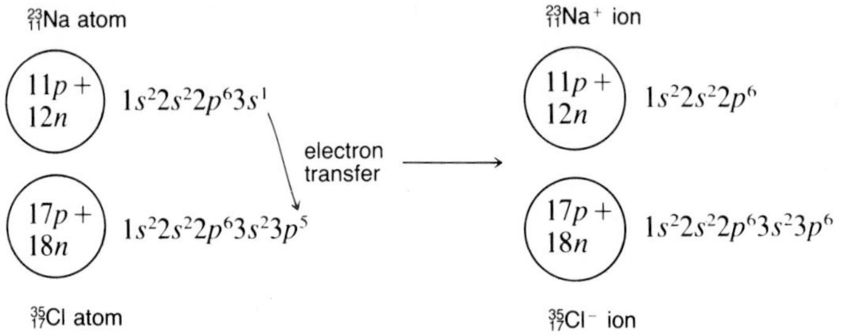
$^{23}_{11}Na$ atom
$11p+$ $12n$
$1s^22s^22p^63s^1$
electron transfer
$17p+$ $18n$
$1s^22s^22p^63s^23p^5$
$^{35}_{17}Cl$ atom
$^{23}_{11}Na^+$ ion
$11p+$ $12n$
$1s^22s^22p^6$
$17p+$ $18n$
$1s^22s^22p^63s^23p^6$
$^{35}_{17}Cl^-$ ion

In the formation of magnesium chloride, each magnesium atom (Group IIA) loses two electrons. But one chlorine atom can gain only one electron. Therefore, two chlorine atoms, each of which gains one electron, are needed to receive the two electrons lost by the magnesium atom. This transfer is seen pictorially in Figure 5.1. Notice that after the transfer each ion has a noble-gas configuration. Mg^{2+} has the same electron configuration as Ne. Cl^- has the Ar electron configuration. It is the attraction between the cations and anions that is the **ionic bond** that holds ionic compounds together.

FIG. 5.1
Mg atoms give up two electrons to two Cl atoms. The magnesium acquires a +2 charge; that is, it becomes the Mg^{2+} ion. Each Cl atom has acquired an extra electron and so acquires a −1 charge; that is, it becomes the Cl^- ion. The attraction between cations and anions holds the compound $MgCl_2$ together.

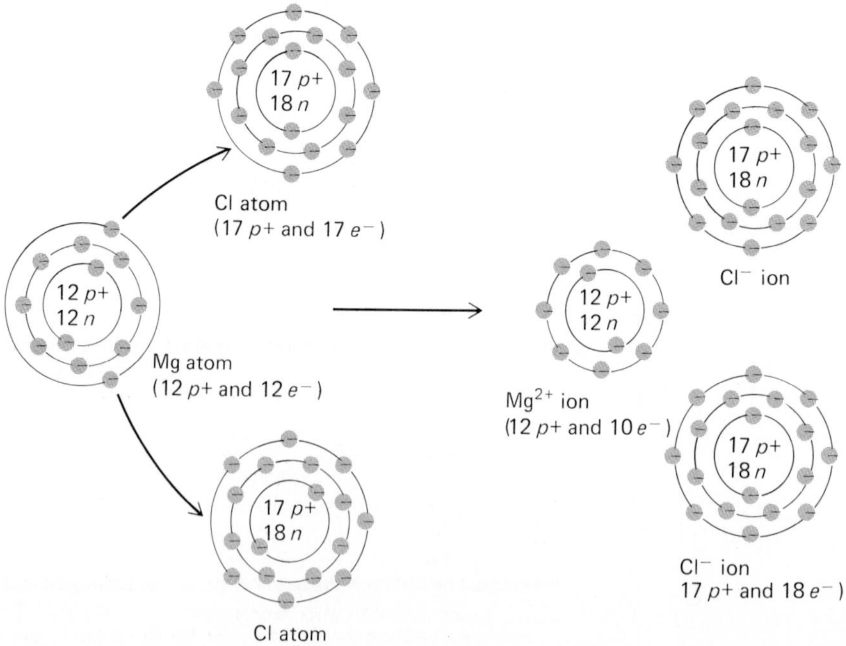
17 p+
18 n
Cl atom
(17 p+ and 17 e−)
12 p+
12 n
Mg atom
(12 p+ and 12 e−)
17 p+
18 n
Cl atom
12 p+
12 n
Mg²⁺ ion
(12 p+ and 10 e−)
17 p+
18 n
Cl− ion
17 p+
18 n
Cl− ion
17 p+ and 18 e−)

General Rules for Ionic Compound Formation

1 **Metals lose** electrons to attain a noble-gas configuration. For A Group metals, the number of electrons lost equals the group number.
2 **Nonmetals gain** electrons to attain a noble-gas configuration. For A Group nonmetals, the number of electrons gained equals the difference between eight and the group number.
3 Total number of electrons **lost** always **equals** total number of electrons **gained.**

SAMPLE EXERCISE 5.3

Describe how the ionic compound composed of potassium and oxygen forms.

Solution Follow the preceding general rules.

1 K is in Group IA. Therefore, each atom tends to lose $1e^-$. and becomes the K^+ ion.
2 O is in Group VIA. Therefore, each atom tends to gain $8 - 6 = 2e^-$ and becomes the O^{2-} ion.
3 Electrons lost must equal electrons gained, so 2K atoms must each lose $1e^-$ to satisfy 1O atom.

This explains why the formula for the compound is K_2O. As mentioned in Section 1.6, subscripts indicate the relative numbers of atoms (or ions) in compounds. Formula writing will be discussed in Section 5.9.

Problem 5.2 Develop a pictorial representation similar to Figure 5.1 for the formation of ionic compounds made from:
a Calcium and oxygen
b Aluminum and chlorine

5.6 OXIDATION-REDUCTION

The two processes that occur simultaneously during electron transfer have special names. **Oxidation** is the process in which there is a *loss of electrons* from the substance oxidized. **Reduction** is the process in which there is a *gain of electrons* by the substance reduced. Electron transfer reactions are called **redox** reactions because **red**uction and **ox**idation necessarily occur simultaneously.

Oxidation can be shown in the following way:

$$Mg \longrightarrow Mg^{2+} + 2e^- \qquad \textbf{Oxidation}$$

We begin with magnesium metal on the left which gives up two electrons. Thus on the right we see the Mg^{2+} ion and the two electrons.

Similarly, we can portray reduction:

$$Cl_2 + 2e^- \longrightarrow 2Cl^- \qquad \textbf{Reduction}$$

Chlorine, which exists as diatomic molecules, i.e., two atoms hooked together, accepts two electrons. The result is two anions. In Section 5.10 you will learn that they are named chloride ions.

Alternatively, we can show the oxidation and reduction coupled together, as indeed they would be if magnesium and chlorine were to react:

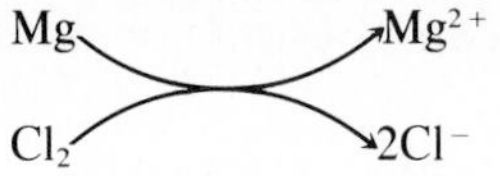
Mg
Mg^{2+}
Cl_2
$2Cl^-$

Notice in this case that the two electrons lost and gained are not written down. In this type of representation with intersecting arrows it is understood that something is transferred at the intersection. In this case it must be two electrons if Mg is becoming Mg^{2+} and two Cl^- ions are forming. We will return to this representation of coupled processes when we discuss biochemical reactions.

5.7 IONIC CHARGES

Because you now know that electron transfer occurs so that ions achieve noble-gas electron configurations, you could predict the correct combining ratio of metals and nonmetals in ionic compounds. That is, for example, the analysis of compound formation in Sample Exercise 5.3 led to the formula K_2O because it was found that two K combine with one O. Another method of predicting combining ratios and writing correct formulas is based simply on the idea that compounds must be electrically neutral and does not require an analysis of electron transfer. Your ability to use this method depends on your knowledge of the ionic charges which the common elements acquire.

For the A Group elements, you learned in Sections 5.3 and 5.4 how to use the periodic table to predict the ionic charge acquired by an element. Table 5.3 summarized the ionic charges for the A Groups.

Problem 5.3 What is the charge on the sulfur ion?

It is not possible to predict the magnitude of the positive charges on B Group metals from the periodic table. Fortunately there are only a few you need to know, and these are listed in Table 5.4. Notice in this table that many so-called transition-metal elements (B Group) and some A Group elements form ions with more than one charge.

5.8 POLYATOMIC IONS

The ions we have discussed so far are *simple* ions, formed from one atom of one element. Simple ions are also called **monatomic ions,** because the prefix *mono-* means "one" (the terminal *o* in *mono-* is dropped before a vowel).

There are several **polyatomic ions,** so-called because the prefix *poly-* means "many." Thus a polyatomic ion is a charged species made from many atoms. Table 5.5 lists the names and formulas of the more common polyatomic ions. Let us look at the formula for sulfate, SO_4^{2-}. This formula tells us that four oxygen atoms and one sulfur atom are combined in such a way that the group of atoms as a whole has a -2 charge. Subscripts always refer to number of atoms in a unit; when no subscript appears, a 1 is understood. The whole sulfate group is the ion; the component sulfur and oxygen atoms are inseparable. In Section 6.7 you will see that covalent bonding holds the sulfur and oxygens together in the sulfate ion. Polyatomic ions are groups of covalently bonded atoms that have a charge.

Many beginning chemistry students have difficulty recognizing polyatomic ions. Please study Table 5.5 and get acquainted with the idea that a total group of atoms has a charge associated with it. For example, one nitrogen and three oxygens have the charge -1 (NO_3^-), and the combination of one hydrogen, one phosphorus, and four oxygens has a -2 charge (HPO_4^{2-}).

TABLE 5.4 Common Transition-Metal Ions and A Group Ions with More than One Charge*

+1 Charge	+2 Charge	+3 Charge	+4 Charge
Copper(I) Cu^{+} (cuprous)	Copper(II) Cu^{2+} (cupric)	Chromium Cr^{3+}	Tin(IV) Sn^{4+} (stannic)
Silver Ag^{+}	Iron(II) Fe^{2+} (ferrous)	Iron(III) Fe^{3+} (ferric)	Lead(IV) Pb^{4+} (plumbic)
	Lead Pb^{2+} (plumbous)		
	Mercury(II) Hg^{2+} (mercuric)		
	Nickel Ni^{2+}		
	Tin(II) Sn^{2+} (stannous)		
	Zinc Zn^{2+}		

* Roman numerals are used to distinguish between different ions from the same metal. The roman numeral is the same as the charge on the ion. This is discussed in Section 5.10. The words in parentheses are archaic names used to distinguish metal ions.

5.9 FORMULAS FOR IONIC COMPOUNDS

Once you know the ionic charges, you can readily write correct formulas for ionic compounds using the principle that cations (+) and anions (−) combine in such a way that the magnitude of total positive charge just equals the magnitude of total negative charge and the compound is thereby electrically neutral. For example, the proper formula for the combination of the elements potassium and bromine in the sedative potassium bromide is KBr because a one-to-one combination of K^+ and Br^- results in electrical neutrality (see Table 5.3).

TABLE 5.5 Polyatomic Ions

+1 Charge	−1 Charge	−2 Charge	−3 Charge
Ammonium, NH_4^{+}	Acetate, $C_2H_3O_2^{-}$	Carbonate, CO_3^{2-}	Phosphate, PO_4^{3-}
	Cyanide, CN^{-}	Dichromate, $Cr_2O_7^{2-}$	
	Dihydrogen phosphate $H_2PO_4^{-}$	Hydrogen phosphate, HPO_4^{2-}	
	Hydrogen carbonate* HCO_3^{-}	Sulfate, SO_4^{2-}	
	Hydrogen sulfate† HSO_4^{-}	Sulfite, SO_3^{2-}	
	Hydroxide, OH^{-}		
	Nitrate, NO_3^{-}		
	Nitrite, NO_2^{-}		
	Permanganate, MnO_4^{-}		

* Commonly called bicarbonate.
† Commonly called bisulfate.

In writing a formula, we write the positive ion first and then the negative ion. Ionic charges are not shown in the formula. Even though the compound KBr is made up of K^+ and Br^- ions, the formula does not show the charges explicitly. We indicate with subscripts, after each ion, the number of that ion present in the compound. For example, in Sample Exercise 5.3, two potassium ions were indicated in the formula, K_2O. If no subscript is written, then it is understood to be 1, as in KBr or for oxygen in K_2O.

SAMPLE EXERCISE 5.4

What is the formula for the compound formed from Ca^{2+} and Br^-?

Solution The combination of one Ca^{2+} and one Br^- ion would leave the compound with a $+1$ charge: $(+2) + (-1) = +1$. Instead, one Ca^{2+} ion and two Br^- ions must combine: $(+2) + 2(-1) = 0$. The formula thus is $CaBr_2$. Note the use of subscript 2 to indicate two bromide ions. $CaBr_2$ is made up of one Ca^{2+} and two Br^- ions.

When polyatomic ions are involved, the rules work very much the same way. What is the formula for the compound formed from Mg^{2+} and OH^-? In order to balance the charges, we will need one magnesium ion combining with two hydroxide ions. But where will we write our subscript to clearly indicate two hydroxide ions? The formula is written $Mg(OH)_2$. Whenever we have more than one of a polyatomic ion, we surround that ion with parentheses and indicate with a subscript after the parentheses the number of polyatomic ions that are present. $Mg(OH)_2$ is made up of one Mg^{2+} and two OH^- ions.

SAMPLE EXERCISE 5.5

What is the formula for the compound formed from Na^+ and OH^-?

Solution To balance the charges, we simply combine one Na^+ and one OH^-. The formula will then be NaOH. We do *not* have to enclose the hydroxide ion within parentheses because there is only one of this polyatomic ion present in NaOH. However, you must be able to recognize the unit OH^- in the absence of parentheses.

SAMPLE EXERCISE 5.6

What is the formula for the compound formed from Al^{3+} and SO_4^{2-}?

Solution How are we to balance the charges $+3$ and -2? Begin by remembering that the common denominator of 3 and 2 is 6. If we take two $+3$ and three -2 ions, the total positive charge is $+6$ and the total negative is -6; the total positive and negative charges are now balanced. Therefore, the correct formula is $Al_2(SO_4)_3$. This formula tells us that the ratio of ions in this ionic compound is two aluminum ions to three sulfate ions.

Problem 5.4 Write formulas for the ionic compounds formed from:

a Na^+ and I^-
b NH_4^+ and S^{2-}
c Al^{3+} and NO_3^-
d Al^{3+} and CO_3^{2-}
e Al^{3+} and PO_4^{3-}

Problem 5.5 What is the ratio of cations to anions present in each of the following ionic compounds?

a CaO
b MgF_2
c Na_3PO_4
d $KC_2H_3O_2$
e $Ca_3(PO_4)_2$

Suppose we ask you to write a formula for the combination of calcium and sulfur. This time we have not indicated the charges as in the previous examples, so you must first ask yourself what ions these elements form and then proceed as usual. Calcium is in Group IIA; therefore, it forms Ca^{2+}. Sulfur is in Group VIA; therefore, it is S^{2-}. The correct formula is CaS.

SAMPLE EXERCISE 5.7 Write the correct formula for the compound formed by combining zinc and bromine.

Solution The compound formed from zinc and bromine must be an ionic compound since zinc is a metal and bromine a nonmetal. Remember or look up the charge of the zinc ion, which is Zn^{2+}. Bromine is in Group VIIA; therefore its charge is -1. To form a neutral compound, the correct combination is one cation and two anions. Thus we get $ZnBr_2$.

If an element can form an ion with more than one charge (Table 5.4), we must be given additional information as to which ion is actually forming in the particular problem at hand.

SAMPLE EXERCISE 5.8 Write a correct formula for the compound formed from iron and chlorine. Assume that iron will form the $+3$ ion.

Solution We are told that in this problem iron forms the Fe^{3+} ion, and we already know that in all ionic compounds chlorine forms the Cl^- ion. The correct combination to form a neutral compound is one cation and three anions. Thus the formula is $FeCl_3$.

General Rules for Writing Formulas for Ionic Compounds

1 Determine the ionic charges of the elements which are combining.
2 Choose a combination of the ions such that the sum of the ionic charges equals zero.
3 Write the symbol for the positive ion to the left and that for the negative ion to the right. Do not include the charges.
4 Use subscripts after each ion to indicate the numbers present in the compound. Polyatomic ions require parentheses where there is more than one. The number 1 is understood.

Problem 5.6 Write a correct formula for the compounds formed from the following:

a Magnesium and sulfur
b Chromium and chlorine
c Calcium and nitrogen
d Lead (assume it forms a $+4$ ion) and oxygen

Shortcut for Writing Ionic Formulas

There is a shortcut method for writing ionic formulas which you should feel free to use. However, it is a good idea to always remember that the basis for ion combination is the formation of a neutral (uncharged) compound. Let us see how this method works for the compound formed from Al^{3+} and S^{2-}.

STEP 1 Write the two ions next to each other with their charges as superscripts:

Al^{3+} S^{2-}

STEP 2 Then bring the 2 of the sulfur down as the subscript for Al and the 3 from the aluminum as the subscript for the S; that is, "crisscross" the charges as subscripts without signs. Always remember to omit the signs in the subscripts.

Al^{3+} ✕ S^{2-} gives Al_2S_3

STEP 3 Now check to see that the subscripts are the smallest possible whole-number ratio. If the subscript ratio can be converted into a smaller whole-number ratio, this must be done. In this example the subscript ratio 2:3 cannot be converted into a smaller whole-number ratio.

Applying the first two rules to the combination of Ca^{2+} and O^{2-},

Ca^{2+} ✕ O^{2-} gives Ca_2O_2

But the subscript ratio 2:2 can be converted into the smaller whole-number ratio 1:1 by dividing both subscripts by 2. This gives the correct formula, CaO. Similarly, for Pb^{4+} combining with O^{2-},

Pb^{4+} ✕ O^{2-} gives Pb_2O_4

which should be written PbO_2.

Do not forget the parentheses around polyatomic ions. For the combination of Ba^{2+} and NO_3^-,

Ba^{2+} ✕ NO_3^- gives $Ba(NO_3)_2$

The subscript ratio of 1:2 cannot be converted into a smaller whole-number ratio.

Problem 5.7 Redo Problem 5.4 using the shortcut method for formula writing.

5.10 NOMENCLATURE

There are over 3 million known compounds today. If each one had a separate arbitrarily assigned name, called a *common name,* we would have a totally impossible situation. To alleviate this, sets of rules for **systematically naming** compounds have been developed. These naming rules are what is known as **nomenclature.** Despite the advantages of using systematic names, some compounds have been in use so long and are so widely used that they are often called by their common names even by chemists. Table 5.6 shows the common names of some familiar ionic compounds. In this chapter you will learn the rules for giving systematic names to ionic compounds. In the later chapters of this book you will learn to name organic (carbon) compounds, the most prevalent and important molecular compounds.

In naming an ionic compound, we simply name the cation first, followed by a separate word for the anion.

Cation	**Anion**
First name	Last name

Cations are named exactly the same as the metals from which they come. Simple monatomic anions are named by adding the suffix *-ide* to **root** names from nonmetals. Table 5.7 shows that the root is a portion of the name of the nonmetal from which the anion comes. Thus, to name NaI, we start with

TABLE 5.6 Commonly Encountered Ionic Compounds

Systematic Name	Formula	Common Name	Occurrence in the Body or Common Use
Aluminum acetate	$Al(C_2H_3O_2)_3$	Burrow's solution	Medical astringent
Ammonium carbonate	$(NH_4)_2CO_3$	Smelling salts	
Ammonium chloride	NH_4Cl		Diuretic, expectorant
Barium sulfate	$BaSO_4$		X-ray analysis aid
Calcium carbonate	$CaCO_3$	Limestone	Antacid, soil sweetener
Calcium oxalate	CaC_2O_4		Urinary stones
Calcium phosphate	$Ca_3(PO_4)_2$		Bone component
Calcium sulfate	$CaSO_4$		Plaster casts
Magnesium hydroxide	$Mg(OH)_2$	Milk of magnesia	Laxative
Magnesium sulfate	$MgSO_4$	Epsom salts	Laxative
Potassium* chloride	KCl		
Potassium* hydrogen carbonate	$KHCO_3$		
Potassium hydrogen tartrate	$KHC_4H_4O_6$	Cream of tartar	Baking staple
Potassium iodide	KI		Expectorant, antifungal
Potassium nitrate	KNO_3	Saltpeter	Diuretic, antiseptic
Potassium permanganate	$KMnO_4$		Anti-infectant
Silver nitrate	$AgNO_3$		Antiseptic
Sodium bromide	$NaBr$		Sedative
Sodium carbonate	Na_2CO_3	Washing soda	General cleanser
Sodium† chloride	$NaCl$	Table salt	Food flavoring
Sodium fluoride	NaF		Dental and bone components
Sodium† hydrogen carbonate (sodium bicarbonate)	$NaHCO_3$	Baking soda	Makes bread rise
Sodium hydroxide	$NaOH$	Lye	Unclogs drains
Sodium iodide	NaI		Treatment of iodine deficiency
Zinc oxide	ZnO		Astringent

* Potassium ions are very prevalent in body cells along with a variety of anions, mostly HPO_4^{2-}.
† Sodium ions are very prevalent in bodily fluids along with a variety of anions, mostly Cl^- and HCO_3^-. See Section 5.13.

the cation name *sodium* and follow it with the word *iodide* [(root = **iod**) + *ide*]. Thus NaI is sodium iodide.

Consider the name of K_2S, potassium sulfide. Note that no mention is made in the name of the subscript 2 after potassium. This is always true for *ionic* compounds. The name contains no mention of subscript numbers.

The names of the polyatomic ions appeared in Table 5.5. In Na_2SO_4, we

TABLE 5.7 Naming Simple Anions

Nonmetal*	Anion name (root + *-ide*)
Oxygen	**Ox**ide
Chlorine	**Chlor**ide
Bromine	**Brom**ide
Iodine	**Iod**ide
Fluorine	**Fluor**ide
Phosphorus	**Phosph**ide
Nitrogen	**Nitr**ide
Sulfur	**Sulf**ide

* Root is indicated by boldface type.

recognize the sodium cation and the SO_4^{2-} sulfate anion, so the name is sodium sulfate. NH_4Cl is ammonium chloride.

If the cation can have a variable charge, we follow the name of the cation by a roman numeral, inside parentheses, indicating the magnitude of the charge. For example, Fe^{2+} combines with Cl^- to give the compound $FeCl_2$. This is named iron(II) chloride to distinguish it from the compound that results from the combination of Fe^{3+} and Cl^-, $FeCl_3$, iron(III) chloride.

SAMPLE EXERCISE 5.9

Name the following ionic compounds: MgO, Na_3PO_4, $CuBr$, $CuBr_2$, $AlCl_3$, $Al_2(SO_4)_3$, $(NH_4)_3PO_4$, and Cu_2CO_3.

Solution

MgO	Magnesium oxide.
Na_3PO_4	Sodium phosphate. Notice there is no mention of the subscript numbers.
$CuBr$	Copper(I) bromide. Note here the need to indicate the magnitude of the charge on copper. We realize that the charge on Cu is +1, since Br is always −1 and there is only one copper ion to balance the −1 charge. Hence the charge on Cu in this compound must be +1.
$CuBr_2$	Copper(II) bromide. There are two Br^- ions, and therefore the one copper ion must have a +2 charge.
$AlCl_3$	Aluminum chloride.
$Al_2(SO_4)_3$	Aluminum sulfate.
$(NH_4)_3PO_4$	Ammonium phosphate.
Cu_2CO_3	Copper(I) carbonate. The copper ion has a +1 charge, and carbonate a −2 charge. The two positive ions therefore have a total positive charge of +2, and since there are two copper ions in the formula, each must be charged +1.

Problem 5.8 Name the following compounds: $NaCl$, NH_4NO_3, $CaBr_2$, K_2CO_3, FeO, Fe_2O_3, $Pb_3(PO_4)_2$, and CuO.

Writing correct formulas from names requires the skills presented in Section 5.9. Review especially Sample Exercises 5.7 and 5.8. From this review you should recognize that the compound calcium chloride, for example, is made up of calcium ions and chloride ions. Write down the symbols for the ions with their charges, and achieve a neutral combination by the crisscross method or otherwise:

Ca^{2+} ⤫ Cl^- gives $CaCl_2$

Given the name copper(I) bromide, notice that the roman numeral tells you that it is Cu^+ that is combined in this compound. Bromide is always Br^-. Thus the correct formula is

Cu^+ ⤫ Br^- gives $CuBr$

SAMPLE EXERCISE 5.10

Write the formula for the compound ammonium sulfate.

Solution The name ammonium sulfate is associated with a compound made up of the ammonium ion NH_4^+ and the sulfate ion SO_4^{2-}. We write down the two ions,

NH_4^+ SO_4^{2-}

and balance the charges by taking two ammonium ions and one sulfate ion, giving

$(NH_4)_2SO_4$

Since the ammonium ion is a polyatomic ion, we must enclose it within parentheses when the subscript is greater than 1. Once again we see that the name of the compound does not directly tell us the numbers of each ion involved. It does tell us what ions are involved, and if we know the charges of these ions (as we should), then we can immediately write a correct formula with the proper subscripts.

5.11 FORMULA WEIGHT

In Chapter 3 we discussed the atomic weights of atoms. It should come as no surprise that the weights assigned to compounds are the sums of the weights of the atoms or ions that make them up. Because ions are formed from atoms by the loss or gain of electrons, and because electrons have essentially no mass, the masses of ions are for all practical purposes the same as those of the atoms from which they were formed.

For ionic compounds we can calculate the **formula weight,** that is, the weight of one **formula unit,** the smallest grouping of ions corresponding to the correct fixed ratio of ions in an ionic compound. Note that one formula unit does not have an independent existence. Figure 5.2 accurately portrays the ionic compound NaCl as a collection of ions. The boxed formula unit merely reflects the fact that the ratio of ions is 1:1.

Let us see how a formula weight is determined, using Na_2SO_4 as our example. Sodium ions are formed from sodium atoms with essentially no mass change. Sulfate ions are formed from sulfur atoms and oxygen atoms with essentially no mass change. Therefore, we can get masses for Na^+, S, and O from the periodic table. Each mass is multiplied by the number of that atom (or ion) present in one formula unit. The products of the multiplications are added together. In summary:

FIG. 5.2
(*a*) Schematic representation of a NaCl crystal. The alternating positive and negative charges maximize positive-negative attraction and minimize negative-negative and positive-positive repulsion. The result is a definite crystalline shape. Also, it is very difficult to pull the ions apart. (*b*) Actual photograph of a sodium chloride crystal. Note the regular cubic appearance. Magnification = 100×. (*Courtesy of G. W. Luther, Kean College.*)

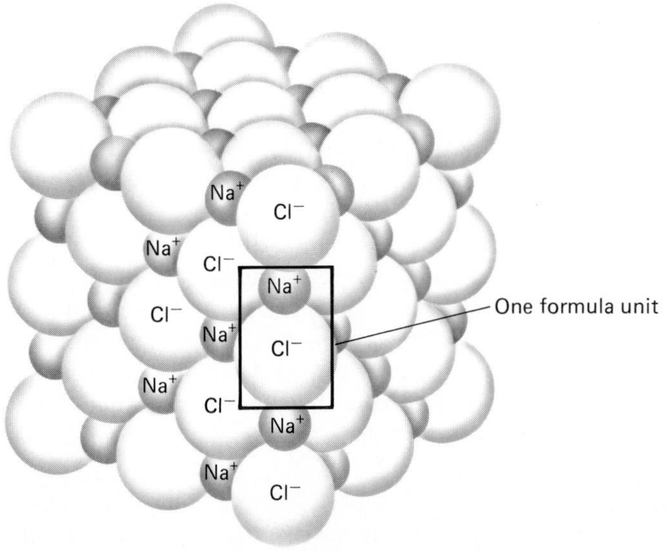

Na+
Cl−
Na+
Cl−
Na+
Cl−
Na+
Cl−
Na+
Cl−
Na+
Na+
Cl−
One formula unit

(a)

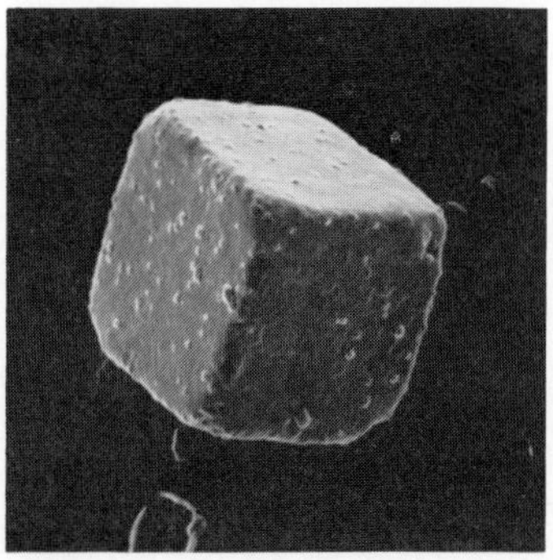

(b)

Na_2SO_4 Contains:		Mass of Each Atom or Ion, amu	×	Number of That Atom or Ion Present	=	Total Mass of Each Element, amu
$2Na^+$ ions		23.0		2		46.0
$1SO_4^{2-}$ ion	1S atom	32.1		1		32.1
	4O atoms	16.0		4		64.0
				Total mass in Na_2SO_4	=	142.1

The formula weight of Na_2SO_4 reported to 3 significant figures is 142 amu.

When an ion is enclosed in parentheses in a formula, remember that the number outside the parentheses multiplies all the atoms inside the parentheses. Look over Sample Exercise 5.11.

SAMPLE EXERCISE 5.11

What is the formula weight of $Mg(OH)_2$?

Solution The method is the same as that shown in the summary of the calculation for Na_2SO_4, but this example points out the multiplication of both atoms O and H within the parentheses.

$Mg(OH)_2$ Contains:		Mass of Each Atom or Ion, amu	×	Number of That Atom or Ion Present	=	Total Mass of Each Element, amu
$1Mg^{2+}$ ion		24.3		1		24.3
$2OH^-$ ions	2O atoms	16.0		2		32.0
	2H atoms	1.01		2		2.02
				Total mass in $Mg(OH)_2$	=	58.3

The formula weight of $Mg(OH)_2$ is 58.3 amu.

Problem 5.9 Determine the formula weights of NaOH, $CaCl_2$, and $Al_2(CO_3)_3$.

5.12 THE NATURE OF THE IONIC BOND

What is called the **ionic bond** is the electrostatic attraction between cations (+) and anions (−) in an ionic compound. As was discussed in Section 4.4, *electrostatic* is the adjective used to describe any attraction between plus and minus charges. Section 4.4 also pointed out that bringing together plus and minus charges results in a lower-energy state. Cation-anion attraction provides the lowering of energy that enables the ionization process and electron transfer to occur.

The strong ionic bond, i.e., the attraction between cations and anions, gives ionic compounds their characteristic properties. For example, ionic compounds typically are solids with high melting points. In an ionic compound the ions are arranged in an alternating cation-anion pattern (Figure 5.2). This gives the material the definite shape characteristic of a solid. Because the cation-anion attractions are so strong, a great deal of energy must be supplied to move them apart. This means that to change a solid to a liquid (melting), a great deal of heat energy must be supplied in order to pull the ions out of their rigid shape and into the shapeless liquid state. Thus, ionic compounds have very high melting temperatures. For example, the melting point of sodium chloride is 801°C.

Because we have just said that a great deal of heat energy must be applied to separate the Na^+ ions and the Cl^- ions in solid sodium chloride, it will probably come as a surprise to you to learn that in salt water the Na^+ ions and Cl^- ions are separated and "swim around" independently of one another. Water has the ability to pull ions out of their orderly array in the solid state. You will need to master some other concepts and wait until Chapter 9 before you can fully understand how water does this. We introduce this idea now because it is one of much critical importance, especially in the clinical laboratory. When you dissolve salt in water to cook vegetables or to make an antiseptic solution for gargling, you make the observation that the solid seems to disappear. In fact, you are witnessing the separation of the ions and their dispersion among water molecules.

5.13 IONS IN BLOOD

Any solution of an ionic compound is a mixture of cations, anions, and solvent, the liquid dissolving medium. The individual content of each ion in the mixture can be determined independently, and such determinations are commonly reported clinically. Figure 2.4, which depicts a typical clinical lab report, offers some examples.

Sodium ion intake must be restricted by persons suffering from hypertension. In fact, prolonged excessive Na^+ ion intake can cause hypertension. Figure 5.3 shows the increase in incidence of hypertension with an increase in dietary salt. Monitoring Na^+ ion intake does not mean just watching the use of table salt (NaCl). Any ionic compound wherein sodium is the cation is a dietary source of Na^+ ion. The flavor enhancer monosodium glutamate (MSG) is Na^+ ions ionically bound to the polyatomic ion glutamate. In baked goods, sodium bicarbonate is another source of Na^+ ions in addition to the table salt that is commonly used in making them.

The normal range of sodium ion content in blood is 3170 to 3330 mg/L. It is unfortunately a common practice in clinical texts to refer to sodium content, but there are never any sodium metal atoms or any other metal atoms in blood—there are only metal ions from ionic compound sources.

If the sodium ion level is elevated, this is described as the condition **hypernatremia.** The prefix *hyper-* means "high"; the suffix *-emia* means "in the blood." The root of the word, *natr,* comes from the Latin word for sodium, *natrium.* Natrium is also the source of the symbol for sodium, Na. A deficiency of sodium ion in blood is termed **hyponatremia.** The prefix *hypo-* means "low."

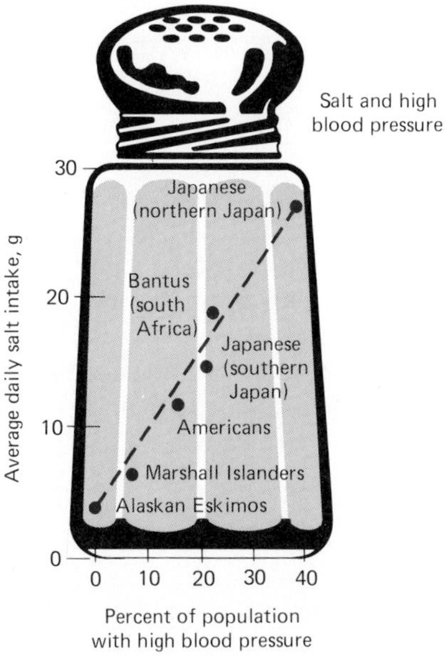
Salt and high blood pressure
Average daily salt intake, g
30
20
10
0
Japanese (northern Japan)
Bantus (south Africa)
Japanese (southern Japan)
Americans
Marshall Islanders
Alaskan Eskimos
0 10 20 30 40
Percent of population with high blood pressure

FIG. 5.3
Evidence that hypertension or high blood pressure occurs more often among heavy salt eaters continually mounts. *(© 1983 by The New York Times Company. Reprinted by permission.)*

TABLE 5.8 Causes and Consequences of Abnormal Blood Ion Concentrations

Condition	Cause	Symptoms
Hyponatremia	Loss of Na^+ ion, e.g., in vomiting or diarrhea; excessive water intake	Slow reduction: mild anorexia, nausea, weakness, giddiness Sudden reduction: agitation, confusion, convulsions, coma
Hypernatremia	Excessive loss of water; excessive intake of sodium	Thirst; dry, sticky mucous membranes; flushed skin; dry, rough tongue; reduced urine output
Hypokalemia	Loss of K^+ ion, in vomiting or diarrhea; insufficient K^+ in diet	Neuromuscular disturbances; cardiac abnormalities, e.g., arrhythmias
Hyperkalemia	Excess retention if renal insufficiency; cell injuries which release K^+ into blood serum	Neuromuscular problems; cardiac manifestations, such as bradycardia
Hypocalcemia	Parathyroid hormone deficiency; Vitamin D deficiency; kidney failure	Tetany, i.e., muscle spasms
Hypercalcemia	Excess parathyroid hormone; excess vitamin D	Emotional instability, psychosis, coma; cardiac failure
Hypomagnesia	Insufficient magnesium in diet; absorption problems	Tremors, mental disorientation, slow heartbeat
Hypermagnesia	Excess intake of magnesium-containing drugs such as antacids and laxatives	Drowsiness, lethargy

Proper potassium ion levels in blood are particularly critical. Either high or low levels can lead to heart failure. The normal range of K^+ ion content in blood is 137 to 195 mg/L. **Hyperkalemia** and **hypokalemia** are the conditions of high and low potassium levels, respectively. The root, *kal,* in this case is from the Latin word for potassium, *kalium.* Kalium is also the source of the symbol for potassium, K. The symptoms of these conditions appear in Table 5.8.

Although most body calcium ions are found in the bones and teeth, a small amount (~1 percent) is present in blood serum. Ca^{2+} is necessary for proper blood clotting (Section 21.10), contraction of muscle, transmission of signals within the nervous system, and good cardiac function. The normal level of serum Ca^{2+} is 90 to 105 mg/L. The causes and symptoms of hypocalcemia and hypercalcemia appear in Table 5.8.

After calcium, potassium, and sodium, magnesium is the fourth most abundant cation in the body. Like calcium, only 1 percent is found in blood; most Mg^{2+} ions are in bone and within cells. Mg^{2+} functions in muscle and nerve action and is also part of many biological catalysts. The normal serum level of Mg^{2+} is 19 to 25 mg/L. The causes and symptoms of hypomagnesia and hypermagnesia appear in Table 5.8.

CHAPTER ACCOMPLISHMENTS

After completing this chapter you should be able to define **key words** and do the following:

5.1 INTRODUCTION

1. Give the general names for the two major types of compounds.

5.2 HOW CAN ATOMS ACHIEVE LOWER-ENERGY STATES?

2. State why most elements interact to form compounds.
3. State the common pattern in the electron configuration of noble gases which is responsible for their stability.
4. State the type of electron configuration atoms strive to obtain when elements combine to form compounds.

5.3 METALS LOSE ELECTRONS

5. Determine the number of electrons an A Group metal will lose to obtain a noble-gas configuration.

5.4 NONMETALS GAIN ELECTRONS

6. Determine the number of electrons an element in Group VA, VIA, or VIIA will gain to obtain a noble-gas configuration.
7. Identify the electron configuration of an A Group ion with that of a noble gas.
8. Relate the terms *cation* and *anion* to the terms *metal* and *nonmetal.*
9. Using a periodic table, predict the charge on an ion formed from a given A Group element.

5.5 ELECTRON TRANSFER

10. Develop a pictorial representation for the formation of ions from atoms.
11. State the relationship between the number of electrons lost by a metal and the number of electrons gained by a nonmetal in the formation of an ionic compound.

5.6 OXIDATION-REDUCTION

12. Explain why electron transfer reactions are called *redox* reactions.

5.7 IONIC CHARGES

13. Write the symbol and charge of each A Group ion, given a periodic table.
14. Given the name or the symbol, state the charge(s) of each ion in Table 5.4.

5.8 POLYATOMIC IONS

15. Distinguish between the terms *monatomic ion* and *polyatomic ion.*
16. Given the name, write the formula and charge of a polyatomic ion, or given the formula, name each polyatomic ion in Table 5.5.

5.9 FORMULAS FOR IONIC COMPOUNDS

17. Write a correct formula for an ionic compound, given the combining elements and a periodic table.
18. Write the formula for an ionic compound, given two combining polyatomic ions or a polyatomic ion and a combining element and a periodic table.

5.10 NOMENCLATURE

19. Name ionic compounds, when given a chemical formula.
20. Write a correct formula for an ionic compound, given the name of that compound.

5.11 FORMULA WEIGHT

21. Given a periodic table and a correct ionic formula, calculate the formula weight of an ionic compound.

5.12 THE NATURE OF THE IONIC BOND

22. Explain why an ionic bond lowers the energy of a compound relative to the individual ions or atoms.
23. Account for the observation that ionic compounds have high melting points.

5.13 IONS IN BLOOD

24. Describe the electronic form of metals found in blood.
25. State what is meant by the conditions hyper- and hyponatremia, hyper- and hypokalemia, hyper- and hypocalcemia, and hyper- and hypomagnesia.

PROBLEMS

5.10 What is a chemical bond?

5.11 List two types of chemical bonds.

5.12 What types (metallic, nonmetallic) of elements combine to form ionic compounds?

5.13 Explain why Group IA metals form a +1 cation.

5.14 Explain why Mg does *not* form a Mg^{3+} ion.

5.15 *a* Write the electron configuration of the Li^+ ion.
b Which inert gas has an electron configuration identical to that of Li^+?

5.16 *a* How many electrons must a barium atom lose to form an ion with a noble-gas configuration?
b Which noble gas has an electron configuration identical to that of barium ion?

5.17 Give the symbols of two cations that have electron configurations identical to that of krypton.

5.18 Explain why nonmetals gain rather than lose electrons in forming ionic bonds.

5.19 Explain why Group VIA nonmetals form a −2 anion.

5.20 *a* Write the electron configuration of the Se^{2-} ion.
b Which inert gas has an electron configuration identical to that of Se^{2-}?

5.21 Give the symbols of three anions that would have electron configurations identical to that of krypton.

5.22 Explain why two sodium atoms are needed to react with one sulfur atom in the formation of the ionic compound sodium sulfide.

5.23 Write the symbol and charge of the ion that could form from each of the following elements:
a Bromine
b Strontium
c Cesium
d Phosphorus
e Aluminum
f Lithium

5.24 Describe each ion in Problem 5.23 as either a cation or an anion.

5.25 *a* Develop a pictorial representation similar to that in Figure 5.1 for the formation of sodium fluoride.
b Explain how your diagram accounts for the formula NaF.

5.26 *a* Develop a pictorial representation similar to that in Figure 5.1 for the formation of potassium nitride.
b Explain how your diagram accounts for the formula K_3N.

5.27 *a* Develop a pictorial representation similar to that in Figure 5.1 for the formation of aluminum oxide.
b Explain how your diagram accounts for the formula Al_2O_3.

5.28 Write the formula for the ionic compound that could form from the elements with atomic numbers 20 and 17.

5.29 Describe the following as oxidation or reduction:
a Cation formation
b Anion formation

5.30 Explain in your own words why electron transfer and *redox* are synonymous.

5.31 Label the following reactions as oxidation or reduction:
a $P + 3e^- \longrightarrow P^{3-}$
b $K \longrightarrow K^+ + 1e^-$
c $S^{2-} \longrightarrow S + 2e^-$
d $Na^+ + 1e^- \longrightarrow Na$

5.32 Write the formula of the ionic compound that could form from elements X and Y if X has two valence electrons and Y five valence electrons.

5.33 Write formulas for the compounds formed from the following ions:
a Cs^+ and F^-
b Ba^{2+} and I^-
c Ba^{2+} and O^{2-}
d NH_4^+ and CO_3^{2-}
e Al^{3+} and SO_4^{2-}
f Na^+ and MnO_4^-
g Li^+ and PO_4^{3-}
h K^+ and $C_2H_3O_2^-$
i Mg^{2+} and N^{3-}
j Zn^{2+} and HPO_4^{2-}
k Pb^{4+} and SO_4^{2-}
l Fe^{3+} and O^{2-}
m Ca^{2+} and HCO_3^-

5.34 Name the compounds formed in Problem 5.33.

5.35 Write a correct formula for the compound formed by combining:
a Lithium and oxygen
b Barium and chlorine
c Sodium and sulfur
d Nickel and nitrogen
e Silver and bromine
f Iron (the +2 ion) and oxygen
g Iron (the +3 ion) and oxygen
h Tin (the +2 ion) and chlorine
i Tin (the +4 ion) and chlorine
j Gallium and sulfur
k Calcium and iodine

5.36 Name the compounds formed in Problem 5.35.

5.37 Indicate the names and numbers of each ion in the following formulas:
a KBr
b $AlPO_4$
c $MgBr_2$
d Li_2CO_3
e $(NH_4)_2SO_4$
f $Ca(OH)_2$
g $Ba(NO_3)_2$
h $NaHCO_3$
i CuO
j CuBr
k $FeCl_3$
l PbO

5.38 Determine the total number of ions in one formula unit of each of the following:
a $CaCl_2$
b $Ca_3(PO_4)_2$
c $Al(HCO_3)_3$
d $(NH_4)_2SO_4$

5.39 Determine the total number of atoms in one formula unit of each of the compounds in Problem 5.38.

5.40 Name the ion common to each compound in each set:
a NaCl, $NaHCO_3$, NaC_2O_4

b KCl, $MgCl_2$, $AlCl_3$
c NH_4Cl, NH_4Br, NH_4I

5.41 Explain why people on a "low-salt" (low-sodium-ion) diet must restrict their intake of MSG ($NaC_5H_8O_4N$) as well as of table salt (NaCl).

5.42 A student suffering from hypertension purchases a bag of pretzels labeled "no salt added." One of the ingredients is baking soda. Will the pretzels be low in sodium ion? Explain.

5.43 Write formulas for the following compounds:
a Silver oxide
b Potassium chloride
c Magnesium hydrogen carbonate
d Aluminum dihydrogen phosphate
e Sodium phosphate
f Barium sulfate
g Copper(II) iodide
h Lead(IV) sulfide
i Copper(I) oxide
j Tin(II) nitride

5.44 Calculate the formula weight of the following ionic compounds:
a K_2S
b NaF
c CaI_2
d Li_2SO_4
e $(NH_4)_3PO_4$
f $MgSO_3$
g $NaC_2H_3O_2$
h $Pb(NO_3)_2$
i K_2CrO_4
j $(NH_4)_2CO_3$

5.45 Explain why ionic solids tend to have high melting points.

5.46 Is hypernatremia a condition due to a high level of sodium metal or sodium ion?

CHAPTER SIX

MOLECULAR COMPOUNDS

6.1 INTRODUCTION

Chapter 5 discussed the structure and properties of ionic compounds, cited the everyday acquaintance you have with them, and mentioned some of their physiological roles. The other type of compound, the molecular compound, which is held together by covalent bonds, has very different properties. Also, it turns out that covalent, or molecular, compounds are much more prevalent. Biological molecules such as the foodstuffs carbohydrates, lipids, and proteins are all held together principally by covalent bonds.

In this chapter we will explore the nature of the covalent bond and covalent, or molecular, compounds. Many of the examples of covalent compounds we'll discuss will be **organic compounds.** The distinction of organic versus inorganic compounds in chemistry arose because prior to about 1830, scientists believed that living matter was distinctly different from nonliving matter. They classified compounds originating from living organisms as **organic** and compounds from nonliving sources as **inorganic.** It was also observed that all organic compounds contained the element carbon.

The belief in a fundamental difference in compounds of living organisms, which was known as the Vital Force Theory, was disproved in 1828 when Friedrich Wöhler converted a nonliving material (ammonium cyanate, classified as inorganic) into urea, an organic compound found in living creatures, including human beings. The classification of compounds as organic or inorganic has been maintained and is based on the observation that most compounds that contain the element carbon are very different in properties from those which do not.

The old classification of organic vs. inorganic which related to living systems does not hold. Most *simple* organic compounds are not involved in physiological chemistry, and many inorganic compounds do play a significant role in body chemistry. For example, the ionic compounds discussed in Chapter 5 are all inorganic. The modern definition of organic chemistry is that it is the study of carbon compounds. Today the field of organic chemistry encompasses all the compounds of carbon, with the major exception of the carbonates and certain metal carbon compounds which we will ignore.

Because organic compounds are covalent, or molecular, compounds, you will find this chapter essential to your subsequent studies of organic and biochemistry and will have frequent occasion to refer to it.

6.2 WHY ARE THERE TWO TYPES OF COMPOUNDS?

Why are there the two categories of compounds, ionic compounds and molecular compounds? Briefly, there are two types of compounds because there are two types of bonds, the forces that hold elements together in compounds.

The different natures of the ionic bond and covalent bond give rise to two classes of compounds with different properties.

The nature of the ionic bond was explored in Chapter 5. The combination of *metals* and *nonmetals* leads to the ionic bond and ionic compound formation through electron transfer. Combining two or more nonmetals will not lead to electron transfer and the formation of an ionic bond because all nonmetals tend to gain electrons. Yet nonmetals do bond to one another. Nonmetals combine to form molecular compounds. Nonmetallic atoms become linked together by forces known as covalent bonds in units called **molecules.** Figure 6.1 depicts molecules of some commonly encountered molecular compounds. Molecules are often portrayed by molecular models which use balls to represent atoms, and sticks or springs to represent the covalent bonds holding the atoms together.

The atoms in molecules are held together by *shared pairs* of electrons. A shared pair of electrons is called a **covalent bond.** As in the case of the formation of ionic bonds, covalent bonds form so that atoms can achieve a noble-gas configuration.

Thus compound formation in general occurs in order that atoms can achieve lower-energy, noble-gas configurations. There are two types of compounds because there are two ways to achieve the noble-gas configuration:

1 **Transfer of electrons** from metals to nonmetals leads to ionic bonds and ionic compounds.
2 **Sharing of electrons** between nonmetals leads to molecules that are held together by covalent bonds. Molecular compounds are also called covalent compounds.

6.3 DIATOMIC MOLECULES

Certain elements (H_2, O_2, N_2, halogens) exist as diatomic molecules. These elements exist as diatomic molecules rather than individual atoms because the atoms share their electrons in order to achieve a noble-gas configuration.

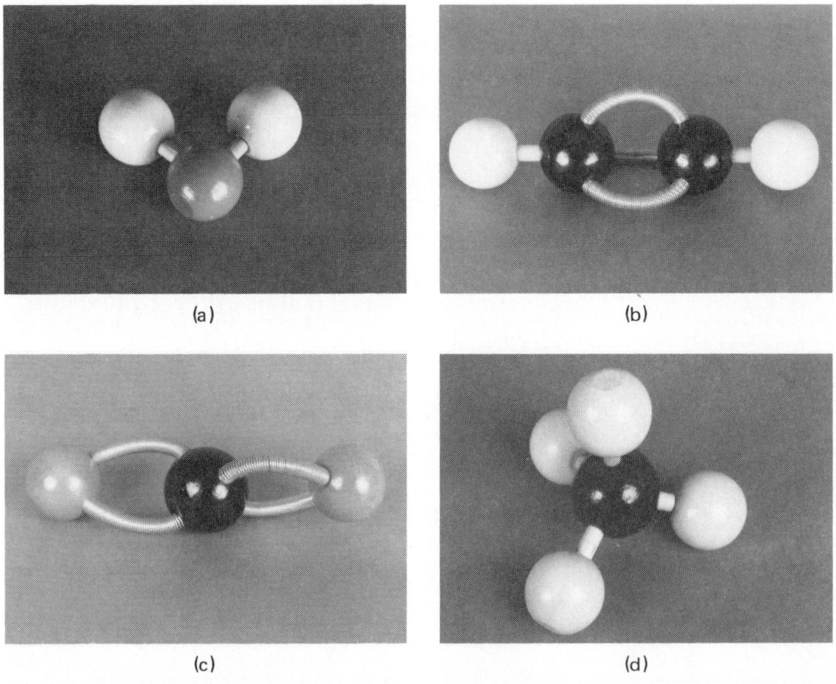
(a)
(b)
(c)
(d)

FIG. 6.1
Molecular models of covalent compounds. Covalent bonds hold atoms together in molecules. (*a*) Bonds hold two hydrogen atoms and one oxygen atom in the order H—O—H in water molecules. Molecular models depict atoms as balls and covalent bonds as sticks or springs, and show the arrangements of the atoms in space. Besides water, this figure shows other covalent compounds with which you are familiar: (*b*) Acetylene, HC≡CH, a gas used in welding torches; (*c*) carbon dioxide, CO_2, which we exhale; (*d*) methane, CH_4, which is natural gas. (*Photograph by Bryan Lees.*)

For example, consider individual chlorine atoms (Group VIIA), for which we would write the Lewis electron dot symbols (Section 4.10):

:C̤̈l· ·C̤̈l:

Seven dots represent the seven valence electrons in the third main level of $_{17}Cl$, $1s^22s^22p^63s^23p^5$. Each atom can acquire eight valence electrons (and consequently greater stability and lower energy) if the two come together to share a pair of electrons. As the two atoms come close together, each atom contributes one unpaired electron to form a shared pair between the atoms. The shared pair belongs equally to both atoms, so each now has a complete octet for its valence electrons, i.e., a noble-gas configuration.

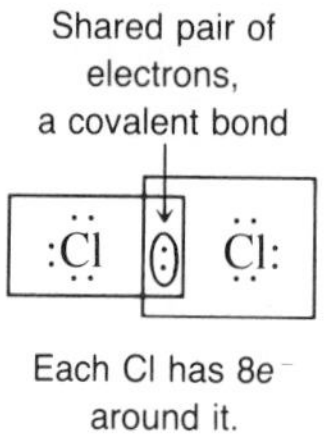
Shared pair of electrons, a covalent bond
Each Cl has 8e⁻ around it.

The shared pair can also be indicated by a dash (—).

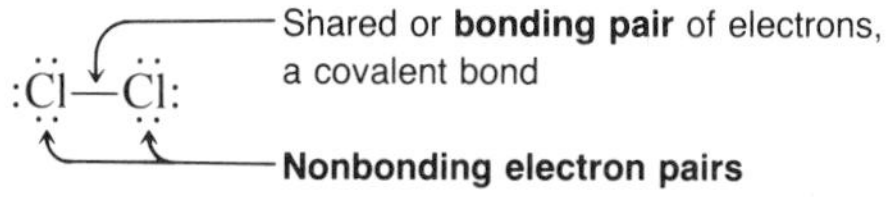
Shared or **bonding pair** of electrons, a covalent bond
Nonbonding electron pairs

The valence electrons not involved in bonding are called **nonbonding electrons.** They are also called **lone pairs.**

All the other halogens (Group VIIA) form covalent bonds in exactly the same way:

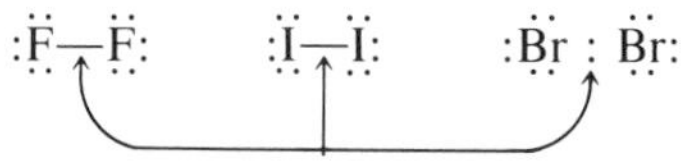

Remember, the dash or shared pair means exactly the same thing.

H_2 molecules are also held together by a shared pair. The two H atoms come together to form H:H, or H—H. Each H atom now has the helium duet, two electrons.

When we try to account for the bonding in O_2 with a single pair of shared electrons, the oxygen atoms do not obtain a complete octet. Oxygen is in Group VIA:

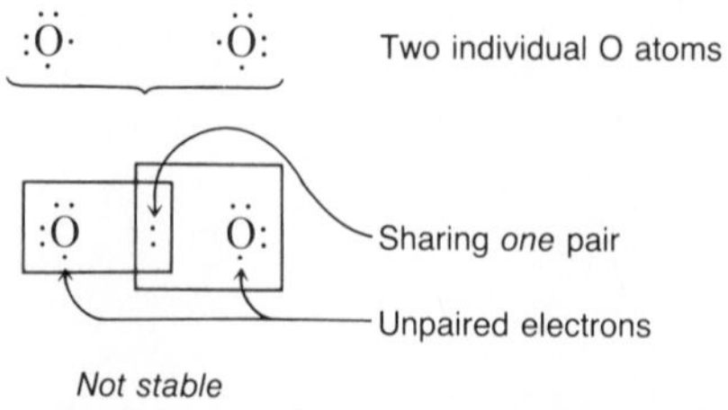
Two individual O atoms
Sharing *one* pair
Unpaired electrons
Not stable

Each O has only seven electrons (count within the boxes around the atoms), but if the oxygens share their remaining unpaired electrons, octets will be attained.[1]

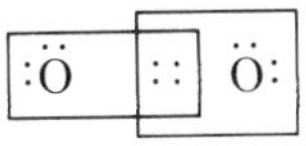

or :Ö=Ö:

Each O has eight electrons. Two shared pairs of electrons are called a *double covalent bond,* or simply a **double bond.**

In order for each nitrogen in N_2 to have a noble-gas configuration, three shared pairs are required. This is known as a **triple bond.**

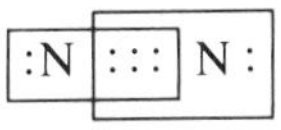

or :N≡N:

There are *no* examples of quadruple bonds.

6.4 THE NATURE OF THE COVALENT BOND

You have seen that it is the electrostatic attraction between cations and anions that lowers energy in the formation of ionic bonds. The formation of covalent bonds leads to a lowering of energy because of the attraction between the electrons (−) of one atom and the nucleus (+) of another atom. Let us see how energy is lowered in the H_2 molecule (relative to two separate H atoms) through covalent bond formation.

In the two separate H atoms the lone valence electron of each isolated atom is attracted to its respective nucleus.

H1
H2

As the atoms come closer together, there are two new attractive forces (electron$_1$ to nucleus$_2$, electron$_2$ to nucleus$_1$) and two repulsive forces (between electron$_1$ and electron$_2$, between nucleus$_1$ and nucleus$_2$). Up to a certain distance apart, called the *bond length,* the new attractive forces are larger than the repulsive forces. This leads to a lowering of energy for the H_2 molecule, i.e., two H atoms held together at a fixed distance compared with two separate H atoms (see Figure 6.2). The atoms cannot move apart to a greater distance than the bond length because this would raise the energy by reducing the attractive forces. They cannot move closer than the bond length because this would also raise the energy by increasing the repulsive forces.

6.5 LEWIS ELECTRON DOT FORMULAS

It is important to know *why* covalent bonds form. It is also necessary to be able to tell *how* electrons are shared in a molecule for which the molecular formula is available. The sharing pattern is shown by writing Lewis electron dot formulas, just as we have done previously for the diatomic molecules.

Another simple example of how to write a Lewis formula can be seen by considering the bonding in HCl. First write the Lewis dot symbols for the atoms:

[1]This electronic structure for oxygen does not account for some of its observed properties and consequently represents a failure of Lewis theory to predict structure. However, it offers a simple example of the concept of a double bond.

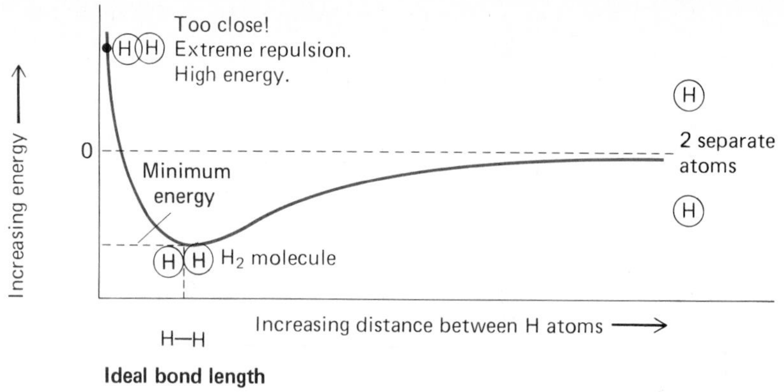
Too close!
Extreme repulsion.
High energy.
Increasing energy
0
Minimum energy
H₂ molecule
2 separate atoms
Increasing distance between H atoms
H—H
Ideal bond length

FIG. 6.2
Minimum energy is achieved when the distance between H atoms is the bond length. Separate atoms have higher energy. Atoms closer together have higher energy.

H· ·C̈l:

Group IA Group VIIA

Now see how the unpaired electrons can be shared so that both H and Cl attain a noble-gas configuration. This is achieved easily for this small molecule:

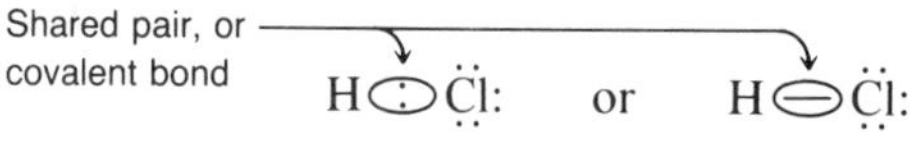
Shared pair, or covalent bond
or

Whether we write the formula horizontally, vertically, or at an angle makes no difference:

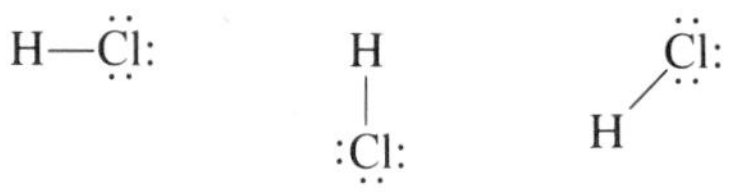

H has a duet, and Cl has an octet.

Lewis electron dot formulas are also called **Lewis structures** or **structural formulas** because they tell how the molecule is *constructed,* or built.

Problem 6.1 Write the Lewis electron dot structures for HBr, HF, and HI.

By "playing around" with the unpaired electrons of the individual atoms of H_2O, you probably could figure out that the Lewis structure is H—Ö—H.

Group IA
Group VIA

However, for more complicated molecules, particularly those with many atoms, it would be helpful to have guidelines for writing Lewis structures instead of having to just "play around" or flounder.

Most Lewis structures that you will be asked to write or interpret will be those for organic molecules. For this reason, we will offer simple guidelines for writing Lewis structures that work almost invariably for organic compounds and for many common inorganic compounds such as water. The simple method does not always work for inorganic compounds; some of these difficulties will be addressed in Section 6.6.

The guidelines we suggest depend on the observation that certain elements characteristically form certain numbers of bonds in order to attain octets. For example, consider carbon. Carbon is in Group IVA; hence carbon atoms have four valence electrons. Each carbon needs four more electrons to achieve an octet. Consequently, we find that in its compounds each carbon must participate in four shared electron pairs (bonds) with other atoms in order to achieve an octet. The sharing can involve single bonds, double bonds, or triple bonds.

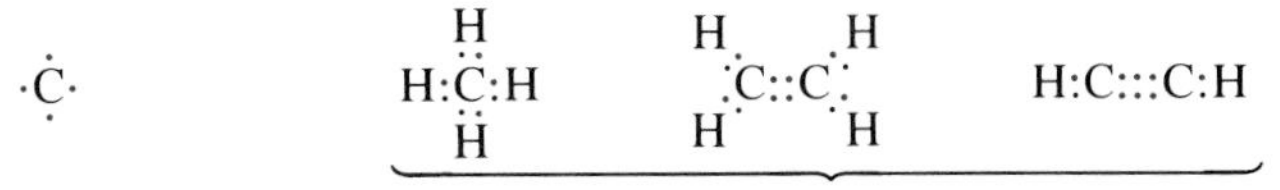

Carbon atom with four valence electrons

Stable molecules containing carbon; each carbon atom has an octet of electrons.

In organic chemistry, the dash (—) is more often used to represent an electron pair. Thus these structures would be:

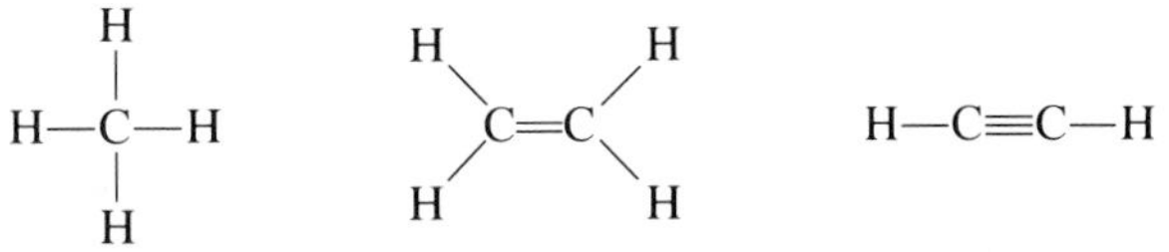

If we apply the same type of analysis to the other elements commonly found in organic molecules, we would find the general bonding patterns shown in Table 6.1. Recognition of these patterns offers an approach to writing Lewis structures, or **structural formulas.**

Given a molecular formula, one writes a correct Lewis, or structural, formula by providing each given atom with the numbers of bonds indicated in Table 6.1. For example, chloroform has the molecular formula $CHCl_3$. Carbon must form four bonds. Each chlorine atom and the H atom can form only one bond each, but there are a total of four such atoms (three Cl's and one H). Therefore, a suitable structure is

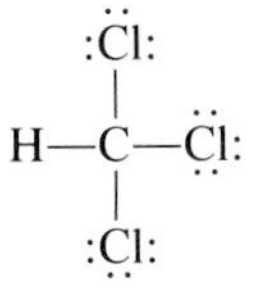

This method of writing Lewis formulas centers on bonding electrons. However, the nonbonding electrons, or lone pairs, should not be ignored because they are often required to assure all atoms in the compound an octet. Table 6.1 shows the lone pairs that surround oxygen, nitrogen, and halogens in their compounds. In writing the structure for chloroform ($CHCl_3$), we simply carried along the lone pairs on chlorine from Table 6.1.

When writing Lewis structures, after you have satisfied the bonding requirements of atoms, check that each atom has an octet (H requires only a duet). If in writing the structure for $CHCl_3$ we had merely formed the required

TABLE 6.1 Typical Covalent Bonding Patterns for the Common Elements

Lewis Dot Structure	Periodic A Group	Number of Bonds Usually Formed*	Bonding Patterns
·Ċ·	IV	4	—C— four single bonds
			—C= two single, one double bond
			—C≡ one single, one triple bond
H·	I	1	—H one single bond
:Ẍ· †	VII	1	:Ẍ— one single bond
:Ö: ‡	VI	2	—Ö— two single bonds
			:Ö= one double bond
·Ṅ· §	V	3	—N— three single bonds
			—N= one single, one double bond
			:N≡ one triple bond

* These patterns do not take into account the possibility of coordinate covalent bonding. Hence, for example, N usually forms three bonds, but occasionally we see it forming four bonds, as in the case of ammonium, NH_4^+. This is discussed in Section 6.6.

† X represents the halogens chlorine, fluorine, bromine, and iodine.

‡ Sulfur is in the same periodic group as oxygen; therefore it follows the same bonding patterns.

§ Phosphorus theoretically can use the same patterns as nitrogen. In fact, P most often forms three single bonds or it is pentavalent (i.e., forms five bonds), as in the phosphates.

number of bonds without regard to the numbers of valence electrons surrounding the atoms, our structure would have been

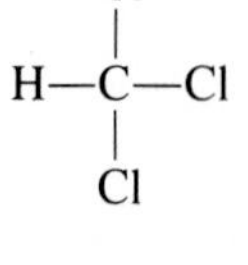

In this incomplete structure, an octet is shown around carbon, but each chlorine lacks an octet. This is rectified and the structure is corrected by placing the appropriate number of electron dots (in this case, six) around each chlorine symbol. These six electrons are in chlorine's valence shell. One of the seven valence electrons is shared in bonding; the other six are lone pairs.

Ammonia, the gas inhaled from smelling salts, has the molecular formula NH_3. Using the information in Table 6.1, we arrive at the Lewis structure for ammonia:

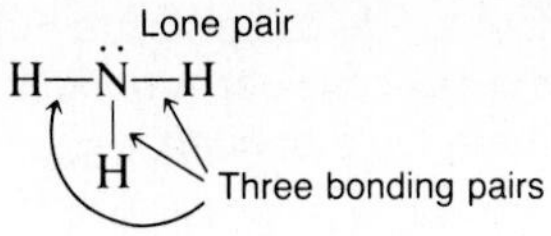
Lone pair
Three bonding pairs

Nitrogen forms three bonds; its octet is completed by the remaining lone pair from its valence electrons.

Sample Exercises 6.1 through 6.3 use the information in Table 6.1 to write Lewis structures for several different molecules. They are varied and point out how you can recognize different bonding patterns as you examine the molecular formula.

SAMPLE EXERCISE 6.1

Write a structural formula for the molecular formula CH_4O.

Solution Begin with the atom that forms the largest number of bonds, namely, carbon:

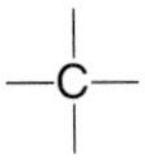

Next, consider oxygen which forms two bonds. "Attach" it (i.e., bond it) to carbon and see how many more bonds must be formed.

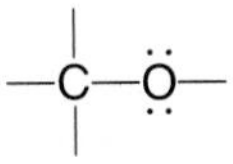

Carbon requires three more bonds and oxygen one more. There are four hydrogens, each of which forms one bond to complete the structure:

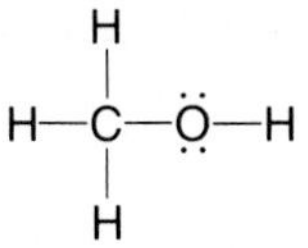

Check that each atom (except H) has an octet. If we had neglected to include the lone pairs on oxygen, this oversight would be revealed in the check.

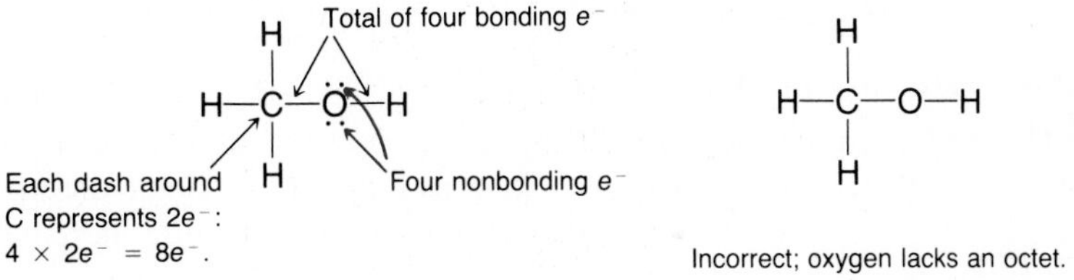
Total of four bonding e^-
Four nonbonding e^-
Each dash around C represents $2e^-$: $4 \times 2e^- = 8e^-$.
Incorrect; oxygen lacks an octet.

SAMPLE EXERCISE 6.2

Construct a structural formula for formaldehyde (CH_2O), a preservative for biological specimens.

Solution Begin as in Sample Exercise 6.1 and arrive at

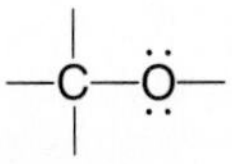

In this case we do not have four atoms to complete the structure as before. We only have two H atoms. This tells us to consider a bonding pattern with multiple bonds; i.e., carbon and oxygen can be double-bonded:

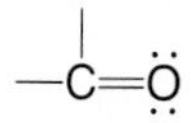

Now we see that only two hydrogens are required to complete the structure:

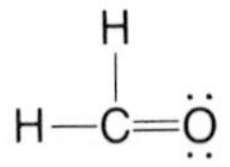

Check that each atom (except H) has an octet.

SAMPLE EXERCISE 6.3

Write a Lewis structure for hydrogen cyanide (HCN), the lethal gas used in gas chambers.

Solution Begin as in the previous examples with carbon, which forms four bonds, and attach it to nitrogen, which forms three bonds.

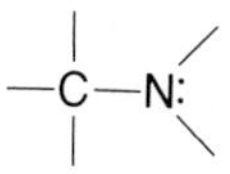

The one H in the molecular formula cannot satisfy the bonding requirements of C and N. Therefore, you conclude that multiple bonding is necessary. A double bond between C and N is not sufficient. There must be a triple bond:

—C≡N:

The one H atom must be bonded to C to satisfy carbon's bonding requirements and complete the structure:

H—C≡N:

Check that each atom (except H) has an octet.

Problem 6.2 Use Table 6.1 to write a structural formula for CH_5N.

6.6 COORDINATE COVALENT BONDS

Most covalent bonds are formed by each bonded atom contributing one electron to the pair. The method of writing Lewis structures suggested in Section 6.5 presupposes this. These single electron contributions from each atom can be seen by looking at diatomic molecules and at NH_3. The N—H

Separate Atoms	Diatomic Molecules
H· ·H	H:H
:Cl· ·Cl:	:Cl:Cl:
H· ·Cl:	H:Cl:

bonds in NH_3 can easily be envisioned as arising from the contribution of one electron from N and one from H.

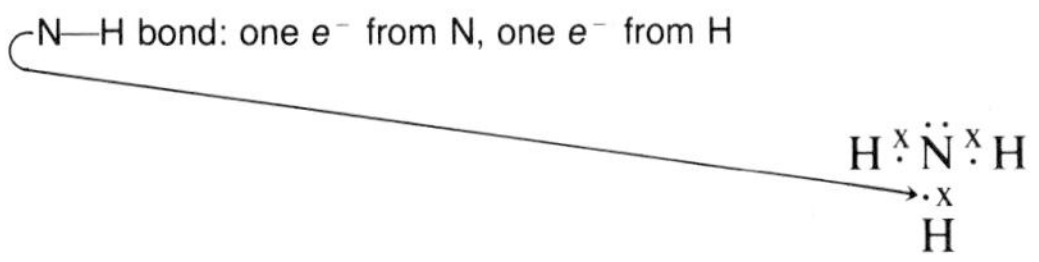
N—H bond: one e⁻ from N, one e⁻ from H

N has five valence electrons (represented here as ·); each H has one valence electron (represented here as x).

It is also possible to form a bond in which one of the bonded atoms furnishes both electrons. A bond where one atom furnishes both electrons is called a **coordinate covalent bond.** For example, the polyatomic ammonium ion NH_4^+ forms if the two nonbonding electrons on NH_3 form a coordinate covalent bond with an H^+ ion which has no electrons to contribute.

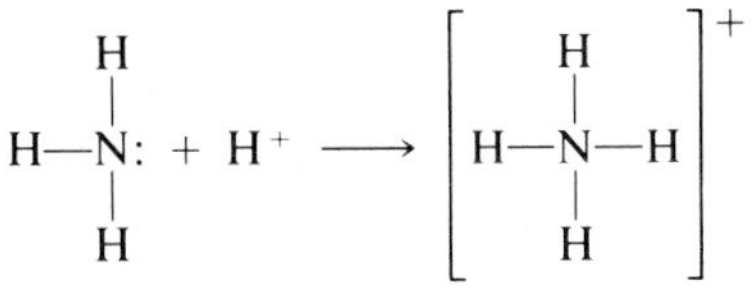

The four N—H bonds in NH_4^+ are identical. Once formed, a coordinate covalent bond has the same properties as any other covalent bond.

Problem 6.3 Show how the ion H_3O^+ can be formed by coordinate covalent bond formation between H_2O and H^+.

A notable example of a coordinate covalent bond is that between oxygen and the iron in the molecule hemoglobin formed in red blood cells. It is this bond that allows hemoglobin to carry oxygen to the cells of the body. Figure 6.3 shows a structure known as *heme,* the iron-containing portion of hemoglobin.

Both bonding electrons for joining oxygen to iron are supplied by oxygen; hence, a coordinate covalent bond is formed.

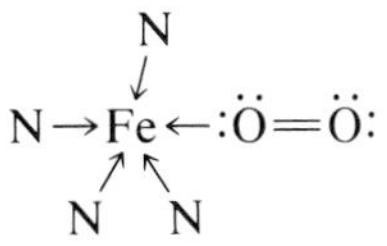

(No geometry is implied.)

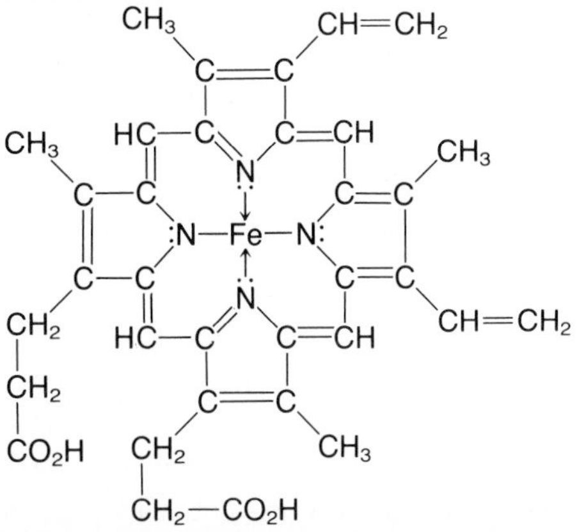

FIG. 6.3 Heme, an important part of the hemoglobin molecule. The iron ion (Fe^{2+}) is bonded to four nitrogen atoms in a complex molecule known as the protoporphyrin IX. Each five-membered, nitrogen-containing ring is called a porphyrin ring. The Fe^{2+} ion is capable of accepting a pair of electrons from an oxygen molecule to form a coordinate covalent bond.

Iron is capable of forming such bonds with other electron pairs. For example, a gas found in automobile exhaust, carbon monoxide, can bond to iron in hemoglobin through its lone pairs (:C≡O:). Because this bond is stronger than the iron-oxygen bond, oxygen is prevented from linking up with iron and is not carried to the body's cells; this is what makes CO toxic. (See Section 19.12 for a fuller discussion of oxygen transport.)

6.7 POLYATOMIC IONS

Ammonium, NH_4^+, which was just discussed in Section 6.6, is one of the polyatomic ions. Like ammonium, all of the polyatomic ions are held together by covalent bonds. Unlike neutral molecules, which have total numbers of bonding and nonbonding electrons equal to the total numbers of valence electrons in the unbonded atoms, polyatomic ions have charges because they have extra electrons, or, as in the case of NH_4^+, are electron deficient. (It was pointed out in Section 6.6 that one hydrogen in NH_4^+ does not bring a valence electron to the structure.)

Most polyatomic ions are negatively charged because they have extra electrons, that is, more electrons than the atoms contribute from their valence electrons. Polyatomic ions form from molecules by bonds breaking so that one of the atoms gets both electrons of the bond. For example, consider the acetate ion which forms from acetic acid:

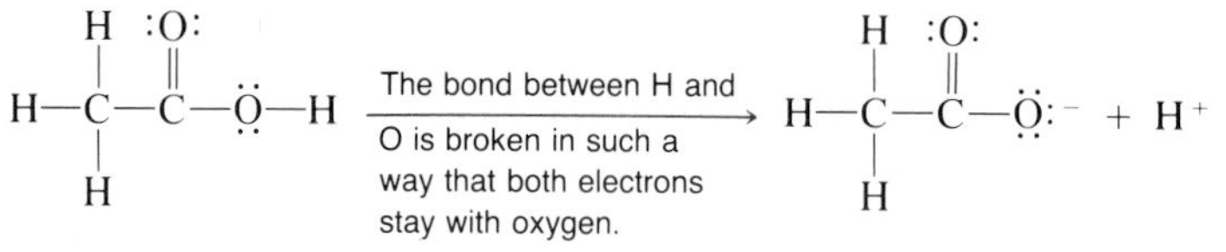
The bond between H and O is broken in such a way that both electrons stay with oxygen.

Acetic acid

Acetate ion

The hydrogen ion is produced; it is a bare proton without the electron found in a hydrogen atom. The acetate ion has an extra electron and hence has a negative charge. The total number of valence electrons from two carbons, two oxygens, and three hydrogens is 23 (four from each C, six from each O, and three from each H). There are 24 electrons depicted in the bonding and nonbonding electrons of acetate ion. This is what is meant by an extra electron.

Sulfate ion arises in an analogous way:

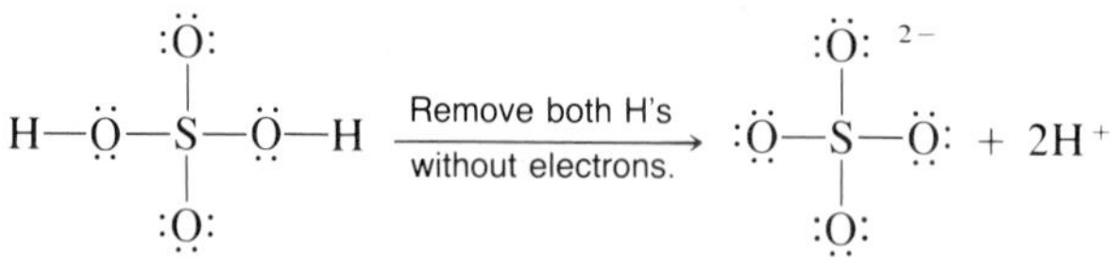
Remove both H's without electrons.

Sulfuric acid, H_2SO_4

Sulfate ion, SO_4^{2-}

The guidelines in Section 6.5 would not permit you to write a Lewis structure for H_2SO_4 or SO_4^{2-}, but the basic Lewis rule of establishing octets around atoms would. You need not be able to write these structures to understand the nature of polyatomic ions. What is essential is that you realize that (1) the atoms in a polyatomic ion are held together by covalent bonds and (2) because they are charged, they can form electrostatic or ionic bonds with oppositely charged species.

Ammonium acetate ($NH_4C_2H_3O_2$) is composed entirely of nonmetals. However, it is an ionic compound because ammonium is a cation and acetate is an anion. The cation and anion are held together by electrostatic attraction, the ionic bond.

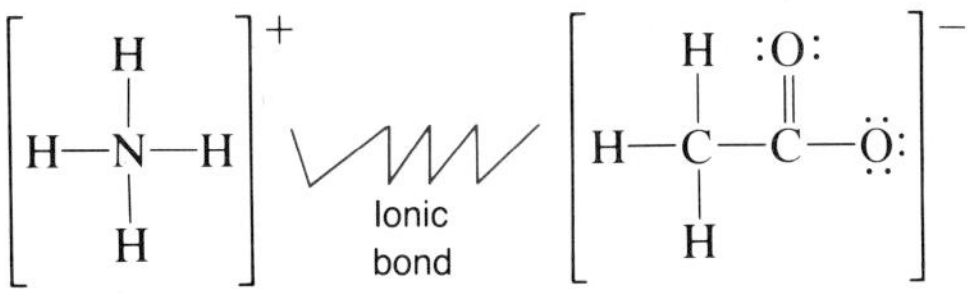
Ionic
bond

6.8 ELECTRONEGATIVITY AND POLARITY

So far, we have discussed the electron pair in a covalent bond as if it were shared equally between the two atoms. This is definitely the case when the two bonded atoms are identical, as with H_2 or Cl_2. However, in most bonds one bonded atom attracts the bonding electrons more strongly. For example, in H—Cl the chlorine has a stronger attraction for electrons than does the hydrogen. This unequal attraction provides the basis for the unequal breaking of bonds wherein one atom gets both electrons, as discussed in Section 6.7.

The attractive force that an atom exerts on shared electrons in a chemical bond is known as its **electronegativity.** A scale of relative electronegativities has been devised by the Nobel Laureate Linus Pauling (Nobel prizes: Chemistry 1954, Peace 1962). Values for the representative elements are given in Table 6.2. When you examine this table, you will see that electronegativity is a periodic property. In general it increases from left to right across a period and up a group from bottom to top. This means that nonmetals have high electronegativities and metals have low electronegativities. Because metals have low electronegativities, they are said to be *electropositive.*

The difference in electronegativities between two bonded atoms offers an indication of the **polarity** in the bond. Polarity is a measure of the inequality in the distribution of bonding electrons. A **nonpolar bond** is one in which there is *equal sharing* of bonding electrons. In a **polar bond** there is *unequal sharing;* the more electronegative atom pulls the electrons closer.

H:H

Nonpolar bond; electrons shared equally

H :Cl

Polar bond; electrons closer to Cl, the more electronegative element

TABLE 6.2 Electronegativities of Some Representative Elements

IA	IIA	IIIA	IVA	VA	VIA	VIIA
H 2.1						
Li 1.0	Be 1.5	B 2.0	C 2.5	N 3.0	O 3.5	F 4.0
Na 0.9	Mg 1.2	Al 1.5	Si 1.8	P 2.1	S 2.5	Cl 3.0
K 0.8	Ca 1.0	Ga 1.6	Ge 1.8	As 2.0	Se 2.4	Br 2.8
Rb 0.8	Sr 0.9	In 1.7	Sn 1.8	Sb 1.9	Te 2.1	I 2.5
Cs 0.7	Ba 0.9	Tl 1.8	Pb 1.8	Bi 1.9	Po 2.0	At 2.2
Fr 0.7	Ra 0.9					

Figure 6.4 shows the sharing pictorially.

To obtain a rough measure of the polarity of a bond, take the mathematical difference between the electronegativities of the bonded atoms. The larger the difference, the more polar the bond.

For example, in the case of H_2 (H—H) the electronegativity difference (2.1 − 2.1) is zero; thus the bond is a *nonpolar covalent bond,* or simply a *covalent bond.* The two electrons in a (nonpolar) covalent bond are equally distributed between the atoms.

In H—Cl the electronegativity difference between H (2.1) and Cl (3.0) is 0.9 (3.0 − 2.1). Because of the electronegativity difference, the H—Cl bond is called a *polar covalent bond.* If we look more closely at the consequences of unequal sharing of electrons, we will find how this word *polar* originates. In H—Cl, because the bonding electrons are closer to Cl, Cl acquires a partial negative charge (remember that electrons are negative). H acquires a partial positive charge because the electrons neutralizing the nuclear charge are being drawn away. This is written as

$$^{\delta+}\text{H} \quad :\text{Cl}^{\delta-}$$

where the symbol δ means "partial." The bond has developed a positive pole and a negative pole, just as we speak of the + and − poles of a battery. That is why the bond is said to be polar.

Because there are *two* poles (one + and one −) one can also say that a **dipole** exists in the bond. (The prefix *di-* always means "two.") A crossed arrow (+⟶) pointing toward the more electronegative element indicates the presence of a dipole.

$$\overset{+\!\!\longrightarrow}{\text{H—Cl}}$$

The greater the electronegativity difference, the more polar the bond and the larger the dipole. As you continue your study of chemistry, bond polarity will play an important role in understanding the properties of molecules.

SAMPLE EXERCISE 6.4

For the bonds indicated below, (1) decide whether they are nonpolar or polar covalent, (2) mark the dipole, and (3) decide which is the most polar:

a H—I
b The N—H bond in NH_3
c The O—H bond in H_2O

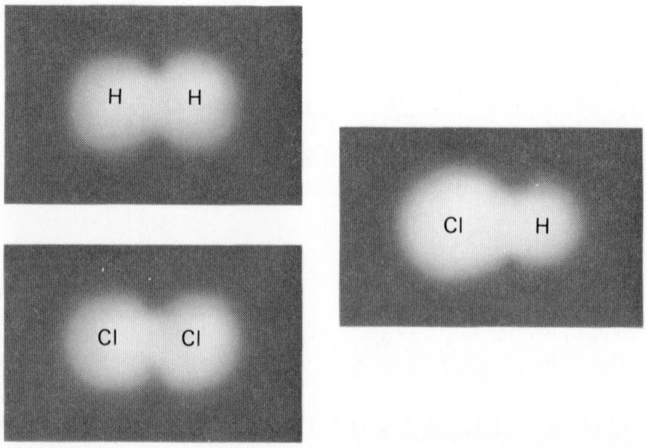
H H
Cl H
Cl Cl

FIG. 6.4
The fuzzy cloud around the atoms is the shared pair of electrons. In H_2 and Cl_2, the cloud is evenly distributed. In HCl the cloud is drawn more toward the more electronegative atom, which is Cl.

Solution

1 Calculate the difference in electronegativities between bonded atoms (see Table 6.2). Subtract the smaller electronegativity from the larger one.

a 2.5 − 2.1 = 0.4
(I) (H)
Although the difference is small, 0.4, we classify this bond as polar based on the discussion so far.

b 3.0 − 2.1 = 0.9
(N) (H)
We classify this bond as polar, because of the electronegativity difference.

c 3.5 − 2.1 = 1.4
(O) (H)
Again, we classify the bond as polar.

2 The dipole is marked by a crossed arrow pointing in the direction of the more electronegative element.

a $\overset{+\!\!\longrightarrow}{\text{H—I}}$

b $\overset{\longleftarrow\!\!+}{\text{N—H}}$

c $\overset{\longleftarrow\!\!+}{\text{O—H}}$

3 The O—H bond is the most polar because the electronegativity difference (1.4) is largest.

6.9 RECOGNIZING IONIC VERSUS MOLECULAR COMPOUNDS

A simple generalization that enables us to readily distinguish between ionic and molecular (covalent) compounds is that *most ionic compounds are formed from metals and nonmetals,* whereas *most molecular compounds are formed from combinations of nonmetals.* This rule is very convenient for purposes of nomenclature and formula writing because metal plus nonmetal combinations are named according to the rules for ionic compounds, and compounds containing exclusively nonmetals usually are not.

However, in considering the properties of a compound, its ionic or covalent (molecular) nature depends on the electronegativity differences between the bonded atoms. See Table 6.3 for a comparison of the properties of ionic vs. covalent compounds. There is a gradual transition from covalent to polar covalent to ionic bonding as the electronegativity differences increase. The properties of the compound depend on the bonding character. The following ranges of electronegativity differences may be used to predict bonding character:

From 0 to 0.5, covalent (to only slightly polar)

From 0.5 to 1.7, polar covalent

Over 1.7, ionic

TABLE 6.3 Comparison of Properties of Ionic and Covalent Compounds

Characteristic	Ionic Compounds	Covalent Compounds
Smallest particles present	Ions.	Molecules.
Physical state at room temperature	Solids.	Gases, liquids, or low-melting-point solids.
Electrical conductivity of the molten state	Conductors.	Nonconductors.
Water solubility	Many are soluble.	Most are not soluble.
Electrical conductivity of water solution	Conductors.	Most are nonconductors (unless they react with water).

Because the transition in properties is gradual, the cut-off points of 0.5 and 1.7 are somewhat arbitrary, and bonds with electronegativity differences near (± 0.2) these values are expected to be intermediate in character.

Notice that the rule that a combination of a metal and nonmetal yields an ionic compound works very well for most examples we encounter. For example, from Table 6.2 we can obtain the electronegativities for Na and Cl as 0.9 and 3.0, respectively. The difference of 2.1 between these two numbers tells us that the bond between Na and Cl is ionic, and of course, we know from experience that sodium chloride (table salt) exhibits the properties of an ionic compound, as indicated in Table 6.3.

In contrast, gasoline exhibits the properties of a covalent compound. Its formula C_8H_{18}, which shows only nonmetals in the compound, and the small (0.4) electronegativity difference between C and H both predict that this should be so. Gasoline should also be recognized as an organic compound. Carbon compounds are organic compounds. Organic compounds are almost exclusively covalent compounds.

6.10 MOLECULAR SHAPE

Atoms are three-dimensional. We have envisioned them as spheres or clouds. Molecules also have three-dimensional shapes; the particular shape depends on how the atoms in the molecule are arranged. Figure 6.5 depicts the typical molecular shapes that we will discuss. Because Lewis structures are written on paper, they have only two dimensions. However, from Lewis structures we can determine the three-dimensional arrangement in space of electron pairs (electron-pair geometry) and then determine the molecular shape.

The Valence Shell Electron-Pair Repulsion theory (VSEPR, pronounced

FIG. 6.5
Molecular shapes. The shape of the molecule is determined by the arrangement of the atoms.

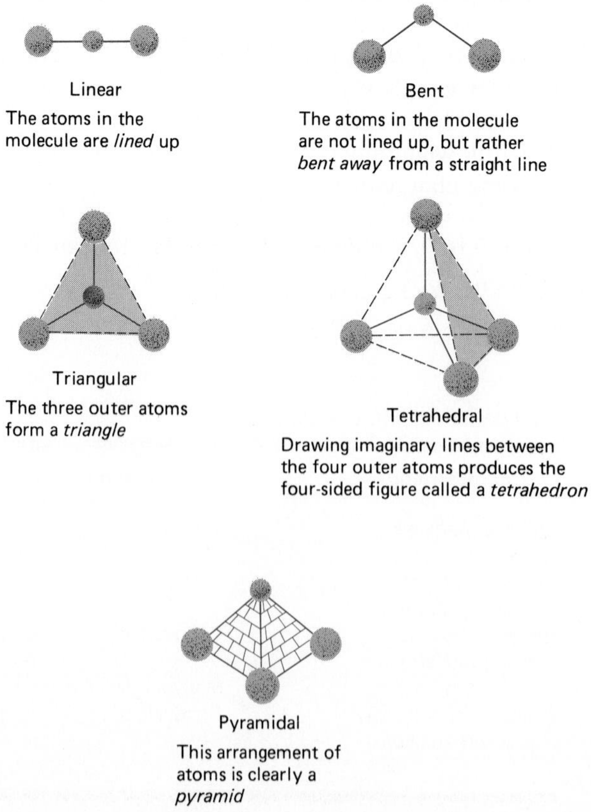
Linear
The atoms in the molecule are *lined* up
Bent
The atoms in the molecule are not lined up, but rather *bent away* from a straight line
Triangular
The three outer atoms form a *triangle*
Tetrahedral
Drawing imaginary lines between the four outer atoms produces the four-sided figure called a *tetrahedron*
Pyramidal
This arrangement of atoms is clearly a *pyramid*

"vesper" for short) allows us to predict the three-dimensional arrangement of electrons around a central atom. The basis for this theory is the idea that electron pairs keep as far away from each other as possible. Table 6.4 shows the geometries associated with the maximum possible separation in space for two, three, and four points around a central point.

The geometry that will keep two points farthest apart is a linear geometry, a separation of 180°. For three points, maximum separation is achieved by a planar triangular geometry, a separation of 120°. For four points, the greatest separation occurs in a tetrahedral geometry, a separation of 109°28′.

Given any Lewis structure, we can determine the number of *sets* of electron pairs around a central atom. For example,

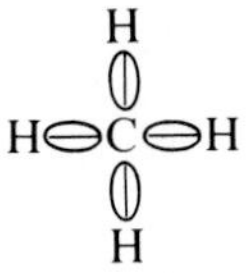
Four sets

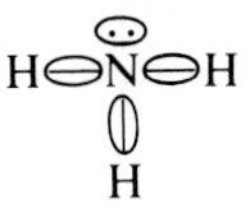
Four sets

Two sets

Two sets

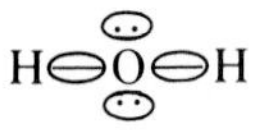
Four sets

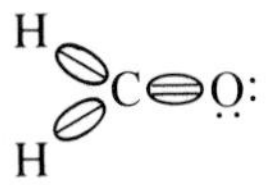
Three sets

Notice that we are not simply counting electron pairs, but rather we are considering how the pairs are grouped in *sets*. A set can consist of one pair (either a single bond or a nonbonding pair), two pairs (a double bond), or three pairs (a triple bond). Let us call this number of sets of electron pairs around a central atom the **set number. The set number can also be determined**

TABLE 6.4 Geometries of Maximum Separation of Points (•) in Space around a Central Point (●)

Number of Points	Arrangement		Angle of Separation
2		Linear, i.e., along a line.	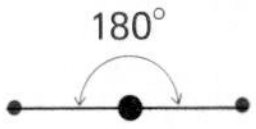180°
3	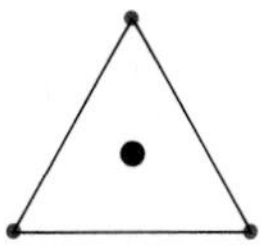	Planar triangular; i.e., in one plane in space, and a triangle is formed by connecting the dots (•).	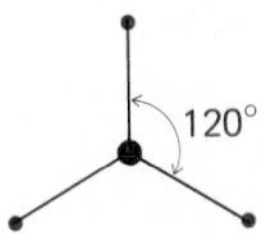120°
4	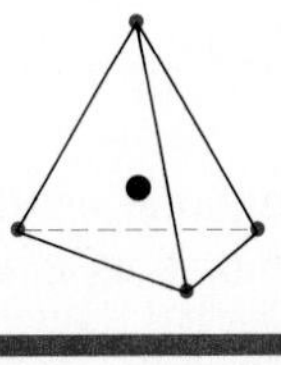	Tetrahedral; i.e., connecting the dots (•) forms the geometric figure called a tetrahedron.	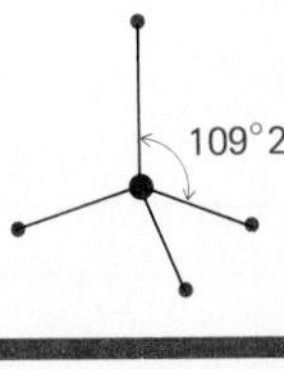109°28′

by counting the number of atoms bonded to the central atom plus the number of nonbonding pairs on the central atom. The electron distributions for the Lewis structures shown above are summarized as follows:

	Set Number	=	Number of Bonded Atoms	+	Number of Nonbonding Pairs
CH_4	4	=	4	+	0
NH_3	4	=	3	+	1
HCN	2	=	2	+	0
CO_2	2	=	2	+	0
H_2O	4	=	2	+	2
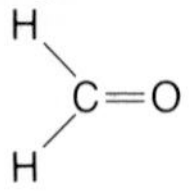	3	=	3	+	0

The **set number** tells you the **electron-pair geometry** directly:

Set number	Geometry
2	Linear
3	Planar triangular
4	Tetrahedral

The molecular shape may or may not be the same as the electron-pair geometry. If there are *no nonbonding electron pairs* on the central atom, the **electron-pair geometry** and **molecular shape** *are identical.* For example,

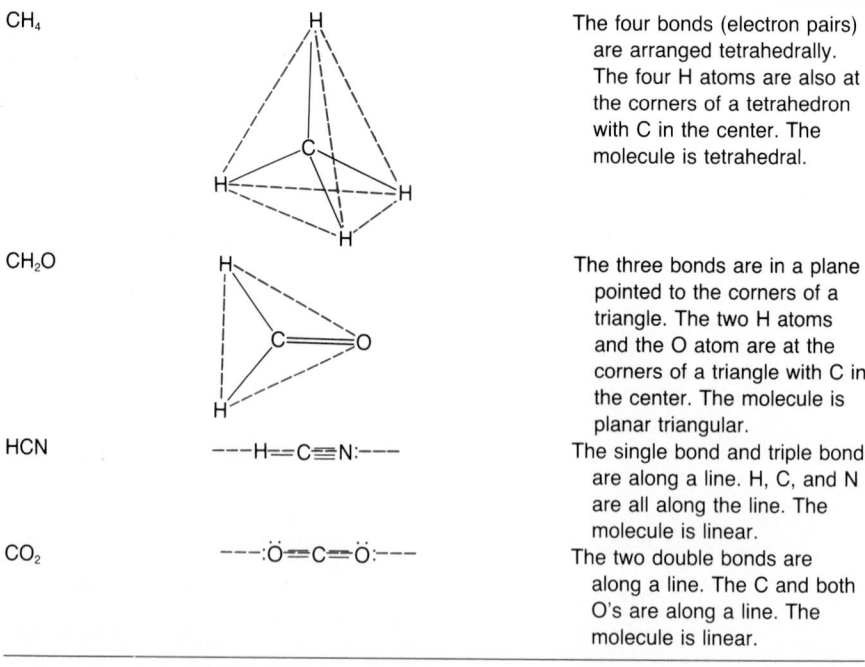

CH_4	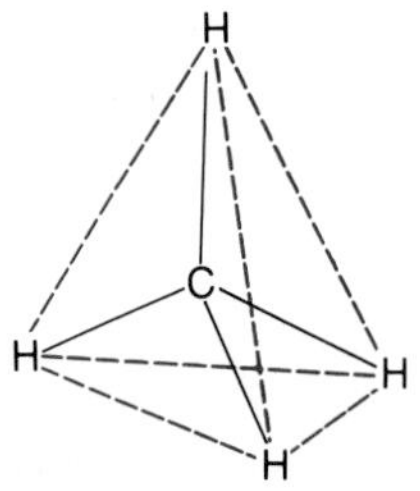	The four bonds (electron pairs) are arranged tetrahedrally. The four H atoms are also at the corners of a tetrahedron with C in the center. The molecule is tetrahedral.
CH_2O	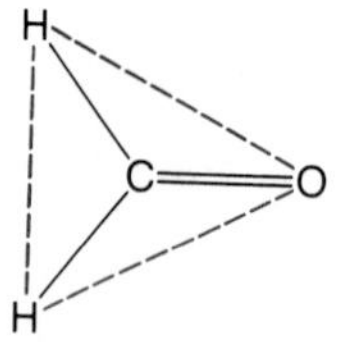	The three bonds are in a plane pointed to the corners of a triangle. The two H atoms and the O atom are at the corners of a triangle with C in the center. The molecule is planar triangular.
HCN	---H—C≡N:---	The single bond and triple bond are along a line. H, C, and N are all along the line. The molecule is linear.
CO_2	---:Ö═C═Ö:---	The two double bonds are along a line. The C and both O's are along a line. The molecule is linear.

If there are nonbonding electron pairs around the central atom, then not every electron pair is associated with a bonded atom and the electron-pair geometry and molecular shape differ. *Only the arrangement of bonded atoms determines* **molecular shape.** For example,

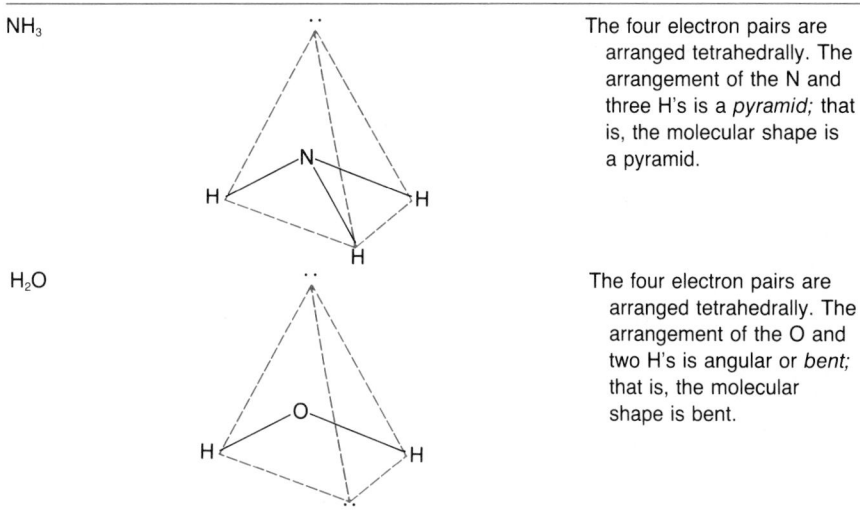

NH_3

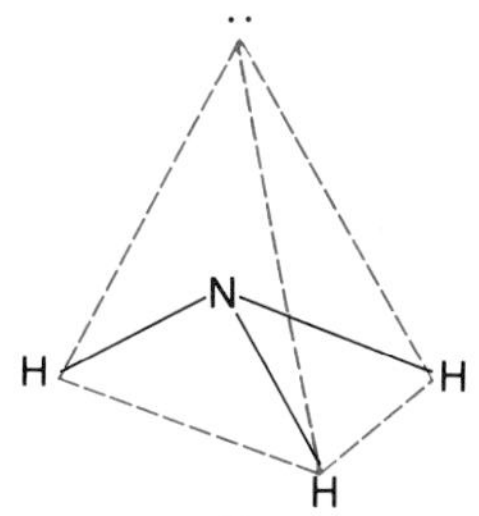
N
H
H
H

The four electron pairs are arranged tetrahedrally. The arrangement of the N and three H's is a *pyramid;* that is, the molecular shape is a pyramid.

H_2O

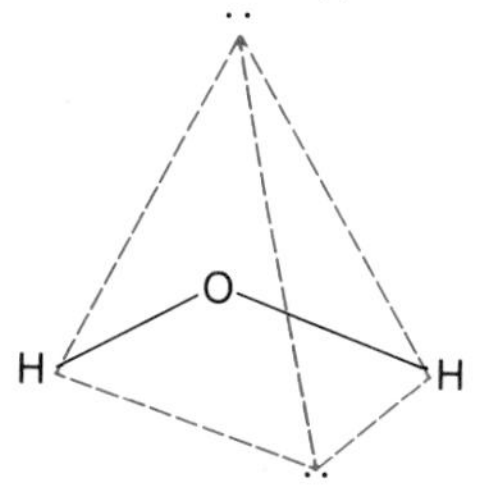
O
H
H

The four electron pairs are arranged tetrahedrally. The arrangement of the O and two H's is angular or *bent;* that is, the molecular shape is bent.

Electron pairs cannot be seen, but electron-pair geometry can be readily predicted theoretically by using VSEPR theory. However, molecular shape, the arrangement of atoms in a molecule, is real and can be determined experimentally. Similarly, when we build molecular models we see the shape defined by the atoms—we cannot see the electron pairs. Look back at Figure 6.1. This is why only the atoms are considered in determining molecular shape. Table 6.5 summarizes some relationships among **set number,** *number of nonbonding electron pairs,* **electron-pair geometry,** and **molecular shape.**

SAMPLE EXERCISE 6.5

What are the electron-pair geometry and molecular shape of the following compounds?

a CH_2Cl_2
b H_2S

Solution Write the Lewis structures according to the guidelines of Table 6.1 and determine the set numbers and number of nonbonding electron pairs around the central atom. From these numbers and from Table 6.5 the geometries can be determined.

TABLE 6.5 Electron-Pair Geometry and Molecular Shape*

Set Number	Number of Nonbonding Pairs	Electron-Pair Geometry	Molecular Shape	Example
2	0	Linear	Linear	CO_2
3	0	Planar triangular	Planar triangular	H_2CO
4	0	Tetrahedral	Tetrahedral	CH_4
4	1	Tetrahedral	Pyramidal	NH_3
4	2	Tetrahedral	Bent	H_2O

* Set number governs electron-pair geometry. Notice that when the number of nonbonding pairs is zero, molecular shape is identical to electron-pair geometry.

a 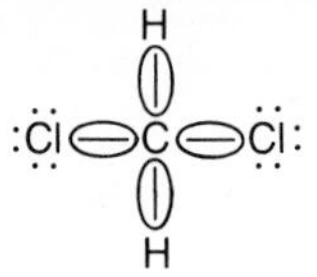

There are four sets of electrons around the central carbon, so the set number is four and the electron-pair geometry is tetrahedral. Because there are no nonbonding electrons, the molecular shape is also tetrahedral.

b 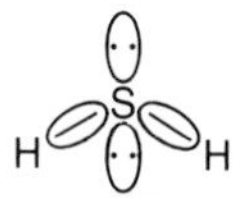

Electron-pair geometry is tetrahedral because once again there are four sets of electrons. However, there are only two bonded atoms, so the molecular shape is defined by the shape 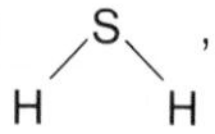, which is bent.

6.11 MOLECULAR POLARITY

In Section 6.8 it was established that many covalent bonds are polar because of electronegativity differences between bonded atoms. When bond dipoles exist in a molecule, the molecule as a whole *may* have a dipole, that is, a + end and a − end; such a molecule is a *polar molecule*.

Whether or not a molecule is polar depends both on the existence of bond dipoles within the molecule and on the molecular shape. For example, in Section 6.8 we established that HCl has the bond dipole $^{\delta+}$H — Cl$^{\delta-}$, or $\overset{+\!\!\longrightarrow}{\text{H—Cl}}$. Because this bond dipole involves the entire molecule, the bond dipole and molecular dipole are the same thing. Therefore, this molecule is polar.

The bond dipoles in water (the molecular shape of which is bent, as we have seen) can be designated as follows:

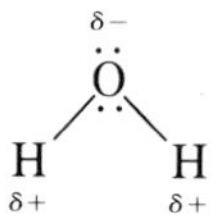

or

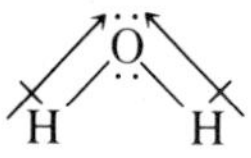

Recall that the more electronegative element draws the electrons toward itself and becomes slightly negative. In water, the fact that both bond dipoles point in the same general direction leads to a molecular dipole that can be marked as:

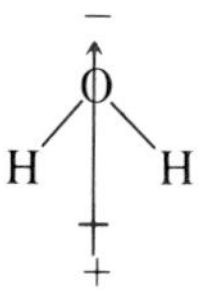

Water molecules are polar because their molecular shape is such that the effects of the two bond dipoles add together and create a permanent molecular dipole. That is, there exists a region of higher electron density (the negative end) and a region of lower electron density (the positive end).

In some molecules, because of the molecular shape, bond dipoles exert their effects in opposite directions and cancel each other out, leaving the molecule as a whole with no molecular dipole. Two common cases in which bond dipoles cancel each other out follow:

1 Z—A—Z

Molecular shape is linear, with two identical atoms Z bonded to a central atom A. For example, carbon dioxide:

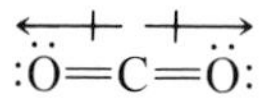

2

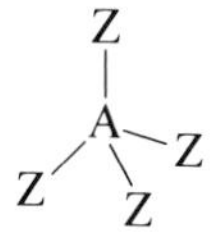

Molecular shape is tetrahedral, with four identical atoms Z tetrahedrally arranged around a central atom A. For example, carbon tetrachloride:

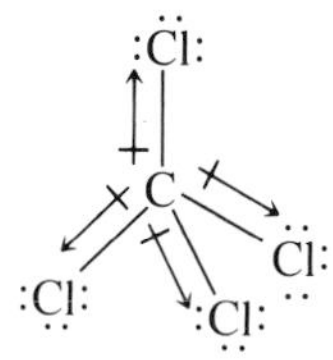

There is no + end and no − end in these molecules. In each example, electrons surround the central atom symmetrically.

SAMPLE EXERCISE 6.6

Which of the following molecules are polar?

a :F̤̈—F̤̈:

b :C̤̈l—B̤̈r:

c :C≡O:

d

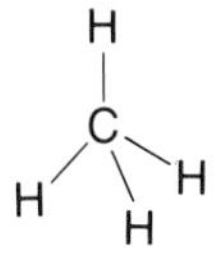

e

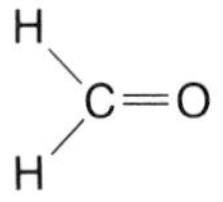

Solution

a Nonpolar. No bond dipole (zero electronegativity difference).

b Essentially nonpolar. Very small bond dipole (electronegativity difference of 0.2).

c Polar. 

d Nonpolar. Very small bond dipoles which cancel.

e Polar.

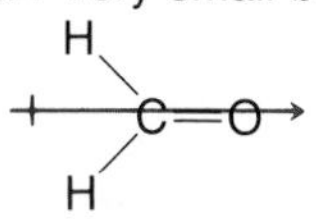

All bond dipoles point in the same general direction. Note that C—H bond polarity is quite small and can be disregarded without changing the conclusion about molecular polarity.

6.12 MOLECULAR WEIGHT

The molecular formula for a covalent compound indicates the kinds and numbers of atoms linked together in one molecule. Not surprisingly, we can determine the **molecular weight,** that is the weight of one molecule, by simply adding the weights of all the atoms in the molecule. For H_2O this weight is 2 × 1.01 amu for hydrogen plus 16.0 amu for oxygen, giving 18.0 amu. If we could weigh one H_2O molecule, it would weigh 18.0/12.0 = 1.50 times as heavy as one carbon-12 atom. *Molecular weights* are relative weights compared with the weight of one carbon-12 atom because they are the sums of relative atomic weights based on that scale.

SAMPLE EXERCISE 6.7

Nitric acid, HNO_3, can be used as a test for the presence of proteins. Determine its molecular weight.

Solution

Atoms in One HNO_3 Molecule	Mass of Each Atom, amu	×	Number of That Atom Present	=	Total Mass for Each Element, amu
1H	1.01		1		1.01
1N	14.0		1		14.0
3O	16.0		3		48.0
			Molecular weight	=	63.0

One HNO_3 molecule has a mass of 63.0 amu. One HNO_3 molecule is 63.0/12.0 = 5.25 times as heavy as one carbon-12 atom.

Problem 6.4 Determine the molecular weights of SO_2, N_2O_4, and H_2SO_4.

6.13 LAW OF DEFINITE COMPOSITION REVISITED

You first encountered the Law of Definite Composition in Section 1.5 as part of the first discussion of the properties of compounds. This law and the idea of fixed combining weight ratios are hard to appreciate before you know about the existence of atoms and how they combine. Now that you have explored bonding, you know that elements always show some definite combining ratio of atoms or ions, and this inevitably leads to some definite combining weight ratio. For water (H_2O) the combining atomic ratio is 2 hydrogens to 1 oxygen because this ratio allows all atoms to have octets (O) or duets (H) of electrons. For sodium chloride the combining ionic ratio is Na:Cl = 1:1.

Because atoms (and ions) have definite atomic weights, a definite combining ratio of atoms (or ions) means a definite combining ratio by weight. Thus the weight ratio or fixed proportion of elements in H_2O is always 2.0:16 = 1.0:8.0 because two hydrogens weigh 2.0 amu and one oxygen weighs 16 amu. Similarly the weight ratio for NaCl is always 23.1:35.5.

Problem 6.5 Acetylene gas, a compound used in welding, has the combining ratio of 2 carbon atoms with 2 hydrogen atoms. What is the weight ratio for acetylene?

CHAPTER ACCOMPLISHMENTS

After completing this chapter you should be able to define **key words** and do the following:

6.1 INTRODUCTION

1. Distinguish between *organic* and *inorganic* compounds.

6.2 WHY ARE THERE TWO TYPES OF COMPOUNDS?

2. Explain why some elements combine to form ionic bonds and others combine to form covalent bonds.

6.3 DIATOMIC MOLECULES

3. Distinguish between single, double, and triple covalent bonds.

6.4 THE NATURE OF THE COVALENT BOND

4. State why there is a lowering of energy relative to the separate atoms when a covalent bond forms.

6.5 LEWIS ELECTRON DOT FORMULAS

5. Given a molecular formula, write a Lewis electron dot structure for a molecular compound.

6.6 COORDINATE COVALENT BONDS

6.7 POLYATOMIC IONS

6. Describe the makeup of polyatomic ions.

6.8 ELECTRONEGATIVITY AND POLARITY

7. Describe the trends in electronegativity across a row and down a column of the periodic table.
8. State the difference between polar and nonpolar bonds.
9. Given a table of electronegativities, indicate whether a bond formed between two atoms will be nonpolar or polar.
10. Use the crossed arrow ($\mapsto$) to label the dipole in a polar covalent bond.

6.9 RECOGNIZING IONIC VERSUS MOLECULAR COMPOUNDS

11. Distinguish between ionic and molecular compounds for purposes of nomenclature and formula writing.
12. Given a table of electronegativities, distinguish between covalent, polar covalent, and ionic bonds.
13. Compare the properties of ionic and molecular compounds.

6.10 MOLECULAR SHAPE

14. State the essential idea of VSEPR theory.
15. State the geometries of maximum separation for two, three, or four sets of electron pairs.
16. Given the Lewis structure of a molecule, predict the electron-pair geometry.
17. Given the Lewis structure of a molecule, predict the molecular shape.

6.11 MOLECULAR POLARITY

18. Given the Lewis structure of a molecule and a table of electronegativities, predict whether a molecule will be polar or nonpolar.

6.12 MOLECULAR WEIGHT

19. Given a periodic table and a correct molecular formula, calculate the molecular weight of a molecular compound.

6.13 LAW OF DEFINITE COMPOSITION REVISITED

20. Predict the weight ratio of elements in a compound from the combining ratio of atoms or ions in the compound.

PROBLEMS

6.6 Name two types of chemical compounds.

6.7 Name two types of chemical bonds.

6.8 Distinguish between ionic and covalent bonds based on the way that they form.

6.9 Write Lewis electron dot structures for the following diatomic molecules:
a ICl *b* BrCl *c* ClF

6.10 Using Table 6.2, predict whether the bonds in Problem 6.9 are polar or nonpolar.

6.11 *a* Write the Lewis symbols for a hydrogen and a fluorine atom.
b Indicate the new attractive and repulsive forces that arise as the two atoms are brought together to form a covalent bond.

6.12 According to Figure 6.2, what is true about the energy content of a covalent bond at a:
a Separation distance greater than the bond length?
b Separation distance less than the bond length?
c Separation distance equal to the bond length?

6.13 Use Table 6.1 to determine how many bonds each atom in the following molecules usually form:
a H_2S *b* CH_2Br_2

6.14 Write the Lewis structures showing all bonding and nonbonding valence electrons for each of the molecules in Problem 6.13.

6.15 Write Lewis structures for the following molecules:
a NF_3 *d* PH_3
b CO_2 *e* OF_2
c CO *f* H_2O_2

6.16 Write structural formulas for:

a C_2H_6 *c* C_2H_3Cl *e* C_2HCl

b $CHBr_3$ *d* CH_4S

6.17 Write Lewis structures for SO_2 and O_3 (ozone).

6.18 *a* Write Lewis structures for H^+, H, and H^-.

b Which two could form a coordinate covalent bond?

6.19 In general, what kinds of substances could act as the electron-pair donor in the formation of a coordinate covalent bond?

6.20 In general, what kinds of substances could act as the electron-pair acceptor in the formation of a coordinate covalent bond?

6.21 What is the origin of the negative charges on many polyatomic ions?

6.22 Define electronegativity.

6.23 Describe the trends in electronegativity found in the periodic table.

6.24 *a* Where in the periodic table do we find elements with the highest electronegativity?

b Where do we find elements with the lowest electronegativity?

6.25 Using Table 6.2 and the guidelines in Section 6.9, classify the following bonds as nonpolar or polar:

a P—Br *b* H—O *c* I—Br *d* N—H

6.26 Using Table 6.2, arrange the bonds given in Problem 6.25 in order of increasing polarity.

6.27 For each polar bond that you found in Problem 6.25, use the crossed-arrow symbol (⟼) to indicate the dipole in the bond.

6.28 *a* Write a Lewis structure for CF_4.

b Use the symbol ⟼ to indicate the polarity in each of the bonds in CF_4.

c Where is the center of the $\delta-$polarity in CF_4?

d Where is the center of the $\delta+$ polarity in CF_4?

6.29 For Problem 6.15, use ⟼ to indicate any bond polarities in the Lewis structures.

6.30 Using Table 6.2 and the guidelines in Section 6.9, indicate whether the following bonds are nonpolar, polar, or ionic:

a H—Br *c* Li—O *e* Al—Cl

b H—F *d* Mg—Cl *f* C—Br

6.31 Indicate whether the following compounds are ionic or molecular, and state whether the smallest basic grouping is a molecule or a formula unit:

a H_2O *e* $PbCrO_4$ *h* FeS

b NaBr *f* HBr *i* $CdCl_2$

c Al_2O_3 *g* K_2CO_3 *j* SiO_2

d CH_4O

6.32 Compound A is a high-melting-point, water-soluble solid. Compound B is a liquid which does not conduct electricity. Indicate the likely ionic or covalent nature of compounds A and B.

6.33 What is the electron-pair geometry of the following compounds?

a CF_4 *b* C_2H_2 *c* PH_3

6.34 What is the molecular shape of each of the compounds in Problem 6.33?

6.35 *a* Why does moving electron pairs far apart lower their energy?

b In the case of a linear arrangement of electron pairs, what prevents their moving apart to an angle of 200°?

6.36 *a* List the relationship between the number of atoms around a given carbon and the geometric arrangement around the carbon atom.

b What is the most common geometric arrangement of covalent bonds around a carbon?

6.37 Give the shape about each carbon atom in each of the following structures:

a 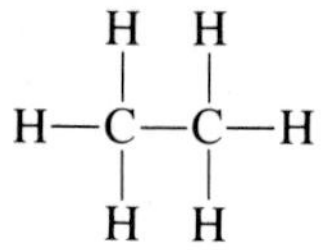

b 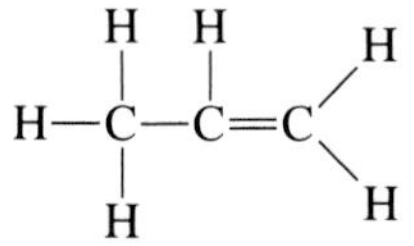

c 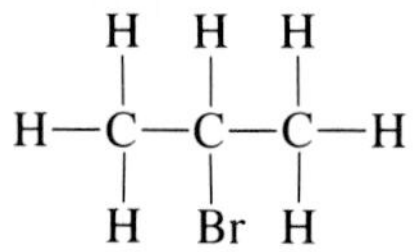

d 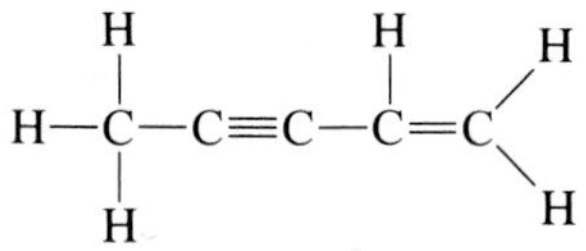

e 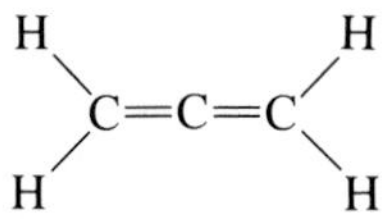

6.38 *a* Write out the Lewis structures of CH_2Cl_2 and CCl_4.

b Compare the molecular polarities of CH_2Cl_2 and CCl_4.

6.39 List HF, HCl, HBr, and HI in order of increasing molecular polarity.

6.40 *a* Draw a Lewis structure for NH_3.

b Mark the bond dipoles in NH_3.

c Mark the molecular dipole in NH_3.

6.41 Pick out the polar molecules from the following list:

a IF *c* NF_3 *e* $SiBr_4$

b Br_2 *d* HCN *f* CS_2

6.42 Calculate the molecular weight of the following compounds:

a N_2 *e* Cl_2O *h* PBr_3

b $CHCl_3$ *f* P_2O_5 *i* H_2SO_4

c H_2S *g* $HC_2H_3O_2$ *j* NH_3

d CI_4

6.43 Propane, a fuel used in mobile homes, has a combining ratio of 3 carbon atoms with 8 hydrogen atoms. What is the weight ratio for propane.

6.44 Describe how Dalton's atomic model provides an explanation for the Law of Definite Composition.

6.45 What is the combining ratio of atoms in each of the following formulas?

a C_2H_6 *c* $CHCl_3$
b SF_4 *d* H_2SO_4

6.46 What is the weight ratio of the elements in each of the compounds indicated in Problem 6.45?

CHAPTER SEVEN

THE MOLE CONCEPT AND STOICHIOMETRY

7.1 INTRODUCTION

By this chapter you should have a mental picture of what individual atoms, ions, and molecules are like. You should also be aware that these particles are extremely tiny—we cannot see, let alone hold, an individual atom, ion, or molecule. Samples of matter that we handle are collections of enormous numbers of these particles. Aluminum foil is a huge collection of Al atoms, a glass of water contains a gigantic number of H_2O molecules, and table salt is a huge collection of Na^+ ions and Cl^- ions.

In this chapter we will see how one can keep track of the numbers of particles in a sample, and, furthermore, how numbers of particles are related to the mass of a sample. Knowledge of this relationship is absolutely essential if one is to develop successful "recipes" for doing chemical reactions in the laboratory. It is not the intention of this text to make you working chemists. However, it is necessary to understand how chemists measure out known quantities of atoms, ions, and molecules. Their units of measurement (moles) for specifying mass-particle relationships are used in other fields besides chemistry, most notably in the clinical laboratory for the specifications of ion concentrations and levels of other components in blood samples, and in the preparation of medicinal solutions. (See especially Chapter 9.)

7.2 INDIVIDUALS VERSUS "PACKAGES"

Historically, the relationship between numbers of particles and mass was worked out in the early nineteenth century, long before there was any detailed knowledge of the nature of particles of matter. The strategy that chemists employed to relate numbers of particles to mass involved defining a unit of matter containing a definite number of particles. This unit of matter is called a **mole** (abbreviated mol). Think of a mole as a "package" containing a definite number of particles (atoms, molecules, etc.) just as a dozen is a "package" containing 12 eggs or 12 doughnuts or 12 anything (see Figure 7.1). Besides the dozen, we daily encounter other "package" concepts:

Package	Number of Items within Package
Dozen	12
Pair	2
Gross	144
Mole	6.02×10^{23}

The number of items in a mole is so large (6.02×10^{23}) that it is hard to

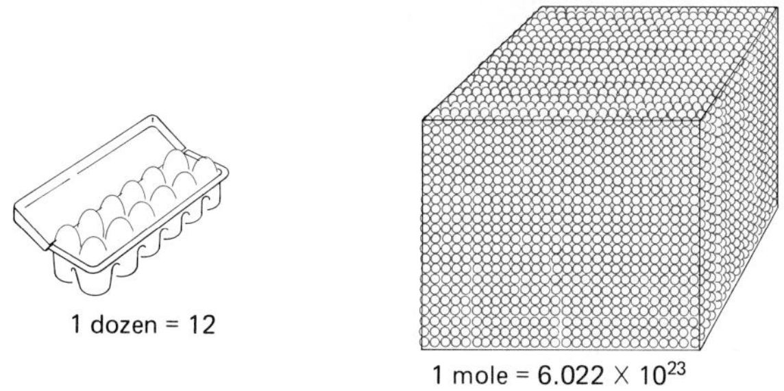
1 dozen = 12
1 mole = 6.022 X 10^23

FIG. 7.1
Eggs come in dozens. Atoms and molecules come in moles.

imagine, but the concept is the same as that of the dozen. Moles are packages of particles; dozens are packages of various items, such as eggs.

The mole concept bridges the gap between the *microscopic* world of atoms, molecules, and formula units which we cannot see and handle and the *macroscopic* world of elements and compounds which we can see and hold.

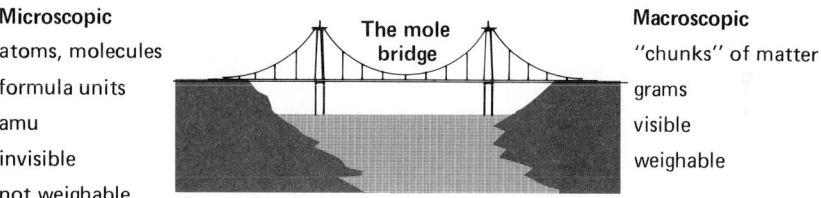

Microscopic	Macroscopic
atoms, molecules	"chunks" of matter
formula units	grams
amu	visible
invisible	weighable
not weighable	

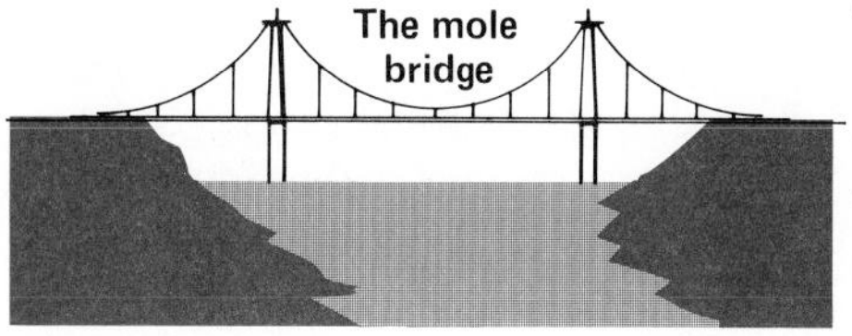
The mole bridge

Before seeing how the mole "package" concept can be related to the mass of a sample, it might be useful to review Sections 3.8, 3.9, 5.11, and 6.12 before proceeding. These sections review atomic, molecular, and formula weight.

7.3 RELATIVE WEIGHTS

The concept of relative weights (Section 3.8) leads directly to the relationship between numbers of particles, packages of particles, and mass in a sample. Let us choose the elements carbon and oxygen to demonstrate the relationship between numbers of atoms and mass of a sample.

From the periodic table we see that the average atomic weights of carbon and oxygen are 12.01 and 16.00 amu, respectively. As a matter of practicality we can accept these numbers as the relative masses of individual atoms:[1]

Atomic ratio   Mass ratio

$$\frac{1 \text{ atom O}}{1 \text{ atom C}} \qquad \frac{16.00 \cancel{\text{amu}}}{12.01 \cancel{\text{amu}}} = 1.332$$

Table 7.1 shows that the mass ratio remains the same so long as the number of oxygen atoms equals the number of carbon atoms. What Table 7.1 establishes is that *for equal numbers of atoms the mass ratio is always the same as the ratio of the atomic weights given in the periodic table.*

The reverse of this statement is also true and is of great importance: **If the mass ratio of two samples corresponds to the ratio of atomic weights, then the two samples must contain equal numbers of atoms.** This is true of all the examples in Table 7.1 and will always be true, regardless of the units of mass, because identical mass units cancel out.

[1]Section 3.9 carefully explains that average atomic weight is a weighted average of the masses of all isotopes of an element, and therefore no single atom actually has that weight. However, because we must deal with atoms "on the average," we routinely use the periodic table of atomic weights for the weights of single atoms.

TABLE 7.1 Mass Ratios for Varying Numbers of Oxygen and Carbon Atoms

Atomic Ratio	Mass Ratio
$\frac{1 \text{ atom O}}{1 \text{ atom C}}$	$\frac{1 \times 16.00\ \cancel{\text{amu}}}{1 \times 12.01\ \cancel{\text{amu}}} = \frac{16.00}{12.01} = 1.332$
$\frac{2 \text{ atoms O}}{2 \text{ atoms C}}$	$\frac{2 \times 16.00\ \cancel{\text{amu}}}{2 \times 12.01\ \cancel{\text{amu}}} = \frac{32.00}{24.02} = 1.332$
$\frac{10 \text{ atoms O}}{10 \text{ atoms C}}$	$\frac{10 \times 16.00\ \cancel{\text{amu}}}{10 \times 12.01\ \cancel{\text{amu}}} = \frac{160.0}{120.1} = 1.332$
$\frac{10^6 \text{ atoms O}}{10^6 \text{ atoms C}}$	$\frac{10^6 \times 16.00\ \cancel{\text{amu}}}{10^6 \times 12.01\ \cancel{\text{amu}}} = \frac{1.600 \times 10^7}{1.201 \times 10^7} = 1.332$
$\frac{6.02 \times 10^{23} \text{ atoms O}}{6.02 \times 10^{23} \text{ atoms C}}$	$\frac{6.02 \times 10^{23} \times 16.00\ \cancel{\text{amu}}}{6.02 \times 10^{23} \times 12.01\ \cancel{\text{amu}}} = \frac{9.635 \times 10^{24}}{7.232 \times 10^{24}} = 1.332$

7.4 HOW MANY PARTICLES IN A MOLE?

Because maintaining equal numbers of atoms in samples of two elements depends on their atomic-weight ratios, and because the unit of mass in the metric system is the gram, it is convenient to consider "packages" of material corresponding to the atomic weight of the material in grams, or the *gram atomic weight* (GAW). The GAW for the element carbon is 12.01 g. For the element oxygen, GAW = 16.00 g. We know from Table 7.1 that $16.00\ \cancel{g}/12.01\ \cancel{g} = 1.332$; therefore the number of atoms in 12.01 g of carbon must be equal to the number of atoms in 16.00 g of oxygen. Or, in general, **the number of atoms in the GAW of one element is always equal to the number of atoms in the GAW of any other element.**

It turns out that the number of atoms in a GAW of any element is 6.02×10^{23}. Unfortunately, the experiments by which this actual number is determined are too complicated to be described in this text. We ask you to accept this number without proof. Thus, the most important concept is

GAW for any element = the same number of atoms as for any other element, namely, 6.02×10^{23} atoms.

This number, 6.02×10^{23}, is called **Avogadro's number** to honor the Italian physicist Amedeo Avogadro, who in the early nineteenth century recognized the importance of being able to relate numbers of particles to measurable quantities of matter.

Now you know what the unit or "package" called the **mole** contains. Because of the reasoning just detailed, a **mole** *is defined as Avogadro's number of constituent units,* and this corresponds to easily assignable weights. For atoms,

$$1 \text{ mol} = 6.02 \times 10^{23} \text{ atoms} = \text{GAW}$$

This gives us equalities and thus conversion factors for relating moles, grams, and numbers of atoms.

SAMPLE EXERCISE 7.1

a How many grams are there in 1 mol of Mg?
b How many grams are there in 2 mol of S?

Solution *a* Consult the periodic table and find that the atomic weight of magnesium is 24.31 amu. Therefore, the GAW is 24.31 g.

b The GAW of S is 32.06 g. Therefore,

1 mol of S = 32.06 g of S

2 mol of S = 64.12 g of S

Using the Unit Conversion Method,

Given quantity × conversion factor = new

$$2\ \cancel{\text{mol of}}\ \text{S} \times \frac{32.06\ \text{g of S}}{1\ \cancel{\text{mol of}}\ \text{S}} = 64.12\ \text{g of S}$$

Problem 7.1 What are the gram atomic weights (GAW) of Al, F, and Br?

Problem 7.2 How many grams are in 3 mol of Al?

7.5 MOLES OF COMPOUNDS

Because the mole is defined as Avogadro's number of consituent units, for compounds,

1 mol of a molecular compound = 6.02×10^{23} **molecules**

1 mol of an ionic compound = 6.02×10^{23} **formula units**

Just as 1 mol = Avogadro's number = GAW for elements, it can be seen that for molecular compounds,

1 **mol** = 6.02×10^{23} molecules = GMW

or for ionic compounds,

1 **mol** = 6.02×10^{23} formula units = GFW

where GMW is the molecular weight expressed in grams, and GFW is the formula weight expressed in grams.

Therefore, to relate moles and grams for any material, look up or calculate the atomic weight, molecular weight, or formula weight, and that number of grams is 1 mol. Figure 7.2 shows the relationship between moles (6.02×10^{23} particles) and mass.

Collections of atoms:	1 mol = GAW
Molecular compounds:	1 mol = GMW
Ionic compounds:	1 mol = GFW

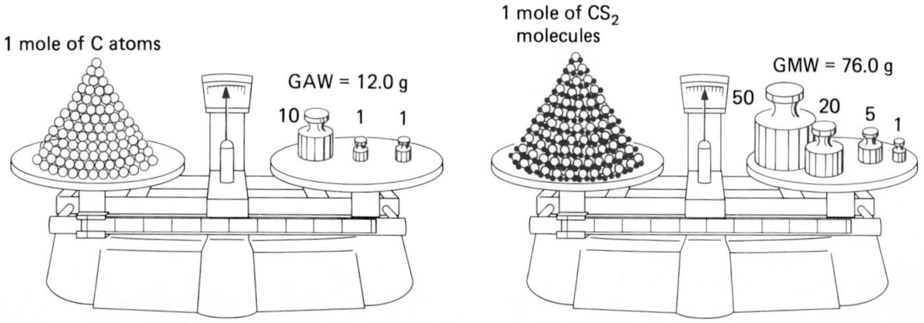
1 mole of C atoms
GAW = 12.0 g
10
1
1
1 mole of CS_2 molecules
GMW = 76.0 g
50
20
5
1

FIG. 7.2
A mole of material weighs exactly the GAW, GMW, or GFW of that material.

SAMPLE EXERCISE 7.2

How many grams are in 1 mol of each of the following?

a N_2O, laughing gas
b $Al(C_2H_3O_2)_3$, the astringent in Burrow's solution

Solution

a To find the molecular weight (Section 6.12), we sum up the atomic weights of all the atoms present. For N_2O this is

$$\begin{aligned} 2 \times N &= 2 \times 14.01 = 28.02 \text{ amu} \\ 1 \times O &= 1 \times 16.00 = \underline{16.00 \text{ amu}} \\ & \qquad\qquad\qquad 44.02 \text{ amu} \end{aligned}$$

The molecular weight of N_2O is 44.02 amu. The molecular weight in grams, or gram molecular weight (GMW), is 44.02 g.

$$1 \text{ mol of } N_2O = 44.02 \text{ g of } N_2O$$

b To find the formula weight (Section 5.11), we sum up the weights of all the ions present in one formula unit. For $Al(C_2H_3O_2)_3$ this is

$$\begin{aligned} 1 \times Al^{3+} &= 1 \times 26.98 \text{ amu} = 26.98 \text{ amu} \\ 3 \times C_2H_3O_2^- &= 3 \times 59.04 \text{ amu} = \underline{177.12 \text{ amu}} \\ & \qquad\qquad\qquad\qquad 204.10 \text{ amu} \end{aligned}$$

The formula weight of $Al(C_2H_3O_2)_3$ is 204.10 amu. The formula weight in grams is 204.10 g.

$$1 \text{ mol of } Al(C_2H_3O_2)_3 = 204.10 \text{ g of } Al(C_2H_3O_2)_3$$

Problem 7.3 How many grams are there in 1 mol of $CaCl_2$?

7.6 GRAM-MOLE-PARTICLE CONVERSIONS

Probably the conversion factors that you will use most often in chemistry will be those which relate moles to grams through the GAW, GMW, or GFW. Occasionally you will also need to remember the relationship of moles to numbers of particles (atoms, molecules, formula units) as well.

Because the GAW, GMW, and GFW all refer to the weight of 1 mol of material, we can use the more general term of **molar weight** in referring to any of the three. The molar weight of a monoatomic element is the element's GAW; the molar weight of a molecular compound is its GMW; and the molar weight of an ionic compound is its GFW.

Molar weight is another name for GAW, GMW, or GFW.

From now on in this text we will use the term **molar weight** instead of GAW, GMW, or GFW. The molar weight is the weight of 1 mol of anything.

SAMPLE EXERCISE 7.3

Perform the conversions indicated.

a How many grams are contained in 1.45 mol of H_2O?
b 14.9 g of Na_2SO_4 is equivalent to how many moles of Na_2SO_4?
c How many molecules of NH_3 are in 7.2 mol of NH_3?

Solution

In all the parts of this exercise we will be using one or more of the conversion factors resulting from the equalities

$$1 \text{ mol} = 6.02 \times 10^{23} \text{ particles} = \textbf{molar weight}$$

and the Unit Conversion Method.

a Use the following stepwise procedure:

STEP 1 The *given* is 1.45 mol of H_2O.

STEP 2 The *new* unit is grams of H_2O.

STEP 3 The equality relating the given and new is

$$1 \text{ mol} = \text{molar weight}$$

$$1 \text{ mol of } H_2O = 18.0 \text{ g of } H_2O$$

STEP 4 Proper format is **given × conversion factor** (new units/given units) = **new quantity and unit.** Therefore,

$$1.45\ \cancel{\text{mol of } H_2O} \times \frac{18.0 \text{ g of } H_2O}{1\ \cancel{\text{mol of } H_2O}} = 26.1 \text{ g of } H_2O$$

b Follow the same procedure:

STEP 1 Given is 14.9 g of Na_2SO_4.

STEP 2 New unit is moles of Na_2SO_4.

STEP 3 Equality is

$$1 \text{ mol} = \text{molar weight}$$

$$1 \text{ mol of } Na_2SO_4 = 142 \text{ g of } Na_2SO_4$$

STEP 4 Proper format is **given × new units/given units = answer.**

$$14.9\ \cancel{\text{g of } Na_2SO_4} \times \frac{1 \text{ mol of } Na_2SO_4}{142\ \cancel{\text{g of } Na_2SO_4}} = 0.105 \text{ mol of } Na_2SO_4$$

c Follow the same procedure:

STEP 1 Given is 7.2 mol of NH_3.

STEP 2 New unit is molecules of NH_3.

STEP 3 Equality is 1 mol = 6.02×10^{23} molecules.

STEP 4 Proper format is **given × conversion factor = answer.**

$$7.2\ \cancel{\text{mol of } NH_3} \times \frac{6.02 \times 10^{23} \text{ molecules of } NH_3}{1\ \cancel{\text{mol of } NH_3}} = 4.3 \times 10^{24} \text{ molecules of } NH_3$$

Problem 7.4 How many moles of lithium sulfide correspond to 75.8 g of lithium sulfide?

7.7 MOLES WITHIN MOLES

The meaning of **subscripts** in chemical formulas unfolded in Chapters 5 and 6. When a **coefficient** is placed before a chemical formula, it *multiplies the entire formula*. For example, $2N_2O_5$ means two molecules of N_2O_5, therefore four N atoms and 10 O atoms must be present.

Coefficients are similarly applied to formulas of ionic compounds. Therefore, 2NaCl indicates two formula units of sodium chloride, hence two Na^+ ions and two Cl^- ions.

When we write 1 mol of NaCl, we now know that this means 6.02×10^{23} formula units of NaCl, and hence 6.02×10^{23} Na^+ ions and 6.02×10^{23} Cl^- ions.

$$6.02 \times 10^{23} \text{ NaCl contains} \begin{cases} 6.02 \times 10^{23} \text{ Na}^+ \text{ ions} \\ \text{and} \\ 6.02 \times 10^{23} \text{ Cl}^- \text{ ions} \end{cases}$$

So we see that 1 mol of NaCl contains 1 mol of Na^+ ions (6.02×10^{23} ions) and 1 mol of Cl^- ions (6.02×10^{23} ions) or 2 mol of ions.

$$1 \text{ mol of NaCl contains} \begin{cases} 1 \text{ mol of Na}^+ \text{ ions} \\ \underline{1 \text{ mol of Cl}^- \text{ ions}} \\ 2 \text{ mol of ions} \end{cases}$$

Some students have difficulty visualizing 2 mol of particles coming from 1 mol. Remembering that the mole concept is like the dozen concept might be helpful in this case. In one dozen eggs, there are one dozen yolks and one dozen whites. So we see two dozen subunits (one dozen yolks and one dozen whites) within one dozen of the whole eggs. As another example, in 1 mol of married couples, there is 1 mol of males and 1 mol of females, that is, 2 mol of human beings within 1 mol of couples.

Just as the subscripts in a chemical formula tell us the number of atoms or ions within a molecule or formula unit, the *subscripts also tell us the numbers of moles of atoms or ions contained within 1 mol of the compound.* For NaCl, the unexpressed subscripts are ones, so 1 mol of NaCl holds within it 1 mol of Na^+ and 1 mol of Cl^-. In 1 mol of Na_2SO_4 there are 2 mol of Na^+ (because the subscript is 2) and 1 mol of SO_4^{-2}. In 1 mol of N_2O_5 there are 2 mol of N and 5 mol of O.

7.8 CHEMICAL REACTIONS AND EQUATIONS

So far this book has dealt with chemicals only one compound or element at a time. The excitement of chemistry builds when you combine materials and watch chemical reactions occur. A **chemical reaction** is the process by which one or more chemical substances are converted into one or more *different* chemical substances.

Within your body chemical reactions occur constantly. Your car hums as a consequence of the chemical reaction between gasoline and air. Batteries energize because of chemical reactions. Nails rust and silver tarnishes as a result of chemical reactions. In all these cases, **reactants** (or *starting materials*) are converted into **products** (new materials). See Figure 7.3. The mole concept will enable us to relate and calculate amounts of products and reactants, but first we must learn how chemists represent chemical reactions by chemical equations, such as those shown in Figure 7.3.

The symbolism of chemistry is like a new language to be mastered. By now you are familiar with chemical **symbols** (letters) and have a vocabulary of chemical **formulas** (words). To describe chemical reactions, we will use chemical **equations** (sentences), which are shorter descriptions than descriptions by English sentences.

For example, if you swallow an antacid tablet like Tums for indigestion, the reaction that occurs would be described in English as follows: Calcium carbonate (Tums) reacts with hydrochloric acid (stomach acid) to yield cal-

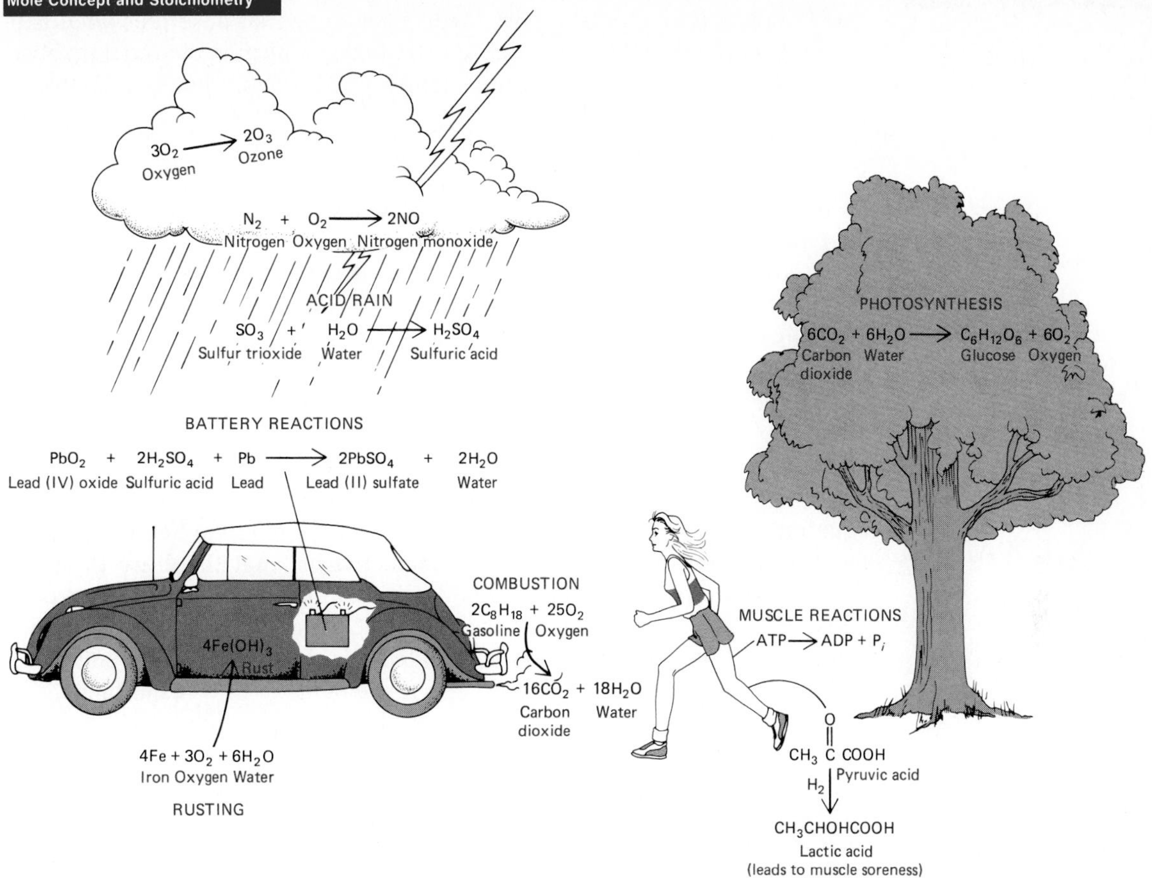

$3O_2 \longrightarrow 2O_3$
Oxygen Ozone
$N_2 + O_2 \longrightarrow 2NO$
Nitrogen Oxygen Nitrogen monoxide
ACID RAIN
$SO_3 + H_2O \longrightarrow H_2SO_4$
Sulfur trioxide Water Sulfuric acid
BATTERY REACTIONS
$PbO_2 + 2H_2SO_4 + Pb \longrightarrow 2PbSO_4 + 2H_2O$
Lead (IV) oxide Sulfuric acid Lead Lead (II) sulfate Water
PHOTOSYNTHESIS
$6CO_2 + 6H_2O \longrightarrow C_6H_{12}O_6 + 6O_2$
Carbon dioxide Water Glucose Oxygen
COMBUSTION
$2C_8H_{18} + 25O_2$
Gasoline Oxygen
$16CO_2 + 18H_2O$
Carbon dioxide Water
MUSCLE REACTIONS
$ATP \longrightarrow ADP + P_i$
$4Fe(OH)_3$
Rust
$4Fe + 3O_2 + 6H_2O$
Iron Oxygen Water
RUSTING
$CH_3COCOOH$
Pyruvic acid
H_2
$CH_3CHOHCOOH$
Lactic acid
(leads to muscle soreness)

FIG. 7.3
Chemical reactions abound in everyday life.

cium chloride, carbon dioxide, and water. Clearly the chemists' way of saying this is shorter:

$$CaCO_3 + 2HCl \longrightarrow CaCl_2 + CO_2 + H_2O$$

In addition to being shorter, as you will discover as you continue to read this chapter, the chemical equation contains a great deal more useful information than the English sentence does.

Chemical equations always have the form:

Reactants $\longrightarrow$ Products

Starting materials Final materials

The arrow means "yields" or "gives." It is important to remember that the arrow can also be regarded as an equals (=) sign. We can translate an English sentence that describes a chemical reaction into a chemical equation that describes the same reaction by:

1 Writing correct formulas that correspond to the names of the chemical substances mentioned
2 Deciding which formulas are reactants and which are products
3 Following the preceding format, writing reactant formulas to the left of the arrow and product formulas to the right of the arrow

For example, let us translate the English sentence that describes what happens when you burn charcoal for your barbecue. Burning charcoal involves carbon (charcoal) reacting with oxygen to yield carbon dioxide. Following the preceding steps, we write:

1 C for carbon, O_2 for oxygen (the element oxygen occurs in air as a diatomic molecule), and CO_2 for carbon dioxide. *Establishing correct formulas is the most important step.*
2 "Carbon reacting with oxygen" tells us that C and O_2 are the reactants, "to yield carbon dioxide" tells us that CO_2 is the product.
3 Thus, $C + O_2 \longrightarrow CO_2$.

To repeat, the most important step in writing chemical equations is to write *correct chemical formulas*.

7.9 THE MEANING OF BALANCED EQUATIONS

There is usually another step required in order to write a correct chemical equation. After following the three-step procedure noted in Section 7.8, one must be certain the equation is *balanced,* that is, that there are the same numbers of atoms of each kind on both sides of the arrow. An equation must be balanced in order to satisfy the Law of Conservation of Matter, which states that "matter can neither be created nor destroyed in a chemical reaction." Chemical reactions are "rearrangements," or "reshufflings," of atoms, but no new atoms are formed, nor old ones lost. Atoms are conserved during a chemical reaction.

Let us look at the preceding example to see what is meant by "balanced":

C	$+$	O_2	$\longrightarrow$	CO_2
One carbon atom		Two oxygen atoms hooked together in a diatomic molecule		One carbon and two oxygen atoms hooked together in one carbon dioxide molecule

This equation is balanced because on each side of the arrow there are one carbon and two oxygen atoms. What has changed is the "hookup" (Figure 7.4). That is, the bonding changed; electrons are shared differently.

The following example is also balanced:

$NaCl$	$+$	$AgNO_3$	$\longrightarrow$	$AgCl$	$+$	$NaNO_3$
1 Na^+ ion 1 Cl^- ion		1 Ag^+ ion 1 NO_3^- ion		1 Ag^+ ion 1 Cl^- ion		1 Na^+ ion 1 NO_3^- ion

Each side of the equation contains one of each ion, but the cations and anions have switched partners in moving from reactants to products.

FIG. 7.4
Chemical reactions are rearrangements of atoms. No atoms are lost or created. In this example, a separate C atom inserts itself between the two atoms of the O_2 molecule. The result is a new molecule, CO_2.

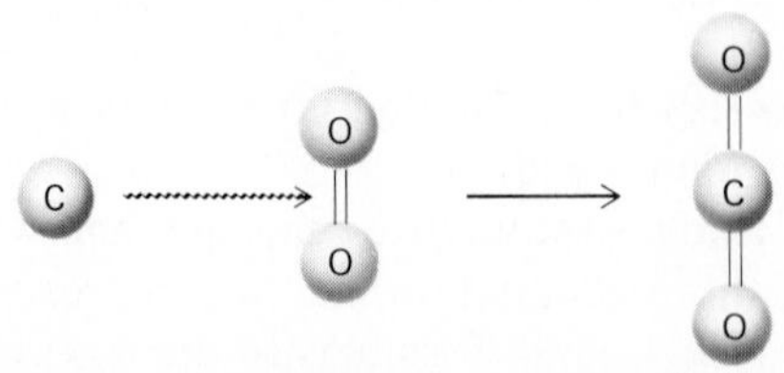
C
O
O
O
C
O

Notice that in the first example of an equation, the Tums equation in Section 7.8, there is a coefficient of 2 before the HCl. This coefficient is necessary in order to have a balanced equation, i.e., in order to have equal numbers of each kind of atom on both sides of the equation.

BALANCED

$$CaCO_3 + 2HCl \longrightarrow CaCl_2 + CO_2 + H_2O$$

1Ca 1C 3O 2H 2Cl 1Ca 2Cl 1C 2H 3O

UNBALANCED

$$CaCO_3 + HCl \longrightarrow CaCl_2 + CO_2 + H_2O$$

1H 1Cl 2Cl 2H

When the translation of an English sentence into a chemical equation by the three steps outlined in Section 7.8 does not lead automatically to a balanced equation, balancing is accomplished by inserting proper coefficients. The coefficients are usually whole numbers.

7.10 BALANCING EQUATIONS

Hydrogen reacts with oxygen to yield water. Let us write a *balanced* chemical equation for this reaction.

1 The correct formulas are H_2, O_2, and H_2O.
2 The reactants are H_2 and O_2. The product is H_2O.
3 Thus, $H_2 + O_2 \longrightarrow H_2O$.

The next step is to determine whether the equation is balanced. This is done by counting the numbers of each kind of atom on each side of the equation and comparing. We determine the number of atoms of a kind in each formula by multiplying for each formula

Coefficient × subscript = number of atoms

Remember that when there is no coefficient, the coefficient is understood to be 1.

$$H_2 + O_2 \longrightarrow H_2O$$

H_2	O_2	H_2O	
1 × 2 = 2H	1 × 2 = 2O	1 × 2 = 2H	1 × 1 = 1O
(Coefficient × Subscript)	(Coefficient × Subscript)	(Coefficient × Subscript)	(Coefficient × Subscript)

We see that this equation is *not* balanced, because whereas there are two oxygen atoms on the left, there is only one oxygen atom on the right. To balance an equation, one *inserts coefficients*. **Never** alter subscripts because that would produce incorrect formulas.

For this equation we begin the balancing procedure by placing the coefficient 2 before H_2O:

$$H_2 + O_2 \longrightarrow 2H_2O$$

2H 2O

$2H_2O$: 2 × 2 = 4H (Coefficient × Subscript); 2 × 1 = 2O (Coefficient × Subscript)

Now there are two oxygens on each side of the equation, but the hydrogen balance is upset. To correct this we put a 2 before H_2:

$$2H_2 + O_2 \longrightarrow 2H_2O$$

4H	2O	4H	2O
(2×2)	(1×2)	(2×2)	(2×1)

Now the equation is balanced.

To repeat, balance an unbalanced equation by inserting coefficients. **Never** alter subscripts. Notice in the preceding equation that $H_2 + O_2 \longrightarrow H_2O$ could be "balanced" by writing $H_2 + O_2 \longrightarrow H_2O_2$. But this would be very **wrong** because H_2O_2 is **not** water.

Problem 7.5 Decide whether the following are balanced:

a $KOH + H_2SO_4 \longrightarrow K_2SO_4 + H_2O$

b $2Na + H_2O \longrightarrow 2NaOH + H_2$

c $H_2 + Cl_2 \longrightarrow 2HCl$

Let us summarize the total method for equation writing and balancing:

STEP 1 Write the equation using the stepwise procedure described in Section 7.8.

STEP 2 Determine whether the equation is balanced by multiplying the coefficient times the subscript for each atom in each formula on each side and comparing.

STEP 3 If the equation is *not* balanced, balance it by inserting the correct coefficients.

STEP 4 When you think the equation is balanced, check it by again multiplying the coefficient times the subscript for each atom on each side.

SAMPLE EXERCISE 7.4 Write a balanced equation for the reaction between nitrogen and hydrogen which yields ammonia (NH_3).

Solution Follow steps 1 through 4 as outlined immediately preceding this exercise.

STEP 1 $N_2 + H_2 \longrightarrow NH_3$

STEP 2

Coefficient × subscript	Coefficient × subscript
$1 \times 2 = 2N$	$1 \times 1 = 1N$
$1 \times 2 = 2H$	$1 \times 3 = 3H$

Neither N nor H is balanced.

STEP 3 Begin by balancing N, which can be accomplished by placing a coefficient of 2 before NH_3.

$$N_2 + H_2 \longrightarrow 2NH_3$$

This coefficient multiplies the entire NH_3 formula, so that we now have $2 \times 3 = 6H$ on the right. We can get 6H on the left side by using the coefficient 3 for H_2:

$$N_2 + 3H_2 \longrightarrow 2NH_3$$

STEP 4	**Coefficient × subscript**	**Coefficient × subscript**
	$1 \times 2 = 2N$	$2 \times 1 = 2N$
	$3 \times 2 = 6H$	$2 \times 3 = 6H$

The equation is balanced because we have exactly 2N and 6H on each side.

Problem 7.6 Write a balanced equation for the reaction between hydrogen and bromine which yields hydrogen bromide.

When polyatomic ions appear on *both* sides of an equation, they can be counted and balanced as a whole. In the following example we can balance SO_4^{2-} rather than balancing S and O individually.

$$Al + H_2SO_4 \longrightarrow Al_2(SO_4)_3 + H_2$$

COUNT

$1 \times 1 = 1Al$	$1 \times 2 = 2H$	$1 \times 2 = 2Al$	$1 \times 2 = 2H$
	$1 \times 1 = 1SO_4$	$1 \times 3 = 3SO_4$	

BALANCE

$$2Al + 3H_2SO_4 \longrightarrow Al_2(SO_4)_3 + 3H_2$$

COUNT

$2 \times 1 = 2Al$	$3 \times 2 = 6H$	$1 \times 2 = 2Al$	$3 \times 2 = 6H$
	$3 \times 1 = 3SO_4$	$1 \times 3 = 3SO_4$	

The only way to learn to balance equations comfortably is through practice. Proceed slowly atom by atom, and keep track by writing down the numbers of atoms on either side of the equation at each step. Always make a final check, and always remember that *correct equations require correct formulas*.

7.11 TYPES OF REACTIONS

In all the previous examples of chemical equations, you have been shown the reactants and the products of a chemical reaction. It is possible to predict the products of a reaction, that is, given just the reactants that are mixed together, tell what the products must be. The ability to predict products takes a great deal of experience, and even professional chemists cannot always make predictions. However, it is useful to see how predictions are made for some simple reactions.

The study of chemical reactions can be greatly simplified by realizing that almost all reactions can be classified into four major types:

1 Combination
2 Decomposition
3 Single replacement
4 Double replacement

Learning to recognize reaction types organizes and simplifies the study of reactions.

Combination Reactions

As the name implies, in combination reactions two reactants *combine* or join together to form one product.

Combination Model Equation

$$A + B \longrightarrow AB$$

Two reactants One product

All reactions in which a compound is formed from its elements are combination reactions. Notice how the following examples fit the model equation.

$$C + O_2 \longrightarrow CO_2$$

$$Mg + Cl_2 \longrightarrow MgCl_2$$

$$2Na + Br_2 \longrightarrow 2NaBr$$

In each example, *two elements* react to give *one product* (a compound). These are the combination reactions we will encounter most frequently. Now, whenever you see two elements reacting, you can predict that the product will be the compound that they form. For the combination of metallic and nonmetallic elements, you can complete the equation by writing a correct formula for the ionic compound formed according to the rules of Section 5.9. For example, given the uncompleted equation,

$$Al + Cl_2 \longrightarrow$$

you should recognize the reactants as the metallic element aluminum and the nonmetallic element chlorine and realize that the product will be the ionic compound that they form, namely, aluminum chloride. Complete the equation by writing the correct formula for aluminum chloride, and then balance the equation.

COMPLETE

$$Al + Cl_2 \longrightarrow AlCl_3 \qquad (Al^{3+} \; Cl^{1-})$$

BALANCE

$$2Al + 3Cl_2 \longrightarrow 2AlCl_3$$

There are many examples of organic molecules reacting in combination reactions. In organic chemistry these are more often called *addition* reactions. For example, the combination of a water molecule with an organic molecule is a frequent occurrence in carbohydrate and fat metabolism. Such a reaction may be written:

$$\underset{\text{Fumaric acid}}{HOOC{-}CH{=}CH{-}COOH} + H{-}OH \longrightarrow \underset{\text{Malic acid}}{HOOC{-}\underset{H}{\underset{|}{CH}}{-}\underset{OH}{\underset{|}{CH}}{-}COOH}$$

Notice that the two reactants have combined into one product. The structural formulas show how the combination has occurred. The water has "added across" the carbon-carbon double bond (H goes to one C; OH to the other C). You will see this reaction discussed fully in Section 13.8.

Decomposition Reactions

As the name implies, in decomposition reactions one reactant *decomposes*, or breaks apart, into two or more products.

Decomposition Model Equation

$$XY \longrightarrow X + Y$$

One reactant Two (or more) products

Examples:

$$2HgO \xrightarrow{\text{heat}} 2Hg + O_2$$

$$2KClO_3 \xrightarrow{\text{heat}} 2KCl + 3O_2$$

Smelling salts, $(NH_4)_2CO_3$, work because of their decomposition to the products shown:

$$(NH_4)_2CO_3 \longrightarrow 2NH_3\uparrow + CO_2\uparrow + H_2O$$

The pungency of the liberated NH_3 gas is what revives the semiconscious person. When gases are produced in reactions, arrows pointing upward are used in equations to show the gases escaping.

In organic chemistry, decomposition reactions are more often called *elimination* reactions. For example, in metabolic reactions in your body several eliminations occur. Two of them are shown here:

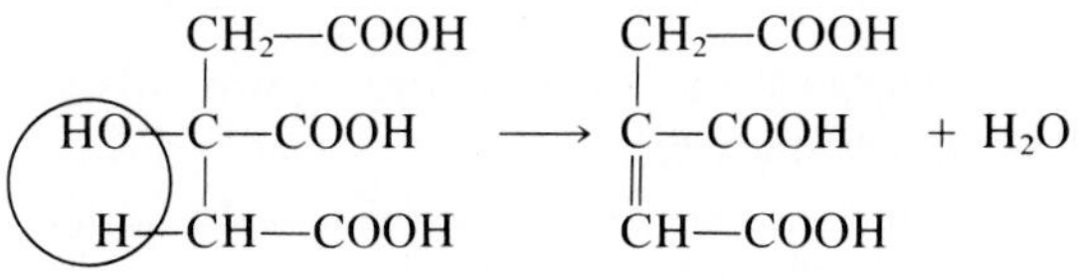

Citric acid Aconitic acid

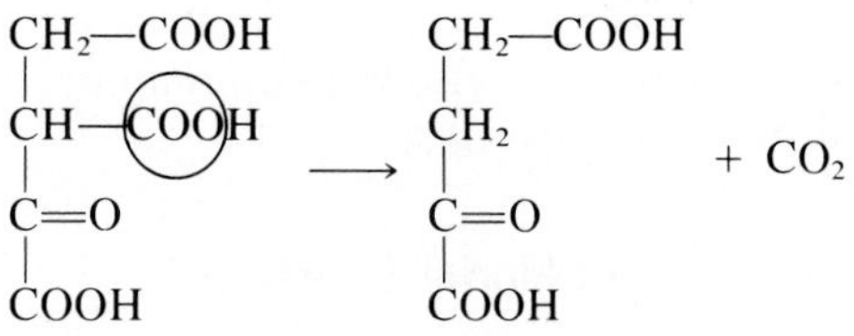

Oxalosuccinic acid α-Ketoglutaric acid

In the first case H_2O is eliminated or "kicked out" of citric acid, and the reactant thereby decomposes. In the second case CO_2 is eliminated. Decarboxylation (loss of CO_2) is common in both organic and inorganic chemistry [see $(NH_4)_2CO_3$ above].

Decomposition reactions are very easy to recognize because they are the only type of reaction in which there is *only one* reactant.

Single-Replacement Reactions

In single-replacement reactions an element reacts with a compound in such a way that the element replaces one of the existing elements in the compound.

Single-Replacement Model Equation

$$E + AB \longrightarrow EB + A$$

Two reactants (one element + one compound) | Two products (one compound + one element)

Some metallic elements replace other metal cations in ionic compounds. Remember that ionic compounds are made up of ions even though the chemical formulas do not show the ionic charges.

$$Zn + CuSO_4 \longrightarrow ZnSO_4 + Cu$$

Zn	$CuSO_4$	$ZnSO_4$	Cu
Metallic element	Ionic compound ($Cu^{2+}SO_4^{2-}$)	Ionic compound ($Zn^{2+}SO_4^{2-}$)	Metallic element

Notice how this example fits the model equation. Notice also that zinc atoms (Zn) are becoming zinc ions (Zn^{2+}) and that the replaced cations (Cu^{2+}) are becoming copper atoms (Cu). Other examples all show these same features, namely, one metallic element is changed from atoms to ions, while the other metallic element undergoes the change from ions to atoms.

$$2K + Pb(NO_3)_2 \longrightarrow 2KNO_3 + Pb$$

(Pb^{2+}) (K^+)

$$Sn + 2AgNO_3 \longrightarrow Sn(NO_3)_2 + 2Ag$$

(Ag^+) (Sn^{2+})

A counterpart reaction from organic chemistry is the *substitution* reaction. An example follows:

$$Cl^- + CH_3I \longrightarrow CH_3Cl + I^-$$

Double-Replacement Reactions

In double-replacement reactions two compounds react with each other to form two different compounds.

Double-Replacement Model Equation

$$AB + XY \longrightarrow AY + XB$$

Two reactants (two compounds) | Two products (two compounds)

There is a double-replacement (two replacements) in the sense that A replaces X in XY and X replaces A in AB. It is perhaps easier to view the reaction as a "switching of partners." The AB and XY partnerships are dissolved and the AY and XB partnerships are formed in their place.

Ionic compounds can react in double-replacement reactions.

$NaCl + AgNO_3 \longrightarrow NaNO_3 + AgCl$

(Na^+, Cl^-) (Ag^+, NO_3^-) (Na^+, NO_3^-) (Ag^+, Cl^-)

Notice how this example fits the model equation. The cations and anions switch partners to form new ionic compounds. Look at some more examples:

$K_2S + MgSO_4 \longrightarrow K_2SO_4 + MgS$

(K^+, S^{2-}) (Mg^{2+}, SO_4^{2-}) (K^+, SO_4^{2-}) (Mg^{2+}, S^{2-})

$Na_2CO_3 + CaCl_2 \longrightarrow 2NaCl + CaCO_3$

(Na^+, CO_3^{2-}) (Ca^{2+}, Cl^-) (Na^+, Cl^-) (Ca^{2+}, CO_3^{2-})

Not all ionic compounds actually react to produce visible products. You cannot be expected to predict which do and do not at this point. In this chapter we will give you examples of ionic compounds which *do react* in double-replacement reactions and ask you to correctly complete the equations for the reactions by switching cation-anion partners. To do this, you must *write correct formulas* for ionic compounds according to the principles in Section 5.9.

SAMPLE EXERCISE 7.5

Complete the following equation:

$MgCl_2 + K_2CO_3 \longrightarrow$

Solution

STEP 1 Recognize that the reactants are two ionic compounds. This signals a double-replacement reaction.

STEP 2 Identify the cation-anion pairs in the reactants. $MgCl_2$ is made up of Mg^{2+} cations and Cl^- anions, and K_2CO_3 is made up of K^+ cations and CO_3^{2-} anions.

STEP 3 Switch the cation-anion partners, and write correct formulas for the new cation-anion pairs. Mg^{2+} teams up with CO_3^{2-} to give $MgCO_3$, and K^+ teams up with Cl^- to give KCl:

$MgCl_2 + K_2CO_3 \longrightarrow MgCO_3 + KCl$

STEP 4 Balance the equation.

$MgCl_2 + K_2CO_3 \longrightarrow MgCO_3 + 2KCl$

7.12 OXIDATION-REDUCTION REACTIONS

You have just seen chemical reactions categorized as combinations (additions), decompositions (eliminations), single replacements (substitutions), or double replacements. Reactions may also be classified as oxidation-reduction reactions. In Section 5.6 the definitions of oxidation, reduction, and redox were given:

Oxidation = loss of electrons

Reduction = gain of electrons

Redox reaction = electron transfer reaction (one substance gains and one substance loses electrons)

The principal simple examples of redox are the electron transfer reactions that occur during the formation of an ionic compound. For example, if magnesium metal and chlorine gas react to form magnesium chloride, electrons are transferred from magnesium to chlorine. That is, magnesium is oxidized and chlorine is reduced.

$Mg \longrightarrow Mg^{2+} + 2e^-$	Oxidation "half" reaction
$Cl_2 + 2e^- \longrightarrow 2Cl^-$	Reducton "half" reaction
$Mg + Cl_2 + \cancel{2e^-} \longrightarrow Mg^{2+} + 2Cl^- + \cancel{2e^-}$	Overall redox reaction

(The electrons are canceled out because an equal number appear on both sides of the arrow, which is really an "equals" sign.)

Single-replacement reactions offer another example of simple redox reactions. Look back at the discussion in Section 7.11 concerning the reaction between Zn and $CuSO_4$. In this case, zinc is being oxidized and copper is being reduced:

$Zn \longrightarrow Zn^{2+} + 2e^-$	Oxidation "half" reaction
$Cu^{2+} + 2e^- \longrightarrow Cu$	Reduction "half" reaction
$Zn + Cu^{2+} + \cancel{2e^-} \longrightarrow Zn^{2+} + Cu + \cancel{2e^-}$	Overall redox reaction

The material that is oxidized is also referred to as the **reducing agent** because it provides the electrons that are gained by the material that is reduced. In the example of zinc in copper sulfate solution, zinc is being oxidized (losing electrons) and is thereby acting as a reducing agent because it provides the electrons which reduce copper. Similarly, the material that is reduced is referred to as the **oxidizing agent** because it accepts the electrons lost in oxidation. Therefore, Cu^{2+} is an oxidizing agent for Zn, from which it gains electrons and is thereby reduced. Remember,

Material **oxidized** = **reducing agent**
(electrons lost)

Material **reduced** = **oxidizing agent**
(electrons gained)

In addition to the simple redox reactions already discussed, there are many complex redox reactions, involving several reactants and products, in which the oxidizing and reducing agents are not obvious as in the simple cases. For example, the following equation represents a complex redox reaction:

$$2KMnO_4 + 6HCl + 5H_2C_2O_4 \longrightarrow 2MnCl_2 + 10CO_2 + 2KCl + 8H_2O$$

Deep purple (for $KMnO_4$); Oxalic acid (for $H_2C_2O_4$); Light pink (for $MnCl_2$); Colorless (for CO_2, KCl, H_2O)

$KMnO_4$ is the oxidizing agent and oxalic acid is the reducing agent. This equation represents the reaction that occurs when oxalic acid is used to remove purple permanganate stains from fabrics or other objects. Whereas the reactant $KMnO_4$ is deep purple, the products of the reaction are colorless with the exception of $MnCl_2$, which is light pink and water-soluble, so it can be washed away. Recognition of the electron transfer that is occurring in this reaction requires special skills which will not be explained in this text. Happily, however, important redox processes in the body, such as the electron transport system (Section 22.6), can be readily understood if one knows simply the basic definitions of oxidation and reduction.

Many oxidizing agents are used medicinally. They kill microorganisms by oxidizing vital bacterial components. Several topical antiseptics, that is, materials which kill or inhibit the growth of microorganisms when applied to living tissue, without harm to the tissue, are oxidizing agents. For example, a 0.01 to 0.2% solution of the oxidizing agent $KMnO_4$ is used to treat bladder and urethra infections. Aqueous solutions of iodine, known as Lugal's solution, are frequently used in surgery because of iodine's antiseptic properties and because its striking color allows nontreated spots to be easily detected. A 3% hydrogen peroxide solution also functions as an antiseptic; the actual oxidizing agent is oxygen. Hydrogen peroxide (H_2O_2) decomposes to oxygen and water:

$$2H_2O_2 \longrightarrow 2H_2O + O_2$$

The turbulent action of the evolved gas may also aid in the mechanical removal of foreign matter from the wound. Hydrogen peroxide can also be used to remove bloodstains from linen or cotton.

Other oxidizing agents also kill microorganisms, but they are also harmful to living tissue, and therefore are applied as disinfectants to nonliving matter. All the halogens are effective oxidizing agents:

$$X_2 + 2e^- \longrightarrow 2X^-$$

X_2 is reduced; therefore it is an oxidizing agent. So, like iodine, chlorine can be used to kill microorganisms, and it is, in drinking water. In water solutions of Cl_2 the actual oxidizing agent is the hypochlorite ion, OCl^-. This ion in the compounds $Ca(OCl)_2$ or NaOCl is added directly to swimming pools as a disinfectant. Cl_2 can cause severe irritation or even burns to skin or mucous membranes. Also, there is some fear that Cl_2 may react with organic matter in the water to form toxic chlorinated organic compounds, and for this reason many municipalities are beginning to use ozone (O_3) as a disinfectant in their water supply system. Calcium hypochlorite is another oxidizing agent; it has been used to disinfect hospital floors, clothing, and bedding.

Reducing agents can also act as antiseptics by reacting with bacterial components. An example is sulfur dioxide (SO_2), which is often used to disinfect rooms of patients with contagious diseases. SO_2 is also used in the food-packing industry as a preservative. It preserves by killing bacteria. SO_2 is added to wine must (crushed grapes and juice prior to fermentation) to destroy undesirable microorganisms present in wild yeasts which grow on the surface of grapes. Wine yeasts are not very susceptible to SO_2. Sulfur dioxide also acts as an antioxidant in wine and prevents the formation of vinegar.

The definitions of oxidation and reduction in terms of electron loss or gain are the most modern and broadest definitions of these processes. Historically, and still to some extent today in organic chemistry, the definitions were framed in terms of gain or loss of oxygen or hydrogen. That is:

Oxidation—gain of oxygen or loss of hydrogen
Reduction—loss of oxygen or gain of hydrogen

You will encounter oxidation and reduction in organic chemistry in Section 14.7.

7.13 WHAT IS STOICHIOMETRY?

The rather forbidding word **stoichiometry** is pronounced "stoy-key-ah′-meh-tree" and is defined as calculations relating the amounts of reactants and products in chemical reactions. Stoichiometry applies the mole concept to the balanced chemical equation. When you finish the study of stoichiometry in this chapter, you will be fully prepared to go into a laboratory, measure out the proper amounts of reactants required for a particular reaction, and predict how much product should be produced.

7.14 MOLAR INTERPRETATION OF THE BALANCED EQUATION

The meaning of a balanced chemical equation was discussed in Section 7.9 in terms of interacting particles, that is, atoms, molecules, and ions. This is certainly one useful interpretation of a balanced equation. For example,

$$C + O_2 \longrightarrow CO_2$$

tells us that one C atom plus one molecule of O_2 yields one molecule of CO_2. Similarly,

$$3H_2 + N_2 \longrightarrow 2NH_3$$

means that three molecules of H_2 plus one molecule of N_2 yields two molecules of NH_3.

If we multiply all the coefficients of a balanced equation by the same number, we still have a balanced equation. For example, the coefficients of the equation $C + O_2 \longrightarrow CO_2$ can be multiplied by 2 to give

$$2C + 2O_2 \longrightarrow 2CO_2$$

This is still a balanced equation. We now have two carbon atoms on each side and four oxygen atoms on each side. Similarly, multiplying $3H_2 + N_2 \longrightarrow 2NH_3$ by 3 gives

$$\underset{18H}{9H_2} + \underset{6N}{3N_2} \longrightarrow \underset{6N\ \ 18H}{6NH_3}$$

In balancing equations we always try to use the smallest set of numbers, just as we try to use the simplest ratio in any mathematical relationship. However, an exact multiple of the simplest set of coefficients still gives a balanced equation.

Let us multiply through by Avogadro's number.

C	+	O_2	$\longrightarrow$	CO_2
One C atom		One molecule of O_2		One molecule of CO_2
6.02×10^{23} C	+	6.02×10^{23} O_2	$\longrightarrow$	6.02×10^{23} CO_2
6.02×10^{23} C atoms 1 mol of C		6.02×10^{23} molecules of O_2 1 mol of O_2		6.02×10^{23} molecules of CO_2 1 mol of CO_2

We see by doing this that the coefficients of a balanced equation not only tell us the ratio of reacting particles, but also tell us the reacting **mole ratio.** We can read the coefficients of balanced equations as the numbers of moles of reactant or product.

Molar amounts are easily related to weighable gram amounts through the molar weight (Section 7.6), and in practice in the laboratory, chemists almost always interpret the coefficients of balanced chemical equations as numbers of moles of reactants or products. Thus, *we can, and should, read chemical equations in terms of moles.*

$$Ca + 2H_2O \longrightarrow Ca(OH)_2 + H_2$$

$$1 \text{ mol of } Ca + 2 \text{ mol of } H_2O \longrightarrow 1 \text{ mol of } Ca(OH)_2 + 1 \text{ mol of } H_2$$

$$2KClO_3 \longrightarrow 2KCl + 3O_2$$

$$2 \text{ mol of } KClO_3 \longrightarrow 2 \text{ mol of } KCl + 3 \text{ mol of } O_2$$

Problem 7.7 Interpret the following equations in terms of the numbers of moles of reactants and products:

a $2Al(OH)_3 + 3H_2SO_4 \longrightarrow Al_2(SO_4)_3 + 6H_2O$

b $4Li + O_2 \longrightarrow 2Li_2O$

7.15 THE MOLE RATIO

Given any balanced chemical equation, you can write mole ratios that relate molar amounts of one reactant to molar amounts of another reactant, or molar amounts of reactant to molar amounts of product. For example, for the balanced equation $N_2 + 3H_2 \longrightarrow 2NH_3$, because 1 mol of N_2 reacts with 3 mol of H_2, 1 mol of N_2 is needed to produce 2 mol of NH_3, and 3 mol of H_2 is required to yield 2 mol of NH_3, we can write the following mole ratios:

$$\frac{1 \text{ mol of } N_2}{3 \text{ mol of } H_2} \qquad \frac{1 \text{ mol of } N_2}{2 \text{ mol of } NH_3} \qquad \frac{3 \text{ mol of } H_2}{2 \text{ mol of } NH_3}$$

Of course, we can also write the reciprocals of these fractions. These are the mole-ratio conversion factors relating *reactant to reactant or reactant to product*. Mole ratios employ the coefficients of the balanced equations.

SAMPLE EXERCISE 7.6

What is the mole ratio of HF to SnF_2 in the following reaction, which is a means of making stannous fluoride, the "fluoride" in toothpaste?

$$Sn + 2HF \longrightarrow SnF_2 + H_2$$

Solution The coefficients give you the ratio of the reacting molar amounts. In this case, 2 mol of HF produces 1 mol of SnF_2. The mole ratio, which can be written and used as a conversion factor is 2 mol of HF/1 mol of SnF_2.

To do stoichiometry one must *always* use an appropriate *mole ratio* that has been obtained from a balanced chemical equation. Mole ratios are used just as any other conversion factor in the Unit Conversion Method is used.

Problem 7.8 What is the mole ratio of Sn to H_2 in the reaction given in Sample Exercise 7.6?

7.16 MOLE-MOLE, MOLE-GRAM, GRAM-GRAM CONVERSIONS

The simplest stoichiometry problems are those in which we are given a balanced chemical equation and a molar amount of one reactant and asked to calculate the molar amounts of the other reactants and products. These problems are handled by the Unit Conversion Method in the usual manner (Section 2.4). We identify the given quantity and extract a conversion factor from the balanced equation.

Mole-mole conversions are simple, one-step unit conversions, and you are probably ready to apply the method even to very complex and unfamiliar reactions.

SAMPLE EXERCISE 7.7 How many moles of $H_2C_2O_4$ are required to react completely with 1.50 mol of $KMnO_4$ according to the following equation:

$$2KMnO_4 + 6HCl + 5H_2C_2O_4 \longrightarrow 2MnCl_2 + 10CO_2 + 2KCl + 8H_2O$$

See Section 7.12 for a discussion of the utility of this reaction.

Solution STEP 1 The *given* is 1.50 mol of $KMnO_4$.

STEP 2 The *new* is moles of $H_2C_2O_4$.

STEP 3 From the balanced equation write the mole ratio which is the conversion factor. The appropriate form of the conversion factor is **new/given,** that is,

$$\frac{5 \text{ mol of } H_2C_2O_4}{2 \text{ mol of } KMnO_4}$$

STEP 4 The format is

$$1.50 \cancel{\text{mol of } KMnO_4} \times \frac{5 \text{ mol of } H_2C_2O_4}{2 \cancel{\text{mol of } KMnO_4}}$$

$$= \frac{1.50 \times 5 \text{ mol of } H_2C_2O_4}{2} = 3.75 \text{ mol of } H_2C_2O_4$$

Notice that the coefficients from the equation (5 and 2) are *exact* numbers, so they do not limit the number of significant figures in the answer.

Problem 7.9 Write a *balanced* equation for the reaction between K and Br_2 that forms the sedative KBr, and calculate the number of moles of KBr produced by the reaction of 7.50 mol of Br_2.

In Section 7.6 we told you that the conversion factors most often used in chemistry are those relating moles and grams through the molar weight. We are going to use these factors extensively in stoichiometry. Using both gram-mole or mole-gram conversion factors and mole-ratio conversion factors from balanced equations enables us to relate gram amounts of one reactant or product to molar amounts of other reactants or products, or vice versa.

Central to all stoichiometry problems is the concept that the *mole ratio* relates the new quantity to the given quantity. Identify the mole ratio first. Then identify the gram-mole conversions necessary to use this mole ratio as a factor. Notice how this order of identifying equalities and conversion factors is used in Sample Exercise 7.8.

SAMPLE EXERCISE 7.8 According to the equation

$$Fe_2O_3 + 2Al \longrightarrow Al_2O_3 + 2Fe$$

how many moles of Al_2O_3 are produced by the reaction of 81 g of Al?

Solution Use the Unit Conversion Method. Two conversion factors will be required.

STEP 1 The *given* is 81 g of Al.

STEP 2 The *new* unit is moles of Al_2O_3.

STEP 3 The *mole ratio* necessary to convert moles of Al to moles of Al_2O_3 is

$$\frac{1 \text{ mol of } Al_2O_3}{2 \text{ mol of Al}} \qquad \frac{\textbf{New} \text{ material}}{\textbf{Given} \text{ material}}$$

To use this factor, grams of Al must be changed to moles of Al through the molar weight.

$$\frac{1 \text{ mol of Al}}{27 \text{ g of Al}} \qquad \frac{\textbf{New} \text{ unit}}{\textbf{Given} \text{ unit}}$$

STEP 4 Use the format beginning with the *given* and using conversion factors of the form **new/given** to cancel out units.

$$81 \not{\text{g of Al}} \times \frac{1 \not{\text{mol of Al}}}{27 \not{\text{g of Al}}} \times \frac{1 \text{ mol of } Al_2O_3}{2 \not{\text{mol of Al}}}$$

$$= \frac{81 \times 1 \times 1 \text{ mol of } Al_2O_3}{27 \times 2} = 1.5 \text{ mol of } Al_2O_3$$

When chemists do chemical reactions in the laboratory, they measure out reactants and products in grams. It is essential to be able to relate gram amounts of reactants to gram amounts of products. As in previous stoichiometry problems, to do gram-gram stoichiometry problems we need a balanced chemical equation to give us a mole-ratio conversion factor that relates the materials involved in the conversion. Once again the mole ratio will be the central focus and should be determined first. The other conversion factors will be gram-mole or mole-gram factors.

SAMPLE EXERCISE 7.9

Sodium iodide can be made in the laboratory by combining sodium metal and iodine. The balanced equation for the reaction is $2Na + I_2 \longrightarrow 2NaI$. How many grams of I_2 are required to produce 225 g of NaI?

Solution Use the Unit Conversion Method. Three conversion factors will be required. Determine the mole ratio first.

STEP 1 The *given* is 225 g of NaI.

STEP 2 The *new* is grams of I_2.

STEP 3 The *mole ratio* necessary to convert NaI to I_2 is

$$\frac{1 \text{ mol of } I_2}{2 \text{ mol of NaI}} \qquad \frac{\textbf{New} \text{ material}}{\textbf{Given} \text{ material}}$$

To use this factor, grams of NaI must be changed to moles of NaI through the molar weight.

$$\frac{\textbf{New} \text{ unit}}{\textbf{Given} \text{ unit}} \qquad \frac{1 \text{ mol of NaI}}{150. \text{ g of NaI}} \qquad \begin{pmatrix} \text{Atomic weights} \\ \text{Na} \quad 23.0 \\ \text{I} \quad \underline{127.} \\ 150. \end{pmatrix}$$

To get the answer in grams of I_2, moles of I_2 must be converted to grams of I_2 through the molar weight.

$$\frac{\textbf{New} \text{ unit}}{\textbf{Given} \text{ unit}} \qquad \frac{254 \text{ g of } I_2}{1 \text{ mol of } I_2} \qquad (2 \times 127 = 254)$$

STEP 4 Use the usual format, beginning with the given, 225 g of NaI. The mole ratio is boxed for emphasis.

$$225\ \cancel{\text{g of NaI}} \times \frac{1\ \cancel{\text{mol of NaI}}}{150.\ \cancel{\text{g of NaI}}} \times \boxed{\frac{1\ \cancel{\text{mol of I}_2}}{2\ \cancel{\text{mol of NaI}}}} \times \frac{254\ \text{g of I}_2}{1\ \cancel{\text{mol of I}_2}} =$$

$$\frac{225 \times 1 \times 1 \times 254\ \text{g of I}_2}{150. \times 2 \times 1} = 191\ \text{g of I}_2$$

Notice that recognizing the mole ratio as the central focus in Sample Exercise 7.9 or any problem helps in setting up the format. All gram-gram conversions have the following "skeleton":

Mole ratio

$$\textbf{Given}\ \text{grams} \times \frac{\quad}{\quad} \times \boxed{\frac{\textbf{new}\ \text{material}}{\textbf{given}\ \text{material}}} \times \frac{\quad}{\quad} = \textbf{new}\ \text{grams}$$

Gram-mole conversion factors arranged to cancel out units

SAMPLE EXERCISE 7.10

The powerful, but ecologically problematic insecticide DDT (*d*ichloro-*d*iphenyl-trichloroethane) can be made by the following reaction:

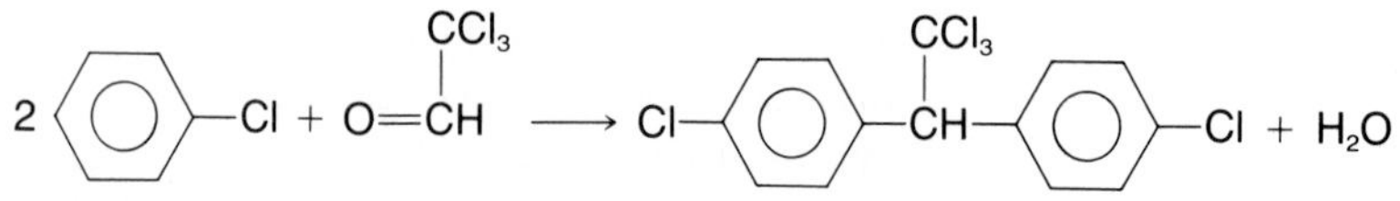

2
Cl
CCl3
O=CH
Cl
CCl3
CH
Cl
H2O

Chlorobenzene (mw = 112.5) Trichloroethanal (mw = 147.5) DDT (mw = 354.5)

How many grams of chlorobenzene are required to react completely with 29.5 g of trichloroethanal to produce DDT?

Solution Work this through as in Sample Exercise 7.8, recognizing 29.5 g of trichloroethanal (TCE) as the given and grams of chlorobenzene as **new,** *or* try fitting the data into the preceding "skeleton."

We begin this by inserting the quantities into the clearly labeled parts of the "skeleton."

$$29.5\ \text{g of TCE} \times \frac{\quad}{\quad} \times$$ 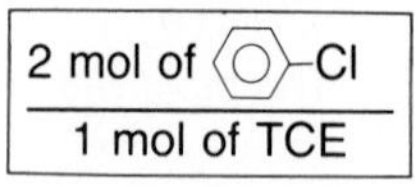 $$\times \frac{\quad}{\quad} = \text{g of chlorobenzene}$$

2 mol of
Cl
1 mol of TCE

Given **Mole ratio** **New**

Now the first conversion factor must cancel grams of TCE. So it will be

$$\frac{1\ \text{mol of TCE}}{147.5\ \text{g of TCE}}$$

The last conversion factor must produce the new unit, grams of chlorobenzene. So it will be

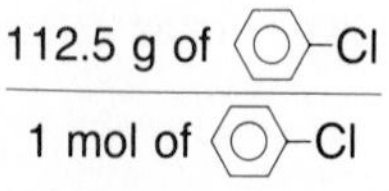

112.5 g of
Cl
1 mol of
Cl

Thus, the setup is

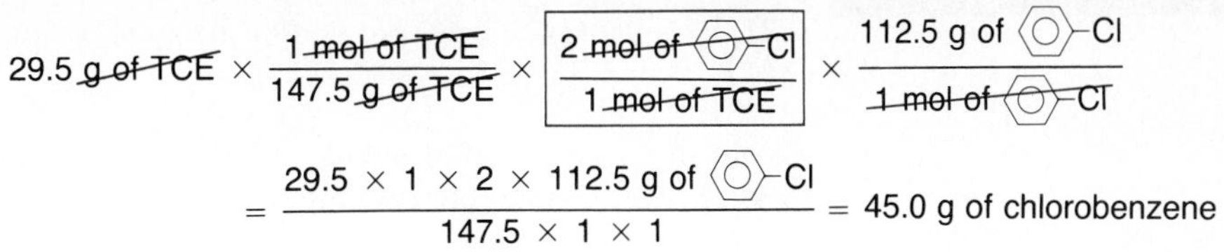
29.5 g of TCE × 1 mol of TCE / 147.5 g of TCE × 2 mol of Cl / 1 mol of TCE × 112.5 g of Cl / 1 mol of Cl
= 29.5 × 1 × 2 × 112.5 g of Cl / 147.5 × 1 × 1 = 45.0 g of chlorobenzene

Problem 7.10 How many grams of DDT are produced by the reaction described in Sample Exercise 7.10?

All stoichiometric conversions require a balanced equation from which one determines an appropriate mole ratio. Because *the mole ratio must be a conversion factor,* gram amounts of material must be converted into molar amounts (or vice versa) in the course of stoichiometry calculations. A general plan of attack for all stoichiometry problems appears below:

1 Determine the mole ratio necessary to convert the given material into the new material. The mole ratio comes from the coefficients of the balanced equation and will have the form

$$\frac{\text{Number of moles of } \textbf{new} \text{ material}}{\text{Number of moles of } \textbf{given} \text{ material}}$$

2 If *both* the given and new units are stated in *moles,* then the problem will be a one-step conversion.

3 If either or both the given and new units are stated in *grams,* then there must be gram-mole or mole-gram conversion factors. See Figure 7.5.

7.17 HEAT AS A REACTANT OR PRODUCT

Whenever a chemical reaction occurs, there is an energy change. Energy was first discussed in Section 1.7 and again in Section 4.2. It is convenient to measure energy changes during chemical reactions in the form of heat energy called **enthalpy** using the units of kilocalories (abbreviated kcal). Heat can be released as the reaction proceeds. When heat is released, we can regard heat as a product. A reaction in which heat is a product is called an **exothermic** reaction. (*Exo* means "out," and *thermic* means "heat." Hence, heat, or enthalpy, comes out or is released.) An example is the reaction of methane, CH_4, the gas in bunsen burners, with O_2. The letters in parentheses in Equation (7.1) specify the physical state (*g* for gas and *l* for liquid) of the reactants and products.

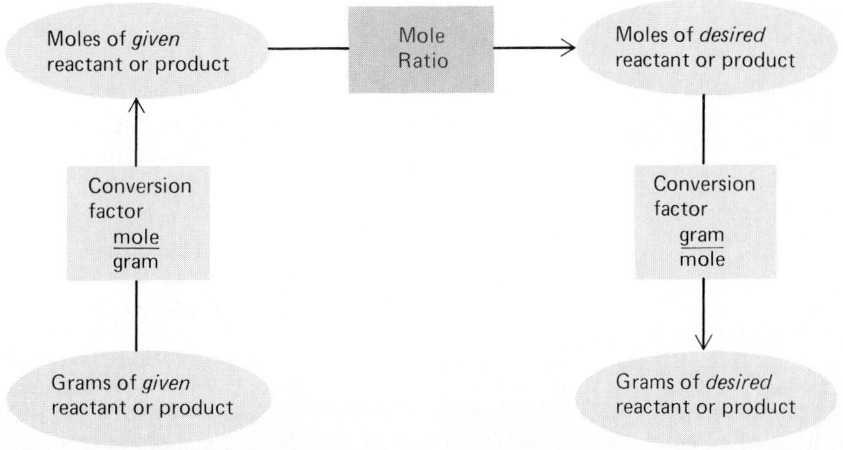
Moles of given reactant or product
Mole Ratio
Moles of desired reactant or product
Conversion factor mole/gram
Conversion factor gram/mole
Grams of given reactant or product
Grams of desired reactant or product

FIG. 7.5
The *big* picture of stoichiometry! Begin with grams or moles given and end with moles or grams desired by proceeding along the part of this pathway that applies.

$$CH_4(g) + 2O_2(g) \rightleftharpoons CO_2(g) + 2H_2O(l) + \underset{\text{Product}}{213 \text{ kcal}} \tag{7.1}$$

Note that the potential energy of the chemical products is less than that of the reactants since energy has been given off in the form of heat.

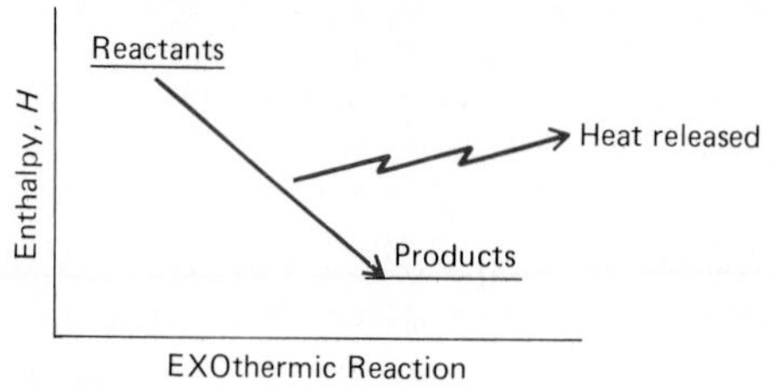
Reactants
Heat released
Products
Enthalpy, H
EXOthermic Reaction

Enthalpy is represented symbolically by the letter *H*. Heat (enthalpy) changes during a chemical reaction are termed **heats of reaction** and are symbolized ΔH.

Sometimes heat must be supplied in order for a reaction to occur. In this case we can regard heat as a reactant. A reaction in which heat (enthalpy) is a reactant is called an **endothermic** reaction. (*Endo* means "in," indicating that heat is put in or is absorbed.) The formation of laughing gas, N_2O, from N_2 and O_2 is an example of an endothermic reaction.

$$2N_2(g) + O_2(g) + \underset{\text{Reactant}}{39.0 \text{ kcal}} \longrightarrow 2N_2O(g) \tag{7.2}$$

In the case of an endothermic reaction the potential energy of the chemical products is higher than that of the reactants since energy has been added.

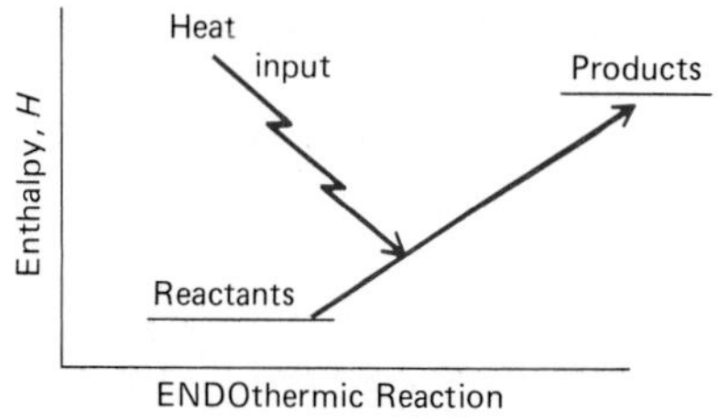
Heat
input
Products
Reactants
Enthalpy, H
ENDOthermic Reaction

Problem 7.11 Identify the following reactions as *exo*thermic or *endo*thermic:

a $2PCl_3 + 146 \text{ kcal} \longrightarrow 2P + 3Cl_2$

b $C + O_2 \longrightarrow CO_2 + 94.1 \text{ kcal}$

c $2O_3 \longrightarrow 3O_2 + 68.1 \text{ kcal}$

Because heat can be regarded as a reactant or product in a chemical reaction, one can obtain from the balanced equation mole-ratio conversion factors which relate the amount of heat that has been supplied or produced to the amounts of reactants used or the amounts of products formed. Following are the mole ratios (or the reciprocals) relating to heat released or absorbed in Equations (7.1) and (7.2):

From Equation (7.1),

$$\frac{213 \text{ kcal}}{1 \text{ mol of } CH_4} \qquad \frac{213 \text{ kcal}}{2 \text{ mol of } O_2} \qquad \frac{213 \text{ kcal}}{1 \text{ mol of } CO_2} \qquad \frac{213 \text{ kcal}}{2 \text{ mol of } H_2O}$$

From Equation (7.2),

$$\frac{39.0 \text{ kcal}}{2 \text{ mol of } N_2} \qquad \frac{39.0 \text{ kcal}}{1 \text{ mol of } O_2} \qquad \frac{39.0 \text{ kcal}}{2 \text{ mol of } N_2O}$$

Using conversion factors and the same general format as for other stoichiometry problems, we can calculate the amount of heat that must be supplied or the amount of heat that has been produced given an amount of some chemical reactant or product.

SAMPLE EXERCISE 7.11

How much heat is released when 8.00 g of CH_4 burns? Burning is the reaction of CH_4 with O_2, as shown in Equation (7.1).

Solution Use the Unit Conversion Method. The equation gives you the mole ratios as shown preceding this sample exercise. The molecular weight of CH_4 is 16.0; thus 1 mol of CH_4 = 16.0 g.

Given × **conversion factors** = **new**

$$8.00\ \cancel{\text{g of CH}_4} \times \frac{1\ \cancel{\text{mol of CH}_4}}{16.0\ \cancel{\text{g of CH}_4}} \times \frac{213\ \text{kcal}}{1\ \cancel{\text{mol of CH}_4}} = 107\ \text{kcal}$$

Problem 7.12 How much heat is required to produce 132 g of laughing gas, N_2O, according to Equation (7.2)?

CHAPTER ACCOMPLISHMENTS

After completing this chapter you should be able to define **key words** and do the following:

7.2 INDIVIDUALS VERSUS "PACKAGES"

1. Compare the use of the mole as a package unit to the use of the dozen as a package unit.

7.3 RELATIVE WEIGHTS

2. Recognize that equal numbers of atoms are present in any two samples of elements if the mass ratio of the samples corresponds to the atomic-weight ratio of the elements.

7.4 HOW MANY PARTICLES ARE IN A MOLE?

3. Recognize that the abbreviation GAW stands for gram atomic weight.
4. State the relationship between the number of atoms in a GAW of one element and the number of atoms in a GAW of any other element.
5. State Avogadro's number.
6. Relate 1 mol of an element to the number of atoms in a mole, and to the mass of the element.

7.5 MOLES OF COMPOUNDS

7. Relate 1 mol of a molecular compound to a number of molecules and 1 mol of an ionic compound to a number of formula units.
8. Recognize that the abbreviation GMW stands for gram molecular weight and that GFW stands for gram formula weight.
9. Relate 1 mol of a compound to the mass of the compound.

7.6 GRAM-MOLE-PARTICLE CONVERSIONS

10. Given the formula of a substance, convert (*a*) a given number of moles of that substance to grams of that substance, (*b*) a given number of grams of that substance to moles of that substance, and (*c*) a given number of moles of that substance to particles of that substance.

7.7 MOLES WITHIN MOLES

11. Correctly interpret coefficients that are placed before chemical formulas.
12. Interpret subscripts of chemical formulas in terms of moles of constituent particles.

7.8 CHEMICAL REACTIONS AND EQUATIONS

13. Translate English sentences that describe chemical reactions into chemical equations.
14. Recognize the importance of correct chemical formulas in chemical equations.

7.9 THE MEANING OF BALANCED EQUATIONS

15. State the Law of Conservation of Matter.
16. State the meaning of the word *balanced* as it refers to a chemical equation.

7.10 BALANCING EQUATIONS

17. Recognize whether an equation is unbalanced or balanced.
18. Balance a chemical equation.

7.11 TYPES OF REACTIONS

19. Name the four major types of reactions.
20. Given chemical equations, recognize combination, decomposition, single-replacement, and double-replacement reactions.
21. Given metal and nonmetal reactants, write a balanced chemical equation for their chemical reaction.
22. Given two ionic compound reactants, write a balanced equation for their reaction.

7.12 OXIDATION-REDUCTION REACTIONS

23. Recognize electron transfer reactions as redox reactions.
24. Name some medicinal oxidizing agents.

7.13 WHAT IS STOICHIOMETRY?

7.14 MOLAR INTERPRETATION OF THE BALANCED EQUATION

25. Interpret the coefficients of a chemical equation as numbers of moles of reactants and products.

7.15 THE MOLE RATIO

26. Given a balanced chemical equation, construct *mole-ratio* conversion factors which relate molar amounts of reactants and/or products.

7.16 MOLE-MOLE, MOLE-GRAM, GRAM-GRAM CONVERSIONS

27. Given a balanced equation and a molar or gram amount of one reactant or product, calculate the molar or gram amounts of other substances reacted or produced.
28. State the importance of the mole ratio in all stoichiometry calculations.

7.17 HEAT AS A REACTANT OR PRODUCT

29. Identify chemical equations as representing either exothermic or endothermic reactions.
30. Recognize the term *enthalpy* (and the symbol *H*) as another word for heat energy.
31. Given a balanced chemical equation, write conversion factors relating heat and the molar amount of reactants or products.
32. Given a balanced equation and an amount of one reactant or product, calculate the quantity of heat consumed or produced as the reaction proceeds.
33. Given a balanced equation and an amount of heat consumed or produced, calculate the amount of reactant consumed or product produced.

PROBLEMS

7.13 What kinds of packages, other than a dozen, are you familiar with?

7.14 Is the number of aluminum atoms in 26.98 g of aluminum equal to the number of fluorine atoms in 21.00 g of fluorine?

7.15 Give the gram atomic weight for the following elements:

a Ar
b S
c I
d Ag
e K
f Si

7.16 What would be the mass of Avogadro's number of calcium atoms?

7.17 How many grams are there in the following molar amounts?

a 2.00 mol of lithium
b 5.93 mol of phosphorus
c 99.5 mol of neon

7.18 How many moles are there in the following gram amounts?

a 19.5 g of sodium

b 78.9 g of boron
c 19.2 g of copper

7.19 *a* How many moles are there in 112 mg of potassium?
b How many millimoles is this?

7.20 Convert the following molar amounts to milligram amounts:
a 2.00 mmol of Li
b 5.93 mmol of P
c 99.5 mmol of Ne

7.21 We have 8.3 mol of sodium. How many moles of potassium should be measured out to ensure that we have an equal number of K atoms and Na atoms?

7.22 You already possess 0.250 mol of magnesium. How many grams of beryllium would ensure a number of beryllium atoms that equals the number of atoms in the magnesium?

7.23 *a* A student has weighed out 11.9 g of nickel; how many grams of copper should the student weigh out to ensure equal numbers of nickel and copper atoms?
b Suppose the student needs twice as many copper atoms as nickel atoms; how many grams of copper should be weighed out?

7.24 Compute the molar weight of each of the following:
a KCl
b H_2O
c H_3PO_4
d NH_4Cl
e $(NH_4)_2CO_3$
f Na_3PO_4

7.25 How many grams are there in 1 mol of each of the following?
a H_2
b Cl_2
c Li_2CO_3
d $Ba_3(PO_4)_2$
e K^+
f SO_4^{2-}

7.26 *a* We wish to combine equal numbers of CH_4 molecules and O_2 molecules. If we have 0.50 mol of CH_4, how many moles of O_2 should we measure out?
b If we had 64.0 g of CH_4, how many moles of O_2 would be required to ensure equal numbers of CH_4 and O_2 molecules? How many grams of O_2 is this?

7.27 Calculate the number of moles in each of the following:
a 4.04 g of H_2
b 85.9 g of Br
c 0.512 g of H_2O
d 1.19×10^2 g of $Ca_3(PO_4)_2$
e 50.9 g of $(NH_4)_2CO_3$
f 39.1 g of K^+

7.28 Calculate the number of grams contained in each of the following:
a 0.750 mol of H_2
b 11.3 mol of N_2
c 7.50×10^{-3} mol of H_2O
d 0.915 mol of $Al_2(SO_4)_3$
e 6.02×10^{23} formula units of NaCl
f 1.50 mol of Na^+

7.29 State the number of moles of each ion contained in each of the following:
a 3.00 mol of LiBr
b 4.25 mol of $CaCl_2$
c 1.10×10^3 mol of $Mg_3(PO_4)_2$
d 0.750 mol of $BaCO_3$

7.30 Calculate the number of moles of each atom contained in each of the following:
a 3.00 mol of H_2
b 4.50 mol of N_2
c 1.25×10^{-1} mol of $C_2H_4Cl_2$
d 0.450 mol of H_3PO_4

7.31 What is the mass of Avogadro's number of NH_4Br formula units?

7.32 Define or explain the following terms:
a Chemical reaction
b Chemical equation
c Reactant
d Product

7.33 Translate the following English sentences into chemical equations:
a Copper reacts with sulfur to yield copper(II) sulfide.
b Potassium reacts with nitrogen to form potassium nitride.
c Calcium hydroxide decomposes to calcium oxide and water.

7.34 Translate these equations into English sentences:
a $MgCl_2 + Pb(NO_3)_2 \longrightarrow PbCl_2 + Mg(NO_3)_2$
b $Si + O_2 \longrightarrow SiO_2$
c $2K + Br_2 \longrightarrow 2KBr$

7.35 What is the relationship between the Law of Conservation of Matter and a balanced chemical equation?

7.36 To balance chemical equations, one should use __________. In balancing equations, __________ in chemical formulas should never be altered.

7.37 Identify each of the following equations as balanced or unbalanced:
a $Br_2 + 2KI \longrightarrow I_2 + KBr$
b $4P + 5O_2 \longrightarrow 2P_2O_5$
c $Al + HBr \longrightarrow AlBr_3 + 3H_2$
d $2NH_3 + H_2SO_4 \longrightarrow (NH_4)_2SO_4$
e $H_2O_2 \longrightarrow H_2O + O_2$

7.38 Balance the equations that you identified as unbalanced in Problem 7.37.

7.39 Balance the following equations:
a $N_2 + O_2 \longrightarrow NO$
b $NaBr + Cl_2 \longrightarrow NaCl + Br_2$
c $P + Cl_2 \longrightarrow PCl_5$
d $BaCl_2 + (NH_4)_2SO_4 \longrightarrow BaSO_4 + NH_4Cl$
e $K_2O + H_2O \longrightarrow KOH$
f $Fe + O_2 \longrightarrow Fe_2O_3$
g $CaC_2 + H_2O \longrightarrow C_2H_2 + Ca(OH)_2$
h $Zn + HNO_3 \longrightarrow Zn(NO_3)_2 + H_2$
i $NH_4NO_2 \longrightarrow N_2 + H_2O$
j $PbO + O_2 \longrightarrow PbO_2$

7.40 Find and correct the translation error in each of the following:
a Sentence: Sodium reacts with iodine to form sodium iodide.
Equation: $Na + I_2 \longrightarrow NaI_2$

b Sentence: Silver oxide decomposes to silver and oxygen.
Equation: $Ag_2O \longrightarrow 2Ag + O$

c Sentence: Lead(II) nitrate reacts with potassium chloride to form lead(II) chloride and potassium nitrate.
Equation: $Pb(NO_3)_2 + KCl \longrightarrow PbCl + K(NO_3)_2$

d Sentence: Silicon reacts with oxygen to form silicon dioxide.
Equation: $S + O_2 \longrightarrow SO_2$

7.41 Balance the following equations:
a $Fe(OH)_3 + H_2SO_4 \longrightarrow Fe_2(SO_4)_3 + H_2O$
b $C_2H_6 + O_2 \longrightarrow CO_2 + H_2O$
c $NaOH + Al(OH)_3 \longrightarrow NaAlO_2 + H_2O$
d $P_4O_{10} + H_2O \longrightarrow H_3PO_4$
e $K_2O + P_4O_{10} \longrightarrow K_3PO_4$
f $MgI_2 + H_2SO_4 \longrightarrow HI + MgSO_4$
g $PCl_5 + H_2O \longrightarrow HCl + H_3PO_4$
h $Al + Sn(NO_3)_2 \longrightarrow Al(NO_3)_3 + Sn$

7.42 Name four types of chemical reactions.

7.43 Look through the equations in Problems 7.33 through 7.41 and find one example of each reaction type.

7.44 Identify each of the following reactions as a combination, decomposition, single replacement, or double replacement:
a $3Na + Al(NO_3)_3 \longrightarrow 3NaNO_3 + Al$
b $Na_3PO_4 + Al(NO_3)_3 \longrightarrow AlPO_4 \downarrow + 3NaNO_3$
c $2H_2O_2 \longrightarrow 2H_2O + O_2 \uparrow$
d $BaO + SO_3 \longrightarrow BaSO_4$
e $2NaClO \longrightarrow 2NaCl + O_2 \uparrow$
f $Cl_2 + 2KI \longrightarrow 2KCl + I_2$

7.45 Complete and balance the following equations:
a $Mg + I_2 \longrightarrow$
b $Li + O_2 \longrightarrow$
c $Al + S \longrightarrow$
d $K + P \longrightarrow$
e $Ca + N_2 \longrightarrow$

7.46 Complete and balance the following equations:
a $Mg + CuSO_4 \longrightarrow$
b $Al + Ni(NO_3)_2 \longrightarrow$
c $Al + SnCl_2 \longrightarrow$

7.47 In Problems 7.45 and 7.46 which elements are losing electrons?

7.48 In Problems 7.45 and 7.46 which elements are being oxidized?

7.49 How does chlorine function as a disinfectant?

7.50 Methyl alcohol's toxic effect in the body arises through its conversion to formaldehyde, the biological specimen preservative. The conversion in brief is

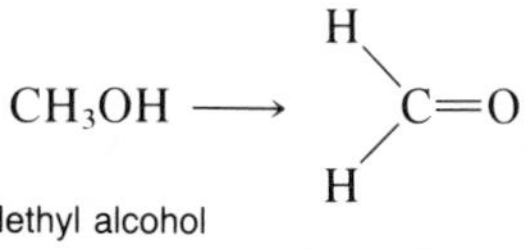

$$CH_3OH \longrightarrow \quad \overset{\displaystyle H}{\underset{\displaystyle H}{\diagdown}} C = O \text{ (Equation } not \text{ balanced)}$$

Methyl alcohol

Formaldehyde

Is the methyl alcohol being oxidized or reduced?

7.51 During the synthesis of fatty acids in the body the following reaction occurs:

$$CH_3(CH_2)_{10}\overset{\displaystyle O}{\overset{\|}{C}}-CH_2-\overset{\displaystyle O}{\overset{\|}{C}}-SCoA + H_2 \longrightarrow$$

$$CH_3(CH_2)_{10}\overset{\displaystyle OH}{\overset{|}{C}H}-CH_2-\overset{\displaystyle O}{\overset{\|}{C}}-SCoA$$

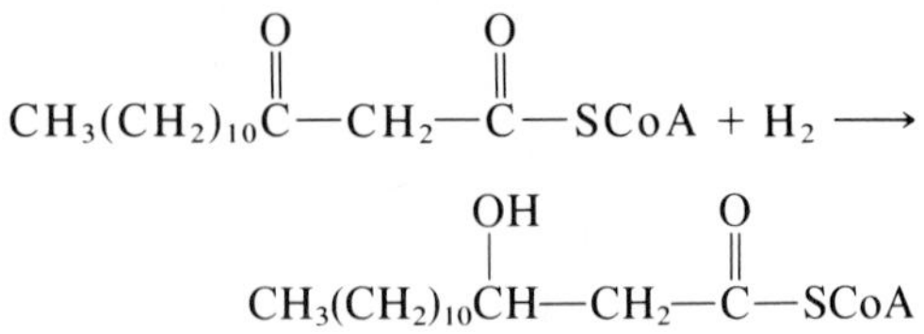

Is the organic reactant being oxidized or reduced?

7.52 Complete and balance the following equations representing double-replacement reactions:
a $K_2SO_4 + Ba(NO_3)_2 \longrightarrow$
b $(NH_4)_2CO_3 + MgCl_2 \longrightarrow$
c $(NH_4)_3PO_4 + Ca(NO_3)_2 \longrightarrow$
d $FeCl_2 + K_3PO_4 \longrightarrow$
e $Na_2S + Ni(NO_3)_2 \longrightarrow$

7.53 Rusting involves the chemical reaction of iron, water, and oxygen to form iron(III) hydroxide. Write a balanced equation for the rusting process.

7.54 The combustion of gasoline is the reaction between octane (C_8H_{18}) and oxygen to form carbon dioxide and water. Write a balanced equation for the combustion.

7.55 Photosynthesis is the combination of carbon dioxide and water in the presence of light to form glucose ($C_6H_{12}O_6$) and oxygen. Write a balanced equation for photosynthesis.

7.56 For the following balanced chemical equations, (1) interpret them in terms of relative numbers of moles of reactants and products, and (2) write some of the mole ratios between reactants and products which the balanced equations indicate.
a $2P(s) + 3H_2(g) \longrightarrow 2PH_3(g)$
b $HBr(g) + KOH(aq) \longrightarrow KBr(aq) + H_2O(l)$
c $2CO(g) + O_2(g) \longrightarrow 2CO_2(g)$
d $C(s) + 2Cl_2(g) \longrightarrow CCl_4(l)$
e $Mg(s) + 2HCl(aq) \longrightarrow MgCl_2(aq) + H_2 \uparrow$

7.57 Mole-mole conversions. Given the balanced equation

$$Na_2Cr_2O_7 + 6HI + 4H_2SO_4 \longrightarrow$$
$$3I_2 + Cr_2(SO_4)_3 + Na_2SO_4 + 7H_2O$$

a How many moles of I_2 form when 2.0 mol of HI react?
b How many moles of HI react with 2.0 mol of H_2SO_4?
c How many moles of water form at the same time that 7.0 mol of I_2 form?
d How many moles of HI are required to produce 0.14 mol of H_2O?

7.58 Mole-gram conversion. Given the equation

$$2Al + 6HCl \longrightarrow 2AlCl_3 + 3H_2 \uparrow$$

a How many grams of Al are needed to release 0.45 mol of H_2?
b How many grams of $AlCl_3$ are obtained from 15 mol of HCl?
c How many grams of H_2 are released from 10. mol of HCl?

7.59 Gram-mole conversion. Given the balanced equation

$$2KOH + H_2SO_4 \longrightarrow K_2SO_4 + 2H_2O$$

a How many moles of H_2SO_4 are needed to make 87 g of K_2SO_4?
b How many moles of H_2O are produced from the reaction of 19.6 g of H_2SO_4?
c How many moles of KOH are needed in order to produce 100. g of water?

7.60 Gram-gram conversions. Given the equation

$$MnO_2 + 4HCl \longrightarrow MnCl_2 + Cl_2 + 2H_2O$$

a How many grams of chlorine are formed from 11 g of MnO_2?
b How many grams of HCl are needed to make 5.0 of $MnCl_2$?
c How many grams of water form when 11 g of HCl react?

7.61 The reaction that occurs in chlorine bleaches that leads to whitening is

$$NaClO + NaCl + H_2O \longrightarrow Cl_2 + 2NaOH$$

In a cup of a typical bleach there is approximately 12 g of NaClO.
a How many moles of Cl_2 are liberated by 12 g of NaClO?
b How many grams of NaOH are produced?

7.62 A propellant used in rocket engines is a mixture of hydrazine (N_2H_4) and hydrogen peroxide (H_2O_2). This mixture reacts spontaneously to yield N_2 and H_2O:

$$N_2H_4 + 2H_2O_2 \longrightarrow N_2 + 4H_2O$$

a How much hydrazine in grams is needed to completely react with 0.30 mol of H_2O_2?
b How many grams of N_2 are produced by the combination of ingredients described in part *a*?

7.63 Write a balanced equation and calculate how many grams of chlorine (Cl_2) are needed to produce 103 g of sodium chloride from sodium metal and chlorine gas. *A balanced equation is* **always** *necessary to do stoichiometry!*

7.64 Old oil paintings are darkened by PbS which forms by the reaction of the Pb in paint with H_2S in air.

$$Pb + H_2S \longrightarrow PbS + H_2$$

Suppose H_2S comes in contact with 2.0 g of Pb. How much PbS forms?

7.65 Old oil paintings can be cleaned by the reaction of H_2O_2 with PbS.

$$PbS + 4H_2O_2 \longrightarrow PbSO_4 + 4H_2O$$

a What is the yield of $PbSO_4$ if 0.60 g of H_2O_2 react?
b How much PbS will be removed?

7.66 Identify the following reactions as exothermic or endothermic:
a $CaCO_3 + 42\ \text{kcal} \longrightarrow CaO + CO_2$
b $MnO_2 + Mn \longrightarrow 2MnO + 59.5\ \text{kcal}$
c $C + H_2O + 31\ \text{kcal} \longrightarrow CO + H_2$

7.67 The propellant reaction given in Problem 7.62 is highly exothermic:

$$N_2H_4(l) + 2H_2O_2(l) \longrightarrow N_2(g) + 4H_2O(g) + 153\ \text{kcal}$$

a How much heat is liberated when 7.00 mol of H_2O_2 reacts?
b How much heat is released as 3.00 mol of water forms?

7.68 The air pollutant SO_3 combines with water to form H_2SO_4, which is destructive to marble and limestone, and the reaction also contributes to thermal pollution (excessive heat in the atmosphere).

$$SO_3(g) + H_2O(l) \longrightarrow H_2SO_4(aq) + 31\ \text{kcal}$$

How much H_2SO_4 in grams and how much heat in kilocalories is produced by 12 g of SO_3?

7.69 How many moles of carbon dioxide can be released from $MgCO_3$ if 8.43 kcal of energy is supplied?

$$MgCO_3 + 28.1\ \text{kcal} \longrightarrow MgO + CO_2$$

7.70 When 5.0 g of nitroglycerin blows up, the reaction can be represented by the equation

$$4C_3H_5O_9N_3 \longrightarrow 12CO_2\uparrow + 6N_2\uparrow + O_2\uparrow + 10H_2O + 1725\ \text{kcal}$$

How much heat is evolved?

CHAPTER EIGHT

GASES, LIQUIDS, SOLIDS

8.1 INTRODUCTION

Even before you began reading this text, the classification of matter into physical states based on definiteness of shape and volume was familiar to you (see Section 1.2). Table 8.1 summarizes the shape and volume characteristics of solids, liquids, and gases. In this chapter we will explore other unique properties of these three physical states.

Understanding body activities such as breathing, cell respiration, and detection of distant aromas requires a knowledge of the properties of the gaseous state. Evaporation and fluidity are properties of liquids that are important in medicine and agriculture. The property of rigidity distinguishes solids and permits their use, especially in structural applications, where the other states are unsuited. The analysis of the physical states in this chapter will enable you to understand the molecular origin of these properties. An understanding on the atomic or molecular level means an understanding of matter from the inside out. This provides a deeper understanding, just as studies of cellular reactions and anatomy provide a better understanding of the external person.

It may seem strange to you, but the structures of liquids and solids, which we can readily see and touch, defy understanding more often than the structure and behavior of gases, which often are invisible and difficult to handle. Because of this greater simplicity of gases, we will begin our discussion with them, giving a brief overview of the unique properties of the gaseous state. The physical properties and gas laws will lead us to a model of gases. Then, a discussion of the intermolecular forces that hold molecules together in collections will show us how to modify our original model to account for the liquid and solid states.

8.2 CHARACTERISTICS OF GASES

Gases have indefinite shapes and volumes. Another way of saying the same thing appears in statements 1 and 2 in the list of gas characteristics that follows; that is, gases are easily *compressed* and readily *expand*. This list presents other properties unique to the gaseous state.

1 *Gases can be easily compressed* by applying pressure to the walls of a flexible container (Figure 8.1*a*). Compression implies a *decrease* in volume. When we try the ''pressing'' experiment on the container filled with water, we find that liquids are not easily compressed (Figure 8.1*b*). Solids are the least compressible of the three states (Figure 8.1*c*).
2 *Gases expand to fill the entire volume of their container.* Expansion implies an *increase* in volume. If a small container of a gas is opened in

TABLE 8.1 Shape and Volume Characteristics of the Three Physical States

Physical state	Shape	Volume	Example
Solid	Definite—does not depend on the shape of the container	Definite—has a fixed volume	A gold bar occupies a clearly defined amount of space.
Liquid	Indefinite—takes on the shape of its container	Definite—has a fixed volume	A glass of milk spilled on the floor spreads out, but not forever.
Gas	Indefinite—takes on the shape of its container	Indefinite—has no fixed volume	A small vial of gas opened in a large auditorium spreads to fill the entire room.

a classroom, the gas escapes and soon expands to fill the entire room. The mass of the gas undergoing expansion remains the same; only the volume changes; that is, the same amount of material is spread out within a larger space.

3 *Gases have indefinite densities.* Because density is the ratio of mass to volume and the volume of a gas varies with its container but the mass remains constant, density also varies as the volume of the container does.

4 *Gases have a low density* compared with liquids and solids (Table 2.3). This is why gas bubbles within a liquid tend to rise.

5 Assuming that they do not react chemically, *gases can diffuse (mix) rapidly through each other* in all directions. Ammonia gas released at the front of a classroom rapidly diffuses through the air (a mixture of gases), as can be verified by a student sitting in the back. The odor of ammonia is quite pungent. This student's observation would also be a demonstration of a gas sample expanding to fill an entire room.

6 *Gases exert a pressure on the wall of any container or surface that they touch.* At sea level, air pushes against every square inch of our bodies with a force of 14.7 lb.

7 *Gases expand in volume when they are heated and contract when they are cooled.* Liquids and solids also expand and contract with temperature, but not to the same extent as gases.

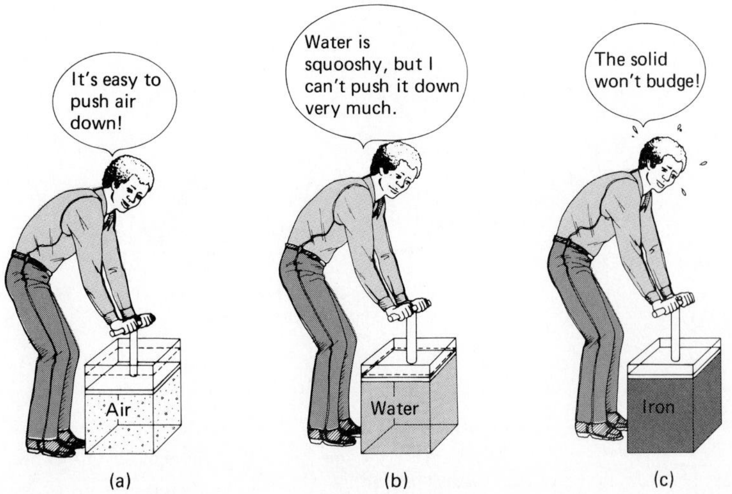
It's easy to push air down!
Air
(a)
Water is squooshy, but I can't push it down very much.
Water
(b)
The solid won't budge!
Iron
(c)

FIG. 8.1 Gases are easily compressed, liquids are difficult to compress, and solids are the least compressible of the three physical states.

Before proposing a molecular model that accounts for these properties, we will explore the gas laws, which tell us the quantitative relationships among the gas variables, i.e., among the number of moles, the volume, the pressure, and the temperature of a gas. In order to understand how the gas laws were developed, it is necessary to know how to measure the gas variables, the number of moles n, the volume V, the pressure P, and the temperature T. The designated letters small n and capitals V, P, and T are the standard symbols for the gas variables.

8.3 MEASUREMENTS OF GAS VARIABLES *n, T, V,* AND *P*

The number of moles n can be determined by measuring the mass of a gas sample. The relationship between mass and moles was well established in Chapter 7. Just as we measure the mass of a liquid by difference—that is, we determine the mass of both empty and filled containers and subtract the mass of the empty container from the mass of the filled container to find the liquid's mass—so do we determine the mass of a gas. The temperature T of a gas can be measured with a thermometer. The volume V of a gas can be measured by placing the gas in a container of known volume. The unit of volume most often employed in gas measurements is the liter.

Pressure P is defined as the force exerted per unit area, or

$$P = \frac{\text{force}}{\text{area}}$$

Some typical forces are pushing, pulling, and collisions. Given the same area, the larger the force, the greater the pressure. Given the same force, pressure is greater when the force operates over a smaller area (see Figure 8.2). Notice that force and pressure are not the same thing. An identical force can lead to very different pressures depending on the area over which it is distributed.

You will see in Section 8.8 that particles in a sample of gas are constantly moving. The pressure of a gas comes from the force exerted on the inside area of the gas's container by the collisions of moving molecules (see Figure

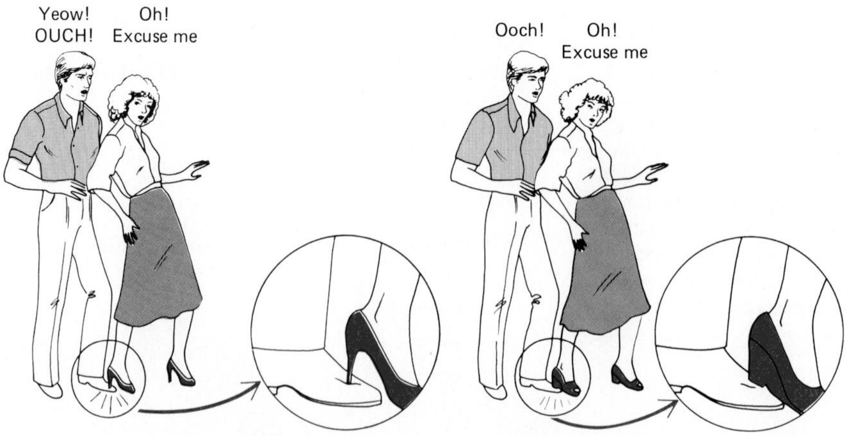
Yeow! OUCH!
Oh! Excuse me
Ooch!
Oh! Excuse me

The woman wearing stiletto-type heels (area = 0.5 cm^2) steps on a man's toes. The pressure exerted is

$$P = \frac{125\ \text{lb}}{0.5\ \text{cm}^2} = 250\ \text{lb/cm}^2$$

It really HURTS!

The same woman wearing flat heels (area = 4 cm^2) steps on man's toes. The pressure exerted is

$$P = \frac{125\ \text{lb}}{4\ \text{cm}^2} = 31\ \text{lb/cm}^2$$

It doesn't hurt nearly as much.

Pressure is eight times greater in the case of the smaller heel, even though the force remained the same.

FIG. 8.2 Pressure varies inversely with area. The same force over a smaller area results in a greater pressure.

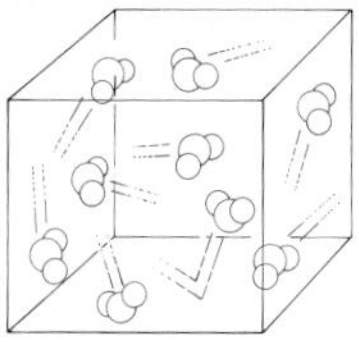

FIG. 8.3
The pressure of a gas comes from the force of collisions of moving molecules exerted on the inside area of the gas's container.

8.3). The pressure of the air around us, which is known as atmospheric pressure, is the pressure exerted when molecules in air collide with our bodies, the ground, or other objects on earth. This pressure is approximately 14.7 lb/in^2. (Pounds per square inch can also be abbreviated psi.)

$$\frac{14.7 \text{ lb}}{1 \text{ in}^2} \qquad \frac{\text{force}}{\text{area}}$$

Air pressure, or atmospheric pressure, can be measured with a barometer (see Figure 8.4). A long (about 80 cm), narrow (diameter = 1 cm) tube, open at one end, is filled with mercury. The tube is stoppered and inverted into a dish containing mercury. When the stopper is removed, some of the mercury empties out of the tube into the dish, but a column of mercury approximately 76 cm (760 mm) high remains in the tube. Because the tube was originally completely filled with mercury, the space that is left above the mercury column when some of the mercury flows out contains *no* air. A vacuum, i.e., a totally empty space, is created, and there is no force pushing down on the top of the mercury column. The pressure from the weight of the mercury in the tube equals the pressure from the atmosphere, that is, the pressure of the weight of air on the mercury in the dish. Examine Figure 8.4 carefully.

The height of the mercury in the tube gives us a measure of the atmospheric pressure. If the atmospheric pressure increases, the height of the mercury in the tube will increase because more mercury is pushed into the tube. If the atmospheric pressure decreases, the height in the tube will decrease as mercury flows into the dish. At sea level on an average day, atmospheric

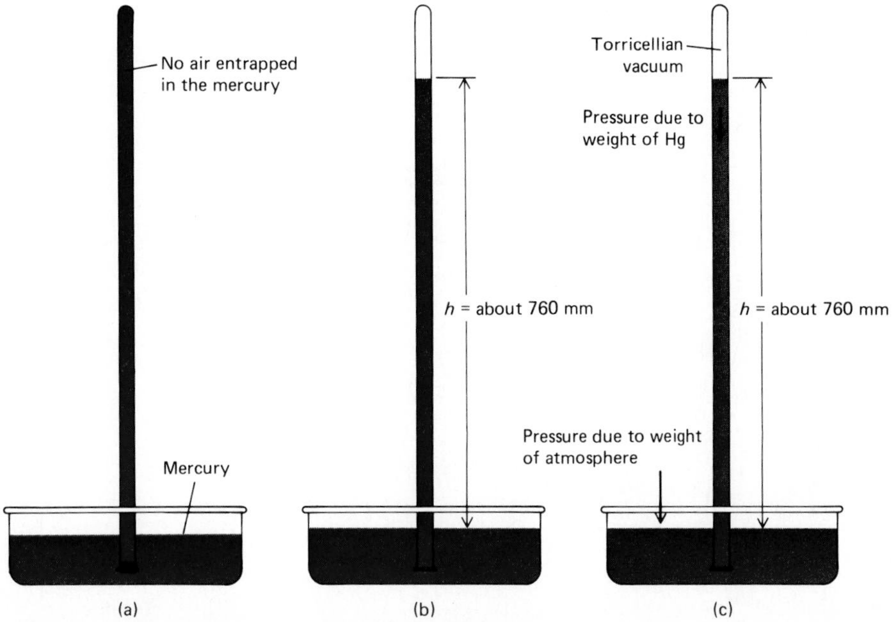
No air entrapped in the mercury
Mercury
(a)
h = about 760 mm
(b)
Torricellian vacuum
Pressure due to weight of Hg
h = about 760 mm
Pressure due to weight of atmosphere
(c)

FIG. 8.4
The construction of a barometer. (*a*) A stoppered tube full of mercury is inverted in a mercury bath. (*b*) The stopper is removed and some mercury falls out of the tube. The remaining height of mercury above the mercury surface exposed to the atmosphere is 760 mm. (*c*) The weight of the atmosphere pushes down on the open bath of mercury and thus holds up the column of mercury in the tube. Atmospheric pressure exactly balances the pressure exerted by the mercury column.

TABLE 8.2 Equalities among Common Units of Pressure

1 atm = 760 mmHg	760 mmHg = 760 torr
1 atm = 760 torr	1 mmHg = 1 torr
1 atm = 14.7 lb/in^2	760 mmHg = 14.7 lb/in^2

pressure can support a column of mercury 760 mm high. This pressure is called one standard atmosphere (1 atm). In honor of the Italian scientist Torricelli, who first developed the barometer, the unit of pressure, millimeters of mercury, has been given the name torr. So 760 mmHg = 760 torr. Table 8.2 relates the common units of pressure.

1 standard atmosphere = 760 mmHg = 760 torr

Problem 8.1 Weather reporters typically say, "Today's barometric pressure is 30 and falling." The 30 is the day's atmospheric pressure in inches of mercury. What is the pressure in torr?

Now that we have explored the measurement of n, T, V, and P, we can pursue the gas laws that govern them.

8.4 DALTON'S LAW OF PARTIAL PRESSURES

Suppose we have three 1.0-L containers. The first one contains a gas a at pressure P_a, the second one contains a gas b at a pressure of P_b.

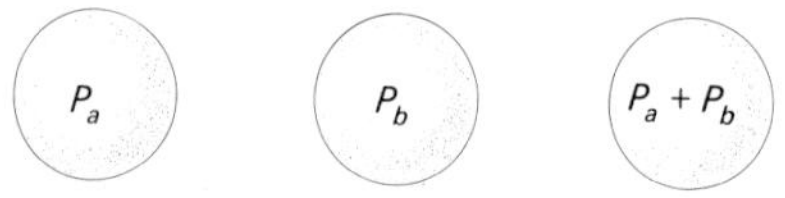
P_a
P_b
$P_a + P_b$

Container 1 Container 2 Container 3

Now we empty the gases from containers 1 and 2 into container 3. The volume of the combined gases is 1.0 L, the same as the volume of the original pure gases. The pressure in the third container is found to be equal to the sum of the pressures that each gas exerted by itself.

$$P_{\text{total}} = P_a + P_b$$

John Dalton (the same scientist who proposed the Atomic Theory of Matter) conducted many experiments of the type just described and found that the total pressure of a mixture of gases is always equal to the sum of the pressures of the pure gases taken separately. Stated another way, each gas in a mixture exerts a pressure as if it were alone in the container. The pressure of each gas in the mixture is known as the partial pressure of that gas. **Dalton's Law of Partial Pressures** states that the total pressure of a mixture of gases is equal to the sum of all the partial pressures of each gas in the mixture:

$$P_{\text{total}} = p_1 + p_2 + p_3 + p_4 + \cdots \tag{8.1}$$

Partial pressures are commonly represented by small p's.

Problem 8.2 A container holds three gases, argon, neon, and krypton, with partial pressures of 1, 0.5, and 1.5 atm, respectively. What is the total gas pressure in the container?

The concept of partial pressures plays an important role in understanding respiration in multicellular organisms. Respiration is the physiological process by which oxygen and carbon dioxide are exchanged between the cells, the blood, the lungs, and the outside atmosphere (Figure 8.5). In this section we will discuss the role that partial pressure plays in gas exchange; in Section 19.12 we will connect this idea with the critical role of hemoglobin in O_2 and CO_2 transport.

Gases flow spontaneously from a region of higher partial pressure to a region of lower partial pressure. As Figure 8.5 shows, atmospheric oxygen flows into the alveoli, small sacs in the lungs, because p_{O_2} is greater in the inspired air (158 mmHg) than in alveolar air (101 mmHg). Carbon dioxide flows in the reverse direction from the alveolar sacs to the outside air because p_{CO_2} in the alveoli (40 mmHg) is much greater than in the atmosphere (0.3 mmHg).

In a second exchange, O_2 diffuses from the alveolar sacs (p_{O_2} = 101 mmHg) across the alveolar capillary membrane into the circulating *venous* blood (p_{O_2} = 45 mmHg). The oxygen in the now arterial blood has a p_{O_2} of about 100 mmHg. Although a small amount is dissolved in blood, oxygen is predominantly transported in chemical combination with hemoglobin, a component of red blood cells (Section 19.12). At the same time that oxygen moves from the lungs to the blood, carbon dioxide migrates from the venous blood (p_{CO_2} = 46 mmHg) into alveolar sacs (p_{CO_2} = 40 mmHg).

The partial pressure of O_2 in tissue is about 35 mmHg, and therefore O_2 migrates from the red blood cells (p_{O_2} = 100 mmHg) into the tissues. The deoxygenated blood (venous blood) has a p_{O_2} of about 45 mmHg and returns

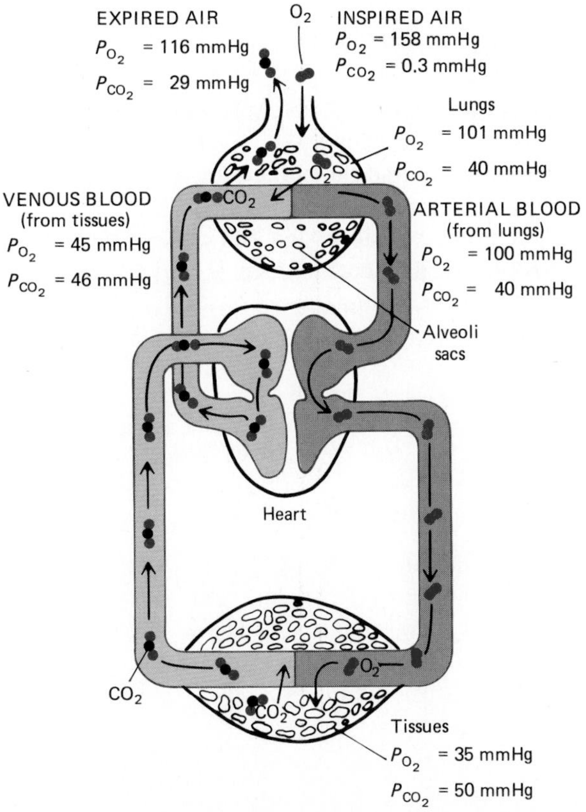
EXPIRED AIR
P_{O_2} = 116 mmHg
P_{CO_2} = 29 mmHg
O_2
INSPIRED AIR
P_{O_2} = 158 mmHg
P_{CO_2} = 0.3 mmHg
Lungs
P_{O_2} = 101 mmHg
P_{CO_2} = 40 mmHg
CO_2
O_2
VENOUS BLOOD
(from tissues)
P_{O_2} = 45 mmHg
P_{CO_2} = 46 mmHg
ARTERIAL BLOOD
(from lungs)
P_{O_2} = 100 mmHg
P_{CO_2} = 40 mmHg
Alveoli
sacs
Heart
CO_2
O_2
CO_2
Tissues
P_{O_2} = 35 mmHg
P_{CO_2} = 50 mmHg

FIG. 8.5
Inspired air, with its relatively high p_{O_2} enriches the lungs' alveoli with oxygen and oxygen then enters the blood stream. This arterial blood then takes oxygen to the tissues. Carbon dioxide from tissues enters the blood. Blood with a depleted oxygen supply and high CO_2 concentration is venous blood, which returns to the lungs, where CO_2 is expired and inspired oxygen "converts" venous blood to arterial blood

to the lungs where the cycle is repeated. Meanwhile CO_2 flows from the tissues (p_{CO_2} = 50 mmHg) into arterial blood (p_{CO_2} = 40 mmHg). The venous blood flowing back to the lungs has a p_{CO_2} = 46 mmHg, which will be reduced at the lung capillaries, where CO_2 will diffuse into the alveoli, repeating the CO_2 cycle.

During periods of breathing difficulty, the partial pressure of O_2 being delivered to a patient can be increased through the use of a nasoinhaler or face mask. The inhaler raises p_{O_2} in the inspired air by a factor of about 2, whereas the face mask, which encloses the nasal area, increases p_{O_2} in inspired air by about 3 to 4 times. Increased p_{O_2} can be dangerous under certain circumstances: oxygen administration to premature infants has resulted in blindness (retrolental fibrosis).

8.5 BOYLE'S LAW

When we increase the pressure on a bubble of gas by squeezing it, its volume gets smaller. When we lessen the pressure, the volume gets larger again. There appears to be an *inverse* relationship between the volume and the pressure of a gas; raising or lowering one has the opposite effect on the other gas variable.

In the seventeenth century Robert Boyle investigated the quantitative relationship between the volume and pressure of a gas. In his experiments Boyle held the temperature and the number of moles of gas at fixed values and observed the change in volume as he varied the pressure on the gas. He found that the relationship was exactly inversely proportional; that is, for example, when the pressure is doubled, the volume is cut in half. When the pressure is quadrupled, the volume is reduced to one quarter the original volume. When the pressure is cut in half, the volume is doubled. These results can be stated mathematically by

$$V \propto \frac{1}{P} \quad \text{at constant } n \text{ and } T \tag{8.2}$$

($\propto$ is the symbol which expresses proportionality.) This expression is read, "The volume of a gas is *inversely* proportional to the pressure when the temperature and the number of moles are held constant." This statement is Boyle's Law.

A proportionality sign can always be replaced by an equals sign and a proportionality constant, thus converting the relationship into an equation. In this case, using k as a constant, we get

$$V = \frac{k}{P} \tag{8.3}$$

Multiplying both sides of the equation by P gives us

$$PV = k \quad \text{at constant } n \text{ and } T \tag{8.4}$$

This equation states that for a fixed number of moles at a given temperature, the product of the pressure and volume of a gas is equal to a constant.

For a fixed quantity of gas at a particular temperature, if we vary the pressure from some value P_1 to a new value P_2, the volume must change from V_1 to V_2, but the product ($P \times V$) will still be equal to the same constant k.

$$\left.\begin{aligned} P_1V_1 &= k \\ P_2V_2 &= k \end{aligned}\right\} \quad \text{at constant } n \text{ and } T$$

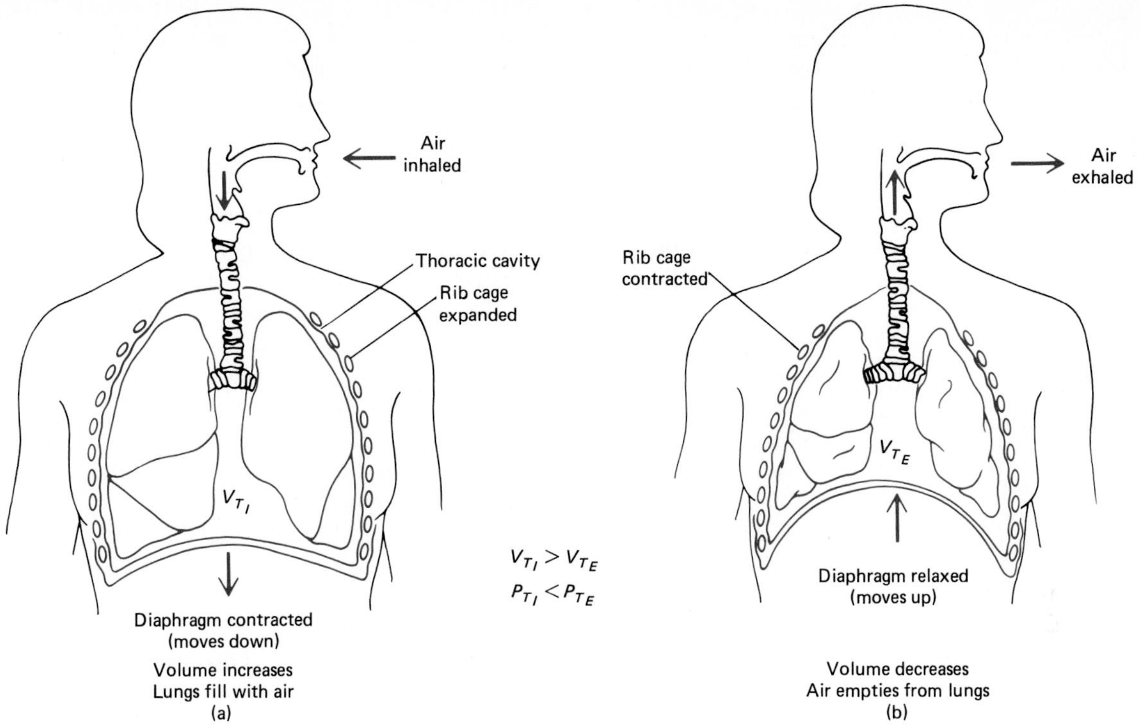
Air
inhaled
Thoracic cavity
Rib cage
expanded
V_{T_I}
Diaphragm contracted
(moves down)
Volume increases
Lungs fill with air
(a)
$V_{T_I} > V_{T_E}$
$P_{T_I} < P_{T_E}$
Air
exhaled
Rib cage
contracted
V_{T_E}
Diaphragm relaxed
(moves up)
Volume decreases
Air empties from lungs
(b)

FIG. 8.6
Breathing is a direct demonstration of the inverse relationship between volume and pressure. Increased volume reduces pressure in the lungs. Decreased volume increases pressure in the lungs. The volume of the thoracic cavity is represented by V_T, V_{T_I} for inhalation, and V_{T_E} for exhalation.

or

$$P_2V_2 = P_1V_1 \tag{8.5}$$

You will see numerical problems involving Boyle's law in Section 8.7.

An application of the inverse relationship between volume and pressure can be seen in the physiological process of breathing. Moving your diaphragm downward (Figure 8.6*a*) expands the rib cage and increases the volume of the thoracic cavity ($V_T \uparrow$). The principles in this section tell us that as V_T increases, the pressure in the thoracic cavity P_T must decrease, that is, fall below atmospheric pressure. The air outside at atmospheric pressure then rushes into the lungs. Moving the diaphragm upward (Figure 8.6*b*) contracts the rib cage, decreasing the volume of the thoracic cavity ($V_T \downarrow$). P_T is increased and air is pushed out of the lungs.

8.6 CHARLES' LAW

If we take a balloon filled with air and place it in boiling water, the volume of the balloon visibly increases (see Figure 8.7). If we take the balloon and place it in ice water, the volume decreases. In both cases, the volume change occurs at a constant pressure (P_{atm}) and with a fixed amount of gas. We can conclude that when the temperature of a gas increases, its volume increases. When the temperature decreases, the volume of the gas decreases. That is, there is a *direct* relationship between the volume and the temperature of a gas. Raising or lowering one has the same effect on the other gas variable. In 1787 Jacques Charles investigated the quantitative relationship between

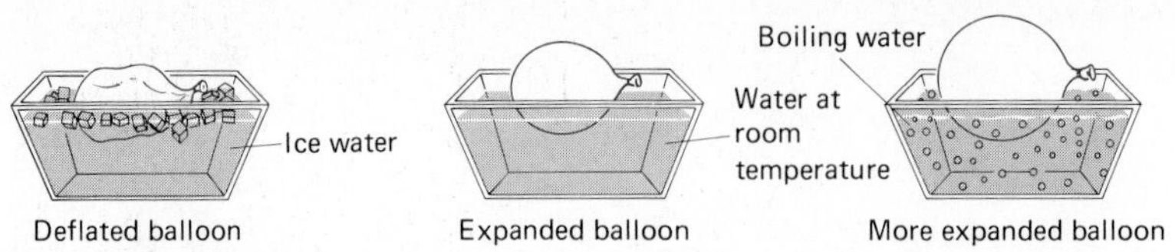
Ice water
Boiling water
Water at room temperature
Deflated balloon
Expanded balloon
More expanded balloon

FIG. 8.7 The volume of the air in the balloon increases as the temperature increases. The pressure is constant and equal to atmospheric *P*.

the volume and the temperature of a fixed quantity of gas held at a constant pressure. A typical set of data from his experiments is shown in Table 8.3. We observe that as the temperature increases, the volume increases, and as the temperature decreases, the volume decreases.

A plot of volume versus temperature (in degrees Celsius) gives the solid straight line shown in Figure 8.8. The line can be extended (broken dashes) to its intercept on the temperature axis. The value is −273°C for all gases subjected to this experiment. In principle this intercept represents the temperature at which the volume of any gas should be zero. In practice, however, all gases liquefy before reaching this temperature.

Mathematically, a straight-line relationship between two variables is simpler if the horizontal intercept is zero, because then the relationship between the variables is a direct proportionality. We can accomplish this for the volume-temperature relationship by establishing a new temperature scale with the zero of this new scale equal to −273°C. This temperature is called absolute zero because it is not possible to achieve lower temperatures than this. We call this new scale the **Kelvin temperature scale** (after Lord Kelvin who constructed it) or the **absolute temperature scale.** The units of this scale are represented by a capital letter K. One does *not* use the degree symbol as in °C and °F. Any Celsius temperature can be converted to Kelvin by using the formula

$$K = °C + 273$$

In the third column of Table 8.3 we list the Kelvin temperature for each Celsius reading in the second column. By convention a capital T designates a Kelvin temperature and a small t designates a Celsius temperature. In Figure 8.9 we plot volume versus Kelvin temperature.

Problem 8.3 Convert the following temperatures to Kelvin temperatures:

a 0°C
b 25°C
c 212°F

TABLE 8.3 Volume-Temperature Data for a Fixed Mass of Gas at a Constant Pressure

Volume, $cm^3 \times 10^2$	Temperature, °C	Temperature, K
9.6	−10.0	263
10.0	0.0	273
10.5	10.0	283
10.7	20.0	293
11.1	30.0	303
11.5	40.0	313
11.9	50.0	323
12.2	60.0	333
12.6	70.0	343
13.0	80.0	353
13.4	90.0	363

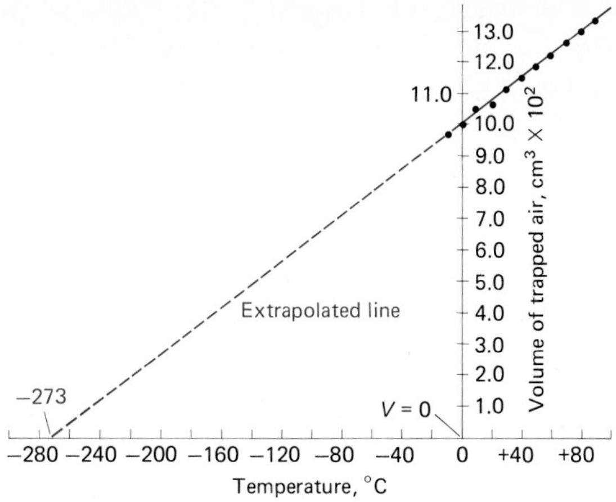
13.0
12.0
11.0
10.0
9.0
8.0
7.0
6.0
5.0
4.0
3.0
2.0
1.0
Volume of trapped air, cm³ × 10²
Extrapolated line
−273
V = 0
−280 −240 −200 −160 −120 −80 −40 0 +40 +80
Temperature, °C

FIG. 8.8
Plot of volume versus temperature (°C) from Table 8.3. Notice that the line extrapolated to zero volume intercepts the temperature axis at $t = -273$°C.

The results of the preceding experiment can be stated mathematically by

$$V \propto T \qquad \text{at constant } n \text{ and } P \tag{8.6}$$

This expression is read, "The volume of a gas is *directly* proportional to the Kelvin temperature when the pressure and number of moles are held constant." This statement is **Charles' Law.**

As we did with Boyle's law, we can convert the proportionality into an equation by using a constant. In this case, we use the constant *m:*

$$V = mT \qquad \text{at constant } n \text{ and } P \tag{8.7}$$

Dividing each side of Equation (8.7) by T gives us

$$\frac{V}{T} = m \qquad \text{at constant } n \text{ and } P \tag{8.8}$$

This equation states that for a fixed number of moles at a given pressure, the quotient of the volume divided by the Kelvin temperature is equal to a constant. Just as in Section 8.5 we rewrote $PV = k$ as $P_2V_2 = P_1V_1$, so too we can rewrite Equation (8.8) in the form

$$\frac{V_2}{T_2} = \frac{V_1}{T_1} \tag{8.9}$$

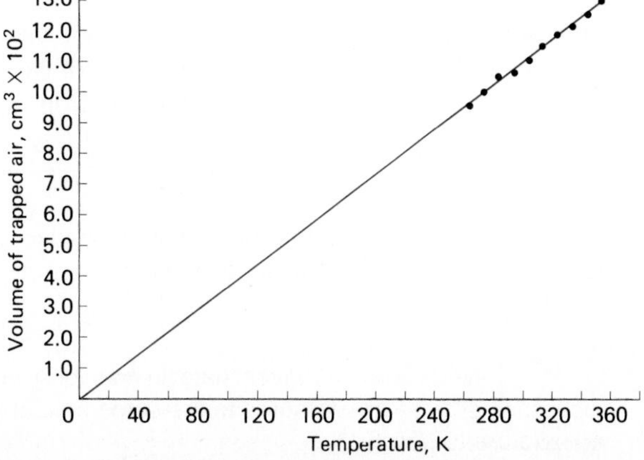
13.0
12.0
11.0
10.0
9.0
8.0
7.0
6.0
5.0
4.0
3.0
2.0
1.0
Volume of trapped air, cm³ × 10²
0 40 80 120 160 200 240 280 320 360
Temperature, K

FIG. 8.9
Plot of volume versus Kelvin temperature from Table 8.3. This graph through the origin ($T = 0$; $V = 0$) clearly shows the direct proportionality between V and Kelvin T.

Remember that the volume is directly proportional only to the Kelvin temperature; *no direct* proportionality exists between volume and Celsius temperature. In gas-law problems, Celsius temperatures must always be converted to Kelvin.

Problem 8.4 *a* Using the data in Table 8.3, show that the ratio V/T is a constant.
b Using the data in Table 8.3, show that the ratio V/t (in °C) is *not* a constant.

You will encounter problems that require Charles' law in Section 8.7.

8.7 COMBINED GAS LAWS

Boyle's law and Charles' law can be combined into a single expression:

$$V \propto \frac{1}{P} \quad \text{at constant } n \text{ and } T \quad \text{Boyle's law} \tag{8.2}$$

$$V \propto T \quad \text{at constant } n \text{ and } P \quad \text{Charles' law} \tag{8.6}$$

Therefore,

$$V \propto \frac{T}{P} \quad \text{at constant } n \quad \text{Combined law} \tag{8.10}$$

This expression is read, "For a fixed mass of gas the volume is directly proportional to the Kelvin temperature and inversely proportional to the pressure." We can write this in an equation form by replacing the proportionality sign by an equals sign and a constant r.

$$V = \frac{rT}{P} \quad \text{at constant } n \tag{8.11}$$

Multiplying both sides of the equation by P and dividing by T gives

$$\frac{PV}{T} = r \quad \text{at constant } n \tag{8.12}$$

If the pressure, volume, and temperature are changed from P_1, V_1, and T_1 to P_2, T_2, and V_2, then

$$\frac{P_2V_2}{T_2} = \frac{P_1V_1}{T_1} \quad \text{at constant } n \tag{8.13}$$

Equation (8.13) can be used to solve problems involving a change in pressure, volume, and temperature for a fixed mass of gas (constant number of moles). The units used on both sides of the equation must be the same.

SAMPLE EXERCISE 8.1 In the laboratory, 4.00 L of helium gas is trapped in a cylinder at a pressure of 7.00 atm. The pressure is decreased, at a constant temperature, to 2.00 atm. What is the new volume?

Solution STEP 1 Write down all given data in the form of initial and final conditions. If necessary, convert all pressures to the same unit, all volumes to the same unit, and temperature to Kelvin:

Initial	Final
$P_1 = 7.00$ atm	$P_2 = 2.00$ atm
$V_1 = 4.00$ L	$V_2 = ?$

Temperature is constant ($T_1 = T_2$).

STEP 2 We recognize this as a problem involving pressure and volume changes at constant temperature and mass. We write down the combined gas law:

$$\frac{P_2V_2}{T_2} = \frac{P_1V_1}{T_1} \qquad (8.13)$$

STEP 3 Omit terms which are identical on both sides. Since in this problem, $T_1 = T_2$, we can omit the temperature term:

$$P_2V_2 = P_1V_1$$

STEP 4 Isolate the unknown variable to one side. We are asked to solve for the new volume, so we divide both sides by P_2, isolating V_2 on one side:

$$\frac{\cancel{P_2}V_2}{\cancel{P_2}} = \frac{P_1V_1}{P_2}$$

$$V_2 = \frac{P_1V_1}{P_2}$$

STEP 5 Fill in the given data, and be careful to have all pressures in the same unit and all volumes in the same unit. Do arithmetic operations and cancel the common units.

$$V_2 = \frac{(7.00\ \cancel{\text{atm}})(4.00\ \text{L})}{(2.00\ \cancel{\text{atm}})}$$

$$V_2 = 14.0\ \text{L}$$

ANSWER CHECK It is a very good idea to check the reasonableness of your answer by thinking through the problem in terms of the qualitative discussions of P, V, and T changes in Sections 8.5 and 8.6. In this problem, because the volume is varying with a change in pressure at constant temperature, a Boyle's law (Section 8.5) relationship should be recognized. Because the pressure is decreasing (from 7 to 2 atm), the volume must increase; i.e., the final volume must be greater than the initial volume of 4 L. The final answer of 14 L is reasonable according to this qualitative analysis.

SAMPLE EXERCISE 8.2

A balloon containing 0.80 L of air at 25.0°C is warmed up to 80°C. What is the new volume of the balloon?

Solution STEP 1

Initial	Final
$V_1 = 0.80$ L	$V_2 = ?$
$T_1 = 25°\text{C} + 273$	$T_2 = 80°\text{C} + 273$
$= 298$ K	$= 353$ K

Pressure is constant ($P_1 = P_2$).

STEP 2 This problem involves volume and temperature changes. The mass of air inside the balloon is fixed, and the pressure P_{atm} is constant.

$$\frac{P_2V_2}{T_2} = \frac{P_1V_1}{T_1} \qquad (8.13)$$

STEP 3 Since $P_1 = P_2$, we have

$$\frac{V_2}{T_2} = \frac{V_1}{T_1}$$

STEP 4 To isolate the wanted variable, V_2, multiply both sides by T_2, giving

$$\frac{V_2\cancel{T_2}}{\cancel{T_2}} = \frac{V_1T_2}{T_1}$$

$$V_2 = \frac{V_1T_2}{T_1}$$

STEP 5

$$V_2 = \frac{(0.80\ \text{L})(353\ \cancel{\text{K}})}{298\ \cancel{\text{K}}}$$

$$V_2 = 0.95\ \text{L}$$

ANSWER CHECK The volume is varying because of a change in temperature at constant pressure. This is a Charles' law (Section 8.6) relationship. Because the temperature is increasing (from 298 to 353 K), the volume must increase; i.e., the final volume must be greater than the initial volume of 0.80 L. The final answer of 0.95 L is reasonable.

SAMPLE EXERCISE 8.3

A sample of helium gas has a volume of 1.25 L at −125°C and 5.00 atm. The gas is compressed at 50.0 atm to a volume of 325 mL. What is the final temperature of the helium gas in degrees Celsius?

Solution STEP 1

Initial	**Final**
$P_1 = 5.00$ atm	$P_2 = 50.0$ atm
$V_1 = 1.25$ L	$V_2 = 325$ mL
$T_1 = -125°\text{C} + 273$	$T_2 = ?$
$= 148$ K	

The volumes must be expressed in the same unit. We will convert 325 mL to liters:

$$V_2 = 325\ \cancel{\text{mL}} \left(\frac{1\ \text{L}}{1000\ \cancel{\text{mL}}}\right) = 0.325\ \text{L}$$

STEP 2 The problem involves a pressure, volume, and temperature change. Mass is fixed.

$$\frac{P_2V_2}{T_2} = \frac{P_1V_1}{T_1} \qquad (8.13)$$

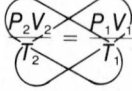

The loops indicate cross-multiplication, as described in Step 4.

STEP 3 Since P, V, and T all change, there are no terms to be omitted.

STEP 4 When two fractions are equal, $\text{numerator}_a \times \text{denominator}_b = \text{numerator}_b \times \text{denominator}_a$, so

$$P_1V_1T_2 = P_2V_2T_1$$

Now divide both sides by P_1V_1, giving

$$\frac{\cancel{P_1}\cancel{V_1}T_2}{\cancel{P_1}\cancel{V_1}} = \frac{P_2V_2T_1}{P_1V_1}$$

$$T_2 = \frac{P_2V_2T_1}{P_1V_1}$$

$$T_2 = \frac{(50.0\ \cancel{\text{atm}})(0.325\ \cancel{\text{L}})(148\ \text{K})}{(5.00\ \cancel{\text{atm}})(1.25\ \cancel{\text{L}})}$$

$$T_2 = 385\ \text{K}$$

We must convert this to degrees Celsius.

$$\text{K} = {}^\circ\text{C} + 273$$

$$385 = {}^\circ\text{C} + 273$$

$$385 - 273 = {}^\circ\text{C}$$

$$112^\circ = {}^\circ\text{C}$$

Problem 8.5 A 738-mL sample of a gas at 0°C and 760 mmHg is cooled to −200°C and compressed to 75.0 atm. What will be the new volume of this gas sample?

A statement about the volume of a gas is meaningless unless we also know the temperature and pressure of the gas, because as we have seen, V changes as T and P change. The volumes of gases can be compared only at the same temperature and pressure. Scientists have agreed on a convenient set of reference conditions. The temperature 273 K (0°C) and the pressure 1 atm (760 mmHg) have been chosen as **standard temperature** and **standard pressure.** The abbreviation **STP** is often used for standard temperature and pressure.

STP = 273 K (0°C) and 1 atm (760 mmHg)

The volume of a gas at non-STP conditions can be converted to STP by the use of the combined gas-law equation [Equation (8.13)].

8.8 IDEAL GAS LAW

In the foregoing sections we have worked with a *fixed* amount of gas and observed quantitative variations in P, V, and T. If we take a balloon (at constant T and P) and simply add more gas to it, the volume of the balloon increases. Of course, it is true for any sample of matter that if we add more mass to the sample, it gets bigger (volume increases). But the remarkable thing about gases is that there is a *direct proportionality* between volume and the amount of material in moles which is the same for all gases. It is remarkable because this proportionality does not hold for the liquid or solid state.

As we have seen previously, a direct proportion is expressed

$$V \propto n \qquad \text{at constant } T \text{ and } P \tag{8.14}$$

We now have proportionality relationships between all four gas variables [Equations (8.13) and (8.14)]. It would be convenient to find a single relationship which would allow the calculation of the fourth variable given any three of the four variables P, V, T, or n at any condition.

Combining $V \propto n$ with Equation (8.10), $V \propto T/P$, gives

$$V \propto \frac{nT}{P} \tag{8.15}$$

Replacing the proportionality by the constant R gives us the so-called **ideal gas law,** Equation (8.17)

$$V = \frac{RnT}{P} \tag{8.16}$$

or

$$PV = nRT \tag{8.17}$$

The gas constant R can be evaluated using known experimental values of P, V, and T. The numerical value of R depends on the units of P. The most common values are

$$R = 0.0821 \frac{(\text{L})(\text{atm})}{(\text{mol})(\text{K})} \tag{8.18}$$

and

$$R = 62.4 \frac{(\text{L})(\text{mmHg})}{(\text{mol})(\text{K})} \tag{8.19}$$

The volume of 1.00 mol of a gas at STP conditions can be calculated using the ideal gas law.

$$V = \frac{nRT}{P}$$

$$V = \frac{(1.00 \cancel{\text{mol}}) \left(0.0821 \dfrac{\text{L}\cdot\cancel{\text{atm}}}{\cancel{\text{mol}}\cdot\cancel{\text{K}}}\right) (273 \cancel{\text{K}})}{1 \cancel{\text{atm}}}$$

$$V = 22.4 \text{ L}$$

This volume is general for any gas and is known as the **molar gas volume.** Equal volumes of gases compared at STP contain equal numbers of moles.

Let us look at some applications of the ideal gas law and the molar gas volume concept.

SAMPLE EXERCISE 8.4

What volume is occupied by 3.25 mol of oxygen gas at 735 mmHg and 25°C?

Solution We recognize that this problem can be solved by the ideal gas law because we are given three variables *n, P,* and *T* and asked to find the fourth gas variable *V.* Therefore, we will use

$$PV = nRT \qquad (8.17)$$

STEP 1 Isolate the variable we are being asked to find. In this problem, the unknown variable is *V,* so we write

$$V = \frac{nRT}{P}$$

STEP 2 Choose the value of *R* that you will use based on the units of *P* given. In this case, because *P* is given in millimeters of mercury, choose the *R* of Equation (8.19), 62.4 L·mmHg/mol·K.

STEP 3 Perform any necessary conversion of given units to conform to the units of *R.* In this case,

$$25°\text{C} = 298\ \text{K} \qquad (25°\text{C} + 273)$$

STEP 4 Substitute the numerical values with their proper units in the equation shown in step 1. Cancel out common units and perform the indicated arithmetic.

$$V = \frac{(3.25\ \cancel{\text{mol}})\ (62.4\ \text{L}\cdot\cancel{\text{mmHg}}/\cancel{\text{mol}}\cdot\cancel{\text{K}})\ (298\ \cancel{\text{K}})}{735\ \cancel{\text{mmHg}}}$$

$$V = 82.2\ \text{L}$$

STEP 5 Check to be sure that the units you are left with, after cancellation of units, is the proper unit for the unknown variable. In this case, liters is the proper unit for volume.

SAMPLE EXERCISE 8.5

What is the volume of 11.9 g of nitrogen gas at STP?

Solution This problem can be solved by the standard Unit Conversion Method.

STEP 1 The **given** is 11.9 g of N_2 at STP.

STEP 2 The **new** quantity to be determined is number of liters at STP.

STEP 3 We do not have a conversion factor to directly convert grams to liters, but at STP we know that 1 mol of N_2 = 22.4 L of N_2 and that the molar weight relates moles to grams. In this case, 1 mol of N_2 = 28.0 g of N_2.

STEP 4 Set up in the proper format:

Given × conversion factors = answer

$$11.9\ \cancel{\text{g of N}_2} \times \frac{1\ \cancel{\text{mol of N}_2}}{28.0\ \cancel{\text{g of N}_2}} \times \frac{22.4\ \text{L of N}_2}{1\ \cancel{\text{mol of N}_2}} = 9.52\ \text{L of N}_2$$

Problem 8.6 How many moles of gas are there in a 400-mL aerosol can at a temperature of 20°C and a pressure of 3.00 atm?

Problem 8.7 Calculate the volume of 7.50 g of hydrogen gas at STP.

8.9 KINETIC THEORY OF GASES

The Kinetic Theory of Gases was developed to account for the physical properties and gas laws that we have discussed in this chapter. The theory proposes a model for gases that is based on the following assumptions:

1 Gases are made up of small particles (either atoms or molecules) which are constantly moving in random, straight-line motion.
2 The distance between particles is very large compared with the size of the particles. A gas is mostly empty space.
3 There are no attractive forces between particles. The particles move independently of each other.

TABLE 8.4 Explanation of Physical Properties and Gas Laws Based on the Kinetic Theory of Gases

Observation	Explanation
1 Gases can be compressed easily.	1 Because gas particles are far apart, they easily can be squeezed closer together by an outside force.
2 Gases expand to fill the volume of their container.	2 Gas particles are constantly moving with no attractive forces between particles, so they will expand until they meet an outside force, namely, the wall of the container.
3 Gases have a low density.	3 Because a gas is mostly empty space, there are few particles (low mass) per unit volume.
4 Gases can diffuse through each other.	4 Gas particles are constantly moving and are separated by large distances. This leads to freedom for particles of one gas to move through the empty space of another gas.
5 Gases exert a pressure on container walls.	5 Moving gas particles collide with container walls, thus exerting a force on every square inch.
6 Boyle's law: $V \propto \frac{1}{P}$	6 When the volume of gas is *decreased,* the particles collide with the walls more often, leading to a *greater pressure.* When volume of gas is *increased,* particles collide less often, leading to a *decreased pressure.*
7 Charles' law: $V \propto T$	7 Increased temperature causes particles to move faster, leading to more and "harder" collisions with walls. Pressure inside the walls is increased until the volume expands to the point where the pressure inside the walls is again equal to the pressure outside.
8 Dalton's law of partial pressure: $P_{total} = p_1 + p_2 + p_3 + \cdots$	8 Since the particles move independently of one another, each gas in a mixture will exert a pressure independent of the pressure of the other gases. The total pressure will be the sum of the individual pressures.

4 The particles collide with each other and with the walls of the container without incurring a loss of energy.
5 The average kinetic energy of the particles is directly proportional to the Kelvin temperature of the gas. (When the average kinetic energy increases, the particles move faster.)

Table 8.4 illustrates how the Kinetic Theory of Gases explains some of our observations about gases.

A gas that has no attractive forces between its particles is an "ideal" gas. The Kinetic Theory of Gases gives us a model of an ideal gas. The gas laws described in this chapter hold exactly only for ideal gases. Most "real" gases show approximately ideal behavior except at low temperature and high pressure. At low temperature, the motion of gas particles becomes so slow that attractive forces begin to play an important role. At high pressure, particles are squeezed close enough together for attractive forces to be important.

8.10 INTERMOLECULAR FORCES

Why isn't all matter in the gaseous state? Why do liquids and solids exist? The answer lies in the hypothesis that there are attractive forces between molecules known as intermolecular forces. Whereas the attractive forces between molecules of gases are so small that we can consider each molecule to be moving independently of others (Section 8.9), the intermolecular forces of liquids and solids are much stronger and the behavior of individual molecules is greatly influenced by surrounding molecules.

In addition to their role as the "glue" holding together collections of molecules in liquids and solids, intermolecular forces have an enormous impact on the chemical and physiological properties of large biochemical molecules. Their importance in chemistry, biology, and life cannot be overstated. Before beginning the study of intermolecular forces, you will find it very helpful to review Section 6.11 on molecular polarity.

The three types of intermolecular forces are:

1 Dipole-dipole interactions
2 Dispersion (London) forces
3 Hydrogen bonds

Dipole-Dipole Interactions

These interactions occur between molecules that have a permanent dipole moment, as shown in Figure 8.10. Notice that the molecules arrange themselves so that the + poles of the molecules are attracted to the − poles of other molecules. All polar molecules are attracted to one another through this *dipole-dipole interaction*. The strength of the dipole-dipole attraction depends on the magnitude of the dipole moment. Nonpolar molecules do not exhibit this interaction because they do not have permanent dipoles.

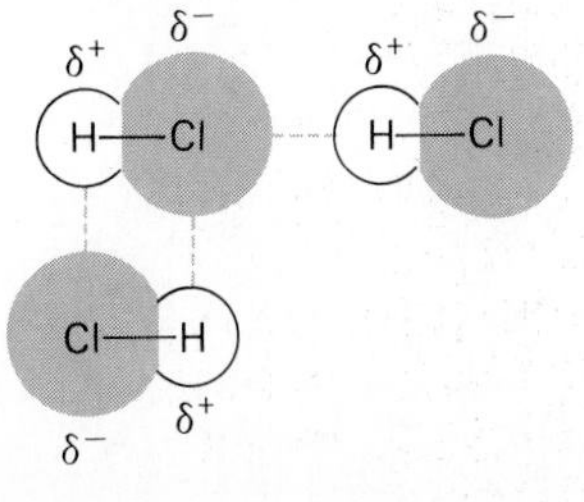
δ+
δ−
H—Cl
δ+
δ−
H—Cl
Cl—H
δ−
δ+

FIG. 8.10 Dipole-dipole interaction is the mutual attraction between the + end of one polar molecule and the − end of another polar molecule. The dipole-dipole interactions are represented by dashed lines.

Dispersion (London) Forces

Figure 8.11 illustrates the development of these intermolecular forces. Electrons in molecules are in constant motion. At any instant the electron distribution within a molecule may be unsymmetrical, so that an "instantaneous dipole" develops (Figure 8.11*b*). This instantaneous dipole induces an unsymmetrical charge distribution in a nearby molecule, causing a net attractive force to develop between the two molecules (Figure 8.11*c*). An instant later the electron distributions shift, setting up instantaneous dipoles with other molecules (Figure 8.11*d*). These are called *dispersion forces* or *London forces* for Fritz London who first offered a mathematical understanding of these interactions. Dispersion forces exist between all kinds of molecules, whether polar or nonpolar.

The strength of dispersion forces depends on the number of electrons in a molecule and the ease with which their distribution can be disturbed. Molecules with greater numbers of atoms and hence greater molecular weights have larger numbers of electrons. Therefore, we see an increase in dispersion forces as molecular weight (and number of electrons) increases.

Hydrogen Bonds

The intermolecular force that is called the *hydrogen bond* occurs between a hydrogen covalently bonded to N, O, or F in one molecule and a nonbonding pair of electrons on N, O, or F in another molecule. Water, H_2O, is the most commonly encountered molecule that shows hydrogen bonding. The electron pair in the covalent bond between O and H is shifted toward the more electronegative oxygen, leaving the hydrogen partially positive in charge. Because of the small size of the hydrogen atom, its positive charge is concentrated, and this leads to a strong intermolecular attraction with a nonbonding pair of electrons on the oxygen of a nearby water molecule.

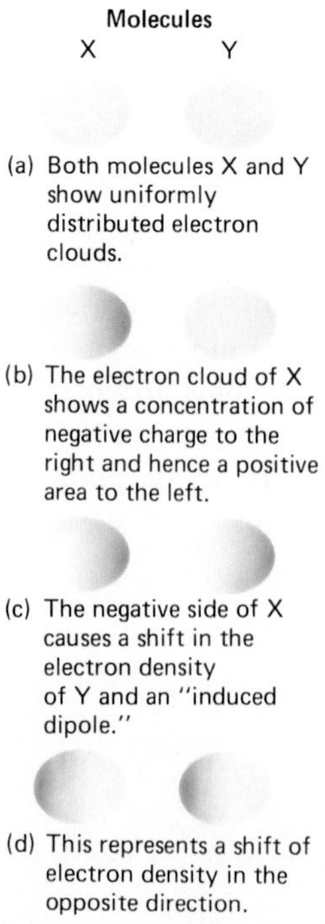
Molecules
X
Y

(a) Both molecules X and Y show uniformly distributed electron clouds.

(b) The electron cloud of X shows a concentration of negative charge to the right and hence a positive area to the left.

(c) The negative side of X causes a shift in the electron density of Y and an "induced dipole."

(d) This represents a shift of electron density in the opposite direction.

FIG. 8.11 Dispersion forces arise through distortions of the electron distributions in molecules which approach one another. The molecules in this figure are pictured as "blobs" of electron density which is initially uniformly distributed and which becomes distorted. Distortions create "instantaneous dipoles," that is, + and − ends to the molecules which exist only for a fleeting moment. All of the events described in (*a*) through (*d*) happen in a small fraction of a second.

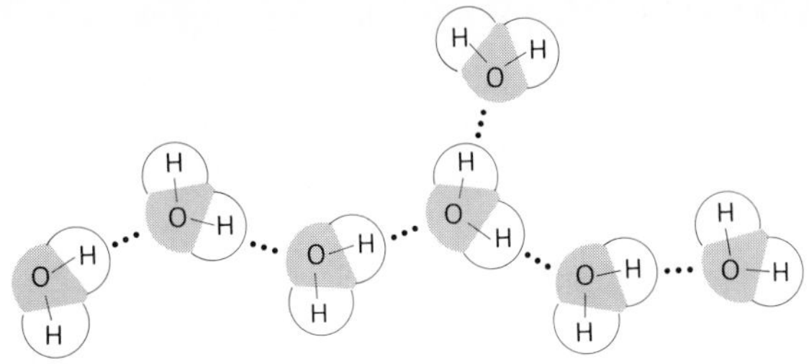
H
O
H
H
O
H
O
H
H
O
H
H
O
H
H
O
H
H
O
H

FIG. 8.12
Hydrogen bonding in water. Intermolecular hydrogen bonds (represented by · · ·) form between the slightly positive hydrogen of an O—H bond in one molecule and the electronegative oxygen of another molecule.

This attraction is known as the hydrogen bond.

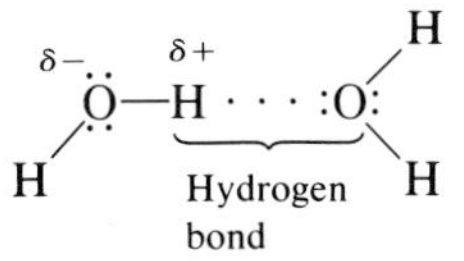
δ−
δ+
O—H · · · :O:
H
H
H
Hydrogen
bond

Each oxygen offers two nonbonding pairs of electrons for hydrogen bonding with hydrogens of two nearby water molecules. The hydrogen bond is stronger than other dipole-dipole attractions or London forces, but much weaker than the normal covalent bond. In Figure 8.12 the covalent bonds are shown by the usual dash (—) and the hydrogen bonds by dots (· · ·).

Many of the special physical properties of water (compared with molecules of similar molecular weight and polarity) arise from the strong hydrogen bonding between water molecules. Water can climb upward in a narrow tube, a phenomenon known as capillary action. Glass tubes or the networks of pores in soil contain oxygen atoms as part of their molecular structure. Hydrogen bonds form between the water molecules and these oxygen atoms in the narrow area, and this helps to pull the water up the tube or pore. A blood sample is drawn up into a capillary tube by this same action.

Hydrogen bonds can also form between two different kinds of molecules, as you will see in Chapters 9, 14, 16, and elsewhere. Whether a molecule or a small part of a large molecule can hydrogen-bond with water has important consequences which you will encounter in subsequent chapters (e.g., see Section 9.4). This property also affects the nature of the three-dimensional shape of large molecules such as proteins, as you will see in Section 19.10. You will see in Section 20.4 that hydrogen bonds act as the "hooks and eyes" that hold together the two chains of the genetic material DNA.

SAMPLE EXERCISE 8.6

Which of the following molecules are capable of forming hydrogen bonds with one another?

a NH_3 *b* CH_4 *c* HF *d* H—C(H)=O

Solution Molecules with N—H, O—H, or F—H bonds show hydrogen bonding.

a NH_3, because it has a polar N—H bond, will show hydrogen bonding.
b CH_4 has only a nonpolar C—H bond and will *not* show hydrogen bonding.
c HF has a polar H—F bond and will show hydrogen bonding.
d In this case, although an oxygen with its nonbonding pair of electrons is present, the hydrogen is not bonded to N, O, or F, and so hydrogen bonding is not possible.

Problem 8.8 Which of the following molecules are capable of forming hydrogen bonds with one another?

a H_2S *b* $CH_3—\overset{\overset{\displaystyle O}{\|}}{C}—NH_2$ 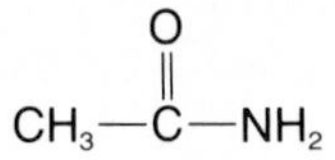   *c* $CH_3—NH_2$ *d* $CH_3—\overset{\overset{\displaystyle CH_3}{|}}{N}—CH_3$ 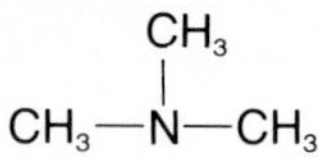

8.11 CONDENSATION OF GASES

Intermolecular forces cause gas molecules to be pulled closer together into the liquid state as a gas is cooled or as the pressure on a gas is increased. The stronger the intermolecular forces that are present, the less cooling or compression is required to cause liquefaction.

Although an "ideal" gas would have no attractive forces between its molecules, all real gases have attractive intermolecular forces. When a gas is cooled, the motion of its molecules is slowed, and eventually after sufficient cooling the attractive forces are able to "pull" the molecules into a liquid. The temperature at which the gas begins to condense into a liquid could be called the *condensation point,* but it is more common to call it the **boiling point.** For any substance the change in state from gas to liquid (condensation) occurs at the same temperature as the change in state from liquid to gas (boiling). Boiling point depends on pressure. When the pressure on a gas is increased, the distance between molecules is decreased, and less cooling is necessary before the intermolecular attractions can force the molecule into a liquid.

Just as for the case of increased pressure pushing molecules together, molecules with strong intermolecular forces will need *less cooling* before they liquefy. The boiling points of such substances are higher than the boiling points of those with weaker intermolecular forces. Water has a boiling point of + 100°C and methane (CH_4) has one of − 162°C because, in addition to the always present dispersion forces, water molecules are held together by hydrogen bonds, the strongest of the intermolecular attractions, and dipole-dipole interactions, whereas methane molecules are attracted to one another only by dispersion forces.

SAMPLE EXERCISE 8.7

Predict the relative values of the boiling points of the following compounds, based on the intermolecular forces present in each:

a $\overset{\displaystyle H}{\underset{\displaystyle H}{>}}C=O$ 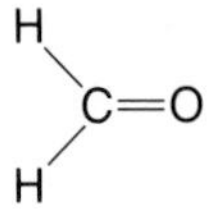

Formaldehyde

b $H—\overset{\overset{\displaystyle H}{|}}{\underset{\underset{\displaystyle H}{|}}{C}}—O—H$ 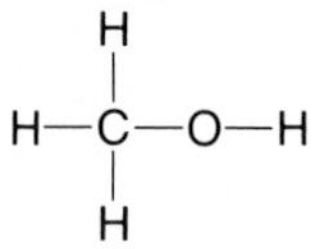

Methyl alcohol

c $H—C\equiv C—H$

Acetylene

Solution

All compounds have dispersion forces. Acetylene has only dispersion forces, and we therefore expect it to have the lowest boiling point (− 82°C); i.e., it must be cooled the most in order for the weak attractions to pull the molecules together. The dispersion forces in acetylene and formaldehyde are about equal because of the similar molecular weights. However, formaldehyde is polar because of the C=O bond dipole. Dipole-dipole interactions will cause it to liquefy at a higher temperature (− 21°C) than acetylene.

Methyl alcohol has an O—H bond, which signals the existence of the strongest intermolecular force, the hydrogen bond; its boiling point is 65°C.

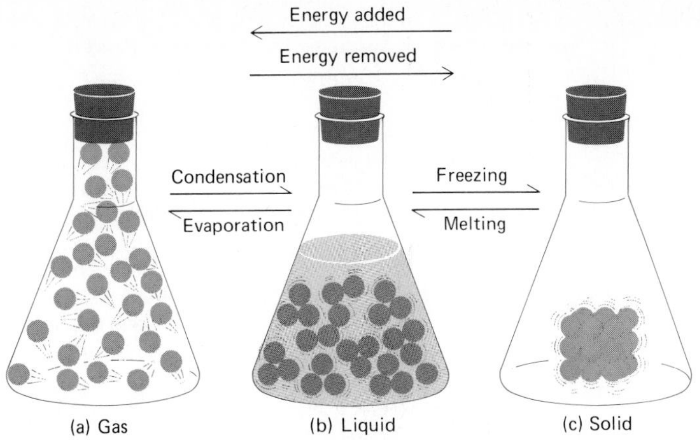
Energy added
Energy removed
Condensation
Evaporation
Freezing
Melting
(a) Gas
(b) Liquid
(c) Solid

FIG. 8.13
Comparision of the behavior of particles in the (*a*) gaseous, (*b*) liquid, and (*c*) solid states of matter. Also see Table 8.5. (*a*) Molecules are far apart; attractive forces between molecules are very small or nonexistent. Molecules are free to move in random motion. (*b*) Molecules are closer together; attractive forces are larger. Molecules can freely move, but cannot separate from each other. (*c*) Molecules or ions are touching; attractive forces are very large and hold particles in a fixed orderly arrangement. Particles can vibrate only about fixed positions. Of course, these molecules are drawn oversized so that we can see them.

Problem 8.9 In each pair pick the compound with the higher boiling point:

a NH_3 or NF_3
b H_2O or H_2S
c Cl_2 or Br_2

When a liquid is sufficiently cooled, eventually the molecular motion slows down to the point where the substance can no longer flow and the liquid is transformed into a solid. The temperature at which this occurs is known as the *freezing point,* or **melting point.** The change in state from liquid to solid (freezing) occurs at the same temperature as the change in state from solid to liquid (melting).

8.12 A MODEL OF LIQUIDS AND SOLIDS

Comparative molecular models of gases, liquids, and solids are shown in Figure 8.13. These models are based on the distinguishing properties of gases, liquids, and solids outlined in Table 8.1 and on the property of compressibility. Table 8.5 describes how these models of liquids and solids explain their characteristic properties.

8.13 PHYSICAL PROPERTIES OF LIQUIDS

Among the more important physical properties of liquids are vapor pressure, boiling point, heat of vaporization, and surface tension.

Vapor Pressure

Liquids left in an open container slowly evaporate or vaporize; i.e., the liquid's molecules escape into the gas phase. The model of liquids can aid in the understanding of this phenomenon. Some molecules near the surface of the liquid have more than the average amount of kinetic energy, which they have gained through collisions with other molecules. These more energetic molecules can overcome the attractive forces of neighboring molecules and escape into the gas phase. Since it is the more energetic molecules

TABLE 8.5 Observable Properties of Liquids and Solids and Their Molecular Explanation

Observable Property	Explanation from Model
LIQUIDS	
Liquids have a definite volume.	Forces of attraction are strong enough to hold molecules together in a definite volume.
Liquids have an indefinite shape.	Particles in a liquid are free to move (but not independently of each other), so that a liquid takes the shape of its container.
Liquids have low compressibility.	Particles in a liquid lie close together with very little space between them.
SOLIDS	
Solids have a definite volume.	Forces of attraction are very strong, holding particles in a definite volume.
Solids have a definite shape.	Particles are held in fixed positions by strong attractive forces.
Solids have very low compressibility.	Particles are touching; increasing pressure cannot squeeze particles closer together.

that are leaving, the average kinetic energy, and therefore the temperature, decreases. It is for this reason that evaporation is said to be a cooling process. Perspiration evaporates from your skin, absorbing heat from the body. Sweating followed by evaporation is one of the body's cooling mechanisms. Liquids with low intermolecular attractions vaporize more readily at the same temperature than liquids with high intermolecular attractions. For example, ether, which does not show hydrogen bonding, evaporates much more readily than water, which has the strong hydrogen-bonding intermolecular attractions. Alcohol, a substance with a vapor pressure higher than water, is rubbed on the skin of a patient with a high fever to introduce an additional evaporative cooling mechanism.

When a liquid is placed into a closed container, the molecules that escape into the gas phase find themselves enclosed in a fixed space. As more and more molecules enter the gas phase, a pressure develops in the closed container (Figure 8.14). Since they cannot escape from the enclosed volume, the gas molecules eventually strike the liquid surface and a few are ''recaptured'' into the liquid phase. After a period of time the rate of condensation becomes equal to the rate of vaporization. Vaporization and condensation continue, but there is no *net* change in the number of molecules leaving the liquid for the gas phase. The pressure of the gas at this point is known as the **vapor pressure** of the liquid.

FIG. 8.14
The liquid is placed in the closed container and begins to evaporate (↑). As evaporation proceeds, more and more gas molecules (•) accumulate, and this leads to an increase in pressure. Gas molecules also begin to condense (↓). When the rates of evaporation and condensation are equal, the pressure remains constant because the net number of molecules in the gas phase remains constant. This constant pressure is called the vapor pressure of the liquid at the specified temperature.

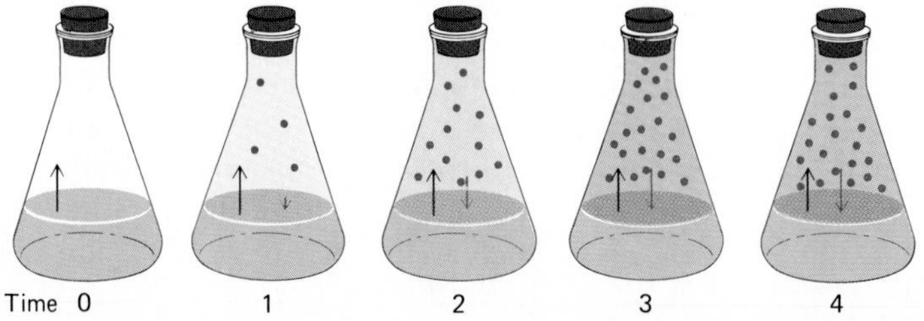

Time 0
1
2
3
4

The value of the vapor pressure depends on the nature of the liquid and on temperature. Liquids with low intermolecular forces, such as ether and gasoline, have high vapor pressures at room temperature. Liquids with high intermolecular forces, such as water and alcohols, have low vapor pressures at room temperature.

Vapor pressure increases with increasing temperature, as shown for water in Figure 8.15. Kinetic energy and temperature are proportional. An increase in temperature gives rise to a corresponding increase in kinetic energy, and hence more molecules near the surface of the liquid are able to overcome the attractive forces of their neighbors and escape into the gas phase. Also at higher temperatures, molecules in the gas phase, having greater kinetic energy, are less likely to be recaptured into the liquid. The result is a greater number of gas molecules constrained within the same volume and hence a higher vapor pressure.

Boiling

A liquid *boils* when its vapor pressure is equal to the external pressure on the surface of the liquid. This condition exists at the **boiling point** (boiling temperature) of the liquid. During evaporation only molecules at the surface escape into the vapor phase, but at the boiling point some molecules *within* the liquid have sufficient energy to overcome the intermolecular attractive forces of their neighbors, so that bubbles of vapor form within the liquid. The bubbles rise in the liquid, and the vapor is released at the surface. It is

FIG. 8.15
The "normal boiling point" of water is 100°C because at that temperature the vapor pressure of water is 760 mmHg, the same value as standard atmospheric pressure. When the vapor pressure of a liquid equals the external (atmospheric) pressure, a liquid boils. (*a*) If the external pressure is lowered, then a liquid can boil at a lower temperature. If the external pressure is only 350 mmHg, then water will boil at 80°C because at that temperature its vapor pressure equals 350 mmHg. (*b*) If the external presure is raised, then a liquid must be heated to a higher temperature before it boils. If the external pressure is 900 mmHg, then water's temperature must be raised to 103°C in order for its vapor pressure to equal 900 mmHg.

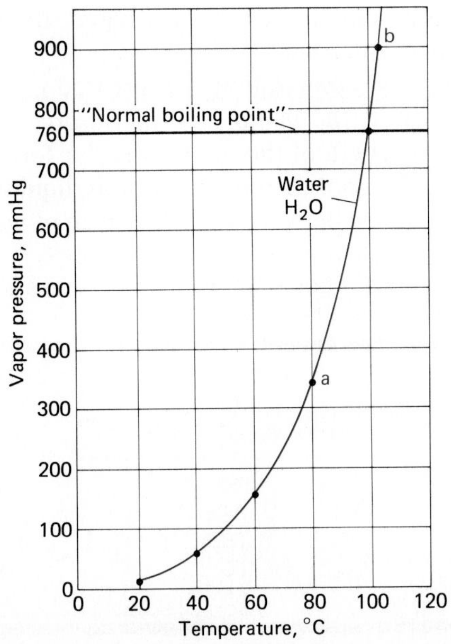
Vapor pressure, mmHg
0
100
200
300
400
500
600
700
760
800
900
"Normal boiling point"
Water
H_2O
a
b
0
20
40
60
80
100
120
Temperature, °C

the formation of vapor bubbles within the liquid itself that characterizes boiling and distinguishes it from evaporation.

Liquids with high intermolecular attractions, such as water, require a relatively high temperature before their vapor pressure equals the external pressure; hence these liquids are found to have a high boiling point. Liquids with low intermolecular attractions, such as ethers, have a lower boiling point.

The so-called normal boiling point is the temperature at which the vapor pressure is equal to an external pressure of 1 atm. For example, water boils at 100°C when the external (atmospheric) pressure is 1 atm, or 760 mmHg, because the vapor pressure of water is 760 mmHg at 100°C.

The boiling point of a liquid can be reduced by lowering the external pressure because then the vapor pressure of the liquid is equal to the external pressure at a lower temperature (Figure 8.15*a*). One application of this principle is in the food industry, where water is removed from such substances as coffee by boiling the liquid under a reduced pressure at a temperature lower than the normal boiling point, where the product might decompose.

The boiling point of a liquid can be increased by raising the external pressure, because then the vapor pressure of the liquid is equal to the external pressure at a higher temperature (Figure 8.15*b*). A home pressure cooker works on this principle. By maintaining a pressure above 1 atm inside the pressure cooker, the temperature of the liquid can rise above 100°C, thus allowing the food to cook in a shorter time.

Although heat must be continuously supplied, the temperature of a boiling liquid remains constant. If we add more heat (raise the flame) to an uncovered pot of boiling water, we find that the water boils faster but the temperature remains constant. If we remove the heat, the boiling process slows down and eventually stops.

Heat of Vaporization

The quantity of heat needed to convert a fixed mass of liquid at a fixed temperature to the gaseous state is known as the *heat of vaporization*. Common units for heat of vaporization are calories per gram (cal/g) and kilocalories per mole (kcal/mol). When an amount of heat equal at least to the heat of vaporization is supplied to a boiling liquid, the liquid continues to boil at a constant temperature.

When the heat of vaporization is expressed in the units of kilocalorie per mole, it is known as the *molar heat of vaporization*. This quantity gives a measure of the strength of the intermolecular forces in the liquid. Table 8.6 lists the heats of vaporization for various liquids at their normal boiling points. The difference between the molar heat of vaporization for water (9.72

TABLE 8.6 Heats of Vaporization for Some Common Substances at Their Normal Boiling Points

		Heat of Vaporization	
Substance	Normal Boiling Point, °C	cal/g	kcal/mol
Methane	−161	138	2.21
Ethyl ether	34.6	89.8	6.64
Ethyl alcohol	78.3	204	9.38
Water	100	540	9.72
Sodium chloride	1465	698	40.8

kcal/mol) and that for methane (2.21 kcal/mol) reflects the strong hydrogen-bonding attractions between water molecules in contrast to the weaker dispersion forces between methane molecules. A greater amount of heat is required to evaporate a gram of water than the same mass of any other common liquid. This property adds to the idealness of water as our body liquid because the evaporation of only a small amount of perspiration leads to significant cooling. The excess heat of an "overheated" person goes into the evaporation process and body temperature is maintained.

When a gas condenses to the liquid state, heat called the *heat of condensation* is given off in an amount exactly equal to the heat of vaporization. The heat of condensation must be removed for the gas to condense to a liquid at a constant temperature. A steam burn is severe because of the large amount of heat that is liberated when the steam vapor condenses to liquid water. Surgical and laboratory equipment is sterilized in an atmosphere of steam rather than boiling water because steam can be heated to temperatures above 100°C and because of the large amount of heat emitted when steam condenses.

SAMPLE EXERCISE 8.8 How much heat is required to vaporize 2.5 mol of ethyl alcohol at its normal boiling point?

Solution The heat needed to convert a liquid to a gas at its boiling point is the heat of vaporization. Since we are given moles of ethyl alcohol, we will use the molar heat of vaporization (9.38 kcal/mol) given in Table 8.6.

$$2.5 \text{ \sout{mol of ethyl alcohol}} \times \frac{9.38 \text{ kcal}}{\text{\sout{mol of ethyl alcohol}}} = 23 \text{ kcal}$$

Problem 8.10 How much heat is liberated when 22.5 g of steam is condensed to liquid water at 100°C?

Surface Tension

Intermolecular attractions lead to some interesting properties at the surface of a liquid. The molecules in the interior of a liquid, which are completely surrounded by other molecules, feel a balanced intermolecular attraction. However, molecules at the surface feel an unbalanced force because they are not attracted to other liquid molecules on one side (Figure 8.16. As a

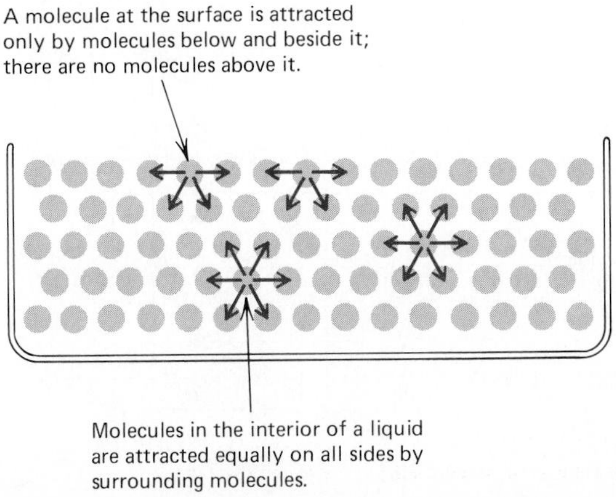
A molecule at the surface is attracted only by molecules below and beside it; there are no molecules above it.
Molecules in the interior of a liquid are attracted equally on all sides by surrounding molecules.

FIG. 8.16 Molecules at the surface boundary of a liquid experience an unbalanced attractive force toward the interior of the liquid. This leads to a net pull inward and a smaller surface area. The result is the phenomenon of **surface tension**.

TABLE 8.7 The Relationship between Intermolecular Forces and the Physical Properties of Liquids

	Rate of Evaporation	Vapor Pressure	Boiling Point	Heat of Vaporization	Surface Tension
Strong intermolecular forces	Low	Low	High	High	High
Weak intermolecular forces	High	High	Low	Low	Low

result the surface molecules experience a pull into the liquid known as surface tension. The surface tension is especially strong in water, where the hydrogen-bonding attractive force is present. The tension at the surface can balance a steel needle if it is carefully placed on the water surface. If the needle is pushed below the surface it will immediately sink because of the greater density of the steel needle compared with water.

An insect such as a water strider can walk on water balanced by the surface tension. Water tends to form spherical droplets because a sphere has the lowest surface area for a given volume, thereby maximizing the inner attractive forces.

Table 8.7 summarizes the relationship of intermolecular forces to the physical properties of liquids.

8.14 CLASSES OF CRYSTALLINE SOLIDS

Solids have a definite volume and shape because the particles making up the solid are held in a fixed arrangement by the attractive forces. Solids made up of a highly regular repeating pattern of particles are known as **crystalline solids.** Most common solids are crystalline or are said to be made up of crystals. If you have ever looked at salt crystals carefully, you probably have noticed that they appear to be in the shape of little cubes with very smooth surfaces. The visible geometry in a crystalline solid arises from the arrangement of the particles (atoms, ions, molecules) within the solid. The particular three-dimensional arrangement of particles in a crystalline solid is known as the **crystal lattice,** and the particles are said to occupy *lattice points*. Figure 8.17 shows a representation of the crystal lattice in sodium chloride. Every crystalline solid has a definite crystal lattice.

Solids which have an irregular packing are known as **amorphous solids;** examples of amorphous solids are glass and many plastics.

Crystalline solids are divided into classes depending on the nature of the particles that make up the crystal lattice. The particles can be ions, atoms, or molecules.

FIG. 8.17 The crystal lattice of sodium chloride. Also see Figure 5.2.

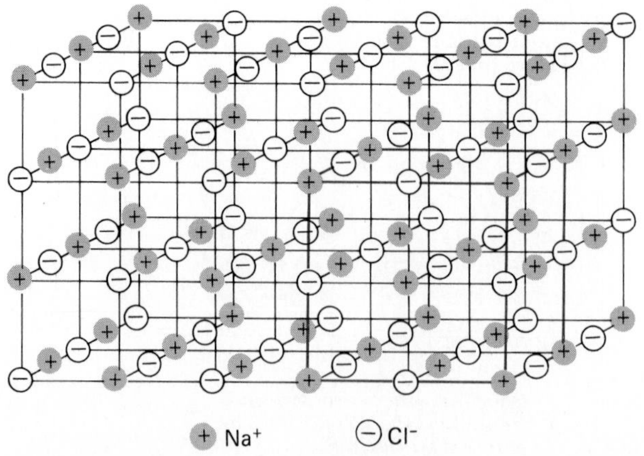

+ Na+
− Cl−

Ionic Crystals

A crystalline solid such as NaCl in which the lattice is made up of oppositely charged ions is said to be an *ionic crystal* (Figure 8.17). All ionic compounds fit into this class (review Section 5.12). The very strong electrostatic forces which exist between oppositely charged particles make it difficult for the individual ions to slide past one another. Therefore, ionic solids are very hard.

Molecular Crystals

Crystalline solids in which molecules occupy the lattice points are known as *molecular crystals*. Examples of molecular crystals include ice, in which an H_2O molecule occupies each lattice point, and "dry ice" (solid carbon dioxide), in which a CO_2 molecule is at each lattice point. The attractive forces in these solids are the intermolecular forces discussed in Section 8.10. Even the strongest intermolecular forces are considerably weaker than the electrostatic attraction which is the ionic bond. Because of the weak attractions between molecules which enable one molecule to slide past another, molecular crystals tend to be soft.

Covalent Crystals

Crystalline solids in which the lattice points are occupied by atoms covalently bonded to one another are called *covalent crystals*. Each atom in a covalent crystal is covalently bonded to at least two other atoms. The entire crystal is actually one large molecule. Because of the strength of covalent bonds, covalent crystals are very hard. Examples of covalent crystals include diamond, quartz (silicon dioxide), and carborundum.

Metallic Crystals

Crystalline solids in which the lattice points are occupied by metallic cations surrounded by moving electrons are called *metallic crystals*. See Figure 8.18. This picture of a metallic crystal arises from the fact that metal atoms lose one or more of their valence electrons, and the electrons are then "free" to move through the entire lattice. The attractive forces in these crystals are the attractions between the positive metal ions and the mobile electrons. The strength of the attractive forces varies according to the size of the ion and number of electrons ionized from each atom. The excellent electrical conductivity of a metal can be accounted for by the freely moving electrons in the lattice of the metallic solid. Electricity is the movement of electrons.

Table 8.8 summarizes this section and also the melting behavior of the

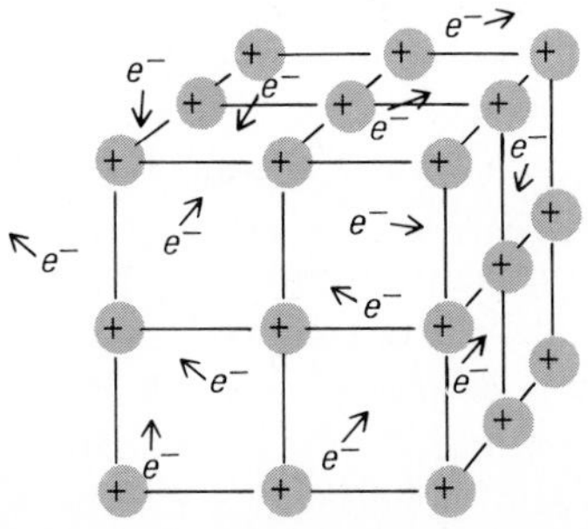

FIG. 8.18
A metallic crystal. Metal cations occupy the lattice points. The valence electrons move freely in an "electron sea."

TABLE 8.8 A Comparison of Crystalline Solids

Type of Crystal	Example	Particles Occupying Lattice Sites	Attractive Forces	Melting Points
Ionic	Sodium chloride, NaCl	Cations and anions	Ionic bond	High
Molecular	Ice, $H_2O(s)$	Molecules	Hydrogen bonds, dipole-dipole attractions, and dispersion forces	Low
Covalent	Diamond, C	Atoms	Covalent bonds	Very high
Metallic	Iron, Fe	Cations	Attraction between the cations and "free" valence electrons	Ranges from low to high

classes of crystalline solids. The melting phenomenon is addressed in Section 8.15.

8.15 PROPERTIES OF SOLIDS

The temperature at which a crystalline solid is converted to a liquid is known as the **melting point.** The melting point of a crystalline solid is the same as the *freezing point* of the liquid.

$$\text{Solid} \underset{\text{freezing}}{\overset{\text{melting}}{\rightleftharpoons}} \text{liquid}$$

In contrast to boiling points, the melting points of most solids do not change greatly with a change in external pressure.

If we heat a solid that is initially at a temperature below its melting point, the temperature increases until it reaches the melting point. As the solid is heated, the particles in the lattice vibrate more rapidly about their fixed positions, though they remain in their same relative positions, as if they were "marching in place." At the melting point the vibrations become so rapid that the particles begin to move apart, and the lattice begins to break down. Though the temperature remains constant, heat must be added continuously to the system to feed the process whereby the lattice is broken down and the solid is converted to a liquid. Solids in which the forces between particles are stronger have higher melting points. More heat must be added to pull particles apart. In Table 8.8 you can see a reflection of the strength of bonds in the relative melting points of the classes of solids.

The amount of heat needed to convert 1 g of a solid to a liquid at the melting point is called the **heat of fusion.** The heat needed for the liquefaction of 1 mol of solid is known as the *molar heat of fusion* (Table 8.9). This is the amount of energy needed to break down the attractive forces in the solid at the melting point. Each gram of ice in an icepack will absorb 80 cal before melting to liquid water at 0°C.

During the reverse process of solidification, an amount of heat that is equal in magnitude to the heat of fusion must be removed or liberated. This quantity of heat liberated, which is exactly equivalent to the heat of fusion, may be called the *heat of solidification*. Vineyard owners protect their grapes by using water's high heat of solidification (80 cal/g). When a nighttime freeze

TABLE 8.9 Heats of Fusion of Various Substances at Their Melting Points

Solid	Melting Point, °C	Heat of Fusion	
		cal/g	kcal/mol
Ice	0.0	80	1.44
Ethyl alcohol	−117	24.9	1.15
Methane	−183	14	0.22
Benzene	6	30.1	2.65
Sodium chloride	804	124	7.19
Cu	1083	49	3.11
Ni	1453	74	4.34

is expected in the fall, a mist of water is sprayed over the grape vines. As the water freezes, 80 cal/g of heat is released and this prevents the temperature of the grapes from falling below 0°C.

Although the particles in a solid are more restricted than those in a liquid, many solids have a measurable vapor pressure and can evaporate directly from the solid to the vapor state without passing through the liquid state. This process is called *sublimation*.

$$\text{Solid} \underset{\substack{\text{deposition} \\ \text{(condensation)}}}{\overset{\text{sublimation}}{\rightleftharpoons}} \text{vapor}$$

The odor of mothballs arises because of the vapor pressure of the solid white napthalene. Napthalene molecules are able to escape into the gaseous state and enter our nostrils. The purplish vapor in a closed bottle of solid iodine crystals presents direct evidence for the sublimation of iodine. The slow

FIG. 8.19
Plot of temperature versus added heat for the conversions solid ⇌ liquid ⇌ gas. (*a*) Temperature rises steadily as the solid is heated. The slope of the line is related to the specific heat of the solid. (*b*) At the melting point, temperature remains constant until enough heat is absorbed to melt the solid completely. (*c*) Temperature rises steadily as the liquid is heated. The slope of the line is related to the specific heat of the liquid. (*d*) At the boiling point, temperature remains constant until enough heat is absorbed to vaporize the liquid completely. (*e*) Temperature rises steadily as the gas is heated. The slope of the line is related to the specific heat of the gas.

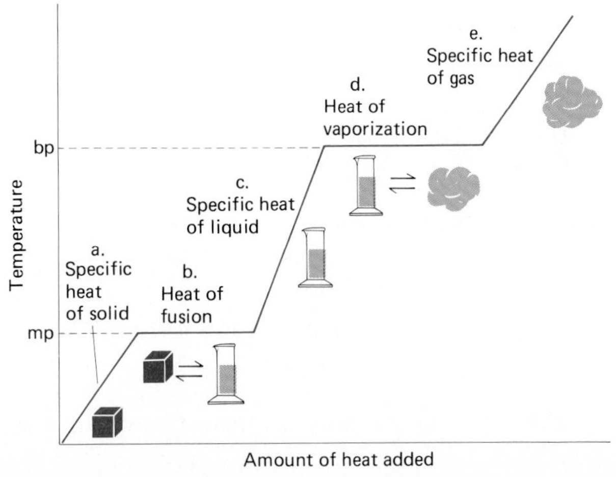

e.
Specific heat
of gas
d.
Heat of
vaporization
bp
c.
Specific heat
of liquid
Temperature
a.
Specific
heat
of solid
b.
Heat of
fusion
mp
Amount of heat added

disappearance of snow in the winter, although the temperature remains below the melting point of ice, visibly demonstrates the sublimation of ice.

Crystalline solids that are molecular crystals provide the most likely candidates for sublimation because they generally contain the weakest attractive forces between their particles.

8.16 HEAT CHANGES AND PHASE CHANGES

When a solid is heated, its temperature rises until it reaches the melting point, and the temperature remains constant until the solid is totally converted to a liquid. If heating continues, the temperature of the liquid increases until it reaches the boiling point, and then the temperature remains constant until the liquid is totally converted to a gas. If the gas continues to be heated, its temperature rises. Figure 8.19 illustrates the changes in temperature of a pure substance as it is heated, beginning with a solid and continuing to the gaseous state, as has just been described. If we are given the specific heat (review Section 2.8) and the heats of fusion and vaporization for a particular substance, we can calculate the amount of heat required for a fixed mass of that substance to undergo any given temperature change.

SAMPLE EXERCISE 8.9

Calculate the amount of heat required to convert 25.0 g of ice initially at −15°C to steam at 125°C.
Given:

Specific heat of ice = 0.500 cal/g·°C

Specific heat of water = 1.00 cal/g·°C

Specific heat of steam = 0.480 cal/g·°C

Also see Tables 8.6 and 8.9.

Solution

STEP 1 Calculate the amount of heat required to raise the temperature of ice from −15°C to its melting point of 0°C (review Section 2.8).

Heat = mass × temperature change × specific heat

$$\text{Heat} = 25.0\ \cancel{g} \times [0\cancel{°C} - (-15\cancel{°C})] \times 0.500\ \frac{\text{cal}}{\cancel{g} \times \cancel{°C}} = 188\ \text{cal}$$

STEP 2 Use the heat of fusion of ice to calculate the amount of heat necessary to melt 25.0 g of ice.

$$25.0\ \cancel{g} \times 80.0\ \frac{\text{cal}}{\cancel{g}} = 2{,}000\ \text{cal}$$

STEP 3 In a manner similar to step 1, calculate the heat required to raise the temperature of water from 0°C to its boiling point (100°C).

$$\text{Heat} = 25.0\ \cancel{g} \times (100\cancel{°C} - 0\cancel{°C}) \times 1.00\ \frac{\text{cal}}{\cancel{g} \times \cancel{°C}} = 2{,}500\ \text{cal}$$

STEP 4 Use the heat of vaporization of water to calculate the amount of heat necessary to vaporize 25.0 g of water.

$$25.0\ \not{g} \times \frac{540\ \text{cal}}{\not{g}} = 13{,}500\ \text{cal}$$

STEP 5 As in steps 1 and 3, calculate the heat required for the indicated temperature change.

$$\text{Heat} = 25.0\ \not{g} \times (125°\not{C} - 100°\not{C}) \times 0.480\ \frac{\text{cal}}{\not{g} \times °\not{C}} = 300\ \text{cal}$$

The total heat input for steps 1 through 5 is:

Step		Heat Input, cal
1	Heating ice	188
2	Ice → water	2,000
3	Heating water	2,500
4	Water → steam	13,500
5	Heating steam	300
	Total	18,488

Notice that the largest portion of the added heat is necessary to convert the liquid water to steam at the boiling point.

CHAPTER ACCOMPLISHMENTS

After completing this chapter you should be able to define **key words** and do the following:

8.1 INTRODUCTION
1. State the shape and volume characteristics of the three physical states.

8.2 CHARACTERISTICS OF GASES
2. State the characteristic properties of gases, and name the four measurable gas variables.

8.3 MEASUREMENTS OF GAS VARIABLES *n, T, V,* AND *P*
3. Describe how the number of moles, volume, and temperature of a gas and atmospheric pressure are measured.
4. State the molecular basis of gas pressure.

8.4 DALTON'S LAW OF PARTIAL PRESSURES
5. State Dalton's law of partial pressures.
6. Given the pressure of each gas in a mixture of gases, calculate the total pressure.
7. Explain the relationship between partial pressure and gas exchange among the lungs, blood, and tissues.

8.5 BOYLE'S LAW
8. State Boyle's law in words and mathematically.

8.6 CHARLES' LAW
9. State Charles' law in words and mathematically.
10. Convert a Celsius temperature to Kelvin.

8.7 COMBINED GAS LAWS
11. Considering the three variables, volume, temperature, and pressure and holding one variable constant (for example, *P*), calculate the effect of a change in the second variable (for example, *T*) on a given amount of the third variable (for example, *V*).

12. Calculate the effect of a change in two variables (for example, P and T) on a given amount of the third variable (for example, V).
13. Given the volume of a fixed amount of gas at a given temperature and pressure, calculate the volume at STP.

8.8 IDEAL GAS LAW

14. State the ideal gas equation that relates the four variables of a gas.
15. Choose the proper value of R, the ideal gas constant, for a given value of the pressure.
16. Given three (for example, P, T, and n) of the gas variables, calculate the fourth (for example, V).
17. State the volume of 1 mol of any gas at STP.
18. Calculate the volume of a given mass of gas at STP.

8.9 KINETIC THEORY OF GASES

19. State the assumptions of the Kinetic Theory of Gases and explain the characteristic properties of gases, Boyle's law, Charles' law, and Dalton's law of partial pressure in terms of the Kinetic Theory of Gases.
20. State the conditions under which gas ideality (zero attractive forces) is most nearly approached.

8.10 INTERMOLECULAR FORCES

21. Describe and distinguish among the three intermolecular forces.
22. State the relationship between the strength of the dipole-dipole force and the molecular polarity in a molecule.
23. State the relationship between the strength of the dispersion forces and the molecular weight of a molecule.
24. Given the Lewis structure of a molecule, predict the existence of hydrogen bonding, dipole-dipole interactions, and dispersion intermolecular forces.

8.11 CONDENSATION OF GASES

25. State the relationship between the boiling point of a liquid and the strength of the intermolecular forces within the liquid.

8.12 A MODEL OF LIQUIDS AND SOLIDS

26. Explain the observable distinguishing properties of liquids and solids by using the molecular model of liquids and solids.

8.13 PHYSICAL PROPERTIES OF LIQUIDS

27. Describe the process by which a constant vapor pressure for a liquid is attained in a closed container.
28. State the relationship between the temperature of a liquid and its vapor pressure.
29. Describe the relationship between the boiling point of a liquid and the external pressure.
30. State the relationships between each of the following properties of liquids and the strength of the intermolecular forces in the liquid: vapor pressure, boiling point, heat of vaporization.
31. Given the heat of vaporization (condensation), calculate the amount of heat needed (released) to vaporize (condense) a given amount of liquid (gas).
32. Explain how the phenomenon of surface tension arises.

8.14 CLASSES OF CRYSTALLINE SOLIDS

33. Name the four classes of crystalline solids and distinguish among the four classes with respect to the nature of the particles occupying their lattice points and the nature of the attractive forces between these particles.
34. State the relationship between the melting point of a solid and the strength of the interparticle attractions within the solid.

8.15 PROPERTIES OF SOLIDS

35. Given the heat of fusion (solidification), calculate the amount of heat needed (released) to melt (solidify) a given amount of solid (liquid).

8.16 HEAT CHANGES AND PHASE CHANGES

36. Given appropriate specific heats and the heats of fusion and vaporization, calculate the amount of heat needed for a specified mass of a substance to undergo a given temperature change.

PROBLEMS

8.11 State three physical properties of gases in general.

8.12 Explain each of the properties you cited in Problem 8.11 on the basis of the Kinetic Theory of Gases.

8.13 Originally 0.96 g of a gas at 20°C occupies 0.50 L. The gas is allowed to expand to fill 12 L at 20°C. Calculate the original and final densities.

8.14 Convert:
a 735 mmHg to atmospheres
b 1.75×10^{-2} atm to millimeters of mercury

8.15 Oxygen and chlorine gas are mixed in a container with partial pressures of 351 mmHg and 0.783 atm, respectively. What is the total pressure inside the container?

8.16 Two mixtures of gases *A* and *B* are separated by a barrier that gases can cross. In the *A* mixture, gas 1 has a partial pressure of 80 mmHg and gas 2 has 138 mmHg. In the *B* mixture, p_1 = 98 mmHg and p_2 = 120 mmHg. Will there be any gas flow between *A* and *B;* if so in which direction? Explain.

8.17 *a* State Boyle's law in terms of a proportionality.
b State Boyle's law in the form of an equation.

8.18 *a* State Charles' law in the form of a proportionality.
b State Charles' law in the form of an equation.

8.19 Give common examples that demonstrate Charles' law and Boyle's Law.

8.20 Convert the following:
a 19°C to Kelvin
b −193°C to Kelvin
c 452 K to degrees Celsius
d 98°F to Kelvin

8.21 *a* A gas is at a temperature of 25°C. Assuming the pressure and mass are held constant, will the volume of the gas double if the temperature is increased to 50°C?
b To what temperature must the gas be heated to double its volume?

8.22 The pressure on a gas is tripled at constant temperature.
a Will the volume increase or decrease?
b By what factor will the volume change?

8.23 A 1.0-L balloon is released at sea level where the pressure is 753 mmHg. Assuming the temperature remains constant, what will be the volume of the balloon at an altitude where the pressure is 351 mmHg?

8.24 At constant pressure, 1.5 L of exhaled gas undergoes a change in temperature from 98°F to 20°C. What is the new volume of the gas?

8.25 A student collects 18.3 mL of argon gas at a temperature of 18°C and a pressure of 758 mmHg. What is the volume of argon at STP?

8.26 A 5.00-L balloon of air is contained within a rocket ship at a temperature of 20°C and 1.00 atm. The balloon is released into space at a temperature of −250°C and 2 mmHg pressure. What is the new volume of the balloon?

8.27 There are fewer air molecules per unit volume at the top of a mountain than at sea level.
a Qualitatively compare the total air pressure and p_{O_2} on the mountain top and at sea level.
b Describe the effect your answers to part *a* have on the breathing and respiration processes.

8.28 Why is it necessary to compare volumes of gas only at the same temperature and pressure?

8.29 Calculate the volume of 5.0 mol of nitrogen gas at STP.

8.30 Calculate the volume of 5.0 mol of nitrogen gas at 745 mmHg and 75°C.

8.31 Calculate the volume of a cylinder that contains 7.65 g of hydrogen gas at STP.

8.32 Calculate the density of nitrogen gas at STP.

8.33 What volume will be occupied by the following weights of each gas at STP:
a 14.5 g of O_2 *b* 6.75 g of HCl
c 539 mg of CO *d* 11.8 g of NH_3

8.34 State the assumptions of the Kinetic Theory of Gases.

8.35 *a* What is meant by an ideal gas?
b Under what conditions of temperature and pressure does a real gas most approach ideal behavior?
c Under what conditions of temperature and pressure does a real gas deviate most from ideal behavior?

8.36 Using the Kinetic Theory of Gases, explain why the pressure in an automobile tire increases after the car has been driven at high speed.

8.37 Using the Kinetic Theory of Gases, explain why an aerosol can should not be heated above the warning temperature listed on the can.

8.38 Using Charles' law, explain why a given gas decreases in density as its temperature rises.

8.39 *a* Does 1.00 mol of a gas always occupy 22.4 L?
b What are the restrictions?

8.40 *a* In which physical state are the attractive forces the weakest?
b Could an ideal gas be condensed to a liquid?

8.41 *a* State the three intermolecular attractive forces.
b Describe the molecular conditions necessary for the existence of each of these forces in molecules.

8.42 State the intermolecular attractive forces present in samples of the following substances:
  a F_2 b H_2O c CH_2Cl_2

8.43 For each example in Problem 8.42 determine which type of attractive force will be most significant in determining the physical properties (such as boiling point) of the sample.

8.44 In which types of molecules do we find dispersion forces?

8.45 Explain why the boiling point of the inert gases increases as we go down the periodic group.

8.46 Which of the following molecules are capable of hydrogen bonding between themselves?

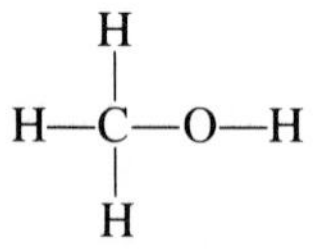

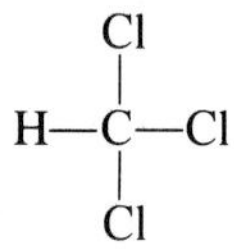

  c NF_3
  d H_2S

  a
$$H-\overset{\displaystyle H}{\underset{\displaystyle H}{C}}-O-H$$

  e 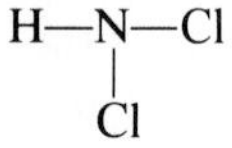
$$H-\overset{}{\underset{\displaystyle Cl}{N}}-Cl$$

  b
$$H-\overset{\displaystyle Cl}{\underset{\displaystyle Cl}{C}}-Cl$$

8.47 Which of the following substances have dipole-dipole attractive forces between their molecules?
  a HBr d $H-C{\equiv}N$
  b $S{=}C{=}S$ e PF_3
  c CF_4 f Br_2

8.48 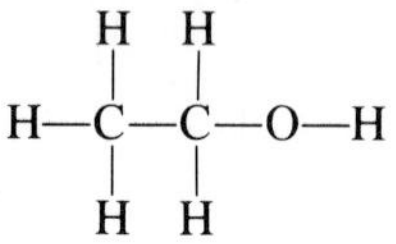
$$H-\overset{\displaystyle H}{\underset{\displaystyle H}{C}}-\overset{\displaystyle H}{\underset{\displaystyle H}{C}}-O-H$$
Ethyl alcohol

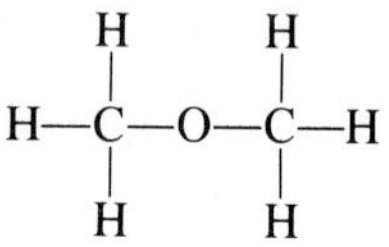

$$H-\overset{\displaystyle H}{\underset{\displaystyle H}{C}}-O-\overset{\displaystyle H}{\underset{\displaystyle H}{C}}-H$$
Methyl ether

Ethyl alcohol and methyl ether have the same molecular weight, but the boiling point of the alcohol (80°C) is considerably higher than that of the ether (-24°C). Explain why this is so.

8.49 Explain why an increase in intermolecular attractions leads to an increase in the boiling point of a liquid.

8.50 Use the models of liquids and solids developed in Figure 8.13 to explain why liquids but not solids can flow.

8.51 Using the model of a liquid developed in Figure 8.13 and the model of a gas developed in Section 8.9, explain why diffusion is much more rapid in a gas than in a liquid.

8.52 Why do liquids with low intermolecular attractive forces have high vapor pressures?

8.53 Why is the process by which perspiration evaporates from your skin a cooling process?

8.54 Indicate which substance in each pair you would expect to have the higher vapor pressure at the same temperature:
  a CH_4 or H_2O

  b
$$H-\overset{\displaystyle H}{\underset{\displaystyle H}{C}}-OH$$
or
$$\overset{\displaystyle H}{\underset{\displaystyle H}{{>}}}C{=}O$$

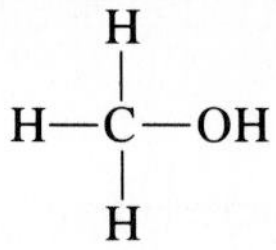

  c
$$H-\overset{\displaystyle H}{\underset{\displaystyle H}{C}}-\overset{\displaystyle H}{\underset{\displaystyle H}{C}}-\overset{\displaystyle H}{\underset{\displaystyle H}{C}}-H$$
or
$$H-\overset{\displaystyle H}{\underset{\displaystyle H}{C}}-\overset{\displaystyle H}{\underset{\displaystyle H}{C}}-\overset{\displaystyle H}{\underset{\displaystyle O}{C}}-H$$

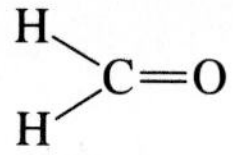

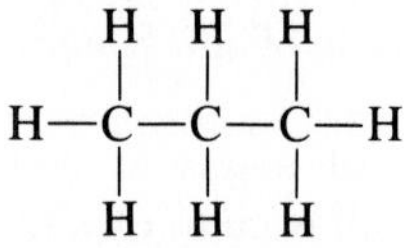

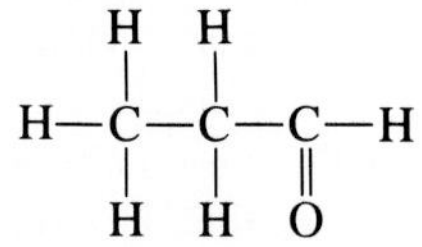

8.55 Why does increasing the temperature of a liquid increase its vapor pressure?

8.56 At 20°C the vapor pressure of substance X is 700 mmHg and that of substance Y is 335 mmHg.
  a Which substance has the higher vapor pressure?
  b Which substance will have the lower normal boiling point?
  c Which substance has the stronger intermolecular forces?

8.57 Describe at least two differences between the processes of evaporation and boiling.

8.58 The vapor pressure of a substance at 25°C is 435 mmHg. If the atmospheric pressure is 758 mmHg, will the substance be a liquid or gas at 25°C?

8.59 a What is meant by the statement that water has a *normal* boiling point of 100°C?
  b Can water be made to boil at temperatures other than 100°C? Explain.

8.60 Liquid A has a molar heat of vaporization of 17.3 kcal/mol. Liquid B has a molar heat of evaporization of 87 kcal/mol.
  a Which liquid has the larger intermolecular forces?
  b Which liquid would you predict to have the lower vapor pressure at 25°C?
  c Which liquid would you predict to have the lower boiling point?

8.61 Predict what effect an increased temperature would have on the surface tension of a liquid.

8.62 Igniting a warmed dish containing alcohol gives a much more vigorous display of flames than igniting the cold dish. Explain.

8.63 Cheese boards typically have covering domes. Why should cheese not be left uncovered?

8.64 Using Table 8.6, calculate how much heat is required to vaporize 11.3 g of H_2O at 100°C.

8.65 Using Table 8.6, calculate how much heat is liberated when 4.9 mol of steam is condensed to liquid water at 100°C.

8.66 Indicate the species that occupy the lattice points for each of the following types of crystalline solids:
  a Ionic
  b Molecular
  c Covalent
  d Metallic

8.67 a Give specific examples of each type of crystalline solid.

b Indicate the particles that would occupy the lattice points for each of your examples.

8.68 Ionic solids and metallic solids are made up of charged particles. However, metallic solids conduct electricity, whereas ionic solids do not conduct electricity in the solid state. Explain.

8.69 Using Table 8.9, calculate the amount of heat liberated by the solidification of 11.4 g of water at 0.0°C.

8.70 When we detect the odor of a solid, what are we actually smelling?

8.71 How many calories are required to convert 50.0 g of solid ethyl alcohol at −163°C to vapor at 79°C? The melting point of the alcohol is −117°C and the boiling point is 79°C. The specific heat of the liquid alcohol is 0.535 cal/(g·°C) and for the solid alcohol 0.232 cal/(g·°C). See Tables 8.6 and 8.9 for heats of vaporization and fusion.

8.72 Use the data in this chapter to calculate the number of kilocalories of heat required to convert 1.45 mol of ice initially at −5.0°C to vapor at 110°C.

CHAPTER NINE

SOLUTIONS

9.1 INTRODUCTION

Up until now we have been discussing *pure substances,* one of the major categories of matter. We will now discuss solutions, probably the most important *mixtures,* the other major category of matter. Indeed the importance of solutions, especially water solutions, cannot be overemphasized. Most chemical reactions in the laboratory are carried out in solution. Most body chemistry occurs in solution. Body fluids are water solutions or colloidal dispersions, both of which will be examined in this chapter.

Solutions are *homogeneous* mixtures of two or more substances. In order for the mixture to be truly homogeneous, i.e., uniform throughout, the particles of the intermixed substances must be ionic or molecular in size. When intermixed particles are larger than this, a true solution cannot be achieved. Rather, we get *colloidal dispersions* or *suspensions* (Section 9.13).

In describing solutions, we identify and give different names to the substance dissolved and the substance doing the dissolving. The substance dissolved is called the **solute.** The substance doing the dissolving is the **solvent.** In general, the solute is the component that is present in lesser amount in the solution. We can envision the solute particles as surrounded by the solvent particles.

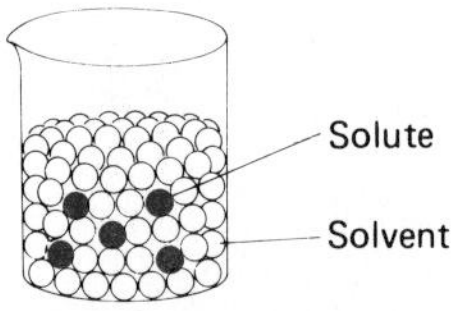
Solute
Solvent

Solute + solvent = solution

Because there are three physical states of matter and each state in principle can be either the solute or solvent in a solution, there are conceivably nine types of solutions. Table 9.1 shows the nine types and gives a common example of each type that actually exists. There are no common examples of solids or liquids truly dissolved in a gas. The most common solutions are those in which the solvent is the liquid water, and those are the solutions which will concern us in this chapter. In discussing mixtures of gases in Chapter 8, we dealt with the other commonly encountered solutions, gases dissolved in gases.

Now let us consider properties of solutions in which the solvent is a liquid.

1 As in all mixtures,
   a The composition is variable (Section 9.8).
   b The solute and solvent may be separated by physical means.

TABLE 9.1 The Nine Solution Types Based on the Physical State of Solute and Solvent, with Examples

	Solvent		
Solute	**Gas**	**Liquid***	**Solid**
Gas	$O_2(g)$ in $N_2(g)$ **Air**	$CO_2(g)$ in $H_2O(l)$ **Soda**	$H_2(g)$ in Pd(*s*) **Hydrogenation catalyst†**
Liquid	No examples exist	Alcohol(*l*) in $H_2O(l)$ **Martini**	Hg(*l*) in Ag(*s*) **Dental fillings** (amalgam)
Solid	No examples exist	NaCl(*s*) in $H_2O(l)$ **Salt water** (saline solution)	Zn(*s*) in Cu(*s*) **Brass‡**

* This section is placed in a box because the most common solutions are those in which the solvent is a liquid.
† This catalyst is used in hydrogenation reactions in organic chemistry (Chapter 13).
‡ All metal alloys like brass are solid-in-solid solutions.

2 Because solutions are homogeneous mixtures,
  a The solute particles must be uniformly distributed among the solvent particles throughout the solution.
  b The solute particles will not settle out.
  c The solution has the same chemical and physical properties in every part.
3 A true solution is transparent, though it may be colored (see Figure 9.1).

The properties cited under items 2 and 3 are a consequence of the small solute particle size. In Section 9.13 you will see that colloids and suspensions have different properties because of larger particle sizes.

9.2 SOLUTION TERMINOLOGY

In addition to the terms *solute* and *solvent,* there are several other words that are commonly used to describe solutions.

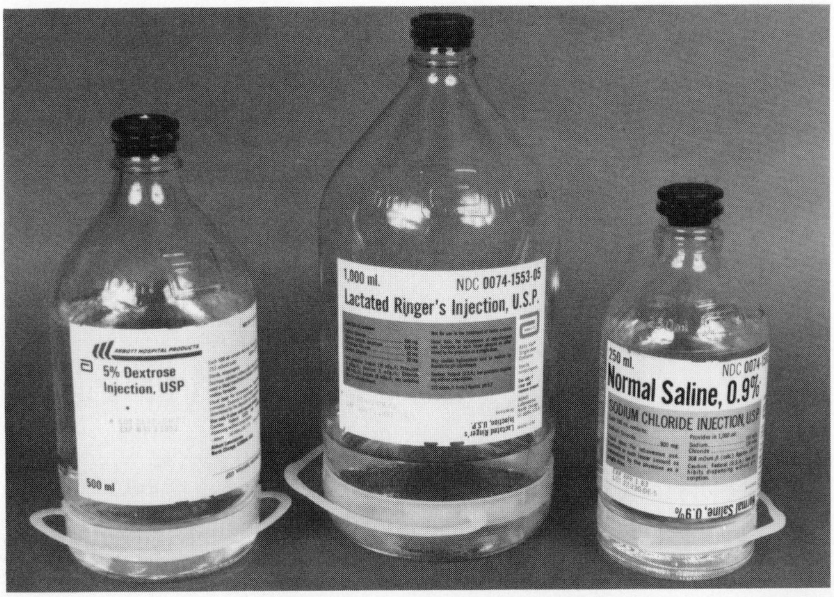
5% Dextrose
Injection, USP
500 ml
1,000 ml.
NDC 0074-1553-05
Lactated Ringer's Injection, U.S.P.
250 ml.
Normal Saline, 0.9%

FIG. 9.1 Some commonly encountered medical solutions displaying the properties described in Section 9.1. (*Photograph by Bryan Lees.*)

Solubility

Solubility is a statement of how much solute can dissolve in some given amount of solvent at a given temperature. A quantitative statement of solubility is a ratio, $\frac{\text{maximum amount of solute}}{\text{amount of solvent}}$, and usually has the units grams of solute per 100 g of solvent. Qualitatively we speak of solutes as **soluble** (some unspecified amount dissolves) or **insoluble** (the solute does *not* dissolve). The modifiers, such as *slightly* or *very* soluble, loosely describe smaller or larger amounts of solute dissolving in a given amount of solvent.

Saturated Solutions

A **saturated solution** is one in which no more solute can dissolve in a given amount of solvent at a given temperature. Table 9.2 lists the solubility limits of some saturated solutions. To prepare a saturated solution of NaCl, one dissolves at least 36.0 g of NaCl in 100 g of water at 20°C. At 20°C, 36.0 g is the maximum amount of NaCl that will dissolve; any excess NaCl will settle to the bottom of the container (see Figure 9.2).

Although a saturated solution containing excess solid appears to the eye to be static (exhibiting lack of activity), the solute particles are actually involved in a dynamic (active) process. The solid solute is dissolving (going into) solution, and at the same time some solute particles are coming out of solution to join the pure solid. For a solution in which water is the solvent, we can represent this process as

$$\text{Undissolved solute } (s) \rightleftharpoons \text{dissolved solute } (aq)$$

where (s) indicates a solid and (aq) means an *aqueous* or water solution. The two half-headed arrows ($\rightleftharpoons$) indicate that the process is reversible and can proceed in either direction. Usually one can actually see any undissolved solute in a saturated solution. Exactly at the solubility limit (for NaCl, 36.0 g of NaCl per 100 g of H_2O at 20°C), the amount of undissolved solute is too small to be seen (Figure 9.2*b*).

Unsaturated Solutions

If the amount of solute dissolved is *less than* the solubility limit, then the solution is **unsaturated.**

TABLE 9.2 Solubility Limits of Some Saturated Solutions

	Solubility, $\frac{\text{max. grams of solute}}{100\text{ g of } H_2O}$	
Solute	**At 20°C**	**At 60°C**
NaCl	36.0	37.3
KBr	65.2	85.5
KNO_3	31.6	110.0
$AgC_2H_3O_2$	1.04	1.89
$K_2Cr_2O_7$	13.1	50.5
$KMnO_4$	6.4	22.2
$AgNO_3$	222	525
$BaSO_4$	0.00023	0.00036

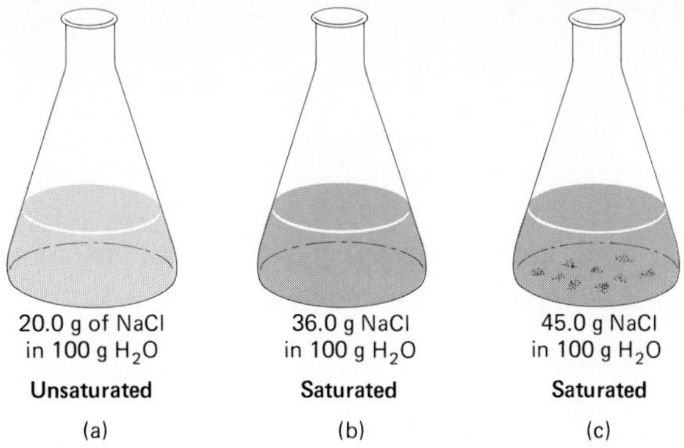
20.0 g of NaCl in 100 g H_2O
Unsaturated
(a)
36.0 g NaCl in 100 g H_2O
Saturated
(b)
45.0 g NaCl in 100 g H_2O
Saturated
(c)

FIG. 9.2
Unsaturated and saturated solutions of sodium chloride at 20°C. (*a*) In this solution 20.0 g is less than the maximum amount of NaCl that can dissolve in 100 g of H_2O at 20°C. (*b*) In this solution 36.0 g is exactly the solubility limit. A small speck of undissolved solid in equilibrium with the dissolved solute may or may not be visible. (c) In this solution 36.0 g of NaCl is in solution. The extra 9.0 g appears as undissolved solid on the bottom of the flask.

Problem 9.1 Consult Table 9.2 and decide whether the following solutions are saturated or unsaturated:

a 15 g of KBr in 100 g of H_2O at 20°C
b 15 g of $KMnO_4$ in 100 g of H_2O at 60°C
c 19 g of NaCl in 50 g of H_2O at 20°C
d 135 g of KNO_3 in 150 g of H_2O at 60°C

Dilute versus Concentrated Solutions

In comparing unsaturated solutions that have identical components, the terms *dilute* and *concentrated* can be used to relate the amounts of solute in a given amount of solvent qualitatively. A **dilute solution** has a comparatively small amount of solute; a **concentrated solution** has a relatively large amount. For example, 1 or 2 g of NaCl in 100 g of water would be a dilute solution, whereas 30 g of NaCl in 100 g of water is concentrated because 30 g is approaching the solubility limit.

Problem 9.2 In comparison with a solution in which 3.0 g of KBr is dissolved in 5.0 g of H_2O, is the solution in Problem 9.1*a* dilute or concentrated?

Miscible and Immiscible

The terms *miscible* and *immiscible* are used in describing mixtures of liquids with liquids. **Miscible** liquids dissolve in each other in all proportions. For example, grain alcohol and water are totally miscible. They dissolve in one another not only on a 50:50 basis but even when one exceeds the other greatly. This is what is meant by "in all proportions." In contrast, oil and water are totally **immiscible;** i.e., they are completely insoluble in one another regardless of the proportions involved.

Hydrophilic versus Hydrophobic

Especially in biological and medical studies the terms **hydrophilic** (water-loving) and **hydrophobic** (water-fearing) are used to describe water-soluble and water-insoluble materials, respectively. The prefix *hydro-* designates "water." The suffix *-philic* means "love," and the suffix *-phobic* denotes "fear."

9.3 SOLUTION FORMATION

You can visualize how a typical ionic compound like NaCl dissolves in water by remembering several ideas that have already been encountered.

1 Ionic compounds are constructed of ions that are held together in orderly arrays in the solid state.
2 Water molecules are freely moving in the liquid state.
3 Water molecules are dipoles with the oxygen end negative and the hydrogen end positive.
4 Cation-anion (positive-negative) attraction is a low-energy (stable) condition (Section 4.2).

Figure 9.3*a* represents solid NaCl just at the moment it has been added to a beaker of water. The positive cations and negative anions are held together by electrostatic attraction.

At the surface of the solid, the moving water molecules can collide with the ions; these collisions can gradually "chip" away the ions from the crystal. The collisions are particularly effective when the negatively charged end of the water dipole collides with a positively charged ion:

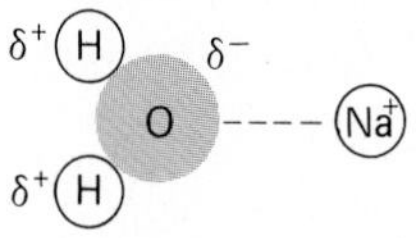

δ+ H
δ−
O
Na+
δ+ H

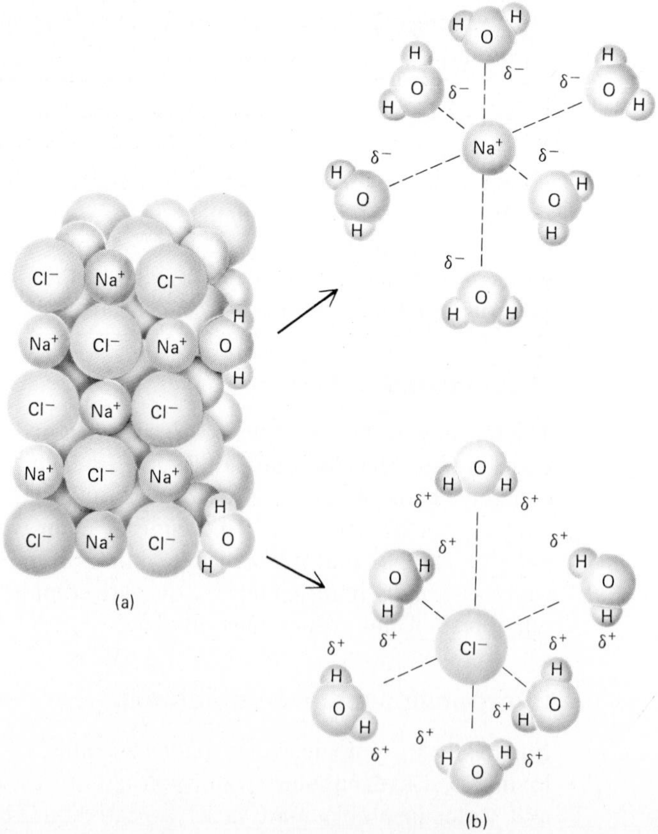

(a)
(b)
Cl−
Na+
H
O
δ−
δ+

FIG. 9.3
Dissolving of an ionic solute such as NaCl in water. (*a*) Solid NaCl is a collection of ions tightly bound together. Water molecules collide with the solid when it is put into water. (*b*) Water molecules have succeeded in "chipping away" ions from the solid and have surrounded the ions.

or the positive end collides with a negative ion:

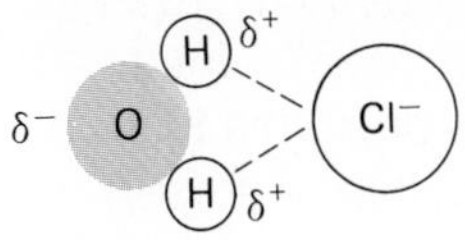
δ⁺
H
δ⁻
O
Cl⁻
H
δ⁺

In these cases the water molecules exert an ion-dipole attraction which helps to pull apart the array of ions.

When the ions are separated from the crystal, the water molecules surround them (see Figure 9.3*b*). Notice that the positively charged ends of the water dipoles point toward Cl^- and the negatively charged ends of the water dipoles point toward the Na^+. This process is known as **hydration** when the solvent is water. Hydration energy is the energy that is released through ion-dipole attractions. **Solvation** is the general name of the process by which solvent particles surround solute particles for any solvent.

The dissolving and hydration process can be represented by a chemical equation of the form

$$NaCl(s) \rightleftharpoons Na^+(aq) + Cl^-(aq)$$

The double-headed arrow is used because the process is reversible (Figure 9.4). As the solution becomes more concentrated, the attraction between the ions overcomes the ion-dipole attraction because there are not enough water molecules for efficient hydration, and some solid comes out of solution.

When an ionic compound dissolves in water, it does so by the process just described. However, different ionic compounds dissolve to different extents. Table 9.2 shows quite large differences in solubility, ranging from the very soluble silver nitrate to the only slightly soluble silver acetate to the practically insoluble barium sulfate. When ionic compounds dissolve only to the very small extent that $BaSO_4$ does (about 2.3×10^{-4} g per 100 g of H_2O), the compound for all practical purposes is classified as insoluble. That is, in the representation of the reversible dissolving process, $BaSO_4(s) \rightleftharpoons Ba^{2+}(aq) + SO_4^{2-}(aq)$, there is almost no tendency for the ions to separate; the process from left to right (→) essentially does not occur.

FIG. 9.4
The dissolving process is reversible: $NaCl(s) \rightleftharpoons Na^+(aq) + Cl^-(aq)$. (*a*) When solid NaCl is first placed in water, only separation of the ions occurs. (*b*) When some ions begin to accumulate in solution, they begin to clump together again into the solid. (*c*) When the rates of the two processes are equal, there is no net change in the number of ions in solution, and the solution is saturated.

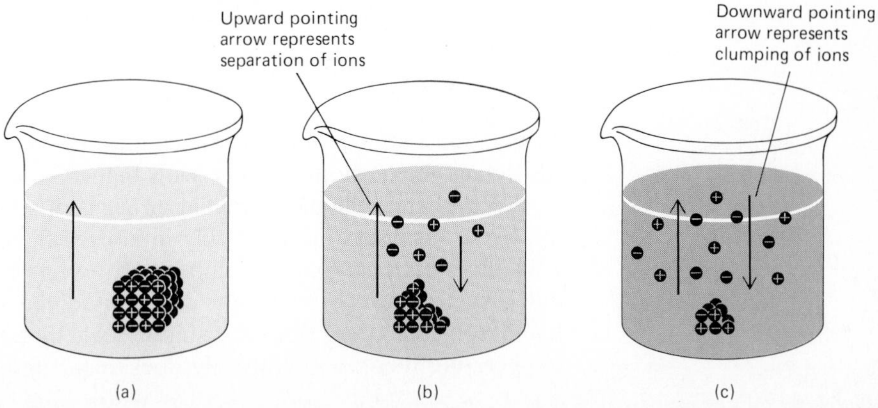
Upward pointing arrow represents separation of ions
Downward pointing arrow represents clumping of ions
(a)
(b)
(c)

It is this property of insolubility that allows the preparation of barium sulfate *suspensions* (Section 9.13), "barium cocktails," which are useful aids to x-ray diagnosis in the gastrointestinal tract. Not only is the opaqueness of the suspended $BaSO_4$ solid a desirable feature, but also Ba^{2+} ions are quite toxic. A solution of barium sulfate, wherein free Ba^{2+} ions were present, would be useless radiographically because it would be clear and the x-rays would pass through unreflected, and it would be toxic.

One of the factors that determines just how soluble an ionic compound will be is the magnitude of its hydration energy compared with the energy associated with the attraction between its ions in the crystalline solid. In many ionic solids, the ions are held together too tightly in the crystal for water molecules to "entice" them away through ion-dipole attractions.

9.4 FACTORS INFLUENCING SOLUBILITY

There are several factors which influence the extent to which a given solute dissolves in a given solvent.

Nature of Solute and Solvent

We have examined the solution process for an ionic compound in water. This process depends on the ionic nature of the solute and the dipole nature of the solvent, which together make interaction and intermingling of particles possible.

Solvation in general involves the solvent molecules "sneaking in between" the solute particles. For this reason, substances which have similar intermolecular forces, either both polar or both nonpolar, tend to dissolve in one another. Substances with dissimilar intermolecular forces do not tend to form solutions. For example, water and grain alcohol are completely miscible because of the similarity of their intermolecular forces. Both are polar and both form hydrogen bonds. Figure 9.5*a* shows how they are able to mix intimately. Figure 9.5*b* shows the hydrogen bonding between the sugar dextrose and water which makes possible the preparation of dextrose solutions which are commonly given intravenously, e.g., in the case of shock.

Gasoline, which can be represented as C_8H_{18}, will not dissolve in either water or alcohol because gasoline is nonpolar. It will, however, dissolve in pentane, C_5H_{12}, or hexane, C_6H_{14}, both of which, like gasoline, display weak dispersion forces (Section 8.10) as their principal intermolecular forces.

This generalization about the way in which similarity of intermolecular forces in solute and solvent leads to solubility is often referred to as the principle of "like dissolves like." Though the principle provides a somewhat useful guideline, the solution process does not depend on relative polarities alone. It is fortunate that it is easy to test for solubility in the laboratory.

Temperature

Temperature affects both the speed and the extent of the solution process. Molecules move more quickly with higher kinetic energies at higher temperatures; hence the intermingling process of solution formation is speeded up. Sugar dissolves more quickly in hot tea than in cold tea.

In the solution process, temperature exerts its effect on the reversible interchange between undissolved and dissolved solute. For solid solutes in liquid solvents, the effect is usually such that the solubility increases as temperature increases. Not only does sugar dissolve *faster* in hot tea than

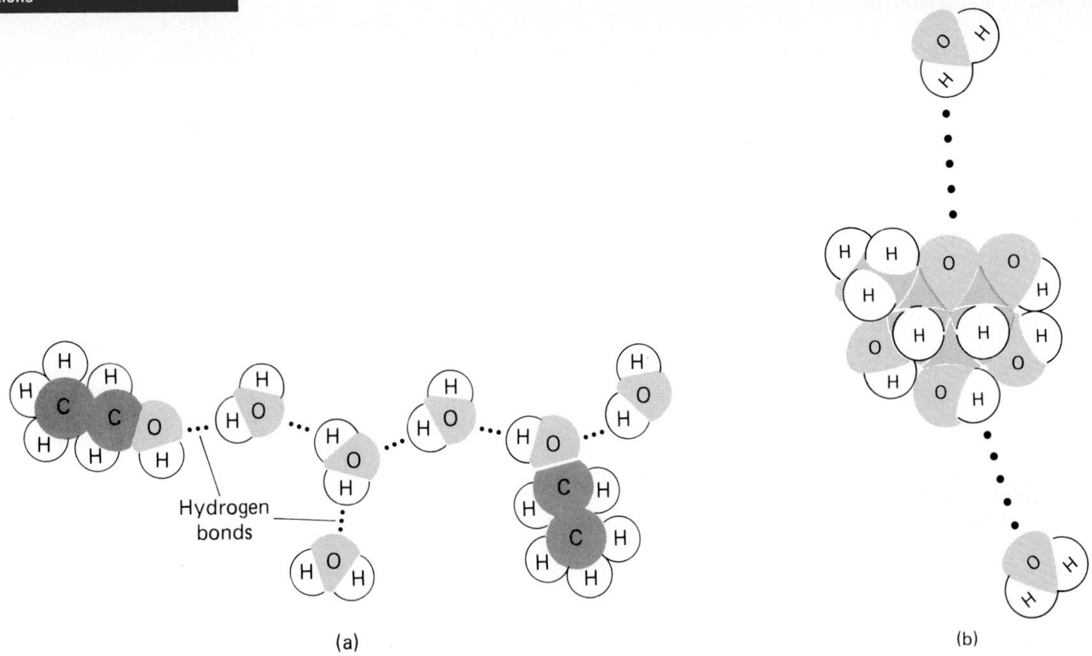
Hydrogen bonds
(a)
(b)

FIG. 9.5
(*a*) Hydrogen bonding between 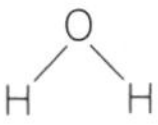 (water) and 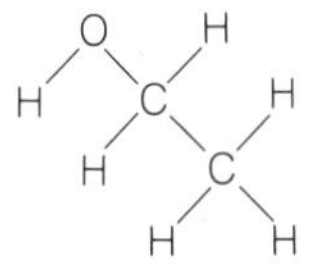 (ethyl alcohol) molecules. Compare this figure with Figure 8.12, which shows hydrogen bonding in pure water. The carbon chain in ethyl alcohol is easily accommodated within the hydrogen-bonding network. (*b*) Hydrogen bonding also leads to water solubility of larger, more complex molecules, such as dextrose.

in cold tea, but also more sugar dissolves in a given amount of hot tea than in the same amount of iced tea.

For gaseous solutes in liquid solvents, increasing temperature decreases solubility. Cold soda and cold beer maintain their fizz (CO_2 gaseous solute) more readily than do warm soda and warm beer.

Pressure

Pressure has almost no effect on the solubility of solids in liquids. The solubility of a gas in a liquid, on the other hand, is directly proportional to the partial pressure of the gas above the solution. "Carbonated" beverages are bottled in such a way that the partial pressure of CO_2 above the solution is greater than 1 atm. This increases the solubility of CO_2 in H_2O. When the bottle is opened, CO_2 gas escapes, thus lowering the partial pressure of CO_2 above the solution. Hence the solubility of CO_2 in the soda is reduced.

The external pressure on the body affects the solubility of gases in the blood. Deep sea divers are subjected to high pressures at low depths, and this increases the amount of nitrogen (from the air they breathe) dissolved in their blood. If they come back to the lower surface pressure too rapidly, the nitrogen gas leaves solution in much the same way CO_2 gas bubbles leave an opened soda. Gas bubbles in the bloodstream can cause pain and even death. The bubbles act as embolisms (clots) and block blood circulation. This condition is know as the "bends."

The effect of pressure on gas solubility is put to good use in **hyperbaric therapy.** Patients are placed in a chamber in which the pressure can be maintained at more than twice atmospheric pressure (Figure 9.6). This chamber can be used to slowly reduce the pressure on divers, thus avoiding the "bends."

Often the hyperbaric chamber is filled with pure oxygen. The increased partial pressure of oxygen leads to greater solubility of oxygen in blood. This is helpful in some cases of hypoxia or oxygen deficiency, such as in carbon monoxide poisoning. Hyperbaric chambers have also been used to treat infections, such as gangrene, caused by anaerobic bacteria. Oxygen is toxic to these bacteria. X-ray radiation of cancer cells is sometimes carried out in a hyperbaric chamber because under high oxygen pressure the difference in sensitivity of cancer and normal cells to the radiation is increased.

Particle Size of Solid Solutes

Because the solution process occurs at the surface of a solid solute, increasing the surface area speeds up the process. Finely pulverized solids offer more surface area per unit mass than big chunks of solid. Particle size affects only the speed of solution formation. The total amount of mass that dissolves is set by other factors and does not depend on whether a fine powder quickly dissolves or a large chunk dissolves slowly.

Stirring

Stirring also affects only the speed of the solution process. It does so by mechanically bringing the solute and solvent into more intimate contact.

9.5 ELECTROLYTES

In Section 9.3 and Figure 9.3 we developed a picture of what a solution of an ionic compound is really like. For example, a solution of NaCl is actually Na^+ ions and Cl^- ions intermingled between (and hydrated by) H_2O molecules. Experimentally, it is found that a solution of an ionic compound conducts electricity. In fact, it is the presence of ions in a solution that allows that solution to conduct electricity in the first place. Solutions that conduct electricity are classified as **electrolyte solutions.**

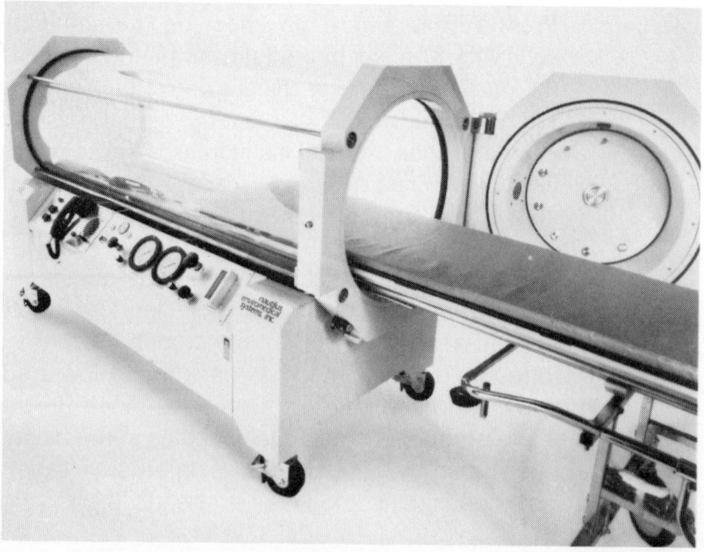

FIG. 9.6
A hyperbaric chamber. The chamber is sealed and the pressure can be increased to over 2 atm. The concentration of oxygen in the air can also be controlled. (*Courtesy of Nautilus Enviromedical Systems, Inc.*)

Electric current is a movement of electric charge. In metal wires electrons carry the charge (see Figure 9.7). If we attach electrodes (metal rods) to the poles of a battery and immerse the electrodes in a solution containing ions, the ions carry the electric charge from one pole to the other to complete the circuit and make the light glow. If there are no ions, no current flows and the light does not glow. **A solution that conducts electricity definitely contains ions.**

Soluble ionic compounds are **strong electrolytes** because they completely *dissociate* in water. **Dissociation** is the name given to the process described in Section 9.3 whereby ions that are closely "associated" in the solid crystal disassociate, or break apart, and become independent in solution. The formula for the ionic compound tells how many ions are present per formula unit or how many moles of ions there are per mole of compound (you might wish to review Section 7.7 at this point). For example,

$$KOH(s) \rightleftharpoons K^+(aq) + OH^-(aq)$$

$$Na_2SO_4(s) \rightleftharpoons 2Na^+(aq) + SO_4^{2-}(aq)$$

$$Al(NO_3)_3 \rightleftharpoons Al^{3+}(aq) + 3NO_3^-(aq)$$

The breaking apart or dissociation of ions is complete (⟶). The tendency to recombine (⟵) can be ignored unless the solubility limit is exceeded.

Most molecular compounds are *non*electrolytes; their solutions do *not* conduct electricity. This is so because molecular compounds are not composed of ions, and thus there can be no dissociation in water. If there are no ions, then the solution is a nonelectrolyte. However, some molecular compounds react with water in such a way as to produce ions. This process is called **ionization.** If by reaction with water a molecular compound produces many ions per mole of compound, it is a *strong* electrolyte. If only a few ions are produced per mole of compound, the compound is a *weak* electrolyte (Figure 9.8).

One particularly important class of molecular compounds that undergoes ionization in water is *acids*. Acids are molecular compounds in which hydrogen is bound to an electronegative element. Their water solutions exhibit certain special properties which will be discussed fully in Chapter 11. The water solutions of acids are electrolytes because acids react in water to form ions, i.e., *acids ionize in water*. For example, in water HCl molecules form H^+ ions and Cl^- ions:

$$HCl(g) \xrightarrow{H_2O} H^+(aq) + Cl^-(aq)$$

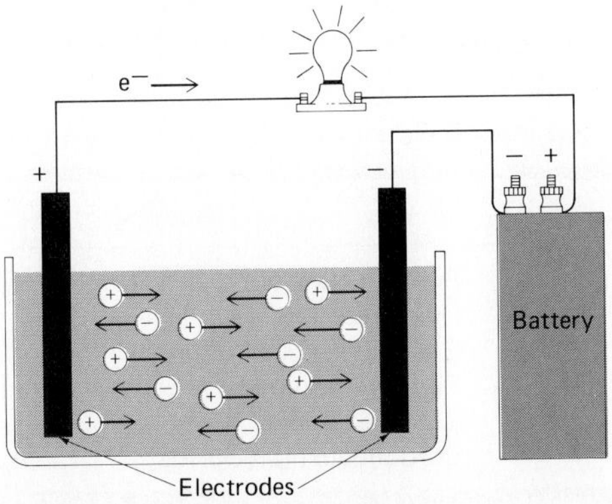
e⁻
+
−
− +
Battery
Electrodes

FIG. 9.7
Ions in solution carry the current (electrons) from one electrode to the other. If there are no ions in solution, then the circuit is open; that is, there is no way for current to span the space between the electrodes.

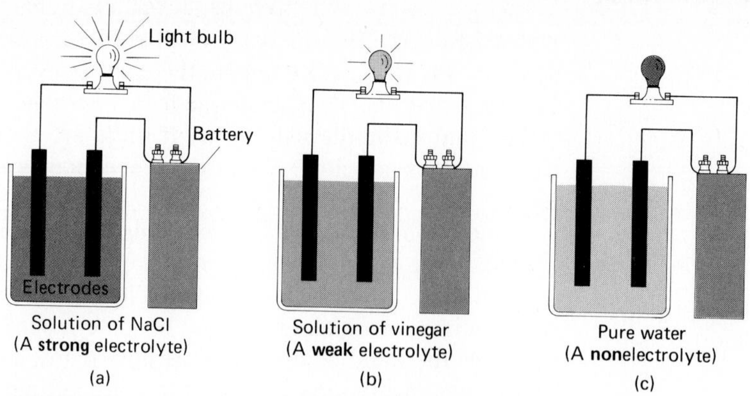
Light bulb
Battery
Electrodes
Solution of NaCl
(A strong electrolyte)
(a)
Solution of vinegar
(A weak electrolyte)
(b)
Pure water
(A nonelectrolyte)
(c)

FIG. 9.8
Strong and weak electrolytes and nonelectrolytes. (*a*) NaCl is a strong electrolyte, as shown by the brightly glowing light bulb. (*b*) Vinegar has only a very small number of ions in solution and hence is a weak electrolyte, as witnessed by the dimly glowing bulb. (*c*) Pure water is a nonelectrolyte, and therefore the light bulb does not glow.

We can ignore the tendency to recombine because it is so small in this case. Thus we say that HCl (hydrochloric acid) ionizes 100 percent. Every molecule in solution ionizes, and HCl therefore is a strong electrolyte. HCl is also termed a **strong acid** because of its 100 percent ionization. There are six common strong acids (strong electrolytes): HCl, HBr, HI, HNO_3, H_2SO_4, and $HClO_4$. All other common acids are weak.

Vinegar is a solution of acetic acid, CH_3COOH. Acetic acid ionizes only to a very limited extent:

$$CH_3COOH \rightleftharpoons H^+(aq) + CH_3COO^-(aq)$$

This representation ($\rightleftharpoons$) indicates that a few molecules of CH_3COOH will form ions in water, but most molecules of CH_3COOH do not react. Because only a small number of ions are produced, CH_3COOH is a weak electrolyte (see Figure 9.8*b*). Most acids are weak electrolytes and thus weak acids. Learn the preceding list of the six strong acids. All others, then, are weak. Aside from acids, almost all other molecular compounds are nonelectrolytes. One notable exception is ammonia, NH_3, which is a weak electrolyte.

9.6 PARTICLES IN SOLUTION

Most chemical reactions are conducted in solution. The reactants are solutes. Products may be solutes, or they may leave the solution if they are either insoluble solids or gases. To paint an accurate picture of a reaction in solution, it is necessary to know the nature of the solute particles as well as the solubilities of the reactants and products. Table 9.3 summarizes the symbols commonly used to show the physical states of reactants and prod-

TABLE 9.3 Symbols Used in Chemical Equations to Denote Physical State

Symbol	Meaning
(*s*)	Solid reactant or product
(*l*)	Liquid reactant or product
(*g*)	Gaseous reactant or product
(*aq*)	Reactant or product in water (aqueous) solution
↑	Gaseous product
↓	Solid product (precipitate)

ucts. You have previously encountered (*s*), (*l*), (*g*), ↑, and (*aq*). The designation ↓, which implies a solid precipitating from (falling out of) solution, is used more often than (*s*) to denote a solid product.

From the preceding discussion about electrolyte solutions we can obtain a summary of the nature of solute particles.

Summary of the Nature of Solute Particles

TYPE 1 The solute particles of *ionic compounds* are *ions*. Therefore, an ionic compound in solution should be represented as the sum of its hydrated ions. For example, NaCl in solution is represented as $Na^+(aq) + Cl^-(aq)$ because the ions are separate and independent and surrounded by water molecules (*aq*).

TYPE 2 The solute particles of *molecular compounds that are nonelectrolytes* are molecules. Sugar is a nonelectrolyte. The solid is composed of molecules $C_{12}H_{22}O_{11}$. A sugar solution is a mixture of water molecules and sugar molecules. The process of dissolving the nonelectrolyte sugar in water can be represented as $C_{12}H_{22}O_{11}(s) \rightarrow C_{12}H_{22}O_{11}(aq)$.

TYPE 3 The solute particles of *molecular compounds that are strong electrolytes* are *ions*. Therefore, such a compound should be represented as the sum of the ions which it forms. HCl, for example, should be expressed as $H^+(aq) + Cl^-(aq)$ if it is in solution.

TYPE 4 The solute particles of *molecular compounds that are weak electrolytes* are mixtures of *molecules* and *ions*. Because the number of molecules is much greater than the number of ions, we usually represent weak electrolytes by their molecular formula.

9.7 IONIC EQUATIONS

Equipped with the knowledge of the nature of solute particles, we can now represent chemical reactions in solution by ionic equations rather than by what we shall call the "traditional" equations that you learned to use in Chapter 7.

For example, a single replacement reaction occurs between magnesium metal and hydrochloric acid to yield magnesium chloride and hydrogen gas. From your experience in Chapter 7 you would write

Traditional equation: $Mg + 2HCl \longrightarrow MgCl_2 + H_2\uparrow$

If you drop Mg metal into HCl solution, bubbles of H_2 gas appear, and a clear solution of $MgCl_2$ forms. The physical picture of this reaction is much better described by an ionic equation in which we are careful to indicate appropriate solute particles for materials in solution. Thus the ionic equation for the reaction would be

Ionic equation:

$$\underbrace{Mg(s)}_{\text{(a)}} + \underbrace{2H^+(aq) + 2Cl^-(aq)}_{\text{(b)}} \longrightarrow \underbrace{Mg^{2+}(aq) + 2Cl^-(aq)}_{\text{(c)}} + \underbrace{H_2\uparrow}_{\text{(d)}}$$

(a) Solid does not dissolve: Mg "disappears" because it reacts.
(b) See Summary, Type 3 in Section 9.6.
(c) See Summary, Type 1.
(d) Gas escapes; it is not in solution.

Notice that this ionic equation also tells us that magnesium and hydrogen were involved in chemical changes, but that chloride ion was unchanged. Mg metal becomes Mg^{2+} ions; H^+ ions become H_2 gas; Cl^- ions remain Cl^- ions. Ions that are unchanged in chemical reactions are called **spectator ions.** Spectators simply "watch" the other ions react.

We can totally represent the chemical changes that occur in the preceding reaction by a so-called net ionic equation, which shows only the reacting species, not spectators.

Net ionic equation: $Mg(s) + 2H^+(aq) \longrightarrow Mg^{2+}(aq) + H_2$

We have previously represented the double-displacement reaction between solutions of NaCl and $AgNO_3$ in the following way:

Traditional equation: $NaCl + AgNO_3 \longrightarrow AgCl + NaNO_3$

In order to write the corresponding ionic equation and show the nature of the solute particles in solution, you need to know the solubility properties of the compounds. In the example, AgCl is insoluble and so comes out of solution (AgCl ↓). The other compounds are soluble.

Ionic equation:

$$\underbrace{Na^+(aq) + Cl^-(aq) + Ag^+(aq) + NO_3^-(aq)}_{\text{See Summary, Type 1.}} \longrightarrow \underbrace{AgCl\downarrow}_{\text{This solid is } not \text{ in solution.}} + \underbrace{Na^+(aq) + NO_3^-(aq)}_{\text{See Summary, Type 1.}}$$

The *net* ionic equation is written by identifying the spectator ions and realizing that because they appear on both sides of the equation they can be canceled out. The spectators are Na^+ ions and NO_3^- ions. Thus,

Net ionic equation: $Cl^-(aq) + Ag^+(aq) \longrightarrow AgCl\downarrow$

In the preceding discussion of the Mg + HCl reaction, the information that the product $MgCl_2$ is soluble, coupled with the recognition of $MgCl_2$ as an ionic compound, suggested the use of Summary, Type 1, to represent it properly. In the other example it was noted that the AgCl was insoluble. If you are not told whether products are soluble, you must consult tables of solubility data to decide whether or not a given ionic compound is soluble.

9.8 CONCENTRATION EXPRESSIONS

Because they are mixtures, solutions have a variable composition. When we state what the composition of a solution is, we call that statement the concentration of the solution. The **concentration** of a solution is the amount of solute that is present in some given amount of solvent or given amount of solution.

Concentration is a ratio of the form:

$$\frac{\text{Amount of solute}}{\text{Amount of solvent}} \quad \text{or more often} \quad \frac{\text{amount of solute}}{\text{amount of solution}}$$

Previously (Section 9.2), we have seen solubility given as a ratio. Notice the difference between the solubility, which is the "maximum" amount of solute per amount of solvent at a given temperature, and the concentration, in which the amount of solute can take on any value up to the solubility limit. Also, the most often used concentration expressions are those in which the amount of solute is given per amount of *solution* rather than per amount of *solvent*.

The various concentration expressions differ in the *units* used for the amounts of solute and solution. The molarity expression used most often by chemists employs the units of moles for the solute and liters for solution. In percentage expressions the units for the solute and solution may be either grams (weight) or milliliters (volume). This gives rise to three possible percentage expressions: percentage weight per weight (% w/w), percentage volume per volume (% v/v), and percentage weight per volume (% w/v).

Percentage Weight/Volume (% w/v)

Weight/volume percentage is the one used most often in medicine because the most common solutions are those in which the solute is a solid, most easily measured in grams, and the solution is a liquid, most easily measured in volume units, e.g., milliliters.

The numerical value expressed in a weight per volume percent is equal to the number of grams of solute present in 100 mL of solution. For example, a 10% w/v dextrose solution has 10 g of dextrose per 100 mL of solution. A 0.9% w/v salt solution (physiological saline) contains 0.9 g of salt per 100 mL of solution. This value can be readily calculated through the equation

$$\% \text{ w/v of solute} = \frac{\text{grams of solute}}{\text{milliliters of solution}} \times 100\% \tag{9.1}$$

SAMPLE EXERCISE 9.1 What is the percent weight/volume of KOH in 60 mL of a solution which contains 5.0 g of KOH?

Solution Use Equation (9.1):

$$\% \text{ w/v of KOH} = \frac{5.0 \text{ g of KOH}}{60 \text{ mL of solution}} \times 100\%$$

$$= 8.3\% \text{ w/v}$$

There is 8.3 g of KOH per 100 mL of solution.

In many problems it is convenient to recognize and use as a conversion factor the ratio of grams of solute per 100 mL of solution which the percent weight/volume gives you. That is, every % w/v value gives you a conversion factor relating grams of solute to milliliters of solution, e.g.,

$$\frac{5 \text{ g of solute}}{100 \text{ mL of solution}} \quad \text{for } 5\% \text{ w/v}$$

SAMPLE EXERCISE 9.2 Physiological saline solution is a 0.900% w/v NaCl solution used for storing red blood cells as well as for other medical purposes. How much salt (in grams) is needed to prepare 250 mL of physiological saline solution?

Solution Use the Unit Conversion Method.

STEP 1 The *given* is 250 mL of solution.

STEP 2 The *new* unit is grams of NaCl.

STEP 3 The percent weight/volume provides the conversion factor 0.9 g of NaCl/100 mL of solution.

STEP 4 Set up the usual format:

Given × conversion factor = new quantity and unit

$$250\ \cancel{\text{mL of soln}} \times \frac{0.9\ \text{g of NaCl}}{100\ \cancel{\text{mL of soln}}} = 2.25\ \text{g of NaCl}$$

Problem 9.3 How many milliliters of 7.0% w/v glucose solution would contain 49 g of glucose?

A closely related concentration expression used in clinical laboratories is the **milligram percent** (mg %). The statement that the concentration of glucose in blood serum is 75 mg % indicates that there are 75 mg of glucose per 100 mL of serum. In general, the numerical value of mg % equals the number of milligrams of solute present in 100 mL of solution.

Percentage Volume/Volume (% v/v)

This expression is encountered only for solutions which are mixtures of liquids. For example, a water solution of alcohol might be labeled 45% v/v. This indicates 45 mL of alcohol per 100 mL of solution. In general, the numerical value of percent volume/volume equals the number of milliliters of solute present in 100 mL of solution.

Percentage Weight/Weight (% w/w)

This expression is rarely encountered. The numerical value of percent weight per weight equals the number of grams of solute present in 100 g of solution.

Parts per Million (ppm)

To understand the expression *parts per million,* it is useful to realize that a statement of percentage is actually a statement of parts per hundred. A 5% solution indicates 5 parts of solute per 100 parts of solution. A solution which has the concentration 5 ppm contains 5 parts of solute per 1 million parts of solution.

The parts per million expression is most useful for very dilute solutions wherein percentage expressions would involve very small numbers. For example, drinking water typically contains only 1 mg of fluoride ion (F^-) per liter of solution. In percent weight/volume this would be

$$\frac{0.001\ \text{g of F}^-}{1000\ \text{mL of soln}} \times 100\% = 0.0001\ \%\ \text{w/v}$$

This concentration in parts per million is 1 ppm:

$$\frac{0.001\ \text{g of F}^-}{1000\ \text{mL of soln}} = \frac{1\ \text{g of F}^-}{1{,}000{,}000\ \text{mL of soln}} = \frac{1\ \text{part}}{1\ \text{million parts}} = 1\ \text{ppm}$$

(Multiply the left-hand term by 1000/1000 = 1 to get the right-hand term.)

For even more dilute solutions the expression *parts per billion* (ppb) is employed.

Molarity

Molarity is the concentration expression used most frequently by chemists because moles are the units of matter most conveniently manipulated in the laboratory in doing chemical reactions (see Chapter 7). Of course, moles of material are always readily related to grams of material through the molar weight (see Section 7.6). **Molarity** (abbreviated *M*) is defined as moles of solute per liter of solution:

$$\text{Molarity } (M) = \frac{\text{moles of solute}}{\text{liters of } \textit{solution}}$$

Whenever one is told the number of moles of solute or the amount of solute in grams that is dissolved in a specified volume of solution, **the molarity can be calculated by dividing the number of moles of solute by the numbers of liters of solution.**

SAMPLE EXERCISE 9.3

Determine the molarity of the following solutions:

a 4 mol of NaCl is dissolved in enough water to make 2 L of solution.

b 58.8 g of KOH is dissolved in enough water to make 700.0 mL of solution.

Solution

a Because the units given are moles of solute and liters of solution, the problem is completely straightforward and we can immediately divide.

$$\frac{\text{4 mol of NaCl}}{\text{2 L of soln}} = \frac{\text{2 mol}}{\text{1 L}} = \text{molarity}$$

b In this case grams of solute must be converted to moles of solute and milliliters of solution to liters of solution before we divide.

$$58.8\ \cancel{\text{g of KOH}} \times \frac{\text{1 mol of KOH}}{56.0\ \cancel{\text{g of KOH}}} = \text{1.05 mol of KOH}$$

$$700.0\ \cancel{\text{mL}} \times \frac{\text{1 L}}{1000\ \cancel{\text{mL}}} = \text{0.7000 L}$$

$$\frac{\text{1.05 mol of KOH}}{\text{0.7000 L}} = \frac{\text{1.50 mol}}{\text{1 L}} = \text{molarity}$$

Note carefully the distinction between moles and molarity. The number of moles of material is an absolute amount and can be directly converted to grams. Molarity is a ratio which tells the number of moles distributed through a volume of solution. It is convenient to recognize that the total number of moles in a given solution can be obtained by multiplying $M \times V$ (in liters). This relationship comes about by rearranging the equation by which molarity is calculated:

$$M = \frac{\text{moles of solute}}{\text{volume (in liters)}} \qquad (9.2)$$

Multiplying each side of the equation by V yields

$$M \times V \text{ (in liters)} = \text{moles of solute} \quad (9.3)$$

We speak of 1 M, 2 M, 6 M solutions (read "1 molar," "2 molar," etc.). The statement of molarity provides a conversion factor between moles of solute and liters of solution. For example,

Molarity:	1 M	2 M	6 M
Conversion factor:	$\frac{1\ \text{mol}}{1\ \text{L}}$	$\frac{2\ \text{mol}}{1\ \text{L}}$	$\frac{6\ \text{mol}}{1\ \text{L}}$

Of course, the reciprocals are also conversion factors. Recognizing the molarity as a conversion factor enables us to handle many different types of concentration problems with the Unit Conversion Method.

SAMPLE EXERCISE 9.4

How many moles of KBr are there in 35.8 mL of a 0.172 M solution?

Solution Use the Unit Conversion Method, with the molarity as a conversion factor.

STEP 1 The *given* is 35.8 mL of solution (0.172 M is also given).

STEP 2 The *new* is moles of KBr.

STEP 3 The necessary conversion factors are

$$\frac{1\ \text{L}}{1000\ \text{mL}}$$

(to use M, the volume of the solution must be in liters) and

$$\frac{0.172\ \text{mol of KBr}}{1\ \text{L of soln}}$$

STEP 4 Set up the usual format:

Given × conversion factors = new quantity and unit

$$35.8\ \cancel{\text{mL}} \times \frac{1\ \cancel{\text{L}}}{1000\ \cancel{\text{mL}}} \times \frac{0.172\ \text{mol of KBr}}{1\ \cancel{\text{L}}} = 0.00616\ \text{mol of KBr}$$

This problem can also be solved by remembering that

$$M \times V \text{ (in liters)} = \text{moles of solute}$$

The Unit Conversion Method is stressed because it has more diverse applications than the $M \times V$ formula.

Of course, conversion factors of the type

$$\frac{\text{Molar weight}}{1\ \text{mol}}$$

continue to be important and are often used in conjunction with the molarity conversion factor.

SAMPLE EXERCISE 9.5 How many grams of NaOH are there in 2.50 L of 0.343 *M* solution?

Solution STEP 1 The *given* is 2.50 L of solution (0.343 *M* is also given).

STEP 2 The *new* is grams of NaOH.

STEP 3 The conversion factors are

$$\frac{0.343 \text{ mol of NaOH}}{1 \text{ L}} \quad \text{and} \quad \frac{40.0 \text{ g of NaOH}}{1 \text{ mol of NaOH}}$$

STEP 4

$$2.50 \cancel{\text{L}} \times \frac{0.343 \cancel{\text{mol of NaOH}}}{1 \cancel{\text{L}}} \times \frac{40.0 \text{ g of NaOH}}{1 \cancel{\text{mol of NaOH}}} = 34.3 \text{ g of NaOH}$$

For solutions in which the concentration of solute is very low, the concentration may be expressed more conveniently as a millimolar solution. For example, a solution which contains 40 mg of NaOH per liter of solution contains 1 mmol per liter. Therefore, it is more conveniently described as 1 m*M* than as 0.001 *M*.

$$40 \cancel{\text{mg of NaOH}} \times \frac{1 \cancel{\text{g}}}{1000 \cancel{\text{mg}}} \times \frac{1 \cancel{\text{mol of NaOH}}}{40 \cancel{\text{g of NaOH}}} \times \frac{1000 \text{ mmol}}{1 \cancel{\text{mol}}} = 1 \text{ mmol of NaOH}$$

or

$$40 \cancel{\text{mg of NaOH}} \times \frac{1 \text{ mmol of NaOH}}{40 \cancel{\text{mg of NaOH}}} = 1 \text{ mmol of NaOH}$$

9.9 DILUTION

Molar solutions of some specified concentration can be prepared directly by dissolving the appropriate amount of solute in the proper volume of solution. Another method of preparing solutions of some desired molarity is through the process of dilution. **Dilution** involves mixing a solvent (usually water) with a concentrated solution. The additional water produces a more dilute solution because the same amount of solute is distributed through a larger amount of solvent or solution. For example, if 1 L of water is mixed with 1 L of a 6 *M* solution, the solution becomes a 3 *M* solution:[1]

Original *Final*

$$6\,M = \frac{6 \text{ mol}}{1 \text{ L of soln}} + 1 \text{ L of water} \longrightarrow \frac{6 \text{ mol}}{2 \text{ L of soln}} = 3\,M$$

Calculations involving dilution problems center on the fact that the same amount of solute is present in both the original solution and the final diluted solution.

$$\textbf{Moles solute}_{\textbf{original solution}} = \textbf{moles solute}_{\textbf{diluted solution}} \qquad (9.4)$$

[1]Combining 1 L of solution with 1 L of water does not produce exactly 2 L of solution. However, this assumption is sufficiently accurate to 2 significant figures.

Furthermore, recall from Equation (9.3) that the number of moles of solute present in any solution can always be determined by multiplying $M \times V$ (in liters):

$$M \times V = \text{moles of solute}$$

$$\frac{\text{moles of solute}}{1\ \cancel{\text{L of soln}}} \times 1\ \cancel{\text{L of soln}} = \text{moles of solute}$$

Therefore, we can rewrite Equation (9.4) as

$$M_o \times V_o = M_d \times V_d \qquad (9.5)$$

where *o* stands for the original solution and *d* stands for the diluted solution.

SAMPLE EXERCISE 9.6

What is the molarity of the solution prepared by mixing 2 L of water with 1.5 L of 0.5 M KOH solution?

Solution Whenever water (solvent) is added to a solution, the problem should be recognized as a dilution problem. Then use Equation (9.5).

Original Solution		Diluted Solution
$M_o = 0.5\ M$		$M_d = ?$
$V_o = 1.5$ L		$V_d = 1.5\text{ L} + 2\text{ L} = 3.5$ L
$M_o \times V_o$	=	$M_d \times V_d$
$0.5\ M \times 1.5$ L	=	$M_d \times 3.5$ L

Divide each side of the equation by 3.5 L to isolate M_d.

$$M_d = \frac{0.5\ M \times 1.5\ \cancel{L}}{3.5\ \cancel{L}}$$

$$M_d = 0.2\ M$$

[Because volume appears on both sides of Equation (9.5), the units of volume must be consistent, but they do not necessarily have to be liters. In this case, you could have used 1500 mL and 3500 mL and obtained the same result. However, one cannot use liters on one side and milliliters on the other.]

SAMPLE EXERCISE 9.7

What volume of a 0.34 M $MgCl_2$ solution is required to make 450 mL of a 0.10 M solution by dilution?

Solution The word *dilution* signals the use of Equation (9.5).

Original Solution		Diluted Solution
$M_o = 0.34\ M$		$M_d = 0.10\ M$
$V_o = ?$		$V_d = 450$ mL
$M_o \times V_o$	=	$M_d \times V_d$
$0.34\ M \times V_o$	=	$0.10\ M \times 450$ mL

Divide each side by 0.34 M to isolate V_o.

$$V_o = \frac{0.10\ \cancel{M} \times 450\ \text{mL}}{0.34\ \cancel{M}}$$

$$V_o = 130\ \text{mL}$$

Problem 9.4 In a laboratory, 500. mL of water is mixed with 250 mL of 0.75 M NaOH solution. What is the molarity of the resulting solution?

9.10 EQUIVALENTS AND MILLIEQUIVALENTS

Moles and millimoles are units for "packages" of particles (Section 7.2). For example, 1 mmol of Ca^{2+} is a package of 6.02×10^{20} calcium ions. Molar units do not take into account the charges on ions. Because in physiological fluids it is often necessary to know about the charge content, the package units called **equivalents** and **milliequivalents** have been defined in medicinal chemistry. Equivalents represent the total units of charge contributed by a collection of ions, because the number of equivalents in a sample is calculated by multiplying the number of moles of ions times the magnitude of the charge on the ion.[2] The sign of the charge is disregarded.

$$\text{Number of equivalents of an ion} = \text{number of moles} \times \text{magnitude of charge on the ion}$$

$$\text{Milliequivalents} = \text{millimoles} \times \text{ionic charge}$$

Later on in this chapter, Table 9.6 records the ionic concentrations in bodily fluids in the units milligrams per liter, millimoles per liter, and milliequivalents per liter. For example, the plasma concentration of Ca^{2+} is given as 100 mg/L, 2.5 mmol/L, and 5 meq/L. Given milligrams per liter, you can convert to millimoles per liter by using the molar weight as a conversion factor:

$$100\ \cancel{\text{mg of Ca}^{2+}} \times \frac{1\ \text{mmol of Ca}^{2+}}{40\ \cancel{\text{mg of Ca}^{2+}}} = 2.5\ \text{mmol of Ca}^{2+}$$

To calculate the number of milliequivalents, simply multiply by the magnitude of the ionic charge:

$$\text{Millimoles} \times \text{magnitude of ionic charge} = \text{milliequivalents}$$

$$2.5\ \text{mmol of Ca}^{2+} \times 2 = 5\ \text{meq of Ca}^{2+}$$

SAMPLE EXERCISE 9.8 What is the intracellular concentration of magnesium ion in milliequivalents per liter if the Mg^{2+} is 42.5 mg % w/v?

Solution The given concentration tells us that there are 42.5 mg in 100 mL of solution. This is the definition of milligram percent weight/volume. Using this ratio as a conversion

[2]This definition of equivalents in terms of ionic charge is not the actual definition. Equivalency is actually based on the number of electrons transferred in reactions of sample, but because number of electrons transferred is related to ionic charge, this simple rule-of-thumb definition works.

factor enables the determination of milligrams per liter.

$$\frac{1000\ \cancel{\text{mL}}}{1\ \text{L}} \times \frac{42.5\ \text{mg of Mg}^{2+}}{100\ \cancel{\text{mL}}} = \frac{425\ \text{mg of Mg}^{2+}}{1\ \text{L}}$$

Then,

$$425\ \cancel{\text{mg of Mg}^{2+}} \times \frac{1\ \text{mmol of Mg}^{2+}}{24.3\ \cancel{\text{mg of Mg}^{2+}}} = 17.5\ \text{mmol of Mg}^{2+}$$

$$\text{meq} = \text{mmoles} \times \text{magnitude of charge}$$

$$\text{meq of Mg}^{2+} = 17.5\ \text{mmol of Mg}^{2+} \times 2$$

$$= 35\ \text{meq of Mg}^{2+}$$

The concentration of Mg^{2+} is 35 meq/L.

9.11 COLLIGATIVE PROPERTIES OF SOLUTIONS

Section 8.13 discussed the physical properties of liquids. The physical properties of liquids are altered in predictable ways by the presence of dissolved solutes. Thus, for example, boiling points of solutions are typically higher than boiling points of the pure solvent, and solution freezing points are typically lower. Both of these effects stem from the fact that the vapor pressure of a solution is lower than that of a pure solvent. Furthermore, solutions demonstrate a property, osmotic pressure, which is not present in a pure liquid. The numerical values of the vapor pressure lowering, the boiling point elevation, the freezing point depression, and the osmotic pressure depend *only* on the concentration of solute particles in solution, and not on the identity of the solute particles.

Such properties, which do not depend on the nature of the dissolved species, but only on the *number of dissolved particles*, are called **colligative properties.** Glance back at Section 9.6 to remind yourself that ionic compounds and some ionizable molecular compounds produce more particles in solution than do molecular nonelectrolytes.

Vapor Pressure of a Solution

Recall from Section 8.13 and Figure 8.15 that vapor pressure increases with temperature. Figure 9.9 shows the typical variation of the vapor pressure of a pure solvent and a solution. Notice that the vapor pressure of the solution is always *lower* than that of the pure solvent.

The lowering of vapor pressure by the presence of solute particles is readily explained by remembering that the development of vapor pressure depends on the ability of molecules to escape from the surface of a liquid. As Figure 9.10 shows, in a solution solute particles literally get in the way of solvent molecules and interfere with their escape. The greater the number of solute particles, the greater the interference effect, and therefore, the greater is the lowering of vapor pressure.

Boiling Point Elevation of a Solution

Lowering of vapor pressure leads directly to the elevation of the normal boiling point of a solution compared with that of the pure solvent. Recall that the normal boiling point is defined as the temperature at which the vapor pressure is equal to an external pressure of 760 mmHg. Because the vapor pressure of a solution is lower at all temperatures than that of the pure

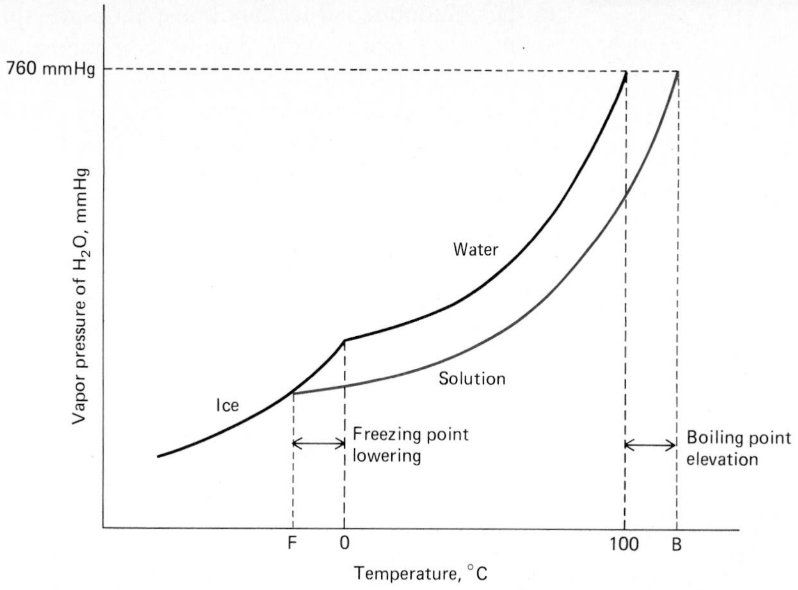
760 mmHg
Vapor pressure of H_2O, mmHg
Water
Solution
Ice
Freezing point lowering
Boiling point elevation
F
0
100
B
Temperature, °C

FIG. 9.9
The black line shows the change in vapor pressure of pure water as temperature varies. The color line plots vapor pressure of a water *solution* versus temperature. Vapor pressure of the solution is lower than that for the pure solvent at every temperature. Because the normal boiling temperature is that at which vapor pressure equals 760 mmHg, solutions boil at higher temperatures; the intersection of 760 mmHg and the solution curve occurs at a higher temperature (*B* in the figure). Freezing points are lowered because the solution curve and vapor pressure curve for ice intersect at a lower temperature (*F* in the figure).

solvent, a higher temperature is required to reach a vapor pressure of 760 mmHg. This idea is shown graphically in Figure 9.9.

Adding salt to a pot of boiling water leads to a slightly increased boiling temperature because the salt lowers the vapor pressure of the solution and a higher temperature is necessary in order to reestablish the vapor pressure

FIG. 9.10
(*a*) In the pure solvent every space on the surface is occupied by a solvent molecule. (*b*) In a solution some spaces are occupied by solute particles. Thus there is less surface area from which solvent molecules can escape, and vapor pressure is lowered.

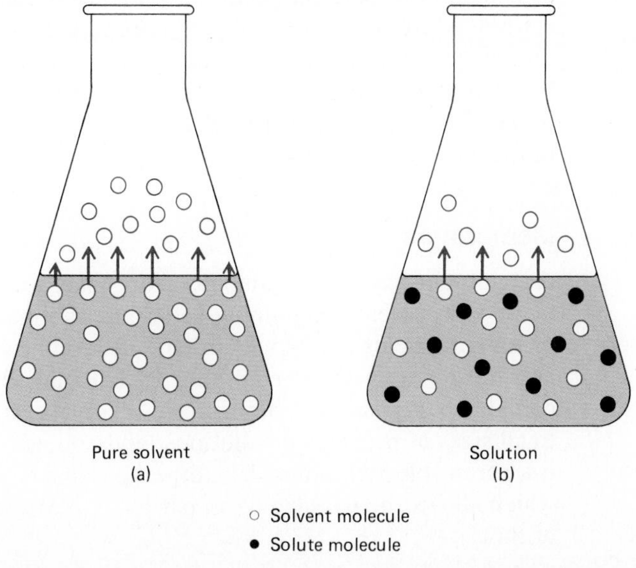
Pure solvent
(a)
Solution
(b)
Solvent molecule
Solute molecule

of the solution equal to the atmospheric pressure. Adding more salt to the solution will increase the boiling temperature still further. Mole for mole, salt has twice the effect on boiling point as sugar because sugar is a molecular solid and produces 1 mol of sugar molecules in solution, whereas salt produces 2 mol of ions (1 mol of Na^+ and 1 mol of Cl^-) per mole of compound. Variations in colligative properties depend on numbers of particles.

Freezing Point Depression of a Solution

A liquid freezes at the temperature at which the vapor pressure of the liquid equals the vapor pressure of the solid state of the substance. For water the vapor pressure of liquid water and the vapor pressure of ice are equal at 0°C. Figure 9.9 shows that the lowering of the vapor pressure curve for a liquid solution results in the intersection of that curve with the solid state curve (for pure solvent) at a lower temperature. This means that solutions freeze at lower temperatures than pure solvents.

This colligative property of freezing point depression is put to good practical uses in cold winter weather. Pure water in an automobile radiator would freeze in winter and damage the engine. "Antifreeze" is simply a solution. The solute (usually ethylene glycol, $HOCH_2CH_2OH$) depresses the freezing point of the solution to temperatures far below 0°C.

The same principle is employed to melt ice on roadways. Remember melting and freezing occur at the same temperature. Whereas pure water freezes or ice melts at 0°C, a solution undergoes these transitions at lower temperatures. Thus "salt" sprinkled on icy roads forms a solution with the ice that melts at a lower temperature. The principle that the magnitude of effects on colligative properties depends on numbers of particles is employed by using $CaCl_2$ rather than NaCl. $CaCl_2$ produces 3 mol of particles (1Ca^{2+} and 2Cl^-) per mole of compound, while NaCl forms 2 mol. Thus $CaCl_2$ is $1\frac{1}{2}$ times more effective at melting ice on streets and roadways than NaCl.

9.12 OSMOTIC PRESSURE OF SOLUTIONS

Osmotic pressure is a property that you may have not encountered previously. Because of this probable unfamiliarity and because the development of osmotic pressure has many physiological consequences, we will devote a whole section to this colligative property. Osmotic pressure depends on the phenomenon of *osmosis* and the more general concept of particle flow through *membranes,* so let us begin with the question, "What is a membrane?" Membranes surround cells and are of vital importance to bodily function.

Membranes

A membrane can be thought of as a thin sheet with holes in it. Placed between two compartments containing solutions, the membrane allows the passage of some particles and blocks others, depending on the sizes of the particles and the holes in the membrane. (See Figure 9.11.) A membrane can be completely *permeable,* that is, have relatively large holes and allow passage to all sizes of particles in solutions (and colloids, Section 9.13). Filter paper is a permeable membrane. The other extreme is an *impermeable* membrane, which allows the passage of no particles. A stretched balloon is an example of this.

FIG. 9.11
A semipermeable membrane can separate particles based on size. The semipermeable membrane shown is an osmotic membrane. Its "holes" allow the passage of H_2O molecules in both directions, but the membrane prevents the passage of glucose molecules, Na^+ ions, and Cl^- ions, which are larger than water molecules.

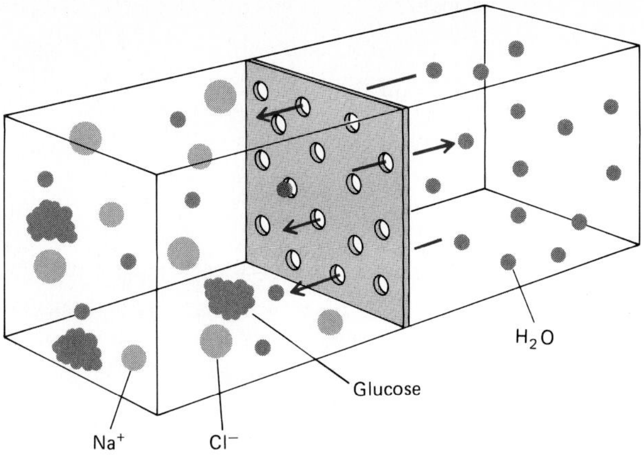
H2O
Glucose
Na+
Cl−

Biological membranes are *semipermeable;* that is, they allow the passage of some small particles, but hold back larger ones. Semipermeable biological membranes are categorized as osmotic or dialyzing membranes. This is a subtle distinction that depends only on the size of the particles which they allow to pass through their pores. The consequences of semipermeability are the same for both osmotic and dialyzing membranes. **Osmotic membranes** allow the passage of *water* and other small molecules, the dimensions of which are not significantly larger than water, e.g., dissolved gases such as CO_2, O_2, and N_2, and some small organic molecules.

Osmosis

Osmosis is defined as the flow of water across an osmotic membrane from a more dilute solution (or pure water) into a more concentrated solution. This represents flow from a region where there are relatively more water molecules to where there are fewer. It is important to realize that water flows across the osmotic membrane in both directions, as shown in Figure 9.11. However, there is a *net* flow of water from a solution with a larger relative number of water molecules (dilute solution or pure solvent) to one with relatively fewer water molecules (concentrated solution).

Once again we can use the idea of solute particles "getting in the way" to explain this net flow of water. Figure 9.12*a* shows solute particles blocking some water molecules from passing through a semipermeable membrane. This blockage is less effective in a dilute solution than in a concentrated solution because there are fewer solute particles in a dilute solution. The net flow of water from the dilute to the concentrated chamber increases the volume of the more concentrated solution, as shown in Figure 9.12*b*. The net movement of water cannot continue indefinitely because the increased height of the more concentrated solution causes a downward pressure from the added weight of liquid, which eventually balances the opposing tendency of water to flow into the more concentrated solution. When this balance of forces occurs, there is no further net movement of water. At this point the concentrated solution has become more dilute and the dilute solution more concentrated. Another way of saying this is that the concentration gradient has been diminished. A **concentration gradient** is a *difference* in concentration between solutions. The greater the difference or gradient, the greater the potential for a net flow of water in the direction dilute (more water) $\longrightarrow$ concentrated (less water).

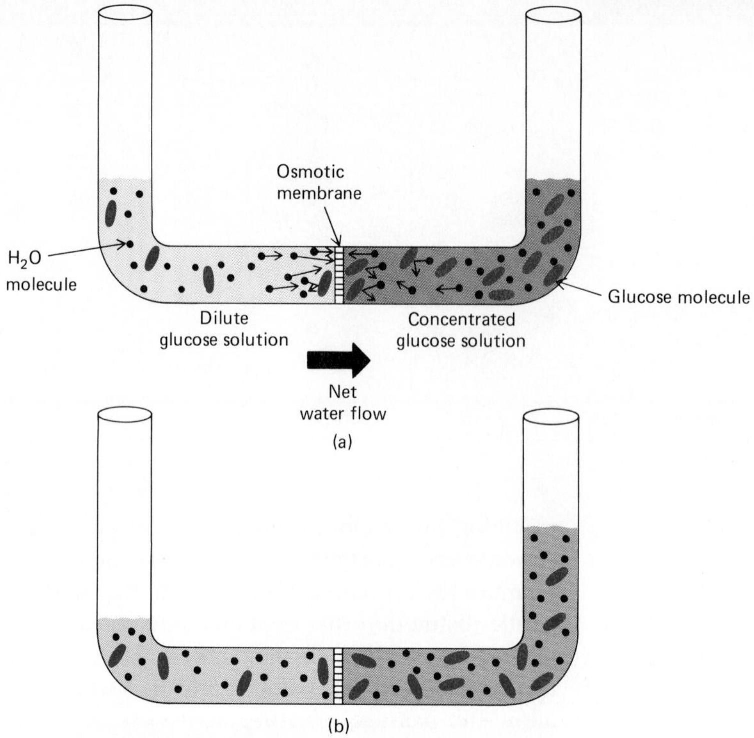
Osmotic membrane
H_2O molecule
Glucose molecule
Dilute glucose solution
Concentrated glucose solution
Net water flow
(a)
(b)

FIG. 9.12
(*a*) Two glucose solutions (one dilute and one concentrated) are placed in compartments separated by an osmotic membrane. Only water molecules can fit through the "holes" of the osmotic membrane. Larger glucose molecules cannot pass through. Also, glucose molecules block the passage of water molecules by deflecting them. The more glucose, the more the interference. Thus the net flow is from the dilute solution where there is more water and less blocking glucose into the more concentrated solution. (*b*) This shows the compartments after time has elapsed. The direction of net flow is obvious from the altered water levels.

Osmotic Pressure

When a solution is separated from pure water by an osmotic membrane, the pressure that develops from the excess height of the solution side as the water flows in is called the **osmotic pressure** of the solution. The osmotic pressure of a solution can also be defined as the external pressure which must be applied to prevent a net flow of water across a semipermeable membrane from pure water to an aqueous solution. Figure 9.13 shows this application of pressure. The osmotic pressure measures the tendency for water to flow from pure water across a semipermeable membrane into a solution. This tendency depends only on the concentration of solute particles in the solution; that is, it is a colligative property. The greater the concentration of solute particles, the larger the solution's osmotic pressure and the greater the tendency for water to flow into it.

It is the difference in osmotic pressure of two solutions with different solute concentrations that causes the net flow of water that we've been discussing. In comparing solutions, a more dilute solution with a lower osmotic pressure is said to be **hypotonic** compared with a more concentrated solution with a higher osmotic pressure. In turn, the more concentrated solution is called **hypertonic** compared with the dilute solution. Solutions of exactly the same particle concentration have identical osmotic pressures and

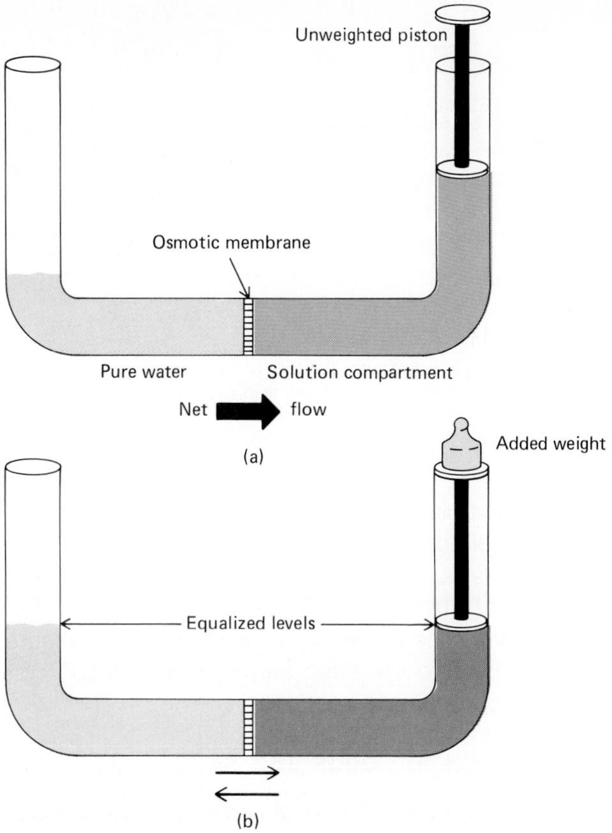
Unweighted piston
Osmotic membrane
Pure water
Solution compartment
Net flow
(a)
Added weight
Equalized levels
(b)

FIG. 9.13
Measurement of osmotic pressure. (*a*) A solution is separated from pure water by an osmotic membrane, and the water flows into the solution, raising the level in the solution compartment. (*b*) Pressure is applied by adding weights to the piston mechanism until there is no net flow into the solution compartment. The pressure that must be applied to accomplish this is the **osmotic pressure**.

are said to be **isotonic.** We can summarize these ideas schematically:

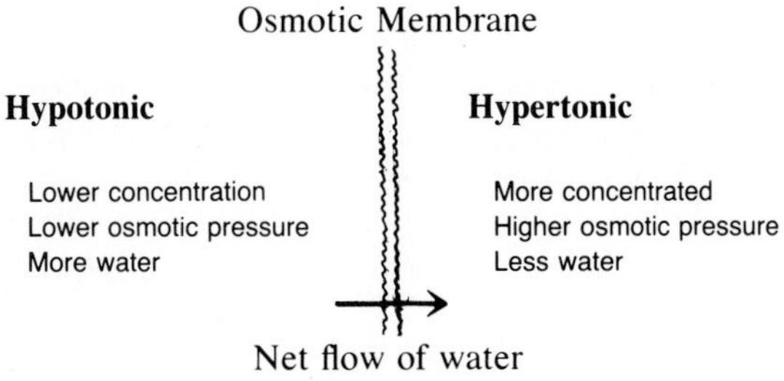
Osmotic Membrane
Hypotonic
Lower concentration
Lower osmotic pressure
More water
Hypertonic
More concentrated
Higher osmotic pressure
Less water
Net flow of water

Red Blood Cells

The membranes of red blood cells allow the passage of water molecules. Red blood cells are stored safely in physiological saline solution, which is 0.9% NaCl solution, because the physiological saline solution is isotonic with the red blood cells; that is, the solute concentrations within the two solutions are the same. The nature of the particles within the two solutions is different, but the concentration of particles within the two solutions is the same. There is no concentration gradient.

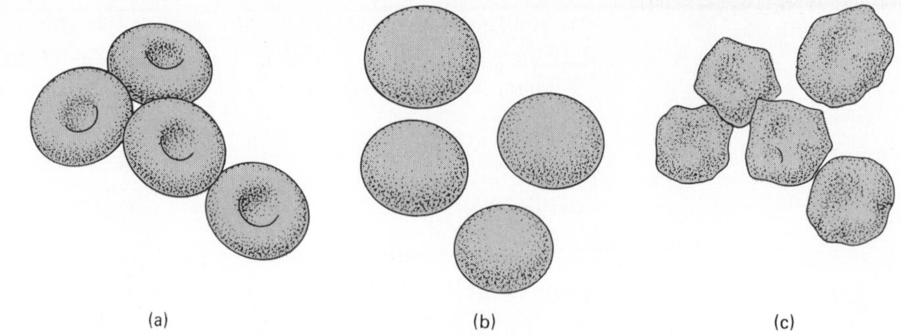
(a)
(b)
(c)

FIG. 9.14
(*a*) Red blood cells in an isotonic solution are plump, but properly formed. (*b*) In a hypotonic solution red blood cells swell and may burst. (*c*) In a hypertonic solution red blood cells shrink.

If red blood cells are placed in pure water or a hypotonic solution, there is a net flow of water into the red blood cell. The cell swells and may burst or undergo *hemolysis,* the breaking apart of red blood cells. When red blood cells are placed in a hypertonic solution, water leaves the red blood cells, and the cells shrink or undergo *crenation.* Figure 9.14 shows red blood cells subjected to the three surrounding solution conditions: isotonic, hypotonic, and hypertonic.

Osmosis in Foods and Agriculture

We witness the effects of osmosis in many foods. (See Figure 9.15.) Prunes swell up if placed in water (a hypotonic medium to the prune), cucumbers shrink and become pickles in brine (a hypertonic medium to the cucumber), eggplants give off their water when sprinkled with salt, and hams are dehydrated and therefore preserved by "cooking" them in a brine solution. The

FIG. 9.15
The skins around fruits and vegetables act as osmotic membranes. Prunes swell in water because water enters the prune to dilute the concentrated sugar solution within. Cucumbers shrink into pickles in brine because water leaves the comparatively dilute solution within the cucumber.

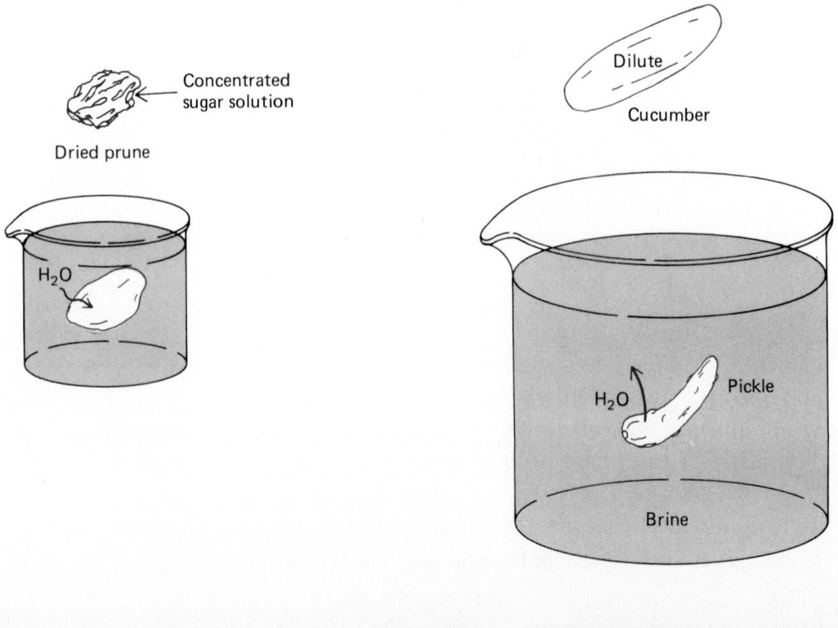
Concentrated
sugar solution
Dried prune
H_2O
Dilute
Cucumber
H_2O
Pickle
Brine

movement of water in each case is through cell membranes. In agriculture osmosis helps one understand how water flows from the ground root system (hypotonic) to the concentrated sap cells (hypertonic) in the upper regions of a tree.

Osmolarity

Because quantitatively colligative properties such as osmosis depend on the concentration of *particles* in solution, a concentration expression which fully reflects this concentration is sometimes used. The expression is called **osmolarity.** It is most useful for electrolyte solutions in which there are a greater number of moles of particles than moles of solute compound.

An ionic compound such as NaCl completely dissociates in solution (see Section 9.6).

$$NaCl(s) \longrightarrow Na^+(aq) + Cl^-(aq)$$

Consequently 1 L of a 1 *M* NaCl solution contains 2 mol of particles. In 1 mol of NaCl there is 1 mol of Na^+ and 1 mol of Cl^- (see Section 7.7). A 1 *M* NaCl solution may be described as being 2 osmolar (2 *osM*). The 2 reflects the fact that there are 2 mol of particles per mole of solute.

In general,

$$\text{Osmolarity} = \text{molarity} \times \frac{\text{moles of particles}}{\text{1 mol of solute}}$$

For examples, 1 *M* $MgCl_2$ is 3 *osM* because there are 3 mol of particles (Mg^{2+} and $2Cl^-$) per mole of solute ($1 \times 3 = 3$); and 2 *M* KNO_3 is 4 *osM* because there are 2 mol of particles (K^+ and NO_3^-) per mole of solute ($2 \times 2 = 4$).

The osmolarity of a nonelectrolyte solution is always the same as the molarity because the number of moles of particles per mole of solute is *one*.

Problem 9.5 What is the osmolarity of each of the following solutions:

a 1.5 *M* KBr
b 2 *M* glucose
c 2 *M* K_2SO_4

Referring to a 1 *M* NaCl solution as 2 *osM* and a 1 *M* glucose solution as 1 *osM* tells immediately that the equimolar salt solution has twice the effect on colligative properties. That is, the magnitudes of the boiling point elevation, freezing point depression, and osmotic pressure are twice as great for the 2 *osM* NaCl as for the 1 *osM* glucose.

9.13 COLLOIDS AND SUSPENSIONS

It was mentioned in Section 9.1 that a true solution can be achieved only if the solute particles are very small. To quantify this, the diameters of the particles must be in the range 0.05–0.25 nanometers ($1 \text{ nm} = 1 \times 10^{-9}$ m). This is often described as "ionic" or "molecular" in size. Such particles cannot be seen even by the most powerful microscopes.

When larger particles are interspersed within a medium such as water, mixtures with properties and names different from those of solutions arise: colloids and suspensions. Table 9.4 summarizes the properties of solutions, colloids, and suspensions. Muddy water is a suspension of dirt particles in water. We can actually see the large dispersed particles, the particles even-

TABLE 9.4 Comparative Properties of Solutions, Colloids, and Suspensions

Property	Solution	Colloid	Suspension
Particle size (diameter)	0.05–0.25 nm	1–100 nm	100 nm
Particle visibility	Invisible	Visible with electron microscope	Visible to naked eye
Particle settling on standing	No settling	No settling	Settling
Particle separation on filter paper	Nonfilterable	Nonfilterable	Filterable
Behavior to light	Transparent	Tyndall effect	Tyndall effect
Particle passage through dialyzing membrane	Pass	Do not pass	Do not pass

tually settle out of solution, and the particles do not pass through any membrane. Even filter paper will prevent suspended particles from passing through. A **suspension** then is a heterogeneous mixture which settles on standing and its components can be separated by filtration. Common suspensions encountered medically are the "barium cocktail" (Section 9.3) and milk of magnesia, which is a suspension of relatively insoluble $Mg(OH)_2$ in water.

The properties of colloids or *colloidal dispersions* are probably least familiar to you, though colloids are encountered daily and most bodily fluids are colloidal in nature. **Colloids** are homogeneous mixtures which do not settle nor are their components filterable, but their solute particles are visible under high magnification and will not pass through a semipermeable membrane. Table 9.5 gives examples of colloids classified according to the physical states of the colloidal dispersed particles and the dispersing medium. The particle size in colloids is such that they pass through holes of filter paper, but not through a semipermeable membrane. These particles can be seen with an electron microscope. Many individual protein molecules and other large biological molecules are colloidal in size. Other colloid particles are aggregates of many small molecules. For example, in whipped cream, aggregates of many gas molecules are dispersed through the liquid cream, thus giving the system an "airiness."

Some colloids look like solutions; that is, they appear to be transparent. However, light does not pass through a colloidal dispersion unreflected as it does through a true solution (see Figure 9.16). Solute particles are too small to reflect light; colloidal particles can reflect light and consequently

TABLE 9.5 Types of Colloidal Dispersions

Dispersed Phase	Dispersing Medium	Name of Type	Examples
Liquid	Gas	Liquid aerosol	Fog, clouds, aerosol sprays, smoke
Solid	Gas	Solid aerosol	Smoke
Gas	Liquid	Foam	Whipped cream, suds,
Liquid	Liquid	Liquid emulsion	Milk, mayonnaise
Solid	Liquid	Sol or gel	Jellies, paint, protoplasm, blood plasma, beer
Gas	Solid	Solid foam	Marshmallows, foam rubber
Liquid	Solid	Solid emulsion	Butter, cheese
Solid	Solid	Solid sol	Opals, pearls, some alloys

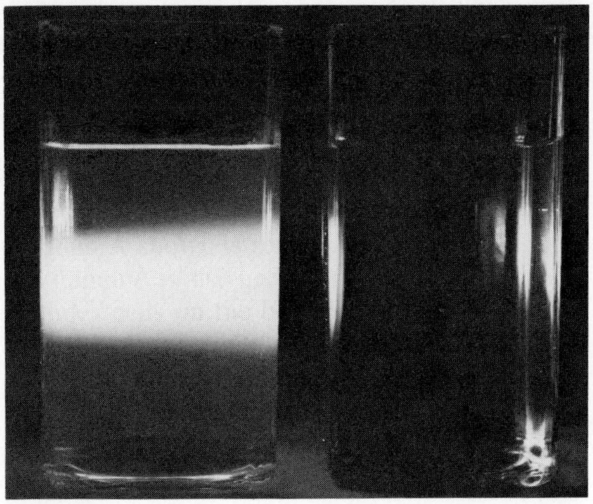

FIG. 9.16
Solutions are truly transparent because solute particle size is too small to allow deflection of light. Particles in colloidal dispersions deflect light, and the result is that the beam of light can be seen in the dispersion. This phenomenon is called the *Tyndall effect.* (*Photograph by Bryan Lees.*)

the beam of light can be seen in the colloidal dispersion. This is known as the **Tyndall effect.** The visibility of the light beam from a movie projector is another example of the Tyndall effect. Dust particles in air reflect the light and make it visible.

9.14 DIALYSIS

All cell membranes demonstrate osmosis in that they allow the movement of water molecules from a region of low solute concentration (more dilute solution) to a region of high solute concentration (more concentrated solution). Dissolved gases (CO_2, O_2) and some molecules such as alcohol or ammonia can also readily pass through an osmotic cell membrane; ions and larger molecular solutes cannot readily cross these membranes without transport assistance (Section 18.10).

However, cells in the kidneys, blood capillaries, and intestinal walls have a type of semipermeable membrane, known as a dialyzing membrane, which has pores sufficiently large to allow the passage of ions and certain molecules along with water and gases. **Dialysis** then is the movement of ions and small molecules, including water, across a dialyzing membrane. Colloidal-sized particles, such as protein molecules, cannot pass through a dialyzing membrane.

The direction of net transfer of materials in dialysis is governed by the same principles as in osmosis. Dissolved ions or molecules move from the region of their higher concentration to a region of lower concentration.

Dialysis can be used to separate colloidal-sized particles, such as protein and polysaccharide molecules, from ions and smaller molecules. Figure 9.17*a* shows a cellophane bag containing a mixture of water, salt, and colloidally dispersed starch. Water continuously circulates around the bag. Cellophane is a dialyzing membrane. Na^+ and Cl^- ions move out of the bag into the circulating water where they are removed. Starch is colloidal; it cannot cross the membrane and therefore remains in the bag. The final result is that the starch and any other colloidal material are separated from the dissolved ionic and small molecular substances.

The principles of dialysis govern the functioning of the kidneys whereby they separate waste molecules of metabolism, such as urea and excess ions, from the blood. This function is performed for people with kidney failure

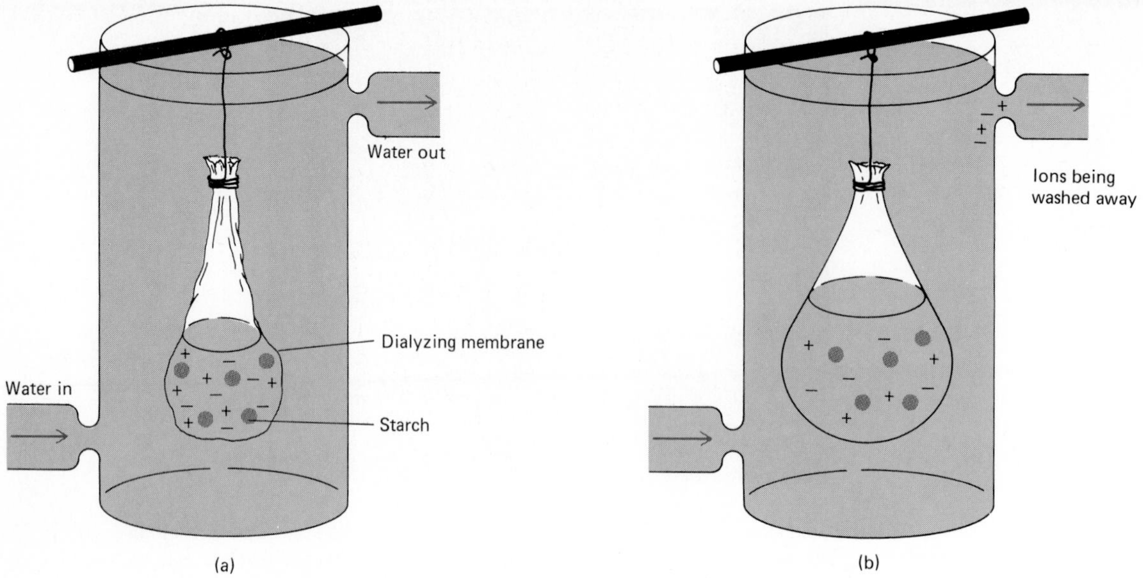
Water out
Ions being washed away
Dialyzing membrane
Water in
Starch
(a)
(b)

FIG. 9.17
(*a*) A cellophane bag containing water, starch (colloidal-sized particles), and sodium chloride is immersed in a water bath. The water is constantly being flushed and replenished. (*b*) After some time elapses, we see fewer ions inside the bag because Na^+ and Cl^- ions can cross the dialyzing membrane into the bath and be washed away. The starch cannot cross the membrane. Separation of particles is thereby effected.

by the artificial kidney machine, which employs the process of **hemodialysis,** or blood dialysis. The patient's blood is pumped through a cellophane coil that winds through a solution containing adjusted concentrations of Na^+, K^+, Cl^-, and HCO_3^- ions (Figure 9.18). The concentration is adjusted so that excess ions will leave the blood via dialysis across the cellophane mem-

FIG. 9.18
Hemodialysis. The artificial kidney uses dialysis to remove wastes from blood. The patient's blood is pumped through a long, coiled cellophane dialysis tube that is immersed in a solution called the dialysate. The dialysate has the proper physiological concentration of essential ions. Excess ion concentration in the blood and other small-molecule metabolic wastes pass through the dialysis membrane from the blood into the dialysate and are thus removed from the blood.

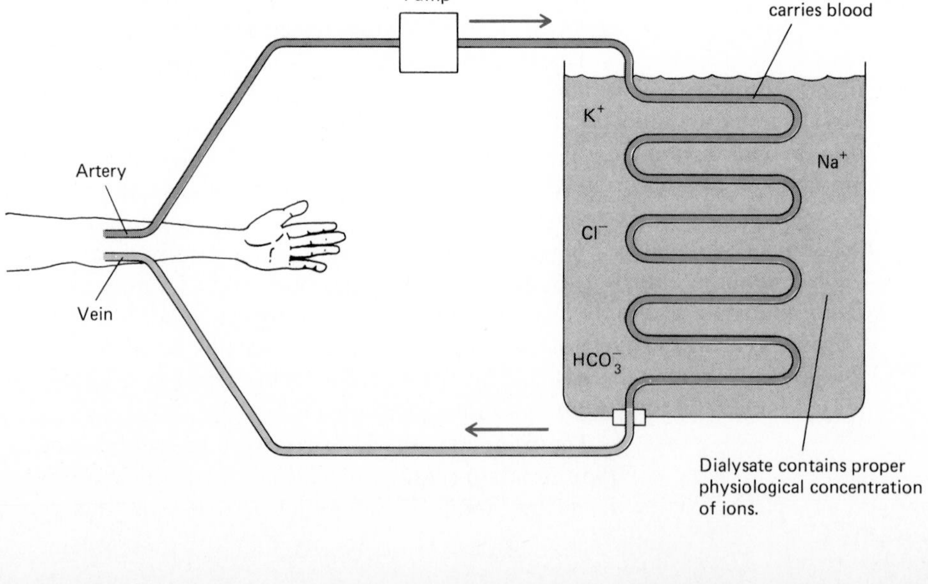
Pump
Cellophane dialyzing coil carries blood
K⁺
Na⁺
Artery
Cl⁻
Vein
HCO₃⁻
Dialysate contains proper physiological concentration of ions.

brane. Proper healthy ion concentrations are maintained within the blood by maintaining an appropriate concentration gradient between blood and the dialysate bath solution. There is no danger of loss of colloidal-sized blood proteins across the membrane.

9.15 ACTIVE TRANSPORT

Osmosis and dialysis are spontaneous events in the presence of a concentration gradient across a membrane. There is a natural tendency for a net transfer from a high concentration area to a low concentration area. Sometimes it is necessary for ions or other particles to cross the cell membrane against the concentration gradient. This process is called **active transport** and requires energy input. We will examine this phenomenon more closely after studying some concepts about energy in Chapter 10 and after further exploring the nature of cell membranes. (Section 18.10).

9.16 ELECTROLYTE BALANCE

There are three so-called bodily fluid compartments: the fluid within cells, or **intracellular** compartment; the **interstitial** fluid between tissue cells; and the **plasma** fluid compartment within the blood vessels. Figure 9.19 shows the three compartments. Table 9.6 gives the normal concentrations of the common ions found in the intra- and extracellular fluids. The extracellular fluids are those *outside* the cell. Interstitial fluid and plasma are grouped together as extracellular because for the most part their ion concentrations are very similar. However, the protein content is quite different, as you will see in Section 9.17. These concentrations must be maintained for good health.

This maintenance of proper ion concentrations is known as proper **electrolyte balance** in the body. One critical example of the importance of electrolyte concentration is the role it plays in nerve transmission in nerve cells. Proper nerve transmission depends on the correct Na^+ ion concentration outside the cell and the proper K^+ ion concentration inside the cell. Some other consequences of incorrect cation concentrations have already been discussed in Section 5.13. The role of chloride, the most abundant anion in plasma, is to maintain electrical neutrality as other ions move across mem-

FIG. 9.19

Approximately 60 percent of body weight is water. There are various compartments in the body. Most water is contained intracellularly, i.e., inside cells (40 percent). There is also water between cells in the so-called interstitial fluid compartment (15 percent). The remainder of the water is in blood plasma (5 percent).

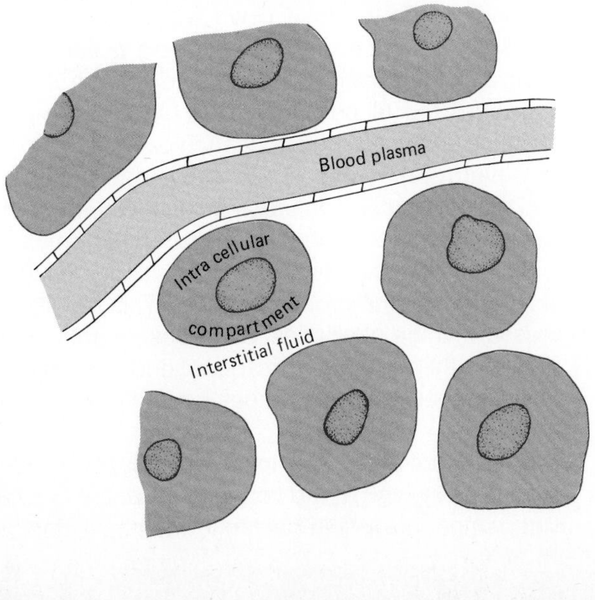
Blood plasma
Intra cellular
compartment
Interstitial fluid

TABLE 9.6 Typical Ionic Concentrations in the Intracellular and Extracellular Fluid Compartments*

Intracellular				Extracellular (plasma)			
Ion	mg/L	mmol/L	meq/L	Ion	mg/L	mmol/L	meq/L
CATIONS							
K^+	5850	150	150	Na^+	3270	142	142
Mg^{2+}	480	20	40	K^+	156	4	4
Na^+	230	10	10	Ca^{2+}	100	2.5	5
				Mg^{2+}	24	1	2
ANIONS							
HPO_4^{2-}	5952	62	124	Cl^-	3550	100	100
HCO_3^-	610	10	10	HCO_3^-	1525	25	25
Cl^-	70	2	2	HPO_4^{2-}	95	1	2
SO_4^{2-}	1152	12	24	SO_4^{2-}	50	0.5	1

* Total cationic concentration shown is greater than total anionic concentration shown because negative ion contributions by proteins have not been included.

branes. It plays a part in acid-base balance in blood, which is regulated in part by bicarbonate anion shifts across membranes (see Section 11.13). Electrolyte balance is regulated by the kidneys. The adjustment of ion concentrations is accomplished in the body through the processes of osmosis, dialysis, and active transport.

9.17 OSMOTIC PRESSURE AND FLUID TRANSPORT

Blood plasma and interstitial fluid, the liquid medium between cells, have approximately the same total concentration of electrolytes. However, blood plasma has a much higher concentration of protein. This results in a higher osmotic pressure in blood plasma, by about 25 torr, than in the interstitial fluid. This osmotic pressure difference is called the **colloid osmotic pressure** or the **oncotic pressure.** Along with blood pressure, oncotic pressure governs water and nutrient transport into and out of the blood capillaries.

Oncotic pressure alone would dictate movement across the membrane from the interstitial area (region of lower osmotic pressure) into the blood plasma (region of higher osmotic pressure). However, blood pressure can oppose this flow. Blood flows within the capillaries because of the blood pressure developed by the pumping action of the heart. Blood capillaries have an arterial and a venous end (Figure 9.20). At the arterial end of the capillaries the blood pressure is higher than the colloid osmotic pressure, and fluid and nutrients are pushed out into the interstitial fluid around tissues. However, at the venous end of the capillaries, the blood pressure is lower and the colloid osmotic pressure is now greater than the blood pressure, and fluids and metabolic waste products flow from the tissues into the capillaries.

Physiological conditions such as shock or the malnutrition disease kwashiorkor lower the protein in blood. This results in a lowered colloid osmotic pressure in the capillaries, and consequently more fluid flows from the capillaries to the tissues at the arterial end and less fluid returns to the capillaries at the venous end. The tissues become swollen with fluid, a condition known as edema.

Edema can also result in cases of hypertension where the blood pressure is sufficiently high at the venous end of the capillaries that the osmotic pressure cannot draw fluid back into the capillaries.

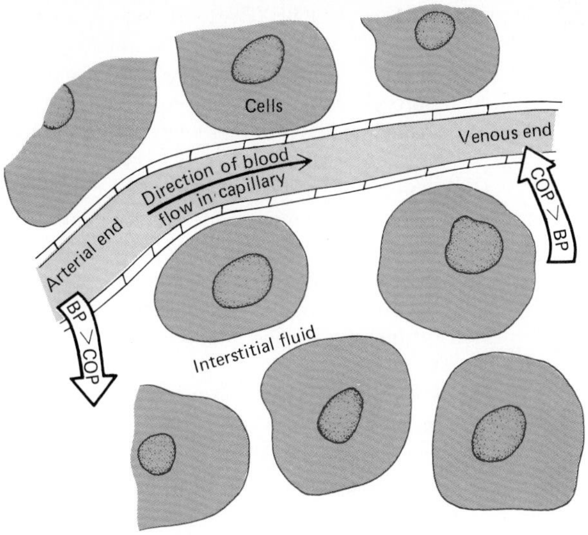
Cells
Venous end
Direction of blood flow in capillary
Arterial end
COP > BP
BP > COP
Interstitial fluid
BP = blood pressure
COP = colloid osmotic pressure

FIG. 9.20
Direction of flow of fluids between capillary and interstitial fluid. At the arterial end of the capillary, blood pressure (BP) is approximately 35 mmHg, while colloid osmotic pressure (COP) is 25 mmHg. Consequently the flow is in the direction shown. The higher blood pressure forces flow in that direction. At the venous end, blood pressure is only about 15 mmHg, that is, less than COP = 25 mmHg, and the net flow is thereby from the interstitial fluid into the capillary.

CHAPTER ACCOMPLISHMENTS

After completing this chapter you should be able to define **key words** and do the following:

9.1 INTRODUCTION
1. Name and distinguish between the two components of a solution.

9.2 SOLUTION TERMINOLOGY
2. Given solubility data, state whether a solution is saturated or unsaturated.
3. Indicate the type of mixtures to which the terms *miscible* and *immiscible* are applied.

9.3 SOLUTION FORMATION
4. Describe the formation of a water solution of an ionic solid.

9.4 FACTORS INFLUENCING SOLUBILITY
5. State the factors which affect the extent of solubility of a given solute in a given solvent.
6. Given the Lewis structures of two molecular compounds, predict whether one will dissolve in the other.
7. Describe the relationship between the partial pressure of a gas and its solubility in a liquid.

9.5 ELECTROLYTES
8. Give specific examples of strong and weak electrolytes and nonelectrolytes.
9. Distinguish between the processes of dissociation and ionization.

9.6 PARTICLES IN SOLUTION
10. Describe the basic particles found in solutions of ionic compounds and in solutions of molecular electrolytes and nonelectrolytes.

9.7 IONIC EQUATIONS
11. Given a traditional equation, write a net ionic equation for a reaction in which a gas or precipitate is formed.

9.8 CONCENTRATION EXPRESSIONS

12. Distinguish between concentration and solubility.
13. Given the grams of solute and the volume of a solution, calculate the percentage weight/volume of the solution.
14. Calculate the number of grams of solute needed to prepare a given volume of solution of some given percentage weight/volume.
15. Given the moles (or grams) of solute and the volume of the solution, calculate the molarity of the solution.
16. Given the volume and molarity of a solution, calculate the number of moles of solute in that solution.
17. Given the volume and molarity of a solution, calculate the number of grams of specified solute in that solution.

9.9 DILUTION

18. Given the volume and molarity of an original solution, calculate the molarity of a new solution prepared by adding a given amount of solvent to the original solution.
19. Calculate the volume of a more concentrated solution of given molarity which can be diluted to prepare a specified volume of a new solution of lower molarity.

9.10 EQUIVALENTS AND MILLIEQUIVALENTS

20. Given some number of moles (or millimoles) of an ion, indicate the corresponding number of equivalents (or milliequivalents).
21. Given some concentration of an ion in solution, express that concentration in milliequivalents per liter.

9.11 COLLIGATIVE PROPERTIES OF SOLUTIONS

22. Name four colligative properties.
23. Describe the effect of added solute on the vapor pressure, boiling point, and freezing point of a liquid solvent.

9.12 OSMOTIC PRESSURE OF SOLUTIONS

24. Distinguish among permeable, impermeable, and semipermeable membranes.
25. Given the concentrations of two solutions separated by an osmotic membrane, tell the direction in which the net flow of water will occur.
26. Distinguish among isotonic, hypotonic, and hypertonic solutions.
27. Given the molarity of a solution, calculate the osmolarity of that solution.

9.13 COLLOIDS AND SUSPENSIONS

28. Distinguish among solutions, colloids, and suspensions in terms of particle size.
29. Describe the Tyndall effect.

9.14 DIALYSIS

30. Given the concentrations of two solutions separated by a dialyzing membrane, tell the direction of net flow of solution components.

9.15 ACTIVE TRANSPORT

9.16 ELECTROLYTE BALANCE

9.17 OSMOTIC PRESSURE AND FLUID TRANSPORT

31. Describe the relationship between blood pressure and oncotic pressure in capillaries that accounts for nutrients entering and waste material leaving tissues.

PROBLEMS

9.6 Define in your own words what is meant by a solution.

9.7 A compound, like a solution, is homogeneous. Describe how a solution differs from a compound.

9.8 Give specific examples, other than those in Table 9.1, for each of the three types of solution in which the solvent is a liquid.

9.9 *a* Are all solutions mixtures?
b Are all mixtures solutions? Explain.

9.10 Which of the following examples would you classify as solutions?
a Helium gas dispersed in air
b Scotch whiskey
c Urine

d Fog
e Soda water
f Bronze (tin homogeneously dispersed in copper)
g Blood
h Milk
i Milk of magnesia

9.11 Indicate a possible solute and a possible solvent for each of the examples you classified as a solution in Problem 9.10.

9.12 Why is dusty air not considered a solution?

9.13 *a* Does a saturated solution necessarily have to be a concentrated solution? Explain.
b Give a specific example from Table 9.2 of a solution that is saturated, yet dilute.

9.14 Use Table 9.2 to decide whether the following solutions are saturated or unsaturated:
a 21 g of NaCl in 100 g of H_2O at 20°C
b 18 g of $KMnO_4$ in 100 g of H_2O at 60°C
c 2.1 g of $AgC_2H_3O_2$ in 100 g of H_2O at 60°C
d 62 g of KBr in 100 g of H_2O at 20°C

9.15 Consult Table 9.2 and decide whether the following solutions are saturated or unsaturated:
a 17 g of KBr in 25 g of H_2O at 20°C
b 7.5 g of $KMnO_4$ in 50 g of H_2O at 60°C
c 45 g of NaCl in 150 g of H_2O at 60°C
d 41 g of KNO_3 in 125 g of H_2O at 20°C

9.16 Other than by adding more solute, describe how a dilute solution of NaCl can be made more concentrated.

9.17 *a* To which type of mixture do the terms *miscible* and *immiscible* apply?
b Give an example of two miscible substances other than alcohol and water.
c Give an example of two immiscible substances other than oil and water.

9.18 Protein molecules are very large. Some segments of these molecules are water soluble and some are not. The segments can be described as ________ (water-loving) or ________ (water-fearing).

9.19 *a* What attractive forces must be broken down in the dissolving of an ionic solid?
b What new attractive forces are formed in the dissolving of an ionic solid?

9.20 Explain why NaCl does not dissolve in a nonpolar solvent such as gasoline.

9.21 *a* How does the nature of a solute affect whether it will dissolve in a given solvent?
b What type of solutes will dissolve best in water solution?
c What type of solutes will dissolve best in gasoline?

9.22 Explain why a solute such as sugar dissolves faster in hot water than in cold water.

9.23 *a* Explain why a bottle of soda goes "flat" if left open to the air.
b Would keeping the bottle warm help the soda to stay fizzier longer? Explain.

9.24 What effect does changing external (atmosphere) pressure have on the solubility of gases in blood?

9.25 *a* Write out Lewis structures for methyl alcohol, CH_3OH, and hydrogen chloride, HCl.
b Predict whether HCl will dissolve in CH_3OH. Explain your answer.

9.26 Examples of the three categories of biological molecules could be represented by the structures:

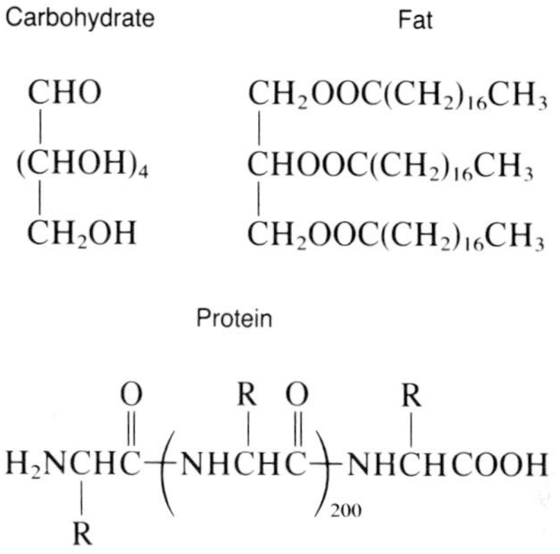

Carbohydrate
Fat
CHO
(CHOH)4
CH2OH
CH2OOC(CH2)16CH3
CHOOC(CH2)16CH3
CH2OOC(CH2)16CH3
Protein
H2NCHC(NHCHC)200NHCHCOOH

Which class might be expected to form true water solutions? Explain.

9.27 State the evidence for the existence of ions in a solution of NaCl.

9.28 Describe how you would experimentally show that HCl gas ionizes when dissolved in water.

9.29 HBr is a covalent substance which acts as a strong electrolyte in aqueous solution. CH_3COOH is a covalent substance which acts as a weak electrolyte in aqueous solution. Explain how one covalent substance can be a strong electrolyte and the other weak.

9.30 Using NaCl and HCl as specific examples, describe the differences between the processes of dissociation and ionization.

9.31 An aqueous solution of HCl conducts electricity. However, a solution of HCl in benzene does not conduct electricity. Explain what is happening in the two solutions.

9.32 Give the names and formulas of three strong and three weak acids.

9.33 What solute particles will be found in aqueous solutions of the following?
a KBr
b $Ca(OH)_2$
c HNO_3
d CH_3COOH
e H_2SO_3
f Glucose, $C_6H_{12}O_6$
g Methyl alcohol, CH_3OH

9.34 Write balanced net ionic equations for the following reactions:
a $K_2SO_4(aq) + Ba(NO_3)_2(aq) \longrightarrow KNO_3(aq) + BaSO_4\downarrow$
b $Zn(s) + H_2SO_4(aq) \longrightarrow ZnSO_4(aq) + H_2\uparrow$
c $NaCl(aq) + AgNO_3(aq) \longrightarrow AgCl\downarrow + NaNO_3(aq)$

d $(NH_4)_2S(aq)$ + $Pb(NO_3)_2(aq)$ ⟶ $NH_4NO_3(aq)$ + PbS ↓

9.35 Describe and illustrate by example the difference between solubility and concentration.

9.36 A solution was prepared by dissolving 13.8 g of KNO_3 in 75.0 mL of water at 20°C. What is the percentage weight/volume of this solution?

9.37 In a laboratory, 179 mL of a silver nitrate solution is evaporated to dryness. After all the water is evaporated, 69.3 g of silver nitrate remains. What is the percentage weight/volume of $AgNO_3$ in the original solution?

9.38 How many grams of sugar would you use to prepare 20.0 mL of a 0.50% w/v solution?

9.39 How many milliliters of 4.0% w/v $Ba(NO_3)_2$ solution would contain 1.25 g of barium nitrate?

9.40 Using Table 9.2, calculate the weight percentage of a saturated aqueous $KMnO_4$ solution at 60°C.

9.41 How much NaCl would be present in 800. mL of a 7.0 mg % saline solution?

9.42 What mass of sugar would be needed to prepare 80.0 mL of a 15.0 mg % sugar solution?

9.43 An antifreeze solution was prepared by dissolving 250. mL of methyl alcohol in 700. mL of water. What is the concentration of this solution in % v/v?

9.44 Which solution is more concentrated: a 0.01% w/v glucose solution or a 50-ppm solution?

9.45 Calculate the molarity of a solution which contains 2.50 mol of KNO_3 dissolved in 2.50 L.

9.46 How many moles of NaCl are present in 50.0 mL of a 0.125 *M* solution?

9.47 How many grams of glucose, $C_6H_{12}O_6$, are present in 1.50 L of a 0.500 *M* solution?

9.48 Find the number of grams of NaOH present in 375 mL of a 0.800 *M* solution.

9.49 What is the molarity of 50.0 mL of a 0.50 *M* NaOH solution after it has been diluted to 450. mL?

9.50 300.0 mL of water is added to 200.0 mL of a 0.250 *M* Na_2SO_4 solution. What is the molarity of the resulting solution?

9.51 What volume of a 1.50 *M* KNO_3 solution is required to make 1.00 L of a 0.100 *M* solution by dilution?

9.52 To what volume must you dilute 75.0 mL of 3.0 *M* $CuSO_4$ to have a 0.50 *M* solution?

9.53 A student needs to prepare 100.0 mL of a 4.0 *M* HCl solution. The student has available a 12.0 *M* HCl solution. What volume of the concentrated solution must be measured out to prepare the 4.0 *M* solution by dilution?

9.54 How many equivalents of each ion are represented by the following?

a 1 mol of Ca^{2+}
b 0.5 mol of Na^+
c 0.25 mol of Mg^{2+}
d 1 mol of SO_4^{2-}
e 0.5 mol of F^-

9.55 How many milliequivalents of each ion are present in 1 L of solution with the following concentrations?

a 0.5 m*M* Ca^{2+}
b 3 m*M* K^+
c 1 m*M* Cl^-
d 0.25 m*M* SO_4^{2-}

9.56 What is the concentration of each ion in milliequivalents per liter for the following solutions?

a 2 m*M* K^+
b 1.5 m*M* Ca^{2+}
c 23 mg % Na^+
d 10 mg % Ca^{2+}

9.57 Name four colligative properties.

9.58 Comparing pure water and a 10% aqueous glucose solution, which has the higher (*a*) vapor pressure? (*b*) Boiling point? (*c*) Freezing point? (*d*) Osmotic pressure?

9.59 With respect to the same properties, how would a 10% NaCl solution compare with the two solutions mentioned in Problem 9.58?

9.60 What is meant by the term *semipermeable membrane?*

9.61 Compartments A and B are separated by an osmotic membrane. Given the following concentrations of solutions in the compartments, tell if there will be a net flow of water, and if so, in which direction.

a A contains pure water; B contains 1% glucose.
b A contains 1 *M* glucose; B contains 1 *M* NaCl.
c A contains 1 *M* NaCl; B contains 0.5 *M* KBr.
d A contains 0.5 *M* KBr; B contains 1 *M* glucose.

9.62 Red blood cells are safely stored in 0.9% NaCl solution because this solution is *isotonic* with the interior of the cells. What does this mean?

9.63 Describe each of the following solutions as *hypertonic* or *hypotonic* with respect to 0.9% NaCl (0.15 *M* NaCl):

a 2% NaCl
b 0.15 *M* glucose
c 0.35 *M* glucose
d 0.12 *M* $CaCl_2$

9.64 Indicate the osmolarity of each solution described in Problem 9.61*b*, *c*, and *d*.

9.65 Describe how solutions, colloids, and suspensions differ.

9.66 How are solutions and colloids alike?

9.67 What is the Tyndall effect and how could you use it to distinguish between solutions and colloids?

9.68 A mixture of starch (colloidal-sized particles), table salt, and water is placed in a cellophane bag. Cellophane is a dialyzing membrane. The bag is placed in a beaker of pure water, and 3 h elapse.

a Is the water in the beaker still pure? Explain.
b Has the bag shrunk or swollen? Explain.

9.69 Compartments X and Y are separated by a dialyzing membrane. Given the following concentrations of solutions in the compartments, indicate whether there will be a net flow of water, and if so, in which direction. Also tell the direction of flow of any other solution components.

a X contains 0.1 *M* NaCl; Y contains 1 *M* glucose and 1% starch.

b X contains 1 *M* NaCl; Y contains 1 *M* KCl.

c X contains 2 *M* KCl and 0.1% protein; Y contains 0.1 *M* KCl.

d X contains 1% starch; Y contains pure water.

9.70 Describe how the artificial kidney machine works.

9.71 Ions can enter cells against a concentration gradient. What is this process called?

9.72 What is meant by electrolyte balance in the body?

9.73 State a physiological consequence of extremely low blood pressure.

9.74 Give one symptom of high blood pressure.

CHAPTER TEN

KINETICS AND EQUILIBRIUM

10.1 INTRODUCTION

Because it is such an important concept, the idea of nature spontaneously tending toward minimum energy has been noted repeatedly: in Chapter 4 you discovered the low-energy electron arrangements, in Chapters 5 and 6 compound formation was described as an energy-lowering event, and in Chapter 7 you learned that in exothermic reactions energy is released so that the products have a lower energy content. Of course, nonspontaneous processes which lead to higher-energy states can be made to occur by pumping in energy. For example, electrons can be excited, and endothermic reactions can be made to occur.

Schematically:

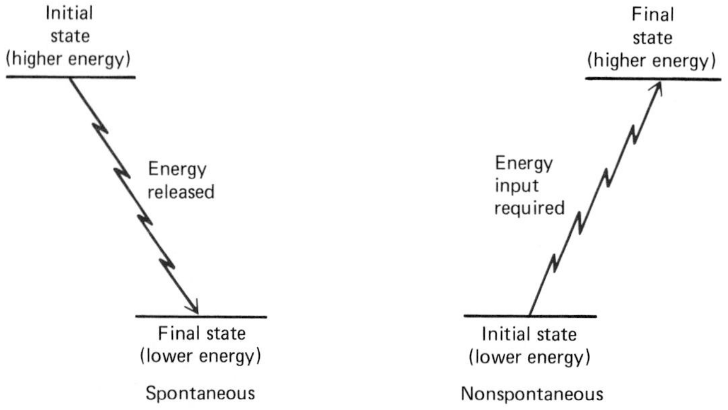
Initial
state
(higher energy)
Energy
released
Final state
(lower energy)
Spontaneous
Final
state
(higher energy)
Energy
input
required
Initial state
(lower energy)
Nonspontaneous

As chemists use the term, a **spontaneous** event is defined as one that leads to lower-energy products. *Spontaneous* as a scientific term is not meant to imply anything about the speed of an event. Unfortunately, our everyday usage of the term connotes quickness, and we must be careful to avoid this everyday idea about spontaneity when we encounter it in a chemistry textbook. Considerations of energy and spontaneity which take into account only the initial and final states are called **equilibrium** or **thermodynamic** considerations. These considerations are important and tell us many things about processes and reactions. For example, the products of the metabolism of carbohydrates and fats are CO_2 and H_2O. Just from the knowledge of the initial food we eat and the final waste products, we can predict the total amount of energy we might get from food. We need not know about the intermediate processing steps.

Problem 10.1 Do the waste products of the human body have more or less energy than the original starting foodstuffs? Explain your answer.

Thermodynamic considerations do not tell us how fast a reaction will occur. The study of the speed of a chemical reaction is called **kinetics.** Speed or *rate* of reaction is related to the pathway by which an initial state is transformed into a final state. Cellular reactions take place at much faster rates than the same reactions done in test tubes. To understand this physiological acceleration, you must know the basic concepts of kinetics. The final picture of physiological rates of reaction will emerge in Chapter 21, Enzymes.

10.2 SPEED VERSUS SPONTANEITY

In order to explore the important distinctions between spontaneity and speed, that is, between thermodynamics and kinetics, let us begin by examining the reaction between hydrogen and oxygen molecules to yield water.

$$2H_2(g) + O_2(g) \longrightarrow 2H_2O(g) + 117 \text{ kcal} \qquad (10.1)$$

The release of 117 kcal as a product indicates that this reaction is very exothermic. Therefore we would expect the reaction to occur spontaneously in order to arrive at the lower-energy state of the product. But this reaction is very slow, to the point that no reaction occurs if H_2 and O_2 gas are merely mixed together. There appears to be no spontaneous event. However, if a spark is provided or if a little finely divided metal is added to the gas mixture, the reaction immediately proceeds with explosive speed, giving off the expected amount of heat energy.

In Section 10.3 we will explore the origin of the heat energy (**enthalpy**) (Section 7.17) difference between the reactant gases and the product water that makes the reaction spontaneous. Later, in Section 10.7, you will learn the reason for the varying speed of the reaction under different conditions.

10.3 THE ORIGIN OF HEATS OF REACTION (ΔH)

Every covalent bond has a characteristic bond dissociation enthalpy, that is, the heat energy required to break that bond. This amount of energy is equal to the energy released when that bond is made. Figure 10.1 shows graphically the idea that bond breaking requires energy input and bond making is accompanied by energy release. Also see Figure 6.2.

The enthalpy difference between products and reactants, i.e., the heat of reaction, ΔH arises from the difference in bond dissociation enthalpies of bonds broken and formed during the reaction. Let's look again at the reaction to form water and this time write the reactants and products so that their bonds are shown:

$$2\ H{-}H(g) + O{=}O(g) \longrightarrow 2\ H{-}O{-}H(g)$$

From this equation you can see that in the reaction to form water (H—O—H), on the reactant side two H—H bonds must be broken and one O═O bond must be cleaved, while on the product side two O—H bonds per water molecule are formed.

Because we have already been told in Section 10.2 that this reaction is exothermic, it must be that more energy is released in bond making than is consumed in bond breaking. The actual numbers are 444 kcal released in forming the bonds in 2 mol of water and 327 kcal consumed in breaking the bonds in 2 mol of H_2 and 1 mol of O_2.

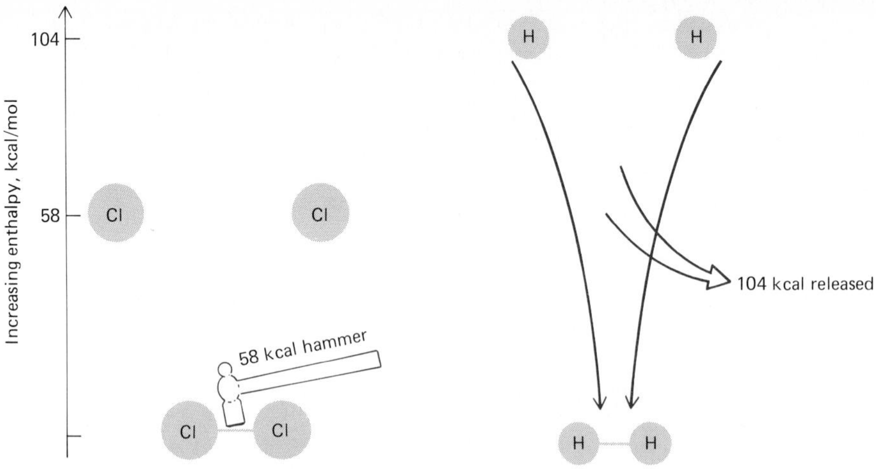
104
58
Increasing enthalpy, kcal/mol
Cl
Cl
58 kcal hammer
Cl
Cl
H
H
104 kcal released
H
H

FIG. 10.1
Energy must be put in to break a bond. The separated atoms are of higher energy. In this case we see that it requires 58 kcal/mol to break Cl—Cl bonds. When bonds are made, energy is released. Just as it takes 104 kcal/mol to break H—H bonds, this much energy is released when bonds form. See also Figure 6.2.

By convention ΔH is calculated from these data by doing the following subtraction:

$$\Delta H = \text{energy input for bond breaking} - \text{energy released in bond making}$$

Consequently, for an exothermic reaction where more energy is released than consumed, ΔH is always negative:

$$\Delta H = 327 \text{ kcal} - 444 \text{ kcal} = -117 \text{ kcal}$$

In general, if more energy is released in bond making than is required for bond breaking, the reaction is exothermic and ΔH is negative. If, on the other hand, more energy is required for bond breaking than is released in bond making, the reaction is endothermic and ΔH is positive.

Problem 10.2 Is ΔH for the conversion of food to waste products + or −?

10.4 GIBBS FREE ENERGY

Thus far in Chapter 7 and in the foregoing sections we have discussed only one type of energy, heat content, or enthalpy. Whereas it is certainly true that a process or reaction that leads to a lowered enthalpy is generally spontaneous, there are also spontaneous processes in which there is no enthalpy change (or even an increased heat content). For example, there is no enthalpy change when a gas expands to fill any available volume, and ice above 0°C spontaneously changes to liquid water although the enthalpy (heat) content of the liquid water is higher than that of the solid ice.

The spontaneity in these cases arises through a different energy factor. In both of them there is a change in the randomness of the system. The spreading out of the gas molecules results in an increased randomness of the gaseous system, and as you saw in Section 8.12, the molecules of liquid water are more randomly arranged than in solid ice. **Entropy** (symbolized S) is the scientific term for randomness or disorder.

You are probably familiar with many common spontaneous events that occur with an increase in entropy, i.e., with a tendency toward increasing disorder. A new deck of playing cards will fall into a randomized mixture, increasing the entropy of the cards; only by providing your energy input can you once again order the cards. A clean, organized room has a definite tendency to become disorganized after a period of use. These are specific examples of a general phenomenon, the tendency of increasing entropy in natural processes. An increase in entropy (randomness) is described as a favorable or positive change; a decrease in entropy (less randomness, more order) is an unfavorable or negative change.

Thus there are two energy factors that determine spontaneity: changes in enthalpy (ΔH) and changes in entropy (ΔS). The American scientist Josiah Willard Gibbs (1839–1903) combined these ideas into the concept of free energy. Free energy (symbolized G) is defined as the energy available to do useful work. Knowledge about ΔG, the change in free energy, is totally sufficient to predict the spontaneity of a reaction. The mathematical relationship between ΔG, ΔH, and ΔS is as follows:

$$\Delta G = \Delta H - T\Delta S \tag{10.2}$$

where ΔG = change in free energy
ΔH = change in enthalpy
T = absolute temperature
ΔS = change in entropy for a particular process or chemical reaction at a fixed temperature

If a reaction or process leads to a lowered free energy, the reaction is spontaneous; i.e., nature tends toward minimum energy. Such a process is termed *exergonic*.

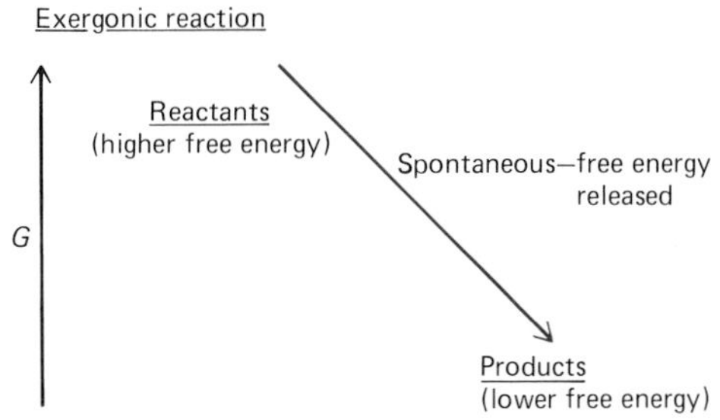
Exergonic reaction
Reactants
(higher free energy)
Spontaneous—free energy released
G
Products
(lower free energy)

In the opposite case, where the products of a reaction have a higher free-energy content than the reactants, the reaction is termed *endergonic*.

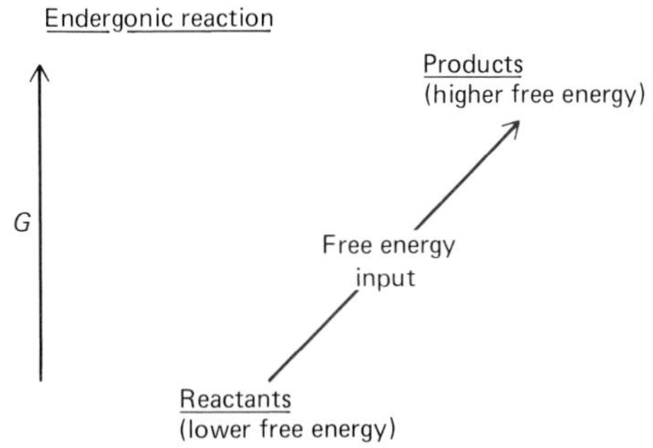
Endergonic reaction
Products
(higher free energy)
G
Free energy input
Reactants
(lower free energy)

You may ask, since free energy gives the total picture of spontaneity, why have we previously considered only enthalpy, which does not always tell the whole story? The answer is that heats of reaction (ΔH) are much more easily determined experimentally than are ΔG values. Thus enthalpy data are much more readily available. Also, in many cases entropy changes (ΔS)

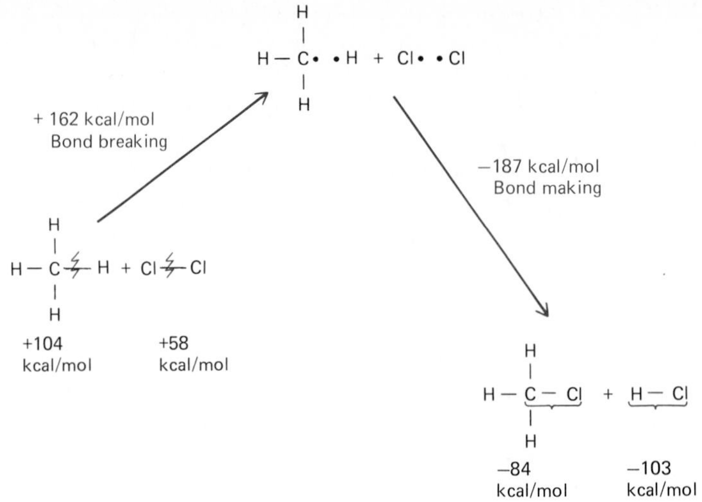
H—C• •H + Cl• •Cl
+ 162 kcal/mol
Bond breaking
−187 kcal/mol
Bond making
H—C—H + Cl—Cl
+104
kcal/mol
+58
kcal/mol
H—C—Cl + H—Cl
−84
kcal/mol
−103
kcal/mol

FIG. 10.2
In the relation between methane (CH_4) and chlorine to form methyl chloride (CH_3Cl) and HCl ($CH_4 + Cl_2 \rightarrow CH_3Cl + HCl$), there must be energy input to break C—H and Cl—Cl bonds before C—Cl and H—Cl bonds can be made. However, bond breaking need not be complete before bond making begins.

are small or are such as to enhance the effect of the enthalpy change (ΔH). Examine Equation 10.2 again and you will see that if $\Delta S = 0$, then $\Delta G = \Delta H$:

$$\Delta G = \Delta H - T\Delta S \qquad (10.2)$$

Substituting 0 for ΔS:

$$\Delta G = \Delta H - T(0)$$

$$\Delta G = \Delta H$$

ΔS is usually small. Thus, more often than not, trends in ΔG and ΔH are identical. Consequently, exergonic reactions are usually exothermic reactions. In the case of the formation of water from hydrogen and oxygen, we have seen this reaction to be exothermic ($\Delta H = -117$ kcal). It is also exergonic ($\Delta G = -110$ kcal) despite a slightly unfavorable (negative) entropy factor. More order is imposed in going from 3 mol of reactant molecules to 2 mol of product molecules.

For the remainder of this text when we refer to energy changes we will mean Gibbs free-energy changes (ΔG) unless we specify otherwise. Remember, knowledge of ΔG enables a foolproof prediction of spontaneity. **When ΔG is negative, the reaction is spontaneous. If ΔG is positive, the reaction is nonspontaneous.**

10.5 ACTIVATION ENERGY

Now let us explore the questions of why the spontaneous, highly exergonic reaction in which water is formed from its elements is very slow, but can be greatly speeded up by a spark or finely divided metal particles. This exploration will answer the more general question of what determines reaction rate or kinetics.

The answers lie in the concept of a **reaction pathway.** A proposed pathway gives all the intimate details of how reactants are converted into products.

Which bonds are broken, which bonds are formed, and the timing of these processes are all postulated in a proposed pathway.

The question of the timing of bond breaking and bond making is crucial to an understanding of how the energy content of a system varies as a reaction occurs. Before new bonds can be formed, releasing energy, some bond(s) of the original reactant must be broken; this requires energy input. Therefore, even in an exergonic reaction (where ultimately there is energy *output*) there will be some amount of energy *input* necessary before a reaction can take place. Figure 10.2 shows the input-output idea.

The amount of energy input varies for different reactions because of differences in the bonds being broken and formed and the timing involved. The minimum energy that must be added to the reactants to initiate a reaction is called the **activation energy.** This is often represented E_{act}. Every reaction, except one in which no bonds are broken, has an activation energy. A reaction with a large activation energy is slow; one with a small activation energy is fast.

10.6 REACTION PROGRESS DIAGRAMS

Chemists pictorially describe the energy changes of chemical reactions in reaction progress diagrams such as those in Figure 10.3. The vertical axis of such a diagram plots potential free energy. The horizontal axis measures how far the reaction has proceeded from reactants written on the left of the diagram to the products written on the right, in the same manner as chemical equations are usually written. Notice in Figure 10.3 that the shape of the reaction progress diagram is always like a road over a hill. One begins at reactants, goes over the top of a hill, and ends up at products. The products are lower in energy than the reactants in an exergonic reaction and higher in energy in the endergonic case.

These diagrams show both the free energy ($\boldsymbol{\Delta G}$) of the reaction and the activation energy ($\boldsymbol{E_{act}}$). The plotted difference in energy between the products and reactants is the free energy of the reaction ($\boldsymbol{\Delta G}$). In an exergonic reaction the products appear below the reactants on the potential energy scale because the products have less energy. In an endergonic reaction the products appear above the reactants on the potential energy scale because the products have more energy than the reactants.

The curved line that extends from reactants to products shows the energy change as the system progresses from reactants to products. The highest point on this curve represents the maximum energy on the pathway from

FIG. 10.3

Typical reaction progress diagram. The *y* axis plots free energy and the *x* axis represents progress along the path from reactants to products. ΔG and E_{act} are labeled in each case. E_{act} is always positive (energy input). ΔG is negative for exergonic reactions (energy released) and positive for endergonic reactions (energy absorbed).

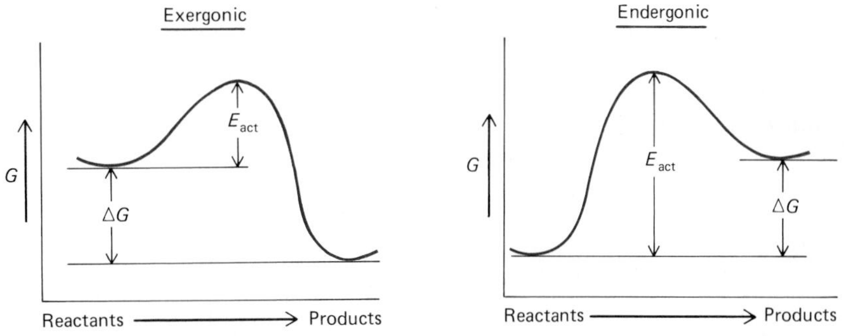
Exergonic
Endergonic
G
E_{act}
ΔG
G
E_{act}
ΔG
Reactants
Products
Reactants
Products

reactants to products. **The energy difference between this highest-energy point and the reactants is the activation energy (E_{act}),** that is, the amount of energy input needed to get over the "hill" or energy barrier between reactants and products.

10.7 PATHWAYS, E_{act}, AND KINETICS

The reaction progress diagram gives a graphical picture of the concept of reaction pathway. Keeping this in mind enables us to see the relationship between pathway, activation energy, and kinetics, or rate of reaction. If the pathway from reactants to products involves going over a low "hill", i.e., if the E_{act} is small, then the reaction is relatively fast. If, on the other hand, a high "hill" separates reactants and products, i.e., a large E_{act}, then the reaction is slow. As shown in Figure 10.4, activation energy and reaction rate are related in exactly the same way that the height of a hill relates to the speed with which you can surmount the hill.

Now let us again consider the speed of the reaction between H_2 and O_2 to form H_2O. Figure 10.5 shows the reaction progress diagram for this reaction both in the absence and in the presence of metal particles. The pathway that must be traversed in the absence of metal involves quite a large E_{act} or "high hill." Activation energy is provided by the energy of motion, i.e., the kinetic energy of molecules. In Section 8.9 you learned that molecular motion and kinetic energy are directly related to temperature. At room temperature, if H_2 and O_2 gases are simply mixed together, the molecules are moving too slowly to provide the activation energy needed to surmount the barrier. A spark allows the reaction to occur at an observable rate by providing the necessary activation energy to a few molecules. When these reactant molecules get over the barrier and form product molecules, the energy released as water forms continues to supply the needed activation energy.

In the presence of finely divided metal particles the reactant molecules can react by a different reaction pathway. This new pathway has a lower E_{act} and hence the reaction occurs more quickly. The metal remains unchanged in the course of the reaction. A material that speeds a reaction by lowering E_{act} and is itself unchanged by the reaction is called a **catalyst.** Notice in Figure 10.5 that the catalyst does not affect the free-energy difference between the reactants and products. ΔG is identical for the catalyzed and uncatalyzed reactions.

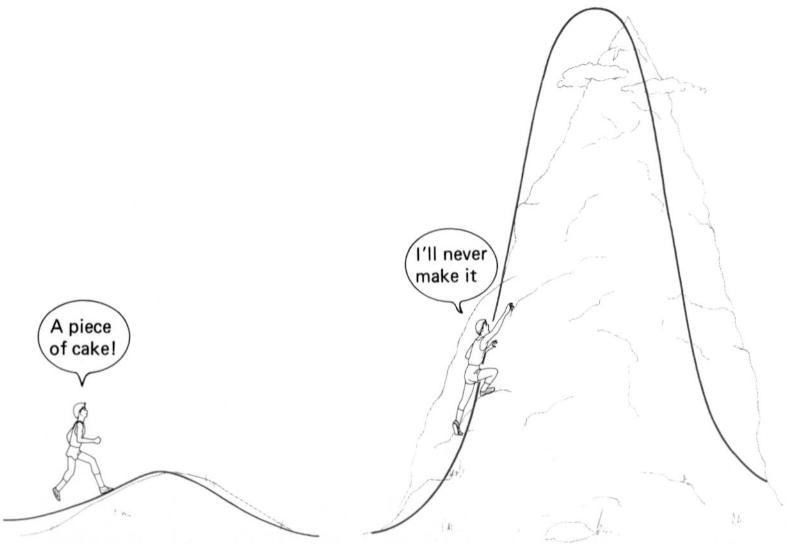
I'll never make it
A piece of cake!

FIG. 10.4
Activation energy is like a hill which reactant molecules must climb over in order to become products. A low hill allows a fast trip. A high hill means a slow one.

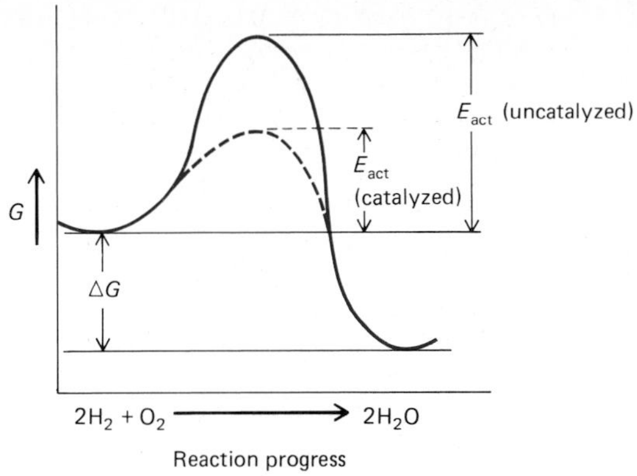

E_{act} (uncatalyzed)
E_{act}
(catalyzed)
G
ΔG
$2H_2 + O_2$ → $2H_2O$
Reaction progress

FIG. 10.5
This reaction progress diagram demonstrates the difference in energies of the pathways for an uncatalyzed (solid line) versus a catalyzed (dotted line) reaction. In the catalyzed case the activation energy is lower and hence the reaction proceeds more quickly. Notice that ΔG is the same for both reactions.

Physiological reactions are greatly speeded up by biological catalysts called enzymes. For example, the enzyme urease increases the rate of decomposition of urea (a waste product of proteins and other nitrogen-containing compounds) by a factor of 10^{13} compared with the reaction done in a test tube. Like other catalysts, enzymes lower the activation energy of reactions in which they participate compared with uncatalyzed reactions. Enzymes will be discussed further in Chapter 21.

SAMPLE EXERCISE 10.1

Figure 10.6 shows the reaction progress diagrams for the decomposition of the antiseptic hydrogen peroxide (H_2O_2) under different conditions of catalysis. H_2O_2 works as an antiseptic by decomposing and releasing O_2; the bubbles seen when H_2O_2 is applied to a wound are the evolved oxygen. Catalase, an enzyme present in blood, accelerates this decomposition. Read off the following parameters from the diagram:

a ΔG, the free energy of the reaction
b E_{act} (activation energy) of the uncatalyzed reaction
c E_{act} of the Pt-catalyzed reaction
d E_{act} of the enzyme-catalyzed reaction

Solution *a* The difference in energy between the reactants and products is −29 kcal. The sign is negative because the reaction is exergonic, that is, energy is released. Notice that ΔG is identical for all three reaction pathways because it involves only the initial and final states, *not* the pathway.

FIG. 10.6
The diagram is needed to answer the questions in Sample Exercise 10.1. The scale of the *y* axis is 1 mm = 1 kcal/mol.

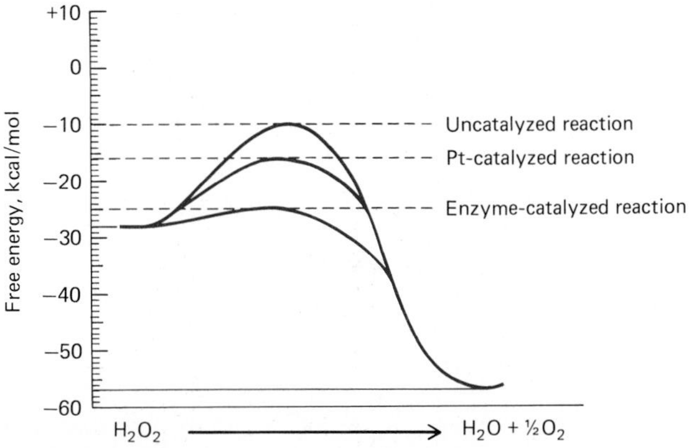

+10
0
−10
−20
−30
−40
−50
−60
Free energy, kcal/mol
Uncatalyzed reaction
Pt-catalyzed reaction
Enzyme-catalyzed reaction
H_2O_2 → $H_2O + \frac{1}{2}O_2$

b The energy difference between reactants and the top of the hill for the uncatalyzed reaction is 18 kcal. E_{act} is always positive.

c The energy barrier in the Pt-catalyzed reaction is 12 kcal.

d The energy barrier in the enzyme-catalyzed case is 3 kcal.

Problem 10.3 In Sample Exercise 10.1, which reaction (one of the catalyzed ones or the uncatalyzed one) is fastest? Explain how you know this from the reaction progress diagram.

10.8 FACTORS INFLUENCING REACTION RATE

So far you have seen that the rate of a reaction can be changed by using a catalyst that provides an alternative pathway with a different activation energy. There are other factors that affect the rate of a reaction without changing the activation energy. For example, **the rate of reaction can be increased by increasing** the molar **concentration** of one or more of the reactants, or by increasing the **temperature** of the reaction.

One simple and useful way to understand these effects is to realize that the activation energy for a reaction is provided principally by the molecular motion and collision of reactant molecules, as mentioned briefly in Section 10.7. In the course of reactants being converted into products, the reactants must collide so that the electron environment around the reactant atoms is changed to yield that of the products. Or more simply, old bonds (electron pairs) must begin to break before new ones form.

In order for products to form, the collision between reactants must be sufficiently energetic so that the collision *complex* is supplied with an amount of energy at least equal to the activation energy. If this is so, the collision complex is known as an **activated complex,** as shown in Figure 10.7. The activated complex is a high-energy, unstable state because the kinetic energy of the colliding reactants has been incorporated into the internal potential energy of the complex. Once formed, this excited state can fall apart back to reactants, or it can go on to the product state.

Not all collisions between reactant molecules are *effective*. In a reaction mixture there may be many soft, low-energy collisions in which the collision complex does not have enough energy to overcome the activation barrier and simply falls back to reactants. Figure 10.8 portrays effective and ineffective collisions. These simple ideas about the energetics of molecular col-

FIG. 10.7
Reactants collide. The kinetic energy of the collision is incorporated into the internal energy of the resulting complex. If that energy is at least as large as E_{act}, then an activated complex is formed. The double-headed arrows ($\rightleftharpoons$) indicate that the unstable (high-energy) activated complex can either go back to reactants or go on to product.

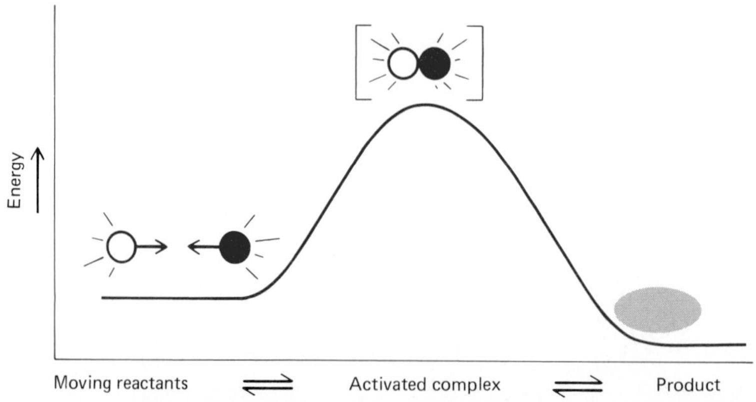
Energy
Moving reactants ⇌ Activated complex ⇌ Product

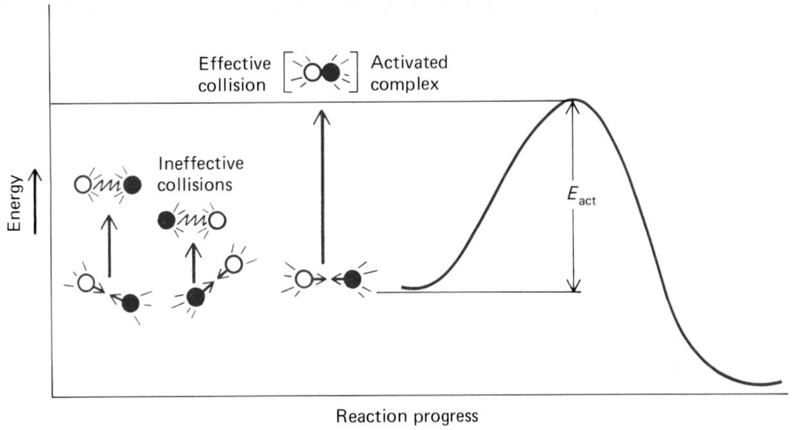
Effective collision
Activated complex
Ineffective collisions
Energy
E_{act}
Reaction progress

FIG. 10.8
Some collisions are effective; that is, the complex has suffficient energy to surmount the E_{act} barrier and go on to product. Many collisions are ineffective. In an ineffective collision the collision complexes do not have sufficient energy to surmount the barrier. They fall apart back to reactant molecules.

lisions are sufficient to explain the effects of increased concentration and temperature on reaction rate.

Concentration Effect

First, let's consider the effect of increasing concentration. An increase in concentration of reactants in a reaction vessel means that moving reactant molecules are crowded more closely together. This leads to a greater collision frequency and therefore an increase in the probability of product formation. An analogy can be made to a game of billiards: the greater the number of billiard balls on the table, the greater the number of possible and probable collisions with a moving ball.

Problem 10.4 Explain how a decrease in reactant concentration leads to a decrease in the rate of product formation.

Temperature Effect

Now let's consider the effect of temperature on reaction rate. An increase in the temperature of a reaction mixture results in an increase in the average kinetic energy of the molecules and thereby an increase in the average molecular speed. This implies that molecules will collide more often and that each collision will be more energetic and thus have a greater chance of overcoming the activation barrier. Therefore, an increase in temperature should yield an increase in the rate of product formation as more colliding reactant molecules reach the activated complex state.

Problem 10.5 Explain how a decrease in temperature leads to a decrease in the rate of product formation.

10.9 REVERSIBLE REACTIONS

So far we have looked at chemical reactions as if they began with an initial set of materials (reactants) and went completely to a final set of materials (products). Actually, in many cases the products can react to re-form the original reactants. Reactions in which the products can themselves react to re-form reactants are called **reversible reactions.**

You have previously seen examples of *reversible* processes. For example, in Section 8.13 you encountered the two opposing proceses, vaporization and condensation, which occur whenever a liquid is in a closed container. You saw that, in a closed container, eventually the rate of condensation equals the rate of vaporization, and the vapor pressure remains constant thereafter (Figure 10.9). It is important to realize that in such a case we are not dealing with a *static* (inactive) mixture of liquid and vapor. Rather, active vaporization and condensation are still taking place, although the processes are difficult to see because they are occurring at the same rate. We call this a state of *dynamic (active) equilibrium.* This term is used whenever two opposing processes coexist at the same rate.

A dynamic equilibrium also exists between a dissolved and undissolved substance in a *saturated* solution.

$$NaCl(s) \rightleftharpoons Na^+(aq) + Cl^-(aq)$$

$$\text{Sugar}(s) \rightleftharpoons \text{sugar}(aq)$$

The two half-headed arrows ($\rightleftharpoons$) indicate that the process is reversible and can proceed in either direction. The rate at which the solid NaCl dissociates into ions is equal to the rate at which ions emerge from solution, and the rate at which sugar dissolves is equal to the rate at which it comes out of solution.

FIG. 10.9
Establishment of liquid-vapor equilibrium.

$$\text{Liquid} \underset{\text{condensation}}{\overset{\text{vaporization}}{\rightleftharpoons}} \text{vapor}$$

At time zero, only vaporization (↑) is occurring. Vaporization continues at a steady rate, as shown by the straight horizontal line on the graph. As vapor forms, condensation slowly starts (↓). At time 1, the condensation rate is R_{c_1} and the rate of vaporization is R_v. At time 2, the condensation rate is R_{c_2} and the vaporization rate remains R_v. Eventually the rate of condensation equals the rate of vaporization and there is a dynamic equilibrium. At equilibrium the liquid and vapor concentration remain fixed.

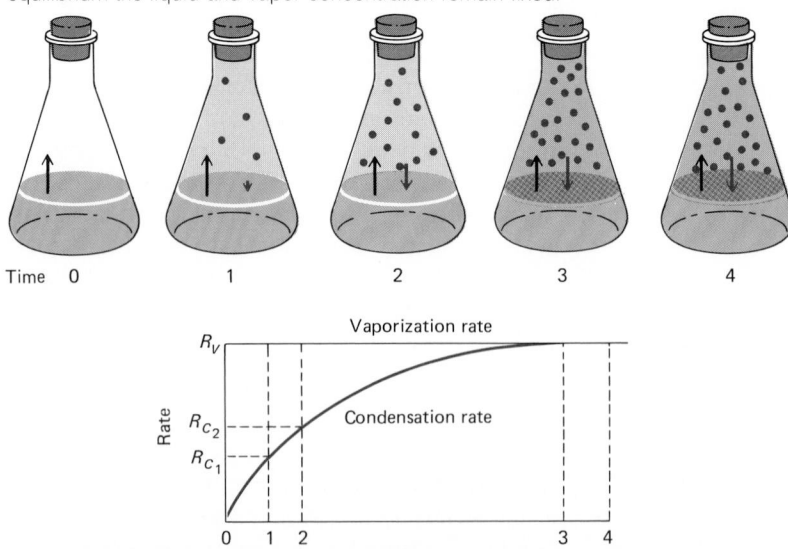

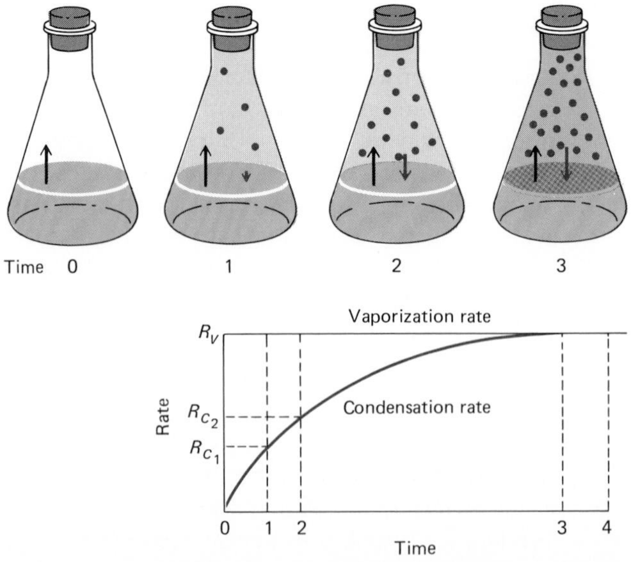
Time 0
1
2
3
4
Vaporization rate
Condensation rate
Rate
R_V
R_{C_2}
R_{C_1}
0 1 2 3 4
Time

With this concept in mind of two opposing processes, one that is exactly the reverse of the other, let us consider an example of a reversible chemical reaction.

If we place 0.1 mol of N_2O_4, a colorless gas, into a sealed 1-L tube and heat it to 100°C, we obtain visible evidence of the formation of a reddish brown gas, which on analysis proves to be NO_2. Thus we have evidence that the reaction $N_2O_4 \longrightarrow 2NO_2$ is occurring. The intensity of the reddish brown color increases rapidly in the beginning, but after a period of time the color reaches a maximum. The contents of the container at this point are found to include both NO_2 and N_2O_4.

If we reverse the experiment and start with 0.1 mol of NO_2, the reddish brown color begins to diminish. This is evidence that the reaction $2NO_2 \longrightarrow N_2O_4$ is occurring. After a period of time the reddish brown color reaches a minimum and does not diminish further. We conclude from these experiments that the reaction is reversible because:

1 If we start with N_2O_4, NO_2 forms, and if we start with NO_2, N_2O_4 forms.
2 The reactant is not totally converted to product in either experiment. Evidence for this lies in the color of the mixture.
   a If we begin with colorless N_2O_4, a reddish brown coloration as dark as pure NO_2 will never develop.
   b If we begin with NO_2, a totally colorless gas indicative of pure N_2O_4 will not be produced.

The reversible reaction between N_2O_4 and NO_2 can be written in the

FIG. 10.10
The $N_2O_4 \rightleftharpoons 2NO_2$ equilibrium. At time zero, beginning with pure, colorless N_2O_4, only the forward reaction occurs at a rate designated by the arrow (↑). (1) As red-brown NO_2 forms, the reverse reaction can occur, but at first the rate is slow because of the small concentration of NO_2. Notice that the rate of the forward reaction has decreased because there is less N_2O_4. (2) The rate of the forward reaction decreases because N_2O_4 decreases. The rate of the reverse reaction increases because NO_2 increases. (3) and (4) Equilibrium is obtained: $rate_{forward} = rate_{reverse}$ and the concentration of N_2O_4 and NO_2 remain fixed. Compare this figure with Figure 10.9.

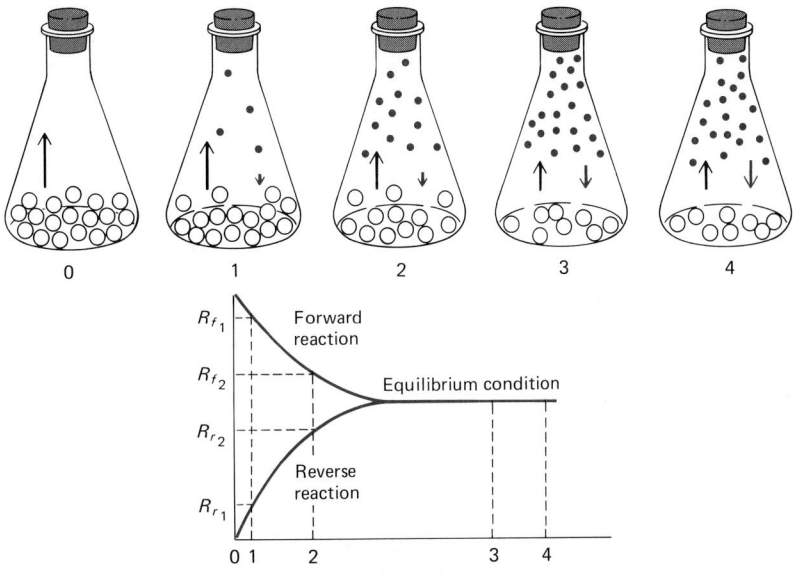

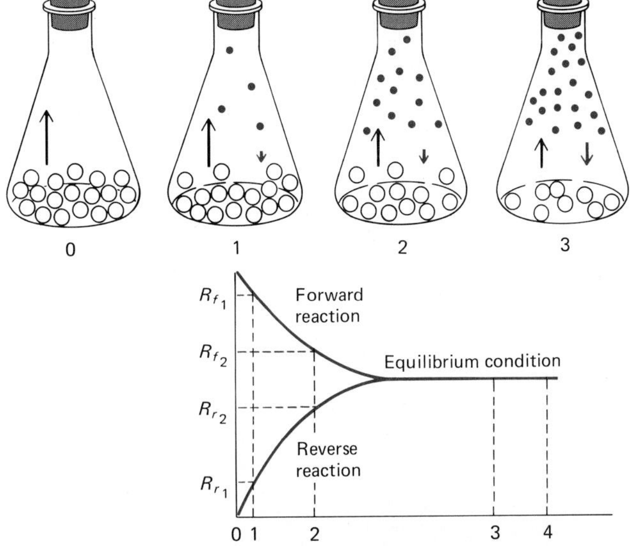
0
1
2
3
4
R_{f_1}
R_{f_2}
R_{r_2}
R_{r_1}
Forward reaction
Equilibrium condition
Reverse reaction
0 1
2
3
4

following way:

$$N_2O_4(g) \rightleftharpoons 2NO_2(g) \tag{10.3}$$

This is an example of dynamic equilibrium for a chemical system, i.e., a **chemical equilibrium.**

Now let us consider the reaction represented by Equation (10.3) in terms of the effect of concentration on reaction rate. The rate of the forward reaction depends on the concentration of N_2O_4, and the rate of the reverse reaction on the NO_2 concentration. If we begin with pure N_2O_4, then initially the concentration of N_2O_4 is large, there are many collisions of N_2O_4 molecules, and so the rate of decomposition to NO_2 is fast. As time passes, the concentration of N_2O_4 decreases, the chances for collisions decrease, and therefore the rate of decomposition decreases. However, the concentration of NO_2 is increasing, and thereby the rate of the reverse reaction increases. At some point the rate of the reverse reaction becomes equal to the rate of the forward reaction and the concentrations of reactants and products no longer change. This process is illustrated diagramatically and graphically in Figure 10.10. When the rate of the forward reaction equals the rate of the reverse reaction, we say that the reaction is at **chemical equilibrium,** and the concentrations of reactants and products at equilibrium are called **equilibrium concentrations.** It is important to remember that the equilibrium condition is a dynamic (active) condition.

At *equilibrium*

$$\textbf{Rate}_{\textbf{forward}} = \textbf{rate}_{\textbf{reverse}}$$

10.10 EQUILIBRIUM CONSTANT

The equality of forward and reverse rates at equilibrium leads to a constancy in the ratio of equilibrium concentrations of products and reactants in any reversible reaction. The constant ratio is known as the **equilibrium constant,** and is symbolized K_{eq}.

$$K_{eq} = \frac{\text{equilibrium product concentration}}{\text{equilibrium reactant concentration}}$$

Equilibrium constant — Constant ratio

In a correct equilibrium-constant expression for a given balanced equation the numerator contains each product concentration raised to a power equal to the coefficient of that product in the balanced equation (1 is understood), and the denominator contains the reactant concentrations raised to their respective powers indicated by their coefficients. For the generalized balanced equation

$$aA + bB \rightleftharpoons cC + dD$$

the equilibrium expression is written

$$K_{eq} = \frac{[C]^c\,[D]^d}{[A]^a\,[B]^b}$$

SAMPLE EXERCISE 10.2

Write an equilibrium-constant expression for the equilibrium

$$N_2O_4(g) \rightleftharpoons 2NO_2(g)$$

Solution STEP 1 In the numerator write the concentration of each product raised to a power equal to its coefficient in the balanced equation:

$$K_{eq} = \frac{[NO_2]^2}{?}$$

STEP 2 In the denominator write the reactant concentrations, each raised to a power equal to its coefficient in the balanced equation:

$$K_{eq} = \frac{[NO_2]^2}{[N_2O_4]}$$

SAMPLE EXERCISE 10.3 When the color in the reaction vessel remained the same, indicating that equilibrium had been established for the gas mixture reaction, a chemist found that the concentration of N_2O_4 in the vessel was 0.04 mol/L and that the concentration of NO_2 was 0.12 mol/L. Calculate the equilibrium constant for the $N_2O_4(g) \rightleftharpoons NO_2(g)$ reaction.

Solution Use the equilibrium-constant expression written in Sample Exercise 10.2:

$$K_{eq} = \frac{[NO_2]^2}{[N_2O_4]}$$

Fill in the appropriate numerical concentrations and do the operations indicated by the exponents and fraction line:

$$K_{eq} = \frac{(0.12)^2}{0.04} = \frac{0.014}{0.04} = 0.36$$

The equilibrium expression should contain the concentration terms for gases and substances in solution only. The concentration term ([solid] or [liquid]) for a pure liquid or solid does not appear in the normal equilibrium expression, because the concentration of a pure liquid or solid is a constant and so it is automatically incorporated into the equilibrium constant. Water acting both as a solvent and as a reactant or product is a commonly encountered pure liquid for which a concentration term does not appear in the equilibrium expression. The water concentration is essentially constant if water is the solvent.

SAMPLE EXERCISE 10.4 Write an equilibrium constant expression for the equilibrium

$$\text{Sucrose}(aq) + H_2O(l) \rightleftharpoons \text{glucose}(aq) + \text{fructose}(aq)$$

Solution In this example water is acting as both the solvent, as indicated by (*aq*) following the other reactant and each of the products, and as a reactant, as can be seen by the fact that water does not appear as a product. The water concentration does not change, and so it should not be included in the equilibrium constant expression.

$$K_{eq} = \frac{[\text{glucose}][\text{fructose}]}{[\text{sucrose}]}$$

Problem 10.6 Write an equilibrium-constant expression for each of the following equilibria:

a Glucose 1-phosphate(*aq*) $\rightleftharpoons$ glucose 6-phosphate(*aq*)

b Maltose(*aq*) + $H_2O \rightleftharpoons$ 2 glucose(*aq*)

10.11 INTERPRETING THE VALUE OF K_{eq}

What is the purpose or usefulness of knowing the value of K_{eq} for a reaction? Consider the simple example

$$R \rightleftharpoons P$$

for which the equilibrium expression is

$$K_{eq} = \frac{[P]}{[R]}$$

Because the equilibrium constant gives us a measure of the concentration of product [*P*] compared with reactant [*R*] concentrations at equilibrium, the value of K_{eq} tells us how far a reaction has proceeded toward product at the equilibrium point.

If the equilibrium constant is large ($>1 \times 10^2$), then the equilibrium system is made up mostly of product. We say that the reaction has proceeded far to the right. If $[P]/[R] = 100$, then clearly the numerator, [*P*], is considerably greater than the denominator, [*R*].

If the equilibrium constant is small ($<1 \times 10^{-2}$), the equilibrium condition consists mainly of reactants. The equilibrium is said to be far to the left. If $[P]/[R] = 0.01$, then clearly the denominator, [*R*], is considerably greater than the numerator, [*P*].

For reactions in which the numerical value of K_{eq} is between 1×10^{-2} and 1×10^2, the equilibrium state is a mixture containing significant quantities of products and reactants. Table 10.1 compares K_{eq} values and product and reactant concentrations at equilibrium, and Sample Exercise 10.5 offers some illustrations.

SAMPLE EXERCISE 10.5

Given K_{eq}, describe qualitatively the equilibrium composition for each of the following reactions:

a Glucose 1-phosphate(*aq*) $\rightleftharpoons$ glucose 6-phosphate(*aq*) K_{eq} (at 25°C) = 17

b Alanine(*aq*) + glycine(*aq*) $\rightleftharpoons$ alanylglycine(*aq*) + $H_2O(l)$
K_{eq} (at 25°C) = 1.3×10^{-3}

c Sucrose(*aq*) + $H_2O(l)$ $\rightleftharpoons$ glucose(*aq*) + fructose(*aq*)
K_{eq} (at 25°C) = 1×10^5

d Glucose(*aq*) + $O_2(g)$ $\rightleftharpoons$ 2 pyruvate$^-$(*aq*) + $2H^+(aq)$ + $2H_2O(l)$
K_{eq} (at 25°C) = 1×10^{88}

TABLE 10.1 Comparison of K_{eq} Value and Reactant and Product Concentrations at Equilibrium

K_{eq}	Description of Equilibrium Condition
1×10^{30} 1×10^{20} 1×10^{10}	Essentially all products at equilibrium. More products than reactants at equilibrium.
1×10^{2} 1 1×10^{-2}	Significant amounts of products and reactants at equilibrium.
1×10^{-10} 1×10^{-20} 1×10^{-30}	More reactants than products at equilibrium. Essentially all reactants at equilibrium. Reaction has not occurred to any significant extent.

Solution

a According to the guidelines given in Table 10.1, the equilibrium composition consists of significant quantities of products and reactants because K_{eq} is between 1×10^{-2} and 1×10^{2}.

b There is a greater amount of reactant than product at equilibrium.

c There is a much greater amount of product than reactant in the equilibrium state.

d The equilibrium state contains essentially only product.

Problem 10.7 Given K_{eq} and the guidelines in Table 10.1, describe qualitatively the equilibrium composition for each of the following reactions:

a Fructose 6-phosphate(*aq*) $\rightleftharpoons$ glucose 6-phosphate(*aq*) $K_{eq} = 2.0$

b ADP + phosphate $\rightleftharpoons$ ATP K_{eq} (at 37°C) $= 1.3 \times 10^{6}$

10.12 LE CHATELIER'S PRINCIPLE

It is possible to change the point at which equilibrium is achieved, which is called the position of equilibrium, and therefore the equilibrium concentrations of products and reactants by altering the conditions of concentration, pressure, or temperature that a system in equilibrium experiences. This means that we can sometimes "shift" a reaction in such a way as to produce more of a desired product.

Shifts in equilibrium are governed by **Le Chatelier's principle,** which states that if a stress is applied to a system in equilibrium, the equilibrium will shift so as to relieve that stress. Whenever a physiological reaction or process is reversible, which is often the case, Le Chatelier's principle explains how a change in condition leads to the particular observed consequences. For example, transport of oxygen by hemoglobin in red blood cells from the lungs, through the bloodstream, to the tissues depends on the reversible reaction

$$\underset{\text{Hemoglobin}}{Hb{-}H^+(aq)} + O_2(g) \rightleftharpoons \underset{\text{Oxygenated hemoglobin}}{HbO_2(aq)} + H^+(aq)$$

As you are about to see, exchange of oxygen occurs by equilibrium shifts on the basis of Le Chatelier's principle.

Effect of a Change in Concentration

Let us examine how Le Chatelier's principle applies to changing concentrations, the most important stress under physiological conditions, and how, in particular, it affects the yield of oxygenated hemoglobin (HbO_2) in the reaction

$$Hb{-}H^+(aq) + O_2(g) \rightleftharpoons HbO_2(aq) + H^+(aq)$$

Oxygenated hemoglobin (HbO_2) is the carrier of oxygen in blood (Section 19.12). $Hb{-}H^+$ is the symbolic form of (unoxygenated) hemoglobin which upon the binding of an oxygen molecule loses H^+, the hydrogen ion.

The starting point in this discussion is always the equilibrium condition; i.e., we start with equilibrium concentrations of $Hb{-}H^+$, O_2, HbO_2, and H^+. Increasing the concentration of either of the reactants, $Hb{-}H^+$ or O_2, for example, by the administration of pure oxygen gas, will cause the equilibrium to shift to the product side away from the stress of added reactant concentration. The equilibrium will reestablish itself with a higher concentration of oxygenated hemoglobin and a lower concentration of the reactant that was not added.

Equilibrium: $Hb—H^+ + O_2 \rightleftharpoons HbO_2 + H^+$
Stress by adding O_2
Shift $\rightarrow$ to consume O_2 and relieve stress

Such shifting also consumes $Hb—H^+$ and produces more HbO_2 and H^+.

If the equilibrium is disturbed by the removal of hydrogen ion, which occurs in the small blood vessels of the lungs, the system reacts to this stress of lowered product concentration by shifting to the product side. The result will be an increase in the concentration of oxygenated hemoglobin. $Hb—H^+$ and O_2 will be consumed.

Equilibrium: $Hb—H^+ + O_2 \rightleftharpoons HbO_2 + H^+$
Stress by removing H^+
Shift $\rightarrow$ to replace H^+ and relieve stress

Such shifting also consumes $Hb—H^+$ and O_2 and produces HbO_2.

If the equilibrium is disturbed by increased hydrogen ion, a condition known as acidosis which can arise in untreated diabetes or starvation, the system reacts to this stress of increased product concentration by shifting to the reactant side. The result is an increase in the equilibrium concentrations of $Hb—H^+$ and O_2.

Equilibrium: $Hb—H^+ + O_2 \rightleftharpoons HbO_2 + H^+$
Stress by adding H^+
Shift $\leftarrow$ to consume H^+ and relieve stress

Such shifting also increases $Hb—H^+$ and O_2 and decreases HbO_2.

Effect of Temperature

You have previously seen (Section 7.17) that heat can be treated as if it were a product in an exothermic reaction or a reactant in an endothermic reaction. Increasing the temperature can thus be thought of as producing more heat for the reaction, and decreasing the temperature as providing less heat for the reaction.

The reaction of hydrogen and nitrogen to produce ammonia is exothermic (It is also exergonic at room temperature or below.):

$$N_2 + 3H_2 \rightleftharpoons 2NH_3 + \text{heat}$$

Increasing the temperature of this reaction causes the equilibrium to shift to the reactant side, thereby decreasing the yield of ammonia. On the other hand, the reaction responds to the stress of a decreased temperature by shifting the equilibrium to the product side, thus increasing the yield of ammonia.

Equilibrium: $N_2 + 3NH_2 \rightleftharpoons 2NH_3 + \text{heat}$

Stress by increased T means more heat

Shift ← to consume heat

Such shifting also consumes NH_3.

Stress by decreased T means less heat

Shift → to produce heat

Such shifting also produces NH_3.

Effect of Pressure

A change in pressure affects the equilibrium of only those reactions which undergo a change in volume in proceeding from reactant to product. In practice, this means that pressure changes affect those reactions which show a change in the number of moles of gaseous materials. The stress of an increased pressure (decreased V) is relieved by a shift in the equilibrium to the side having fewer gas molecules. A decrease in pressure (increased V) shifts the equilibrium to the side having a greater number of gas molecules. In our ammonia example,

$$N_2(g) + 3H_2(g) \rightleftharpoons 2NH_3(g)$$

1 mol of N_2 + 3 mol of H_2 2 mol of NH_3

4 mol of gaseous reactants are converted into 2 mol of gaseous products. An increase in pressure P will shift the equilibrium to increase the yield of ammonia. Decreased P has the opposite effect.

Equilibrium: $N_2(g) + 3H_2(g) \rightleftharpoons 2NH_3(g)$

Stress by increasing P and decreasing V

Shift → to produce fewer moles of gas and relieve stress

Such shifting also produces NH_3.

Stress by decreasing P and increasing V

Shift ← to produce more moles of gas and relieve stress

Such shifting also consumes NH_3.

Le Chatelier's principle indicates that the best equilibrium yield of ammonia should be obtained from using an excess concentration of one of the reactants at a low temperature and a high pressure. Fritz Haber applied these ideas to the commercial synthesis of ammonia, a very important chemical in the manufacture of farm fertilizers.

CHAPTER ACCOMPLISHMENTS

After completing this chapter you should be able to define **key words** and do the following:

10.1 INTRODUCTION

1. Cite the two states with which equilibrium or thermodynamics is concerned.
2. Distinguish between thermodynamic considerations and kinetic considerations with respect to chemical reactions.

10.2 SPEED VERSUS SPONTANEITY

3. Distinguish between spontaneity and speed.

10.3 THE ORIGIN OF HEATS OF REACTION (ΔH)

4. Describe how heats of reaction are related to bond dissociation energies.
5. State the relationship between the algebraic sign of ΔH and the exothermic or endothermic nature of a reaction.

10.4 GIBBS FREE ENERGY

6. Give specific examples of processes involving increasing and decreasing entropy changes.
7. State the criterion for spontaneity in a chemical reaction.
8. State the relationship between the algebraic sign of the free-energy change and exergonic and endergonic reactions.

10.5 ACTIVATION ENERGY

9. Describe what is meant by a reaction pathway.
10. Tell why there is a necessary energy input into an exergonic reaction.

10.6 REACTION PROGRESS DIAGRAMS

11. Describe what is being plotted in a reaction progress diagram.
12. Describe the general shapes of reaction progress diagrams for exergonic and endergonic reactions.

10.7 PATHWAYS, E_{act}, AND KINETICS

13. Describe the relationship between reaction rate and activation energy.
14. Describe the relationship between a catalyst and the pathway of a reaction.
15. Modify a given reaction progress diagram to show the effect of a catalyst.
16. Determine the free-energy change of a reaction from a given reaction progress diagram.
17. Determine the activation energies from a given reaction progress diagram showing an uncatalyzed and a catalyzed reaction.

10.8 FACTORS INFLUENCING REACTION RATE

18. State three factors which influence the rate of a reaction.
19. Describe the collision model of a chemical reaction.
20. Describe how the collision model accounts for the concentration and temperature effects on reaction rate.

10.9 REVERSIBLE REACTIONS

21. Relate the terms *reversible reaction* and *dynamic equilibrium.*

10.10 EQUILIBRIUM CONSTANT

22. Write the equilibrium-constant expression for a reaction given a balanced chemical equation.

10.11 INTERPRETING THE VALUE OF K_{eq}

23. Given a value of K_{eq}, qualitatively describe the equilibrium composition of a given reaction.

10.12 LE CHATELIER'S PRINCIPLE

24. Given a chemical equation, describe the effect on the equilibrium concentrations of a change in concentration of one species.
25. Given an exothermic or endothermic chemical reaction, describe the effect of a change in temperature on the equilibrium concentrations.
26. Given a chemical equation involving one or more gases, describe the effect on the equilibrium concentrations of a change in total pressure on the reaction.

PROBLEMS

10.8 The reaction of hydrogen and oxygen to yield water is highly exothermic.

$$2H_2(g) + O_2(g) \rightleftharpoons 2H_2O(g) + \text{heat}$$

Which has a lower enthapy content: 2 mol of water, or the combination of 2 mol of H_2 and 1 mol of O_2?

10.9 Explain in your own words what is meant by a spontaneous reaction.

10.10 Is a spontaneous reaction necessarily a fast reaction? Explain and give an example to justify your answer.

10.11 Based on your everyday knowledge of the two substances, does gasoline or water have a higher amount of chemical energy? Explain your answer.

10.12 Is the reaction $Br_2 \longrightarrow 2Br$ exo or endothermic? Explain how you can tell.

10.13 Consider the reaction:
$C + O{=}O \longrightarrow O{=}C{=}O$.
a What bonds are broken on the reactant side?
b What bonds are formed on the product side?

10.14 Define the term *entropy* in your own words.

10.15 Events in nature have a tendency to proceed toward greater entropy. Comment on this statement, using specific examples to justify your answer.

10.16 In each of the following equations, which side has the greater entropy?
a Stones + mortar ⟶ stone wall
b Liquid water ⟶ ice
c An initial telephone conversation ⟶ the conversation after 10 repetitions
d Sun + water + soil + fertilizer + seeds ⟶ a tomato
e A living, functioning cell ⟶ a dead cell

10.17 State the determining criterion for a spontaneous process.

10.18 *a* Give a definition for the terms *exergonic* and *endergonic*.
b A chemical reaction is endothermic; *must* it also be endergonic? Explain.
c A particular process is exergonic; must it also be exothermic? Explain.

10.19 Indicate which of the following changes must be spontaneous:
a An exothermic reaction with a positive entropy change
b A reaction in which the enthalpy content decreases with a positive entropy change
c An endothermic reaction with a positive entropy change
d An endothermic reaction with a negative entropy change
e An exothermic reaction which takes place at absolute zero (−273°C)

10.20 Glucose can be formed from starch and water. Does this statement describe the pathway of the reaction? Explain.

10.21 Define the term *activation energy*.

10.22 Explain why even an exergonic reaction has an activation energy.

10.23 *a* Draw and label a general reaction progress diagram for an exergonic reaction.
b Draw and label a general reaction progress diagram for an endergonic reaction.

10.24

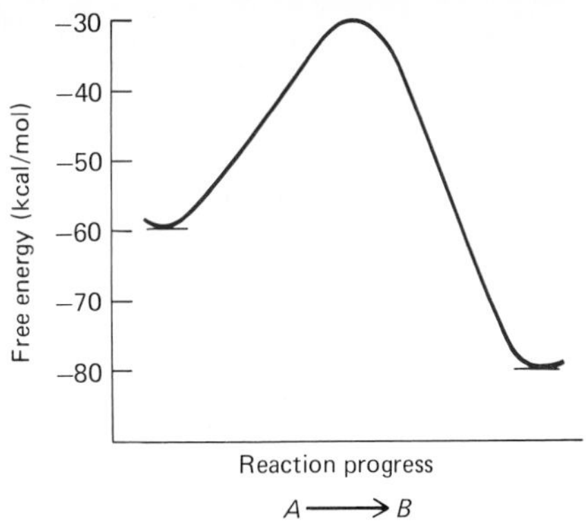
Free energy (kcal/mol)
−30
−40
−50
−60
−70
−80
Reaction progress
A⟶B

a Calculate the free-energy change for the reaction shown in the figure.
b Is the reaction exergonic or endergonic?
c Calculate the activation energy for the change *A→B*.

10.25 Describe how an experimenter can provide the required activation energy for a reaction.

10.26 Explain why a highly exothermic reaction with a low activation energy might tend to become explosive as it proceeds.

10.27 Explain the general method by which a catalyst speeds up a chemical reaction.

10.28 When an enzyme is added to the reaction of Problem 10.24, the reaction progress diagram becomes:

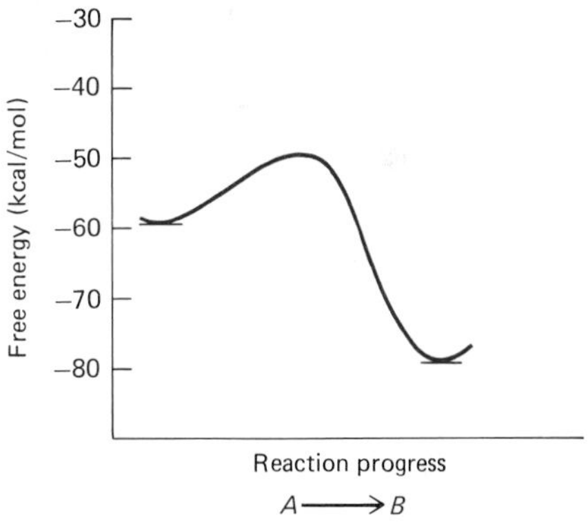
Free energy (kcal/mol)
−30
−40
−50
−60
−70
−80
Reaction progress
A⟶B

a Is the enzyme-catalyzed reaction faster or slower than the uncatalyzed reaction? How do you know?
b Compare the free-energy change for the catalyzed reaction to the uncatalyzed reaction. Is there a difference? Explain.

10.29 Describe the relationship between the magnitude of the activation energy and the rate of a chemical reaction.

10.30 State one way to affect the value of the activation energy.

10.31 State two factors, other than activation energy, which affect the rate of reaction.

10.32 *a* Describe a chemical reaction on the molecular level. Use the concepts of collision theory and activated complex in your model.

b Show how your model accounts for an increased rate of reaction with increasing reactant concentrations.

c Describe how your model accounts for an increased rate of reaction with increased temperature.

10.33 Does every collision between two species lead to a chemical reaction? What is the key determining factor as to whether there is a reaction?

10.34 Describe two nonchemical examples of a dynamic equilibrium.

10.35 A dynamic equilibrium exists according to the following equation:

$$A(g) + B(g) \rightleftharpoons C(g) + D(g)$$

A is a yellowish gas, *D* is a bluish gas, and *B* and *C* are colorless gases.

a Predict the observable changes that occur when 0.1 mol of *A* and 0.1 mol of *B* are placed in an empty container.

b Predict the observable changes that occur if the reaction is begun with 0.1 mol of *C* and 0.1 mol of *D*.

c Describe the appearance of the system if the reaction is left at equilibrium.

10.36 *a* Describe how the evaporation of water in a closed container leads to a dynamic-equilibrium condition.

b Would the evaporation of water in an open container lead to a dynamic equilibrium? Explain.

10.37 What statement can be made concerning the rates of the forward and reverse reactions at the equilibrium condition?

10.38 Write an equilibrium-constant expression for each of the following equilibria:

a α-D-Glucose(*aq*) $\rightleftharpoons$ β-D-glucose(*aq*)

b $CH_3OH(aq) + CH_3COOH(aq) \rightleftharpoons CH_3COOCH_3(aq) + H_2O(l)$

c Hemoglobin(*aq*) + oxygen(*g*) $\rightleftharpoons$ oxyhemoglobin(*aq*)

d Double-strand DNA $\rightleftharpoons$ 2 single-strand DNA

10.39 Write an equilibrium-constant expression for each of the following equilibria:

a $H_2(g) + Cl_2(g) \rightleftharpoons 2HCl(g)$

b $PCl_5(g) \rightleftharpoons PCl_3(g) + Cl_2(g)$

c $CO(g) + H_2O(g) \rightleftharpoons CO_2(g) + H_2(g)$

d $4NH_3(g) + 5O_2(g) \rightleftharpoons 4NO(g) + 6H_2O(g)$

10.40 What is the equilibrium-constant expression for each of the following:

a $H_2(g) + Br_2(g) \rightleftharpoons 2HBr(g)$

b $CH_4(g) + Cl_2(g) \rightleftharpoons CH_3Cl(g) + HCl(g)$

c $2NO_2(g) \rightleftharpoons N_2(g) + 2O_2(g)$

d $4NH_3(g) + 3O_2(g) \rightleftharpoons 2N_2(g) + 6H_2O(g)$

10.41 Use Table 10.1 and the given K_{eq} to qualitatively describe the equilibrium composition:

a ATP + $H_2O \rightleftharpoons$ ADP + HPO_4^{2-}
K_{eq} (at 25°) = 1.3×10^5

b Arginine + $HPO_4^{2-} \rightleftharpoons$ arginine-phosphate + H_2O K_{eq} (at 25°) = 7.9×10^{-6}

c Glucose-6-phosphate + arginine $\rightleftharpoons$ glucose + arginine-phosphate
K_{eq} (at 37°) = 1.5×10^{-3}

d ATP + CH_3COOH + coenzyme A $\rightleftharpoons$ AMP + pyrophosphate + acetylcoenzyme A
K_{eq} (at 37°) = 1.0

10.42 Use Table 10.1 and the given K_{eq} to qualitatively describe the equilibrium composition of each of the following reactions:

a $2NO_2(g) \rightleftharpoons N_2(g) + 2O_2(g)$
K_{eq} (at 25°C) = 6.7×10^{16}

b $CH_3OH(g) \rightleftharpoons CO(g) + 2H_2(g)$
K_{eq} (at 100°C) = 7.37×10^{-8}

c $N_2(g) + O_2(g) \rightleftharpoons 2NO(g)$
K_{eq} (at 2400°C) = 3.4×10^{-3}

d $2SiO(g) \rightleftharpoons 2Si(l) + O_2(g)$
K_{eq} (at 1727°C) = 9.62×10^{-13}

10.43 State Le Chatelier's principle in your own words.

10.44 The equilibrium between carbonic acid (H_2CO_3) and bicarbonate ion is a very important one in the body's buffer system. State the effect, if any, of the following changes in the position of the equilibrium

$$H_2CO_3(aq) \rightleftharpoons H^+(aq) + HCO_3^-(aq)$$

a Adding H_2CO_3 to the equilibrium mixture

b Removing H^+ as it is formed

c Adding H^+ to the equilibrium mixture

d Adding HCO_3^- to the equilibrium mixture

e Removing H_2CO_3 as it is formed

10.45 Reconsider Problem 10.44*c*. What happens to the bicarbonate (HCO_3^-) concentration upon addition of H^+ and the consequent shift?

10.46 An equilibrium that exists in solutions of carbon dioxide such as soda and beer is

$$H_2CO_3(aq) \rightleftharpoons CO_2(g) + H_2O(l)$$

Use Le Chatelier's principle to explain why an open bottle of beer goes flat.

10.47 State the effect, if any, of the following changes on the position of equilibrium for the reaction

$$4NH_3(g) + 3O_2(g) \rightleftharpoons 2N_2(g) + 6H_2O(g) + 366 \text{ kcal}$$

a Increasing the concentration of oxygen

b Adding N_2 to the equilibrium mixture

c Removing H_2O as it is formed

d Increasing the reaction temperature

e Decreasing the total pressure

10.48 Would you heat or chill the equilibrium mixture in Problem 10.47 to achieve a best yield of N_2?

10.49 State the effect, if any, of the following changes on the position of equilibrium for the reaction

$$2SO_3(g) + 47\ \text{kcal} \rightleftharpoons 2SO_2(g) + O_2(g)$$

a Increasing the concentration of SO_3
b Adding O_2 to the equilibrium mixture
c Removing O_2 as it is formed
d Increasing the reaction temperature
e Decreasing the total pressure

10.50 What is the best combination of temperature and pressure conditions to achieve the decomposition of SO_3? (See Problem 10.49.)

10.51 *a* Describe the effect of a catalyst on the activation energy of a chemical reaction.
b Describe the effect of a catalyst on the free-energy change of a chemical reaction.

10.52 Criticize the following statement: "A reaction with a positive free-energy change can be made spontaneous by the use of an enzyme."

CHAPTER ELEVEN

ACIDS AND BASES

11.1 INTRODUCTION

We have daily contacts with acids and bases. Acids are necessary for digestion of proteins in the stomach (hydrochloric acid); they are found on salads in vinegar (acetic acid), in fruits such as lemons (citric acid), in battery acid (sulfuric acid); and the drug, aspirin, is acetylsalicylic acid. Bases occur in lye (sodium hydroxide), window cleaner (ammonia solution), soaps and detergents (mixtures of various bases), and antacids (aluminum hydroxide, magnesium hydroxide, and sodium bicarbonate). Figure 11.1 depicts some familiar acid and base products.

Table 11.1 lists the names and formulas of some acids and bases and cites their occurrence or use in food, medicine, agriculture, or industry. The name of the acid actually applies to the aqueous solution. Hydrochloric acid, for example, is formed by dissolving hydrogen chloride, a gas, in water. Notice that with the exception of the nitrogen bases, ammonia and methylamine, all the bases in the table are ionic compounds, which is to say, combinations of metal ions and hydroxide ions.

Practitioners in the health-related, nutritional, and agricultural professions should be able to recognize acids and bases and be aware of the physical and chemical properties of acidic and basic solutions. An understanding of buffers, sweet and sour soils, acid indigestion, acidosis, and alkalosis requires a knowledge of acidity and basicity. Table 11.2 lists some general properties of acids and bases. Although we will discuss some of these properties in more detail as this chapter unfolds, you can begin to recognize various foods, drugs, agricultural products, and other commonly encountered substances as acidic or basic by using this table.

We have delayed a discussion of acids and bases to this point because it required a knowledge of solutions and ionic equations (Sections 9.6 and 9.7) and equilibrium concepts (Chapter 10). As you proceed in this chapter and whenever you encounter acids and bases, you will find it helpful to form the following associations:

Associate **acid with H^+** (hydrogen ion or proton, that is, an H atom without its one electron).

Associate **base with OH^-** (hydroxide ion).

Let us see how these associations hold within common definitions of acids and bases.

11.2 THE ARRHENIUS DEFINITION

As scientific study of acids and bases proceeded through the nineteenth and twentieth centuries, various definitions of these substances were formulated. These different definitions do not contradict one another. The newer defi-

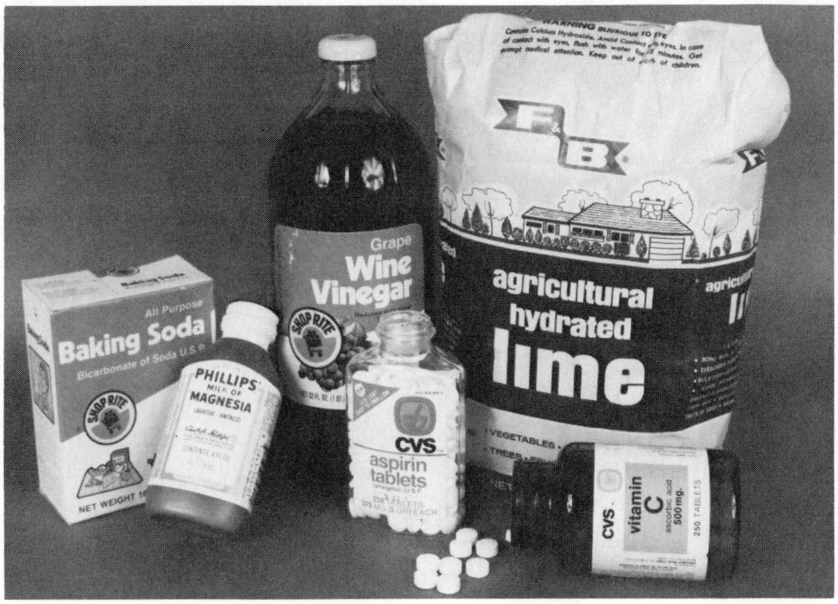
Baking Soda
PHILLIPS' MILK OF MAGNESIA
Grape Wine Vinegar
agricultural hydrated lime
CVS aspirin tablets
CVS vitamin C

FIG. 11.1 Some commonly encountered acids and bases. **Acids**: Vinegar is acetic acid; aspirin is acetylsalicylic acid; vitamin C is ascorbic acid. **Bases**: Soil sweetener is lime, which is essentially $Ca(OH)_2$; milk of magnesia is $Mg(OH)_2$; the bicarbonate anion (HCO_3^-) of baking soda acts as a Brønsted-Lowry base. *(Photograph by Bryan Lees.)*

nitions, such as that of Brønsted and Lowry (Section 11.3), are broader and more general definitions that include the older ones, such as that of Arrhenius. We will examine both definitions, because sometimes one offers a clearer explanation of some observed phenomenon, and sometimes the other definition provides more clarity.

TABLE 11.1 Common Acids and Bases

Name	Formula	Use or Occurrence
ACIDS, HA		
Boric acid	H_3BO_3	Eyewash
Carbonic acid	H_2CO_3	Carbonated water
Hydrochloric acid	HCl	Gastric juice
Nitric acid	HNO_3	Test for proteins
Phosphoric acid	H_3PO_4	Soft drink additive
Sulfuric acid	H_2SO_4	Manufacture of fertilizers
ORGANIC ACIDS, RCOOH*		
Acetic acid	CH_3COOH	Vinegar
Acetylsalicylic acid	$C_6H_4(OOCCH_3)COOH$	Aspirin
Benzoic acid	C_6H_5COOH	Preservative in foods
Lactic acid	$CH_3CH(OH)COOH$	Formed in muscles
Formic acid	$HCOOH$	Insect bites
BASES, METAL HYDROXIDES		
Aluminum hydroxide	$Al(OH)_3$	Antiperspirant
Calcium hydroxide	$Ca(OH)_2$	In cement
Magnesium hydroxide	$Mg(OH)_2$	Milk of magnesia
Sodium hydroxide	$NaOH$	Lye
BASES, NITROGEN-CONTAINING		
Ammonia	NH_3	Respiratory stimulant
Methylamine	CH_3NH_2	Tanning of leather

* Organic acids contain the structural fragment, $-\overset{\overset{O}{\|}}{C}-OH$, which is often written —COOH. The rest of the acid molecule is represented by R—.

TABLE 11.2 Properties of Acids and Bases

	Property
Acids	Turn litmus (a dye) red Taste sour (not a recommended general test) React with base in a neutralization reaction
Bases	Turn litmus blue (remember *b* for blue, *b* for base) Taste bitter Feel slippery (soapy) React with an acid in a neutralization reaction

In 1884 Svante Arrhenius defined acids and bases in terms of the ions which they release in aqueous solution:

Acids release H^+ into water.

Bases release OH^- into water.

Considering the metal hydroxide (MOH) bases, the origin of OH^- ions in solutions of these compounds is obvious. These are ionic compounds made up of metal cations and hydroxide anions in the solid state. When ionic compounds dissolve, the ions separate and are solvated by water (see Section 9.3). We can represent this **dissociation** by using a general equation for the solubility equilibrium in a saturated base solution:

$$MOH(s) \rightleftharpoons M^+(aq) + OH^-(aq)$$

Problem 11.1 Use equations to show how KOH(*s*) and $Ca(OH)_2(s)$ release OH^- ions into solution.

Acids are molecular compounds which undergo ionization in aqueous solution; indeed, as you saw in Section 9.5, they are the principal examples of molecular electrolytes. **Ionization** is the reaction of a molecular compound with water to produce ions. In the case of the ionization of molecular compounds of the general formula HA, one of the ions produced is H^+; therefore we classify HA as an **Arrhenius acid.**

$$HA(g \text{ or } l) \rightleftharpoons H^+(aq) + A^-(aq)$$

This reaction is more correctly represented as

$$HA(g \text{ or } l) + H_2O \rightleftharpoons \underbrace{H_3O^+(aq)}_{\text{Hydronium ion}} + A^-(aq)$$

which shows that HA ionizes by reaction with H_2O. For simplicity, we often just write $H^+(aq)$, as before. However, you should remember that an H^+ in aqueous solution will always form a coordinate covalent bond with a lone pair of electrons on water's oxygen. This arrangement is the hydronium ion.

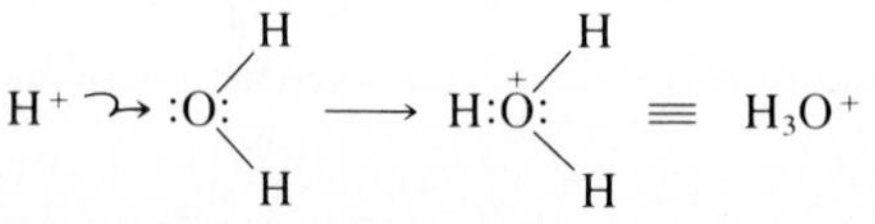
H
H⁺
:O:
H
H:O:
H
H
≡ H₃O⁺

Organic compounds that have as part of their structural formulas the group, $\overset{\displaystyle O}{\overset{\|}{—C—}}OH$, are classified as acids because the H bonded to O is ionizable. That is, these compounds undergo an ionization in aqueous solution which can be represented by

$$\underbrace{CH_3—\overset{\displaystyle O}{\overset{\|}{C}}—OH}_{\substack{\text{A—H} \\ \text{Acetic acid}}}(s \text{ or } l) \rightleftharpoons \underbrace{CH_3—\overset{\displaystyle O}{\overset{\|}{C}}—O^-}_{\substack{\text{A}^- \\ \text{Acetate ion}}}(aq) + H^+(aq)$$

Thus, such compounds also fit the Arrhenius definition of an acid. The example in this case is acetic acid. The ionization produces $H^+(aq)$ and the acetate anion.

11.3 BRØNSTED-LOWRY DEFINITION

If you examine Table 11.1 again, you will find that all the acids and bases there easily fit the Arrhenius definitions except for ammonia and methylamine. That is, the bases are of the form $M(OH)_n$ and the acids are of the form HA or contain the group—COOH. Because of NH_3 and some other materials with acidic or basic properties that do not obviously fit the Arrhenius definitions, J. N. Brønsted and T. M. Lowry offered new definitions in 1923:

Acids are H^+ (proton) donors.

Bases are H^+ (proton) acceptors.

The Brønsted-Lowry definition of acids is essentially identical to the Arrhenius definition. An acid donates H^+ or releases H^+ into the solution.

In the Brønsted-Lowry theory, for compounds of the type MOH, M^+ is regarded strictly as a spectator ion (see Section 9.7). The true basic species is the hydroxide anion (OH^-), because it is OH^- that will quite readily accept H^+ (a proton) to form water.

$$\underset{\text{Hydroxide}}{OH^-} + \underset{\text{Proton}}{H^+} \longrightarrow \underset{\text{Water}}{H_2O}$$

Hydroxide ion (OH^-) is clearly a base because it accepts H^+. MOH is a base because in water it dissociates into M^+ and OH^- ions. It is convenient to ignore M^+ when writing Brønsted-Lowry acid-base expressions that involve metal hydroxide bases. The Brønsted-Lowry definition of bases allows us to understand ammonia's (NH_3) basicity by recognizing that the lone pair on nitrogen is available for bonding with H^+. NH_3 can accept H^+,

$$\ddot{N}H_3 + H^+ \rightleftharpoons \underset{+}{\overset{H}{\ddot{N}}}H_3 \equiv NH_4^+$$

There is a group of organic compounds, called amines, which have properties similar to ammonia. Methylamine, CH_3NH_2, is the simplest example. Methylamine and all other organic amines are bases by the Brønsted-Lowry definition because the lone pair on their nitrogen can accept protons.

$$CH_3\ddot{N}H_2 + H^+ \rightleftharpoons CH_3\underset{+}{\overset{H}{\ddot{N}}}H_2 \equiv CH_3NH_3^+$$

Many drugs such as quinine, morphine, and caffeine (see Chapter 16) are amines and have the bitter taste typical of bases. Histamine is an amine involved in allergic reactions, and it is basic.

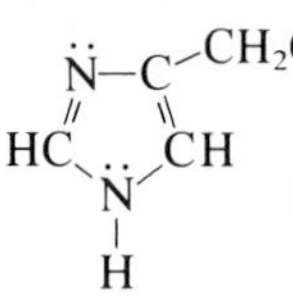

Histamine

For an aqueous solution of NH_3, water is the source of H^+,

$$H_3N\colon + H\colon\ddot{\underset{..}{O}}—H \rightleftharpoons H_3\overset{+}{N}\colon H + {}^-\colon\ddot{\underset{..}{O}}—H$$

This equation shows us that NH_3 is a base in the Arrhenius sense also because the reaction of NH_3 with H_2O releases OH^- ions in solution.

Notice in the ammonia equilibrium

$$NH_3(g) + H_2O(l) \rightleftharpoons NH_4^+(aq) + OH^-(aq)$$

that water acts as a Brønsted-Lowry acid because it donates H^+ to NH_3. When we examine the preceding equilibrium from right to left (←), we observe that the ammonium ion (NH_4^+) acts as an acid (i.e., it gives up H^+) and hydroxide (OH^-) acts as a base (i.e., it accepts H^+). The NH_4^+ ion is said to be the **conjugate acid** of the base NH_3. The hydroxide ion (OH^-) is the **conjugate base** of the acid H_2O. Every acid has a conjugate base, the species remaining after the acid has donated a hydrogen ion. Every base has a conjugate acid, the species formed after the base has accepted a hydrogen ion.

$$\underset{\text{Acid}}{HA} + \underset{\text{Base}}{B} \rightleftharpoons \underset{\substack{\text{Conjugate}\\\text{acid}}}{HB^+} + \underset{\substack{\text{Conjugate}\\\text{base}}}{A^-}$$

Water, which we saw act as a Brønsted-Lowry acid in the preceding example, can also act as a Brønsted-Lowry base. Water acts as a base when it accepts H^+ to form the hydronium ion. For example, when hydrogen chloride gas is dissolved in water, the following reaction occurs:

$$HCl(g) + H_2O \rightleftharpoons H_3O^+(aq) + Cl^-(aq)$$

Here HCl is an acid because it donates a hydrogen ion. H_2O is a base in this reaction because it accepts a hydrogen ion. H_3O^+ is the conjugate acid of the base H_2O, and Cl^- is the conjugate base of the acid HCl. A substance such as H_2O which can act as both an acid and a base is called **amphoteric.**

SAMPLE EXERCISE 11.1

Identify the acid-conjugate base and base-conjugate acid pairs for the equilibrium

$$H_2O + CH_3NH_2 \rightleftharpoons CH_3NH_3^+ + OH^-$$

Solution

We recognize *H_2O as an acid,* in this equation, by noticing that it has given up H^+ as the reaction proceeds from left to right. *OH^- is the conjugate base,* the species left when H^+ is released from the acid H_2O. We can recognize *CH_3NH_2 as a base* from Table 11.1 or by noticing that it accepts H^+ as the reaction proceeds from left to right. *$CH_3NH_3^+$ is the conjugate acid,* the species formed when the base CH_3NH_2 accepts H^+.

Problem 11.2 Label the acid, base, conjugate acid, and conjugate base for the equilibria:

a $HCl + NH_3 \rightleftharpoons NH_4^+ + Cl^-$

b $OH^- + CH_3{-}\underset{\underset{O}{\Vert}}{C}{-}OH \rightleftharpoons CH_3{-}\underset{\underset{O}{\Vert}}{C}{-}O^- + H_2O$

Clearly our experience is that water does not display the characteristics of either an acid or a base as they are described in Table 11.2. This brings us to the important point that any substance that donates H^+ can be classified *theoretically* as an acid. However, unless the concentration of H^+ produced reaches some lower limit (10^{-6} *M*), we do not witness the properties in Table 11.2. It is easy to recognize the acids and bases listed in Table 11.1 by their properties because they produce sufficiently large concentrations of H^+ or OH^- in solution. Let us now examine the idea of acid and base strength and see how it is related to the amount of H^+ or OH^- in solution.

11.4 ACID AND BASE STRENGTH

In Section 9.5 you saw that a strong electrolyte is one that has many ions in solution. Similarly, the criterion for a strong acid is that it ionizes to a great extent and produces many ions, specifically H^+ ions (or, more correctly, H_3O^+). The six strong acids you encountered in Section 9.5 essentially ionize 100 percent. For example, because the equilibrium

$$HCl \rightleftharpoons H^+ + Cl^-$$

lies so far to the right, we may write

$$HCl \longrightarrow H^+ + Cl^-$$

Weak acids, on the other hand, ionize to only a limited extent. All the common organic acids are weak acids. For example, acetic acid ionizes less than 5 percent:

$$CH_3COOH \rightleftharpoons H^+ + CH_3COO^-$$

Because the equilibrium lies to the left and only a small number of H^+ ions (H_3O^+) are produced, acetic acid is a weak acid.

The equilibrium constant corresponding to the ionization of an acid is called the **ionization constant** and is designated $\boldsymbol{K_a}$. For $HA \rightleftharpoons H^+ + A^-$,

$$K_a = \frac{[H^+][A^-]}{[HA]}$$

Table 11.3, which lists the relative strengths of selected acids and bases, also gives some selected K_a values. Notice that it is arranged in order of decreasing acid strength reading from top to bottom. As with other equilibrium constants, a large value for K_a indicates that the equilibrium lies far to the right, and a small value indicates that the equilibrium position is to the left. Thus strong acids have high K_a values and weak acids, low ones. In particular, compare the K_a values of HCl (strong) and acetic acid (weak).

Table 11.3 also shows that if acids have more than one ionizable H^+, we consider the equilibria for removing one H^+ at a time. For example, H_2SO_4

TABLE 11.3 Relative Strengths of Acids and Bases

K_a	Acid Name	Acid Formula	Base Formula	Base Name	K_b
1×10^{7}	Hydrochloric	HCl	$\rightleftharpoons H^+ + Cl^-$	Chloride ion	1.4×10^{-21}
	Nitric	HNO_3	$\rightleftharpoons H^+ + NO_3^-$	Nitrate ion	
	Sulfuric	H_2SO_4	$\rightleftharpoons H^+ + HSO_4^-$	Hydrogen sulfate ion	
	Hydronium ion	H_3O^+	$\rightleftharpoons H^+ + H_2O$	Water	
1.2×10^{-2}	Hydrogen sulfate ion	HSO_4^-	$\rightleftharpoons H^+ + SO_4^{2-}$	Sulfate ion	8.3×10^{-13}
	Phosphoric	H_3PO_4	$\rightleftharpoons H^+ + H_2PO_4^-$	Dihydrogen phosphate ion	
	Lactic	$CH_3CH(OH)COOH$	$\rightleftharpoons H^+ + CH_3CH(OH)COO^-$	Lactate ion	
	Acetylsalicylic	$C_6H_4(OOCCH_3)COOH$	$\rightleftharpoons H^+ + C_6H_4(OOCCH_3)COO^-$	Acetylsalicylate ion	
	Benzoic	C_6H_5COOH	$\rightleftharpoons H^+ + C_6H_5COO^-$	Benzoate ion	
1.8×10^{-5}	Acetic	CH_3COOH	$\rightleftharpoons H^+ + CH_3COO^-$	Acetate ion	5.6×10^{-10}
	Carbonic	H_2CO_3	$\rightleftharpoons H^+ + HCO_3^-$	Bicarbonate ion	
	Dihydrogen phosphate ion	$H_2PO_4^-$	$\rightleftharpoons H^+ + HPO_4^{2-}$	Hydrogen phosphate ion	
	Boric	H_3BO_3	$\rightleftharpoons H^+ + H_2BO_3^-$	Dihydrogen borate ion	
5.6×10^{-10}	Ammonium ion	NH_4^+	$\rightleftharpoons H^+ + NH_3$	Ammonia	
	Bicarbonate	HCO_3^-	$\rightleftharpoons H^+ + CO_3^{2-}$	Carbonate	
	Methyl ammonium ion	$CH_3NH_3^+$	$\rightleftharpoons H^+ + CH_3NH_2$	Methylamine	
	Hydrogen phosphate ion	HPO_4^{2-}	$\rightleftharpoons H^+ + PO_4^{3-}$	Phosphate ion	
1.0×10^{-14}	Water	H_2O	$\rightleftharpoons H^+ + OH^-$	Hydroxide ion	5.5×10^{1}

Increasing ↑ Acid Strength ↓ Decreasing

Decreasing ↑ Base Strength ↓ Increasing

is **diprotic,** which means it has two ionizable hydrogens. We represent the equilibria for their ionization

$$H_2SO_4 \rightleftharpoons H^+ + HSO_4^-$$

$$HSO_4^- \rightleftharpoons H^+ + SO_4^{2-}$$

There is a greater tendency for the first H^+ to ionize than the second. Therefore, we see that H_2SO_4 is a stronger acid than HSO_4^-. In general, there is always a greater tendency for the first H^+ to ionize in any acid with more than one ionizable H^+.

Base strength follows the same principle as electrolyte strength or acid strength; i.e., extensive dissociation or ionization producing many ions means that the base is strong, and little dissociation or ionization yielding few ions means it is weak. In the case of bases, it is the hydroxide ion (OH^-) concentration that is of interest. Therefore, soluble metal hydroxides, such as the Group IA hydroxides, are strong bases because these ionic compounds dissociate 100 percent as they dissolve.

Magnesium hydroxide is classified as a strong base because all the formula units that do enter solution do so by completely dissociating by the process

$$Mg(OH)_2 \rightleftharpoons Mg^{2+} + 2OH^-$$

However, the Group IIA hydroxides, such as $Mg(OH)_2$, have limited solubilities in aqueous solution. So, in contrast to soluble strong bases, few formula units dissolve and a saturated solution of magnesium hydroxide contains only a few hydroxide ions. Consequently, unlike a metal hydroxide base such as NaOH (lye), $Mg(OH)_2$ solutions can be ingested and are used to treat acidic stomach conditions and peptic ulcers. Bases neutralize acids, as was pointed out in Table 11.2 and will be discussed in Section 11.8.

The equilibrium

$$NH_3(aq) + H_2O \rightleftharpoons NH_4^+(aq) + OH^-(aq)$$

lies to the left. Therefore, because there is little ionization and only a small number of OH^- ions are produced, ammonia is a weak base. The equilibrium constant corresponding to this reaction is the so-called K_b (basicity constant) of NH_3 and equals 1.8×10^{-5}. Notice that this equilibrium constant gives a measure not only of the extent to which OH^- ions are produced (the Arrhenius criterion for basicity), but also of how well a base *accepts* H^+ (The Brønsted-Lowry criterion). All the organic amine bases are weak bases.

Table 11.3 also gives us a measure of base strength. In this case, the measure of base strength is in terms of the Brønsted-Lowry test of how well the base accepts H^+. Recall that there is a conjugate base corresponding to every acid.

$$HA \rightleftharpoons H^+ + A^-$$

Acid Conjugate base

Therefore, because Table 11.3 ranks the tendency of acids to give up H^+ (tendency for the reaction to proceed →), it also tells us the tendency for the conjugate base to accept H^+ (tendency for the reaction to proceed ←).

$$HA \underset{\text{base reaction}}{\overset{\text{acid reaction}}{\rightleftharpoons}} H^+ + A^-$$

Strong acids have weak conjugate bases. For example, because HCl completely ionizes to $H^+ + Cl^-$, Cl^- has essentially no tendency to accept H^+ and therefore is a very weak base. Conversely, very weak acids have strong conjugate bases. Water has only a very slight tendency to ionize:

$$HOH \rightleftharpoons H^+ + OH^-$$

The conjugate base of water is the strong base OH^-.

Problem 11.3 Arrange the following acids in order of decreasing acid strength (use Table 11.3): HNO_3, H_3BO_3, $CH_3CH(OH)COOH$ (lactic acid), H_2CO_3, H_2O.

Problem 11.4 Arrange the conjugate bases of the acids in Problem 11.3 in order of increasing base strength.

11.5 IONIZATION OF WATER

Water is the solvent in bodily solutions, and its properties thereby influence many bodily processes. A consideration of its acid-base properties is essential to understanding the maintenance of the proper acidity conditions in blood and other fluids.

Although water is a molecular compound, it contains a very small concentration of ions. In viewing water as an acid (HA), we say that it undergoes the ionization reaction

$$\underset{\text{(HA)}}{HOH(l)} \rightleftharpoons H^+(aq) + \underset{\text{(A}^-\text{)}}{OH^-(aq)}$$

As for any acid, this equilibrium is more correctly shown as

$$\underset{\text{Acid}}{HOH(l)} + \underset{\text{Base}}{H_2O(l)} \rightleftharpoons \underset{\text{Acid}}{H_3O^+(aq)} + \underset{\text{Base}}{OH^-(aq)}$$

This representation shows us that the equilibrium is possible because of the amphoteric nature of water. Reference to Table 11.3 tells us that the equilibrium lies to the left because H_3O^+ and ^-OH are the stronger acid-base pair. Therefore, in pure water the concentrations of hydronium and hydroxide ions are very small.

As we proceed in our discussion, we will use the simpler expression for the water-ionization equilibrium, that is,

$$HOH(l) \rightleftharpoons H^+(aq) + OH^-(aq)$$

The equilibrium-constant expression for this reaction is

$$K_{eq} = [H^+][OH^-]$$

(Remember, pure liquids and solids do not appear in K_{eq} expressions.) Because this equilibrium expression is so important in acid-base chemistry, it is given the special symbol $\boldsymbol{K_w}$. Also, the product of the concentrations, $[H^+]$ $[OH^-]$, is sometimes called the **ion product** of water.

$$K_w = [H^+][OH^-]$$

In pure water at 25°C, $[H^+] = [OH^-]$, and each is found to have the concentration 1.0×10^{-7} *M*. Thus the value for K_w can be found by substitution:

$$K_w = (1.0 \times 10^{-7})\,(1.0 \times 10^{-7})$$

$$K_w = 1.0 \times 10^{-14}$$

The addition of acids to water will raise the concentration of H^+ above 1.0×10^{-7} *M*. The addition of bases will raise the concentration of OH^- above 1.0×10^{-7} *M*. But the *product of the concentrations* of H^+ and OH^- ions in water must remain constant and equal to K_w. This is so because the water equilibrium $HOH(l) \rightleftharpoons H^+(aq) + OH^-(aq)$ is present in all aqueous solutions and the equilibrium-constant expression ($K_w = [H^+][OH^-] = 1.0 \times 10^{-14}$) for this equilibrium must always be satisfied. In accord with Le Chatelier's principle, when $[H^+]$ increases, $[OH^-]$ decreases, and when $[OH^-]$ increases, $[H^+]$ decreases, and the ion product, $[H^+][OH^-]$, always remains 1.0×10^{-14}.

Neutral solution:

$$[H^+] = [OH^-] = 1.0 \times 10^{-7}\ M$$

Acidic solution:

$[H^+] > 1.0 \times 10^{-7}\ M$

$[OH^-] < 1.0 \times 10^{-7}\ M$

Basic solution:

$[OH^-] > 1.0 \times 10^{-7}\ M$

$[H^+] < 1.0 \times 10^{-7}\ M$

SAMPLE EXERCISE 11.2 An aqueous solution is found to have $[H^+] = 1.0 \times 10^{-4}\ M$. Is this solution acidic, basic, or neutral?

Solution Since $1 \times 10^{-4}\ M$ is greater than $1 \times 10^{-7}\ M$, the solution is acidic.

SAMPLE EXERCISE 11.3 What is $[OH^-]$ in Sample Exercise 11.2?

Solution If we are given $[H^+]$ or $[OH^-]$ and asked to determine $[OH^-]$ or $[H^+]$, we make use of the fact that $K_w = [H^+][OH^-] = 1.0 \times 10^{-14}$. In this example, $[H^+]$ is given as $1.0 \times 10^{-4}\ M$ and the $[OH^-]$ is unknown. Substituting in the K_w expression,

$$K_w = [H^+]\,[OH^-] = 1.0 \times 10^{-14}$$

$$[OH^-] = \frac{1.0 \times 10^{-14}}{[H^+]}$$

$$[OH^-] = \frac{1.0 \times 10^{-14}}{1.0 \times 10^{-4}}$$

$$[OH^-] = 1.0 \times 10^{-10}\ M$$

Consistent with this being an acidic solution, $[OH^-]$ turns out to be less than $1.0 \times 10^{-7}\ M$.

Problem 11.5 What is $[OH^-]$ in a solution in which $[H^+] = 3.2 \times 10^{-9}\ M$? Is this solution acidic, basic, or neutral?

11.6 pH

The concentrations of H^+ and OH^- mentioned or calculated in the previous section were somewhat unwieldy exponential numbers. In order to be able to express the acidity or basicity of solutions without the use of exponential numbers, chemists devised the pH scale of acidities. The pH of a solution is defined as

$$pH = -\log [H^+]$$

which is said, "pH equals the negative logarithm of the hydrogen ion concentration in moles per liter." The pH is very easy to calculate for solutions in which $[H^+]$ is an exact power of 10, that is, $1 \times$ the power of 10. In this case,

$$[H^+] = 1.0 \times 10^{-x}$$

$$pH = x$$

The logarithm of 1.0×10^{-x} is $-x$. The negative logarithm is $-(-x) = x$. For $[H^+] = 1.0 \times 10^{-7}$, the logarithm of $1.0 \times 10^{-7} = -7$. The negative log is $-(-7) = 7$, and pH = 7.

SAMPLE EXERCISE 11.4

Determine the pH of the following solutions, in which

a $[H^+] = 1.0 \times 10^{-11}$

b $[H^+] = 0.001$

Solution The negative of the exponent is the pH.

a pH = 11

b First write 0.001 in scientific notation, $0.001 = 1 \times 10^{-3}$. pH = 3.

Problem 11.6 Determine the pH of the following solutions, in which

a $[H^+] = 1.0 \times 10^{-5}$

b $[H^+] = 0.000010$

Let us summarize the pH characteristics of neutral, acidic, and basic solutions:

Neutral solution:

$[H^+] = 1.0 \times 10^{-7}\ M$

pH = 7.00

Acidic solution:

$[H^+] > 1.0 \times 10^{-7}\ M$

pH < 7.00

Basic solution:

$[H^+] < 1.0 \times 10^{-7}$

pH > 7.00

Thus the pH scale extends from 0 to 14, with values less than 7 indicating an acidic solution, and values greater than 7, a basic solution.

For most solutions, the hydrogen ion concentration is *not* an exact power of 10. That is, in most solutions $[H^+]$ is not equal to 1.0×10^{-x}, but rather $[H^+] = N \times 10^{-x}$, where N is some number other than 1. In order to determine the pH in this case, you must be able to manipulate logarithms. There are two ways: with a calculator and using tables.

Those of you who have a logarithm button on your calculator can perform these manipulations quite simply. Most calculators take "logs" by two steps: First, enter the number. Second, press the log button. The logarithm is then displayed on the screen. Change the sign to determine the pH. For example, given

$[H^+] = 4.0 \times 10^{-3}$

enter 4.0×10^{-3} (or 0.0040). Next, press the log button. Display is −2.3979. Change the sign and round off to two decimal places. pH = 2.40.

If your calculator is not equipped to do logarithms, then you will need a log table to calculate pH. Table 11.4 is a two-place log table adequate for

TABLE 11.4 Two-Place Logarithms

	Tenths									
Ones	**0.0**	**0.1**	**0.2**	**0.3**	**0.4**	**0.5**	**0.6**	**0.7**	**0.8**	**0.9**
1	0.00	0.04	0.08	0.11	0.15	0.18	0.20	0.23	0.26	0.28
2	0.30	0.32	0.34	0.36	0.38	0.40	0.41	0.43	0.45	0.46
3	0.48	0.49	0.51	0.52	0.53	0.54	0.56	0.57	0.58	0.59
4	0.60	0.61	0.62	0.63	0.64	0.65	0.66	0.67	0.68	0.69
5	0.70	0.71	0.72	0.72	0.73	0.74	0.75	0.76	0.76	0.77
6	0.78	0.79	0.79	0.80	0.81	0.81	0.82	0.83	0.83	0.84
7	0.85	0.85	0.86	0.86	0.87	0.88	0.88	0.89	0.89	0.90
8	0.90	0.91	0.91	0.92	0.92	0.93	0.93	0.94	0.94	0.95
9	0.95	0.96	0.96	0.97	0.97	0.98	0.98	0.99	0.99	1.00

our needs. Every number written in scientific notation has two parts, the decimal number (between 1 and 10) and the power of 10 (Section 2.13). For example,

$$3.9 \times 10^{-6}$$

Decimal number   Power of 10

The logarithm of this number is the sum of the log of the decimal number, which we get from Table 11.4, and the log of the power of 10. *The log of a power of 10 is the exponent*. Thus, if

$$[H^+] = N \times 10^{-x}$$

$$\log [H^+] = \log N + \log 10^{-x}$$

$$\log [H^+] = \log N + (-x)$$

From log table

$$pH = -\log [H^+]$$

For the specific example $[H^+] = 3.9 \times 10^{-6}$ we need to look up the log of 3.9 and add it to -6. To use the table to determine the log of 3.9, we locate 3 in the vertical (ones) column and 0.9 in the horizontal (tenths) row: 0.59 appears in the box at the intersection of the 3 in the ones column and 0.9 in the tenths row. Thus the log of 3.9 is 0.59; $\log 3.9 \times 10^{-6} = 0.59 + -6 = -5.41$. So for $[H^+] = 3.9 \times 10^{-6}\ M$, the $\log [H^+] = -5.41$, and the pH equals 5.41.

Check that the table gives you the same answer as a calculator by using the table to determine the pH of a solution where $[H^+] = 4.0 \times 10^{-3}\ M$, the example treated previously by the calculator method. In the table 0.60 appears at the intersection of the 4 in the ones column and 0.0 in the tenths row. $0.60 + -3 = -2.40$. Thus we obtain pH = 2.40, as before.

SAMPLE EXERCISE 11.5

Determine the pH of a solution in which $[H^+] = 2.3 \times 10^{-5}\ M$.

Solution STEP 1 Recognize that this $[H^+]$ concentration is of the form $[H^+] = N \times 10^{-x}$. Use a calculator with a log button or work out as follows:

$$\log [H^+] = \log N + (-x)$$

STEP 2 Look up the log of 2.3 in Table 11.4.

$\log 2.3 = 0.36$

STEP 3

$\log [H^+] = 0.36 + (-5) = -4.64$

STEP 4

$pH = -\log [H^+] = -(-4.64) = 4.64$

Problem 11.7 Determine the pH of a solution in which $[H^+] = 0.032$.

11.7 MEASUREMENT OF pH

Most bodily fluids such as blood and urine have a normal pH range (Table 11.5). Continued deviation from a particular range usually indicates some pathological condition and therefore offers a means of diagnosis. Medical laboratories require a simple and quick method of measuring pH.

Plants grow best in soil with a specific pH range and, as Figure 11.2 shows, do not show maximum growth or fruit production outside this range. Optimal soil pH ranges for several common plants are shown in Table 11.6. Farmers must be able to measure the pH of the soil (actually the suspension formed from soil and water) in which each crop is planted, so they can adjust it if necessary.

The pH of a colorless or lightly colored solution or suspension is often determined by the use of an **indicator,** a dye that changes color as pH changes. Frequently the dye is impregnated on paper to which a solution to be tested can easily be applied. Indicators are weak acids which have one color in the acid form (symbolized HIn) and another color in the conjugate base form (In^-).

$$HIn \rightleftharpoons H^+ + In^-$$

Color at lower pH   Color at higher pH

When H^+ is removed from the equilibrium (by reaction with some base), there is a shift in the position of equilibrium to the In^- side (Le Chatelier's principle). When the concentration of In^- exceeds that of HIn, the color is

TABLE 11.5 Normal pH Range for Some Bodily Fluids

Fluid	Normal pH Range
Liver bile	7.4–8.0
Blood	7.35–7.45
Gastric juice	1–2
Pancreatic juice	7–8
Saliva	6.4–7.0
Sweat	4.5–7.5
Tears	7.0–7.4
Urine	4.5–7.5

TABLE 11.6 pH Range for Optimum Plant Growth

Plant	pH Required
Apples	5.0–6.5
Cherries	6.0–7.5
Strawberries	5.0–6.5
Peas	6.0–7.5
Snap beans	6.0–7.5
Tomatoes	5.5–7.5
Tulips	6.0–7.0
Roses	6.0–8.0
Azaleas	4.5–6.0
Hydrangeas	5.0–6.0

FIG. 11.2
The importance of proper soil pH in growing crops. Spinach was planted throughout the field. In the foreground, the soil was not treated with lime [$Ca(OH)_2$] to adjust pH to conditions favorable for spinach; only stunted plants or nothing at all grow in this area. In the back, lime was applied and you can see the spinach growing in dark rows. *(Courtesy of Cook College, Rutgers University.)*

that of the high pH range. For the dye named litmus, a commonly used indicator, this is blue. If the solution is now made acidic (that is, H^+ is added), the equilibrium shifts to the HIn side, giving the color of the low pH range which, for litmus, is red.

Litmus changes color at pH 5–8, but other indicators change at other pH values. Each changes over a unique pH range depending on its acid ionization constant.

The use of litmus paper, which is red at any pH less than 5 and blue at any pH greater than 8, enables us to determine only whether a solution is acidic or basic. A universal pH paper can be prepared by impregnating paper with several dyes that change color at narrow and different pH ranges (Figure 11.3). The pH of an unknown solution is then determined by comparing the color developed on pH paper from a drop of the solution with that on a chart which relates color to pH.

A pH meter (Figure 11.4) is used to determine pH values more accurately and to determine those of highly colored solutions. The pH meters work by accurately measuring H^+ ion concentration. Table 11.7 lists the pH values of some commonly encountered materials.

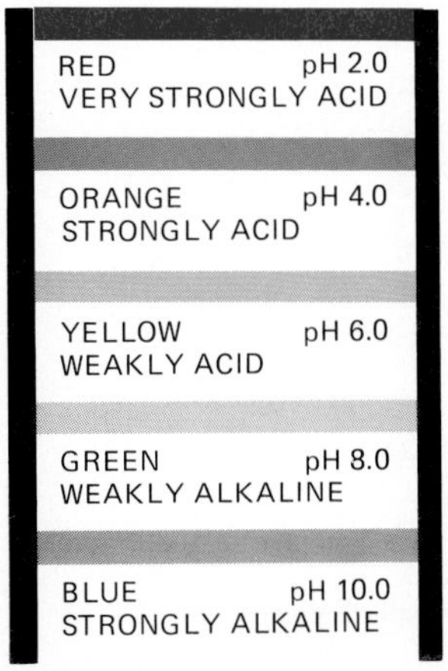
RED pH 2.0
VERY STRONGLY ACID
ORANGE pH 4.0
STRONGLY ACID
YELLOW pH 6.0
WEAKLY ACID
GREEN pH 8.0
WEAKLY ALKALINE
BLUE pH 10.0
STRONGLY ALKALINE

FIG. 11.3
Color chart for universal indicator paper. Solutions of different pH turn the paper different colors.

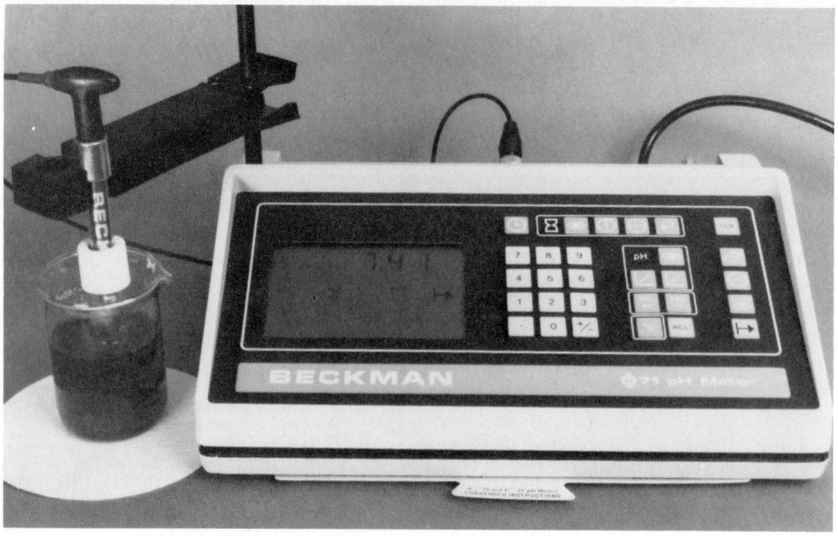
BECKMAN
pH

FIG. 11.4
A pH meter. The electrodes are immersed in blood, so that the pH displayed is that of blood (7.41). *(Photograph by Bryan Lees.)*

11.8 REACTIONS OF ACIDS AND BASES

The reactions of acids and bases occur as parts of natural processes. In addition, some drugs or other medical remedies rely on the reactions of acids and bases. Probably the most common and important reaction of acids and bases is their reaction with one another, the *neutralization reaction,* which we will discuss first.

Neutralization

When acids and bases react with one another, ionic compounds, called **salts,** form. Table salt, NaCl, is only one example of the class of compounds known as salts. **Salts** are ionic compounds formed from the reaction of an acid with a base: the cation of the salt comes from the base, and the anion comes from the acid.

TABLE 11.7 Approximate pH Values of Some Common Materials

pH	Solution
0	Battery acid
1	Stomach acid
2	Lemon juice, lime juice
3	Vinegar, wine, soft drinks, beer, orange juice, pickles
4	Tomatoes, grapes
5	Black coffee, rainwater
6	Urine, milk, saliva
7	Pure water, blood
8	Seawater
9	Clorox, phosphate detergent
10	Soap, milk of magnesia
11	Household ammonia
12	Hair remover
13	Oven cleaner

If the base reacting in the neutralization reaction is a hydroxide base, then water is also formed. Consider the neutralization reaction between HCl and NaOH:

$$\underset{\text{Acid}}{HCl(aq)} + \underset{\text{Base}}{NaOH(aq)} \longrightarrow \underset{\text{Salt}}{NaCl(aq)} + \underset{\text{Water}}{H_2O}$$

The salt gets its cation (Na^+) from the base and its anion (Cl^-) from the acid. Water forms from the hydroxide of the base and H^+ from the acid.

Strong Acid–Strong Base Neutralization Reactions

For a strong acid–strong base reactant pair, which HCl and NaOH are, the reaction can be viewed as a double displacement (see Section 7.11), in which cations and anions switch partners.

$$H\text{—}Cl + Na^+OH^- \longrightarrow Na^+Cl^- + H\text{—}OH$$

Other examples of strong acid–strong base neutralization reactions are

$$HNO_3(aq) + LiOH(aq) \longrightarrow LiNO_3(aq) + H_2O$$

$$2HCl(aq) + Ca(OH)_2(aq) \longrightarrow CaCl_2 + 2H_2O$$

$$\underset{\text{Acid}}{H_2SO_4(aq)} + \underset{\text{Base}}{2NaOH(aq)} \longrightarrow \underset{\text{Salt}}{Na_2SO_4} + \underset{\text{Water}}{2H_2O}$$

Because the strong acid is completely ionized, and the strong base completely dissociates in aqueous solution, the following ionic equations (review Section 9.7) apply:

Traditional equation: $\underset{\text{Strong acid}}{HA} + \underset{\text{Strong base}}{MOH} \longrightarrow \underset{\text{Salt (base cation + acid anion)}}{MA + H_2O}$

Full ionic:

$$H^+(aq) + \cancel{A}^-(aq) + \cancel{M}^+(aq) + OH^-(aq) \longrightarrow \cancel{M}^+(aq) + \cancel{A}^-(aq) + H_2O(l)$$

Net ionic: $H^+ + OH^- \longrightarrow H_2O$

The net ionic equation for any strong acid–strong base combination is simply $H^+ + OH^- \longrightarrow H_2O$. The pH at neutralization of a strong acid–strong base pair is 7.0.

In terms of Brønsted-Lowry theory this strong acid–strong base reaction is a proton transfer reaction in which the equilibrium lies far to the right.

$$\underset{\text{Strong acid}}{HCl} + \underset{\text{Strong base}}{OH^-} \rightleftharpoons \underset{\text{Weak base}}{Cl^-} + \underset{\text{Weak acid}}{HOH}$$

A single-headed arrow is typically used for the strong acid–strong base reaction because the equilibrium is so far to the right.

$$HCl + OH^- \longrightarrow Cl^- + HOH$$

Notice again that in this Brønsted-Lowry equation NaOH is represented simply as OH^- because Na^+ is merely a spectator.

Neutralization Reactions of Weak Acids or Weak Bases

The neutralization reactions of weak acids or weak bases are best considered in terms of Brønsted-Lowry theory because the weak species is not fully ionized in solution. For example, consider the reaction between the weak acid, acetic acid, and a strong base:

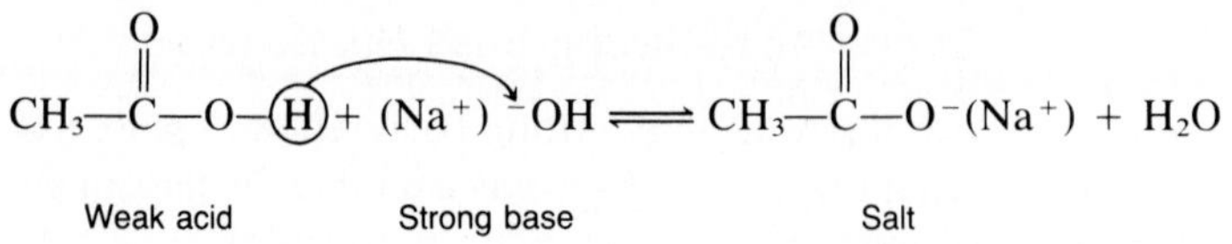
CH₃—C(=O)—O—(H) + (Na⁺) ⁻OH ⇌ CH₃—C(=O)—O⁻(Na⁺) + H₂O
Weak acid
Strong base
Salt

The acid donates H^+ and the base accepts it; water forms because the base is hydroxide. The salt, sodium acetate, forms in the usual manner, cation from base, anion from acid. The sodium ion is shown here in parentheses because ordinarily the metal ion of MOH is not represented in Brønsted-Lowry reactions.

Now consider the neutralization reaction of a weak acid and weak base:

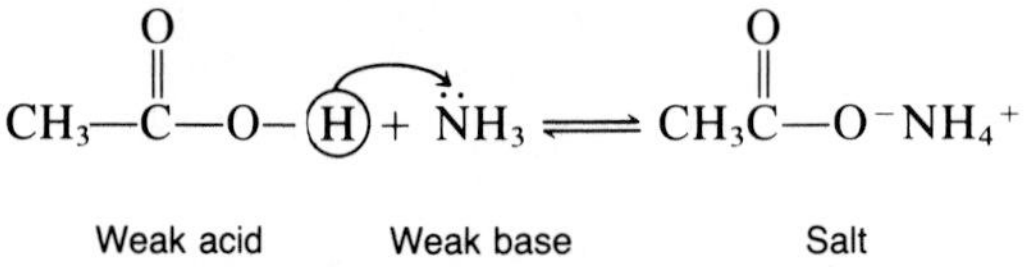
CH₃—C(=O)—O—(H) + N̈H₃ ⇌ CH₃C(=O)—O⁻NH₄⁺
Weak acid
Weak base
Salt

Again a salt forms wherein the cation comes from the base and the anion from the acid. No water forms because the base is not a hydroxide base. The reaction between a strong acid, such as HCl, and NH_3 gives a similar result:

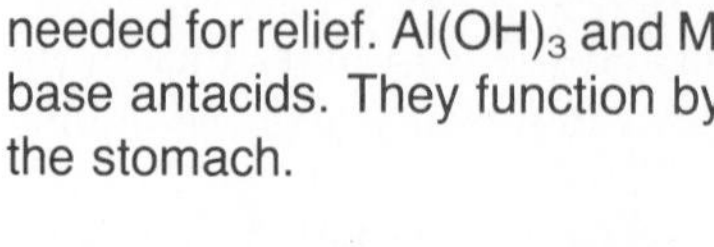
(H)Cl + N̈H₃ ⇌ NH₄⁺Cl⁻
Strong acid
Weak base
Salt

Neutralization reactions are the relief of indiscreet eaters and stomach ulcer sufferers. Food overindulgence or stressful situations can lead to an oversecretion of hydrochloric acid in the stomach, for which an antacid is needed for relief. $Al(OH)_3$ and $Mg(OH)_2$ are the two most common hydroxide-base antacids. They function by neutralizing some of the excess H^+ ions in the stomach.

$$2HCl + Mg(OH)_2 \longrightarrow MgCl_2 + 2H_2O$$

$$3HCl + Al(OH)_3 \longrightarrow AlCl_3 + 3H_2O$$

A mixture of the aluminum and magnesium compounds is often desirable because aluminum compounds tend to cause constipation, whereas magnesium compounds have a laxative effect. Figure 11.5 depicts a variety of common antacid products.

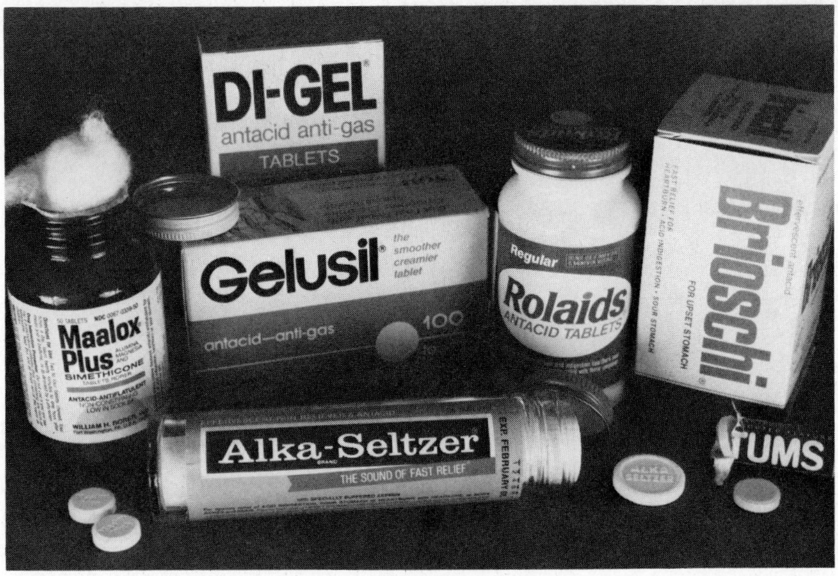
DI-GEL
antacid anti-gas
TABLETS
Gelusil
the smoother creamier tablet
antacid—anti-gas
Maalox Plus
SIMETHICONE
Regular
Rolaids
ANTACID TABLETS
Brioschi
effervescent antacid
FOR UPSET STOMACH
Alka-Seltzer
THE SOUND OF FAST RELIEF
TUMS

FIG. 11.5
Commercial antacid products.

Reactions of Acids with Carbonates and Bicarbonates

Other antacid medications employ the reactions of acids with basic bicarbonate and carbonate anions. Acids stronger than H_2CO_3 (see Table 11.3) react with carbonates and bicarbonates to form carbonic acid, H_2CO_3, most of which (because of the instability of H_2CO_3) decomposes to CO_2 and H_2O. We can write an equation for the reaction between hydrochloric acid and sodium bicarbonate which also shows the carbonic acid equilibrium leading to water and carbon dioxide.

$$\begin{array}{l} HCl(aq) + NaHCO_3(aq) \longrightarrow NaCl(aq) + H_2CO_3(aq) \\ \qquad\qquad\qquad\qquad\qquad\qquad\qquad\quad \updownarrow \\ \qquad\qquad\qquad\qquad\qquad\qquad\qquad H_2O + CO_2(g) \end{array}$$

Keeping in mind that $Na^+(aq)$ and $Cl^-(aq)$ are spectator ions and that the position of the carbonic acid equilibrium is far to the side of $H_2O + CO_2$, we can write the net ionic equation for the reaction of any acid with a bicarbonate as

$$H^+(aq) + HCO_3^-(aq) \longrightarrow H_2O + CO_2(g)$$

Similarly, for a carbonate the reaction would be

$$2H^+(aq) + CO_3^{2-}(aq) \longrightarrow H_2O + CO_2(g)$$

Many popular antacids remove excess H^+ through a bicarbonate (Alka-Seltzer, Bromo Seltzer) or carbonate (Tums, Pepto-Bismol) ingredient (Figure 11.5). They make you burp because CO_2 gas is released in the stomach.

Acid-base reactions are also important in baking. In baking powder, sodium bicarbonate (baking soda) is combined with acidic ingredients. When wet, this mixture releases carbon dioxide, causing the rising of such baked goods as biscuits and certain breads.

Reactions of Acids with Active Metals

Another example of an acid-base reaction which impacts on food handling is the reaction of acid with some metals. Acids react with Group IA, IIA, and IIIA metals to form hydrogen gas and a salt of the metallic ion. For example,

$$2Al(s) + 6HCl(aq) \longrightarrow 2AlCl_3(aq) + 3H_2(g)$$

Metal | Acid | Salt of the metal | Hydrogen

Some transition metals such as iron and zinc react in a similar way, whereas others such as gold, silver, copper, and mercury do not react with acids to form hydrogen gas.

Acidic foods such as wine or tomatoes should not be brought into contact with cast iron or carbon steel utensils because the acid in these foods will react with the iron. In stainless steel, iron is alloyed in such a way that it no longer reacts with acid. There is also increasing concern that the weak acid H_2CO_3 in soda is reacting with aluminum metal soda cans, albeit slightly, to produce undesirable concentrations of aluminum ions in soda.

11.9 DETERMINATION OF ACID-BASE CONCENTRATIONS—TITRATION

The neutralization reaction discussed in Section 11.8 can be used to determine the unknown concentration of an acid or base given a known concentration of base or acid. The procedure is called **titration.** Consider the neutralization reaction between HCl and NaOH again:

$$HCl + NaOH \longrightarrow NaCl + H_2O$$

or in ionic equation form:

$$H^+(aq) + Cl^-(aq) + Na^+(aq) + OH^-(aq) \longrightarrow Na^+(aq) + Cl^-(aq) + H_2O(l)$$

At the point of neutralization, called the **equivalence point,** the number of moles of H^+ from the acid exactly equals the number of moles of OH^- from the base. This is particularly apparent from the net ionic equations for the reactions of any strong acid and strong base:

$$H^+ + OH^- \longrightarrow H_2O$$

In doing titration calculations you must always remember that at the equivalence point

$$\text{Moles of } H^+ = \text{moles of } OH^-$$

In doing a titration experiment, a certain volume of the unknown acid or base is placed in the flask with an indicator (and/or it is attached to a pH meter). The indicator must be one that undergoes a color change at the equivalence point. A burette is then used for the addition of base to unknown acids or for the addition of acid to unknown bases until the equivalence point is reached. The burette enables one to deliver very accurately measured volumes of solutions.

Figure 11.6 shows the measured addition of a volume of base of known concentration until the known volume of unknown acid is neutralized, as witnessed by the change in indicator color. We know the volume and concentration of added base; then we can calculate the moles of OH^-, and this

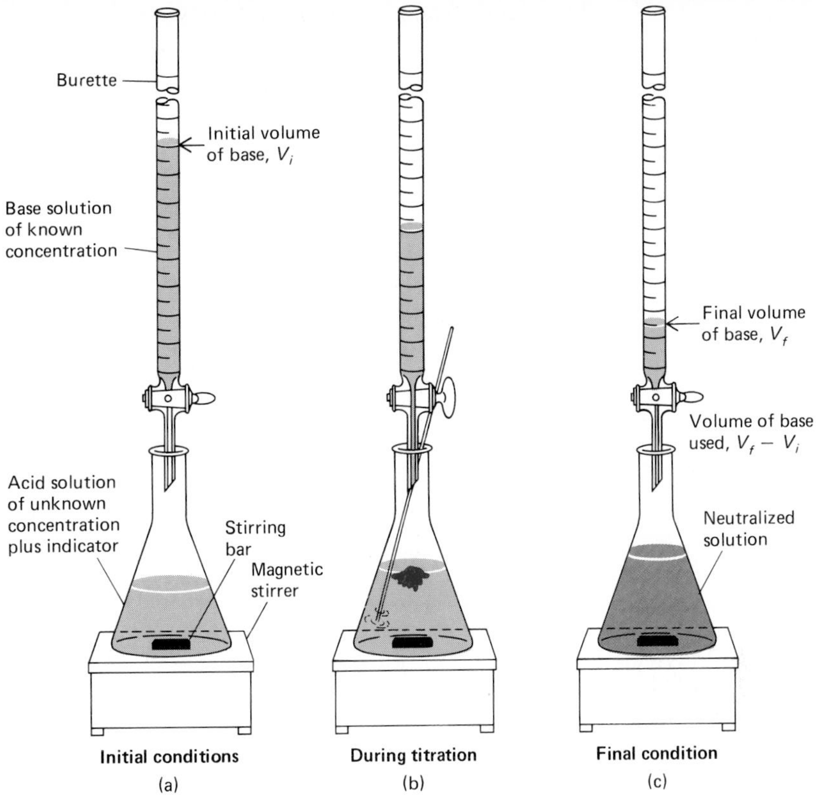
Burette
Initial volume of base, V_i
Base solution of known concentration
Final volume of base, V_f
Volume of base used, $V_f - V_i$
Acid solution of unknown concentration plus indicator
Stirring bar
Magnetic stirrer
Neutralized solution
Initial conditions
(a)
During titration
(b)
Final condition
(c)

FIG. 11.6 Acid-base titration procedure. A base (or acid) of known concentration is added drop by drop to an acid (or base) of unknown concentration until the indicator just changes color. From the magnitude of the added volume of known concentration, the unknown concentration can be calculated. (*a*) Initial conditions. (*b*) During titration a color change may be noted as a drop of base hits the solution. It is temporary and disappears with stirring. (*c*) This is the point at which the color change becomes permanent.

is equal to the number of moles of H^+ neutralized. Titration calculations always involve a calculation of the number of moles of H^+ or OH^- neutralized and then a calculation of the molarity of H^+ or OH^-.

Calculation of the number of moles of added OH^- from the known concentration (M) and volume is done by remembering from Section 9.8 that $M \times V$ (in liters) = moles of OH^-, and at the equivalence point moles of OH^- = moles of H^+. Then the molarity of H^+ is calculated by dividing the number of moles by the volume (in liters) of the unknown acid. See Sample Exercise 11.6.

SAMPLE EXERCISE 11.6

Using a pH meter a chemist finds that 42.50 mL of a 0.100 *M* NaOH solution is required to titrate (neutralize) 31.00 mL of a hydrochloric acid solution of unknown concentration. What is the molarity of the hydrochloric acid?

Solution

STEP 1 Calculate the moles of added OH^- from M of $OH^- \times V$ (in liters) = moles of OH^-.

$$\frac{0.100 \text{ mol}}{1 \text{ L}} \times 0.0425 \text{ L} = 0.00425 \text{ mol of } OH^-$$

(0.100 *M* NaOH is 0.100 *M* in OH^-; $42.5 \cancel{\text{mL}} \times 1 \text{ L}/1000 \cancel{\text{mL}} = 0.0425$ L.)

STEP 2 At the equivalence point, moles of OH^- = moles of H^+. Therefore, 0.00425 mol of H^+ was neutralized.

STEP 3 Calculate the acid molarity by dividing moles by liters of solution. There is

0.00425 mol of H^+ in 0.00425 mol of HCl.

$$31.00 \text{ mL of HCl} \times \frac{1 \text{ L}}{1000 \text{ mL}} = 0.03100 \text{ L of HCl solution}$$

$$M \text{ of HCl} = \frac{\text{moles of HCl}}{\text{liter of solution}}$$

$$M \text{ of HCl} = \frac{0.00425 \text{ mol of HCl}}{0.03100 \text{ L of HCl}} = 0.137\ M$$

In doing titration calculations you must sometimes take into account the fact that the molarity of the acid is not always equal to the molarity of the H^+ ion and the molarity of the base is not always equal to the molarity of OH^- ion. This lack of equality occurs for diprotic and triprotic acids, for example, H_2SO_4 and H_3PO_4, and for metal hydroxides which have more than one hydroxide anion, for example, $Mg(OH)_2$. The subscript of the hydrogen ion or hydroxide ion relates the molarity of the ion to the molarity of the acid or base. Thus, 1 *M* H_2SO_4 is 2 *M* in H^+ because there are 2 mol of H^+ per 1 mol of H_2SO_4, and 1 *M* H_3PO_4 is 3 *M* in H^+. Multiply the acid molarity by the subscript of hydrogen to determine the molarity of H^+. Similarly, 0.1 *M* $Mg(OH)_2$ is 0.2 *M* in OH^- because there are 2 mol of OH^- per 1 mol of $Mg(OH)_2$.

Problem 11.8 What is the molarity of H^+ in each of the following acid solutions?

a 0.5 *M* H_3PO_4
b 0.2 *M* H_2SO_4
c 0.45 *M* HBr

SAMPLE EXERCISE 11.7

A student neutralizes 29.10 mL of a potassium hydroxide solution with 15.30 mL of a 0.500 *M* H_2SO_4 solution. What is the molarity of the KOH solution?

Solution Note that in this case an acid of known concentration is used to titrate a base of unknown concentration. The principles are just the same as those of the reverse type of titration.

STEP 1 As before, we want to calculate moles of added known, in this case, H^+ from *M* of $H^+ \times V$ (in liters) = moles H^+. 0.500 *M* H_2SO_4 is 1 *M* in H^+ because there are 2 mol of H^+ per 1 mol of H_2SO_4:

$$\frac{0.500 \text{ mol of } H_2SO_4}{\text{L}} \times \frac{2 \text{ mol of } H^+}{1 \text{ mol of } H_2SO_4} = \frac{1 \text{ mol of } H^+}{\text{L}}$$

$$15.30 \text{ mL of soln} \times \frac{1 \text{ L}}{1000 \text{ mL}} = 0.0153 \text{ L of soln}$$

$$M \times V = \frac{1 \text{ mol of } H^+}{\text{L}} \times 0.0153 \text{ L} = 0.0153 \text{ mol of } H^+$$

STEP 2 At the equivalence point, moles of H^+ = moles of OH^-. Therefore, 0.0153 mol of OH^- was neutralized.

STEP 3 Calculate the base molarity by dividing moles by liters of solution. There is 0.0153 mol of OH^- in 0.0153 mol of KOH.

$$29.10 \text{ mL of KOH} \times \frac{1 \text{ L}}{1000 \text{ mL}} = 0.02910 \text{ L of KOH}$$

$$M = \frac{0.0153 \text{ mol KOH}}{0.02910 \text{ L KOH}} = 0.526\ M \text{ KOH}$$

SAMPLE EXERCISE 11.8 A student neutralized 14.40 mL of a H_2SO_4 solution with 35.2 mL of an 0.200 *M* NaOH solution. What is the molarity of the H_2SO_4 solution?

Solution STEP 1 Calculate the moles of added known:

$$M \text{ of } OH^- \times V \text{ (in liters)} = \text{moles of } OH^-$$

$$\frac{0.200 \text{ mol of } OH^-}{\text{L}} \times 0.0352 \text{ L} = 0.00704 \text{ mol of } OH^-$$

STEP 2 At the equivalence point, moles of OH^- = moles of H^+. Therefore, 0.00704 mol of H^+ was neutralized.

STEP 3 Calculate the acid molarity by dividing moles of acid by liters of solution. First determine moles of acid from moles of H^+.

$$0.00704 \text{ mol of } H^+ \times \frac{1 \text{ mol of } H_2SO_4}{2 \text{ mol of } H^+} = 0.00352 \text{ mol of } H_2SO_4$$

$$14.40 \text{ mL of soln} \times \frac{1 \text{ L}}{1000 \text{ mL}} = 0.0144 \text{ L of soln}$$

$$M = \frac{0.00352 \text{ mol of } H_2SO_4}{0.0144 \text{ L of soln}} = 0.244\ M\ H_2SO_4$$

11.10 WHAT IS A BUFFER?

Although, as Table 11.5 shows, the pH values of various bodily fluids can be very different, for example, pH 1–2 in gastric juice versus pH 7–8 in pancreatic juice, the normal range within any particular fluid is quite narrow. This is especially true in blood plasma, where the pH of a healthy individual must remain between 7.35 and 7.45. Should the blood pH fall to pH = 7.2, oxygenated hemoglobin (HbO_2), the carrier of O_2 from the lungs to all cells in the body, releases its oxygen, which leads to cell starvation and eventually bodily death. On a more subtle level, as you will see when we look at proteins in Section 19.15, very small changes in pH affect the concentrations of the protonated and unprotonated forms of biological molecules and in so doing affect the solubility and three-dimensional shape of these molecules.

The body uses a system of buffers to maintain the proper pH of bodily fluids within necessary narrow ranges. A **buffer** is a weak acid–weak base pair that by reacting with added amounts of a base or acid can resist large changes in the solution's pH. For example, if 1 mL of 1.0 *M* HCl is added to 100 mL of pure water, the pH of the water plummets 5 whole units from 7 to 2. If the same 1 mL of 1.0 *M* HCl is added to 100 mL of bicarbonate buffer (we will discuss the composition of this buffer in the next section), the pH of the buffer solution decreases from 7.4 to 6.9, a difference of only 0.5 pH units.

Let us now look at the different types of buffers and learn how they work. Buffers are of two general types:

1 Weak conjugate acid-base pairs. Either (*a*) a weak acid (e.g., acetic acid) and its conjugate base (the acetate ion) or (*b*) a weak base (e.g., ammonia) and its conjugate acid (the ammonium ion).
2 Protein buffers.

Proteins have both weak acid and weak base sites in the same molecule (Sections 19.3 and 19.8).

Many bodily fluid buffers are of the weak acid-conjugate base type and so we will concentrate our attention on these buffers, using the acetic acid–acetate ion system as a general example.

11.11 HOW A BUFFER FUNCTIONS

Before we look at how buffers work, let us see first how we prepare a buffer of the weak acid-conjugate base type. A solution of acetic acid contains both the molecular (un-ionized) acid (CH_3COOH) and the conjugate base, the anion CH_3COO^-.

$$CH_3COOH(aq) \rightleftharpoons CH_3COO^-(aq) + H^+(aq) \qquad (11.1)$$

The anion (base) concentration is very small because the equilibrium lies to the left. We can increase the concentration of anion by adding to the solution an acetate salt which completely dissociates. For example, sodium acetate:

$$CH_3COO^-Na^+ \longrightarrow CH_3COO^-(aq) + Na^+(aq)$$

A buffer is made by combining a solution of a weak acid of known concentration with a solution of known concentration of the salt (anion) of that acid. The H^+ concentration, and hence the pH, of the solution depends on the ratio of the acid to the anion. This can be seen easily by an examination of the equilibrium-constant expression for the acetic acid ionization shown in Equation (11.1).

$$K_a = \frac{[CH_3COO^-][H^+]}{[CH_3COOH]} \qquad (11.2)$$

We isolate (H^+) by multiplying both sides by $[CH_3COOH]$ and dividing by $[CH_3COO^-]$:

$$\frac{K_a\,[CH_3COOH]}{[CH_3COO^-]} = [H^+] \qquad (11.3)$$

Because K_a is a constant, the $[H^+]$ (and pH) changes only as the ratio of the concentrations change.

Let us now suppose that we have made a buffer solution such that the acetic acid and acetate ion concentrations are equal and quite high (~0.1 *M*). We can use Le Chatelier's principle to see the effect of added acid or base on the position of equilibrium and then qualitatively predict the effect on the acid/anion ratio and hence $[H^+]$.

Equilibrium: $CH_3COOH(aq) \rightleftharpoons CH_3COO^-(aq) + H^+(aq)$

Stress by adding H^+

The equilibrium shifts to the left to relieve the stress by consuming the added $H^+(aq)$. Acetate ion is also consumed and acetic acid is formed. However, since the concentrations of acid and acetate ion were initially large, the ratio after equilibrium is reestablished is not much affected and the hydrogen ion concentration after equilibrium is reestablished is essentially the same as it was before acid was added.

Initial equilibrium	New equilibrium after H^+ addition
$[H^+] = K_a \dfrac{[CH_3COOH]}{[CH_3COO^-]}$	$(H^+) = K_a \dfrac{[CH_3COOH + x]}{[CH_3COO^- - x]}$

x = concentration of H^+ added as stress to original solution

As an actual example, suppose sufficient H^+ is added to the original solution (0.1 M in both acetic acid and acetate ion) to make the H^+ concentration = 0.001 M. Then after the added H^+ reacts with CH_3COO^-,

$$[CH_3COO^-] = 0.1 - 0.001 = 0.09901$$

because the acetate ion is diminished by combination with H^+, and

$$[CH_3COOH] = 0.1 + 0.001 = 0.101\ M$$

because additional CH_3COOH is produced. Therefore, when equilibrium is reestablished,

$$[H^+] = \frac{K_a\,[0.101]}{[0.099]} = K_a\,(1.02)$$

The H^+ concentration has increased only very slightly (2 percent) compared with the original value of $[H^+] = K_a\,[0.1]/[0.1] = K_a$ in the buffer solution.

Now let us consider the effect of added base.

Equilibrium: $CH_3COOH(aq) \rightleftharpoons CH_3COO^-(aq) + H^+(aq)$
Stress by adding OH^-

The added OH^- reacts with H^+ to form water,

$$OH^-(aq) + H^+(aq) \longrightarrow H_2O$$

thus decreasing the $H^+(aq)$ concentration. Adding base provides the stress of removing H^+ from the equilibrium. The original acetic acid equilibrium shifts to the right to relieve the stress by forming $H^+(aq)$. The concentration of acetic acid is also decreased and that of acetate ion is increased. However, the new ratio of acid to anion is relatively unaffected, and the hydrogen ion concentration after equilibrium is reestablished is essentially the same as it was before hydroxide was added.

Initial equilibrium	New equilibrium after OH^- addition
$[H^+] = K_a \dfrac{[CH_3COOH]}{[CH_3COO^-]}$	$[H^+] = K_a \dfrac{[CH_3COOH - x]}{[CH_3COO^- + x]}$

x = small change in concentration of acetic acid and acetate ion that occurs as a consequence of addition of base

If 0.001 M OH^- is added so that 0.001 M H^+ is consumed, then the equilibrium shifts produce new equilibrium values of

$$[CH_3COOH] = 0.1 - 0.001 = 0.099$$

and

$$[CH_3COO^-] = 0.1 + 0.001 = 0.101$$

and

$$[H^+] = K_a \frac{0.099}{0.101} = K_a\,(0.98)$$

which is very close to the original value for $[H^+]$.

The rearranged equilibrium-constant expression, in which $[H^+]$ is isolated, readily shows us the hydrogen ion concentration of a buffer solution as a function of acid and anion concentrations. The application of some simple math to this expression yields an equation that readily shows us the pH of a buffer solution. The new equation is simply the negative logarithmic form of

$$[H^+] = K_a \frac{[CH_3COOH]}{[CH_3COO^-]} \qquad (11.3)$$

which reads:

$$-\log\,[H^+] = -\log K_a - \log \frac{[CH_3COOH]}{[CH_3COO^-]}$$

or

$$pH = pK_a + \log \frac{[CH_3COO^-]}{[CH_3COOH]} \qquad (11.4)$$

where $pK_a = -\log K_a$

Equation (11.4) can be written for any weak acid in the form

$$pH = pK_a + \log \frac{\text{anion of the weak acid}}{\text{weak acid}} \qquad (11.5)$$

This equation is known as the Henderson-Hasselbach equation for the biochemists who developed it.

If we know the identity of the acid, and thereby the pK_a, and the ratio of concentrations of anion to weak acid, the pH of the solution can be calculated. The pK_a or K_a must either be given or be calculated from the K_a in Table 11.3. You will see these data given and this equation used in Section 11.12.

11.12 BLOOD BUFFERS

Now we can discuss the body's mechanisms for protecting blood from dramatic pH changes. Three interconnected systems maintain the pH in blood: (1) the blood buffers, which actually serve to neutralize the added hydrogen and hydroxide ions which form from the body's metabolic reactions; (2) the lungs, which are involved in the excretion and inhalation of carbon dioxide, and thereby maintain the concentration of carbonic acid in blood; and (3) the kidneys, which excrete hydrogen ions and bicarbonate ions from the blood. In this section we will concentrate on the role of the blood buffers and witness the supporting role of the other two systems.

There are three major body buffers: (1) the H_2CO_3/HCO_3^- buffer, (2) the $H_2PO_4^-/HPO_4^{2-}$ buffer, and (3) the protein buffers. The carbonic acid–bicarbonate mixture is the major buffer in blood. As we have seen before, carbonic acid is an unstable weak acid which in aqueous solution, in this case blood, is always in equilibrium with $CO_2(aq)$.

$$H_2CO_3(aq) \rightleftharpoons CO_2(aq) + H_2O \tag{11.6}$$

The position of this equilibrium is to the right. Dissolved $CO_2(aq)$ is also in equilibrium with $CO_2(g)$ in the lungs.

$$CO_2(aq) \rightleftharpoons CO_2(g) \tag{11.7}$$

Blood Lungs

A moment's thought (and perhaps some help from Le Chatelier) should convince you that the concentration of $H_2CO_3(aq)$ can be directly affected by that of $CO_2(g)$.

Carbonic acid is also in equilibrium with the other half of the buffer mixture, i.e., the bicarbonate anion, through the acid dissociation:

$$H_2CO_3(aq) \rightleftharpoons HCO_3^-(aq) + H^+(aq) \tag{11.8}$$

The operation of the buffering action here is completely analogous to the acetic acid–acetate ion system discussed in Section 11.11. Acidic by-products (H^+) of the metabolic cycles can be neutralized by $HCO_3^-(aq)$ forming $H_2CO_3(aq)$ [equilibrium of Equation (11.8) shifts left]. When in excess, H_2CO_3 is removed from the body as $CO_2(g)$ [equilibria of Equations (11.6) and (11.7) shift right]. Excess base is neutralized by $H_2CO_3(aq)$ forming $HCO_3^-(aq)$ [equilibrium of Equation (11.8) shifts right]. Figure 11.7 summarizes the equilibrium shifts in the buffer system caused by additions of acid or base.

The concentrations of $HCO_3^-(aq)$ and $H_2CO_3(aq)$ in the blood of a healthy individual are 2.5×10^{-2} *M* and 1.25×10^{-3} *M*, respectively. The pK_a of carbonic acid is 6.1. Using these data and the Henderson-Hasselbach relationship [Equation (11.5)],

$$\text{pH} = \text{p}K_a + \log \frac{[HCO_3^-]}{[H_2CO_3]}$$

$$\text{pH} = 6.1 + \log \frac{2.5 \times 10^{-2}}{1.25 \times 10^{-3}}$$

$$\text{pH} = 6.1 + \log 20$$

$$\text{pH} = 6.1 + 1.3 = 7.4$$

The pH we should expect to find in normal blood is 7.4.

The capacity of a particular buffer to resist changes in pH, called the **buffering capacity,** depends on the ratio of anion to acid remaining fairly constant. For the carbonic acid–bicarbonate buffer this means a relative constancy in the ratio $[HCO_3^-]/[H_2CO_3]$ of about 20:1.

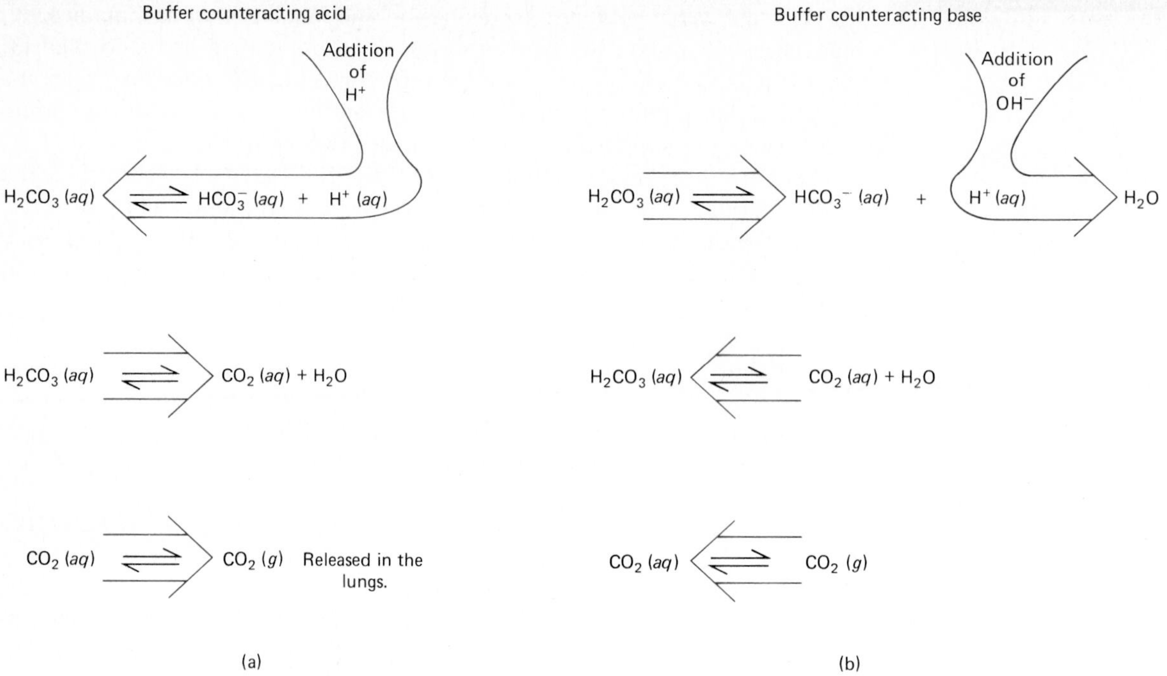
Buffer counteracting acid
Addition of H+
H2CO3 (aq) ⇌ HCO3− (aq) + H+ (aq)
H2CO3 (aq) ⇌ CO2 (aq) + H2O
CO2 (aq) ⇌ CO2 (g) Released in the lungs.
(a)
Buffer counteracting base
Addition of OH−
H2CO3 (aq) ⇌ HCO3− (aq) + H+ (aq) H2O
H2CO3 (aq) ⇌ CO2 (aq) + H2O
CO2 (aq) ⇌ CO2 (g)
(b)

FIG. 11.7
Action of the carbonic acid–bicarbonate buffer upon addition of acid (*a*) or base (*b*). (*a*) Addition of H^+ shifts (⇐) the equilibrium to the left, producing H_2CO_3. This new H_2CO_3 constitutes a stress on the second equilibrium, shifting it to the right. $CO_2(aq)$ is thus produced, and this shifts the third equilibrium to the right. (*b*) Addition of OH^- consumes H^+ and shifts the first equilibrium to the right, thus diminishing H_2CO_3. To compensate, the second equilibrium shifts left to replace H_2CO_3. The third equilibrium shifts left to replace $CO_2(aq)$.

A key factor in the maintenance of this 20:1 ratio is the coupling of the H_2CO_3 concentration to the partial pressure of CO_2 in the lungs, as was shown in Equations (11.6) and (11.7). The CO_2 in our lungs provides a ready and essentially limitless source of more H_2CO_3 so that a shortage is unlikely. Furthermore, excess H_2CO_3 can be disposed of by shifting the equilibrium of Equations (11.6) and (11.7) to the right in the direction of $CO_2(g)$. Increased $CO_2(g)$ exhalation accomplishes this.

The concentration of HCO_3^- is regulated by the kidneys. If the HCO_3^- concentration drops, the kidneys remove H^+ from the blood and the H^+ concentration in urine increases. Removal of H^+ shifts the equilibrium, $H_2CO_3 \rightleftharpoons HCO_3^- + H^+$, to the right, thus replenishing $[HCO_3^-]$. As we discussed, the H_2CO_3 lost by this shift can be replaced by CO_2. The kidneys promote the excretion of excess concentrations of HCO_3^-.

The carbonic acid–bicarbonate system in blood has a high buffering capacity because of the coupled equilibria between H_2CO_3 and the unlimited supply of CO_2 in the lungs and the regulation of $[HCO_3^-]$ by the kidneys. If both $[HCO_3^-]$ and $[H_2CO_3]$ are maintained at nearly constant values, then the ratio $[HCO_3^-]/[H_2CO_3]$ remains constant and so does the pH of the solution.

The other major buffer system is the $H_2PO_4^-/HPO_4^{2-}$ weak acid/conjugate base system. This is the major buffer in intracellular fluids. This buffer mixture functions through the equilibrium

$$H_2PO_4^-(aq) \rightleftharpoons HPO_4^{2-}(aq) + H^+(aq)$$

Weak acid neutralizes added base	Weak base neutralizes added acid

Problem 11.9 pK_a of $H_2PO_4^-$ is 7.21; for normal cell concentrations $[HPO_4^{2-}] = 2.4 \times 10^{-3}\ M$ and $[H_2PO_4^-] = 1.5 \times 10^{-3}\ M$. Use the Henderson-Hasselbach equation,

$$pH = pK_a + \log \frac{[HPO_4^{2-}]}{[H_2PO_4^-]}$$

to show that a pH of 7.4 is maintained within cells by this phosphate buffer system.

11.13 ACIDOSIS AND ALKALOSIS

Blood buffers usually maintain the blood pH in the normal range pH 7.35–7.45. However, disorders in either the respiratory or metabolic systems may lead to a decrease in the blood pH below 7.35, a condition given the medical term **acidosis;** or, to an increase in blood pH above 7.45, a condition known as **alkalosis.** Respiratory problems primarily affect the pH through abnormalities in the concentration of $CO_2(g)$ and hence $H_2CO_3(aq)$. Metabolic imbalances chiefly affect the pH through abnormalities in the concentration of $HCO_3^-(aq)$. We will look separately at respiratory and metabolic acidosis and alkalosis. Table 11.8 summarizes the discussion that follows.

If there is an excess in the blood carbonic acid concentration caused by hypoventilation (reduced respiration or too little "fresh air"), the blood pH

TABLE 11.8 A Look at Acidosis and Alkalosis

	Respiratory Acidosis	Respiratory Alkalosis	Metabolic Acidosis	Metabolic Alkalosis
pH level	Low	High	Low	High
Primary cause	Excess H_2CO_3	Deficiency in H_2CO_3	Deficiency in HCO_3^-	Excess HCO_3^-
Plasma concentrations:				
$[HCO_3^-]$	High	Low	Low	High
$[H_2CO_3]$	High	Low	Low	High
Possible underlying disorder	Pulmonary edema, pneumonia, emphysema, asthma, heart failure	Rapid breathing due to insufficient oxygen, prolonged crying, nervousness, high fever, central nervous disorder, improper respirator treatment	Diabetes mellitus, starvation, kidney failure, severe diarrhea	Overdose of antacid, excessive loss of gastric juice
Typical symptoms	Hypoventilation, irregular pulse, headache, vertigo, disorientation	Light-headedness, profuse perspiration, irregular pulse, hyperventilation	Deep rapid breathing, increased and more acidic urine, restlessness, disorientation, drowsiness, coma	Slow respiration, numbness, tingling, headache, possible convulsions
Patients likely to have this condition	Long-term heavy smokers, who have diminished lung capacity	Very apprehensive, anxious individuals, who tend to hyperventilate	Diabetics, whose metabolism produces excess H^+	Persistent vomiters, who lose H^+ in gastric juice

becomes abnormally low and a person suffers **respiratory acidosis.** In this condition, the lungs cannot exhale $CO_2(g)$ effectively and its partial pressure increases, leading to increased $[H_2CO_3]$.

$$CO_2(aq) \rightleftharpoons CO_2(g) \quad \text{increase } CO_2(g) \text{ and shift} \longleftarrow \tag{11.7}$$

$$H_2CO_3(aq) \rightleftharpoons CO_2(aq) + H_2O \quad \text{increase } CO_2(aq) \text{ and shift} \longleftarrow \tag{11.6}$$

Respiratory acidosis may arise from any condition causing slow shallow breathing, such as pulmonary edema, pneumonia, asthma, or heart failure. Typical symptoms of respiratory acidosis are irregular pulse, headache, vertigo, and disorientation. One may suffer mild respiratory acidosis in a stuffy room.

In **respiratory alkalosis** the blood pH is abnormally high because of a deficiency in blood carbonic acid concentration. Hyperventilation is the cause. Too much carbon dioxide is exhaled, leading to a lowering of the carbonic acid concentration through equilibrium shifts (to the right) in the equilibria in Equations (11.6) and (11.7). Respiratory alkalosis may result from an increased breathing rate caused by prolonged crying, nervousness, high fever, central nervous system disorders, and improper respirator treatment. Typical symptoms of respiratory alkalosis are light-headedness, profuse perspiration, and irregular pulse. Babies are particularly prone to this condition if they cry excessively.

In **metabolic acidosis** a low blood pH is caused by a deficiency in bicarbonate concentration. The deficiency arises either from a loss of bicarbonate ion or from excess metabolic formation of H^+. Excess H^+ diminishes HCO_3^- by reacting with it. The excess formation of H^+ may result from diabetes mellitus, starvation, or kidney insufficiency. Severe diarrhea can result in the loss of bicarbonate. Hyperventilation is a response to this condition because hyperventilation lowers $[H_2CO_3]$, which helps to maintain the normal 20:1 ratio for $[HCO_3^-]/[H_2CO_3]$ in the face of decreased $[HCO_3^-]$. The kidneys also help to compensate by sending more H^+ into the excreted urine. Typical symptoms of metabolic acidosis are deep, rapid breathing, increased urine output with an increased acidity, restlessness, disorientation, drowsiness, and eventually coma.

In **metabolic alkalosis** there is an increase in bicarbonate concentration and a high blood pH. The increase in $[HCO_3^-]$ may be direct as in the case of overingestion of antacid medicines or in kidney disease where the kidneys fail to remove HCO_3^- from the blood. A decrease in $[H^+]$ also increases the HCO_3^- concentration because of increased ionization of H_2CO_3 to replace H^+:

$$H_2CO_3 \xrightleftharpoons{\text{shift}} HCO_3^- + H^+$$

Loss of H^+ from blood may arise from excessive loss of gastric juice through prolonged vomiting or from intracellular loss of K^+ from diuretic therapy. H^+ from blood migrates into cells in this case.

Hypoventilation is a physiological compensation for metabolic alkalosis because it tends to increase $[H_2CO_3]$ and thereby maintains the $[HCO_3^-]/[H_2CO_3]$ ratio in the face of increased $[HCO_3^-]$. If the kidneys are functioning normally, they respond to metabolic alkalosis by eliminating excess bicarbonate and reducing the excretion of H^+. Typical symptoms are slow respiration, numbness, tingling, headache, and possibly convulsions.

Proper buffer function to maintain pH and avoid acidosis and alkalosis is especially critical to good health. The ability of hemoglobin to transport oxygen depends on pH, and cells rapidly die if deprived of oxygen. The rates of metabolic reactions are extremely sensitive to pH because the catalytic activity of enzymes depends on pH, as you will see in Section 21.7. Large, short-term fluctuation in pH or smaller, but pronounced, prolonged deviations can lead to death. Even very small pH deviations can cause discomfort because of the disruption of metabolic activity.

CHAPTER ACCOMPLISHMENTS

After completing this chapter you should be able to define **key words** and do the following:

11.1 INTRODUCTION

1. List some common acids and bases and their application in everyday life.
2. State the physical and chemical properties which distinguish acids from bases.

11.2 THE ARRHENIUS DEFINITION

3. Recognize acids and bases according to the Arrhenius definition.
4. State the ions present in an aqueous solution of a given metallic hydroxide or a given acid.

11.3 BRØNSTED-LOWRY DEFINITION

5. Recognize acid-base pairs and their conjugate acid-base pairs in a given equation.
6. Describe the requirement for an acid or base showing the properties of Table 11.2.

11.4 ACID AND BASE STRENGTH

7. Describe the relationship between the classification of acids and bases as strong or weak and the extent of their ionization.
8. Describe what is meant by the acid and base ionization constants.
9. Given a table of acid and base ionization constants, arrange a list of acids and bases in order of increasing or decreasing acid and base strength.

11.5 IONIZATION OF WATER

10. Write the equilibrium-constant expression corresponding to K_w.
11. State the value of $[H^+]$ and $[OH^-]$ in pure water.
12. State the value of K_w.
13. Given a value of $[H^+]$ or $[OH^-]$, state whether the solution is acidic or basic.
14. Given a value for $[H^+]$, calculate $[OH^-]$, or given a value of $[OH^-]$, calculate $[H^+]$.

11.6 pH

15. Given the pH of a solution, state whether the solution is acidic, basic, or neutral.
16. Calculate the pH of a solution given the value of $[H^+]$ or $[OH^-]$.

11.7 MEASUREMENT OF pH

17. State the normal pH range for blood.
18. State two methods by which the pH of a solution can be determined experimentally.

11.8 REACTIONS OF ACIDS AND BASES

19. Write an equation for the neutralization reaction between a given acid and base.
20. Write an equation for the reaction between a given acid and a bicarbonate or carbonate salt.

11.9 TITRATION

21. Describe how a titration is performed.
22. Given the volume of one solution (acidic or basic) and the volume and molarity of a second solution (basic or acidic) needed to titrate the first, calculate the molarity of the first solution.

11.10 WHAT IS A BUFFER?

23. State the two general types of biological buffers.

11.11 HOW A BUFFER FUNCTIONS

24. Describe how a buffer of the weak acid-conjugate base type is prepared.
25. State what the $[H^+]$ concentration of a given buffer depends on.
26. Describe the changes in equilibrium that occur when a small amount of acid is added to a given weak acid-conjugate base buffer.
27. Given concentrations of weak acid and conjugate base and an appropriate pK_a, use the Henderson-Hasselbach equation to calculate the pH of a given weak acid-conjugate base buffer.

11.12 BLOOD BUFFERS

28. State the three systems that serve to regulate the pH in blood.
29. State the three major blood buffers.
30. State the three equilibria that influence the concentration of H_2CO_3 in blood plasma.
31. Describe how excess acid and excess base are neutralized by the carbonic acid–bicarbonate buffer.
32. Describe why the carbonic acid–bicarbonate system in blood has a high buffering capacity.

11.13 ACIDOSIS AND ALKALOSIS

33. State the pH, the primary cause, typical symptoms, and possible underlying disorders for each of the following: respiratory acidosis, respiratory alkalosis, metabolic acidosis, and metabolic alkalosis.
34. Describe how respiratory disorders may be distinguished from metabolic disorders.

PROBLEMS

11.10 *a* Give two examples of an Arrhenius acid.
b Give two examples of an Arrhenius base.

11.11 Show by an equation how the ionization of HNO_3 in aqueous solution leads to the formation of the hydronium ion.

11.12 *a* Distinguish between a Brønsted-Lowry acid and a Brønsted-Lowry base.
b Distinguish between a Brønsted-Lowry acid and an Arrhenius acid.
c Distinguish between a Brønsted-Lowry base and an Arrhenius base.

11.13 Show by equations how the ion HSO_4^- can act as an amphoteric substance.

11.14 Show by an equation how the ion NH_4^+ can act as a Brønsted-Lowry acid.

11.15 Identify the acid-conjugate base and the base-conjugate acid pairs in the following reactions:
a $HBr + H_2O \rightleftharpoons H_3O^+ + Br^-$
b $NH_3 + H_3O^+ \rightleftharpoons NH_4^+ + H_2O$
c $HSO_4^- + OH^- \rightleftharpoons SO_4^{2-} + H_2O$
d $HCO_3^- + H_3O^+ \rightleftharpoons H_2CO_3 + H_2O$
e $NH_3 + H_3PO_4 \rightleftharpoons NH_4^+ + H_2PO_4^-$

11.16 Give the conjugate bases of the following substances:
a H_2O *c* $H_2PO_4^-$
b HF *d* NH_4^+

11.17 Give the conjugate acids of the following substances:
a H_2O *c* $H_2PO_4^-$
b F^- *d* NH_3

11.18 Give the names of two acids and two bases found in commercial products in your household.

11.19 Describe two procedures you would use in distinguishing between an acidic and basic substance.

11.20 Arrange the following substances in order of increasing acid strength: H_3BO_3, HSO_4^-, $CH_3CH(OH)COOH$, H_2O, and HNO_3.

11.21 Arrange the conjugate bases of the acids in Problem 11.20 in order of increasing base strength.

11.22 *a* Write an equilibrium-constant expression for the acid ionization of CH_3COOH.
b What quantities would have to be deter-

mined experimentally to evaluate this equilibrium constant?

11.23 Identify the Brønsted-Lowry acid and the Brønsted-Lowry base in each case and complete the following equations by transferring a proton from the acid to the base:

a $NH_4^+ + NH_3 \rightleftharpoons$
b $H_2SO_4 + H_2O \rightleftharpoons$
c $HCl + CN^- \rightleftharpoons$
d $H_2CO_3 + NO_3^- \rightleftharpoons$
e $HSO_4^- + OH^- \rightleftharpoons$
f $HSO_4^- + NO_3^- \rightleftharpoons$

11.24 *a* Write an equilibrium-constant expression for the reaction of H_2O as a Brønsted-Lowry acid in pure water.
b Write an equilibrium-constant expression for the reaction of H_2O as a Brønsted-Lowry base in pure water.
c What is the relationship between the equilibrium constants in parts *a* and *b*?
d What are the numerical values of these equilibrium constants?

11.25 State two pieces of evidence that indicate water can be only very slightly ionized.

11.26 Why must aqueous solutions with a high $[OH^-]$ concentration have a low $[H^+]$ concentration?

11.27 State the importance of pH measurements to farmers.

11.28 Indicate whether the following solutions are acidic, basic, or neutral:

a $[H^+] = 9.3 \times 10^{-11}\ M$
b $[H^+] = 6.2 \times 10^{-6}\ M$
c $[OH^-] = 3.2 \times 10^{-11}\ M$
d $[OH^-] = 5.0 \times 10^{-5}\ M$
e $[H^+] = 0.0013\ M$
f $[H^+] = 0.0000001\ M$

11.29 Determine the pH of the following solutions, and indicate whether the solution is acidic, basic, or neutral:

a $[H^+] = 1.0 \times 10^{-7}\ M$
b $[H^+] = 1.0 \times 10^{-2}\ M$
c $[H^+] = 1.0 \times 10^{-10}\ M$
d $[H^+] = 0.0001\ M$

11.30 Determine the pH of the following solutions, and indicate whether the solution is acidic, basic, or neutral:

a $[H^+] = 3.0 \times 10^{-5}\ M$
b $[H^+] = 8.7 \times 10^{-9}\ M$
c $[H^+] = 0.074\ M$
d $[H^+] = 0.19\ M$
e $[OH^-] = 6.3 \times 10^{-4}\ M$
f $[OH^-] = 4.9 \times 10^{-12}\ M$
g $[OH^-] = 0.013\ M$
h $[OH^-] = 0.000020\ M$

11.31 Indicate whether the following solutions are acidic, basic, or neutral:

a Seawater: pH = 8.1
b Vinegar: $[H^+] = 1.6 \times 10^{-3}\ M$
c Coffee: pH = 5.42
d Orange juice: $[H^+] = 2.0 \times 10^{-4}\ M$
e Blood: pH = 7.4
f Soda water: pH = 2.8
g Stomach fluid: pH = 1.8
h Ammonia water: $[H^+] = 1.9 \times 10^{-11}\ M$

11.32 Describe how a pH meter could be used to indicate the equivalence point of a titration reaction between a strong acid and a strong base.

11.33 Complete and balance the following neutralization reactions:

a $KOH + HCl \longrightarrow$
b $Ba(OH)_2 + HNO_3 \longrightarrow$
c $NaOH + H_3PO_4 \longrightarrow$
d $Ca(OH)_2 + H_2SO_4 \longrightarrow$
e $Ca(OH)_2 + H_3PO_4 \longrightarrow$

11.34 Indicate the acid and base needed to prepare the following salts:

a $NaNO_3$
b K_2SO_4
c $CaCl_2$
d $AlPO_4$
e $(CH_3COO)_2Ca$
f C_6H_5COOLi

11.35 Complete and balance the following equations:

a $HNO_3 + KHCO_3 \longrightarrow$
b $HCl + Li_2CO_3 \longrightarrow$
c $Zn + HCl \longrightarrow$
d $Na_2CO_3 + CH_3COOH \longrightarrow$
e $Mg + H_2SO_4 \longrightarrow$

11.36 A student found that 86.30 mL of an HCl solution was required to completely neutralize 31.75 mL of 0.150 *M* NaOH. What is the molarity of the acid solution?

11.37 The active sour ingredient in vinegar is acetic acid, CH_3COOH. In order to test the acid molarity of a vinegar solution, a laboratory technician titrated 24.50 mL of vinegar with 0.4789 *M* NaOH. She found that it took 46.09 mL of NaOH to completely neutralize the vinegar. What is the acid molarity of the vinegar? Assume that the only acid present is CH_3COOH.

11.38 How many milliliters of 0.250 *M* H_2SO_4 are required to neutralize 125 mL of 0.100 *M* NaOH?

$$2NaOH + H_2SO_4 \longrightarrow Na_2SO_4 + 2H_2O$$

11.39 Explain why a 0.30 *M* H_2SO_4 solution is 0.60 *M* in H^+.

11.40 What ions are actually reacting in a neutralization reaction between a strong acid and strong base? Write the net ionic equation.

11.41 Gastric juice, found in the stomach, contains hydrochloric acid. A 30.10-mL sample of gastric juice is titrated with 0.1050 *M* NaOH. The chemist finds that 17.20 mL of NaOH is needed to neutralize the gastric juice acid. What is the molarity of acid in gastric juice? Assume that HCl is the only substance in gastric juice that reacts with NaOH.

11.42 In an experiment, 25.0 mL of 1.00 *M* HCl is required to titrate a Drano (active basic ingredient NaOH) solution. How many moles of NaOH are present in the Drano solution?

11.43 In another experiment, 10.0 g of vinegar (active acid ingredient is acetic acid, CH_3COOH) is titrated with 65.40 mL of 0.150 *M* NaOH.

$$CH_3COOH + NaOH \longrightarrow CH_3COONa + H_2O$$

a How many moles of CH_3COOH are present in 10.0 g of the vinegar?
b How many grams of CH_3COOH are present in 10.0 g of vinegar?
c What is the percentage by weight of acetic acid in the vinegar solution?

11.44 *a* Use Table 11.3 to indicate the components of two weak acid–conjugate base buffers.
b Use Table 11.3 to indicate the components of two weak base–conjugate acid buffers.

11.45 Define a buffer solution.

11.46 What substance would you need to add to benzoic acid to prepare a buffer solution?

11.47 Using the appropriate equilibrium equations, describe how a benzoic acid–sodium benzoate (the anion of benzoic acid) mixture can act as a buffer.

11.48 Describe how the $[H^+]$ concentration of a given buffer varies when the acid/anion concentration is:
a Doubled
b Decreased to one-third of its original value

11.49 Qualitatively describe the changes in the concentrations of $[H^+]$, $[H_2CO_3]$, and $[HCO_3^-]$ when a small amount of hydrochloric acid is added to a carbonic acid–bicarbonate buffer.

11.50 Calculate the pH of a acetic acid–acetate ion buffer that contains $[CH_3COOH] = 0.15\ M$ and $[CH_3COO^-] = 0.10\ M$ (pK_a for acetic acid is 4.7).

11.51 Calculate the pH of a blood sample in which the bicarbonate concentration is twice the carbonic acid concentration (pK_a for H_2CO_3 in blood plasma $= 6.1$).

11.52 Explain how the $[H_2CO_3]$ concentration in blood plasma would vary if:
a The concentration of $CO_2(aq)$ increased.
b The partial pressure of $CO_2(g)$ in the lungs increased.
c The concentration of bicarbonate decreased.

11.53 What would be the pH of blood if $[H_2CO_3]$ and $[HCO_3^-]$ were present in equal concentrations?

11.54 *a* Describe the role of the lungs in regulating blood pH.
b Describe the role of the kidneys in regulating blood pH.

11.55 Although the concentration of $[H_2CO_3]$ in blood plasma is only one-twentieth of the $[HCO_3^-]$ concentration, blood has a high buffering capacity toward base. Explain.

11.56 Why would breathing into a paper bag help a patient with respiratory alkalosis?

11.57 Compare the $[H_2CO_3]$ and $[HCO_3^-]$ concentrations in respiratory acidosis and metabolic acidosis. Could they be used to distinguish between these two disorders?

CHAPTER TWELVE

INTRODUCTION TO ORGANIC CHEMISTRY

12.1 INTRODUCTION

Living matter is made up of carbon compounds, called organic compounds. The origin of the term *organic* was discussed in Section 6.1. Organic compounds always contain carbon and hydrogen, frequently contain oxygen and/or nitrogen, and sometimes contain phosphorus or sulfur. The halogens and even metals can be incorporated into organic structures. Carbon compounds make up the food that we eat, the muscles that allow us motion, the brain that allows us thought, and the sperm or egg cells within us that allow our hereditary character to be passed on to our offspring. All living cells contain thousands of these compounds, ranging in chemical structure from comparatively simply molecules such as ethyl alcohol and glucose to more complex molecules like cholesterol and still more complicated molecules such as hemoglobin.

The modern definition of **organic chemistry** is the chemistry of carbon compounds. Not all carbon compounds are associated with living systems, but those are the ones in which we are most interested. In this chapter we will begin the study of the general concepts of organic chemistry which apply to small, simple carbon compounds and to more complex molecules such as carbohydrates, fats, and proteins, which are the ultimate carbon compounds of interest.

As you approach the study of organic chemistry, please realize that you will need to remember many concepts from general chemistry. The same principles of bonding, thermodynamics, polarity, solubility, and acid-base behavior that you have already learned apply to organic as well as inorganic compounds. You will see that this is so if we consider some general properties of organic compounds.

12.2 GENERAL PROPERTIES OF ORGANIC COMPOUNDS

Before pursuing more theoretical ideas of bonding and geometry, let's take a look at the tangible physical properties of organic compounds. The differences in the physical properties of organic versus inorganic compounds follow from differences in the nature of their bonding. The covalent bond is predominant in organic compounds, whereas inorganic compounds more often display ionic bonding. (Review Section 6.9.)

Organic compounds, owing to their weaker intermolecular forces, generally have lower melting points than inorganic compounds. Because there are no ions present, organic compounds are nonconductors of electricity. Many organic compounds can be combined with oxygen in a highly exothermic reaction to form carbon dioxide and water. We often call this light and heat producing reaction "burning." Most inorganic compounds do not react with oxygen in a combustion reaction.

Problem 12.1 Why are the properties of table salt (NaCl) and gasoline (C_8H_{18}) so very different?

The water solubility of an organic compound depends first on the presence of atoms that can hydrogen-bond with water, and then on the size of the interacting group of atoms compared with the rest of the molecule. For example, the presence of the grouping —OH within an organic molecule gives the molecule the potential of hydrogen bonding with water. Organic compounds with —OH groups are called *alcohols*. The —OH fragment may be part of a small organic molecule, that is, one with only a small number of carbons, in which case the OH would represent a substantial fraction of the molecular makeup. Or, it may represent only a small fraction of the atoms in a large molecule. The alcohol of alcoholic beverages, ethyl alcohol, exemplifies the first case, and everyday experience with mixed drinks tells us it is water soluble.

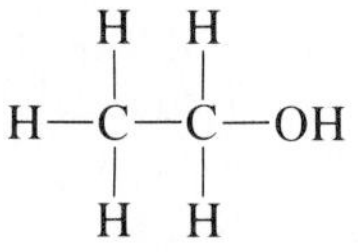

Ethyl alcohol

However, an alcohol such as 1-octanol is insoluble. The character of this alcohol is determined more by the large carbon chain than by the small —OH group.

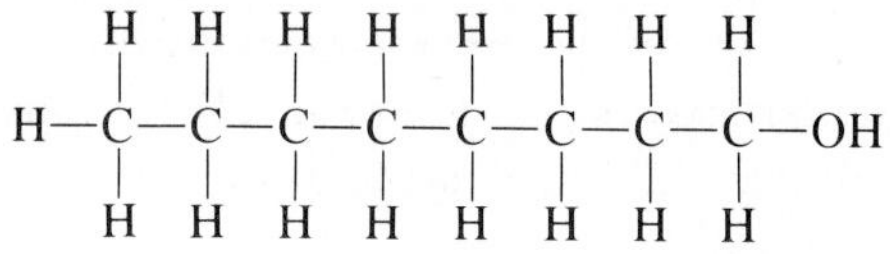

1-Octanol

In general, alcohols of four carbons or fewer are soluble in water, whereas those of six carbons or more, containing only one —OH group, are insoluble in water. Compounds that do not contain atoms capable of hydrogen bonding with water, such as O or N, are insoluble in water. As a general rule of thumb, most organic compounds are insoluble in water; however, they are soluble in nonpolar solvents such as carbon tetrachloride and benzene.

Table 12.1 compares the general properties of organic and inorganic compounds.

Problem 12.2 Blood sugar (glucose) is water-soluble although it has six carbons. Its Lewis structure is

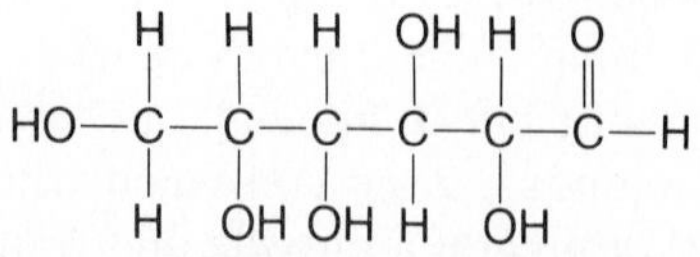

Why is glucose water-soluble?

TABLE 12.1 A General Comparison of Organic and Inorganic Compounds

Type of Compound	Melting and Boiling Points	Electrical Conductivity	Combustion	Water Solubility
Organic	Low	Nonconductors	Many can burn.	Those which have a group capable of hydrogen bonding with water and four or fewer carbons are soluble; the remainder are usually not soluble in water.
Inorganic	High	Conduct in molten state or solution	Usually not flammable.	Generally water soluble.

12.3 BONDING IN CARBON COMPOUNDS

In contrast to only a few hundred thousand known inorganic compounds, there are well over 3 million known organic compounds. Many occur naturally, but a large number have been prepared synthetically in the laboratory. Students often ask why there are so many organic compounds. The answer can be found in the nature of the chemical bonds that carbon forms.

The predominant type of bonding between carbon atoms and between carbon and other elements in organic compounds is covalent or polar covalent (Chapter 6). Carbon is unique among the elements in its ability to form strong bonds *to itself* as well as to other elements such as hydrogen, oxygen, and nitrogen.

The fact that carbon-carbon bonds are strong means that chains of carbon atoms can form and will be stable. The chains can be of varying length, leading to literally an infinite number of compounds differing in chain length. Figure 12.1 shows carbon compounds with three carbons and with six carbons and a segment of polyethylene. The common plastic polyethylene is constructed of huge molecules in which thousands of carbon atoms are joined in a chain. In addition, carbon atoms can also be joined together to form rings. Furthermore, both chains and rings can have "branches" of one or more carbons.

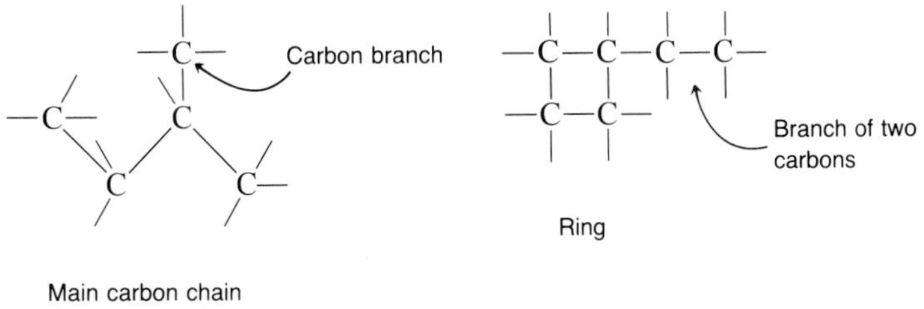
Carbon branch
Main carbon chain
Branch of two carbons
Ring

Thus, because of the chain- and ring-forming abilities of carbon, we see that a great variety of organic compounds arise just through different arrangements of different numbers of carbon and hydrogen atoms. However, carbon can also bond to other elements. Most notably, in organic compounds of physiological interest, we find the elements oxygen, nitrogen, phosphorus, and sulfur in addition to carbon and hydrogen. Each different structural arrangement of the atoms, which can be represented by a structural formula (Lewis structure), corresponds to a different compound with a unique set of physical and chemical properties. All the foregoing leads to a very large number of possible compounds with a wide spectrum of properties and reactivities.

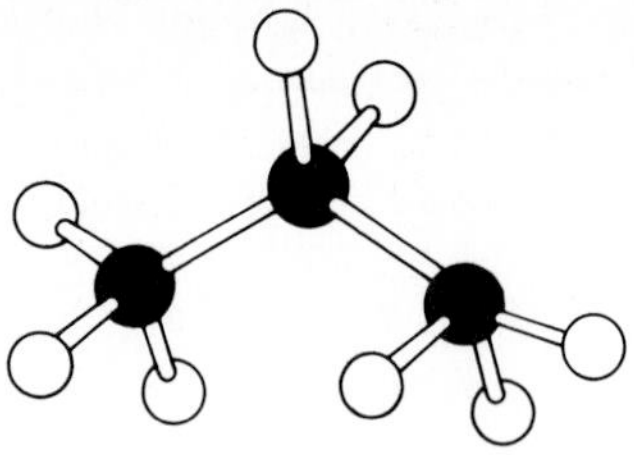

Three-carbon chain (propane)

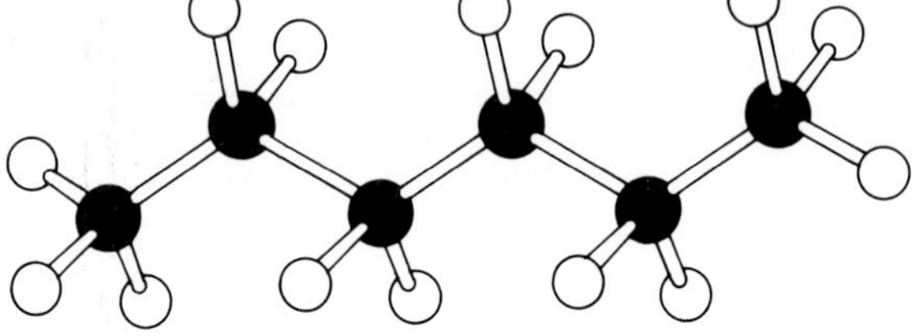

Six-carbon chain (hexane)

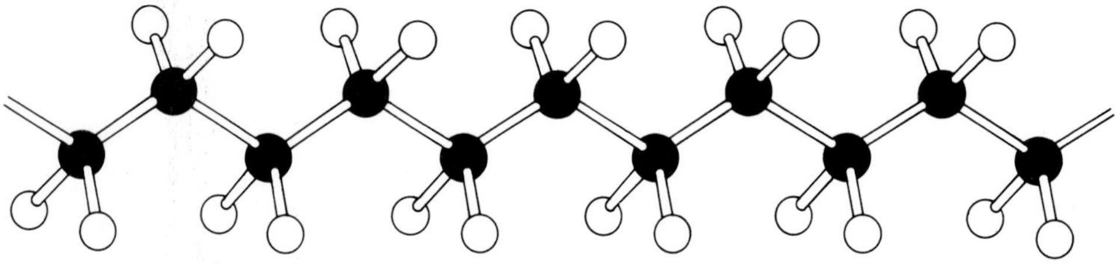

Polyethylene

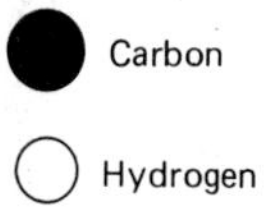
Carbon
Hydrogen

FIG. 12.1
Carbon chains can be any length, from very short to literally infinite. We can show only a short segment of polyethylene because it is a chain of thousands of carbon atoms. The zigzag relationship of adjacent carbon-carbon bonds arises from the tetrahedral geometry around each carbon.

12.4 STRUCTURAL FORMULAS FOR ORGANIC MOLECULES

Organic molecules are almost always represented by structural formulas rather than molecular formulas. A structural formula shows the kinds and numbers of atoms in a molecule and the way in which they are bonded together. In Section 6.5 you learned that elements typically form certain numbers of covalent bonds: carbon, four; hydrogen, one; oxygen, two; and nitrogen, three. The bonding patterns were summarized in Table 6.1. From their molecular formulas and these data you were able to write Lewis structures or structural formulas for covalent compounds. Typical examples of structural formulas follow.

Carbon forms four bonds:

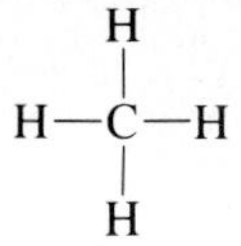

Four single bonds to C

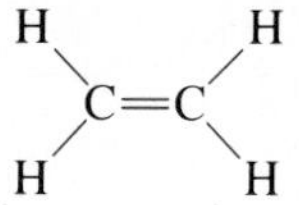

One double bond and two single bonds to each C

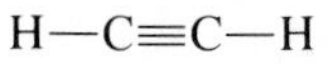

One triple bond and one single bond to each C

Hydrogen forms only one bond, as shown in all examples.

Oxygen forms two bonds:

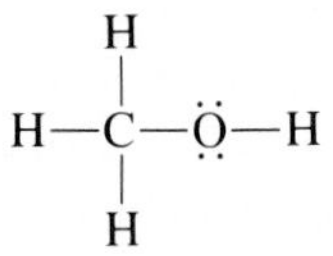

Two single bonds to O

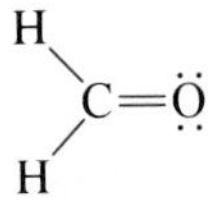

One double bond to O

Nitrogen forms three bonds:

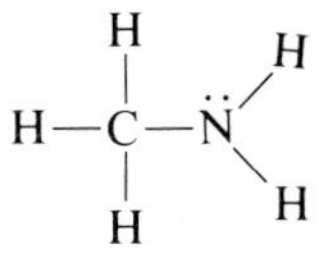

Three single bonds to N

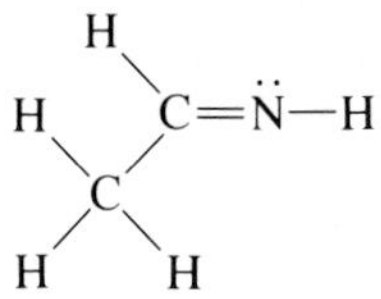

One double bond and one single bond to N

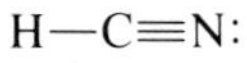

One triple bond to N

SAMPLE EXERCISE 12.1

What is the structural formula for propane, C_3H_8?

Solution For compounds with many carbon atoms, always begin by linking the given numbers of carbons together:

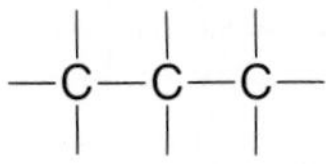

Then count the number of bonding slots to be filled. In this case, there are eight slots and eight hydrogens to fill them:

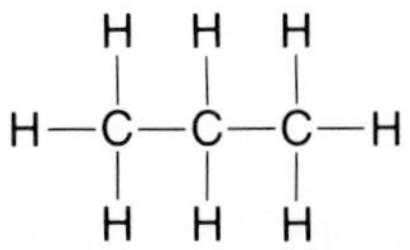

If there were too few atoms to fill the slots, then that would be an indicator that a multiple-bonding pattern was necessary.

Review Section 6.5, especially Sample Exercises 6.1 to 6.3, carefully if you do not remember how to write structural formulas.

12.5 CONDENSED STRUCTURAL FORMULAS

So far we have used only *full* structural formulas which show individual symbols for each atom and all connecting bonds. For example, in the case of propane, C_3H_8, in Sample Exercise 12.1, we came to the structure

$$\begin{array}{cccc} & H & H & H \\ & | & | & | \\ H- & C- & C- & C-H \\ & | & | & | \\ & H & H & H \end{array}$$

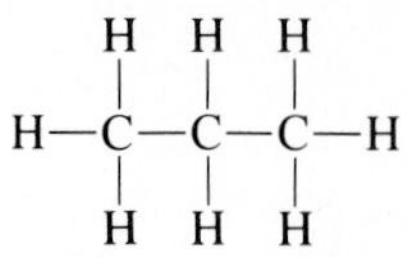

The three carbons, eight hydrogens, and all bonds are shown.

To save space and time of writing, *condensed* structural formulas are more often employed. In these formulas most carbons are individually represented, but hydrogens on the same carbon are added up, and the bonding dashes are not shown. For example, the condensed structural formula for propane is $CH_3CH_2CH_3$. Note carefully in the full structural formula that the three carbons in propane are bonded in a chain with no intervening hydrogens. It is merely the convention employed in writing condensed structures to list atoms bonded to each carbon after each carbon.

$$CH_3CH_2CH_3$$

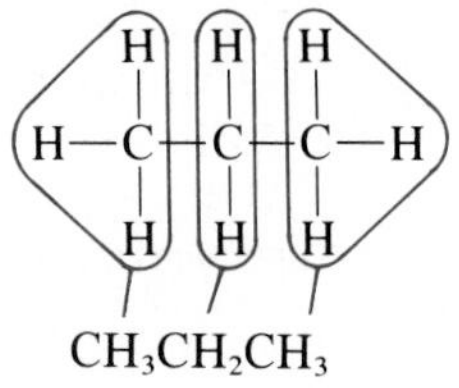

Branches off of a main carbon chain are handled by parentheses in condensed structural formulas:

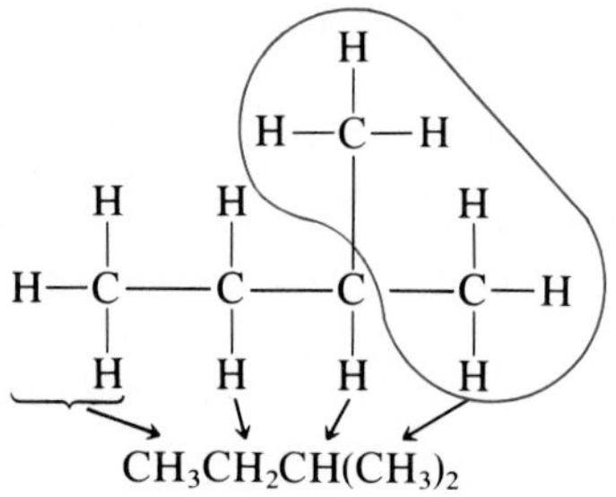

$$CH_3CH_2CH(CH_3)_2$$

Whenever you are in doubt about the meaning of a condensed structure, write out the *full* structure by arranging the atoms indicated after a particular C atom around that C atom. For example,

Condensed	Full
$CH_3CH_2CH_2CH_3$	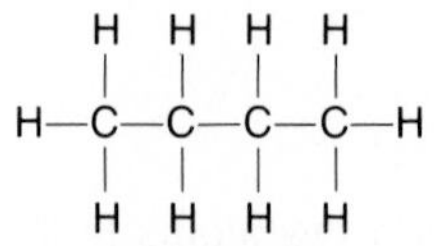 $\begin{array}{cccc} H & H & H & H \\ | & | & | & | \\ H-C-C-C-C-H \\ | & | & | & | \\ H & H & H & H \end{array}$
$CH_3CHBrCH_3$	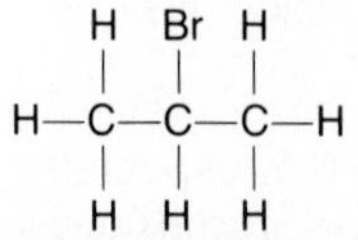 $\begin{array}{ccc} H & Br & H \\ | & | & | \\ H-C-C-C-H \\ | & | & | \\ H & H & H \end{array}$

(Continued)

Condensed	Full
$CH_3CH_2C(CH_3)_3$	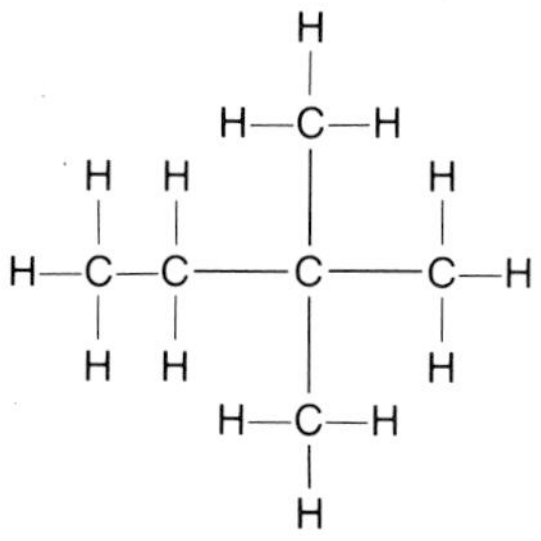

Problem 12.3 Write full structural formulas from the following condensed structures:
a CH_2ClCH_3
b $CH_3CH_2CH(CH_3)CH_2CH_3$
c CH_3CH_2OH

Compounds with rings, i.e., *cyclic* compounds, are represented by regular polygons. Each apex of the polygon represents a carbon atom. Though H's are not written, it is understood that a number of H's are attached to each carbon atom such that each carbon forms four bonds.

For example, compare the full and condensed formulas for rings:

Full	Condensed
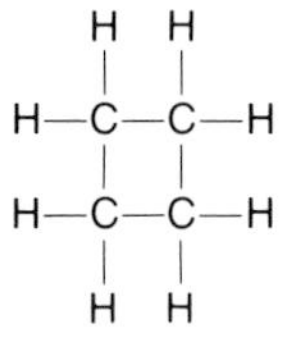	 Each apex is a carbon bonded to two other carbons. Therefore, there must be two H's bonded to each carbon.
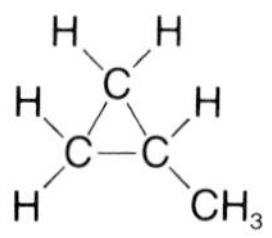	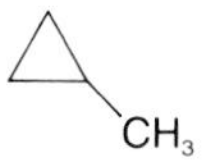 Attachments other than H atoms are always shown.
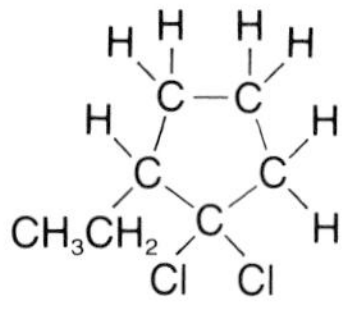	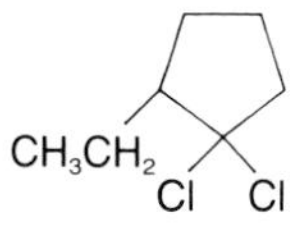

12.6 FUNCTIONAL-GROUP CONCEPT

Because of the large number of organic compounds, a classification system for organizing them is needed. Organic compounds are classified according to the presence of so-called functional groups in the compound. A **functional group** is a specific group of atoms, such as the —OH group or the $—NH_2$ group, that gives a compound a particular set of physical and chemical properties. Table 12.2 lists the common functional groups.

TABLE 12.2 Common Functional Groups

Structural Feature*	Name of Class
$>C<$	Alkane
$>C{=}C<$	Alkene
$-C{\equiv}C-$	Alkyne
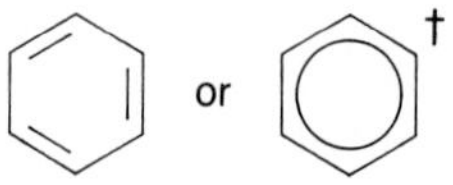	Aromatic
$-C-OH$	Alcohol
$-C-NH_2$	Amine
$-C-O-C-$	Ether
$-C-X$	Alkyl halide
$(H)C{=}O$ (—CHO)	Aldehyde
$-C-C({=}O)-C-$	Ketone
$-C({=}O)-OH$ (—COOH)	Carboxylic acid
$-C({=}O)-NH_2$ (—$CONH_2$)	Amide
$-C({=}O)-O-C-$ (—COOC—)	Ester
$-C({=}O)-X$	Acyl halide
$-C({=}O)-O-C({=}O)-$	Acid anhydride

* Where other atoms are not shown on the bonds to carbon, hydrogens or carbons are normally found bonded at these positions. Forms in parentheses are the common abbreviations of these structural features. Halogens (F, Cl, Br, I) are represented by X.

† The meaning of this representation will be clarified in Section 13.11.

For example, a compound with an —OH group is an alcohol. An alcohol with one carbon (methanol) and an alcohol with two carbons (ethanol) have very similar chemistry. The functional group, and not the number of carbons and hydrogens, is usually the key factor in determining the chemical properties of an organic compound.

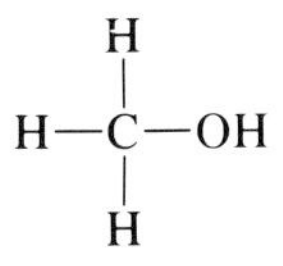
H
H—C—OH
H

One-carbon alcohol

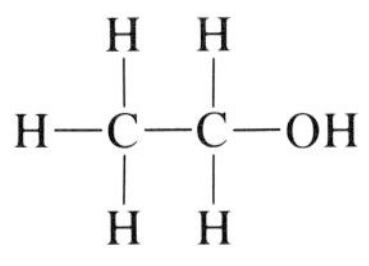
H H
H—C—C—OH
H H

Two-carbon alcohol

The carbons and hydrogens act as a backbone or skeleton on which the functional groups are placed. In order to make predictions concerning the properties and reactions of specified organic compounds, it is necessary to be able to identify the functional group(s) present in the compound. Your instructor will probably want you to memorize the structures and names in Table 12.2, so that given a structure such as in Sample Exercise 12.2 or 12.3, you can identify the functional group(s) present without referring to the table.

SAMPLE EXERCISE 12.2

Indicate the functional groups in the following structures:

a 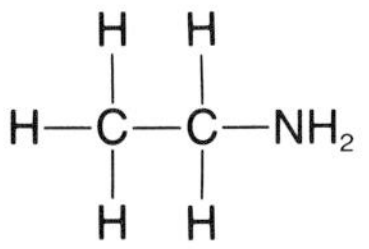
H H
H—C—C—NH2
H H

c 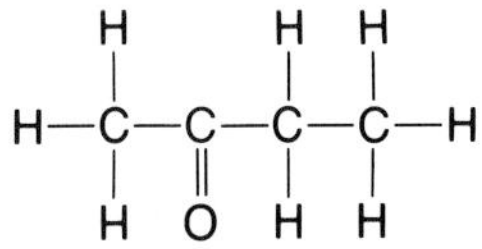
H H H
H—C—C—C—C—H
H O H H

b 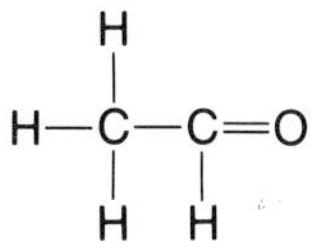
H
H—C—C=O
H H

d 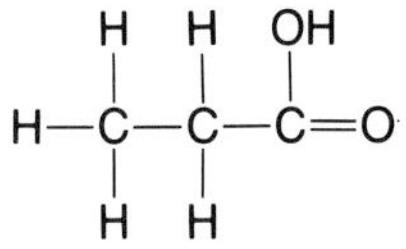
H H OH
H—C—C—C=O
H H

Solution Refer to Table 12.2 to identify the functional groups.

a The presence of the $—NH_2$ group indicates an *amine.*

b The presence of the —C=O (with —H on the C) group indicates an *aldehyde.*

c The C—C—C (with =O on the middle C) indicates a *ketone.* Notice the difference between this group and the aldehyde group, which has an H attached directly to the C that is double-bonded to O.

d —C=O (with —OH on the C) indicates the carboxylic *acid* group. Do not confuse this group with the alcohol functional group. Alcohols have —O—H and no C=O; acids have —OH attached to C=O. Also compare amines with amides, ethers with esters, and alkyl with acyl halides.

Many important biochemical compounds have more than one functional group. In most cases you may think of each group contributing its set of physical and chemical properties unaffected by the presence of other groups.

SAMPLE EXERCISE 12.3

Circle and label the functional groups present in the compound benzocaine.

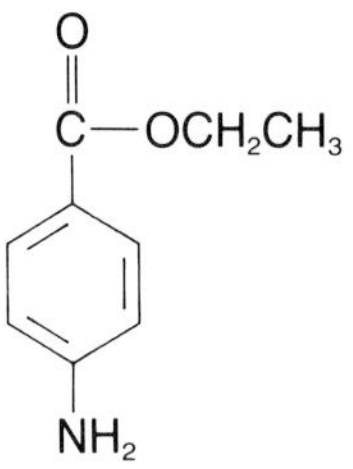
O
C—OCH₂CH₃
NH₂

Solution

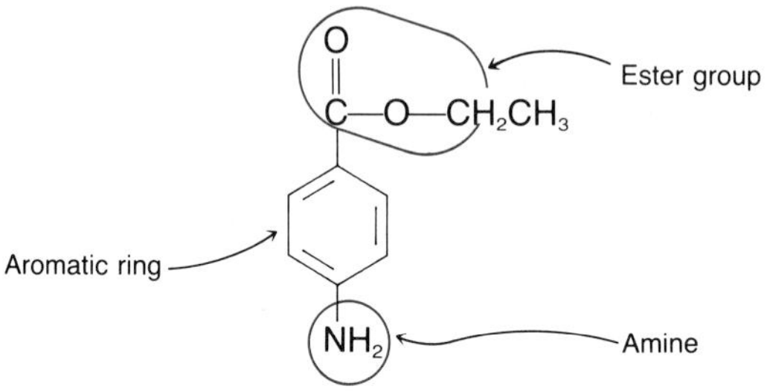
O
C—O—CH₂CH₃
Ester group
Aromatic ring
NH₂
Amine

Whereas maintaining the same functional group but changing the number of carbons, for example from methanol to ethanol, does not drastically alter chemical reactivity, it can have very dramatic physiological consequences. Methanol (methyl alcohol or wood alcohol), when ingested, inhaled, or absorbed through the skin, is a toxic substance which can lead to visual impairment or in acute cases, blindness and death. Ethanol (ethyl alcohol or grain alcohol), at least in moderate amounts, can be taken internally.

You will see why these two chemically very similar compounds lead to such different biological results when you study the oxidation of alcohols in Section 14.7.

12.7 HOMOLOGOUS SERIES

The functional-group concept is the common way of classifying organic compounds. Sets of compounds containing the same functional groups belong to the same families. For example, consider the sets of compounds below:

Alkanes	Alcohols
CH_4	CH_3OH
CH_3CH_3	CH_3CH_2OH
$CH_3CH_2CH_3$	$CH_3CH_2CH_2OH$
$CH_3CH_2CH_2CH_3$	$CH_3CH_2CH_2CH_2OH$
$CH_3CH_2CH_2CH_2CH_3$	$CH_3CH_2CH_2CH_2CH_2OH$

The alkanes are the family characterized by having exclusively single bonds between carbons and containing only the elements carbon and hydrogen. In the other set all the compounds have the same OH functional group and belong to the alcohol family. Furthermore, going down a list, each member of the family, for example $CH_3CH_2CH_2OH$, can be derived from the one before it, in this example CH_3CH_2OH, by the addition of a $—CH_2—$ grouping.

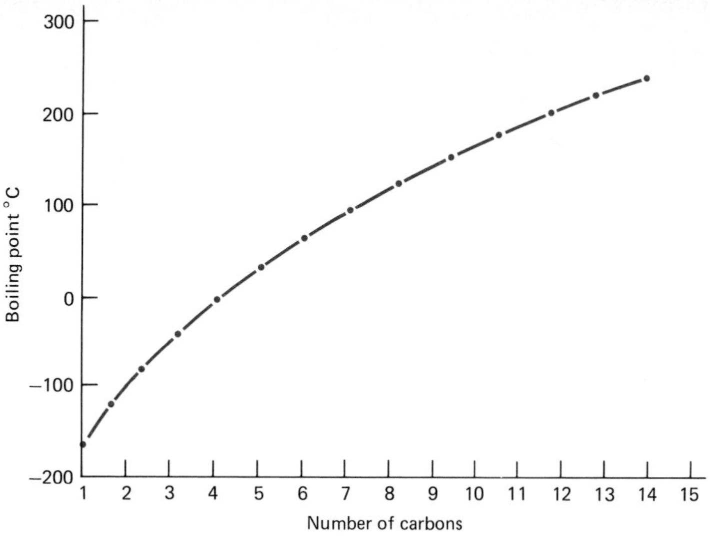
300
200
100
0
−100
−200
Boiling point °C
1 2 3 4 5 6 7 8 9 10 11 12 13 14 15
Number of carbons

FIG. 12.2 Boiling points of straight-chain alkanes. Boiling points increase as chain length increases because additional atoms in the chain increase dispersion forces between chains.

A series of compounds which have the same functional group and differ only in the number of $—CH_2—$ groups present constitutes a **homologous series.** Each of the functional-group families forms a homologous series.

Members of a homologous series have similar chemical properties, because they all have the same functional group which determines chemical reactivity. However, physical properties, such as boiling point and melting point, show a general increase with increasing numbers of $—CH_2—$ groups in the structural backbone. The added CH_2 groups increase the molecular weight, which increases the always present dispersion forces (Section 8.10), and thereby a higher temperature is required to separate the molecules. A heavier molecule also requires a higher temperature to reach the same speed as a lighter molecule. Figure 12.2 shows the variation in boiling point for the homologous series of straight-chain alkanes; the first five members of this series were shown at the beginning of this section. For a homologous series, such as the alcohols, in which the lower members are water-soluble (Section 12.2), the water solubility decreases as the chain of $—CH_2—$ groups increases.

12.8 THE R-GROUP CONCEPT

In considering organic reactions you will find that the hydrocarbon chain length usually is not important because all reactivity occurs at the functional group. Because of the usual *un*importance of the carbon chain in an organic compound compared with the functional group, it is a common practice to represent the chain in a general, nonspecific way. For example, ROH represents any alcohol. The R may stand for one carbon (with attached hydrogens), two carbons, or a seven-carbon chain, or any group of C and H atoms.

Similarly, RCOOH can be used to represent a carboxylic acid. RNH_2, RX, and ROR are amines, alkyl halides, and ethers, respectively. Alternatively, we can say that ROH represents a member of the homologous series of alcohols. RX represents the alkyl halide homologous series. R clearly does not specify chain length.

12.9 THE "ACTION" IS AT THE FUNCTIONAL GROUP

In examining an organic compound to predict its chemical reactivity, you should focus on the functional group present because that is where the reaction will occur. The hydrocarbon chain is usually unchanged during reactions. For example, consider the following reaction:

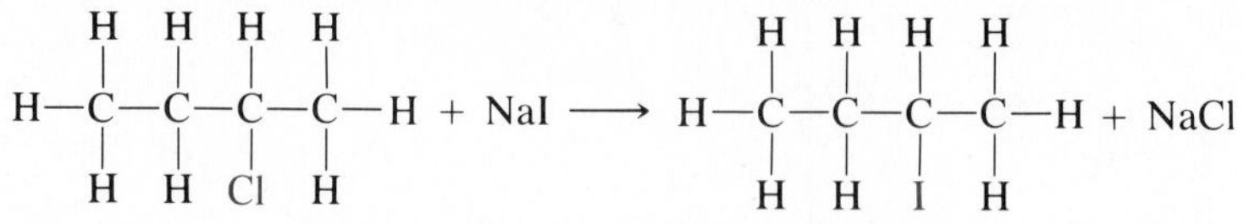

The Cl is the functional group and identifies the class known as alkyl chlorides. Iodide displaces and replaces chloride. All the action is at the functional-group carbon; none of the other carbons is changed in the reaction. Organic chemists sometimes represent such reactions very simply, using R groups:

$$R{-}Cl + NaI \longrightarrow R{-}I + NaCl$$

In Chapters 13, 14, 16, and 17 we will discuss the basic chemistry of the functional groups. A knowledge of the chemistry of each functional group will be as important to your understanding of organic and biochemistry as was a knowledge of the periodic-group concept to studying the chemistry of the elements.

12.10 ISOMERISM

Throughout the foregoing we have been using *structural* formulas to represent organic molecules and will continue to do so in all succeeding chapters. It is essential that we use full or condensed structural formulas in organic chemistry rather than molecular formulas, which simply give the numbers of each type of atom in the compound. This is so because often more than one structure can be written for a particular molecular formula. For example, consider the molecular formula C_2H_6O. Two correct Lewis structures can be written for C_2H_6O:

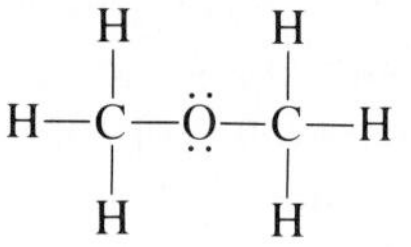

Structure 1

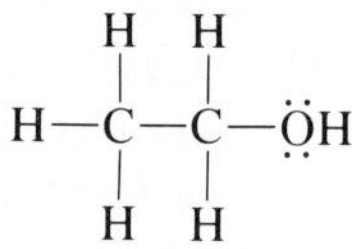

Structure 2

The fact that two structural formulas can be written for C_2H_6O tells us that there are two distinctly different compounds with the same molecular formula. Compounds which have the same molecular formula but different structural formulas are called **isomers.**

Isomers are completely different compounds, each with its own unique set of properties. In the example just given, compound 1 is an ether and has the characteristic properties of an ether, whereas compound 2 is an alcohol with the properties characteristic of that functional group. Compound 1 is a gas at room temperature (bp = −23°C), whereas compound 2 is a liquid which boils at 78°C.

Whenever you are given a molecular formula and asked to write a structural formula, you should remember that there may be more than one structure, which is consistent with the Lewis rules for writing formulas, that can be written. For example, suppose you were asked to write a structural formula for the molecular formula $C_2H_4Br_2$. If you begin in the usual manner of writing structures, you would connect the two carbons and find six available slots for the H's and Br's:

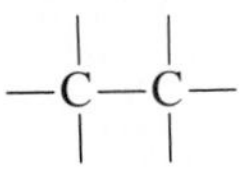

This allows for two unique arrangements of hydrogen and bromine atoms. The two bromines can be on the same carbon:

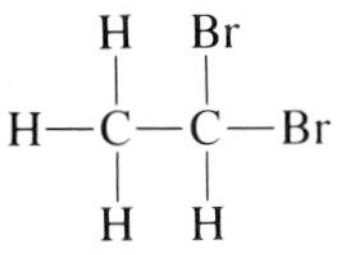 or CH_3CHBr_2

or each C can bear one Br:

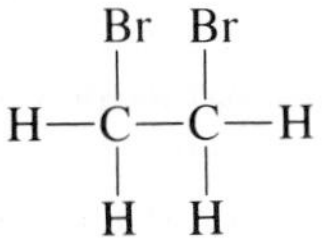 or CH_2BrCH_2Br

These are the only two unique arrangements; i.e., there are only two isomers corresponding to the formula $C_2H_4Br_2$. The arrangements

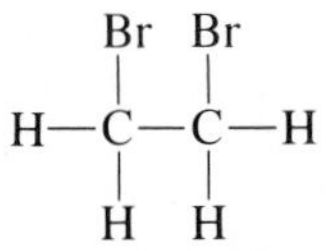 and 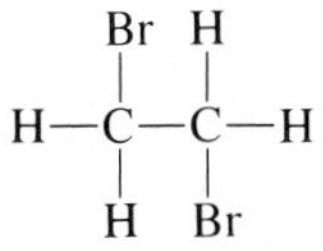

are identical even though they might look different to you written in two dimensions on paper. But remember that molecules are three-dimensional, as we have seen in Sections 6.10 and 6.11 and are about to review in Section 12.11. Rotation about carbon-carbon bonds also assures the identity of these two representations, as we will see in Section 12.12.

Often it is necessary to be able to recognize whether two given structures are related as isomers. This can be done according to the following guidelines.

Guidelines for Recognizing Structures as Identical, Isomeric, or Unrelated

STEP 1 Determine the molecular formula of the given structures. Isomers (and identical compounds) must have the same molecular formula. If the molecular formulas of the given structures are *not* the same, the two structures are neither identical nor isomers. If the molecular formulas are the same, proceed to step 2.

STEP 2 Examine the bonding arrangement of the atoms in the structure. Isomers will have their atoms arranged differently, while identical compounds will exhibit the same bonding arrangement.

SAMPLE EXERCISE 12.4 For each of the following pairs of structures, decide whether the two structures are identical, isomeric, or unrelated:

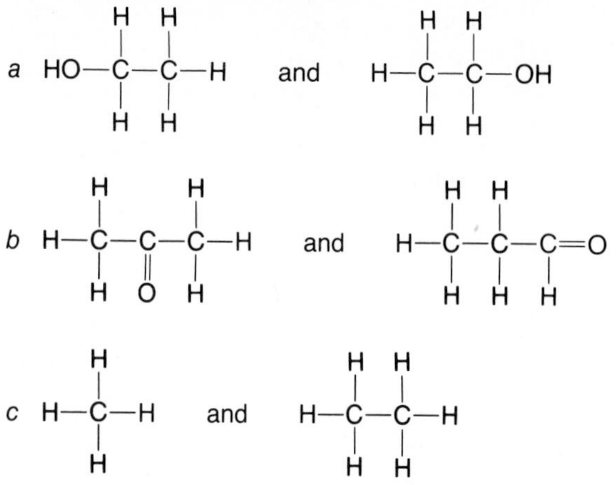

Solution

a *Identical.* The two structures have the same molecular formula. C_2H_6O. Although one is written as the reverse of the other, a careful examination will show that the atoms are bonded together in exactly the same arrangement in both structures; the basic skeleton is 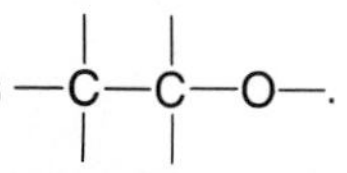.

b *Isomers.* The two structures have the same molecular formula, C_3H_6O. However, an examination of the bonding arrangement shows that the structural formulas are different. The location of the carbon-oxygen double bond differs, as does the placement of H's on C's. In this case, recognizing one compound as a ketone and the other as an aldehyde emphasizes the difference in structure.

c *Unrelated.* The molecular formula for the compound on the left is CH_4. The one on the right has the formula C_2H_6. Therefore, these are neither identical nor isomeric.

Isomerism is another phenomenon of organic chemistry which leads to the huge numbers of organic compounds that exist. For a given molecular formula, even one with only a small number of atoms, there may be a large number of structural isomers. Not only can there be any number of carbon atoms bonded in chains or rings and a variety of different attached elements, but also the atoms can be arranged in different ways, still consistent with the rules for writing Lewis structures. Structural isomers are different compounds which have different physical properties and often different chemical reactivities.

12.11 GEOMETRY AROUND CARBON ATOMS

Before continuing the discussion of isomerism, let us review the three-dimensional arrangements around carbon atoms in organic molecules. The spatial arrangement around a particular carbon atom follows directly from its Lewis structure and VSEPR theory (see Section 6.10). The geometry, or arrangement, of bonded atoms around a particular carbon atom can readily be determined by the carbon's bonding pattern. Four single bonds are arranged tetrahedrally; when a double bond is present, the array is planar

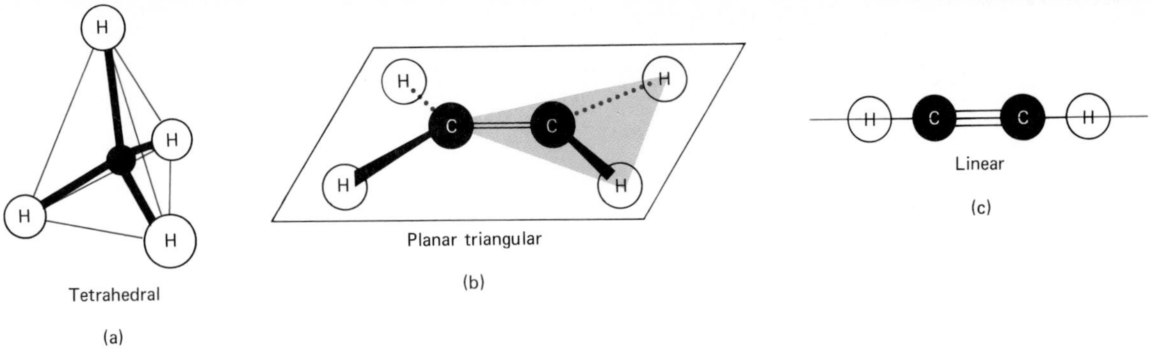
H
H
H
H
Tetrahedral
(a)
H
H
C
C
H
H
Planar triangular
(b)
H
C
C
H
Linear
(c)

FIG. 12.3
Geometry around carbon atoms. (*a*) four single bonds to carbon are arranged in three dimensions so that they point to the corners of a tetrahedron. CH_4 offers a simple example. (*b*) When a carbon is involved in a double bond, the carbon and all of the atoms bonded to it lie in the same plane. The full description of the geometry is planar triangular because the atoms bonded to the doubly bonded carbon can be conceived to form the apexes of a triangle. (*c*) When a carbon is involved in a triple bond, the carbon and all of the atoms bonded to it lie along a line, a linear array.

triangular; the presence of a triple bond produces a linear array. Figure 12.3 shows these arrangements, and they can be summarized as follows:

Bonding Pattern		Geometry
Four singles		Tetrahedral
One double and two singles	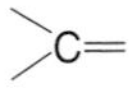	Planar triangular
One triple and one single	—C≡	Linear

In this chapter the discussion will be confined to the tetrahedral arrangement around carbon. Section 13.7 discusses the planar triangular arrangement, and Section 13.9, the linear array. The tetrahedral arrangement around carbon is the most commonly encountered geometry because carbon atoms that form four single bonds are most prevalent. In a connecting chain of all single-bonded carbons, the tetrahedral shape about each carbon leads to a zigzag array for the chain, as you have already seen in Figure 12.1. A chain of four carbons (butane) looks like

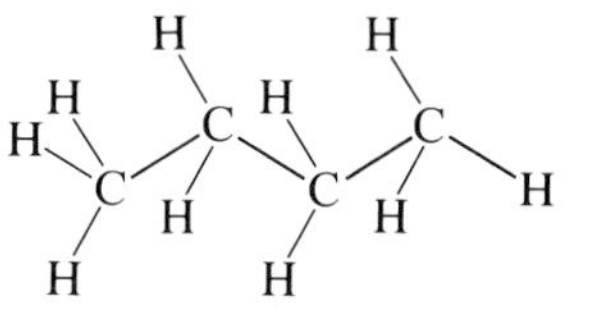

12.12 ROTATION ABOUT SINGLE BONDS

The nature of the carbon-carbon single bond allows rotation of bonded atoms around the bond without significantly affecting the strength of the bond. Such rotation was previously referred to in Section 12.10 in comparing the

two representations

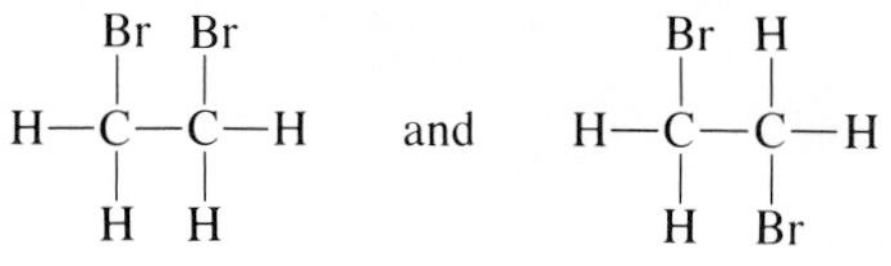

Using perspective drawings to show the tetrahedral arrangement of carbon bonds and showing rotation about the C—C bond, the identity of these two representations can be demonstrated:

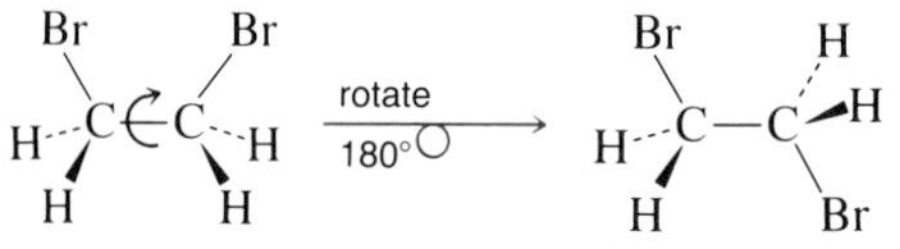

This rotation is rapid in molecules of this compound.

For the molecule butane ($CH_3CH_2CH_2CH_3$), a rotation about the bond connecting the second and third carbons can be represented:

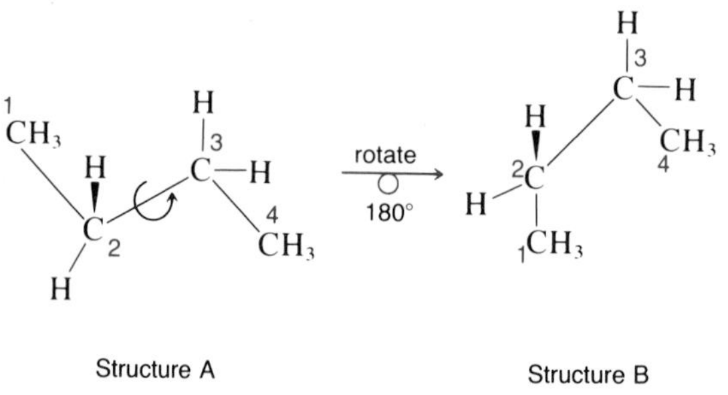

The "rotated" structure B still has tetrahedral geometry around all the carbons. If we continue rotating another 180° around the 2, 3 bond, so that altogether the rotation has been 360°, the "rotated" form B returns to the original structure A.

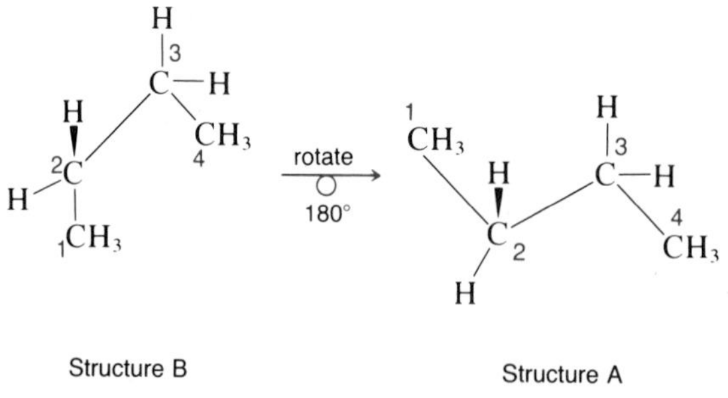

In a butane lighter or other container, butane molecules are going through these rotations all the time. Structures A and B above are different **conformations** of the molecule butane. Two structures are considered to be **conformers** if by rotating about a single bond we can convert one into the other. The two representations of CH_2BrCH_2Br previously discussed are also conformers.

In this text, and most other texts, organic structures are written in a straight line for convenience. That is, for butane we write

$$\begin{array}{c} \text{H} \quad \text{H} \quad \text{H} \quad \text{H} \\ | \quad\; | \quad\; | \quad\; | \\ \text{H}-\text{C}-\text{C}-\text{C}-\text{C}-\text{H} \\ | \quad\; | \quad\; | \quad\; | \\ \text{H} \quad \text{H} \quad \text{H} \quad \text{H} \end{array}$$

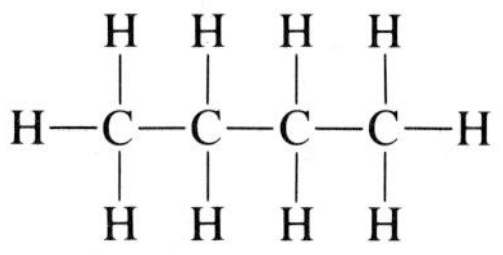

or use the condensed formula, $CH_3CH_2CH_2CH_3$. However, recognizing that bond rotation is possible tells us that this bent or twisted representation

$$\begin{array}{c} \text{H} \quad \text{H} \quad \text{H} \\ | \quad\; | \quad\; | \\ \text{H}-\text{C}-\text{C}-\text{C}-\text{H} \\ | \quad\; | \quad\; | \\ \text{H} \quad \text{H} \quad \text{H} \\ \quad\; | \\ \quad \text{H}-\text{C}-\text{H} \\ \qquad | \\ \qquad \text{H} \end{array}$$

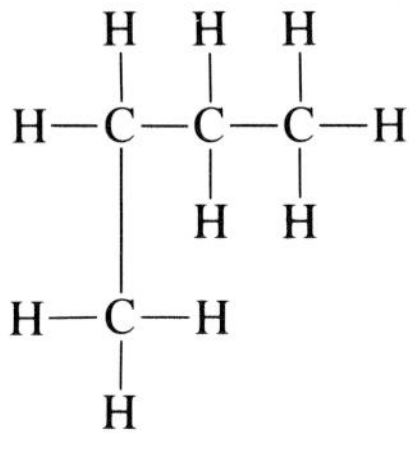

is also butane. This second representation still has four continuously connected carbon atoms, though they do not all appear on the same line—one is rotated below. As you begin to write and read structural formulas, you will need to recognize conformers as different representations of the same compound.

Consider:

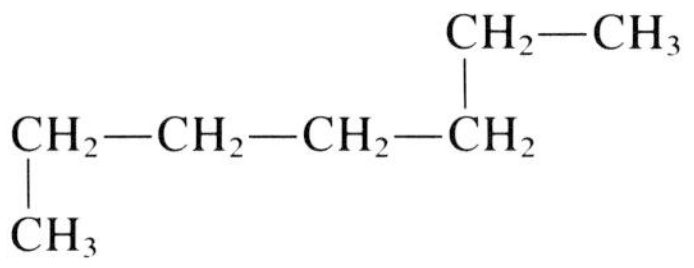

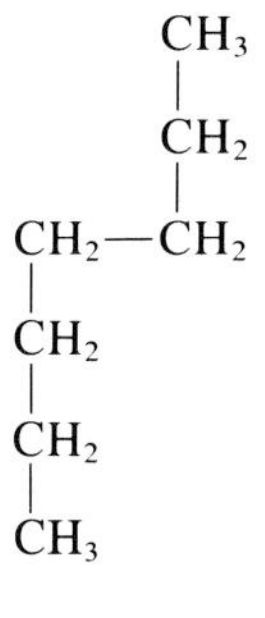

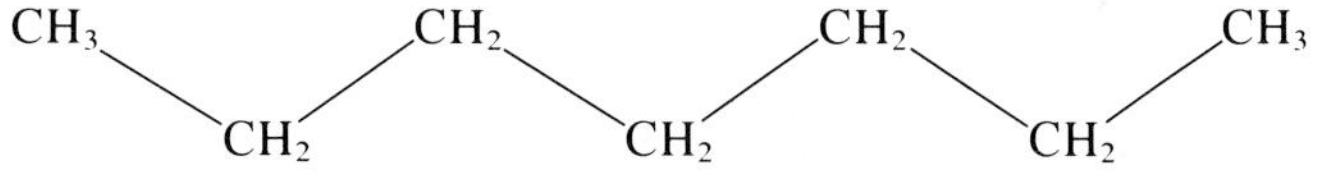

The above are all acceptable ways of writing structures for the same compound of seven carbons connected in a continuous chain.

12.13 ISOMERISM REVISITED

As you learned in Section 12.10, **isomers** are compounds which have the same molecular formula but different structural formulas. These structural isomers differ in the way in which the atoms are bonded, or connected together. Because of the various ways of writing structures, we must be wary in examining given structures in order to identify truly different arrangements. There may be different functional groups in isomers, in which case the different arrangements are easy to spot. Alternatively it might be the carbon chain that is arranged differently. This section discusses spotting isomers arising from different arrangements of the carbon chain.

The two structures for molecular formula C_4H_{10}, *n*-butane (structure 1) and isobutane (structure 2), are isomers

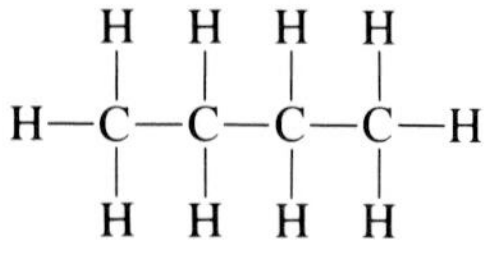

Structure 1

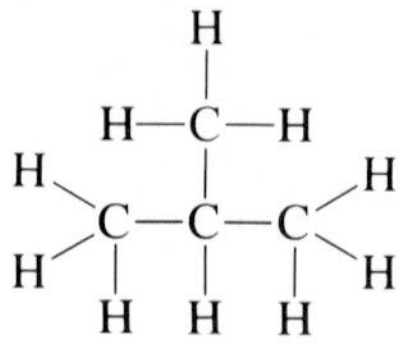

Structure 2

because the atoms are connected together differently in the two structures. In structure 1 there is a continuous, unbranched chain of four carbons, whereas in structure 2 there is a chain of three carbons with one branch. Identifying the longest continuous chain in any organic structure is an important skill to acquire for the purpose of recognizing and naming compounds. One way to find the longest continuous chain of carbons in a structure is to use a procedure similar to a "connect the dots" game. Mark through a carbon chain with a pencil, going from carbon to carbon along the bonds, and observe how far you can proceed without lifting the pencil off the paper. See Figure 12.4.

Structures 2 and 3 are not structural isomers; they represent different conformations of the compound isobutane. Structure 2 can be converted into structure 3 by a rotation about the bond indicated (↻).

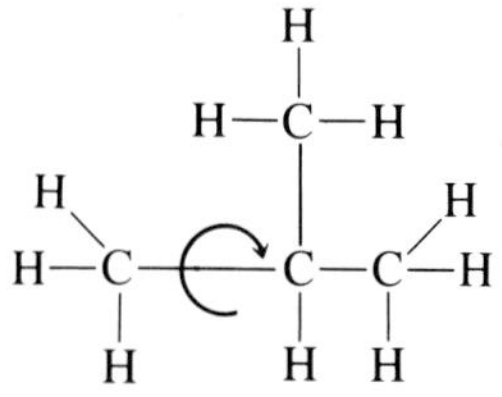

Structure 2

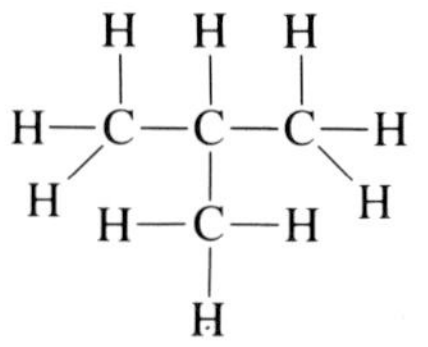

Structure 3

FIG. 12.4
(*a*) *n*-Butane can be identified by recognizing that there are four carbons in a continuous chain. The "connect-the-dots" game demonstrates the continuity whether the four carbons are written across the horizontal or in some twisted fashion. (*b*) In the branched chain, isobutane, only three carbons can be traced in a continuous fashion no matter where we start tracing or how the structure is drawn.

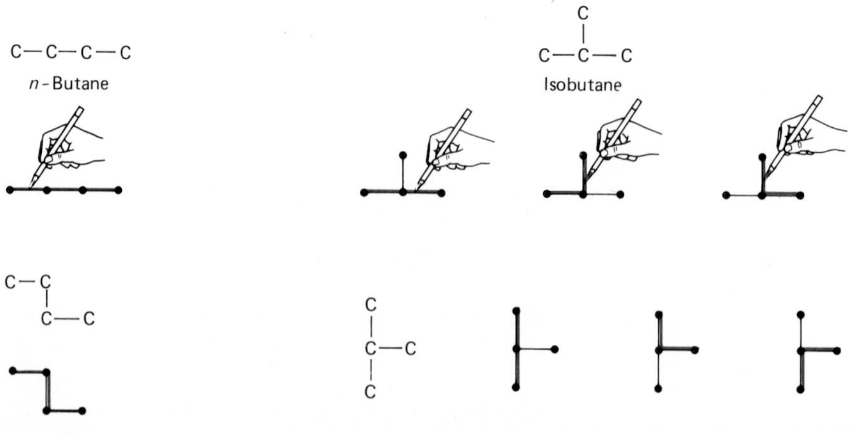
n-Butane
Isobutane
(a)
(b)

Two structures which can be interconverted by a bond rotation are not structural isomers, but instead are **conformers.** Alternatively, one can recognize structures 2 and 3 as identical by turning either one upside down; i.e., rotate the whole structure 180° in space. A third way to recognize the identical nature of structures 2 and 3 is by using the connect-the-dots test. You will find three carbons in a row as the maximum length of the chain no matter how you check. There is a branch on the middle of the three-carbon chain. We will further discuss and use the connect-the-dots test when we discuss nomenclature in Section 12.14.

Problem 12.4 Indicate in each pair whether the representations given are structural isomers or the same molecule or unrelated.

a $H-\overset{\overset{\displaystyle H}{|}}{\underset{\underset{\displaystyle H}{|}}{C}}-\overset{\overset{\displaystyle Cl}{|}}{\underset{\underset{\displaystyle H}{|}}{C}}-H$ and $H-\overset{\overset{\displaystyle H}{|}}{\underset{\underset{\displaystyle H}{|}}{C}}-\overset{\overset{\displaystyle H}{|}}{\underset{\underset{\displaystyle H}{|}}{C}}-Cl$

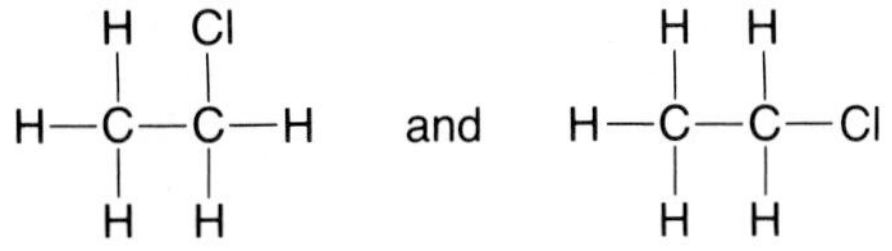

b $H-\overset{\overset{\displaystyle H}{|}}{\underset{\underset{\displaystyle H}{|}}{C}}-\overset{\overset{\displaystyle Cl}{|}}{\underset{\underset{\displaystyle H}{|}}{C}}-Cl$ and $Cl-\overset{\overset{\displaystyle H}{|}}{\underset{\underset{\displaystyle H}{|}}{C}}-\overset{\overset{\displaystyle H}{|}}{\underset{\underset{\displaystyle H}{|}}{C}}-Cl$

c $CH_3CH_2-\overset{\overset{\displaystyle CH_3}{|}}{\underset{}{\overset{\displaystyle CH_2}{|}}}CH_2$ and $\overset{\overset{\displaystyle CH_3}{|}}{\underset{}{\overset{\displaystyle CH_2}{|}}}CH_2-CH_2-\underset{\underset{\displaystyle CH_3}{|}}{CH_2}$

d $H-\overset{\overset{\displaystyle H}{|}}{\underset{\underset{\displaystyle H}{|}}{C}}-\overset{\overset{\displaystyle OH}{|}}{\underset{\underset{\displaystyle H}{|}}{C}}-\overset{\overset{\displaystyle H}{|}}{\underset{\underset{\displaystyle H}{|}}{C}}-H$ and $H-\overset{\overset{\displaystyle H}{|}}{\underset{\underset{\displaystyle H}{|}}{C}}-\overset{\overset{\displaystyle H}{|}}{\underset{\underset{\displaystyle OH}{|}}{C}}-\overset{\overset{\displaystyle H}{|}}{\underset{\underset{\displaystyle H}{|}}{C}}-H$

e $H_3C-\overset{\overset{\displaystyle CH_3}{|}}{\underset{\underset{\displaystyle CH_3}{|}}{C}}-CH_3$ and $CH_3CH_2CH_2CH_2CH_3$

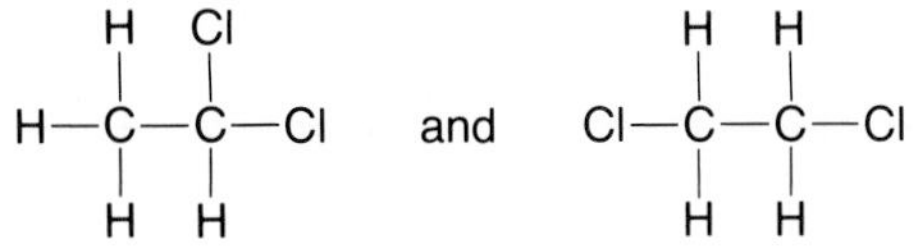

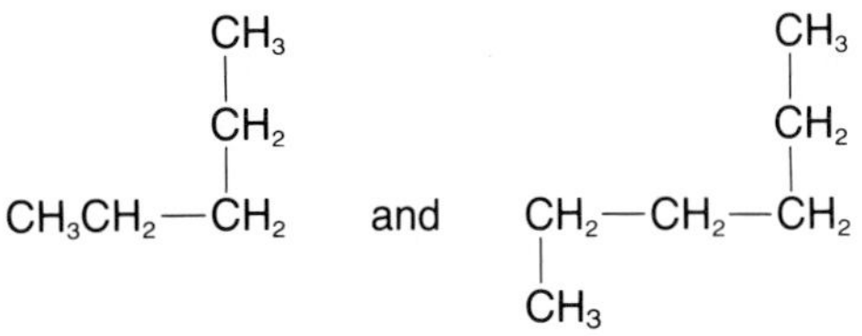

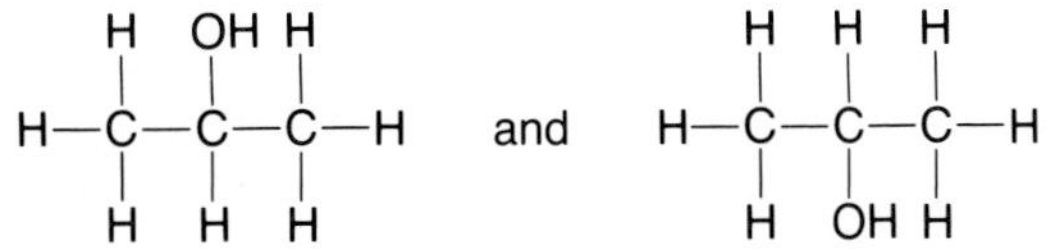

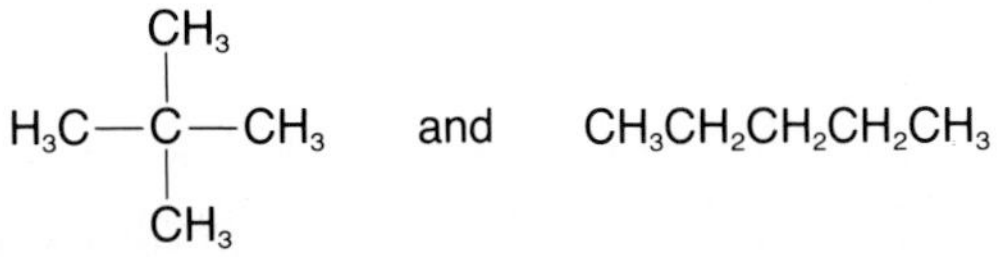

12.14 NOMENCLATURE

The method or set of rules for naming compounds is called **nomenclature.** In Section 5.10 you learned the rules for naming ionic compounds. The name of an organic compound must specify a unique structure; that is, from the written name for a compound it should be possible to construct a structural formula. A systematic nomenclature provides a means of constructing names so that there is a unique correspondence between name and structure.

The International Union of Pure and Applied Chemistry (IUPAC) has developed a systematic approach for organic nomenclature known as the **IUPAC** system of rules. In the IUPAC system the root of the name for a given compound is based on the name of the longest continuous chain of carbons in the compound. A suffix is added to the root to indicate the functional-group family. Other connected substituent groups are named as branches on the longest chain.

Before you can proceed further, you will need to learn the root names of the first 10 straight carbon chains in Table 12.3. This table also includes the

TABLE 12.3 Root Names for Continuous Carbon Chains and the Names of the First 10 Straight-Chain Alkanes

Number of Carbons in a Continuous Chain	Root Name	Alkane Structure	Alkane Name
1	Meth-	CH_4	Methane
2	Eth-	CH_3CH_3	Ethane
3	Prop-	$CH_3CH_2CH_3$	Propane
4	But-	$CH_3CH_2CH_2CH_3$	Butane
5	Pent-	$CH_3CH_2CH_2CH_2CH_3$	Pentane
6	Hex-	$CH_3CH_2CH_2CH_2CH_2CH_3$	Hexane
7	Hept-	$CH_3(CH_2)_5CH_3$	Heptane
8	Oct-	$CH_3(CH_2)_6CH_3$	Octane
9	Non-	$CH_3CH_2)_7CH_3$	Nonane
10	Dec-	$CH_3(CH_2)_8CH_3$	Decane

names of the first 10 *alkanes*. As you learned in Section 12.7, the alkane family is characterized by the fact that its members contain only hydrogen and carbon atoms connected by single bonds. The list of alkane structures in Table 12.3 is an extension of the alkane homologous series you have already seen (Section 12.7). The suffix which denotes the alkane family is *-ane*. Notice that the names in the table are constructed by combining the root name (indicating the number of carbons in the chain) with *-ane*.

In order to name alkanes with branched chains, you must learn the names of some alkyl groups. An **alkyl group** (R group) is the fragment of a molecule that remains if one atom is removed from a carbon. For example, when one of the hydrogens of ethane is removed, the fragment that remains is called the ethyl group.

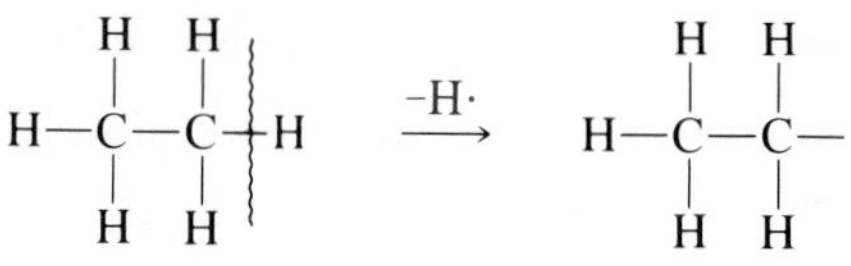

Ethane Ethyl

The grouping CH_3CH_2— is not a compound (one carbon has only three bonds) and must be bonded to another grouping or atom. You will find such alkyl groups as substituents on the longest (root) continuous chain of carbons. For example,

$CH_3—CH_2—CH_2—CH_2—CH_2—CH_3$

Hexane

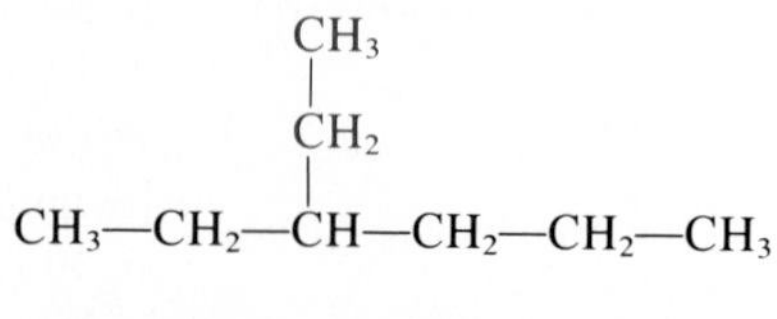

Hexane with an ethyl substituent or branch

Table 12.4 lists the structure and names of some alkyl groups. Note that the name of an alkyl group is constructed:

Root name + suffix *-yl*

The two-carbon fragment is ethyl; the one-carbon fragment CH_3— is methyl.

The alkane propane yields two different alkyl groups depending on which

TABLE 12.4 Names of Common Alkyl Groups

CH_3—	Methyl
CH_3CH_2—	Ethyl
$CH_3CH_2CH_2$—	*n*-Propyl
CH_3CHCH_3 (bond from central C)	Isopropyl
$CH_3CH_2CH_2CH_2$—	*n*-Butyl
$CH_3CH_2CHCH_3$ (bond from CH)	*sec*-Butyl
$CH_3CH(CH_3)CH_2$—	Isobutyl
$CH_3C(CH_3)CH_3$ (bond from central C)	*t*-Butyl

hydrogen is removed. Removal of a hydrogen from either of the end carbons gives a straight-chain alkyl group, called normal propyl, *n*-propyl.

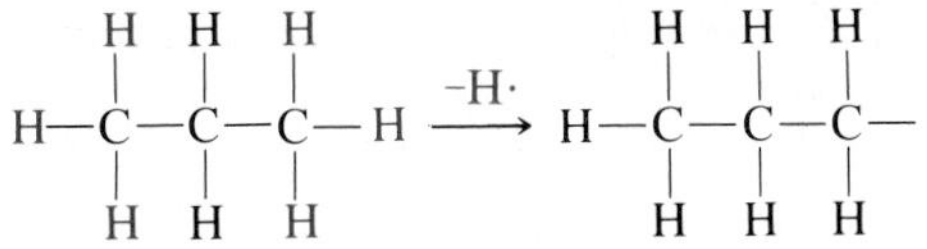

Removal of any one of the six colored hydrogens gives the same group. *n*-Propyl

Removal of a hydrogen from the middle carbon gives the isopropyl group.

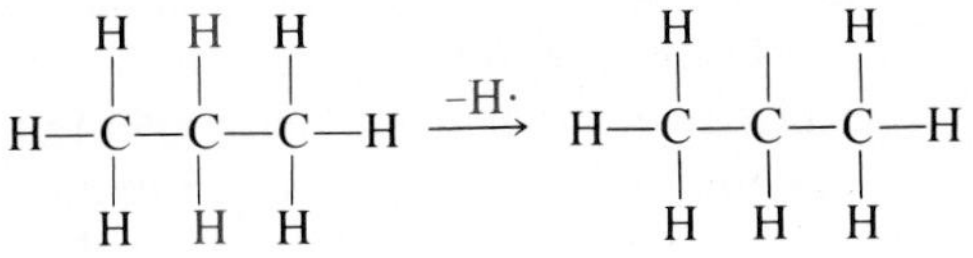

Removal of one of the two colored hydrogens gives the same group. Isopropyl

The basic idea of how root names and alkyl group names are used can be demonstrated by a simple example: CH_3—$\underset{}{CH}(CH_3)$—CH_2CH_3 is named 2-methylbutane. The longest continuous chain of four carbons provides the root name; hence the parent chain is a butane. There is a methyl substituent or branch on the chain. Methyl is connected to the *second* carbon along the butane chain, hence the number 2 to designate its position. As structures become more complex, so must the name, but the basic idea always involves finding a parent chain and recognizing the attached substituents.

Let us lay out a stepwise procedure for naming alkanes and demonstrate the procedure by naming

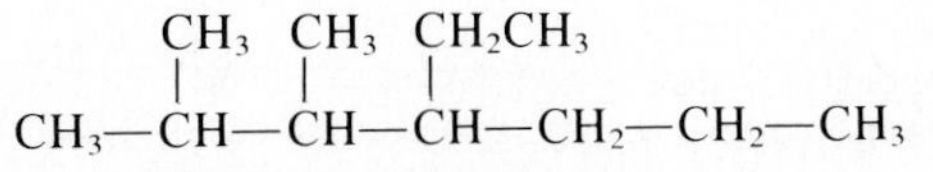

STEP 1 **Find the longest continuous chain of carbons to establish the parent name.**

Seeing how many carbons you can pencil through without lifting the pencil off the paper (the connect-the-dots system discussed in Section 12.13) is one way of accomplishing this. In this case the longest continuous chain happens to occur straight across the horizontal, but you should be aware that this is not always the case. The longest continuous chain may go ''around corners.'' See, for example, Sample Exercise 12.5.

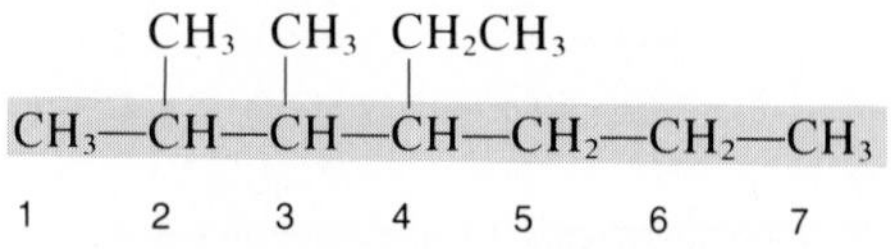

Because the longest chain is seven carbons, this is a heptane.

STEP 2 **Identify the substituents on the longest chain.**

In our example the branches are two methyl groups and one ethyl group:

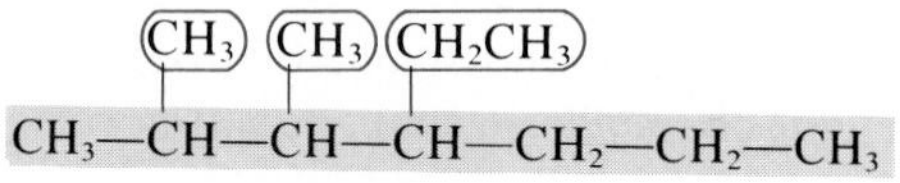

STEP 3 **Use appropriate prefixes to indicate more than one of a particular substituent.**

In the example there are two methyl groups. This can be indicated by saying *dimethyl*. The prefix *di-* means ''two.'' Table 12.5 lists other numerical prefixes. Only *di-* and *tri-* are used commonly. There is only one ethyl group, so no prefix is needed.

STEP 4 **Number the chain so that the lowest possible numbers are assigned to the substituent positions.**

In the example, number the chain from left to right because that gives the substituent positions the numbers 2, 3, and 4, rather than 4, 5, and 6, which would be assigned if numbering began at the right end of the chain. Thus we have 2, 3-dimethyl and 4-ethyl substituents.

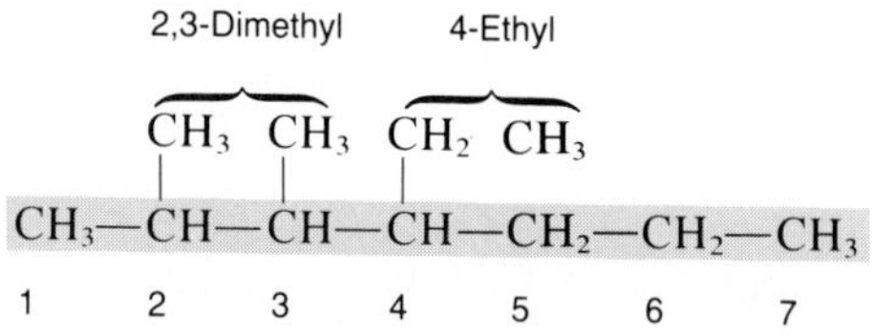
2,3-Dimethyl
4-Ethyl

STEP 5 **Use commas between substituent numbers and dashes between numbers and prefixes.**

Unlike substituent groups or different branching positions are separated by a dash. The final substituent group and the base name are written together as one word. It is preferred that the substituent names be written in alphabetical order without consideration of numerical prefixes. The assembled name for the sample compound is

4-ethyl-2,3-dimethylheptane

4-ethyl	2,3-dimethyl	heptane
An ethyl on carbon 4	Two methyl groups on carbons 2 and 3	A 7-carbon chain

TABLE 12.5 Numerical Prefixes in Organic Nomenclature

Number of Substituents	Prefix
1	Mono-*
2	Di-
3	Tri-
4	Tetra-
5	Penta-
6	Hexa-
7	Hepta-
8	Octa-
9	Nona-
10	Deca-

* Rarely encountered.

SAMPLE EXERCISE 12.5

Give the IUPAC name for

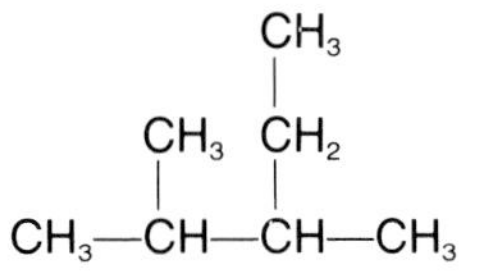

Solution The longest chain, as shown, is five carbons, giving the parent name pentane.

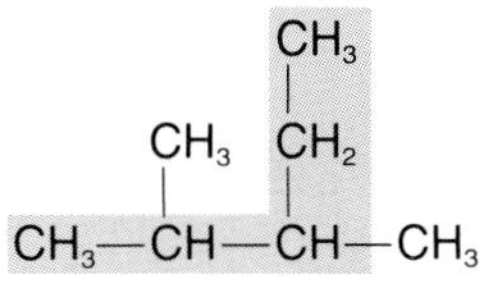

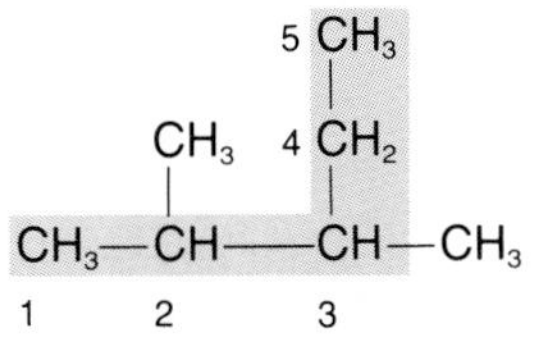

There are methyl groups branched on the chain in the 2 and 3 positions. The full name is 2,3-dimethylpentane.

Problem 12.5 Give IUPAC names for

a 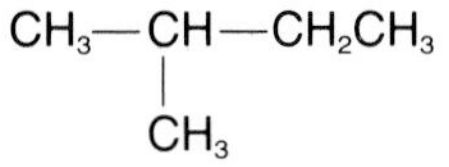

b 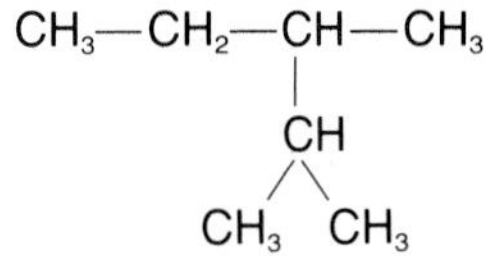

Alkyl Halides

The stepwise procedure for nomenclature is basically the same for all organic compounds. For example, in the IUPAC system alkyl halides are named as hydrocarbons with halogens as substituents on the longest chain. A halogen substituent is designated by the root of the halogen with the suffix *-o:*

F- = fluoro

Cl- = chloro

Br- = bromo

I- = iodo

So, $CH_3—CH(Cl)—CH_2—CH_3$ is 2-chlorobutane. When a carbon chain has both alkyl and halogen branches, an alphabetical listing in the name is preferred. As examples:

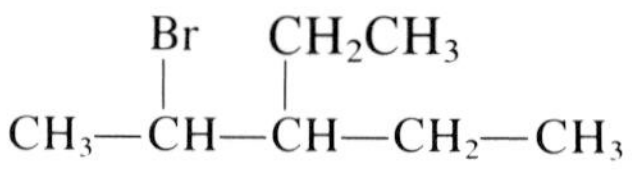
2-Bromo-3-ethylpentane

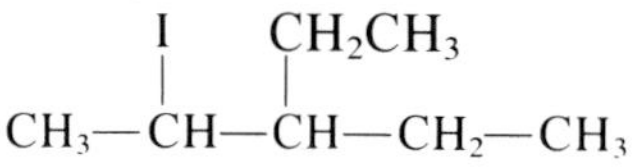
3-Ethyl-2-iodopentane

SAMPLE EXERCISE 12.6

Give an IUPAC name for

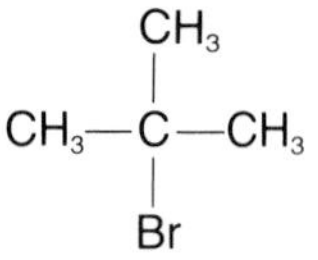

Solution The longest chain is three carbons, giving propane as the parent name.

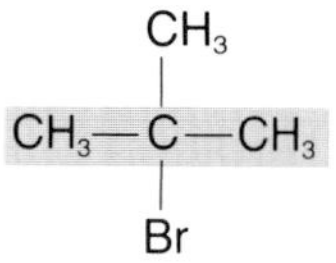

The substituents are a methyl group and a bromine in the 2 position. The full name is 2-bromo-2-methylpropane.

Problem 12.6 Give IUPAC names for

a 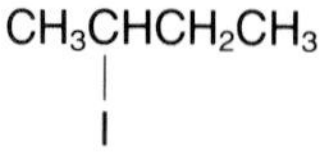

b 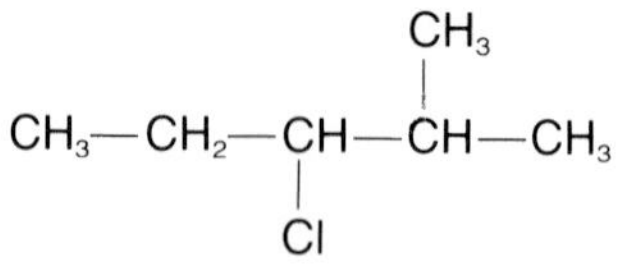

12.15 COMMON NOMENCLATURE

Although it is probably true that life would be easier if everyone, including textbook authors, held to one systematic approach to nomenclature, many compounds are referred to by their historical *common* names. Many common names were developed prior to the establishment of the IUPAC system and remain entrenched in the chemical community. For commonly used but very complicated molecules you may agree that the common name is highly preferable to the systematic name. For example, the IUPAC name for the penicillin V molecule is 3,3-dimethyl-7-oxo-6-[(phenoxyacetyl) amino]-4-thia-1-azabicyclo[3,2,0]heptane-2-carboxylic acid. Do you agree that the common name is easier to say and write?

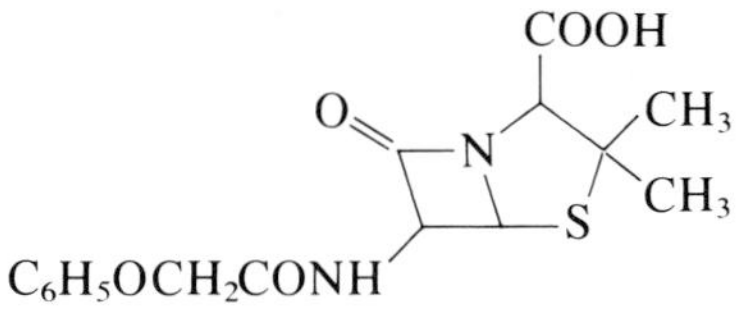

Penicillin V

Let us look at one common approach to naming that finds wide usage. This approach depends on perceiving a molecule as a combination of an alkyl group and a functional group.

R — Z

Alkyl group   Functional group

The method is to name the alkyl group and follow this with the name of the functional group. Consider the following examples:

CH_3Br = methyl bromide

CH_3CH_2OH = ethyl alcohol

$CH_3CH_2CH_2Cl$ = *n*-propyl chloride

$CH_3CH(OH)CH_3$ = isopropyl alcohol

CH_3NH_2 = methyl amine

Other widely used common names will be introduced when we require them in their appropriate functional-group setting.

12.16 WRITING STRUCTURES FROM NAMES

We have been concentrating on writing a name for a given structure. Let us now consider the reverse process: given a name, write the corresponding structural formula. This is easily accomplished by reading the name "backward," i.e., consider the name reading from right to left.

Let us apply this idea to the name

2,5-dimethyl-3-isopropylhexane

- 2,5-dimethyl: Two methyl groups at the 2 and 5 positions
- 3-: Position of isopropyl chain
- hex-: Six carbons as the longest continuous chain
- -ane: Alkane family

From this analysis of the name, write the structure by following the steps:

STEP 1 **Write down the longest continuous carbon chain and number it.** In the hexane example,

C — C — C — C — C — C

1 2 3 4 5 6

The numbers need not be written down.

STEP 2 **Attach the specified substituents at the correct positions.** In this case, methyls should be attached at carbons 2 and 5, and isopropyl at position 3.

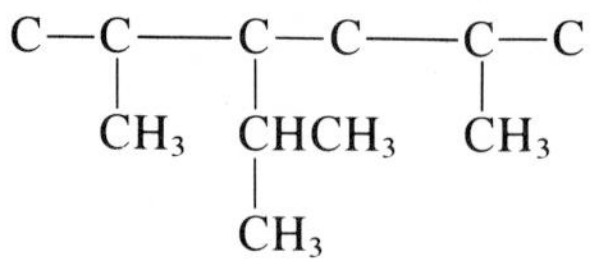

STEP 3 **Complete the structure by supplying each carbon with enough hydrogens so that each has four bonds.**

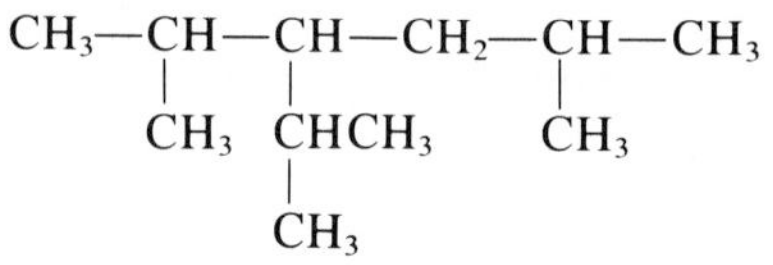

Problem 12.7 Write a structure for 2,2,3-trimethylbutane.

12.17 NOMENCLATURE AND ISOMERISM

The logic involved in assigning systematic names to organic molecules can be used to distinguish between sets of isomers and identical compounds. For example, suppose you were trying to decide whether the following two structures were identical, isomeric, or unrelated.

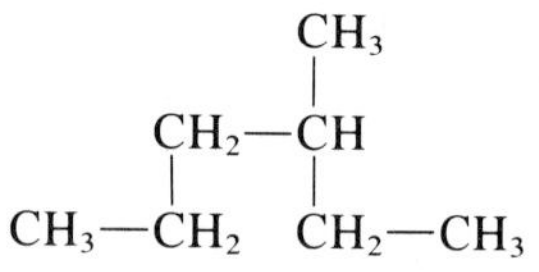

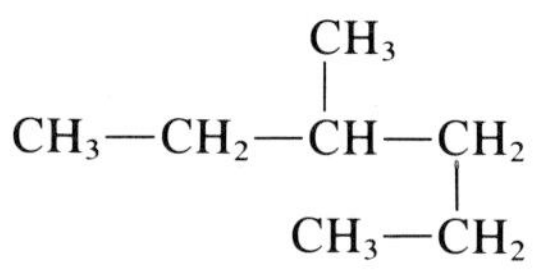

Counting up carbons and hydrogens tells you that they must either be identical or isomers because both structures have the molecular formula C_7H_{16}. If you locate the longest chain in each structure, you'll find that these are both hexanes, six-carbon chains;

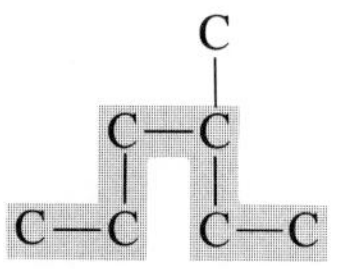

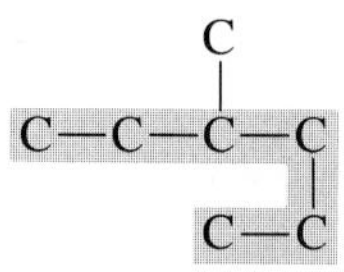

These structures are identical because not only do they have the same parent chain, but also they have an identical branch (a methyl group) at the same chain position (carbon 3). These are two representations of the same compound, 3-methylhexane.

If you name the structure

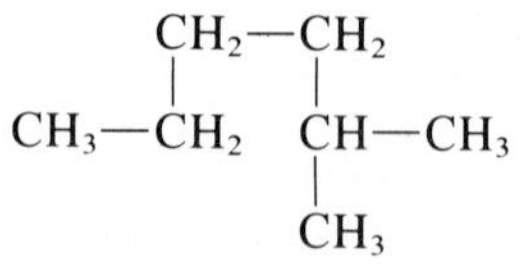

you'll find that it is an isomer of the previous two; i.e., it is 2-methylhexane. Application of the systematic nomenclature rules offers another method of coping with problems of isomer identification.

CHAPTER ACCOMPLISHMENTS

After completing this chapter you should be able to define **key words** and do the following:

12.1 INTRODUCTION

12.2 GENERAL PROPERTIES OF ORGANIC COMPOUNDS

1. State the general differences in properties between organic and inorganic compounds.

12.3 BONDING IN CARBON COMPOUNDS

2. State the predominant type of bonding in organic compounds.
3. Give at least three reasons why there is a very large number of organic compounds.
4. State the elements most commonly found in organic compounds.

12.4 STRUCTURAL FORMULAS FOR ORGANIC MOLECULES

5. Given a molecular formula, write a structural formula using the covalent bonding patterns of Table 6.1.

12.5 CONDENSED STRUCTURAL FORMULAS

6. Draw full structural formulas from a given condensed structural formula and vice versa.

12.6 FUNCTIONAL-GROUP CONCEPT

7. Explain why the functional-group concept is useful.
8. Identify the functional groups present in an organic compound given the structural formula.

12.7 HOMOLOGOUS SERIES

9. State whether a given set of compounds constitutes a homologous series.
10. Describe the general trend in boiling points and melting points in a homologous series.

12.8 THE R-GROUP CONCEPT

11. Recognize the meaning of R in a given organic structure.

12.9 THE "ACTION" IS AT THE FUNCTIONAL GROUP

12.10 ISOMERISM

12. State the relationship between compounds with the same molecular formula but different structural formulas.
13. Given a set of structural formulas, distinguish among isomers, identical compounds, and unrelated compounds.

12.11 GEOMETRY AROUND CARBON ATOMS

14. Given the bonding pattern at a particular carbon, state the geometry at that carbon.

12.12 ROTATION ABOUT SINGLE BONDS

15. Recognize conformers as different representations of the same compound.

12.13 ISOMERISM REVISITED

12.14 NOMENCLATURE

16. State the names of the first 10 alkanes.
17. Give the names and structures of the common alkyl groups.
18. Write an IUPAC name for a given alkane structural formula.
19. State an IUPAC name for a given structural formula of an alkyl halide.

12.15 COMMON NOMENCLATURE

20. Write a structure for an alkyl halide, given a common name.

12.16 WRITING STRUCTURES FROM NAMES

21. Given an IUPAC name, draw the structure of an alkane or alkyl halide.

12.17 NOMENCLATURE AND ISOMERISM

22. Use the nomenclature rules to help you identify isomers.

PROBLEMS

12.8 Explain why the melting points of organic compounds are generally lower than those of inorganic compounds.

12.9 Methyl alcohol dissolves in water; however, the solution does not conduct electricity. Explain.

12.10 Predict which of the following compounds will be the most water-soluble:

a CH_3CH_2OH *c* CH_3CH_3
b $CH_3(CH_2)_5CH_2OH$

12.11 When methane burns, CO_2 and H_2O form. Which has a lower free energy, methane and O_2 or the combination of CO_2 + H_2O? What evidence do you have for your answer?

12.12 State whether each of the following descriptions best applies to an organic or inorganic compound:

a The compound melts at 945°C and dissolves in water. It conducts electricity.
b The compound is a liquid which does not dissolve in water and does not conduct electricity.
c The compound melts at 115°C and burns in air.
d The compound is a colorless gas which dissolves in water. Its solution conducts electricity.
e The compound is a colorless gas which burns in air and does not dissolve in water.

12.13 *a* Describe the major compositional difference between organic and inorganic compounds.
b What element besides carbon is common to almost all organic compounds?

12.14 Give reasons for the large number of organic compounds in existence.

12.15 What other element in the periodic table would you expect to show bonding properties similar to those of carbon and hence possibly form a large number of compounds?

12.16 Write structural formulas for the following molecular formulas:

a C_2H_6
b $CHBr_3$
c C_2H_3Cl
d CH_4S
e C_2HCl

12.17 Write full structural formulas corresponding to the following condensed structures:

a $CH_3CH(OH)CH_3$
b $CH_3CH_2CH(CH_2CH_3)CH_3$
c $CH(CH_3)_3$
d $CH(CH_2CH_3)_3$

12.18 Write a condensed structural formula for each of the following full structural formulas:

a
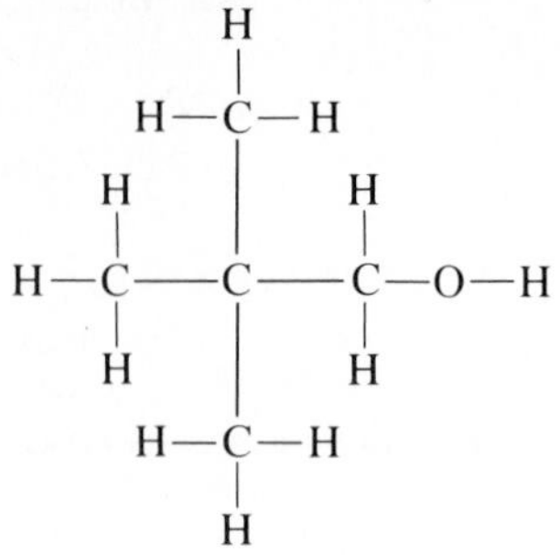

b
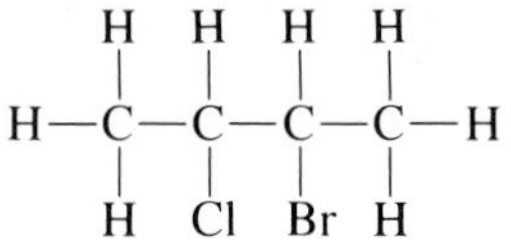

12.19 Classify the following compounds by functional group:

a
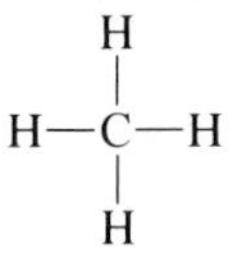

b
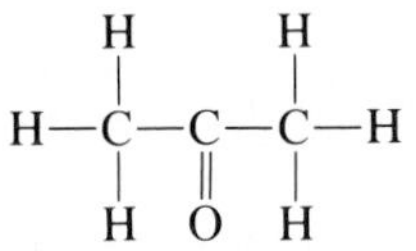

c
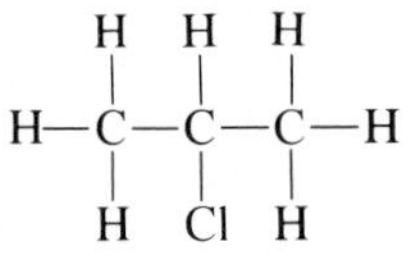

d
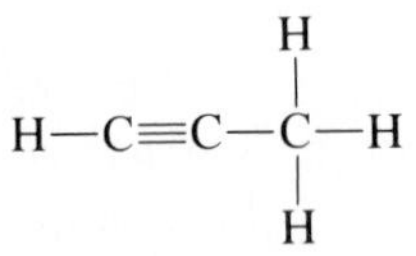

e
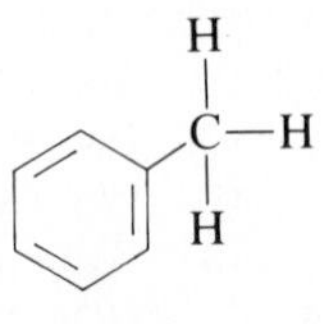

f
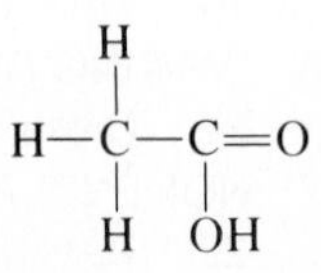

g
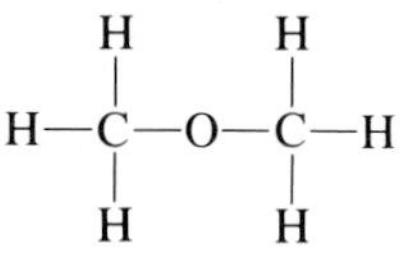

h
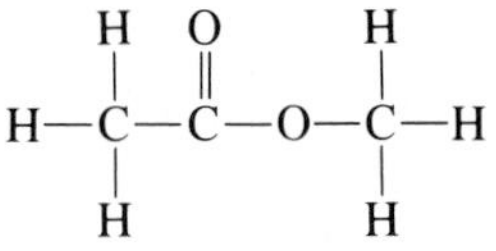

i
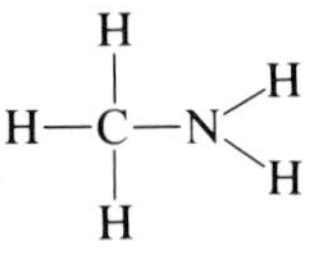

j
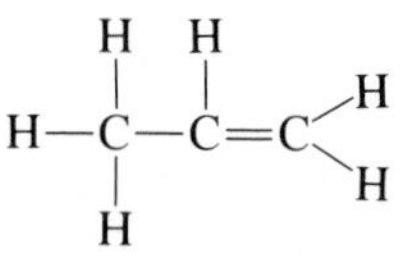

k
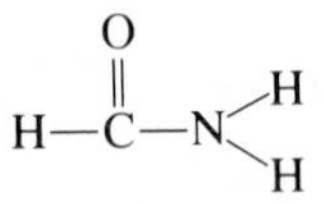

l
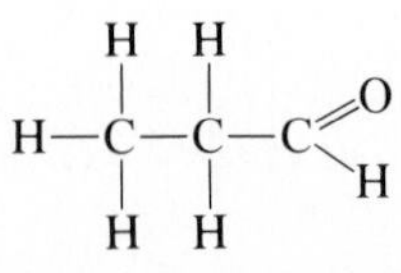

12.20 Circle and label the functional groups present in each compound:

a

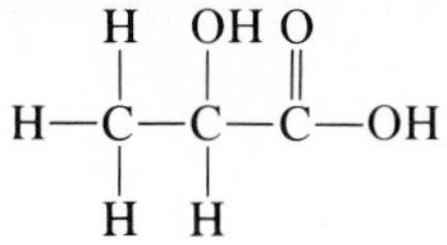

b

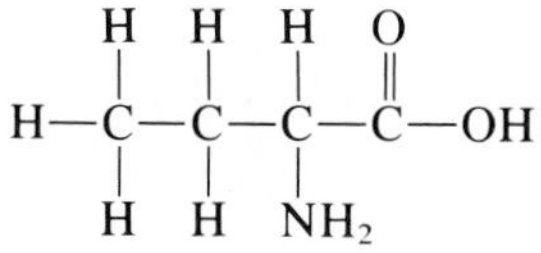

c

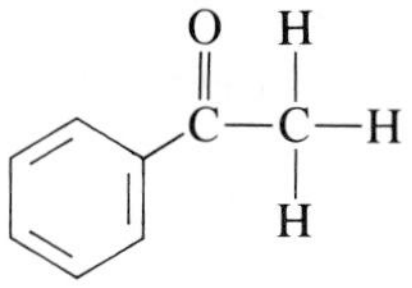

d

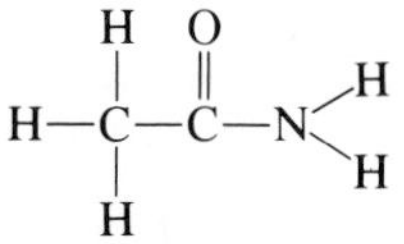

e

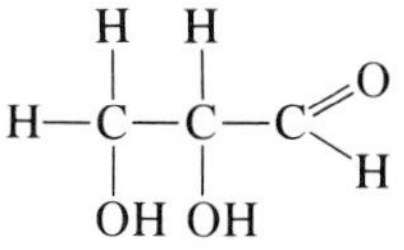

f

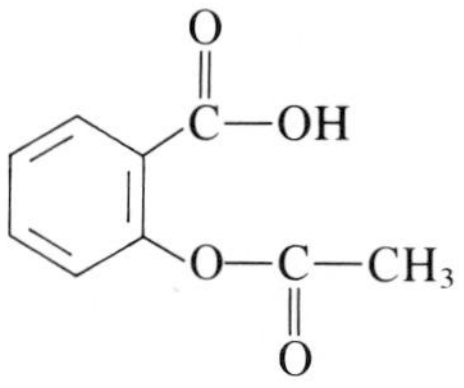

12.21 The two compounds below belong to a homologous series. Fill in the structures of the missing members between these two compounds:

CH_3CH_2Cl __________ $CH_3(CH_2)_7CH_2Cl$

12.22 Give two reasons why the boiling point of $CH_3(CH_2)_7CH_2Cl$ will be higher than that of CH_3CH_2Cl.

12.23 Give a meaning to R as used in the following structure:

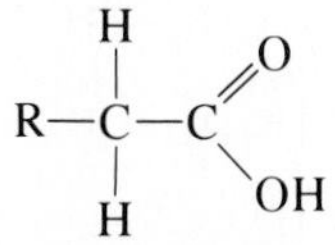

12.24 In your own words, carefully distinguish among isomers, identical compounds, and unrelated compounds.

12.25 For each of the following pairs of structures, indicate whether the two are identical, isomeric, or unrelated:

a

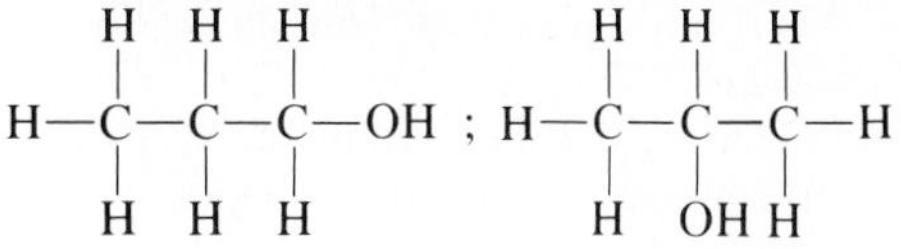

b

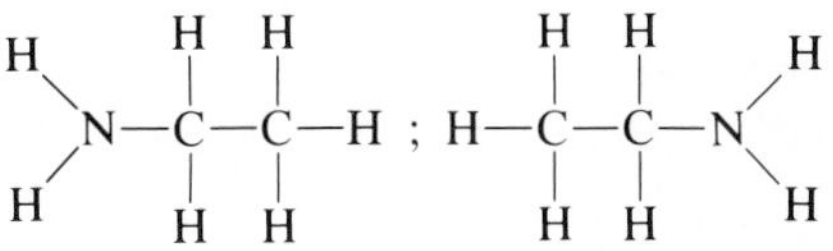

c

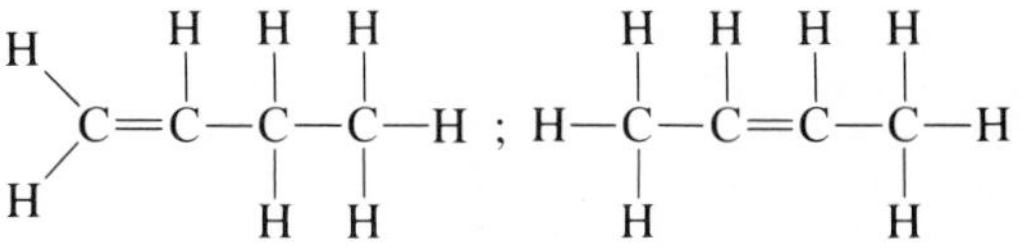

d

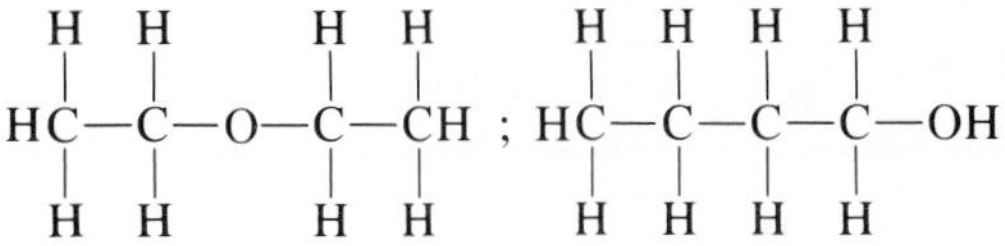

e

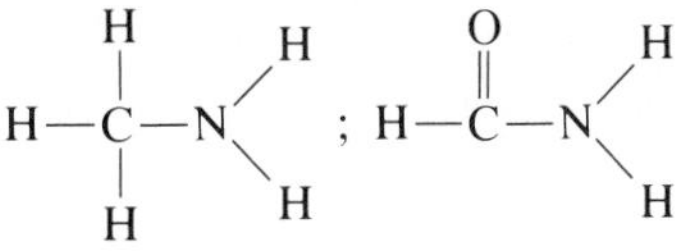

f

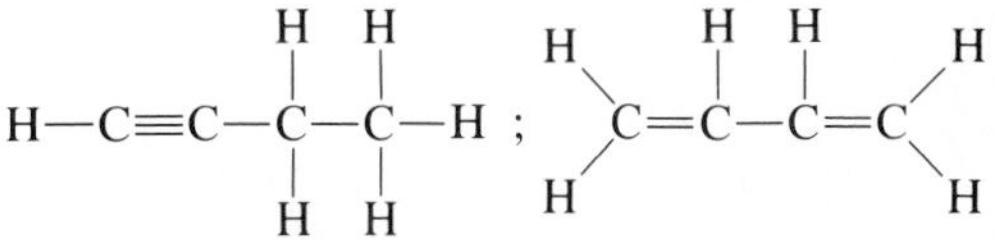

g

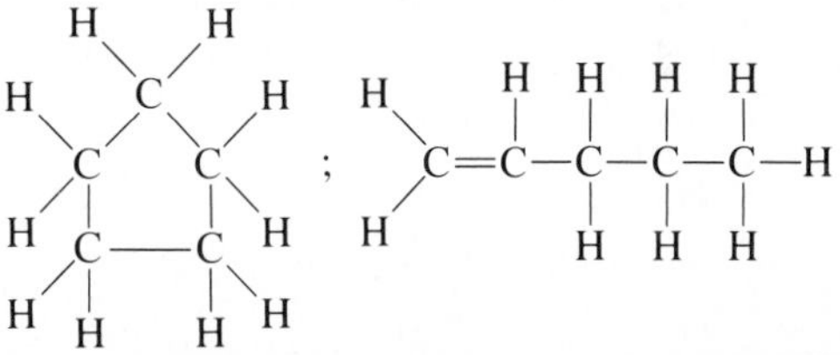

12.26 Write structural formulas for isomers of molecular formula C_5H_{12}. (You should find three.)

12.27 Write structural formulas for isomers with molecular formula C_3H_7Cl. (There are two.)

12.28 Write structural formulas for isomers with molecular formula C_4H_9Br. (You should find four.)

12.29 Give the geometry around each of the circled carbons:

a

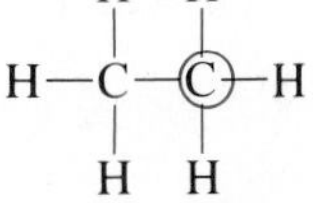

c

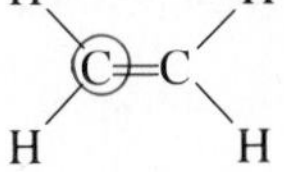

b

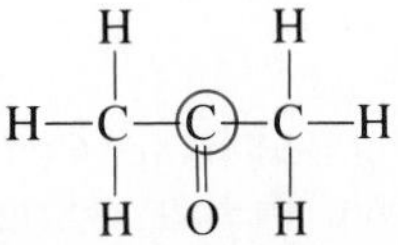

d

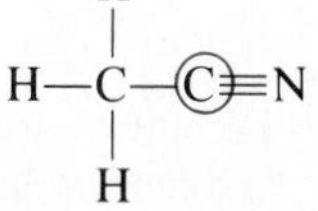

12.30 *a* List the relationship between the number of atoms around a given carbon and the geometric arrangement around the carbon atom.

b What is the most common geometric arrangement of covalent bonds around a carbon?

12.31 Give the shape about each carbon atom in each of the following structures:

a CH_3-CH_3

b

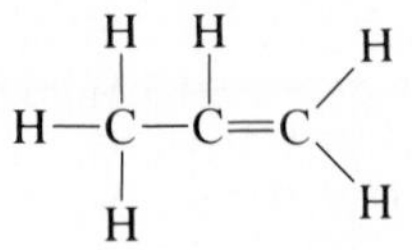

c

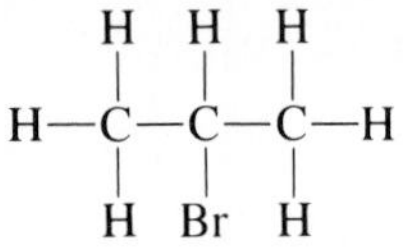

d

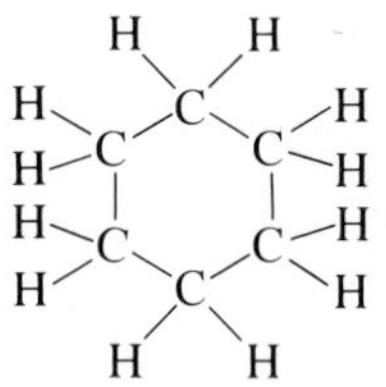

e

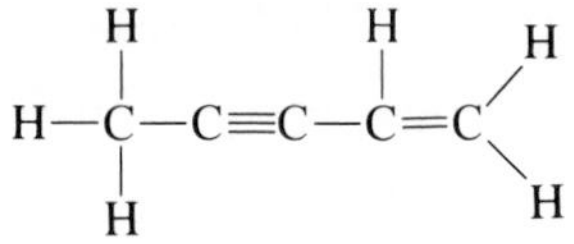

f

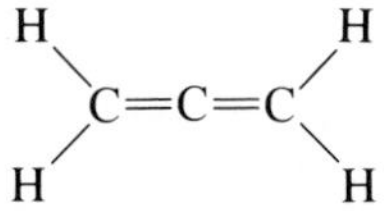

12.32 Write the structures of two conformers of the following structure:

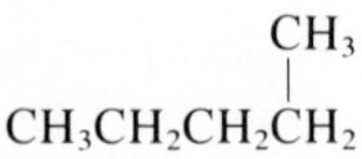

12.33 Do the following two structures represent different compounds? Explain your answer.

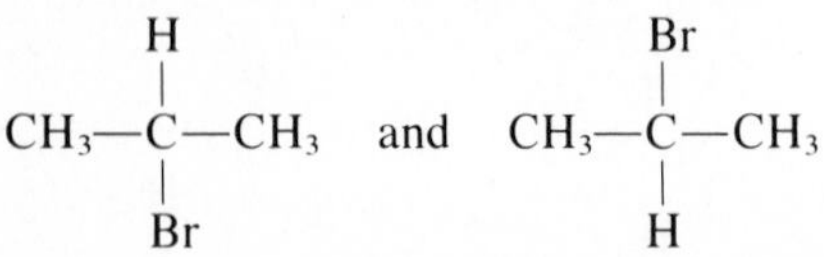

12.34 Draw two structural isomers for C_3H_7Br.

12.35 Indicate in each pair whether the representations given are structural isomers, the same molecule, or unrelated:

a

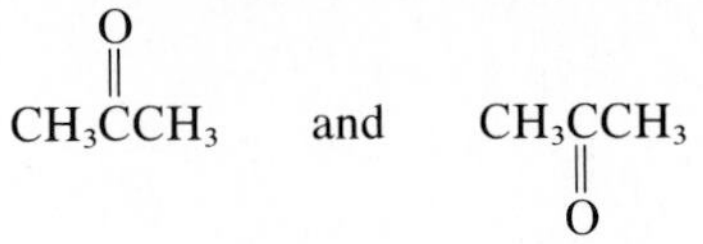

b

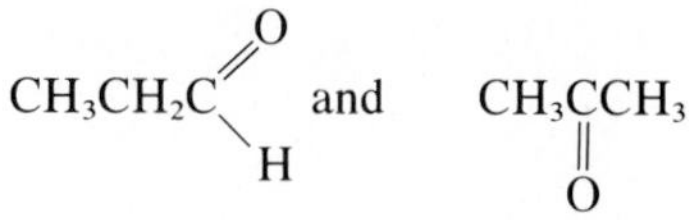

c

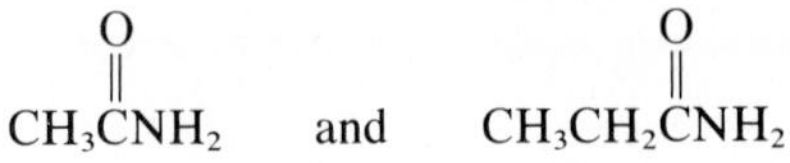

d

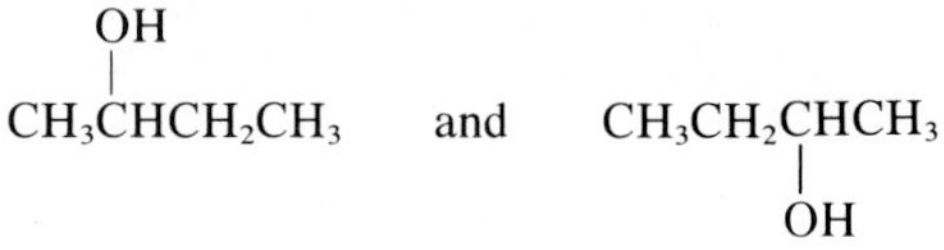

e 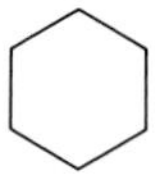 and $CH_3CH_2CH_2CH_2CH_2CH_3$

f 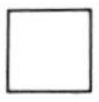 and $CH_3CH{=}CHCH_3$

g

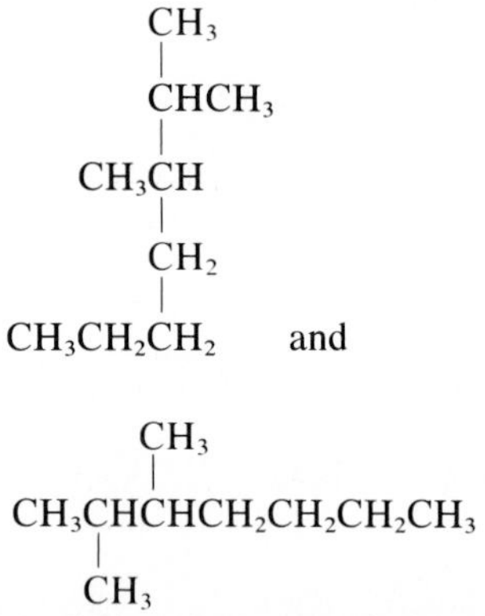

12.36 Write the roots of the names of the first 10 continuous carbon chains.

12.37 Write the structures and names of the two alkyl groups that can be derived from propane, $CH_3CH_2CH_3$.

12.38 Give an unambiguous name for each of the following:

a

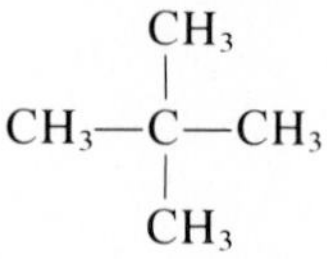

b

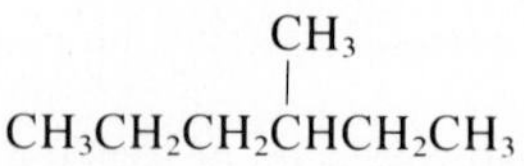

c

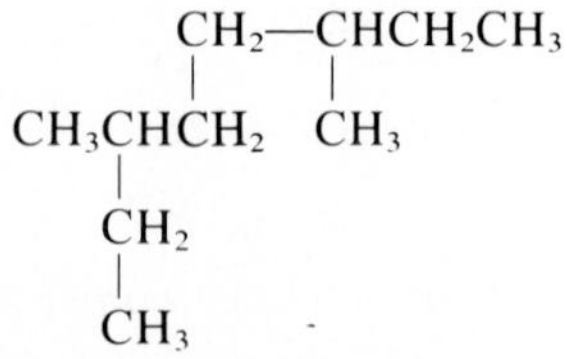

d 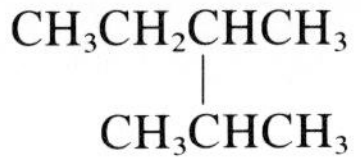

e 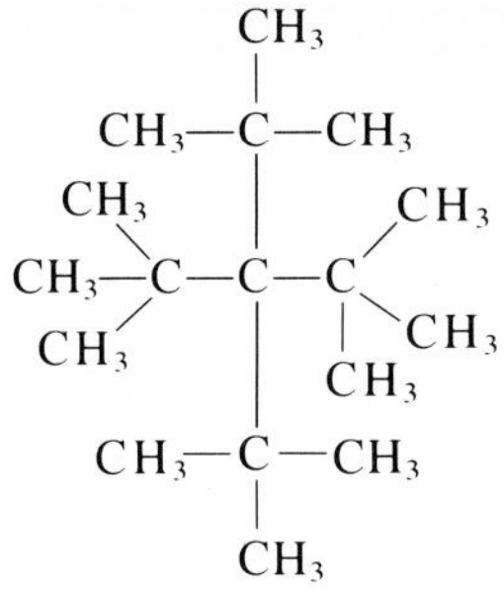

12.39 Give an unambiguous name for each of the following alkyl halides:

a 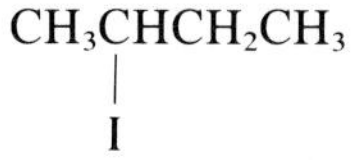

b 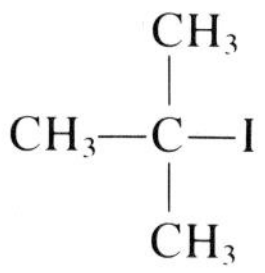

c 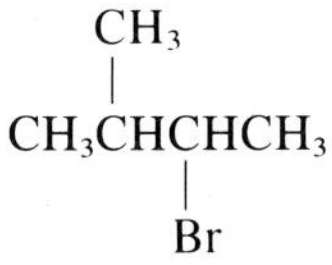

d 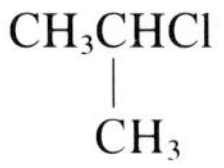

e 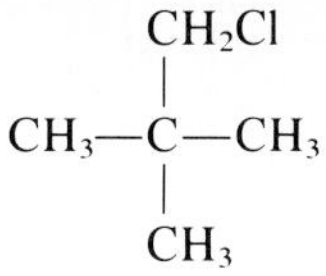

f 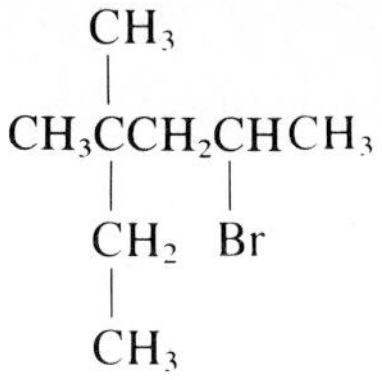

12.40 Write a structure corresponding to each of the following names:

a 2-Methylpentane

b 2,2-Dimethyl-5-ethyloctane

c 4-Isopropylheptane

d 1-Bromobutane

e Ethyl chloride

f 2,4-Dimethyl-3-chloropentane

CHAPTER THIRTEEN

HYDROCARBONS

13.1 INTRODUCTION

The simplest organic compounds in terms of elemental composition are the hydrocarbons. **Hydrocarbons** are the families of organic compounds that contain only the elements hydrogen and carbon. Industrially they are extremely significant. They are fuels obtained directly from natural gas, petroleum, and coal sources, and they serve in the first line of synthesis of a variety of organic materials and commercially important products such as plastics, fibers, etc.

The hydrocarbons are subdivided into classes depending on the type of carbon-carbon bonding and whether the particular compound is an open chain or contains a ring of carbons. Figure 13.1 summarizes this classification scheme and offers examples of all the hydrocarbons we will examine in this chapter. Another classification scheme divides hydrocarbons into two sets: saturated and unsaturated hydrocarbons. The **alkanes**, which you have already encountered in Chapter 12, are also known as the *saturated* hydrocarbons, because each carbon has completely filled (or saturated) its bonding capacity with bonds to four other atoms. All the other hydrocarbons are unsaturated.

Let us begin our discussion of hydrocarbons by seeing how these basic raw materials, or reactants in organic synthesis, are claimed from natural sources.

13.2 SOURCE AND USE OF HYDROCARBONS

The basic sources of all the hydrocarbons and, indirectly, most organic chemicals, are petroleum, natural gas, and coal. Gas, oil, and coal are the decayed remains of animal and vegetable matter that have been subjected to high pressures and temperatures in underground cavities under anaerobic conditions, i.e., without oxygen. Petroleum, or crude oil, is a mixture of hydrocarbons which is refined for use by a process called fractional distillation. The crude oil is heated and vaporized, the vapor moves up a large refinery column, and different components of the vapor condense at different points to give various separated liquid fractions or "cuts" of the original mixture. The higher-boiling-point, least volatile components condense at the bottom of the column, whereas the low-boiling-point fractions appear at the top of the column. Some of the lowest-boiling-point components are expensive to recover and so are burned at the top of the column.

Table 13.1 shows the results of a typical crude oil distillation. The various fractions or cuts are designated by the boiling-point range and the number of carbons within molecules of a given fraction. The boiling-point range of

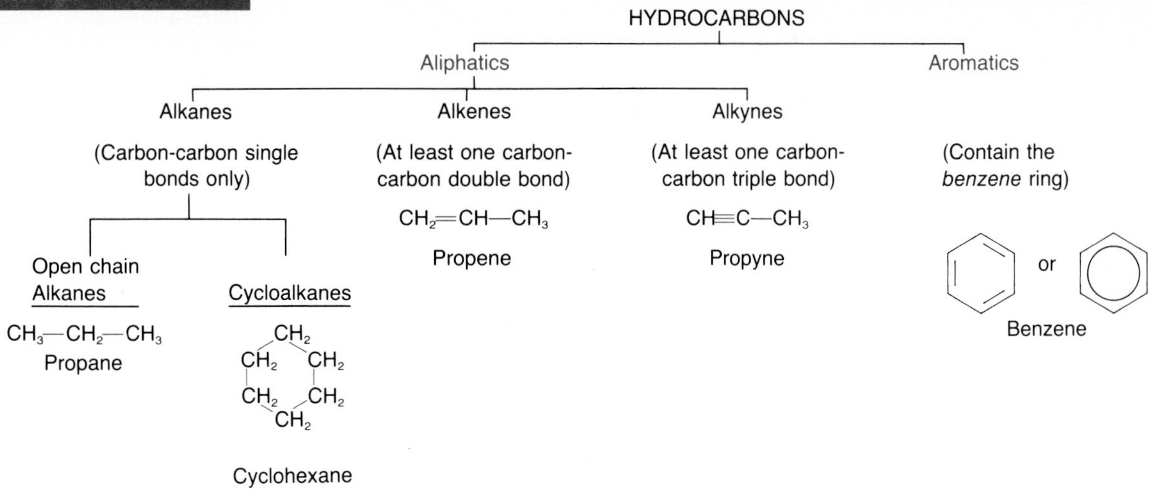
HYDROCARBONS
Aliphatics
Aromatics
Alkanes
Alkenes
Alkynes
(Carbon-carbon single bonds only)
(At least one carbon-carbon double bond)
(At least one carbon-carbon triple bond)
(Contain the *benzene* ring)
CH2=CH—CH3
Propene
CH≡C—CH3
Propyne
or
Benzene
Open chain Alkanes
Cycloalkanes
CH3—CH2—CH3
Propane
CH2
CH2 CH2
CH2 CH2
CH2
Cyclohexane

FIG. 13.1
Hydrocarbons are compounds containing only the elements carbon and hydrogen. They are classified according to bond type and the presence or absence of rings. A representative example is shown for each class.

a fraction increases with the length of the carbon chains. You saw this behavior previously in Section 12.7.

The major use of alkanes is as an energy source. Methane, CH_4, is the major component of natural gas which is piped into homes and businesses for use as a space heating and cooking fuel. The lower-weight liquid fractions of petroleum are used in gasoline to keep our automobiles running. The higher-weight fractions are used as fuel to heat homes; run electrical generating plants; and power ships, planes, and diesel engines.

Alkanes can also serve as a source of other classes of hydrocarbons, most notably alkenes. Alkanes are converted to alkenes by a process known as

TABLE 13.1 Crude Oil Distillation

Fractions Obtained	Boiling-Point Range, °C	Hydrocarbon Size	Comments
Gases: CH_4, C_2H_6, C_3H_8, C_4H_{10}	<20	C_1–C_4	Similar to natural gas, useful for fuel and chemicals. Much of it, however, is flamed because of expense of recovery.
Light naptha	20–150	C_4–C_9	Both light and heavy naptha are the base for gasoline and useful for fuel and chemicals.
Heavy naptha	150–200	C_9–C_{11}	
Kerosene	175–275	C_9–C_{16}	Compounds which find use as jet, tractor, and heating fuel.
Gas oil	200–400	C_{15}–C_{25}	Useful for diesel and heating fuel. Becoming a major source of lower-weight alkanes through thermal conversion.
Lubricating oil	>350	C_{20}–C_{30}	Used for lubrication. May be chemically converted to lower-molecular-weight fractions.
Heavy fuel oil	>350	C_{22}–C_{40}	Boiler fuel. Petroleum jelly may be chemically converted to lower-molecular-weight fractions.
Asphalt	Solid residue	C_{25}–C_{50}	Paving, coating, and structural uses.

cracking, which will be discussed in Section 13.5. The alkenes are more chemically reactive than alkanes and can thereby serve as raw materials for the synthesis of a wide variety of other organic chemicals and larger molecules which eventually become the synthetic fabrics and plastics that we are so familiar with in our everyday life. About 75 percent of all organic chemicals are prepared from three hydrocarbons: the alkenes ethene and propene and the aromatic benzene. The alkenes with four carbon atoms, methyl-substituted benzenes, and methyl alcohol are also important raw materials.

13.3 PHYSICAL PROPERTIES OF HYDROCARBONS

Because their elemental composition is the same (C and H), the physical properties of all classes of hydrocarbons are very similar. Consequently, we can discuss trends in physical properties of hydrocarbons in general. The trends apply equally well to alkanes, alkenes, alkynes, and aromatics.

As you saw in Figure 12.2, the boiling points in any homologous series of hydrocarbons increase with increasing molecular weight. The only intermolecular force present in hydrocarbons is the dispersion force, and this attractive force increases with increasing molecular weight (Section 8.10).

If we examine a group of hydrocarbons, all of which have the *same* molecular weight, the boiling point decreases with increasing branching on the main carbon chain. Table 13.2 offers an example of this phenomenon; all the examples are isomers of C_6H_{14} (molecular weight = 86). This decrease in boiling point with branching arises as a consequence of the nature of dispersion forces which were first discussed in Section 8.10. Electron distributions in molecules are more easily disturbed by neighboring molecules, and dispersion forces are stronger, if the "surfaces" of the electron clouds of the molecules can more nearly touch. For any given molecular weight, a molecule that has the shape of a spaghetti strand will have greater dispersion forces than a molecule that is ball-shaped, as shown in Figure 13.2.

Continuous- or straight-chain molecules can be viewed as spaghetti strands which can line up and touch along their entire length. Branched structures are more ball-like in shape, and their surface interactions are thus less effective.

All the hydrocarbons have densities less than that of water (1.0 g/mL). As a result, they float on water. Unfortunately, this fact has been verified too often by oil spills from tankers at sea; the floating oil despoils our bays and beaches.

Hydrocarbons are not soluble in water. This can be deduced from the observation that they float on water and do not diffuse through water. As we saw in Section 6.8, there is basically no electronegativity difference between carbon and hydrogen; thus the hydrocarbons are nonpolar and tend to dissolve only in nonpolar solvents. This is another example of the general principle that "like dissolves like."

TABLE 13.2 Comparison of Boiling Point and Branching in Six-Carbon Alkanes

Name	Structure	Boiling Point, °C
Hexane	$CH_3CH_2CH_2CH_2CH_2CH_3$	68.7
2-Methylpentane	CH_3 | $CH_3CHCH_2CH_2CH_3$	60.3
2,3-Dimethylbutane	CH_3 CH_3 | | $CH_3CH—CHCH_3$	49.7

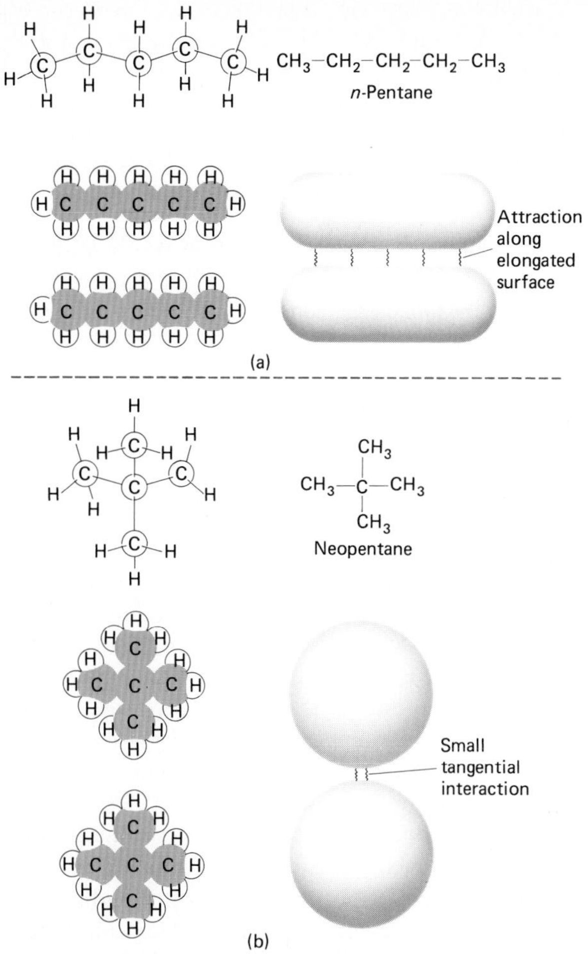
$CH_3-CH_2-CH_2-CH_2-CH_3$
n-Pentane
Attraction along elongated surface
(a)
Neopentane
Small tangential interaction
(b)

FIG. 13.2 The effect of molecular shape on the extent of dispersion forces. (*a*) We can envision molecules of *n*-pentane "touching" and interacting along their entire elongated surface. (*b*) Neopentane is more like a sphere and there is less "contact." The difference in intermolecular attraction is reflected in the boiling points of these two compounds. The boiling point of *n*-pentane is 36.2°C, whereas that of neopentane is lower, 9.5°C, although both have the same molecular weight.

Organic and biochemical molecules that have a large alkyl-group portion relative to a small polar functional group, as in a 10-carbon alcohol or aldehyde, have many of the same physical properties (density < 1, insolubility in water) as the hydrocarbons. The relationship between the relative sizes of the hydrocarbon fragment and the polar (hydrogen-bonding) fragment and water solubility was discussed in Section 12.2.

13.4 CHEMICAL PROPERTIES OF ALKANES

Now let us discuss the chemistry of the classes of hydrocarbons one at a time, beginning with the unique saturated class, the alkanes. Alkanes are inert toward most chemical reactants and conditions. However, they can react with oxygen and with the halogens fluorine, chlorine, and bromine.

The reaction of any alkane with excess oxygen yields carbon dioxide, water, and energy given off as heat and light. This chemical reaction is termed *combustion,* but is commonly called "burning." Combustion is a highly exothermic reaction, but it does not occur until the required activation energy is supplied by a match or spark. That is, combustion is thermodynamically favorable, but kinetically unfavorable in the sense that there is a high activation energy. (Review Section 10.2.) Burning is, of course, of enormous commercial importance because this is the exothermic reaction which makes alkanes excellent fuels. Most of the energy we use for heating, running our automobiles, cooking food, and generating electricity comes from combustion of alkanes.

Sufficient oxygen to produce CO_2 from the alkane is necessary for complete combustion. The balanced equation for the complete combustion of methane is

$$CH_4(g) + 2O_2(g) \longrightarrow CO_2(g) + 2H_2O(g) \qquad (13.1)$$

Methane — Carbon dioxide

The heat evolved in this reaction is 192 kcal/mol of CH_4.

Problem 13.1 Write a balanced equation for the complete combustion of butane (C_4H_{10}).

If insufficient oxygen is present, a competing reaction that forms carbon monoxide occurs:

$$CH_4(g) + \tfrac{3}{2}\,O_2(g) \longrightarrow CO(g) + 2H_2O(g) \qquad (13.2)$$

Methane — Carbon monoxide

This reaction produces only 127 kcal/mol of CH_4. The conversion to carbon monoxide produces less energy per mole of CH_4 than the complete combustion to carbon dioxide. A similar result is found for any other alkane, indeed any other organic substance undergoing combustion. On an energy savings basis alone you should keep your oil or gas furnace and automobile engine in good "tune" with a proper proportion of air to fuel to assure complete combustion to CO_2.

Furthermore, carbon monoxide production is undesirable because it is a highly toxic gas. CO can bind to hemoglobin, the oxygen-carrying molecules in blood, thereby preventing it from bonding oxygen and carrying it through the circulatory system. Carbon monoxide is colorless, odorless, and tasteless, so that an individual does not have any warning of its presence until he or she begins to experience the symptoms of headache, mental dullness, nausea, and decreased pulse and respiratory rates. If exposure continues, collapse and death can be the result. Many automobile accidents may result from the impaired reflexes of a driver who has experienced some CO poisoning from sitting in an urban traffic jam. Carbon monoxide is one of the toxic substances found in cigarette smoke, making cigarette smoking dangerous to your health.

Alkanes can react with the halogens (except iodine) to form alkyl halides or halogenated alkanes. The reaction is an example of the basic reaction type in organic chemistry known as **substitution.** A halogen atom substitutes for a hydrogen atom. For example, in the reaction of ethane with chlorine gas

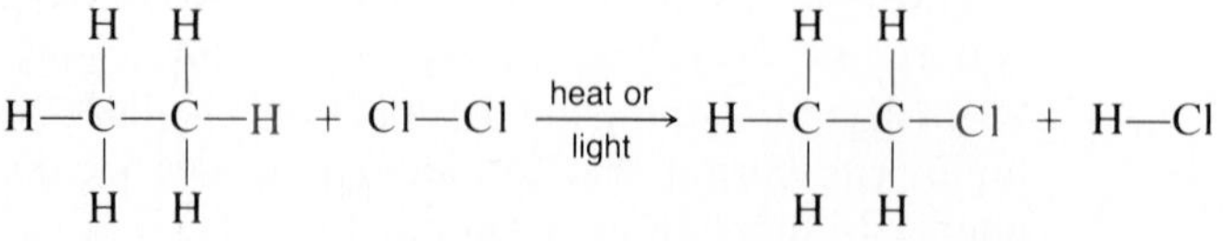
heat or
light

(13.3)

Ethane — Chlorine — Chloroethane or ethyl chloride — Hydrogen chloride

Or in condensed form,

$$CH_3CH_3 + Cl_2 \longrightarrow CH_3CH_2Cl + HCl$$

A chlorine atom has substituted for a hydrogen in the reacting alkane. More than one hydrogen can be removed from an alkane, yielding a *poly*haloalkane, an alkane with *many* halogen substituents. For example, methane reacts with chlorine to form chloroform.

$$\underset{\text{Methane}}{CH_4} + 3Cl_2 \longrightarrow \underset{\text{Chloroform}}{CHCl_3} + 3HCl \qquad (13.4)$$

Most alkanes can yield more than one monosubstitution product with halogen. Propane, for example, reacts with chlorine to form both 1-chloropropane and 2-chloropropane.

$$\underset{\text{Propane}}{CH_3CH_2CH_3} + \underset{\text{Chlorine}}{Cl_2} \longrightarrow \underset{\text{1-Chloropropane}}{CH_3CH_2CH_2Cl} + HCl$$

$$\underset{\text{Propane}}{CH_3CH_2CH_3} + \underset{\text{Chlorine}}{Cl_2} \longrightarrow \underset{\text{2-Chloropropane}}{CH_3\underset{\displaystyle Cl}{\underset{|}{C}}HCH_3} + HCl$$

Problem 13.2 How many isomeric products are obtained by monochlorination of butane?

13.5 ALKENES

Hydrocarbons with at least one double bond are called **alkenes.** There are both open-chain and cyclic alkenes, but we will discuss only open-chain alkenes in this section. Alkenes are termed *unsaturated* because each doubly bonded carbon is connected to only three other atoms and could be converted by a chemical change to a saturated carbon, bonded to four atoms. We will see such reactions in Section 13.8.

The need for alkenes in the chemical industry far exceeds their natural occurrence in petroleum and natural gas. Thus, most commercial alkenes are formed from the saturated alkanes found in crude oil and natural gas by a process called "cracking." Alkanes, when heated strongly in the absence of air, "crack" by splitting off hydrogen molecules and by fragmentation into smaller carbon chains. In the chemical laboratory, and in many physiological processes, double bonds are built into molecules through elimination reactions (Section 7.11) involving either alcohol or alkyl halide reactants. When heated with sulfuric acid as a catalyst, alcohols eliminate a molecule of water, as you will see in Section 14.6. Alkyl halides undergo elimination of HX when the alkyl halide is heated with a strong base solution. The H and X are eliminated from adjacent carbons and a double bond forms between these carbons.

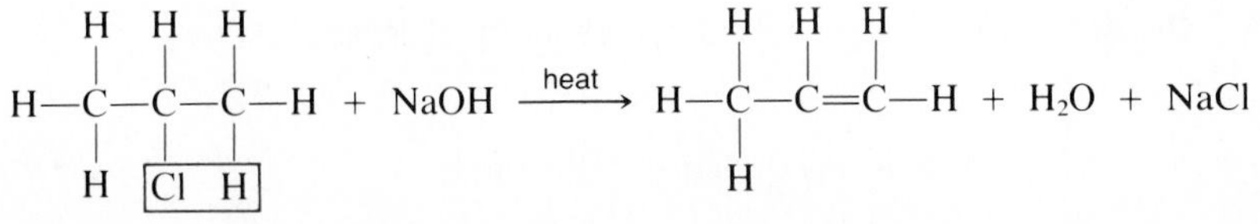

H H H
H—C—C—C—H + NaOH
H Cl H
heat
H H H
H—C—C=C—H + H2O + NaCl
H

(13.5)

2-Chloropropane Propene

Problem 13.3 What alkene is formed when 1-chlorobutane is reacted with strong base?

$$\text{Natural gas and crude oil} \xrightarrow[\text{thermal cracking}]{500^\circ\text{C}} \underset{\text{Ethene}}{CH_2{=}CH_2} + \underset{\text{Propene}}{CH_3{-}CH{=}CH_2} + H_2 \quad (13.6)$$

Of the 300 billion lb of organic chemicals that are produced yearly, about 50 percent require ethene as a raw material in their production.

13.6 NOMENCLATURE OF ALKENES

The IUPAC rules for naming the open-chain alkenes follow very closely the general rules for alkanes given in Section 12.14 except that the suffix *-ene* is used instead of *-ane*. Examine the rules below to see how we would name

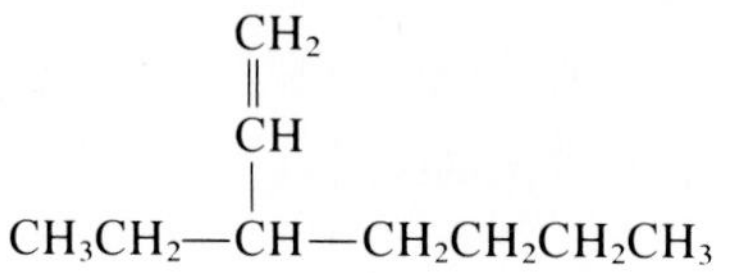

STEP 1 **Find the longest chain of carbons that contains the double bond to establish the parent name.** In the example the chain would be traced:

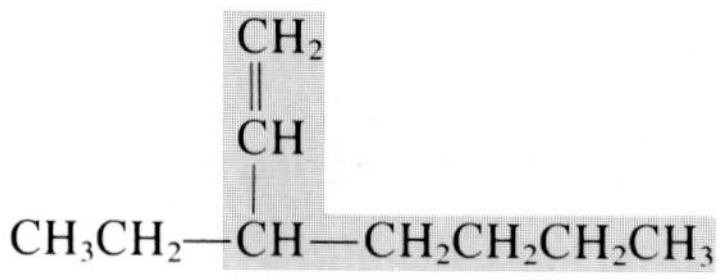

The longest chain is seven carbons; therefore this is a heptene.

STEP 2 **Number the chain so that the first carbon of the double bond is given the lowest possible number.** The position of the double bond is indicated by a number before the parent name.

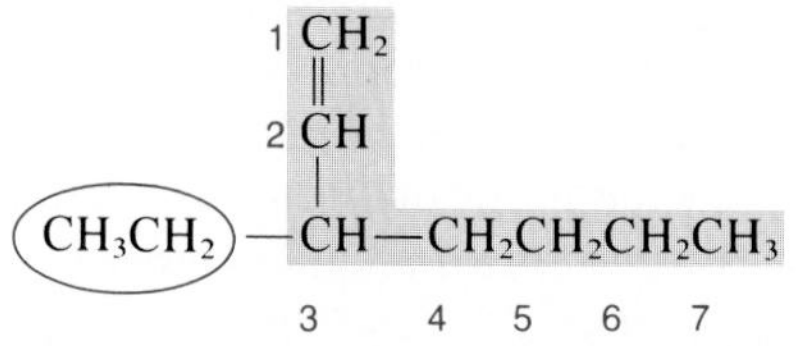

This is a 1-heptene.

STEP 3 **Locate and number the branched alkyl groups on the main chain.** Look back and see the ethyl group at position 3. The complete name of the example is 3-ethyl-1-heptene.

Some of the most common, industrially important alkenes are often named in a non-IUPAC way by adding *-ene* to the parent alkyl group name with the same number of carbons. See Table 13.3 for examples of this.

Problem 13.4 Give IUPAC names for the following alkenes:

a $CH_3{-}CH{=}CH{-}CH_3$

b 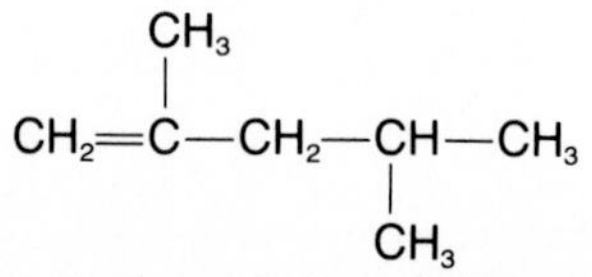

TABLE 13.3 A Comparison of IUPAC and Common Names for Alkenes

Structure	IUPAC Name*	Common Name
$CH_2=CH_2$	Ethene	Ethylene
$CH_3—CH=CH_2$	Propene	Propylene
$CH_3—C(CH_3)=CH_2$	2-Methylpropene	Isobutylene

* Notice that the double-bond position number is omitted in unambiguous cases, such as for two- or three-carbon chains in which the double bond *must* be at carbon 1.

13.7 GEOMETRIC ISOMERISM

As you first saw in Section 12.10, for most molecular formulas several isomeric structures can be written by using different carbon skeletons. An example of this type of isomerism among alkenes is offered by the following structures (straight chain vs. branched chain) corresponding to molecular formula C_4H_8:

$CH_2=CHCH_2CH_3$ (1-Butene) and $CH_2=C(CH_3)—CH_3$ (Isobutylene)

In alkenes, for a given length carbon chain there can be many structural isomers, just depending on the position of the double bond in that chain. For example, for a chain of four carbons and one double bond there are two structural isomers:

$CH_2=CHCH_2CH_3$ (1-Butene) and $CH_3CH=CHCH_3$ (2-Butene)

One isomer has the double bond in the 1,2 position and the other between carbons 2 and 3. Of course, 2-butene is also an isomer of isobutylene. These isomers are all described as *structural isomers;* the atoms are bonded together differently in the isomers shown. Other types of isomerism are possible, as you are about to see in this section and in later chapters.

The geometry around carbons in double bonds is planar triangular, as you have seen in Section 6.10 and again in Section 12.11. Consequently, the double-bonded carbons and the four atoms directly bonded to them all lie in the same plane. Furthermore, unlike rotation of a single bond, rotation about the carbon-carbon double bond is restricted. Restricted rotation introduces the possibility of a new type of isomerism called **geometric isomerism.** Full structural formulas for 2-butene reveal an example of this type of isomerism. The structures can be drawn in two ways:

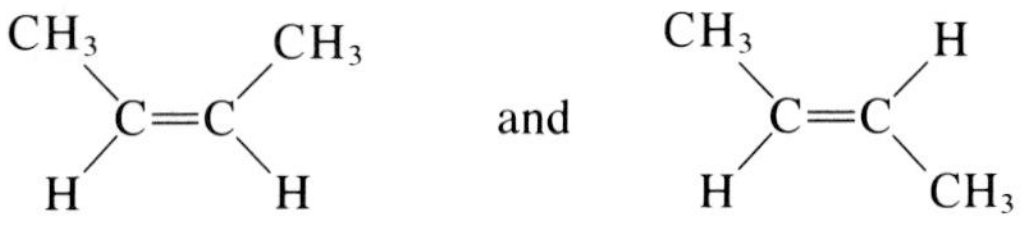

Structure 1
cis-2-Butene
(bp = 4°C;
mp = −139°C)

Structure 2
trans-2-Butene
(bp = 1°C;
mp = −106°C)

Figure 13.3*a* shows molecular models of these two structures in three dimensions.

Geometric isomers differ in the placement of substituents relative to a defined plane in space. In this case the carbon-carbon double bond defines the plane. The geometric isomer with the two methyl groups on the *same side* of the plane perpendicular to the molecular plane (Figure 13.3*b*) is known as the cis isomer (structure 1 above), whereas the one with the methyl groups on *opposite sides* of the plane is the trans isomer (structure 2 above). Because of the restricted rotation about double bonds, cis-trans isomers cannot be interconverted under normal conditions.

Each isomer of a cis-trans pair has its own unique set of physical and chemical properties. The two distinct isomers of 2-butene, *cis*-2-butene (1), and *trans*-2-butene (2), may be formed separately from one another in ordinary chemical reactions. If mixed, they can be separated by fractional distillation because of their different boiling points.

Whereas there are two geometric (cis-trans) isomers of 2-butene, 1-butene does not display geometric isomerism. Nor is there a cis-trans pair for 2-methyl-2-butene. How can one tell which alkenes will display geometric isomerism? The answer lies in an examination of the groups attached to the

FIG. 13.3
(*a*) Molecular models of *cis*- and *trans*-2-butene. The two methyl groups are on the same side of the plane defined by the double bond (the two springs connecting C-2 and C-3) in the cis isomer. The methyls are on opposite sides of the plane in the trans isomer. (*b*) This depiction emphasizes the constraint of atoms attached to doubly bonded carbons in the same plane. It also shows that the cis-trans (same-opposite) relationship is with respect to a plane perpendicular to that plane. (*Photograph by Bryan Lees.*)

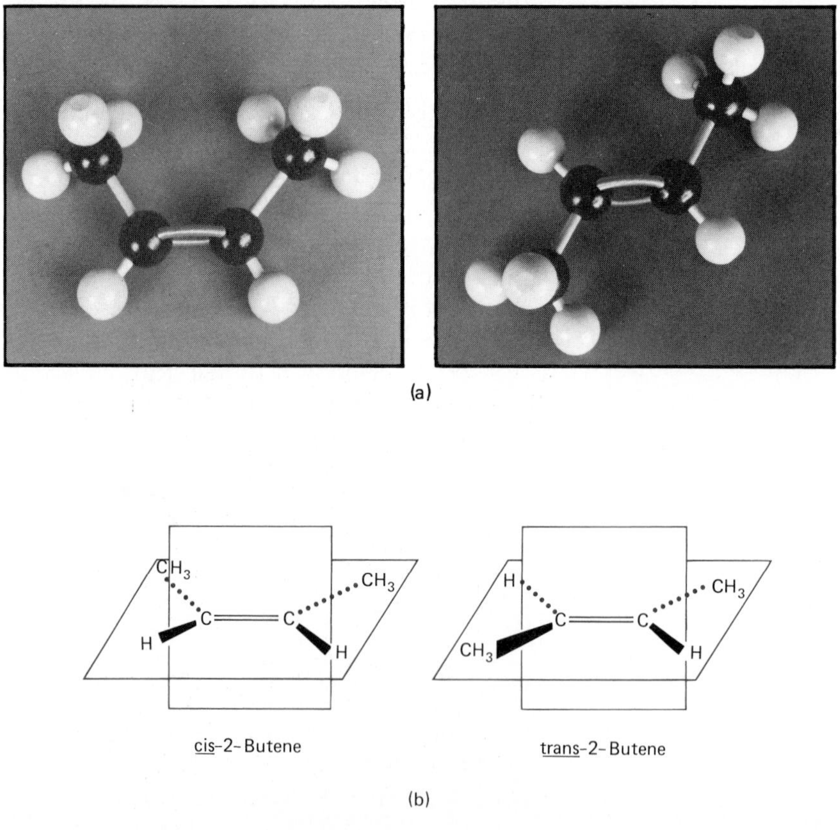
(a)
CH3
CH3
C
C
H
H
cis-2-Butene
H
CH3
C
C
CH3
H
trans-2-Butene
(b)

doubly bonded carbons:

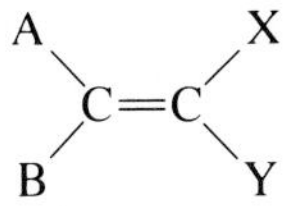

For unique cis-trans isomers to exist, A must be different from B, and X must be different from Y; the relationship between A,B and X,Y is not significant. Look back at structures 1 and 2, that is, *cis*- and *trans*-2-butene. In this case the condition is met; A and B are CH_3 and H, and X and Y are H and CH_3. When we consider 1-butene, however, we find identical groups on carbon in position 1:

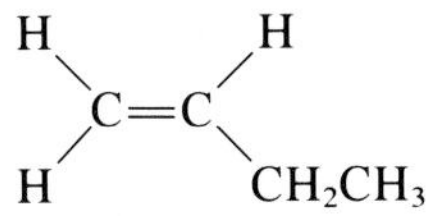

There can be no cis-trans isomerism if one of the doubly bonded carbons bears identical substituents. Thus there is only one isomer for 2-methyl-2-butene ($CH_3{-}\overset{\overset{\displaystyle CH_3}{|}}{C}{=}CH{-}CH_3$).The two different representations that could be drawn are not isomeric, but identical:

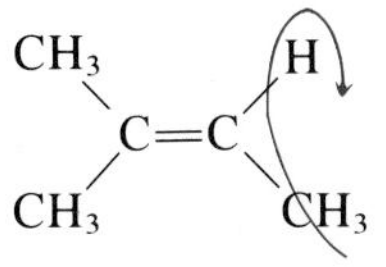

Structure 3

and

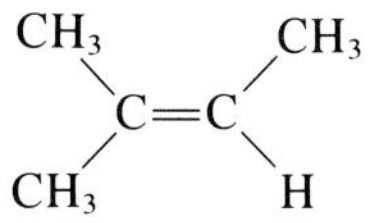

Structure 4

Rotation of structure 3 in space 180° in the manner shown by the arrow shows it to be identical to structure 4; that is, flipping structure 3 over gives

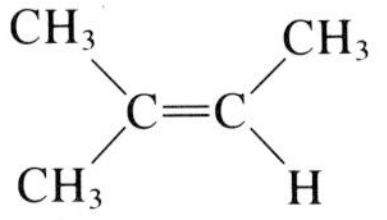

Structure 3 "flipped"

Problem 13.5 Identify the alkenes that can display cis-trans isomerism:

a $CH_3CH_2CH{=}C(CH_3)_2$

b $CH_3CH_2CH{=}CHCH_3$

c $CH_2{=}\overset{\overset{\displaystyle CH_3}{|}}{C}CH_2CH_3$

d $CH_3CH_2\underset{\underset{\displaystyle CH_3CH_2}{|}}{C}{=}CHCH_3$

Cis-trans isomerism plays an important role in the chemistry of vision. A receptor cell, known as a rod, in the retina of the human eye contains a compound rhodopsin. Rhodopsin is made up of a protein, opsin, linked to a light-absorbing compound, 11-*cis*-retinal. As you can see in Figure 13.4, the compound has several double bonds. This 11-*cis*-retinal has a shape that

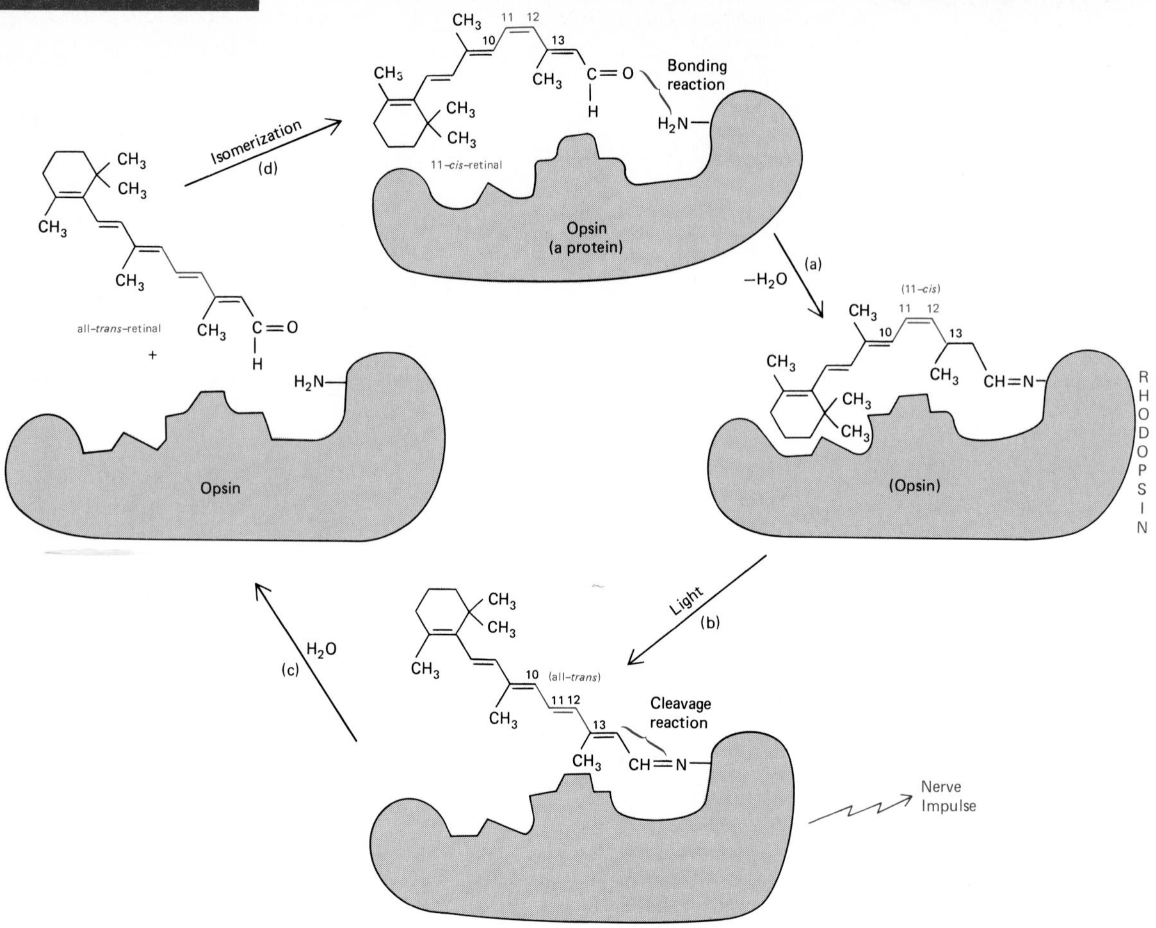

Isomerization
(d)
11-cis-retinal
Bonding reaction
Opsin (a protein)
$-H_2O$
(a)
all-trans-retinal
+
Opsin
(11-cis)
(Opsin)
RHODOPSIN
Light
(b)
H_2O
(c)
(all-trans)
Cleavage reaction
Nerve Impulse

FIG. 13.4

The visual cycle. (*a*) 11-*cis*-Retinal reacts with and bonds to opsin to form rhodopsin. (*b*) Light energy isomerizes the double bond at the 11-12 position from the cis to the trans configuration. A nerve impulse is generated as a result of isomerization and a change in shape of the rhodopsin. The nerve impulse is generated as a result of isomerization and a change in shape of the rhodopsin. The nerve impulse is perceived as light. (*c*) 11-*trans*-Retinal has a different shape and does not fit so nicely onto the surface of opsin. Consequently, it is cleaved off. (*d*) By a series of reactions all *trans*-retinal is isomerized back to 11-*cis*-retinal, which can then form rhodopsin and the visual cycle can be repeated. If 11-*cis*-retinal and rhodopsin were not restored, we would lose the ability to see light.

fits exactly on the surface of the opsin (Figure 13.4*a*). When light energy is absorbed by the rod cells, the cis double bond at the 11 position in retinal is broken and then re-formed as the trans isomer (Figure 13.4*b*). The trans isomer has a different shape from that of the cis isomer, and this leads to a change in the shape of the entire rhodopsin molecule. In a way that is not completely understood, the structurally altered rhodopsin causes a nerve impulse to be generated at the optic nerve and travel to the brain, where it is perceived as light. Finally, the 11-*trans*-retinal is released from opsin (Figure 13.4*c*) and chemically converted back to 11-*cis*-retinal in the retina and in the liver (Figure 13.4*d*). Rhodopsin is then regenerated through the combination of opsin and 11-*cis*-retinal (Figure 13.4*a*). This regeneration must occur or we would no longer be able to perceive light coming into the eye.

Although trans double bonds are usually more stable than cis double bonds, there is an activation energy (a kinetic barrier) of about 25 kcal/mol for conversion of cis to trans alkenes. In the vision process described, light supplies the energy required for conversion.

13.8 ALKENE REACTIONS

As in all organic molecules, the main reactive site is the functional group, that is, for alkenes the double bond. Hydrogen, the halogens, hydrogen halides, and water are typical reagents which can add to the double bond and thereby "saturate" the bond. Using X — Y as a representation for the attacking reagent, the general addition reaction can be shown:

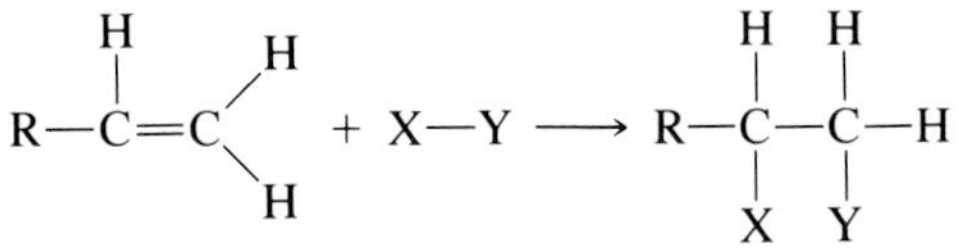

(13.7)

In the reaction the double bond is converted into a single bond and two new single bonds are formed, one to each of the original doubly bonded carbons.

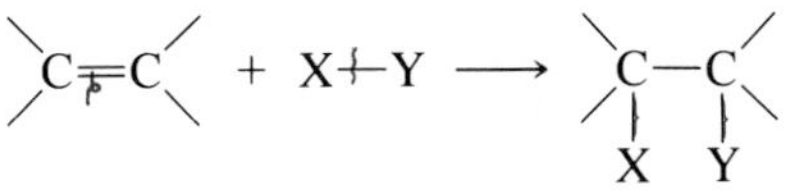

Bonds broken   Bonds made

The reaction is called addition because two reactants have added together to form one product; the one product is always the exact sum of the reactants. Let us turn from the general reagent X—Y to the specific, adding reagents mentioned at the beginning of this paragraph.

Hydrogenation

Alkenes add hydrogen to form alkanes. This addition reaction is known as hydrogenation. One H bonds to each of the originally doubly bonded carbons, and the double bond is converted to a single bond. For example,

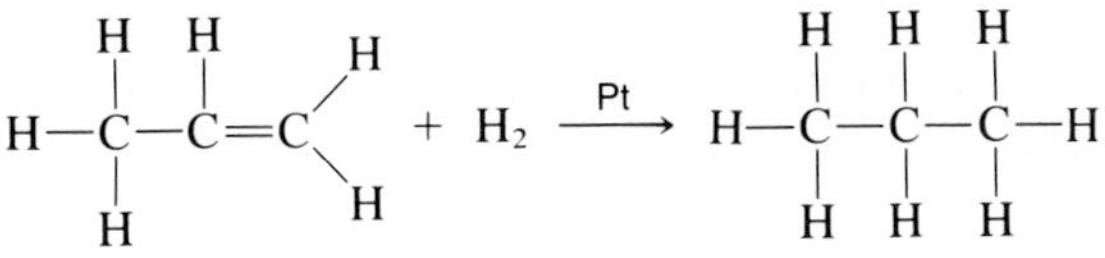

(13.8)

Propene   Propane

A metal catalyst such as platinum (Pt) is necessary for all hydrogenation reactions. As with any catalyzed reaction, the purpose of the catalyst is to lower the activation energy to allow the reaction to proceed at a feasible rate.

Hydrogenation is used commercially to convert inexpensive liquid oils, such as soybean oil and cottonseed oil, into saturated fats of higher demand, such as margarine and cooking shortenings (Section 18.4).

Halogenation

The halogens chlorine and bromine rapidly react with alkenes at room temperature. The product of the reaction is a dihaloalkane. Again, the double bond is converted to a single bond, and one halogen bonds to each of the originally doubly bonded carbons. For example,

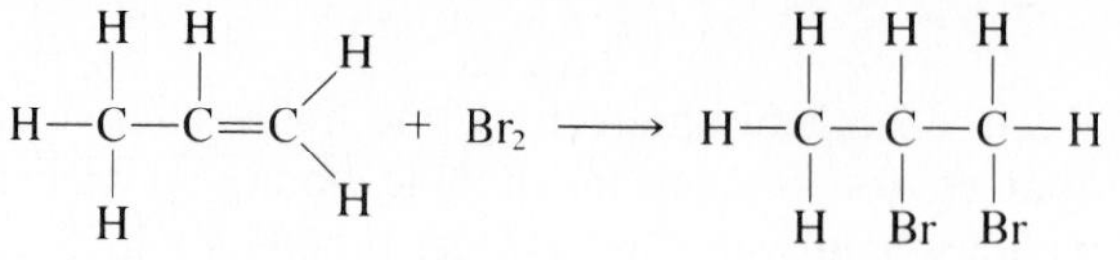

$$\begin{matrix} & & & & \text{H} \\ & & & & | \\ \text{H} & \text{H} & & \text{H} & \\ | & | & & & \\ \text{H}-\text{C}-\text{C}=\text{C} & & + \text{ Br}_2 \longrightarrow \\ | & & & \text{H} \\ \text{H} & & & \end{matrix} \qquad \begin{matrix} \text{H} & \text{H} & \text{H} \\ | & | & | \\ \text{H}-\text{C}-\text{C}-\text{C}-\text{H} \\ | & | & | \\ \text{H} & \text{Br} & \text{Br} \end{matrix}$$ (13.9)

Propene	Bromine	1,2-Dibromopropane
(colorless)	(red-brown)	(colorless)

The bromine addition reaction is often used as a quick chemical test for the presence or absence of double bonds. Bromine solutions are reddish brown; alkenes and the product dibromides are colorless. When bromine is added to a substance with double (or triple) bonds, the reddish brown disappears rapidly. If the substance tested does not contain a double (or triple) bond, the color will persist.

Problem 13.6   Draw a structural formula for the product of the addition of Br_2 to 2-methyl-2-butene.

Hydrogen Halide Addition

The hydrogen halides (HCl, HBr, HI) react with alkenes to form alkyl halides. When HCl reacts with an alkene, a C—H bond is formed to one of the doubly bonded carbons and a C—Cl bond to the other, and as in all additions, the double bond becomes a single bond.

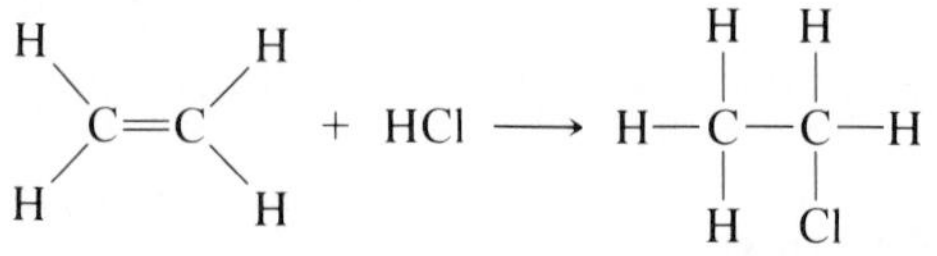

$$\begin{matrix} \text{H} & & \text{H} \\ \diagdown & & \diagup \\ & \text{C}=\text{C} & \\ \diagup & & \diagdown \\ \text{H} & & \text{H} \end{matrix} \quad + \text{ HCl } \longrightarrow \quad \begin{matrix} \text{H} & \text{H} \\ | & | \\ \text{H}-\text{C}-\text{C}-\text{H} \\ | & | \\ \text{H} & \text{Cl} \end{matrix}$$ (13.10)

Ethene   Ethyl chloride

In the reaction of ethene shown, the direction of addition does not matter. That is, the Cl could have been added to the C on the left and the product would still have been ethyl chloride. However, for most alkenes two different products (which will be structural isomers of one another) might be expected, depending on which carbon takes on the H and which the halide.

$$\begin{matrix} \text{H} & \text{H} & & \text{H} \\ | & | & & \diagup \\ \text{H}-\text{C}-\text{C}=\text{C} & & + \text{ HCl } \longrightarrow \\ | & & & \text{H} \\ \text{H} & & \end{matrix} \qquad \begin{matrix} \text{H} & \text{H} & \text{H} \\ | & | & | \\ \text{H}-\text{C}-\text{C}-\text{C}-\text{H} \\ | & | & | \\ \text{H} & \text{Cl} & \text{H} \end{matrix}$$

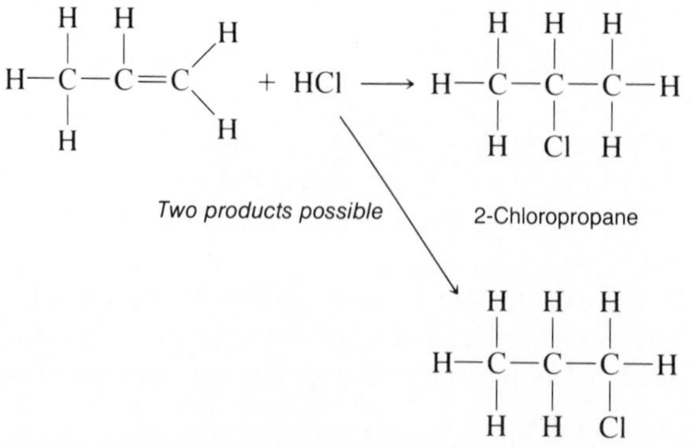

Two products possible   2-Chloropropane

$$\begin{matrix} \text{H} & \text{H} & \text{H} \\ | & | & | \\ \text{H}-\text{C}-\text{C}-\text{C}-\text{H} \\ | & | & | \\ \text{H} & \text{H} & \text{Cl} \end{matrix}$$

1-Chloropropane

It turns out that when there are two possible products, one of the products always forms in much greater amount than the other. In the example given above, 2-chloropropane is the major product and 1-chloropropane is formed in only a minor amount. How can we predict what the major product will be? In 1870 Vladimir Markovnikov, a Russian chemist, formulated a simple rule based on many observed HX additions to alkenes. Markovnikov's rule states that **in the addition of H—X to a carbon-carbon double bond, H bonds to that carbon which already has the greater number of hydrogens.** The rule is sometimes paraphrased, "The rich get richer," regarding hydrogen atoms as riches.

Observe how the following two examples, for which only the major products are shown, fit the rule.

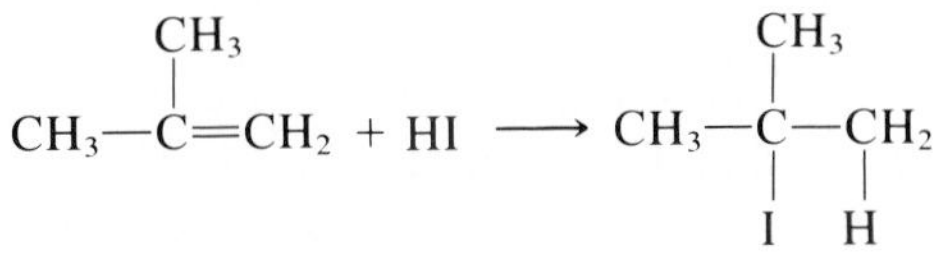

2-Methylpropene 2-Iodo-2-methylpropane

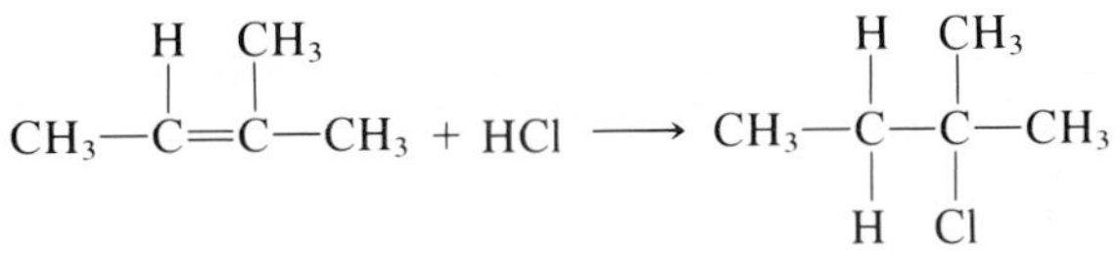

2-Methyl-2-butene 2-Chloro-2-methylbutane

Problem 13.7 Write the structural formula for the major product of the following reaction:

$CH_3—CH_2—CH═CH_2 + HCl \longrightarrow$

Hydration

Alkenes add water (HOH) in acid-catalyzed reactions called hydration reactions. The addition products are alcohols. The reactions follow Markovnikov's rule; the —OH of HOH corresponds to X.

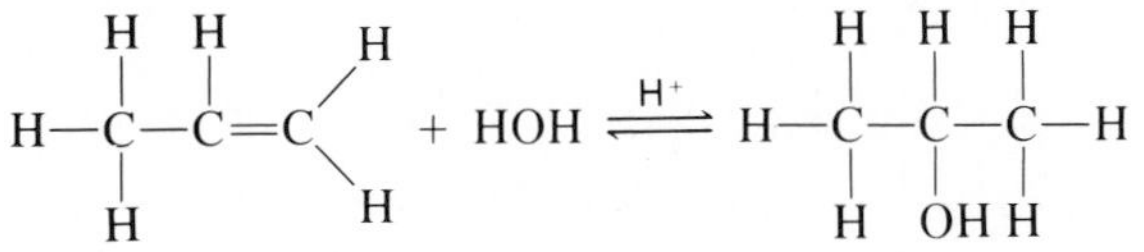

(13.11)

Propene 2-Propanol

A biological example of the alkene hydration reaction is the conversion of fumaric acid to (−) malic acid, one of the reactions of the Krebs citric acid cycle, an important metabolic cycle that you will study in Section 23.4. Don't worry for now that these structures and their names are unfamiliar; just examine the chemical change at the carbon-carbon double bond.

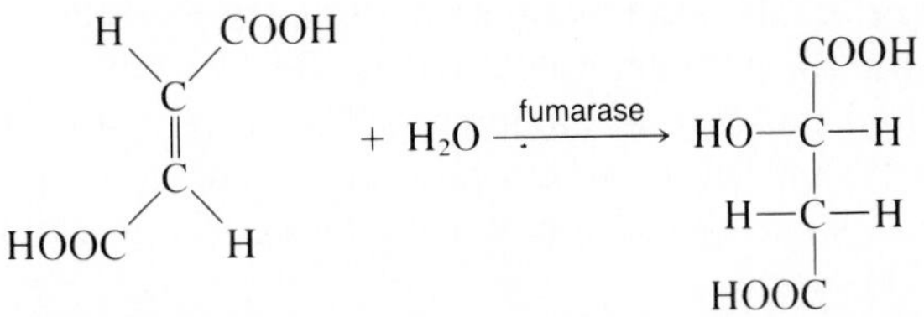

Fumaric acid

(−) Malic acid

There are some very interesting differences between the laboratory reaction and the biological reaction. For one, the presence of the biological catalyst (enzyme) fumarase permits this reaction to occur under physiological pH conditions (7.4) rather than the acid conditions required in the test tube reaction. Also, the enzyme directs the —OH to take its place only as shown.

Problem 13.8 Write the structural formula for the *major* product of the following reaction:

$$CH_3CH_2CH{=}CH_2 + H_2O \xrightarrow{H^+}$$

Polymerization

Under certain catalyzed conditions alkene molecules can add to one another to form long chains known as polymers. For example, ethene, commonly called ethylene, can form the polymer *polyethylene:*

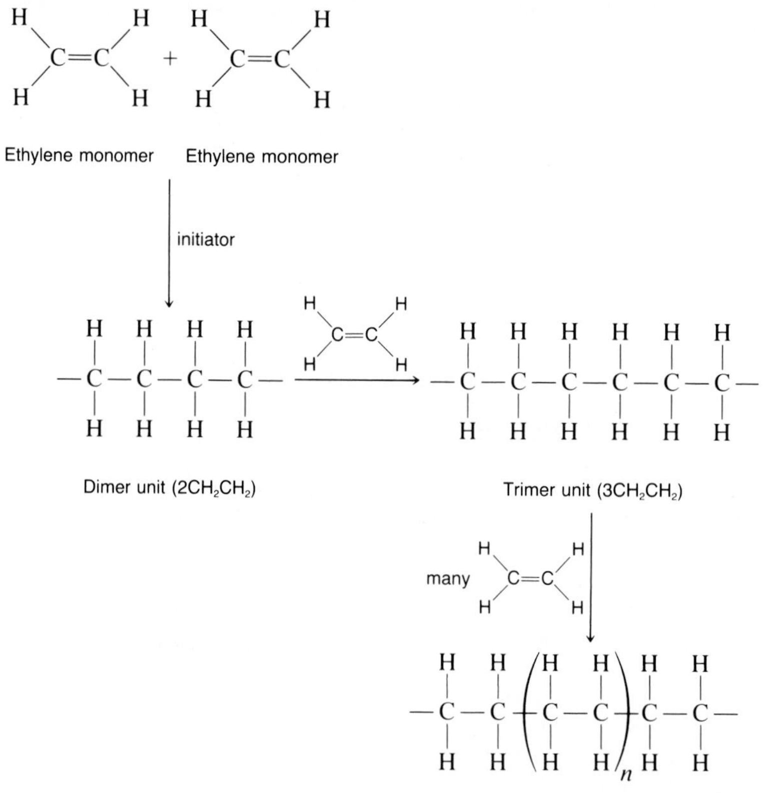
Ethylene monomer
Ethylene monomer
initiator
Dimer unit (2CH2CH2)
Trimer unit (3CH2CH2)
many

Polyethylene

This polymer is a chain of the repeating unit $—CH_2—CH_2—$. The value of *n*, the number of repeating units, can vary from 100 to 1000 and is controlled by the reaction conditions. The chemical unit that repeats to form the polymer is called the monomer. Ethene ($CH_2{=}CH_2$) monomer units combine to form the polymer polyethylene. Besides ethylene itself, many derivatives of ethylene serve as monomers. The variety of monomers creates a variety of polymers with varying properties and therefore different uses. Table 13.4 lists some common polymers derived from additions of various monomers.

You come into contact with polymers every day in a variety of ways. The proteins and carbohydrates in your food are biopolymers (polymers of bio-

TABLE 13.4 Alkene Addition Polymers

Monomer (Common Name in Parentheses)	Polymer	Use
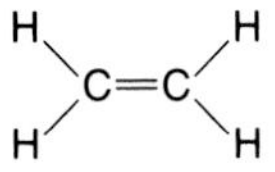 Ethene (ethylene)	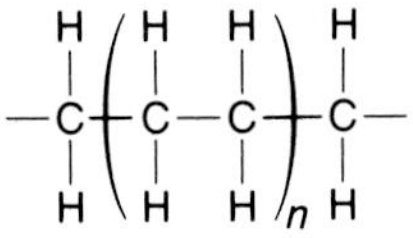 Polyethylene	Film, tubing, toys, containers, electrical insulation
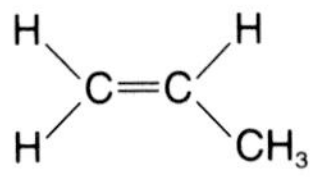 Propene (propylene)	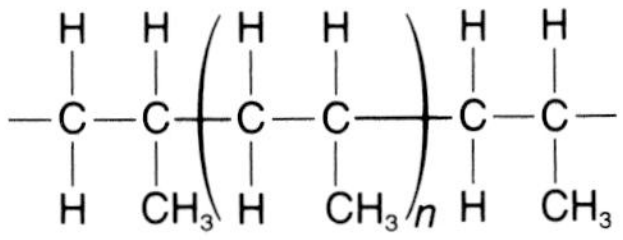 Polypropylene	Carpet and clothing fibers, steering wheels, synthetic bone joints
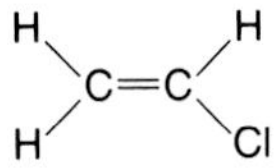 Chloroethene (vinyl chloride)	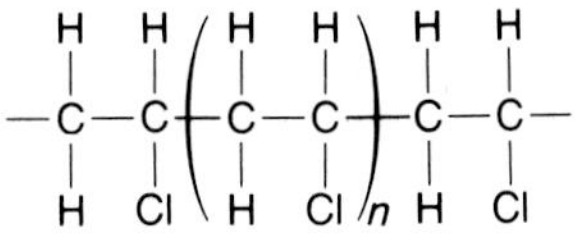 Poly(vinyl chloride), PVC	Phonograph records, house siding, electrical insulation, floor coverings, garden hose, synthetic leather
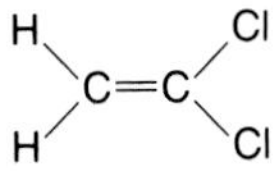 1,1-Dichloroethene (vinylidene chloride)	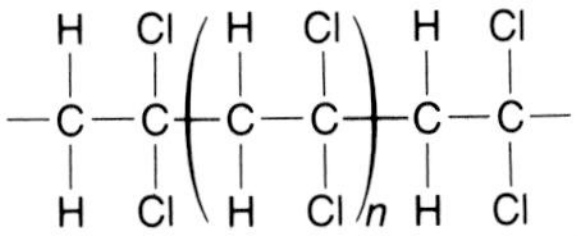 Poly(vinylidene chloride)	Saran Wrap
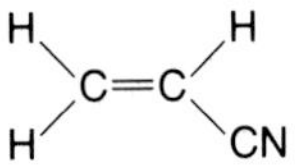 Propenenitrile (acrylonitrile)	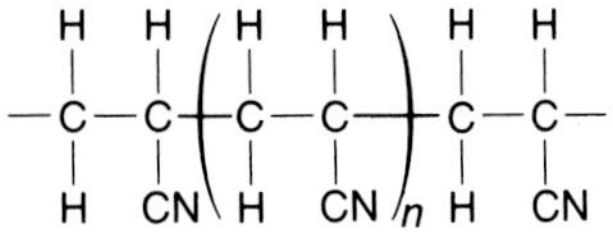 Polyacrylonitrile	Orlon and Acrilan fibers for rugs and clothing
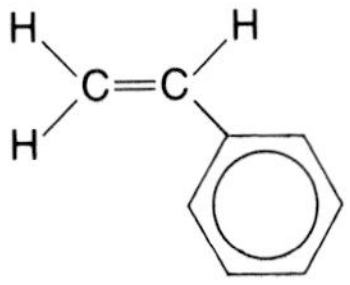 Phenylethene (styrene)	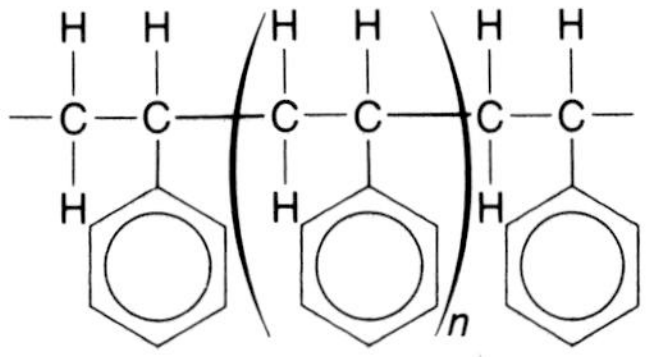 Polystyrene	Styrofoam insulation, hot-drink cups
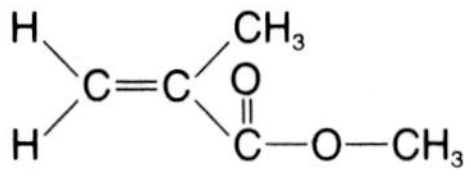 Methyl 2-methylpropenoate (methyl methacrylate)	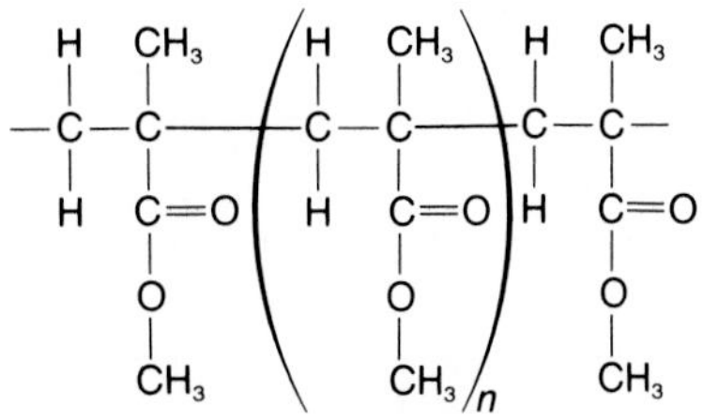 Poly(methyl methacrylate)	Plexiglas, Lucite, hard contact lenses

TABLE 13.4 *(Continued)*

Monomer (Common Name in Parentheses)	Polymer	Use
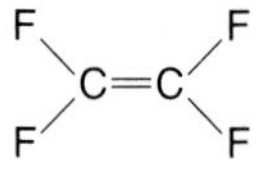 Tetrafluoroethene (tetrafluoroethylene)	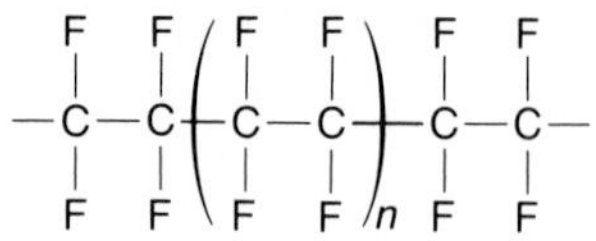 Polytetrafluoroethylene, PTFE (Teflon)	Stopcocks, nonstick pots and pans, artificial heart valves, pacemaker housings

logical origin). Polymers are the major components of the ubiquitous plastics we encounter; synthetic rubber is polymeric, as are synthetic fibers. New polymeric materials are constantly being developed to replace metals and other traditional construction materials. The new materials are designed to have superior properties and lower costs.

In industry, products from polymers are classified as plastics, elastomers, foams, or fibers, depending on their properties. Polyethylene, polypropylene, and poly(methyl methacrylate) are examples from Table 13.4 which can be formed into familiar solid plastic objects. Elastomers are materials which show elasticity; that is, they can be stretched and return to their original shape. Rubber is an elastomer; the most common synthetic rubbers are polymers of dienes (alkenes with *two* double bonds), with four carbons as the longest continuous chain. Foams are made by blowing gases into molten polymeric material. Styrofoam cups or insulation offer examples of this kind of material. Many polymers can be drawn into fibers; indoor-outdoor carpeting is often made from polypropylene fibers. In Chapter 17 you will encounter other polymeric fibers, polyesters, and polyamides, which form by a different polymerization process than the addition polymerization described for ethylene.

13.9 ALKYNES

Alkynes are hydrocarbons which contain a carbon-carbon triple bond, —C≡C—. Like the alkenes, alkynes are unsaturated. IUPAC naming of alkynes follows closely the rules for alkenes given in Section 13.6 with the exception that the suffix for alkynes is *-yne*. For example,

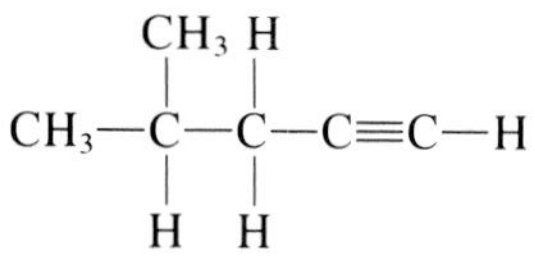

is called 4-methyl-1-pentyne. Because the triple bond has a linear structure, there is no possibility of geometric isomerism. The simplest alkyne, H—C≡C—H, has the IUPAC name ethyne but is more commonly known as acetylene. Acetylene gives a very hot flame when burned with oxygen because the combustion reaction is very exothermic; it is therefore widely used in welding torches. Acetylene can be prepared in the laboratory from the reaction of water with the inorganic substance calcium carbide (CaC_2).

$$\underset{\text{Calcium carbide}}{CaC_2} + 2H_2O \longrightarrow \underset{\text{Acetylene}}{C_2H_2} + Ca(OH)_2$$

Industrially, it is made from natural gas.

Because they are unsaturated, the alkynes, like the alkenes, undergo addition as their most important reactions. Hydrogen, halogens, hydrogen halides, and water can all be added to alkynes. Either 1 or 2 mol of the adding reagent can react with 1 mol of alkyne, depending upon the stoichiometry and catalyst employed. If 1 mol is added, an alkene forms, whereas if 2 mol is added, an alkane is formed. For example, propyne can be hydrogenated to form propene or propane.

$$\underset{\text{Propyne}}{CH_3{-}C{\equiv}C{-}H} + H_2 \xrightarrow{\text{poisoned Pd}} \underset{\text{Propene}}{CH_3{-}\overset{\overset{\large H}{|}}{C}{=}CH_2}$$

Poisoned palladium is a specific catalyst for stopping the reaction at the alkene stage.

$$CH_3{-}C{\equiv}C{-}H + 2H_2 \xrightarrow{\text{Ni}} \underset{\text{Propane}}{CH_3{-}CH_2{-}CH_3}$$

Addition follows Markovnikov's rule; that is, when HX adds, the H bonds to that carbon already richer in H's.

$$\underset{\text{Propyne}}{CH_3{-}C{\equiv}C{-}H} + HCl \longrightarrow \underset{\text{2-Chloropropene}}{CH_3{-}\overset{\overset{\large Cl}{|}}{C}{=}CH_2}$$

$$\underset{\text{Propyne}}{CH_3C{\equiv}C{-}H} + 2HCl \longrightarrow \underset{\text{2,2-Dichloropropane}}{CH_3{-}\overset{\overset{\large Cl}{|}}{\underset{\underset{\large Cl}{|}}{C}}{-}CH_3}$$

13.10 CYCLIC HYDROCARBONS

Nomenclature

Cyclic hydrocarbons were introduced in Section 12.5, and their representation by regular polygons was explained at that time. Cyclic compounds occur in nature in various ring sizes from the smallest rings of just three atoms to very large rings containing as many as 30 atoms. Saturated rings are named as cycloalkanes, i.e., by placing the prefix *cyclo-* before the parent name indicating the number of carbons in the ring. Thus,  is cyclobutane and 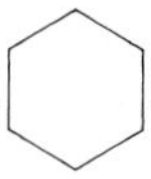 is cyclohexane. Rings containing carbon-carbon double bonds are named as cycloalkenes, for example, 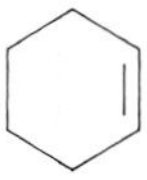 is called cyclohexene.

A single substituent on a cycloalkane ring need not be numbered since it is defined as position 1. Two or more substituents are given the lowest set of numbers possible circling around the ring in either a clockwise or a counterclockwise direction, whichever way gives the lower set. The following examples illustrate these rules.

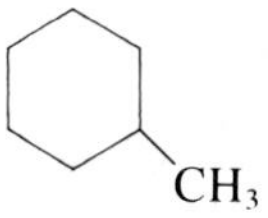
CH3

Methylcyclo-
hexane

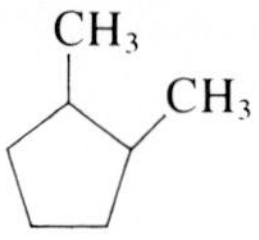
CH3
CH3

1,2-Dimethylcyclo-
pentane

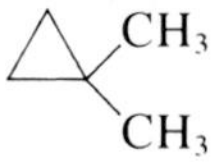
CH3
CH3

1,1-Dimethylcyclo-
propane

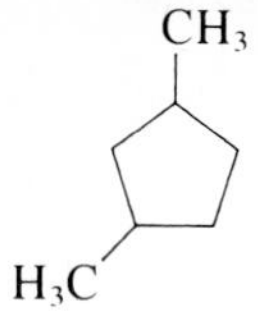
CH3
H3C

1,3-Dimethylcyclo-
pentane (*not* 1,4-)

In cycloalkenes, the double-bonded carbons are assigned positions 1 and 2 and substituents in the ring are given the lowest possible number relative to the double bond.

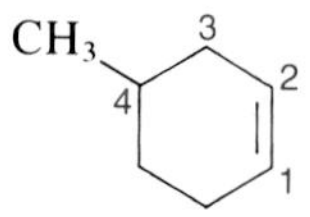
CH3
3
4
2
1

4-Methylcyclohexene

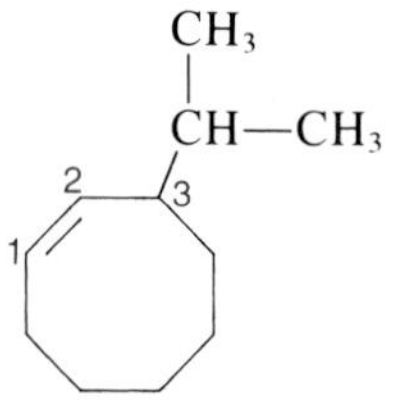
CH3
CH—CH3
2
3
1

3-Isopropylcyclooctene

Disubstituted cycloalkanes can exist as pairs of cis-trans isomers, because unlike their open-chain counterparts, rotation is restricted about the carbon-carbon bonds in a ring. Two substituents on different carbons of a ring can lie either on the same side (cis isomer) or on opposite sides (trans isomer) of the plane of the ring. The appropriate designation is included in the name. For example:

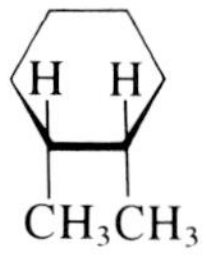
H H
CH3CH3

cis-1,2-Dimethylcyclohexane

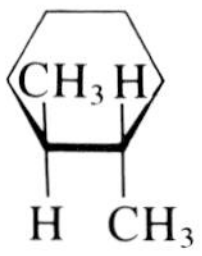
CH3 H
H CH3

trans-1,2-Dimethylcyclohexane

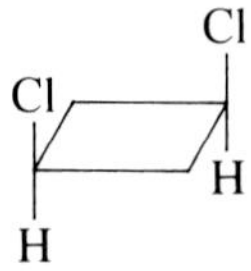
Cl
Cl
H
H

cis-1,3-Dichlorocyclobutane

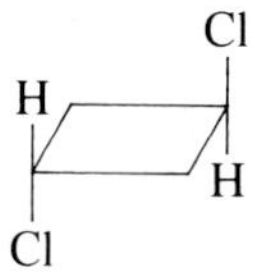
Cl
H
H
Cl

trans-1,3-Dichlorocyclobutane

Stability

Hooking carbons together in rings can, in some cases, weaken the carbon-carbon bonds and lower the stability of the cyclic compound. One cause of this phenomenon can be seen by examining the bond angles in ring and open-chain compounds. A carbon with four single bonds has a tetrahedral geometry and in principle bond angles of 109°28′. That is, the angle between any two bonds to the same saturated carbon is 109°28′.

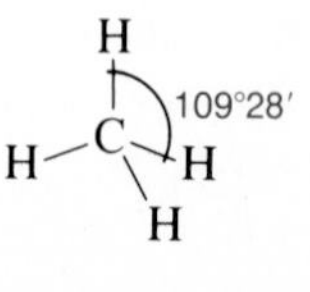
H
109°28′
H
C
H
H

In open-chain compounds there is no restraint on carbon's achieving these angles. In cyclic compounds, ring geometry may make it impossible for normal tetrahedral bond angles to exist.

For example, three- and four-membered rings require interior bond angles of 60° and 90°, respectively.

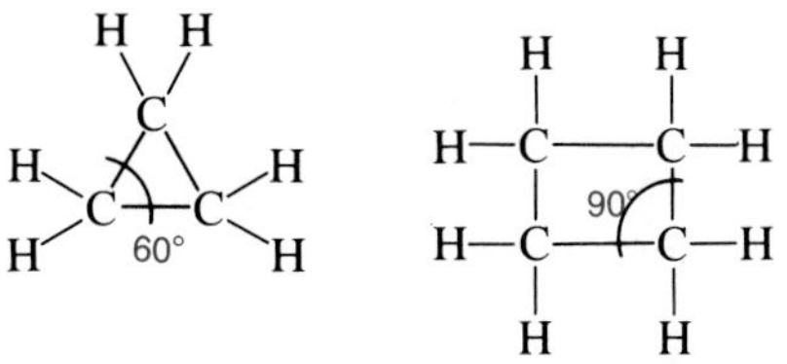
60°
90°

This difference between the actual angle and desirable 109°28′ creates angle strain; that is, these small rings are unstable and susceptible to ring-opening reactions which would relieve the strain. For example, cyclopropane readily reacts with bromine to yield 1,3-dibromopropane,

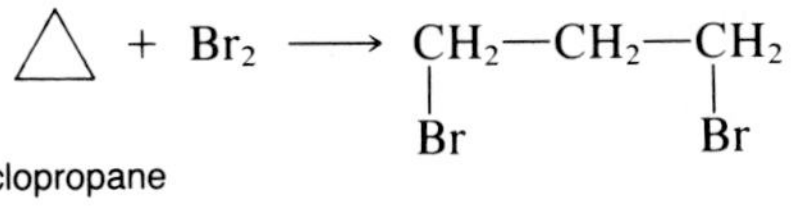
Cyclopropane
1,3-Dibromopropane

A regular pentagon has interior angles of 108°, and consequently cyclopentane displays little angle strain and does not open readily. A flat ring of six atoms, i.e., a regular hexagon, has interior angles of 120°. Thus if the six carbons of cyclohexane exist in the same plane, cyclohexane should show angle strain because of the distortion of bond angles from 109°28′. But cyclohexane is found to be strain free. This is because the six carbons of the ring are not all in the same plane. As Figure 13.5 shows, the carbons are arranged: four carbons in a plane, one carbon above that plane, and one carbon below that plane. This arrangement is described as the cyclohexane "chair." The arrangement allows for tetrahedral bond angles around each carbon. For convenience in writing, cyclohexane is frequently shown as a planar hexagon, but it actually exists as the three-dimensional chair. Probably because of their greater stability, cyclohexane and cyclopentane rings are much more common in physiological compounds then either cyclopropane or cyclobutane.

FIG. 13.5
The cyclohexane chair. (*a*) A flat ring (all six carbons in one plane) would have strained bond angles of 120°. The nonplanar ring is strain free; the bond angles have the tetrahedral value, 109°28′. (*b*) The cyclohexane chair has four carbons in one plane (circled in black) and one carbon above and one carbon below the plane (circled in color). The molecular model shows this more clearly; most hydrogens are not shown on the model so that the carbons are not obscured.

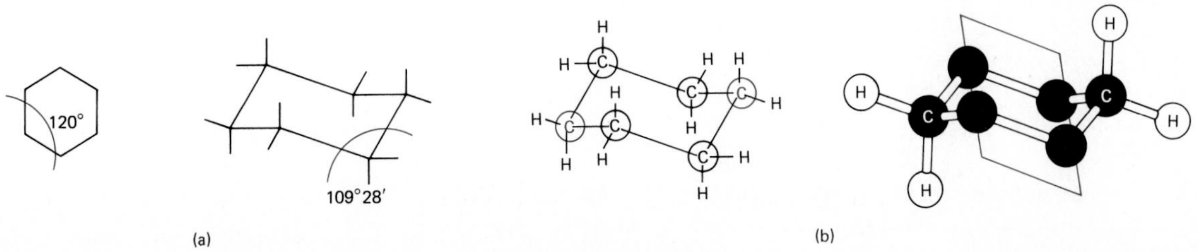
120°
109°28′
(a)
(b)

Reactions

Other than the ring-opening reactions of small strained rings, the chemical and physical properties of the cycloalkanes and cycloalkenes are very similar to those of the open-chain alkanes and alkenes. That is, cycloalkanes undergo monochlorination reactions (Section 13.4), and cycloalkenes undergo addition reactions (Section 13.8).

13.11 AROMATIC HYDROCARBONS

The hydrocarbons that have been discussed so far, alkanes, cycloalkanes, alkenes, and alkynes, are all classified as being **aliphatic,** an archaic designation which signifies their source in fats (the Greek *aleiphar*). There is another category of hydrocarbons known as **aromatic,** originally so designated because of their characteristic aromas. Today the aliphatic-aromatic classification is based on the absence (aliphatic) or presence (aromatic) of the benzene ring system and its unique properties. Many familiar compounds contain the benzene ring and are classified as aromatic.

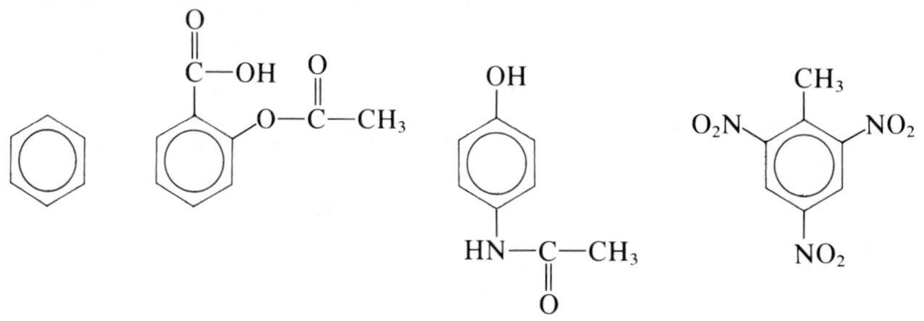

Benzene | Acetylsalicylic acid (aspirin) | Acetaminophen (Tylenol) | 2,4,6-Trinitrotoluene (TNT)

Benzene has the molecular formula C_6H_6. If your class were assigned to write a correct Lewis structure for C_6H_6, and given the hint that benzene is a six-membered ring, some of you might arrive at the structure

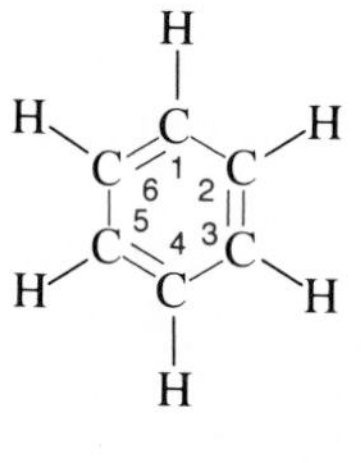

Structure 1

whereas others of you might write

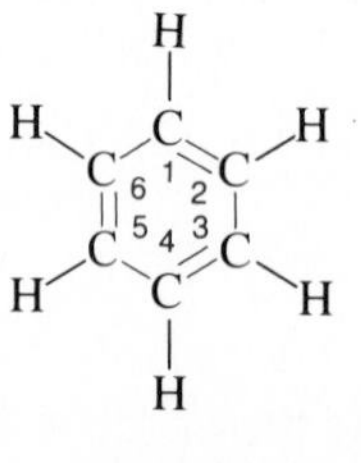

Structure 2

In structures 1 and 2 the location of atoms is identical. The difference between the two structures is in the placement of electrons; in structure 1 the double bonds appear between carbons at positions 2 and 3, 4 and 5, and 6 and 1, whereas in structure 2 the electrons of the double bonds appear between carbons at positions 1 and 2, 3 and 4, and 5 and 6.

Both structures 1 and 2 are correct Lewis structures. However, neither one by itself can account for the properties of benzene and therefore neither structure 1 nor structure 2 can accurately represent the benzene molecule. The unique properties of benzene that we must account for are:

1 All carbon-carbon bonds are identical in properties such as length and strength, rather than there being three single bonds and three double bonds.
2 Benzene does not undergo the addition reactions characteristic of double bonds.
3 Benzene is unusually stable. That is, its total energy is less than what would be predicted based on structures 1 or 2. These properties can be explained only by realizing that there is a unique bonding arrangement in the benzene ring.

In Section 6.4 you saw that the energy lowering accompanying covalent bond formation arises from the attraction between the electrons of one atom and the nucleus of another atom. That is, energetically, it is preferable for electrons to be shared between atoms. The six electrons of the "second bonds" in the double bonds of benzene are not restricted to three localized double bonds between specific carbon atoms as structures 1 and 2 imply. Rather, apparently they can circulate over all six carbons in the ring; consequently all six electrons are attracted by all six carbon nuclei. This lowers the energy of the benzene molecule compared with that of a structure in which each of the six electrons is attracted to only two nuclei, as would be the case in structures 1 and 2. The six "double bond" electrons in benzene are said to be **delocalized** because they are free to circulate over the entire benzene ring and are not restricted to three localized double bonds.

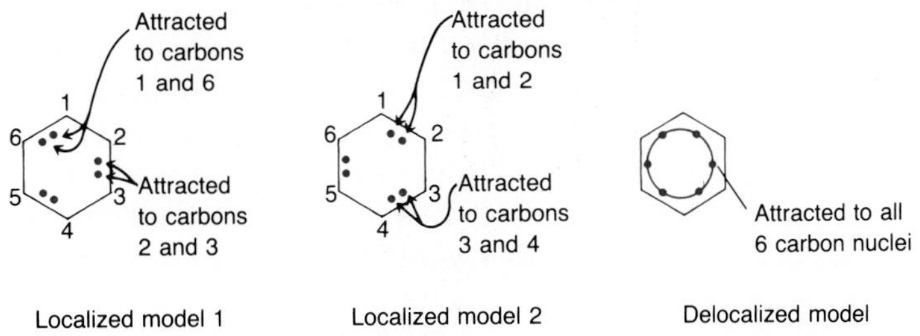
Attracted to carbons 1 and 6
1
2
3
4
5
6
Attracted to carbons 2 and 3
Localized model 1
Attracted to carbons 1 and 2
Attracted to carbons 3 and 4
Localized model 2
Attracted to all 6 carbon nuclei
Delocalized model

Figure 13.6 attempts to give an accurate picture of the delocalized electrons in benzene. There is electron density above and below the plane of the six carbon atoms. Aromatic rings are represented by hexagons with circles inscribed within to represent the circulating electrons.

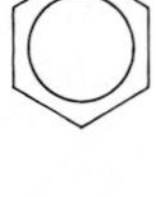

Benzene

Let us see how the delocalized model of benzene explains the unique properties of benzene:

1 All six carbon-carbon bonds are identical because each is composed of a single bond and an equal contribution from the six circulating electrons.

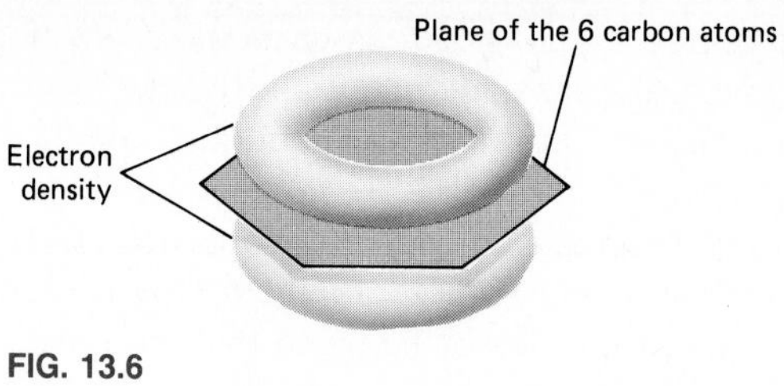
Plane of the 6 carbon atoms
Electron
density

FIG. 13.6
The delocalized electrons in benzene produce electron density in the shape of doughnut clouds above and below the plane of the six carbon atoms. The important idea is the recognition of this unique arrangement of benzene's electrons which gives rise to unique properties and to the representation 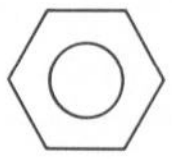

2 Benzene does not undergo addition reactions because there are no localized double bonds as there are in alkenes, which do undergo addition.
3 Benzene gains stability from the attractions between each of the six circulating electrons and each of the six carbon nuclei in the benzene ring.

A complete understanding of the extra stability of a delocalized electron system compared with a localized one requires a quantum mechanical treatment beyond the scope of this text.

We will use the representation 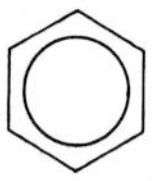 throughout this text. In other books you will still encounter the structures 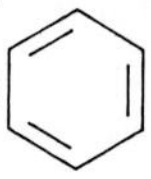 and 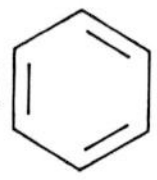, both because of former usage and because these forms are sometimes useful in discussions of organic reaction mechanisms.

We have said that benzene does not undergo addition because it does not have localized double bonds. A more accurate explanation lies in the fact that addition would destroy the aromaticity; that is, the stabilizing electron doughnut would be destroyed. For example, if chlorine is added to benzene, the aromatic ring is replaced by an aliphatic one. We must use structure 1 or 2 to show this addition reaction,

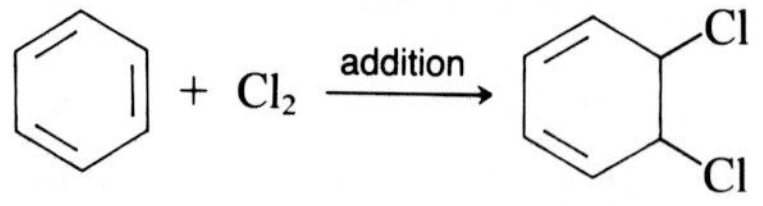
+ Cl2
addition
Cl
Cl

Aromatic Aliphatic

This reaction does not occur. Rather, benzene resists addition and reacts instead by substitution for one of the hydrogens.

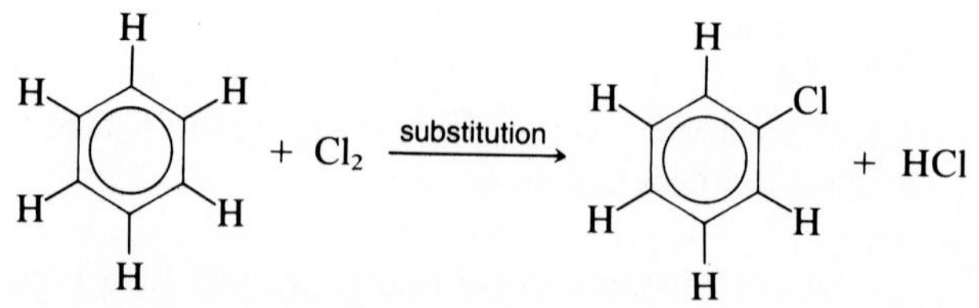
H
H
H
H
H
H
+ Cl2
substitution
H
H
Cl
H
H
H
+ HCl

Thus the electron circulation and stability are preserved. Usually a catalyst is required. Because all six hydrogens in benzene are equivalent, it makes no difference which one is removed in the substitution reaction. In this example of substitution, the H's on the ring are shown for clarity. As you know, representations of rings do not ordinarily show H's. In general, the benzene substitution reaction can be thought of as occurring according to the following scheme:

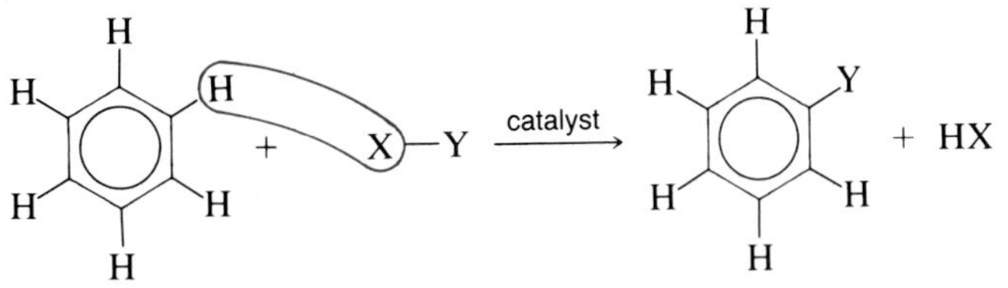
catalyst
+ HX

The substitution of Cl for H was shown. Other atoms or groups of atoms that can be placed on the aromatic ring by substitution for H are —Br, $—NO_2$, $—SO_3H$, and alkyl groups. More than one substituent can be placed on the ring.

The IUPAC rules for naming monosubstituted benzenes simply prefix the word *benzene* with the substituent name. No numbering system is needed. For example,

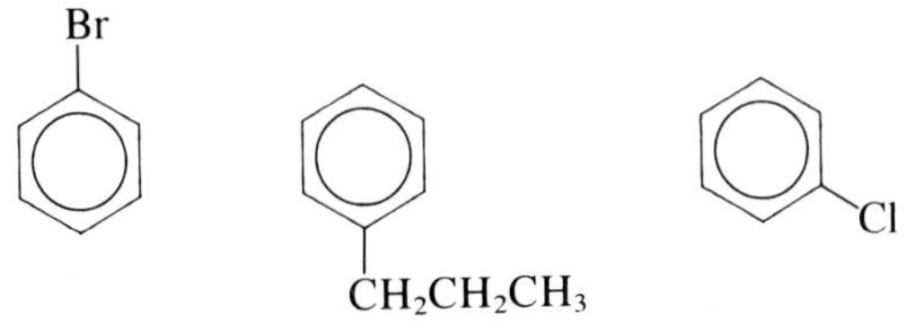
Br
$CH_2CH_2CH_3$
Cl

Bromobenzene *n*-Propylbenzene Chlorobenzene

However, many frequently encountered monosubstituted benzenes are known exclusively by their common names. Some examples are:

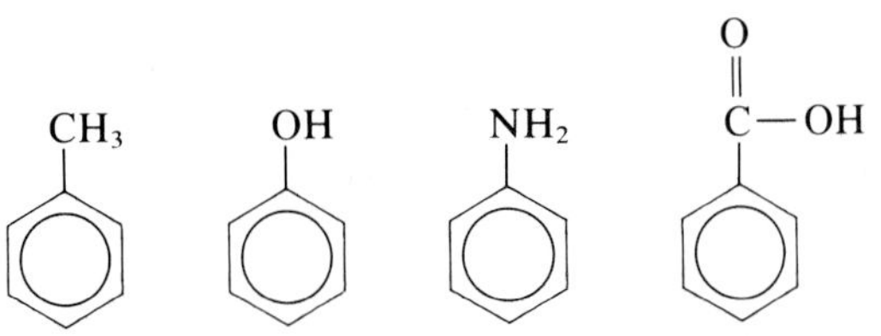
CH_3
OH
NH_2
O
C—OH

Toluene Phenol Aniline Benzoic acid

You will see these examples again in the chapters dealing with the functional groups attached to the ring.

An unambiguous name for a *di*substituted benzene derivative must provide information about the relative positions of the two groups on the ring. The prefix *ortho-* is used when the substituents are on adjacent carbons (i.e., numbered 1, 2), *meta-* is employed if the substituted carbons are separated by one carbon (i.e., numbered 1, 3), and *para-* is used if they are separated by two carbons (i.e., numbered 1, 4). The designations *ortho-*, *meta-*, and *para*-, abbreviated *o*-, *m*-, and *p*-, are only used for disubstituted benzenes.

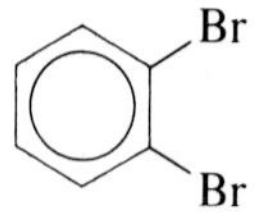
Br
Br

ortho-Dibromobenzene

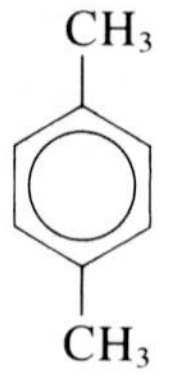
CH_3
CH_3

para-Dimethylbenzene
(commonly called *p*-xylene)

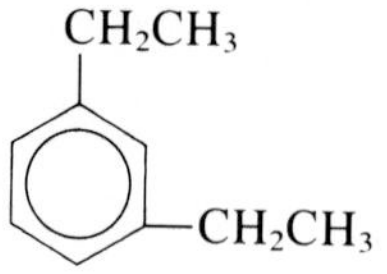
CH_2CH_3
CH_2CH_3

meta-Diethylbenzene

Often when there are two different substituents, one substituent is named as part of the common base names, toluene, phenol, aniline, or benzoic acid. For example,

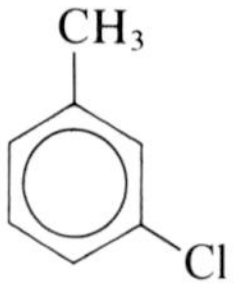
CH_3
Cl

meta-Chlorotoluene

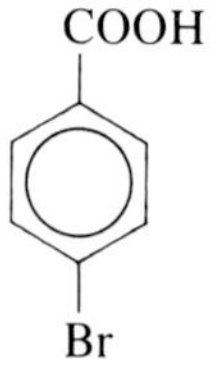
COOH
Br

para-Bromobenzoic acid

The designation of positions of groups on benzene rings with more than two substituents is accomplished with numbers.

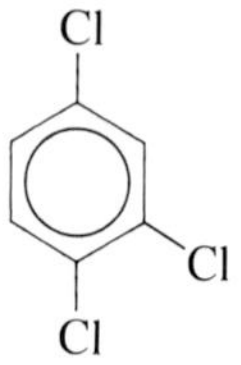
Cl
Cl
Cl

1,2-4-Trichlorobenzene

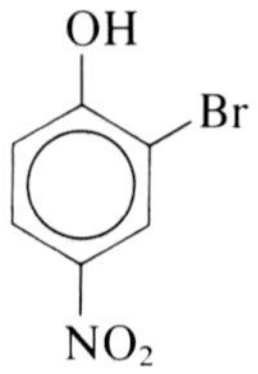
OH
Br
NO_2

2-Bromo-4-nitrophenol

Problem 13.9 *a* Write structural formulas for three isomers of chlorotoluene.
b Name each isomer in part *a*.

Aromatic compounds are easily recognized because of the distinguishing representation, 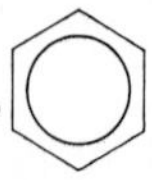. Some aromatic compounds have two or more benzene rings joined together and sharing common carbon atoms. These compounds are known as the polycyclic aromatic hydrocarbons. A few examples are

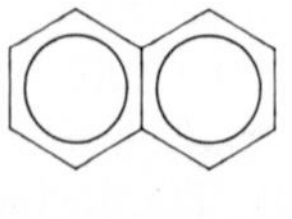

Naphthalene

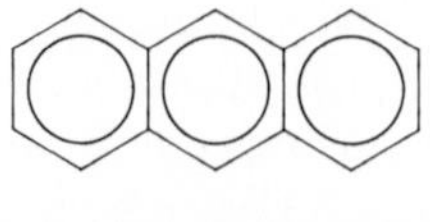

Anthracene

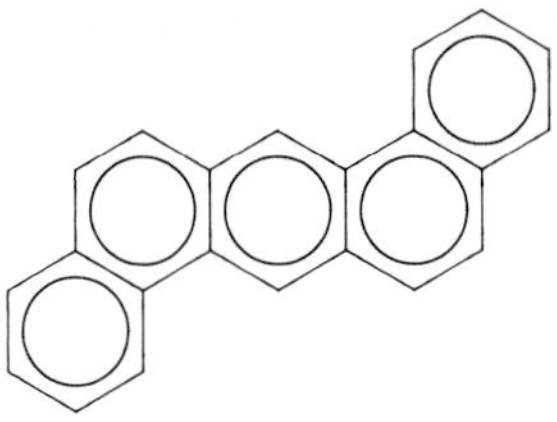

Dibenzoanthracene

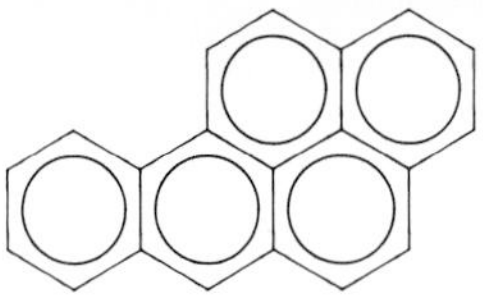

Benzopyrene

All the polycyclic aromatics are solids. However, napthalene sublimes easily, and this volatility leads to its use in mothballs. Moths are repelled by this aromatic. Dibenzoanthracene and benzopyrene are known carcinogens, i.e., chemical substances that cause cancer. They can form in cigarette smoke, during the barbecuing of steaks, and, indeed, during the combustion of any organic matter. Recent research indicates that the polycyclic aromatic hydrocarbons are not themselves carcinogens. Rather, organisms such as human beings, in an attempt to convert the hydrocarbons to water-soluble excretion products, break down the material in a series of steps involving several intermediates. Some of the intermediates are the actual carcinogens. An organism's susceptibility to the carcinogenic effects of polycyclic aromatic hydrocarbons may depend on the activity of its detoxifying enzymes which form and break down the intermediates on the pathway from polycyclic aromatic hydrocarbon to the final excretion product.

13.12 BIOLOGICALLY SIGNIFICANT HYDROCARBONS

Although the hydrocarbons play an enormously important role as a fuel source in the industrial world, they do not act as an energy source in the physiological world. The molecules that provide fuel for the body all contain oxygen as well as carbon and hydrogen.

Nonetheless, hydrocarbons do have biological and medical significance. For example, cyclopropane has been used as an anesthetic. Mineral oil, which is actually a mixture of hydrocarbons, is useful as a skin lubricant. When taken internally it can act as a laxative, probably by retarding the reabsorption of water in the large intestine and thus softening the stools. However, adverse side effects may accompany its use as a laxative. For example, if mineral oil is accidently aspirated while swallowing, a type of pneumonia may develop, and mineral oil might interfere with absorption of fat-soluble vitamins (Section 26.9) in the intestine. For these reasons, use of mineral oil as a laxative should probably be discontinued.

Insects use some hydrocarbons as chemical communicators. To communicate with one another, insects secrete these chemicals, called pheromones, as airborne messages to members of the same species. The messages can attract a mate, sound an alarm, or indicate a source of food. For example, the common female housefly emits the pheromone *cis*-9-tricosene as a sex attractant.

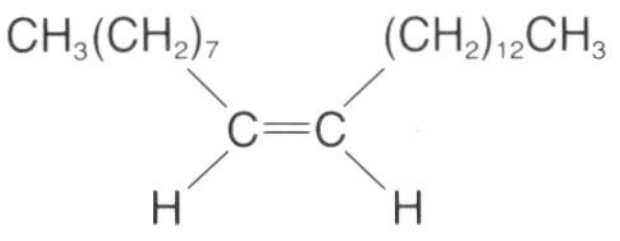
$CH_3(CH_2)_7$
$(CH_2)_{12}CH_3$
C=C
H
H

cis-9-Tricosene

Interestingly, the trans isomer is physiologically inactive. This is another example of the importance of three-dimensional shape in living chemistry.

Ants, when disturbed, release alarm pheromones which have been identified as the alkanes undecane and tridecane.

$CH_3(CH_2)_9CH_3$ — Undecane

$CH_3(CH_2)_{11}CH_3$ — Tridecane

The pheromones may serve as an ecologically safe method of insect control. Some commercial traps baited with sex pheromones for the gypsy moth and Japanese beetle are already available. Because the traps require only localized, minute amounts of natural products, insect populations can be controlled without depositing insecticides in the environment.

13.13 PROPERTIES OF THE HALOGENATED HYDROCARBONS

Halogenated hydrocarbons serve, among other uses, as dry-cleaning solvents, refrigerants, and insecticides.

All four hydrogens of methane can potentially be replaced by chlorine in the chlorination substitution reaction (Section 13.4). Schematically,

$$\underset{\text{Methane}}{CH_4} \xrightarrow[\text{uv light}]{Cl_2} \underset{\text{Methyl chloride}}{\overset{HCl}{\underset{CH_3Cl}{+}}} \xrightarrow[\text{uv light}]{Cl_2} \underset{\text{Methylene chloride}}{\overset{HCl}{\underset{CH_2Cl_2}{+}}} \xrightarrow[\text{uv light}]{Cl_2} \underset{\text{Chloroform}}{\overset{HCl}{\underset{CHCl_3}{+}}} \xrightarrow[\text{uv light}]{Cl_2} \underset{\text{Carbon tetrachloride}}{\overset{HCl}{\underset{CCl_4}{+}}}$$

These halogen compounds are all known by common names. Methyl chloride is a gas at room temperature. The other three methane derivatives have boiling points above room temperature because of their greater molecular weight and are therefore liquids at room temperature. CH_2Cl_2 and $CHCl_3$ are only slightly polar, and CCl_4, because of its symmetry, is completely nonpolar. All three are good solvents for nonpolar hydrocarbons and therefore have been used as cleaning fluids. However, chloroform is an acute toxin to the respiratory and cardiovascular systems, and chronic carbon tetrachloride exposure leads to kidney and liver damage. Both $CHCl_3$ and CCl_4 are also suspected of being carcinogenic.

CCl_4 and $CHCl_3$ have previously been used in clinical environments. CCl_4 was used to rid human beings of intestinal hookworm parasites, and $CHCl_3$ was used as an anesthetic. Our increased knowledge of their toxicity has led to a discontinuance in use, at least for human beings. Trichloroethylene and tetrachloroethylene have largely replaced carbon tetrachloride in the dry-cleaning industry.

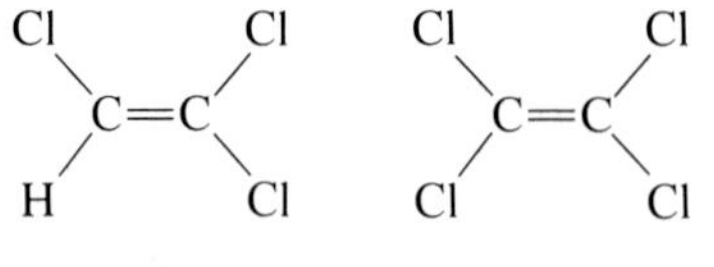

Trichloroethylene   Tetrachloroethylene

However, there is some evidence from experiments with mice that trichloroethylene is also carcinogenic. All chlorinated hydrocarbons are probably toxic to some extent and should be used only in well-ventilated areas.

Freons are halogenated derivatives of methane which contain both fluorine and either chlorine or bromine.

CCl_2F_2

Dichlorodifluoromethane
(Freon-12)
(bp = −30°C)

These compounds are nontoxic, nonflammable, and noncorrosive. Because of these properties and their low boiling points, the freons have found wide use as the refrigerants in refrigerators, air conditioners, and freezers. Heat is exchanged through vaporization and condensation of the freons. The freons were formerly used as the propellant in aerosol sprays. However, this practice has been discontinued because freons react with the ozone layer (O_3) found in the upper atmosphere. The ozone layer acts as a screen that prevents excessive ultraviolet radiation from the sun from reaching the earth. A depletion of the ozone layer which would result from reactions with freon could lead to an increase in skin cancer and other undesirable effects of exposure to ultraviolet radiation. The federal government has banned the use of freons in aerosol propellants.

Other fluorinated hydrocarbons have been used as blood substitutes. Colloidal dispersions of perfluorinated hydrocarbons in water act in this way. The description *perfluorinated* means that all H's in a compound are replaced by F's, so that the compound is actually a fluorocarbon rather than a hydrocarbon. One example of a fluorocarbon blood substitute is perfluorobutyltetrahydrofuran:

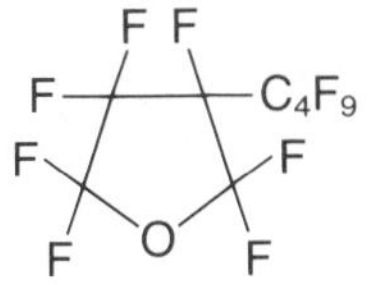
F F
F
C_4F_9
F
F
F
O
F

This compound has an ether functional group, but it is the fluorinated carbon skeleton that produces its physiological usefulness. The fluorocarbon must be dispersed in water with other ingredients to achieve miscibility with normal blood.

Fluorocarbon dispersions can dissolve almost 3 times the quantity of oxygen per unit volume as natural whole blood. "Bloodless" rats have survived with such mixtures replacing natural blood in their veins. Indeed, human beings have survived when fluorocarbons were transfused rather than normal blood. These cases occurred when patients refused human blood transfusions for religious reasons, but were willing to accept artificial blood to replace life-threatening losses.

Animals (or human beings) that have received artificial blood continue to live and produce normal blood. The artificial blood is displaced and slowly excreted from the body over a period of weeks. The solubility capabilities of fluorocarbons allow them to perform the oxygen transport function of blood. Their inertness prevents them from doing any bodily harm.

Artificial blood even has some advantages over normal blood. No blood typing is required before transfusion, and it is not susceptible to some poisons. For example, carbon monoxide can kill because it prevents hemoglobin in normal blood from transporting oxygen. CO does not interfere with oxygen transport by fluorocarbon dispersions.

In the future artificial blood will probably be used in medical emergencies, for short-term replacement for rare blood types, and as a preservation fluid for transplantable organs.

Another familiar and useful perfluorinated compound is Teflon. Teflon is a polymer of tetrafluoroethylene; the monomer and polymer are shown in Table 13.4. Its unique antistick properties arise from the same source as its physiological inertness.

Two halogenated ethanes serve as anesthetics. Ethyl chloride is a local anesthetic. It functions by "freezing" the area on which it is sprayed. The "freezing," or numbing sensation, occurs because of the high volatility of ethyl chloride which leads to rapid evaporation and a consequent heat loss from the sprayed spot (Section 8.13). You may have noticed the use of some spray in a baseball game when a player is hit by a fast-moving ball. Ethyl chloride in the spray numbs the pain. Halothane (trade name Fluothane) is a widely used inhalation general anesthetic. In contrast to many other inhalation anesthetics, it is nonflammable and nonexplosive.

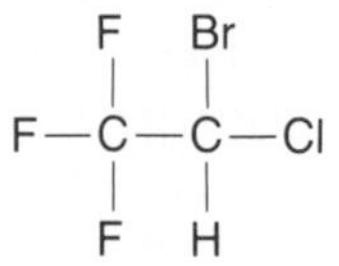

Halothane

An insecticide is a chemical substance that kills insects. Chlorinated hydrocarbons have been the most widely used insecticides; the most well known of them is DDT, which kills insects by acting on their nervous systems.

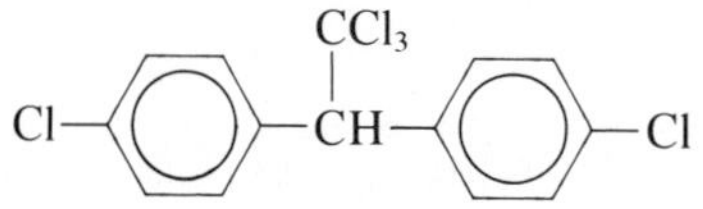

1,1′-(2,2,2-Trichloroethylidene)bis[4-chlorobenzene]
(Dichloro-diphenyl-trichloroethane)

The insecticidal properties of this compound were discovered in 1939 by the Swiss scientist Paul Mueller, and it came into wide use shortly thereafter. During World War II, DDT was used by the Allies to control malaria and typhus by destroying the insects which carry the diseases. After World War II, DDT use escalated agriculturally and domestically. Insect-carried diseases such as yellow fever (mosquito), typhus (flea), and malaria (mosquito) were eliminated from Europe and the United States and diminished in incidence throughout the rest of the world partly through the effective use of DDT. Besides its disease-control applications, DDT helped to dramatically improve crop yields.

However, in the 1960s the adverse aspects of DDT became apparent. DDT is not biodegradable, so that it persists in the environment, and it is not water-soluble, so that when it is absorbed by animals it remains and accumulates in their fatty tissue. Many other chlorinated hydrocarbons have been prepared to serve as insecticides, but all have the same negative properties as DDT.

Organophosphates and carbamates have now largely replaced the chlorinated hydrocarbons as insecticides for horticultural use. These compounds can be hydrolyzed, i.e., broken down by water. Therefore, neither of these types of compounds is as persistent in the environment as the chlorinated hydrocarbons.

CHAPTER ACCOMPLISHMENTS

After completing this chapter you should be able to define **key words** and do the following:

13.1 INTRODUCTION

1. Recognize and distinguish a hydrocarbon from other organic compounds.
2. Classify the structural formula of a given hydrocarbon as an alkane, alkene, alkyne, cycloalkane, or aromatic hydrocarbon.
3. Classify the structural formula of a given hydrocarbon as saturated or unsaturated.

13.2 SOURCE AND USE OF HYDROCARBONS

4. State the basic source of all hydrocarbons.
5. State commercial uses of hydrocarbons.

13.3 PHYSICAL PROPERTIES OF HYDROCARBONS

6. Arrange a given set of homologous compounds in order of increasing boiling point.
7. Describe the polarity and water solubility of hydrocarbons.

13.4 CHEMICAL PROPERTIES OF ALKANES

8. State the general chemical properties of an alkane.
9. Write a balanced equation for the complete combustion of a given hydrocarbon.
10. Using structural formulas, write an equation for the reaction of an alkane with halogen to yield a monohalogenation product or products.

13.5 ALKENES

11. State the name of the industrial reaction by which alkenes are formed from alkanes.

13.6 NOMENCLATURE OF ALKENES

12. Write an unambiguous name for a given alkene structure.
13. Draw the structure of an alkene, given its name.

13.7 GEOMETRIC ISOMERISM

14. Indicate whether a given alkene can display cis-trans isomerism.
15. Distinguish between structural isomers and cis-trans isomers for a given alkene.
16. Describe the relationship of cis-trans isomerism to the chemistry of vision.

13.8 ALKENE REACTIONS

17. Write the structure of the product produced when a given alkene is hydrogenated.
18. Write the structure of the product obtained when a given alkene is halogenated with either Cl_2 or Br_2.
19. State Markovnikov's rule.
20. Write the structure of the major product formed when HCl, HBr, HI, or H_2O is added to a given alkene.
21. Determine the structure of a monomer necessary to form a given polymer.

13.9 ALKYNES

22. Write an unambiguous name for a given alkyne structure.
23. Draw the structure of an alkyne, given its name.
24. Write a structure for the major product formed when either 1 or 2 mol of H_2, Cl_2, Br_2, HCl, or HBr is added to 1 mol of a given alkyne.

13.10 CYCLIC HYDROCARBONS

25. Name a given cycloalkane structure.
26. Draw a cycloalkane, given its name.

27. Explain why three- and four-membered rings are unusually reactive.
28. Explain why cyclohexane rings take on a nonplanar shape.

13.11 AROMATIC HYDROCARBONS

29. Distinguish between the terms *aliphatic* and *aromatic.*
30. Give an unambiguous name for a substituted benzene.
31. Write a structure for an aromatic hydrocarbon, given its name.
32. Draw the ortho, meta, and para isomers of a given disubstituted benzene.

13.12 BIOLOGICALLY SIGNIFICANT HYDROCARBONS

33. Give examples of the significance of hydrocarbons in the biological world.

13.13 PROPERTIES OF THE HALOGENATED HYDROCARBONS

34. State the general uses of halogenated hydrocarbons.
35. State the properties of DDT which make its use undesirable.

PROBLEMS

•13.10 Indicate whether the following compounds are saturated or unsaturated:

a $CH_3(CH_2)_5CH_3$

b $CH_3—C≡C—CH_3$

c $CH_3C(H)=C(H)CH_2CH_3$ (H H drawn above the double-bonded carbons)

d 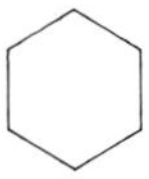

e 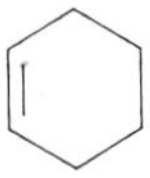

f 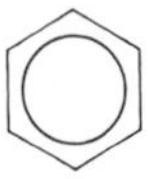

•13.11 Classify the following substances according to the scheme of Figure 13.1:

a $CH_3(CH_2)_5CH_3$

b $CH_3—C≡C—H$

c 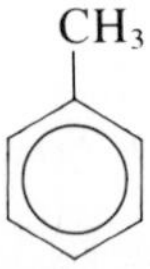

d 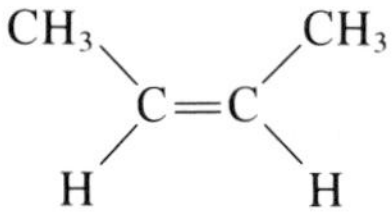

e CH_4

f 

13.12 What elements are present in a hydrocarbon?

13.13 What products might animal and vegetable matter be converted to if they were aerobically decayed?

13.14 Why are the highest-molecular-weight hydrocarbons found at the bottom of a refinery column?

13.15 Explain why the lighter-weight hydrocarbons are found at the top of a refinery column.

13.16 *a* Give two uses of alkanes as fuels.
b Give one nonfuel use of alkanes.

13.17 Arrange the following set of compounds in order of increasing boiling point:

a CH_3CH_3
b $CH_3(CH_2)_5CH_3$
c $CH_3CH_2CH_2CH_3$

13.18 Predict the compound with the highest boiling point in the following set:

a $CH_3C(CH_3)(H)CH_2CH_3$ (drawn as $CH_3CCH_2CH_3$ with CH_3 above and H below the second C)

b $CH_3CH_2CH_2CH_2CH_3$

c $CH_3—C(CH_3)_2—CH_3$ (drawn as $CH_3—C—CH_3$ with CH_3 above and below the central C)

13.19 Explain why it is possible to have an oil fire on the ocean.

13.20 How could you tell whether a colorless, odorless liquid is a water solution or a hydrocarbon?

13.21 Predict the solubility of sodium chloride in pentane. Briefly explain your answer.

13.22 Predict whether methane will be more soluble in water or in pentane.

13.23 Compare the thermodynamic stability of an alkane with that of its combustion products.

•13.24 Write a balanced chemical equation for the complete combustion of ethane, C_2H_6.

13.25 Write a balanced chemical equation for the complete combustion of the octane (C_8H_{18}) in gasoline.

13.26 Write structures for the monobromination products of the following:
a Propane, $CH_3CH_2CH_3$
b 2-Methylpropane
c 2,2-Dimethylpropane

•13.27 Give an unambiguous name for the following alkenes. Ignore cis-trans isomerism in the naming.

a $H_2C=CH_2$

b $H_2C=CHCH_2CH_3$

c $CH_3—C(CH_3)_2—CH=CHCH_3$ (drawn as $CH_3—C—CH=CHCH_3$ with CH_3 above and below the C) *d* $CH_3CH_2C(=CH_2)CH_2CH_3$

•13.28 Which of the following alkenes can display cis-trans isomerism?
a $CH_3CH{=}CCl_2$
b $CH_3CH{=}C(CH_3)Cl$
c $CH_3CH_2CH{=}CHCH_2CH_3$

13.29 Draw the structures of the cis and trans isomers for each alkene in Problem 13.28 that you identified as a candidate for cis-trans isomerism.

•13.30 Write a structure for the organic product of the following reaction:

$$CH_3CH_2Cl \xrightarrow[\text{heat}]{\text{NaOH}}$$

•13.31 Write a structure for the major organic product of each of the following reactions:
a $CH_3CH{=}CHCH_3 + Br_2 \longrightarrow$
b $CH_3CH{=}C(CH_2CH_3)CH_3 + HCl \longrightarrow$
c $CH_3CH{=}C(CH_2CH_3)CH_3 + H_2 \xrightarrow{\text{Pt}}$
d $CH_3C(CH_3){=}C(CH_3)CH_3 + H_2O \xrightarrow{H^+}$
e $CH_3CH_2C(=CH_2)CH_2CH_3 + HBr \longrightarrow$

13.32 What is the monomer unit in the following partial structure of a polymer chain?

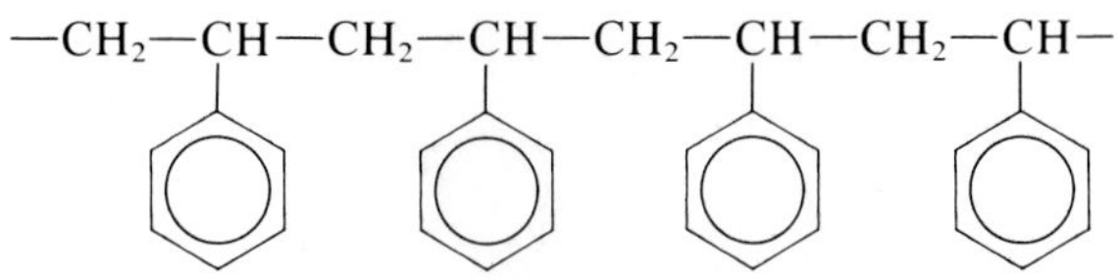

13.33 Write a structure for the major organic product of each of the following reactions:
a $CH_3C{\equiv}CH + 2HCl \longrightarrow$
b $CH_3C{\equiv}CCH_3 + H_2 \xrightarrow[\text{Pd}]{\text{poisoned}}$

13.34 Give an unambiguous name to each of the following alkynes:
a $HC{\equiv}CCH_2CH_3$
b $CH_3C{\equiv}CCH_3$
c $CH_3CH_2CH(CH_3)C{\equiv}CH$

13.35 Draw a structure corresponding to each of the following names:
a Acetylene
b 2-Pentyne
c 4,4-Dimethyl-2-pentyne

13.36 Give an unambiguous name to each of the following:

a 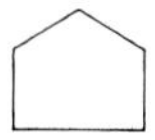

b 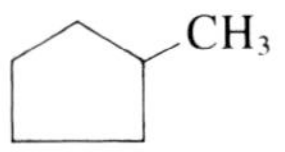

c 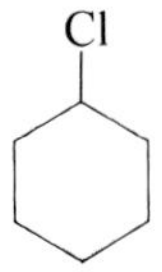

d 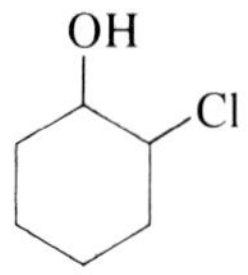

•13.37 Draw a structure corresponding to each of the following names:
a 1,1-Dimethylcyclohexane
b Cyclopentene
c 4-Bromocyclohexanol

•13.38 Indicate which of the following can show cis-trans isomerism and write structures to represent these cis-trans isomers:

a 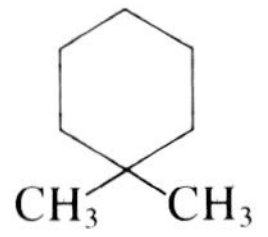

b 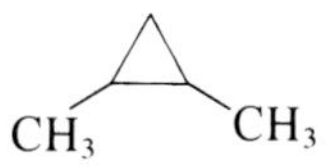

c 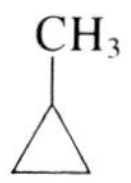

d 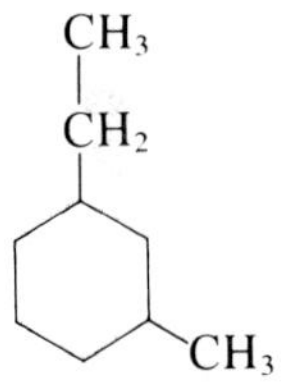

13.39 What feature in the structure of cyclopropane and cyclobutane makes them highly reactive?

13.40 *a* How many 1,2-dimethylbenzenes could exist if benzene contained three double bonds in its ring structure?
b There is only one known 1,2-dimethylbenzene. What does this say about the presence of double bonds in the benzene ring?

•13.41 *a* What type of chemical reaction should occur if benzene contained double bonds?
b What type of reaction actually does take place on the benzene ring?

13.42 Give an unambiguous name to each of the following aromatic compounds:

a 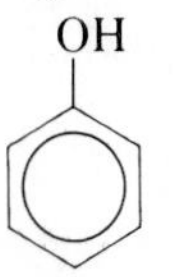

b 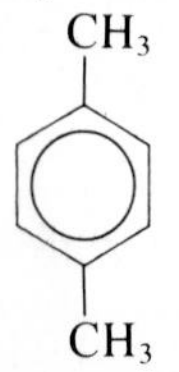

c 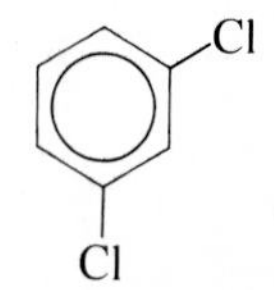

d 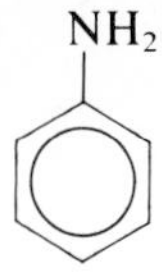 *e* 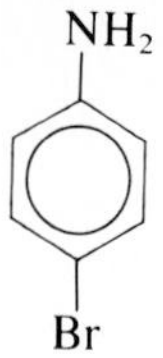

13.43 Draw a structure corresponding to each of the following names:
a *ortho*-Dimethylbenzene
b 2,4-Dibromotoluene
c *meta*-Nitrobenzoic acid

13.44 Draw and name all the isomers of dibromobenzene.

•13.45 Classify each of the following reactions as substitution, elimination, or addition:
a 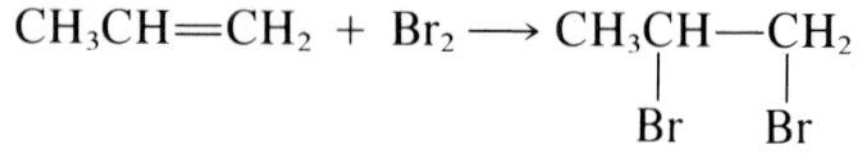
b 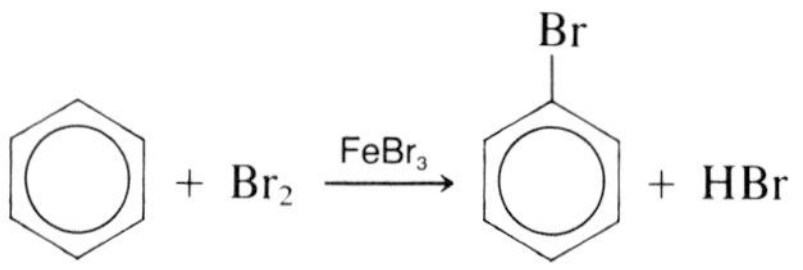
c 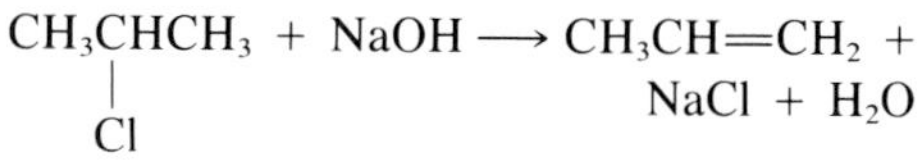
d $CH_3CH_3 + Br_2 \xrightarrow{\text{uv light}} CH_3CH_2Br + HBr$

13.46 Give three uses of hydrocarbons in the home.

13.47 Give at least one medical use of a hydrocarbon.

13.48 Describe how a pheromone may be used for insect control.

13.49 Give an unambiguous name to each of the following compounds. Indicate cis or trans where necessary to designate the isomer drawn.
a $CH_3CH_2CH_2CH_2Br$
b 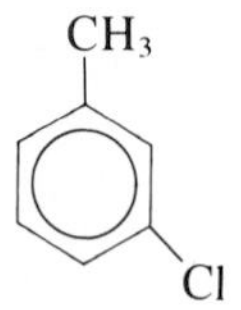
c $CH_3CH_2CH_2CH{=}CH_2$
d 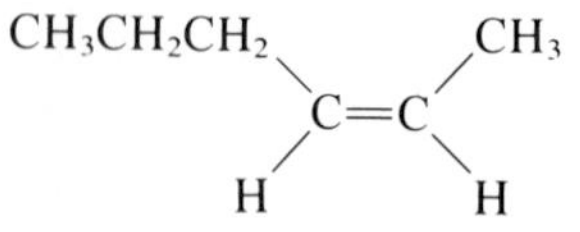
e 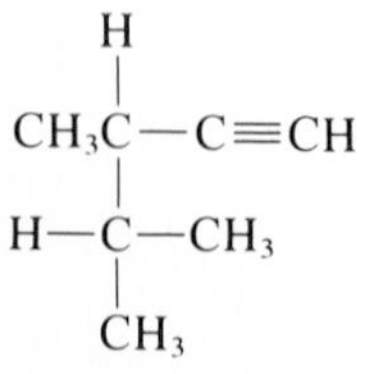
f 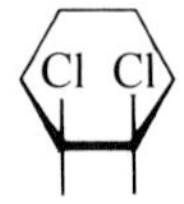
g 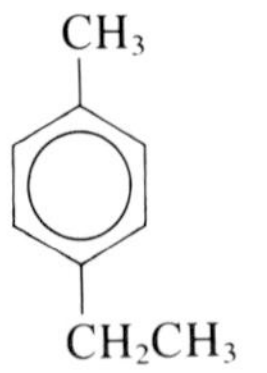
h 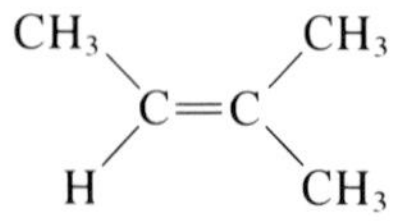

13.50 Write the structure of the major organic product formed in the following reactions:
a 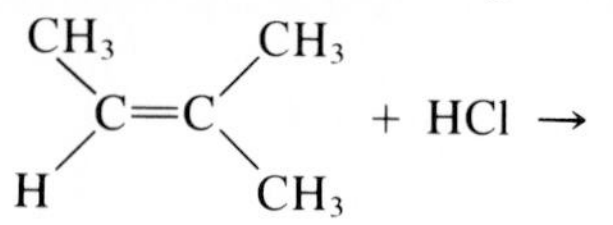
b $CH_3CH_3 + Cl_2 \xrightarrow{\text{uv light}}$
c $HC{\equiv}CCH_2CH_3 + 2H_2 \xrightarrow{\text{Pt}}$
d $CH_3CH_2CH{=}CH_2 + H_2O \xrightarrow{H^+}$
e $HC{\equiv}CCH_2CH_2CH_3 + HCl \longrightarrow$

13.51 Draw structural formulas for the following compounds:
a 2-Methylbutane
b 3-Isopropyl-2-methylhexane
c 1,4-Dimethylbenzene
d 3-Methyl-1-butene
e *trans*-1,2-Dichlorocyclopropane
f 2-Methyl-3-hexyne

13.52 Which of the following structures represent identical compounds?
a 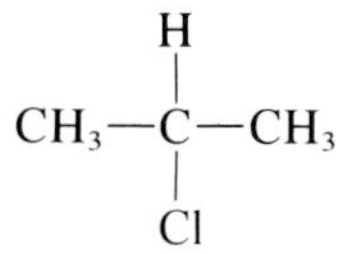
d 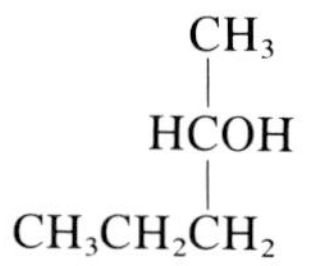
b $CH_3CH_2CH_2Cl$
c 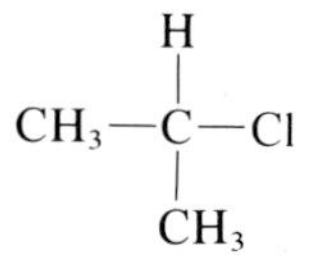
e 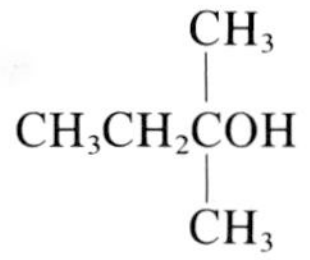
f 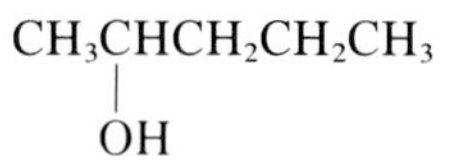

13.53 *a* Describe how you would use a bromine solution to decide whether an unknown colorless liquid is an alkene or water.
b Is there any other type of hydrocarbon that would react with bromine at room temperature?

13.54 Chloroform, $CHCl_3$, has been used as an anesthetic. Would you recommend its use today? Why or why not?

13.55 A hydrocarbon of formula C_4H_8 reacts with 1 mol of Br_2 to yield 2,3-dibromobutane. What was the structure of the original hydrocarbon?

13.56 Examine the following structures:

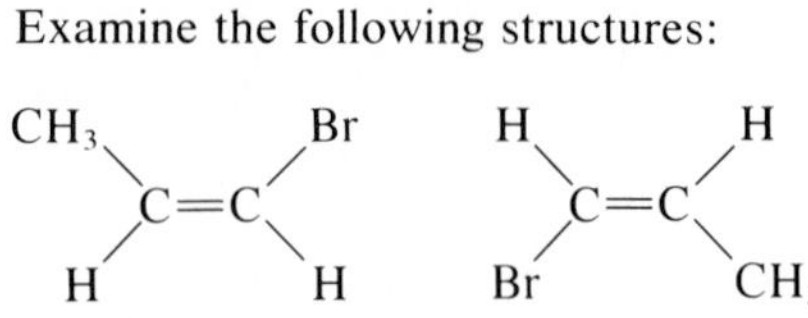

1 2

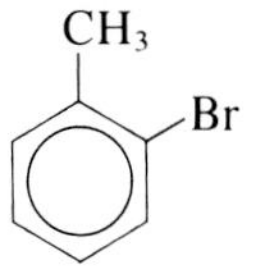

3

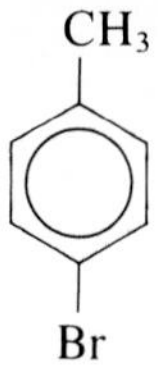

4

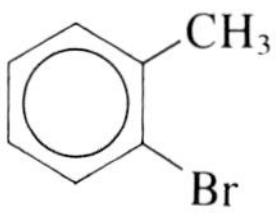

5

6

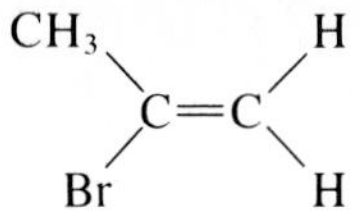

7

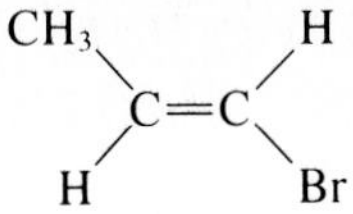

8

a Which structure(s), if any, is (are) identical to structure 1?

b Which structure(s), if any, is (are) a structural isomer of structure 1?

c Which structure(s), if any, is (are) a geometric isomer of structure 1?

d Which structure(s), if any, is (are) identical with structure 3?

e Which structure(s), if any, is (are) an isomer of structure 3?

CHAPTER FOURTEEN

ALCOHOLS AND THEIR OXIDATION PRODUCTS

14.1 INTRODUCTION

Hydrocarbons are rather dull chemically, especially the saturated hydrocarbons. The hydrocarbon chain is often regarded as the skeleton to carry around the interesting, active part of the molecule, the functional group. Physiological activity, as well as chemical interest, is usually associated with functional groups.

If you review the functional groups which you first encountered in Table 12.2, you'll realize that these fragments center around elements other than carbon or hydrogen. Most often oxygen is part of a functional group. In this chapter we will consider alcohols, phenols, ethers, and the oxidation products of alcohols, the aldehydes and ketones. These compounds all have just one oxygen in the functional group.

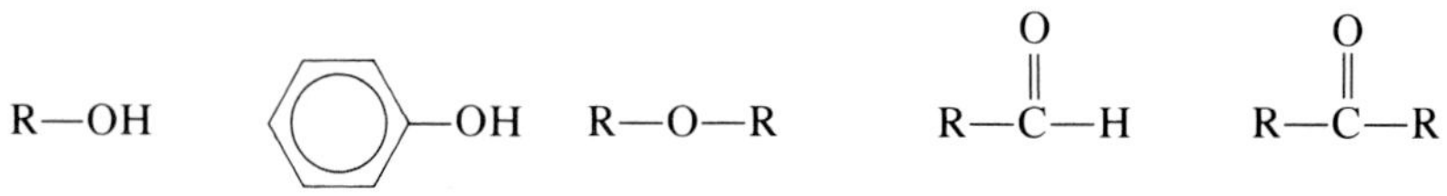

Alcohol — Phenol — Ether — Aldehyde — Ketone

We will also discuss thiols, R—SH, sulfur analogs of alcohols; S replaces O in the functional group.

Alcohols are the most prevalent of these compounds, and the most important to chemists because they are frequently starting materials for synthesizing other compounds. With the exception of phenols, alcohols can be converted into all of the other functional-group categories discussed in this chapter. Several alcohols are probably familiar to you: grain alcohol, wood alcohol, rubbing alcohol (Section 14.5).

Ethers are anesthetics and solvents, phenols have germicidal properties, and aldehydes and ketones occur in spices and perfumes and as products of physiological reactions. Let us begin our discussion of these varied but related compounds with the largest class, the alcohols.

14.2 THE ALCOHOL FUNCTIONAL GROUP

As you know, alcohols are distinguished by the —OH group (hydroxyl) covalently bonded to a saturated carbon atom. The covalently bonded organic alcohol group (—OH) behaves very differently from the hydroxide ion (OH^-), as you will see in the study of reactions of alcohols in Sections 14.6 and 14.7.

Alcohols can be thought of as derivatives of water in which one of water's hydrogens has been replaced by an alkyl group. Alcohol molecules and water molecules have the same shape around the oxygen atom because in both cases the oxygen atom is bonded to two other groups and has two nonbonded pairs of electrons. This leads to a tetrahedral electron-pair geometry (Section 6.10) and a bent molecular shape.

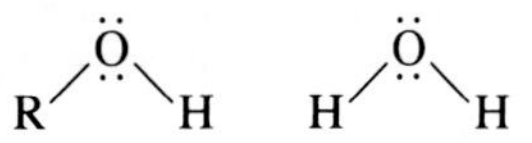

Alcohols and water also share the ability to engage in hydrogen bonding, as will be discussed in more detail in the next section.

The alcohols can be divided into three classes depending on how many alkyl groups are connected to the carbon to which the hydroxyl group is bonded.

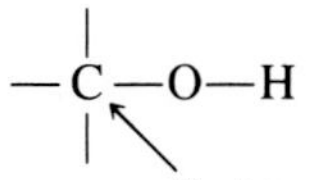

To determine alcohol class, count the number of alkyl groups bonded to this carbon.

If one or no (only methanol) alkyl group is bonded to the —OH-bearing carbon, the alcohol is called *primary*, symbolized 1°. If two alkyl groups are directly attached to the specified carbon, the alcohol is *secondary* (2°), and if three alkyl groups are attached, it is a *tertiary* alcohol (3°). You will find it helpful to be able to recognize alcohols as 1°, 2°, or 3°, particularly when we discuss the oxidation reactions of alcohols.

SAMPLE EXERCISE 14.1

Classify each alcohol below as primary, secondary, or tertiary.

a $CH_3CH_2CHOHCH_3$

b 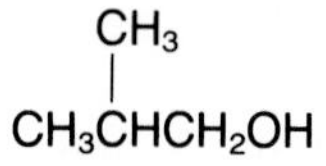

c CH_3OH

d 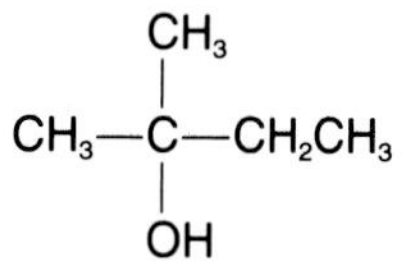

Solution

a *Secondary*—the —OH-bearing carbon is bonded to two alkyl groups. Write out the full structure to see that this is so.

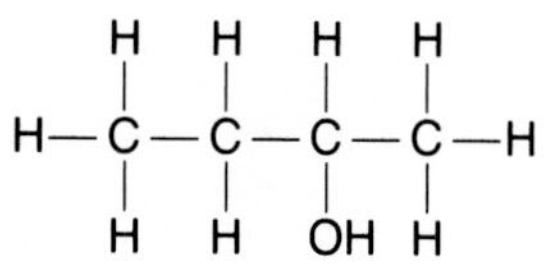

b *Primary*—the functional-group carbon is bonded to only one alkyl group. The full structure shows this better.

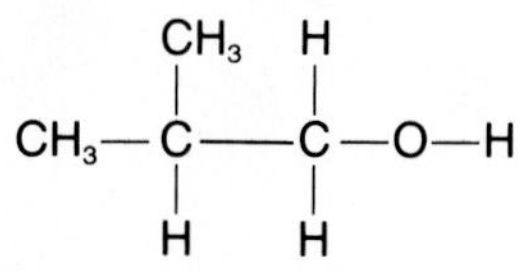

c *Primary*—there is only one carbon in methanol, and in its oxidation reaction methanol behaves as a primary alcohol.

d *Tertiary*—the carbon to which —OH is attached is also connected to three alkyl groups.

14.3 HYDROGEN BONDING IN ALCOHOLS

Alcohols, like water, have a polar O—H bond in which the shared electron pair is shifted closer to the oxygen, and thus the oxygen acquires a partial negative charge (δ^-). The hydrogen acquires a partial positive charge (δ^+) because electrons have moved away from it. Section 6.8 discusses polarity. This δ^+ hydrogen is attracted to one of the nonbonding electron pairs of another oxygen in a separate molecule (Figure 14.1). This weak attractive force is known as hydrogen bonding, as we saw for water in Section 8.10. Hydrogen bonding exists between separate alcohol molecules and holds them together in the liquid state.

Hydrogen bonding has a marked effect on the physical properties of alcohols. Energy in the form of heat must be added to an alcohol in the liquid state in order to break down the intermolecular hydrogen bonds so that the alcohol molecules can independently enter the gas phase. Thus intermolecular hydrogen bonding leads to a higher boiling point for alcohols than for compounds, such as the hydrocarbons, which cannot engage in hydrogen bonding. For example, although ethanol (molecular weight 46) and propane (molecular weight 44) have similar molecular weights (and thus the dispersion forces between their molecules are very similar), the boiling point of ethanol is 79°C whereas that of propane is −42°C.

When an alcohol is mixed with water, the alcohol molecules can hydrogen-bond with water molecules, as was shown in Figure 9.5*a*.

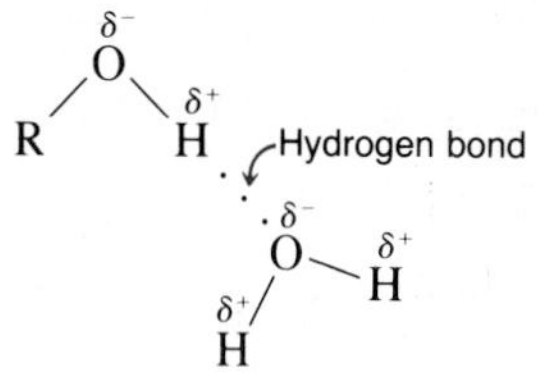
Hydrogen bond

In order for an alcohol to dissolve in water, the alcohol molecules must be able to get between the water molecules. This interpenetration disrupts the hydrogen bonding between water molecules. However, the broken water-

FIG. 14.1
Hydrogen bonding between alcohol molecules. This is the intermolecular hydrogen bond between —OH groups of different molecules.

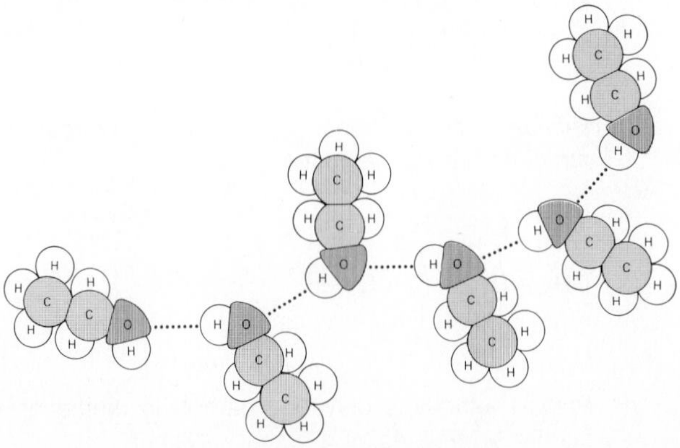

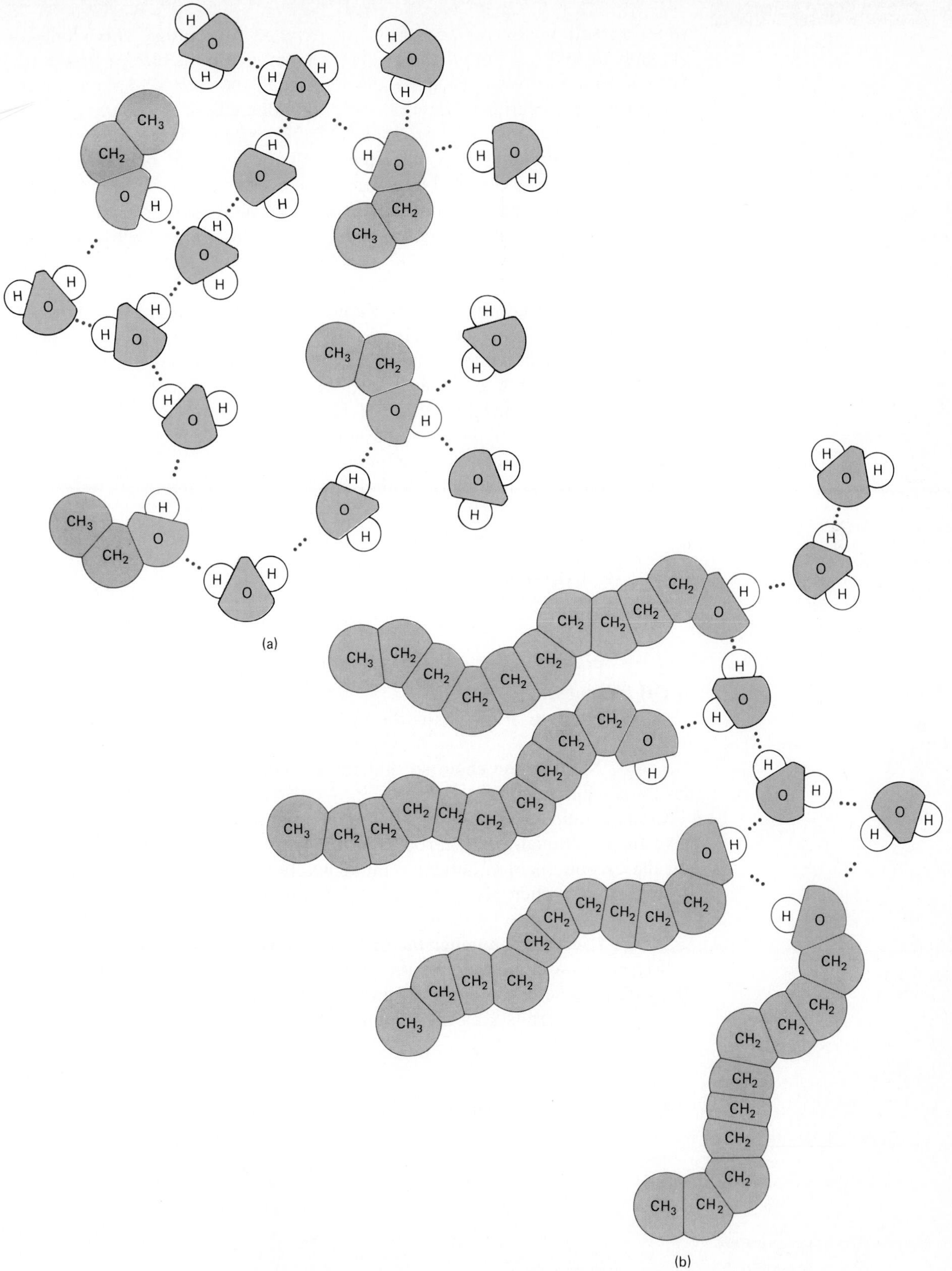

(a)
(b)

FIG. 14.2
(*a*) The carbon chain in ethanol is sufficiently short to allow effective mingling of alcohol and water molecules and hydrogen bonding between them. Thus, ethanol is soluble in water. (*b*) Water molecules are unable to surround molecules of decanol because of the bulky carbon chain. Thus, this alcohol is not water-soluble.

water attractions are replaced by alcohol-water attractions. This ability of alcohols to hydrogen-bond with water is largely responsible for their solubility in aqueous media. However, as the size of the carbon chain increases in a homologous series of alcohols, solubility decreases. As you first saw in Section 12.2, alcohols with six or more carbons are insoluble in water. The large alkyl chain physically interferes with the close fit needed between alcohol and water molecules in order to achieve the hydrogen bond (Figure 14.2). In general, as the alkyl chain of an alcohol increases in size, the physical properties of the alcohol become more like that of a hydrocarbon and less alcohol-like.

14.4 NOMENCLATURE OF ALCOHOLS

The stepwise procedure for nomenclature is basically the same for all organic compounds. This procedure was first discussed in Section 12.14. The major difference between names is the suffix, which indicates the functional group in the molecule. For example, alcohols are named based on the longest continuous chain which bears the —OH group. They are named as alkanols, that is, *-ol* replaces *-e* of the corresponding alkane chain name; an alcohol with a five-carbon chain is a pentanol.

Let us name the following alcohol in a step-by-step manner:

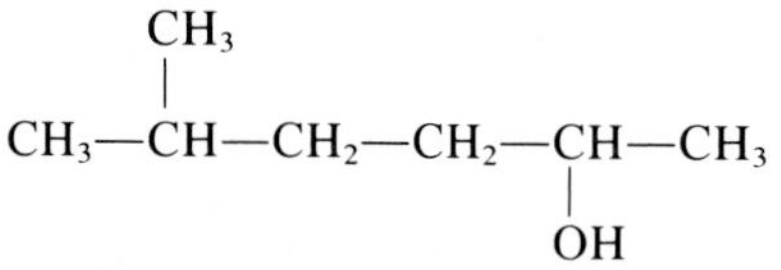

STEP 1 **Find the longest continuous chain that contains the carbon to which —OH is bonded.**

In this example, it is a six-carbon chain. This is a hexanol.

STEP 2 **Number the chain so that the carbon bonded to OH is given the lowest possible number.**

In the example the chain must be numbered from right to left in order to give the functional-group carbon the number 2. The position of the —OH on the carbon chain is indicated immediately before the parent name with a separating hyphen. The example is 2-hexanol.

STEP 3 **Locate and number the branched alkyl groups on the main chain.**

The names and numbers of the alkyl groups are then written in front of the parent name of the alcohol following the same rules as for the alkanes. In this case, there is a methyl group at the carbon in position 5:

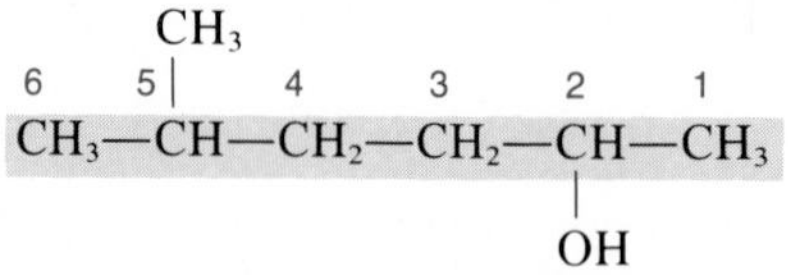

The complete name of the example is 5-methyl-2-hexanol.

SAMPLE EXERCISE 14.2

Give the IUPAC name for

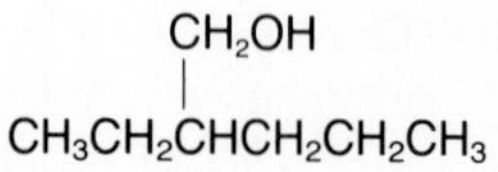

Solution The longest chain containing the —OH group is five carbons. (Note that there is a six-carbon chain but it does *not* contain the —OH group.)

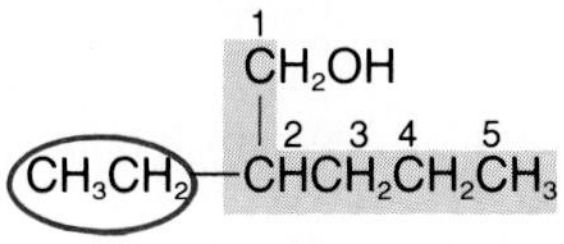

The parent name is 1-pentanol. An ethyl group branches at the second carbon of the chain. The complete name is 2-ethyl-1-pentanol.

Problem 14.1 Give IUPAC names for the following:

a 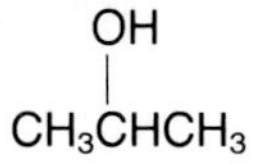

b 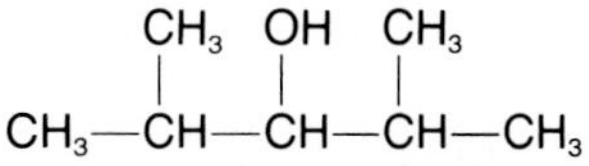

There may be more than one hydroxyl group attached to a carbon chain. *Diols* are compounds with two (*di-*) —OH (*-ol*) groups; triols have three hydroxyls. Compounds with two —OH groups attached to the *same* carbon are usually unstable and will not be discussed here. Compounds with two hydroxyl groups on different carbons are stable and are named in the IUPAC system using the rules for simple alcohols except that the suffix is *-diol,* rather than *-ol,* and this suffix is attached to the alkane chain name. Two numbers, separated by a comma, must be used as a prefix to indicate the positions of the —OH groups on the longest continuous chain. Triols are named in a similar way. Study the examples below:

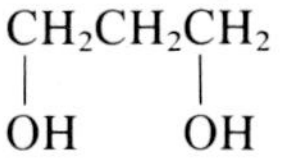

1,3-Propanediol

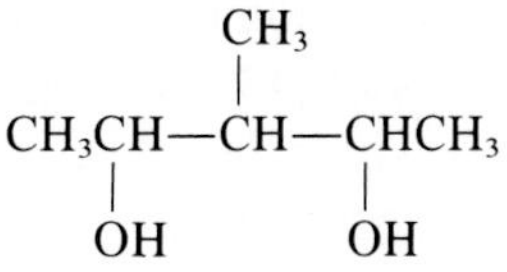

3-Methyl-2,4-pentanediol

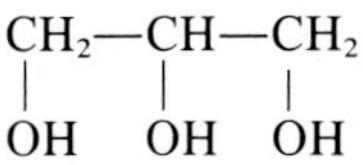

1,2,3-Propanetriol
(glycerol)

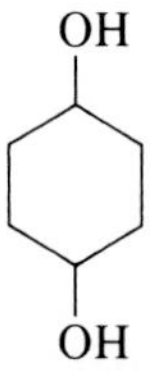

1,4-Cyclohexanediol

Common names were introduced in Section 12.15. Alcohols can be named as combinations of alkyl group names and the word *alcohol* as the name of the —OH group. Thus,

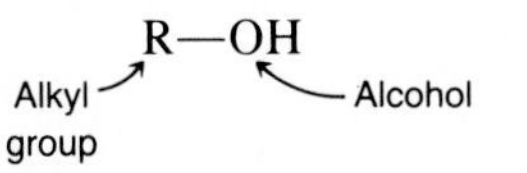
R—OH
Alkyl group
Alcohol

CH_3—OH Methyl alcohol

CH_3CH_2—OH Ethyl alcohol

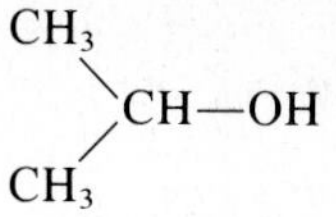 Isopropyl alcohol

This common naming system is used only when the alkyl groups are small and easily named.

There are some common names for diols and triols worth knowing. One is ethylene glycol for 1,2-ethanediol $\left(\begin{array}{cc} OH & OH \\ | & | \\ CH_2 & CH_2 \end{array}\right)$. Ethylene glycol is the main component in antifreeze solutions in car radiators. The method of action of antifreeze solutions was explained in Section 9.11. Prestone and Xerex antifreeze solutions are essentially ethylene glycol with different colored additives. The compound 1,2,3-propanetriol $\left(\begin{array}{ccc} OH & OH & OH \\ | & | & | \\ CH_2 & CH & CH_2 \end{array}\right)$ is commonly named glycerol (or glycerine). Glycerol is formed as a digestion product of fats and oils and is used as a moisture-retaining ingredient in hand lotions.

14.5 INDUSTRIAL SOURCE AND USE OF SOME ALCOHOLS

Alcohols are important industrially, both in their own right and as raw materials for the synthesis of other organic compounds. Alcohols also frequently have medicinal uses. Some of the major industrial volume alcohols are discussed in the following.

Methanol

Methanol, also called wood alcohol because of its original manufacture from the destructive distillation of wood, is commerically prepared from the reaction of carbon monoxide and hydrogen gases passed over a metal catalyst at elevated temperatures:

$$CO + 2H_2 \xrightarrow[\text{heat}]{\text{catalyst}} CH_3OH$$

The combination CO and H_2 is known as water gas because it is formed in the reaction of hot coke (from coal) and steam:

$$C + H_2O \longrightarrow CO + H_2$$

Methanol is important as a raw material in the manufacture of plastics and as an industrial solvent. It is very useful to be able to prepare it from coal, which is in abundant supply in the United States, rather than from petroleum, which has become very expensive as domestic supplies dwindle.

Methanol is highly toxic. Its acute effects range from headache, fatigue, nausea, and visual impairment to complete and permanent blindness, acidosis, coma, and death. Death can result from ingestion of as little as 30 mL; the typical fatal dose is 100 to 250 mL.

Ethanol

Ethanol or ethyl alcohol, is the alcohol found in all alcoholic beverages. The ethanol in these beverages is produced by the action of yeast on sugars found in barley and hops (beer), grapes (wine), and grains such as rye and corn (whiskey). In the chemical process known as fermentation, the biological catalyst or the enzyme called zymase, found in yeast, catalyzes the conversion of the sugar glucose into ethanol and carbon dioxide.

$$\text{Complex sugars} \xrightarrow{\text{hydrolysis}} C_6H_{12}O_6 \xrightarrow{\text{zymase}} 2CH_3CH_2OH + 2CO_2 \qquad (14.1)$$

Fermentation ceases when the alcohol concentration reaches about 16 percent by volume because the alcohol interferes with the ability of the enzyme to catalyze the reaction. Spirits such as whiskey, vodka, and brandy, which usually have alcohol concentrations of 40 to 50 percent, are produced by distilling the original fermentation mixture. Because the boiling point of ethanol (79°C) is lower than that of water, the distillation produces a solution which is higher in alcohol concentration.

Alcohol concentrations on commercial spirit bottles are given by the term *proof,* which is twice the percentage of alcohol by volume. Thus, a Scotch whiskey labeled 86 proof has an ethanol concentration of 43 percent. The individual tastes of wines and spirits arise from the small concentrations of higher-chain alcohols, aldehydes, organic acids, and esters that form in the fermentation process and during the aging period of the particular wine or spirit.

Internal consumption of ethyl alcohol at low concentration leads to increased respiration and a feeling of mild euphoria. However, continued intake and higher concentrations result in a general depression of the central nervous system and a reduction in mental and motor abilities, drowsiness, and eventually, unconsciousness. Because it is rapidly absorbed from the gastrointestinal tract and therefore available as a quick source of energy (7 kcal/g), ethanol is sometimes used to overcome the effects of shock.

Much industrial production of ethanol (and other alcohols) for nonconsumable purposes, is produced by the hydration of alkenes (Section 13.8).

$$\underset{\text{Ethylene}}{CH_2{=}CH_2} + H_2O \xrightarrow{H_2SO_4} \underset{\text{Ethanol}}{CH_3CH_2OH} \qquad (14.2)$$

Besides its most common presence in alcoholic beverages, ethanol is widely found in most homes as a solvent for pharmaceuticals (tinctures), colognes, perfumes, mouthwashes, and hairsprays. A 70 percent by volume solution of ethanol in water is used as a local antiseptic. This solution penetrates the cell walls of microorganisms and kills them; you will see the molecular basis of this action in Section 19.15.

2-Propanol

Isopropyl alcohol is commercially prepared by the addition of water to propene.

$$\underset{\text{Propene}}{CH_3{-}CH{=}CH_2} + H_2O \xrightarrow{H^+} \underset{\substack{\text{2-Propanol} \\ \text{(isopropyl alcohol)}}}{CH_3{-}\underset{\displaystyle OH}{\underset{|}{CH}}{-}CH_3} \qquad (14.3)$$

It is also known as rubbing alcohol, which is used in sponge baths and as an external astringent. It is toxic if taken internally, but it is not absorbed through the skin. Because of its toxic character and similarity in physical and chemical properties to ethanol, 2-propanol can be added to ethyl alcohol to render it unfit for internal consumption. This is known as "denaturing" the ethyl alcohol. Although unfit for beverage use, the denatured ethyl alcohol is usable for industrial purposes and is sold on a tax-free basis for that use only.

Glycerol

Glycerol, or glycerine [$CH_2(OH)CH(OH)CH_2OH$], is a triol which is obtained from fats and oils. It is a syrupy liquid which absorbs moisture from the air, and it helps to maintain moisture when applied to the skin. It is used for this purpose in moisturizing lotions. Also, most rectal suppositories contain glycerine as the moisturizing, lubricating ingredient.

14.6 DEHYDRATION OF ALCOHOLS

Now that you are familiar with the sources and uses of some common alcohols, we can move to a discussion of the reactions of alcohols. Two important reactions of alcohols, dehydration and oxidation, have relevant applications in physiological chemistry. We will consider dehydration first.

1- methyl (Methane)
2 - di
3. - pro
4. - but
5. pent
6. - hex

The prefix *de-* before a word means "to remove or take away," so that the word *dehydration* means to remove "hydration," that is, to remove water. When an alcohol is heated in the presence of sulfuric acid, the components of water (H and OH) are removed to form an alkene and water. This loss of water occurs from within one alcohol molecule and is therefore known as an *intra*molecular (*within* one molecule) dehydration. Dehydration can also occur between *two* alcohol molecules, in which case an ether and water are the products. The ether-yielding reaction is termed *inter*molecular (*between* two molecules) dehydration. Let us take a closer look at each of these dehydrations.

Intramolecular Dehydration

Intramolecular dehydration is carried out at a relatively high temperature (180°C for 1° alcohols) in the presence of an acid catalyst (usually H_2SO_4 or H_3PO_4). This reaction is an example of an elimination reaction. In an elimination reaction a small molecule, in this case water, is removed from a single reactant molecule in such a way as to introduce a carbon-carbon double bond into the carbon chain. The H and OH of water are removed from adjacent carbons of the reactant molecule. For many alcohols there is more than one way in which this can be accomplished, and therefore the double bond may be introduced in different positions. The major product, in such cases, is found to be the compound in which the double bond has the greater number of alkyl-group substituents and the lesser number of hydrogens.

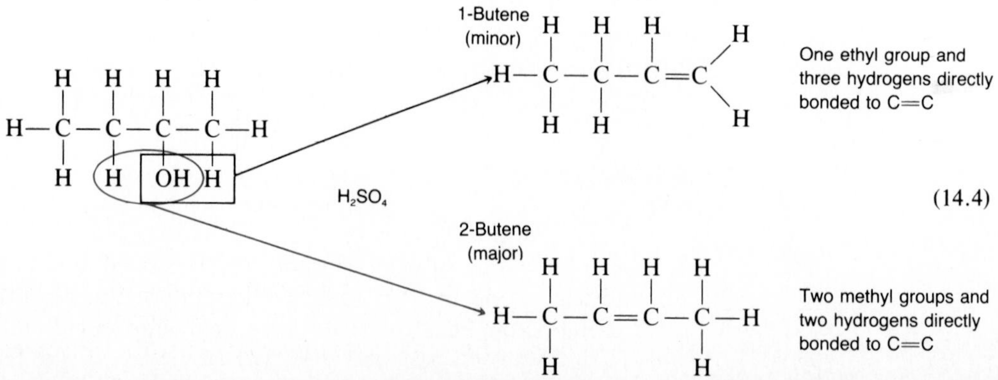
1-Butene
(minor)
One ethyl group and three hydrogens directly bonded to C=C
H₂SO₄
2-Butene
(major)
Two methyl groups and two hydrogens directly bonded to C=C

(14.4)

You may find it useful to have a set of guidelines for writing the structure of the alkene formed in a dehydration reaction.

Guidelines for Intramolecular Dehydration

1 Remove H and OH from adjacent carbons and form a carbon-carbon double bond between these adjacent carbons.

2 If there is more than one hydrogen-bearing carbon neighboring the hydroxyl-bonded carbon, remove the hydrogen from the carbon bearing the lesser number of hydrogens. This rule should also be followed for the elimination of HX from alkyl halides, which was discussed in Section 13.5. Chemists often refer to this guideline as Saytzeff's rule.

SAMPLE EXERCISE 14.3

Write the structure of the *major* product formed in the following dehydrations:

a 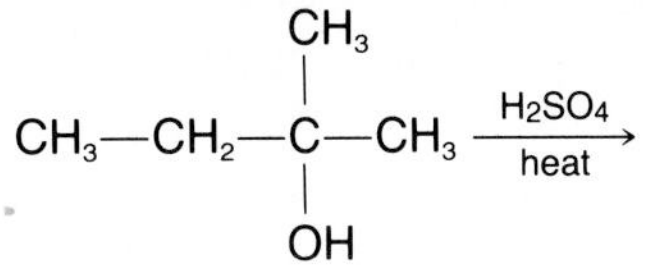
CH3
CH3—CH2—C—CH3
OH
H2SO4
heat

b 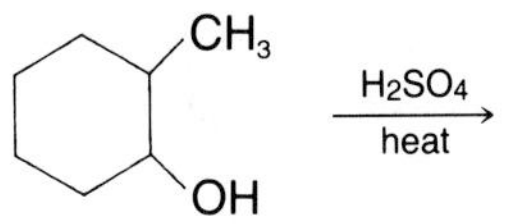
CH3
OH
H2SO4
heat

Solution *a* 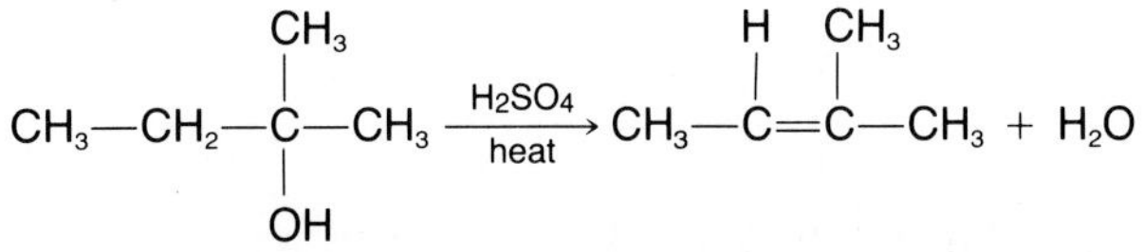
CH3
CH3—CH2—C—CH3
OH
H2SO4
heat
H CH3
CH3—C=C—CH3 + H2O

Loss of H from the $—CH_2$ group rather than from the $CH_3—$ group leads to the more highly substituted double bond.

b 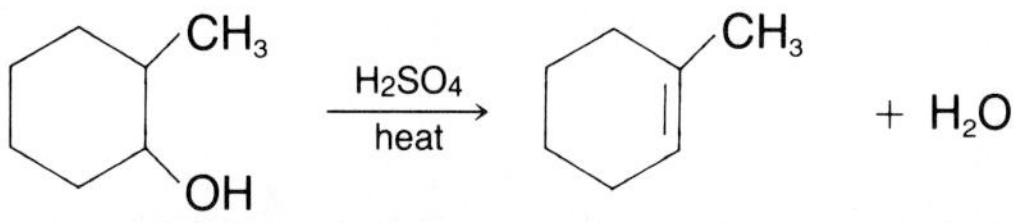
CH3
OH
H2SO4
heat
CH3
+ H2O

You may need to fill in the hydrogens on the cyclohexane ring carbons to see this solution.

Problem 14.2 Write the structure of the major product formed in the reactions of the following with H_2SO_4 in the presence of heat:

a $CH_3CH_2CH_2OH \longrightarrow$

b 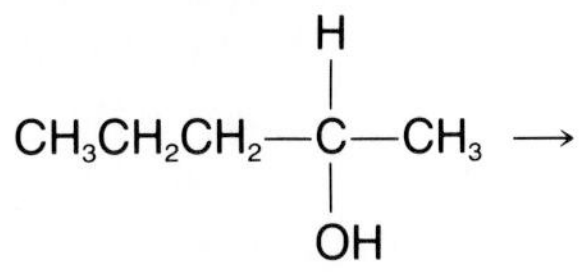
H
CH3CH2CH2—C—CH3
OH

c 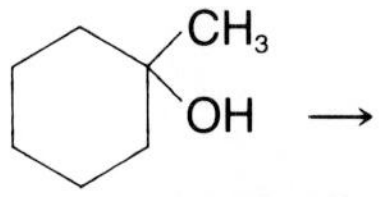
CH3
OH

The dehydration of alcohols to form alkenes is the exact reverse of the hydration reaction of alkenes which you saw in Section 13.8. However, most organic reactions do not go to completion, but rather are reversible equilibria.

Following Le Chatelier's principle (Section 10.12), the position of equilibrium can be shifted by varying the concentrations of reactants and products. To form the alkene in a dehydration reaction, we should remove the water as it forms; to form the alcohol in a hydration reaction, a large excess of water is the preferred condition.

$$CH_2{=}CH_2 + H_2O \underset{\text{dehydration}}{\overset{\text{hydration}}{\rightleftharpoons}} CH_3CH_2OH \qquad (14.5)$$

Intermolecular Dehydration

At a lower temperature than that required for the elimination reaction, two alcohol molecules can react together and split out a water molecule between them. The product of this reaction is an ether. A reaction in which two molecules combine together by splitting out a small molecule is known as **condensation.** The intermolecular dehydration of two alcohol molecules is an example of a condensation reaction.

$$2CH_3CH_2OH \xrightarrow[140^\circ]{H_2SO_4} CH_3CH_2{-}O{-}CH_2CH_3 + H_2O \qquad (14.6)$$

If we write out the two reactant alcohol molecules so that the two OH groups "face one another," you can see that HOH is lost as H from one molecule and OH from the other.

$$CH_3CH_2O\boxed{H + HO}CH_2CH_3 \xrightarrow[140^\circ]{H_2SO_4} CH_3CH_2{-}O{-}CH_2CH_3 \qquad (14.7)$$

The remaining oxygen bonds to what was the hydroxyl carbon of the molecule that lost OH. The ether-yielding dehydration is signaled by the presence of two molecules of alcohol as reactants, whereas the dehydration leading to an alkene involves only a *single* alcohol molecule as reactant.

The condensation of alcohols is sometimes useful in preparing ethers, but does not find simple application in physiological chemistry.

14.7 OXIDATION OF ALCOHOLS

Besides dehydration reactions, alcohols also undergo oxidation. You first encountered oxidation reactions in Section 7.12. Oxidation reactions in organic and biochemistry can be viewed most simply as the loss of hydrogen or gain of oxygen by the organic reactant. Primary alcohols can be oxidized to form aldehydes; secondary alcohols form ketones; tertiary alcohols do not react by oxidation under normal conditions. The common oxidizing agents that are used in this reaction are $K_2Cr_2O_7/H_2SO_4$, hot $KMnO_4$, or CrO_3-pyridine. The last reagent is specific for the conversion of 1° alcohols to aldehydes. However, because the chemistry of these inorganic oxidizing agents is quite complex and not essential to understanding the alcohol reactivity, we shall indicate oxidizing conditions in laboratory reactions by the symbol [O].

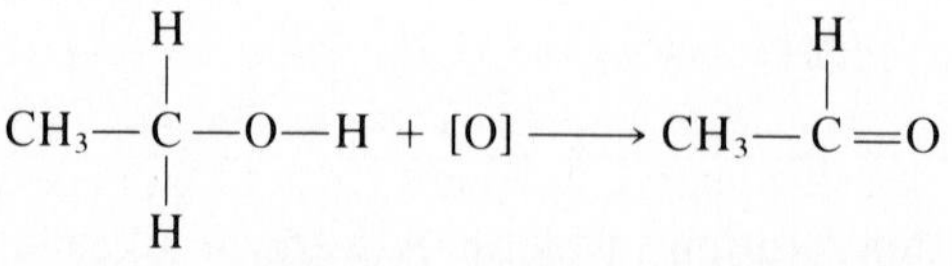
H
CH3—C—O—H + [O] ——→ CH3—C=O
H
H

(14.8)

Primary alcohol Aldehyde

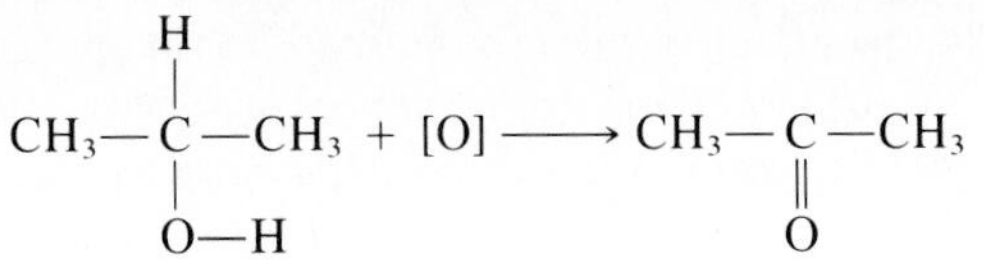
(14.9)

Secondary alcohol — Ketone

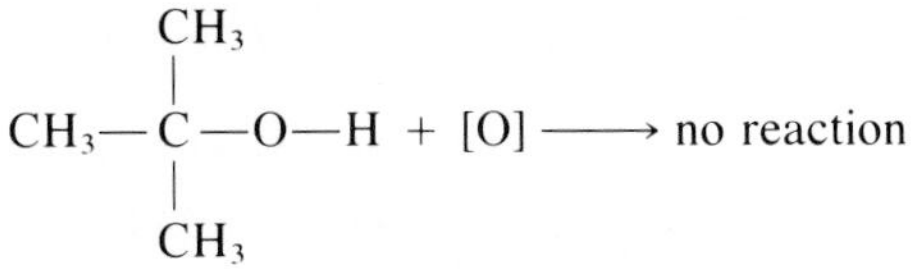

Tertiary alcohol

To write the structure for the oxidation product of any alcohol, remove the H from the O and the H from the C connected to the O and form a double bond between the C and O. Note the bond changes below:

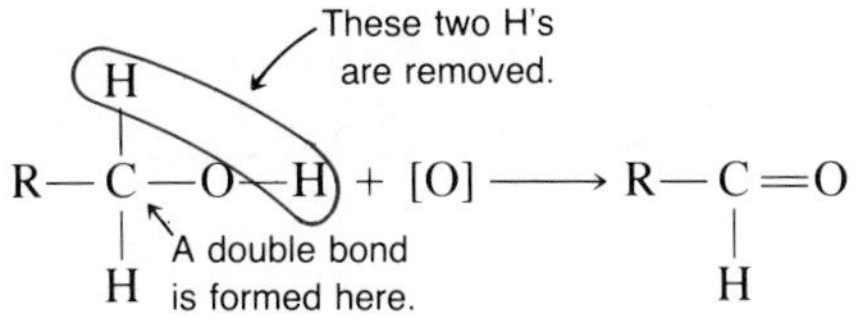
These two H's are removed.
A double bond is formed here.

The conversion from alcohol to aldehyde or ketone can be seen as an oxidation reaction because hydrogen is lost in the reaction. Furthermore, since H must be lost from both O and C, you can see that a tertiary alcohol does not give an oxidation product because there is no H on the C connected to O.

Problem 14.3 Write the structure for the product of each reaction below. If no reaction occurs, write NR.

a 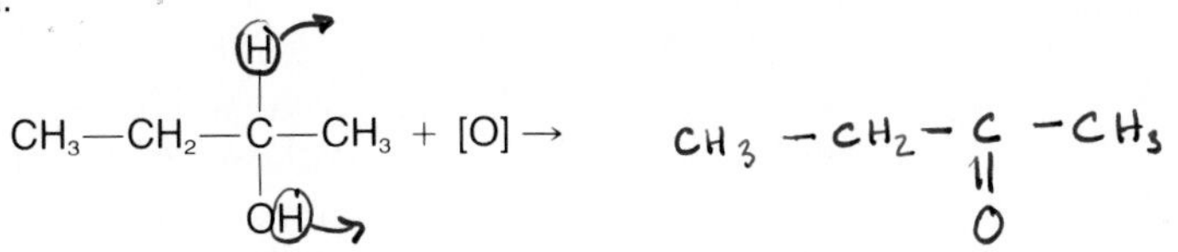

b $CH_3CH_2CH_2CH_2OH + [O] \rightarrow$ 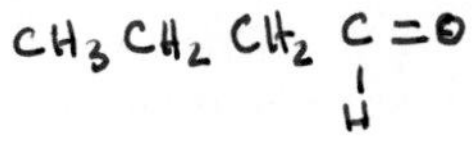

c 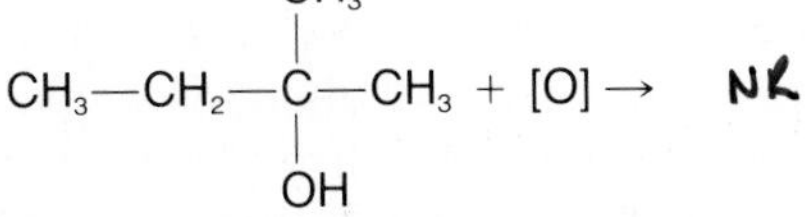

There are many examples in the physiological breakdown of foodstuffs in which an alcohol undergoes oxidation. Indeed, the first step in the breakdown of ethyl alcohol is its oxidation in the liver to an aldehyde called acetaldehyde. Sustained concentrations of acetaldehyde, from the consumption of large amounts of alcohol, raise the blood pressure, cause headaches, and cause a general feeling of malaise, in short, a *hangover.*

The toxicity of methyl alcohol is actually due to its oxidation product, formaldehyde. Formaldehyde is highly injurious to proteins and upsets their role as catalytic enzymes. Furthermore, the presence of formaldehyde has a degenerating effect on the optic nerve. Ethanol is an immediate antidote for the accidental ingestion of methanol because the key oxidation step of both alcohols involves the same enzyme. The enzyme catalyzes the oxidation of ethanol more effectively than that of methanol, allowing the unreacted methanol to be excreted in the urine. See also Section 21.8.

14.8 ETHERS

You have seen that ethers can be prepared by the intermolecular dehydration of alcohols. We will now briefly look at this class of compounds, some members of which have served as important anesthetics.

Ethers are organic compounds that contain the —C—O—C— functional group. Many compounds of biochemical interest have the ether linkage in a three-membered ring known as an epoxide.

$CH_3CH_2—O—CH_2CH_3$

Typical open-chain ether

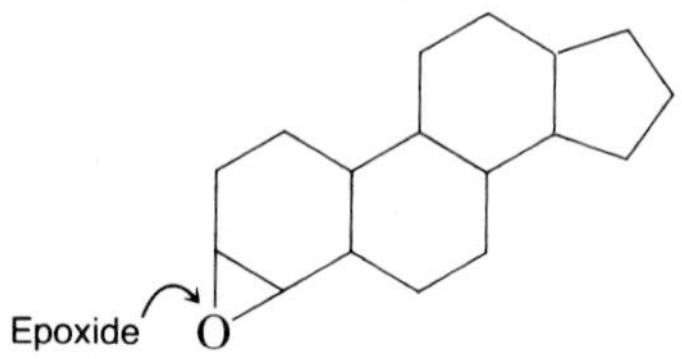
Epoxide
O

Ether names are commonly formed by naming the two alkyl or aromatic groups attached to the oxygen. Look at the examples below:

$CH_3CH_2—O—CH_2CH_3$

Diethyl ether (often shortened to ethyl ether)

$CH_3—O—CH_3$

Dimethyl ether

$CH_3CH_2—O—CH_3$

Ethyl methyl ether

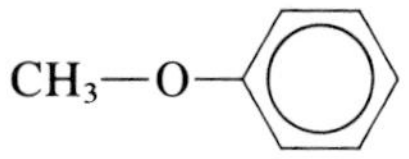
CH3—O—

Methyl phenyl ether (also known as anisole)

Ethers lack the required type of hydrogen (O—H, N—H, or F—H) for hydrogen bonding, and consequently, as you can see from Table 14.1, the boiling points of ethers are comparable to those of alkanes of similar molecular weight rather than to those of alcohols. However, ethers, like alcohols, do have a nonbonding pair of electrons on an oxygen and can act as a hydrogen-bond acceptor with water molecules, as shown in Figure 14.3. This ether-water attraction can replace the water-water hydrogen bonding and accounts for the point shown in Table 14.1 that ethers of six carbons or fewer are soluble or partially soluble in water.

The open-chain ethers are about as unreactive as alkanes and hence are described as being inert. But they are able to dissolve many types of organic substances and thus are important as inert solvents. In using ethers one

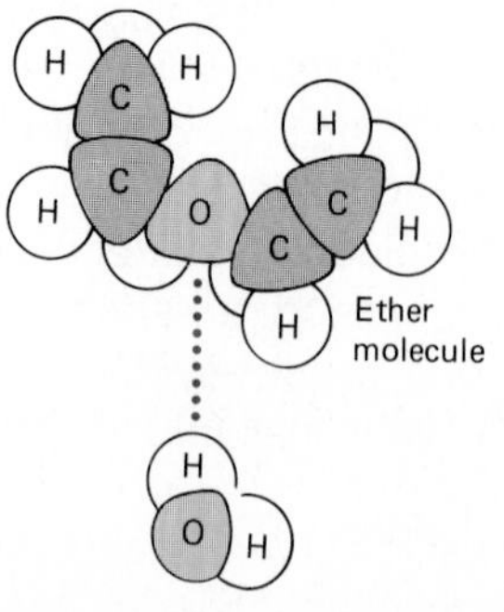
H
C
H
H
C
O
C
C
H
H
H
Ether
molecule
H
O
H

FIG. 14.3
Hydrogen bonding between an ether molecule and a water molecule. Notice that the ether molecule cannot form hydrogen bonds with itself because it does not have an H covalently bonded to oxygen. For the same reason the hydrogen of the hydrogen bond in this system must always be supplied by water.

TABLE 14.1 Comparative Boiling Points and Solubilities of Alkanes, Ethers, and Alcohols

Name	Structure	Molecular Weight	Boiling Point, °C	Water Solubility
Propane	$CH_3CH_2CH_3$	44	−42	Insoluble
Dimethyl ether	$CH_3—O—CH_3$	46	−23	Soluble
Ethyl alcohol	CH_3CH_2OH	46	78	Very soluble
Pentane	$CH_3CH_2CH_2CH_2CH_3$	72	36	Insoluble
Diethyl ether	$CH_3CH_2—O—CH_2CH_3$	74	34	Slightly soluble
Butyl alcohol	$CH_3CH_2CH_2CH_2OH$	74	117	Soluble

chemical property must always be kept in mind. They are highly flammable. This is particularly the case for the most common ether, diethyl ether.

Ether as an Anesthetic

Ethyl ether was introduced to the medical community as a general anesthetic by a dentist, Dr. William Morton, in 1846. So began the era of modern anesthesiology and "painless" surgery. It seems impossible to imagine in a medical world of open-heart surgery and organ transplants, an era only 140 years ago when there was no general anesthetic available.

Diethyl ether is a central nervous system depressant which in small doses produces lessened perception of pain (analgesia) and loss of consciousness and, in larger doses, eventually causes muscle relaxation. Although it does not appear to be harmful to other bodily functions, ether often leads to post-operative nausea and in some cases inflammation of the upper respiratory tract. Along with the always present danger of flammability, these problems have led medical and dental surgeons to search for other general anesthetics.

General anesthesia can be accomplished with the halogenated hydrocarbon halothane (see Section 13.13) or the halogenated ethers methoxyflurane (Penthrane) and enflurane (Enthrane).

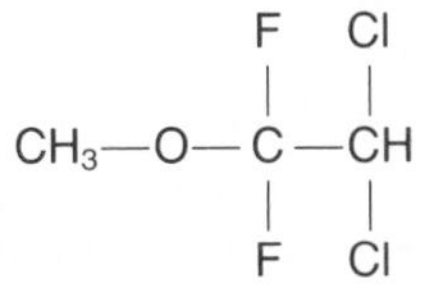
F Cl
CH3—O—C—CH
F Cl

Methoxyflurane

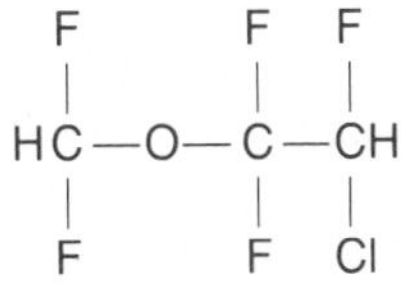
F F F
HC—O—C—CH
F F Cl

Enflurane

These substances are nonflammable and do not irritate the respiratory passages. However, their use requires even greater skill on the part of the anesthesiologist than that needed with diethyl ether because the difference between a safe and a lethal dose is small. Lower doses are effective if these halogenated compounds are used in combination with nitrous oxide; this is common practice.

Ether and other anesthetics have a high degree of solubility in lipids (fat tissue). The mechanism of their central nervous system depression seems to depend on their solubility in the lipid material of nerve cells, which in some way apparently blocks transmission of nerve impulses. When the high concentration of anesthetic is removed, the anesthetic flows out of the lipid material; and the patient quickly resumes consciousness. Many questions regarding this mechanism remain to be answered.

14.9 THE CARBONYL GROUP

Alcohols, as you saw in Section 14.7, can be oxidized to aldehydes and ketones. The key to the chemistry of both these classes of compounds is the carbonyl group, $\rangle C{=}\ddot{\underset{..}{O}}$. Aldehydes differ from ketones in that they have at least one hydrogen bonded to the carbonyl carbon,

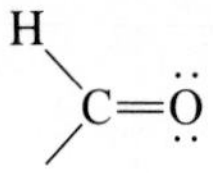

This means that the aldehyde group is necessarily *terminal,* i.e., at the end of the carbon chain. Ketones, on the other hand, have two carbon groups bonded to the carbonyl carbon,

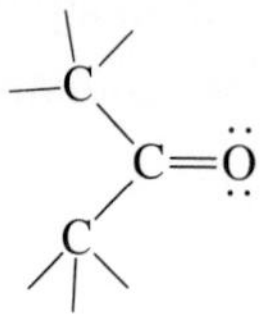

The ketone functional group is therefore necessarily located in an *interior* chain position.

Because of the significant difference in electronegativity between carbon and oxygen, the carbonyl group is polarized in such a way that the oxygen is slightly negative ($\delta-$) and the carbon is slightly positive ($\delta+$). Thus

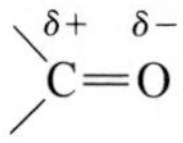

Keeping this polarity in mind will often be helpful in understanding the chemical reactivity of carbonyl-containing compounds.

The aldehyde and ketone functional groups are found in a variety of physiological compounds from small simple ones such as acetaldehyde (CH_3CHO) and acetone $\left(CH_3\overset{\overset{O}{\|}}{C}CH_3\right)$ to complex carbohydrates. Indeed, as you will see in Chapter 15, either the aldehyde or ketone group appears along with alcohol groups in all carbohydrates. In the next few sections we will discuss the nomenclature, physical properties, and chemical reactivity of aldehydes and ketones. You will later see these reactions in physiological settings.

14.10 ALDEHYDE AND KETONE NOMENCLATURE

It is always convenient to know the names of structures that you study.

Aldehydes

The usual stepwise procedure of the IUPAC nomenclature system is followed. The suffix which characterizes aldehydes is *-al*. To name an aldehyde:

STEP 1 **Find the longest continuous chain which includes the aldehyde $\left(-\underset{\underset{O}{\|}}{C}-H\right)$ functional group.**

STEP 2 **Number the chain so that the terminal aldehyde carbon is always position 1.**

STEP 3 **Locate and number the branched alkyl groups on the main chain.**

SAMPLE EXERCISE 14.4

Name the compound $CH_3-CH_2-CH(CH_3)-C(=O)-H$.

Solution

STEP 1 The longest chain of carbons including the aldehyde functional group is four. Thus the root name must be the same as for butane. Remove the *-e* from *butane* and add on the aldehyde suffix *-al*. This compound is a butanal.

STEP 2 The $-C(=O)-H$ group defines position 1.

STEP 3 Numbering the chain beginning with the aldehyde carbon, we find a methyl group at position 2. The full name of the compound is 2-methylbutanal.

Problem 14.4 Name the compound

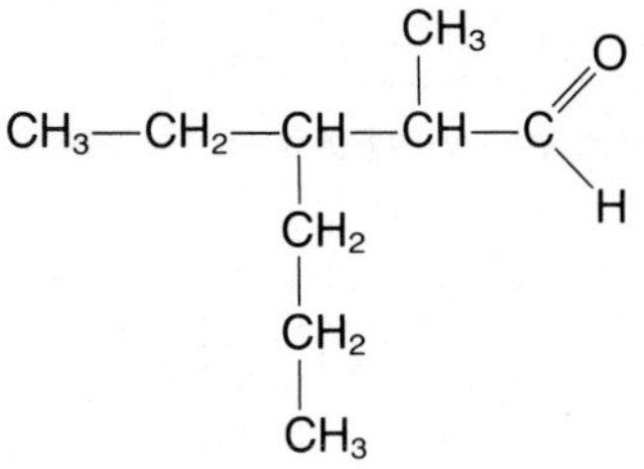

Aldehydes with four or fewer carbons are often called by common names. Their names are easily recognized as names of aldehydes because they end with the suffix *-aldehyde*. Because they are in common use in biochemistry, you should learn to recognize these names and their corresponding structures.

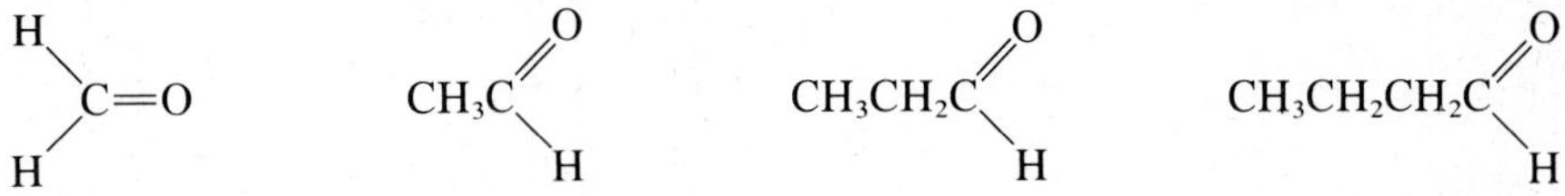

Formaldehyde Acetaldehyde Propionaldehyde Butyraldehyde

Aromatic aldehydes are named as derivatives of benzaldehyde.

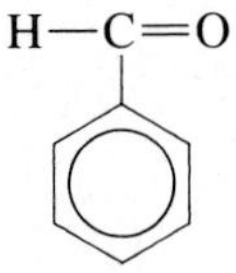

Benzaldehyde

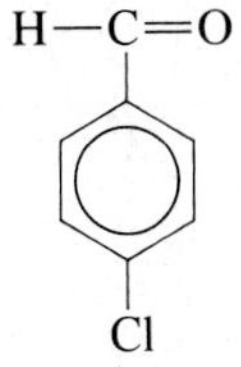

4-Chlorobenzaldehyde

Ketones

To name a ketone by the IUPAC system:

STEP 1 **Find the longest continuous chain which includes the carbonyl group.** The appropriate suffix for ketones is *-one*.

STEP 2 **Number the chain so that the carbonyl group has the lowest possible number.** The position of the carbonyl group is noted before the base name.

STEP 3 **Locate and number the branched alkyl groups on the main chain.**

SAMPLE EXERCISE 14.5

Name the compound

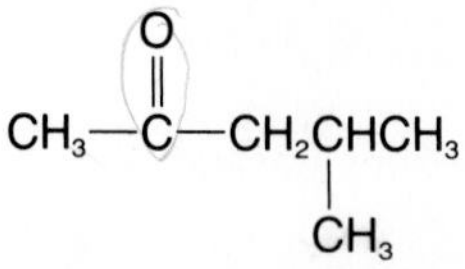

Solution STEP 1 Recognize the ketone functional group,

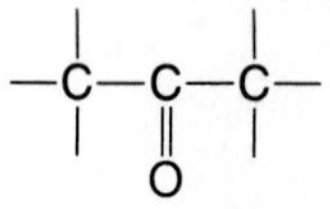

The longest chain of carbons including the carbonyl group is five. The alkane name would therefore be pentane. Remove the *-e* ending and add on the ketone suffix, *-one*. This compound is a pentanone.

STEP 2 Number the chain. The lowest number that can be assigned to the carbonyl group is 2. This is a 2-pentanone.

STEP 3 Find and number the branched alkyl group. A methyl substituent is at position 4. The full name of the structure is 4-methyl-2-pentanone.

Problem 14.5 Name the compound

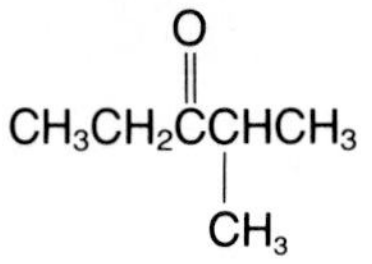

Common names for ketones are constructed by naming the alkyl groups attached to the carbonyl function and adding the word *ketone;* this method may remind you of the scheme for naming ethers.

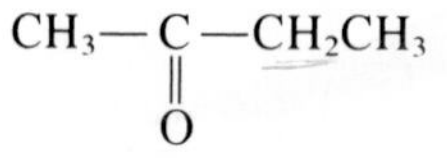

Ethyl methyl ketone

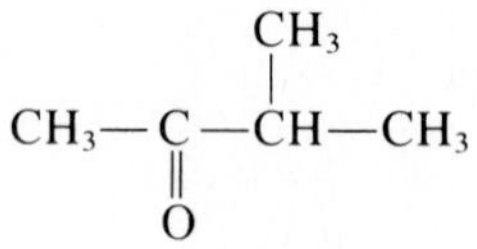

Methyl isopropyl ketone

The most common ketone, propanone or $CH_3\overset{\overset{O}{\|}}{C}CH_3$, is called acetone.

14.11 PHYSICAL PROPERTIES OF ALDEHYDES AND KETONES

In general, aldehydes and ketones boil at higher temperatures than alkanes and ethers and at lower temperatures than alcohols of similar molecular weight. Aldehydes and ketones of five or fewer carbons are soluble in water. Table 14.2 compares the boiling points and water solubilities of an aldehyde, a ketone, an alkane, an ether, and an alcohol of similar molecular weight.

TABLE 14.2 Comparative Boiling Points and Water Solubilities of an Alkane, an Ether, an Aldehyde, a Ketone, and an Alcohol of Similar Molecular Weight

Name	Structure	Molecular Weight	Boiling Point, °C	Water Solubility
Pentane	$CH_3CH_2CH_2CH_2CH_3$	72	36	Insoluble
Diethyl ether	$CH_3CH_2—O—CH_2CH_3$	74	34	Slightly soluble
Butanal	$CH_3CH_2CH_2—C(=O)H$	72	76	Soluble
2-Butanone	$CH_3—C(=O)—CH_2CH_3$	72	79	Soluble
1-Butanol	$CH_3CH_2CH_2CH_2OH$	74	117	Soluble

Aldehydes and ketones boil at a higher temperature than alkanes and ethers of similar weight because of the high polarity of the carbonyl group present in aldehydes and ketones. However, the boiling point of an aldehyde or ketone is lower than that of an alcohol of similar size because there is no hydrogen bonding between molecules of aldehydes or ketones.

Low-molecular-weight aldehydes and ketones are more soluble in water than ethers are. This solubility is a result of the attractive hydrogen-bonding interaction between the lone pair on the carbonyl oxygen and the $\delta+$ hydrogen on a water molecule (Figure 14.4).

14.12 OXIDATION AND REDUCTION OF ALDEHYDES AND KETONES

The oxidation of aldehydes forms the basis of one clinical test for abnormal sugar levels in urine. The basic principles of these oxidations appear here. Similarly, this section discusses the reduction of carbonyl groups, a frequent metabolic reaction, as you will see in Chapters 22 through 25.

Oxidation

Aldehydes are very easily oxidized to carboxylic acids. In contrast, ketones are usually resistant to oxidation. Like tertiary alcohols, which do not undergo oxidation, ketones have no hydrogen attached to the functional-group carbon.

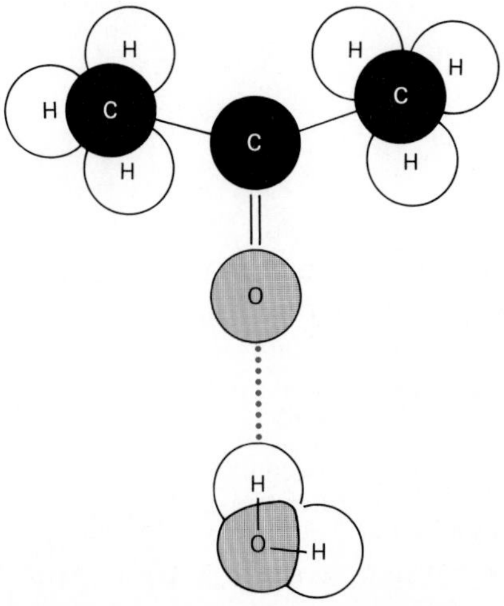
H
H
H
C
H
C
H
C
H
H
O
H
O
H

FIG. 14.4
It is possible for a hydrogen bond to form between the oxygen of a carbonyl group and the hydrogen of a water molecule. This interaction accounts for the water solubility of low-molecular-weight aldehydes and ketones, such as acetone, which is shown in this illustration.

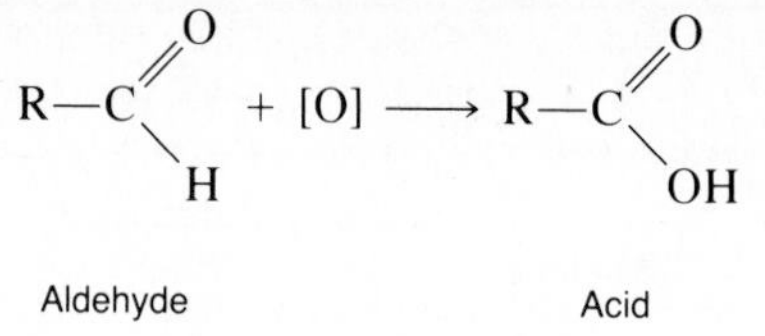
R—C(=O)H + [O] ⟶ R—C(=O)OH
Aldehyde
Acid

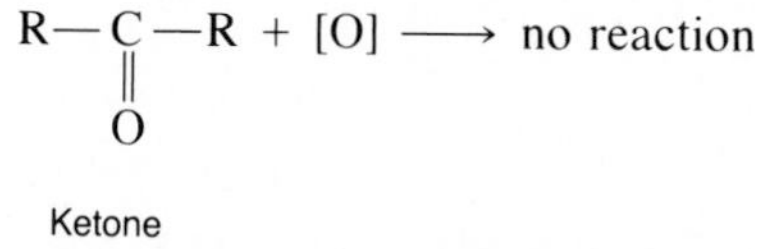
R—C(=O)—R + [O] ⟶ no reaction
Ketone

Aldehydes are oxidized by mild oxidizing agents which do not affect other functional groups. An example of such an oxidizing agent is the silver ammonia complex, $Ag(NH_3)_2^+$, in basic solution, known as *Tollens' reagent.*

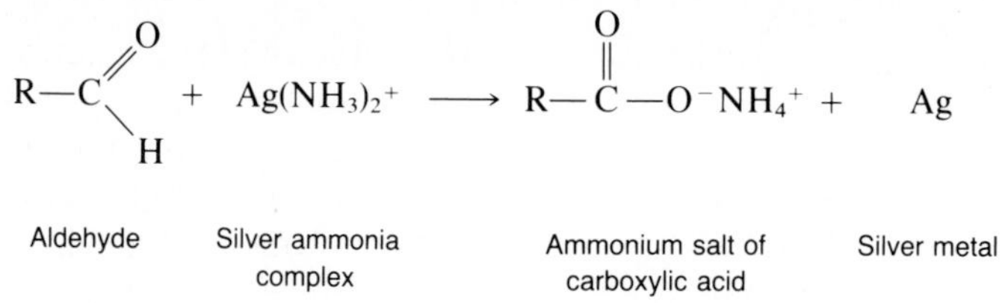
R—C(=O)H + Ag(NH3)2+ ⟶ R—C(=O)—O⁻NH4+ + Ag
Aldehyde
Silver ammonia complex
Ammonium salt of carboxylic acid
Silver metal

The basic condition of the reaction neutralizes the carboxylic acid, and the result is that the anion of the acid is a product. The silver, which is a positive ion (Ag^+) in Tollens' reagent, gains an electron to yield Ag, silver metal. In a clean test tube this silver plates out on the glass as a *silver mirror*. Indeed this oxidation reaction (done with an inexpensive aldehyde) has been used to make silver mirrors. This reaction occurs only for the aldehyde functional group. Consequently the Tollens test is used to distinguish between aldehydes and ketones and, in general, to provide evidence for the presence or absence of the aldehyde group. The silver mirror formed in a positive test provides a convenient visual marker.

Benedict's reagent is a basic solution containing Cu^{2+} ions complexed with polyatomic citrate ions ($C_6H_6O_7^{2-}$). Solutions of Cu^{2+} ions are characteristically blue. Benedict's reagent selectively oxidizes aldehydes and so-called α-hydroxy ketones. The carbon adjacent to the carbonyl group is called the α-carbon.

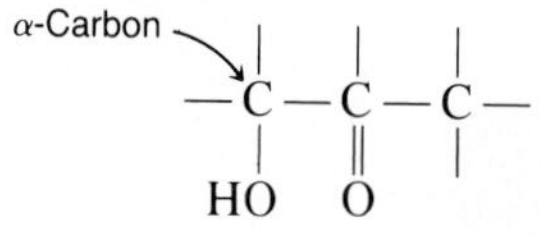
α-Carbon
HO
O

α-Hydroxy ketone

You will see in the next chapter that aldehyde and α-hydroxy ketone functional groups are present in many simple carbohydrates, and consequently, these carbohydrates can be oxidized by Benedict's reagent and the reaction can be seen visually as a color change.

Carbohydrate containing aldehyde or α-hydroxy ketone	+	Cu^{2+} (complex) Benedict's reagent	⟶	$Cu_2O \downarrow$	+	oxidation products
		(*Blue color*)		(*Brick-red* precipitate)		

This reaction is the basis for a qualitative and quantitative test for carbohydrates such as glucose in urine and blood samples. A positive test is indicated by the color change (blue $\longrightarrow$ green $\longrightarrow$ brick red) as the copper ion is reduced from Cu^{2+} to Cu^+. Clinitest tablets are a commercial form of the solid Benedict's reagent.

Reduction

As you saw in Section 7.12, reduction reactions are the reverse of oxidation reactions. Aldehydes can be oxidized to carboxylic acids, and acids can theoretically be reduced to aldehydes, although it is not easy to do in the laboratory.

In Section 14.7 you learned that aldehydes and ketones can be prepared by the oxidation of primary and secondary alcohols, respectively. The opposite reaction can be carried out; that is, aldehydes and ketones can be reduced to 1° and 2° alcohols. This reaction is easily done in the laboratory. The overall result of the reduction is addition of hydrogen, one H bonds to the carbonyl carbon and the other H to the carbonyl oxygen.

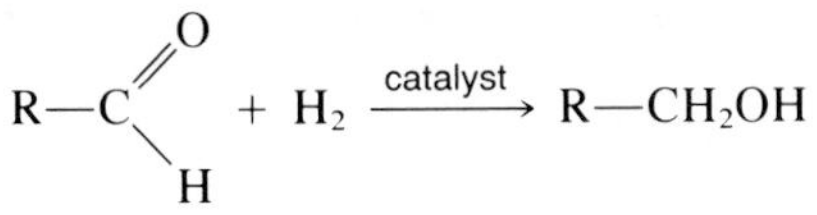

$$R-C\overset{O}{\underset{H}{\diagup}} + H_2 \xrightarrow{\text{catalyst}} R-CH_2OH$$

$$R-\underset{\underset{O}{\|}}{C}-R + H_2 \xrightarrow{\text{catalyst}} R-\underset{\underset{OH}{|}}{\overset{\overset{H}{|}}{C}}-R$$

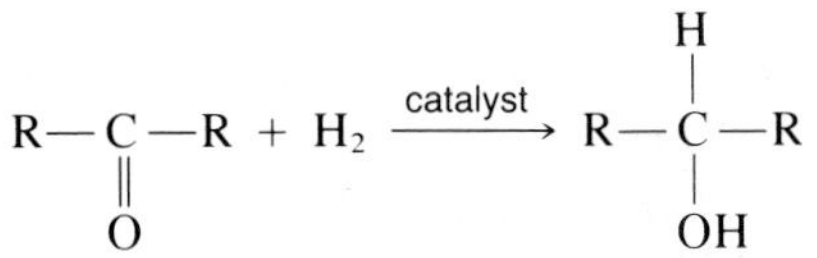

Such reductions occur in the body. The hydrogen is provided by complex molecules called coenzymes (Chapter 21). $NADH/H^+$, the reduced form of the coenzyme *n*icotinamide *a*denine *d*inucleotide (NAD^+), is such a biological reducing agent. In living systems, $NADH/H^+$, as part of an enzyme structure, reduces carbonyl groups to alcohol groups. For example, the ketone functional group in pyruvic acid is reduced to —OH, and the product is called lactic acid. This reaction occurs in our muscles during vigorous exercise.

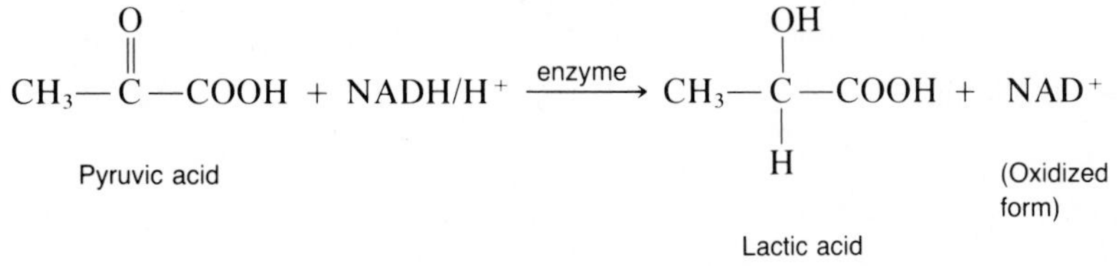

$$\underset{\text{Pyruvic acid}}{CH_3-\underset{\underset{O}{\|}}{C}-COOH} + NADH/H^+ \xrightarrow{\text{enzyme}} \underset{\underset{\text{Lactic acid}}{}}{CH_3-\underset{\underset{H}{|}}{\overset{\overset{OH}{|}}{C}}-COOH} + \underset{\substack{\text{(Oxidized} \\ \text{form)}}}{NAD^+}$$

Oxidizing and reducing capabilities of NAD^+ and $NADH/H^+$ are discussed more fully in Section 22.6.

14.13 ALCOHOL ADDITION TO ALDEHYDES AND KETONES

Like the $\diagdown C = C \diagup$ double bond, the $\diagdown C = O$ double bond can undergo addition reactions. Indeed, the reduction just discussed was an addition reaction. Alcohol molecules add to the carbonyl group to form hemiacetals and hemiketals, new functional groups which have importance in carbohydrate chemistry. These new functional groups are characterized by a carbon which bears an —OH group and is also involved in an ether linkage:

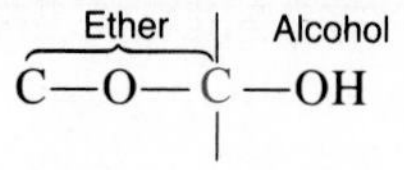
Ether
Alcohol

Hemiacetal or hemiketal

Hemiacetal and Hemiketal Formation

Aldehydes and ketones react with alcohols to form hemiacetals and hemiketals, respectively. The course of their reaction can be followed more readily if we keep in mind the polarities of the carbonyl and alcohol functional groups. Using hemiacetal formation as an example and marking the polarities, we see that we have another case of "opposites attract." That is, new bonds form (as indicated by arrows) between positive centers (H and C) and negative centers (O's).

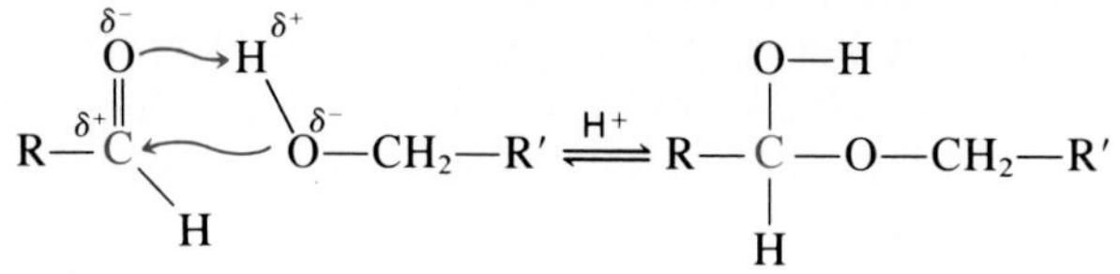

Aldehyde Alcohol Hemiacetal

The negative alcohol oxygen is attracted to the positive carbonyl carbon, and the negative carbonyl oxygen attracts the positive hydrogen. The carbonyl double bond and O—H bond must break to complete product formation. The highlighted carbon is the carbonyl carbon in the reactant and the center of the hemiacetal functional group. Hemiacetal formation can be summarized in the following steps:

STEP 1 The carbonyl carbon–oxygen double bond is converted to a single bond. In essence, the electrons of the double bond move onto the carbonyl oxygen. This produces a negative oxygen (extra electrons) and an electron-deficient positive carbon.

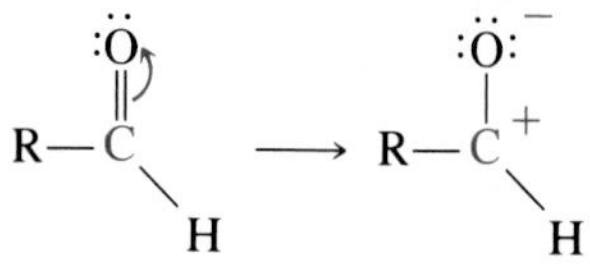

STEP 2 The alcohol oxygen is attached to the carbonyl carbon.

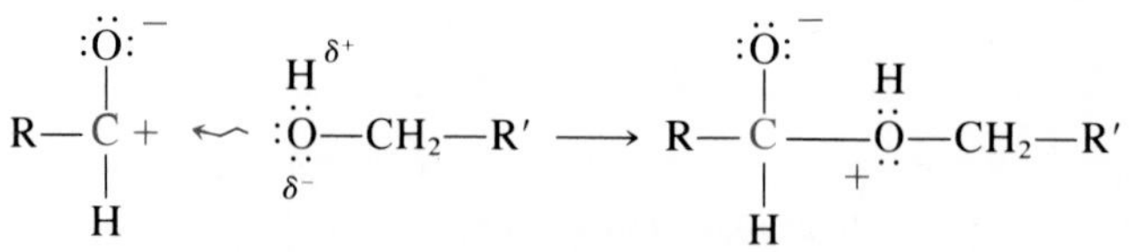

STEP 3 The old alcohol O—H bond is broken and a new one, between the H and the carbonyl oxygen, is formed.

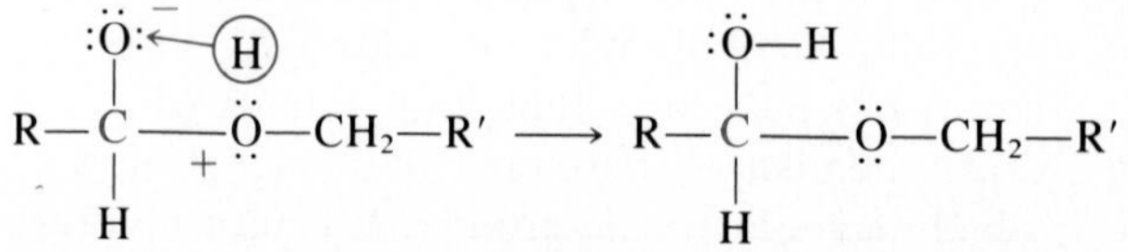

A hemiketal is formed in a similar sequence when the reacting materials are a ketone and an alcohol:

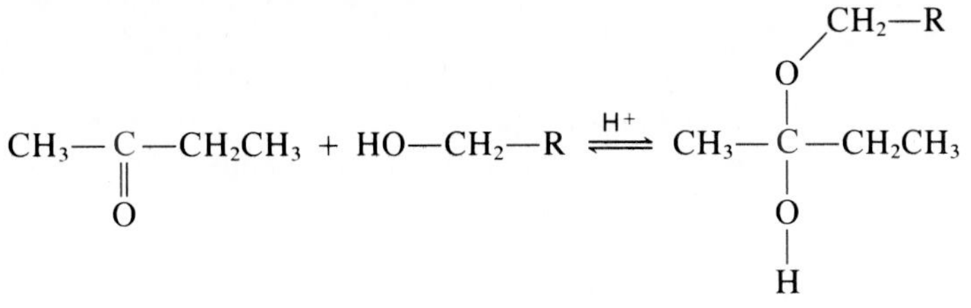

Ketone Alcohol Hemiketal

Carbohydrates, as you will see in the next chapter, have both alcohol groups and aldehyde or ketone groups as part of their molecules. These functional groups can react intramolecularly, producing a cyclic hemiacetal or hemiketal. The reaction proceeds by the same guidelines noted previously. In the following example the carbons are numbered so that you may more easily see the relationship between the cyclic and open-chain structures. The oxygen in the ring was originally the alcohol oxygen in 5-hydroxypentanal.

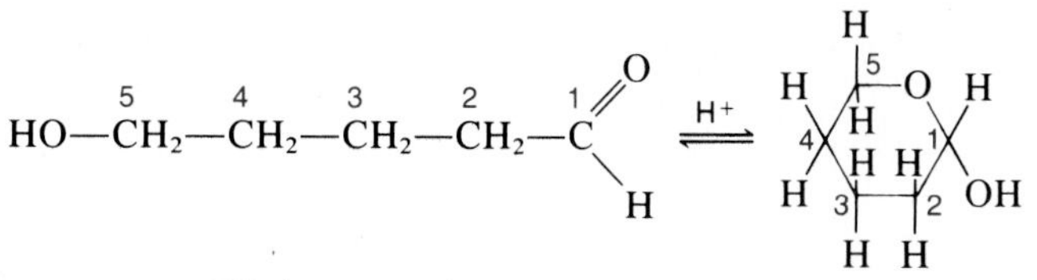

5-Hydroxypentanal

Cyclic hemiacetal

You can recognize the typical bonding pattern of hemiacetal and hemiketal functional groups in the reaction products of alcohols and carbonyls by noting that there is an alcohol group and ether linkage bonded to a common carbon. This common carbon was originally the carbonyl carbon in the starting aldehyde or ketone. This carbon is highlighted in the examples given. Identifying hemiacetals (or hemiketals), especially cyclic ones, will be a useful skill in Chapter 15.

Problem 14.6 Indicate which of the following structures contain a hemiacetal or hemiketal functional group:

a 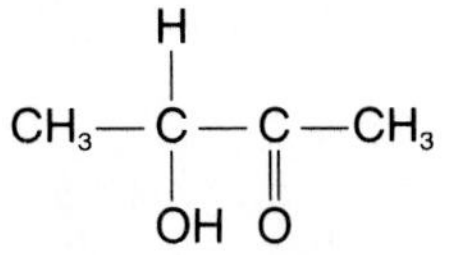

d 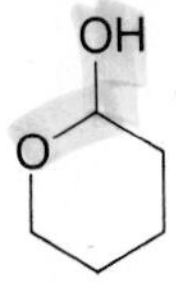

b 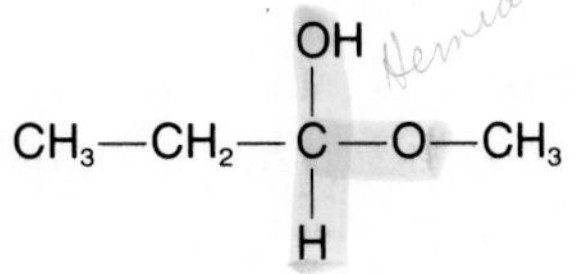

e 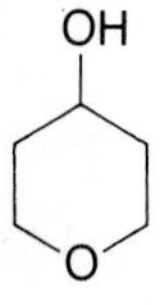

c 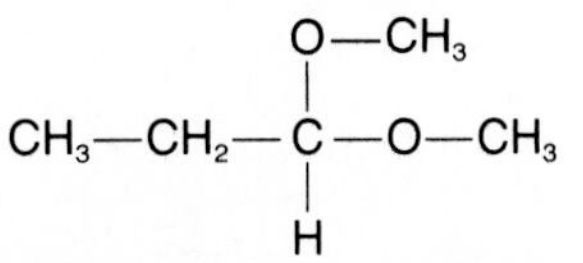

In the laboratory flask, hemiacetal (or hemiketal) formation can take place with either an acid or base catalyst. As the equations above show, the reaction is an equilibrium. With the important exception of carbohydrate structures, the mixture of separate alcohol and aldehyde (or ketone) groups is more stable and hence predominates in the equilibrium mix. The equilibrium lies to the left; there is little of the hemiacetal (or hemiketal). **The real importance of hemiacetals and hemiketals will be seen in carbohydrate chemistry (Section 15.9).**

Acetal and Ketal Formation

As the prefix *hemi-,* meaning "half," suggests, hemiacetals and hemiketals are halfway on the route to acetals and ketals, starting with carbonyl compounds and alcohols. In acid solution the hemiacetal product from the reaction of an aldehyde and alcohol can react with additional alcohol to form an acetal group. The acetal functional group is characterized by two ether groups bonded to a common carbon.

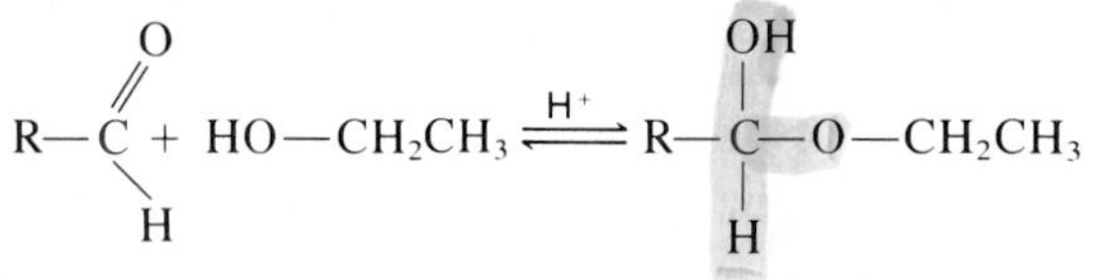

$$
\underset{\text{Aldehyde}}{R-\overset{\displaystyle O}{\overset{\|}{C}}-H} + \underset{\text{Alcohol}}{HO-CH_2CH_3} \underset{}{\overset{H^+}{\rightleftharpoons}} \underset{\text{Hemiacetal}}{R-\overset{\displaystyle OH}{\underset{\displaystyle H}{C}}-O-CH_2CH_3}
$$

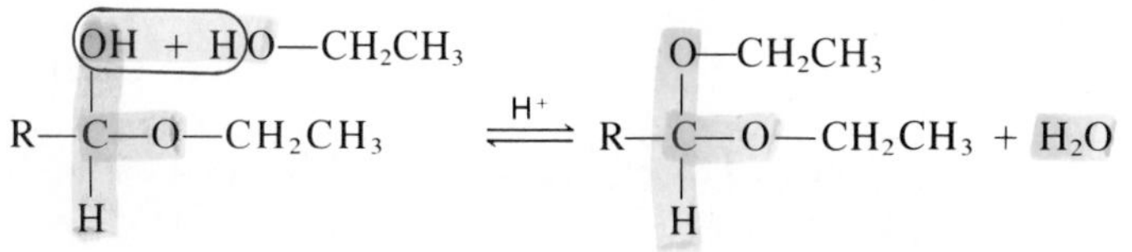

$$
\underset{\text{Hemiacetal}}{R-\overset{\displaystyle OH}{\underset{\displaystyle H}{C}}-O-CH_2CH_3} + \underset{\text{Alcohol}}{HO-CH_2CH_3} \overset{H^+}{\rightleftharpoons} \underset{\text{Acetal}}{R-\overset{\displaystyle O-CH_2CH_3}{\underset{\displaystyle H}{C}}-O-CH_2CH_3} + H_2O \qquad (14.10)
$$

Acetal formation [Equation (14.10)] is a condensation reaction between a hemiacetal and alcohol in which a water molecule is split out between them. A similar reaction occurs with ketones leading to ketals.

The acetal and ketal reactions are reversible, but occur only in acid solution. Therefore, unlike a hemiacetal, a compound containing an acetal group will be stable in basic or neutral solution. In examining a given structure for the presence of an acetal or ketal, look for the bonding arrangement

$$
-\overset{|}{\underset{|}{C}}-O-\overset{|}{\underset{|}{C}}-O-\overset{|}{\underset{|}{C}}-.
$$

The highlighted C was originally the carbonyl C of an aldehyde or ketone.

You will see in Section 15.12 that the reaction of cyclic hemiacetals with alcohols to yield acetals is significant in the formation of di- and polysaccharides.

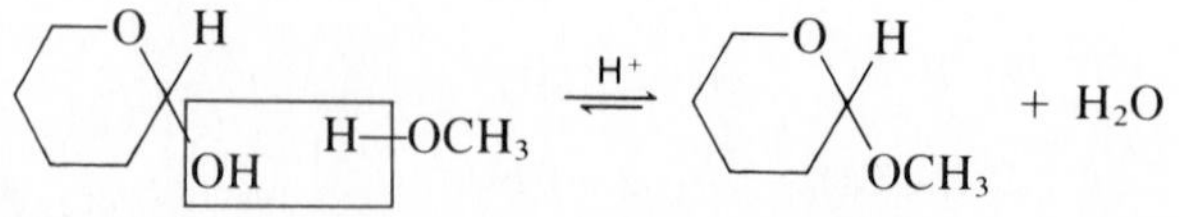

Cyclic hemiacetal   Alcohol   Acetal

14.14 ALDOL ADDITION OF ALDEHYDES AND KETONES

The aldol addition reaction is the addition of one aldehyde (or ketone) molecule to another aldehyde molecule. It occurs in stepwise fashion and we will discuss it one step at a time.

STEP 1 **Formation of anion**

Unlike other carbon-hydrogen bonds which we have examined in this text, the hydrogens on the carbon adjacent to a carbonyl group, i.e., the α-carbon position, are sufficiently acidic that in a strongly basic solution the following reaction produces a small amount of aldehyde species with a negative charge.

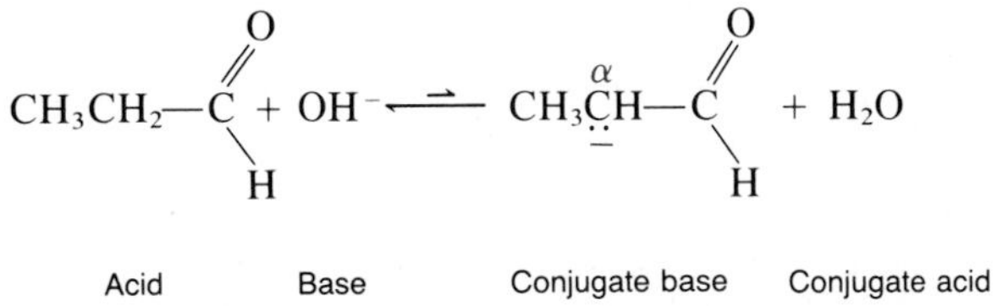
Acid
Base
Conjugate base
Conjugate acid

The aldehyde, by giving up H^+, is acting as a very weak acid. Because it is such a weak acid, the equilibrium lies toward the aldehyde side and only a small amount of anion forms.

STEP 2 **Attack of anion on carbonyl group of second molecule**

Because of the negative charge on carbon, the anion formed in step 1 is highly reactive and can attack at the positively polarized carbon in the carbonyl group of another aldehyde molecule. The result is the formation of a new carbon-carbon bond.

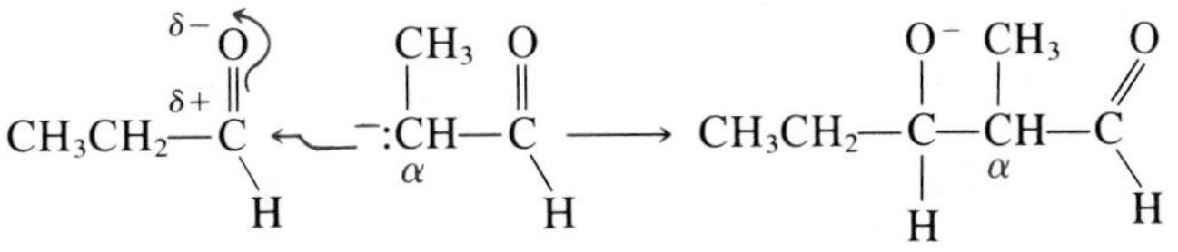

The carbonyl group of the attacked molecule is converted to a single C—O bond, and the oxygen acquires a negative charge.

STEP 3 **H^+ bond to negative oxygen**

The negative oxygen removes an H^+ ion from the solvent by an acid-base reaction. The final neutral product contains both **ald**ehyde and alcoh**ol** functionalities and is therefore dubbed an **aldol.**

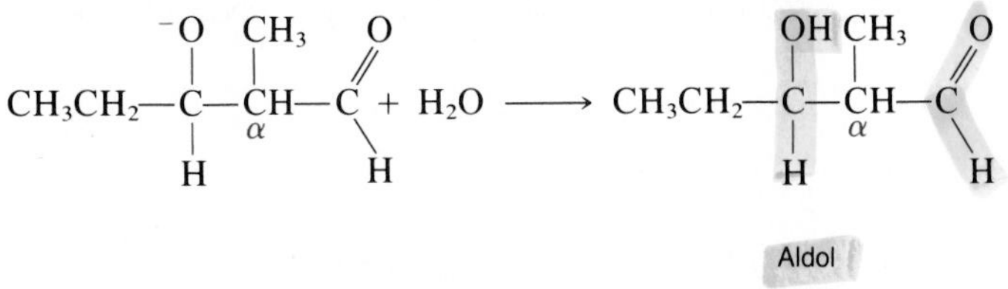
Aldol

The overall equation for the aldol addition between two molecules of the same aldehyde is

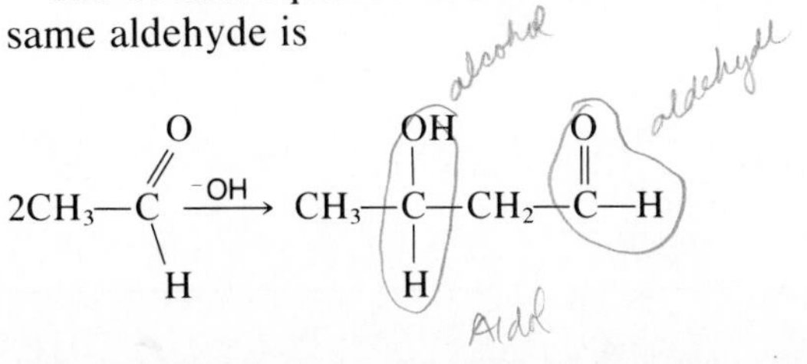
alcohol
aldehyde
Aldol

When you write the structure for the aldol addition product, it is useful to keep in mind the following points:

1. The α-carbon of an aldehyde or ketone bonds to the carbonyl carbon of a second aldehyde.
2. The alcohol group is always on the third carbon, numbering the carbonyl group as position 1 in the aldol product.

Consider the two aldol examples just given:

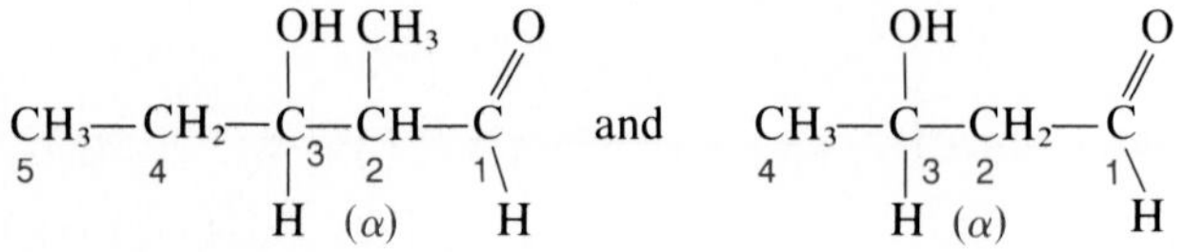

The aldol addition is a very useful reaction in the laboratory because it allows chemists to lengthen the carbon chain and thus prepare larger molecules from smaller ones. Nature uses this reaction in the biological synthesis of large molecules such as collagen in connective tissue. **In Section 23.2 you will see a step in the metabolism of sugars that involves a reverse aldol reaction, known as a retroaldol, in which a sugar is broken down into an aldehyde and a ketone.**

14.15 REACTION OF ALDEHYDES AND KETONES WITH NITROGEN COMPOUNDS

Aldehydes and ketones react with ammonia and other nitrogen compounds to form imines, compounds which have carbon-nitrogen double bonds.

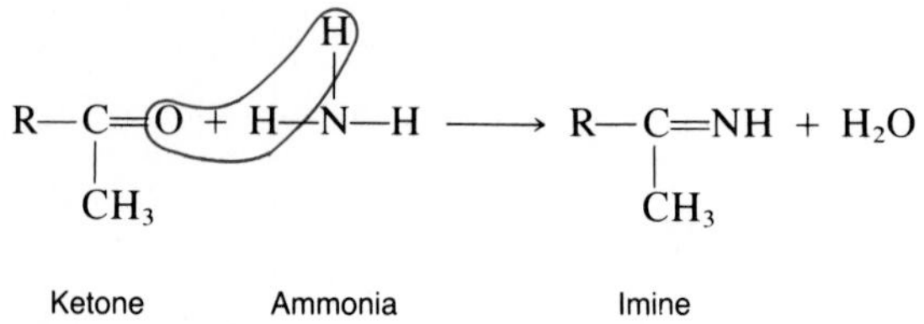
Ketone
Ammonia
Imine

Water is formed in this reaction as a result of the combination of two hydrogens, which were bonded to the nitrogen, and the carbonyl oxygen. This type of reaction is known as *transamination* in living species, which use it in the biological synthesis of amino acids from ketones, and it will be examined in the chapter on protein metabolism (Chapter 25).

Problem 14.7 Write the structural formula for the product in each of the following reactions:

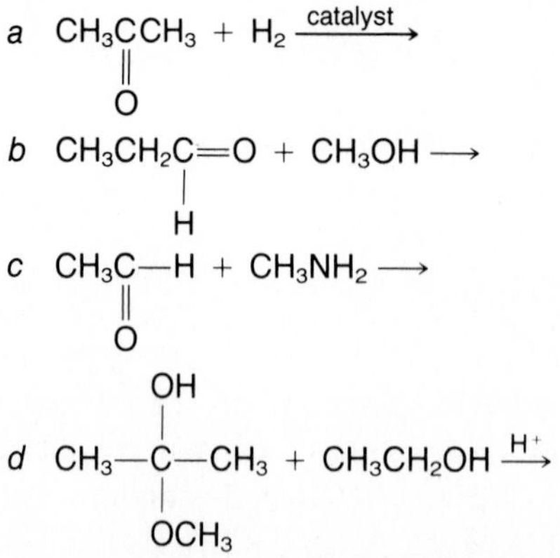

14.16 THE OCCURRENCE AND USE OF A FEW ALDEHYDES AND KETONES

Formaldehyde (methanal) is the simplest aldehyde and the only one without an alkyl group bonded to the carbonyl carbon.

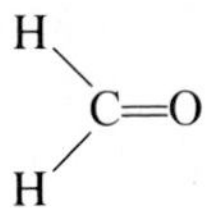

It is a gas at room temperature, but it is rarely encountered in pure form.

Usually formaldehyde is used as a 40 percent solution in water, known as Formalin. Formaldehyde vapors are very irritating to the mucous membranes; ingestion can lead to severe medical problems. Formalin reacts with protein in tissues and causes the tissue to harden. For this reason, it is used as an embalming fluid and a biological specimen preservative. Commercially the most important use of formaldehyde is as a monomer in the manufacture of various plastics such as Formica (counter tops, telephones, etc), Melmac (plastic plates and utensils), and Bakelite (handles on pots and pans).

Acetaldehyde, as we have already mentioned in Section 14.7, is the human body's initial breakdown product for ethyl alcohol.

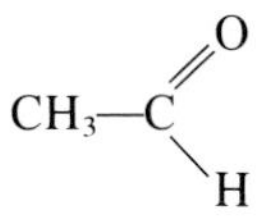

Not unexpectedly, the symptoms for chronic intoxification by acetaldehyde resemble those for chronic alcoholism. Acetaldehyde is used in the manufacture of acetic acid, butanal, and some perfumes and in the silvering of mirrors.

Aromatic Aldehydes

Spices are usually complex mixtures of ingredients. Vanilla and cinnamon owe their characteristic flavors and aromas to the presence of two aromatic aldehydes, vanillin and cinnamaldehyde, respectively.

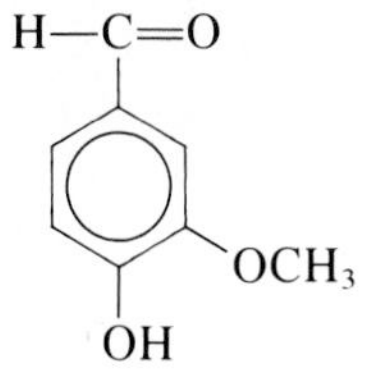

Vanillin

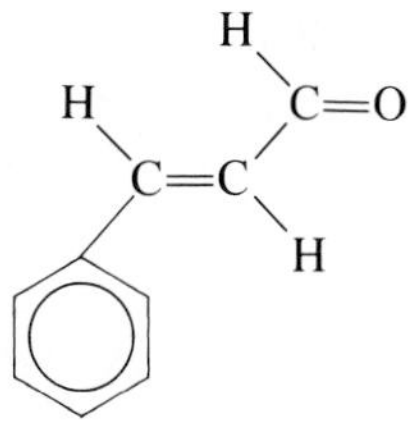

Cinnamaldehyde

Acetone, $CH_3{-}\overset{\overset{\displaystyle O}{\|}}{C}{-}CH_3$, is the simplest ketone and the one of greatest commercial importance because it is miscible with both water and many organic materials. It is the usual solvent in nail polish remover. Acetone is normally found in the blood in very low concentrations (less than 1 mg/100 mL). However, disorders in fat metabolism, such as those that occur in untreated diabetes, lead to much higher blood concentrations of this ketone. The sweetish odor of acetone on the breath and acetone excretion in the urine are symptoms detected in an untreated diabetic patient.

Complex Ketones

Some higher-molecular-weight ketones are useful because of their unique aroma. Camphor, a moth repellent and a preservative in some pharmaceuticals and cosmetics, was formerly extracted from the camphor tree, but is now mostly produced synthetically. Muscone, which has been extracted from the musk glands of the male musk deer, is used to give the musky aroma to some perfumes.

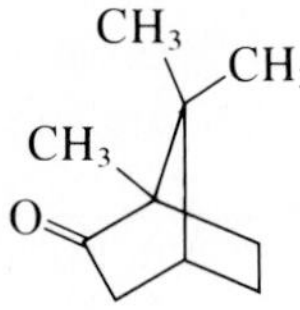

Camphor

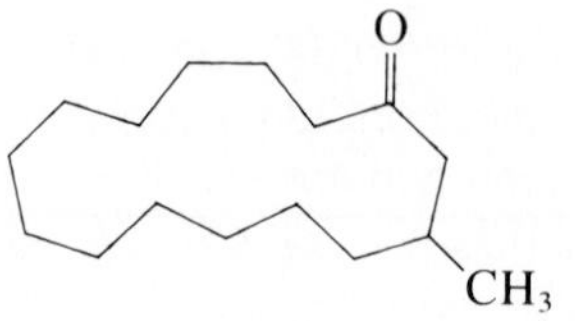

Muscone

14.17 PHENOLS

The phenols are the class of compounds that have the —OH group directly bonded to an aromatic ring. This distinguishes them from the alcohols, in which —OH is bonded to an alkyl group. The properties of the phenols, particularly the chemical properties, are very different from those of the alcohols.

Nomenclature

The simplest member of this class of compounds is simply called phenol. When there are other substituents on the phenol aromatic ring, the compounds are named following the system for aromatic nomenclature discussed in Section 13.11.

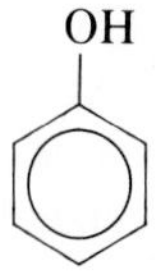

Phenol

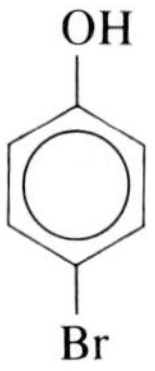

4-Bromophenol

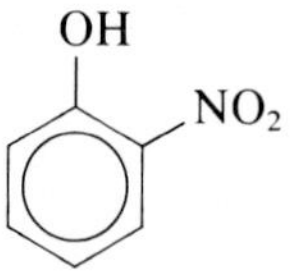

o-Nitrophenol

Physical Properties

Phenols have strong hydrogen-bonding interactions among OH groups, and consequently most phenols are solids at room temperature. Although it contains a large nonpolar group of six carbons, phenol itself is moderately soluble in water (9 g/100 mL). This indicates strong hydrogen bonding between phenol and water molecules. Water solubility can be conferred in otherwise nonsoluble aromatic components by placing one or more hydroxyl groups on the ring.

Chemical Properties

Phenols are weak acids and consequently react with strong bases such as sodium hydroxide to form salts.

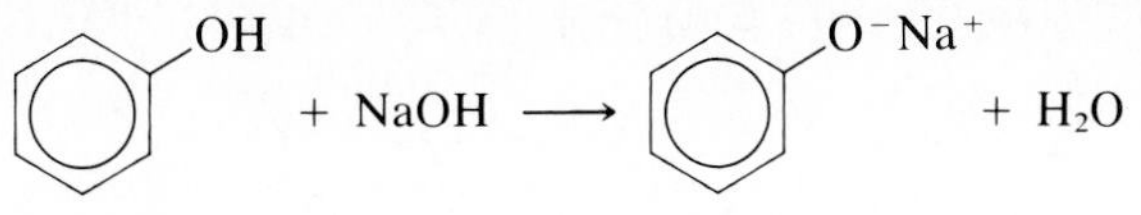

Phenol

Sodium phenoxide, a salt

Phenols do not undergo the dehydration reaction which is common for aliphatic alcohols.

Phenols readily react by oxidation. Phenol itself can be oxidized to a 1,4-diketone known as quinone:

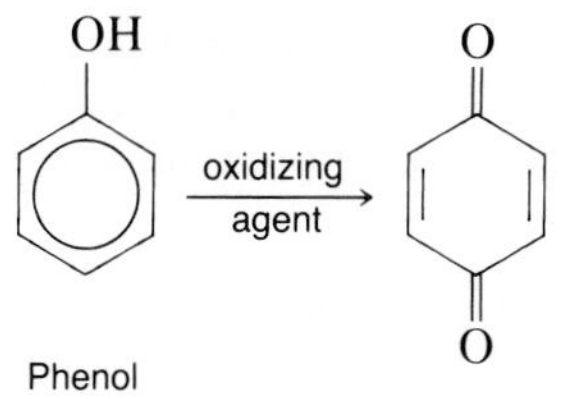

Phenol

Quinone

However, perhaps the most important example of an oxidation of a phenol is that of an aromatic ring with two hydroxy groups in a 1,4 relationship:

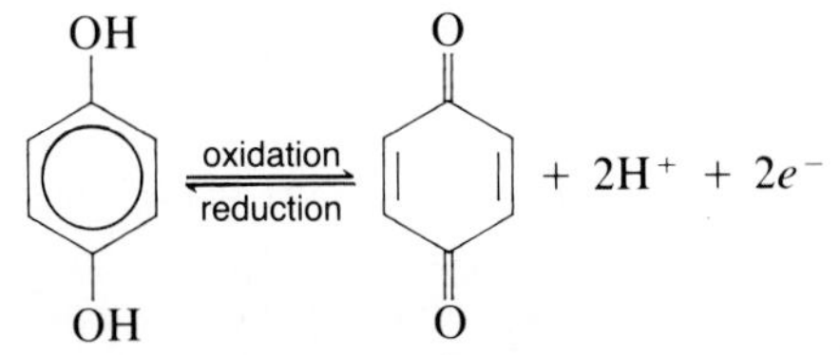

Hydroquinone

Quinone

This reaction occurs in the metabolic process known as the respiratory chain, as you will see in Section 22.6.

Use and Structure of Some Important Phenols

Phenol, which was called carbolic acid in the nineteenth century, was the first antiseptic used in surgery. It was introduced in the 1860s by a surgeon, Joseph Lister, who knew the work of Louis Pasteur and who sought a germ killer which would cut down on postoperative infections. Phenol itself, because of its structural alteration of tissue protein, is no longer used as an antiseptic except in some cases to disinfect surgical instruments.

Compounds containing the phenol group are in wide use in the home as disinfectants and in mouthwashes and throat lozenges.

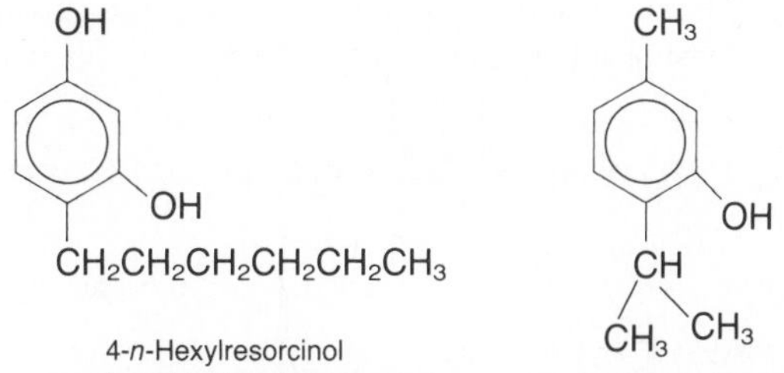

4-*n*-Hexylresorcinol (used in throat lozenges and mouthwashes)

Thymol (gives minty taste; used in mouthwashes)

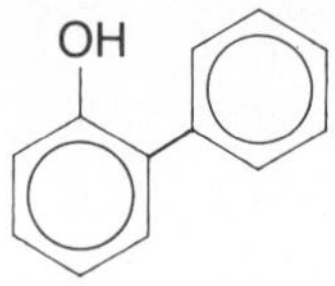

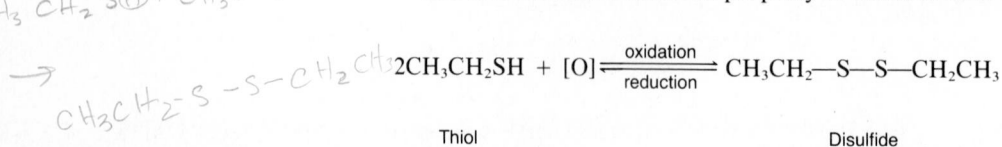

o-Phenylphenol
(found in home
disinfectant sprays)

Because of the ease with which they can be oxidized, some phenols are used as antioxidants in commercial foodstuffs. Antioxidants preserve foodstuffs by protecting them against oxidation. In the presence of air the phenol rather than the food is oxidized. Two antioxidants in common use are butylated hydroxy anisole (BHA) and butylated hydroxy toluene (BHT).

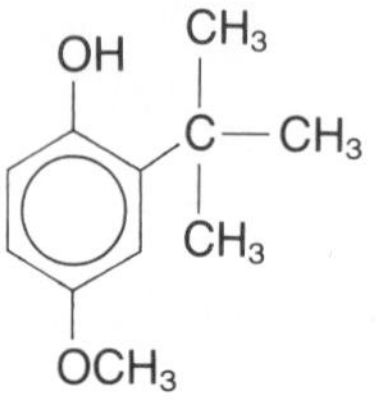

Butylated hydroxy anisole

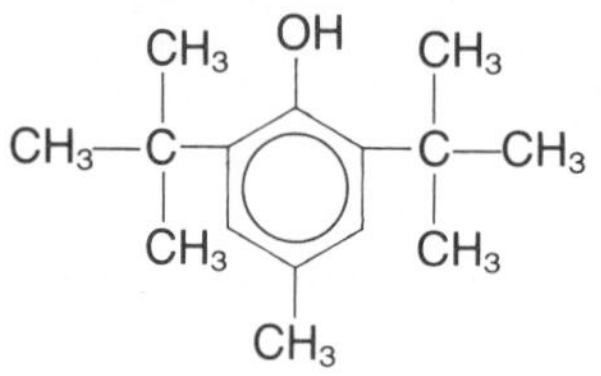

Butylated hydroxy toluene

14.18 THIOLS

Compounds which contain the —SH group are analogs of alcohol; that is, the functional group is similar in form to that of alcohols but has sulfur in place of an oxygen. These —SH-containing compounds are known as thiols or sometimes by an older name, mercaptans. A few examples are

CH_3SH	CH_3CH_2SH	$CH_3CH_2CH_2SH$
Methanethiol (methyl thiol) (bp = 6°C)	Ethanethiol (ethyl thiol) (bp = 36°C)	1-Propanethiol (*n*-propyl thiol) (bp = 68°C)

Although they are higher in molecular weight, thiols boil at lower temperatures than alcohols with the same number of carbons. For example, ethyl alcohol boils at 78°C. These lower boiling points are a consequence of the absence of hydrogen bonding in the thiols.

The most striking physical property of thiols is their foul smell. Butanethiols are the ''active'' component of a frightened skunk's spray. Good use can be made of the bad smell: Natural gas which is odorless, but toxic, is made detectable by the introduction of a very small concentration of methanethiol or ethanethiol. Thiols are responsible for onions' pungency, but not everyone would agree the aroma of onions (propanethiol) is unpleasant.

The most revelant chemical property of thiols is their oxidation to disulfides,

$$2CH_3CH_2SH + [O] \underset{\text{reduction}}{\overset{\text{oxidation}}{\rightleftharpoons}} CH_3CH_2—S—S—CH_2CH_3$$

Thiol   Disulfide

Comparing their reaction with the oxidation of alcohols (Section 14.7), we see that while the carbon is oxidized in alcohol, in thiols it is the sulfur atom that is undergoing oxidation. This can be seen from the loss of hydrogen from each sulfur. Thiol groups are found along protein chains. When these groups are oxidized, one part of the chain is linked to another, as you will see in Section 19.10; this gives characteristic three-dimensional shapes to proteins.

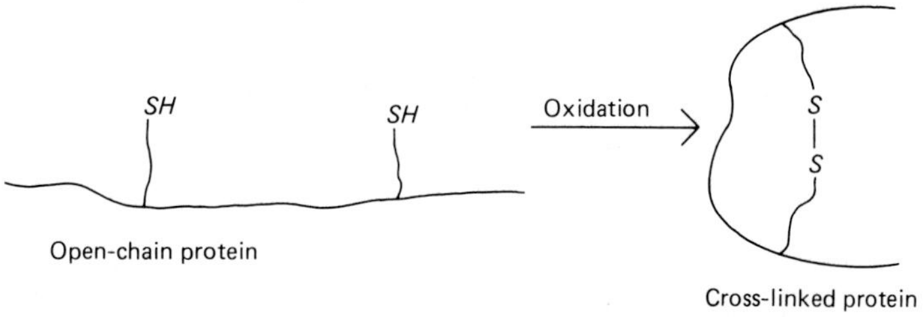
SH
SH
Oxidation
S
S
Open-chain protein
Cross-linked protein

CHAPTER ACCOMPLISHMENTS

After completing this chapter you should be able to define **key words** and do the following:

14.1 INTRODUCTION

1. State what is common to the alcohol, phenol, ether, aldehyde, and ketone functional groups.

14.2 THE ALCOHOL FUNCTIONAL GROUP

2. Recognize the alcohol functional group and classify a given alcohol as primary, secondary, or tertiary.

14.3 HYDROGEN BONDING IN ALCOHOLS

3. Describe in words and use diagrams to show hydrogen-bonding interactions between alcohol molecules and between alcohol and water molecules.
4. State the relationship between the R-group size and the boiling point and water solubility of an alcohol.

14.4 NOMENCLATURE OF ALCOHOLS

5. Give an unambiguous name to a given alcohol, diol, or triol.
6. Given the name of an alcohol, write a structural formula.

14.5 INDUSTRIAL SOURCE AND USE OF SOME ALCOHOLS

7. Describe the commercial preparation and at least one use of methanol, ethanol, and 2-propanol.
8. State the relationship between the proof and the percent concentration of alcohol.

14.6 DEHYDRATION OF ALCOHOLS

9. Recognize a given reaction as dehydration.
10. Write a structure of the major organic product formed in an intramolecular dehydration reaction.
11. Write the structure of the product formed in an intermolecular dehydration.

14.7 OXIDATION OF ALCOHOLS

12. Recognize a given reaction as oxidation.
13. Write the structural formula for the oxidation product of a given alcohol.
14. Describe the role of ethanol as an antidote in methanol poisoning.

14.8 ETHERS

15. Recognize the ether functional group.
16. Give an unambiguous name to a given ether.
17. Given the name of an ether, write a structural formula.
18. Account for the observation that although ethers have a very low boiling point, they are (up to six carbons) at least partially soluble in water.
19. State two contraindications to the use of ether as a general anesthetic.

14.9 THE CARBONYL GROUP

20. Describe the polarized form of the carbonyl group.

14.10 ALDEHYDE AND KETONE NOMENCLATURE

21. Recognize the aldehyde and ketone functional groups in a given compound.
22. Give an unambiguous name for a given aldehyde or ketone.
23. Write the structural formula of an aldehyde or ketone, given a name.

14.11 PHYSICAL PROPERTIES OF ALDEHYDES AND KETONES

24. Order a given set of aldehydes, ketones, ethers, and alcohols by boiling point or water solubility.

14.12 OXIDATION AND REDUCTION OF ALDEHYDES AND KETONES

25. Write the structural formula for the product of an aldehyde oxidation.
26. Write the structural formula for the reduction product of a given aldehyde or ketone.

14.13 ALCOHOL ADDITION TO ALDEHYDES AND KETONES

27. Recognize a hemiacetal, hemiketal, acetal, or ketal functionality in a given structural formula.
28. Write the structural formula for the addition product of an alcohol to either an aldehyde or a ketone.
29. Write the structural formula for the condensation product of an alcohol with either a hemiacetal or a hemiketal.

14.14 ALDOL ADDITION OF ALDEHYDES AND KETONES

30. Write the structural formula for the product of an aldol condensation between two molecules of the same aldehyde.

14.15 REACTION OF ALDEHYDES AND KETONES WITH NITROGEN COMPOUNDS

31. Write the structural formula for the condensation product of an aldehyde or ketone with ammonia or a primary amine.

14.16 THE OCCURRENCE AND USE OF A FEW ALDEHYDES AND KETONES

32. State commercial uses of formaldehyde, acetaldehyde, and acetone.
33. State the physiological condition under which higher than normal concentrations of acetone are found in blood.
34. Describe the physiological condition which may lead to abnormally high acetaldehyde concentrations.

14.17 PHENOLS

35. Recognize the phenol functional group in a given structure, and state two chemical differences between phenols and alcohols.
36. Recognize the hydroquinone-quinone oxidation-reduction in a given chemical transformation.
37. State two commercial uses of phenols.

14.18 THIOLS

38. Recognize the thiol and disulfide groups in a given structure.
39. Account for the differences in boiling point and water solubility of thiols and alcohols.
40. Write the structural formula for the oxidation product of a thiol or the reduction product of a disulfide.

PROBLEMS

• 14.8 Classify each alcohol below as primary, secondary, or tertiary:

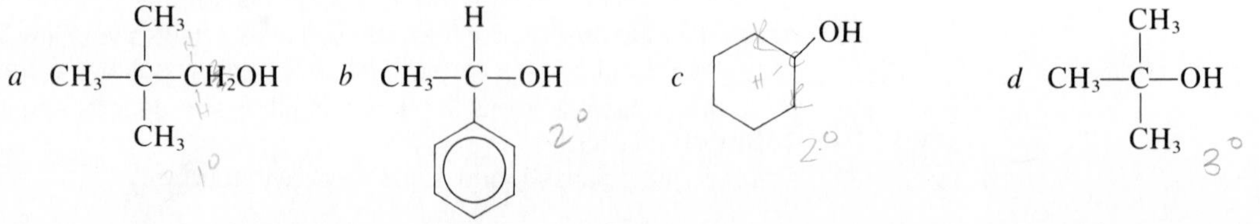

e 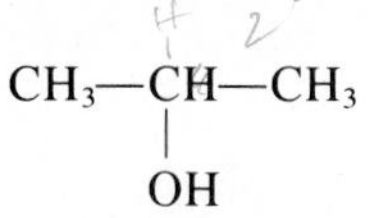

f 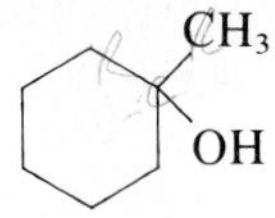

14.9 Ethyl alcohol (CH_3CH_2OH) and dimethyl ether have the same molecular formula and molecular weight (46). However, at room temperature ethyl alcohol is a liquid whereas dimethyl ether is a gas. Explain.

14.10 Arrange the following compounds in order of increasing boiling point from lowest to highest:
a CH_3OH *c* $CH_3CH_2CH_3$
b CH_4 *d* $CH_3CH_2CH_2OH$

14.11 Compare the solubility in water of CH_3OH and $CH_3CH_2CH_2OH$; that is, which is more soluble? Briefly explain your answer.

14.12 *a* Draw a diagram showing the hydrogen-bonding interaction between a molecule of CH_3CH_2OH and CH_3—O—CH_3.
b Would you predict that dimethyl ether is soluble in ethyl alcohol? Why or why not?

•14.13 Give an unambiguous name for each of the following alcohols:

a $CH_3CH_2CH_2OH$

d 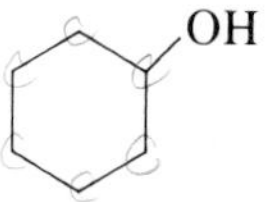

b 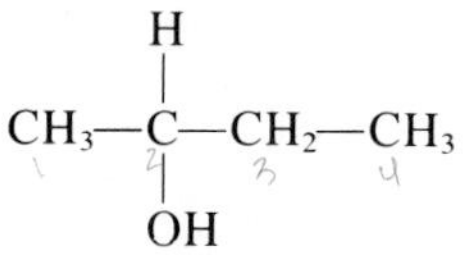

e 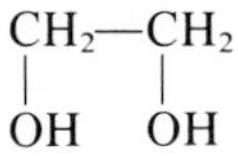

c $HOCH_2CH_2CH_2OH$

f 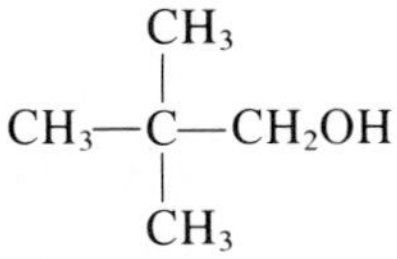

14.14 What is the percentage of alcohol in a bourbon whiskey which is labeled 110 proof?

14.15 List at least three products in your home that have an alcohol as one of the ingredients.

14.16 Calculate the number of kilocalories due to the alcohol in a 12-oz can of beer, assuming the beer to be 4 percent alcohol. (Note: 1 oz = 28 g.)

14.17 Explain what is meant by denatured alcohol.

14.18 Vodka was originally produced by the fermentation of potatoes. What component in the potatoes actually undergoes fermentation?

14.19 What substance would you add to grape juice before fermentation to increase the alcohol concentration in the final product?

• 14.20 Write the structure for the *major* product formed in the following elimination reactions:

a 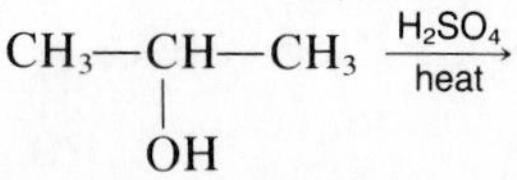

b 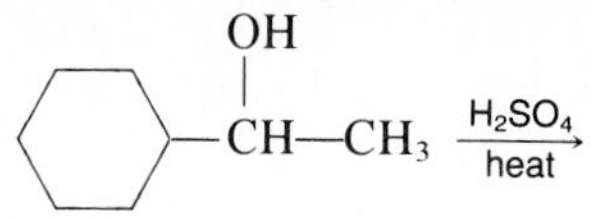

c 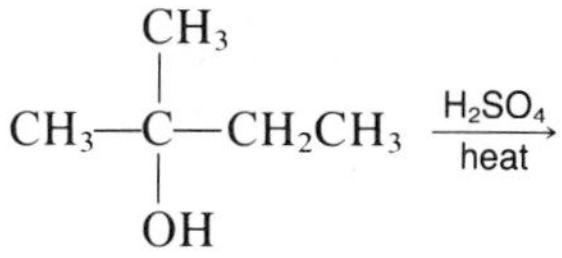

14.21 Write the structure of the ether formed in each of the following condensation reactions:

a $2CH_3OH \xrightarrow[140°C]{H_2SO_4}$

b 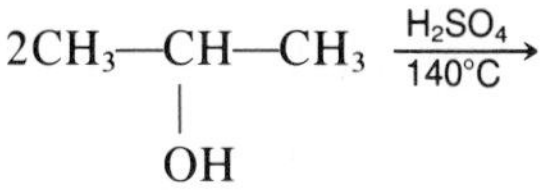

c 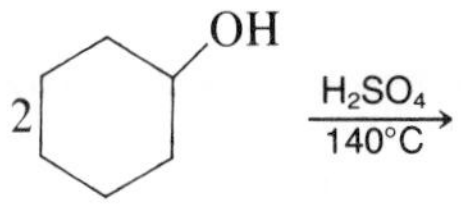

14.22 Write the structure for the product of each oxidation reaction below. If no reaction occurs, write NR.

a H_3C—OH + [O] ⟶

b 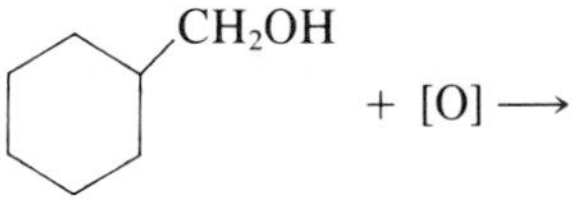

c 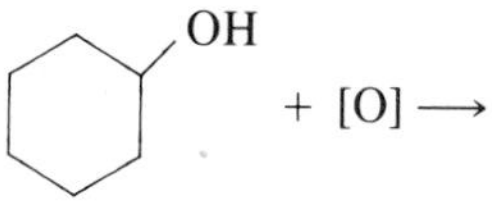

d 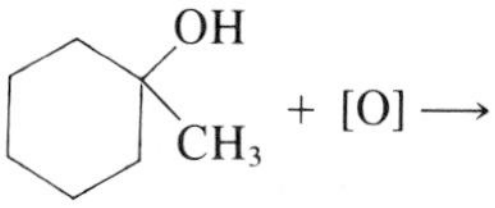

14.23 Explain how ethanol can function as an immediate antidote for methanol poisoning.

14.24 Circle the ether functionality in each of the following natural substances:

a 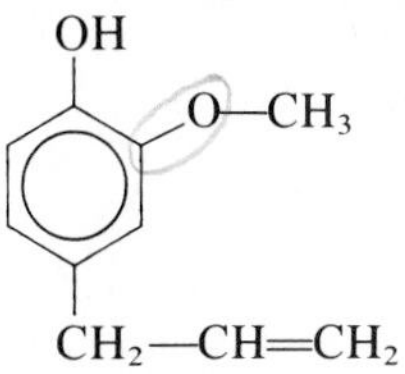

Eugenol
(found in cloves)

b 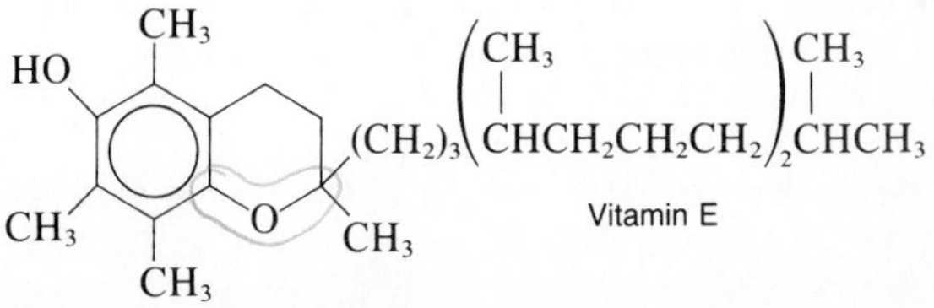

Vitamin E

c 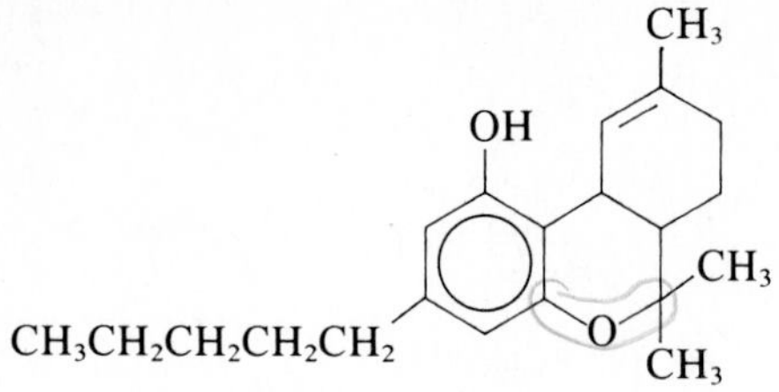

Tetrahydrocannabinol

14.25 Give an unambiguous name for each of the following ethers:

a $CH_3{-}O{-}CH_2CH_3$

b 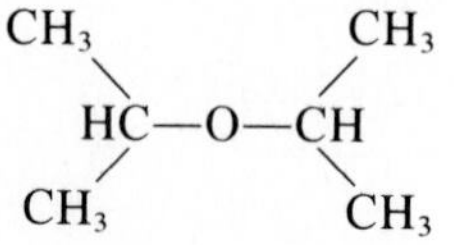

c $CH_3CH_2{-}O{-}CH_2CH_2CH_3$

14.26 State two problems with the use of diethyl ether as a general anesthetic.

• 14.27 Circle and label the aldehyde and ketone functional groups in the following compounds:

a 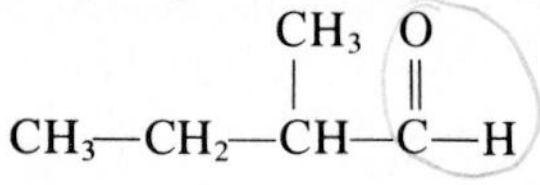

b 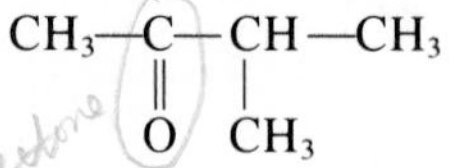

c 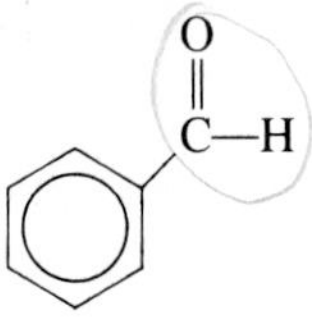

d 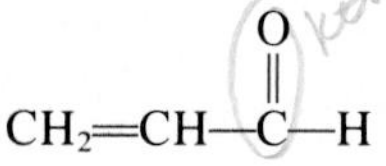

e 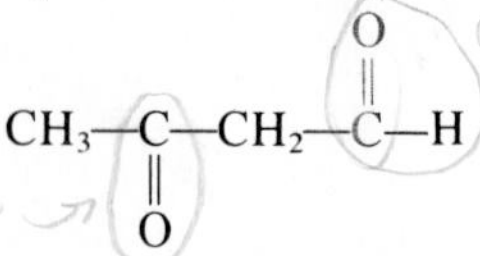

f 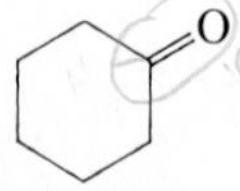

• 14.28 Give a correct chemical name to each of the following aldehydes and ketones:

a 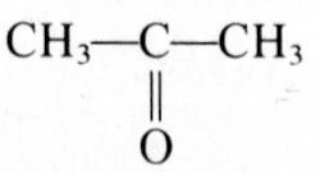

b 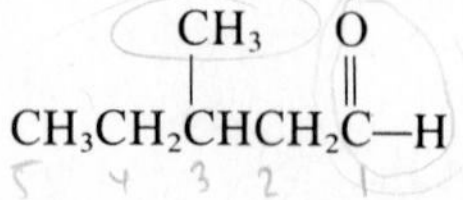

c 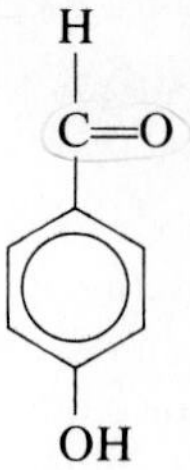

d 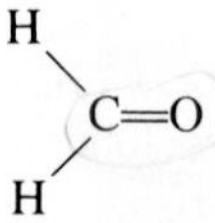

e 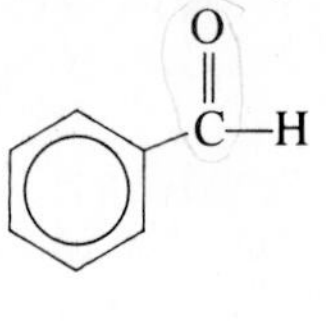

f 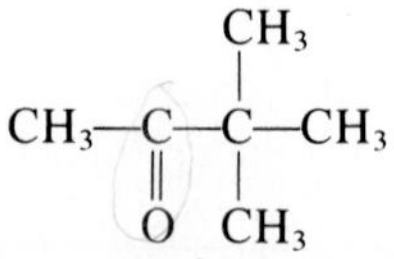

14.29 Aldehydes and ketones boil at a higher temperature than ethers and alkanes of similar molecular weight.

a Is hydrogen bonding responsible for this observation? Explain.

b If your answer in part *a* is no, please provide an alternative explanation of the original observation.

14.30 Aldehydes and ketones boil at a lower temperature than corresponding alcohols of similar molecular weight. Explain.

14.31 Write the structure of the product in the following oxidation and reduction reactions. If no reaction occurs, write NR.

a 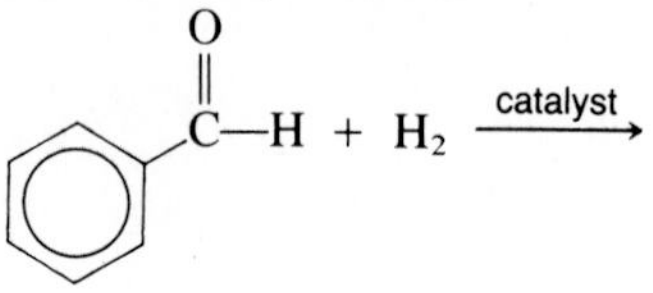

b 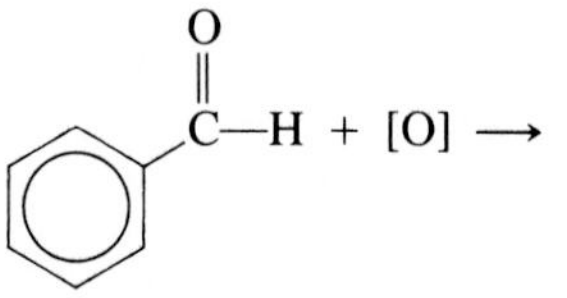

c 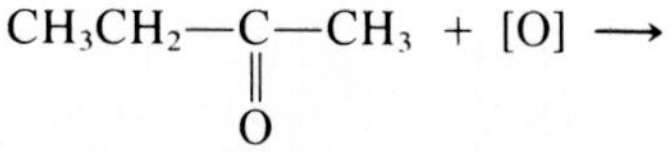

d 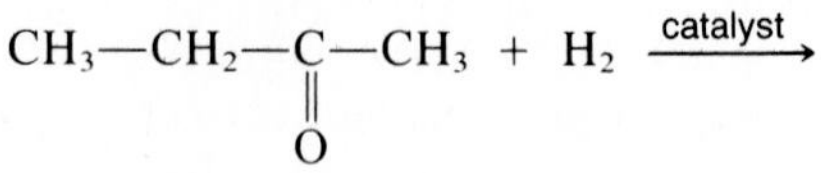

e 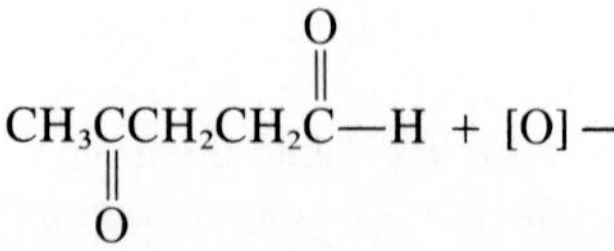

14.32 Explain how Tollens' reagent can be used to distinguish between aldehydes and ketones.

• 14.33 Which of the following substances will give a positive Benedict's test?

a 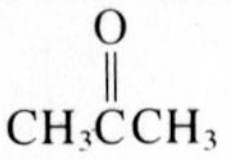

b 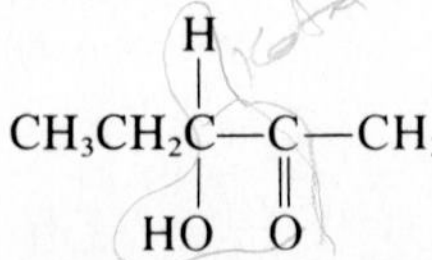

c 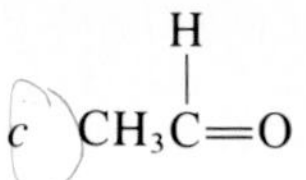

d 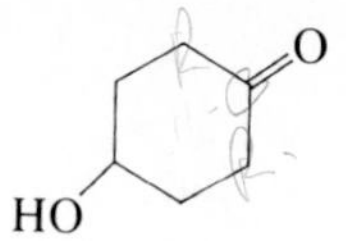

14.34 What chemical element does NADH/H^+ supply when it acts as a reducing agent?

• 14.35 Write the structure of the hemiacetal (or hemiketal) formed in the following addition reactions:

a 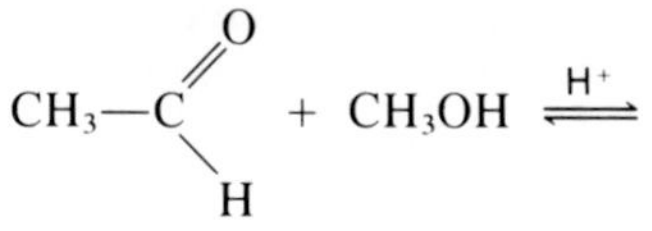

b 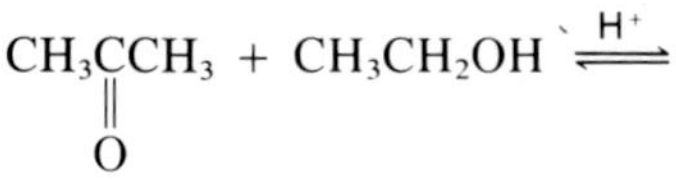

c 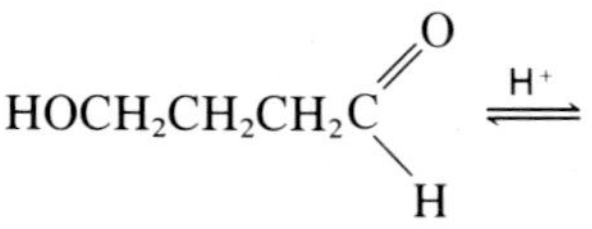

d 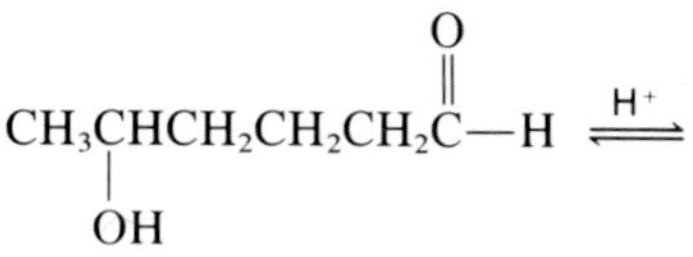

14.36 *a* Write the structure of the acetal formed when the product in Problem 14.35*c* is combined with CH_3OH.

b Write the structure of the acetal formed when the product in Problem 14.35*d* is combined with CH_3OH.

14.37 *a* The chemical change $Cu^{2+} \longrightarrow Cu^+$ is said to be reduction. What definition of reduction is being used here?

b Into what category of substance does Benedict's reagent fit?

14.38 Describe how Benedict's reagent might be used as a quantitative test for carbohydrate in urine samples.

• 14.39 Indicate which of the following structures contain a hemiacetal or hemiketal functional group:

a $CH_3—O—CH_2CH_2OH$

b 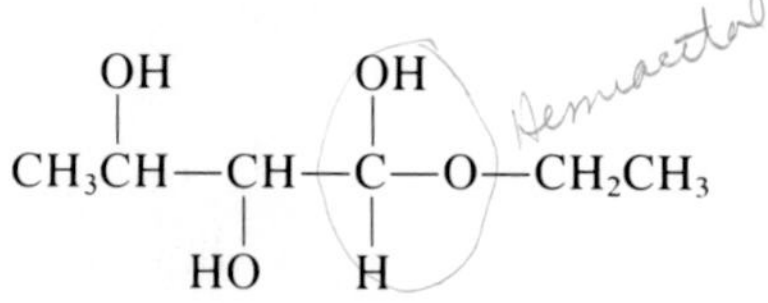

c 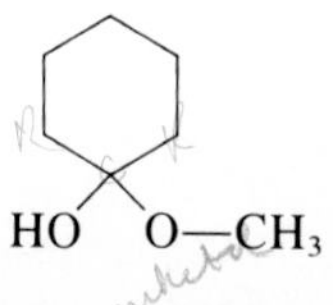

d 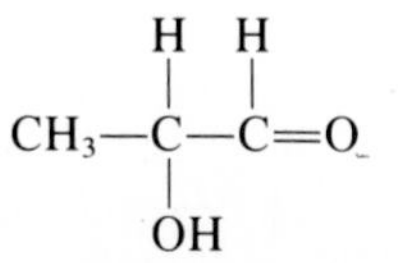

14.40 A carbohydrate, A, when heated in acid solution breaks down into two smaller compounds, B and C. State a possible hypothesis that could explain this observation

• 14.41 Write the structure of the aldol addition product formed in each of the following examples:

a 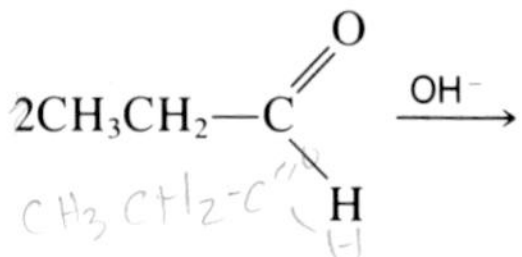

b 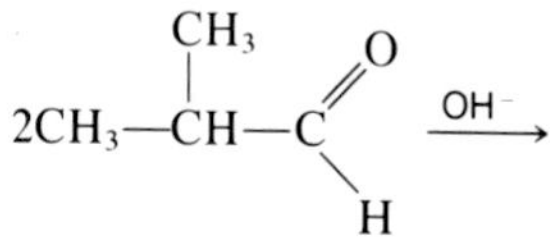

c 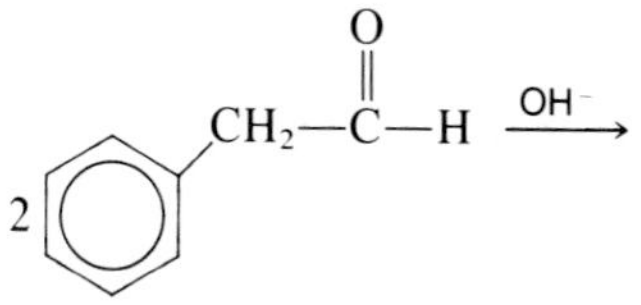

14.42 Write the structure of the imine formed in the following reactions:

a 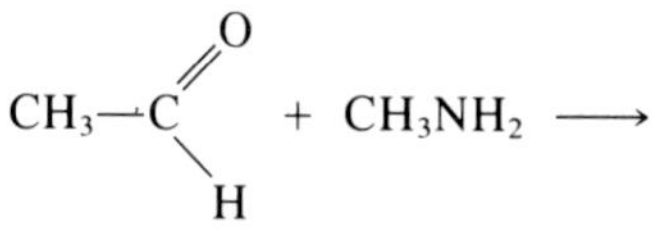

b 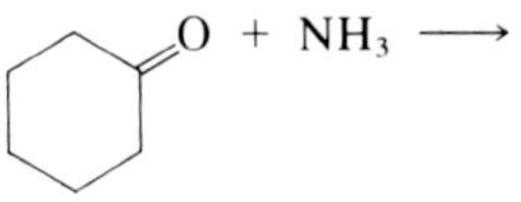

c 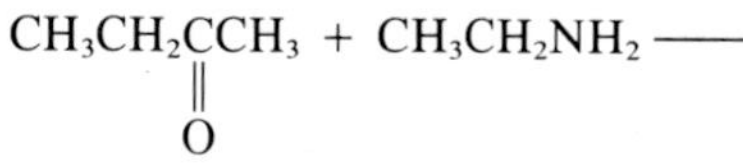

14.43 Circle and label any aldehyde or ketone functional groups which appear in the following physiologically active compounds:

a 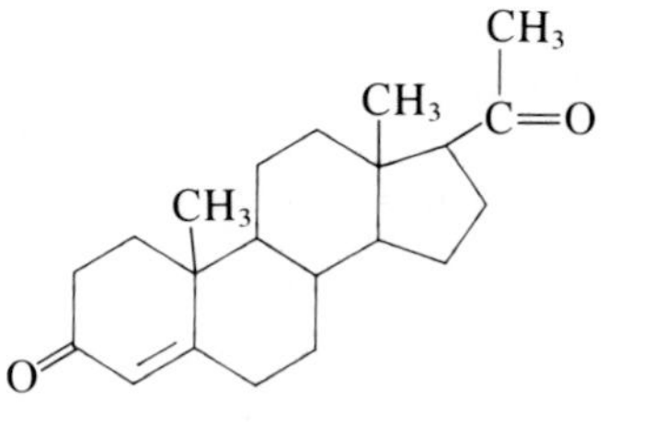

Progesterone

b 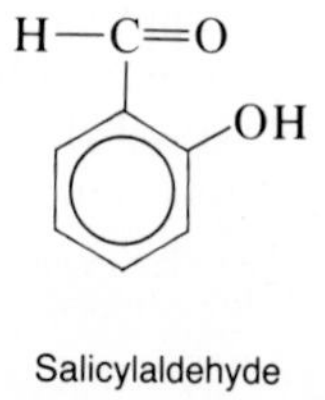

Salicylaldehyde

c 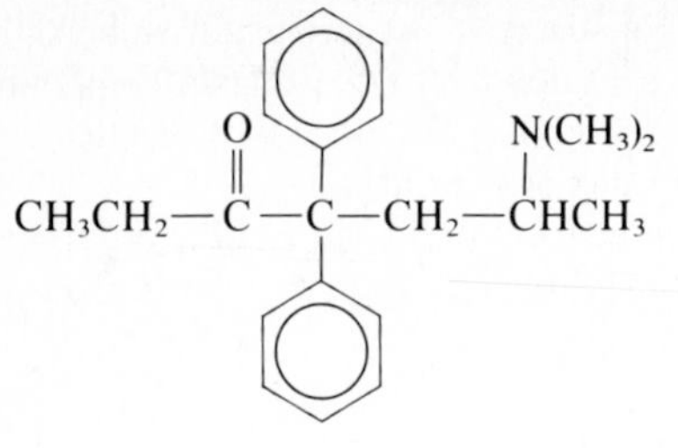

Methadone

d 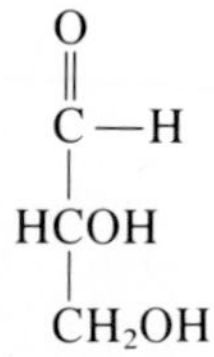

Glyceraldehyde

e 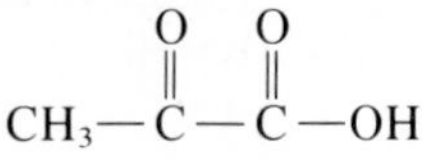

Pyruvic acid

•14.44 If acetaldehyde is the initial breakdown product of ethyl alcohol, what do you predict is the initial breakdown product of methyl alcohol?

$$CH_3CH_2OH \longrightarrow CH_3-\overset{\displaystyle O}{\overset{\|}{C}}-H$$

Ethyl alcohol → Acetaldehyde

$$CH_3OH \longrightarrow ?$$

Methyl alcohol

14.45 How might acetaldehyde be used in the silvering of mirrors?

14.46 Would the presence of acetone in the blood or urine be detected by a Benedict's test?

14.47 What is the difference in the functional-group structure between an alcohol and a phenol?

14.48 In the hydroquinone-to-quinone equilibrium:
a Is hydroquinone a reducing or oxidizing agent?
b Is quinone a reducing or oxidizing agent?

14.49 Phenols are weak acids. However, they are more acidic than alcohols. Which of the following two compounds do you predict exhibits stronger hydrogen bonding? Explain your answer.

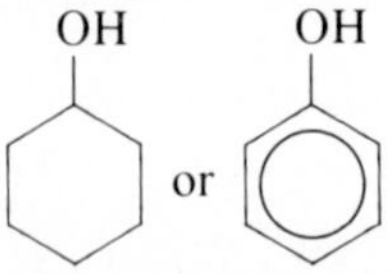

14.50 What chemical reaction would need to be carried out on a disulfide to convert it to a thiol?

14.51 Boiling point normally increases with increased molecular weight. However, ethyl alcohol (MW = 46) boils at 79°C, whereas ethyl thiol (MW = 62) boils at 36°C. Explain.

14.52 Write an equation showing phenol acting as an acid.

14.53 An alcohol upon oxidation gives a product which yields a positive Benedict's test. Was the original alcohol 1°, 2°, or 3°? Explain your reasoning.

14.54 Would you predict phenols to be more soluble in pure water or in aqueous sodium hydroxide? Explain.

14.55 An unknown compound is believed to be either a 10-carbon alcohol or a 10-carbon phenol. The unknown dissolves in sodium hydroxide solution. Which answer is more reasonable?

14.56 Arrange the following compounds in order of increasing boiling point, with the lowest on the left and the highest on the right:

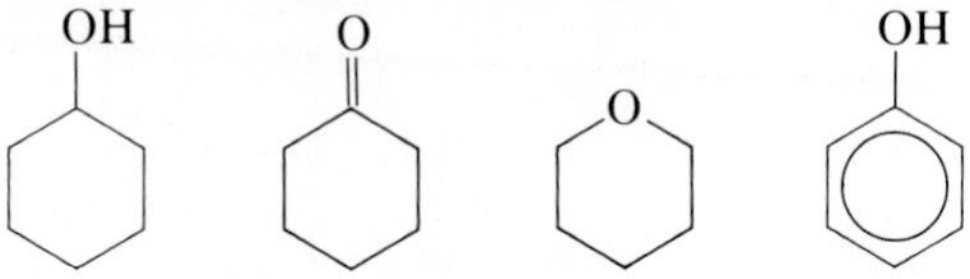

14.57 Given that the carboxylic acid group does not undergo oxidation, write the structure for the oxidation product in the reaction below:

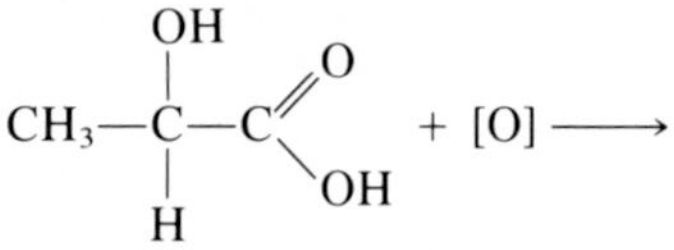

Lactic acid

•14.58 Label the functional group in each of the following compounds as an ether, an alcohol (classify further as 1, 2, or 3°), a phenol, an aldehyde, a ketone, a thiol, or a disulfide:

a 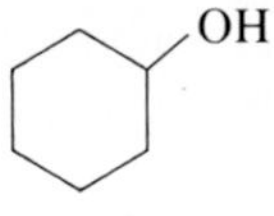

b CH_3SH

c 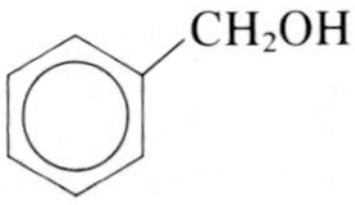

d 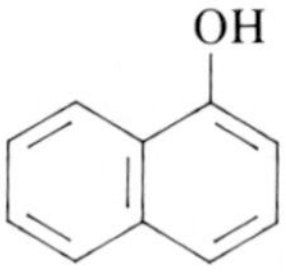

e 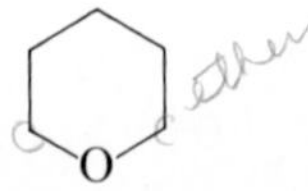

f 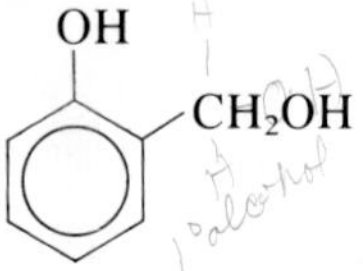

g 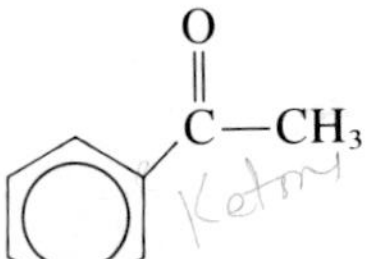

h 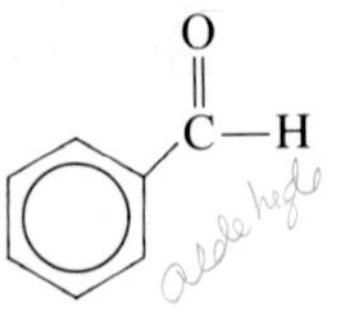

14.59 Write the structures of the alcohols that you would use to form the following aldehydes and ketones in oxidation reactions:

a $CH_3-\underset{\displaystyle O}{\underset{\|}{C}}-CH_3$

b 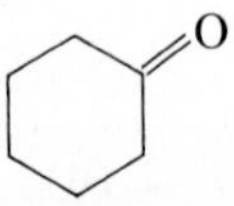

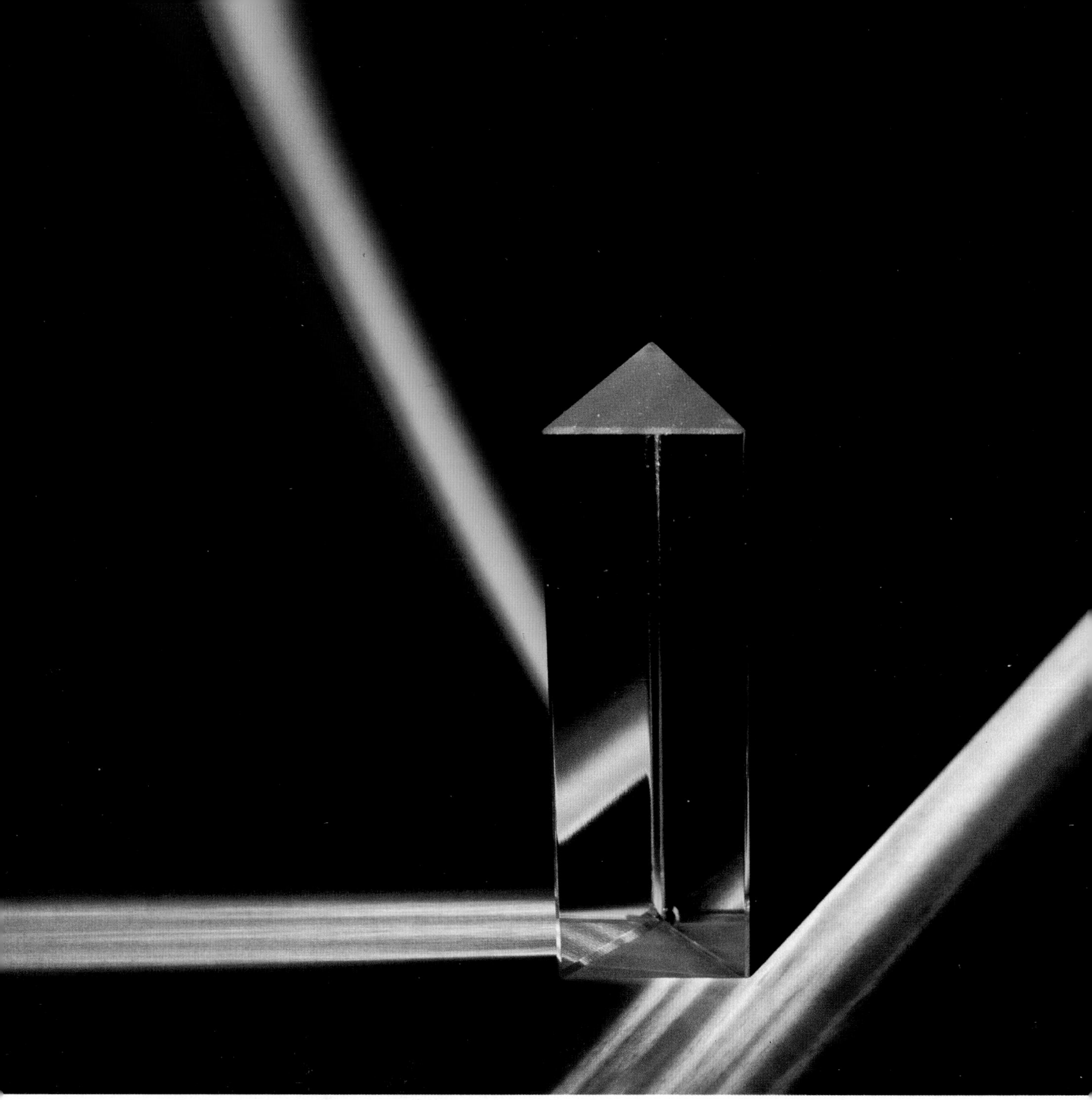

Plate 1

Formation of a spectrum as light passes through a prism. See Section 4.6, Evidence and uses of electron energy levels. *(Richard Megna, Fundamental Photographs)*

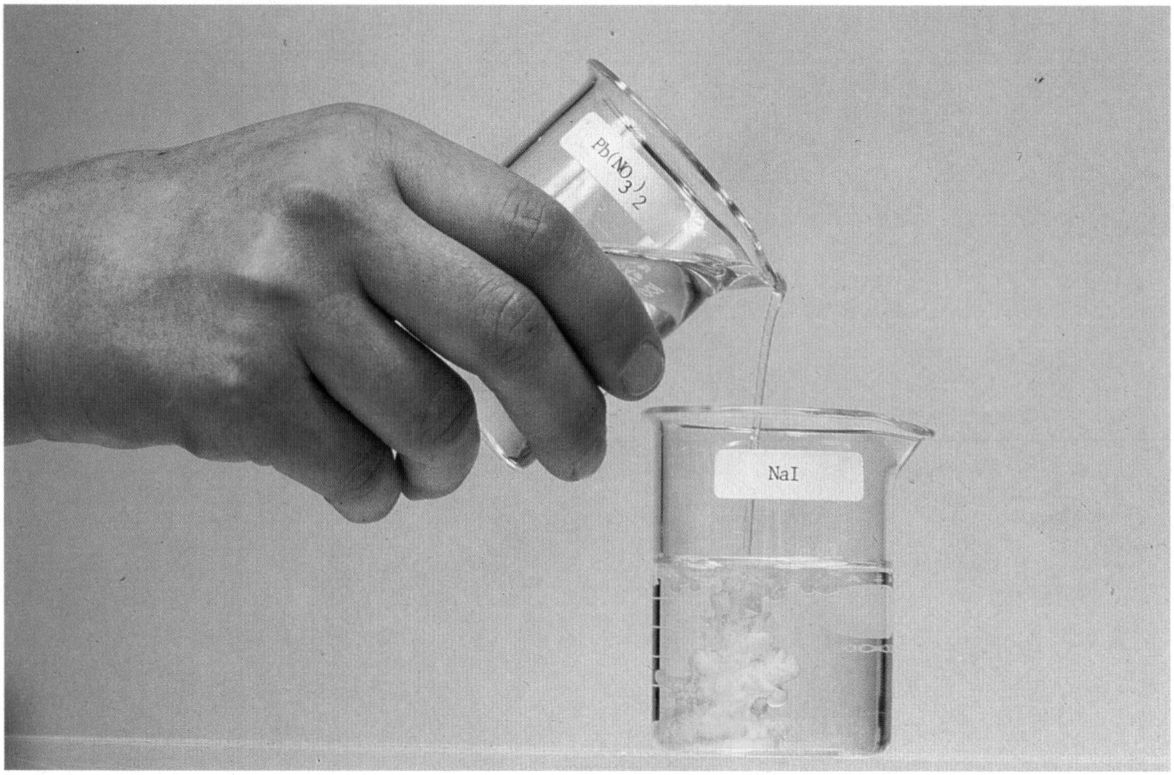
Pb(NO3)2
NaI

Plate 2

Formation of a yellow precipitate (PbI_2) from two clear solutions when $Pb(NO_3)_2$ and NaI are mixed. See Section 7.11, Double replacement reactions, and Section 9.7, Ionic equations. *(Yoav/PhotoTake)*

80
±5%
100 ml
40
20

Plate 3

Test for alkenes by adding bromine to a solution. Full-strength bromine in the bottle is diluted with water in the dropper for safety. When the dilute bromine comes into contact with the cyclohexene in the beaker, its orange-brown color disappears, indicating the presence of an alkene. See Section 13.8, Halogenation. Also note the gaseous bromine above the liquid bromine in the bottle. *(Yoav/PhotoTake)*

Plate 4

Some of the many colors, shapes, and forms of solid chemicals. See Section 8.14, Classes of crystalline solids. (a) Salt (NaCl) and gypsum crystals ($CaSO_4 \cdot 2H_2O$) underwater in the Afar Rift area. *(Vulcain-Explorer/Photo Researchers)* (b) Wulfenite ($PbMoO_4$) from Mexico. *(Tom McHugh/Photo Researchers)* (c)Uranium ore. *(Russ Kinne/Photo Researchers)*

46
47
48
49
50

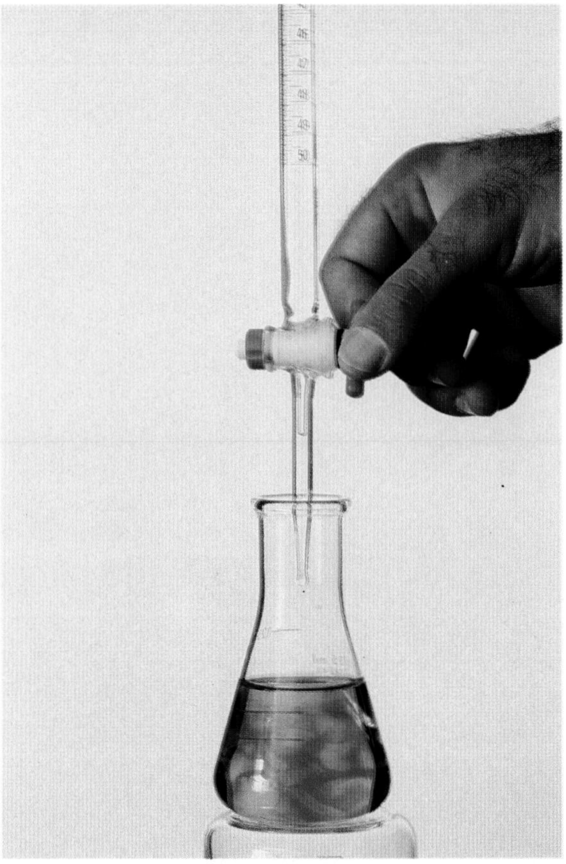

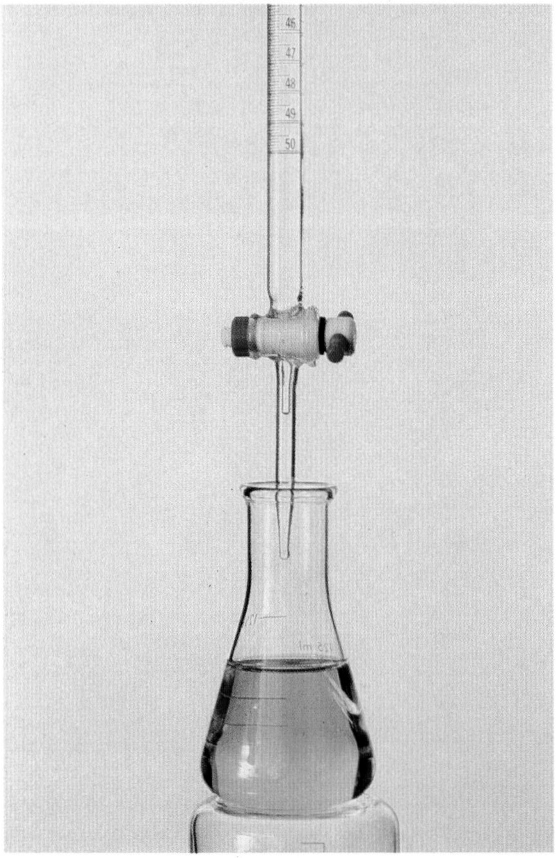
46
47
48
49
50

Plate 5

Titration using bromcresol purple as an indicator. The NaOH in the beaker loses the initial purple color of the indicator as HCl is added from the buret; at approximately pH 6, the color changes to yellow. See Section 11.9, Determination of acid-base concentrations. *(Yoav/PhotoTake)*

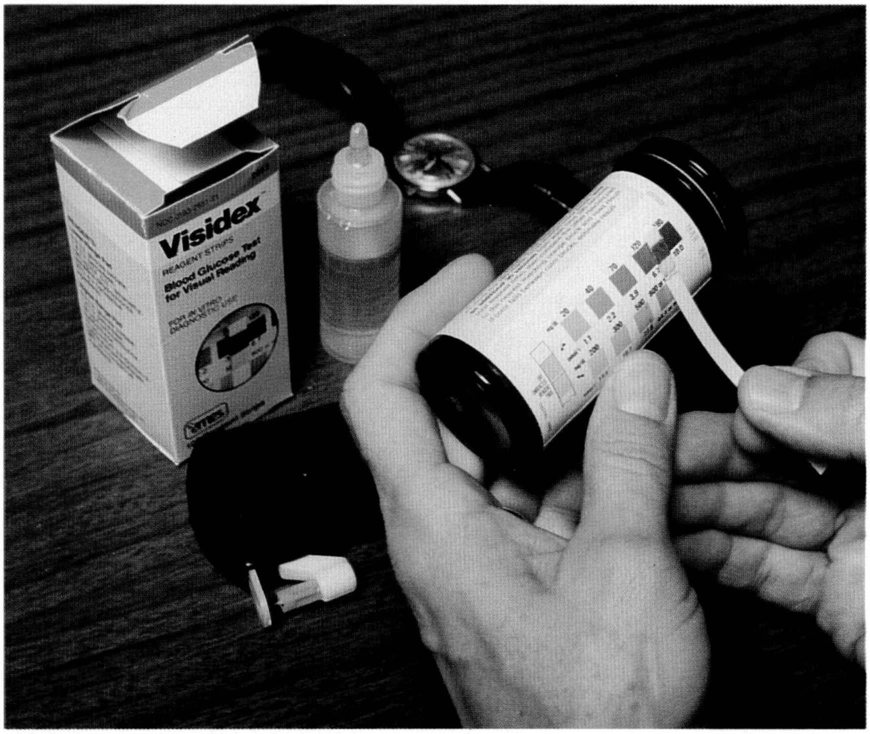
Visidex
REAGENT STRIPS
Blood Glucose Test
for Visual Reading

Plate 6
People with diabetes can use kits like the Visidex system shown here to monitor their blood sugar levels. The reagent strip contains enzymes that react with glucose and indicators that change color depending on the extent of that reaction and thus the concentration of sugar in a blood sample. Patients match the color of the strip to the color key on the side of the container; the green shown here represents a high sugar level for nondiabetics and a good level for diabetics. See Section 21.12, Medical uses of enzymes. *(Ames Division, Miles Laboratories, Inc.)*

Plate 7
A nurse with some of the many chemicals she and other allied-health professionals use in modern hospitals. *(William Hubbell/Woodfin Camp)*

Plate 8

Insoluble azo compounds used as pigments in dyes: toluidine red and dinitroaniline orange. Some other beautifully colored azo compounds are suspected of being carcinogens. See Section 16.11, Amines as carcinogens. *(Ciba-Geigy Plastics and Additives)*

37 ml

Plate 9

Paints admirably display the properties of mixtures. See Section 1.5, Elements, compounds, and mixtures. *(Randy Matusow)*

Plate 10
Effects of nitrogen fertilizer: increased number of roots and greater overall growth. See Section 25.2, Nitrogen fixation. *(Arcadian Corp.)*

Plate 11
Statues are susceptible to damage by the environment. Hazards include oxidation of metals and reactions with acid rain. Marble statues are largely $CaCO_3$ which decomposes through reactions with acids. See Section 11.8, Reactions of acids with carbonates and bicarbonates. *(PhotoTake)*

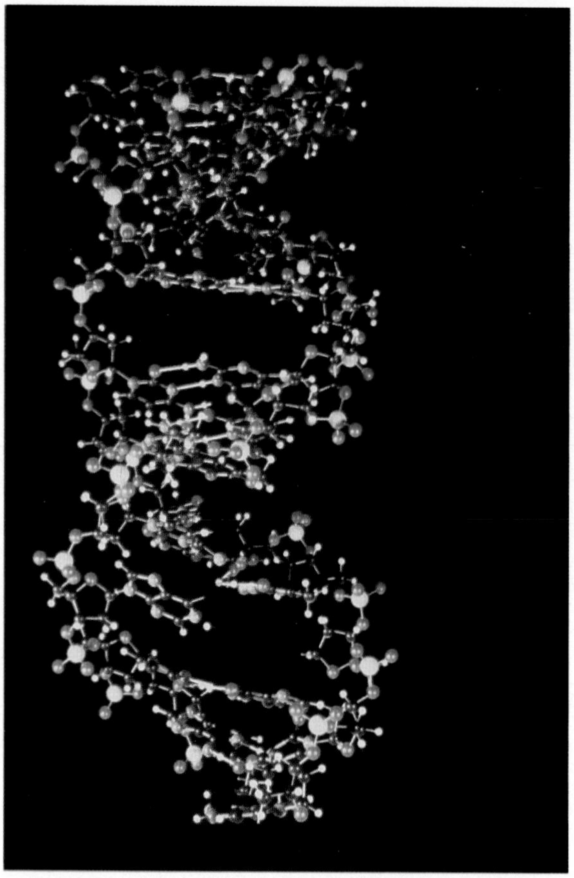

Plate 12
The double-helix of DNA, shown by a molecular model. See Section 20.4, The secondary structure of DNA. *(David Attie/PhotoTake)*

Plate 13
An entire DNA molecule, shown by an electron microscope photograph. *(PhotoTake)*

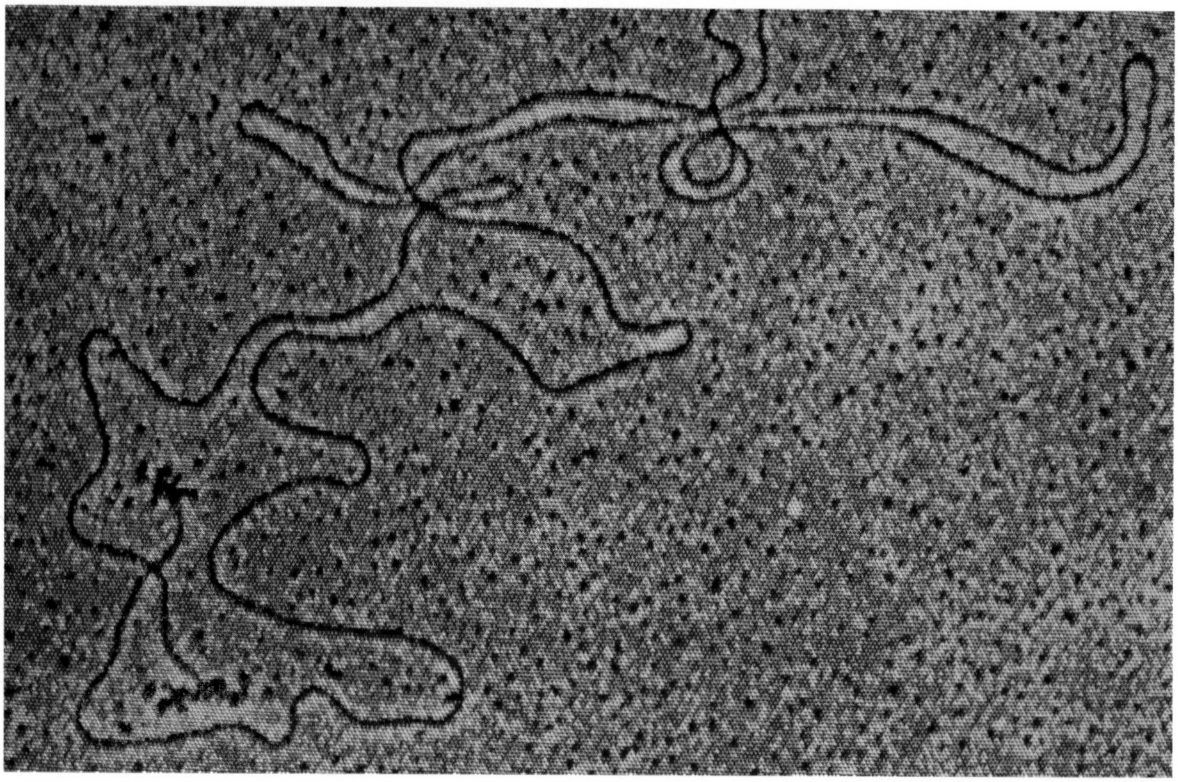

c 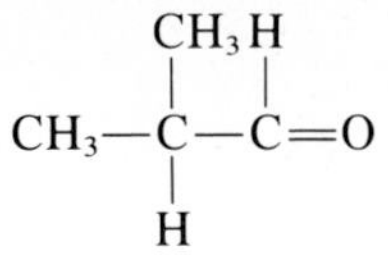

e 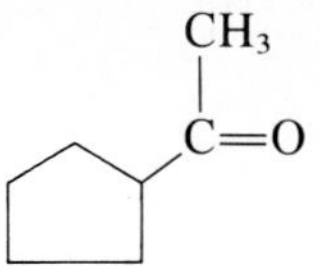

d 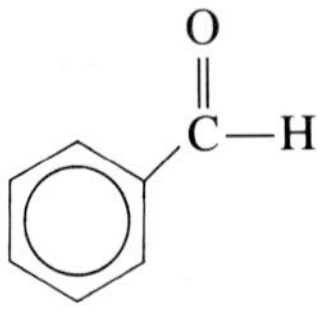

14.60 Write an equation with structural formulas demonstrating each of the following reactions:

a Oxidation of an alcohol to a ketone
b Oxidation of an alcohol to an aldehyde
c Addition of an aldehyde to an alcohol to form a hemiacetal
d Reduction of a ketone to form an alcohol
e Oxidation of an alcohol to form a carboxylic acid

14.61 Identify the following chemical changes as oxidation or reduction of the *organic reactant:*

a 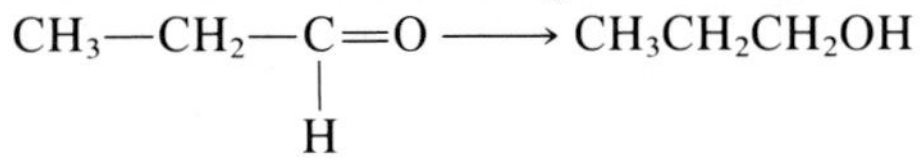

b $2CH_3SH \longrightarrow CH_3-S-S-CH_3$

c 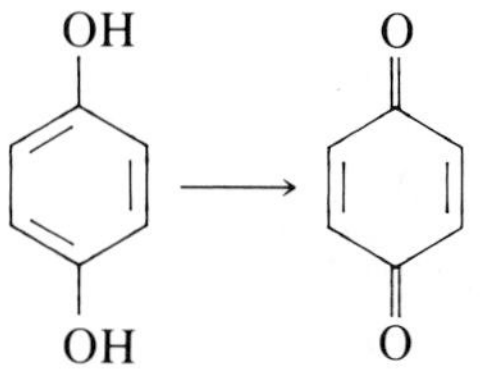

d 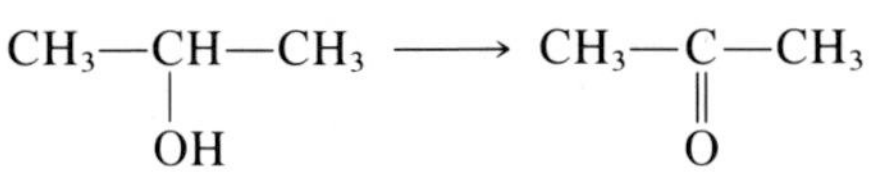

e 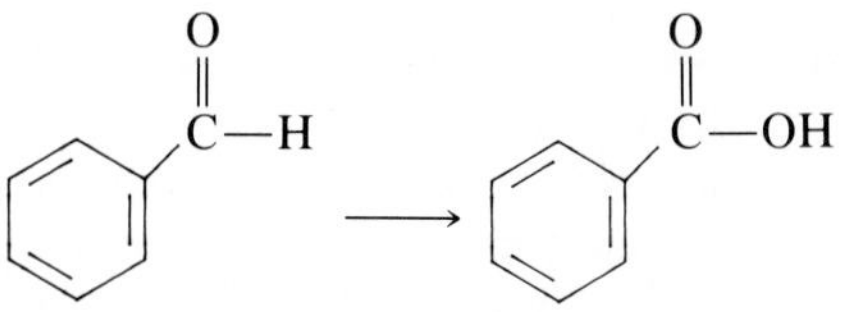

14.62 Write the structural formula for each of the following:

a 3-Methyl-1-pentanol
b 3-Methylpentanal
c 3-Methylpentanone
d Cyclopentanol
e Cyclopentanone
f Isopropyl thiol
g 4-Nitrophenol
h Ethyl methyl ether
i Diphenyl ether
j Glycerol
k Butyraldehyde

14.63 Write structural formulas for X and Y:

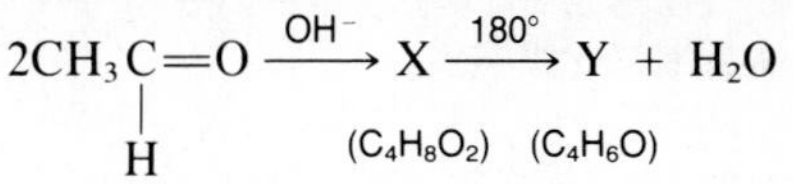

14.64 Write the structural formula for the organic product of each of the following reactions. If no reaction occurs, write NR.

a 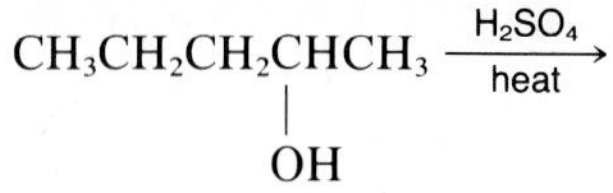

b 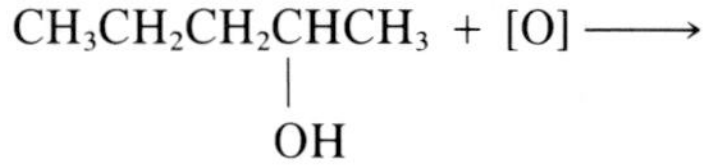

c 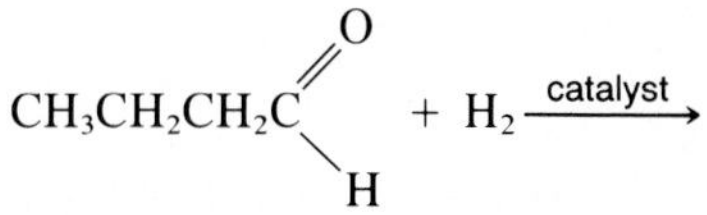

d 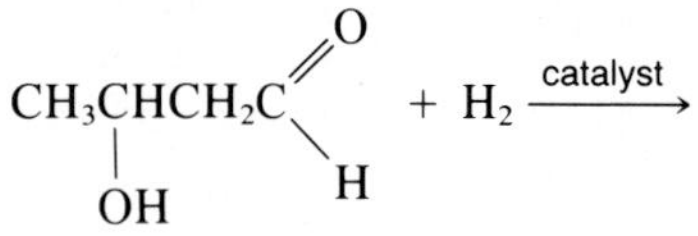

e $2CH_3CH_2OH \xrightarrow{140°}$

f 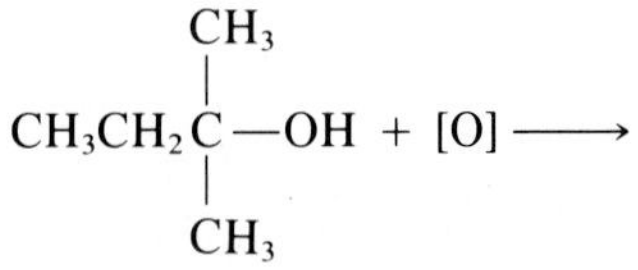

g 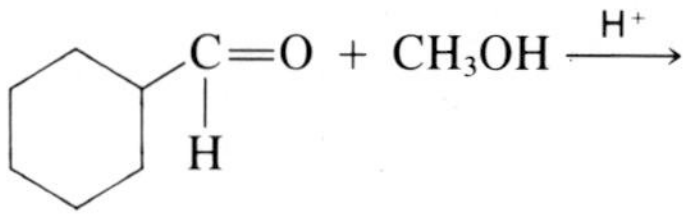

h 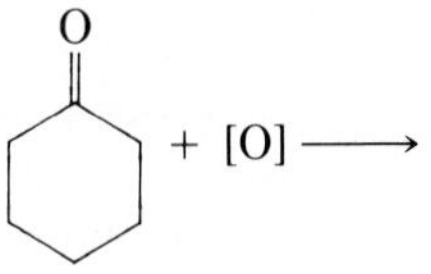

i 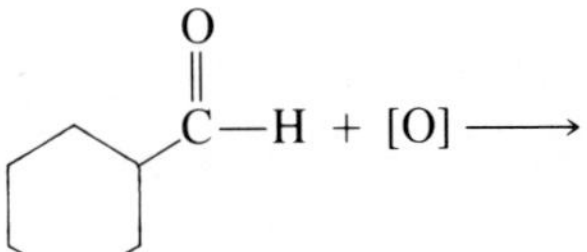

j 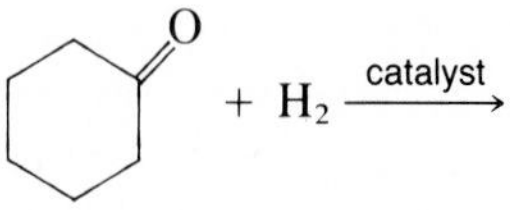

k 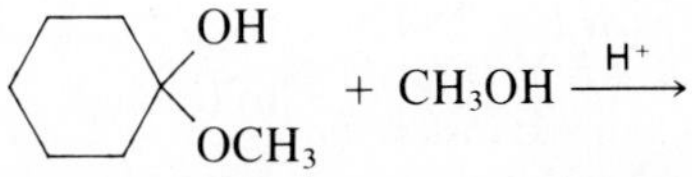

CHAPTER FIFTEEN

CARBOHYDRATES

15.1 INTRODUCTION

Now that you have learned the chemistry of the alcohol, aldehyde, and ketone functional groups, you are ready to consider the structure and reactions of one of the major classes of foodstuffs, the carbohydrates. **Carbohydrates** are defined as polyhydroxyaldehydes and polyhydroxyketones or materials which can be broken down into structures with those functional-group classifications. Glucose, or blood sugar, and starch (corn starch, potato starch) are common examples of carbohydrates. The many hydroxy groups and the aldehyde group are clearly evident in the structure of glucose that follows. Starch is a polymer which can be hydrolyzed into glucose monomers; this is the reaction that occurs when starch is digested.

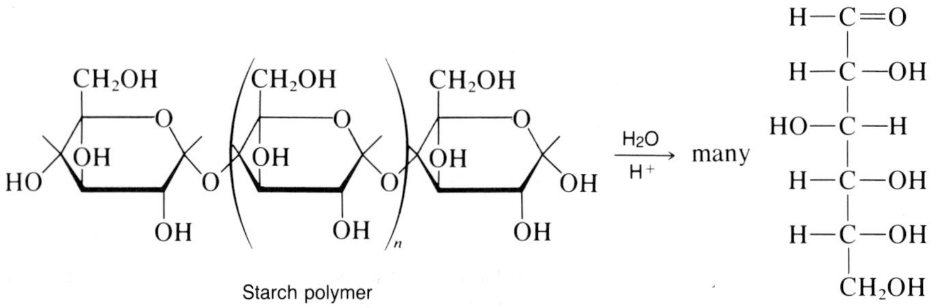

CH2OH
O
OH
HO
OH
n
H2O
H+
many
H—C=O
H—C—OH
HO—C—H
CH2OH

Starch polymer

Glucose monomer

The rings in starch probably don't look like glucose monomer units to you, but they are. The explanation of this will unfold in Section 15.9.

Carbohydrates are also called sugars or saccharides. Because, as the starch example points out, there are small monomeric carbohydrates and large polymeric carbohydrates, you will also see the term *monosaccharide* applied to sugars such as glucose and the term *polysaccharide* applied to starch. Monosaccharides are also called simple sugars.

15.2 CLASSIFICATION: ALDOSES VERSUS KETOSES

There is a systematic nomenclature for carbohydrates, but the parent or base names are all common names. For example, note the names of the following:

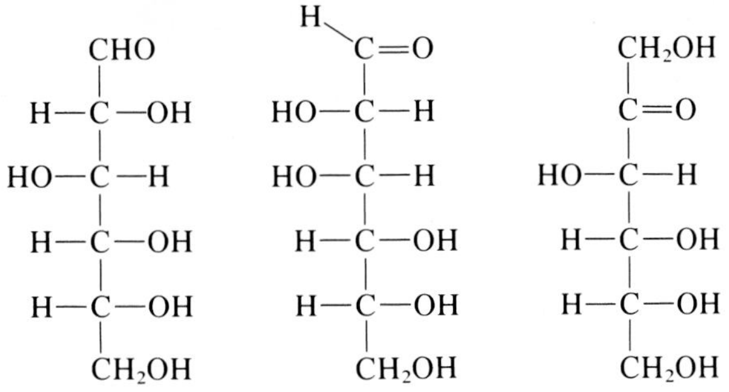

Glucose Mannose Fructose

The suffix *-ose* denotes a simple sugar or monosaccharide, but there is no system based on structure for the parent names; glucose, mannose, and fructose are historical common names. Should some substituent be appended to the parent molecule, then the nomenclature rules you have already seen apply. For example,

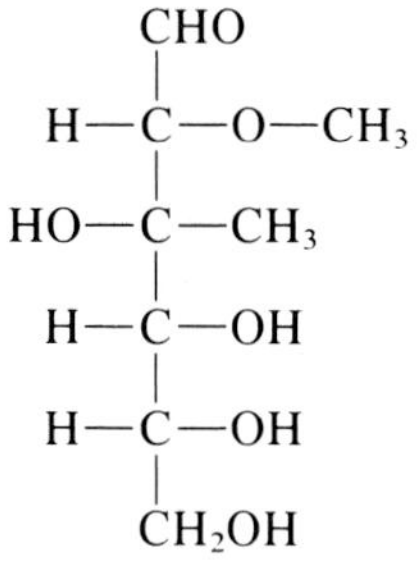

might be named 2-methoxy-3-methylglucose.

In general, we will not be concerned with the intricacies of nomenclature. But it is convenient to be aware of the various ways in which carbohydrates are grouped together or classified. One classification rests on the occurrence of an aldehyde or ketone group in the sugar. If an aldehyde group is present, the sugar is an **aldose.** The presence of a ketone group leads to the classification of a **ketose.** Thus glucose and mannose are aldoses, and fructose is a ketose. Notice that the aldehyde group is commonly written —CHO rather than showing the structural detail, H—C=O.

Problem 15.1 Classify the following as aldoses or ketoses:

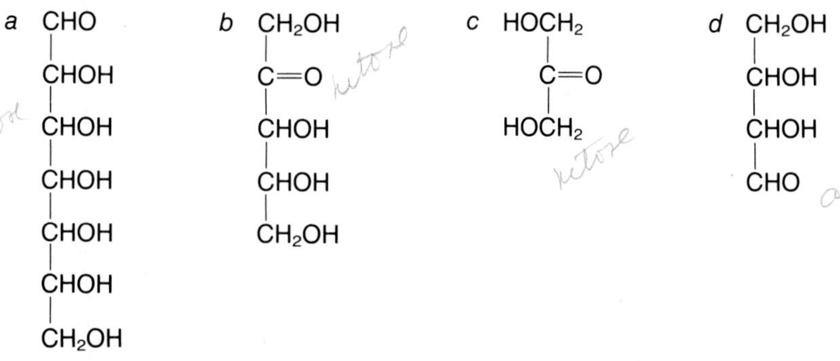

Carbohydrates are frequently grouped together according to chain length. Glucose, mannose, and fructose are all *hex*oses, that is, *six*-carbon sugars.

Similarly *pentoses* are *five*-carbon sugars. For sugars of five or more carbons the familiar numerical roots are used, whereas three-carbon and four-carbon sugars are classified as trioses and tetroses, respectively. Both classifications—functional (aldehyde vs. ketone) and chain length—can be reflected in one name. Thus, glucose and mannose are both aldohexoses, and fructose is a ketohexose.

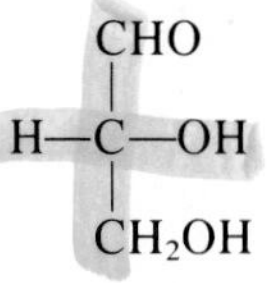

An aldotriose

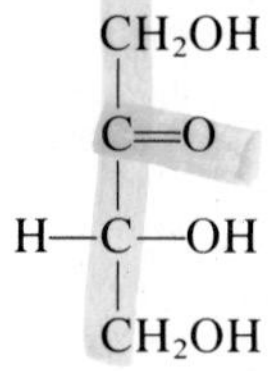

A ketotetrose

Problem 15.2 Add a chain-length classification to the functional classification which you made in Problem 15.1.

15.3 STEREOISOMERISM

Before proceeding with the discussion of carbohydrates, we must digress and consider a kind of isomerism that you have not previously encountered. Many physiologically active organic compounds such as glucose and other carbohydrates exhibit a type of isomerism known as *stereoisomerism*. Biological systems require specific stereoisomers in their reactions and will not function if the wrong isomer is supplied. Thus it is necessary to be aware of what stereoisomers are.

What Are Stereoisomers?

The atoms of two stereoisomers are connected in exactly the same sequence, but differ with respect to their arrangement in three-dimensional space. That is, examining the structural formulas of two stereoisomers written in two dimensions reveals no obvious difference. However, examining three-dimensional representations or molecular models shows that the two isomers have unique, nonidentical arrangements of their atoms, i.e., different constructions in space. The prefix *stereo-* is derived from a word meaning "space."

An example of stereoisomerism can be seen by carefully examining the structural formula for the simplest sugar, glyceraldehyde. In two dimensions one might write structures for this compound such as

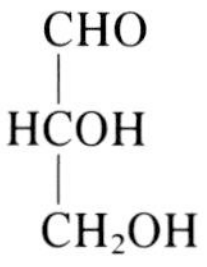

and

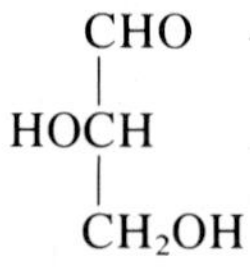

In two dimensions these structures, which differ only in the placement of H and OH on the central carbon, do not appear to represent isomers. However, if you consider the actual three-dimensional array around the central carbon, you will find that there are two unique compounds because there are two unique spatial arrangements of atoms. In the following drawings, the central carbon is in the plane of the paper, the wedges indicate that the atoms attached are coming toward you in front of the paper plane, and the dashed lines extend away from you behind the paper plane.

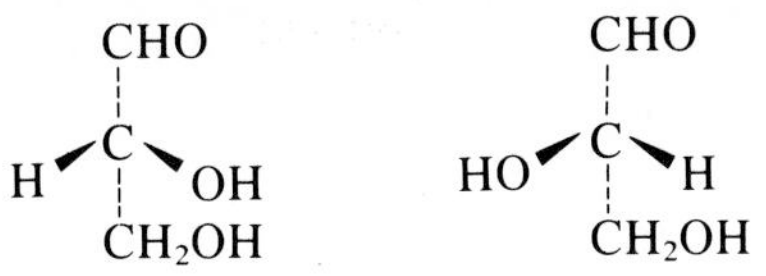

I II

Structures I and II are not identical because they are *not* superimposable. No manipulation will allow you to overlap all atomic centers simultaneously. Try sliding I over II in the plane of the paper. The central carbons in the plane of the paper will overlap, as will the —CHO and —CH_2OH groups which extend behind the paper. But the H and OH which are coming forward will not overlap. Figure 15.1 shows this.

Try another test of superimposition. Turn II 180° in the paper plane to give

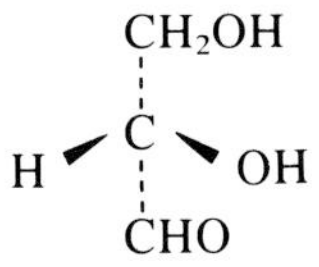

Now try to slide I over it. In this case the central carbons will overlap and the H and —OH groups will overlap, but —CH_2OH and —CHO will *not*. Try other tests and convince yourself that I and II are not superimposable—molecular models are very helpful if they are available to you.

Though not superimposable, I and II do bear a definite relationship to one another. They are related in the same way that your right and left hands are: right and left hands, right and left feet, right and left ears are nonsuperimposable mirror images. Figures 15.2 and 15.3 depict the concept of object and mirror image relationships.

Stereoisomers such as I and II which bear the relationship of being nonsuperimposable mirror images are called **enantiomers.** Note that to be enantiomers two structures must be *both* nonsuperimposable and mirror images. All structures have mirror images but usually the object and mirror image can be superimposed. Consider the representation below of 2-propanol and its mirror image:

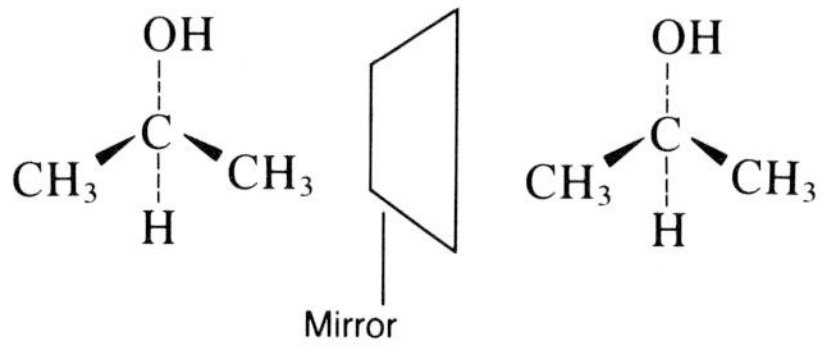
Mirror

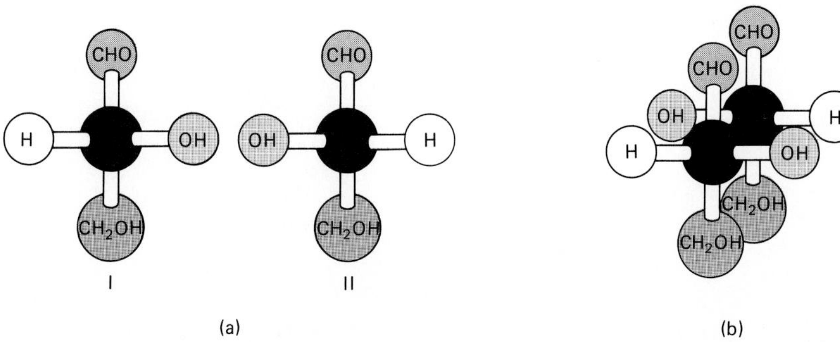
CHO
H
OH
CH2OH
I
II
(a)
(b)

FIG. 15.1
Structures such as I and II are nonsuperimposable. (*a*) Three-dimensional models of I and II. H and OH come forward; CHO and CH_2OH go back. (*b*) Sliding I over II in the plane of the paper shows nonsuperimposability. The H's and OH's do not overlap.

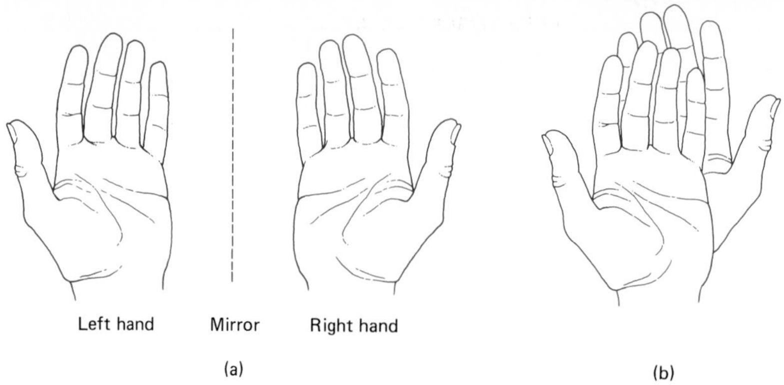
Left hand
Mirror
Right hand
(a)
(b)

FIG. 15.2
(*a*) The mirror reflection of a left hand is a right hand. (*b*) Left and right hands are not superimposable.

2-Propanol and its mirror image are superimposable. Just slide the object into the mirror and all atomic centers overlap.

Chirality

Is there an easy way to tell which compounds like glyceraldehyde will display enantiomerism and which (like 2-propanol) will not? The answer lies in an examination of the central carbon in these structures. Notice that there are four different groups (H, OH, CHO, CH_2OH) around the central carbon in glyceraldehyde. A carbon atom with four different groups bonded to it is

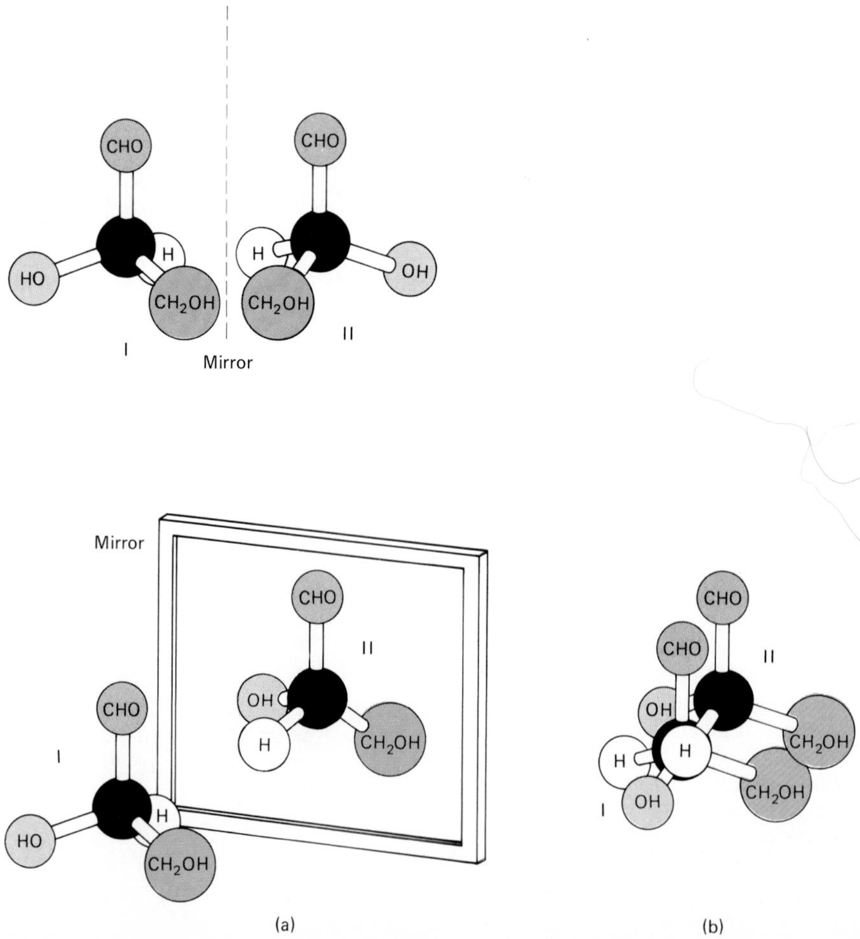
CHO
HO
H
CH2OH
I
Mirror
CHO
H
OH
CH2OH
II
Mirror
CHO
II
OH
H
CH2OH
CHO
I
HO
H
CH2OH
(a)
CHO
CHO
II
OH
H
H
CH2OH
CH2OH
I
OH
(b)

FIG. 15.3
(*a*) The two isomers of glyceraldehyde, I and II, are related as object and mirror image. This can be depicted by actually showing the mirror. More often the mirror is represented by a line drawn between the two structures, as shown in the top illustration. (*b*) Imagine bringing image II out of the mirror and trying to overlap its substituents with those of I. This attempt shows the nonsuperimposability.

called a **chiral carbon.** In general, any atom with four different groups bonded to it is called a chiral center. It turns out that any compound containing a single chiral carbon is a chiral molecule and can exist as a pair of enantiomers. In the case of 2-propanol the central carbon atom has only three different kinds of groups (H, OH, and two CH_3's). There are no chiral carbons in 2-propanol; the molecule is therefore achiral (*not* chiral), and enantiomerism is not possible.

The existence of *one* chiral center *guarantees* that a structure will be chiral and capable of existence in two enantiomeric forms. You will see shortly in Section 15.7 that the occurrence of more than one chiral center in a molecule also leads to stereoisomerism, but the isomers are not necessarily related as object and mirror image.

15.4 PROPERTIES OF ENANTIOMERS

In biological systems, reactions that require or produce compounds with chiral carbons generally require or produce only one enantiomer. This is because, as you will see in Chapter 21, reactions are catalyzed by enzymes in such a way that reacting molecules must be able to "fit together," that is, shape is important. The concept of chirality applies here in the same sense that right feet only fit comfortably into right shoes. The important point is that enantiomers can often be distinguished from one another biologically: one is active and one is not, or, in some cases, one is an essential nutrient or beneficial drug and the other is a deadly poison.

Whereas there may be dramatic differences in physiological properties between enantiomers, their chemical and physical properties are remarkably similar. Enantiomers react the same in most ordinary chemical reactions, and their physical properties such as melting point, boiling point, solubility, etc., are identical. For example, the two enantiomers of glyceraldehyde, designated I and II, have the same experimental melting point. The one notable and easily detected difference between enantiomers is their interaction with plane-polarized light.

Interaction of Enantiomers with Polarized Light

The energy of ordinary light from the sun or a light bulb comes toward us in all planes perpendicular to an imaginary line connecting our eye to the source of light. These planes are represented in Figure 15.4*a* by the many arrows arranged like spokes on a wheel. Some substances, such as calcite (crytalline $CaCO_3$) and the synthetic plastic Polaroid, when placed in a beam of ordinary light, act as a filter and allow only one of the possible planes of light to pass through. This is shown in Figure 15.4*b* by the emergence of only one "spoke" of light. The light passing through the calcite or Polaroid filter is said to be **plane-polarized** and the filter itself, a polarizer.

If plane-polarized light is passed through a solution of achiral molecules, it emerges unaffected; i.e., it remains in the same plane. When plane-polarized light passes through a sample of a chiral compound, the plane of light is rotated, as shown in Figure 15.4*c*. A device called a polarimeter is used to detect the extent of the rotation, which is called **optical rotation.**

Enantiomers rotate the plane of polarized light equal numbers of degrees but in opposite directions. That is, for example, the optical rotation of the enantiomer of glyceraldehyde designated I is $+8.7°$. The $+$ sign indicates rotation in a clockwise fashion. The other enantiomer II rotates light $-8.7°$, that is, in a counterclockwise fashion. Stereoisomers are often identified by citing their optical rotation.

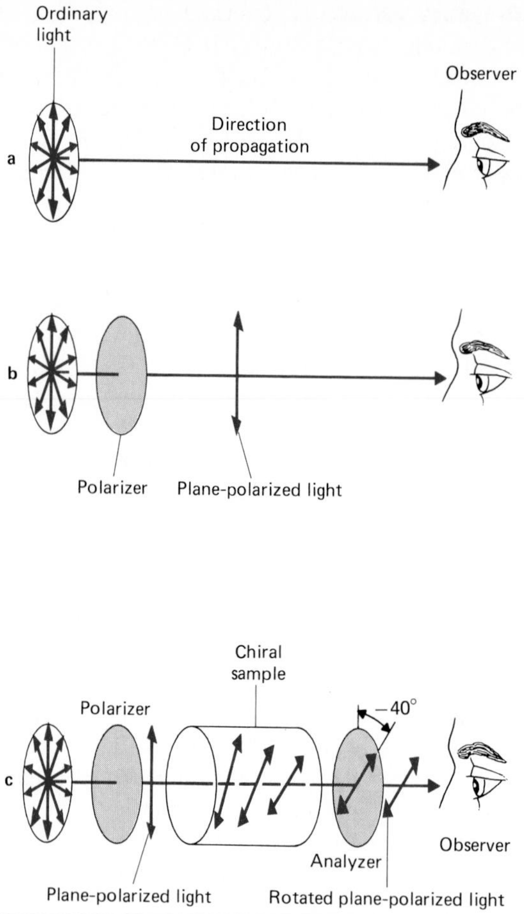
Ordinary
light
Observer
Direction
of propagation
a
b
Polarizer
Plane-polarized light
Chiral
sample
Polarizer
−40°
c
Observer
Analyzer
Plane-polarized light
Rotated plane-polarized light

FIG. 15.4
(*a*) In ordinary light, light energy radiates in all planes perpendicular to the direction of propagation. (*b*) Plane-polarized light is confined to one plane perpendicular to the direction of propagation. (*c*) When plane-polarized light passes through a sample of a chiral compound, the plane of light is rotated.

Problem 15.3 One enantiomer rotates plane-polarized light −41°. What is the sign and magnitude of rotation for a solution of the other enantiomer of the same concentration?

15.5 STEREOISOMERISM AND GLUCOSE

Now let us examine the structure of glucose with the knowledge that chiral carbons lead to stereoisomerism.

Glucose's molecular formula was determined as $C_6H_{12}O_6$ in the nineteenth century, and early in this century it was known that this compound was an aldehyde and also had five alcohol (OH) groups. This then established the structure as

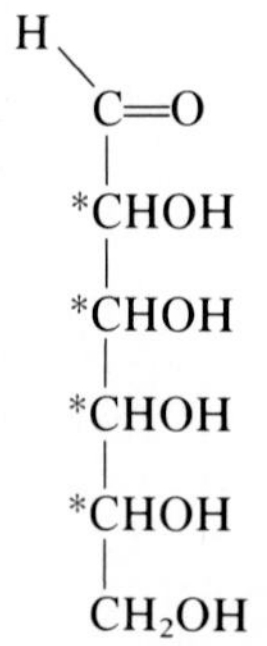
H
C=O
*CHOH
*CHOH
*CHOH
*CHOH
CH2OH

But there are 16 stereoisomers corresponding to this aldohexose structure because there are four chiral carbons (starred on structure). Figure 15.5

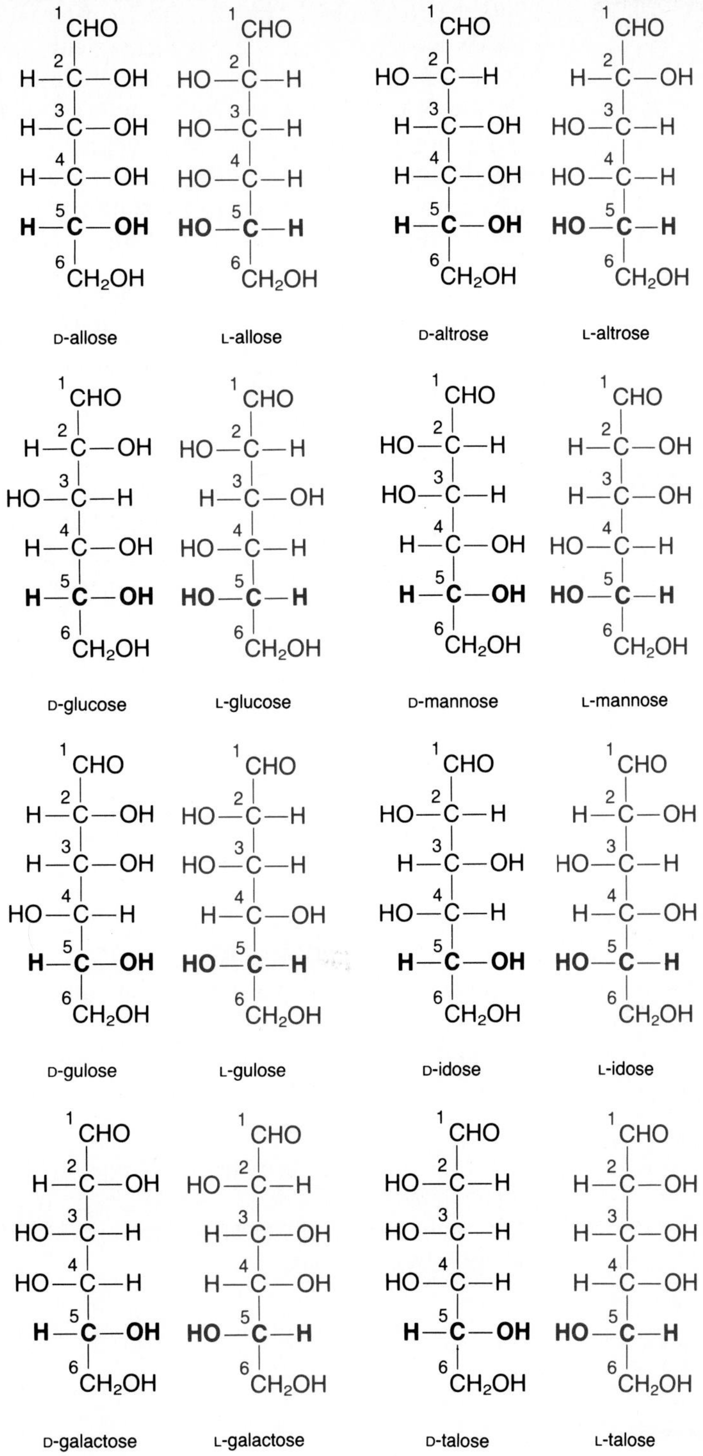
D-allose
L-allose
D-altrose
L-altrose
D-glucose
L-glucose
D-mannose
L-mannose
D-gulose
L-gulose
D-idose
L-idose
D-galactose
L-galactose
D-talose
L-talose

FIG. 15.5
Because there are four chiral carbons in the aldohexose structure, there are 16 stereoisomers, that is, eight pairs of enantiomers. Each enantiomer is classified as D or L depending on the orientation of —OH at the 5 carbon. The D-sugars are shown in black and L-sugars in color.

depicts the 16 isomers. Among these 16 different three-dimensional arrangements, glucose is by far the most prevalent stereoisomer found in nature, and it is of premier importance in carbohydrate metabolism, as you will see in Chapter 23. Only isomers very similar to glucose in structure, such as mannose or galactose, can function metabolically, and these do so often by first undergoing isomerization to the glucose structure.

In Section 15.3 you saw the two enantiomers of the triose, glyceraldehyde, drawn by using wedges and dashed lines. This type of depiction becomes cumbersome as the number of chiral centers increases. Imagine what glucose with four chiral carbons would look like with all those wedges and dashes! For that reason the most famous carbohydrate chemist, Emil Fischer (1852–1919; Nobel laureate, 1902), suggested a representation for stereoisomers called the Fischer projection.

15.6 FISCHER PROJECTIONS

Fischer projections offer an easy way to draw three-dimensional molecules on paper in two dimensions. The atoms are all projected onto one plane. Figure 15.6 shows the projection idea for glyceraldehyde, for which the Fischer projections of the two enantiomers are

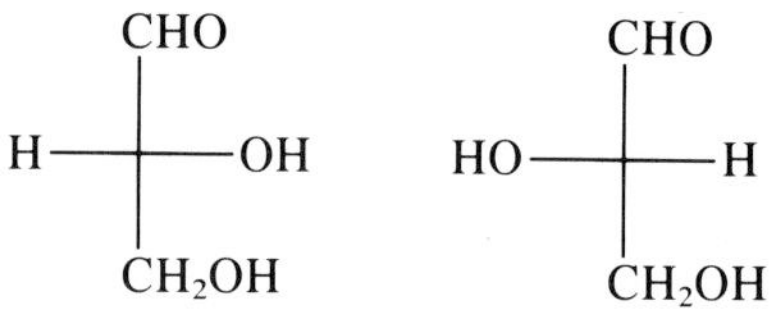
CHO
H
OH
CH2OH
CHO
HO
H
CH2OH

The Fischer rules for showing the array around a chiral center are as follows:

1 Write down or at least envision the carbon chain of the compound written vertically with the carbonyl group at the top.[1]
2 Represent the chiral carbon(s) as the intersection of crossed lines, i.e.,

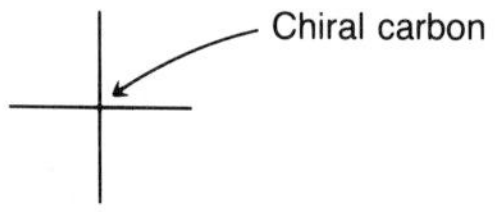
Chiral carbon

3 Substituents on the vertical lines are understood to be going back behind the plane of the paper. The chiral center is in the paper plane.
4 Substituents on the horizontal rungs are understood to be coming forward out of the paper plane.

These rules are applied in Sample Exercise 15.1.

SAMPLE EXERCISE 15.1

Write Fischer projections for the two enantiomers of the structure $HOCH_2—CHOH—CH_2—CHO$.

Solution Follow the rules just given.

[1]Fischer rules were formulated for carbohydrates, all of which have carbonyl groups. A more general statement would say: Put the carbon in position 1 as defined by nomenclature rules at the top.

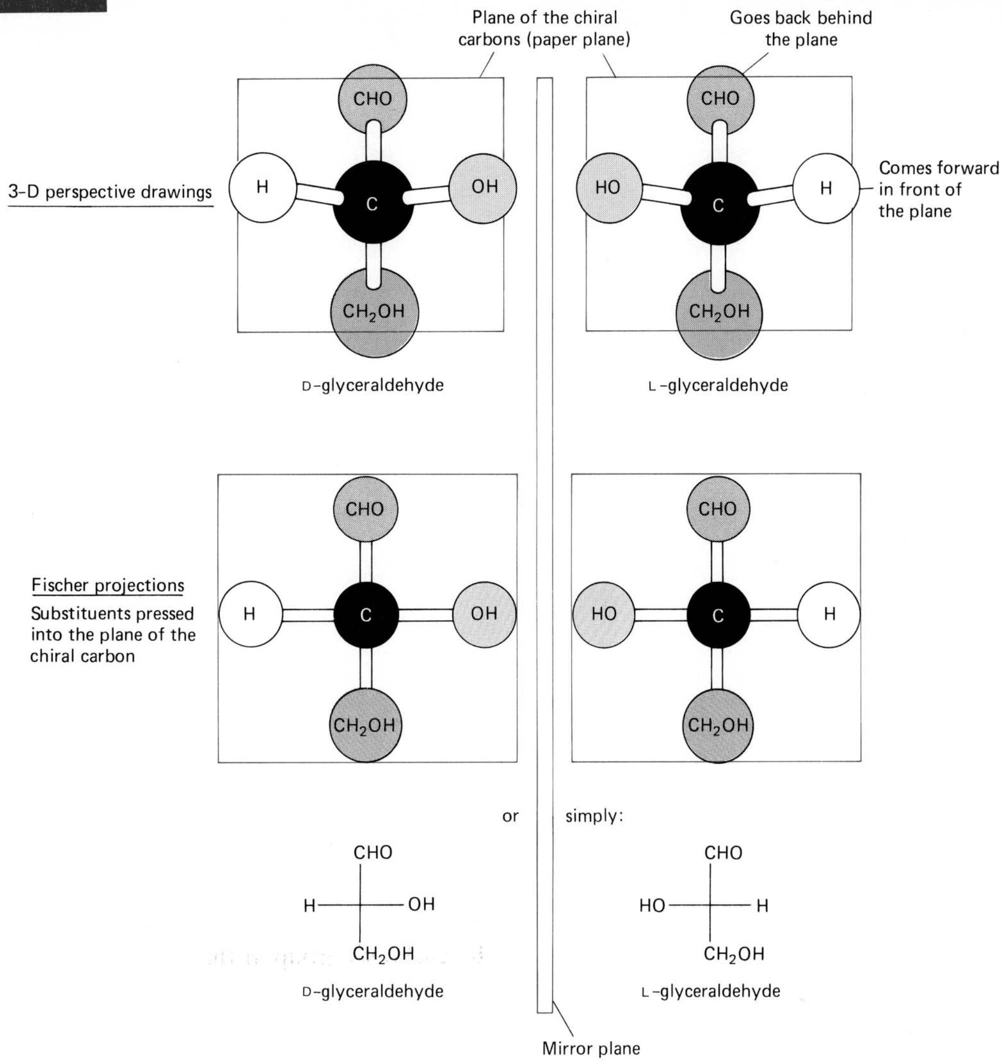
Plane of the chiral carbons (paper plane)
Goes back behind the plane
3-D perspective drawings
CHO
H
C
OH
CH₂OH
HO
Comes forward in front of the plane
D-glyceraldehyde
L-glyceraldehyde
Fischer projections
Substituents pressed into the plane of the chiral carbon
or simply:
Mirror plane

FIG. 15.6
The meaning of Fischer projections. Perspective drawings show that molecules are three-dimensional. If the chiral carbon is in the plane of the paper, the drawing indicates that —CHO and —CH_2OH are going back behind the plane and —H and —OH are coming forward in front of the plane. If we press all substituents onto the paper plane, i.e., bring —CHO and —CH_2OH forward and press —H and —OH back, we have a planar Fischer projection. All substituents are projected onto one plane.

STEP 1 **Rewrite the structure vertically, starting with the aldehyde group at the top:**

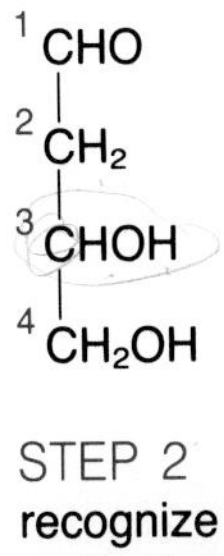
1 CHO
2 CH₂
3 CHOH
4 CH₂OH

STEP 2 **Because it bears four different groups (H, OH, CH_2OH, and CH_2CHO), recognize that the carbon in position 3 is the chiral carbon and represent that portion as the intersection of crossed lines:**

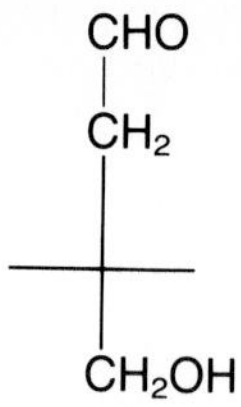

STEP 3 Because the carbon chain is written vertically, recognize that it is going back behind the paper plane.

STEP 4 Attach the horizontal substituents in the two possible arrangements:

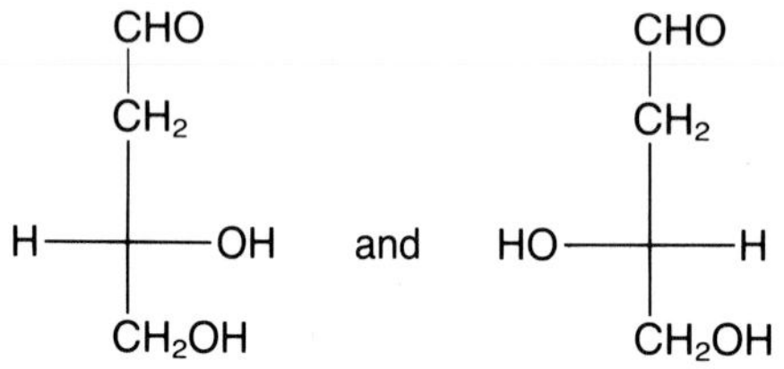

Problem 15.4 Write Fischer projections for the two enantiomers of

CHO—CHOH—$CH_2CH_2CH_3$

15.7 MULTIPLE CHIRAL CENTERS

Fischer projections can be used when writing structures for stereoisomers of molecules with more than one chiral center. The structures of glucose and the other 15 stereoisomers of formula $C_6H_{12}O_6$ in Figure 15.5 are all Fischer forms.

Let's write Fischer projections for the stereoisomeric aldotetroses of formula $C_4H_8O_4$; that is, we'll show the different arrays possible around the chiral carbons in

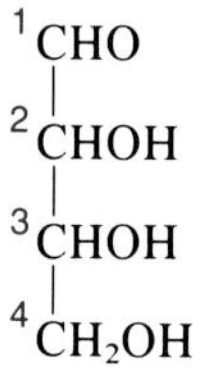

Carbons in positions 2 and 3 are chiral. One Fischer form we can write is

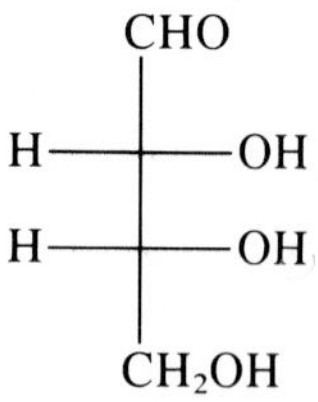

III

The nonsuperimposable mirror image, or enantiomer, of this structure is

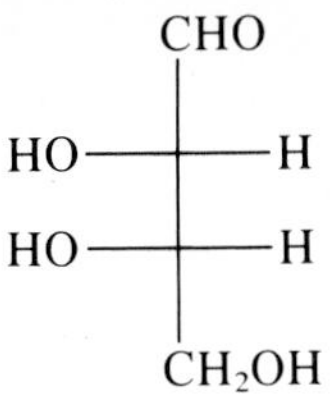

IV

The pair of enantiomers III and IV have in common that both OH's are on the same side of the carbon chain. Recognizing this suggests another arrangement in which the OH groups are on opposite sides of the chain.

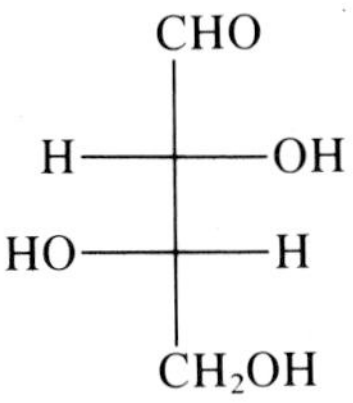

V

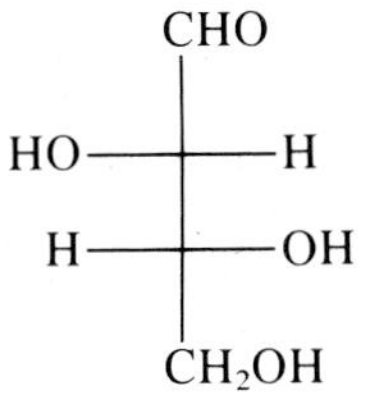

VI

Structures V and VI represent two additional aldotetrose stereoisomers. Structures V and VI are enantiomers; i.e., they are nonsuperimposable mirror images. However, as Figure 15.7 shows, the relationship of III and IV to V and VI is not enantiomeric; i.e., they are not related as object and mirror image. The relationship between isomers in the set III, IV and isomers in the set V, VI is described as **diastereomeric.** Diastereomers are stereoisomers that are not related as object and mirror image.

For an aldotetrose with two chiral centers we have discovered four stereoisomers, two sets of enantiomers. Is there a general rule for predicting the number of stereoisomers based on the number of chiral centers? Yes, there is. The *maximum* number of stereoisomers that can exist is 2^n where n equals the number of chiral carbons. Thus, for example, if there is one chiral carbon, $2^n = 2^1 = 2$ and there are two stereoisomers, i.e., one pair of enantiomers. If there are two chiral carbons, $2^n = 2^2 = 4$ and there are four stereoisomers, i.e., two pairs of enantiomers.

Notice that this rule stipulates the **maximum** number of stereoisomers that can exist for a given formula. For compounds with more than one chiral carbon, it sometimes turns out that there are fewer than this maximum number of stereoisomers. Tartaric acid offers an example of this phenomenon.

Writing Fischer projections for HOOC—CHOH—CHOH—COOH, tartaric acid, in the same manner as we did for the aldotetrose above, we would construct:

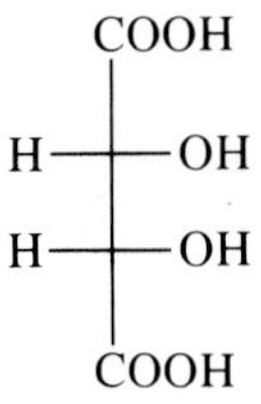

VII

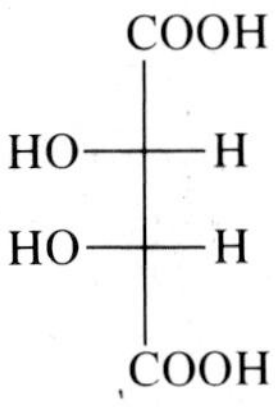

VIII

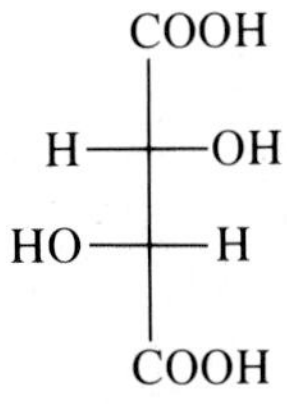

IX

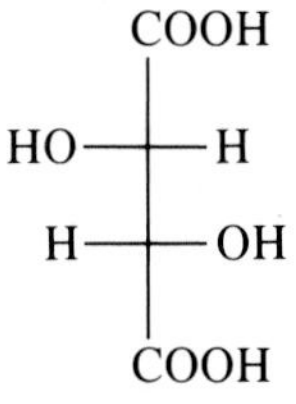

X

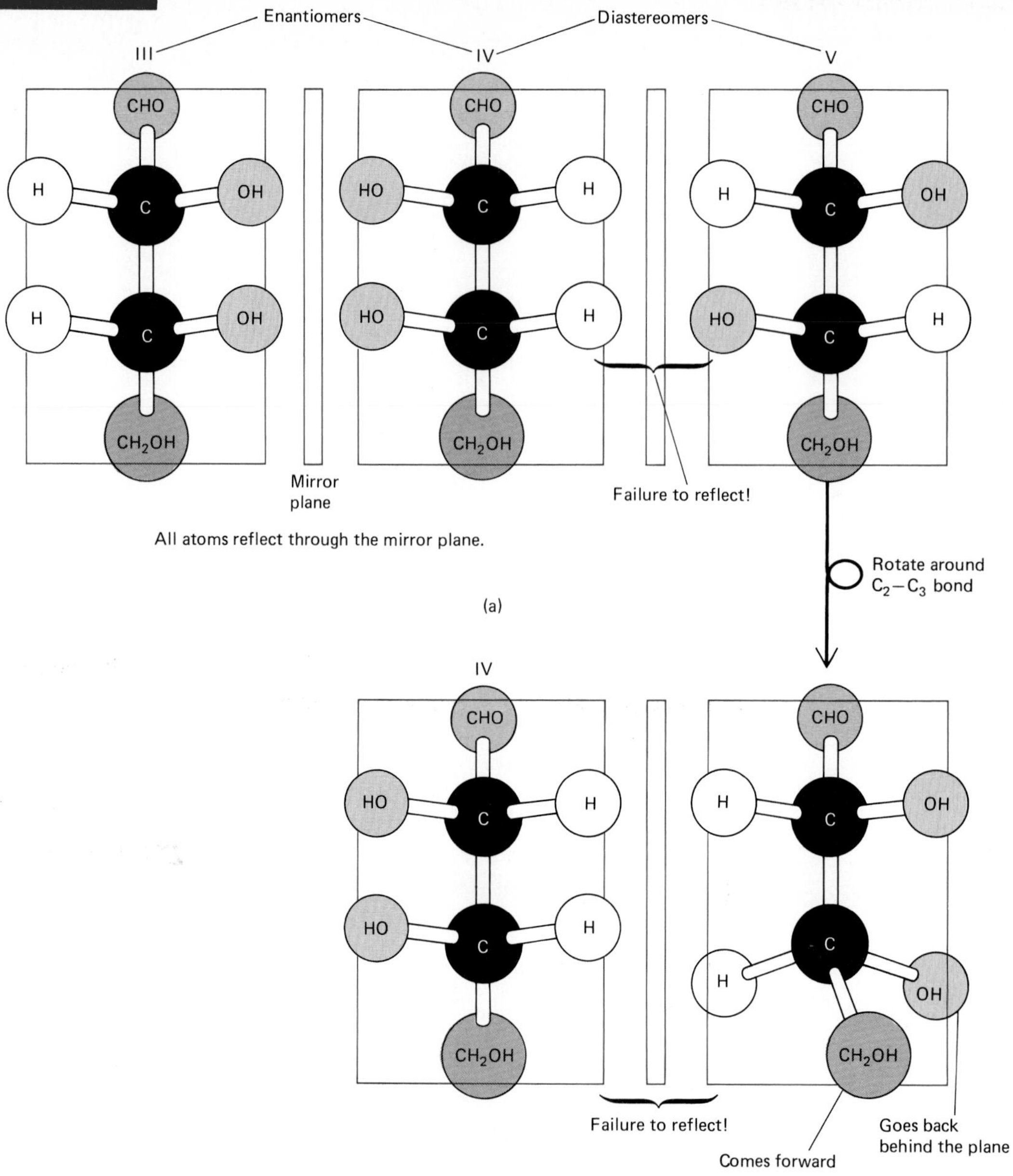
Enantiomers
Diastereomers
III
IV
V
CHO
H
C
OH
HO
CH₂OH
Mirror plane
Failure to reflect!
All atoms reflect through the mirror plane.
Rotate around C_2-C_3 bond
(a)
Failure to reflect!
Comes forward
Goes back behind the plane
(b)

FIG. 15.7
Enantiomers vs. diastereomers. (*a*) Enantiomers are reflected as object and mirror image through the mirror plane. Diastereomers are not related as object and mirror image. (*b*) One or more atoms fail to be reflected through the mirror plane regardless of the way in which bonds are rotated or the molecule is held. Note that the rotated form of V is not a correct Fischer form.

Isomers IX and X are nonsuperimposable mirror images of one another; i.e., they are enantiomers. Isomers VII and VIII appear to be enantiomers, but it happens that they are *identical*. If you rotate Fischer projection VII 180° in the plane of the paper you will find that it superimposes exactly with VIII. If VII ≡ VIII, then there are only three stereoisomers for tartaric acid: the enantiomers IX and X and a diastereomer of these which could be represented by either equivalent structure VII or VIII.

A compound such as this unique isomer of tartaric acid is called a *meso* compound. Meso compounds are characterized by an internal reflection plane, that is, one-half of the molecule reflects the other.

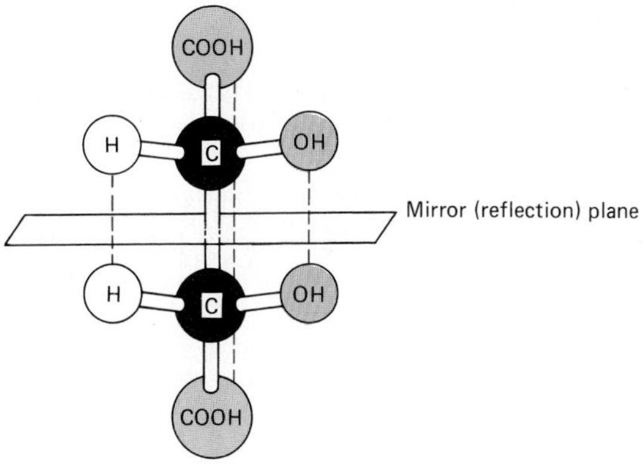
COOH
H
C
OH
Mirror (reflection) plane
H
C
OH
COOH

It is also true in meso compounds that each chiral carbon has the same set of four different substituents. For *meso*-tartaric acid this set is —H, —OH, —COOH, and —CHOH—COOH. Because the common carbohydrates that you will study all have differently substituted chiral carbons, you will not encounter *meso* compounds among them. That is, the 2^n rule for the number of existing isomeric structures will apply.

15.8 CLASSIFICATION: D-FAMILY VERSUS L-FAMILY

How can we describe the specific stereoisomer of a set of stereoisomers that we mean? This is indeed a complex problem because there are so many stereoisomers for structures with multiple chiral centers. We will not discuss a system that uniquely specifies stereoisomers because your further studies do not depend on it. However, you should be aware that carbohydrate isomers are divided into two major categories depending on the arrangement of groups around the chiral carbon which bears the highest number in the carbon chain, i.e., is farthest away from the carbonyl group. The two categories are called the D-family and the L-family.

A sugar isomer is assigned to the **D-family** if the —OH substituent on the highest-numbered chiral carbon appears on the right in the Fischer projection. *D* is used because *dexter* is the Latin word for "right."

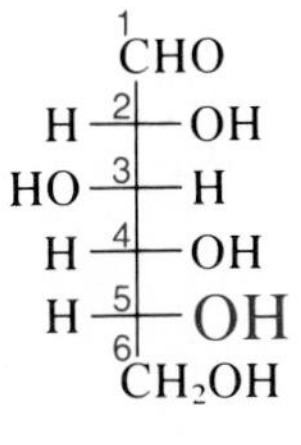
CHO
H—OH
HO—H
H—OH
H—OH
CH2OH

This structure for glucose is designated as D-glucose because of the position of —OH on the carbon in position 5. The orientation of the —OH groups on the other chiral carbons in no way influences the assignment.

Alternatively, a sugar is of the **L-family** if the —OH substituent on the highest-numbered chiral carbon appears on the left in the Fischer projection. *L* is used because *laevus* is the Latin word for "left." In this case, of course, there is also a correspondence to the English word.

L-Arabinose is

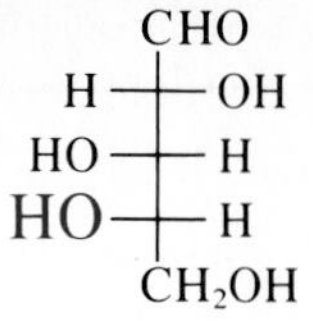

The 16 aldohexose stereoisomers shown in Figure 15.5 have been divided into D- and L-families.

Problem 15.5 Classify the following as D- or L-sugars:

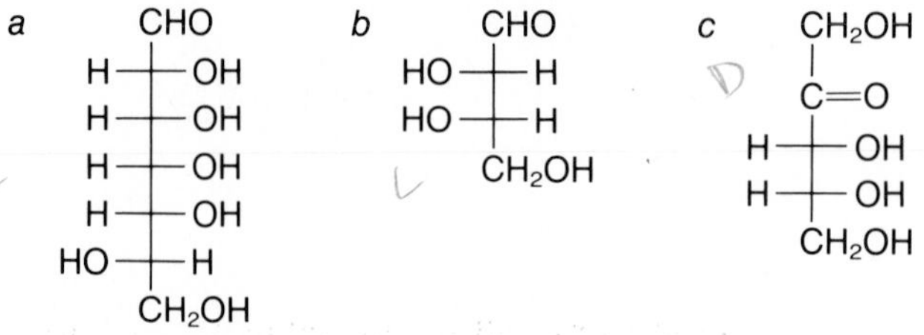

Most biologically active carbohydrates belong to the D-family. We readily metabolize D-glucose, D-mannose, D-fructose, and D-galactose. L-sugars usually cannot be used by the body. That the body requires one family of sugars and rejects the other is a demonstration of the idea that the three-dimensional shape of a molecule is as important as its molecular formula for proper biological activity. This biological demand is based on the requirements of enzymes for "hand-in-glove" fits between reacting molecules. It is also true that far fewer L-sugars exist naturally; they must be synthesized. D-glucose is far and away the most prevalent and the most important monosaccharide.

15.9 INTRAMOLECULAR HEMIACETALS AND HEMIKETALS

Although to explain various classification and stereochemical concepts we have written the sugars as straight or so-called open-chain forms, they do not exist in that fashion in nature. D-glucose and the other monosaccharides exist as cyclic hemiacetals (or hemiketals) because of the ability of the carbonyl group to react with an alcohol group intramolecularly. This possibility was first mentioned in Section 14.13. In that section the reaction of 5-hydroxypentanal was shown:

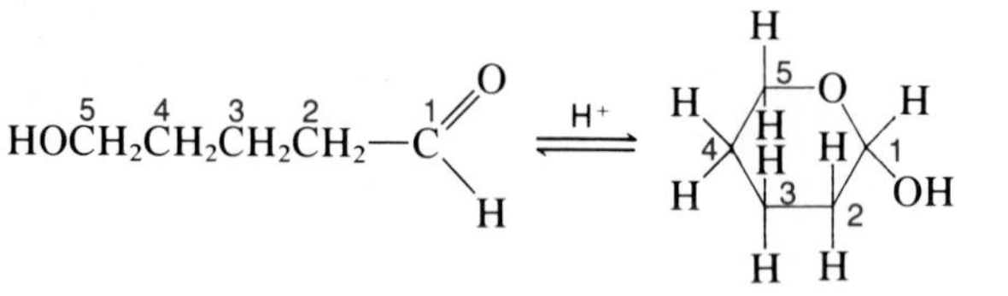

Such a reaction for glucose could be written in the same way:

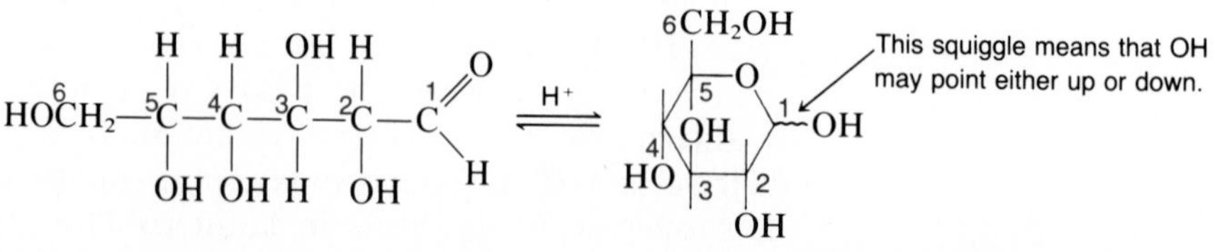
This squiggle means that OH may point either up or down.

(For convenience and clarity, ring hydrogens usually are not shown.)

The fundamental reaction is the same in each case, but prediction of the structure of the product is much more complicated in the case of glucose. There are three complications:

1 How do we know which —OH group of glucose will react with the carbonyl?
2 Ring closure produces a new chiral center at C-1, which was previously achiral. Thus for every open-chain form there are two cyclic stereoisomers.
3 How can we relate the arrangement of ring substituents to the orientation of groups in the open-chain structure?

Ring Size

Let us address these issues one at a time. The first issue was a puzzling question historically when chemists first realized that sugars were cyclic but did not know the ring size. Today we know that there is a great preference in nature for the formation of six-membered rings because of their stability. Five-membered rings are only slightly less favorable energetically.

Consequently, all of the aldohexoses form hemiacetals through the reaction of the aldehyde group with the alcohol group at C-5, as has been shown for D-glucose. The six-membered ring so formed includes five carbons (at positions 1 through 5) and one oxygen (the O at C-5) and is called a *pyranose ring*.

Aldopentoses form five-membered rings; e.g., ribose closes as shown:

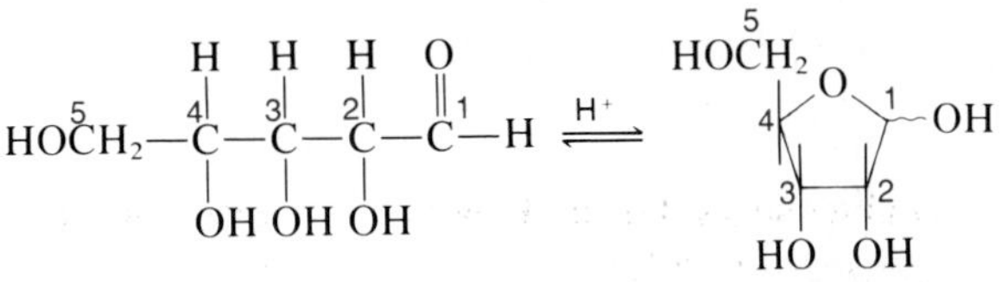

The hexose fructose also forms a five-membered ring:

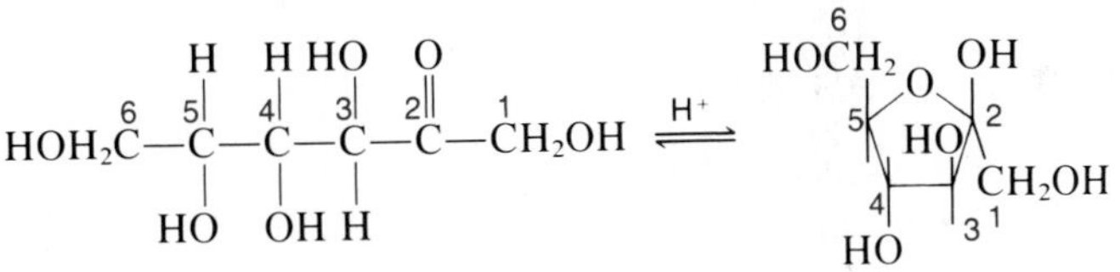

Five-membered ring sugars with four carbons and one oxygen in the ring are called *furanoses*. Biologically active monosaccharides are aldohexopyranoses, aldopentofuranoses, and the one ketohexofuranose, fructose.

C-1 Isomers

The second issue involves the generation of a chiral center at C-1, that is, the carbonyl carbon in the open-chain form. Ring closure can be effected in two ways leading to two isomers. If the OH at C-5 attacks C-1 when the carbonyl oxygen is pointing downward, the new OH at C-1 will point downward.

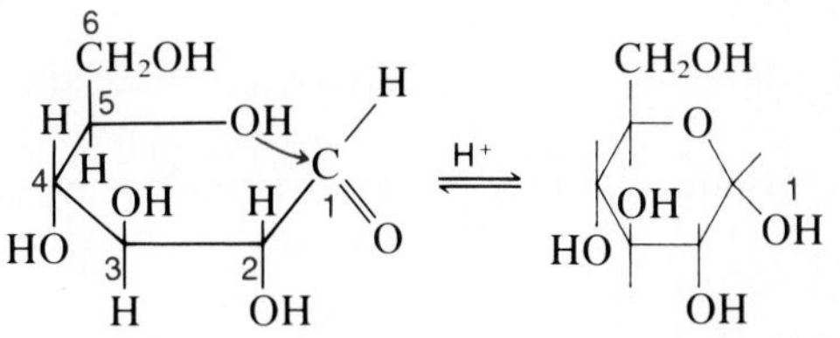

α-D-Glucose

If, on the other hand, the OH at C-5 attacks C-1 when the carbonyl oxygen is pointing upward, the new OH at C-1 will point upward.

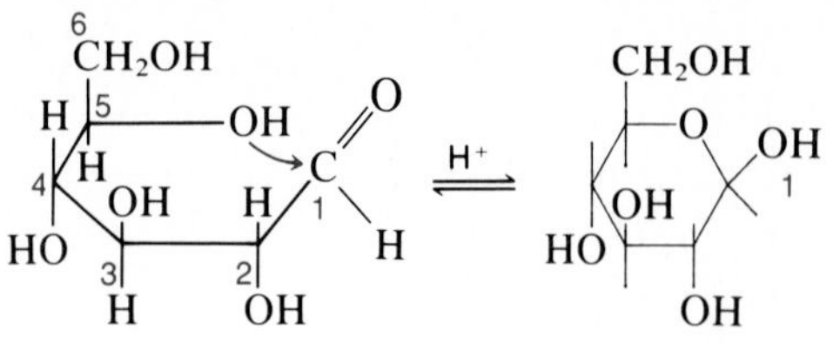

β-D-Glucose

Use the upward-pointing C-6 (the CH_2OH group) as a reference point to distinguish between these two isomers.

The only structural difference between these two isomers is that —OH at C-1 goes down, i.e., is on the opposite side of the ring from C-6, in one structure, whereas it goes up on the same side in the other. By convention, the one with —OH down is designated α and the one with —OH up is designated β. Isomers which differ in this way, only at C-1, are given the special name **anomers.** When open-chain monosaccharides close, they form two anomers, the α-anomer and the β-anomer.

The third issue of relating the stereoisomeric picture shown by the Fischer projection to that shown by the ring form was addressed by the carbohydrate chemist Sir Walter N. Haworth. He won the Nobel Prize in 1939 for his work in carbohydrate chemistry.

15.10 HAWORTH PROJECTIONS

If we attempt to write the hemiacetal forms of glucose in a Fischer projection, they would look like this:

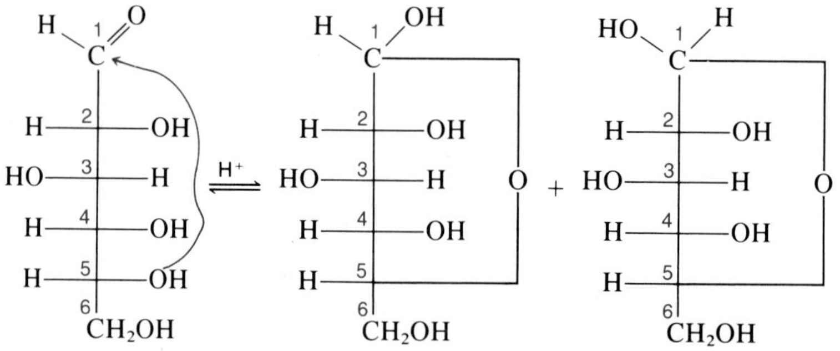

α-D-Glucose β-D-Glucose

These structures are clearly not good representations of a six-membered ring. Consequently, Haworth devised a system for transposing substituents of the Fischer projection onto a more reasonably shaped ring, 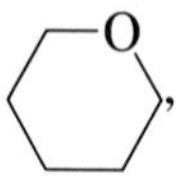, in a correctly oriented fashion.

The system can be summarized by three rules:

1 The hemiacetal ring is represented and numbered as follows.

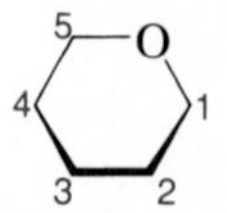

The shading of the ring indicates that it should be viewed as perpendicular to the paper plane. The carbons in positions 1 and 4 are in the plane, the carbons at positions 2 and 3 come toward you, and the carbon at position 5 and the oxygen are behind the plane.

2 For all D-sugars (the only ones we'll consider), the carbon at position 6 is attached above the plane of the ring:

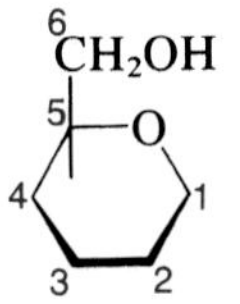
6
CH₂OH
5
O
4
1
3
2

3 Groups on the *right* in the Fischer projection are written down below the ring plane. Groups on the *left* are written *up*. The H's on carbons at positions 2, 3, and 4 have been omitted to avoid clutter in the representation. Thus:

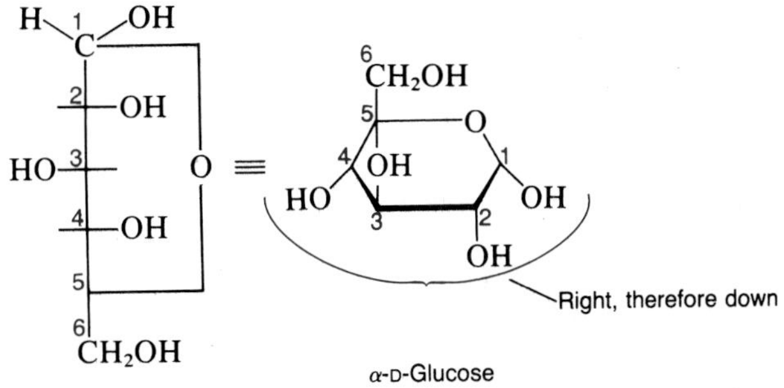
Right, therefore down

α-D-Glucose

α-D-Glucose

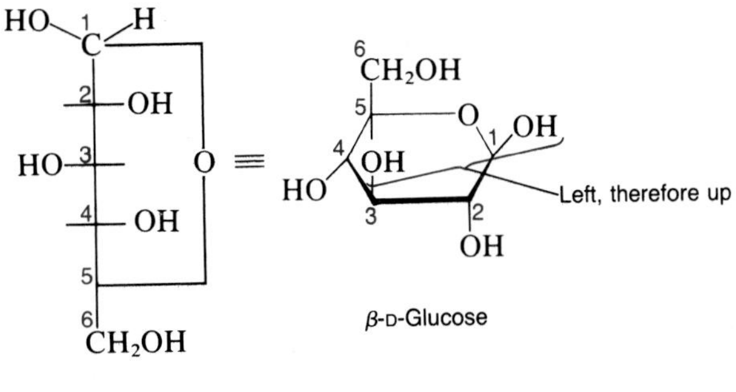
Left, therefore up

β-D-Glucose

β-D-Glucose

SAMPLE EXERCISE 15.2 Write Haworth projection formulas for the two anomers of D-galactose:

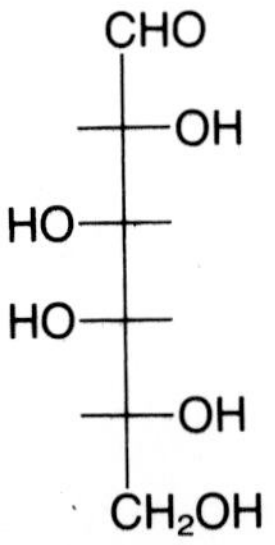

Solution Galactose is an aldohexose. Therefore, from the discussion, the two anomers will be six-membered cyclic hemiacetals formed by the reaction of the hydroxyl group at C-5 with the carbonyl group.

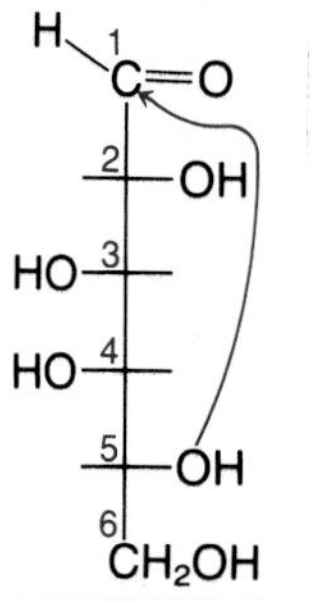
H
1
C=O
2
OH
HO
3
HO
4
5
OH
6
CH₂OH

To write Haworth formulas, follow the rules given. According to rules 1 and 2, the ring is

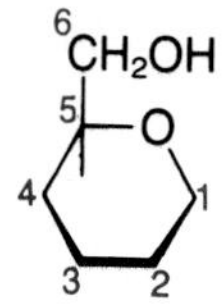
6
CH₂OH
5
O
4
1
3
2

Attach substituents using the convention in rule 3:

Right in Fischer = Down in Haworth

Left in Fischer = Up in Haworth

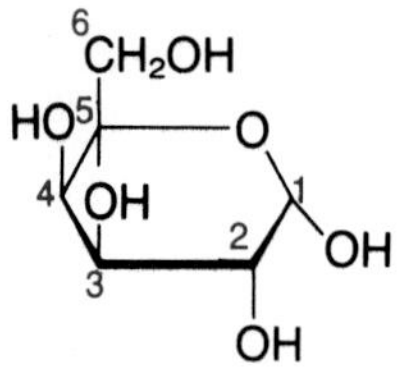
6
CH₂OH
HO
5
O
4
OH
1
2
OH
3
OH

α-D-Galactose

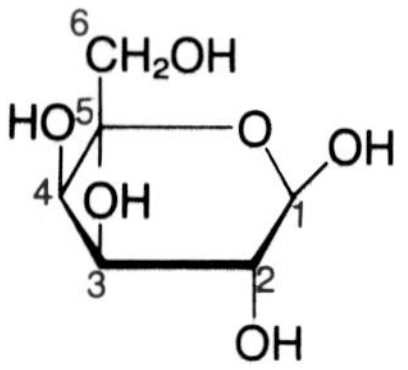
6
CH₂OH
HO
5
O
OH
4
OH
1
3
2
OH

β-D-Galactose

Notice that these structures differ from glucose only at C-4. This is the only difference between glucose and galactose.

Problem 15.6 Write Haworth projection formulas for the two anomers of D-allose:

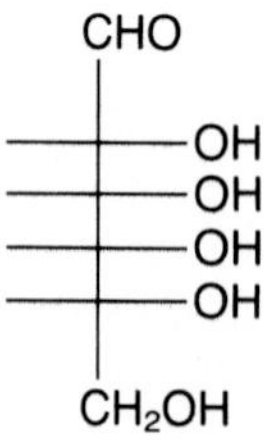
CHO
OH
OH
OH
OH
CH₂OH

15.11 MUTAROTATION

What is D-glucose really like? How does it look and what are its properties? In the solid state, glucose is a white crystalline solid. One might have a sample of pure α-D-glucose (mp = 146°C and optical rotation = +112°), a sample of pure β-D-glucose (mp = 150°C and optical rotation = +19°), or a mixture of the two with intermediate properties. In any case, the solid would look a lot like table sugar.

Problem 15.7 Did you expect the values of the optical rotations of the α- and β-forms to be related? Explain.

When the white solid dissolves in water, a clear solution results and the following equilibrium is found:

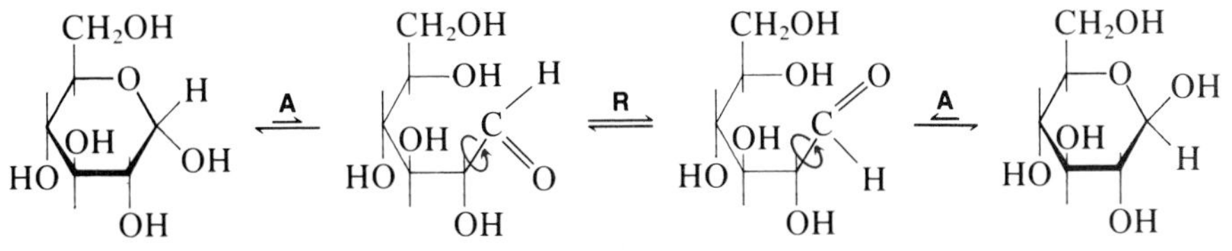
CH₂OH
O
H
OH
HO
OH
OH
A
CH₂OH
OH
H
C
OH
HO
O
OH
R
CH₂OH
OH
O
C
OH
HO
H
OH
A
CH₂OH
O
OH
OH
HO
H
OH

α-D-Glucose (~ 36%) | Open-chain aldehyde (0.1%) | β-D-Glucose (~ 64%)

Because hemiacetal formation is reversible, in solution the glucose ring can open up; i.e., the hemiacetal group is reconverted to the aldehyde and alcohol groups from which it originally formed. This process is marked **A** in the equilibrium equations. The lengths of the equilibrium arrows are intended to reflect the position of the equilibrium which lies far toward the hemiacetal forms.

When the ring has opened up, free rotation around carbon-carbon bonds (which is restricted in the ring) becomes possible. The equilibrium process marked **R** shows rotation around the C-1—C-2 bond, which results in different orientations of the carbonyl oxygen. This rotation provides a route for the conversion of the α-form to the β-form or the reverse process. If ring closure (i.e., hemiacetal formation) occurs when the carbonyl oxygen faces down, the α-form is obtained. If on the other hand, the oxygen points up, one gets the β-form.

Pure α-D-glucose displays an optical rotation of +112° if this is measured the moment the sample solution is prepared. With time, the numerical value of the rotation decreases and eventually reaches a lower limit of +52.7°. Pure β-D-glucose displays an optical rotation of +19° in a freshly prepared sample. As the solution sits, this value increases and eventually reaches an upper limit of +52.7°.

This change in optical rotation which results from ring opening, bond rotation, and ring closing is called **mutarotation.** The prefix *muta-* means "change." Reaching the value of 52.7° in each case indicates that equilibrium has been attained. The value reflects the equilibrium concentrations for the two anomers of 36% α-D-glucose and 64% β-D-glucose. That is, given that the rotation of pure α-glucose is +112° and pure β-glucose +19°, it is possible to calculate that the equilibrium percentages of α- and β-glucose must be 36 and 64 percent respectively to yield a rotation of +52.7° for the mixture of anomers.

15.12 FORMATION OF DI- AND POLYSACCHARIDES

So far we have been discussing only simple sugars or monosaccharides. A buildup of these monomer units into dimers (disaccharides) and polymers (polysaccharides) is possible because of the ability of sugar structures to form acetals as well as hemiacetals. Recall from Section 14.13 that the aldehyde group after reacting with 1 mol of alcohol to form a hemiacetal can go on to react with another mole of alcohol to form an acetal:

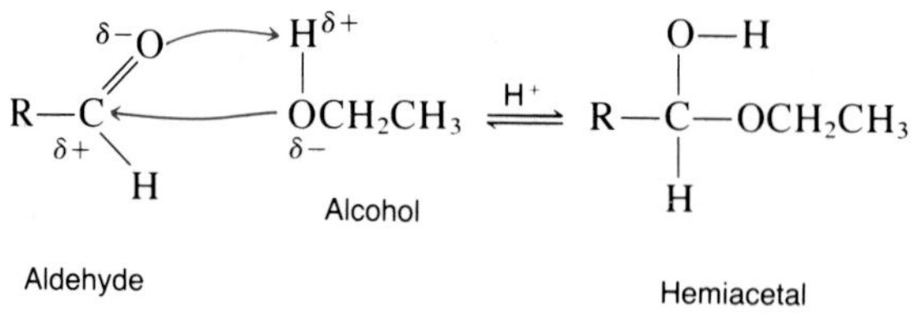
Aldehyde
Alcohol
Hemiacetal

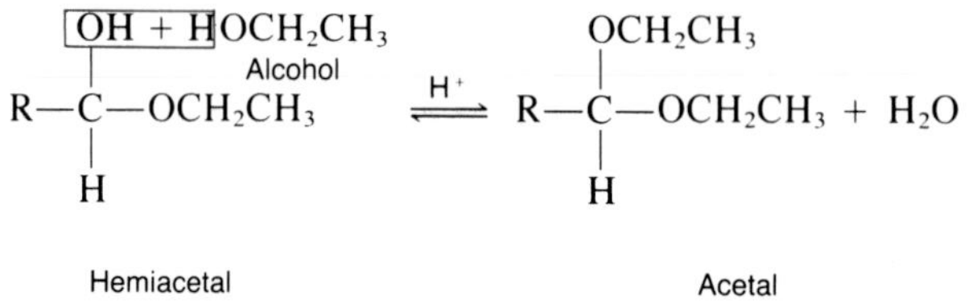
Alcohol
Hemiacetal
Acetal

Similarly, after the open-chain glucose has closed through the reaction of the aldehyde group with one internal alcohol group, the second reaction can occur if an external alcohol group is provided. As a simple example, glucose can react with any alcohol, represented ROH:

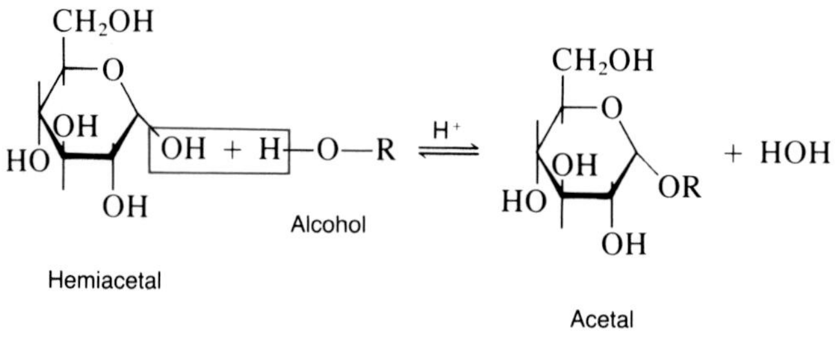
Alcohol
Hemiacetal
Acetal

If instead of a simple alcohol providing —OH, another monosaccharide molecule provides —OH, then a disaccharide forms. That is, the two monosaccharides are joined together

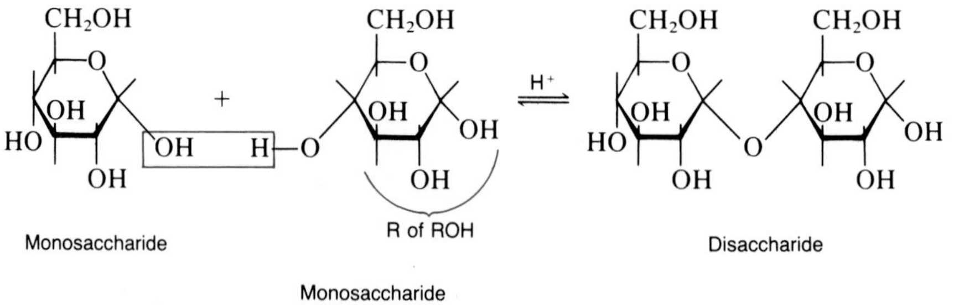
R of ROH
Monosaccharide
Monosaccharide
Disaccharide

The process of linking up monomer units through the anomeric carbon can continue so that trisaccharides, tetrasaccharides, pentasaccharides, etc., and polysaccharides form. Different monosaccharide units can link together, for example, glucose and galactose in lactose. We will discuss specific examples of mono-, di-, and polysaccharides in subsequent sections.

Problem 15.8 Classify the following as mono-, di-, tri-, etc., or polysaccharides:

a

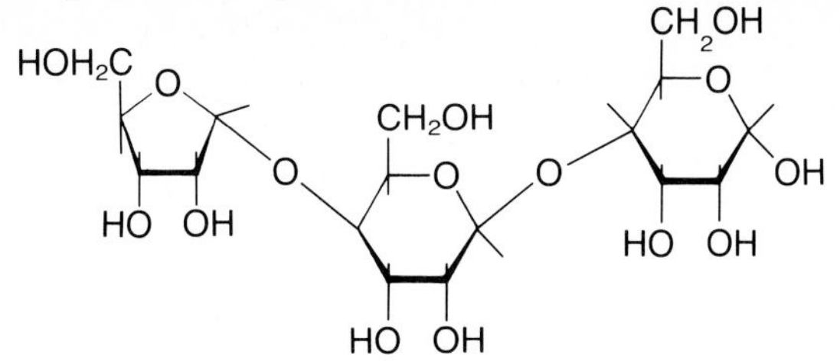

c

b

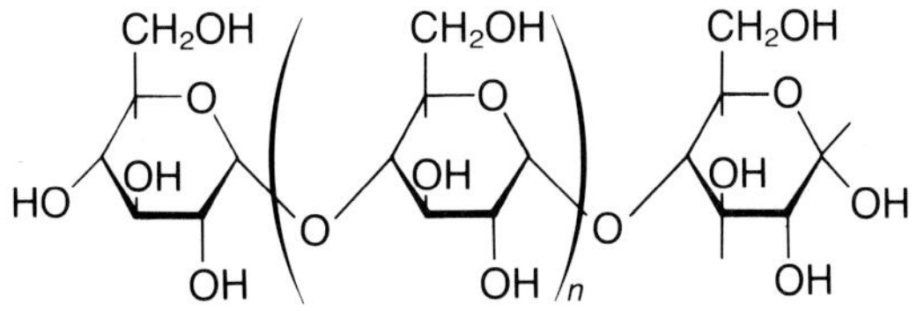

d

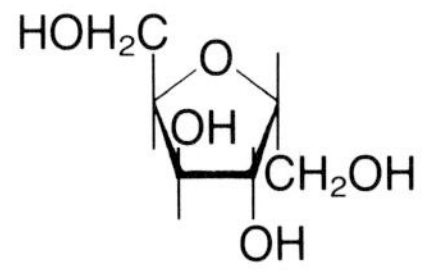

Polysaccharides can be broken down by a hydrolysis process which is just the reverse of acetal formation. This hydrolysis process is carbohydrate digestion, which is discussed in Section 15.17.

15.13 MONOSACCHARIDES

As has been stated, the most important and most prevalent monosaccharide is **glucose.** The other hexoses of biological significance that you will encounter are **fructose, mannose,** and **galactose.**

In your biology courses you will almost certainly be asked to learn the structures of several common monosaccharides, probably the four that have just been named. This is an easy task if you realize the close structural resemblance of all the common monosaccharides to glucose. Notice first of all that all —OH groups are on the right in D-glucose except for the one at C-3. If you know the glucose structure and the relationships, you will know many structures; don't learn the structures individually. For example, D-glucose, D-mannose, and D-fructose are identical from C-3 through C-6:

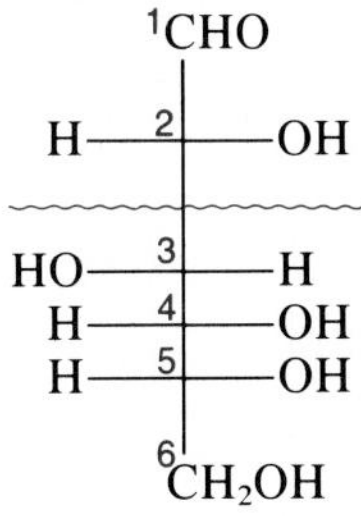

D-Glucose

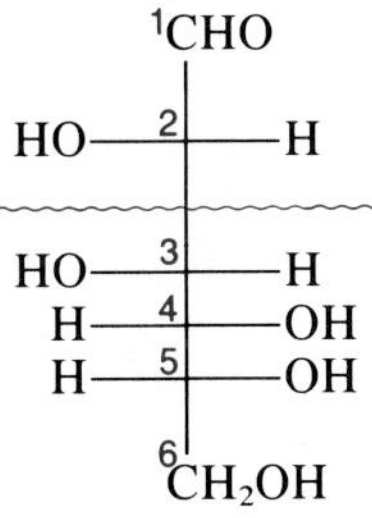

D-Mannose

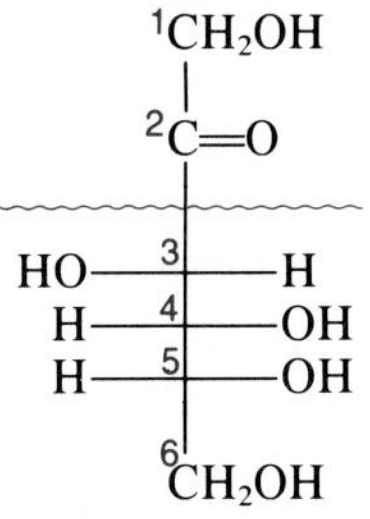

D-Fructose

D-Glucose and D-mannose differ only in the orientation of the —OH group at C-2. The difference between D-glucose and D-fructose is the reversal of the —OH and carbonyl functional groups at C-1 and C-2. Indeed, as you will learn in Section 23.2, this reversal is an important reaction during glycolysis, a carbohydrate metabolic pathway.

D-Glucose and D-galactose differ only in the orientation of —OH at C-4. Knowing this and the structure of glucose, you can immediately write the structure of D-galactose:

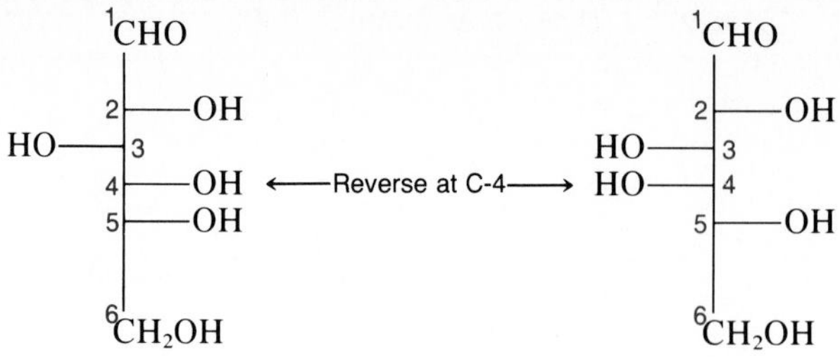

D-Glucose D-Galactose

All these comparisons apply to the cyclic Haworth formulas as well. For the aldohexoses:

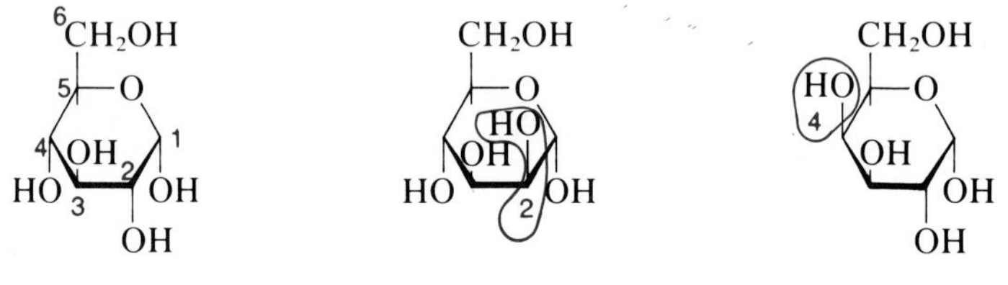

α-D-Glucose α-D-Mannose α-D-Galactose

As in the open-chain forms, D-mannose differs from glucose only at C-2, and D-galactose differs only at C-4.

Glucose is known as *grape sugar* because it occurs in grapes and as *dextrose* because it rotates polarized light in a dextrorotatory sense. The word *dextrorotatory* is used to describe optical rotation in a clockwise sense. Blood sugar is principally glucose; other monosaccharides appear in blood in very small amounts. Blood sugar concentration is approximately 70 to 110 mg/100 mL 2 to 4 h after eating. The properties and behavior of the two anomeric forms, α-D-glucose and β-D-glucose, have been discussed in Section 15.11. As you will see in Chapter 23, carbohydrate metabolism is largely the metabolism of glucose.

Fructose, which is also called *fruit sugar,* is unique among the hexoses we've encountered in several ways. It is the only ketohexose we've seen; it exists in the furanose ring form; and its optical rotation is levorotatory. For this reason it is also known as *levulose*. All other monosaccharides exhibit dextrorotation. Fructose is the sweetest sugar. It is nearly twice as sweet as sucrose or table sugar, and it is three times as sweet as glucose. In Chapter 23 we'll encounter fructose in several of the carbohydrate metabolic pathways.

Galactose is sometimes called brain sugar because it is a component of polymeric glycoproteins (carbohydrate-protein combinations) in brain and nerve tissue. The major dietary sources of galactose are milk products because galactose is part of the disaccharide lactose, or milk sugar. Galactose is metabolized in the body by first being converted to glucose.

Mannose is the least frequently encountered aldohexose of those mentioned. It occurs mostly in nature as the monomer unit of polysaccharides called mannans, components of some berry plants. Mannose is metabolized by conversion into fructose.

Ribose and 2-**deoxyribose** are the most important aldopentoses because they occur in genetic materials DNA (deoxyribonucleic acid) and RNA (ribonucleic acid). The prefix *2-deoxy-* means "without oxygen at position 2."

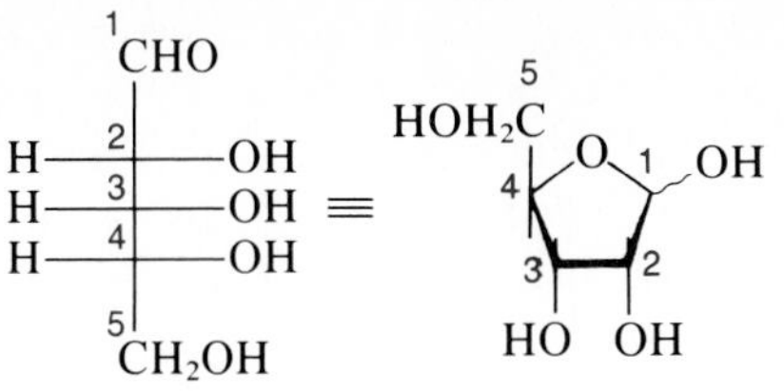

Ribose

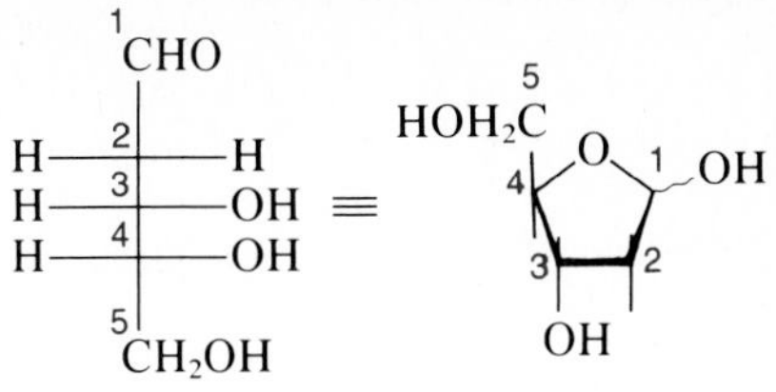

2-Deoxyribose

15.14 CLASSIFICATION: REDUCING AND NONREDUCING SUGARS

All monosaccharides are classified as reducing sugars. What does this mean? The term *reducing* in this case comes from the oxidation-reduction reaction that occurs when an aldehyde (or α-hydroxy ketone) is oxidized by Tollens' or Benedict's reagent (see Section 14.12). In Tollens' or Benedict's solution a monosaccharide such as glucose opens to the free aldehyde form, the aldehyde group is oxidized, and the oxidizing agent is reduced. The monosaccharide is thereby acting as a *reducing* agent.

Tollens' Test

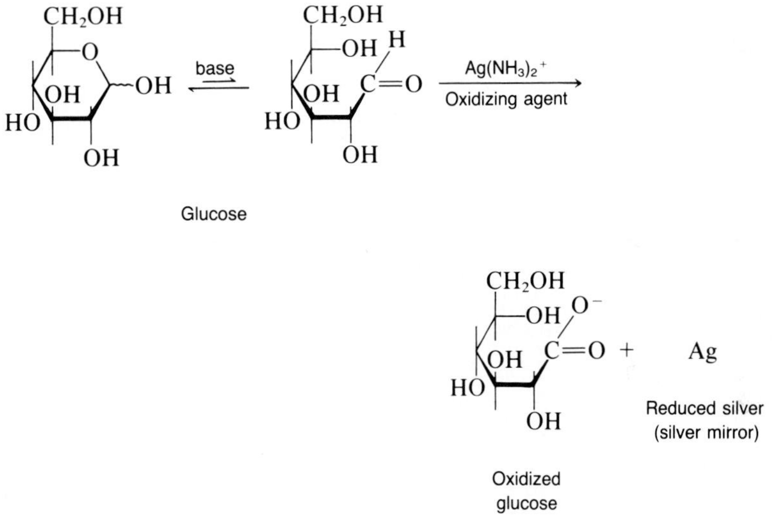
base
$Ag(NH_3)_2^+$
Oxidizing agent
Glucose
+ Ag
Reduced silver
(silver mirror)
Oxidized
glucose

(15.1)

Benedict's Test

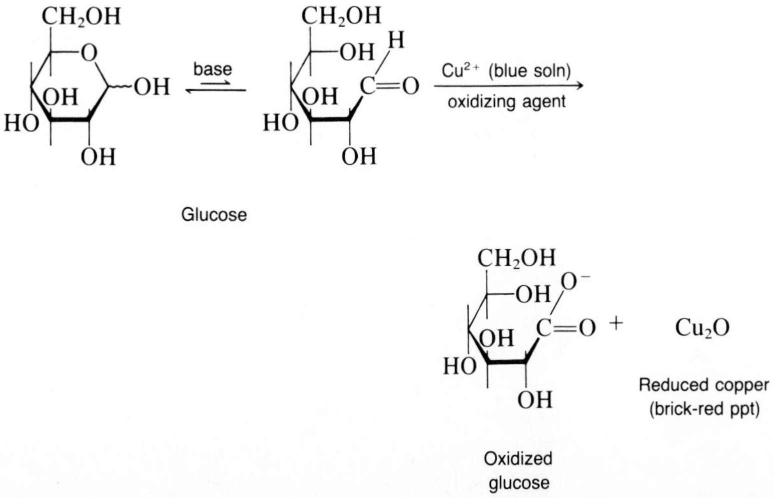
base
Cu^{2+} (blue soln)
oxidizing agent
Glucose
+ Cu_2O
Reduced copper
(brick-red ppt)
Oxidized
glucose

(15.2)

These tests, especially the Benedict's test, are used clinically to detect the presence of glucose in the urine, a condition known as **glucosuria.** Clinitest tablets are solid Benedict's reagent. A positive test is a change in color from blue to red. The presence of glucose in urine is a symptom of hyperglycemia (high blood sugar) and diabetes mellitis, conditions which we will discuss further in Section 23.9.

The equilibrium shown in Equations (15.1) and (15.2) lies toward the ring form, but shifts toward the free aldehyde form as it reacts with the oxidizing agent. This is an equilibrium shift caused by a concentration-change stress; free aldehyde is removed when it is oxidized.

All sugars with hemiacetal functional groups are reducing sugars because they can undergo the reactions shown in Equations (15.1) and (15.2). Thus, the disaccharides, maltose, lactose, and cellobiose are reducing sugars.

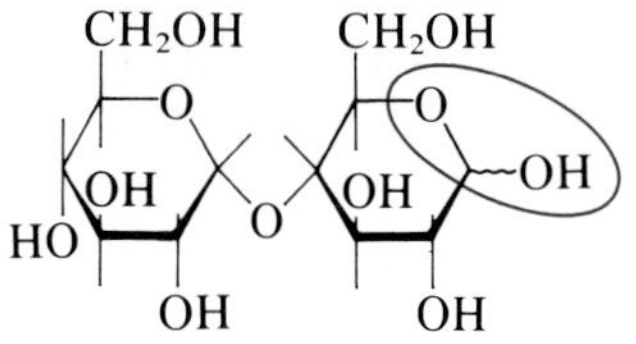

Maltose

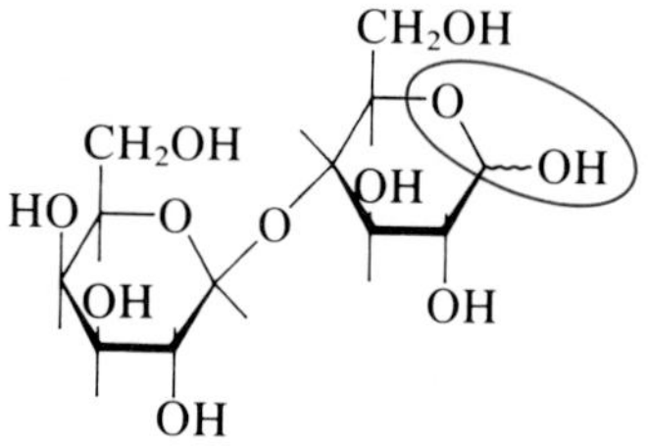

Lactose

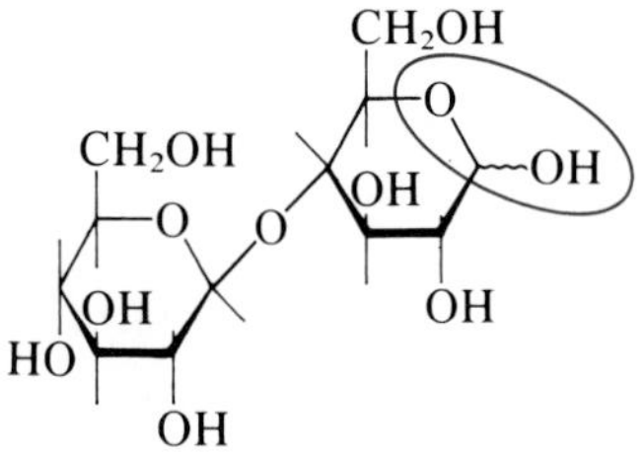

Cellobiose

The circled positions are, of course, the hemiacetal functional group. In the basic solution of the test the rings open, the aldehyde group is restored and oxidized, and the silver or copper is reduced.

Sucrose is a *nonreducing* sugar.

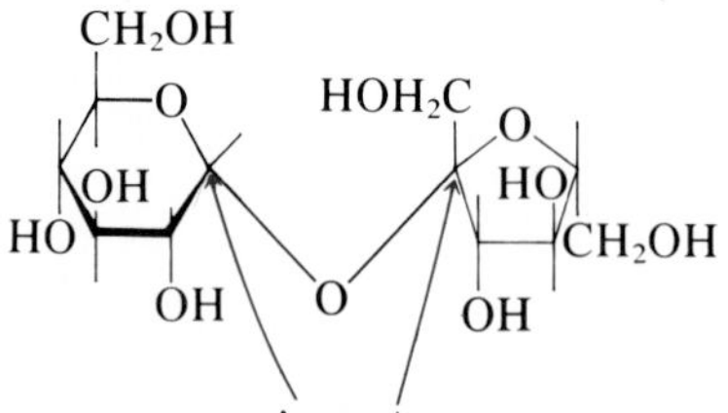
Anomeric
carbons

Notice that there is no hemiacetal group. This is because the glucose and fructose units are joined through their anomeric positions, and both anomeric carbons are involved in acetal linkages. Acetal groups are stable in basic solution. Therefore, under the conditions of the Tollens' or Benedict's test the acetal linkage will not revert back to the free aldehyde form and there is no oxidation-reduction reaction.

Recognizing a sugar as reducing or nonreducing is based on determining the presence or absence of a hemiacetal (or hemiketal) group.

Free aldehyde or hemiacetal (hemiketal) group *present* = *reducing sugar*

Free aldehyde or hemicetal (hemiketal) group *absent* = *nonreducing sugar*

All polysaccharides are nonreducing sugars. It is true that polysaccharides usually have a terminal hemiacetal carbon, for example,

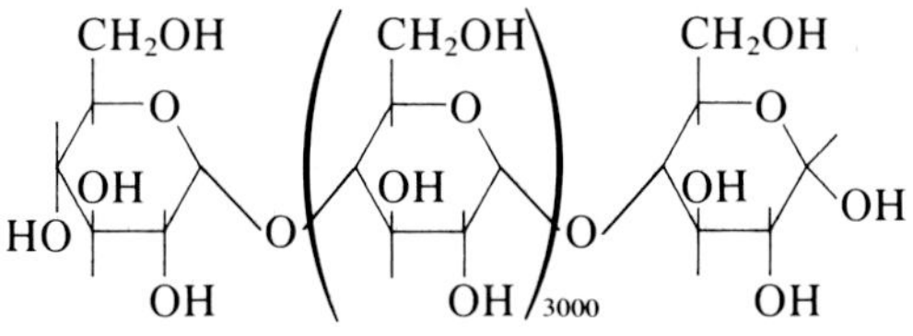

However, polysaccharides are only sparingly soluble, and the one unit with a hemiacetal group is only one out of thousands. Thus even if the group were to react, the concentration change would be so small that a color change would be imperceptible.

SAMPLE EXERCISE 15.3

Identify the following as reducing or nonreducing sugars:

a

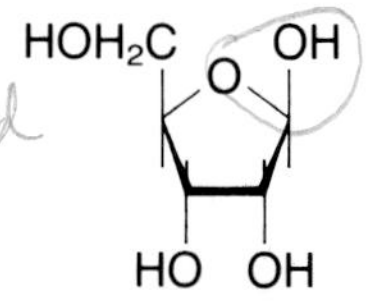

b

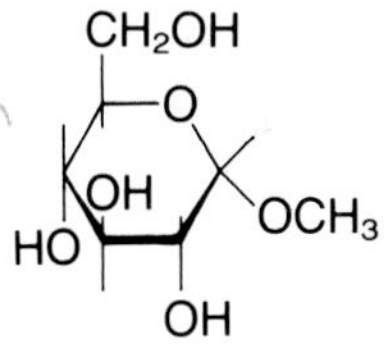

c

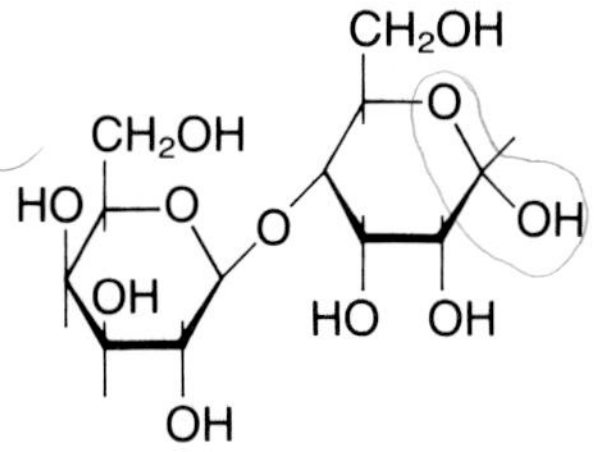

d

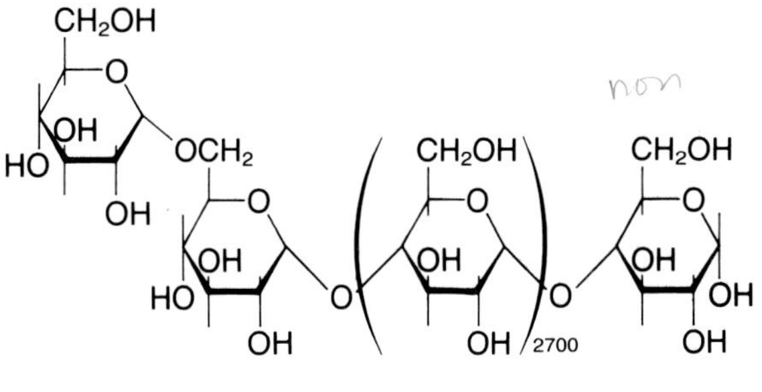

e

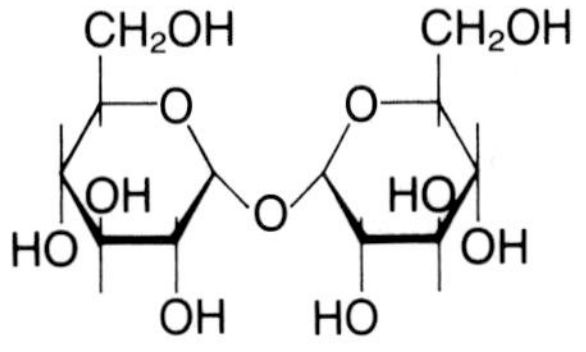

Solution The identification depends on the presence or absence of hemiacetal groups.

a All monosaccharides are *reducing* sugars because of the presence of the hemiacetal carbon.

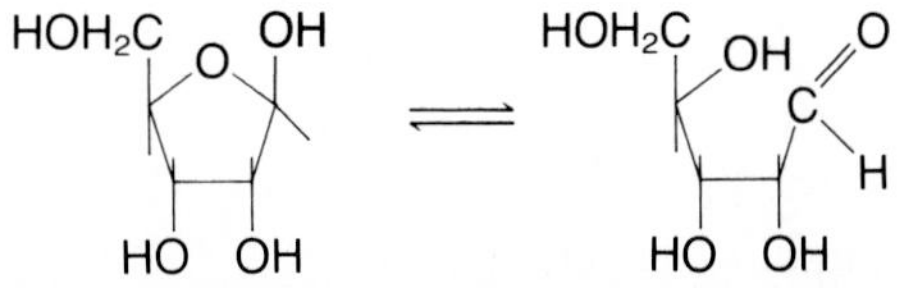

b This is a *nonreducing* structure. The monosaccharide has reacted with methanol to form a methyl acetal. There is no hemiacetal position.
c *Reducing.* Hemiacetal group is present.
d All polysaccharides are *nonreducing.*
e *Nonreducing.* Hemiacetal group is absent.

Benedict's and Tollens' tests are not specific for glucose, but rather show a positive result for all monosaccharides, some disaccharides, and other organic substances capable of being readily oxidized. Lactose, often found in the urine of lactating women, or arabinose, from a diet containing large amounts of fruit or fruit juices, or vitamin C in urine from excess dietary intake of C, might give a positive Benedict's test which could be mistaken for an indication of glucosuria. To avoid this misdiagnosis, a test employing an enzyme, glucose oxidase, that specifically oxidizes and changes color *only* for glucose is used. The test reagent is commercially available impregnated on a dipstick. See Section 21.12.

15.15 DISACCHARIDES

When two monosaccharides are joined into a disaccharide, the linkage is always through C-1 (the anomeric carbon) of one of them. Any —OH group of the other monosaccharide could potentially react and thereby link up with the anomeric carbon. Most often, however, it is the —OH at C-4 that reacts. Thus:

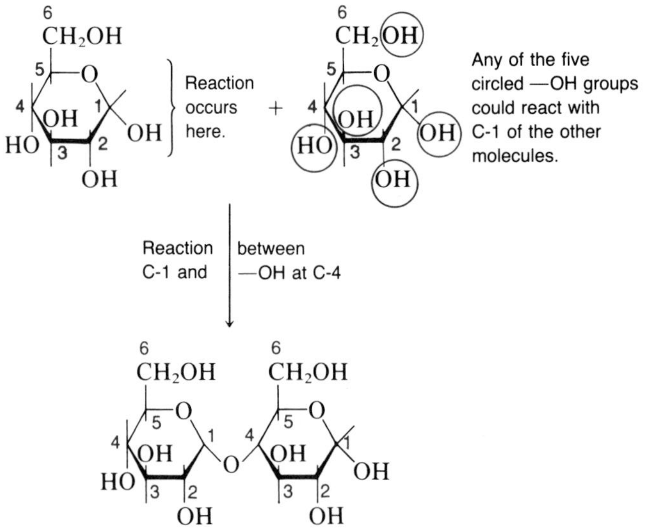
Reaction
occurs
here.
Any of the five
circled —OH groups
could react with
C-1 of the other
molecules.
Reaction
C-1 and
between
—OH at C-4

Linkages between monosaccharides are always described by numbers which denote the ring carbons joined by the linkage. Thus, the linkage shown above is a 1,4-linkage.

These links or bonds between monomer units could be called *acetal* linkages because that is the functional group that is formed. More often biochemists refer to them as *glycosidic* linkages. Glycosidic is a general designation for any sugar in which the hemiacetal functional group has been replaced by another group such as an acetal linkage. The term *glucosidic* refers specifically to glucose.

Glycosidic linkages are not only characterized numerically as previously described, they are also classified according to the α- or β-orientation of the oxygen link at the anomeric carbon. Thus the example used to show a 1,4-

linkage is more specifically an α-1,4-linkage because of the orientation of the linkage. In the α-glycosidic linkage the oxygen link from the anomeric carbon goes down on the opposite side of the ring from C-6. The β-linkage goes up on the same side as C-6.

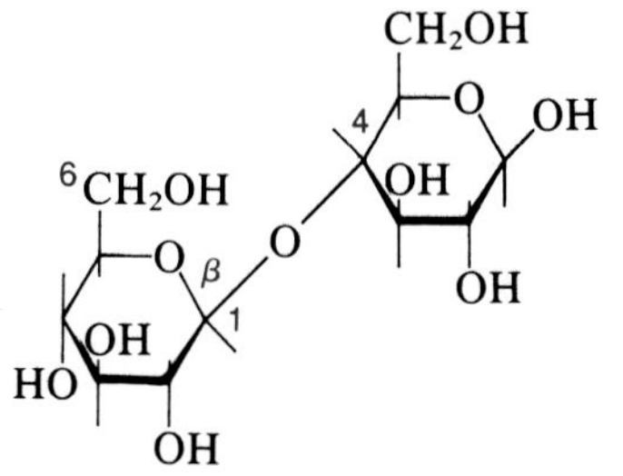
CH₂OH
4
O
OH
⁶CH₂OH
OH
O
O
β
OH
1
OH
HO
OH

Differently oriented linkages produce different isomers with different properties. Two glucose units joined α-1,4 constitute the disaccharide maltose. A β-1,4-glycosidic linkage between glucose units characterizes cellobiose. These two structures, maltose and cellobiose, are given in the discussion of individual disaccharides that follows.

Problem 15.9 Describe the glycosidic linkages in the following disaccharides numerically and according to α- and β-orientation.

a

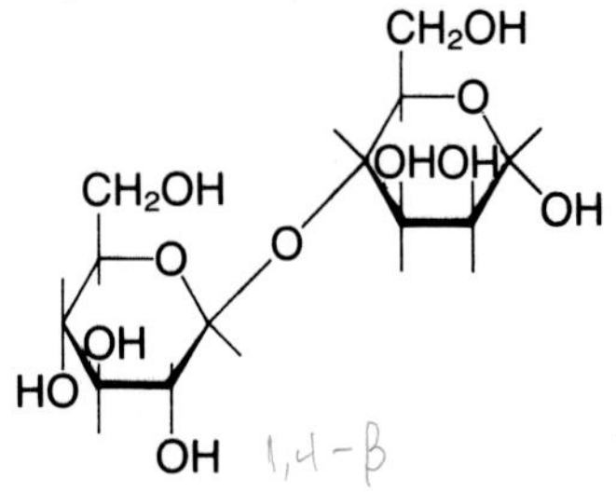
CH₂OH
O
OHOH
CH₂OH
OH
O
O
OH
HO
OH

b

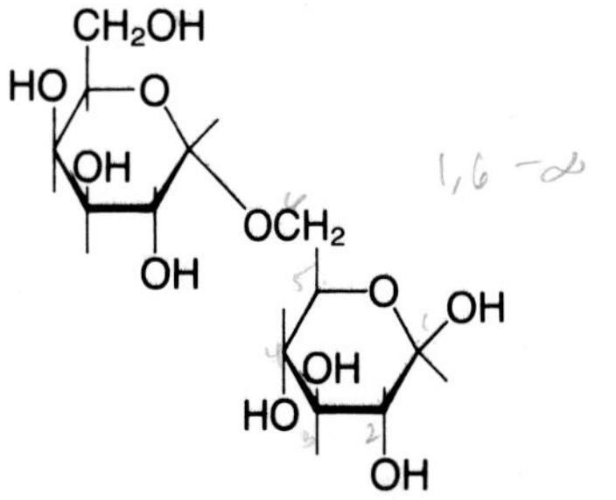
CH₂OH
HO
O
OH
OCH₂
OH
O
OH
OH
HO
OH

Maltose, or malt sugar, is not naturally prevalent. It is encountered most often as an intermediate product in the digestive breakdown of starch.

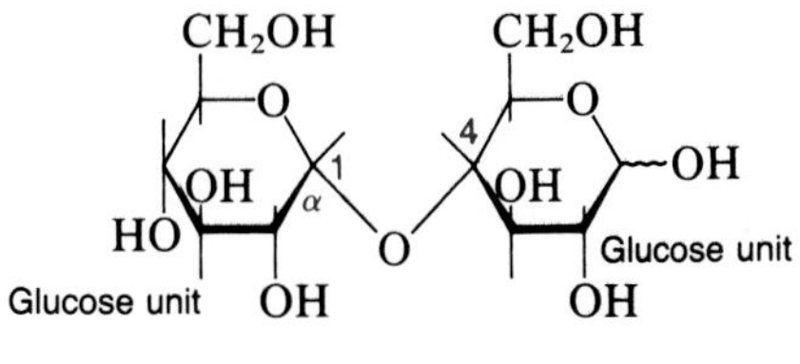
CH₂OH
CH₂OH
O
O
4
OH
1
OH
OH
α
HO
O
Glucose unit
Glucose unit
OH
OH

Maltose

It is a reducing sugar and capable of mutarotation; i.e., the —OH at C-1 in the right-hand ring can be α or β.

Problem 15.10 Explain why maltose is described as a reducing sugar and capable of mutarotation.

Cellobiose is a disaccharide made up of glucose units joined β-1,4. It is also a reducing sugar. The partial hydrolysis of plant cellulose yields cellobiose.

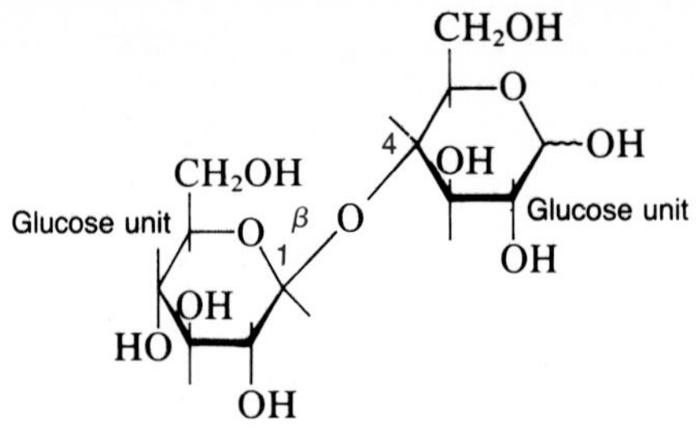
Glucose unit
Glucose unit

Cellobiose

Lactose, or milk sugar, is a disaccharide composed of glucose and galactose joined in the following way:

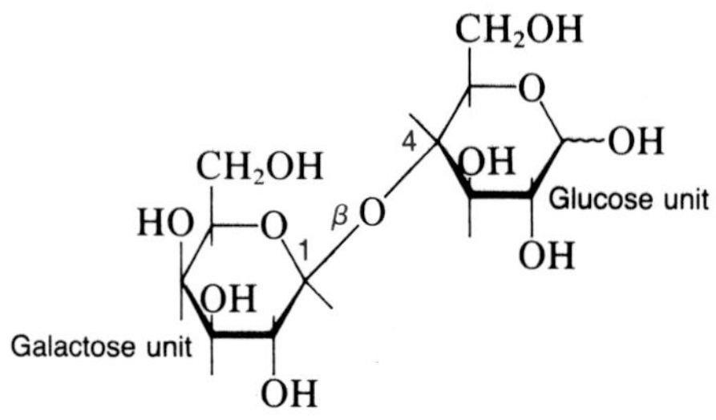
Galactose unit
Glucose unit

Lactose

Galactose provides the anomeric carbon of the glycosidic linkage in a β-orientation. Glucose is hooked on through C-4. Lactose occurs in the milk of mammals at concentrations ranging from 2 to 8 percent. It is a reducing sugar.

Sucrose is the sugar we encounter daily as table sugar. It is a nonreducing sugar because the constituent monomers, glucose and fructose, are both linked through their anomeric carbons. The glycosidic linkage in sucrose is 1,2 and is α with respect to the glucose ring and β with respect to the fructose ring.

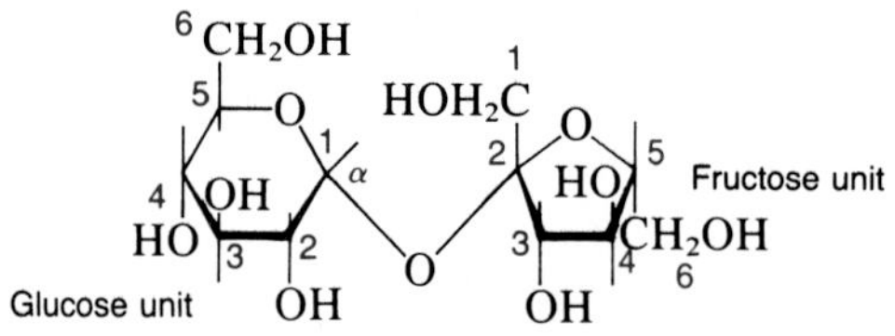
Glucose unit
Fructose unit

Sucrose

Sucrose is obtained from sugar cane or from sugar beets.

Invert sugar is the name given to the 50:50 mixture of glucose and fructose obtained when the disaccharide sucrose is cleaved by hydrolysis.

$$\text{Sucrose} \xrightarrow{\text{hydrolysis}} \underbrace{\text{glucose} + \text{fructose}}_{\text{Invert sugar}}$$

(optical rotation = +66.5°)

Invert sugar
(optical rotation = −39°)

The name arises because of the change or inversion of sign of optical rotation that occurs. Sucrose is dextrorotatory; the mixture of monosaccharides is levorotatory. Invert sugar is sweeter than table sugar because of the free fructose units. Honey is mostly invert sugar, as are many "liquid sugars," i.e., solutions of the crystalline solid sugars.

Problem 15.11 Is invert sugar reducing or nonreducing?

15.16 POLYSACCHARIDES

Whereas polysaccharides could be made up of any monosaccharides, all important and abundant polysaccharides are made of glucose monomer units. The differences between the polymers lies in the nature of the glycosidic linkages.

Cellulose is the most abundant polysaccharide. It is the structural material in plants. Wood, natural fibers, and paper are all mostly cellulose. The glycosidic linkages between glucose units in cellulose are β-1,4. This type of linkage produces a linear polymer, as seen in Figure 15.8. Typically the chains are 2000 to 13,000 glucose units long. These long chains remain straight in nature and line up in parallel arrays held together by hydrogen bonding between chains. This gives the fibrous quality to cellulose materials.

Starch, plant nutrient material, is actually a mixture of two polysaccharides, **amylose** and **amylopectin.** Amylose (Figure 15.9*a*) is characterized by α-1,4 glycosidic linkages between glucose monomers, which lead to a linear chain of glucose units. Amylopectin, on the other hand, is a branched polymer because it contains α-1,6 as well as α-1,4 glycosidic links. Figure 15.9*b* shows the branched structure that results from the inclusion of α-1,6 links.

Hydrolysis of starch, which is carried out commercially for various reasons and which is the process of digestion, usually does not proceed directly from polysaccharide to monomer units. Rather, partial hydrolysis products of intermediate size are obtained. Figure 15.10 shows a commercial process using enzymes from malt to break down amylopectin into **dextrins,** which are still quite large molecules, and maltose disaccharide units. Dextrins are "sticky" and are used in library paste. Another name given to small sugar polymers is **oligosaccharides.** These are usually defined as consisting of from 2 to 10 monomer units.

Glycogen is sometimes called *animal starch* because like plant starch material it is characterized by α-glycosidic links. It has a branched structure similar to that of amylopectin; however, glycogen molecules are more highly branched with shorter branching chains. Also, in general, glycogen molecules have a greater number of glucose units than amylopectin. The particular

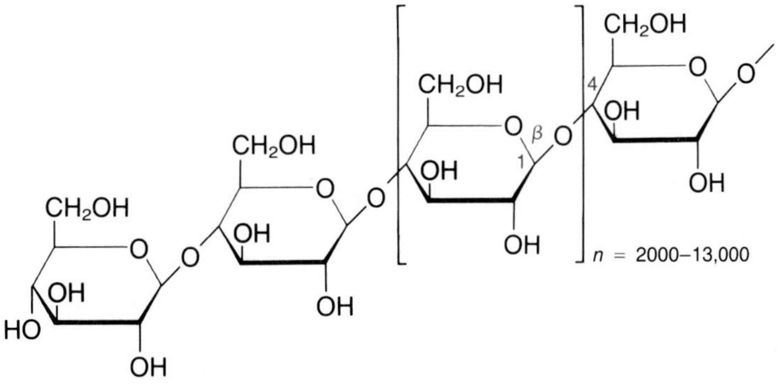
CH₂OH
O
OH
HO
OH
β
1
4
n = 2000–13,000

FIG. 15.8
Cellulose. β-1,4-Glycosidic linkages produce a linear array of monomer units.

number of glucose units in the polymer depends on the species producing the glycogen or amylopectin. The term *animal starch* for glycogen comes from the fact that it is made by animals. Blood glucose monomers are assembled into glycogen polymers for storage in muscles and in the liver.

15.17 CARBOHYDRATE DIGESTION

Carbohydrate digestion is the hydrolysis of disaccharides and polysaccharides into monosaccharide units, mostly glucose units. As you will see in Chapter 23, glucose is directly usable by cells in metabolism; it need not be digested since it is already a usable form. The whole point of digestion is to produce molecules that are the right size and shape to cross membranes, enter the blood and cells, and be used by metabolic cycles. Glucose is suitable for intravenous feeding because it is directly usable by cells.

Equation (15.3) represents the digestion of sucrose. Equation (15.4) shows the complete digestion of amylose.

FIG. 15.9
(*a*) Amylose. α-1,4-Glycosidic linkages produce a linear array of monomer units.
(*b*) Amylopectin. When there are α-1,6-glycosidic linkages as well as α-1,4-linkages, the monomer units cannot all be linear. The structure is described as branched.

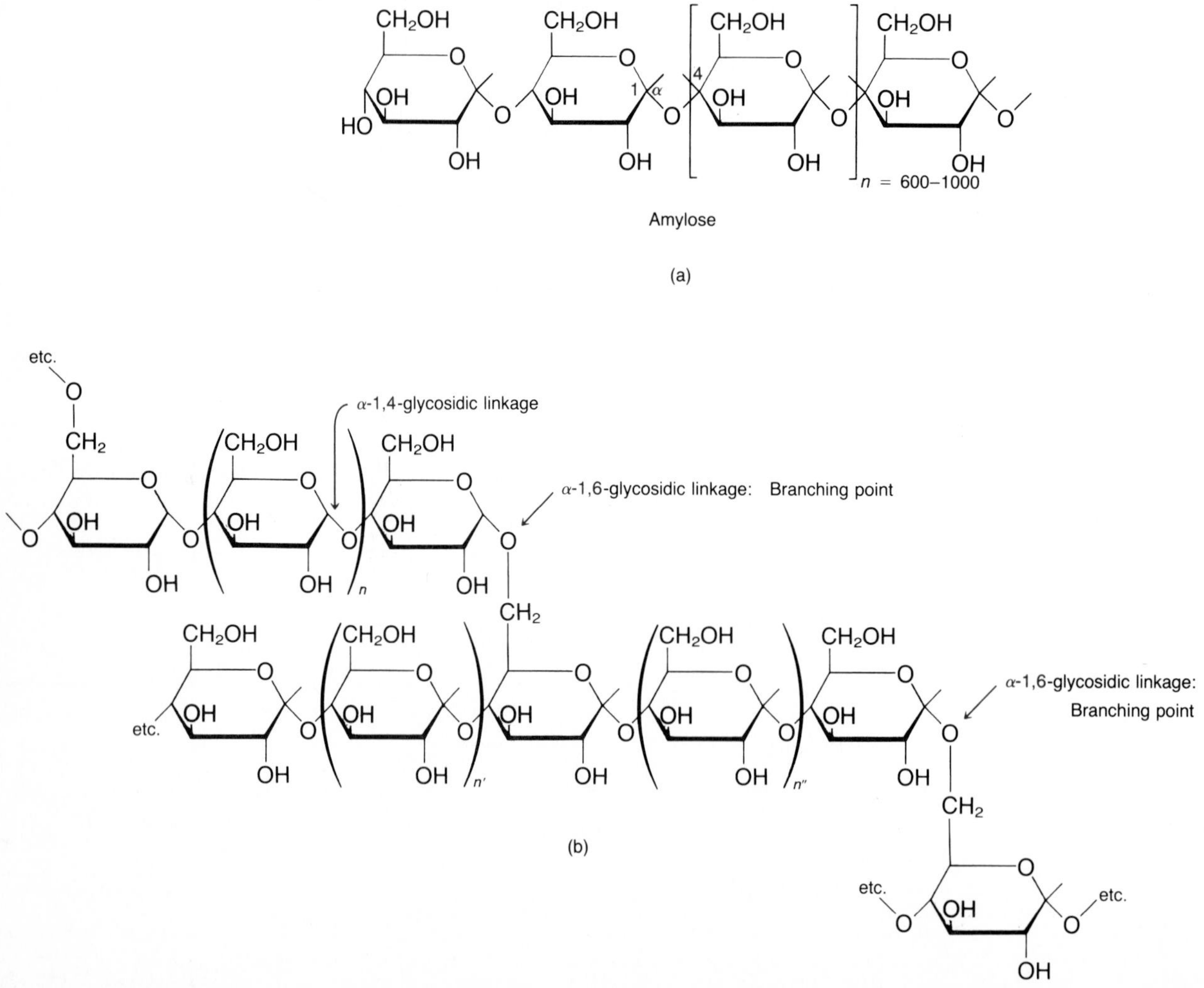

CH2OH
OH
HO
O
1
α
4
n = 600–1000
Amylose
(a)
etc.
CH2
α-1,4-glycosidic linkage
α-1,6-glycosidic linkage: Branching point
n
n′
n″
α-1,6-glycosidic linkage:
Branching point
(b)

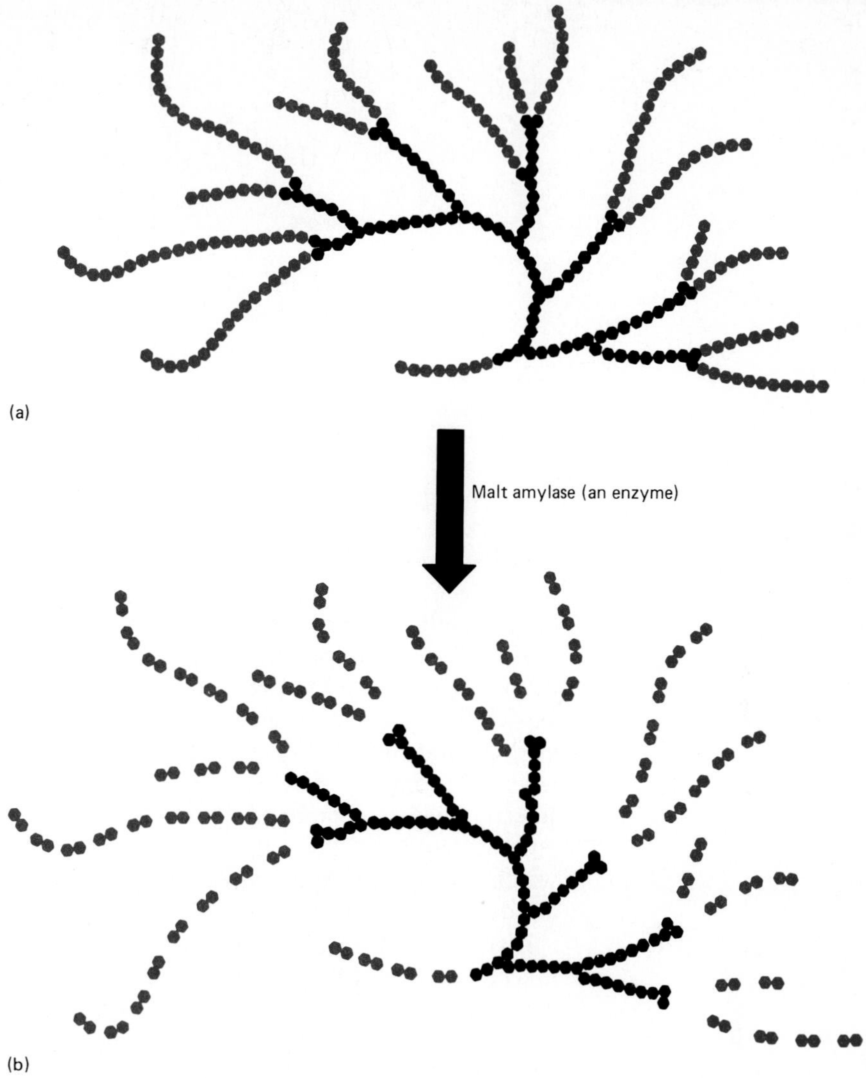
(a)
Malt amylase (an enzyme)
(b)

FIG. 15.10
Hydrolysis of amylopectin by enzymes in malt. (*a*) A schematic view of branched amylopectin. Each hexagon is a glucose unit. The color chain portions are more exposed and vulnerable to attack. The malt enzyme cleaves chains at every other glycosidic linkage, so that disaccharide maltose units are released. The enzyme cannot cleave at branched positions. (*b*) When the maltose units are broken off, smaller polymers called *dextrins*, represented here by the black chain, are left.

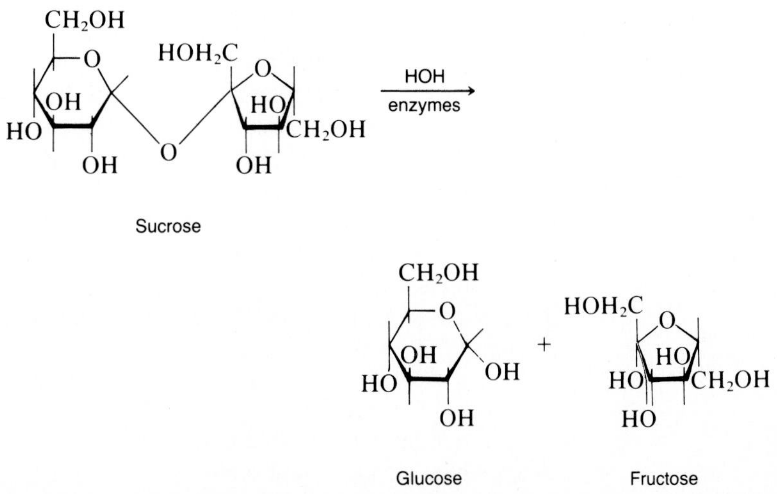
HOH
enzymes
Sucrose
Glucose
Fructose

(15.3)

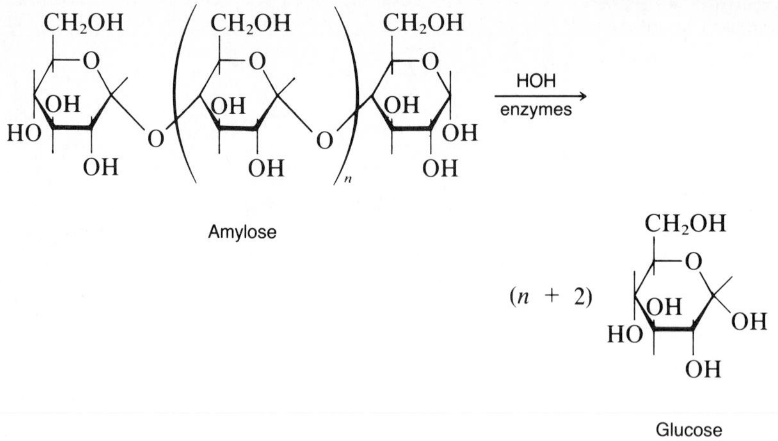
HOH
enzymes
Amylose
Glucose

(15.4)

Carbohydrate digestion begins in the mouth. Saliva contains a lubricant, mucin, and an enzyme which catalyzes the partial hydrolysis of starch to dextrins and disaccharides. The enzyme works best under the slightly acidic to neutral conditions of the mouth ($pH = 5.8 \sim 7.1$). The extent of digestion in the mouth depends on how long the food is chewed.

There is no carbohydrate digestion in the stomach, where there is no further source of carbohydrate digestive enzymes and where any enzyme carried from the mouth with swallowed food is deactivated by the low pH. Carbohydrate hydrolysis (i.e., digestion) continues in the small intestine, where the influx of pancreatic juice reestablishes the pH at higher values. Intestinal juices contain amylase enzymes that continue the breakdown of polysaccharides. Other enzymes present are sucrase, lactase, and maltase, which catalyze the hydrolysis of the disaccharides from which their names are derived.

Digestive enzymes in human beings only allow for the hydrolysis of α-acetal or α-glycosidic linkages. Because of this stereochemical requirement, polysaccharides such as cellulose, which has β-linkages between glucose units, are not digestible. We can digest plant starch (amylose and amylopectin), but not the cellulose which makes up more than 50 percent of the organic matter in plants. Undigested cellulosic fiber from vegetables and whole-grain cereals makes up most of the bulk of feces.

The undigested plant food in our diet, of which cellulose forms a large proportion, is called **dietary fiber.** Some studies suggest that adequate dietary fiber intake is necessary to reduce the risk of diseases of the large intestine, heart disease, diabetes, and obesity. One theory is that fiber may dilute the concentration of chemical carcinogens in the intestine and thereby lower their cancer-causing potential. Fiber may also aid in the excretion of cholesterol and other fats into the feces and thereby possibly result in a lower incidence of heart disease. However, the association between fiber and these conditions is not completely established. The importance of dietary fiber and general considerations of carbohydrate nutrition will be further explored in Section 26.5.

Although human beings do not, termites possess an enzyme which cleaves β-glycosidic linkages and enables them to digest cellulose and thereby destroy wood. Cows have microorganisms in their stomachs which can also hydrolyze β-links, allowing the cow to digest cellulose-containing grass.

CHAPTER ACCOMPLISHMENTS

After completing this chapter, you should be able to define **key words** and do the following:

15.1 INTRODUCTION
1. Cite two alternative names for carbohydrates.

15.2 CLASSIFICATION: ALDOSES VERSUS KETOSES
2. Given a structural formula, identify a sugar as an aldose or ketose and classify it according to chain length.

15.3 STEREOISOMERISM
3. Describe the difference between two stereoisomers.
4. State the necessary criteria for two structures to be enantiomers.
5. Recognize a chiral carbon in a given structural formula.

15.4 PROPERTIES OF ENANTIOMERS
6. Describe the differences and similarities in properties between two enantiomers.
7. Given the optical rotation for one enantiomer, state the value of that property for the other enantiomer.

15.5 STEREOISOMERISM AND GLUCOSE
8. State the significance of stereoisomerism to carbohydrates.

15.6 FISCHER PROJECTIONS
9. Given a structural formula of a compound with one chiral carbon, write Fischer projections for enantiomers of that formula.

15.7 MULTIPLE CHIRAL CENTERS
10. Write Fischer projections for structural formulas with more than one chiral carbon.
11. Distinguish between enantiomers and diastereomers.

15.8 CLASSIFICATION: D-FAMILY VERSUS L-FAMILY
12. Given a Fischer projection for a sugar isomer, classify the carbohydrate as a D-sugar or an L-sugar.
13. State the more prevalent and metabolically important sugar family.

15.9 INTRAMOLECULAR HEMIACETALS AND HEMIKETALS
14. State the reason that carbohydrates exist in nature as cyclic structures.
15. State the ring size for the aldohexoses and aldopentoses and for fructose.

15.10 HAWORTH PROJECTIONS
16. Given a Fischer projection for a sugar, write Haworth projections for the two anomers of that compound.

15.11 MUTAROTATION

15.12 FORMATION OF DI- AND POLYSACCHARIDES
17. Describe the reaction by which monosaccharides are joined to form disaccharides and polysaccharides.
18. Given a structural formula, identify it as a mono-, di-, or polysaccharide.

15.13 MONOSACCHARIDES
19. Name the most prevalent and important monosaccharide.
20. Describe the similarities and differences in structure, properties, and occurrence among glucose, mannose, galactose, and fructose.
21. Name the aldopentoses found in DNA and RNA.

15.14 CLASSIFICATION: REDUCING AND NONREDUCING SUGARS
22. Describe the chemical reactions used to classify sugars as reducing or nonreducing.
23. Given a structural formula, identify a sugar as reducing or nonreducing.

15.15 DISACCHARIDES
24. Given the structural formula of a disaccharide, classify the glycosidic linkage numerically and according to α- or β-orientation.

15.16 POLYSACCHARIDES
25. Name the monomer unit of all common polysaccharides.
26. Describe the nature of the glycosidic linkages in cellulose, starch, and glycogen.

15.17 CARBOHYDRATE DIGESTION

27. Describe the chemical reaction that is the carbohydrate digestion process and name the location of carbohydrate digestion in the body.
28. Describe the stereochemical requirement of polysaccharide glycosidic linkages for digestion to occur.

PROBLEMS

15.12 Define carbohydrates in terms of the functional groups in their structures.

15.13 Carbohydrates are known by other names. Give two other names.

15.14 Classify the following as aldoses or ketoses:

a 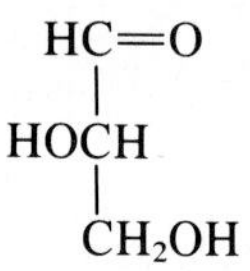

b 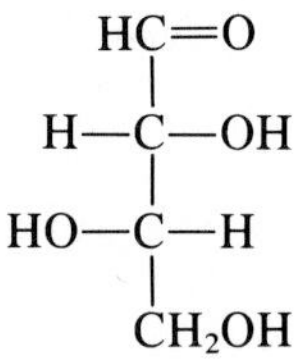

c 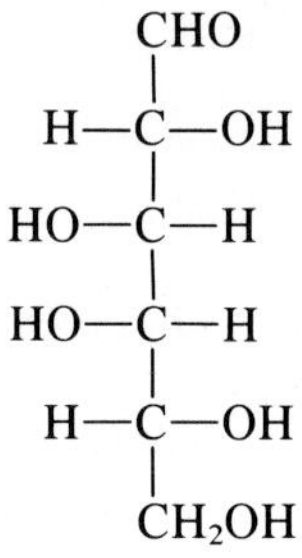

d 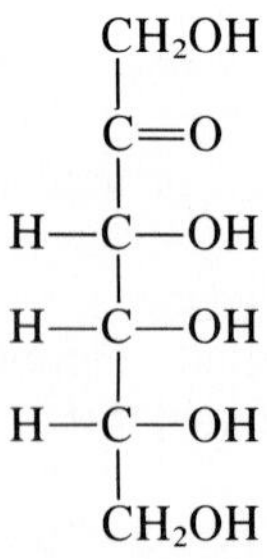

e 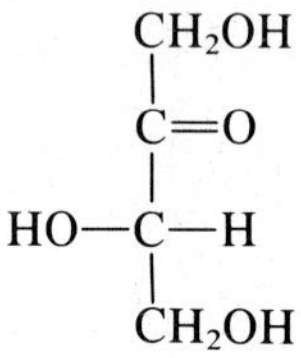

f 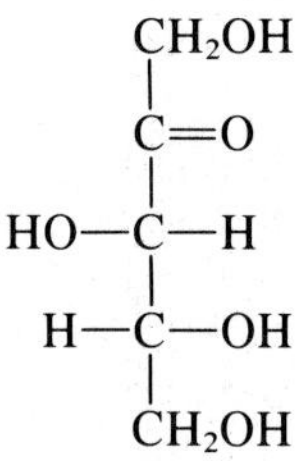

g 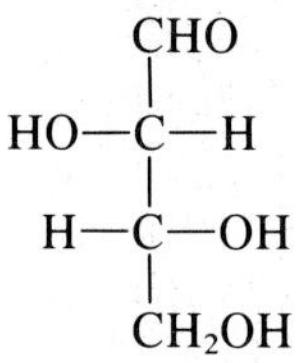

h 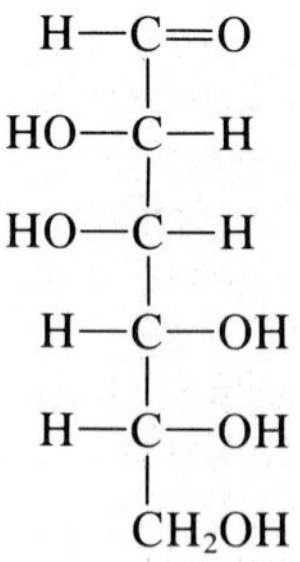

i 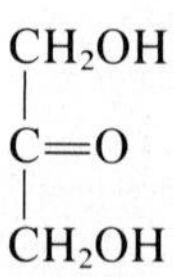

15.15 Give a name for each structure in Problem 15.14 that reflects the chain length as well as the aldose or ketose nature of the sugar.

15.16 Assign each sugar in Problem 15.14 to the D-family or L-family.

15.17 Write Fischer projections for the structures given in Problem 15.14*a*, *e*, and *g* and for their enantiomers.

15.18 Use structural formulas in the Fischer projection to write examples of the following:

a An aldotetrose
b A D-ketoheptose
c A D-ketopentose
d An aldotriose
e A D-aldohexose

15.19 Circle the chiral carbons in each structure you wrote in Problem 15.18.

15.20 Use a Fischer projection to write a structure for the enantiomer of D-ribose:

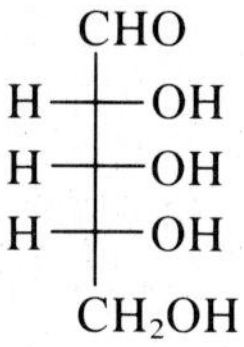

D-Ribose

15.21 Use a Fischer projection to write a structure for a diastereomer of D-ribose.

15.22 What is the maximum number of stereoisomers that can be written for the aldopentoses?

15.23 How many stereoisomeric ketopentoses are there?

15.24 Which so-called optical family, i.e., the D or the L, predominates in nature?

15.25 Are any of the following compounds meso compounds? Which one(s)?

a 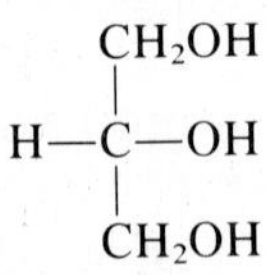

b 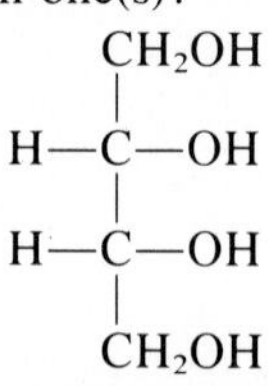

c 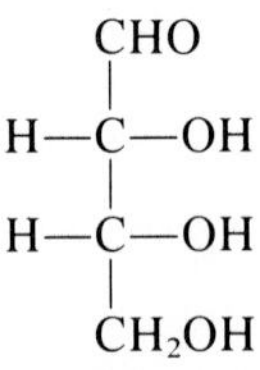

15.26 Does invert sugar belong to the D- or L-family?

15.27 Write the structure for the open-chain form of glucose.

15.28 Write a formula for glucose that accurately reflects the structure in which it exists.

15.29 Why do monosaccharides form cyclic structures?

15.30 Draw a pyranose ring.

15.31 Draw Haworth projection formulas for each of the following:

a

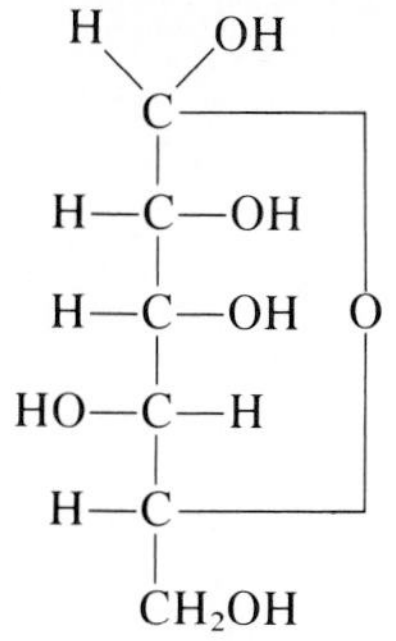

b

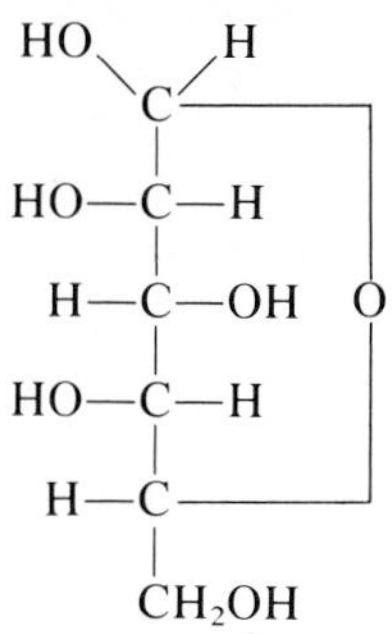

15.32 Designate the structures in Problem 15.31 as either α or β, as appropriate.

15.33 Write Haworth projection formulas for the two anomers of D-mannose.

15.34 Are anomers enantiomers or diastereomers or neither?

15.35 Write structures of the following molecules in the β-D-pyranose ring form (Haworth projection):

a

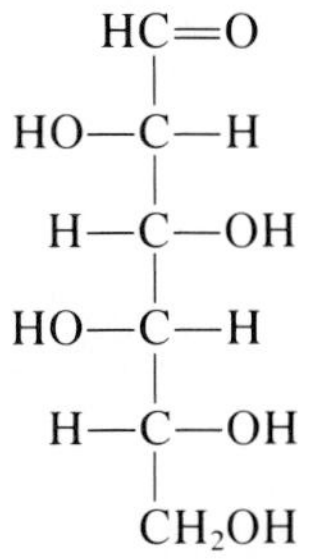

b

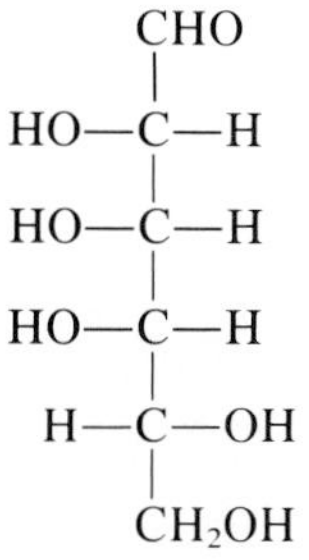

15.36 Draw a furanose ring.

15.37 Name the ketohexose that is a furanose.

15.38 Give the name of a specific aldopentose furanose.

15.39 Complete the following equations:

a

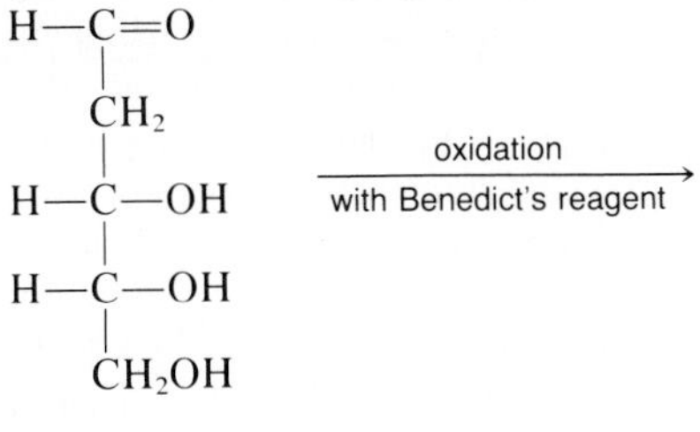
oxidation
with Benedict's reagent

b

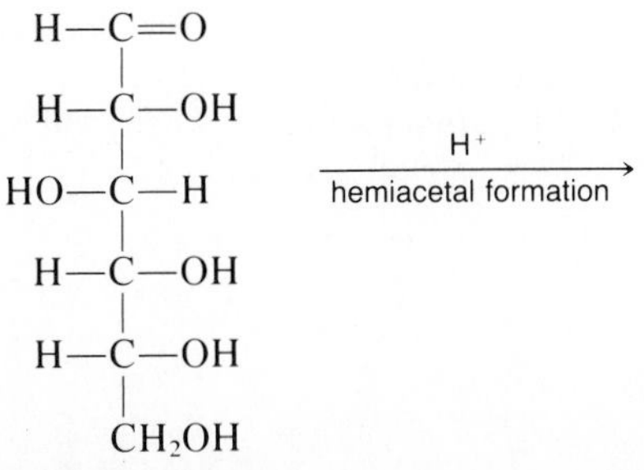
H+
hemiacetal formation

c

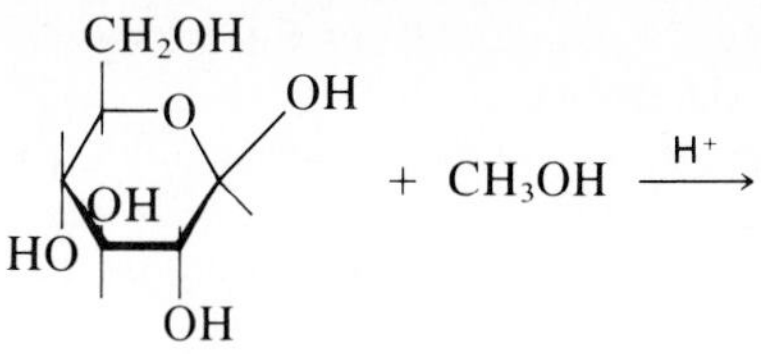

15.40 Identify the following as reducing or nonreducing sugars:

a

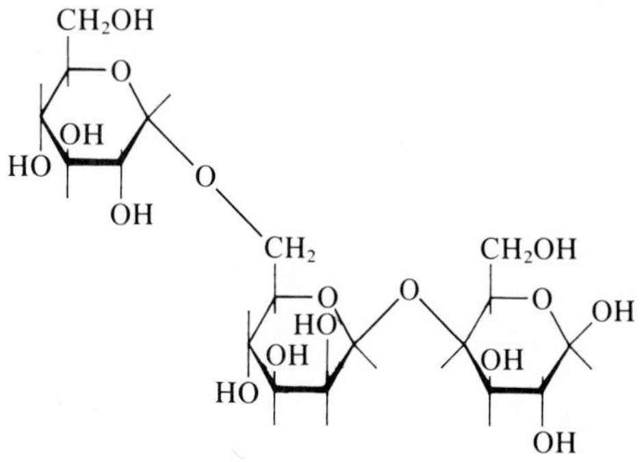

b

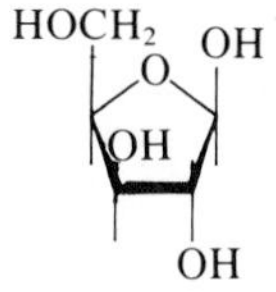

c

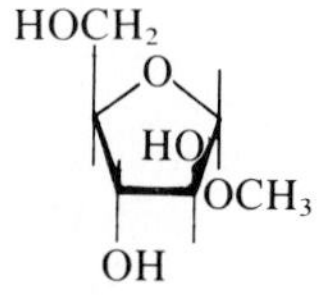

d

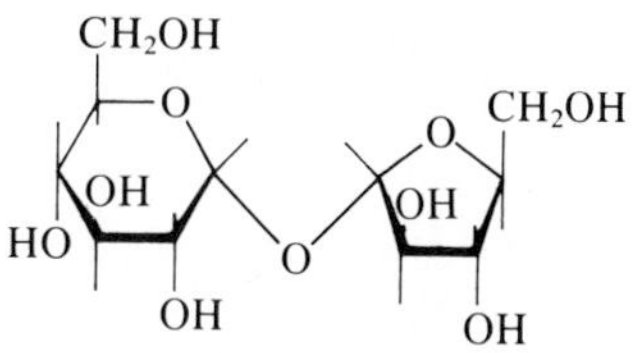

e

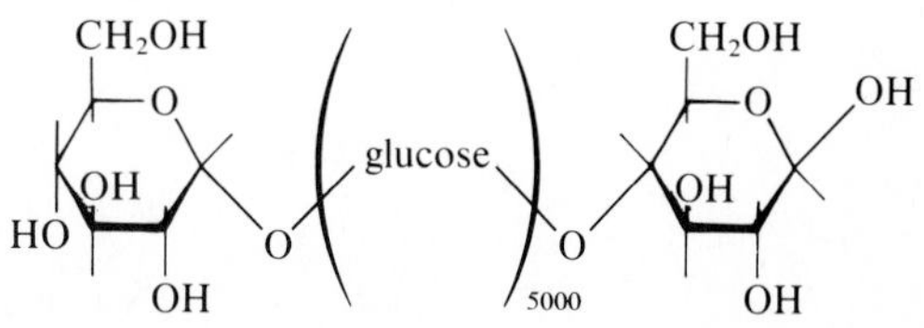
glucose
5000

15.41 Describe the glycosidic linkages in the structure in Problem 15.40*a* numerically and according to α- or β-orientation.

15.42 Write the structures of the monosaccharides that are formed from the trisaccharide shown in Problem 15.40*a* upon hydrolysis.

15.43 All monosaccharides are reducing sugars. It is also true that all monosaccharides mutarotate. How are these two facts related?

15.44 Write equilibrium equations for the mutarotation of galactose.

15.45 Gentiobiose is a reducing disaccharide found in the roots of gentians. Trehalose is a nonreducing disaccharide found in mushrooms. Their structures are shown below:

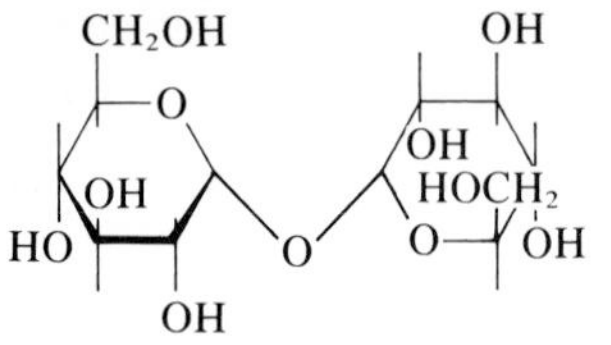

CH2OH
OH
O
OH
OH
HOCH2
HO
O
O
OH
OH

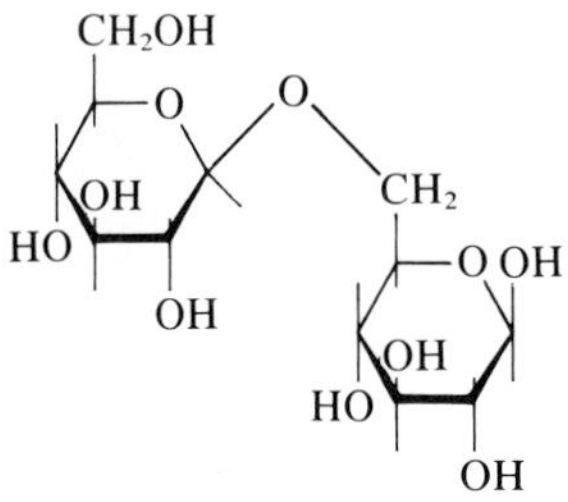

CH2OH
O
O
OH
CH2
HO
O OH
OH
OH
HO
OH

a Label the structures as gentiobiose or trehalose, as is appropriate.

b What monosaccharide(s) would be obtained upon hydrolysis of these disaccharides?

15.46 Describe the glycosidic linkages of the disaccharides in Problem 15.45 numerically and in terms of α- or β-orientation.

15.47 Isotrehalose is an isomer of trehalose that has a 1α-1β-glycosidic linkage. Neotrehalose is an isomer characterized by a 1β-1β-link. Label the structures below appropriately:

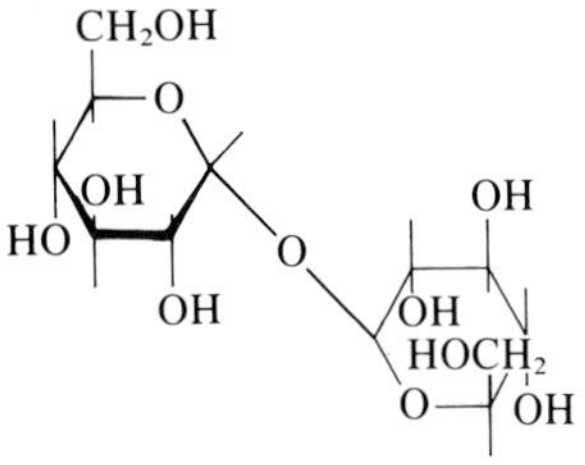

CH2OH
O
OH
OH
HO
O
OH
OH
HOCH2
O
OH

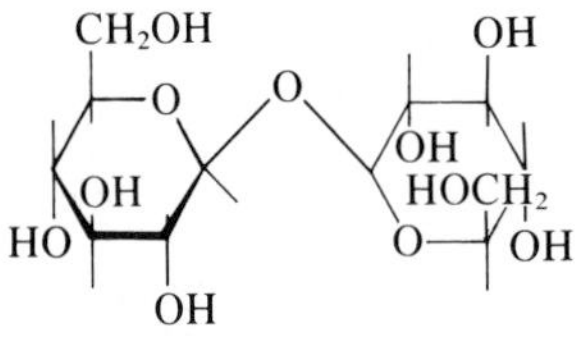

CH2OH
OH
O
O
OH
OH
HOCH2
HO
O
OH
OH

15.48 See Problem 15.47. Are isotrehalose and neotrehalose reducing or nonreducing sugars?

15.49 Write structures for disaccharides in which glucose units are hooked together:

a α-1,4
b β-1,4
c α-1,6
d β-1,6

15.50 Which of the disaccharides in Problem 15.49*a* can be digested by human beings?

15.51 What is the principal structural difference between amylose and cellulose?

15.52 What is the principal structural difference between amylose and amylopectin?

15.53 Is glycogen most like cellulose, amylose, or amylopectin? Explain.

15.54 What is starch?

15.55 Match the following sugar names and descriptions:

a Glucose	*1* Table sugar
b Galactose	*2* Milk sugar
c Sucrose	*3* Blood sugar
d Lactose	*4* Brain sugar
e Mannose	*5* Undigestible disaccharide
f Fructose	*6* Component of RNA
g Cellobiose	*7* Sweetest sugar
h Ribose	*8* Least common sugar of those listed

15.56 The chemical reaction of digestion is ______.

15.57 Cellulose would be a good diet food if it were tasty. Explain.

15.58 Where in the body are carbohydrates digested?

15.59 What sugar can be given in intravenous solutions?

15.60 Dilute solutions can be made of starch, which is a polymer of glucose. Why can't a starch solution be given intravenously?

15.61 State the necessary criteria for two structures to be enantiomers.

15.62 *a* In what way are stereoisomers similar?

b How do stereoisomers differ?

15.63 Which of the following objects are chiral, i.e., nonsuperimposable on its mirror image?

a A hand
b A shoe
c A sock
d A pullover sweater
e A cardigan sweater
f A screw
g A glove
h A mitten

15.64 *a* State two properties in which enantiomers may differ.

b State two properties which will be identical in a pair of enantiomers.

CHAPTER SIXTEEN

AMINES

16.1 INTRODUCTION

Carbon, hydrogen, oxygen, and nitrogen are the predominant elements in compounds in living systems. Functional groups you have encountered up to this point have contained only the first three (C, H, O). In this chapter nitrogen is encountered for the first time in the amine functional group.

Amines and compounds derived therefrom are among the most essential compounds in living systems. Proteins (Chapter 19) are built up from compounds containing the amine group. DNA and RNA (Chapter 20), the hereditary and information molecules, contain the amine functional group. Communication between the brain and other nerve cells in the central nervous system occurs via molecules called neurotransmitters which contain the amine group. Most medicinal drugs have an amine nitrogen as part of their structure. Before pursuing the physiological activity of amines, we will, as usual, consider structure, nomenclature, and chemical reactivity of the functional group.

16.2 CLASSIFICATION OF AMINES

When we introduced alcohols, we mentioned that they can be considered to be derived from water by replacing an H of water by an alkyl group. (H—O—H ⟶ R—O—H) In the same sense, amines can be derived from ammonia (NH_3) by replacing one or more of the hydrogens of ammonia by an alkyl group or aromatic ring.

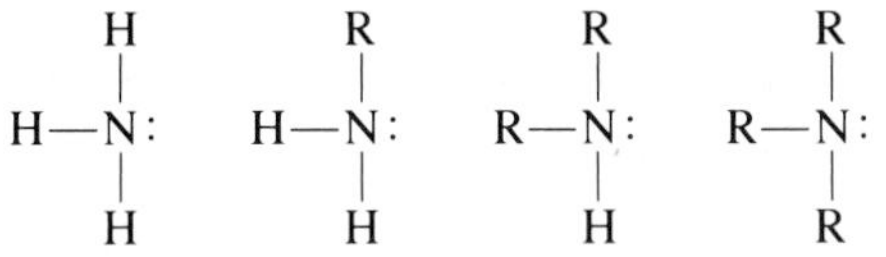

Ammonia   1° Amine   2° Amine   3° Amine

Amines are classified into the categories primary (1°), secondary (2°), and tertiary (3°) depending on the number of organic groups bonded to the amine nitrogen. Primary amines have one organic group, secondary amines have two, and tertiary amines have three organic groups. Because all classes of amines 1°, 2°, and 3°, have a nonbonding pair of electrons on the nitrogen, you shall see that a fourth bond can form to the nitrogen in some chemical reactions. A compound with four carbon bonds to nitrogen is termed a quarternary (4°) salt; these salts are ionic compounds made up of cations with a positive charge on the nitrogen and anions. Tetramethylammonium chloride is a simple example:

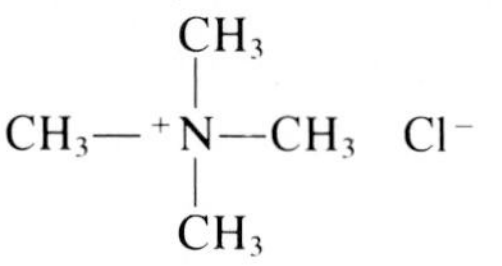

You will see more of quarternary ammonium salts in Section 16.6.

16.3 NOMENCLATURE

Amines with only alkyl groups bonded to the nitrogen are known as aliphatic amines. Such amines are most often named by simply listing the names of the alkyl groups attached to the nitrogen and then adding the word *amine*. The prefixes *di-* and *tri-* are used to indicate the multiple presence of any alkyl group. The alkyl group names can be listed either alphabetically or in order of increasing size. (Prefixes such as *di-*, *tri-*, and *iso-* are ignored when alphabetizing.) The following examples illustrate this common nomenclature system:

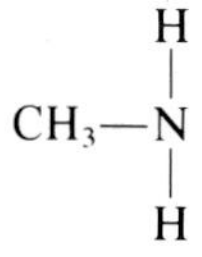

Methylamine

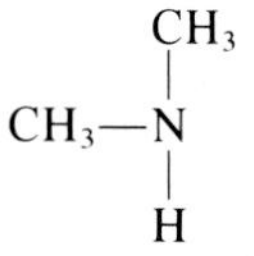

Dimethylamine

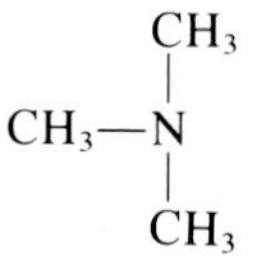

Trimethylamine

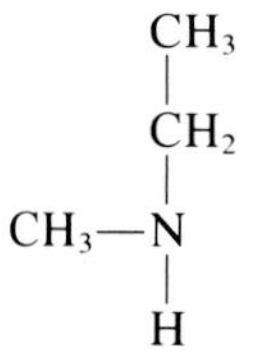

Ethylmethylamine

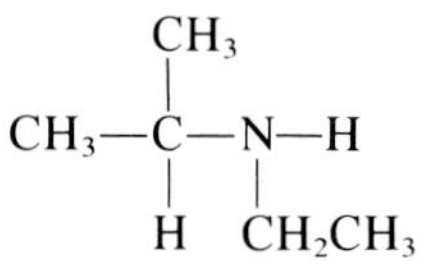

Ethylisopropylamine

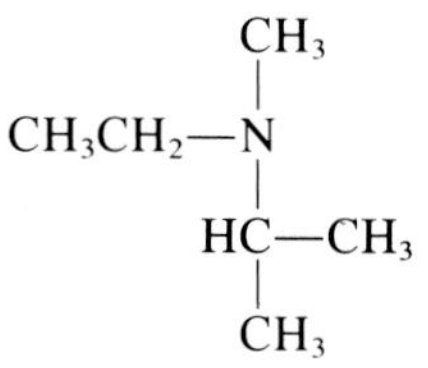

Ethylmethylisopropylamine

Aromatic amines have an aromatic ring directly bonded to nitrogen. They are named as derivatives of aniline.

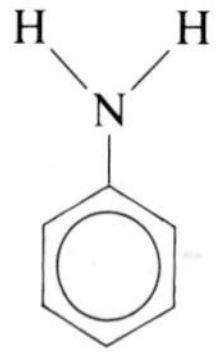

Aniline

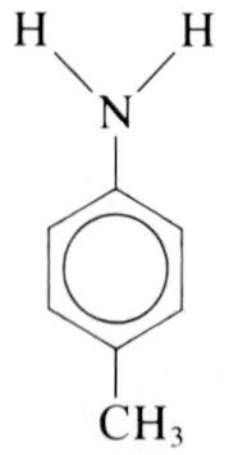

4-Methylaniline

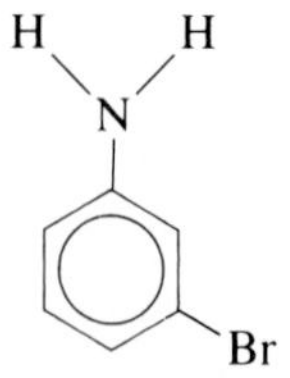

3-Bromoaniline

In the case where an alkyl group is bonded to the nitrogen of an aromatic amine, a capital *N* (rather than a numerical locant) is used to indicate the location of the group. The following examples employ this device to indicate that methyl is bonded to the nitrogen and not to the ring.

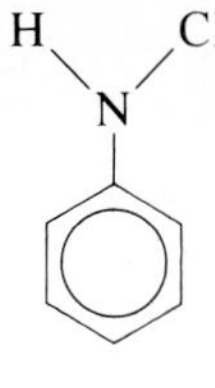

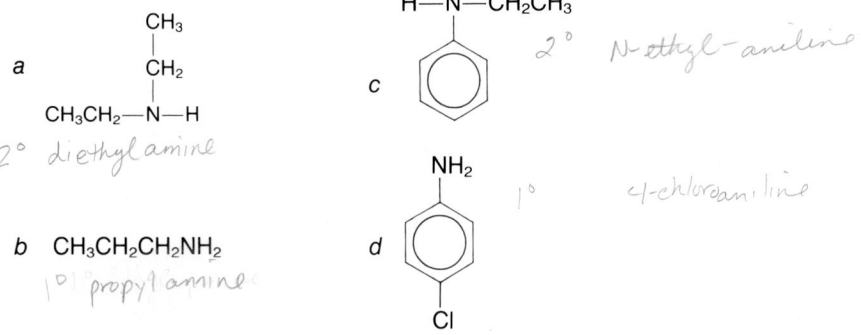

N-Methylaniline

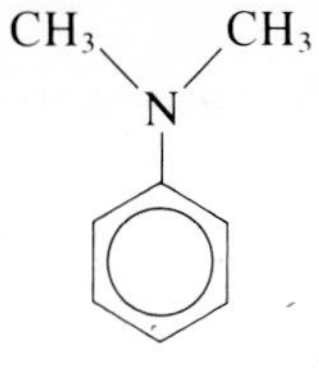

N,N-Dimethylaniline

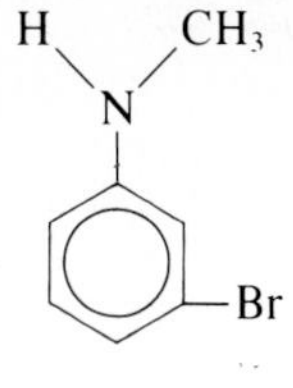

N-Methyl-3-bromoaniline

Although there is an IUPAC system for naming amines, it is rarely used for the simpler amines and we shall not use this nomenclature in this text.

Problem 16.1 Classify as primary, secondary, or tertiary and name each of the following amines:

a

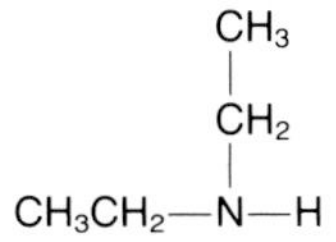

2° diethyl amine

b $CH_3CH_2CH_2NH_2$

1° propyl amine

c

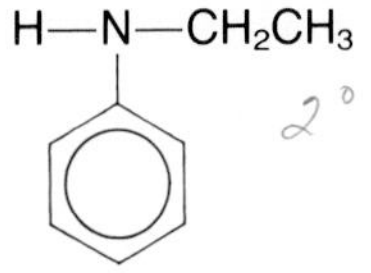

2° N-ethyl-aniline

d

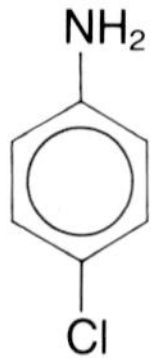

1° 4-chloroaniline

16.4 PHYSICAL PROPERTIES OF AMINES

The boiling point of an amine depends first on its molecular weight and then on its classification as 1°, 2°, or 3°. Primary and secondary amines boil at temperatures higher than hydrocarbons, but lower than alcohols, of similar molecular weight. Table 16.1 shows a comparison of some of these boiling points. However, tertiary amines have boiling points much closer to those of hydrocarbons of similar molecular weight.

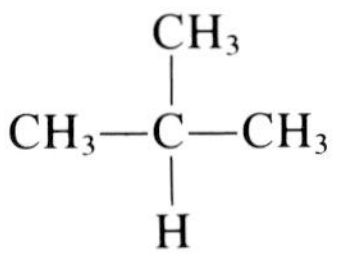

2-Methylpropane
(MW = 58; bp = −11°C)

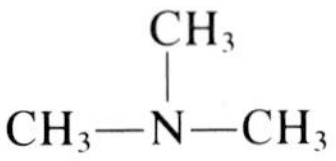

Trimethylamine
(MW = 59; bp = 3°C)

Once again, you can see that elevated boiling points arise when hydrogen bonding is possible. In the absence of hydrogen bonding, dispersion force trends operate as in hydrocarbons so that 3° amines and hydrocarbons of similar molecular weights have comparable boiling points.

As Figure 16.1 shows, 1° and 2° amines have a polar N—H bond which allows for hydrogen bonding between amine molecules in these two categories, and these amines boil at higher temperatures than hydrocarbons of similar molecular weight. However, because the electronegativity of N (3.0) is less than that of O (3.5), the N—H bond is not as polarized as O—H, and hydrogen bonding in amines is weaker than that in alcohols. Both *n*-propylamine and ethylmethylamine boil below 1-propanol. The higher boiling point of *n*-propylamine compared with ethylmethylamine follows from its having two N—H bonds. Tertiary amines do not have an N—H bond, and as a consequence there is no hydrogen bonding at all between tertiary amine molecules.

TABLE 16.1 Comparative Boiling Points of Hydrocarbons, Amines, and Alcohols

Functional Group	Example	Molecular Weight	Boiling Point, °C
Hydrocarbon	$CH_3CH_2CH_2CH_3$ Butane	58	−0.5
1° amine	$CH_3CH_2CH_2NH_2$ *n*-Propylamine	59	48
2° amine	$CH_3CH_2{-}N(CH_3)H$ Ethylmethylamine	59	37
Alcohol	$CH_3CH_2CH_2OH$ 1-Propanol	60	97

Amines with five or fewer carbons are soluble in water. Again hydrogen bonding plays a key role. Amines, even tertiary amines, can form hydrogen bonds with water molecules (Figure 16.2) and thus intermix with water molecules in solution.

One often detects unpleasant odors wafting up from the surface of amines. Remember that amines are structurally similar to ammonia. The origin of the pungent amine odors is the same as that of the strong ammonia smell that you have probably encountered in household products. This is a good example of similar structures producing similar properties and activity.

Amines form in the decay of fish and animal matter and lend the characteristic odor to these processes. Putrescine ($H_2NCH_2CH_2CH_2CH_2NH_2$) and cadaverine ($H_2NCH_2CH_2CH_2CH_2CH_2NH_2$) are formed in the decay of human flesh; their common names have been most appropriately assigned.

16.5 AMINE BASICITY

Like ammonia, amines are weak bases. They can accept a hydrogen ion (proton) from any acid, including water. Review Sections 11.3 and 11.8.

$$RNH_2(aq) + H_2O(l) \rightleftharpoons RNH_3^+(aq) + OH^-(aq) \quad (16.1)$$

SAMPLE EXERCISE 16.1

Write the structure of the product that forms when the following amines react with the given acid:

a $CH_3CH_2NH_2 + HCl \longrightarrow$

b $(CH_3)_3N + H_2SO_4 \longrightarrow$

Solution Using Brønsted-Lowry theory, transfer H^+ from the acid, the proton donor, to the amine base, the proton acceptor. The lone pair on nitrogen accepts and bonds to H^+; the nitrogen develops a positive charge. The anion from the acid balances the charge:

a $CH_3CH_2\ddot{N}H_2 + HCl \longrightarrow CH_3CH_2\overset{+}{N}H_2{-}H\ Cl^-$

b $(CH_3)_3N{:} + H_2SO_4 \longrightarrow (CH_3)_3\overset{+}{N}{-}H\quad HSO_4^-$

(Only one proton from the diprotic acid is transferred to the amine.)

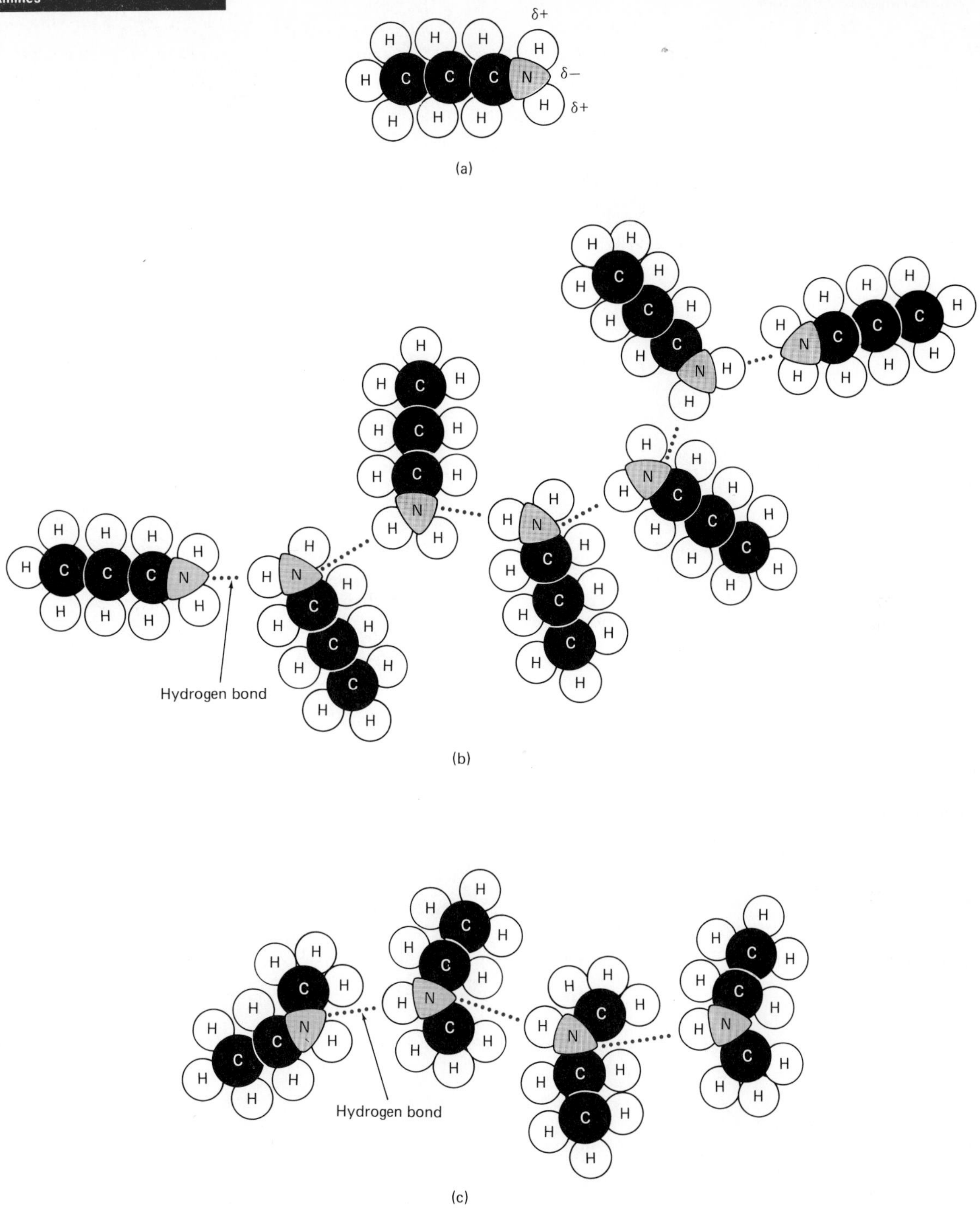
δ+
δ−
δ+
(a)
Hydrogen bond
(b)
Hydrogen bond
(c)

FIG. 16.1

Hydrogen bonding in primary and secondary amines. (*a*) The electronegativity difference between N (3.0) and H (2.1) results in a polar N—H bond. (*b*) The primary amine, *n*-propylamine, has two N—H bonds, thus two H's which can potentially form hydrogen bonds with an electronegative nitrogen on another molecule. However, steric crowding often prohibits formation of both. The boiling point of *n*-propylamine (molecular weight = 59) is 48°C. (*c*) The secondary amine, ethylmethylamine, has only one N—H bond, thus only one H, which can hydrogen-bond with another molecule's nitrogen. The boiling point of ethylmethylamine (molecular weight = 59) is 37°C.

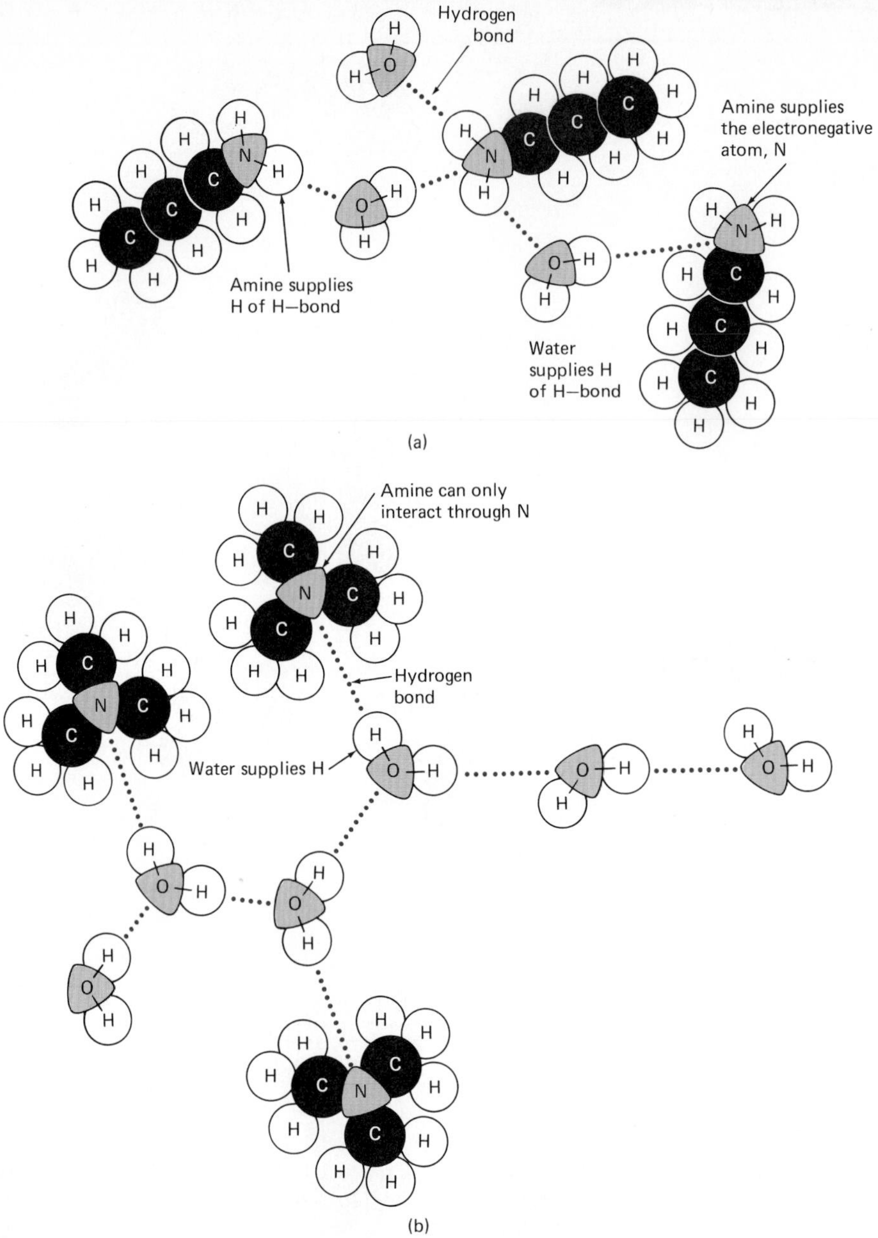
Hydrogen bond
Amine supplies the electronegative atom, N
Amine supplies H of H—bond
Water supplies H of H—bond
(a)
Amine can only interact through N
Hydrogen bond
Water supplies H
(b)

FIG. 16.2
Amines are water-soluble through hydrogen bonding. (*a*) Primary and secondary amines can supply both H and the electronegative atom N to the hydrogen bond. (*b*) A tertiary amine can participate in hydrogen bonding only through the interaction of its N with H from water molecules.

Problem 16.2 Write the structure of the amine salt formed by the following reaction:

$$CH_3CH_2\text{—}\underset{\cdot\cdot}{\overset{\displaystyle CH_3}{\overset{|}{N}}}H + HBr \longrightarrow$$

The equilibrium constant for the reaction shown in Equation (16.1), called K_b, is a measure of base strength, or how far the position of equilibrium is to the right. For example, the K_b of ethylamine is small, 5.6×10^{-4}, and for a 0.1 M solution of this amine, only 7 out of every 100 amine molecules are protonated; i.e., the equilibrium lies to the left. (See Section 11.4 for a review of base strength.)

On the other hand, in a strong acid solution such as the hydrochloric acid medium of the stomach, the equilibrium position lies completely to the product side of Equation (16.2), and all the amine molecules are in the protonated form.

$$RNH_2(aq) + HCl(aq) \longrightarrow R\overset{+}{N}H_3(aq) + Cl^-(aq) \qquad (16.2)$$

The protonated amine cation and its associated anion form an amine salt.

In a manner comparable with the case of ammonia (NH_3) and ammonium (NH_4^+), amine salts are named as organic ammonium salts. That is, when there are only three bonds to nitrogen, the alkyl groups are named and followed by the word *amine*. When there are four bonds to nitrogen, the alkyl groups are again named but are followed by the word *ammonium*. This names the cation. The name is completed by naming the anion. Study the pattern of naming in the following examples:

$$CH_3NH_2 + HCl \longrightarrow CH_3NH_3^+\ Cl^-$$

Methylamine — Methylammonium chloride

$$(CH_3CH_2)_3N + HBr \longrightarrow (CH_3CH_2)_3\overset{+}{N}H\ Br^-$$

Triethylamine — Triethylammonium bromide

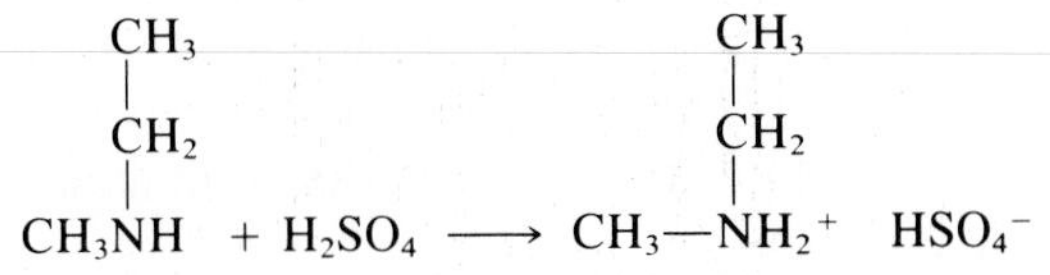

Ethylmethylamine — Ethylmethylammonium hydrogen sulfate

An alternative naming scheme used in the pharmaceutical industry employs the words *amine hydrochloride* in place of *ammonium* and the anion name. For example, methylammonium chloride would be called methylamine hydrochloride. The term *amine hydrochloride* reflects the fact that the salt is made from an amine + hydrogen chloride.

Amine salts are ionic compounds. Consequently, they are solids at room temperature and have much greater water solubility than the amines from which they are derived by reaction with acid. In contrast to gaseous and liquid amines, the solid amine salts, because of their very low vapor pressure, emit no odor. We use this chemistry when we sprinkle our fish or seafood with acidic lemon juice or vinegar or cook it in an oriental style with ginger root. Amines responsible for the sometimes unpleasant ("fishy") aroma emanating from fish are converted, by the acid, into the nonodorous amine salts.

The fact that amine salts have greater water solubility than amines has important consequences in the pharmaceutical industry. Many drugs which contain the amine group, such as morphine and Librium, are prescribed as the hydrochloride, because of the greater water solubility of the salt form of the amine drug. This is particularly important because most drugs have large R groups.

$$R{-}NH_2 + HCl \longrightarrow R{-}\overset{+}{N}H_3Cl^-$$

Amine (smelly liquid; not water-soluble) | Amine salt (odorless solid; water-soluble)

Look at drug labels in your medicine chest. You'll find the word *hydrochloride* on antihistamines and many other preparations. This indicates that the amine is in the water-soluble, odorless salt form.

The equilibrium whereby an amine salt forms is reversible; i.e., the free amine can be reclaimed by reaction with a base.

$$\text{Amine} + HCl \rightleftharpoons \text{amine cation} + Cl^- \qquad (16.3)$$

Base | Acid | Conjugate weak acid | Conjugate weak base

A strong base shifts this equilibrium to the left; that is, amine salts react with NaOH to form the free amine. For example,

H⁺ transfer

$$CH_3\overset{+}{N}H_3\ Cl^- + Na^{+}\ {}^{-}OH \longrightarrow CH_3NH_2 + H_2O + NaCl \qquad (16.4)$$

Acid | Strong base | Weak base | Weak acid

Amines are weak bases and accept H^+ from acids. Amine cations (salts) are weak acids and give up H^+ to bases. This acid-base behavior of amines and their salts is an important aspect in all amine chemistry.

SAMPLE EXERCISE 16.2

Write the structure of the products that form when the following amine salt reacts with the given strong base:

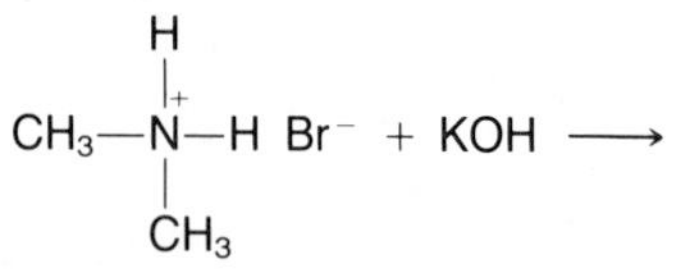

Solution Using Brønsted-Lowry theory, transfer H^+ from the cation of the amine salt, which is the proton donor, to the hydroxide ion, which is the proton acceptor in strong bases. The products are the free amine, water, and a salt formed from the cation of the base and the anion of the amine salt.

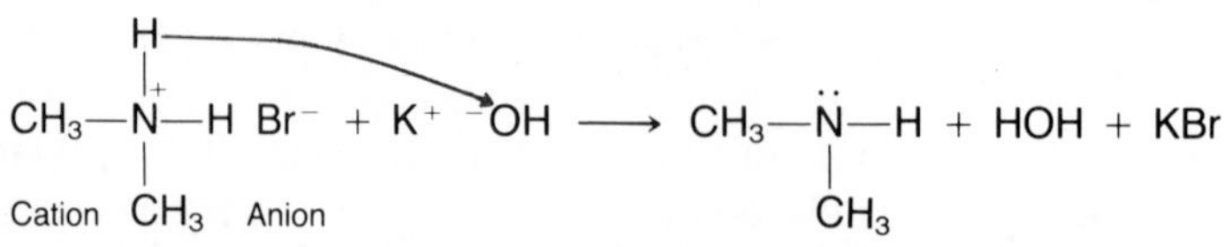
Cation
Anion

Amine salt | Strong base | Amine | Water | Salt

Problem 16.3 Write structures for the products of the following reactions:

a $CH_3CH_2CH_2NH_2 + HBr \longrightarrow$

b $CH_3CH_2CH_2\overset{+}{N}H_3Br^- + NaOH \longrightarrow$

16.6 REACTIONS WITH ALKYL HALIDES

Just as the nitrogen lone pair in amines can bond to H^+, so too can it form a bond to carbon. This happens when an amine reacts with an alkyl halide. Amines use the lone pair of electrons on the nitrogen to form a bond to the carbon of an alkyl halide and thereby displace the halide ion. This is a substitution reaction in which nitrogen replaces halogen. Recall that two electrons can be represented either by two dots (:) or one dash (—).

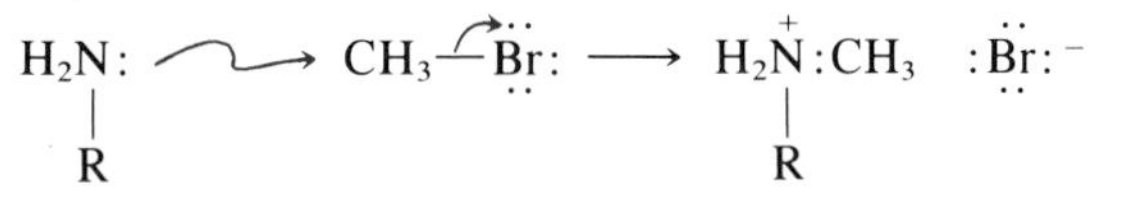

(16.5)

Original amine — Alkyl halide — Alkylamine salt

The product is an amine salt. The halogen is no longer covalently bonded to carbon; the C—Br bond has broken in such a way that Br gets both electrons that were involved in the bond. The bromide anion is ionically bonded to the ammonium cation.

Like any other amine salt, the product of this reaction can react with base to yield the amine [see Equation (16.4)]. The base could be an amine molecule in the original reaction mixture:

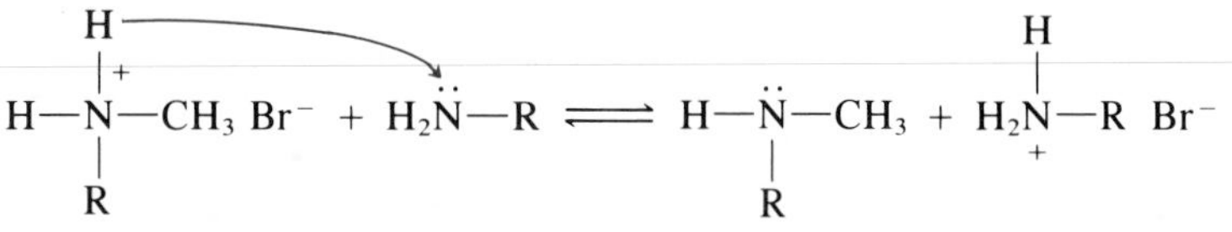

(16.6)

Amine salt — Original amine (acting as a base) — Alkylated amine — Protonated amine

The amine that reacts with an alkyl halide is said to be alkylated because it now has a new alkyl group bonded to the nitrogen. In the example given, the amine R—NH_2 has been methylated; a methyl group is attached to N in the product, $R{-}\overset{CH_3}{\overset{|}{N}}H$. The reaction is useful for preparing new amines. To predict the product of the reaction of an amine and an alkyl halide, recognize that the alkyl group of the alkyl halide will be attached to the N of the amine in the final product. The reaction of ethylamine with isopropyl bromide places an isopropyl group on N. The sequence of events is

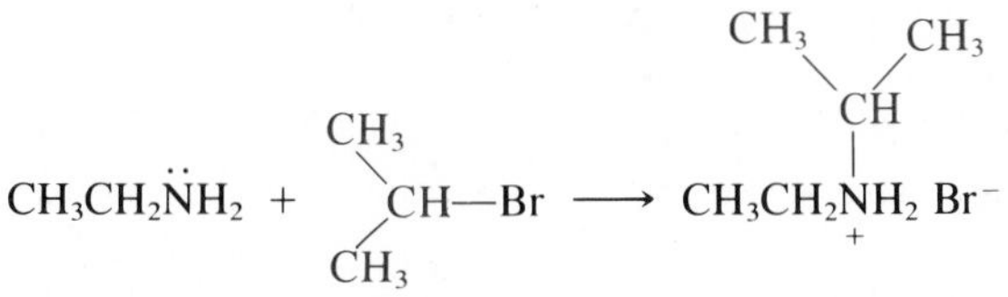

Ethylamine (amine) — Isopropyl bromide (alkyl halide) — Ethylisopropylammonium bromide (amine salt)

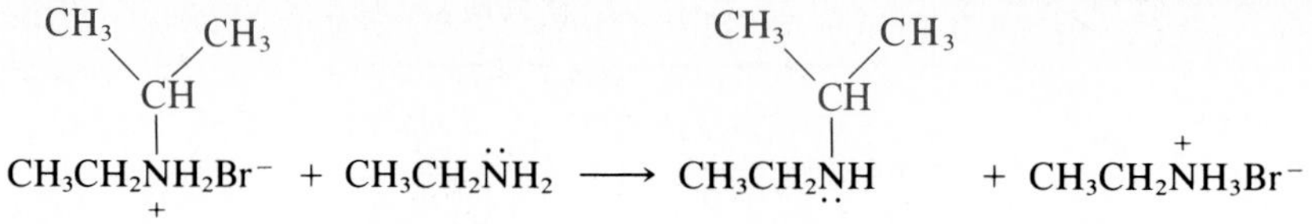

Ethylisopropylammonium bromide (amine salt)	Ethylamine (amine base)	Ethylisopropyl amine (alkylated amine)	Ethylammonium bromide (protonated amine)

The essential event is the placement of another alkyl group (in this case isopropyl) on the amine nitrogen.

Problem 16.4 *a* Starting with aniline, what alkyl halide would be required to prepare

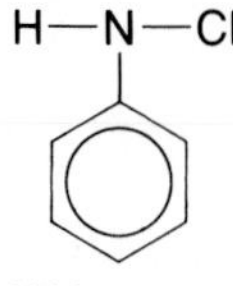

, *N*-methylaniline?

b Write equations for the sequence of steps in the preparation.

When a tertiary amine, R_3N:, reacts with an alkyl halide, an amine salt is formed which has four alkyl groups bonded to the nitrogen. For example, $R_3N: + CH_3Br \longrightarrow R_3\overset{+}{N}:CH_3\ Br^-$. This is a quaternary (4°) ammonium salt. The term *quaternary* indicates that the nitrogen is bonded to four organic groups. Because they do not have an H^+ to yield up to a base from their nitrogen, quaternary ammonium salts are stable in basic solution.

Choline and acetylcholine are two quaternary ammonium salts important in human biochemistry.

$$HOCH_2CH_2\overset{+}{N}(CH_3)_3\ \overset{-}{O}H \qquad CH_3\overset{O}{\overset{\|}{C}}—OCH_2CH_2\overset{+}{N}(CH_3)_3\ \overset{-}{O}H$$

Choline — Acetylcholine

As you will see, choline is part of the structure of compounds making up cell membranes (Sections 18.9 and 18.10), and acetylcholine is a neurotransmitter, a substance involved in passing information in the nervous system (Section 16.9). The quarternary nitrogen confers on both compounds a high degree of polarity and water solubility necessary for their physiological function.

The amine group is present in the structure of deoxyribonucleic acid (DNA) molecules, which have, among their other functions, the responsibility for the proper control of cell division. Uncontrolled cell division is part of the pathological condition known as cancer. Some antitumor drugs employ alkyl halides as alkylating agents for amine nitrogens. The strategy is to alkylate amine groups in DNA of cancerous cells and thereby interfere with cell division. The drug must show at least a slight preference for the DNA of cancer cells as compared with that of normal cells or cell division of normal cells would also stop. One such drug is mechlorethamine hydrochloride (Mustine hydrochloride), which has been used against Hodgkin's disease.

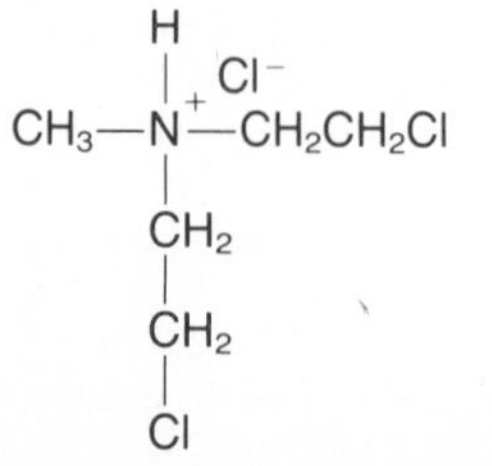

Mechlorethamine hydrochloride (chemotherapeutic compound)

As you can see, there are two alkyl halide groups in the structure of mechlorethamine. It is believed that these two groups alkylate amine groups on separate DNA molecules, thereby cross-linking them together. This prevents effective cell division in cancer cells.

Other important reactions of amines are those with carboxylic acids and their derivatives, but we will defer a discussion of this amine chemistry until Chapter 17. Let us now take a look at some individual amines and classes of amines and the role they play in biological chemistry.

16.7 HETEROCYCLIC AMINES

The concept of cyclic, or ring, compounds was first introduced in Section 12.5, in which we considered only **carbocyclic** structures, that is, rings in which all the atoms connected in the ring are carbons. But, atoms other than carbon can be included in a ring; for example, in Chapter 15 you saw pyranose and furanose rings which include an oxygen atom in the ring forms of carbohydrates. When an atom other than carbon occupies a ring position, the ring is called a **heterocycle.** Many physiologically active compounds contain a nitrogen atom as an amine group in a heterocyclic ring.

The nitrogen heterocycles are conveniently discussed in categories according to ring size. Commonly encountered heterocyclic amines are all either five- or six-membered-ring compounds.

Five-Membered Nitrogen Heterocycles

Pyrrolidine is structurally the simplest example of a five-membered nitrogen heterocycle. It is a saturated ring of four carbons and one nitrogen:

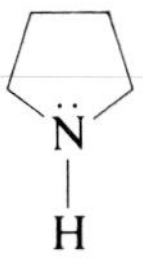
N
H

Pyrrolidine

The pyrrolidine ring can be found in compounds such as proline, an amino acid, that is, a compound that has both an amine functional group and a carboxylic acid group, and in nicotine, a component of tobacco leaves. Nicotine also contains a six-atom heterocycle, which we will encounter in the next section.

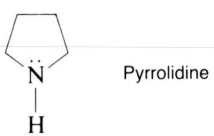

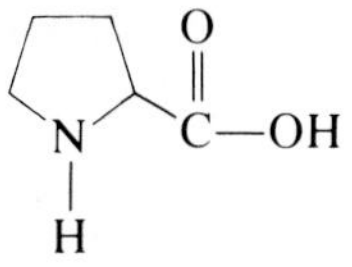
O
N
C—OH
H

Proline

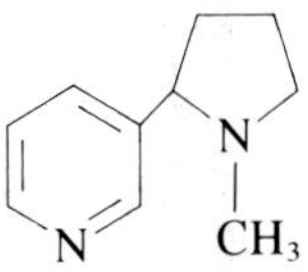
N
N
CH₃

Nicotine

Notice that the usual convention of not writing the carbons and their attached hydrogens when drawing rings is followed in a heterocycle. However, nitrogen and its hydrogens are explicitly shown.

Pyrrole is an unsaturated five-atom ring containing a single nitrogen.

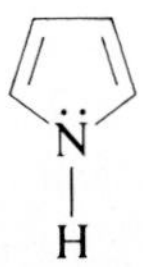
N
H

It is interesting to note that pyrrole displays aromatic character; that is, it acts like benzene. For example, pyrrole does not react by addition, as would a typical alkene, but instead undergoes the typical substitution reactions common to a benzene ring. As in benzene (see Section 13.12), the aromaticity in pyrrole arises from six electrons circulating over the ring. The two apparent double bonds in pyrrole each supply two electrons, thus accounting for four of the six electrons; the other two come from the lone pair on nitrogen. Because the lone pair is not localized on the nitrogen, pyrrole is much less basic than a typical amine; that is, its lone pair is much less willing to accept an H^+. Four pyrrole rings make up the porphyrin ring structure which forms the basis of the heme unit in hemoglobin, the carrier of O_2 in the blood, and in myoglobin, the carrier of O_2 within muscle. The pyrrole rings are highlighted in the structures that follow.

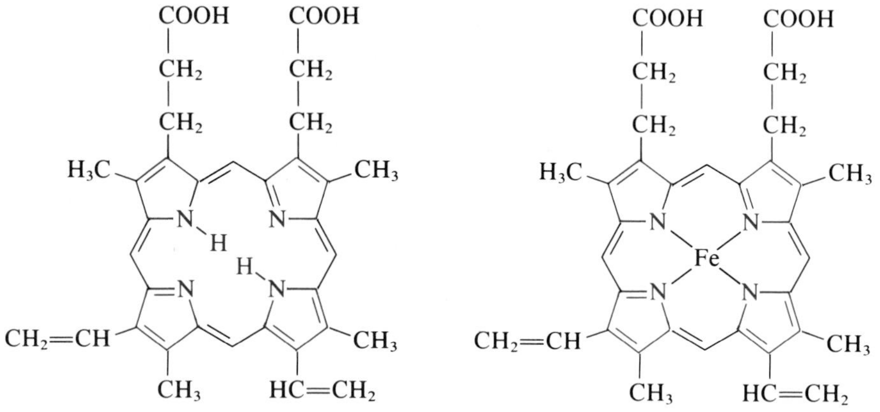
COOH
COOH
CH2
CH2
CH2
CH2
H3C
CH3
N
H
H
N
N
N
CH2=CH
CH3
CH3
HC=CH2
COOH
COOH
CH2
CH2
CH2
CH2
H3C
CH3
N
N
Fe
N
N
CH2=CH
CH3
CH3
HC=CH2

Protoporphyrin IX

Heme group of hemoglobin and myoglobin

In the heme group the four nitrogens bond to a central iron atom. It is this Fe atom which binds O_2 to hemoglobin and myoglobin.

The *indole* structure consists of a benzene ring fused to a five-atom nitrogen heterocycle. Fused means two rings share two atoms in common. The indole system is found in the amino acid tryptophan; in serotonin, a neurotransmitter in the brain (Section 16.9); and in a number of physiologically active amines of plant origin called alkaloids, which we shall examine in Section 16.10.

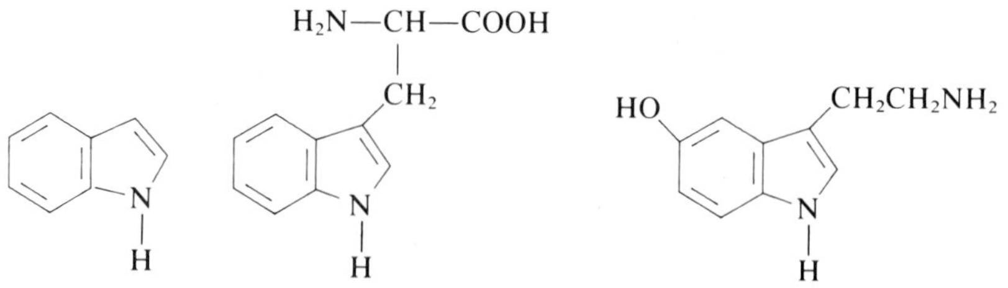
N
H
H2N—CH—COOH
CH2
N
H
HO
CH2CH2NH2
N
H

Indole

Tryptophan

Serotonin

Imidazole is a five-atom ring containing two nitrogens. This system is found in the amino acid histidine and the compound histamine, which is responsible for the unpleasant effects felt by individuals susceptible to nasal and other allergies.

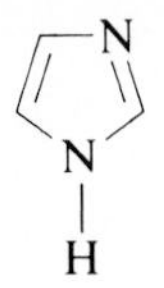

Imidazole

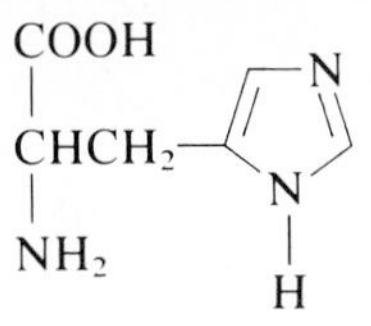

Histidine

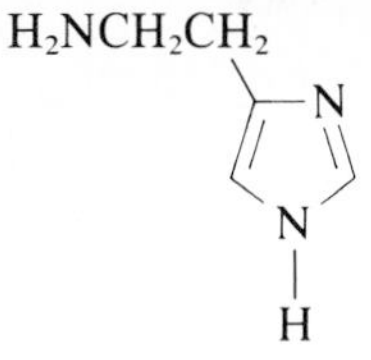

Histamine

Six-Membered Nitrogen Heterocycles

Pyridine is one of the most common of the nitrogen heterocycles; it is a six-membered aromatic ring containing one nitrogen. You saw pyridine earlier in this section as part of the nicotine skelton. The pyridine ring is also found in the structure of nicotinamide, which is also known as the vitamin niacin. The B_6 vitamins, pyridoxine, pyridoxamine, and pyridoxal, incorporate the pyridine ring.

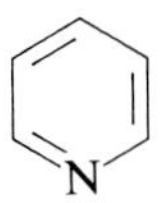

Pyridine

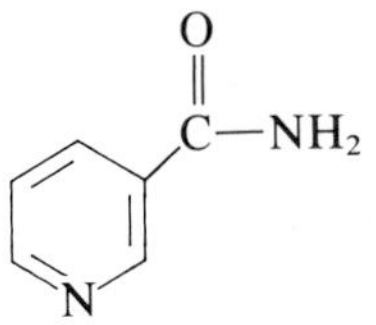

Nicotinamide (niacin)

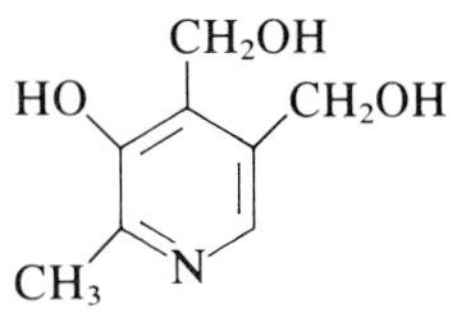

Pyridoxine (one of the B_6 vitamins)

Pyrimidine is a six-atom ring incorporating two nitrogen atoms. The genetic code (Chapter 20) relies on heterocyclic amines, which are usually referred to as nitrogen bases, for its operation. Three of these bases, cytosine, uracil, and thymine, contain the pyrimidine ring. Thiamine (vitamin B_1) also incorporates the pyrimidine ring in its structure.

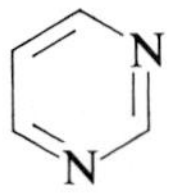

Pyrimidine

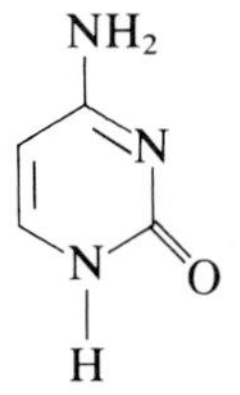

Cytosine (DNA and RNA)

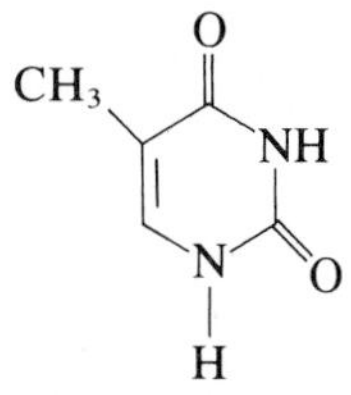

Thymine (DNA only)

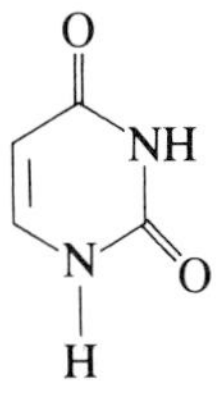

Uracil (RNA only)

In the *purine* structure a pyrimidine ring is fused with an imidazole ring. The purine structure is represented in genetic material (DNA and RNA) by the nitrogen bases, adenine and guanine. The base adenine also makes up a part of the structure of the energy storage molecule ATP and coenzymes which we will encounter in Chapters 22 through 25 on metabolism. The purine ring is also found in caffeine, the stimulating component of tea, coffee, and cola beverages.

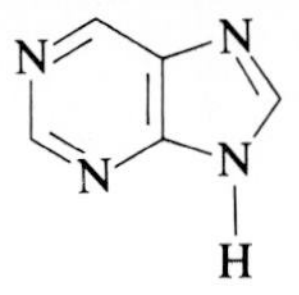

Purine

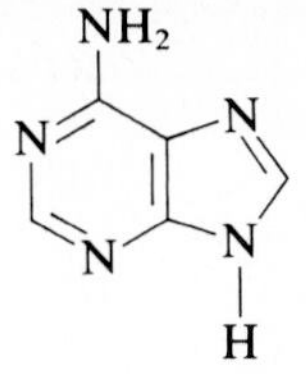

Adenine (DNA and RNA)

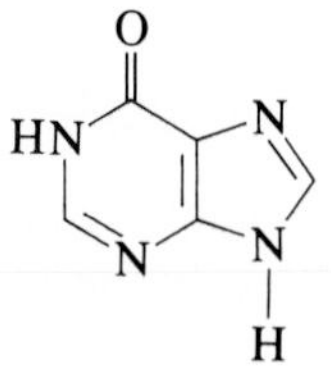

Guanine (DNA and RNA)

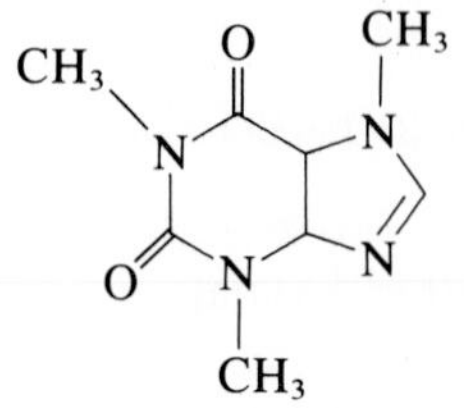

Caffeine

16.8 AMINES AND ALLERGIC REACTIONS

Amines are involved in the body's immune system and the occurrence of allergic reactions in some individuals. The secretion in the body of the amine histamine, which was mentioned in Section 16.7 as an imidazole heterocycle, produces the typical allergy symptoms. Chemists have developed antiallergy drugs through studies of the structure and chemistry of histamine. To understand histamine and antihistamine drugs, it is necessary to know a little about the body's immune or defense system. The body makes antibodies as a defense against certain foreign matter such as pollen, dust, and bacteria that might enter the body. The substance provoking the production of antibodies is termed an *antigen*. The interaction of antigens and antibodies is accomplished through their complementary shapes, or stereochemistry, as shown in Figure 16.3. Their fitting together results in the formation of an antibody-antigen complex and the neutralization of the otherwise possibly dangerous foreign agent.

In some cases antibody-antigen complexes go on to interact with tissues and cells of the body in a manner which is not completely understood. However, it is known that this interaction causes the release from these cells of the heterocyclic amine, histamine.

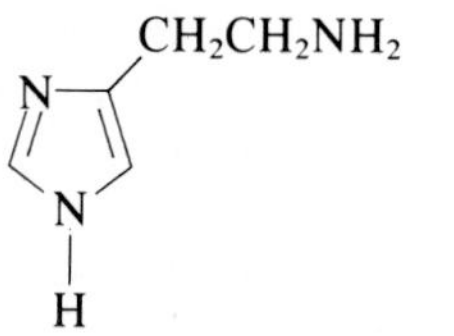

Histamine

The released histamine combines with what are known as H_1 receptor sites, locations on cell surfaces that can bond to and thus recognize other molecules. It is this attachment of histamine to the H_1 receptor site that elicits the typical allergy symptoms of runny eyes and nose, itching, and rashes. The response can be more serious if histamine affects the cardiovascular system and causes edema and/or a fall in blood pressure.

Antihistamines are drugs which block H_1 receptor sites by binding to them. Antihistamines thereby prevent the fit of histamine to these sites and relieve

FIG. 16.3
Antibodies form complexes with antigens by their complementary shapes fitting together. These complexes are then surrounded and destroyed by white blood cells.

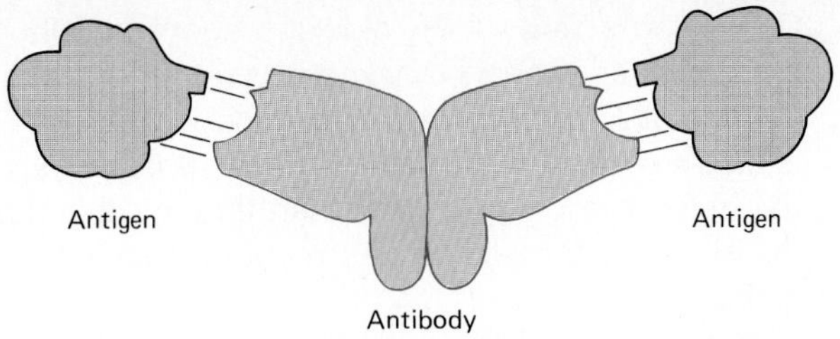
Antigen
Antigen
Antibody

the symptoms of the allergy. Apparently the fit of histamine to the H_1 sites is through the $—CH_2CH_2N<$ side chain because the antihistamines all have this feature in their structure.

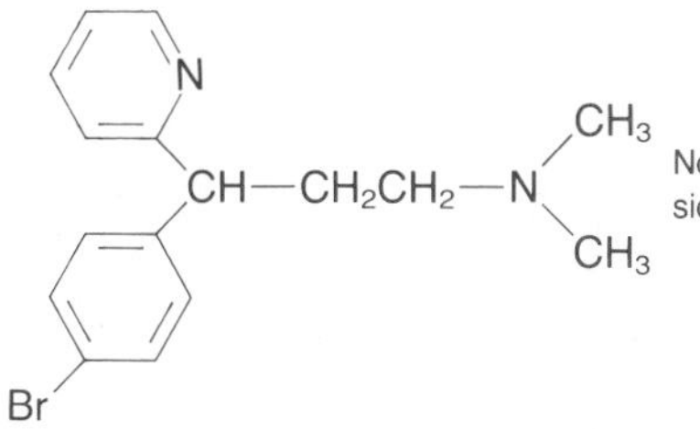

Note the similiarity to side chain of histamine.

Brompheniramine (Dimetane and Dimetapp)

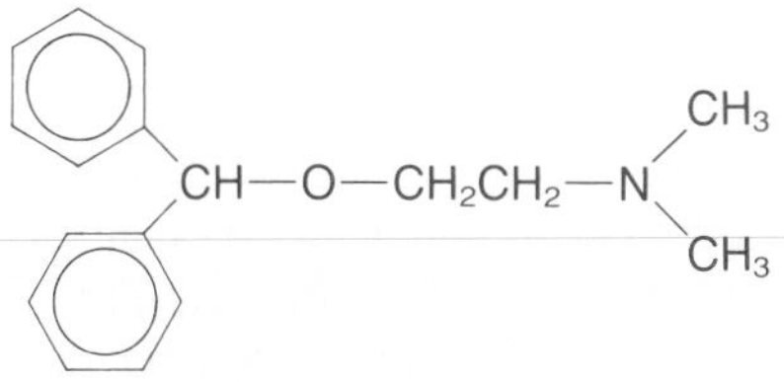

Diphenhydramine (Benadryl)

The antihistamines do not provide a cure, but they do offer relief for the allergy victim.

Because some antihistamines have the side effects of drowsiness and antiemetic activity, that is, they relieve nausea and vomiting, they are sometimes incorporated into sleeping pills and drugs for motion sickness.

Histamine also stimulates the production of stomach acid. This is a serious problem for individuals suffering from stomach ulcers. The usual antihistamines do not relieve this excess acid secretion. It was recently discovered that there are special receptor sites for histamine on gastric membranes. These have been designated H_2 sites. The usual antihistamines do not block the H_2 sites; therefore, they have no effect on the histamine—stomach acid–ulcer problem.

After much research based on the idea of stereochemical fit of molecules, scientists have synthesized a new drug, cimetidine (Tagamet).

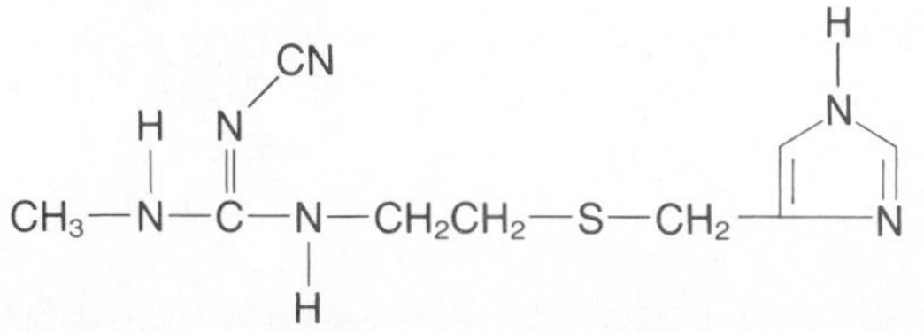

Cimetidine

Cimetidine acts as a H_2 blocking agent and thus suppresses excess acid secretion. It has been very successful in the treatment of peptic ulcers. Notice that cimetidine resembles histamine in that both have an imidazole ring. It is highly probable that both of these substances bind to the H_2 sites through the heterocyclic ring, rather than through the side chain, as in the case of the H_1 sites. Cimetidine has recently become one of the most widely prescribed drugs.

16.9 AMINES AS NEUROTRANSMITTERS

Acetylcholine, norepinephrine, dopamine, and serotonin are amines which have been identified as neurotransmitters. That is, they transmit an impulse signal from one nerve cell, called a neuron, to a connected neuron.

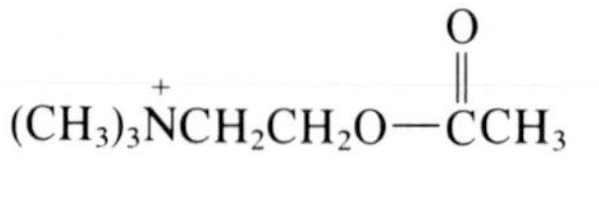

Acetylcholine

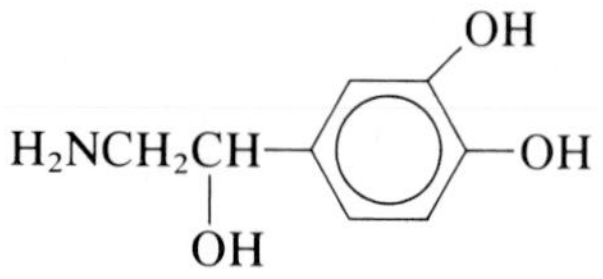

Norepinephrine

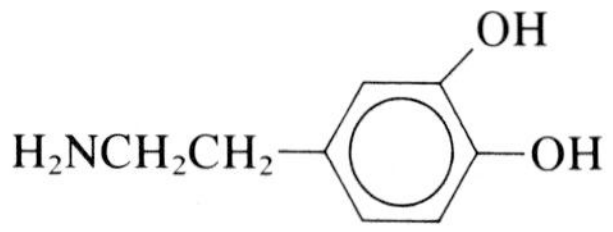

Dopamine

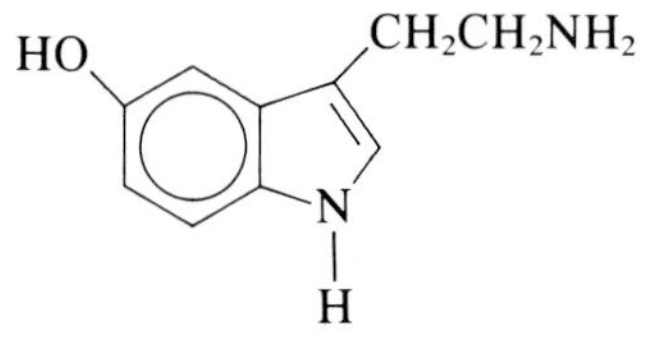

Serotonin

Neurons are separated by a fluid-filled gap known as a synapse (Figure 16.4). When an impulse signal, which is electrical in nature, is received at a presynaptic neuron, neurotransmitters are released. They travel across the gap and fit in at specific receptor sites on the postsynaptic neuron. The receptor sites are precisely tailored to fit the shape of the incoming transmitter. The shape of the receptor is altered after its interaction with the neurotransmitter. The alteration leads to a specific physiological response. This response may be excitatory; for example, it may be that the heart is

FIG. 16.4
A stimulus of the presynaptic (*pre-* = "before") neuron causes the release of a chemical neurotransmitter. The neurotransmitter crosses the synapse and attaches to the postsynaptic (*post-* = "after") neuron. Thus the message of the stimulus, painful or pleasant, is carried from the peripheral sensory neuron to the spinal cord and brain.

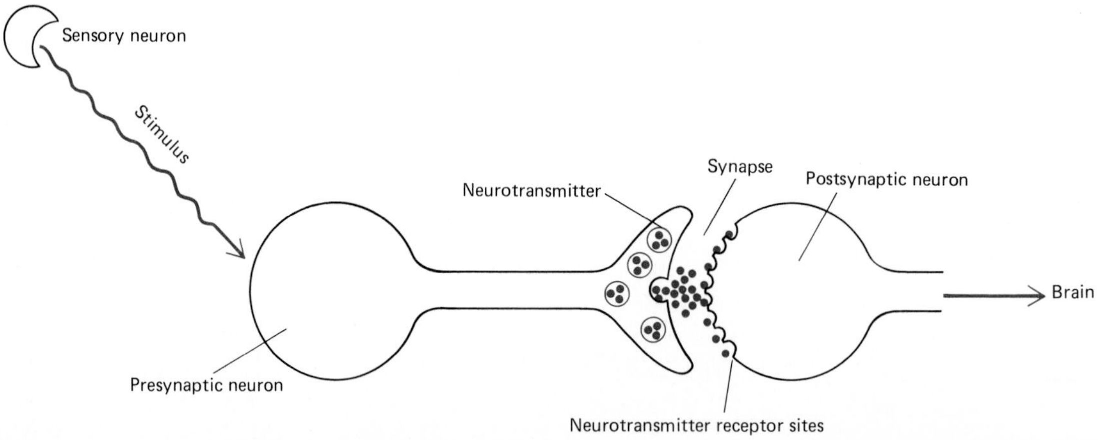
Sensory neuron
Stimulus
Neurotransmitter
Synapse
Postsynaptic neuron
Brain
Presynaptic neuron
Neurotransmitter receptor sites

induced to beat faster. Or, the response might be inhibitory and result in drowsiness and the desire to sleep.

Some drugs exert their effect by either enhancing or suppressing the release of a neurotransmitter from a presynaptic neuron. For example, the abuse of the drug amphetamine leads to delusions of persecution, halluci-

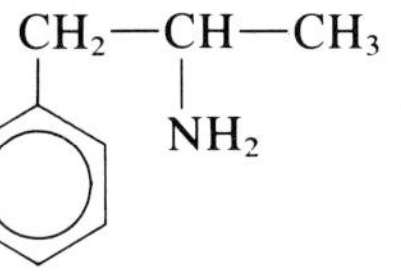

Amphetamine

nations, and a general disruption of thought processes. These are symptoms of schizophrenia, a mental condition which has been found to be associated with high levels of the neurotransmitter dopamine. Apparently amphetamine stimulates dopamine release. Other drugs, such as chlorpromazine (Thorazine), may function as an antipsychotic, reducing the symptoms of schizophrenia, by binding to dopamine receptors and thereby decreasing transmissions in the dopamine pathway. Finally, as we shall see in later sections of this chapter, many psychoactive drugs, particularly the psychedelics, closely resemble the amine neurotransmitters in chemical structure. They may substitute for natural neurotransmitters, such as dopamine, and thereby induce psychotic symptoms.

16.10 ALKALOIDS AND RELATED COMPOUNDS

A large number of nitrogen heterocyclic compounds are extracted from plants. These naturally occurring amines of plant origin are known as alkaloids because they are all basic. *Alkali* is another name for a base. Most of the alkaloids have dramatic physiological activity, and they are classified according to this activity. The following discussion will consider three classes: narcotic analgesics, central nervous system stimulants, and psychedelics. You will find that the often complicated alkaloid structures incorporate the simpler heterocyclic rings described in Section 16.7.

Narcotic Analgesics

The narcotic analgesics are a class of drugs which relieve pain (analgesic property) and induce sleep (sedative property). All the drugs meeting this definition were originally obtained from opium, and hence the narcotic analgesics are also known as opiates.

Opium is a syrupy liquid secreted by the oriental poppy plant. Morphine, which is the most powerful analgesic drug available, is extracted from opium. Unfortunately, it is also addictive. That is, its continued use leads to a physiological need for the drug and then its absence causes withdrawal symptoms. Morphine sulfate, a water-soluble salt of morphine, is used medically for the relief of severe pain in terminal cancer patients, and sometimes for preanesthetic sedation and postoperative analgesia.

Codeine is also extracted from opium. It is much less effective as an analgesic than morphine and is also less addictive. However, it has had a high misuse potential because of its availability in a number of over-the-counter cough preparations. In hospitals it has been used for the relief of moderate pain and as an antitussive to control coughing.

Heroin is synthesized from morphine by the reaction of the phenol groups of morphine with acetic anhydride, which forms two acetyl groups. (You will study this reaction, called esterification, in Section 17.6.) Heroin has more than 3 times the analgesic effect of morphine, but it is so addictive that it is classified as a controlled drug by the federal government, and thus has no accepted medical use in the United States.

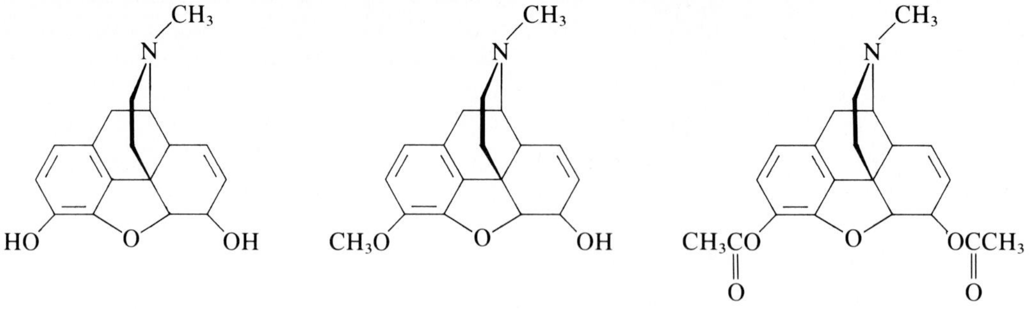
CH3
N
HO
O
OH
CH3
N
CH3O
O
OH
CH3
N
CH3CO
O
O
OCCH3
O

Morphine Codeine Heroin

A number of morphinelike drugs have been synthesized in an effort to find analgesic drugs that are not addictive. Because these synthesized compounds are not derived from plants, they are not actually alkaloids. However, they are included here in Table 16.2 because of their similarity in function and structure to the opiates.

Central Nervous System Stimulants

Stimulants which increase the activity of the central nervous system (abbreviated CNS) are a second class of alkaloids. Some observable symptoms of this stimulation are restlessness, increased motor activity, greater excitement, and a lessened sense of fatigue. Caffeine, nicotine, and cocaine are three drugs in this category. The structures of caffeine and nicotine were given in Section 16.7. Caffeine is a purine, and nicotine has both pyrrolidine and pyridine rings.

Doses as low as 100 to 200 mg of caffeine, which would be found in one to two cups of coffee, increase mental alertness and wakefulness without affecting other CNS functions. There seems to be no physiological addiction to caffeine. However, as many of us are aware through our need for a morning cup of coffee, there is probably a psychological dependence. The physiological mechanism by which caffeine exerts its stimulation will be discussed in Section 21.14.

Nicotine is found in tobacco leaves and is taken into the body either by chewing or smoking tobacco. Nicotine appears to stimulate all levels of the CNS and can lead to tremors and convulsions if consumed in large doses. In addition, nicotine increases the heart rate and blood pressure. Cigarette smoking has an adverse effect on the cardiovascular system because it subjects the body to two hazards: both carbon monoxide and nicotine are found in inhaled cigarette smoke. CO decreases the amount of oxygen available to the heart, while nicotine increases the work being done by the heart.

Cocaine is a powerful CNS stimulant.

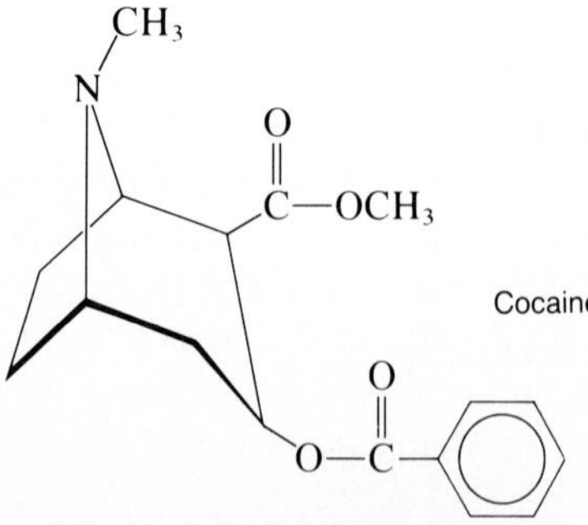
CH3
N
O
C—OCH3
O
O—C—

Cocaine

TABLE 16.2 Synthetic Morphinelike Analgesics

Chemical Name (Trade Name)	Structure	Use and Description
Meperidine (Demerol)	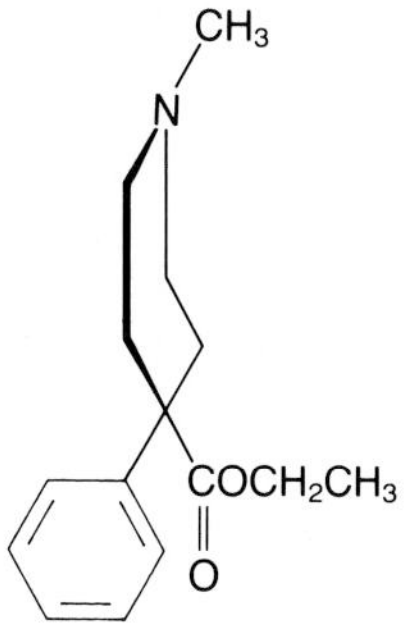	It is widely used for the relief of moderate to severe pain. Its pharmacological effects of analgesia, sedation, euphoria, and respiratory depression are similar to the effects of morphine. Like morphine, it is addictive.
Propoxyphene (Darvon)	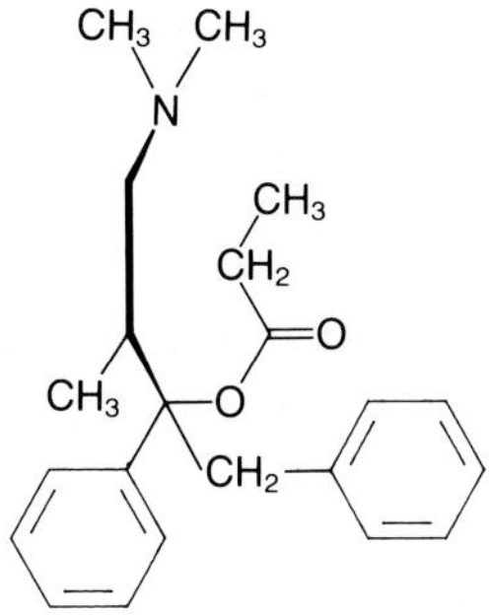	Its analgesic action is about equal to that of aspirin, and it does not appear to be addictive when taken orally.
Methadone (Dolophine)	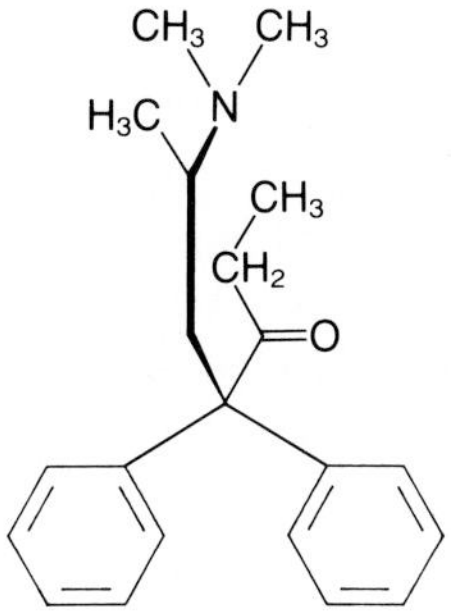	It is used for the controlled withdrawal of heroin addicts. Although it is itself addictive, methadone can be administered orally and does not induce euphoria.
Pentazocine (Talwin)	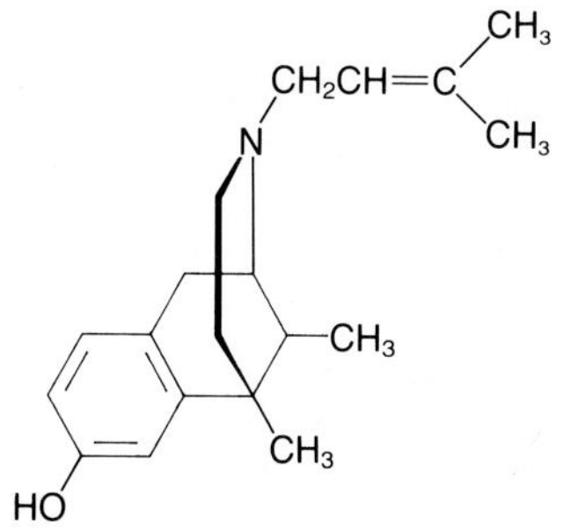	It relieves moderate to severe pain and seems to be less addicting than codeine, Demerol, or morphine.
Butorphanol (Stadol)	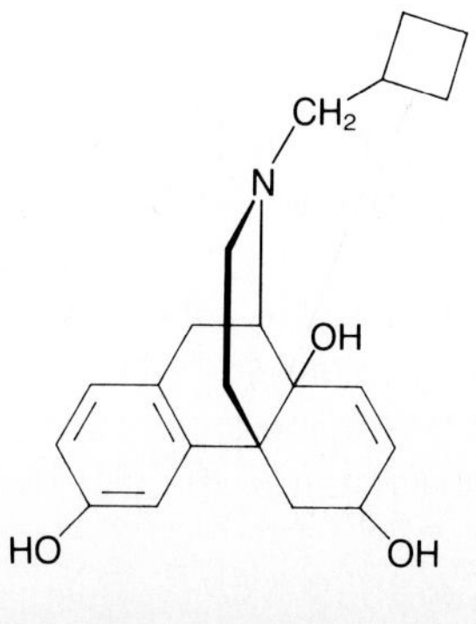	The two synthetic narcotic analgesics, butorphanol and nalbuphine, have recently been introduced into the medical field. They show promise for relief of moderate to severe pain without being addictive, and, unlike the others in this table, are not classified as controlled substances by the government.

TABLE 16.2 *(Continued)*

Chemical Name (Trade Name)	Structure	Use and Description
Nalbuphine (Nubian)	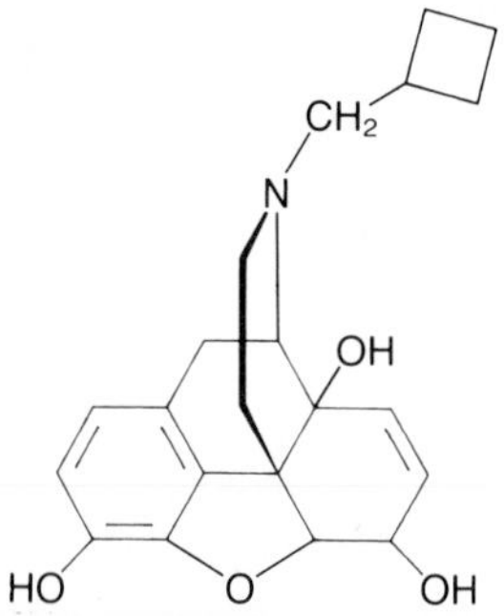	See butorphanol.

It is extracted from the leaves of the cocoa plant. Mental awareness is increased by cocaine while fatigue is decreased. However, the stimulating effect is short-lived and is often followed by a period of deep depression. Cocaine, unlike the morphine opiates, does not appear to be physiologically addictive, but it does lead to a psychological dependence.

Psychedelic Drugs

The psychedelic drugs alter an individual's sensory perception of reality and time. They are sometimes said to "enhance reality" or "expand the mind." Many of the psychedelics exhibit close structural similarity to neurotransmitters such as acetylcholine, norepinephrine, and serotonin (see Section 16.9) and are believed to function by mimicking the actions of the neurotransmitters and thereby altering the flow of information from the sense receptors (eyes and ears) and muscular system to the brain. Table 16.3 shows the structures of a few psychedelics and their similarity to the structures of the neurotransmitters.

16.11 AMINES AS CARCINOGENS

Secondary amines react with nitrous acid to produce compounds known as *N*-nitrosoamines. These compounds are characterized by a nitroso group (—N═O) bonded to an amino nitrogen.

$$(CH_3)_2N\text{—}H + HO\text{—}N\text{=}O \longrightarrow (CH_3)_2N\text{—}N\text{=}O + HOH$$

Nitrous acid

Dimethylamine (a typical 2° amine)

N-Nitroso-dimethylamine (a typical *N*-nitrosoamine)

N-nitrosoamines have been found to be carcinogens which cause kidney, lung, bladder, and brain cancers in a variety of species including monkeys, rodents, birds, and fish. It is possible that *N*-nitrosoamines form in the stomachs of human beings.

Salts containing nitrite ions are used as food additives in curing meats and smoking fish. These additives help maintain the red color of meat, contribute to its flavor, and prevent the growth of botulism bacteria. Unfortunately, nitrite ions in the acidic medium of the stomach will form nitrous acid:

TABLE 16.3 Structural Relationships between Natural Neurotransmitters and Psychedelics

Natural Neurotransmitters	Structural Resemblence	Psychedelics	Source and Effects
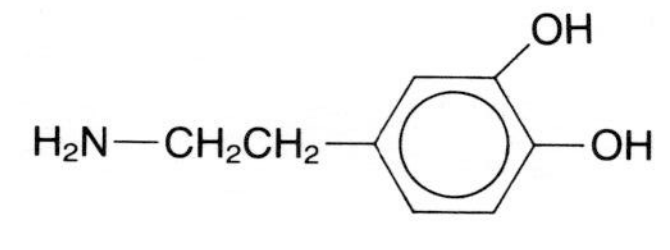 Dopamine	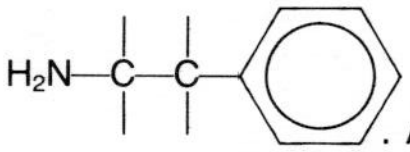 . A two carbon chain attached to an aromatic ring with a terminal amine group is very prevalent in psychoactive drugs. It is called a β-phenylethylamine fragment.	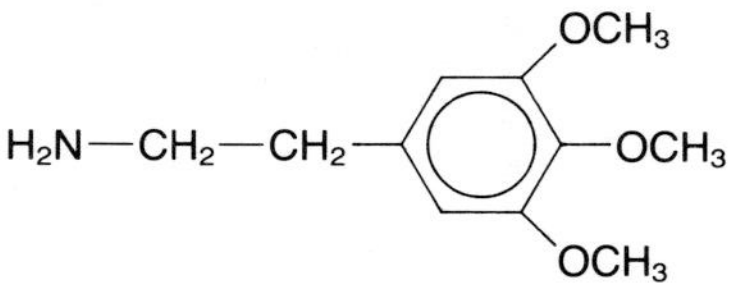 Mescaline	Mescaline is isolated from the peyote plant, a cactus of the southwestern United States. Psychic effects are visual hallucinations and general behavior excitation.
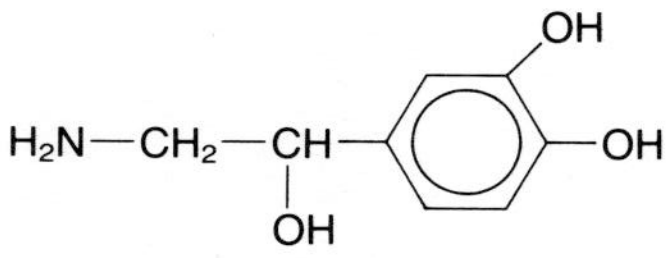 Norepinephrine		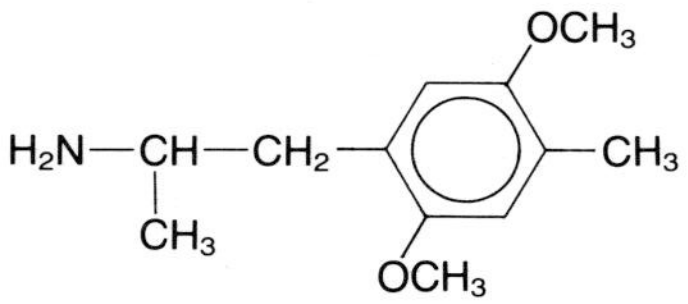 DOM (STP)	This mescaline derivative has similar behavioral effects, but is much more toxic than mescaline.
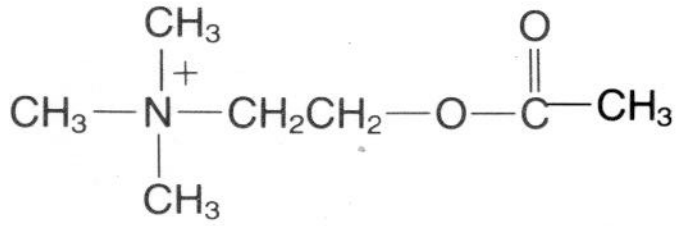 Acetylcholine	Heavily alkylated nitrogen and an ester group. Similarities are highlighted.	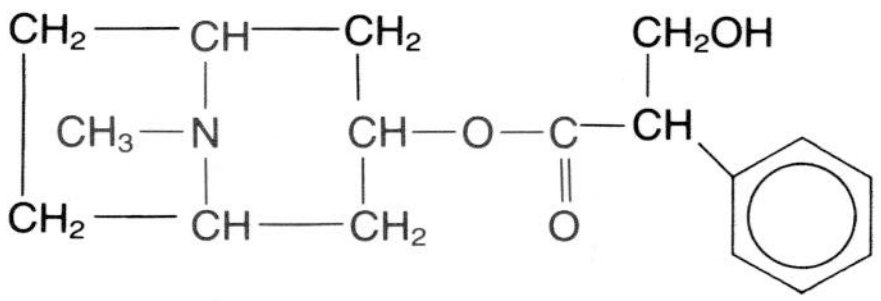 Atropine	Atropine is isolated from the belladonna plant. Its many effects include dilation of pupils in the eye, behavioral depression, mental clouding, and loss of memory. It has been used clinically for eye treatment and for problems in the gastrointestinal tract.
		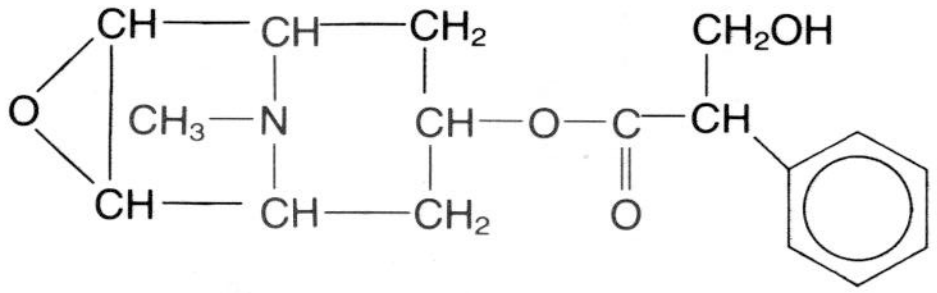 Scopolamine	Scopolamine is also isolated from belladonna and from the shrub henbane. It produces behavioral effects similar to those of atropine. Basic medical use has been preanesthetic medication.

TABLE 16.3 *(Continued)*

Natural Neurotransmitters	Structural Resemblence	Psychedelics	Source and Effects
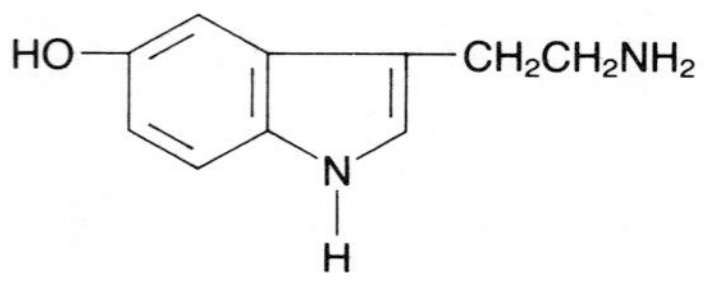 Serotonin	Indole fused rings system and ethylamine group (highlighted in LSD).	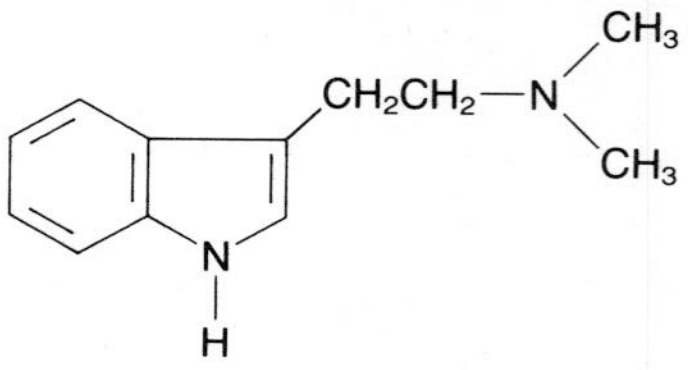 Dimethyltryptamine (DMT) 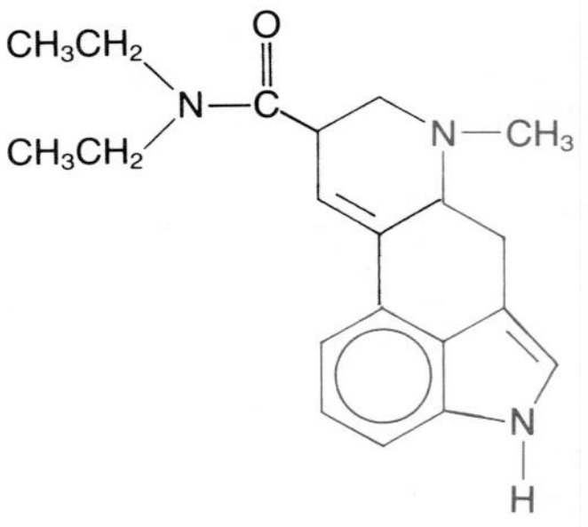 Lysergic acid diethylamide (LSD)	DMT is extracted from the South American plants cohoba and yopo. It produces visual hallucinations, euphoria, and behavioral excitability. The duration of its action is short (1 to 2 h). LSD is a derivative of lysergic acid, which is found in the rye and wheat fungus ergot. It is active in very minute doses (less than 0.0001 g). The physiological changes are remarkably small compared with the psychological alterations. It has been used clinically in the past to provide insight into the mechanisms of mental illness. LSD exists as a pair of enantiomers, and the importance of stereochemistry and critical fit with the receptor site is exemplified by the fact that only the (−) enantiomer is psychoactive.

$$NO_2^- + HCl \rightleftharpoons HNO_2 + Cl^-$$

Nitrite ion Stomach acid Nitrous acid Chloride ion

Then this weak acid can combine with secondary amines, produced from the digestion of protein, to form *N*-nitrosoamines. Nitrosoamines have also been found in smoked fish and cured meats. The highest concentrations are found in fried bacon. The USDA has recently limited the amount of nitrites that can be added to bacon. Whether nitrite ions or nitrosoamines contribute to stomach cancer in human beings is not known with certainty. However, it is now possible to buy, at a higher price, nitrite-free meat products.

Some aromatic amines such as 2-aminonapthalene and benzidine have also been found to be carcinogens. Chemical workers or others in the dye or textile industries occupationally exposed to these aromatic amines are at greater risk of developing bladder cancer than the general population. Aromatic azo compounds such as the dye "butter yellow" are also suspected of being carcinogenic. "Butter yellow" is so named because at one time it was used as the coloring agent for artificial butter, i.e., margarine.

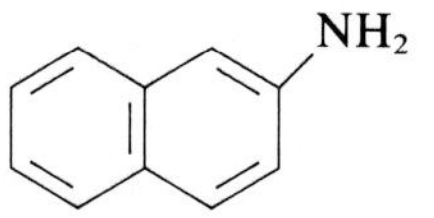
NH2

2-Aminonaphthalene

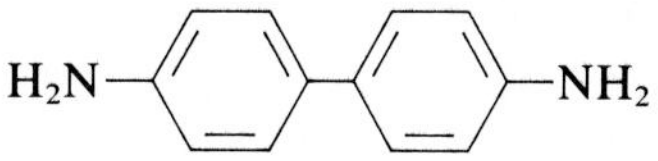
H2N
NH2

Benzidine

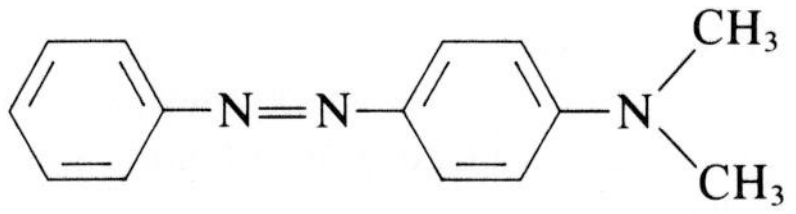
N=N
N
CH3
CH3

4-(*N,N*-Dimethylamino)azobenzene
("butter yellow")

16.12 NUCLEOSIDES

Pyrimidine and purine amine bases react with the monosaccharides ribose and deoxyribose in a condensation reaction to form structures called nucleosides.

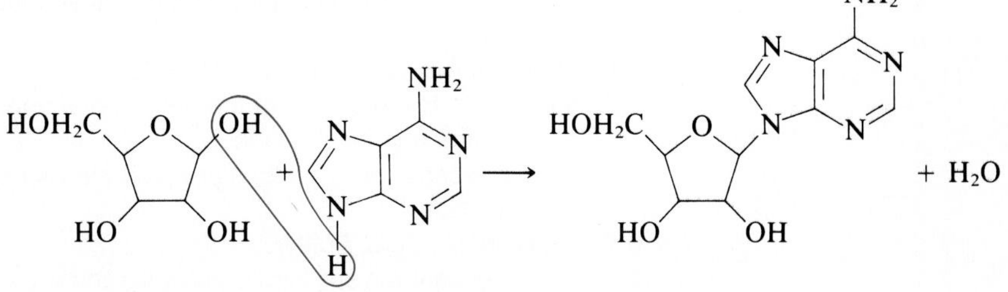
HOH2C
O
OH
HO
OH
+
NH2
N
N
N
N
H
HOH2C
O
HO
OH
NH2
N
N
N
N
+ H2O

β-Ribose Adenine (a purine base) Adenosine (a nucleoside)

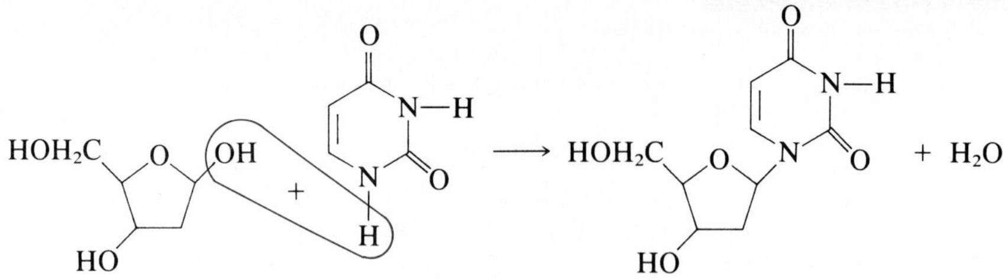
HOH2C
O
OH
+
N—H
N
O
H
HO
HOH2C
N
O
+ H2O

β-Deoxyribose Uracil (a pyrimidine base) Deoxyuridine (a nucleoside)

Nucleosides are a part of the structure of DNA, RNA, and a number of other important biochemical molecules. You will encounter this carbohydrate-amine combination again in Chapter 20.

CHAPTER ACCOMPLISHMENTS

After completing this chapter you should be able to define **key words** and do the following:

16.1 INTRODUCTION

1. Give examples of the occurrence of the amine group in physiological compounds.

16.2 CLASSIFICATION OF AMINES

2. Classify a given amine as primary, secondary, or tertiary.
3. Recognize a quaternary ammonium salt.

16.3 NOMENCLATURE

4. Name a given aliphatic or aromatic amine.

16.4 PHYSICAL PROPERTIES OF AMINES

5. Recognize the possibility of hydrogen bonding in a given amine.
6. Discuss the extent of water solubility for amines.

16.5 AMINE BASICITY

7. Write an equation showing a given amine acting as a Brønsted-Lowry base.
8. Name a given quarternary ammonium salt in two ways.
9. Explain why an amine salt is generally a solid with greater water solubility than the original amine.
10. Explain why many pharmaceutical amines are prescribed in the hydrochloride form.

16.6 REACTIONS WITH ALKYL HALIDES

11. Write the structure of the alkylated amine formed from a given reactant amine and alkyl halide.
12. Describe, on a molecular level, the use of alkylating agents in cancer therapy.

16.7 HETEROCYCLIC AMINES

13. Recognize a specific heterocyclic amine class in a given structural formula.

16.8 AMINES AND ALLERGIC REACTIONS

14. State the possible physiological responses to histamine production.
15. Describe the mode of action of antihistamines.

16.9 AMINES AS NEUROTRANSMITTERS

16. Describe the role of three-dimensional shape in the function of a neurotransmitter.

16.10 ALKALOIDS AND RELATED COMPOUNDS
17. Name three subclasses of alkaloids and their general physiological effects.
16.11 AMINES AS CARCINOGENS
18. State the type of amine that forms *N*-nitrosoamines.
16.12 NUCLEOSIDES
19. State the components of a nucleoside.

PROBLEMS

16.5 *a* State one general class of physiological substances the molecules of which all contain nitrogen.
b Give three specific examples of physiological compounds that contain the amine group.

16.6 Classify each of the following amines as primary, secondary, or tertiary:
a $CH_3CH_2NH_2$ 1°
b $CH_3—N—CH_3$ (with CH_3 on N) 3°
c $CH_3—CH—CH_3$ (with NH_2 on CH) 1°
d 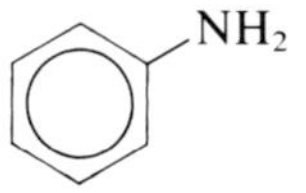 1°
e 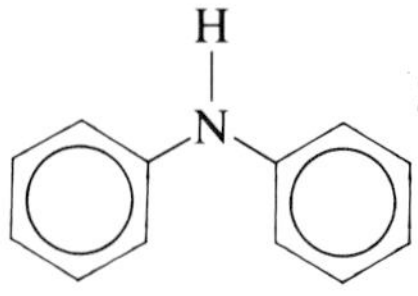 2°
f 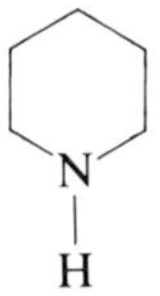 2°
g 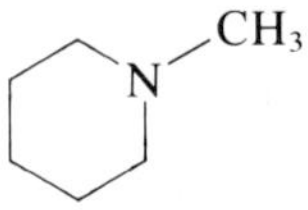 3°

16.7 Give an unambiguous name to each of the following amines:
a CH_3NH_2
b $CH_3—CH—CH_3$ (with NH_2 on CH)
c $(CH_3CH_2)_3N$
d 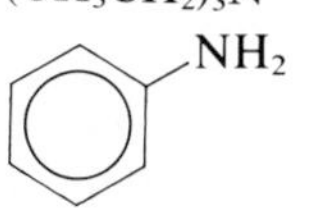
e 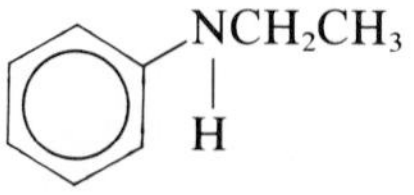
f 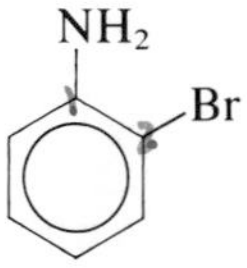
g $CH_3—CH—N—CH_2CH_3$ (with CH_3 on CH and CH_3 on N)
h 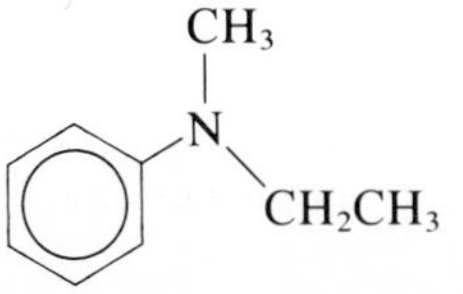

16.8 Draw the structural formula corresponding to each of the following amines:
a Ethylamine
b Diethylamine
c 3-Chloroaniline
d Ethylmethylisopropylamine
e *N*-*n*-Propylaniline

16.9 Although they have similar molecular weights (73 and 72), the boiling point of *n*-butylamine is much higher (78°C) than that of pentane (36°C). Explain.

16.10 Although both have the same molecular weight (73) and a polarized N—H bond, the boiling point of *n*-butylamine (78°) is considerably higher than that of diethylamine (56°C). Explain.

16.11 Although tertiary amines cannot hydrogen-bond to themselves, they can hydrogen-bond with water molecules. Explain this using a diagram to illustrate your point.

16.12 Examine the formulas of putrescine ($H_2NCH_2CH_2CH_2CH_2NH_2$) and cadaverine ($H_2NCH_2CH_2CH_2CH_2CH_2NH_2$). Do you think that the decay process leading to them is reducing or oxidizing?

16.13 *a* Write an equation showing the equilibrium that occurs when ethylamine acts as a base in water solution.
b The K_b for ethylamine is 5.6×10^{-4}. Which side, products or reactants, is favored in this equilibrium?
c What will be the color of blue litmus paper in this solution? Red litmus paper?

16.14 Chlortetracycline hydrochloride (trade name Aureomycin) is an antibiotic. What functional group do you think might be present in this compound?

16.15 Complete the following equations and name the product:
a $CH_3CH_2NH_2 + HBr \rightarrow$
b $CH_3—CH—CH_3 + HNO_3 \rightarrow$ (with NH_2 on CH)
c 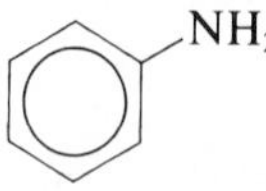 $+ HCl \rightarrow$

16.16 After eating fish or cleaning and packaging it, people often clean their hands with a lemon juice product. What function does the lemon juice serve?

16.17 Amines do not contain hydroxide in their structures, yet they are basic. What definition of basicity do amines fit into?

16.18 Write an equation for an acid-base reaction between methylammonium chloride and ethylamine:

$$CH_3NH_3{}^+Cl^- + CH_3CH_2NH_2 \longrightarrow ?$$

Methylammonium chloride   Ethylamine

16.19 Write the structure of the *amine salt* formed in each of the following reactions:

a $CH_3NH_2 + CH_3CH_2Br \rightarrow$

b

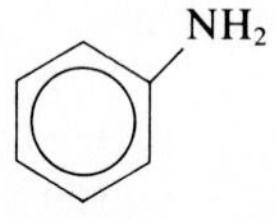
NH₂

$+ CH_3Cl \rightarrow$

16.20 *a* Write the structure of the alkylated amine formed when the amine salt of Problem 16.19*a* is reacted with another molecule of CH_3NH_2.

b Do the same for the amine salt of Problem 16.19*b*.

16.21 Section 16.7 enumerated the classes of heterocyclic nitrogen compounds. Assign each of the following examples to the correct class based on the heterocyclic ring(s) present:

a

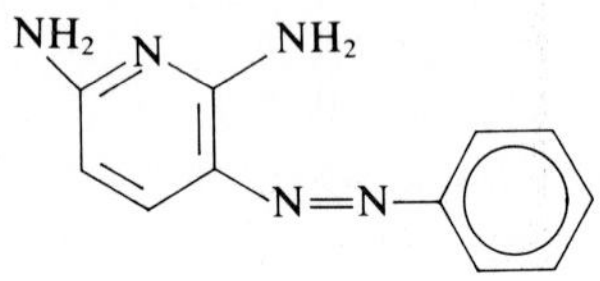
NH₂
N
NH₂
N=N

Azodine (a urinary analgesic)

b

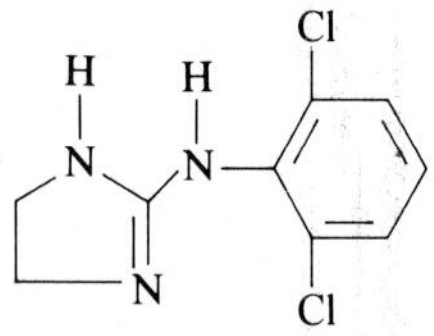
Cl
H
H
N
N
N
Cl

Clonidine
(an antihypertensive; trade name Catapres)

c

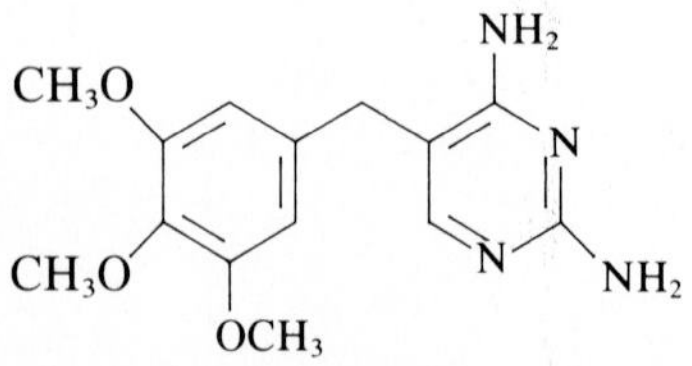
NH₂
CH₃O
N
CH₃O
N
NH₂
OCH₃

Trimethoprim
(an anti-infective; trade name Praloprim)

d

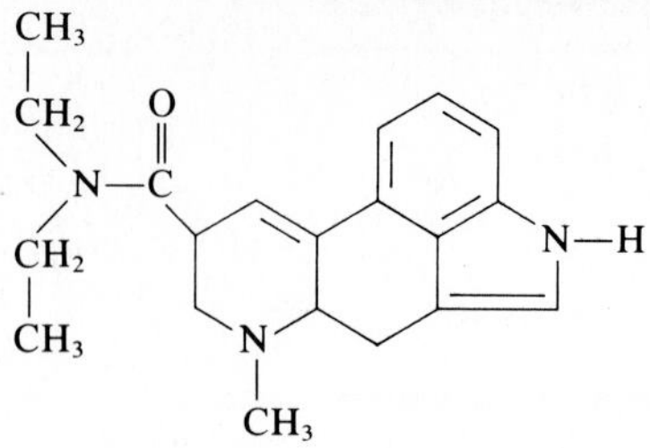
CH₃
CH₂
O
N—C
CH₂
CH₃
N—H
N
CH₃

Lysergic acid diethylamide
(a psychedelic)

e

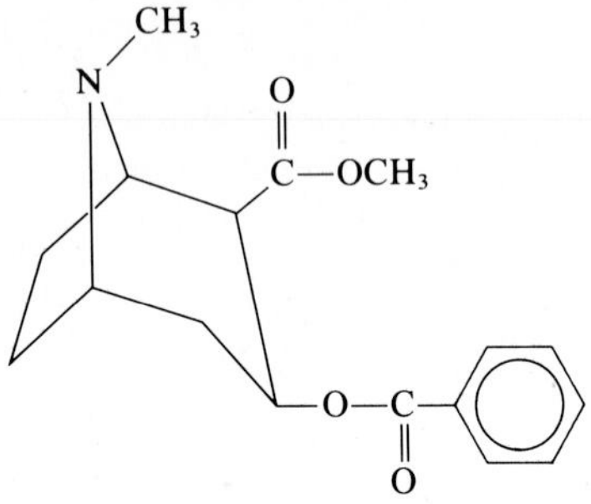
CH₃
N
O
C—OCH₃
O—C
O

Cocaine

16.22 List all the functional groups in the following structures. For alcohols and amines include their classification (1°, 2°, 3°) or heterocyclic ring class.

a

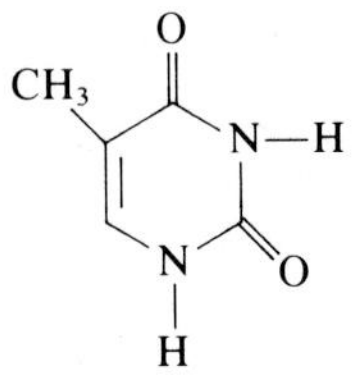
O
CH₃
N—H
N
O
H

Thymine

c

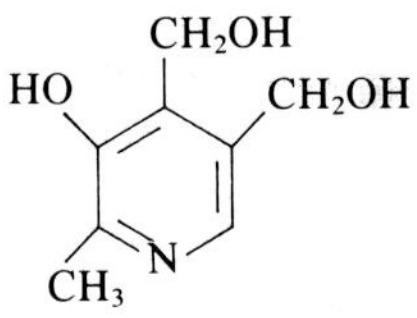
CH₂OH
HO
CH₂OH
CH₃
N

Pyridoxine

b

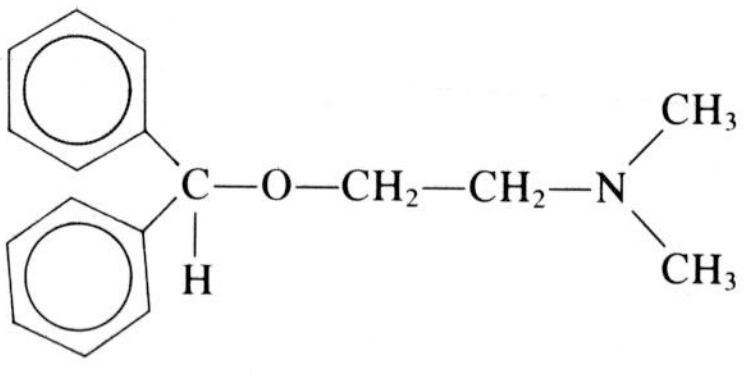
CH₃
C—O—CH₂—CH₂—N
H
CH₃

Benadryl

16.23 Predict whether methylamine or ethylammonium chloride will be more soluble in water. Give your reasoning.

16.24 Which substance would you more likely find as a smelling-salt ingredient, methylamine or methylammonium chloride? Explain your reasoning.

16.25 Some alkyl halides, such as 2-chlorobutane, are suspected carcinogens. How might an alkyl halide disrupt cellular processes, i.e., act as a carcinogen?

16.26 Describe the series of events which occur in individuals allergic to pollen which leads them to "break out" with the typical symptoms of runny eyes and nose upon exposure to pollen.

16.27 Describe how antihistamine drugs may function on a molecular level.

16.28 In what structural features does amphetamine resemble dopamine?

16.29 What functional group accounts for the basicity of alkaloids?

16.30 What common feature can you find in the amine functional group of *all* the narcotic analgesics?

16.31 What common structural feature(s) can you find in the structures of dopamine, norepinephrine, mescaline, and DOM?

16.32 Which compound would you predict to be more water-soluble, norepinephrine or mescaline? Give your reasoning.

16.33 What common structural feature can you find in serotonin, dimethyltryptamine, and lysergic acid diethylamide?

16.34 Identify the chiral carbon(s) in the lysergic acid diethylamide structure.

16.35 Which nitrogen in the structure of LSD could form a *N*-nitrosoamine upon reaction with nitrous acid?

16.36 Which of the following compounds might most likely be suspected of being carcinogenic?

a 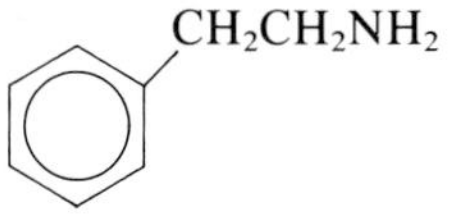

b 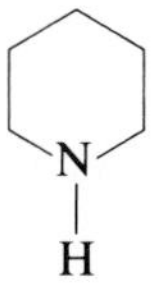

c 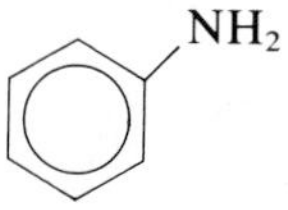

16.37 Write the structures of the organic product(s) that form in the following reactions. If no reaction occurs, write NR.

a 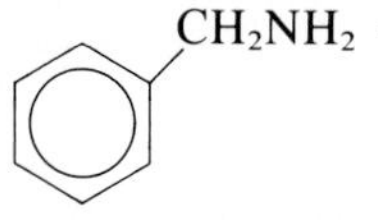 $+ HBr \rightarrow$

b $CH_3—CH(NH_2)—CH_3 + CH_3Br \rightarrow$

c $CH_3—CH(NH_2)—CH_3 + NaOH \rightarrow$

d 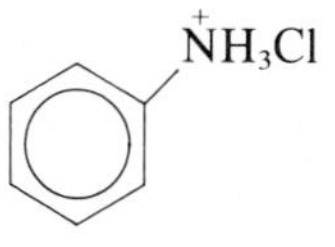 $+ NaOH \rightarrow$

e 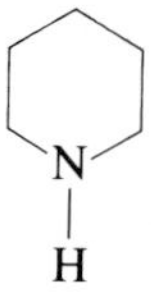 $+ HNO_2 \rightarrow$

f $(CH_3)_3\overset{+}{N}—CH_3\ Br^- + CH_3CH_2Br \rightarrow$

16.38 Name a physiological effect associated with each of the following:

a Morphine
b Nicotine
c Amphetamine
d Caffeine
e Nalbuphine
f Mescaline

16.39 Give the name of the compounds that fit the following descriptions:

a A highly addictive substance prepared from morphine
b Has the simplest possible amine structure
c A quaternary ammonium salt that functions as a neurotransmitter
d Makes up four rings found in the hemoglobin and myoglobin molecules
e An amine implicated in allergic reactions and stomach ulcers
f A noncontrolled (apparently nonaddictive) analgesic with a morphinelike structure
g A stimulant extracted from the leaves of the cocoa plant
h An alkaloid stimulant found in coffee
i A psychedelic compound active in minute doses
j A narcotic analgesic which has been used as an antitussive
k A neurotransmitter containing the indole ring
l Blocks an H_2 receptor site

CHAPTER SEVENTEEN

CARBOXYLIC ACIDS AND THEIR DERIVATIVES

17.1 INTRODUCTION

Compounds which have a carbonyl (>C=O) fragment are traditionally subdivided for study into two categories: (1) aldehydes and ketones, which you have already encountered in Chapter 14, and (2) carboxylic acids and their derivatives, the subject of this chapter. This division arose historically, but it also has a sound basis in the manner of their reactions; that is, there is a similarity in the pathway of reaction among members of a category and a difference between the two categories.

Notice the structural distinction between the two categories: Aldehydes and ketones have only hydrocarbon fragments (i.e., only H or C) attached to the carbonyl carbon.

Aldehyde   Ketone

Carboxylic acids and their derivatives are characterized by the attachment of some element other than H or C to the carbonyl carbon.

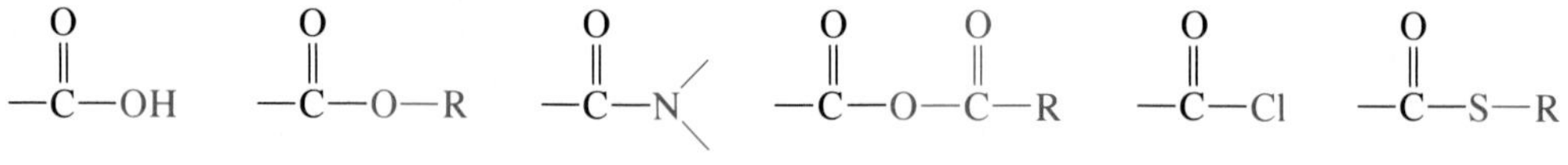

Carboxylic acid   Ester   Amide   Acid anhydride   Acid chloride   Thioester

The derivatives of carboxylic acids are so called because they can be thought of as arising by replacing the —OH of the acid by the appropriate fragment (shown in color). And, indeed, such replacement can be done in the laboratory. For example, esters are made from carboxylic acids by substituting —OR for —OH.

Carboxylic acids are important physiologically both because of the acidic property of the functional group and because of certain reactions of specific acids in such metabolic cycles as the citric acid cycle (Section 23.4). The most significant acid derivatives in biological systems are the triacylglycerols (fats and oils), which are esters, and proteins, which are polyamides. This chapter will deal with carboxylic acids themselves and with the formation reactions of esters and amides. Subsequent chapters will deal with fats and oils (Chapter 18) and proteins (Chapter 19).

17.2 NOMENCLATURE OF CARBOXYLIC ACIDS

There is a systematic IUPAC scheme of nomenclature for carboxylic acids. As usual, it is based on the longest continuous chain dictating the base name and a characteristic suffix for the functional group. The suffix for carboxylic acids is *-oic* followed by the word *acid*. Thus $CH_3CH_2CH_2CH_2—\overset{\overset{O}{\|}}{C}—OH$ is pentanoic acid because there is a five-carbon chain in which is included the acid functional group. Note that the $—\overset{\overset{O}{\|}}{C}—OH$ must be terminal; therefore, it is always in position 1. Notice also that for convenience in typing, the $—\overset{\overset{O}{\|}}{C}—OH$ functional group is frequently represented as —COOH.

$$R—\overset{\overset{O}{\|}}{C}—OH \equiv RCOOH$$

Many physiologically important acids are known almost exclusively by their common names. Common names are usually based on the natural source from which the compound was originally isolated and are unrelated to structure. Table 17.1 lists the structures and common names of important carboxylic and dicarboxylic acids, that is, acids with two carboxylic (—COOH) groups. The systematic names are also given; notice how these follow the usual IUPAC rules.

SAMPLE EXERCISE 17.1

Name the following acid by the IUPAC system:

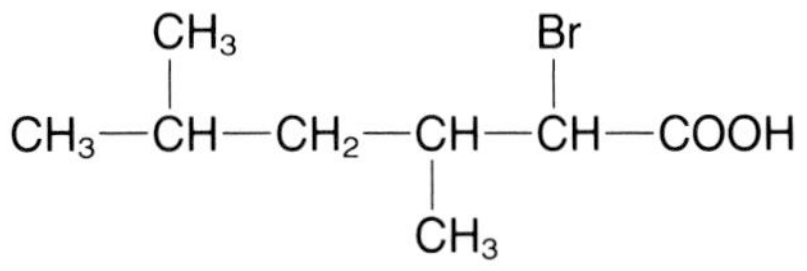
CH3
Br
CH3—CH—CH2—CH—CH—COOH
CH3

Solution Use the usual stepwise procedure.

STEP 1 Identify the longest continuous chain as *six* carbons. This is a *hexanoic* acid.

STEP 2 Number the chain. The terminal carboxylic acid group always defines position 1.

STEP 3 Locate and identify the substituents. These occur at positions 2, 3, and 5.

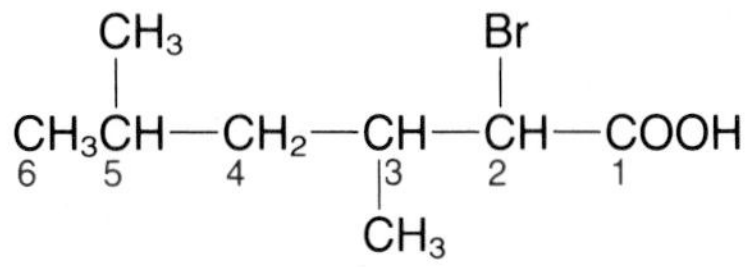
CH3
Br
CH3CH—CH2—CH—CH—COOH
6 5 4 3 2 1
CH3

The full name of the acid is therefore 2-bromo-3,5-dimethylhexanoic acid.

TABLE 17.1 Common Names of Carboxylic Acids

IUPAC Name	Structure	Common Name	Source of Common Name
Methanoic acid	$HCOOH$	Formic acid	*Formica* ("ant"*)
Ethanoic acid	CH_3COOH	Acetic acid	*Acetum* ("vinegar")
Propanoic acid	CH_3CH_2COOH	Propionic acid	*Pro* ("first") + *pion* ("fat"); first fatty acid
Butanoic acid	$CH_3CH_2CH_2COOH$	Butyric acid	*Butyrum* ("butter")
Benzoic acid	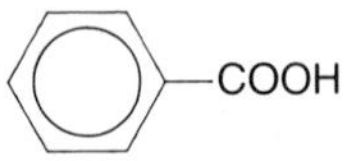	Benzoic acid	Complicated etymology
2-Hydroxybenzoic acid	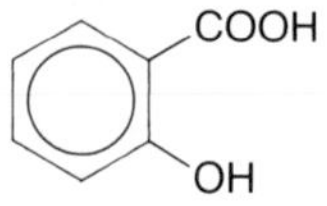	Salicylic acid	*Salix* ("willow")
2-Hydroxypropanoic acid	$CH_3-CH(OH)-COOH$	Lactic acid	*Lactic* ("milk")
2-Ketopropanoic acid	$CH_3-C(=O)-COOH$	Pyruvic acid	*Pyr* ("fire") + *uva* ("grape"); made by heating tartaric acid present in grapes
DICARBOXYLIC ACIDS			
Ethanedioic acid	$HOOC-COOH$	Oxalic acid	*Oxalis* ("sorrel")
Propanedioic acid	$HOOCCH_2COOH$	Malonic acid	*Malum* ("apple"); derived from malic acid found in apples
Butanedioic acid	$HOOCCH_2CH_2COOH$	Succinic acid	*Succinum* ("amber")
Pentanedioic acid	$HOOCCH_2CH_2CH_2COOH$	Glutaric acid	Uncertain etymology

* The poet Ogden Nash showed his knowledge of organic nomenclature when he wrote the poem "The Ant," which goes:
"The ant has made himself illustrious
Through constant industry industrious.
So what?
Would you be calm and placid
If you were full of *formic acid*?"

Source: Verses from 1929 On by Ogden Nash, Copyright 1935 by The Curtis Publishing Company. First appeared in *The Saturday Evening Post.* By permission of Little, Brown and Company.

In keeping with common practice we will use the names *formic acid* and *acetic acid* exclusively for the one-carbon and two-carbon acids, respectively. The terms *propionic* and *propanoic,* and *butyric* and *butanoic,* tend to be used arbitrarily and indiscrimately. Fortunately this practice causes little difficulty because of the great similarity of the common and IUPAC names. Dicarboxylic acids are always called by their common names.

An alternative scheme for labeling the carbon chain of carbonyl compounds is used for many biologically important molecules. In this system, labeling of the chain begins with the carbon attached to —COOH, that is, the 2 position. Lowercase letters of the Greek alphabet are used rather than arabic numbers.

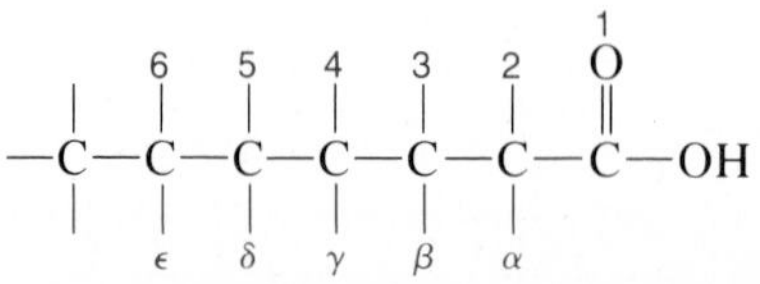

α = alpha
β = beta
γ = gamma
δ = delta
ϵ = epsilon

By this system the compound named in Sample Exercise 17.1 would be called α-bromo-β,δ-dimethylhexanoic acid. One rarely encounters this numbering scheme beyond the β-position. You will see the use of α,β-labeling again in the discussion of amino acids (Section 19.2) and the metabolism of fats (Section 24.3).

17.3 PHYSICAL PROPERTIES OF CARBOXYLIC ACIDS

Carboxylic acids tend to have exceptionally high boiling points and are thereby all liquids or solids at room temperature. As Table 17.2 shows, acids have even higher boiling points than alcohols of similar molecular weight. This is a consequence of the excellent opportunity for hydrogen bonding between acid molecules. Because of hydrogen bonding, a greater amount of heat energy must be supplied in order for molecules to move independently of one another and enter the gas phase (review Sections 8.11 and 14.3). Because of the "double" H-bond formation between the acidic hydrogens and carbonyl

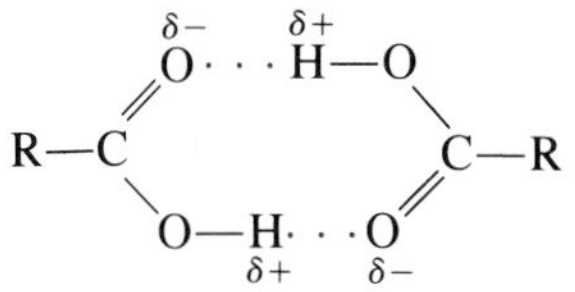
δ−
δ+
O· · · H—O
R—C
C—R
O—H· · ·O
δ+
δ−

oxygens, many carboxylic acids essentially exist as dimers; that is, they move together as two associated molecules. Because of this, their boiling-point behavior is like that of molecules that are twice as large as their apparent molecular weight would indicate.

Carboxylic acids with chains of less than five carbons are water-soluble because of H bonding between the acid and water molecules.

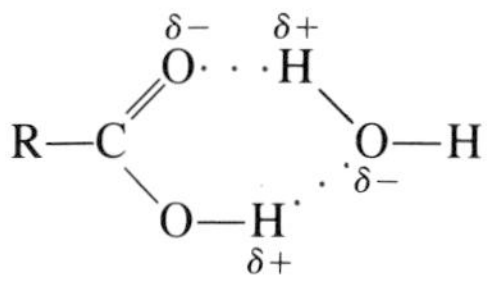
δ−
δ+
O· · ·H
R—C
O—H
δ−
O—H
δ+

In Section 11.1 you learned that sourness is a characteristic property of

TABLE 17.2 Comparative Properties of Carboxylic Acids and Alcohols of Similar Molecular Weights

Number of Carbons	Structure	Name	Molecular Weight	Melting Point, °C	Boiling Point, °C
2	CH_3CH_2OH	Ethanol	46	−115	78
1	HCOOH	Formic acid	46	8	101
3	$CH_3CH_2CH_2OH$	1-Propanol	60	−126	97
2	CH_3COOH	Acetic acid	60	16	118
4	$CH_3CH_2CH_2CH_2OH$	1-Butanol	74	− 90	118
3	CH_3CH_2COOH	Propionic acid	74	− 21	141
5	$CH_3(CH_2)_3CH_2OH$	1-Pentanol	88	− 79	138
4	$CH_3CH_2CH_2COOH$	Butyric acid	88	− 6	164
2	HOOC—COOH	Oxalic acid	90	189*	Decomposes

* Note the particularly striking effect of the presence of two carboxylic acid groups.

acids. You are probably familiar with this property in vinegar, which is a 5% aqueous solution of acetic acid. Oxalic acid imparts tartness to rhubarb, and malic acid (HOOC—CHOH—CH_2COOH) to apples. Sour milk is sour because lactic acid has formed in the milk through bacterial degradation of sugars.

A rank, disagreeable odor accompanies the sourness in longer-chain carboxylic acids. Butter becomes rancid when it has developed a high concentration of butyric acid ($CH_3CH_2CH_2COOH$). Butyric acid is also part of body odor in sweat. Goats have given names to several carboxylic acids which contribute to their antisocial aroma. The Latin word for goat is *caper*. Caproic [$CH_3(CH_2)_4COOH$], caprylic [$CH_3(CH_2)_6COOH$], and capric acids [$CH_3(CH_2)_8COOH$] all occur in goat perspiration.

17.4 ACIDITY AND SALT FORMATION

Carboxylic acids are weak acids. Because acidity was noted and discussed in Chapter 11, you may find it useful to review Sections 11.3, 11.4, and 11.8 before continuing with this section.

The acid equilibrium can be expressed in general as

$$RCOOH + H_2O \rightleftharpoons H_3O^+ + RCOO^-$$

which shows the Brønsted-Lowry proton donation from RCOOH to H_2O. The acid ionization can be represented more simply as

$$RCOOH \rightleftharpoons RCOO^- + H^+$$

Either representation demonstrates the increase in hydrogen (hydronium) ion concentration that is produced in a solution of a carboxylic acid. The anion ($RCOO^-$) is called a *carboxylate* ion.

Like other acids, organic carboxylic acids participate in acid-base neutralization reactions to yield salts. The reacting base might be a strong metal hydroxide base or a weak base such as ammonia or an amine. Look at the acid-base reaction in the Brønsted-Lowry sense of H^+ transfer from acid to base:

$$RCOOH + Na^+(aq) + {}^-OH(aq) \longrightarrow RCOO^-Na^+ + HOH$$

Carboxylic acid — Base — Carboxylate salt

$$RCOOH + H_2\ddot{N}CH_3 \longrightarrow RCOO^-H_3\overset{+}{N}CH_3$$

Acid — Base — Salt

Just as amine salts are water-soluble because of their ionic nature, so too acid salts are likely to be more water-soluble than the acids themselves. Special solubility properties of carboxylic acid salts with long hydrocarbon chains make them useful as soaps. Long-carbon-chain acids are called fatty acids for historical reasons that will unfold in this chapter and the next.

17.5 FATTY ACID SALTS AS SOAPS

The soap action of a fatty acid salt can be explained on a molecular level. Sodium stearate is a good example of a fatty acid salt. It has a straight chain of 17 carbons attached to the carboxylate carbon.

$$CH_3(CH_2)_{16}COOH + NaOH \longrightarrow CH_3(CH_2)_{16}COO^-Na^+ + H_2O$$

Stearic acid — Sodium stearate

Examples of other fatty acids can be found in Table 18.2. Acid salts are named by replacing the *-ic* ending of the acid by *-ate* and naming the cation first. Hence salts of stear*ic* acid are stear*ates*.

The combination within one compound of a polar or hydrophilic ionic end, which is water-soluble, and a long nonpolar or hydrophobic hydrocarbon chain, which is not water-soluble, gives fatty acid salts their soapy properties.

$$CH_3CH_2CH_2CH_2CH_2CH_2CH_2CH_2CH_2CH_2CH_2CH_2CH_2CH_2CH_2CH_2CH_2\overset{\overset{\large O}{\|}}{C}-O^-Na^+$$

Hydrophobic hydrocarbon end (soluble in nonpolar solvents) — Hydrophilic ionic end (water-soluble)

Soaps function by promoting effective mixing of nonpolar dirt or grease with polar water. Grease is hydrocarbon or hydrocarbonlike in nature and binds dirt to surfaces or fabrics. The intimate mixing of essentially noncompatible grease and water which soap promotes is called **emulsification.** Figure 17.1 shows soap function. The nonpolar hydrocarbon ends of sodium stearate dissolve in nonpolar hydrocarbonlike grease and leave the ionic carboxylate ends hanging out. Electrostatic repulsion promotes the breakdown of a big grease blob into tiny droplets. These grease droplets with their outer shell of $—COO^-$ anions are called micelles. The micelles can easily intermingle with polar water molecules because of their ionic exterior. Actually, as Figure 17.1 shows, the soap molecules themselves travel around as micelles in order to shield the hydrophobic hydrocarbon ends from water.

Most solid bars of soap that we encounter are sodium salts of stearic acid or other fatty acids. Liquid soaps are often potassium salts. "Hard" water inactivates soap action because it converts soluble sodium or potassium soaps into insoluble salts. Hard water contains Mg^{2+}, Fe^{3+}, or Ca^{2+} ions. These ions displace Na^+ ions from soap, and the resulting insoluble calcium and magnesium salts precipitate out of water. Bathtub ring is a precipitate of insoluble soaps.

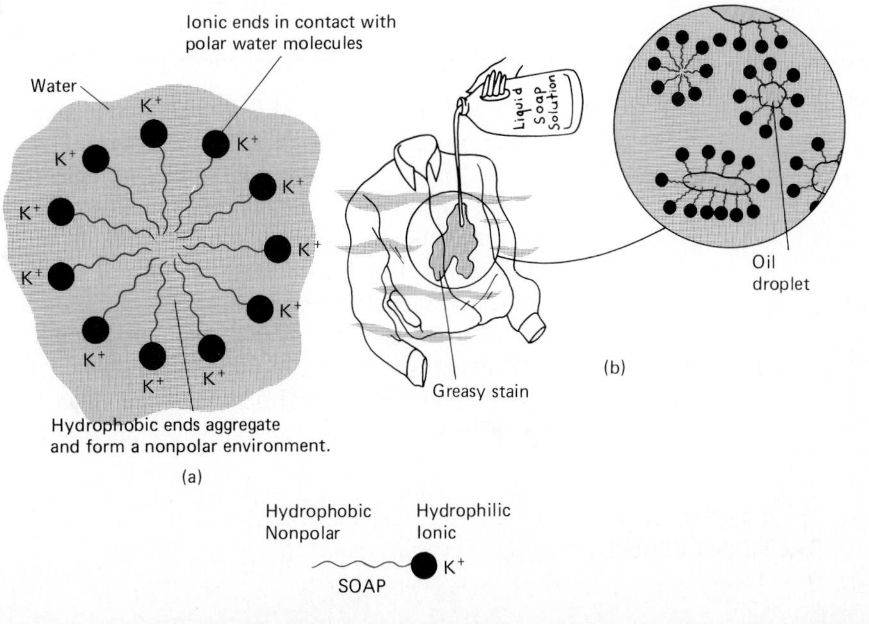
Ionic ends in contact with polar water molecules
Water
K+
Hydrophobic ends aggregate and form a nonpolar environment.
(a)
Liquid Soap Solution
Greasy stain
Oil droplet
(b)
Hydrophobic Nonpolar
Hydrophilic Ionic
SOAP

FIG. 17.1
Molecular soap action. (*a*) Soap particles aggregate into micelles in order to shield the hydrophobic hydrocarbon chains from water. Potassium ions are water-soluble. (*b*) Soap particles break up big blobs of grease, such as those found on dirty clothes, by packing small oil droplets into the center of the micelle structure. Thus the grease "dissolves" and can be washed away.

$$2CH_3(CH_2)_{16}COO^-Na^+\,(aq) + Ca^{2+}\,(aq) \longrightarrow (CH_3(CH_2)_{16}COO^-)_2Ca^{2+}\downarrow + 2Na^+(aq)$$

Soluble soap

Insoluble soap

The excretion of calcium soaps by individuals with celiac disease leads to calcium deficiency in these patients. Celiac disease is characterized by the inability to absorb dietary fatty acids from the small intestine. They therefore pass into the large intestine for elimination. Conversion of the acid to a calcium salt means that calcium ions are lost also as the fatty acids leave the body. Such patients are treated with calcium supplements and put on low-fat diets to reduce the amount of fatty acids produced by digestion of fats.

Soaps are natural products in the sense that they are made from naturally occurring fats, as you will see in Section 18.7. Detergents are synthetic materials made by chemists but based on the molecular design of soaps. Their advantage over soaps is that they function in hard water; calcium and magnesium salts of detergents are water-soluble. The most common detergent is sodium lauryl sulfate:

$$CH_3CH_2CH_2CH_2CH_2CH_2CH_2CH_2CH_2CH_2CH_2CH_2{-}OSO_3^-Na^+$$

Sodium lauryl sulfate

Detergents emulsify, that is, allow water and grease to mix, in the same way that soaps do; the ionic end is hydrophilic, and the hydrocarbon end is hydrophobic. The synthesis of detergents involves the esterification of a long-chain alcohol, such as lauryl alcohol $CH_3(CH_2)_{10}CH_2OH$, with sulfuric acid. We will now look at esterification, the reaction between alcohols and acids.

17.6 ESTERIFICATION

The reaction between a carboxylic acid and an alcohol results in the formation of an ester, a class of compounds characterized by the group, $-\overset{\overset{O}{\|}}{C}-O-C-$. The reaction is therefore known by the name **esterification.**

$$R-\overset{\overset{O}{\|}}{C}-\boxed{O-H + H}-O-CH_3 \longrightarrow R-\overset{\overset{O}{\|}}{C}-O-CH_3 + H_2O \qquad (17.1)$$

Acid Alcohol Ester

Esterification offers another example of a condensation reaction in which a small molecule (usually water) is split out between two molecules which are thereby joined (or condensed) together. Notice how the water is "boxed out" in Equation (17.1) and the remaining pieces of acid and alcohol are joined to make an ester.

Whenever a molecule with an —OH group and one with an H attached to an electronegative element such as O, N, or S, are brought together, water is likely to be split out. You have already seen two examples of this wherein, as in esterification, both condensing molecules have —OH groups. One —OH supplies the —OH of water; the other supplies the H.
Recall the condensation of two alcohols (Section 14.6),

$$CH_3\boxed{OH + H}OCH_3 \longrightarrow CH_3-O-CH_3 + HOH$$

Alcohol Alcohol Ether

and acetal formation (Section 14.13),

$$CH_3-\underset{\underset{OCH_2CH_3}{|}}{\overset{\overset{H}{|}}{C}}-\boxed{OH}\ +\ \boxed{H}OCH_2CH_3\ \longrightarrow\ CH_3-\underset{\underset{OCH_2CH_3}{|}}{\overset{\overset{H}{|}}{C}}-OCH_2CH_3\ +\ HOH$$

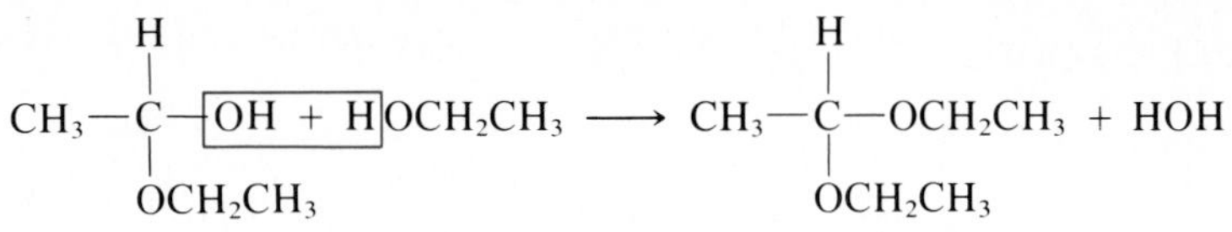

      Hemiacetal   Alcohol   Acetal

Of the examples of condensation reactions presented thus far, esterification is far and away the most common and important one in chemistry laboratories and industrial applications. You will see this in Sections 17.11 and 17.15.

SAMPLE EXERCISE 17.2

Write a structure for the ester that forms when acetic acid reacts with ethanol.

Solution

The structure of the product ester of any acid-plus-alcohol reaction can be determined by following the stepwise procedure:

Guidelines for Ester Formation

STEP 1   Write down the structures of the acid and alcohol so that the —OH groups of these reactants "face one another":

$$CH_3-\overset{\overset{O}{\|}}{C}-OH\ +\ HOCH_2CH_3\ \longrightarrow$$

    Acetic acid   Ethanol

STEP 2   Box out the elements of water:

$$CH_3-\overset{\overset{O}{\|}}{C}-\boxed{OH\ +\ H}OCH_2CH_3\ \longrightarrow$$

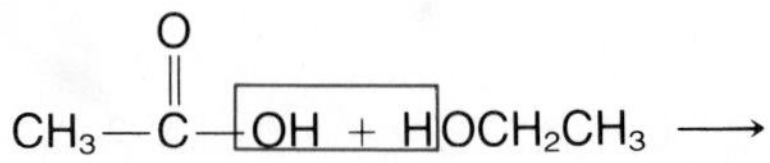

STEP 3   Attach the fragment left from the acid to the fragment of the alcohol to form the ester:

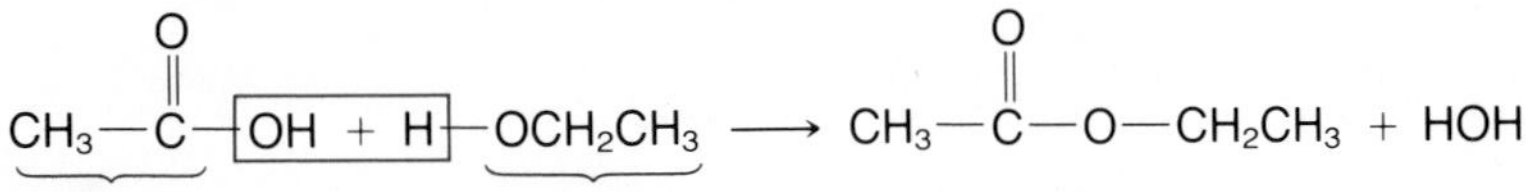

$$CH_3-\overset{\overset{O}{\|}}{C}-\boxed{OH\ +\ H}-OCH_2CH_3\ \longrightarrow\ CH_3-\overset{\overset{O}{\|}}{C}-O-CH_2CH_3\ +\ HOH$$

    Acid   Alcohol   Ester
    fragment   fragment

The ester in this case is ethyl acetate, which is the solvent in airplane glue. You will see how to name esters in the next section.

Problem 17.1   Write a structure for the ester that forms when butyric acid reacts with methanol.

17.7 A CLOSE EXAMINATION OF THE ESTER FUNCTIONAL GROUP

Sample Exercise 17.2 pointed out that in every ester one can identify an acid fragment and an alcohol fragment because esters are constructed by joining these fragments together as water leaves.

$$CH_3-\overset{\overset{\displaystyle O}{\|}}{C}-\boxed{OH \ + \ H}OCH_2CH_3 \longrightarrow CH_3-\overset{\overset{\displaystyle O}{\|}}{C}-O-CH_2CH_3 \ + \ HOH$$

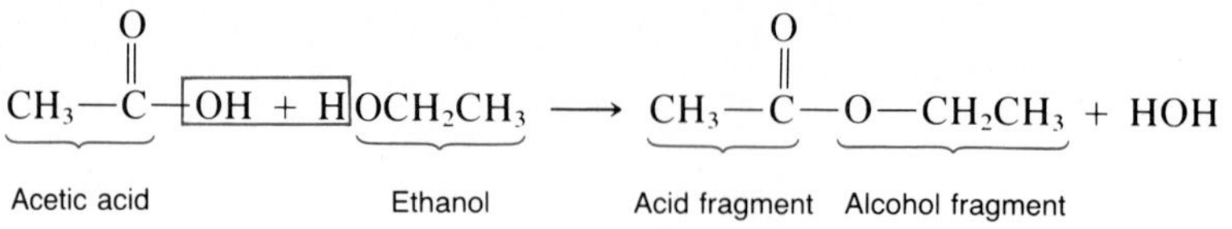

Acetic acid   Ethanol   Acid fragment   Alcohol fragment

The acid fragment always terminates in the carbonyl, $\overset{\overset{\displaystyle O}{\|}}{\underset{\diagup \ \diagdown}{C}}$ 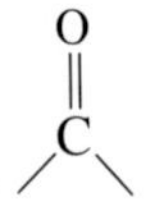, and the alcohol fragment is the group attached to the oxygen.

Esters are named by identifying and naming the alkyl group from the alcohol and then naming the acid fragment. The *-ic* ending of the acid name is changed to *-ate*. Thus the ester in the example would be ethyl (from ethyl alcohol) acetate (from acetic acid). If we had used the IUPAC name, ethanoic acid, then the ester name would be ethyl ethanoate. An understanding of ester nomenclature will make it easier for you to identify the acid and alcohol parts of an ester. Table 17.3 has further examples of ester nomenclature.

SAMPLE EXERCISE 17.3

From what reacting acids and alcohols are the following esters formed?

a $\text{C}_6\text{H}_5-\overset{\overset{\displaystyle O}{\|}}{C}-O-CH_3$

b $CH_3-O-\overset{\overset{\displaystyle O}{\|}}{C}-CH_2CH_3$

Solution

The acid is identified by noting the group attached to the carbonyl. The alcohol is identified by the group attached to oxygen. The alcohol group is named first.

a   This ester is methyl benzoate. It is made from benzoic acid and methanol.

$$C_6H_5-\overset{\overset{\displaystyle O}{\|}}{C}-\boxed{OH \ + \ H}-O-CH_3 \longrightarrow C_6H_5-\overset{\overset{\displaystyle O}{\|}}{C}-O-CH_3 \ + \ HOH$$

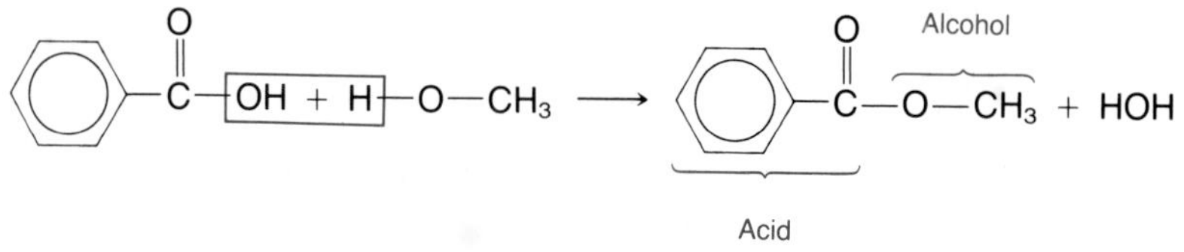

Acid

b   This ester is methyl propionate (IUPAC propanoate), made from methanol and propionic acid (IUPAC propanoic acid).

$$CH_3O\boxed{H \ + \ HO}-\overset{\overset{\displaystyle O}{\|}}{C}-CH_2CH_3 \longrightarrow CH_3O-\overset{\overset{\displaystyle O}{\|}}{C}-CH_2CH_3 \ + \ HOH$$

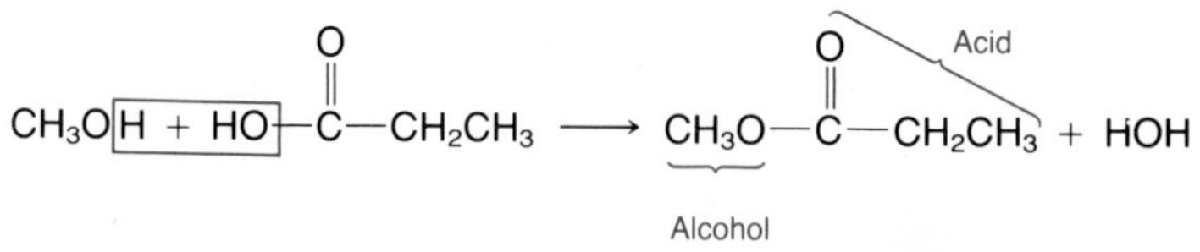

Alcohol

These two examples point out that ester structures can be written from left to right, either acid part then alcohol part, as in $C_6H_5-\overset{\overset{\displaystyle O}{\|}}{C}-O-CH_3$ 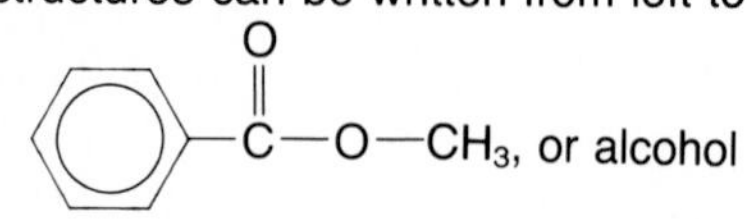, or alcohol part then acid part, as in $CH_3O-\overset{\overset{\displaystyle O}{\|}}{C}-CH_2CH_3$. Not recognizing this can be a pitfall in later applications.

Problem 17.2 What reacting acids and alcohols formed the esters below?

a CH_3CH_2—O—$\overset{\overset{\displaystyle O}{\|}}{C}$—$CH_3$

b CH_3—$\overset{\overset{\displaystyle O}{\|}}{C}$—O—$CH_2$—⬡

17.8 ESTERS FROM PHOSPHORIC ACIDS

An examination of the Lewis structure of many inorganic, oxygen-containing acids reveals a certain structural similarity to carboxylic acids. Note, for example,

Line formula: RCOOH H_3PO_4

Structural formula: R—$\overset{\overset{\displaystyle O}{\|}}{C}$—OH   HO—$\overset{\overset{\displaystyle O}{\|}}{\underset{\underset{\displaystyle OH}{|}}{P}}$—OH

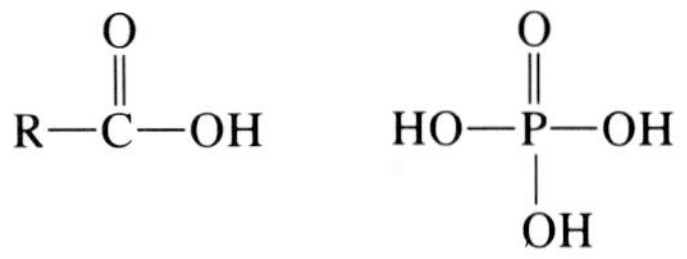

                      Carboxylic acid   Phosphoric acid

In both cases an —OH fragment is attached to an element forming a double bond to oxygen,

$$\overset{\overset{\displaystyle O}{\|}}{C} \quad \text{and} \quad \overset{\overset{\displaystyle O}{\|}}{P}$$

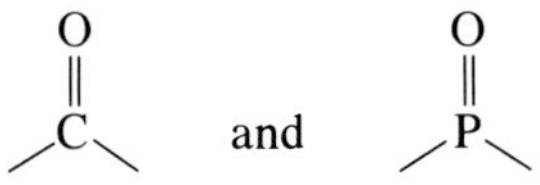

This structural similarity allows phosphoric acid to react with alcohols just as carboxylic acids do. The product of such a reaction is a **phosphate ester:**

HO—$\overset{\overset{\displaystyle O}{\|}}{\underset{\underset{\displaystyle OH}{|}}{P}}$—[OH + H]OCH₃ ⟶ HO—$\overset{\overset{\displaystyle O}{\|}}{\underset{\underset{\displaystyle OH}{|}}{P}}$—O—CH₃ + HOH

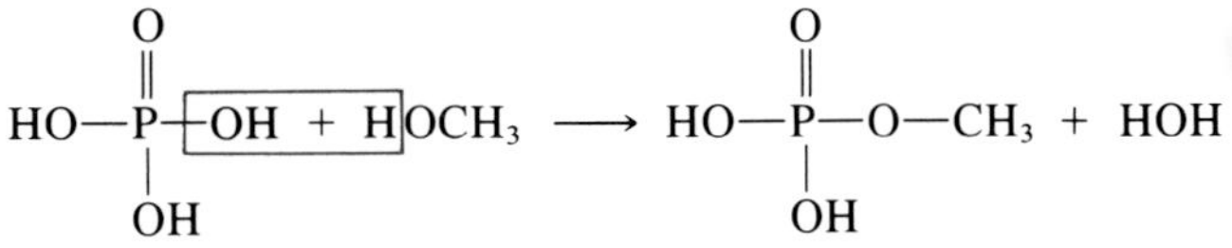

Phosphoric acid   Methanol   Methyl phosphate

Many phosphate esters are found in physiological processes. For example, a first step in carbohydrate metabolism, as you will see in Section 23.2, is the esterification of the alcohol group at position 6 of glucose. When an alcohol is esterified by a phosphoric acid, the process is commonly called **phosphorylation.**

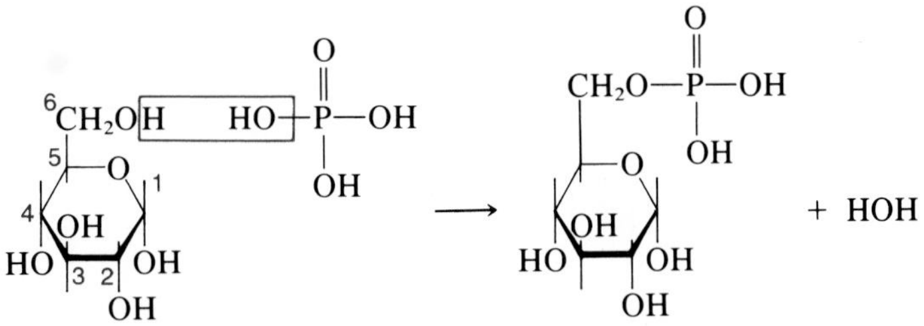

Glucose   Glucose-6-phosphate

Because phosphate esters retain an acidic —OH attached to $-\overset{\overset{\displaystyle O}{\|}}{P}-$, they can form diesters if the acid —OH reacts with another alcohol group. For example, look at the structure of methyl phosphate which was formed above by the reaction of methanol and phosphoric acid. The two remaining —OH groups attached to P could react with other alcohols. The structures of phospholipids, key components of cell membranes, as you will see in Section 18.9, show two phosphate ester linkages.

$$CH_2-O-\overset{\overset{\displaystyle O}{\|}}{C}-R$$

$$HC-O-\overset{\overset{\displaystyle O}{\|}}{C}-R'$$

$$H_2C-O-\overset{\overset{\displaystyle O}{\|}}{\underset{\underset{\displaystyle OH}{|}}{P}}-O-CH_2CH_2\overset{+}{N}(CH_3)_3$$

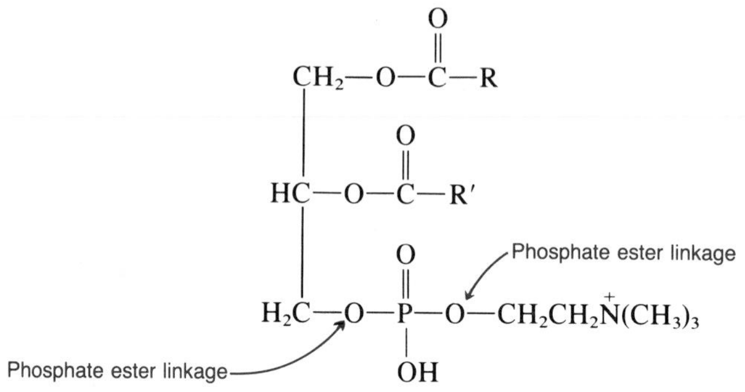

Phosphate ester linkage

Phosphate ester linkage

A lecithin (a phospholipid)

Bringing two molecules of H_3PO_4 together fulfills the requirements for a condensation reaction; that is, one molecule can supply H and the other, OH. Thus, pyrophosphoric (or diphosphoric acid) can form.

$$HO-\overset{\overset{\displaystyle O}{\|}}{\underset{\underset{\displaystyle OH}{|}}{P}}-O\boxed{-H \; + \; HO}-\overset{\overset{\displaystyle O}{\|}}{\underset{\underset{\displaystyle OH}{|}}{P}}-OH \longrightarrow HO-\overset{\overset{\displaystyle O}{\|}}{\underset{\underset{\displaystyle OH}{|}}{P}}-O-\overset{\overset{\displaystyle O}{\|}}{\underset{\underset{\displaystyle OH}{|}}{P}}-OH \; + \; HOH$$

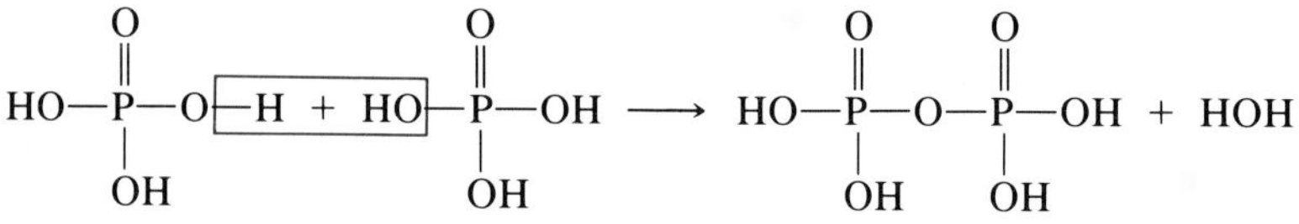

Phosphoric acid   Phosphoric acid   Pyrophosphoric acid

Notice that the product structure is determined in the same way as for other condensation reactions. The linkage, $-\overset{\overset{\displaystyle O}{\|}}{P}-O-\overset{\overset{\displaystyle O}{\|}}{P}-$, formed by splitting out water from phosphoric acid molecules, is called a phosphoric acid anhydride linkage. (*Acid anhydride* means "acid without water.") Esters of pyrophosphoric acid are known as pyrophosphates or diphosphates, which indicates the presence of two $-\overset{\overset{\displaystyle O}{\|}}{P}-$ centers.

Another H_3PO_4 molecule can be condensed onto pyrophosphoric acid to form triphosphoric acid,

$$HO-\overset{\overset{\displaystyle O}{\|}}{\underset{\underset{\displaystyle OH}{|}}{P}}-O-\overset{\overset{\displaystyle O}{\|}}{\underset{\underset{\displaystyle OH}{|}}{P}}-O-\overset{\overset{\displaystyle O}{\|}}{\underset{\underset{\displaystyle OH}{|}}{P}}-OH$$

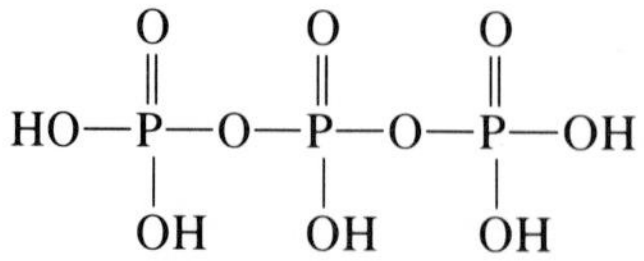

As you might guess, the phosphate esters of triphosphoric acid are called triphosphates. The triphosphate that you will see most often is adenosine

triphosphate (ATP), the triphosphate ester of the nucleoside adenosine, which was shown in Section 16.12.

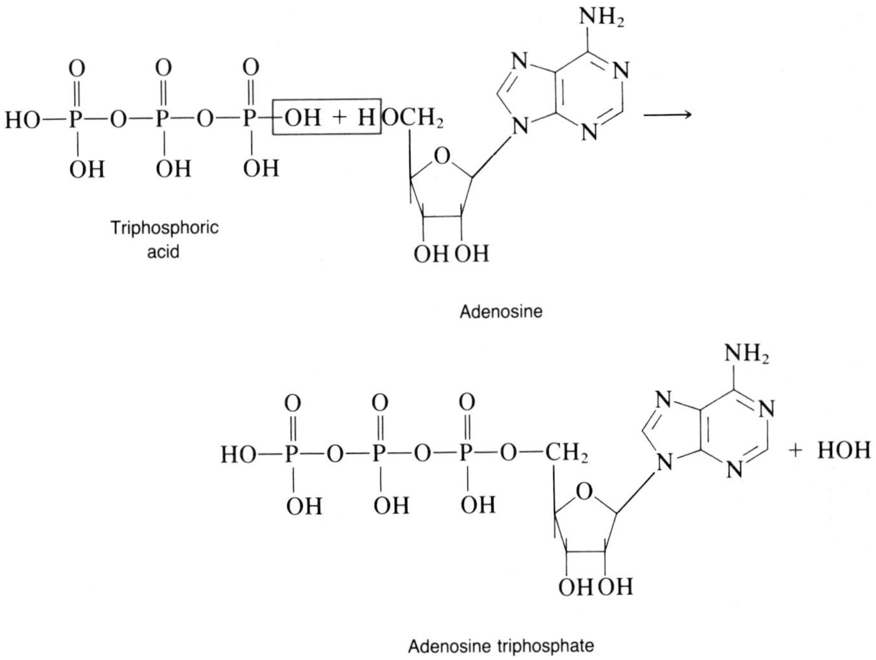
Triphosphoric acid
Adenosine
Adenosine triphosphate

ATP is the carrier of free energy in cells, as you will see in Section 22.4. This energy is transferred by the making and breaking of the $-\overset{\overset{\displaystyle O}{\|}}{P}-O-\overset{\overset{\displaystyle O}{\|}}{P}-$ anhydride bonds in the triphosphate.

17.9 THIOESTERS

Alcohols (ROH) react with acids to form esters. Thiols (RSH) react with acids to form thioesters. The course of the reaction is identical; condensation occurs between an acid, which supplies —OH of water, and the thiol, which supplies —H.

$$\underset{\text{Acid}}{CH_3-\overset{\overset{\displaystyle O}{\|}}{C}-}\boxed{OH + H}\underset{\text{Thiol}}{S-R} \longrightarrow \underset{\text{Thioester}}{CH_3-\overset{\overset{\displaystyle O}{\|}}{C}-S-R} + HOH$$

Thioester formation is essential to fat metabolism, as you will see in Section 24.4. And, acetyl coenzyme A, which plays a central role in all metabolic cycles (see Section 22.11) is a thioester.

$$\underset{\text{Acetic acid}}{CH_3-\overset{\overset{\displaystyle O}{\|}}{C}-}\boxed{OH + H}\underset{\substack{\text{Coenzyme A}^1\\ \text{(a thiol)}}}{-S-CoA} \longrightarrow \underset{\substack{\text{Acetyl CoA}\\ \text{(a thioester)}}}{CH_3-\overset{\overset{\displaystyle O}{\|}}{C}-S-CoA} + HOH$$

[1]Coenzyme A is a large complex molecule which for simplicity is often abbreviated CoA—S—H to indicate the reacting thiol group. For the full structure, see Section 22.11.

17.10 ANHYDRIDES OF CARBOXYLIC ACIDS

Just as two molecules of phosphoric acid can condense to form pyrophosphoric acid by the formation of a phosphoric anhydride linkage, two molecules of a carboxylic acid can condense by splitting out water and form a carboxylic acid derivative called a carboxylic **acid anhydride.**

$$\underset{\text{Carboxylic acid}}{R{-}\overset{O}{\overset{\|}{C}}{-}\boxed{OH + H}{-}}\underset{\text{Carboxylic acid}}{O{-}\overset{O}{\overset{\|}{C}}{-}R} \longrightarrow \underset{\text{Acid anhydride}}{R{-}\overset{O}{\overset{\|}{C}}{-}O{-}\overset{O}{\overset{\|}{C}}{-}R} + H_2O \qquad (17.2)$$

The class name, acid anhydride, literally means acid without the elements of water. The outstanding structural characteristic of an acid anhydride group is an oxygen between two carbonyl groups, $-\overset{O}{\overset{\|}{C}}{-}O{-}\overset{O}{\overset{\|}{C}}-$, in a manner analogous to the phosphoric acid anhydride characterized by an oxygen between two $-\overset{O}{\overset{\|}{P}}-$ groups, $-\overset{O}{\overset{\|}{P}}{-}O{-}\overset{O}{\overset{\|}{P}}-$. The reverse of the reaction in Equation (17.2), that is, the reaction of an acid anhydride with water, produces two carboxylic acid molecules.

$$R{-}\overset{O}{\overset{\|}{C}}{-}O{-}\overset{O}{\overset{\|}{C}}{-}R + H_2O \longrightarrow 2\ R{-}\overset{O}{\overset{\|}{C}}{-}OH$$

Carboxylic acid anhydrides are not commonly found in biological systems, but they are very useful to the laboratory chemist because they are much more reactive than carboxylic acids and esters. The most common and synthetically important acid anhydride is acetic anhydride, that is, the anhydride that forms when two molecules of acetic acid condense.

$$\underset{\text{Acetic acid}}{CH_3\overset{O}{\overset{\|}{C}}{-}\boxed{OH + H}{-}}\underset{\text{Acetic acid}}{O{-}\overset{O}{\overset{\|}{C}}{-}CH_3} \longrightarrow \underset{\text{Acetic anhydride}}{CH_3\overset{O}{\overset{\|}{C}}{-}O{-}\overset{O}{\overset{\|}{C}}{-}CH_3}$$

You will see the synthetic utility of carboxylic acid anhydrides in Sections 17.11 and 17.15.

17.11 PROPERTIES OF ESTERS

Common esters are colorless liquids with pleasant odors. Their pleasant fragrances are in marked contrast to the vile odors of some of the acids from which they are derived. Esters supply the pleasant odors of fruits and flowers. They are isolated from natural sources or synthesized for use in perfumes, cosmetics, and flavorings. Table 17.3 shows the esters responsible for many fruit and spice flavorings.

Esters also act as pheromones (Section 13.13). The aroma of methyl *p*-hydroxybenzoate,

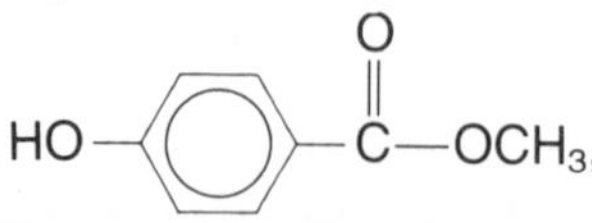

acts as a kind of perfume in the canine species. This ester is secreted in the vagina of dogs in heat and evokes attraction and sexual arousal in male canines.

TABLE 17.3 Flavors or Fragrances of Some Esters

Name	Structure	Flavor or Fragrance
Ethyl formate	$H{-}C({=}O){-}O{-}CH_2CH_3$	Rum
Isobutyl formate	$H{-}C({=}O){-}O{-}CH_2{-}CH(CH_3){-}CH_3$	Raspberries
Methyl butyrate	$CH_3CH_2CH_2{-}C({=}O){-}O{-}CH_3$	Apple
Ethyl butyrate	$CH_3CH_2CH_2{-}C({=}O){-}O{-}CH_2CH_3$	Pineapple
Amyl* acetate	$CH_3{-}C({=}O){-}O{-}CH_2CH_2CH_2CH_2CH_3$	Banana
Isoamyl acetate	$CH_3{-}C({=}O){-}OCH_2CH_2{-}CH(CH_3){-}CH_3$	Pear
Amyl* butyrate	$CH_3CH_2CH_2{-}C({=}O){-}O{-}CH_2CH_2CH_2CH_2CH_3$	Apricot
Octyl acetate	$CH_3{-}C({=}O){-}OCH_2CH_2CH_2CH_2CH_2CH_2CH_2CH_3$	Orange
Methyl salicylate	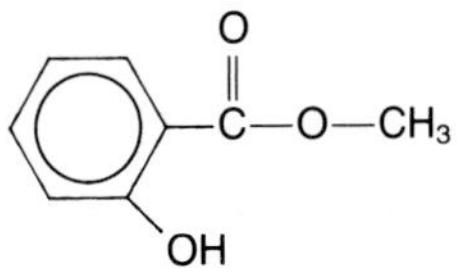	Wintergreen
Benzyl acetate	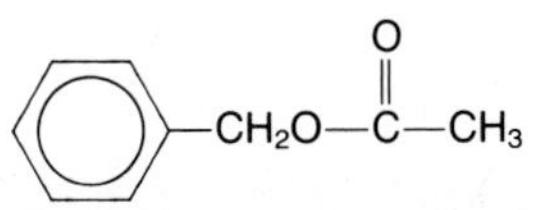	Jasmine

* Amyl = pentyl.

Because they cannot form hydrogen bonds among themselves, esters have lower boiling points than alcohols or acids of comparable molecular weights. Esters are more like ethers in their physical properties than any of the other classes of carbonyl compounds. Consider the molecular weight and boiling point data for various functional group classes shown in Table 17.4.

One of the last entries in Table 17.3 is methyl salicylate, known commonly as oil of wintergreen. It is an ingredient in Ben-Gay and other rub-in preparations for local heat treatment of muscle soreness. Methyl salicylate is a liquid (bp = 228°C). This ester is made through the reaction of salicylic acid and methanol:

TABLE 17.4 Comparative Molecular Weights and Boiling Points of Ethers, Esters, Alcohols, Aldehydes, and Acids

Structure	Name	Molecular Weight	Boiling Point, °C	Functional Group Class
$CH_3CH_2OCH_2CH_3$	Diethyl ether	74	34	Ether
$HCOOCH_2CH_3$	Ethyl formate	74	54	Ester
CH_3COOCH_3	Methyl acetate	74	57	Ester
$CH_3CH_2CH_2CHO$	Butanal	72	76	Aldehyde
$CH_3CH_2CH_2CH_2OH$	1-Butanol	74	118	Alcohol
CH_3CH_2COOH	Propionic acid	74	141	Acid
$CH_3CH_2OCH_2CH_2CH_3$	Ethyl propyl ether	88	64	Ether
$CH_3COOCH_2CH_3$	Ethyl acetate	88	77	Ester
$CH_3(CH_2)_3CHO$	Pentanal	86	103	Aldehyde
$CH_3(CH_2)_3CH_2OH$	1-Pentanol	88	138	Alcohol
$CH_3CH_2CH_2COOH$	Butyric acid	88	164	Acid

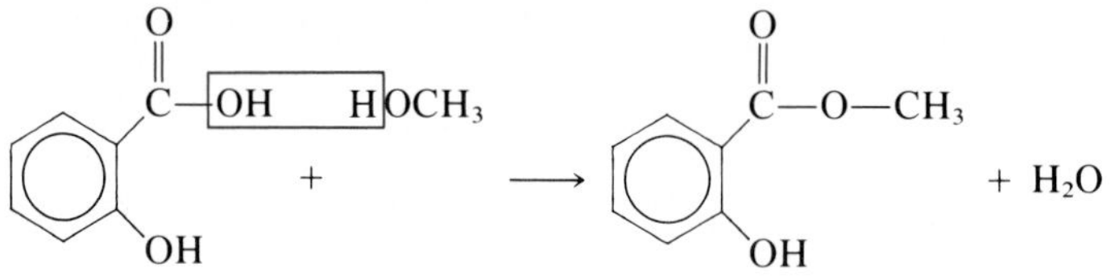

Salicylic acid — Methanol — Methyl salicylate

Aspirin, the common analgesic (pain reliever) and antipyretic (fever reducer), is also an ester of which the salicylic acid molecule provides a fragment. The esterification reaction by which aspirin is produced involves acetic acid and the phenol group of salicylic acid. Phenols react like alcohols in esterification.

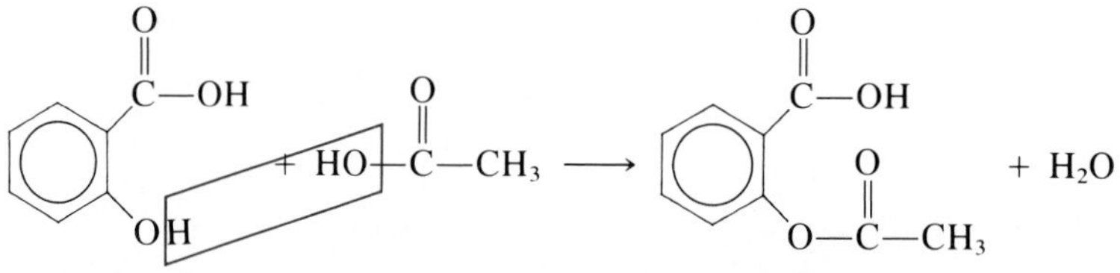

Salicylic acid — Acetic acid — Acetylsalicylic acid (aspirin)

Aspirin has both a carboxylic acid functional group from the salicylic acid and an ester group formed in the reaction shown. In practice, acetic anhydride would be used as a reactant rather than acetic acid. Acetic anhydride is more reactive. A molecule of acetic acid is split out as the condensation reaction occurs.

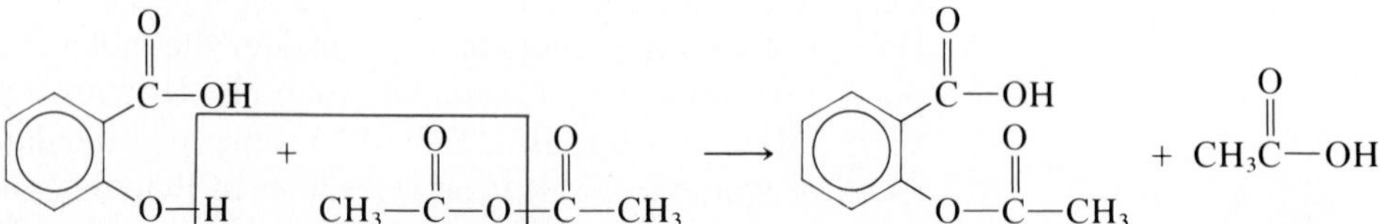

Salicylic acid — Acetic anhydride — Aspirin — Acetic acid

Aspirin is a white crystalline solid (mp = 135°C). Typically there is only 325 mg of aspirin in an aspirin tablet. The rest of the solid material is starch.

17.12 POLYESTERS

The familiar synthetic polyester fabrics, also called by trade names such as Dacron or Fortrel, are chemically exactly what the name *polyester* proclaims; they are polymeric esters. Monomers are joined by many ester linkages formed in condensation reactions. Difunctional compounds, usually diacids and dialcohols, are used to form polymers rather than the monofunctional compounds used to make simple esters. Condensation polymerization requires two reactive sites in each monomer to enable continued reactions and chain formation. Many polyester fibers are made from terephthalic acid, a dicarboxylic acid, and ethylene glycol, a diol:

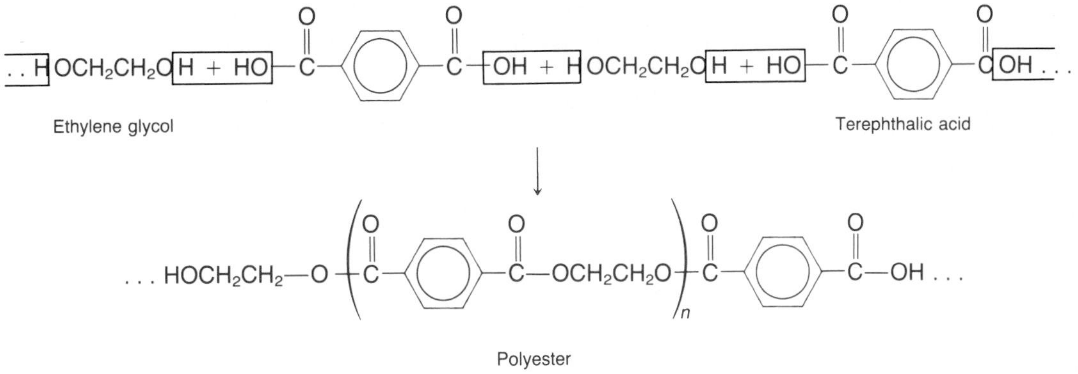
Ethylene glycol
Terephthalic acid
Polyester

Clearly the chain can be extended indefinitely because as each monomer is tacked on by condensation, it still has another functional group to react with another monomer unit. The polyester forms as a molten mass which is spun into fibers. Figure 17.2 shows the formation of synthetic fibers. The fibers are then woven into fabrics or formed into tire cord or otherwise applied. Figure 17.3 shows the surgical use of Dacron fabric as a replacement for a major blood vessel.

FIG. 17.2
Hot molten polymer is spun through a device called a spinneret and emerges as fibers. The spinneret is very much like a shower head, as the figure shows. The synthetic fibers can then be processed and woven as any natural fiber would be. (*Photo courtesy of E. I. du Pont de Nemours & Co.*)

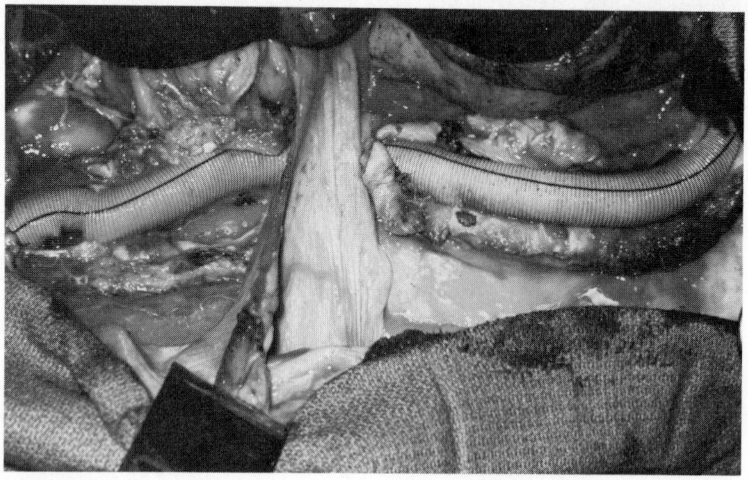

FIG. 17.3
Dacron fabric surgical graft. This photograph shows a patient (head to your right) with an extensive thoracic and abdominal incision to expose the descending thoracic aorta. This major blood vessel has been replaced with a woven polyester fabric known as the Cooley Meadow Low Porosity Aortic Graft. All of the visceral vessels in the abdomen have been reimplanted into the graft. *(Photo courtesy of Denton A. Cooley, M.D., Texas Heart Institute.)*

17.13 HYDROLYSIS OF ESTERS

The most important reaction of esters is their hydrolysis, which is exactly the reverse of the condensation reaction by which they are made. As you have been seeing, condensation involves "splitting water out" and thereby joining together two molecules. Hydrolysis involves "putting water in" and thereby breaking apart one molecule to form two.

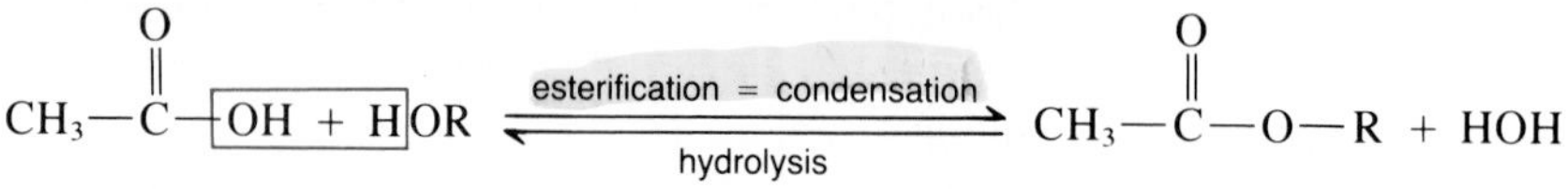
esterification = condensation
hydrolysis

Hydrolysis means breaking by water (*hydro*, "water"; *lysis*, "breaking").

The breakdown of an ester always occurs at the same bond. Water breaks the bond between the carbonyl carbon and the oxygen to which it is singly bonded:

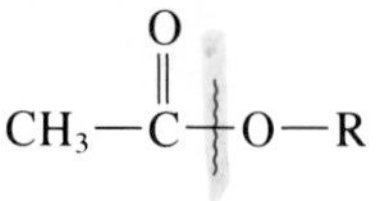

The hydrolysis product structures can be written by attaching the —OH of water to the carbonyl C and the —H of water to the oxygen:

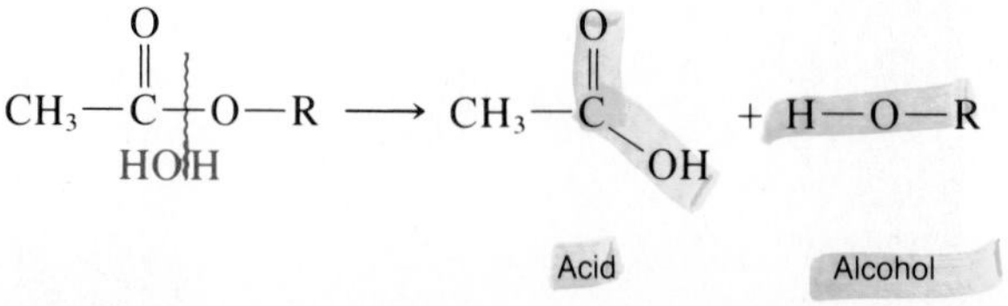
Acid
Alcohol

The products of ester hydrolysis are always the acid and alcohol from which the ester was constructed.

SAMPLE EXERCISE 17.4

Write equations for the hydrolysis reactions of the following esters:

a 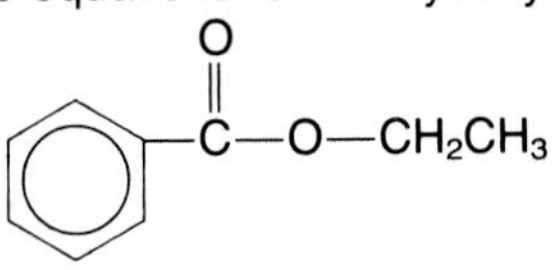 *b* 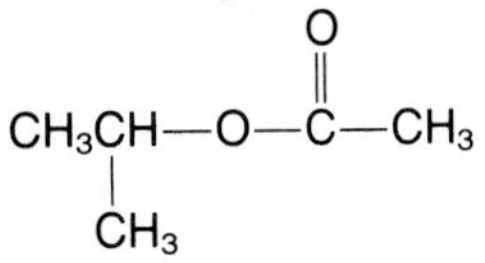

Solution The hydrolysis reaction is the reaction of an ester and H_2O to yield an acid and an alcohol. Follow a stepwise procedure, shown first for part *a*.

Guidelines for Ester Hydrolysis

a STEP 1 Write down the ester structure and slash it at the breaking point, i.e., between —C(=O)— and —O—:

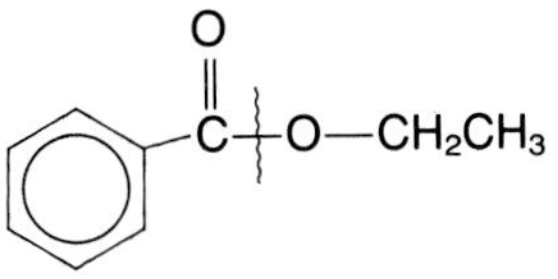

STEP 2 Attach —OH to the carbonyl to determine the acid product,

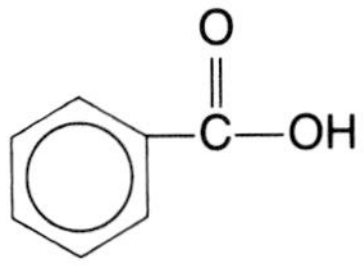

and attach H to the oxygen to determine the alcohol, H—O—CH_2CH_3.

STEP 3 Assemble in equation form:

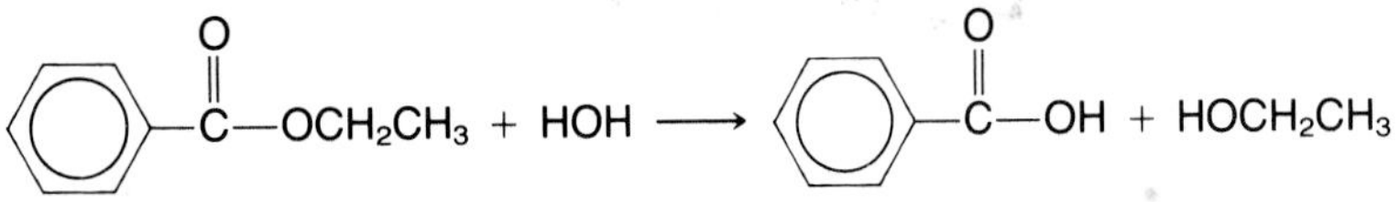

b STEP 1

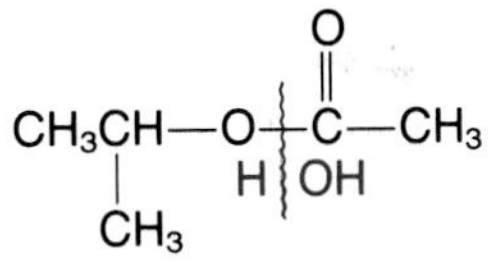

Don't be misled by various ways of writing esters. The ester linkage may be written —C—O—C(=O)— or —C(=O)—O—C—, that is, with —O—C— either to the left or to the right of the carbonyl. Hydrolytic cleavage is *always* between the carbonyl carbon and the oxygen of the ester linkage.

STEP 2

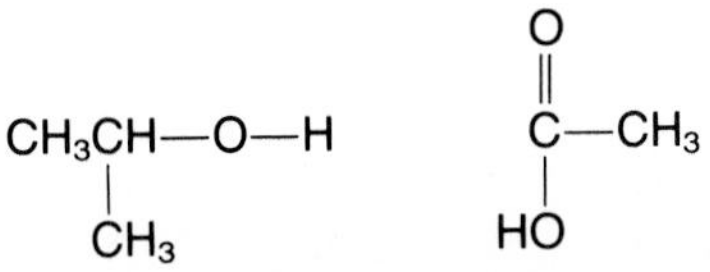

STEP 3

$$CH_3CH-O-\overset{\overset{\displaystyle O}{\|}}{C}-CH_3 + HOH \longrightarrow CH_3CH-OH + HO-\overset{\overset{\displaystyle O}{\|}}{C}-CH_3$$
$$\overset{|}{CH_3} \qquad\qquad\qquad\qquad \overset{|}{CH_3}$$

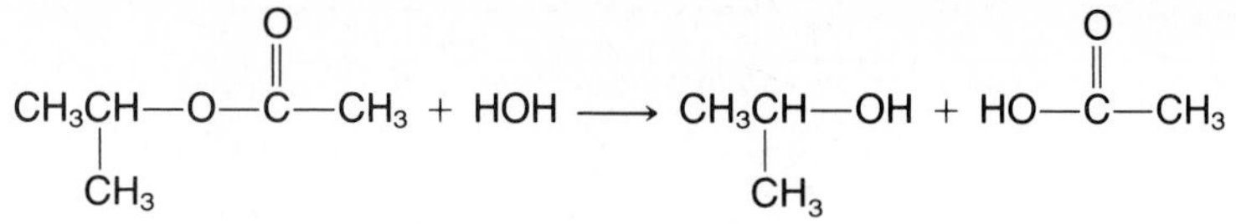

Alternatively, one can answer the question by applying the skill acquired in Sample Exercise 17.3. That is, ask yourself from what acid and alcohol the ester was made. These are the hydrolysis products.

Problem 17.3 Write an equation for the hydrolysis of $\langle\!\bigcirc\!\rangle-CH_2-O-\overset{\overset{\displaystyle O}{\|}}{C}-CH_2CH_2CH_3$. 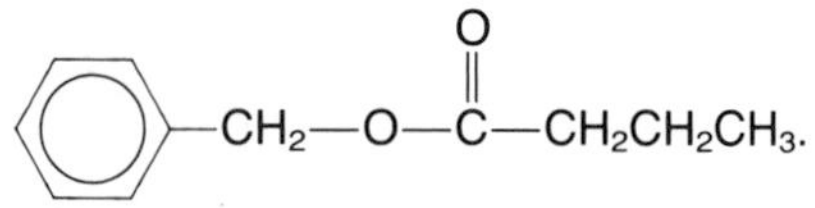

Hydrolysis occurs at best slowly if an ester and water are simply mixed together. Catalysis is required. Both acids and bases catalyze hydrolysis reactions. Acid catalysis explains how acids can "eat holes" in polyester clothing. Clearly water alone does not harm such fabrics; they are washable. But an acidic solution spilled on a polyester fabric can create a hole by hydrolysis of the ester linkages which hold the polymeric chains of the fibers together.

$$\ldots OCH_2CH_2O\!-\!\!\overset{\overset{\displaystyle O}{\|}}{C}\!-\!\langle\!\bigcirc\!\rangle\!-\!\overset{\overset{\displaystyle O}{\|}}{C}\!-\!OCH_2CH_2O\!-\!\overset{\overset{\displaystyle O}{\|}}{C}\!-\!\langle\!\bigcirc\!\rangle\!-\!\overset{\overset{\displaystyle O}{\|}}{C}\!-\!O\ldots$$

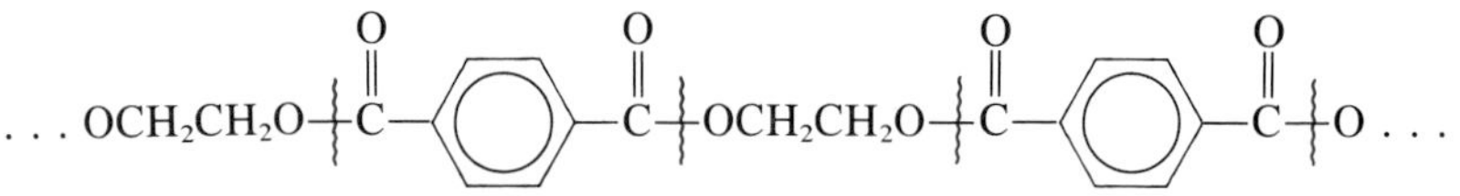

Hydrolysis of a polyester

The natural fiber cotton, which is a form of cellulose (Section 15.16), also suffers hydrolysis in acid solution.

17.14 SAPONIFICATION Saponification is hydrolysis done under conditions of basic catalysis. Because of the presence of a base and the reactivity between acids and bases, the products of basic hydrolysis of an ester are the expected alcohol and the *salt of the acid*.

$$R-\overset{\overset{\displaystyle O}{\|}}{C}-O-CH_3 + NaOH(aq) \longrightarrow R-\overset{\overset{\displaystyle O}{\|}}{C}O^-Na^+ + H-O-CH_3$$

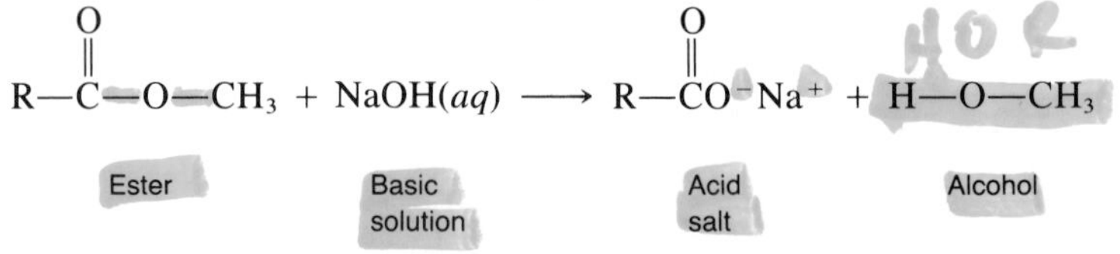

Ester   Basic solution   Acid salt   Alcohol

In basic solution, to predict the hydrolysis products, consider the products formed as water cleaves the ester. Then remember that the product acid reacts with the base to form a salt.

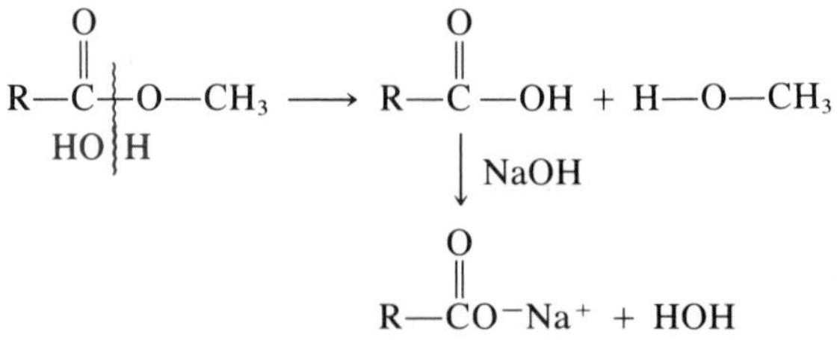

$$R-\overset{\overset{\displaystyle O}{\|}}{C}\!-\!O-CH_3 \longrightarrow R-\overset{\overset{\displaystyle O}{\|}}{C}-OH + H-O-CH_3$$
$$HO\,|\,H$$
$$\Big\downarrow NaOH$$
$$R-\overset{\overset{\displaystyle O}{\|}}{C}O^-Na^+ + HOH$$

Problem 17.4 What are the products of saponification of the ester given in Problem 17.3?

We'll see in Section 18.7 that the saponification of fats and oils, which are esters of long-chain fatty acids, produce soaps. Indeed, the words *soap* and *saponification* are derived from the same root.

17.15 AMIDE FORMATION

Amides are derivatives of carboxylic acids formed by condensation reactions between acids and amines. The amide formation reaction is very similar to esterification.

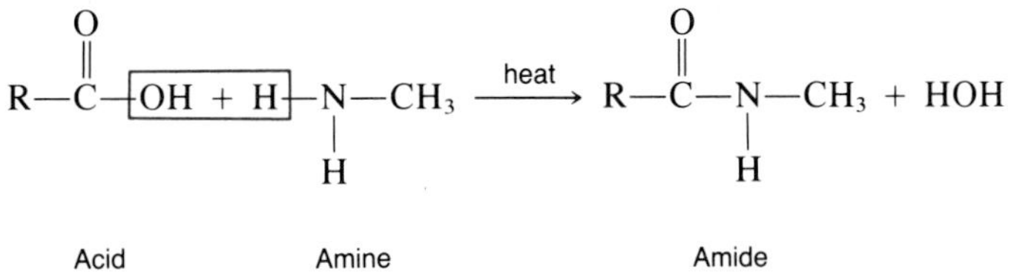

$$\underset{\text{Acid}}{R-\overset{\overset{\textstyle O}{\|}}{C}-\boxed{OH} + \boxed{H}} \underset{\text{Amine}}{-\overset{\textstyle }{\underset{\textstyle H}{N}}-CH_3} \xrightarrow{\text{heat}} \underset{\text{Amide}}{R-\overset{\overset{\textstyle O}{\|}}{C}-\overset{\textstyle }{\underset{\textstyle H}{N}}-CH_3} + HOH$$

Once again water is "boxed out," and the remaining pieces of the reactants are joined to make the amide. The actual course of the reaction is more complex than the equation shows. First, an amine salt,

$$R-\overset{\overset{\textstyle O}{\|}}{C}-O^- \overset{\overset{\textstyle H}{|}}{\underset{\textstyle H}{H-\overset{+}{N}CH_3}}$$

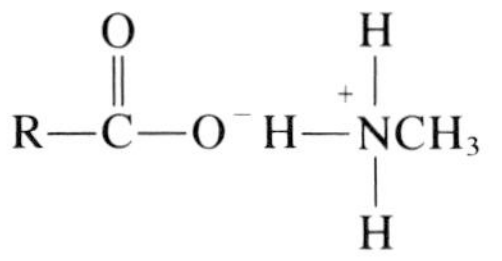

forms from the neutralization reaction of the acid and base. When the amine salt is heated above 100°C, it loses water and forms the amide.

A faster reaction and better yields of product are obtained by using the acid anhydride in place of the carboxylic acid to form the amide.

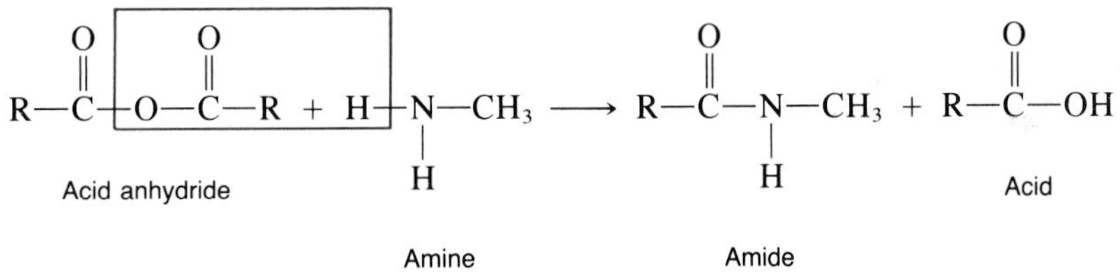

$$\underset{\text{Acid anhydride}}{R-\overset{\overset{\textstyle O}{\|}}{C}-\boxed{O-\overset{\overset{\textstyle O}{\|}}{C}-R} + \boxed{H}} \underset{\underset{\textstyle H}{\text{Amine}}}{-\overset{\textstyle }{\underset{\textstyle }{N}}-CH_3} \longrightarrow \underset{\underset{\textstyle H}{\text{Amide}}}{R-\overset{\overset{\textstyle O}{\|}}{C}-\overset{\textstyle }{N}-CH_3} + \underset{\text{Acid}}{R-\overset{\overset{\textstyle O}{\|}}{C}-OH}$$

In this condensation an acid molecule is split out from between the two reactants.

The acid anhydride, like a carboxylic acid, transfers group $R-\overset{\overset{\textstyle O}{\|}}{C}-$ to the amine nitrogen. The group $R-\overset{\overset{\textstyle O}{\|}}{C}-$ is called an acyl group, and an acid anhydride is a better acyl group transfer agent, or acylating agent, than a carboxylic acid. Another very effective acylating agent used in the laboratory is the acyl or acid halide. For example,

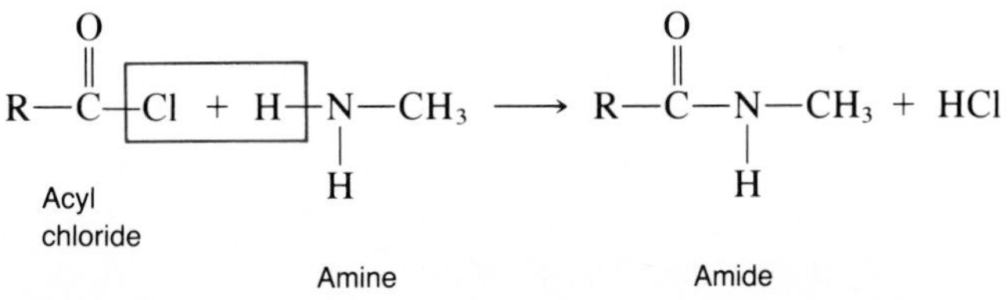

$$\underset{\substack{\text{Acyl}\\\text{chloride}}}{R-\overset{\overset{\textstyle O}{\|}}{C}-\boxed{Cl} + \boxed{H}} \underset{\underset{\textstyle H}{\text{Amine}}}{-\overset{\textstyle }{N}-CH_3} \longrightarrow \underset{\underset{\textstyle H}{\text{Amide}}}{R-\overset{\overset{\textstyle O}{\|}}{C}-\overset{\textstyle }{N}-CH_3} + HCl$$

In the body, the thioester, acetyl coenzyme A, $CH_3-\overset{O}{\overset{\|}{C}}-SCoA$, which was mentioned in Section 17.9, is an excellent and common transfer agent of the specific acyl group, $CH_3-\overset{O}{\overset{\|}{C}}-$, the acetyl group.

Amide structures have multiple opportunities to form hydrogen bonds.

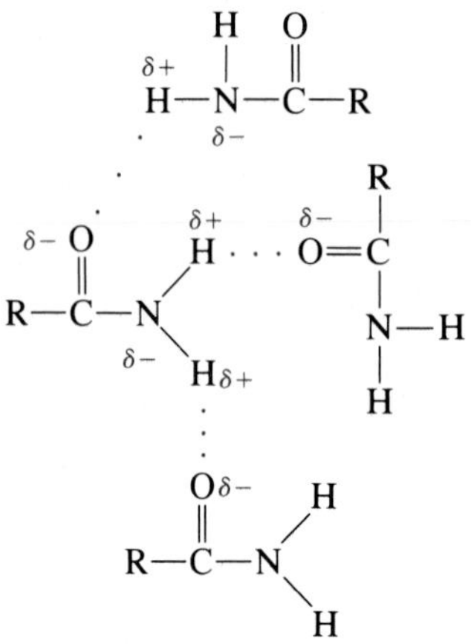

Consequently they have exceptionally high melting points and boiling points. For example, acetamide, $CH_3-\overset{O}{\overset{\|}{C}}-NH_2$ (MW = 59), melts at 81°C, whereas acetic acid, $CH_3-\overset{O}{\overset{\|}{C}}-O-H$ (MW = 60), is a liquid at room temperature (mp = 16°C). Amides are also water-soluble because of hydrogen-bonding possibilities.

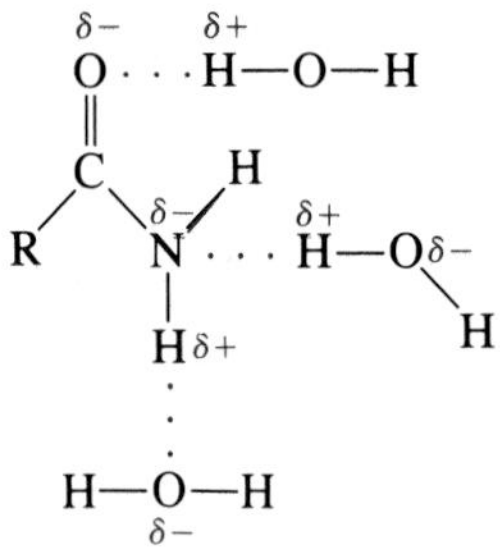

Table 17.5 shows the structures, names, and importance of some useful amides.

SAMPLE EXERCISE 17.5

Write a structure for the amide that forms when acetic acid reacts with methyl ethylamine.

Solution The structure of the product amide of any acid-plus-amine reaction can be determined by following the stepwise procedure:

STEP 1 Write down the structures of the acid and amine so that the —OH of the acid faces the H attached to the amine N:

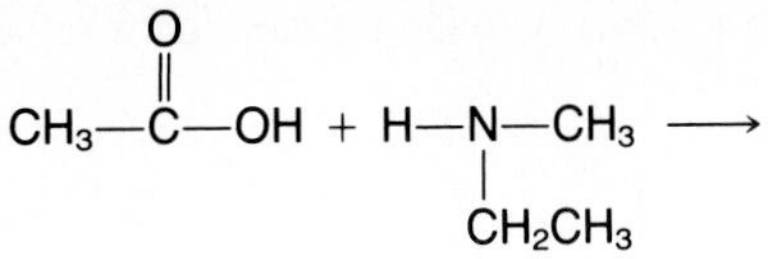

Acetic acid Methyl ethylamine

STEP 2 Box out the elements of water:

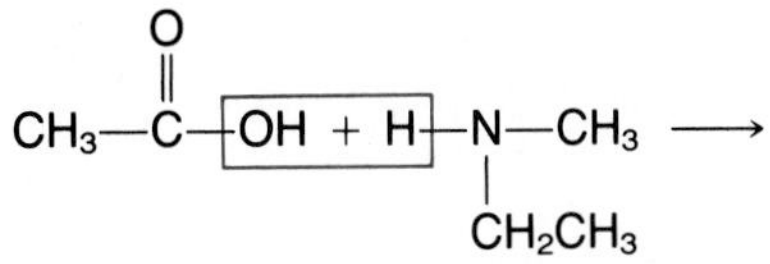

TABLE 17.5 Some Useful Amides

Structure	Name	Importance
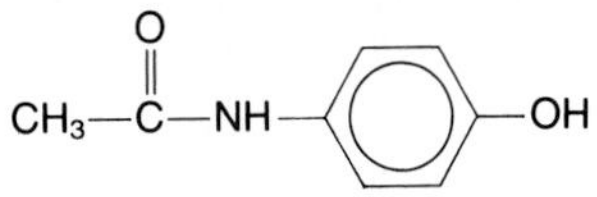	Acetaminophen (Tylenol)	Analgesic
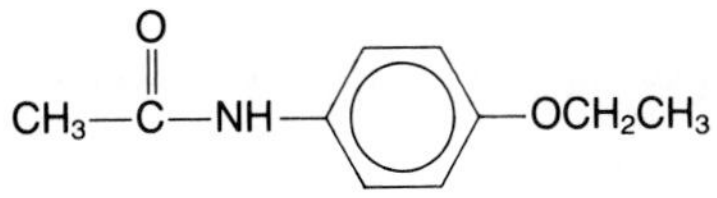	Phenacetin	Analgesic and antipyretic
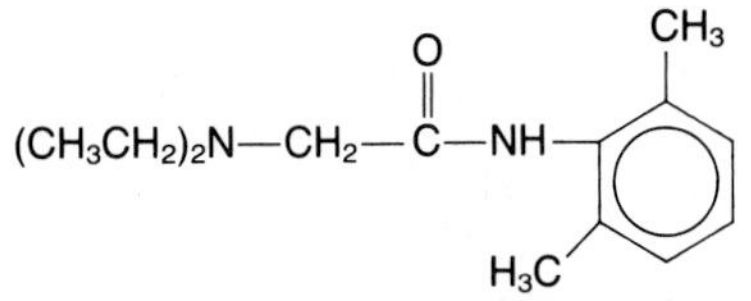	Xylocaine (Lidocaine)	Local anesthetic in sting and itch ointments
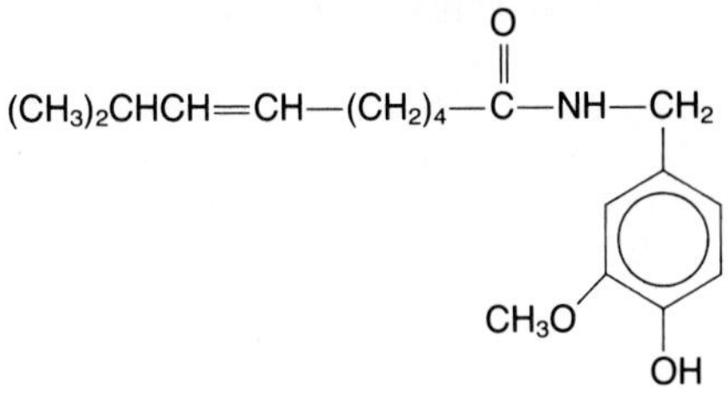	Capsaicin	Pungent spice in paprika and Tabasco sauce
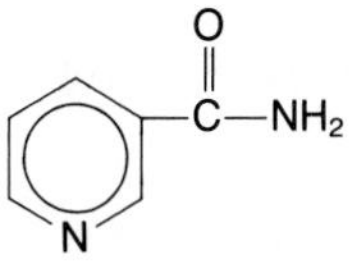	Nicotinamide (Niacinamide)	B vitamin (B_3)
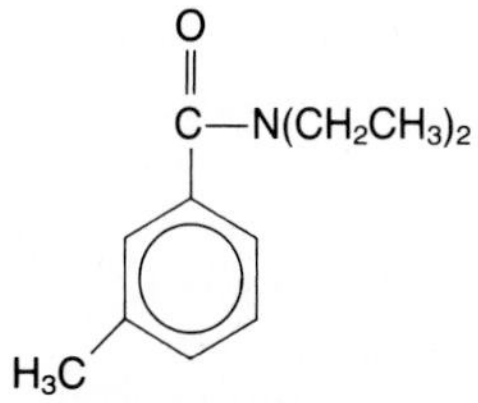	*N,N*-Diethyltoluamide	Mosquito repellent
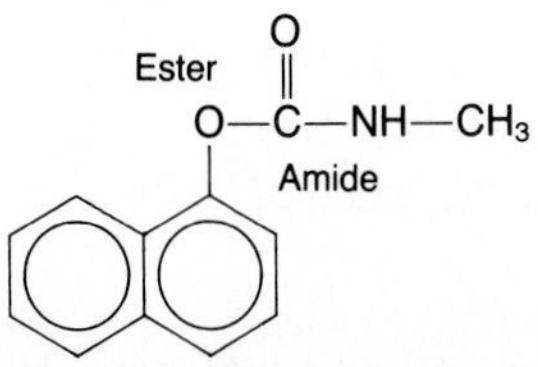 Ester Amide	Carbaryl (Sevin)	Insecticide

STEP 3   Attach the fragment left from the acid to the fragment of the amine to form the amide:

$$CH_3-\overset{\overset{\displaystyle O}{\|}}{C}\boxed{OH + H}-\overset{\overset{\displaystyle}{\underset{CH_2CH_3}{|}}}{N}-CH_3 \longrightarrow CH_3-\overset{\overset{\displaystyle O}{\|}}{C}-\overset{\overset{\displaystyle}{\underset{CH_2CH_3}{|}}}{N}-CH_3 + HOH$$

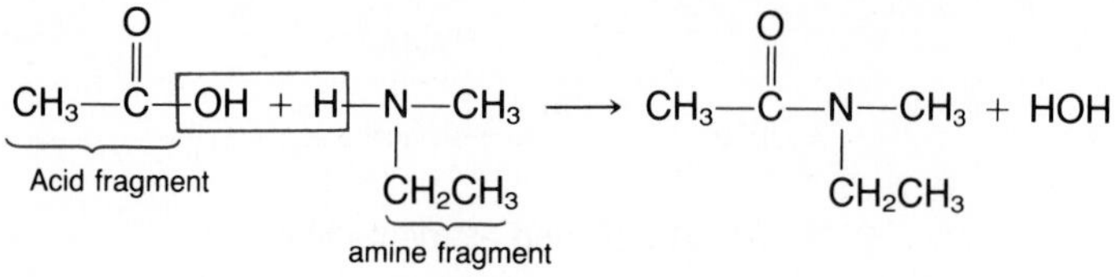

$$\underbrace{\qquad}_{\text{Acid fragment}} \qquad \underbrace{\qquad}_{\text{amine fragment}}$$

Alternatively, the product can be written by realizing that the acyl group, $R-\overset{\overset{\displaystyle O}{\|}}{C}-$, of an acid or acid derivative becomes attached to the N of the amine.

Notice that tertiary amines cannot form amides by reacting with carboxylic acids because tertiary amines lack an H attached to N to team up with the acid —OH to make water.

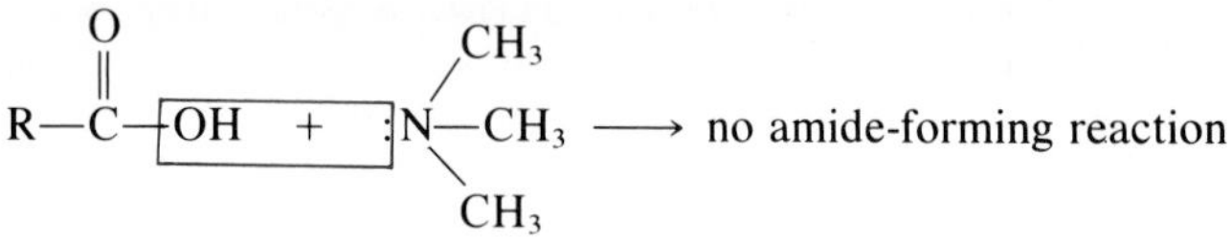

$$R-\overset{\overset{\displaystyle O}{\|}}{C}\boxed{OH} + :\overset{\overset{\displaystyle CH_3}{|}}{\underset{CH_3}{N}}-CH_3 \longrightarrow \text{ no amide-forming reaction}$$

But remember from Section 16.5 that all classes of amines, including tertiary amines, can react as bases with acids to form salts.

$$R-\overset{\overset{\displaystyle O}{\|}}{C}-O\textcircled{H} + :\overset{\overset{\displaystyle CH_3}{\diagup}}{\underset{CH_3}{N}}-CH_3 \xrightarrow[\text{formation}]{\text{salt}} R-\overset{\overset{\displaystyle O}{\|}}{C}O^-\overset{+}{H}\overset{\diagup CH_3}{\underset{\diagdown CH_3}{N}}-CH_3$$

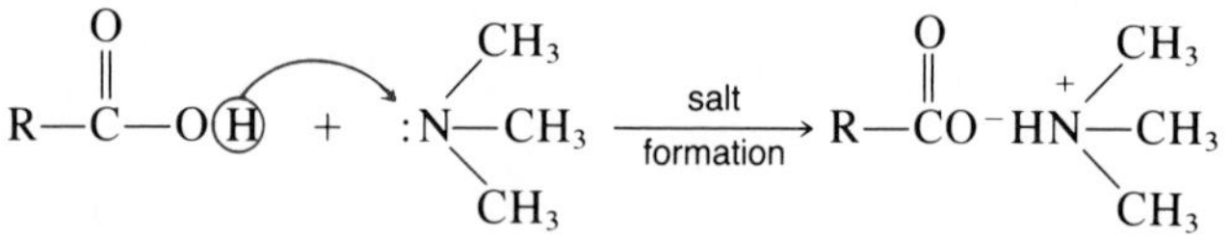

Section 17.4 discussed this type of salt formation.

17.16 POLYAMIDES   Another type of condensation polymer is the polyamide formed by the reaction of compounds with more than one acid or amine functional group. Nylons are polyamides formed by the reaction of diacids and diamines.

$$\ldots\boxed{H}-\overset{\overset{\displaystyle}{\underset{H}{|}}}{N}(CH_2)_6\overset{\overset{\displaystyle}{\underset{H}{|}}}{N}-\boxed{H} + \boxed{HO}\overset{\overset{\displaystyle O}{\|}}{C}(CH_2)_4\overset{\overset{\displaystyle O}{\|}}{C}\boxed{OH} + \boxed{H}\overset{\overset{\displaystyle}{\underset{H}{|}}}{N}(CH_2)_6\overset{\overset{\displaystyle}{\underset{H}{|}}}{N}-\boxed{H} + \boxed{HO}\overset{\overset{\displaystyle O}{\|}}{C}(CH_2)_4\overset{\overset{\displaystyle O}{\|}}{C}-\boxed{OH}\ldots$$

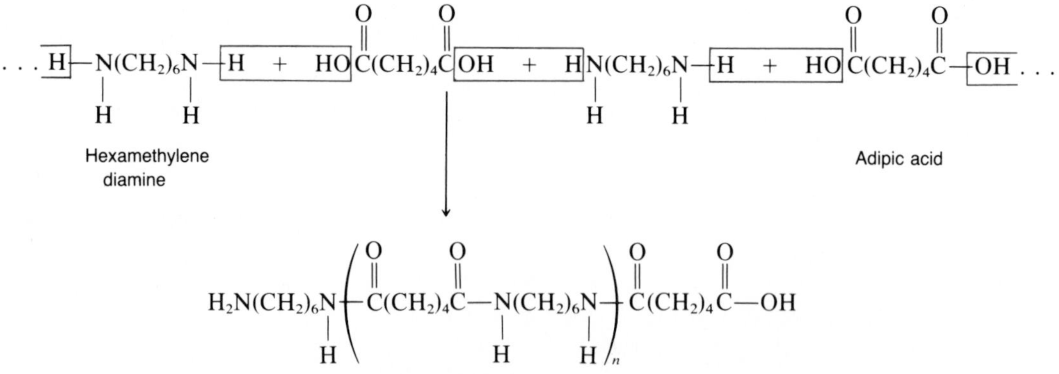

Hexamethylene diamine   Adipic acid

$$H_2N(CH_2)_6\overset{\overset{\displaystyle}{\underset{H}{|}}}{N}\left(\overset{\overset{\displaystyle O}{\|}}{C}(CH_2)_4\overset{\overset{\displaystyle O}{\|}}{C}-\overset{\overset{\displaystyle}{\underset{H}{|}}}{N}(CH_2)_6\overset{\overset{\displaystyle}{\underset{H}{|}}}{N}\right)_n\overset{\overset{\displaystyle O}{\|}}{C}(CH_2)_4\overset{\overset{\displaystyle O}{\|}}{C}-OH$$

Nylon-6,6

In the laboratory, diacid chlorides, such as adipoyl chloride,

$$Cl-\overset{\overset{\displaystyle O}{\|}}{C}(CH_2)_4\overset{\overset{\displaystyle O}{\|}}{C}-Cl,$$ are actually used rather than the diacids themselves

because acid chlorides are more reactive. The condensation idea is the same; HCl is split out.

Nylon-6,6 is so-called because both the amine and the acid have six carbon atoms. The first number codes for the diamine, the second for the acid. Thus nylon-6,10 would be constructed of hexamethylene diamine and the 10-carbon dicarboxylic acid, $HOOC(CH_2)_8COOH$, sebacic acid.

In Section 19.6 you will see the formation of polyamides through the condensation of amino acids, compounds with both the acid and amine functional groups. These polyamides are the biopolymers known as polypeptides or proteins.

17.17 HYDROLYSIS OF AMIDES

As was the case for esters, the most important reaction of amides is their hydrolysis, which is exactly the reverse of the condensation reaction by which they are made. The water of hydrolysis is "put in," and the amide is cleaved between the carbonyl carbon and the nitrogen. The hydrolysis product structures can be written by attaching the —OH of water to the carbonyl and the —H of water to the nitrogen:

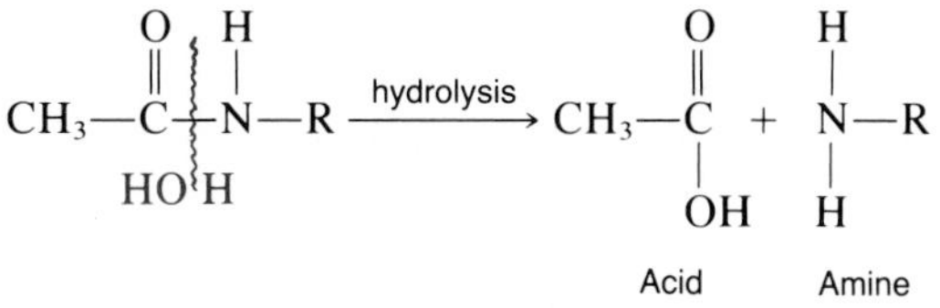
hydrolysis
Acid
Amine

The acidic and basic products will react to form a salt. The acid donates H^+ to the amine so that the actual product is

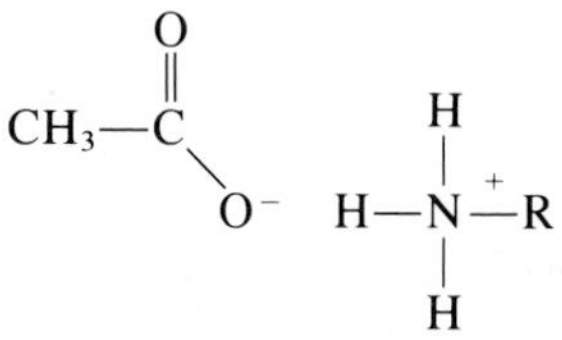

The products of amide hydrolysis are always the acid and amine from which the amide was constructed. As shown, these products form salts, but we will frequently ignore this salt-forming reaction. Like ester hydrolysis, amide hydrolysis is catalyzed by either acids or bases.

SAMPLE EXERCISE 17.6

Write an equation for the hydrolysis of $CH_3-CH(CH_3)-N(H)-C(=O)-CH_2CH_3$.

Solution The hydrolysis reaction is the reaction of amide + H_2O to yield an acid and an amine. Follow a stepwise procedure.

STEP 1 Write down the amide structure and slash it at the breaking point, i.e., between $-C(=O)-$ and $-N(H)-$:

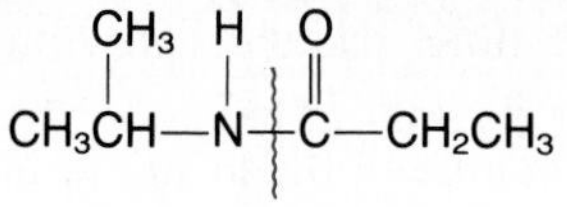

STEP 2 Attach the —OH to the carbonyl to determine the acid product,

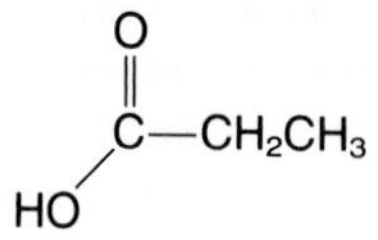

and attach H to the nitrogen to determine the amine,

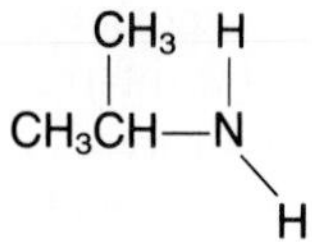

STEP 3 Assemble in equation form:

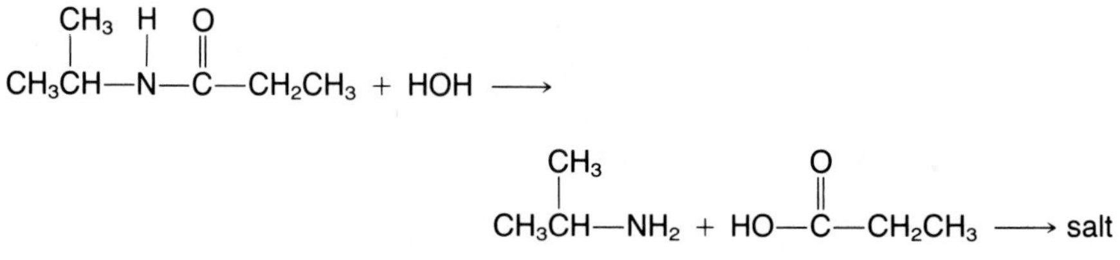

Amine Acid

Problem 17.5 Write an equation for the hydrolysis of

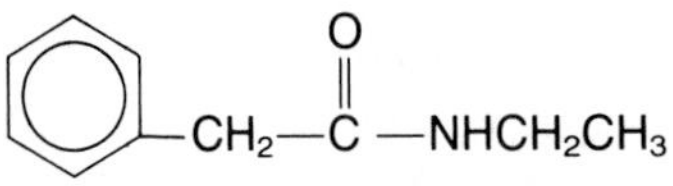

Protein digestion is simply the hydrolysis of the amide links between amino acids, as you will see in Section 19.16. This process occurs in the stomach, where the acid conditions are just right for the hydrolytic enzymes necessary for protein digestion.

17.18 UREA AND THE BARBITURATES

Urea is a unique amide in many ways.

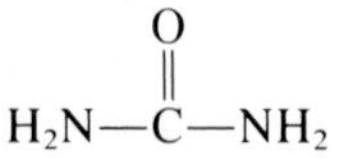

Urea

For example, unlike other amides which are essentially inert except for their hydrolysis reactions, urea undergoes condensation reactions. The products of the condensation of urea with derivatives of malonic acid are barbiturates. Before discussing barbiturates, let's briefy consider urea itself.

Urea is the vehicle by which the body disposes of nitrogen. Through a series of metabolic reactions, carbon dioxide and ammonia combine to form urea.

$$CO_2 + 2NH_3 \rightleftharpoons H_2N{-}\overset{\displaystyle O}{\overset{\|}{C}}{-}NH_2 + H_2O$$

Urea

Urea is a white solid with a melting point of 133°C. It is water-soluble and thus is excreted from the body as the water solution urine. You will learn more about nitrogen metabolism in Chapter 25.

The condensation reaction between urea and malonic acid proceeds as do other condensation reactions.

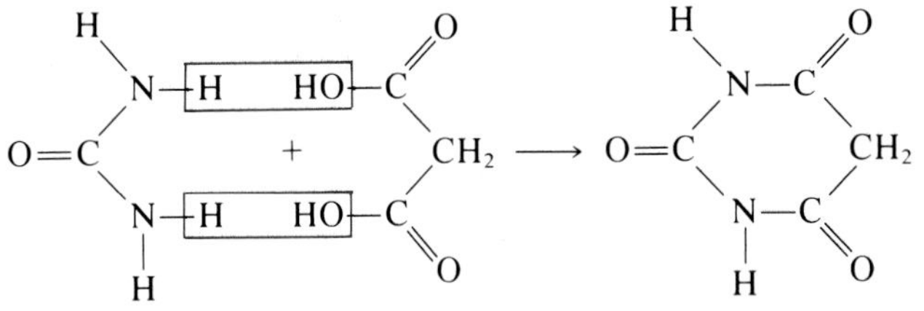

Urea Malonic acid Barbituric acid

The product, barbituric acid, is not physiologically active. However, when derivatives of malonic acid, in which one or both of the alpha hydrogens are replaced by alkyl (or aryl) groups, react with urea, the products are hypnotics and sedatives called barbiturates.

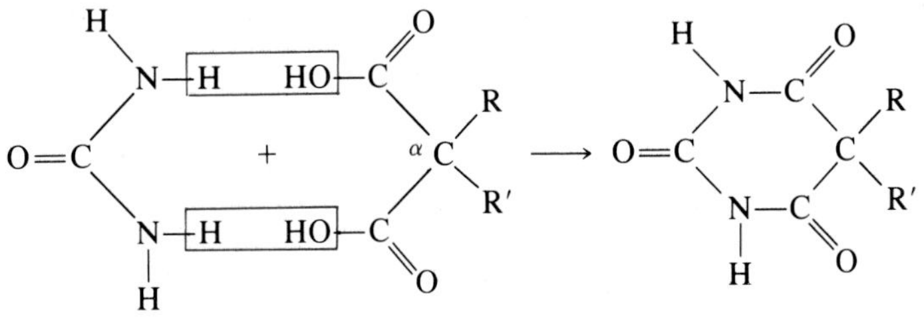

Urea Derivative of malonic acid A barbiturate

Table 17.6 shows the structures, trade names, and uses of some common barbiturates. These compounds are often used pharmaceutically in their salt form. A generalized salt structure is

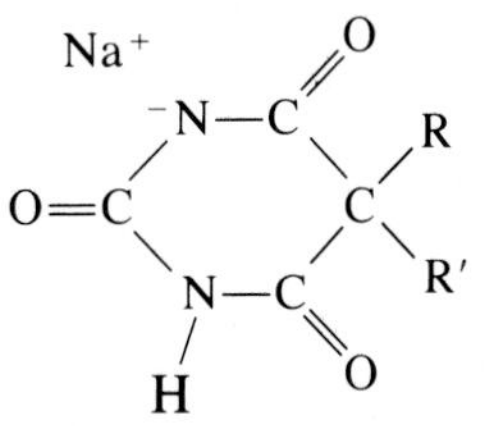

TABLE 17.6 Some Common Barbiturates

Structure	Name (Trade Name)	Medical Use
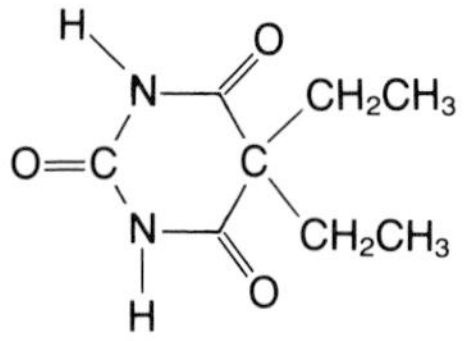	Barbital (Veronal)	Prolonged sedation (10 to 12 h); first synthesized active barbiturate
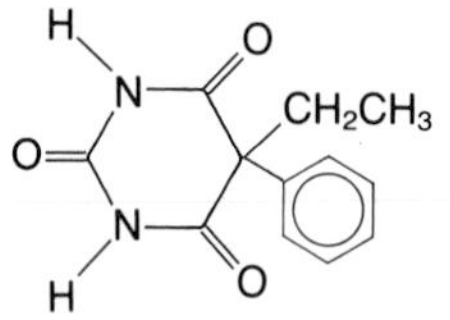	Phenobarbital (Luminal)	Anticonvulsant for epileptics, hypnotic, and sedative
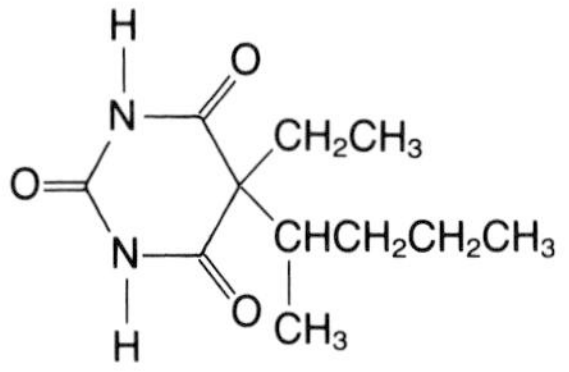	Pentobarbital (Nembutal)	Short-term preanesthetic sedative and hypnotic
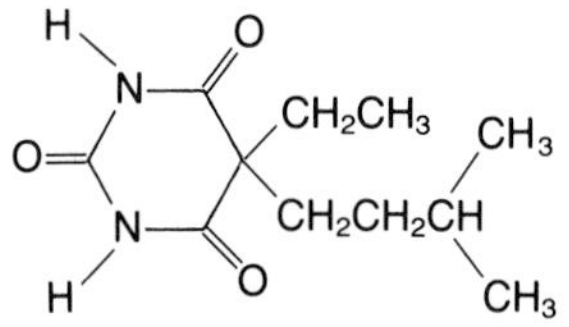	Amobarbital (Amytal)	Antianxiety drug and preanesthetic sedative
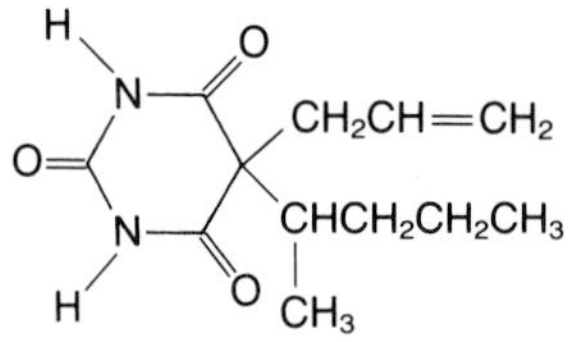	Secobarbital (Seconal)	Short-acting anticonvulsant, hypnotic, and sedative
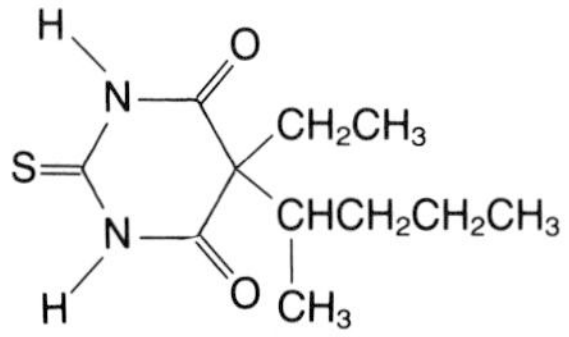	Thiopental* (Pentothal)	Hypnotic and short-term anesthetic; used in psychiatry to release inhibitions; so-called truth serum

* Thiourea, wherein S replaces O in urea, is used to make this compound.

CHAPTER ACCOMPLISHMENTS

After completing this chapter you should be able to define **key words** and do the following:

17.1 INTRODUCTION

1. Name the common functional groups that are considered to be derivatives of carboxylic acids.

17.2 NOMENCLATURE OF CARBOXYLIC ACIDS

2. Give an IUPAC name for a given structural formula of a carboxylic acid.
3. Employ the common names, formic and acetic for the one- and two-carbon carboxylic acids.
4. Label the α- and β-positions of a carboxylic acid or derivative.

17.3 PHYSICAL PROPERTIES OF CARBOXYLIC ACIDS

5. State the origin of the relatively high boiling and melting points of carboxylic acids and the water solubility of the small members of the class.
6. State the taste characteristic of acids.

17.4 ACIDITY AND SALT FORMATION

7. Write an equation for the ionization of a carboxylic acid.
8. Use Brønsted-Lowry theory to complete an equation for the reaction of a given carboxylic acid and base.

17.5 FATTY ACID SALTS AS SOAPS

9. Write a representative formula for a soap.
10. Explain how soap functions on a molecular level.

17.6 ESTERIFICATION

11. Given a reacting carboxylic acid and alcohol, write a structural formula for the product ester.

17.7 A CLOSE EXAMINATION OF THE ESTER FUNCTIONAL GROUP

12. Given the structural formula of an ester, identify the acid and alcohol from which it was made.

17.8 ESTERS FROM PHOSPHORIC ACIDS

13. Write structural formulas for phosphate esters, given the reacting alcohol.

17.9 THIOESTERS

14. Given a reacting acid and thiol, write a structural formula for the product thioester.

17.10 ANHYDRIDES OF CARBOXYLIC ACIDS

15. Recognize acid anhydride functional groups.

17.11 PROPERTIES OF ESTERS

16. Name the property of esters by which they are commonly identified.
17. Compare ester boiling points to those of ethers, alcohols, aldehydes, and acids of similar molecular weight.

17.12 POLYESTERS

18. Describe the chemical reaction by which polyesters are synthesized.

17.13 HYDROLYSIS OF ESTERS

19. Given an ester, write structural formulas for its hydrolysis products.

17.14 SAPONIFICATION

20. Distinguish between hydrolysis and saponification.
21. Given an ester, write structural formulas for its saponification products.

17.15 AMIDE FORMATION

22. Given a reacting carboxylic acid and a primary or secondary amine, write a structural formula for the amide product.
23. Explain the origin of the high melting points and water solubility of the amides.

17.16 POLYAMIDES

24. Describe the chemical reaction by which nylons are synthesized.

17.17 HYDROLYSIS OF AMIDES

25. Given an amide, write structural formulas for its hydrolysis products.

17.18 UREA AND THE BARBITURATES

26. Name the reactants which condense to form the barbiturates.

PROBLEMS

17.6 Name the functional group class for each of the following acid derivatives:

a 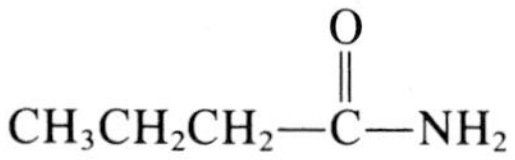

b 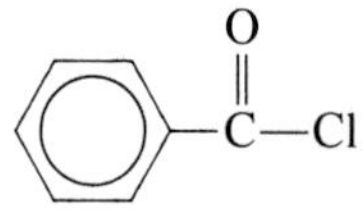

c 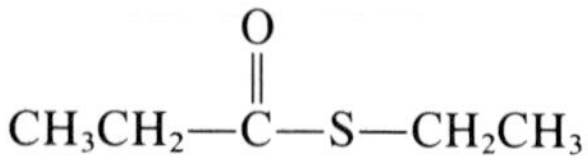

d 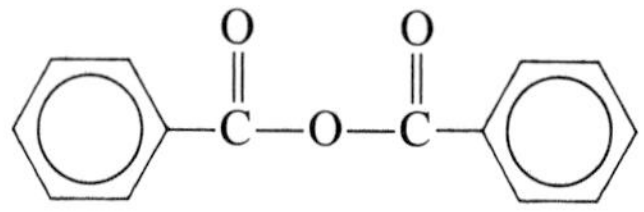

e 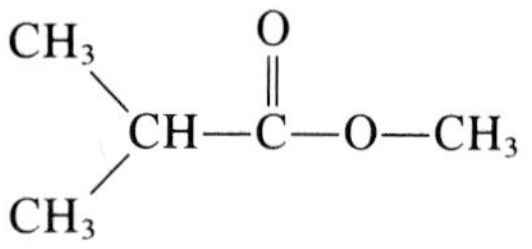

f 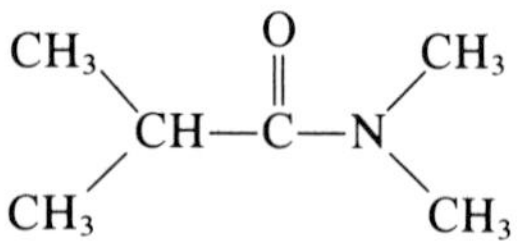

17.7 Write a structural formula for propanoic acid and then write structures for the following derivatives of this acid:

a Acid chloride
b Methyl ester
c An amide
d A thioester

17.8 Name the following acids by the IUPAC system:

a 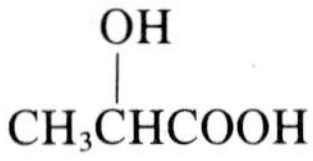 *d* 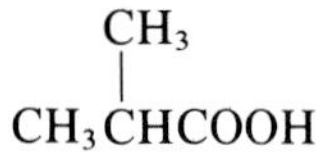

b 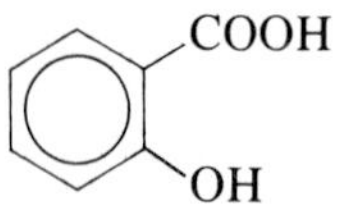 *e* 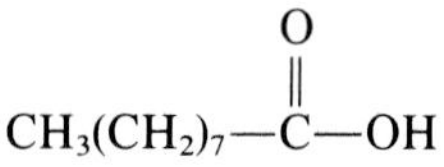

c 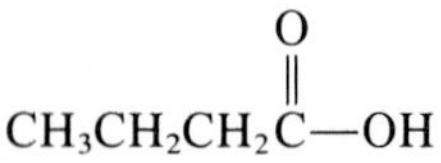

17.9 What are the common names for the structures in Problem 17.8 *a*, *b*, and *c*?

17.10 Write structural formulas for each of the following acids:

a Formic acid *c* Malonic acid
b Acetic acid *d* Pyruvic acid

17.11 The carbon chain of butanoic (butyric) acid could be correctly numbered:

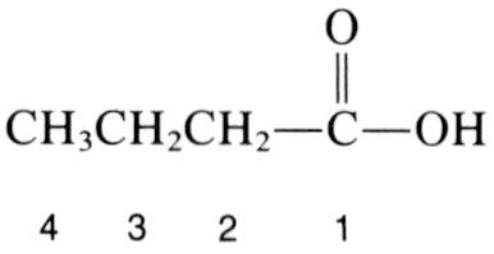

What is another common way of designating the 2 position and the 3 position?

17.12 Dimers of acetic acid form because of the excellent hydrogen-bonding opportunities between acid molecules. Diagram the H bonding.

17.13 Arrange the letters corresponding to the following compounds so that the arrangement reflects decreasing boiling point reading from left to right:

a CH_3CH_2OH *c* $HCOOH$
b CH_3COOH *d* $CH_3—O—CH_3$

17.14 Acetic acid is very water-soluble. What intermolecular interaction accounts for this?

17.15 What is the chemical origin of the tartness of many fruits and vegetables?

17.16 Write a chemical equation which shows the acid ionization of lactic acid.

17.17 Write a structure for the product of the reaction of acetic acid with each of the following reagents:

a NaOH *c* CH_3CH_2OH
b $(CH_3)_3N$: *d* KOH

17.18 Why are acid salts more soluble than the acids themselves?

17.19 The reaction between lignoceric acid and NaOH yields sodium lignocerate:

$$\underset{\text{Lignoceric acid}}{CH_3(CH_2)_{22}COOH} + NaOH \longrightarrow$$

a Write out the structure for sodium lignocerate.
b What household product might you find this material in?

17.20 Why do soaps form micelles in water?

17.21 Phosphoglycerides, which you will encounter in Chapter 18, can also act as soaps or emulsifiers. Explain why this is so, given the following representation for a phosphoglyceride:

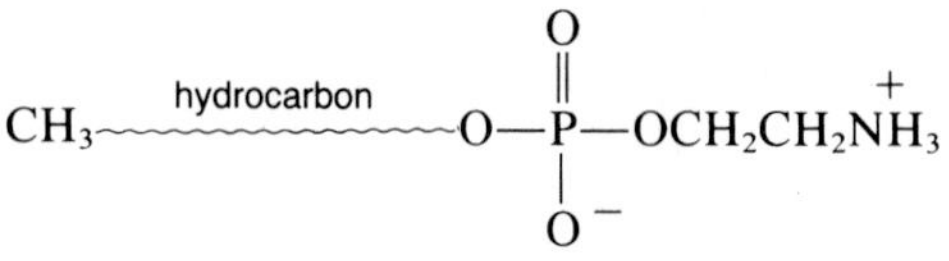
hydrocarbon

17.22 What is the difference between a soap and a detergent?

17.23 Why is it preferable to use a detergent in hard water?

17.24 Write structures for the esters that form when the following acids and alcohols react:
a Pentanoic acid and ethanol
b Acetic acid and 1-pentanol
c Formic acid and 2-propanol
d Propionic acid and methanol

17.25 Are there any isomeric esters produced in Problem 17.24? If so, identify them by their corresponding letters.

17.26 Show the products of the reactions of 1 mol of H_3PO_4 with 1, 2, and 3 mol of ethanol.

17.27 A key reaction of the body's energy transfer system is the reaction between adenosine diphosphate (ADP) and phosphoric acid (and derivatives) (P_i) Briefly,

ADP + P_i ⟶⟵ ATP (adenosine triphosphate)

In more structural detail:

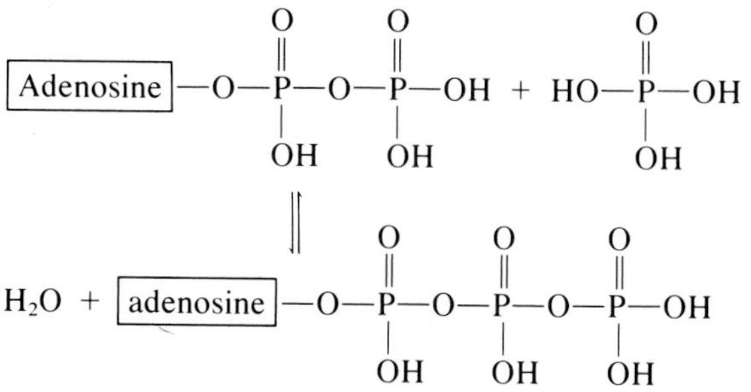

$$\boxed{\text{Adenosine}}-O-\overset{\overset{O}{\|}}{\underset{\underset{OH}{|}}{P}}-O-\overset{\overset{O}{\|}}{\underset{\underset{OH}{|}}{P}}-OH \;+\; HO-\overset{\overset{O}{\|}}{\underset{\underset{OH}{|}}{P}}-OH$$

$$H_2O \;+\; \boxed{\text{adenosine}}-O-\overset{\overset{O}{\|}}{\underset{\underset{OH}{|}}{P}}-O-\overset{\overset{O}{\|}}{\underset{\underset{OH}{|}}{P}}-O-\overset{\overset{O}{\|}}{\underset{\underset{OH}{|}}{P}}-OH$$

What kind of reaction is this?

17.28 Nitroglycerin is misnamed because it is actually an ester (triester) formed by the reaction of glycerol with 3 mol of nitric acid. Write a structural formula for glycerol trinitrate:

$$\begin{array}{c}CH_2OH\\|\\CHOH\\|\\CH_2OH\end{array} \;+\; 3\; HO-N\overset{O}{\underset{O}{\diagup}} \longrightarrow$$

Glycerol   Nitric acid

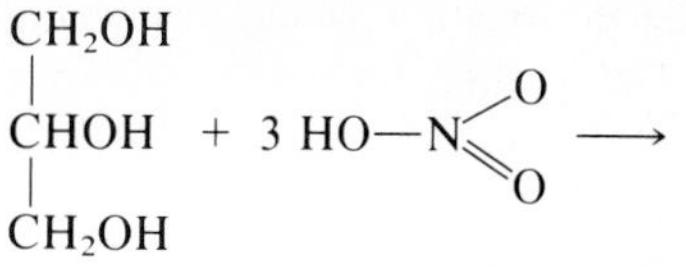

17.29 Thiols behave like alcohols in esterification reactions. Write structures for the thioesters that would form if the corresponding thiols were substituted for the named alcohols in Problem 17.24 (e.g., ethanethiol or CH_3CH_2SH for ethanol).

17.30 The flavors and fragrances of pineapple and oranges come from the following esters:

a $CH_3CH_2CH_2-\overset{\overset{O}{\|}}{C}-OCH_2CH_3$ b $CH_3-\overset{\overset{O}{\|}}{C}-O(CH_2)_7CH_3$

Pineapple   Orange

What acids and alcohols are needed to prepare these esters?

17.31 Label the acid fragment and the alcohol fragment in each of the following esters:
a CH_3COOCH_3 d $CH_3COOCH_2CH_2CH_3$

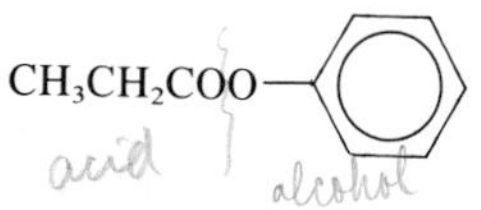

b $HCOOCH_3$

e $CH_3COOCH\overset{\overset{CH_3}{|}}{}CH_3$

c $CH_3CH_2COO-\bigcirc$

f $\bigcirc-COOCH_2CH_3$ 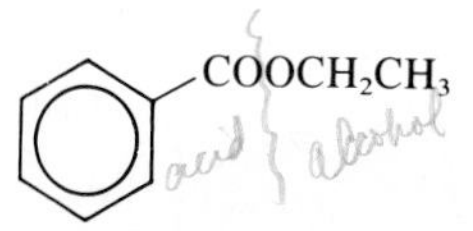

17.32 Wine connoisseurs sniff and enjoy the bouquet of wine. What kind of compound (what functional group) is most likely responsible for the aroma of wine?

17.33 Arrange the letters of the following compounds in order of increasing boiling point from left to

17.35 The essential metabolite acetyl CoA can be assembled by condensation reactions between an adenosine phosphate, pantothenic acid, thioethanolamine, and acetic acid. The order of assembly is reflected in the placement of reactants. Write the structure of acetyl CoA.

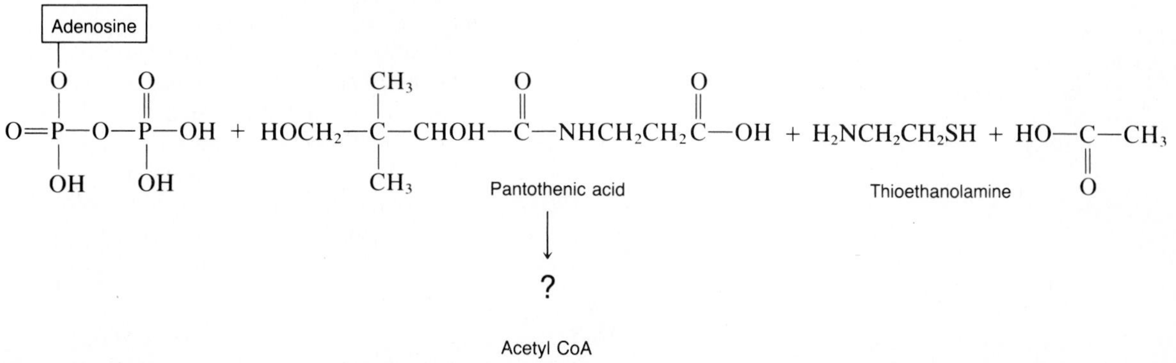

Acetyl CoA

right. The molecular weight of every compound is either 58 or 60.

a CH_3CH_2CHO
b CH_3COOH
c $CH_3CH_2CH_2CH_3$
d $CH_3CH_2CH_2OH$
e $H-\overset{\overset{\displaystyle O}{\|}}{C}OOCH_3$

17.34 Kodel is a polyester fiber used in many garments. It is made from the acid and alcohol shown. Show the repeating feature in Kodel polyester molecules.

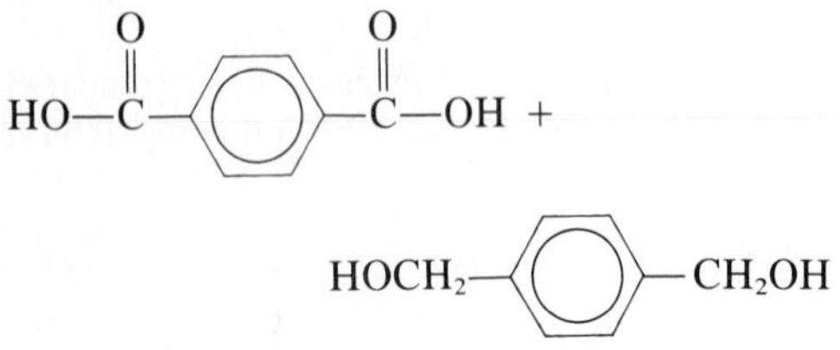

17.35 See bottom page 489.

17.36 What are the names of the reactions that are represented by each of the equilibrium arrows?

$$CH_3-\overset{\overset{\displaystyle O}{\|}}{C}-O-R + HOH \rightleftharpoons CH_3-\overset{\overset{\displaystyle O}{\|}}{C}-OH + HOR$$

17.37 Indicate the hydrolysis products of the following esters:

a 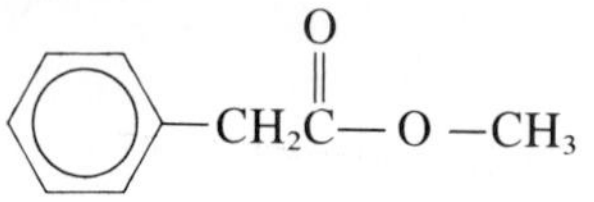

b $CH_3\underset{}{\overset{\overset{\displaystyle CH_3}{|}}{C}}HCH_2O-\overset{\overset{\displaystyle O}{\|}}{C}-CH_3$

c $CH_3CH_2O-\overset{\overset{\displaystyle O}{\|}}{\underset{\underset{\displaystyle O-CH_3}{|}}{P}}-OH$

17.38 Aspirin tablets often smell vinegary after standing for an extended period of time. Write a chemical equation to explain this occurence.

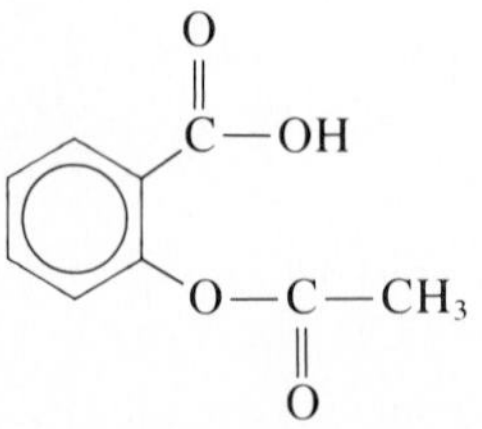

Aspirin

17.39 Distinguish between the terms *hydrolysis* and *saponification*. Use $CH_3CH_2COOCH_3$ as an example, and write an equation to illustrate each term.

17.40 Saponification is done in base. Simple hydrolysis is usually acid-catalyzed. Complete the following equations:

a $CH_3COOCH_3 + H_2O \xrightarrow{HCl}$

b 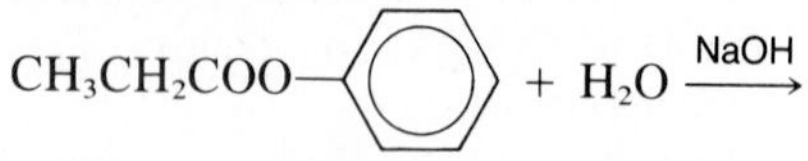

c 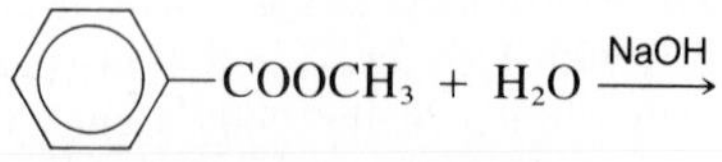

d $CH_3CH_2CH_2COOCH_2CH_2CH_3 + H_2O \xrightarrow{NaOH}$

e $HCOOCH_2CH_3 + H_2O \xrightarrow{KOH}$

f 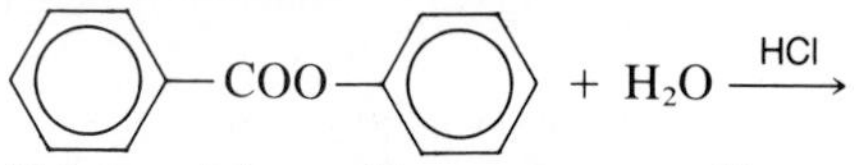

17.41 Tristearylglycerol is an ester you will encounter in Chapter 18. Write the products of the saponification of tristearylglycerol:

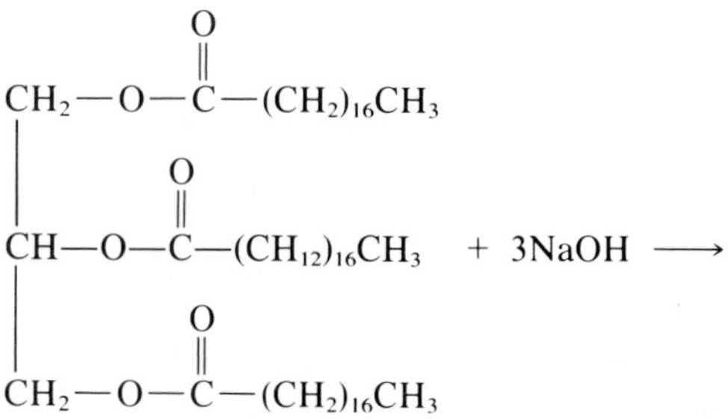

Tristearylglycerol

17.42 Of what practical use are the products of the reaction shown in Problem 17.41?

17.43 Write a structure for the product of the reaction of acetic acid with each of the following amines:
a $CH_3CH_2NH_2$
b $CH_3CH_2-NH-CH_3$
c $CH_3CH_2N(CH_3)_2$

17.44 Write structures for the amides that form when the following acids and amines react:
a Butanoic acid and ethyl amine
b Acetic acid and diethylamine
c Acetic acid and butyl amine
d Propionic acid and dimethylamine

17.45 Are any isomeric amides produced in Problem 17.44? If so, identify them by their corresponding letters.

17.46 Acetamide, $CH_3-\overset{\overset{\displaystyle O}{\|}}{C}-NH_2$, has a molecular weight of 59. Most organic compounds of this size are liquids at room temperature. Acetamide is a solid (mp = 81°C), and furthermore it is more soluble in water than sodium chloride. What phenomenon explains the properties of acetamide?

17.47 Aromatic polyamide fibers such as Kevlar are stronger than steel. A portion of Kevlar polymer can be represented:

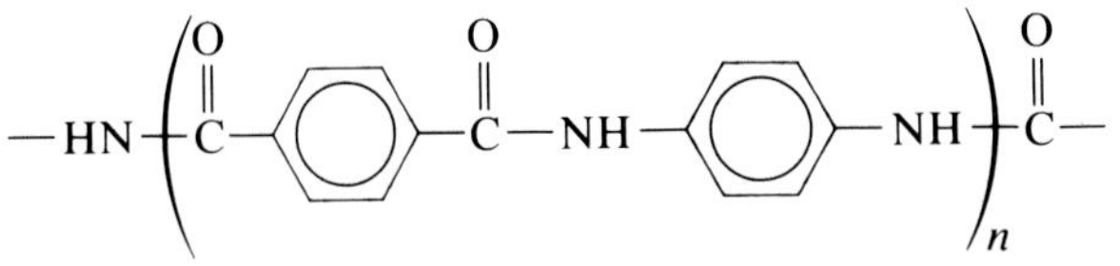

What diacid and diamine condense to form Kevlar?

17.48 Complete the following equations:

a $CH_3-\overset{\overset{\displaystyle O}{\|}}{C}-NHCH_3 + H_2O \longrightarrow$

b

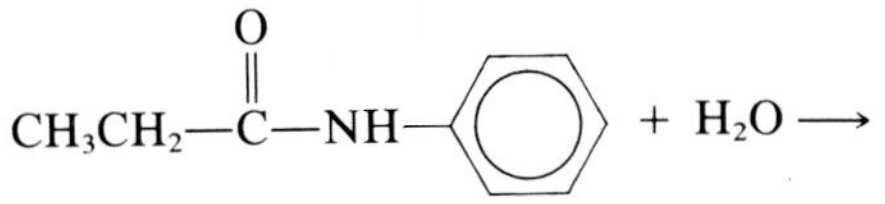

c

d $CH_3CH_2CH_2\overset{\overset{\displaystyle O}{\|}}{C}\underset{\underset{\displaystyle CH_3}{|}}{N}CH_2CH_3 + H_2O \longrightarrow$

e $H-\overset{\overset{\displaystyle O}{\|}}{C}-\underset{\underset{\displaystyle CH_3}{|}}{N}-CH_3 + H_2O \longrightarrow$

f

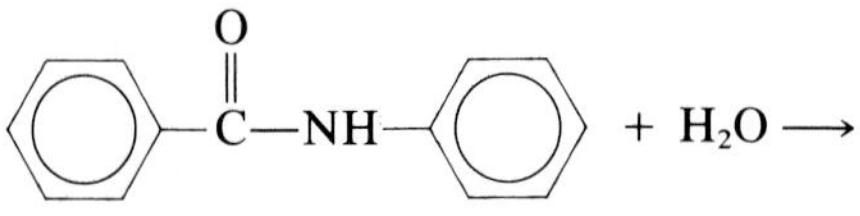

17.49 Hydrolysis of phenobarbital occurs along the jagged line drawn. Name the hydrolysis products of this barbiturate:

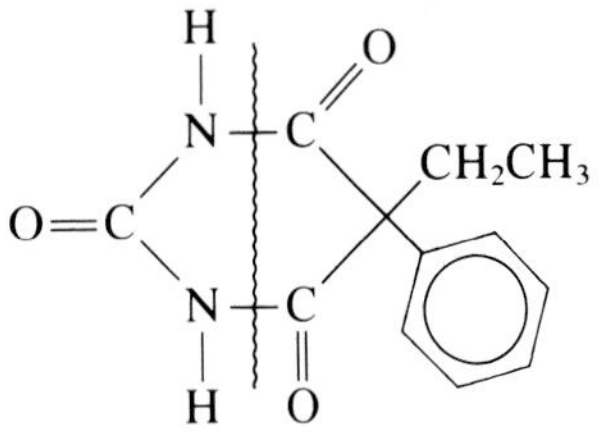

Phenobarbital

17.50 Tell the type of reaction that is occurring and write the product of each of the following:

a

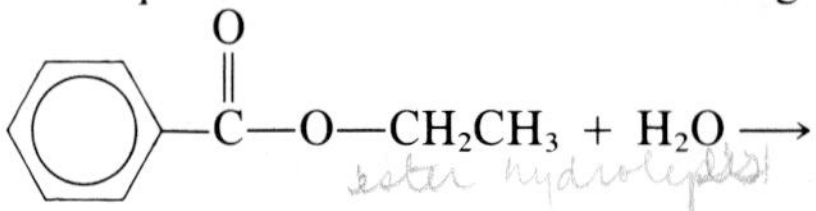

ester hydrolysis

b

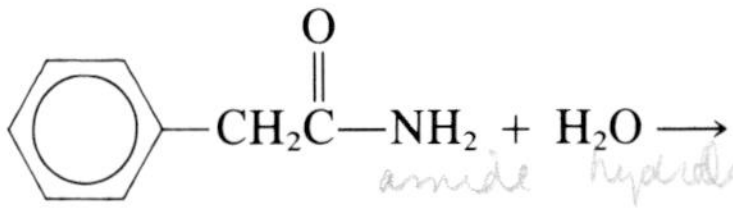

amide hydrolysis

c

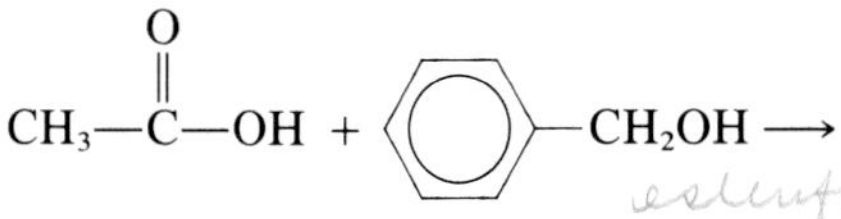

esterification

d

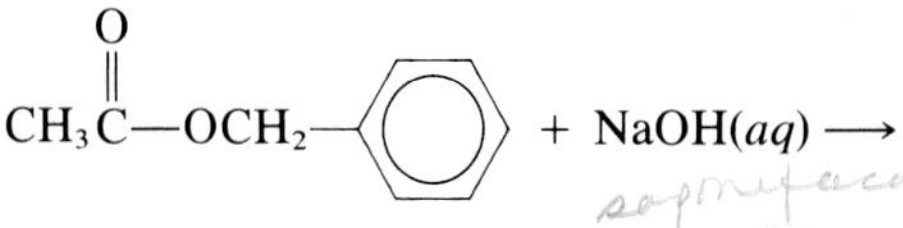

saponification

e

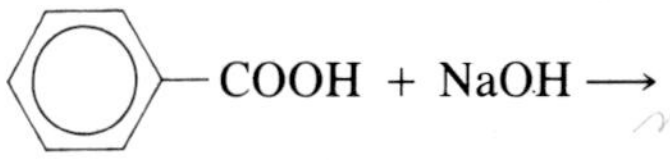

neutralization

f

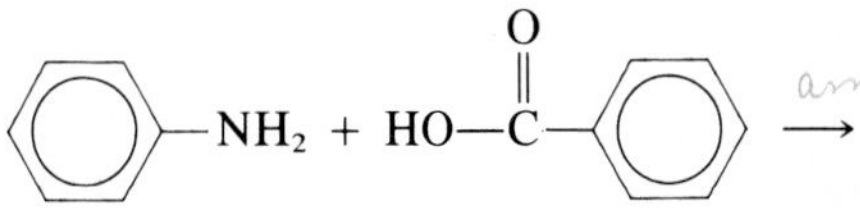

amide formation

17.51 Why do acid solutions make holes in nylon stockings?

CHAPTER EIGHTEEN

LIPIDS

18.1 INTRODUCTION

There are three major classes of biological compounds; that is, there are three components of the *biomass,* the mass of living matter. These three are the carbohydrates, lipids, and proteins. In Chapter 15 you learned the definition or description of carbohydrates as polyhydroxy aldehydes or ketones. You will see in Chapter 19 that proteins are polyamides formed from linking together amino acids.

Lipids, in contrast, cannot be defined structurally. The class called lipids contains a "mixed bag" of compounds with a variety of functional groups and structural features. What the members of the class have in common are their solubility properties. Lipids are not soluble in water; they are hydrophobic. All lipids have extensive hydrocarbon parts in their structures. Lipids are soluble in ether, chloroform, and benzene, that is, in nonpolar solvents.

Their solubility in nonpolar solvents such as ether allows us to separate lipids from carbohydrates and proteins in a sample of biomass: Polar carbohydrates and proteins do not dissolve in nonpolar ether. All the compounds that do dissolve in the nonpolar ether are members of the lipid class.

There is a scheme for classifying the various types of compounds that fall into the lipid category. Many lipids have ester functional groups and are therefore susceptible to saponification, a reaction of esters (Section 17.14). The lipid class can be divided into saponifiable lipids (esters) and nonsaponifiable lipids. The ester, or saponifiable, class is further divided into simple or compound lipids based on the complexity of the ester and the products of their hydrolysis or saponification. Simple lipids yield only alcohols and salts of carboxylic acids upon saponification; compound lipids yield other components, such as phosphoric acid and amines. Figure 18.1 summarizes the lipid classification scheme. As each class is discussed individually, its place in the scheme will be clarified.

Figure 18.2 shows some structures and molecular models of lipids. You can spot the ester linkages in some of them and note that they all have extensive hydrophobic hydrocarbon portions. The figure also points out where you encounter lipid materials: in dietary fats and oils, in vitamins and steroidal drugs, in gallstones, and as cell membranes. Each type of lipid will be discussed in turn.

18.2 WAXES

Simple lipids are esters of alcohols and carboxylic acids. The simplest of these are the waxes, which are mixtures of esters of long-chain, carboxylic acids called fatty acids and long-chain alcohols. Carnauba wax, found in

LIPIDS

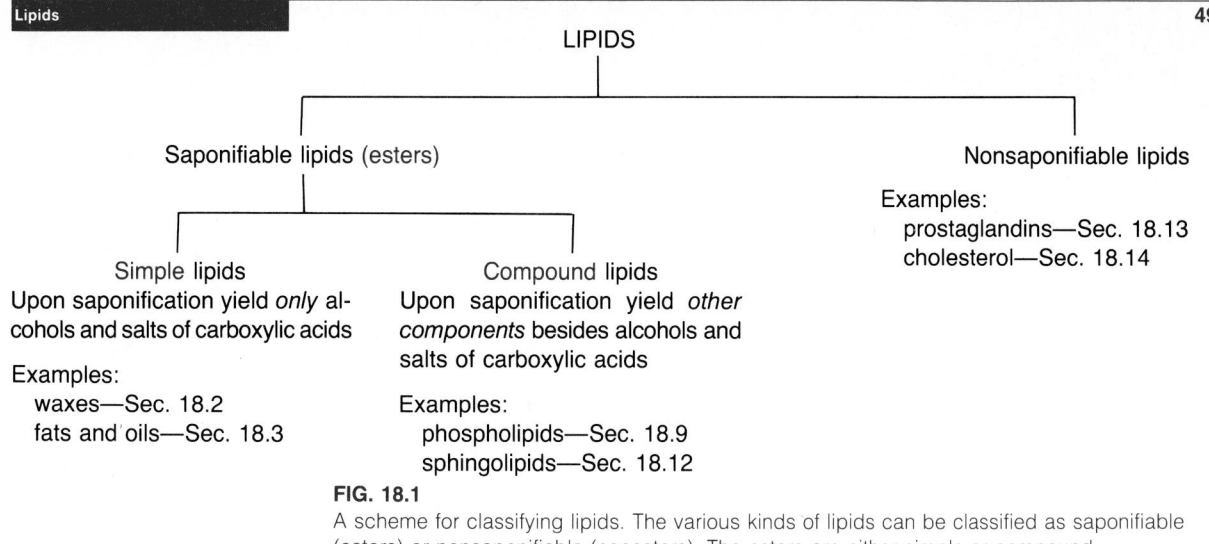

- Saponifiable lipids (esters)
  - Simple lipids
    Upon saponification yield *only* alcohols and salts of carboxylic acids
    Examples:
    waxes—Sec. 18.2
    fats and oils—Sec. 18.3
  - Compound lipids
    Upon saponification yield *other components* besides alcohols and salts of carboxylic acids
    Examples:
    phospholipids—Sec. 18.9
    sphingolipids—Sec. 18.12
- Nonsaponifiable lipids
  Examples:
  prostaglandins—Sec. 18.13
  cholesterol—Sec. 18.14

FIG. 18.1
A scheme for classifying lipids. The various kinds of lipids can be classified as saponifiable (esters) or nonsaponifiable (nonesters). The esters are either simple or compound.

automobile and floor polishes, is an example. The composition of carnauba wax is such that 85 percent of the material can be represented by the formula

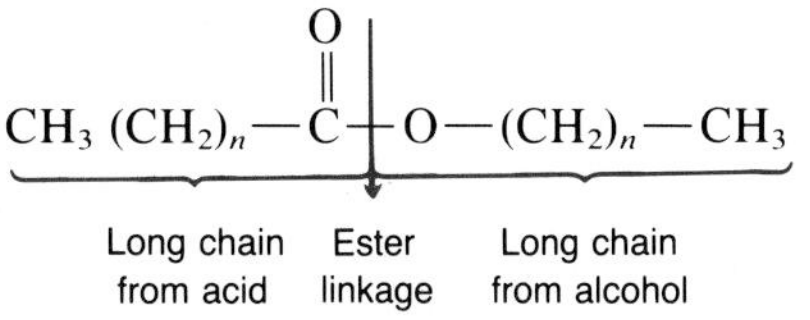
Long chain from acid
Ester linkage
Long chain from alcohol

$n = 17\text{–}33$

That is, 85 percent of the mixture called Carnauba wax is esters of varying chain length specified by the varying value of n. You have seen that the general representation for an ester is $R{-}\overset{\overset{\displaystyle O}{\|}}{C}{-}O{-}R'$. If R and R′ are sufficiently long (about 15 carbons), the ester will acquire the physical properties associated with waxes.

Actually the mixtures known as waxes contain other components besides esters. Typically, free acids and alcohols are present as well as some hydrocarbon material. Some synthetic waxes are essentially all hydrocarbon. The term *wax* is being redefined in terms of the properties we associate with waxes, i.e., malleable solids which melt at only moderately elevated temperatures. Natural waxes are found in insects and whales and on the surfaces of almost all plants. They are protective, water-resistant coatings.

The ester waxes are very important cosmetically and medicinally because they have been designated safe for external application by the Food and Drug Administration and can even be ingested in small amounts. For example, carnauba wax is used to polish candies and pills. Table 18.1 cites the major ester components, the source, and the uses of some common waxes.

18.3 FATS AND OILS

The most prevalent simple lipid compounds are triesters of the triol, glycerol, and fatty acids. **Fatty acids** are carboxylic acids (RCOOH) in which R is a long hydrocarbon chain. These triesters are called variously triglycerides, triacylglycerols, and fats or oils. As an example, consider

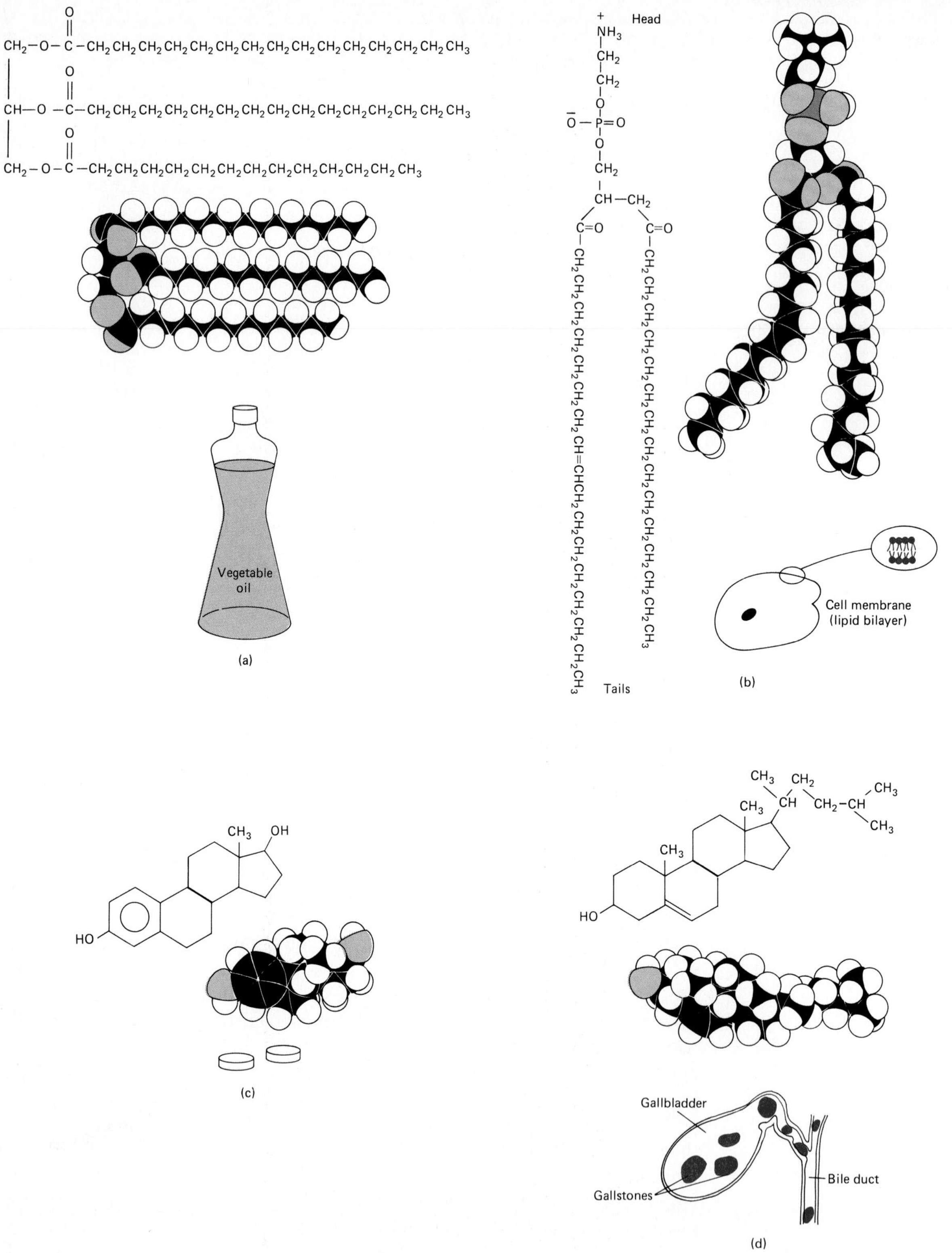
Vegetable oil
(a)
Head
Tails
Cell membrane (lipid bilayer)
(b)
(c)
Gallbladder
Gallstones
Bile duct
(d)

FIG. 18.2
There are a variety of structures among lipids. They have in common extensive hydrocarbon parts that are hydrophobic. Many also have polar, hydrophilic regions. (*a*) A triacylglycerol: cooking fats and oils have this type of structure. (*b*) A phosphoglyceride: cell membranes are constructed of phospholipids. (*c*) Steroids: sex hormones such as 17-β-estradiol are steroids. Some are given medicinally. (*d*) Cholesterol is the most prevalent steroid. Gallstones are mostly cholesterol.

TABLE 18.1 Common Waxes

Wax	Structure of a Principal Ester Component	Source	Uses
Beeswax	$CH_3(CH_2)_{14}{-}\overset{\overset{O}{\|}}{C}{-}O{-}(CH_2)_{29}CH_3$ Myricyl palmitate	Bees	Candles, cosmetics, confections, medicinals, art preservation.
Carnauba wax	$(HO)_aCH_2(CH_2)_b{-}\overset{\overset{O}{\|}}{C}{-}O{-}(CH_2)_cCH_3$ a = 0 or 1 b = 17–29 c = 31 or 33	Brazilian palm trees	Coatings for perishable products; polishing candies and pills; auto and floor waxes.
Spermaceti	$CH_3(CH_2)_{14}{-}\overset{\overset{O}{\|}}{C}{-}O{-}(CH_2)_{15}CH_3$ Cetyl palmitate	Sperm whales	Use in products such as cosmetics, soap, and candles banned in 1976; sperm whales are protected species.
Rice bran wax	$CH_3(CH_2)_m{-}\overset{\overset{O}{\|}}{C}{-}O{-}(CH_2)_n\,CH_3$ m = 20 or 22 n = 21–35	Rice bran	Lipstick base, plastics processing aid.
Bayberry wax	$CH_3(CH_2)_x{-}\overset{\overset{O}{\|}}{C}{-}O{-}(CH_2)_yCH_3$ x = 12–16	Berries of myrtle shrubs	Candles.
Jojoba oil*	$CH_3(CH_2)_7CH{=}CH{-}(CH_2)_n{-}\overset{\overset{O}{\|}}{C}{-}O{-}CH_2(CH_2)_nCH{=}CH(CH_2)_7CH_3$ n = 9 or 11	Jojoba beans	Has replaced sperm whale products.
Jojoba wax	$CH_3(CH_2)_a{-}\overset{\overset{O}{\|}}{C}{-}O{-}CH_2(CH_2)_a{-}CH_3$ a = 18 or 20	Jojoba beans	Cosmetics and candles.

* Section 18.4 shows how jojoba oil can be readily converted to jojoba wax.

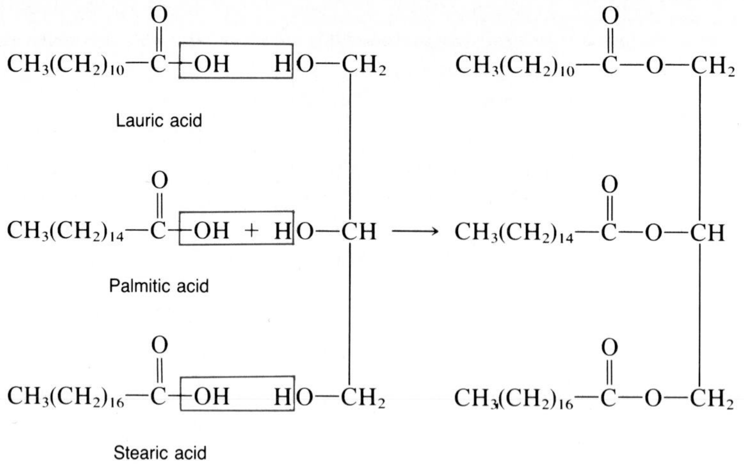
Lauric acid
Palmitic acid
Stearic acid

Fatty acids + Glycerol ⟶ Triacylglycerol (fat)

Notice that the esterification reaction proceeds exactly as we have seen before (Section 17.6). Water is boxed out, and the acid and alcohol fragments are joined together.

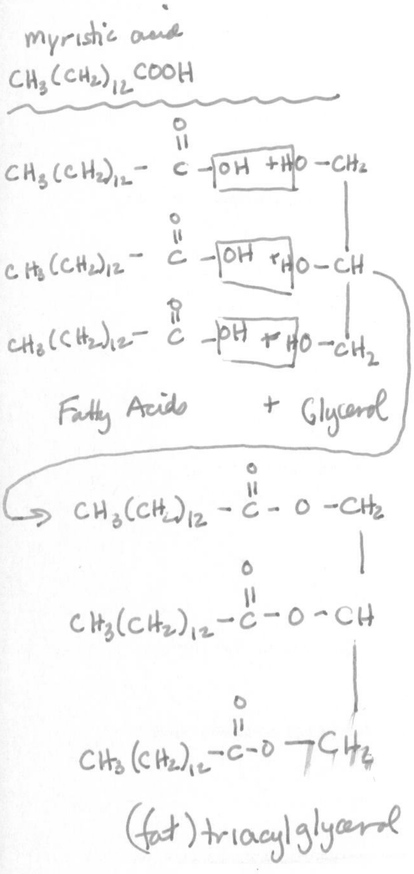

Problem 18.1 Look in Table 18.2 for the acid structures and write a structure for the triester that forms from the reaction of 3 mol of myristic acid with 1 mol of glycerol.

Because glycerol has three alcohol groups, it can react with one, two, or three acids to form mono-, di-, or triesters. These could be called mono-, di-, or triglycerides, or mono-, di-, or triacylglycerols. The triesters, wherein all three alcohol groups have been esterified, are the most common and are the ones that will concern us.

The term *triglyceride,* though in common use, is unsatisfactory from an organic nomenclature point of view. The name is intended to indicate the three ester linkages, but, of course, *-ide* is not the usual ester ending. The term *triacylglycerol* is usefully descriptive of the makeup of these triesters from three fatty acids (*acyl-* is the name of the acid fragment R—C(=O)—) and glycerol. This name corresponds nicely to a representation of these materials that is frequently used:

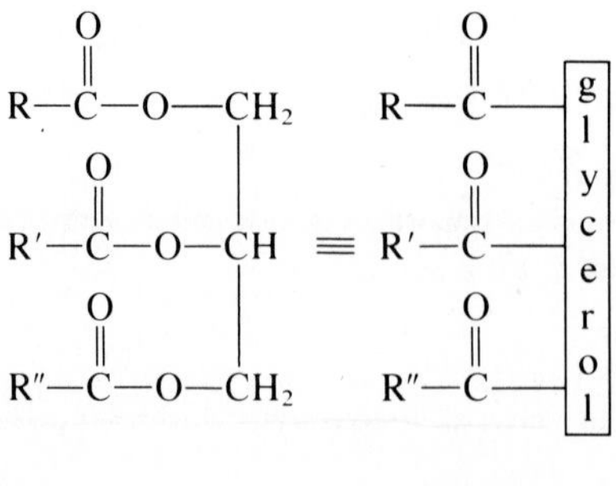
glycerol

Acyl groups

Because all triacylglycerols contain the glycerol backbone, differences between the compounds must arise from the different acyl groups that may be attached. That is, the fatty acids dictate the differing properties of the triacylglycerols.

The terms *fats* and *oils* are totally nondescriptive of structure, but they are the names used most often for the triesters we are discussing. Fatty acids are so-called because they were originally isolated by hydrolyzing (or saponifying) fats. At the time of the naming there was no structural knowledge of fats and oils. Thus the term *fatty acid* simply reflected the source of the materials.

Today we know that long hydrocarbon chains characterize the fatty acids. The total number of carbons in a fatty acid is always an even number because of the way in which they are synthesized biologically, as you will see in Section 24.7. Table 18.2 lists several fatty acids. The properties of the fatty acids are what one would expect for a molecule that has a long hydrophobic hydrocarbon segment terminated by a polar hydrophilic group.

Fatty acids may be saturated, i.e., contain only single bonds between carbons, or they may contain one or more carbon-carbon double bonds, in which case they are unsaturated. The arrangement around the double bonds in unsaturated fatty acids is almost always the cis arrangement. This contributes to lower melting points than for a molecule with the trans array because the cis arrangement produces a molecular shape that is more difficult to pack together in the solid state. Figure 18.3 shows this idea.

What distinguishes a fat from an oil? In terms of physical properties, fats are solids at room temperature, i.e., they have high melting points, and oils are liquids at room temperature. These physical properties are dictated by the structure of the acyl groups from the fatty acids. Saturated acyl chains

TABLE 18.2 Common Fatty Acids*

Number of C Atoms	Number of Double Bonds	Name	Formula	Melting Point, °C
			SATURATED FATTY ACIDS	
4	0	Butyric†	$CH_3(CH_2)_2COOH$	−8
6	0	Caproic	$CH_3(CH_2)_4COOH$	−3
8	0	Caprylic	$CH_3(CH_2)_6COOH$	17
10	0	Capric	$CH_3(CH_2)_8COOH$	32
12	0	Lauric	$CH_3(CH_2)_{10}COOH$	44
14	0	Myristic	$CH_3(CH_2)_{12}COOH$	54
16	0	Palmitic	$CH_3(CH_2)_{14}COOH$	63
18	0	Stearic	$CH_3(CH_2)_{16}COOH$	70
20	0	Arachidic	$CH_3(CH_2)_{18}COOH$	75
22	0	Behenic	$CH_3(CH_2)_{20}COOH$	80
24	0	Lignoceric	$CH_3(CH_2)_{22}COOH$	84
26	0	Cerotic	$CH_3(CH_2)_{24}COOH$	88
			UNSATURATED FATTY ACIDS	
16	1	Palmitoleic	$CH_3(CH_2)_5CH{=}CH(CH_2)_7COOH$	−1
18	1	Oleic	$CH_3(CH_2)_7CH{=}CH(CH_2)_7COOH$	14
18	2	Linoleic	$CH_3(CH_2)_4CH{=}CHCH_2CH{=}CH(CH_2)_7COOH$	−5
18	3	Linolenic	$CH_3CH_2CH{=}CHCH_2CH{=}CHCH_2CH{=}CH(CH_2)_7COOH$	−11
20	4	Arachidonic	$CH_3(CH_2)_4CH{=}CHCH_2CH{=}CHCH_2CH{=}CHCH_2CH{=}CH(CH_2)_3COOH$	−50

* See Table 18.3 for their natural sources in fats and oils.

† Butyric acid is included among long-chain fatty acids because of its occurence in butter and its importance in fatty acid metabolism.

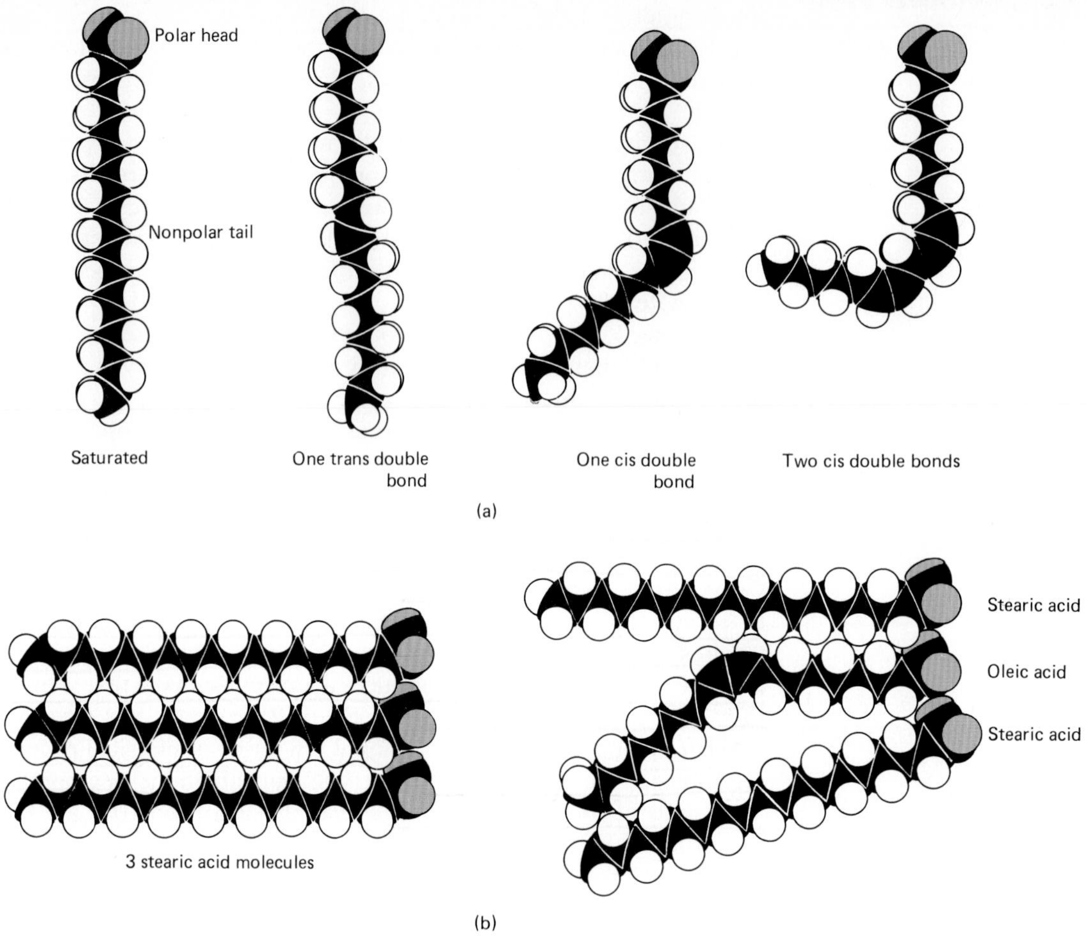
Polar head
Nonpolar tail
Saturated
One trans double bond
One cis double bond
Two cis double bonds
(a)
3 stearic acid molecules
Stearic acid
Oleic acid
Stearic acid
(b)

FIG. 18.3
Influence of fatty acid structure on physical properties. (*a*) The carbon chains of saturated fatty acids and compounds with trans double bonds can be arranged linearly. One cis double bond in the chain produces a bend; two cis double bonds exaggerate the bend. (*b*) Fatty acids with cis double bonds have lower melting points than saturated or trans compounds because it is more difficult to pack the molecules together in the solid state. Compare the ordered packing of three stearic acid molecules to the disrupted array that occurs when an oleic acid molecule is mixed with two stearic acids. Cell membranes are semiliquid because many acyl chains in phospholipids have cis double bonds.

produce the higher melting points of fats; unsaturated chains are more likely to produce the lower melting points of oils.

Fats	Oils
Higher melting point (solids)	Lower melting point (liquids)
Saturated acyl groups	Unsaturated acyl groups

Triacylglycerols in animals tend to have saturated acyl chains and are thus fats, whereas the triacylglycerols in plants typically have unsaturated acyl groups and are usually oils. Polyunsaturated cooking oils from plant sources contain multiple double bonds.

Fats and oils that we encounter in everyday life are mixtures of triacylglycerols. For example, corn oil is a mixture of triacylglycerols each of which can be represented by the general structure:

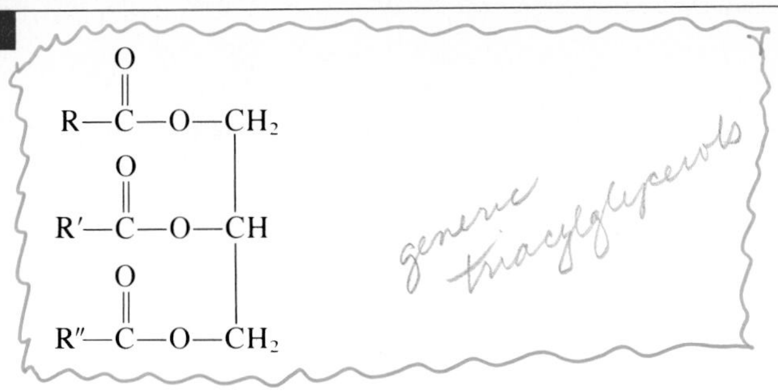
generic triacylglycerols

In corn oil 40 percent of the R groups come from oleic acid or its positional isomers, 40 percent from linoleic acid or its isomers, 5 percent from other unsaturated fatty acids, 4 percent from stearic acid, 10 percent from palmitic acid, and 1 percent from myristic acid. Since fully 85 percent of the R groups are unsaturated, it is not surprising that corn oil is a liquid at room temperature. Table 18.3 shows the composition of other fats and oils.

SAMPLE EXERCISE 18.1

Identify the following as fats or oils:

a 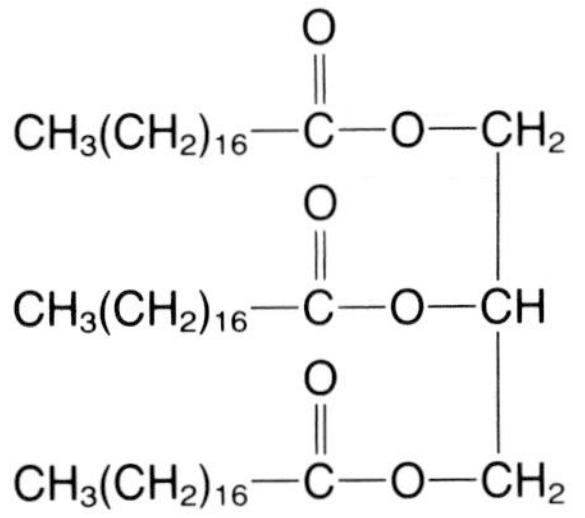

Tristearylglycerol

Fat

b 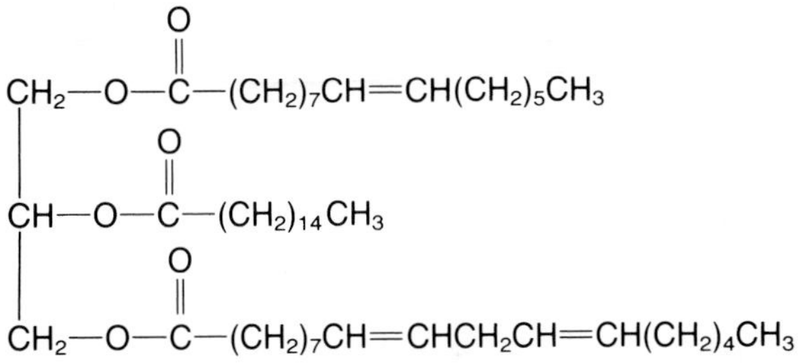

Palmitoleiyl palmityl linoleiyl glycerol

oil

c 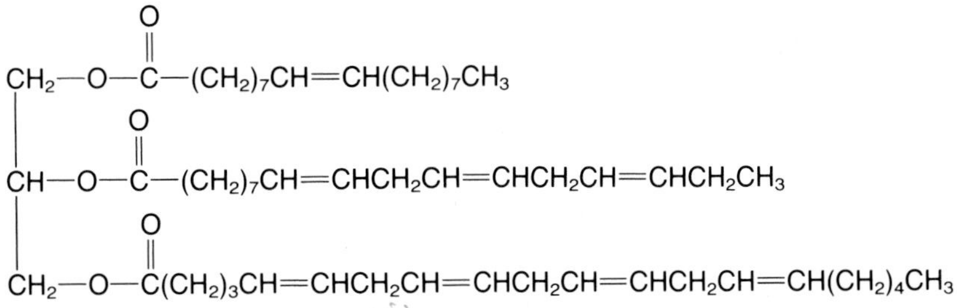

oil

Oleiyl linolenyl arachidonyl glycerol

Solution From given structures we cannot say with certainty which are liquids (oils) and which solids (fats) at room temperature; we would need to observe the actual compounds. However, it is more likely that saturated materials are solids and unsaturated ones liquids. Thus,

a Tristearylglycerol is probably a solid fat because the acyl chains are totally saturated.

b This material would have properties intermediate between those of compounds *a* and *c*.

c This compound is more likely to be an oil because of the many sites of unsaturation in the acyl chains.

TABLE 18.3 Fatty Acid Composition, by Percent, and Iodine Numbers of Common Fats and Oils (Typical Analyses*)

Fatty Acid		Vegetable Oils						
Number of Carbons in Acid Chain	Number of Double Bonds	Castor	Coconut	Corn	Cottonseed	Linseed	Olive	Peanut
				SATURATED				
<8	0							
8	0		8					
10	0		7					
12	0		48			Trace		
14	0		17	1	21	6	Trace	
16	0	2	9	10	2	4	9	7
18	0		2	4	Trace	Trace	2	5
20	0						Trace	4
22	0					Trace		
24	0							3
				UNSATURATED				
16	1			1				
18	1	94	6	40	33	21	83	60
18	2	4	3	40	44	24	6	21
18	3					45		
22	1							
20	Unsat.			4				
22	Unsat.							
Iodine number		86	10	128	100	185	83	93

* The percentages can vary widely in many cases. Environmental factors affect fatty acid composition in plant materials; diet affects animal fat composition.

18.4 HYDROGENATION OF OILS

If you recall the addition reaction of hydrogen to alkenes from Section 13.8, then you can readily see how oils can be converted into fats.

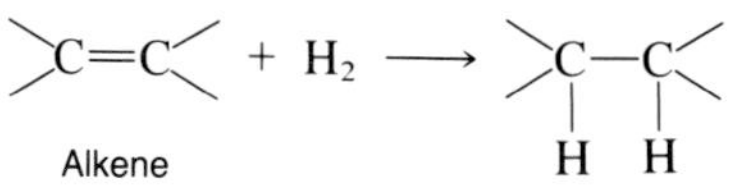

Alkene

Alkane

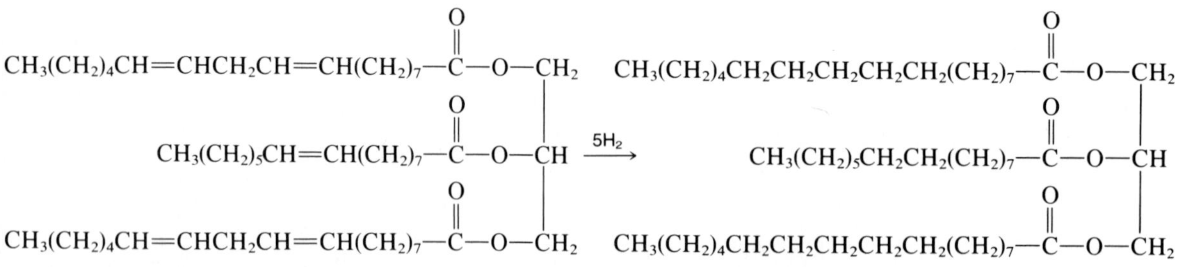

Unsaturated oil with low melting point

Saturated fat with higher melting point

This saturation reaction is important in the margarine industry. Hydrogen gas is bubbled through liquid vegetable oils until the desired semisolid buttery consistency is achieved. Oils are generally not totally saturated to make butter substitutes. Total saturation would produce too hard a margarine, a brittle butter.

TABLE 18.3 *(Continued)*

Vegetable Oils (*cont'd*)				Animal Fats				
Sesame	**Soybean**	**Sunflower**	**Walnut**	**Beef Tallow**	**Butter**	**Human Fat**	**Lard**	**Sardine**
				SATURATED				
					6			
					1			
					3			
			Trace		3			5
	6	4	5	3	11	3	2	14
9	5	3	1	29	29	25	23	3
5	1	1	Trace	19	11	8	12	
Trace								
	Trace	Trace						
				UNSATURATED				
					4			12
47	33	34	18	46	28	46	52	10
39	53	58	73	3	3	14	11	15
	2		3			1		
						1		
					1	1		22
						1		19
110	129	128	150	42	32	68	65	175

18.5 IODINE NUMBER

The addition reaction is also used analytically to determine the degree of unsaturation of oils. In this case, the addition reaction of iodine is observed.

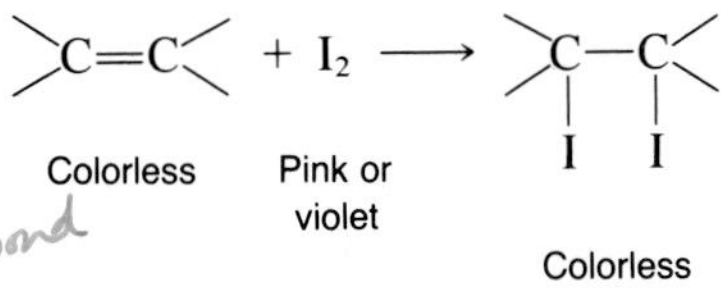
Colorless
Pink or violet
Colorless

The iodine # of an oil or fat is defined as the number of grams of iodine absorbed per 100 g. of fat or oil. Compounds with high iodine numbers are very saturated.

Low iodine #'s correspond with more saturated materials.

The presence of double bonds in a compound can be verified by adding drops of an iodine solution to the compound. If double bonds are present and an addition reaction occurs, the iodine loses its color. That is, elemental iodine is deep purple (pink in dilute solution), but the addition product is colorless.

The test can be used quantitatively. If the concentration of the iodine solution is known and the volume added to a known volume of oil is measured, then one can calculate the number of double bonds in the oil. The iodine is added to the oil in a dropwise fashion until just a very faint color persists. At that point, all double bonds have added iodine, no more iodine can react, and any iodine put into the solution will retain its color.

The *iodine number* of an oil or fat is defined as the number of grams of iodine absorbed per 100 g of fat or oil. Compounds with high iodine numbers are very unsaturated. Low iodine numbers correspond to more saturated materials. Look back at Sample Exercise 18.1; compound *c* would have the highest iodine number because it has the greatest number of double bonds. Animal fats typically have iodine numbers less than 70; for example, for human fat it is 68. On the other hand, the more unsaturated vegetable oils have iodine values that are usually more than 100. Sunflower oil's iodine number is 128. Table 18.3 includes the iodine numbers of each fat and oil listed.

18.6 HYDROLYSIS OF SIMPLE LIPIDS AND DIGESTION

The hydrolysis of a simple lipid is the hydrolysis of an ester. Beeswax, if hydrolyzed, would yield a long-chain acid and a long-chain alcohol:

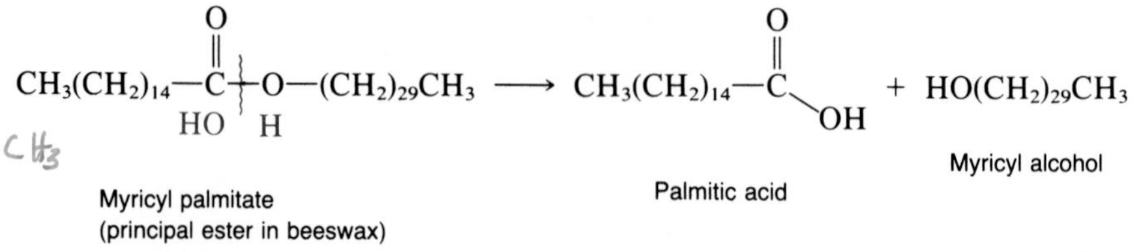
Myricyl palmitate
(principal ester in beeswax)
Palmitic acid
Myricyl alcohol

The hydrolysis of a triacylglycerol requires 3 mol of water per mole of fat or oil:

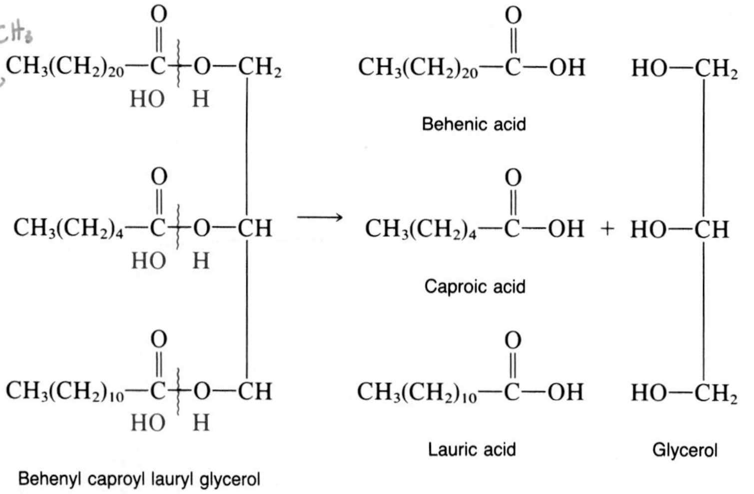
Behenic acid
Caproic acid
Lauric acid
Glycerol
Behenyl caproyl lauryl glycerol

As usual, the ester is cleaved between the carbonyl carbon and the oxygen, and the OH of water goes to the carbonyl carbon. **The products of the hydrolysis of a fat or oil are always glycerol and three fatty acids.**

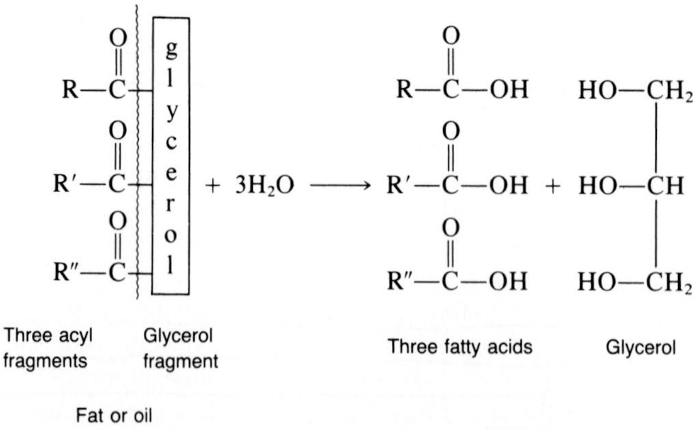
glycerol
Three acyl fragments
Glycerol fragment
Fat or oil
Three fatty acids
Glycerol

Problem 18.2 The major fatty acid components of beef tallow are such that a typical triacylglycerol in it might be the structure which appears below. Write structures for the hydrolysis products of

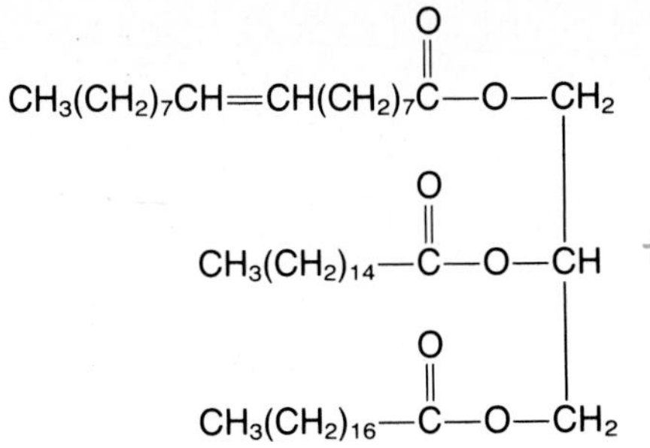

Consult Table 18.2 and name the products.

Rancidity

Unwanted hydrolysis reactions lead to spoilage, called rancidity, in fats and oils. For example, under moist air conditions the triacylglycerols in butter can hydrolyze to form butyric and caproic acids, which have rancid odors. Microorganisms present in the air furnish enzymes which speed this reaction. Rancidity can be prevented by storing butter in a closed container to minimize contact with microorganisms in air and by refrigeration, which slows the rate of all reactions, including hydrolysis reactions.

Polyunsaturated oils are also susceptible to rancidity through oxidation reactions. Oxygen in air reacts with the double bonds and cleaves long chains into shorter-chain fatty acids with rancid odors. To avoid unwanted oxidation, the food industry adds antioxidants such as BHA and BHT, the structures of which were given in Section 14.17. These antioxidants preferentially react with oxygen and prevent the reaction of oxygen with double bonds in polyunsaturated oils. Potato chips are one food product so treated.

Digestion

As was the case for the the digestion of carbohydrates, lipids are also digested by what is essentially a hydrolysis reaction. However, if you'll recall the solubility characteristics of lipids, you'll remember that they are hydrophobic, i.e., not soluble in water. Consequently it is difficult to mix fats and oils intimately with water and achieve the contact necessary for the hydrolysis reaction. This problem is mediated by the alkaline bile and bile salts which function as emulsifiers.

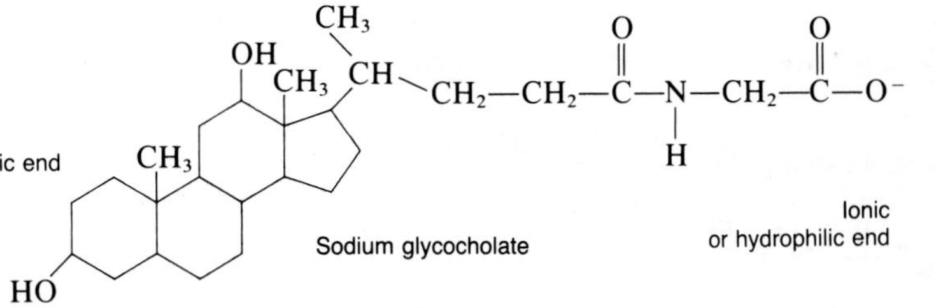
Hydrophobic end
Sodium glycocholate
Ionic
or hydrophilic end

The principal bile salt is sodium glycocholate. Bile salts have both a hydrophilic and a hydrophobic end and act like soap by breaking up the lipid mass into tiny droplets, as described in Section 17.5 and Figure 17.1. These droplets are then susceptible to enzymatic hydrolysis in aqueous media. Because bile is a basic medium, at least some of the fatty acids produced upon hydrolysis are converted into their salts. These salts in turn act as soaps to further aid the emulsification and digestion process. Section 18.15 discusses more details of the digestion and absorption processes.

18.7 SAPONIFICATION

Alkaline hydrolysis of fats and oils or of any ester is called saponification. Section 17.14 discussed saponification of simple esters. The saponification reaction is most readily understood as a two-step process: First, there is a hydrolysis reaction which produces the usual fatty acid and glycerol products from the fat or oil; then because a base is present, the fatty acids are converted into salts. Salts of long-chain fatty acids are *soaps*. The **overall reaction of saponification** is

Fat or oil + aqueous NaOH ⟶ 3 soaps + glycerol

It can be thought of as occurring in the two steps:

STEP 1 **Hydrolysis**

Fat or oil + $3H_2O$ ⟶ 3 fatty acids + glycerol

STEP 2 **Salt formation**

3 fatty acids + 3NaOH ⟶ 3 soaps + $3H_2O$

The products of the saponification of a fat or oil are always glycerol and three soaps, i.e., salts of fatty acids.

SAMPLE EXERCISE 18.2

Write structures for the products of saponification of the following fat:

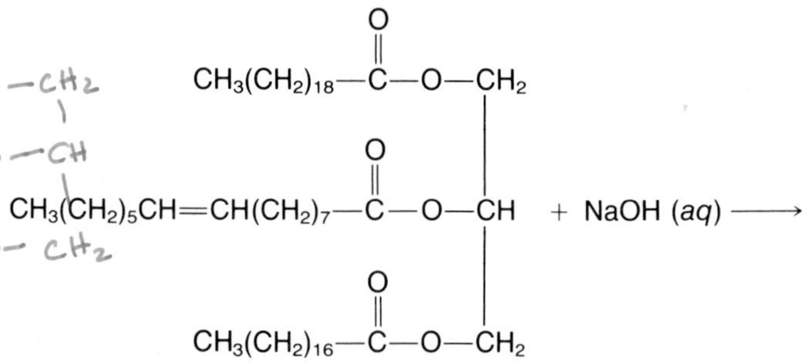

Solution Envision the cleavage of the ester linkages in the usual manner of hydrolysis:

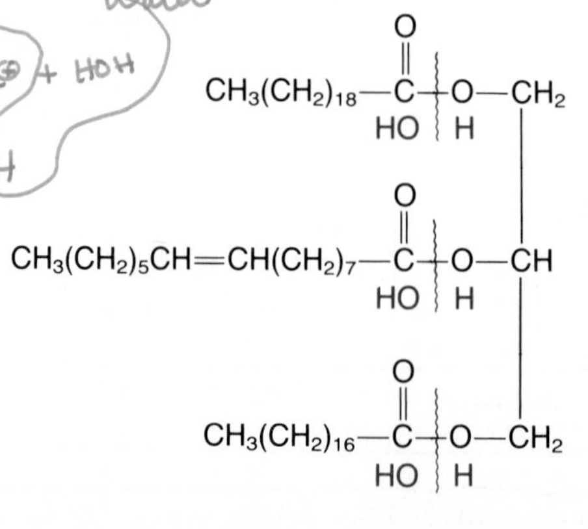

Then realize that each acid will be converted to its sodium salt by the reaction

$$R{-}\overset{\displaystyle O}{\overset{\|}{C}}{-}OH + NaOH \longrightarrow R{-}\overset{\displaystyle O}{\overset{\|}{C}}{-}O^{-}Na^{+} + HOH$$

Therefore, the products are

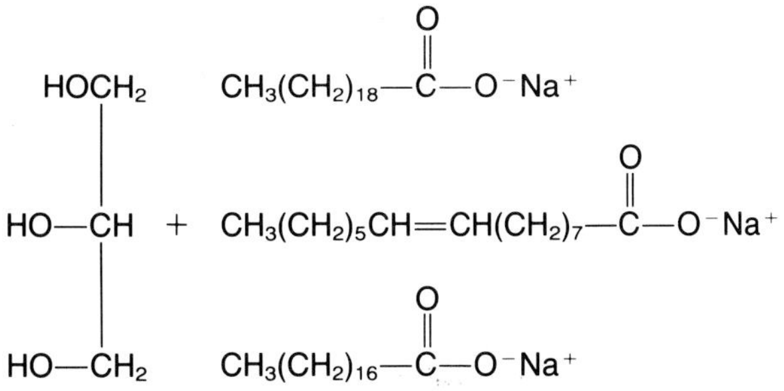

Glycerol   Three soaps

People performed this saponification reaction for centuries without knowing the structures of the materials involved. They knew that if they boiled lard (an animal fat) with lye (NaOH) they got soap. The reaction was witnessed by the change in properties from the ''greasy'' fat to a white solid material with the familiar properties of a bar of soap. The soap had to be washed carefully with water to remove residual caustic lye.

18.8 COMPOUND LIPIDS

Waxes and triacylglycerols, the *simple* lipids, are characterized by the fact that they produce only alcohols and fatty acids upon hydrolysis (or salts upon saponification). That is,

$$\text{Wax} + H_2O \xrightarrow{H^+} \text{fatty acid} + \text{long-chain alcohol}$$

$$\text{Triacylglycerol} + H_2O \xrightarrow{H^+} 3 \text{ fatty acids} + \text{glycerol}$$

Compound lipids, on the other hand, also yield other substances upon hydrolysis (or saponification) because their structures include a greater variety of fragments. These other fragments include phosphoric acid, amines, and sugar molecules.

Whereas the triacylglycerols or fats and oils fulfill nutritional needs, the compound lipids are important structural components of the body. They form the lipid bilayer (Section 18.10) of cell membranes and are found extensively in nervous tissue.

18.9 PHOSPHOLIPIDS

Phospholipids are saponifiable compound lipids which contain a phosphate group. The most prevalent of the phospholipids are the phosphoglycerides, which can be represented as

Phosphoglyceride

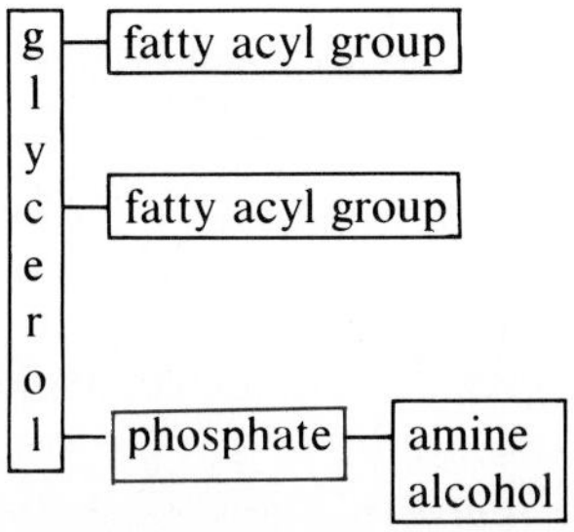
glycerol
fatty acyl group
fatty acyl group
phosphate
amine alcohol

Upon hydrolysis they yield glycerol, two fatty acids, phosphoric acid, and an amino alcohol compound. Using the amino alcohol choline as a specific example, we can envision the construction of a phosphoglyceride in the following way:

$$
\begin{array}{l}
\overset{\displaystyle O}{\underset{\|}{}}\\
CH_2O\!-\!\boxed{H + HO}\!-\!C\!-\!R
\end{array}
$$

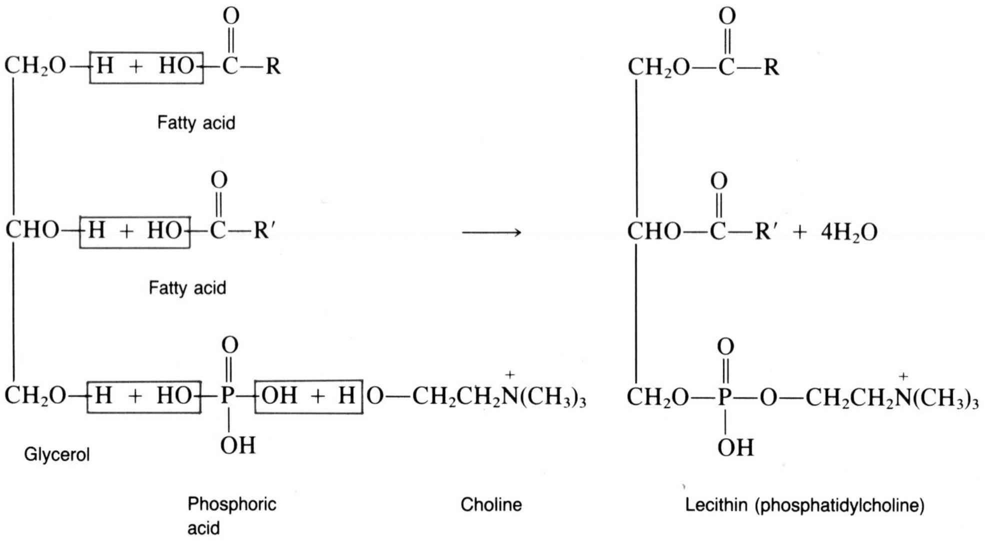

Fatty acid

$$
\begin{array}{l}
\overset{\displaystyle O}{\underset{\|}{}}\\
CHO\!-\!\boxed{H + HO}\!-\!C\!-\!R'
\end{array}
$$

Fatty acid

$$
\begin{array}{l}
\overset{\displaystyle O}{\underset{\|}{}}\\
CH_2O\!-\!\boxed{H + HO}\!-\!P\!-\!\boxed{OH + H}O\!-\!CH_2CH_2\overset{+}{N}(CH_3)_3\\
\hspace{3.5cm}OH
\end{array}
$$

Glycerol   Phosphoric acid   Choline

$\longrightarrow$

$$
\begin{array}{l}
\overset{\displaystyle O}{\underset{\|}{}}\\
CH_2O\!-\!C\!-\!R
\end{array}
$$

$$
\begin{array}{l}
\overset{\displaystyle O}{\underset{\|}{}}\\
CHO\!-\!C\!-\!R' + 4H_2O
\end{array}
$$

$$
\begin{array}{l}
\overset{\displaystyle O}{\underset{\|}{}}\\
CH_2O\!-\!P\!-\!O\!-\!CH_2CH_2\overset{+}{N}(CH_3)_3\\
\hspace{1.5cm}OH
\end{array}
$$

Lecithin (phosphatidylcholine)

When the esterifying amino alcohol is choline, as in the example, the phosphoglyceride formed is identified as a lecithin. Another common amino alcohol component of phosphoglycerides is ethanolamine, $HOCH_2CH_2NH_2$; these phospholipids are called cephalins. Some cephalins incorporate the amino acid serine, $HO\!-\!CH_2\!-\!\underset{\displaystyle COOH}{CH}\!-\!NH_2$, rather than ethanolamine.

Problem 18.3 Write the structure of a cephalin formed from glycerol, oleic acid, palmitic acid, phosphoric acid, and ethanolamine.

At physiological pH (~7), phosphoric acid groups are usually ionized; that is, the following equilibrium lies to the right:

$$
\begin{array}{l}
\overset{\displaystyle O}{\underset{\|}{}}\\
RO\!-\!P\!-\!OH\\
\hspace{1cm}OR
\end{array}
\;\rightleftharpoons\;
\begin{array}{l}
\overset{\displaystyle O}{\underset{\|}{}}\\
RO\!-\!P\!-\!O^- + H^+\\
\hspace{1cm}OR
\end{array}
$$

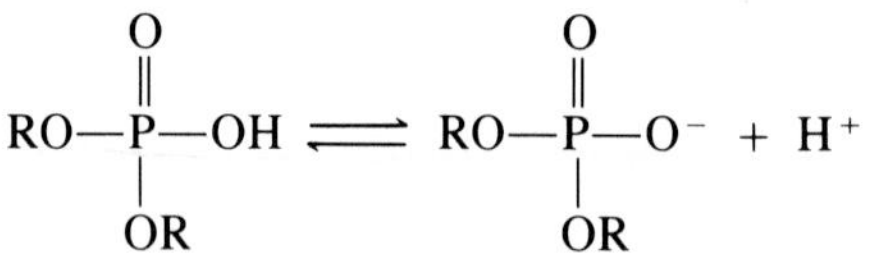

Consequently, nearby amine nitrogens would be protonated.

$$
RNH_2 + H^+ \rightleftharpoons R\overset{+}{N}H_3
$$

At physiological pH, correct structural representations of the lecithins and cephalins would be

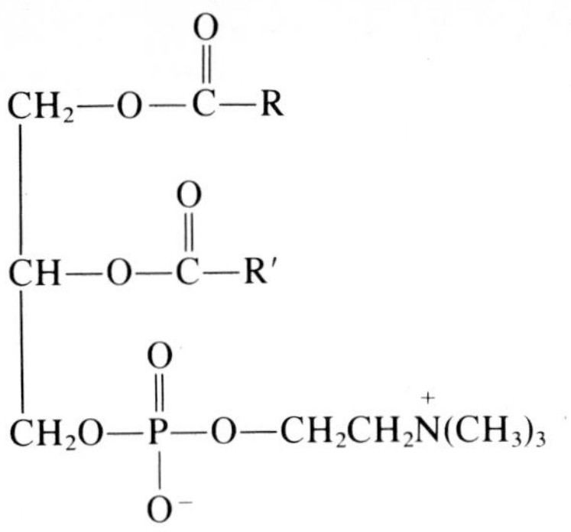

Phosphatidyl choline (a lecithin)

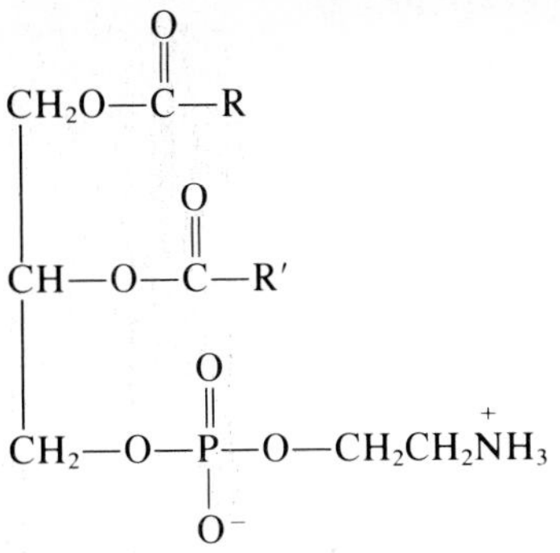

Phosphatidylethanolamine (a cephalin)

The ionic nature of the phosphate and amino portion of the molecule is essential to the structure of the lipid bilayer of cell membranes, which we will explore in the next section.

In order to appreciate the molecular construction of the lipid bilayer, let us begin by writing out the entire structure of a typical phosphoglyceride so that the arrangement of the hydrophilic and hydrophobic portions of the molecule is clear.

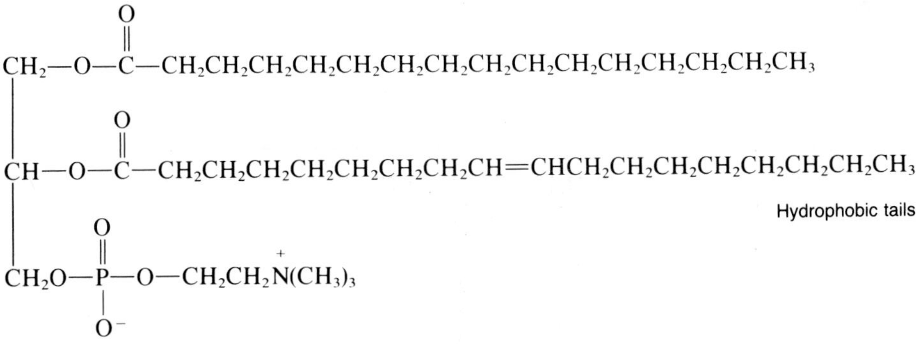
Hydrophobic tails
Hydrophilic head

A molecular model of this compound would look like this:

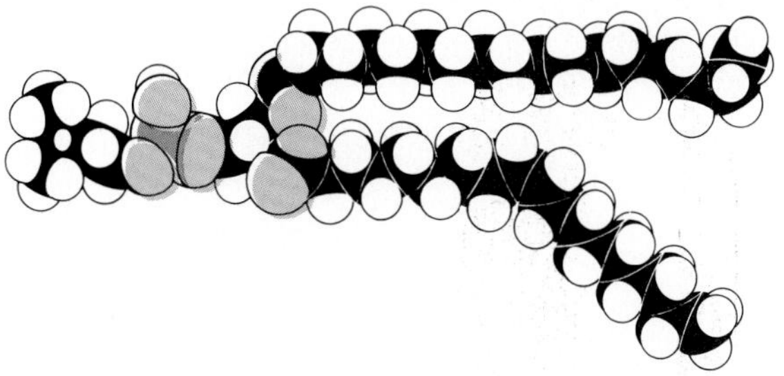

The important thing to notice is that there is an ionic or hydrophilic end, labeled the head, and long hydrophobic hydrocarbon chains, labeled as tails. This allows us to represent these molecules very simply as

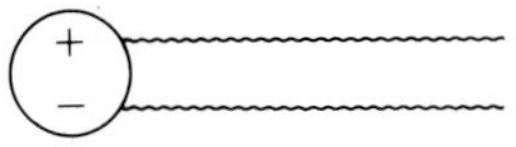
+
−
Hydrophilic
Hydrophobic

In an aqueous medium, these molecules arrange themselves so that their hydrophobic tails intermingle and their hydrophilic heads face the aqueous environment. This is consistent with the idea that "like dissolves like."

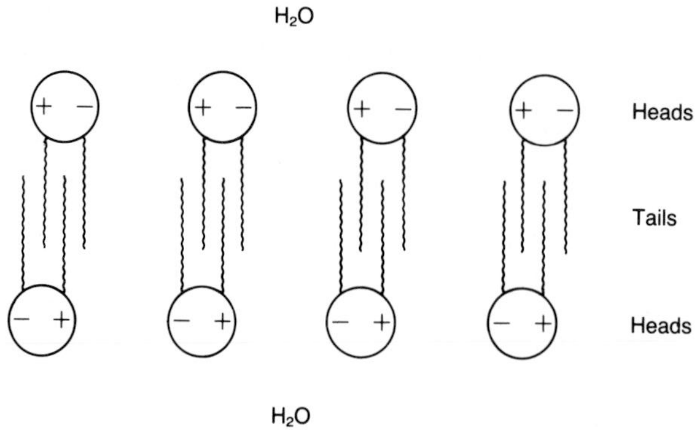
H_2O
Heads
Tails
Heads
H_2O

This creates the **lipid bilayer**—two layers of lipid molecules arranged as shown.

The molecular structure of lecithins also makes them very effective emulsifying agents. That is, they can function like soaps and promote intimate mixing of otherwise immiscible materials. It is the lecithin in egg yolks which allows one to make mayonnaise, ice cream, and custards, which are all emulsions of a fat in an aqueous medium.

18.10 THE LIPID BILAYER OF CELL MEMBRANES

When phospholipids are placed in water, they spontaneously form a lipid bilayer. Such a lipid bilayer forms the protective wrapping of a cell, that is, the cell membrane. Figure 18.4 is a cutaway view of a cell membrane.

FIG. 18.4
The fluid mosaic model of membrane structure. (*a*) Cross section of a bilayer membrane, which shows the interior and the exterior of the cell separated by the membrane. (*b*) Close-up cutaway view of the membrane. The membrane is mostly phospholipid material (the colored heads with two tails). The similar structures with white heads are sphingolipids. Cholesterol molecules move among the hydrophobic tails of phospholipids and sphingolipids. The large structures are embedded proteins. All components can move freely laterally in the membrane. Movement across the membrane is controlled and restricted.

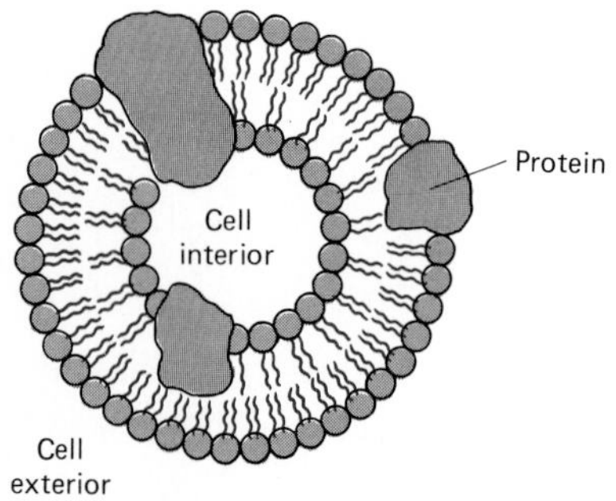
Protein
Cell interior
Cell exterior

(a)

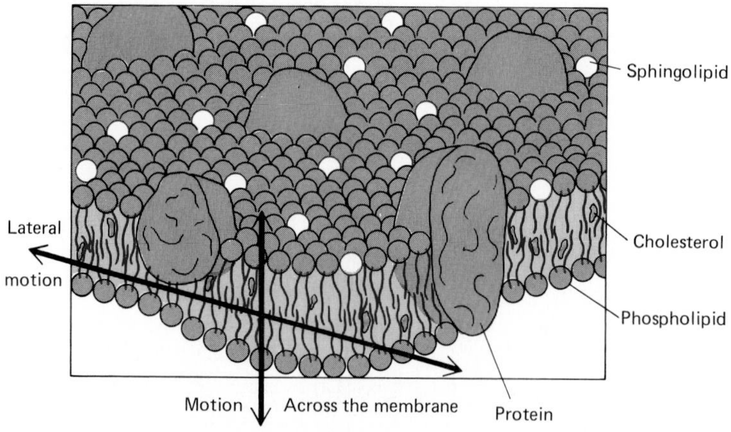
Sphingolipid
Cholesterol
Phospholipid
Lateral motion
Motion
Across the membrane
Protein

(b)

The membrane surrounding a cell is not a rigid solid. Rather, the phospholipid constructive material provides a liquid or jellylike consistency to the membrane. The cis double bonds of acyl chains are one reason for this liquid consistency, as was illustrated in Figure 18.3. Large protein molecules are either embedded in the bilayer or attached to one of the surfaces. This is the so-called fluid-mosaic model of a cell membrane. The protein molecules are able to occupy particular orientations within or on the lipid bilayer through hydrophobic or hydrophilic interactions. In Chapter 19 you will see that protein molecules are huge and have both hydrophilic (polar or ionic) segments and hydrophobic (nonpolar hydrocarbon) sections.

Proteins and lipids can float laterally through the sea of phospholipids which make up the cell membrane. Movement through or across the membrane from the outside to the inside of the cell (or vice versa) is carefully controlled by the proteins in or on the membrane. For example, the proteins on the surface are the receptor sites for neurotransmitters and antigens, which were mentioned in Sections 16.8 and 16.9, and for hormones, the body's chemical messengers, which will be explored in Section 21.14.

The embedded proteins, called intrinsic proteins, allow for passage across the membrane by simple diffusion, facilitated diffusion, or active transport. In the case of simple diffusion, the embedded protein site merely provides a channel or pore through which small molecules and ions such as H_2O, NH_3, O_2, CO_2, and Cl^- can pass. The direction of passage depends on a concentration gradient, as was discussed in connection with dialysis (Section 9.14). Facilitated diffusion involves the protein actually carrying the molecule across the membrane. Glucose crosses the membrane in this way. As in the case of simple diffusion, no energy input is required—the concentration gradient is followed. Active transport requires energy input because in this case molecules and ions are moved against a concentration gradient, that is, they are pumped into or out of the cell (see Section 9.15). The energy source is ATP.

18.11 CELL ORGANIZATION

All living matter is composed of cells. There are two types: eucaryotic and procaryotic cells. Eucaryotic cells are more complex and are the type of cell that makes up animal tissue. Figure 18.5 shows the key structures of a typical *eucaryotic* cell. Eucaryotic cells are organized into compartments within which various activities occur. These compartments are called organelles, meaning little organs. Each organelle has a particular function. For example, in the approximate center of the cell is the nucleus, enclosed by a nuclear membrane. The nucleus controls cell reproduction and is the headquarters for information controlling the synthesis of proteins in that cell, as will be discussed in Chapter 20. The nucleus and the other organelles float in cytoplasm, the jellylike material encased by the cell's membrane. Other organelles include the mitochondria, which are called the "powerhouses" of the cell because most of our biological energy is generated there, as you will see in Chapter 22. Ribosomes, which are not membrane-encased, are the sites of protein synthesis in the body (Chapter 20).

Materials enter and leave the cell compartments or organelles in accord with the important concepts of solubility, diffusion, and active transport because the compartments are separated from their environment by surrounding membranes. The organelles' nutritional needs and waste disposal problems are thus affected by these concepts of membrane chemistry.

There is a simpler type of cell called the procaryotic cell. Bacteria and blue-green algae have this cell design. Procaryotic cells do not have mem-

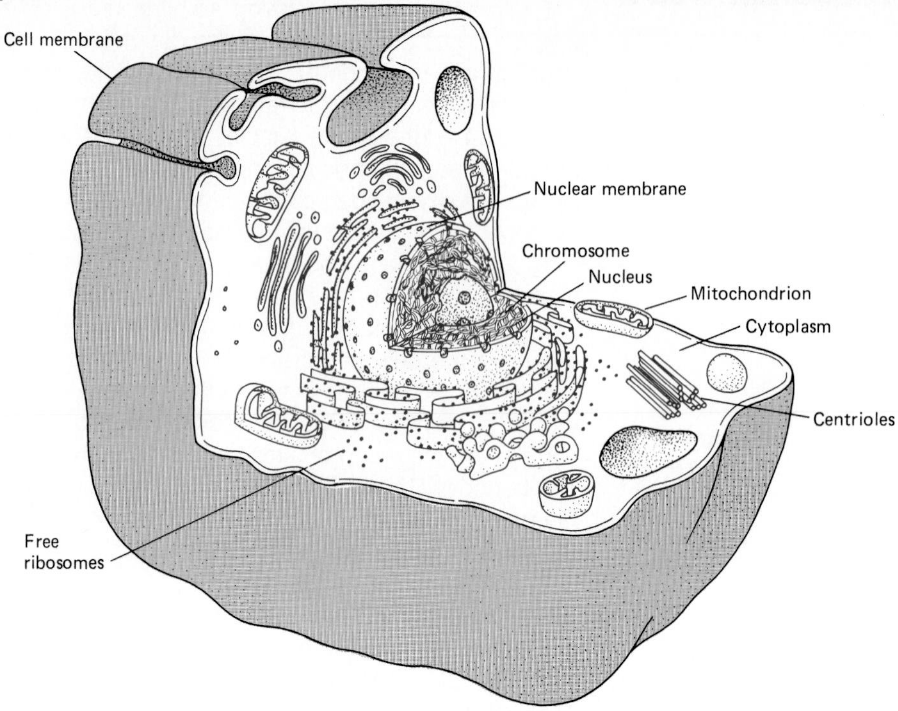
Cell membrane
Nuclear membrane
Chromosome
Nucleus
Mitochondrion
Cytoplasm
Centrioles
Free ribosomes

FIG. 18.5
A typical eucaryotic cell. A membrane surrounds the cytoplasm in which float various organelles ("little organs"). Each organelle has a particular function. For example, mitochondria are sites of energy production, and ribosomes are sites of protein synthesis.

brane-encased compartments or organelles. Most notably, there is no nuclear membrane.

18.12 SPHINGOLIPIDS

Sphingolipids, which are also found in cell membranes, are saponifiable compound lipids characterized by the presence of sphingosine as their backbone, rather than glycerol.

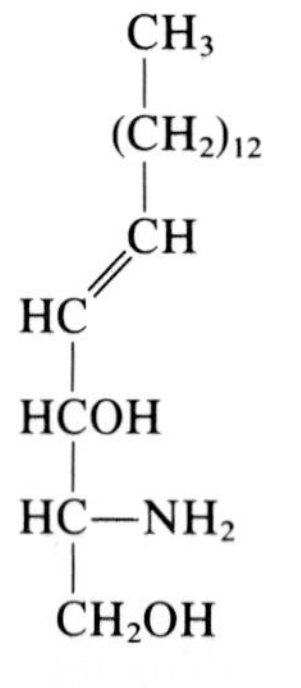
CH3
(CH2)12
CH
HC
HCOH
HC—NH2
CH2OH

Sphingosine

In one type of sphingolipid, called sphingomyelins, the amino group of sphingosine is bonded to a fatty acid by an amide linkage, and the primary alcohol group of sphingosine is esterified with phosphoric acid. The phosphoric acid residue also forms a second ester linkage with an amino alcohol. The general structure of a sphingomyelin can be represented as

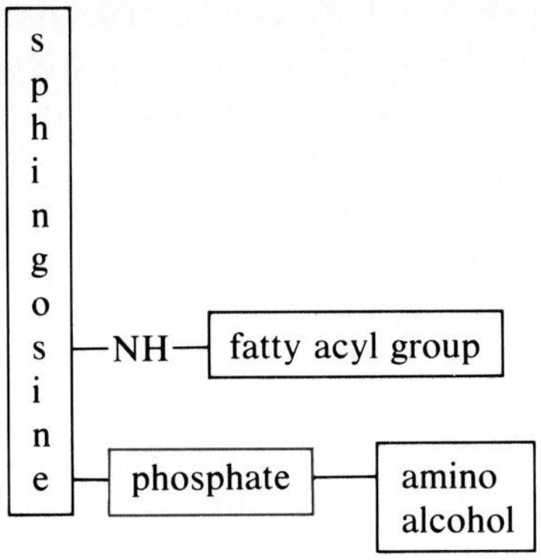
s
p
h
i
n
g
o
s
i
n
e
NH
fatty acyl group
phosphate
amino
alcohol

Sphingomyelin

Because sphingomyelins are saponifiable lipids containing a phosphate group, they can be considered members of the class of phospholipids, as well as sphingolipids.

If we use lignoceric acid and choline to show the formation of a specific sphingomyelin, then

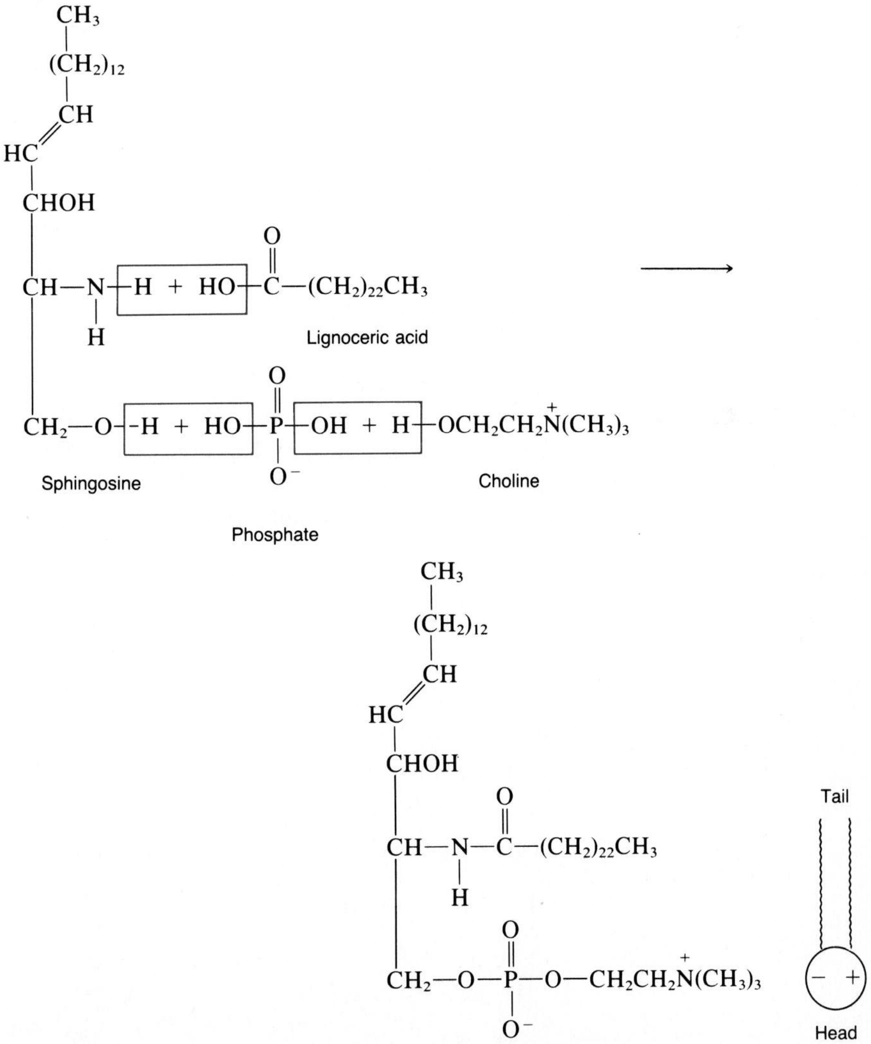
Lignoceric acid
Sphingosine
Phosphate
Choline
Tail
Head

Sphingomyelin

At physiological pH the phosphate is ionized as shown and the sphingomyelins, like the phosphoglycerides, have a hydrophilic head and hydrophobic tail. Therefore, the sphingomyelins can also be represented

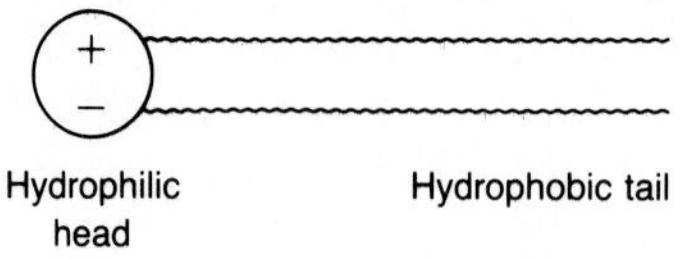
+
−
Hydrophilic head
Hydrophobic tail

as is done next to the full structure in the equation, and they can arrange themselves in the form of a lipid bilayer. Like the phosphoglycerides, sphingomyelins are prevalent in cell membranes, especially in the *myelin sheath* around certain nerve cells.

Problem 18.4 What are the hydrolysis products of a typical sphingomyelin?

Another group of sphingolipids, called glycolipids, contain a carbohydrate component in place of the phosphate fragment of sphingomyelins. *Glyco-* is a general prefix indicating carbohydrate components.

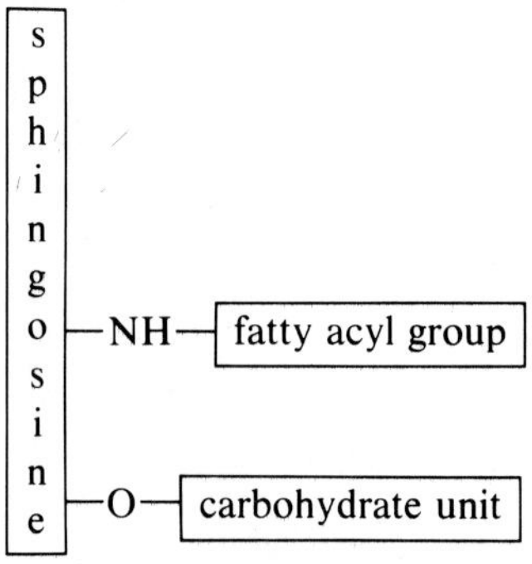
sphingosine
NH
fatty acyl group
O
carbohydrate unit

Glycolipids

Cerebrosides are the simplest glycolipids; the carbohydrate component is a glucose or galactose ring. Gangliosides are more complex; oligosaccharide chains of up to seven units make up the carbohydrate component.

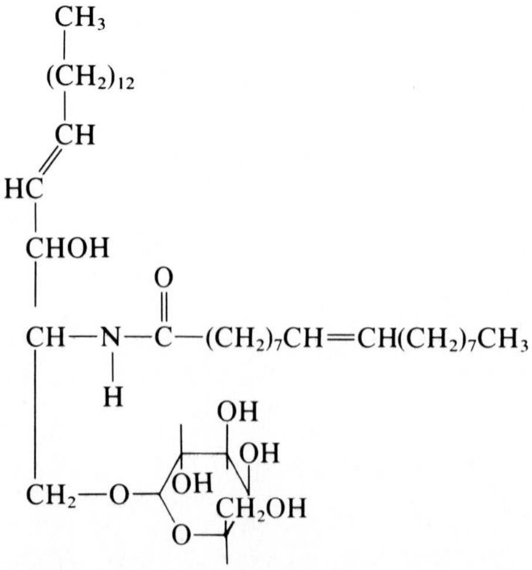

Cerebroside

Cerebrosides which incorporate galactose occur in the cell membranes of the brain. Gangliosides are found at cell surfaces in neural tissue and are often parts of the receptor sites for neurotransmitters (see Section 16.9).

All of the sphingolipids in cell membranes are in a dynamic state, as are many biological structures. This means these molecules are constantly being broken down and replaced by new molecules. Several genetic diseases involve the inability of some individuals to break down sphingolipids. The result is their accumulation in tissues, especially brain tissue, which leads to swelling and disastrous physiological effects. For example, in Niemann-Pick disease the enzyme to break down sphingomyelins is absent. Children with Tay-Sachs disease lack the enzyme needed to degrade gangliosides. In each case, the outcome is mental retardation and eventually death.

18.13 PROSTAGLANDINS

Prostaglandins are 20-carbon fatty acids with a five-carbon ring as part of their structure. For example,

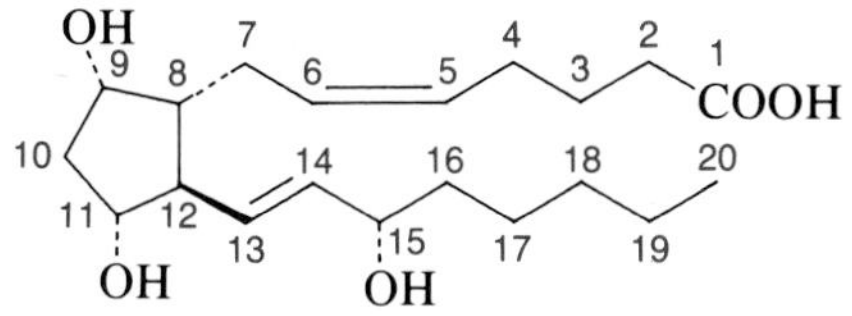
OH
OH
OH
COOH
1
2
3
4
5
6
7
8
9
10
11
12
13
14
15
16
17
18
19
20

Prostaglandin $F_{2\alpha}$

Because these lipids do not contain an ester linkage, they are nonsaponifiable lipids. Prostaglandins were originally isolated from prostate glands, from which they were named, but they are now known to appear in most cells; they are synthesized in cell membranes from phosphoglycerides which have an arachidonic acyl residue at position 2 of the glycerol backbone. The reaction sequence begins with the hydrolysis of the arachidonic ester linkage.

$$^{1}CH_2O{-}\overset{O}{\overset{\|}{C}}{-}R$$

$$^{2}CHO{-}\overset{O}{\overset{\|}{C}}(CH_2)_3CH{=}CHCH_2CH{=}CHCH_2CH{=}CHCH_2CH{=}CH(CH_2)_4CH_3$$

Arachidonic acyl residue

$$^{3}CH_2{-}\boxed{\text{phosphate}}{-}\boxed{\text{amino alcohol}}$$

hydrolysis ↓ phospholipase

$$HO{-}\overset{O}{\overset{\|}{C}}(CH_2)_3CH{=}CHCH_2CH{=}CHCH_2CH{=}CHCH_2CH{=}CH(CH_2)_4CH_3$$

Arachidonic acid

If we rewrite arachidonic acid by bending it around, its relationship to the prostaglandin structure can be seen more readily. The conversion of arachidonic acid to various prostaglandins and related compounds called thromboxanes and leukotrienes is called the **arachidonic acid cascade.**

Arachidonic Acid Cascade

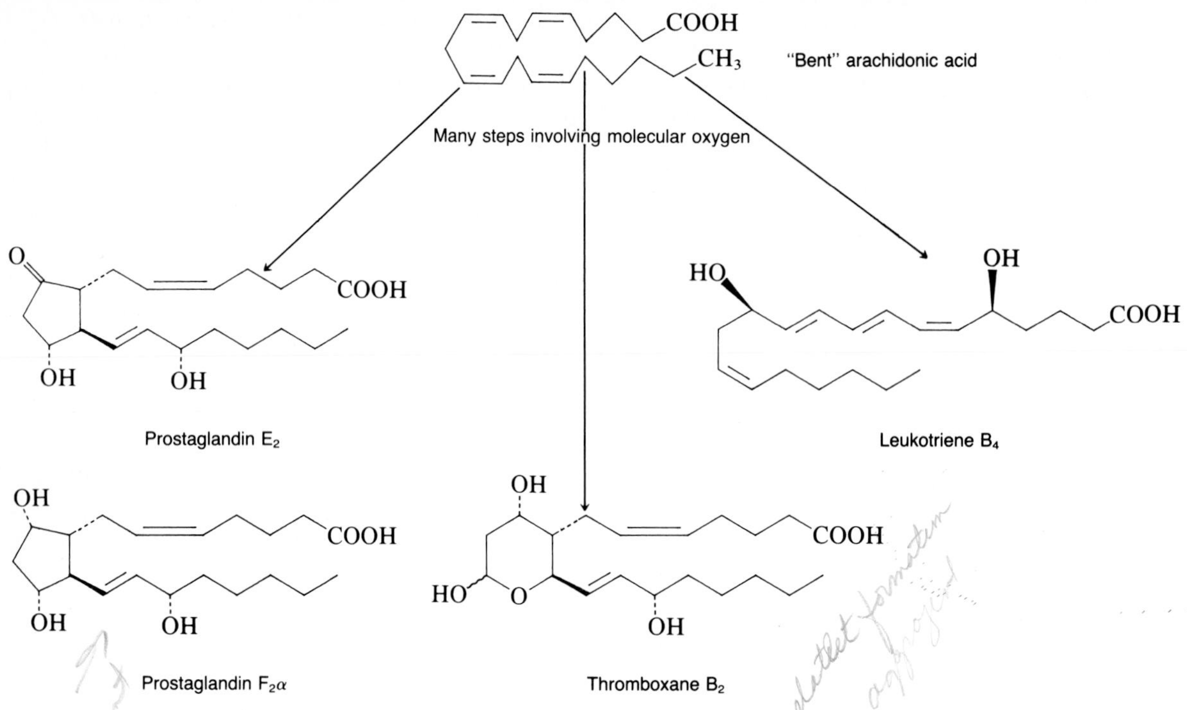
COOH
CH_3
"Bent" arachidonic acid
Many steps involving molecular oxygen
O
COOH
OH
OH
Prostaglandin E_2
HO
OH
COOH
Leukotriene B_4
OH
COOH
OH
OH
Prostaglandin $F_{2\alpha}$
OH
COOH
HO
O
OH
Thromboxane B_2

As these examples show, prostaglandins are characterized by a five-membered ring with oxygen-containing functional groups and two hydrocarbon chains attached at adjacent positions. A carboxylic acid group always terminates one hydrocarbon chain. Writing the arachidonic acid structure bent around emphasizes how easily the cyclopentane ring with two side chains arises:

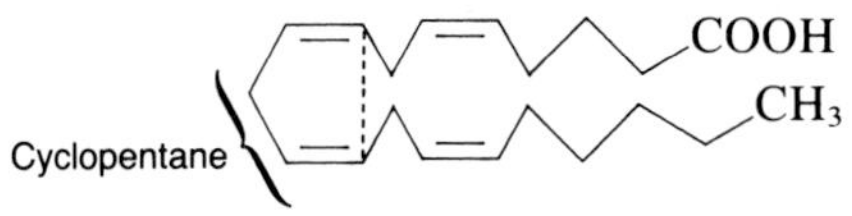
COOH
CH_3
Cyclopentane

The series of reactions leading to prostaglandins from arachidonic acid produces many other related compounds as well as the ones shown.

Prostaglandins produce a variety of physiological effects. They can cause either the contraction or the relaxation of smooth muscle; they lower blood pressure; they stop the flow of gastric juice; and they can induce labor. Some prostaglandins are *pyrogens;* that is, they induce fever, and many are involved in the body's inflammation mechanism. They also influence blood coagulation. Although the exact mechanism of the action of aspirin has not been pinpointed, it is now clear that at least part of its effect is due to the fact that aspirin inhibits prostaglandin synthesis and thereby combats fever and inflammation.

Because of the variety of their physiological activities, it is thought that natural prostaglandins or chemically modified structures might be useful in the treatment of such varied ailments as hypertension, arthritis, asthma, and ulcers. They can also be used to prevent conception, relieve nasal congestion, and prevent blood clots. So far, the prostaglandins have provided no miracle cures, but many pharmaceutical companies are making research efforts in this field.

18.14 STEROIDS

The last lipids that we will consider are the steroids. These compounds, of course, share the solubility characteristics of the lipid class, but they have no ester linkages (i.e., they are nonsaponifiable) and no fatty acid residues in their structures. Steroids are characterized by a tetracyclic ring structure with a side chain at position 17:

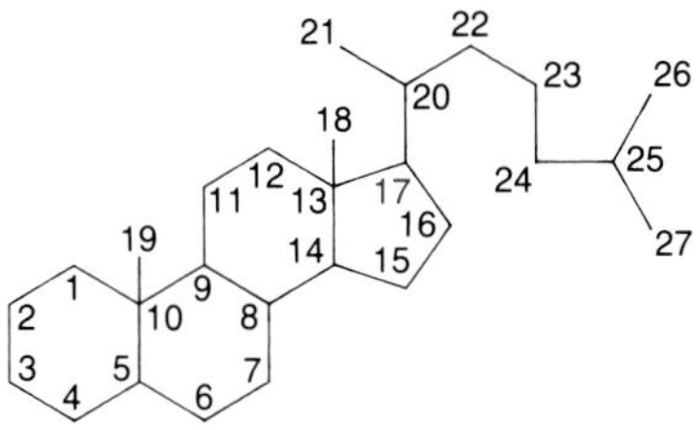

Cholesterol is the most abundant of the steroids and probably the most important because it has its own cellular functions and also serves as the raw material for the synthesis of other steroids.

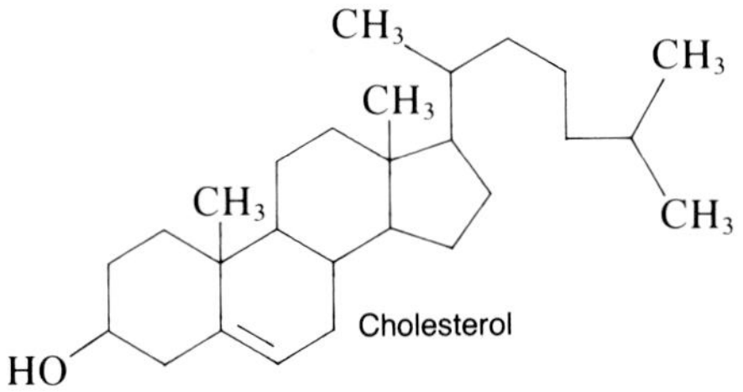
Cholesterol

Cholesterol has the *-ol* ending because it is an alcohol, or more precisely a *sterol*. Cholesterol, which is found dissolved in dietary fats and oils, is also synthesized in the liver from acetyl coenzyme A.

Cholesterol is found in all cell membranes and is essential for proper cell function. However, the presence of excess amounts due to faulty metabolism or inefficient transport can cause problems such as gallstones, which are hard chunks of precipitated cholesterol in the gall bladder or bile duct, or atherosclerosis. Atherosclerosis is a degenerative disease in which lipid deposits build up on the inner walls of arteries and present obstructions, as shown in Figure 18.6.

Many other steroids, including many hormones, are made by the body from cholesterol. Table 18.4 lists the structures and functions of several physiologically active natural and synthetic steroids.

Problem 18.5 Estrone is a female hormone, and androsterone produces male secondary sexual characteristics. Describe the functional-group differences between the two structures.

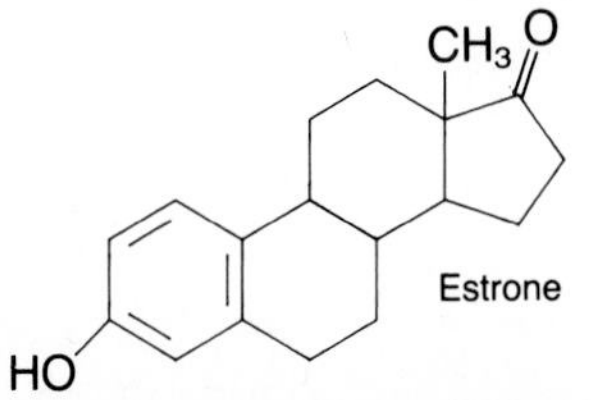
Estrone

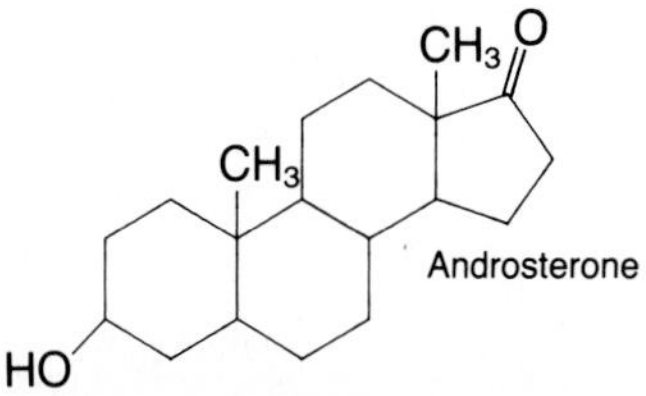
Androsterone

TABLE 18.4 Natural and Synthetic Physiologically Active Steroids

Structure	Function
BILE SALTS	
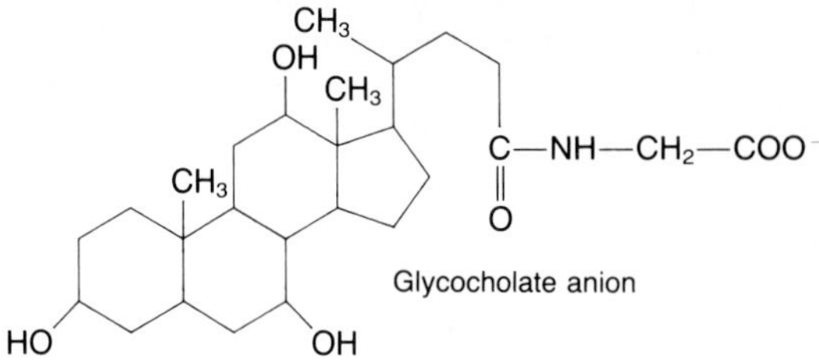 Glycocholate anion	These emulsify fats and aid digestion (Section 18.6).
VITAMIN D	
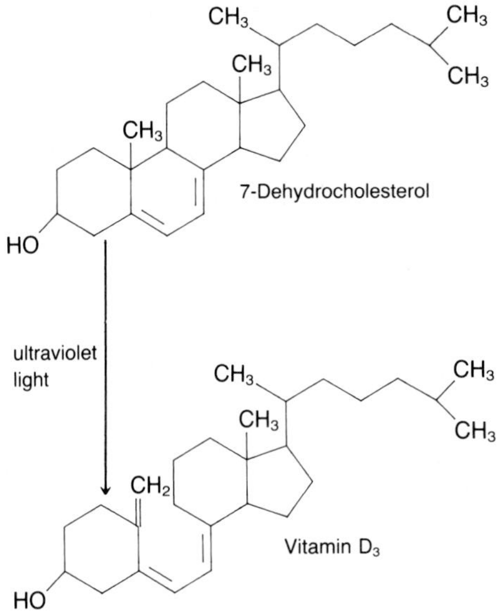 7-Dehydrocholesterol; ultraviolet light; Vitamin D_3	This cholesterol derivative occurs in skin. Sunlight converts the steroid to vitamin D_3, which is needed for healthy bones and teeth.
ADRENAL CORTICORDS	
NATURAL	
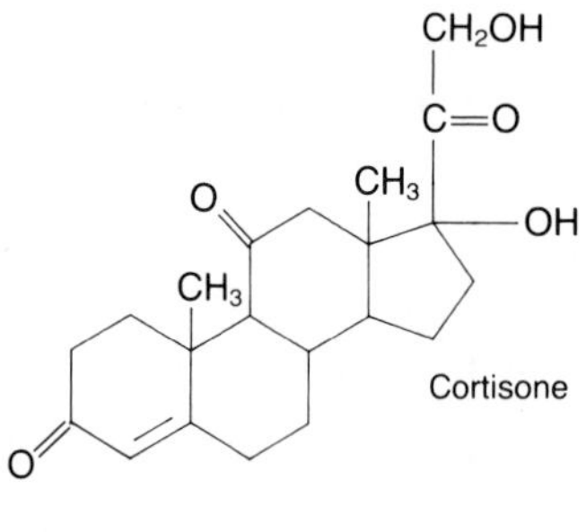 Cortisone	One of the glucocorticords that regulates glucose use. It also reduces the symptoms of arthritis, inflammation, and allergic diseases.
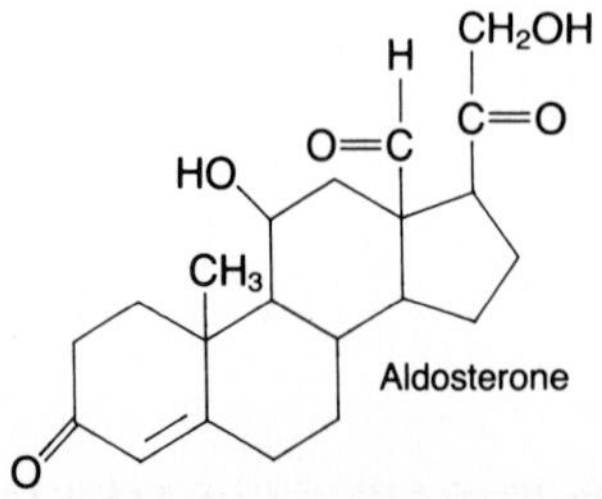 Aldosterone	Regulates sodium chloride and water retention.

TABLE 18.4 *(Continued)*

Structure	Function
SYNTHETIC 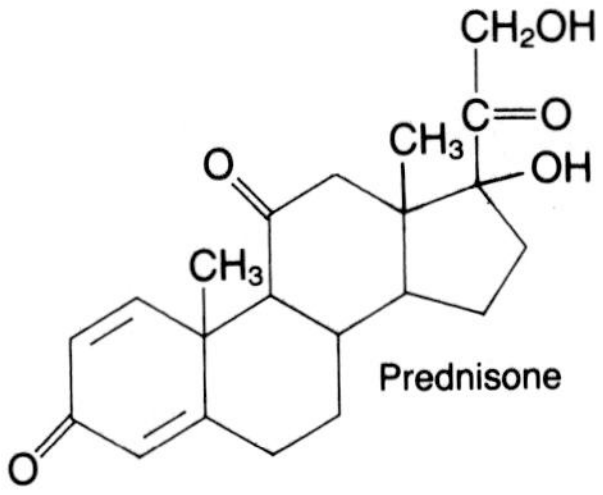 Prednisone	Frequently used clinically as a substitute for cortisone; slight structural variations reduce the adverse side effects of cortisone.
SEX HORMONES **MALE** 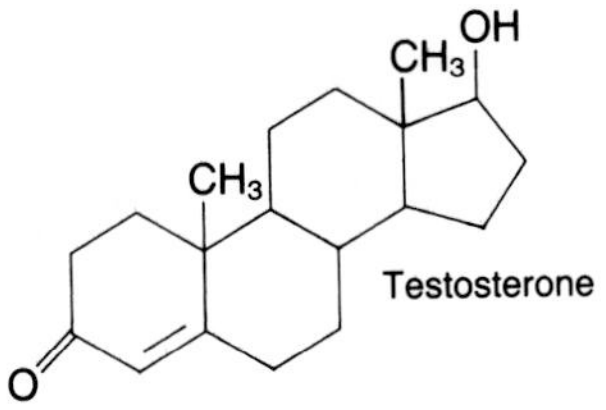 Testosterone	Regulates the development of male reproductive organs and masculine characteristics.
FEMALE 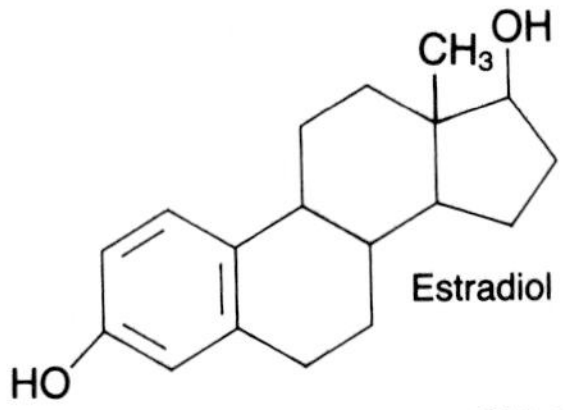 Estradiol	One of the estrogens which regulate female characteristics.
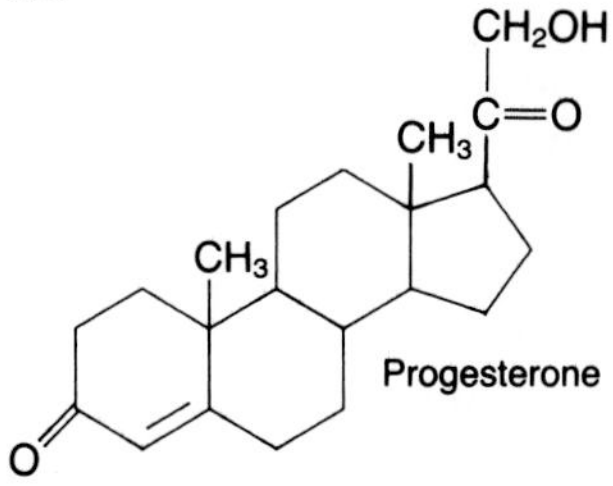 Progesterone	Prepares the uterus wall to accept a fertilized egg and maintain pregnancy.
ORALLY ACTIVE SYNTHETIC FEMALE HORMONES 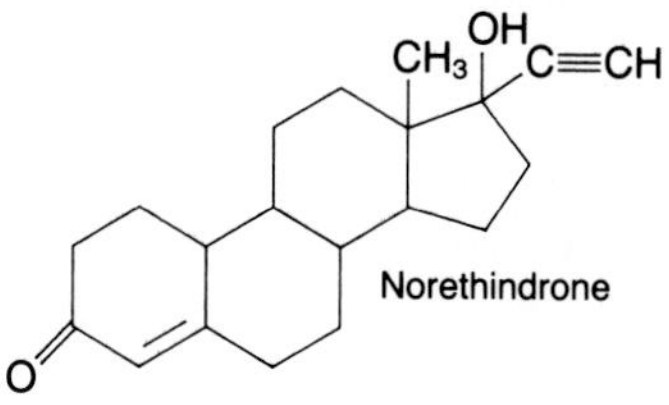 Norethindrone	Like progesterone, this synthetic suppresses release of ova and thus prevents pregnancy. The synthetic must be used because the body would digest orally ingested progesterone.
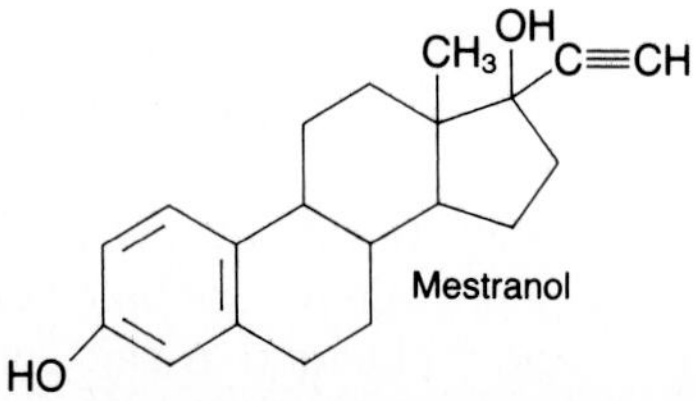 Mestranol	A synthetic estrogen is combined with the actual contraceptive component in birth control pills in order to reduce side effects.

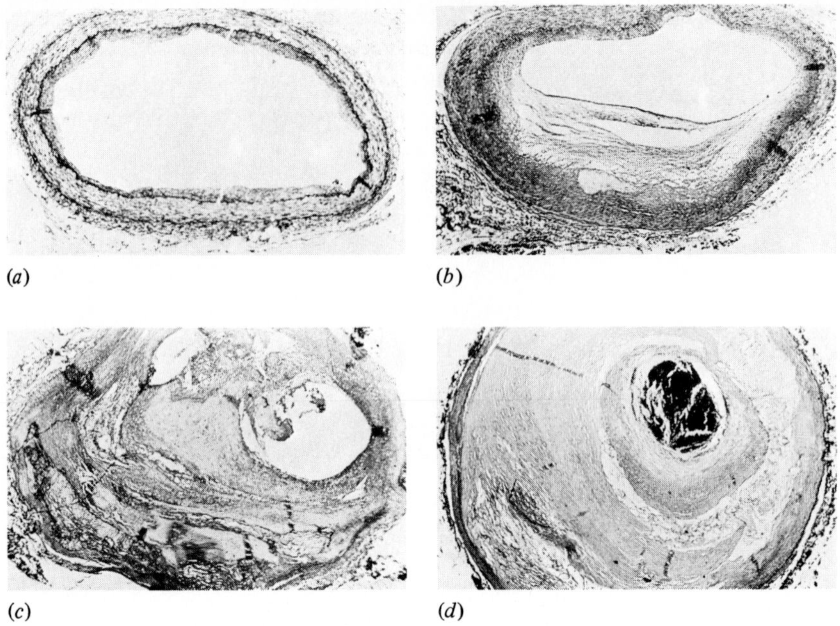

FIG. 18.6
Photomicrographs of cross sections of arterial walls. (*a*) Healthy coronary artery. (*b*) Lipid deposits are starting to form in the inner lining. (*c* Hardened deposits; the channel has become very narrow. (*d*) Narrowed channel is easily blocked by a clot.

18.15 LIPID DIGESTION, ABSORPTION, AND TRANSPORT

Triacylglycerols are emulsified and digested into glycerol and fatty acids and their salts in the small intestine. Biochemists do not agree as to how completely the triacylglycerols are broken down into their component parts. The fatty acids and glycerol cross the intestinal membrane and enter the intestinal mucosa cells, which line the intestine, where they are recombined into triacylglycerols. In order to solubilize these lipids for transport in hydrophilic body fluids, they are complexed with proteins to form lipoproteins. Lipoproteins can be assembled into larger structures called *chylomicrons*. The lipoproteins and chylomicrons carry lipids into the lymph, and then into the bloodstream and to the liver and adipose tissue, where the lipids are again hydrolyzed into their component parts. This process will be explored more fully in Chapter 24.

Phospholipids follow a similar digestion and absorption course. Cholesterol can enter intestinal cells without digestion and be incorporated into lipoproteins and absorbed into the lymph and blood systems. Lipids are always transported through the bloodstream bound to protein material.

In studies of the relationship of cholesterol to heart disease it has been determined that there are several kinds of lipoproteins which can be characterized in terms of density and the relative amounts of lipid and protein material in their structures. Table 18.5 shows the composition, density, and some average values for the blood levels of the three principal categories of lipoproteins.

Notice that the density increases as the percentage of protein increases. The very-low-density lipoproteins (VLDL) and the low-density lipoproteins (LDL) carry lipids, including cholesterol, to cells for metabolism. The high-density lipoproteins (HDL), on the other hand, carry cholesterol and other lipids away from body cells to the liver, where the fats are processed for excretion. Thus the HDL's provide a cleanup mechanism to dispose of excess cholesterol and thus avoid its buildup in blood vessels and atherosclerosis. See Figure 18.6.

TABLE 18.5 Blood Lipoproteins

	Very-Low-Density Lipoproteins (VLDL)	Low-Density Lipoproteins (LDL)	High-Density Lipoproteins (HDL)
Lipid, %	90	75	60
Protein, %	10	25	40
Density, g/mL	0.95–1.01	1.02–1.06	1.06–1.21
Average blood level, mg/100 mL of plasma:			
Men			45
Women			55
Male runners			65
Female runners			75

A distinct correlation between HDL levels and the risk of coronary disease has been established. Assuming constant levels of LDL's, one is less and less likely to have a heart attack as the levels of HDL rise. Or we could say the lower the LDL/HDL ratio, the safer one is from heart disease. One rule of thumb is that one's risk drops 25 percent for each rise of 5 mg/100 mL of HDL above the average values. Women are less susceptible to heart attacks because of their higher HDL concentrations. Runners usually develop higher levels of HDL proportional to the extent to which they exercise. This is thought to be one link between exercise and reduced risk of coronary attacks.

The amount of cholesterol deposits in blood vessels is determined by dietary intake of lipids, the efficiency of cells in accepting and metabolizing lipids, and the amount of high-density lipoprotein available to dispose of excess lipids. Current research indicates the latter two factors are more significant than diet.

CHAPTER ACCOMPLISHMENTS

After completing this chapter you should be able to define **key words** and do the following:

18.1 INTRODUCTION

1. Describe a property which distinguishes lipids from the other components of the biomass.

18.2 WAXES

2. State the distinguishing structural element common to all simple lipids.
3. Describe the structural components common to waxes under the old definition.

18.3 FATS AND OILS

4. Write the structure of a triacylglycerol that could form from the reaction of glycerol with three given carboxylic acids.
5. Distinguish between fats and oils on the basis of physical and structural properties.
6. Given the structure or composition of a triacylglycerol, predict whether the substance is likely to be a fat or an oil.

18.4 HYDROGENATION OF OILS

7. Describe how oils can be converted into fats.

18.5 IODINE NUMBER

8. Given the iodine numbers of two triacylglycerols, compare their degree of unsaturation.

18.6 HYDROLYSIS OF SIMPLE LIPIDS AND DIGESTION

9. Write the structures for the acid-catalyzed hydrolysis products of a given triacylglycerol.
10. Describe the function of the bile salts in the digestion of lipids.

18.7 SAPONIFICATION

11. Write structures for the alkaline hydrolysis products of a given triacylglycerol.

18.8 COMPOUND LIPIDS

12. On the basis of the structural components, distinguish between simple and compound lipids.

18.9 PHOSPHOLIPIDS

13. Write the structure of a lecithin or cephalin formed from a given set of components.
14. Label the hydrophobic tail and hydrophilic head regions of a phosphoglyceride.
15. Draw a pictorial arrangement of phosphoglyceride structures in an aqueous medium.

18.10 THE LIPID BILAYER OF CELL MEMBRANES

16. Describe the structural arrangement of lipid molecules that gives rise to a cell membrane.
17. Compare simple diffusion, facilitated diffusion, and active transport with respect to the presence of a concentration gradient.

18.11 CELL ORGANIZATION

18. Distinguish between eucaryotic and procaryotic cells on the basis of their structural components.

18.12 SPHINGOLIPIDS

19. Describe the structural differences between a phospholipid and a sphingomyelin.
20. Write the structures for the hydrolysis products of a given sphingomyelin.
21. Recognize a cerebroside or a ganglioside in a given structure.
22. State the underlying molecular problem in Tay-Sachs and Niemann-Pick disease.

18.13 PROSTAGLANDINS

23. State the type of lipid from which prostaglandins are produced.
24. State at least two known physiological effects of the prostaglandins.
25. Describe the hypothesis linking the action of aspirin with the prostaglandins.

18.14 STEROIDS

26. Give the general structural formula for a steroid.
27. Describe the functional-group differences between two given steroids.
28. Classify a given lipid structure as a wax, triacylglycerol, phospholipid, sphingolipid, prostaglandin, or steroid.

18.15 LIPID DIGESTION, ABSORPTION, AND TRANSPORT

29. Describe the digestion and absorption process for triacylglyerols.
30. Describe the hypothesis concerning the relationship between the LDL/HDL ratio and the threat of atherosclerosis.

PROBLEMS

18.6 Describe a physical procedure by which lipids can be separated from the other components of the biomass.

18.7 The ester functional group plays a very important role in the chemistry of lipids. Why aren't lipids classified as the components of the biomass that contain an ester functional group?

18.8 Why are many lipids saponifiable?

18.9 Name some lipids you encounter every day.

18.10 Describe a simple lipid.

18.11 Consult Table 18.1 and write the structure of the hydrolysis products from the principal ester component of spermaceti wax.

18.12 *a* State the old definition of a wax.

b State the more recent description of a wax.

c Compare the two definitions. Which is the broader definition?

18.13 Describe a function that may be served by a wax on the surface of a plant leaf.

18.14 Write the structure of the triacylglycerol that could form from glycerol and 3 mol of stearic acid.

18.15 Write structures for the two triacylglycerols that could form from glycerol, 2 mol of lauric acid, and 1 mol of capric acid.

18.16 Write structures for the three triacylglycerols that could form from glycerol and 1 mol each of stearic, lauric, and capric acids.

18.17 Oleic and stearic acids have approximately the same molecular weight (282 and 284, respectively); however, their respective meltings points, 14 and 70°C, are very different. Give a possible explanation for this difference.

18.18 *a* Describe the physical difference between a fat and an oil.

b Describe a structural distinction between a fat and an oil that can account for the physical difference.

18.19 Using Table 18.3, state the most abundant acid in the following:

a Olive oil

b Human fat

c Coconut oil

18.20 What is the difference between a fat and a fatty acid?

18.21 What general type of reaction converts an unsaturated oil into a saturated fat?

18.22 How many moles of hydrogen would be required to form a completely saturated fat from 1 mol of a triacylglycerol containing only linoleic acid?

18.23 Give a definition of iodine number.

18.24 *a* Describe the visual result obtained as iodine is continously added to a solution of an oil.

b What role could the iodine number serve in helping a consumer decide which polyunsaturated oil to purchase?

18.25 *a* A 400-g sample of a fat absorbs 201 g of iodine. What is the iodine number of this fat?

b A 145-g sample of a oil absorbs 198 g of iodine. What is the iodine number of this oil?

18.26 Write structures for the product(s) of reactions between the indicated reagents and the reactant following:

$$H_2C-O-\overset{\overset{\textstyle O}{\|}}{C}-(CH_2)_7CH{=}CH(CH_2)_7CH_3$$
$$HC-O-\overset{\overset{\textstyle O}{\|}}{C}-(CH_2)_8CH_3$$
$$H_2CO-\underset{\underset{\textstyle O}{\|}}{C}-(CH_2)_6(CH_2CH{=}CH)_2(CH_2)_4CH_3$$

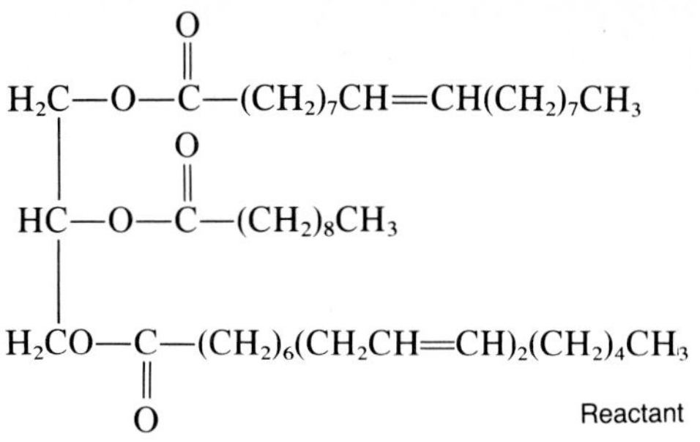

Reactant

a Reactant + 3H$_2$O $\xrightarrow{H^+}$

b Reactant + 3NaOH $\longrightarrow$

c Reactant + 3I$_2$ $\longrightarrow$

18.27 What structural feature of bile salts enables them to function as emulsifiers?

18.28 *a* What is the main physiological functional purpose served by the dietary use of triacylglycerols?

b By the dietary use of compound lipids?

18.29 Describe the structural differences between a triacylglycerol and a phospholipid.

18.30 Describe the structural features shared by lecithins and cephalins.

18.31 Write the structure of the lecithin that could form from glycerol, phosphoric acid, oleic acid, stearic acid, and choline.

18.32 Label the hydrophilic head and the hydrophobic tail regions of the lecithin molecule in Problem 18.31.

18.33 Although triacylglycerols are completely insoluble in water, phospholipids such as the cephalins have some degree of water solubility. Explain why the cephalins might have a greater water solubility than the triacylglycerols.

18.34 A salad dressing of oil and vinegar will only stay "mixed" for a short time; however, if egg yolk is added, a permanent emulsion known as mayonnaise can be obtained. Describe the molecular function of the egg yolk.

18.35 What structural feature of a phospholipid contributes to the "fluid" character in a cell membrane.

18.36 Write structures for all the products that are produced by the saponification of the following lipids:

a 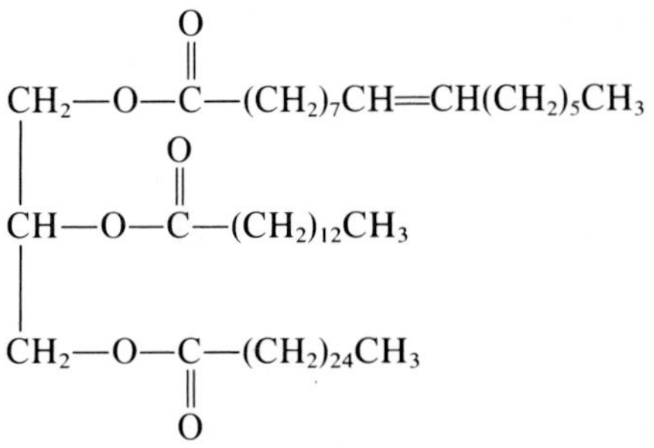
$$CH_2-O-\overset{\overset{\textstyle O}{\|}}{C}-(CH_2)_7CH{=}CH(CH_2)_5CH_3$$
$$CH-O-\overset{\overset{\textstyle O}{\|}}{C}-(CH_2)_{12}CH_3$$
$$CH_2-O-\underset{\underset{\textstyle O}{\|}}{C}-(CH_2)_{24}CH_3$$

b 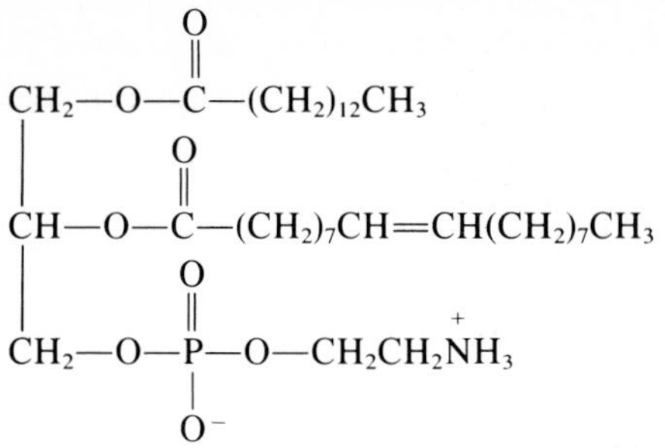

$$CH_2-O-\overset{\displaystyle O}{\overset{\|}{C}}-(CH_2)_{12}CH_3$$

$$CH-O-\overset{\displaystyle O}{\overset{\|}{C}}-(CH_2)_7CH=CH(CH_2)_7CH_3$$

$$CH_2-O-\overset{\displaystyle O}{\overset{\|}{P}}-O-CH_2CH_2\overset{+}{N}H_3$$
$$\underset{\displaystyle O^-}{|}$$

18.37 *a* The concentration of K^+ ions in blood plasma is about 3 millimolar while in the cytoplasm of a red blood cell the concentration of K^+ is 110 millimolar. What type of passage would be needed for K^+ ions to move into the cell? Explain your answer.

b Na^+ concentration in blood plasma is about 140 millimolar, whereas in a red blood cell the concentration is 4 millimolar. What type of passage would be needed for Na^+ ions to move into the cell? To move out into the blood plasma?

18.38 Name the three classes of lipids that are found in cell membranes.

18.39 Describe one difference between a procaryotic and a eucaryotic cell.

18.40 *a* What structural feature allows sphingomyelins to be classified as phospholipids?

b What structural feature allows sphingomyelins to be classified as sphingolipids?

18.41 Write the structure of the sphingomyelin that could form from sphingosine, stearic acid, phosphoric acid, and ethanolamine.

18.42 Label the hydrophilic tail and the hydrophobic head regions of the sphingomyelin structure in Problem 18.41.

18.43 *a* Describe, in general, the hydrolysis products of a sphingomyelin.

b Describe, in general, the hydrolysis products of a glycolipid.

c What differences do you find between the components of parts *a* and *b* of this problem?

d Which structure, a sphingomyelin or a glycolipid, has a more polar head?

18.44 The symptoms of Tay-Sachs disease are generally associated with pathological changes in the nervous system. Describe the underlying molecular basis of this observation.

18.45 What is the molecular disorder in Niemann-Pick disease?

18.46 *a* Write the general structure for a prostaglandin molecule.

b What common features are present in all prostaglandin molecules?

18.47 *a* From what compound lipid are prostaglandins formed?

b What is the name of the fatty acid formed in the route from compound lipid to prostaglandin?

18.48 Describe a hypothesis that can account for aspirin's fever-reducing and anti-inflammatory action.

18.49 *a* Give at least three physiological effects of the prostaglandins.

b What medicinal advantage can be taken of these effects?

18.50 *a* Give the general structure of a steroid.

b Identify position 17 in your structure.

18.51 Cutting back on dietary intake of cholesterol *may* not reduce the level of cholesterol in the blood. Explain.

18.52 What structural differences are there between the male hormonal steroid testosterone and the female hormone estradiol?

18.53 Lipids are not generally soluble in blood plasma. How then are they transported through the body?

18.54 To what class of lipids (wax, triacylglycerol, phospholipid, etc.) do each of the following compounds belong?

a 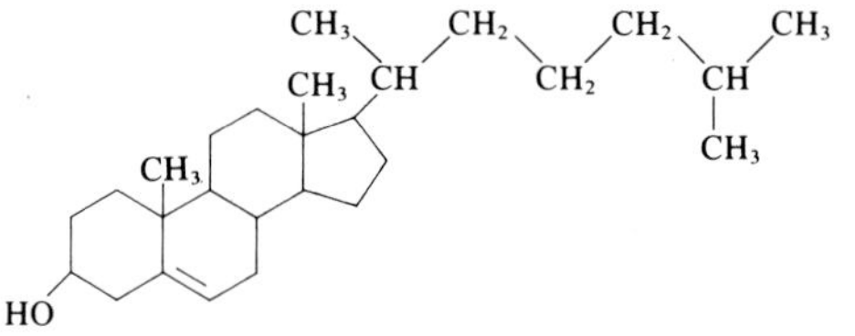

b 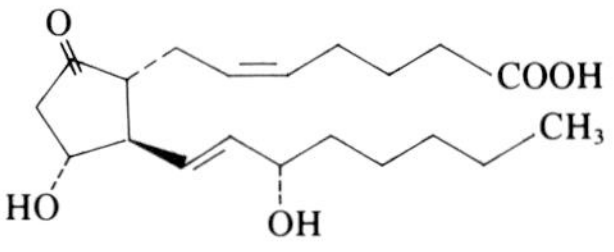

c $CH_3(CH_2)_{20}\overset{\displaystyle O}{\overset{\|}{C}}-O-(CH_2)_{17}CH_3$

d 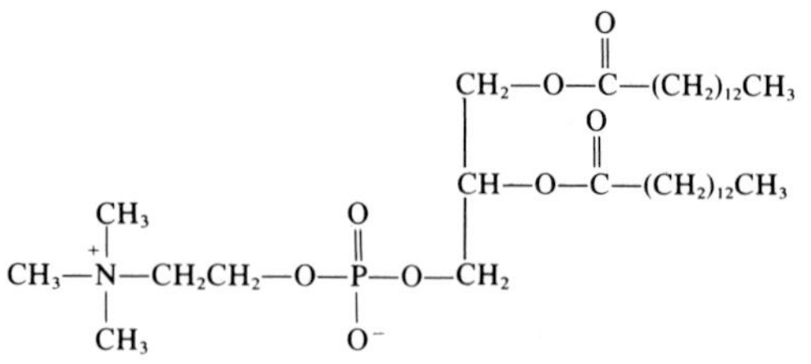

e 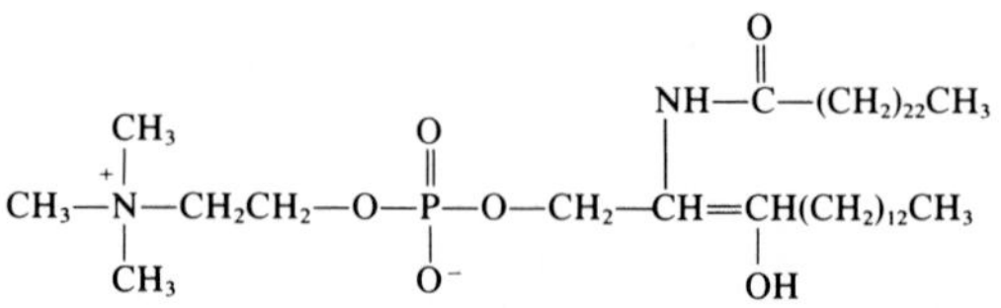

f 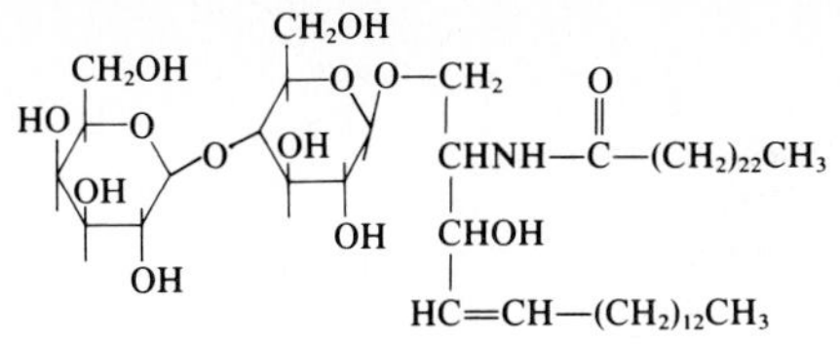
CH₂OH
CH₂OH
HO
O
O
O O—CH₂
OH
OH
O
CHNH—C—(CH₂)₂₂CH₃
OH
CHOH
OH
HC=CH—(CH₂)₁₂CH₃

18.55 Which of the following will not yield a fatty acid upon hydrolysis?
a Cholesterol
b Phosphatidylserine
c Tristearylglycerol
d Cetyl palmitate
e Prostaglandin E_2
f A spingomyelin
g Vitamin D

18.56 *a* Which of the structures given in Problem 18.54 yield glycerol upon hydrolysis?
b Which yield a sphingosine fragment?

18.57 List all the functional groups present in the steroid aldosterone.

18.58 List at least three general functions of lipids.

CHAPTER NINETEEN

AMINO ACIDS AND PROTEINS

19.1 INTRODUCTION

The word *protein* is derived from the Greek word *protos,* meaning "the first." In many ways proteins are first in importance among the three classes of biomolecules: carbohydrates, lipids, and proteins. In most organisms the mass of protein is greater than the combined mass of carbohydrates and lipids. Moreover, proteins serve in a wider variety of biological functions than do the other biomass components.

These functions range from structural support and movement, because proteins are the material of skin, ligaments, tendons, cartilage, and muscle, to biological catalysis, because enzymes are largely proteinaceous. The oxygen transporters, hemoglobin and myoglobin, are proteins, as are the defense and regulatory agents of the body, the antibodies and hormones.

Every cell of the body contains thousands of different proteins, each designed to carry out a specific function. However, each protein is composed from a basic set of 20 amino acids. Proteins, like polysaccharides, are polymers. In polysaccharides the monomeric units, usually glucose, are held together by acetal-type linkages. In proteins the amino acids are linked together by amide-type bonds called peptide bonds. Your knowledge of amines (Chapter 16) and carboxylic acids and amides (Chapter 17) will now be applied to understanding the construction of proteins from amino acids.

19.2 STRUCTURE OF AMINO ACIDS

As the name implies, amino acids contain two functional groups, a carboxylic acid and an amine group. In all amino acids found in proteins the amine functional group is bonded to the alpha (α) carbon, so that these compounds are known as alpha amino acids. In this text we shall simply refer to them as amino acids. (Recall from Section 17.2 that the α-position refers to the carbon in position 2 if one employs the IUPAC numbering system.)

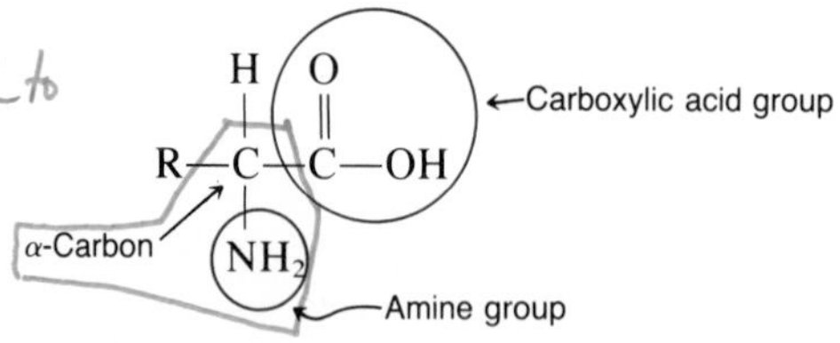
Carboxylic acid group
α-Carbon
Amine group

Because a carboxylic acid group is acidic and an amino group is basic, the structure just given is not really accurate. Rather, a picture of an amino acid containing a single acid and a single basic group at cellular pH (6–7) should show a doubly charged species, wherein the carboxylic acid group has lost a proton and the amino group has gained a proton.

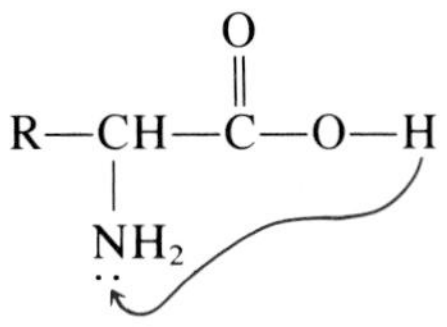

This dipolar ion is often called by the German word *zwitterion*. An amino acid in the zwitterionic form is electrically neutral because the negative charge is balanced by the positive charge.

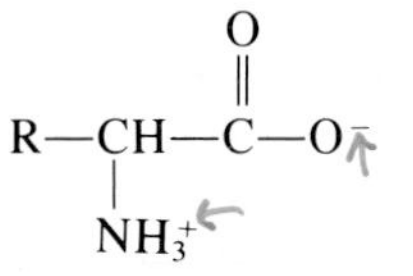

Amino acid in zwitterionic form

The amino acids differ from one another in the nature of the R group. In the simplest amino acid, glycine, R = H and in all other cases R is an organic group. Each amino acid is referred to by its common name or a three-letter abbreviation. These names and standard abbreviations appear in Table 19.1, which appears after its entries are explained in Section 19.3. The differences in physical and chemical properties among the amino acids reflect the differences in the nature of the R group.

Problem 19.1 The R group of a particular amino acid is

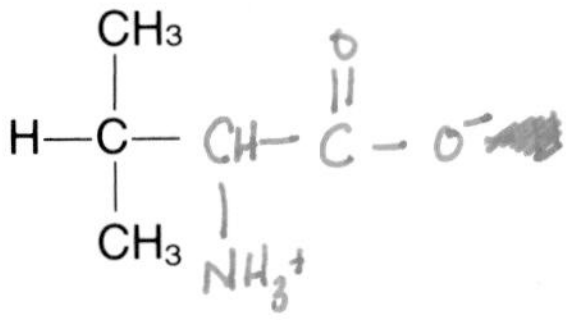

Write the full structure of the amino acid in the zwitterionic form.

19.3 CLASSIFICATION OF AMINO ACIDS

The variety among R groups is such that some interact attractively with water molecules, i.e., they are hydrophilic, and others repel water molecules, i.e., they are hydrophobic. You will see later that the nature of the R group significantly influences protein structure. Nine of the twenty amino acids found in proteins contain R groups which are nonpolar and are therefore hydrophobic. For example, alanine (Ala) has the hydrophobic R group, methyl. The R groups in the remaining 11 amino acids contain functional groups which are hydrophilic because they are able to form hydrogen bonds or they are highly polar. For example, serine's R group is —CH_2OH. The alcohol group can form hydrogen bonds with water molecules.

Furthermore, five of these eleven hydrophilic amino acids contain an additional carboxylic acid group or basic nitrogen group in their R side chain. That is, these amino acids have more than one —COOH group or more than one amino group. At physiological pH, acid groups are ionized and consequently carry a negative charge. That is, at physiological pH, the acid ionization equilibrium (Section 11.4) lies to the right and the anion predominates in solution.

$$RCOOH \rightleftharpoons RCOO^- + H^+$$

Acid Anion

Under these same physiological conditions a basic nitrogen is protonated and thus is positively charged. At a pH of about 7 each of the five amino acids with extra acid or amine functional groups is electrically charged.

Aspartic acid and glutamic acid bear a negative charge because each has a second carboxylic acid group. Their names reflect their classification as acidic amino acids. An amino acid which has more than one carboxylic acid group is classified as an **acidic** amino acid and bears a negative charge under physiological pH conditions.

On the other hand, amino acids that have more than one basic nitrogen are classified as **basic** amino acids, and they bear a positive charge at the body's normal pH. Examples of basic amino acids are lysine, arginine, and histidine. The other hydrophilic amino acids are designated **neutral** amino acids. In Table 19.1 we have classified the 20 amino acids into hydrophobic and hydrophilic categories and have subdivided the hydrophilic grouping into neutral and charged (acidic or basic) subcategories. The numbers called isoelectric points in the table reflect the acidic, basic, or neutral nature of amino acids and will be explained in Section 19.5.

19.4 STEREOISOMERISM IN AMINO ACIDS

With the sole exception of glycine, the alpha carbon in all α-amino acids is a chiral carbon. A pair of nonsuperimposable mirror-image isomers exist for each of these amino acids. Structures for the two enantiomers can be drawn in a similar way to the Fisher projection scheme we saw earlier for carbohydrates. The alpha amino group replaces the hydroxyl (of carbohydrates) at the chiral carbon in the enantiomeric structures.

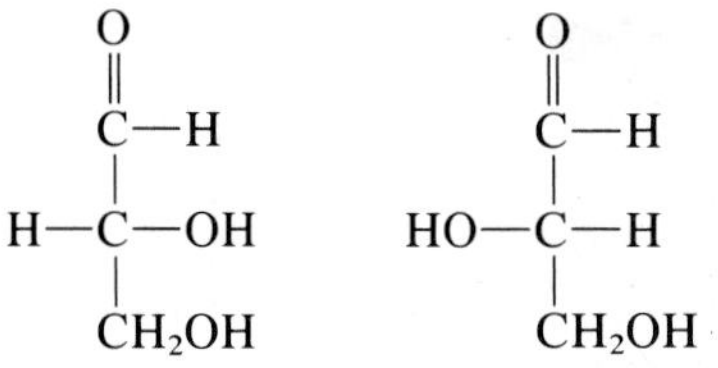

D-Glyceraldehyde L-Glyceraldehyde

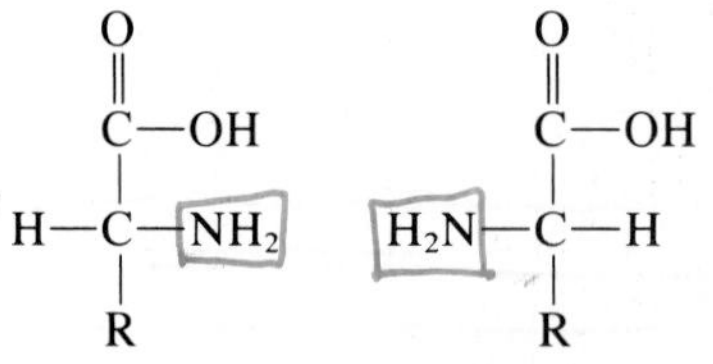

D-Amino acid L-Amino acid

The amino acids from all proteins found in nature belong to the L-series. This particular stereochemistry of the amino acid building blocks fixes a unique spatial orientation on the atoms of a protein molecule. Proteins from living matter on another planet could be made up of D-amino acids. An earthly traveler to that planet could not obtain his or her dietary amino acids from protein matter there because metabolic requirements always specify a definite enantiomer.

TABLE 19.1 Twenty Amino Acids Found in Proteins

Structure	Name	Three-Letter Abbreviation	Isoelectric Point
	I. AMINO ACIDS WITH A NONPOLAR (*HYDROPHOBIC*) R GROUP		
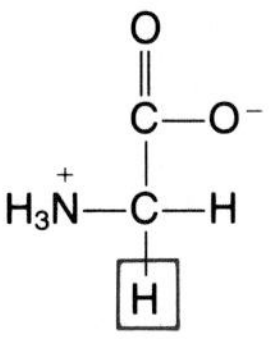	Glycine	Gly	5.97
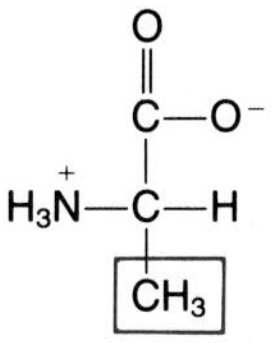	Alanine	Ala	6.02
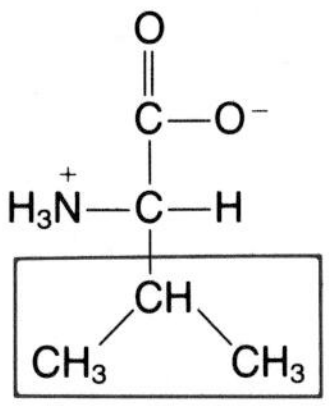	Valine	Val	5.97
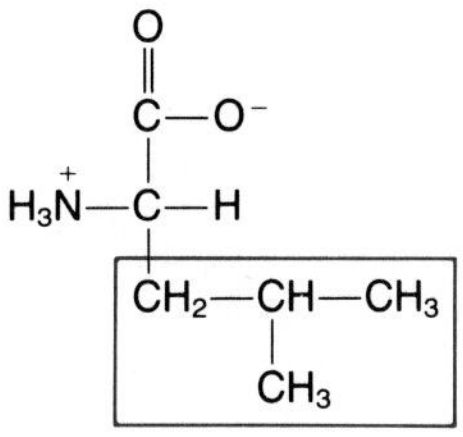	Leucine	Leu	5.98
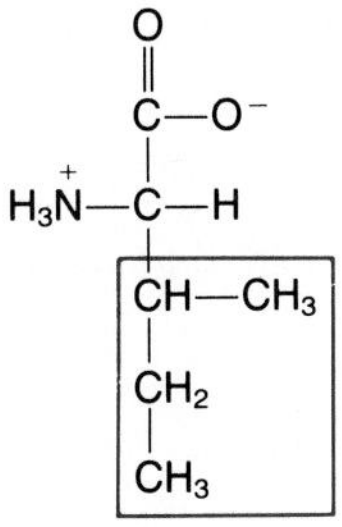	Isoleucine	Ile	6.02
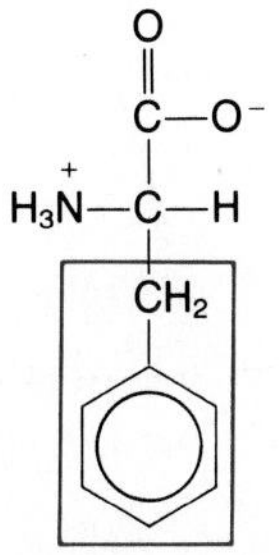	Phenylalanine	Phe	5.48

TABLE 19.1 *(Continued)*

Structure	Name	Three-Letter Abbreviation	Isoelectric Point
I. AMINO ACIDS WITH A NONPOLAR (*HYDROPHOBIC*) R GROUP (*cont'd*)			
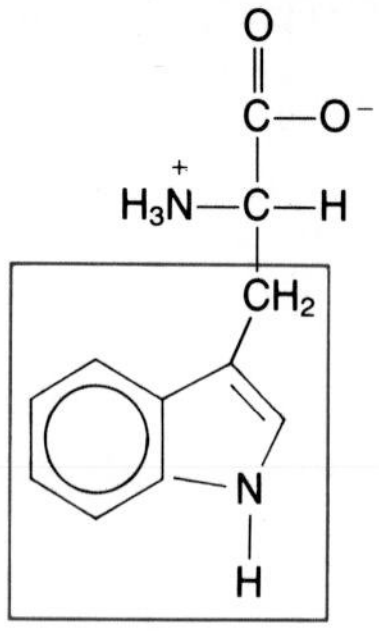	Tryptophan	Trp	5.88
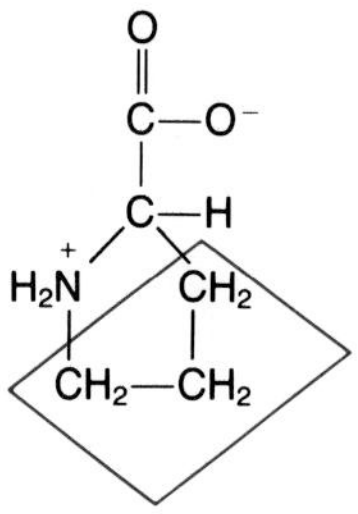	Proline	Pro	6.20
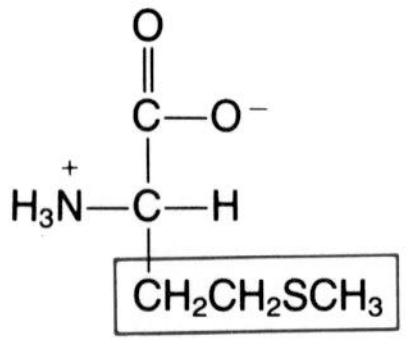	Methionine	Met	5.74
II. AMINO ACIDS WITH A POLAR (*HYDROPHILIC*) R GROUP			
A. ELECTRICALLY NEUTRAL (ZWITTERION) AT PHYSIOLOGICAL pH			
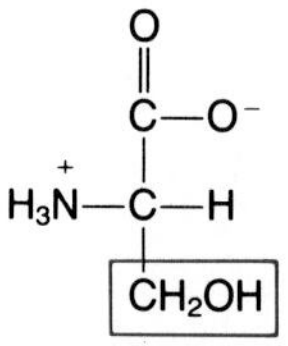	Serine	Ser	5.68
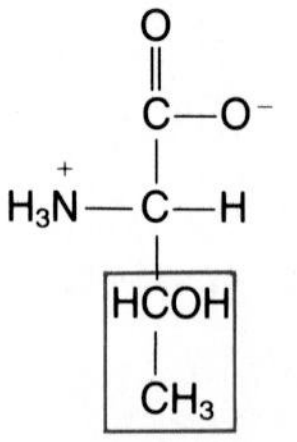	Threonine	Thr	5.60

TABLE 19.1 *(Continued)*

Structure	Name	Three-Letter Abbreviation	Isoelectric Point
II. AMINO ACIDS WITH A POLAR (*HYDROPHILIC*) R GROUP			
A. ELECTRICALLY NEUTRAL (ZWITTERION) AT PHYSIOLOGICAL pH ***(cont'd)***			
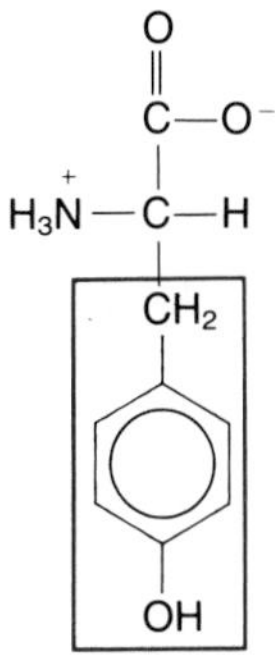	Tyrosine	Tyr	5.66
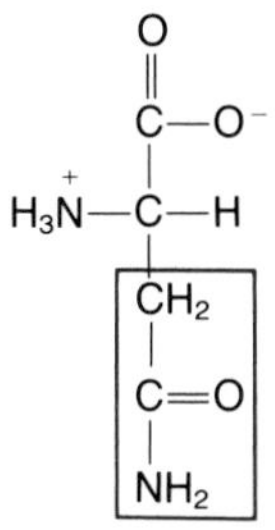	Asparagine	Asn	5.41
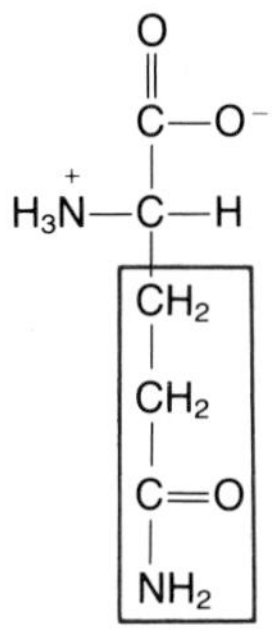	Glutamine	Gln	5.65
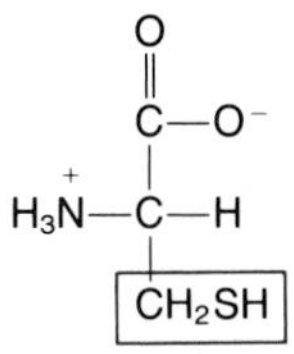	Cysteine	Cys	5.02
B. BASIC AMINO ACIDS—POSITIVELY CHARGED AT PHYSIOLOGICAL pH			
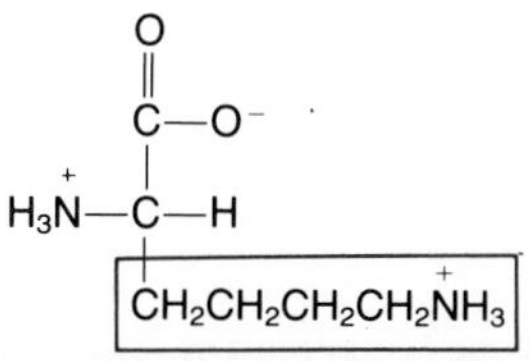 protonated	Lysine	Lys	9.74

TABLE 19.1 *(Continued)*

Structure	Name	Three-Letter Abbreviation	Isoelectric Point
B. BASIC AMINO ACIDS—POSITIVELY CHARGED AT PHYSIOLOGICAL pH ***(cont'd)***			
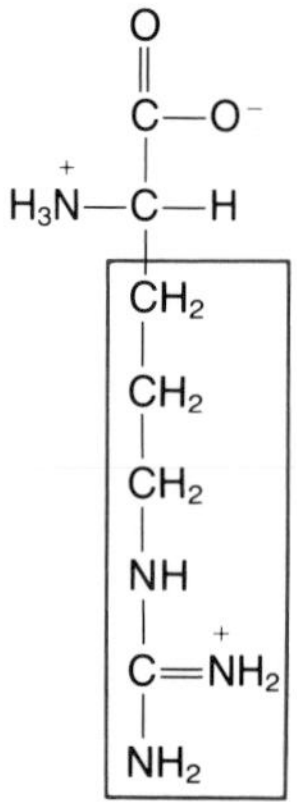	Arginine	Arg	10.76
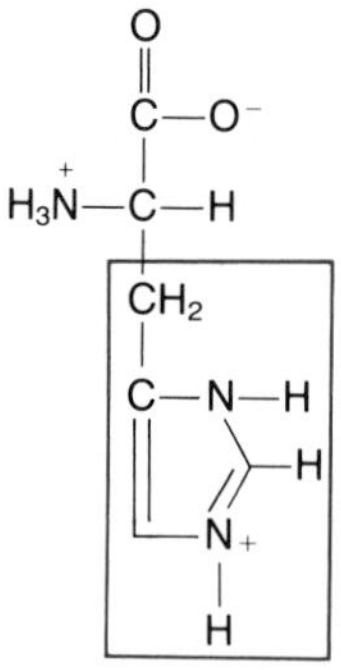	Histidine*	His	7.59
C. ACIDIC AMINO ACIDS—NEGATIVELY CHARGED (ANIONS) AT PHYSIOLOGICAL pH			
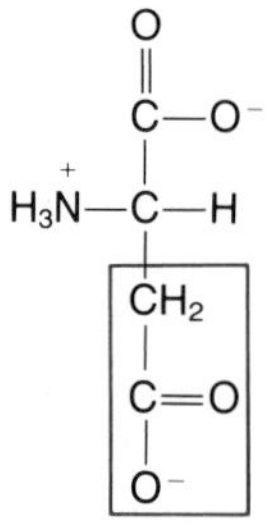	Aspartic acid	Asp	2.98
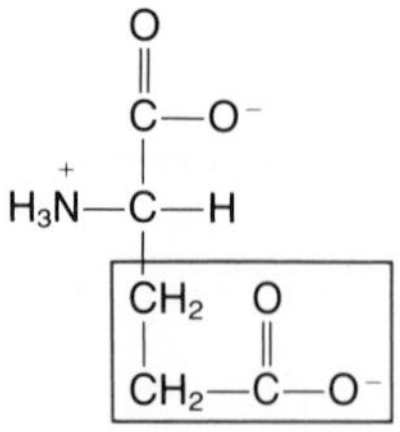	Glutamic acid	Glu	3.22

* Actually the protonated form becomes predominant only at pH values below 6.

19.5 AMINO ACIDS IN SOLUTION

Amino acids are white, crystalline solids with fairly high melting points. They exist as zwitterions in the solid state. Because they are highly polar, as you would expect, amino acids are all at least slightly soluble in water.

Isoelectric Point

In aqueous solution the zwitterionic form of an amino acid is in equilibrium with a positively charged cationic and a negatively charged anionic structure.

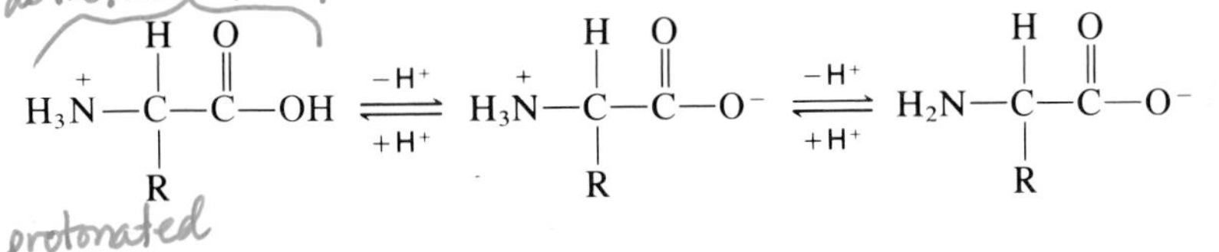
H O
$H_3\overset{+}{N}$—C—C—OH
R
−H⁺
+H⁺
H O
$H_3\overset{+}{N}$—C—C—O⁻
R
−H⁺
+H⁺
H O
H_2N—C—C—O⁻
R

Cation Zwitterion Anion

The exact concentration of each species in the equilibrium mixture is dependent upon the pH of the solution. In strongly acidic solution (pH < 1), such as would be found in the human stomach, the carboxylic acid group is un-ionized, the amine group is protonated, and the amino acid exists predominantly as the cation. If the solution is made highly basic (pH > 11), both the protonated amine and the carboxylic acid group give up H^+ to the basic solution and the predominant species in the equilibrium becomes the anion.

At intermediate pH values such as exist in blood plasma (pH = 7.4) or cellular fluids (pH = 4–8), cations, anions, and zwitterions coexist in equilibrium. At some intermediate pH value the concentration of the zwitterion form is at a maximum and the concentrations of cationic and anionic molecules are equal and at a minimum. This pH value at which there is a maximum amount of zwitterion is called the *isoelectric point*. It is given the symbol pI. The name isoelectric and the associated condition of electrical neutrality derives from the fact that at this pH the sum total of charges on the amino acid molecules is zero. Most molecules are the electrically neutral zwitterion form, and the concentrations of cationic and anionic forms are equal. At pH values below pI the cationic form is favored, and at pH values above pI, the anionic form predominates.

Because the R group can affect the acidity of the carboxylic acid and the basicity of the amino group, each amino acid has a different value of pI. In Table 19.1 you will find the pI values for all 20 amino acids. As you might expect, amino acids with very similar R groups such as leucine and isoleucine have almost identical pI values.

Attraction of Zwitterions

Zwitterion molecules, because they contain a dipole and thus can powerfully attract one another between oppositely charged ends, tend to aggregate together in solution.

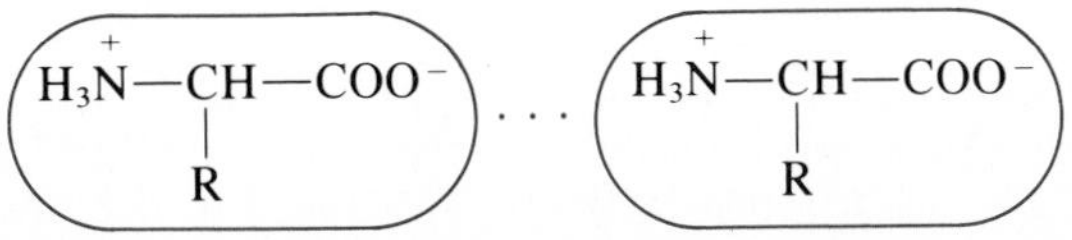
$H_3\overset{+}{N}$—CH—COO⁻
R
. . .
$H_3\overset{+}{N}$—CH—COO⁻
R

or simply,

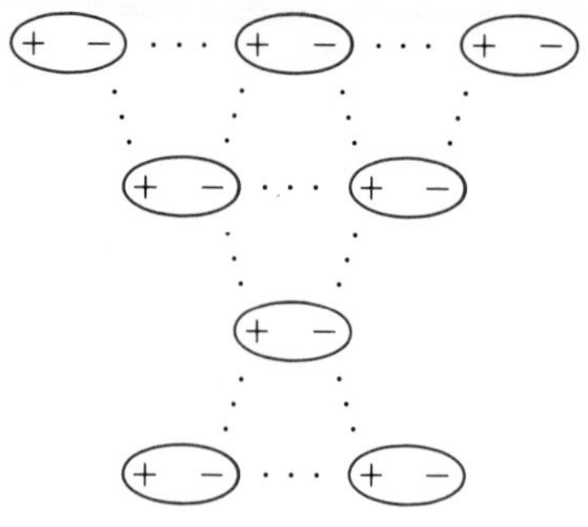

Such aggregation leads to a lowering of solubility. This aggregation is greatest at the isoelectric point because the zwitterion concentration is a maximum at this pH. Thus the solubility of an amino acid is at a minimum at the isoelectric point. This variation of solubility with pH can be used to advantage in separating amino acids and proteins.

Protein molecules also have an isoelectric point at which they exist to a maximum extent in electrically neutral dipolar form. Like amino acids, the solubility of a protein is at a minimum at its isoelectric point. Casein, a protein in milk, precipitates out of solution as the familiar white curds in "sour milk" because the isoelectric point of casein (pI = 4.6) is reached when bacterial action on the milk leads to the production of lactic acid. This acid causes the pH to drop and gives a sour taste to the milk.

Practical Uses of Charge Character

The electrical charge properties and behavior of amino acids in solution are of interest because the same principles apply to proteins, which are polymers of amino acids. Also this behavior is used in the analysis and identification of protein structures, which are common problems in biomedical research. Any protein can be broken down into its constituent amino acids by heating it in a hydrochloric acid solution. As you will see later, this is simply the test tube version of digestion. To separate the amino acids and discover how much of each one is present in the decomposition mixture, we can take advantage of the different ionization characters of the amino acids which affect the proportions of charged and uncharged forms at any particular pH.

For example, in a solution at pH = 6 the predominant forms of the three amino acids, alanine, lysine, and aspartic acid, are

$H_3\overset{+}{N}—CH(CH_3)—COO^-$

Alanine zwitterion
(pI = 6.02)

$H_3\overset{+}{N}—CH((CH_2)_4{}^+NH_3)—COO^-$

Lysine cation
(pI = 9.7)

$H_3\overset{+}{N}—CH(CH_2COO^-)—COO^-$

Aspartic acid anion
(pI = 3.0)

That is, alanine is neutral, lysine is positively charged, and aspartic acid is negatively charged. This permits their separation in an electrical field, as shown in Figure 19.1. The analytical procedure of *electrophoresis* is based on this principle.

In one electrophoresis procedure a mixture of amino acids (e.g., alanine, lysine, and aspartic acid) is embedded in the middle of a strip of cellulose

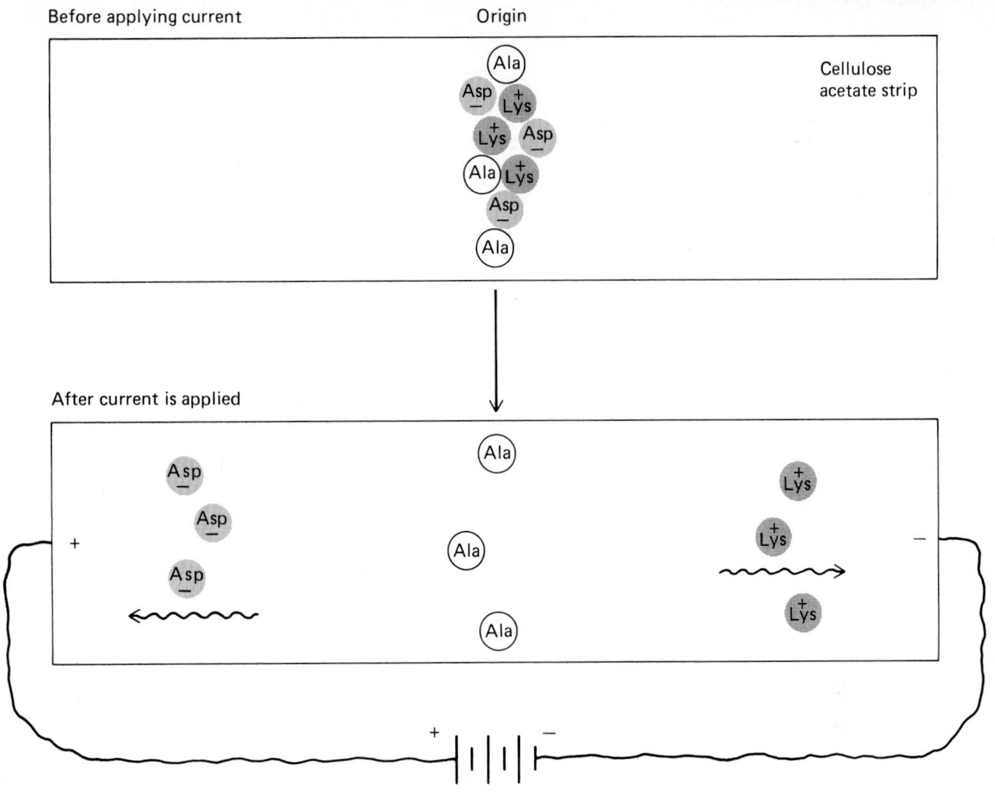

Before applying current
Origin
Cellulose acetate strip
Ala
Asp
Lys
After current is applied
+
−

FIG. 19.1
Electrophoresis. Before current is applied, the amino acids remain mixed together. After current is applied, the amino acids separate by the principle of migration of ions to oppositely charged electrical poles.

acetate soaked with a buffer of fixed pH (in Figure 19.1, pH = 6). Then the ends of the strip are attached to electrodes so that one end acquires a positive and one end acquires a negative charge. The individual amino acids migrate to either the positive or negative pole and at different rates depending on the charge on the amino acid. In this example, the positive lysine goes to the negative pole, the negative aspartic acid goes to the positive pole, and the neutral alanine does not move. At the end of the process the positions of the amino acids are made visible with stains that bond to amino acids. The cellulose acetate strip could be cut into three pieces to reclaim the three individual amino acids.

19.6 THE PEPTIDE BOND

The most important chemical reaction of amino acids is that between the carboxyl group of one and the alpha amino group of a second amino acid to form an amide linkage (Section 17.15). In protein chemistry the amide linkage is called a **peptide bond.** The reaction between glycine and alanine illustrates this reaction and, as shown, gives glycylalanine, a dipeptide.

$$\underset{\text{Glycine}}{H_2N-\underset{H}{\overset{H}{C}}-\overset{O}{\overset{\|}{C}}-OH} + \underset{\text{Alanine}}{H-\overset{H}{N}-\underset{CH_3}{\overset{H}{C}}-\overset{O}{\overset{\|}{C}}-OH} \longrightarrow \underset{\text{Glycylalanine (a dipeptide)}}{H_2N-\underset{H}{\overset{H}{C}}-\overset{O}{\overset{\|}{C}}-\overset{H}{N}-\underset{CH_3}{\overset{H}{C}}-\overset{O}{\overset{\|}{C}}-OH} + H_2O \quad (19.1)$$

water goes out
C bonds to N

The equation for the formation of a peptide bond is written so that the reactant on the left contributes —OH from its carboxyl group while the reactant on the right gives up an H from its alpha amino group to form water. The nitrogen of the reacting alpha amino group bonds to the carboxyl carbon. Despite the fact that amino acids actually exist in solution as ionized structures, we have written the amino acids in the nonionized form to show the loss of water clearly.

Each dipeptide still contains a free carboxyl group on the right end called the **C-terminal end** and a free amino group on the left end called the **N-terminal end.** This allows the condensation of another molecule of amino acid onto the dipeptide at either end. The resulting structure is a tripeptide. In general, the prefix in front of the word *peptide* tells us the number of amino acids combined together in that peptide. Let us condense valine with the C-terminal end of glycylalanine to form the tripeptide, glycylalanylvaline.

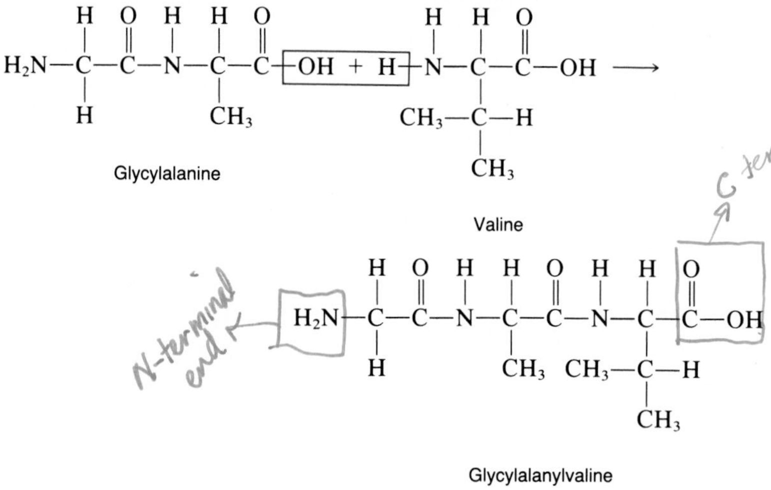
Glycylalanine
Valine
Glycylalanylvaline

(19.2)

Problem 19.2 *a* Write out the structure of the tripeptide that would be obtained by combining valine with the N-terminal end of glycylalanine.
b Name the tripeptide you formed in part *a*.

Additional amino acids can be combined with the tripeptide to form tetrapeptides (four amino acids), pentapeptides (five amino acids), and larger peptides which fall under the general name *polypeptides*.

The two amino acids glycine and alanine, which are shown reacting in Equation (19.1), could have reacted in a reversed manner, that is, employing the carboxyl group of alanine and the amino group of glycine. The resultant product alanylglycine is an isomer of glycylalanine. Place alanine on the left and glycine on the right and form the dipeptide product as was done in Equation (19.1):

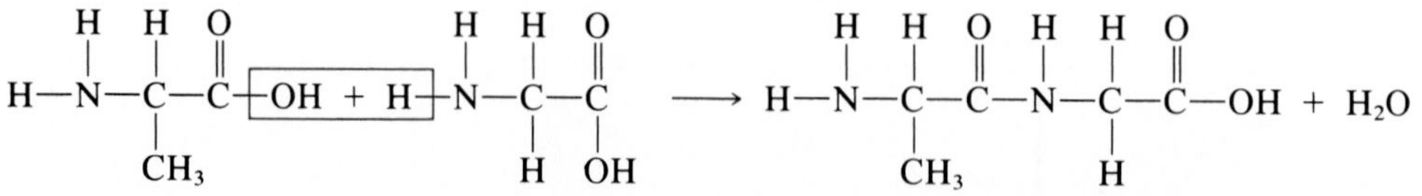

(19.3)

Alanine Glycine Alanylglycine

The structure of glycylalanine, which, for example, has a methyl group on a carbon alpha to a carboxylic acid, is clearly distinct from alanylglycine, which has the methyl group on a carbon alpha to the peptide bond. Glycylalanine and alanylglycine are different isomeric compounds. For every peptide or protein the order of joining together of the amino acids is significant. The unique order or sequence of amino acids in a polypeptide or protein is called its **primary structure.**

As you might imagine, it is inconvenient to use the full amino acid name for large polypeptides. Scientists commonly use the three-letter abbreviations given in Table 19.1 instead. Each amino acid abbreviation is separated by a hyphen and written from left to to right, indicating the sequence of amino acids from the N-terminal to the C-terminal end. Thus, the abbreviated name for glycylalanylvaline is Gly-Ala-Val. We shall refer to polypeptides and later to proteins through their abbreviated names.

SAMPLE EXERCISE 19.1

Write out the complete structure for Met-Asp-Ser.

Solution The amino acid structures may be found in Table 19.1. As shown in Equations (19.1) through (19.3), the organic reaction is most easily shown by using the nonpolar structure for amino acids. That is, the functional groups are shown as —COOH and —NH_2, rather than depicting the proton transfer that has occurred between them.

STEP 1 Write down the structure of the N-terminal amino acid by first writing down the alpha carbon and then attaching the amino group on the left and the carboxyl group on the right and placing H and R above and below the α-carbon:

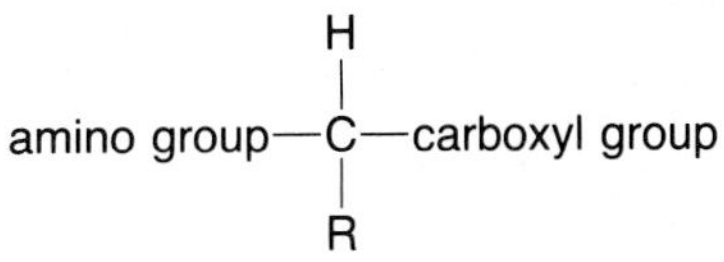
H
amino group—C—carboxyl group
R

For methionine the structure is

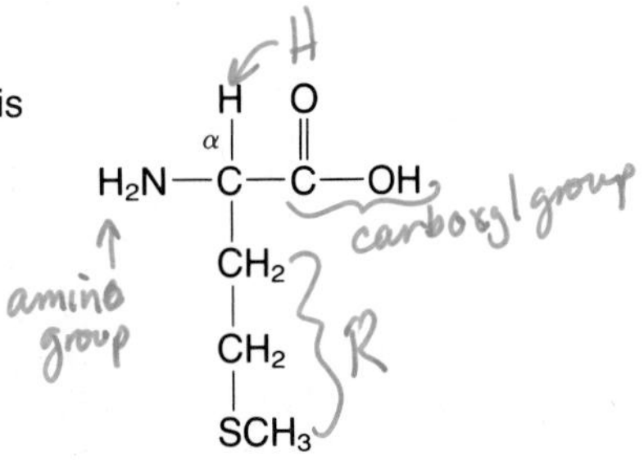
H
H O
α
H2N—C—C—OH
CH2
CH2
SCH3
carboxyl group
amino group
R

Methionine

STEP 2 To the right of the first structure write down the structure of the next amino acid in the chain in the same manner.

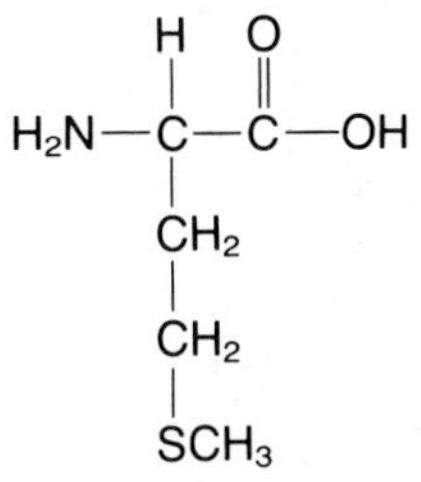
H O
H2N—C—C—OH
CH2
CH2
SCH3

Methionine

STEP 1

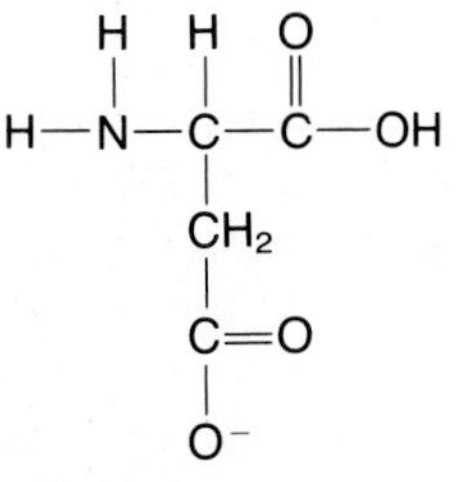
H H O
H—N—C—C—OH
CH2
C=O
O−

Aspartic acid

STEP 2

STEP 3 Remove the elements of H_2O and form a peptide bond by binding the N of aspartic acid to the carboxyl C of methionine.

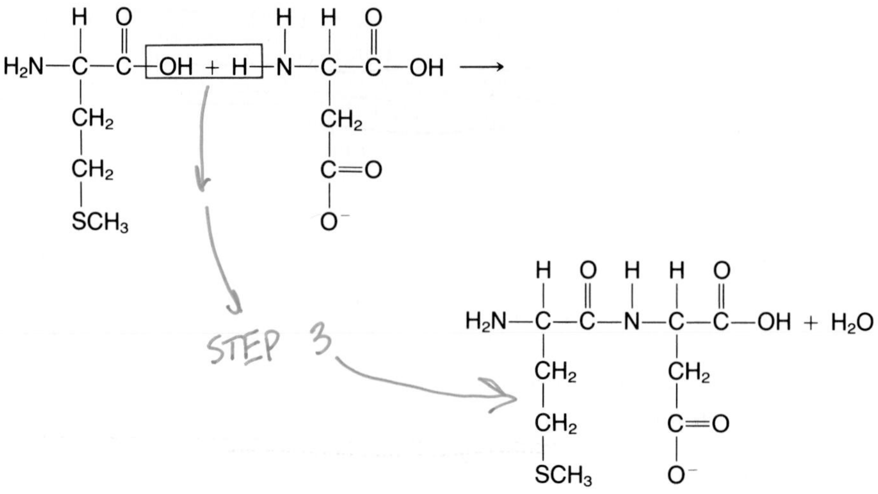

STEP 4 To the right of the just-formed dipeptide, write the structure of the next amino acid and repeat step 3 to form the complete tripeptide.

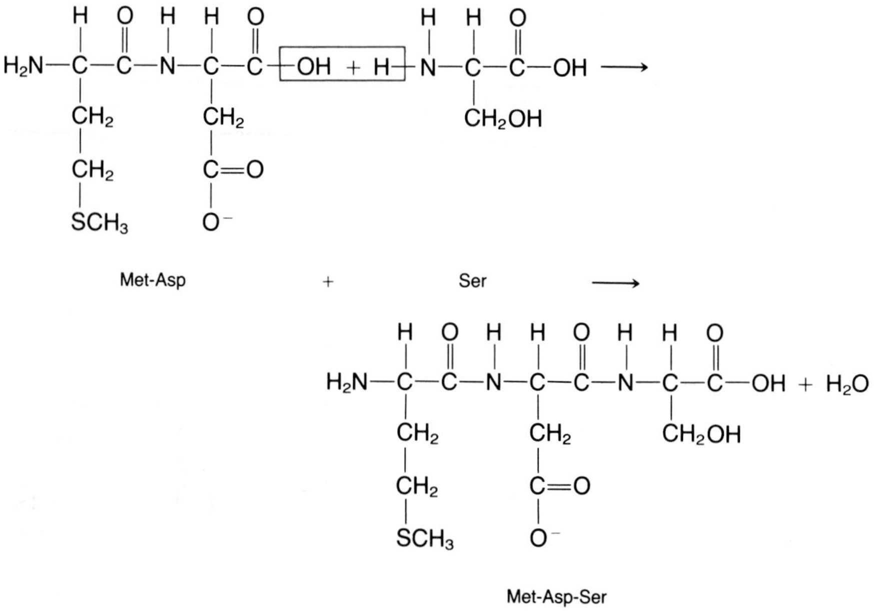

19.7 POLYPEPTIDES

Polypeptides are distinguished from proteins only in that polypeptides contain fewer amino acids. The conventional dividing line is taken as 50. That is, a sequence of 50 or more amino acids is called a protein and one with less than 50, a polypeptide.

A number of polypeptides function as neurotransmitters (Section 16.9) in the central nervous system. Others act as hormones, chemical substances that are released by a gland directly into the blood system and elicit a particular response from a target cell.

Three relatively small polypeptides, substance P, *met*-enkephalin, and *leu*-

enkephalin, appear to be involved in the nerve pathways connected with the body's perception of pain.

Arg-Pro-Lys-Pro-Gln-Gln-Phe-Phe-Gly-Leu-MetNH_2

Substance P

Met is present as the amide derivative instead of as a free carboxylic acid.

Tyr-Gly-Gly-Phe-Met

Met-enkephalin

Tyr-Gly-Gly-Phe-Leu

Leu-enkephalin

Note that the names come from the sole difference in structure, the C-terminal amino acid.

Substance P has been located in peripheral neurons and in spinal cord neurons that respond to pain stimuli. It is released when nerve endings in the skin and viscera are stimulated by mechanical, thermal, or chemical means. This pain stimulus is then transmitted to receiving neurons at the spinal cord (see Figure 19.2*a*).

The enkephalins have also been found in spinal cord neurons adjacent to the synapse between the substance P sensory neuron and the receiving neuron of the spinal cord. Experiments have shown that the enkephalins can suppress the release of substance P from the sensory neurons. By controlling the release of substance P, the enkephalins may regulate the input to the brain of painful stimuli (Figure 19.2*b*). Morphine and other narcotic opiates may have analgesic action because of their ability to mimic the enkephalins and supress the release of substance P. Transcutaneous electrical nerve stimulation (TENS) and acupuncture are used to treat chronic pain. Their function may depend upon the stimulation of neurons to produce enkephalins which act to inhibit the action of neurotransmitters such as substance P.

19.8 PRIMARY STRUCTURE OF PROTEINS

As mentioned in Section 19.6, the sequence of amino acids in a protein is called the primary structure of the proteins. Each protein has a unique primary structure. The concept of amino acid sequence in proteins and its significance bears a resemblance to the concept of arranging letters of the alphabet to form different words. The letters *a, r,* and *t* can form the word *art* or they can be arranged in a different sequence to give *tar,* a word with a very different meaning. They can be arranged yet a third way to give *rat,* a word of still different meaning. Different arrangements of amino acids yield different structures with different functions and activities.

Insulin was the first protein in which the sequence of amino acids was determined. The work, which as you can imagine was quite arduous for a structure with 51 amino acids (a relatively small protein), was accomplished by methods we will not discuss in this text. The analysis was directed by Frederick Sanger of Cambridge University, England, in 1953. The structure of insulin (Figure 19.3) shows two chains of 21 and 30 amino acids, respectively. The chains are connected by disulfide linkages, which we shall examine further in Section 19.10.

Although not structurally identical with human insulin, insulin from cows, horses, pigs, or sheep is biologically active in human beings and can be used for the treatment of diabetes, a disorder characterized by an insulin defi-

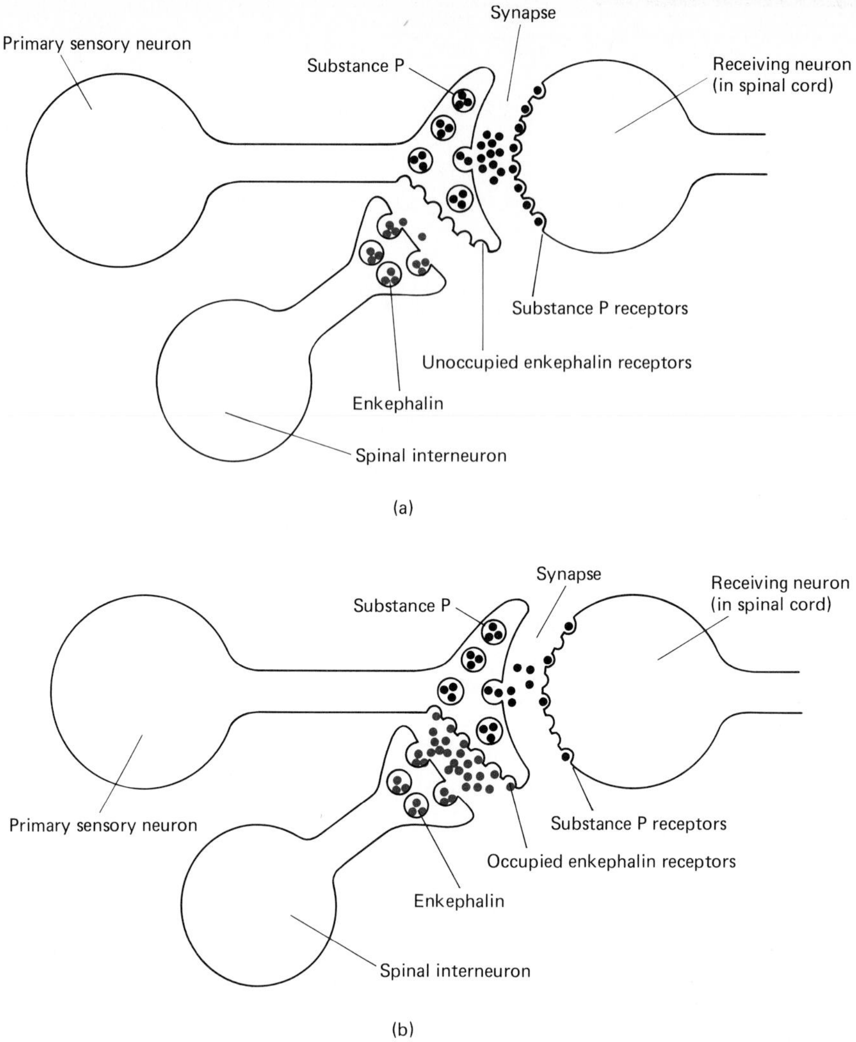
Primary sensory neuron
Synapse
Substance P
Receiving neuron (in spinal cord)
Substance P receptors
Unoccupied enkephalin receptors
Enkephalin
Spinal interneuron
(a)
Synapse
Substance P
Receiving neuron (in spinal cord)
Primary sensory neuron
Substance P receptors
Occupied enkephalin receptors
Enkephalin
Spinal interneuron
(b)

FIG. 19.2
(*a*) If a pain stimulus occurs at the sensory neuron, it may be transmitted by substance P (black dots) across the synapse to the spinal cord neuron. This happens freely and we feel pain if enkephalin (color dots) does not occupy the receptor sites on the primary sensory neuron. (*b*) If enkephalins (color dots) from a spinal interneuron occupy the receptor sites on the primary sensory neuron, they exert an inhibitory effect on substance P. Less substance P (black dots) carries the pain message across the synapse. Note that some of the receptor sites on the spinal cord are unoccupied in this case; thus less pain is felt.

ciency. It is interesting that the amino acids at certain positions in the insulin chain are identical in insulin from sheep, cows, and human beings (see Figure 19.3). This suggests that there are certain critical positions in the chains which are connected with insulin's biological activity.

Since Sanger's pioneering work, the sequences of many other proteins have been determined. Some of these proteins are much larger than insulin. Work on cytochrome c, a protein of 104 amino acids, demonstrates an interesting connection between amino acid sequence and biological evolution. Cyctochrome c is one of the enzymes necessary to the series of metabolic reactions known as the electron transport system, or the respiratory chain (Chapter 22). It is present in all species of plants, animals, and microorganisms that have respiratory chains, and evidence indicates that it was present before the evolutionary divergence of plants and animals.

C-terminal end
Gly
Ile
Val
Glu
5 Gln
Cys
Cys—S—S—Cys
S
S
Ala
Ser
10 Val
Cys
Ser
Leu
Tyr
15 Gln
Leu
Glu
Asn
Tyr
20 Cys—S—S
Asn
A chain
Phe
Val
Asn
Gln
5 His
Leu
Cys
Gly
Ser
10 His
Leu
Val
Glu
Ala
15 Leu
Tyr
Leu
Val
Cys
20 Gly
Glu
Arg
Gly
Phe
25 Phe
Tyr
Thr
Pro
Lys
30 Ala
B chain
N-terminal end

FIG. 19.3
The structure of bovine insulin, showing the amino acid sequence of the two chains and the cross-linkages. The partial numbering of the chains makes it easy to see that there are 21 amino acids in one chain and 30 amino acids in the other. The A chain of the insulins of man, pig, dog, rabbit, and sperm whale are identical. The B chains of the cow, pig, dog, goat, and horse are identical. The amino acid replacements in the A chain usually occur in positions 8, 9, and 10, shown in color.

The amino acid sequence of cytochrome c has been worked out in over 80 species. The difference in amino acid sequence between any two species appears to be related to the evolutionary history of the species. Cyctochrome c in species that are widely different, such as a horse and yeast, differ in the placement of 48 amino acids, whereas there are only two differences between the closely related chicken and duck. A sequence of evolutionary development of plants and animals can be constructed based on information from the amino acid sequence of common proteins such as cytochrome c.

Body proteins are continually being broken down and re-formed from amino acids in the body. How do you and I make protein molecules with the correct amino acid sequence? The enormity of this synthetic problem is brought out by the fact that even for a small tripeptide consisting say of glycine, methionine, and phenylalanine, there are six possible sequences:

Gly-Met-Phe	Met-Gly-Phe	Phe-Gly-Met
Gly-Phe-Met	Met-Phe-Gly	Phe-Met-Gly

For a polypeptide of 20 different amino acids there are more than 2.4 billion possibilities. For a protein of 100 amino acids, the number of possibilities is unimaginably huge. The body's answer to this ordering problem will be discussed in Chapter 20.

19.9 SECONDARY STRUCTURE OF PROTEINS

Primary structure as just discussed might leave us with a view of proteins as being like a string of pearls wherein amino acids are the beads:

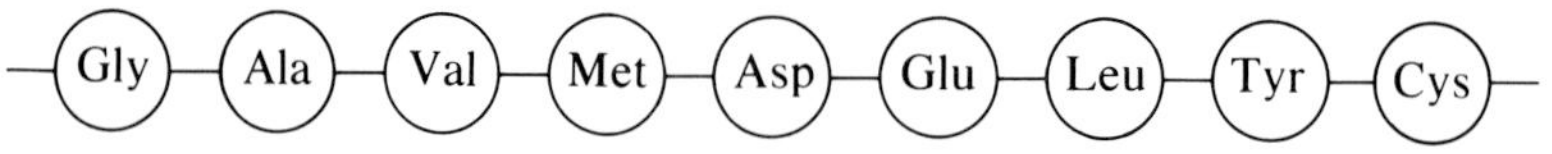
Gly
Ala
Val
Met
Asp
Glu
Leu
Tyr
Cys

This strand might then freely rotate into any shape. For example,

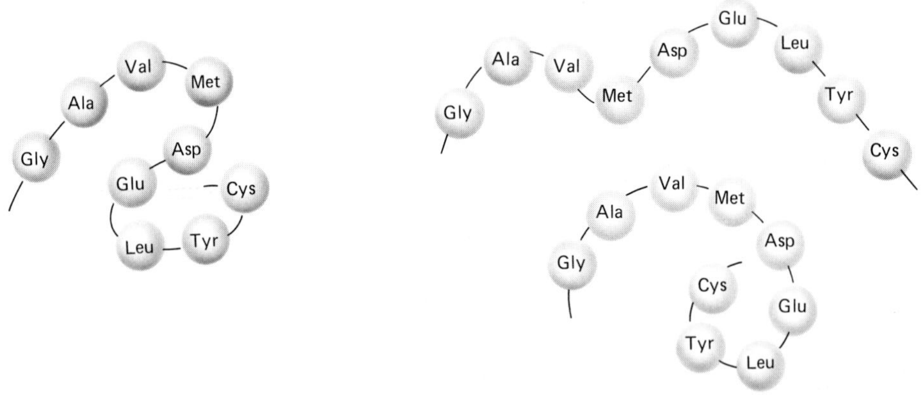
Gly
Ala
Val
Met
Asp
Glu
Cys
Leu
Tyr

However, each protein is known to have a unique three-dimensional shape which is highly important to its function. The regular arrangement or coiling of segments of a protein chain is known as the **secondary structure** of the protein. The α-helix, the β-pleated sheet, and the collagen triple helix are the three types of secondary structure.

α-Helix

In the α-helix structure the protein chain, the string of pearls, coils in the shape of a right-handed helix (Figure 19.4*a*). A helix is the shape of the

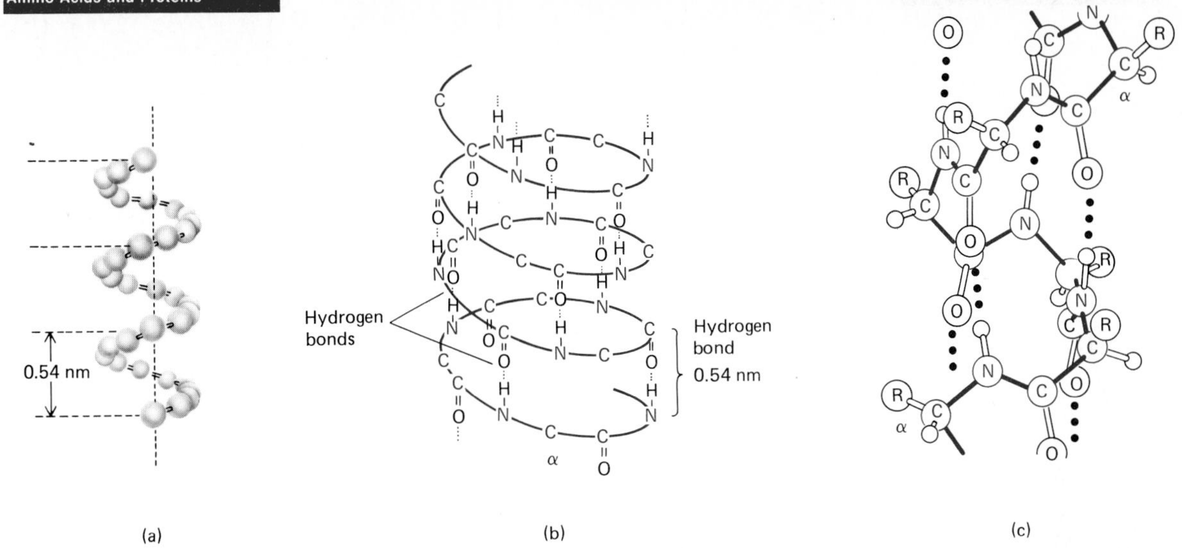

0.54 nm
(a)
Hydrogen bonds
Hydrogen bond
0.54 nm
(b)
(c)

FIG. 19.4
Three representations of the α-helix. (*a*) Each pearl represents an amino acid. The string of pearls (amino acids) coils in space into a right-handed helix. There are 3.6 amino acids per turn of the coil, i.e., within the 0.54 nm distance of one turn. (*b*) Some of the atoms of the amino acid in the helix are shown in order to depict the hydrogen bonding which holds the α-helix in place. The atoms shown in color make up the backbone of the helix. No substituents are shown on the α-carbon in order to avoid clutter in the illustration. (c) A molecular model of the α-helix. The atoms shown in color make up the backbone of the helix.

threads on a screw. The α-helix shape permits a hydrogen bond to form between each H atom attached to a nitrogen atom in a peptide linkage and the carbonyl oxygen in the peptide bond of the fourth amino acid behind and directly above it in the coiled amino acid chain (Figure 19.4*b*). It is the energy lowering, i.e. stabilization, from the large number of hydrogen bonds that is predominantly responsible for the existence of the α-helix structure.

However, the presence of successive groupings of amino acids with R groups that are either electrically charged or bulky tends to prevent formation of the α-helix because of repulsion between these groups. For example, a chain with many aspartic acid residues will not form an α-helix because of repulsions between the charged groups ($\sim COO^-$) on R. A major portion of the protein chain may be in the α-helix, or only a small segment of the chain, or none of the chain at all. The α-keratins, proteins which make up hair and wool fibers, are largely coiled in an α-helix. The stretchability of wool and hair stems from this molecular shape. When we stretch a wool fiber, we are pulling apart the α-helix by breaking the hydrogen bonds. But we do *not* disturb the strong covalent bonds between the "pearls" of the primary structure by the stretching. When we let go, the hydrogen bonds re-form and restore the fiber to its original length.

β-Pleated Sheet

In the β-pleated structure one amino acid chain runs alongside another chain. The chains are each in a fully extended zigzag conformation, and the carbonyl oxygens of one chain are hydrogen-bonded to the hydrogens on peptide nitrogens of the adjacent chain (Figure 19.5). The zigzag arrangement is imposed by the tetrahedral geometry around each carbon and nitrogen atom

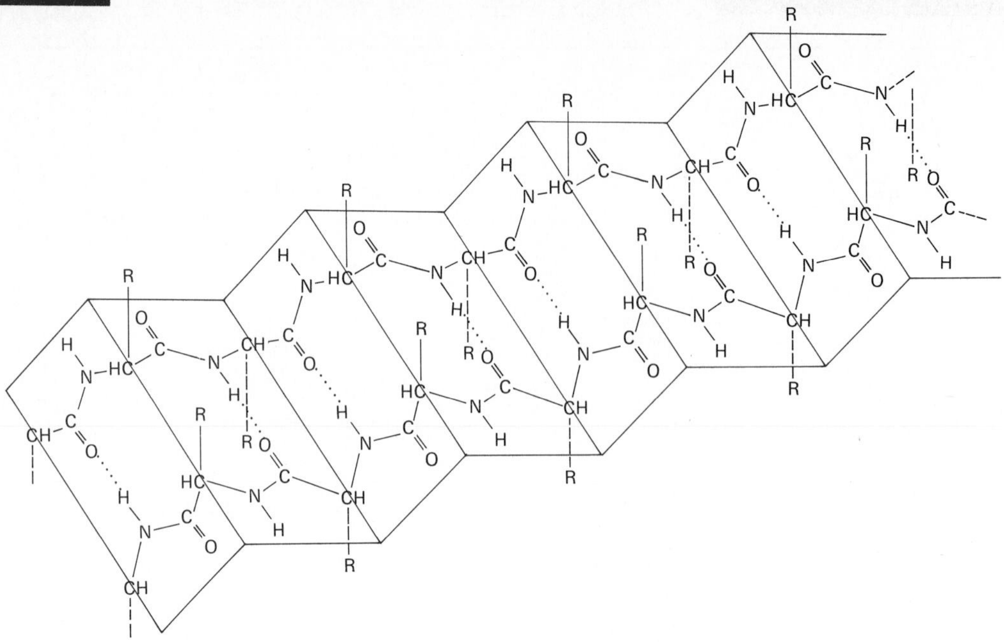

FIG. 19.5
β-Pleated sheet secondary structure. As in the case of the α-helix, hydrogen bonding occurs between —N—H and >C=O of peptide linkages.

in the chain. The zigzag conformation of the chain gives an up-and-down appearance, making the structure resemble a pleated sheet of paper.

Whereas the α-helix is held together by *intra*chain (within the chain) hydrogen bonds, it is *inter*chain (between chains) hydrogen bonding that is present in the β-pleated sheet. β-Keratins, proteins that make up silk fiber, are largely in the β-pleated sheet structure. Proteins in the β-pleated sheet structure are already in the fully extended form and cannot be stretched further without breaking a covalent bond. The nonstretchability of silk fibers can be understood in terms of the molecular structure of silk protein.

Triple Helix

A third type of secondary structure is unique to collagen, the most abundant protein in mammals. Collagen is the basic protein of connective tissue found in tendons, ligaments, blood vessels, skin, and the organic matter of bones and teeth. Collagen is actually a composite of protein material called tropocollagen. Tropocollagen consists of three chains of amino acids wrapped around each other in a tightly coiled triple helix which resembles a rope (Figure 19.6*a*). Each chain in the triple helix is cross-linked to another of the three chains in the triple helix by hydrogen bonding between adjacent peptide bonds. Each tropocollagen triple helix is also cross-linked by covalent bonds to an adjacent triple helix (Figure 19.6*b*). This cross-linked network makes up the collagen fiber. Because of the tight coiling and extensive cross-linking, collagen fibers have very high tensile strength and they have little tendency to stretch. A load of 10 kg is needed to break a fiber 1 mm in diameter. As we grow older, more covalent cross-linking occurs within the collagen fibers. This makes the tissue and bone containing this fiber more

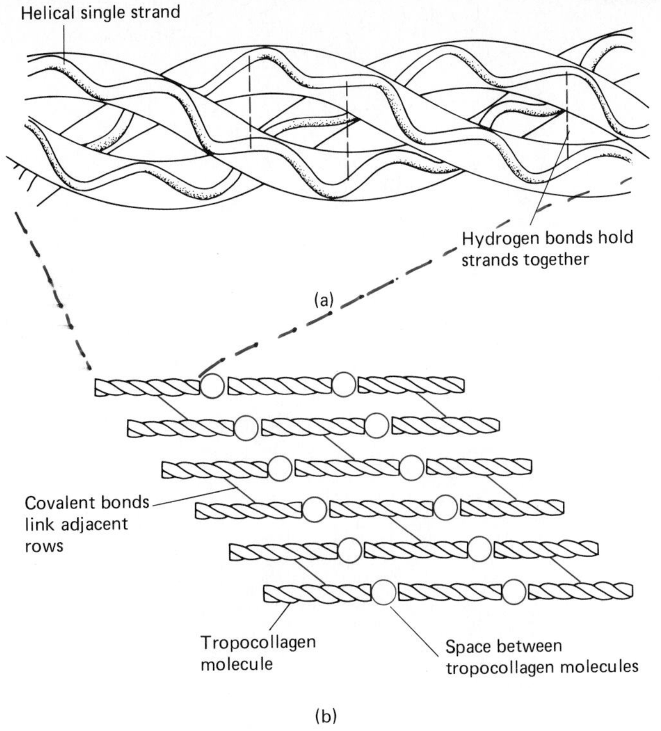
Helical single strand
Hydrogen bonds hold strands together
(a)
Covalent bonds link adjacent rows
Tropocollagen molecule
Space between tropocollagen molecules
(b)

triple helix – found in collagen

FIG. 19.6
The structure of collagen. (*a*) Tropocollagen molecules consist of three intertwined helical structures, a *triple helix*. The strands are held together by hydrogen bonding. (*b*) Collagen fiber is an assembly of tropocollagen triple helices held together by covalent bonds. The circles represent gaps between the tropocollagen molecules. The ropelike tropocollagen molecules are effectively formed into a thicker rope, because the cross-linking shown occurs in three dimensions, not all in one plane, as shown for convenience in illustrations.

rigid, and as a consequence more brittle. Vitamin C is required in the proper formation of collagen. Its absence leads to a condition of weakened blood vessels and skin lesions known as scurvy.

19.10 TERTIARY STRUCTURE OF PROTEINS

The term *fibrous protein* can be applied to the keratins and collagen described in the previous section. Fibrous proteins are composed of long strands or flat sheets. Their shape is determined by their secondary structure. There is another type of protein called *globular protein.* Examples are hemoglobin and cyctochrome c. These proteins have spherical, or globular, shapes which arise because of folding of the protein chain, including segments of the chain which have a particular secondary structure (Figure 19.7). The folding of the chain is referred to as the **tertiary structure** of the protein.

Globular proteins are particularly important because it is in this class that we find the proteins that function as enzymes, antibodies, hormones, transporters, and receptor sites. As we have said before, a key factor in much of this activity is the three-dimensional shape of the protein.

Four types of attractive interactions have been found to contribute to the tertiary structure of a protein. These interactions always arise between the R group side chains and can be between amino acids far apart in sequence. This is the way in which primary structure influences tertiary structure.

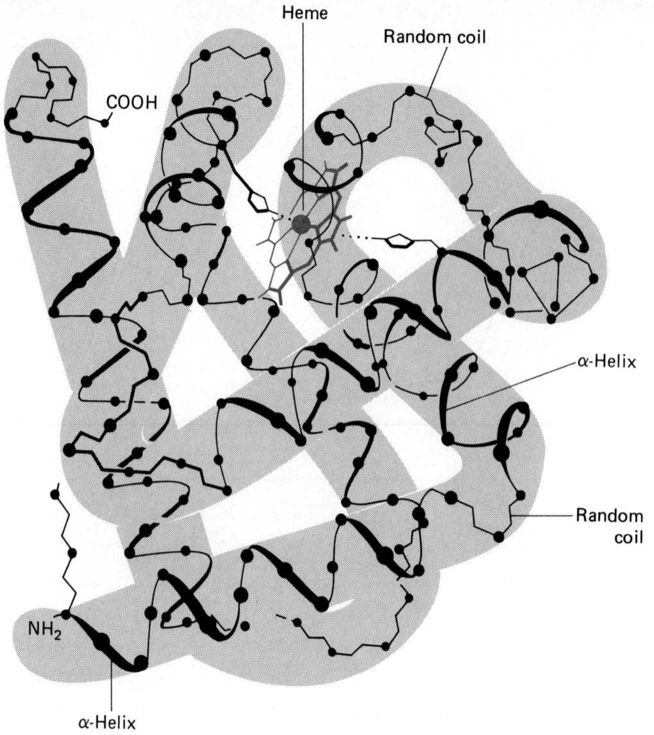
Heme
Random coil
COOH
α-Helix
Random coil
NH2
α-Helix

FIG. 19.7
Myoglobin is a relatively small *globular* protein. It has a single chain of 153 amino acids wrapped around a heme group. The primary structure is known; i.e., the sequence of 153 amino acids in the chain is known. The black dots mark the α-carbons of the amino acids. Most of the chain displays the α-helix secondary structure, though there are unwound or random coil segments. The tertiary structure is the folding pattern shown in this illustration.

1 **Ionic attractions** sometimes called **salt bridges** are interactions between negatively and positively charged substituents on the R groups. For example, the negative charge carried by the carboxylate group ($—COO^-$) on the side chain of aspartic acid may be attracted to the positive charge carried by the ammonium group ($—NH_3^+$) on the side chain of lysine.

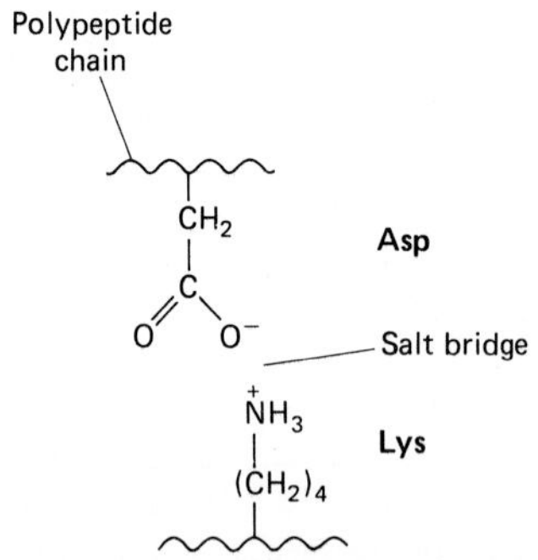
Polypeptide chain
CH2
Asp
C
O
O−
Salt bridge
NH3
Lys
(CH2)4

2 **Hydrogen bonding** can occur between side chains that contain a hydrogen donor, for example, the —O—H of serine, and R groups that contain a hydrogen acceptor, for example, the —C═O of glutamine. In secondary structure, hydrogen bonding occurs between atoms connected to the peptide bond, whereas in tertiary structure, the atoms involved are part of the R side chain. Compare Figure 19.4 with this illustration of tertiary structure H bonding.

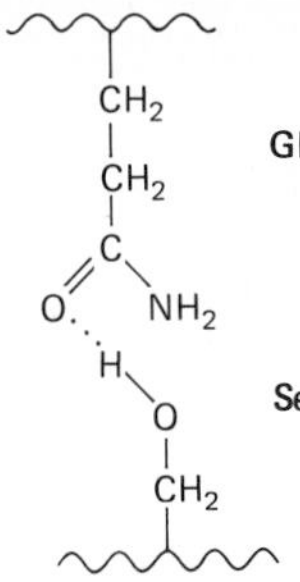
CH2
Gln
CH2
C
O
NH2
H
Ser
O
CH2
Hydrogen bond

strongest →

3 **Disulfide linkages** can form between the —SH groups of two cysteine amino acids. Recall from Section 14.18 that two —SH groups are readily oxidized to give a disulfide bond —S—S—. Because it is a full covalent bond, this linkage is the strongest of the four tertiary interactions and is important in many proteins. You can see it in three places in the insulin molecule shown in Figure 19.3.

4 **Hydrophobic attractions** can exist between the R groups listed as hydrophobic in Table 19.1. In aqueous media the chain folds so that the hydrophobic side chains face toward the interior, away from the outside environment of polar water molecules. The hydrophobic groups associate together, strengthening the attractive dispersion forces and minimizing the possibility of a water molecule coming between them.

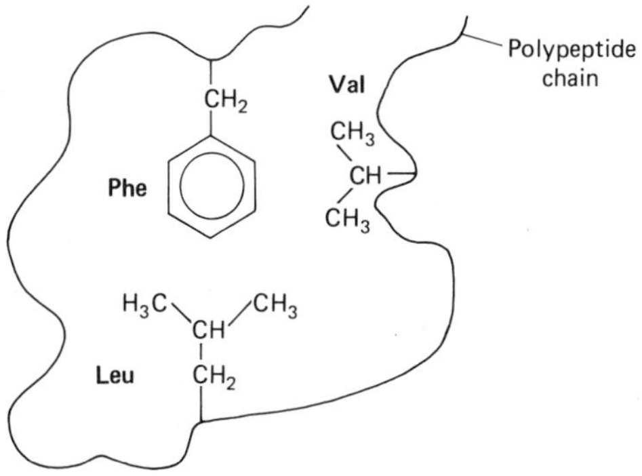
Polypeptide chain
Val
CH2
CH3
CH
CH3
Phe
H3C
CH3
CH
Leu
CH2
Hydrophobic interactions

The primary structure of a protein determines which of these four interactions will occur and where it will occur. Over a short sequence of amino acids (four to six), the primary structure determines the secondary structure. That is, for example, the chain will coil into an α-helix if the R groups of the amino acids don't interfere. Over longer sequences primary structure determines the tertiary structure by allowing intrachain interactions at various points. Given a particular primary structure, the protein establishes a stable shape by maximizing the attractive forces through secondary and tertiary interactions, thereby lowering the energy of the protein. Figure 19.8 schematically shows how the four interactions just described hold a protein chain in its characteristic shape, i.e., tertiary structure.

19.11 QUATERNARY STRUCTURE OF PROTEINS

Some proteins consist of two or more separate, but associated, polypeptide chains and are known as **oligomeric proteins.** The relationship and fit of one chain to another in an oligomeric protein is defined as the **quarternary** (4°) **structure** of that protein. Quaternary structure between chains involves three of the four attractive interactions that lead to the folding of a single chain

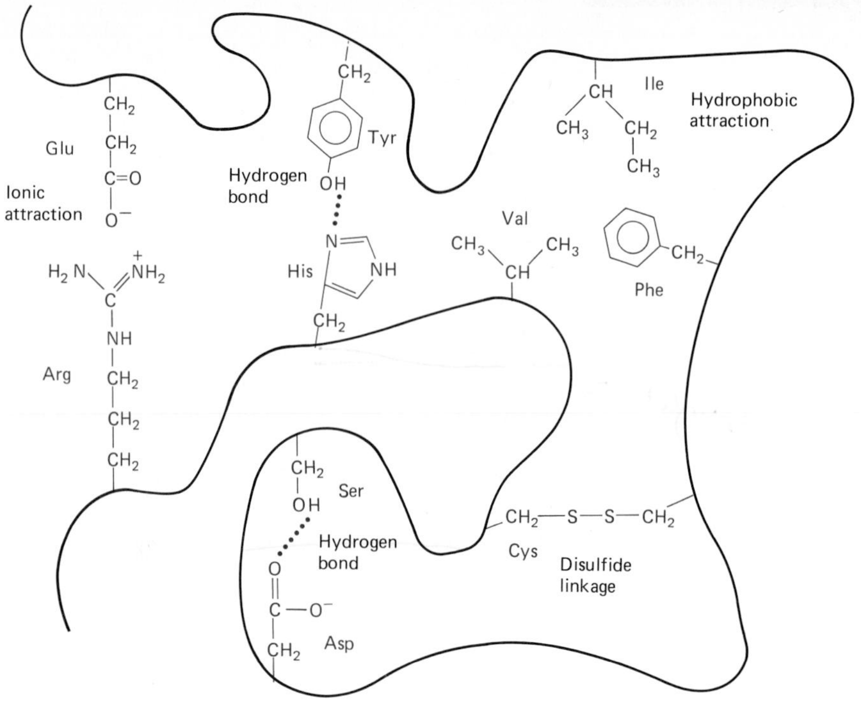
Glu
Ionic attraction
Hydrogen bond
Tyr
His
Arg
Ile
Hydrophobic attraction
Val
Phe
Ser
Hydrogen bond
Asp
Cys
Disulfide linkage

FIG. 19.8
Schematic representation of a protein chain, shown as the wavy, twisted line, held in a characteristic shape by interactions between R group substituents. The R groups are labeled by the abbreviation of the amino acid which they characterize. Most of the R groups on the chain are not shown, nor is the fact that the chain is actually usually in the α-helical array.

into a tertiary structure, i.e., hydrogen bonding, salt bridges, and especially hydrophobic interactions. Each chain within an oligomeric protein is already coiled and folded into its secondary and tertiary structures. These fully arranged and folded chains are then associated with one another through attractive weak interactions. Hemoglobin, the oxygen carrier within red blood cells, is an example of an oligomeric protein whose structure has been extensively studied. See Figure 19.9.

19.12 STRUCTURE AND FUNCTION OF HEMOGLOBIN

Hemoglobin consists of four polypeptide chains: two are identical chains of 141 amino acids and are labeled α, and two are identical chains of 146 amino acids and are labeled β. The quaternary structure of hemoglobin is maintained by noncovalent interactions, predominantly hydrophobic, between an α- and β-chain labeled $\alpha_1\beta_1$ and the other α- and β-chains labeled $\alpha_2\beta_2$ (Figure 19.9). Each of the four chains contains a heme unit (Section 16.7) which is the actual binding site of oxygen. Because there are four chains, a total of four oxygen molecules can be bound to each hemoglobin.

Hemoglobin's Cooperative Effect

Hemoglobin's physiological role is to pick up oxygen in the lungs and deliver it to peripheral cells. Critical to this function is the fact that the affinity of hemoglobin for binding oxygen greatly increases after one oxygen molecule has been attached. A plot of hemoglobin's oxygen saturation vs. the external partial pressure of oxygen shown in Figure 19.10 demonstrates this point graphically. Fractional saturation is the fraction of oxygen bound compared

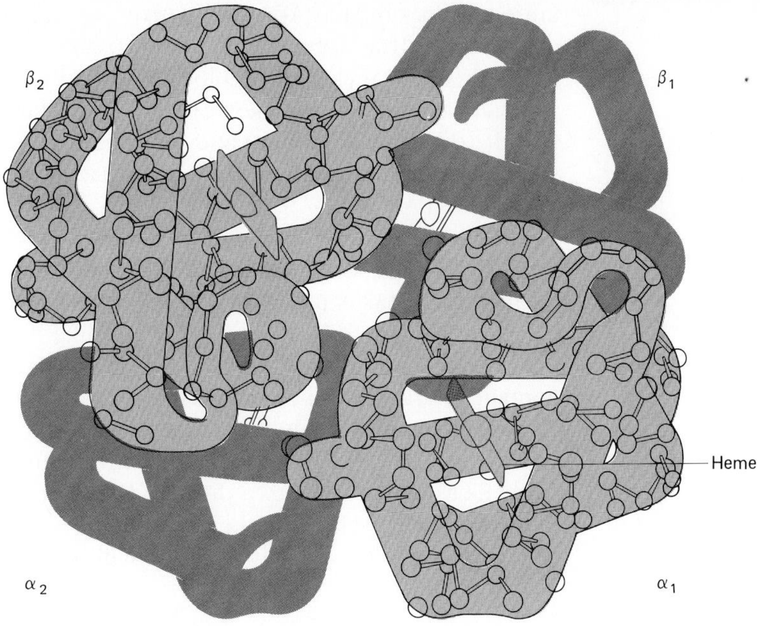

β_2
β_1
Heme
α_2
α_1

FIG. 19.9
Hemoglobin consists of four subunits, each of which is a globular protein chain similar in structure to myoglobin (see Figure 19.7). Quarternary structure is the association of these four units, designated α_1, α_2, β_1, and β_2, in the manner shown. Oxygen binds at the heme groups. This structure is deoxyhemoglobin; i.e., no O_2 is bound. *(Whitaker, Fernandez, and Tsokos: Concepts of General, Organic and Biological Chemistry. © 1981 Houghton Mifflin Company. Used by permission.)*

with the total amount that could be bound. The curve shows that hemoglobin binds very little oxygen at partial pressures between 0 and 10 mmHg. Then there is a sharp increase in oxygen-carrying capacity between p_{O_2} = 10 and 40 mmHg. This increased uptake indicates that in some way the binding of one oxygen molecule to one of hemoglobin's four polypeptide chains "co-operates" in making the other chains more effective oxygen binders.

Notice the contrast with myoglobin's behavior; myoglobin, which is a single polypeptide chain, picks up oxygen rapidly and becomes totally sat-

FIG. 19.10
Cooperative binding of oxygen in hemoglobin: One oxygen binds at low partial pressures and assists the more dramatic uptake that occurs at higher partial pressures (above 10 mmHg). Single-stranded myoglobin takes up oxygen at low pressures and becomes almost saturated at 10 mmHg. Conversely, it will not give up oxygen until p_{O_2} drops below 10 mmHg.

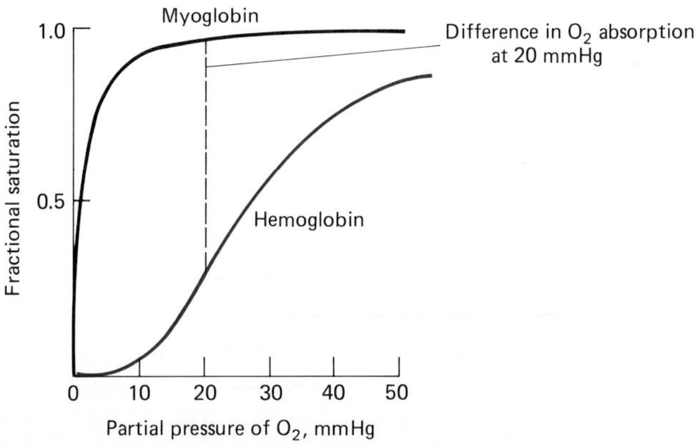

Myoglobin
Difference in O_2 absorption at 20 mmHg
Hemoglobin
Fractional saturation
1.0
0.5
0
10
20
30
40
50
Partial pressure of O_2, mmHg

urated. Experimental x-ray evidence has confirmed that a change in the quarternary structure of hemoglobin occurs when it is oxygenated. Apparently when one oxygen molecule binds to one chain (α_1) of deoxyhemoglobin, a structural shift occurs between the four chains so that the other chains (α_2, β_1, β_2) of hemoglobin have a stronger affinity for oxygen. This effect, sometimes observed in other oligomeric proteins, is known as positive cooperation.

The cooperative binding of oxygen by hemoglobin which gives it the uptake profile of Figure 19.10 makes hemoglobin a very efficient delivery vehicle for oxygen. It takes up oxygen more gradually and also releases it at higher partial pressures than a simpler protein like myoglobin. Consequently, for example, in the lungs where p_{O_2} = 100 mmHg, hemoglobin becomes >97 percent saturated with oxygen. However, in a working muscle where p_{O_2} = 20 mmHg, hemoglobin is only 25 percent saturated with oxygen after contact with a muscle cell. That is, almost three-fourths of hemoglobin's original oxygen has been supplied to the cell. By comparison, myoglobin, the carrier of oxygen in muscles, for the same change in oxygen partial pressure (100 to 20 mmHg) gives up only a very small percent of its bound oxygen.

Effect of H^+ and CO_2 on Hemoglobin

In addition to the partial pressure of oxygen, pH and CO_2 concentration also affect hemoglobin's oxygen-carrying capacity. Hemoglobin can bind to H^+ ions and CO_2. The binding of H^+ or CO_2 to hemoglobin causes oxygen to be released. This process of binding and oxygen release can be summarized by the following two equilibria:

$$\underset{\text{Oxyhemoglobin}}{HbO_2} + H^+ \rightleftharpoons Hb{-}H^+ + O_2 \tag{19.4}$$

$$\underset{\text{Oxyhemoglobin}}{HbO_2} + CO_2 \rightleftharpoons Hb{-}CO_2 + O_2 \tag{19.5}$$

Remembering Le Chatelier's principle, you can see that elevated H^+ or CO_2 concentrations will shift these equilibria to the right and thereby diminish the amount of oxygen bound to hemoglobin.

Deoxygenation of Hemoglobin in Tissues

In active tissue such as muscle cells the partial pressure of CO_2 is high because it is a product of metabolism. The CO_2 diffuses from the tissue into the blood (Figure 19.11*a*) and then into the passing red blood cells, where p_{CO_2} is lower than in the tissues. High CO_2 concentrations produce higher H^+ concentrations in red blood cells through the equilibria shown in Equation (19.6), that is, by combination with H_2O to form H_2CO_3 which then ionizes.

$$CO_2 + H_2O \rightleftharpoons H_2CO_3 \rightleftharpoons H^+ + HCO_3^- \tag{19.6}$$

The high H^+ concentration forces the position of equilibrium in Equation (19.4) to the right. Elevated CO_2 concentrations in the cell also shift the equilibrium in Equation (19.5) to the right, but this is not shown in Figure

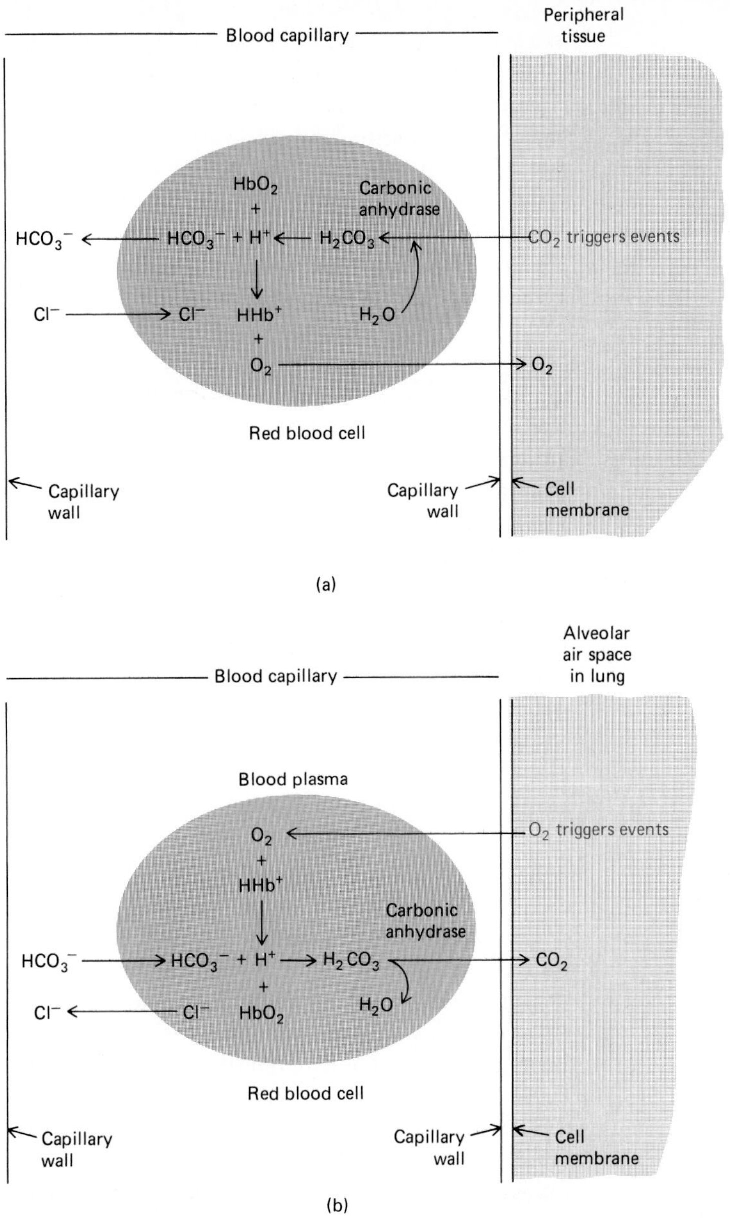
Blood capillary
Peripheral tissue
HbO2
+
Carbonic anhydrase
HCO3−
HCO3− + H+ ← H2CO3
CO2 triggers events
Cl−
Cl−
HHb+
H2O
+
O2
O2
Red blood cell
Capillary wall
Capillary wall
Cell membrane
(a)
Alveolar air space in lung
Blood capillary
Blood plasma
O2
O2 triggers events
+
HHb+
Carbonic anhydrase
HCO3−
HCO3− + H+ → H2CO3
CO2
+
Cl−
Cl−
HbO2
H2O
Red blood cell
Capillary wall
Capillary wall
Cell membrane
(b)

FIG. 19.11
Oxygen–carbon dioxide exchange between blood and tissues and blood and lungs. (*a*) In the tissues partial pressure of CO_2 is high so that it crosses the capillary wall and enters the blood. CO_2 combines with H_2O to form H_2CO_3, which ionizes to HCO_3^- and H^+. H^+ reacts with HbO_2 to form HHb^+ and O_2, which then enters the tissues. HCO_3^- ions leave the cell and chloride ions enter the cell to replace them and preserve electrical neutrality. (*b*) In the lungs partial pressure of oxygen is high so that it crosses the capillary wall and enters the blood. This triggers the reactions shown. O_2 + HHb produces H^+ which combines with HCO_3^- to form H_2CO_3. H_2CO_3 decomposes to H_2O and CO_2, which then exits from blood to lungs. HCO_3^- ions enter the cell and chloride ions shift out of the cell to maintain electrical neutrality both inside the red blood cell and in plasma.

19.11*a*. Thus the cell condition of high CO_2 and H^+ concentration induces the oxygen-saturated hemoglobin in contact with the tissue to deliver its oxygen to the cell: the oxygen diffuses from the region of higher partial pressure within the cell to the plasma and then into the oxygen-starved tissue cell, as shown in Figure 19.11*a*. As a result O_2, a necessary reactant for the continuation of metabolism, is delivered to the cell and two products of metabolism, CO_2 and H^+, are carried off by hemoglobin in the bloodstream to the lungs.

Although some CO_2 is transported as Hb—CO_2 by the shift of the equilibrium in Equation (19.5) to the right, most CO_2 (75 percent) is carried to the lungs as HCO_3^- formed by the equilibrium shown in Equation (19.6) in cells, as shown in Figure 19.11*a*. As the HCO_3^- concentration increases within the red blood cell, HCO_3^- diffuses out of the cell into the blood plasma. To preserve electrical neutrality, Cl^- enters the cell from the plasma, a phenomenon called the **chloride shift.**

Oxygenation of Hemoglobin in the Lungs

Red blood cells which have interacted with tissue as just described move along with plasma through the veins to the lungs. In the lungs O_2 diffuses from a region of high partial pressure, the lung alveoli, into the blood cells, where p_{O_2} is low (Figure 19.11*b*). The increased O_2 concentration in the cells shifts the equilibria in Equations (19.4) and (19.5) to the left, the hemoglobin becomes saturated with oxygen, and CO_2 and H^+ are released. The H^+ ions released combine with HCO_3^- to form H_2CO_3, which then decomposes to H_2O and CO_2, as shown in Figure 19.11*b*. That is, the equilibrium in Equation (19.6) is shifted to the left. The p_{CO_2} within the red blood cell becomes higher than p_{CO_2} in the lungs, and therefore CO_2 diffuses from the cell into the plasma and then into the lungs, where it is exhaled to the external atmosphere. As the HCO_3^- concentration within red blood cells decreases, HCO_3^- enters from the plasma and Cl^- ions leave the cell in a reversal of the **chloride shift.** The net result in the lungs is that hemoglobin has picked up oxygen which can be carried through the arterial system to the tissue, and the metabolic waste product CO_2 has been exhaled into the atmosphere.

The equilibrium of Equation (19.4) points out one importance of the blood's buffering system (Section 11.12). At low pH (high H^+) HbO_2 gives up its oxygen and binds H^+, making hemoglobin a less effective carrier of oxygen. In untreated acidosis (Section 11.13) the blood pH is lower than normal, hemoglobin is not as efficient in carrying oxygen to the cells, and the patient begins to suffer the symptoms of fatigue, confusion, and eventually coma.

The effects of CO_2 and H^+ concentrations on the equilibria in Equations (19.4) and (19.5) and the binding and release of oxygen by hemoglobin are known as the *Bohr effect*. This phenomenon was named after the Danish physiologist Christian Bohr, who made many of the early discoveries in this field.

19.13 SICKLE-CELL ANEMIA: A HEMOGLOBIN DISEASE

Patients suffering from the disease sickle-cell anemia have a low concentration of red blood cells. Under the conditions of low oxygen concentration their red blood cells take on a crescent or sickle shape, as shown in Figure 19.12. These sickle-shaped cells tend to block small blood vessels and lead to failure of the kidney or cardiovascular system.

Examination of the hemoglobin from sickle-cell patients reveals a difference in the primary sequence of amino acids in each of the two β-chains compared with that found in normal hemoglobin. In sickle-cell hemoglobin the amino acid valine replaces the glutamic acid which should be in position 6 of each β-chain. Although the change might appear to be numerically small because there are 574 amino acids in hemoglobin and only two amino acids are affected in sickle-cell hemoglobin, the nonpolar character of valine is so different from that of the polar glutamic acid that the effect is dramatic. Position

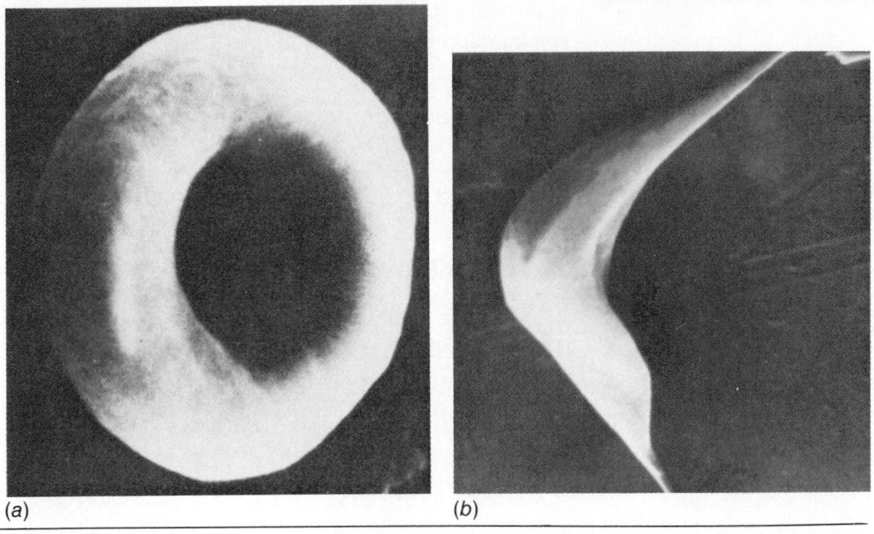

FIG. 19.12 Scanning electron micrographs of red blood cells. (*a*) A normally shaped red blood cell; (*b*) a sickle-shaped red blood cell.

6 of the β-chain is on the outer surface of the hemoglobin molecule, and it is believed that the nonpolar nature of valine creates a hydrophobic interaction between hemoglobin molecules. This leads the molecules to associate together. This association is not present in normal hemoglobin molecules because of the repulsion between negatively charged glutamic acid groups. It is the aggregation of the sickle-cell hemoglobin molecules that apparently is responsible for the sicklelike shape of the red blood cells. A single change in the primary structure leads to a very significant change in the biological properties of the molecule.

Why does a valine instead of a glutamic acid appear at position 6? Why is sickle-cell anemia an inherited disease? These are questions which will be answered in the protein biosynthesis section of the nucleic acids chapter (Chapter 20).

19.14 CLASSIFICATION OF PROTEINS

Although all proteins have in common an amino acid chain, there is a vast variety of different structures and functions among proteins. Therefore, it is convenient to classify these important materials. Such classification is usually done in one of three ways: (1) on the basis of function, (2) on the basis of chemical composition, or (3) on the basis of shape. Let us first take up function, perhaps the most important and common classification.

Classification by Function

The varied role of proteins in the biological world is clearly brought out by the functional classification.

Enzymes

Essentially all physiological reactions are catalyzed by the biological catalysts known as enzymes. All enzymes are proteins. The importance of this group of proteins is such that a whole chapter will be devoted to the study of enzymes (Chapter 21).

Transport

An example of this group is hemoglobin, the oxygen transporter in blood. Many other molecules and ions are transported through blood plasma bound to a protein molecule. We have already seen that lipids such as cholesterol are carried in blood as lipoproteins, that is, lipid bound to protein.

Contractile Proteins

Proteins such as actin and myosin in muscle cells have the ability to contract and expand. This gives these cells the property of motion.

Structural Proteins

The protein collagen is the major component of tendons, cartilage, and skin. These tissues have a high tensile strength because of the triple helix structure of collagen. Other structural proteins are keratin in hair, fingernails, and feathers, and elastin in ligaments.

Defense Proteins

In vertebrates specific proteins serve as antibodies in the immune system. Antibodies recognize, complex with, and thus neutralize foreign proteins in other organisms such as viruses or bacteria. Toxic proteins such as snake venoms serve as protective devices for the organisms producing them.

Regulatory Proteins

Proteins are widely involved in the regulation and control of metabolism, enzymatic biosynthesis, and nerve transmission. Metabolism is mediated by protein hormones such as insulin and parathyroid hormone. In bacteria, repressor proteins are believed to be responsible for turning on and off the mechanisms for enzyme synthesis. Receptor sites at nerve synapses are proteins. Also, polypeptides such as the enkephalins play a role in nerve impulse transmission.

Nutrient Proteins

Some proteins serve as storage forms of nutrients for a developing organism. Examples of nutrient proteins are seed proteins of grain plants; ovalbumin, the protein of egg white; and casein, the major protein in milk.

Classification by Composition

Based on chemical composition, proteins are classified as either simple proteins or conjugated proteins.

Simple Proteins

Simple proteins are made up of amino acids only. Examples in this category are the albumins found in egg white and blood serum, the globulins which are antibodies, and the albuminoids such as keratin, collagen, and elastin.

Conjugated Proteins

Conjugated proteins incorporate in their structures other chemical components in addition to amino acids. Thus there is an amino acid chain portion and some other portion in a conjugated protein. The non-amino-acid-containing portion of the conjugated protein is called a **prosthetic group.** The prosthetic group may be a heme unit as in the case of hemoglobin, a car-

TABLE 19.2 Classification of Conjugated Proteins

Class	Prosthetic Group	Specific Example	Function of Example
Hemoproteins	Heme unit	Hemoglobin Myoglobin	Carrier of O_2 in blood Oxygen binder in muscles
Lipoproteins	Lipid molecule	Low-density lipoprotein High-density lipoprotein	Lipid carrier Lipid carrier
Glycoproteins	Carbohydrate	Gamma globulin Mucin	Antibody Lubricant in mucous secretions
Phosphoproteins	Phosphate group	Casein	Major protein in milk
Nucleoproteins	Nucleic acids	Ribosomes Viruses	Location of protein synthesis in cells Self-replicating, infectious complex
Metalloproteins	Metal ion	Iron—ferritin Zinc—alcohol dehydrogenase	Storage complex of iron Enzyme in alcohol oxidation
Flavoproteins	Flavin nucleotides	Succinate dehydrogenase	Oxidation enzyme

bohydrate derivative, a lipid molecule, or a metal ion, among other possibilities. Conjugated proteins are further classified according to the nature of the prosthetic group (Table 19.2).

Classification by Shape

Classification according to shape has been alluded to before. The two categories are fibrous proteins and globular proteins.

Fibrous Proteins

Fibrous proteins have a long stringy shape and are water-insoluble. They generally serve a structural or covering role. Examples in this category are the collagens, keratins, elastins, myosins, actins, and fibrin.

Globular Proteins

Globular proteins have a tertiary structure which gives them a spherical, or globular, shape. They are often soluble in water solutions and typically function as enzymes, antibodies, transport proteins, and nutrient storage proteins. Common examples in this category are hemoglobin and ovalbumin.

19.15 DENATURATION

When a protein undergoes a change in its normal three-dimensional shape, called its native shape, without breakage of the peptide bonds, it is said to be denatured. Thus, the quaternary, tertiary, and secondary structures are altered by denaturation, but the primary structure remains intact. This is a structural definition of **denaturation.** The biological or functional definition of denaturation is that it is the loss of physiological activity of the protein. The change in structure causes the loss of activity. In the denatured random shape, most proteins are less soluble in water than in their active native state and so precipitate or coagulate out of solution.

Many toxins, such as heavy metal ions, are poisonous because they denature cell proteins. Let us examine some conditions and chemicals that can lead to protein denaturation.

Heat and Ultraviolet Radiation

The input of energy causes atoms to vibrate more rapidly, thereby weakening and eventually breaking the hydrogen bonds which hold 2°, 3°, and 4° structures together. Thus, cooking any food protein denatures the protein. Egg is largely protein material and you are undoubtedly familiar with the dramatic change in an egg produced by denaturing its proteins in a frying pan.

One advantage of denaturing food protein is increased digestibility. Enzymes act on denatured meat protein more easily, and this makes the meat easier to digest. The protein of microorganisms present in food is also denatured by heat; thus cooking kills any unwanted microorganisms in food.

Changes in pH

By adding or removing H^+, acids and bases change the charge character of acidic and basic substituents on the R groups and thus disrupt salt linkages.

Organic Solvents

Some organic solvents such as ethyl alcohol form hydrogen bonds to the protein molecule and thus disrupt hydrogen bonds already existing within the secondary and tertiary structure of the protein. A 70% alcohol solution is used as a skin disinfectant. It penetrates bacterial cell walls and coagulates or clumps together the protein within the bacterium. A more concentrated alcohol solution coagulates the protein at the cell wall and prevents the alcohol from entering the cell. This diminishes its disinfectant properties.

Heavy Metal Ions such as Ag^+, Hg^{2+}, and Pb^{2+}

These cations are electrostatically attracted to the carboxylate anions of acidic amino acids and can also become covalently bonded to the sulfur atom of thiol (—SH) groups. This interaction of the heavy metal ions with the protein disrupts salt bridges and disulfide linkages, thus denaturing the protein. It is the ion and not the elemental metal that is the toxin.

Lead poisoning can occur through the ingestion of lead paint found in older homes or from the absorption of lead in leaded gasoline. Toxicity symptoms of lead poisoning include anemia, loss of memory, and mental deterioration.

Egg white or milk can serve as first aid treatment for heavy metal poisoning if given immediately after ingestion of the metal, before the metal ion has been absorbed into the blood system. The protein in the egg white or milk is itself denatured by the heavy metal ion. This saves the organism's protein. The coagulated egg white or milk-metal complex must then be removed from the stomach by a stomach pump or an emetic before the acidic stomach juices hydrolyze the protein and free the heavy metal ion.

In low concentrations, solutions of heavy metal ions can be used externally as antibacterials. For example a 1% solution of silver nitrate is used for the prevention of gonococcal infections in the eyes of newborn infants.

Violent Agitation

Globular proteins such as those in egg white rapidly denature under the input of mechanical energy. For example, the spherical egg albumin protein is elongated when it is beaten, thus allowing us to make meringues and soufflés.

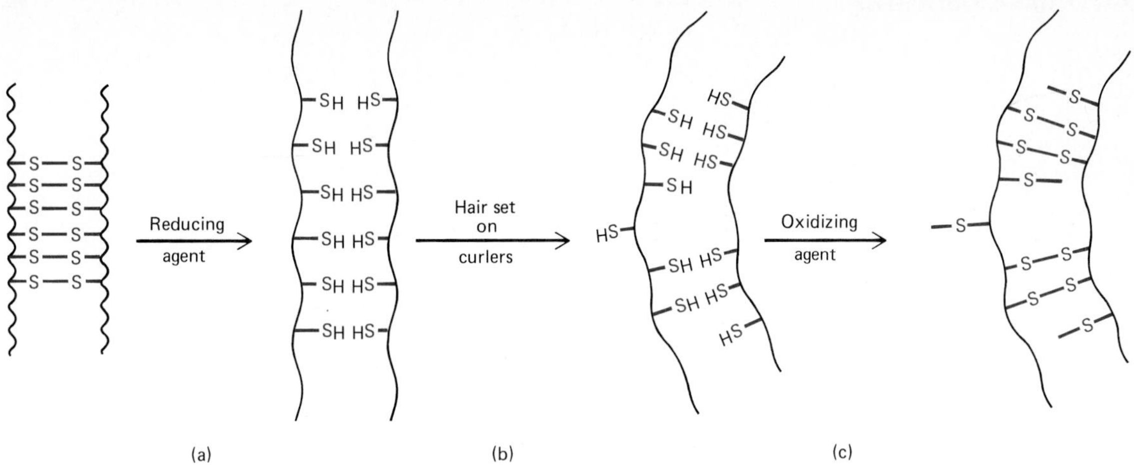

S—S
Reducing agent
SH HS
Hair set on curlers
Oxidizing agent
(a)
(b)
(c)

FIG. 19.13
Steps in permanent waving. (*a*) Cross-links holding the hair in a straight configuration are broken by a reducing agent. (*b*) As the hair is set on curlers, the separate chains are bent and turned so that the original —S—S— links no longer line up. (*c*) New links are formed by use of an oxidizing agent in order to hold the hair in the new shape.

Reduction and Oxidation of Disulfide Linkages

The waviness or straightness of hair arises from the particular low-energy shape of the protein molecules (mainly α-keratin) making up the hair. Hair protein has a high percentage of disulfide linkages maintaining a 3° structure, and it is these linkages that give the overall shape to the hair protein. A "permanent" wave is a method of changing the natural shape of hair. The chemistry of a permanent wave starts with treating the hair with a reducing agent which breaks the disulfide linkages (Figure 19.13*a*). Then the unraveled (denatured) hair is put onto curlers, in the desired shape (Figure 19.13*b*). An oxidizing agent is then added to form *new* disulfide bonds between different pairs of cysteine amino acids. This then holds the hair in the new desired shape (Figure 19.13*c*).

Some denatured proteins return to their native form if the denaturing conditions or chemical are removed. For example, ribonuclease, a key enzyme in nucleic acid chemistry, is denatured and completely loses its biological activity when acted on by a solution of urea and a reducing agent. When the denaturing agent is removed, ribonuclease spontaneously refolds itself into its native shape and regains complete biological activity. This type of experiment verifies that the primary structure is not broken during the denaturation of a protein. Furthermore, it emphasizes that a particular secondary, tertiary, and quaternary structure is set by the nature of the primary structure and is an energy minimum.

19.16 HYDROLYSIS AND DIGESTION OF PROTEINS

When a polypeptide or protein is heated in an acid or base solution, the peptide bonds are hydrolyzed and free amino acids are the products. This hydrolysis of protein is identical to the hydrolysis of a simple amide which you saw in Section 17.17. We can easily write an illustrative equation for this reaction if we use a tripeptide as a reactant:

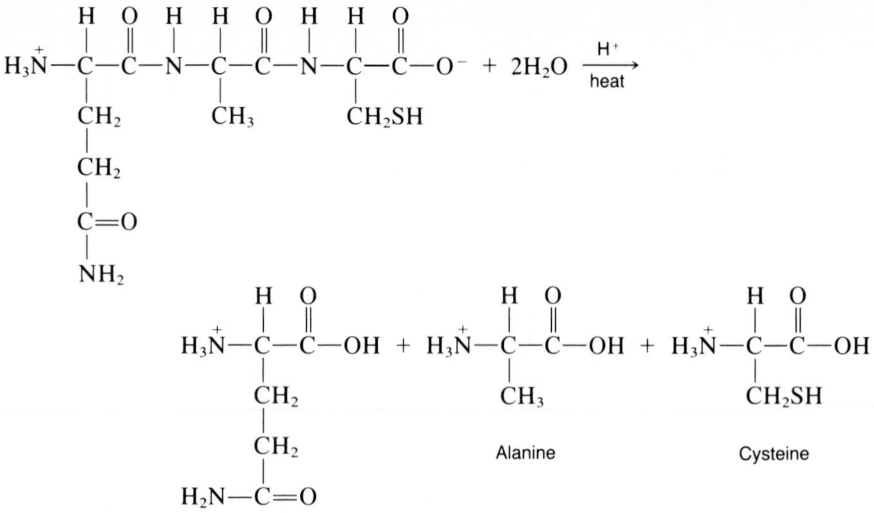

$$\underset{\substack{\text{CH}_2 \\ | \\ \text{CH}_2 \\ | \\ \text{C}=\text{O} \\ | \\ \text{NH}_2}}{\overset{\text{H} \quad \text{O} \quad \text{H} \quad \text{H} \quad \text{O} \quad \text{H} \quad \text{H} \quad \text{O}}{\text{H}_3\text{N}^+\!-\!\text{C}\!-\!\text{C}\!-\!\text{N}\!-\!\text{C}\!-\!\text{C}\!-\!\text{N}\!-\!\text{C}\!-\!\text{C}\!-\!\text{O}^-}} + 2\text{H}_2\text{O} \xrightarrow[\text{heat}]{\text{H}^+}$$

$$\underset{\substack{\text{CH}_2 \\ | \\ \text{CH}_2 \\ | \\ \text{H}_2\text{N}-\text{C}=\text{O}}}{\overset{\text{H} \quad \text{O}}{\text{H}_3\text{N}^+\!-\!\text{C}\!-\!\text{C}\!-\!\text{OH}}} + \underset{\substack{\text{CH}_3}}{\overset{\text{H} \quad \text{O}}{\text{H}_3\text{N}^+\!-\!\text{C}\!-\!\text{C}\!-\!\text{OH}}} + \underset{\substack{\text{CH}_2\text{SH}}}{\overset{\text{H} \quad \text{O}}{\text{H}_3\text{N}^+\!-\!\text{C}\!-\!\text{C}\!-\!\text{OH}}}$$

Glutamine   Alanine   Cysteine

The cation is obtained in each case because of the acidic reaction conditions. Note that other functional groups such as the amide linkage within glutamine are not hydrolyzed under the typical protein hydrolysis conditions.

Sample Exercise 19.2 provides a stepwise approach to writing the products of protein hydrolysis, given a peptide structure.

SAMPLE EXERCISE 19.2

Write structures for the amino acids produced by hydrolysis of

$$\underset{\substack{\text{CH}_3 \qquad\qquad \text{CH}_2\text{OH} \qquad\qquad \text{CH}_2-}}{\overset{\text{O} \quad \text{H} \qquad\qquad \text{O} \quad \text{H} \qquad\qquad \text{O}}{\text{H}_3\text{N}^+\!-\!\text{CH}\!-\!\text{C}\!-\!\text{N}\!-\!\text{CH}\!-\!\text{C}\!-\!\text{N}\!-\!\text{CH}\!-\!\text{C}\!-\!\text{O}^-}}$$

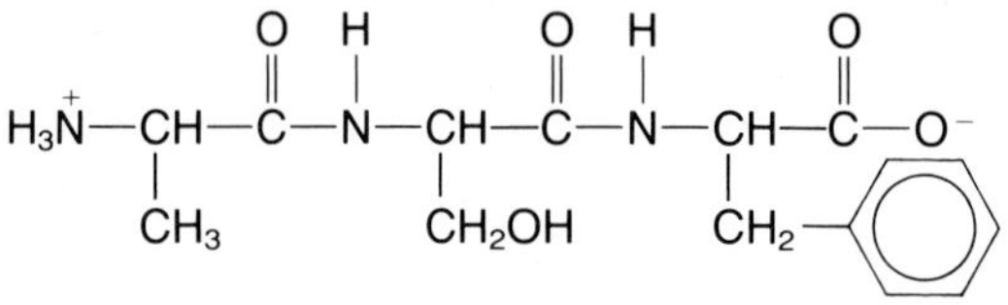

Solution Cleavage occurs at the peptide (amide) bond through the attack of water molecules.

STEP 1 Locate the peptide bonds $\left(\overset{\overset{\displaystyle O}{\|}}{-\text{C}-\text{N}-}\right)$ and draw a slash or wiggly line at the cleavage point:

$$\underset{\substack{\text{CH}_3 \qquad\qquad \text{CH}_2\text{OH} \qquad\qquad \text{CH}_2-}}{\overset{\text{O} \quad \text{H} \qquad\qquad \text{O} \quad \text{H} \qquad\qquad \text{O}}{\text{H}_3\text{N}^+\!-\!\text{CH}\!-\!\text{C}\!+\!\text{N}\!-\!\text{CH}\!-\!\text{C}\!+\!\text{N}\!-\!\text{CH}\!-\!\text{C}\!-\!\text{O}^-}}$$

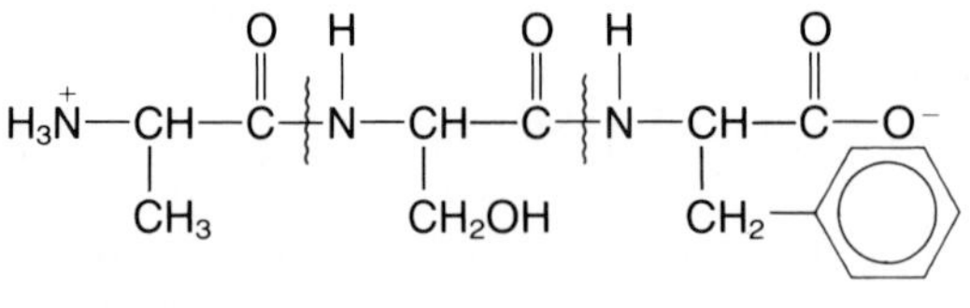

STEP 2 Place water molecules at the cleavage points so that OH is on the carbonyl side and H of water is on the nitrogen side:

$$\underset{\substack{\text{CH}_3 \qquad\qquad \text{CH}_2\text{OH} \qquad\qquad \text{CH}_2-}}{\overset{\text{O} \quad \text{H} \qquad\qquad \text{O} \quad \text{H} \qquad\qquad \text{O}}{\text{H}_3\text{N}^+\!-\!\text{CH}\!-\!\underset{\text{HO}\ \text{H}}{\text{C}}\!+\!\text{N}\!-\!\text{CH}\!-\!\underset{\text{HO}\ \text{H}}{\text{C}}\!+\!\text{N}\!-\!\text{CH}\!-\!\text{C}\!-\!\text{O}^-}}$$

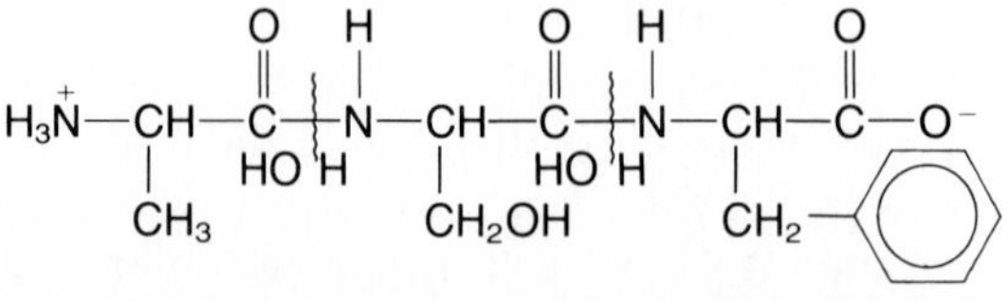

STEP 3 Break the C—N bonds and make new bonds between carbonyl and OH and between N and H. Doing this results in structures for the individual amino acids:

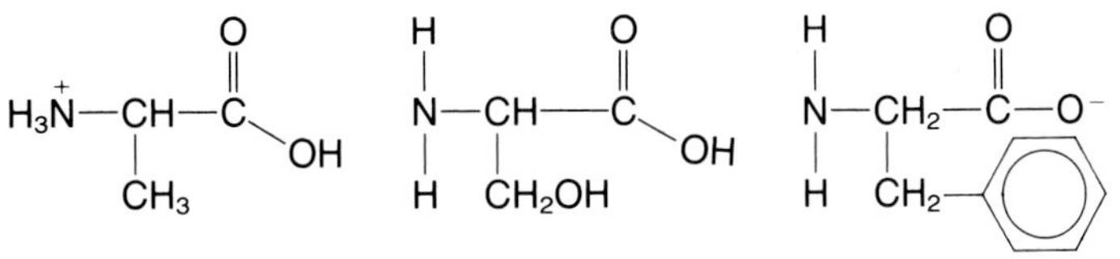

The amino acid structures obtained by this procedure do not necessarily reflect the pH conditions used in the hydrolysis. Instead of the three structures obtained in this example, the cationic forms would actually be produced under acidic conditions and the anionic forms under basic conditions.

Problem 19.3 Write an equation for the acidic hydrolysis of the following tetrapeptide:

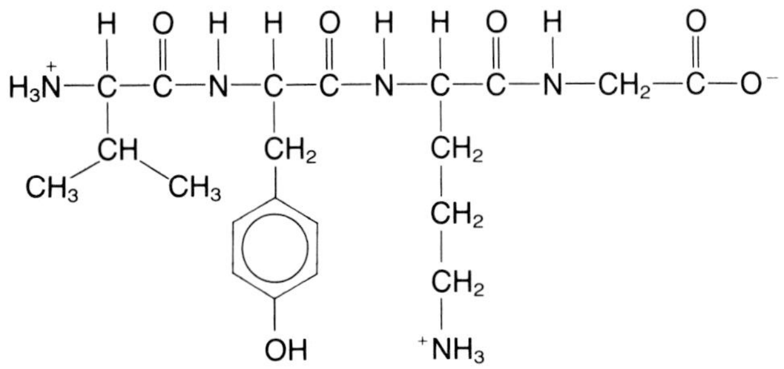

Protein hydrolysis can also be catalyzed by enzymes under milder pH conditions. Digestion is simply the enzyme-catalyzed hydrolysis of the ingested protein in the gastrointestinal tract. The amino acids released by hydrolysis are absorbed through the intestinal wall into the bloodstream and transported to the liver. Large proteins cannot pass through the intestinal wall; small amino acid molecules do so easily.

The amino acids become part of the nitrogen pool which provides raw materials for synthesis of new protein tissue. Excess amino acids are not stored and so must be replenished by a continuous dietary input of protein matter. Chapter 25 elaborates on the fates of amino acids.

CHAPTER ACCOMPLISHMENTS

After completing this chapter you should be able to define **key words** and do the following:

19.1 INTRODUCTION

1. State the functional-group classification of the monomer units making up proteins.

19.2 STRUCTURE OF AMINO ACIDS

2. Give the name and relative position of the functional groups common to all α-amino acids.
3. Given the R group of an amino acid, write the structure of the amino acid in the zwitterionic form.

19.3 CLASSIFICATION OF AMINO ACIDS

4. Classify a given amino acid as hydrophobic or hydrophilic, and further classify a hydrophilic amino acid as acidic, basic, or neutral.

19.4 STEREOISOMERISM IN AMINO ACIDS

5. Recognize a given structure of an amino acid enantiomer as D or L.

19.5 AMINO ACIDS IN SOLUTION

6. Write the structures of a given neutral amino acid in the zwitterionic, cationic, and anionic form, and indicate the general pH conditions in which each is most likely to be found in solution.
7. Describe the nature of the amino acid at its isoelectric point.
8. Describe the relationship between isoelectric point and solubility.

19.6 THE PEPTIDE BOND

9. Write the structure of a di-, tri-, or polypeptide, given the ordered sequence of the component amino acids.

19.7 POLYPEPTIDES

10. State at least two physiological functions of polypeptides.

19.8 PRIMARY STRUCTURE OF PROTEINS

11. Write out all the possible sequences for a tripeptide, given the component amino acids.

19.9 SECONDARY STRUCTURE OF PROTEINS

12. Recognize the α-helix, β-pleated sheet, or collagen triple helix structure within a given diagram of a protein.
13. State the type of attractive interaction in the secondary structure of a protein.

19.10 TERTIARY STRUCTURE OF PROTEINS

14. State and recognize within a given structure the four types of attractive interactions which contribute to the tertiary structure of a protein.
15. Describe how the primary structure determines the tertiary structure.

19.11 QUATERNARY STRUCTURE OF PROTEINS

19.12 STRUCTURE AND FUNCTION OF HEMOGLOBIN

16. Distinguish between the oxygen-carrying capacity of hemoglobin and myoglobin, and describe a hypothesis which could account for the cooperative effect in hemoglobin.
17. State the factors which affect the oxygen-carrying capacity of hemoglobin.

19.13 SICKLE-CELL ANEMIA: A HEMOGLOBIN DISEASE

18. Describe the visually observable abnormality in red blood cells of sickle-cell anemia victims.
19. Describe the molecular difference between sickle-cell hemoglobin and normal hemoglobin, and tell why sickle-cell hemoglobin molecules have a tendency to associate.

19.14 CLASSIFICATION OF PROTEINS

20. State three ways of classifying proteins.
21. State at least four functions of proteins.
22. State the difference between a simple and a conjugated protein and give two specific examples of conjugated proteins and their prosthetic groups.
23. Distinguish between fibrous and globular proteins.

19.15 DENATURATION

24. Describe on a molecular level how heat, ultraviolet radiation, pH changes, organic solvents, heavy metal ions, violent agitation, and reducing agents may denature a protein.
25. Discuss the chemistry involved in a "permanent" wave treatment.

19.16 HYDROLYSIS AND DIGESTION OF PROTEINS

26. Write the structure for the hydrolysis products of a given polypeptide.
27. Distinguish between hydrolysis and denaturation.
28. Describe the digestion process for proteins.

PROBLEMS

19.4 *a* List the three classes of biomolecules.
b Which of these are polymers?

19.5 Name two specific proteins found in the human body and give their function.

19.6 *a* What general monomer unit makes up proteins?
b What type of monomer unit makes up the other biomolecule polymer?

19.7 *a* Draw a structure using R to represent a general alkyl group for an α-amino acid in the zwitterion form.
b Draw the same amino acid in the anionic form.
c Draw the same amino acid in the cationic form.

19.8 Write a zwitterionic structure for an amino acid with the following R groups:

a 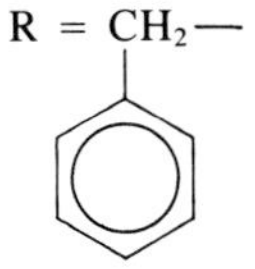

b 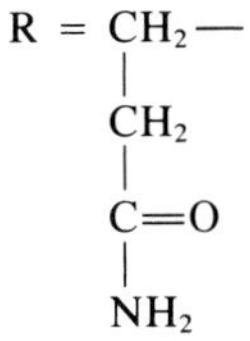

19.9 Without using Table 19.1, indicate whether each of the following amino acids is hydrophobic or hydrophilic:

a 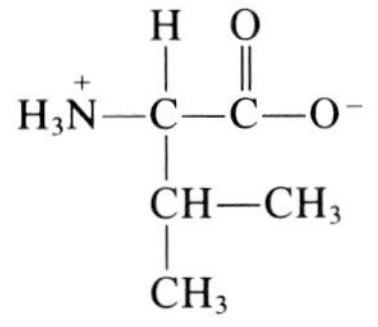

b 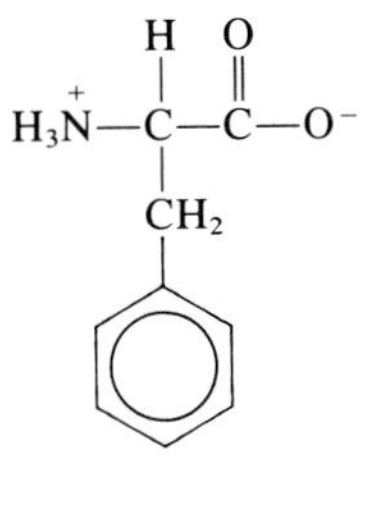

c 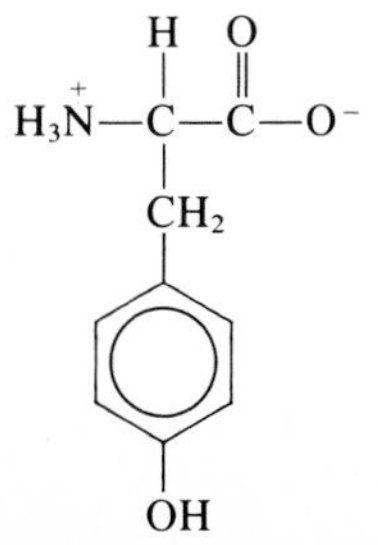

d 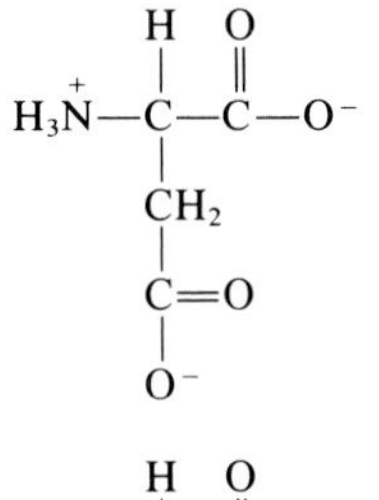

e 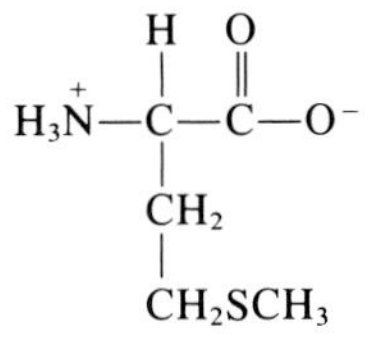

f 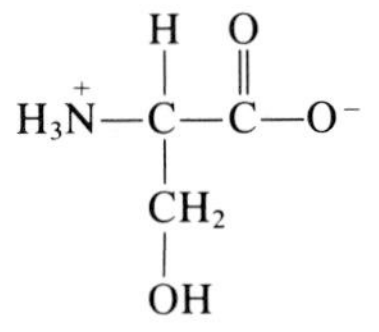

g 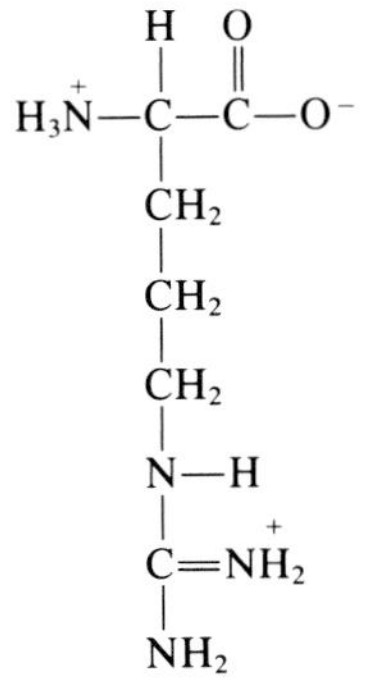

19.10 Classify each hydrophilic amino acid from Problem 19.9 as acidic, basic, or neutral.

19.11 Indicate whether the following structures represent D- or L-amino acids:

a 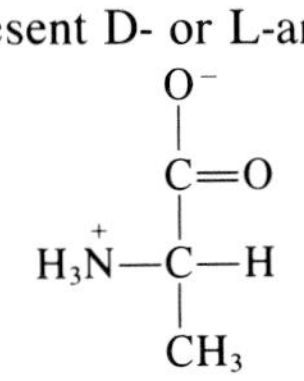

b 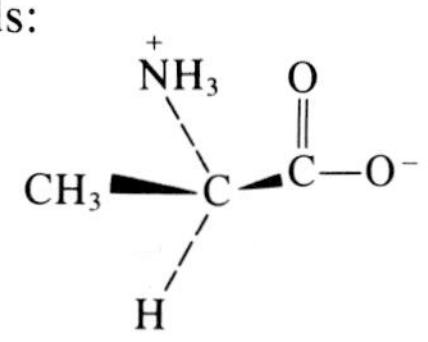

19.12 Which amino acid could an earthling metabolize on a planet which consists of D-amino acids? Why?

19.13 Using Table 19.1, write the structure of threonine that would predominate in a solution at the following pH values:
a pH = 2.5
b pH = 11.0
c pH = 5.60

19.14 Using Table 19.1, write the structure of aspartic acid that would predominate at the following pH values:
a pH = 1.0
b pH = 3.0
c pH = 7.0
d pH = 12.0

19.15 Is the predominant structure of lysine the zwitterionic form at physiological pH (7.4)? If not, at what pH would the zwitterionic form predominate?

19.16 Describe the relationship between the isoelectric point of an amino acid and its solubility.

19.17 What property of an amino acid in its zwitterionic form accounts for the change in solubility as compared with the cationic or anionic form? Explain.

19.18 At what pH would arginine most likely precipitate out of solution?

19.19 A mixture of glycine and glutamic acid at pH = 6.0 is placed in an electrophoresis apparatus. Which amino acid if either, will move to the anode (positive plate)?

19.20 *a* Write the full structural formulas for the dipeptides that could result from combining valine with phenylalanine.

b Write the full structural formula of all the tripeptides that could result from combining the same dipeptides with serine.

19.21 Using the three-letter abbreviations as a convenience, write the structures of all the tripeptides which could result from the combination of valine, leucine, and isoleucine.

19.22 *a* Write out the full structural formula for the tetrapeptide Tyr-Gly-Ser-Cys.

b Write out the full structural formula for the pentapeptide Ala-Val-Leu-Phe-Gly.

c Which of the above two polypeptides has the greater degree of hydrophilic character?

19.23 The hexapeptide Gly-Cys-Phe-Val-Cys-Val can form a cyclic structure. Show how this is possible.

19.24 Which amino acid in the protein fragment Ala-Phe-Asp-Gly would most likely be found in the surface of the protein containing this fragment? Explain your answer.

19.25 Oxytocin is a polypeptide hormone found in the pituitary gland. Its normal physiological function is in the contraction of smooth muscle, but it is sometimes given in obstetrics to induce labor and to stimulate lactation. Write out the full structural formula of oxytocin.

Cys-Tyr-Ile-Gln-Asn-Cys-Pro-Leu-Gly

19.26 What change may occur in the structure of oxytocin (Problem 19.25) if a mild reducing agent is added to a solution of oxytocin?

19.27 List two specific polypeptides and their physiological role in the body.

19.28 Can a polypeptide be a protein? Can a protein be a polypeptide?

19.29 Describe what is meant by the primary structure of a protein?

19.30 Describe how the primary sequence of amino acids in a protein common to two species gives information about the evolutionary development of those species.

19.31 Pick out and label the α-helix and β-pleated structures within the following protein fragment:

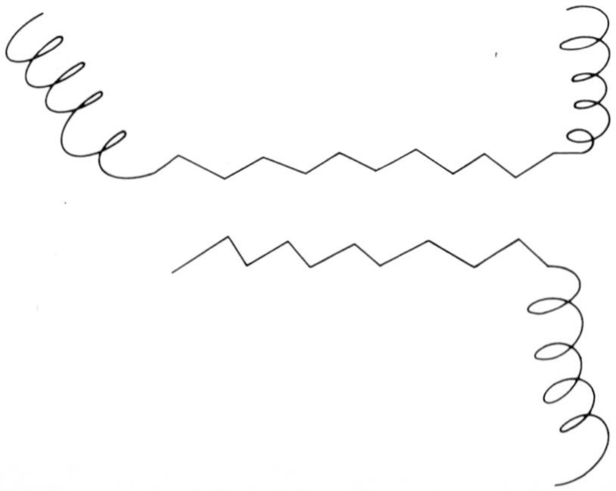

19.32 *a* What is meant by the secondary structure of a protein?

b Can there be a secondary structure without a primary structure?

c Can there be a primary structure without a secondary structure?

19.33 What type of attractive interaction holds together the secondary structure of a protein?

19.34 What common feature exists in the hydrogen bonding found in the α-helix, β-pleat, and collagen triple helix?

19.35 What molecular structure gives rise to the very high tensile strength in the collagen triple helix?

19.36 Fibrous proteins which are composed predominantly in the α-helix are insoluble in water. Hair is a fibrous protein. What types of amino acids would you expect to find in hair protein?

19.37 An absence of Vitamin C in the diet leads to scurvy, a condition of weakened blood vessels and skin lesions. What molecular function does Vitamin C serve in preventing this condition?

19.38 *a* Describe what is meant by the tertiary structure of a protein.

b Can there be a tertiary structure without a primary structure?

c Can there be a tertiary structure without a secondary structure?

19.39 *a* State the four types of attractive interaction that give rise to tertiary structure.

b Distinguish between the origin of the attractive interactions in the tertiary as compared with those in the secondary structure of a protein.

19.40 Hydrogen bonding is an attractive interaction which occurs in the secondary as well as the tertiary structure of a protein. Distinguish any difference between the hydrogen bonding in the secondary as compared with that in the tertiary structure.

19.41 From Table 19.1 pick two amino acids (other than ones illustrated within the text) which when part of a protein contribute to tertiary structure by:

a Ionic attraction (salt bridge)

b Hydrogen bonding

c Hydrophobic interaction

19.42 What type of attractive interaction could occur between the following pairs of amino acids:

a Tyrosine and aspartic acid

b Glutamic acid and histidine

c Alanine and isoleucine

d Cysteine and cysteine

e Serine and serine

19.43 Why does a protein fold or refold into its tertiary structure?

19.44 *a* Define the term *oligomeric protein.*

b Discuss what is meant by the quaternary structure of a protein.

c Give a specific example of an oligomeric protein.

19.45 What heterocyclic amine is found in the heme unit?

19.46 Explain what is meant by the cooperative binding of oxygen by hemoglobin.

19.47 *a* How do hemoglobin and myoglobin differ structurally?

b How does this structural difference play a role in their oxygen-carrying capacity?

19.48 *a* Discuss the effect on hemoglobin's oxygen-carrying capacity of a low blood pH and high CO_2 partial pressure.

b Discuss the effect of high pH and low CO_2 partial pressure on hemoglobin's oxygen-carrying capacity.

c Between what two bodily "stations" does hemoglobin deliver its oxygen?

d At which one of these stations do we find the conditions of part *a*? What function does hemoglobin serve here?

e At which one of the stations do we find the conditions of part *b*? What role does hemoglobin play here?

19.49 Acidosis, a lowering of the blood's normal pH range, can arise from a variety of causes. What effect does acidosis have on hemoglobin's oxygen-carrying capacity.

19.50 What hypothesis has been put forward to explain hemoglobin's cooperative oxygen binding?

19.51 Describe the two ways in which CO_2, which would by itself be quite insoluble in blood, is carried from an active cell to the lungs.

19.52 What is the visible (under a microscope) difference between normal red blood cells and the red blood cells of a sickle-cell anemia patient?

19.53 *a* What is the molecular difference between normal and sickle-cell hemoglobin?

b What is the physiological result of this molecular difference? How may it come about?

19.54 Give the classification category (function, chemical composition, or shape) for proteins described by each of the following:

a A globular protein

b An enzyme

c A lipoprotein

d A tendon

e A nucleoprotein

19.55 Describe the difference between a simple and a conjugated protein and give one specific example of each.

19.56 Which structure of a protein is affected by denaturation?

19.57 An autoclave is used to sterilize surgical instruments. What protein is being denatured in this process?

19.58 In what way is the protein in a cooked egg the same as in a raw egg?

19.59 When we say lead is a toxic metal, what chemical form of the lead do we actually mean?

19.60 *a* Explain how egg white or milk can serve as a emergency antidote for heavy metal poisoning.

b Why is it important to give an emetic after this emergency treatment?

19.61 Describe the reduction and oxidation steps of a permanent wave treatment in terms of their molecular effect on hair protein.

19.62 What is the difference on a molecular level between denaturation and hydrolysis?

19.63 Write structures for the amino acids produced by the acidic hydrolysis of the following pentapeptide:

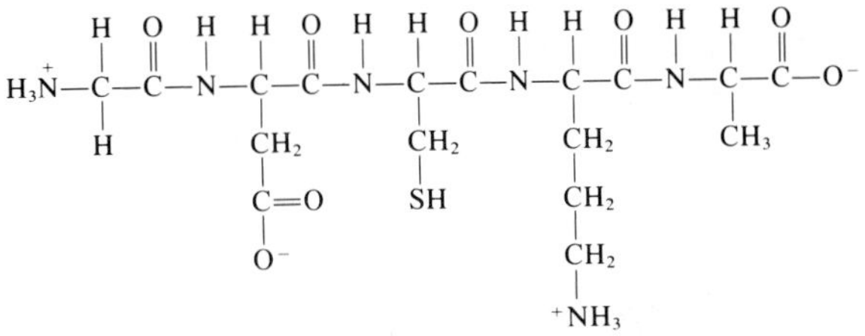

19.64 What is the relationship between hydrolysis under test-tube conditions and digestion under physiological conditions.

19.65 Are the proteins we absorb in our food supply the identical proteins we use in our body? Explain.

19.66 The proteins in seafood such as scallops are sometimes "cooked" by marinating in lemon or lime juice. Explain what is happening.

19.67 Drugs which are proteins, such as insulin, cannot be taken by mouth but must be injected. Explain.

19.68 Give the name of a protein that has the following function or occurrence:

a Is present in milk

b Is a neurotransmitter

c Stores oxygen in muscle cells

d Transports oxygen in blood

e Is an enzyme

f Is found in tendons

19.69 Show by use of equations how a polypeptide or a protein could act as a buffer. Use as an example the dipeptide

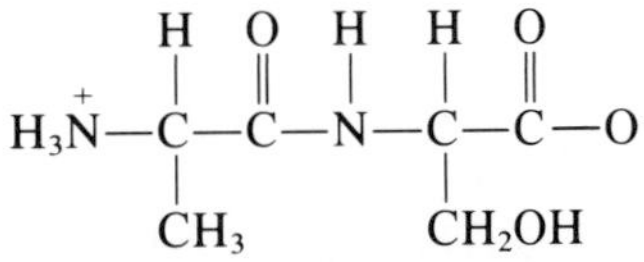

19.70 *a* Could a protein which has been denatured by an acid solution become physiologically active again after the solution is neutralized?

b Could a hydrolyzed protein become active again after neutralization?

CHAPTER TWENTY

NUCLEIC ACIDS

20.1 INTRODUCTION

How is it that we inherit some characteristic features from our parents? Indeed, why is the child of human parents always a human being rather than a lion, an oak tree, or some new species? How is a protein molecule formed in the body with the precise amino acid sequence so that it is physiologically active? These are questions of information transfer. Information is transferred from parents to the first cell of their offspring at the moment of conception. The information is duplicated as the fetus grows from one cell to the approximately 10^{14} (100 trillion) cells in the fully developed baby.

The amount of information is staggering, its nature is highly varied, and it must be very specific. For example, if the child is to have normal glucose metabolism (i.e., be nondiabetic), instructions must be provided to the cells in the pancreas for the correct synthesis of the protein insulin.

In 1945, deoxyribonucleic acid (DNA) was identified as the bearer of hereditary information and the controller of protein synthesis in the cell. The DNA molecule appeared to contain one of the molecular secrets of life. However, the mechanism by which it stored and transferred information was unknown.

In this chapter we will explore the structure of DNA and its close relative ribonucleic acid (RNA) and you will see how the discovery of DNA's three-dimensional structure immediately suggested a hypothesis about the method of storage of hereditary information. The unique structures of DNA and RNA allow them to control protein biosynthesis, and the details of this mechanism are now well known. After you learn the basics of nucleic acid chemistry, you can appreciate the new technology of genetic engineering. Results from this new discipline will have great benefit in medicinal chemistry and many other fields of human endeavor. For example, a herpes vaccine is being developed through techniques of genetic engineering. Let us begin on the road toward understanding the new field by studying the chemical composition of DNA.

20.2 CHEMICAL COMPOSITION OF DNA

Deoxyribonucleic acid got its name because it contains the sugar deoxyribose, was originally found in the nucleus of cells (hence *nucleic*), and chemically is an acid. In multicellulared organisms, such as human beings, DNA is complexed with proteins and forms threadlike structures called **chromosomes.** There are 23 pairs of chromosomes in human cells. When a cell is in the process of dividing, the chromosomes are visible under a microscope (Figure 20.1). Each chromosome thread is composed of a single DNA molecule.

Like proteins and polysaccharides, DNA molecules are polymers. When

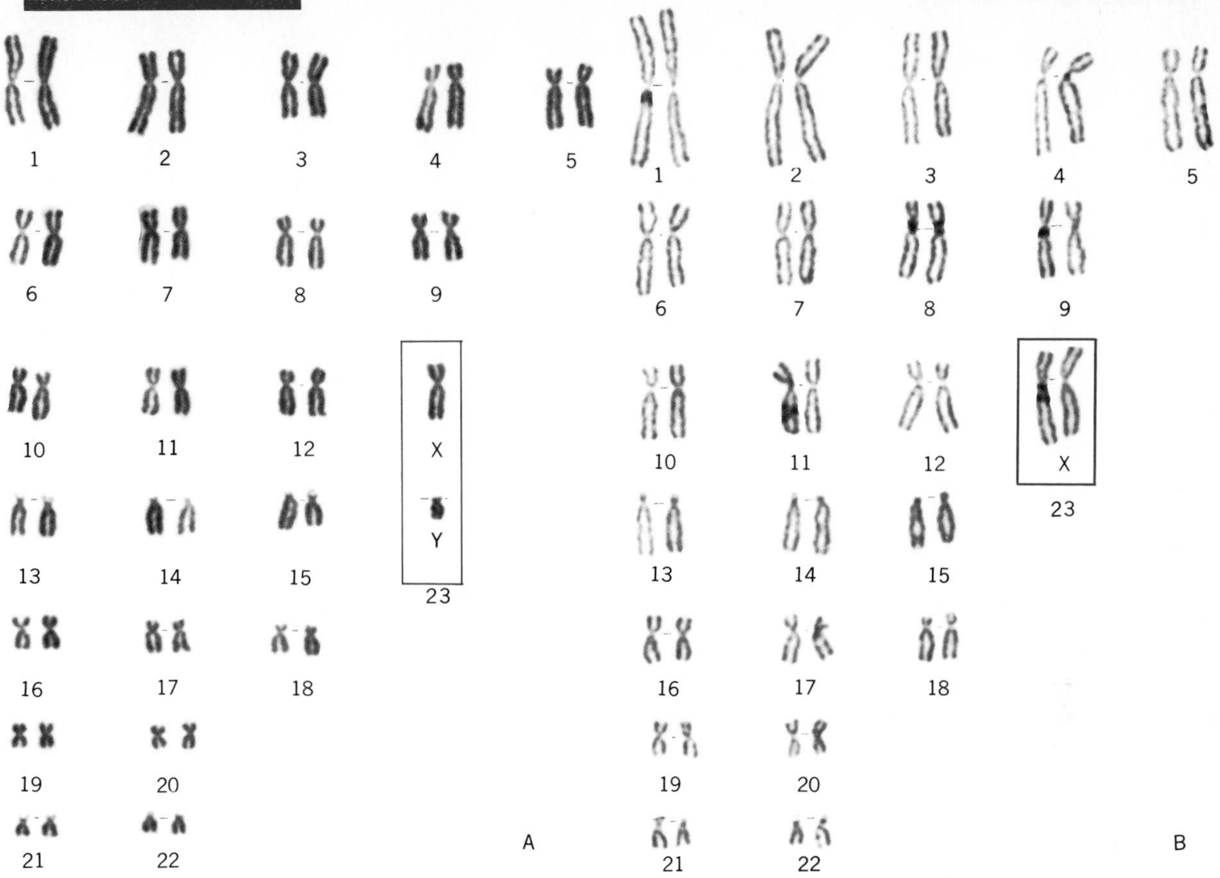
1 2 3 4 5
6 7 8 9
10 11 12
13 14 15
16 17 18
19 20
21 22
X
Y
23
A
1 2 3 4 5
6 7 8 9
10 11 12
13 14 15
16 17 18
19 20
21 22
X
23
B

FIG. 20.1
Chromosomes are combinations of protein and nucleic acids. The molecules are large enough to be seen under an electron microscope. The 46 (23 pairs) human chromosomes have been assigned identifying numbers as shown here for male (A) and female (B) cells. There are 22 pairs of somatic chromosomes and one pair of sex chromosomes: an XY pair for males and an XX pair for females.

DNA is hydrolyzed, monomer units known as **nucleotides** are isolated. Each DNA nucleotide contains residues of a phosphoric acid molecule, a deoxyribose sugar molecule, and a nitrogen heterocyclic base. There are four heterocyclic bases; therefore we can distinguish four DNA nucleotides. The four bases are *a*denine, *c*ytosine, *g*uanine, and *t*hymine. The names of nucleotides are often abbreviated by using just the first letter of the nitrogen base name, A, C, G, or T. Figure 20.2 summarizes the breakdown of chromosomes to sugar, phosphate, and base.

You first encountered the heterocyclic bases in Section 16.7 and learned that two of the nitrogen bases are purines (adenine and guanine) whereas two are members of the pyrimidine class (cytosine and thymine).

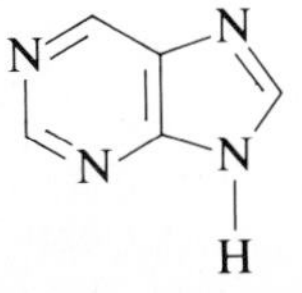
Purine

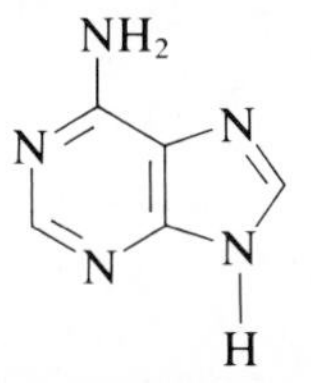
Adenine

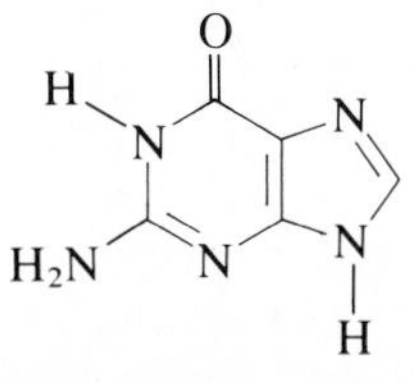
Guanine

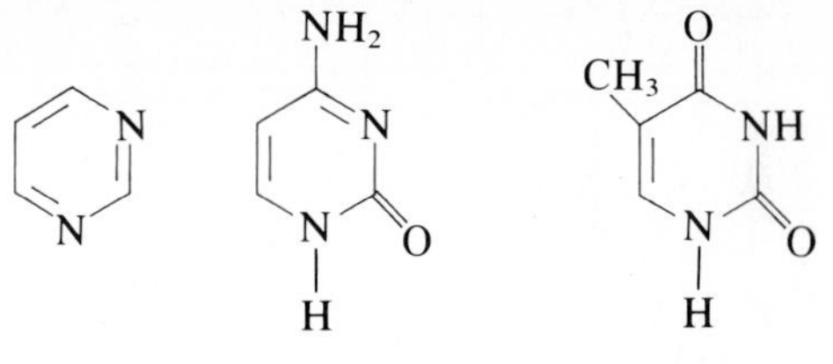

NH2
N
N
H
O
CH3
O
NH
N
O
H

Pyrimidine Cytosine Thymine

In Section 16.12 you encountered **nucleosides,** compounds in which a nitrogen base is bonded to a sugar molecule. For example, deoxyadenosine is the combination of deoxyribose and adenine:

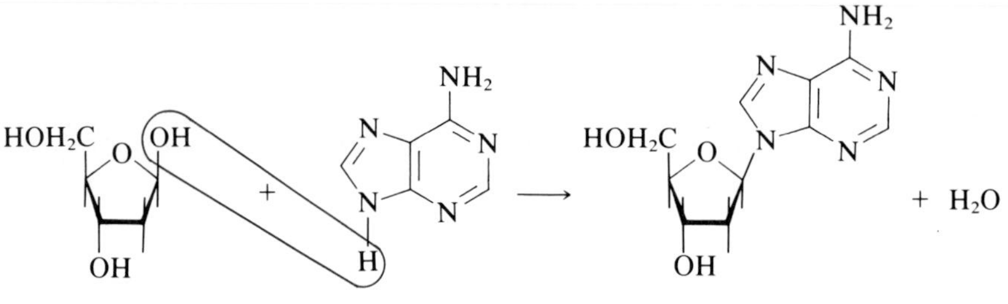

HOH2C
O
OH
OH
+
NH2
N
N
N
N
H
HOH2C
O
OH
NH2
N
N
N
N
+ H2O

Deoxyribose Adenine Deoxyadenosine

A nucleotide is a phosphorylated nucleoside, that is, a molecule in which a phosphoric acid molecule forms a phosphate ester linkage with the 5′ alcohol of the sugar group of a nucleoside. [The numbers of the sugar ring positions have primes (′) to distinguish them from numbers assigned to purine or pyrimidine ring positions.] Schematically,

---

FIG. 20.2

The stepwise breakdown of chromosomes. Chromosomes are nucleoproteins, i.e., combinations of nucleic acids and proteins. Components of nucleic acids are shown in color. **1** Protein is separated from the deoxynucleic acid. Both proteins and nucleic acids are polymers. **2** Nucleic acids are broken down into nucleotide monomer units. **3** Nucleotides are three-component systems; phosphoric acid (phosphate) can be hydrolyzed off, leaving behind a nucleoside. **4** Nucleosides are sugar-base combinations; the sugar and base can be cleaved by hydrolysis.

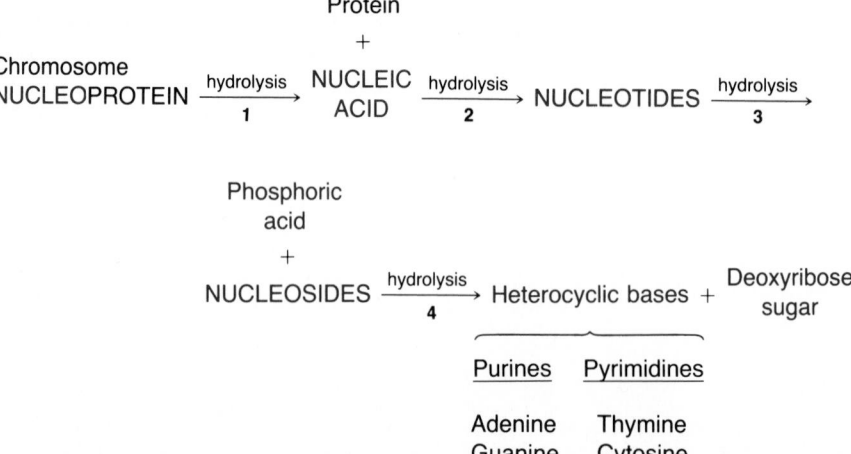

Chromosome NUCLEOPROTEIN $\xrightarrow[1]{\text{hydrolysis}}$ Protein + NUCLEIC ACID $\xrightarrow[2]{\text{hydrolysis}}$ NUCLEOTIDES $\xrightarrow[3]{\text{hydrolysis}}$

Phosphoric acid + NUCLEOSIDES $\xrightarrow[4]{\text{hydrolysis}}$ Heterocyclic bases + Deoxyribose sugar

Purines	Pyrimidines
Adenine	Thymine
Guanine	Cytosine

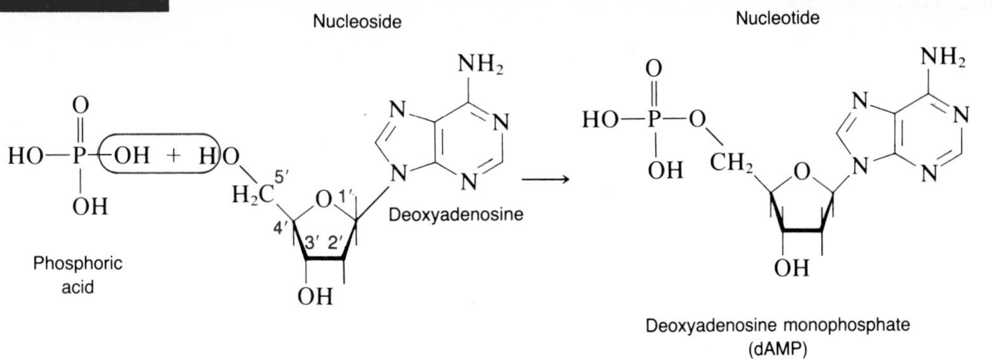
Nucleoside
Nucleotide
Phosphoric acid
Deoxyadenosine
Deoxyadenosine monophosphate (dAMP)

(20.1)

TABLE 20.1 DNA Nucleotides and Their Component Parts

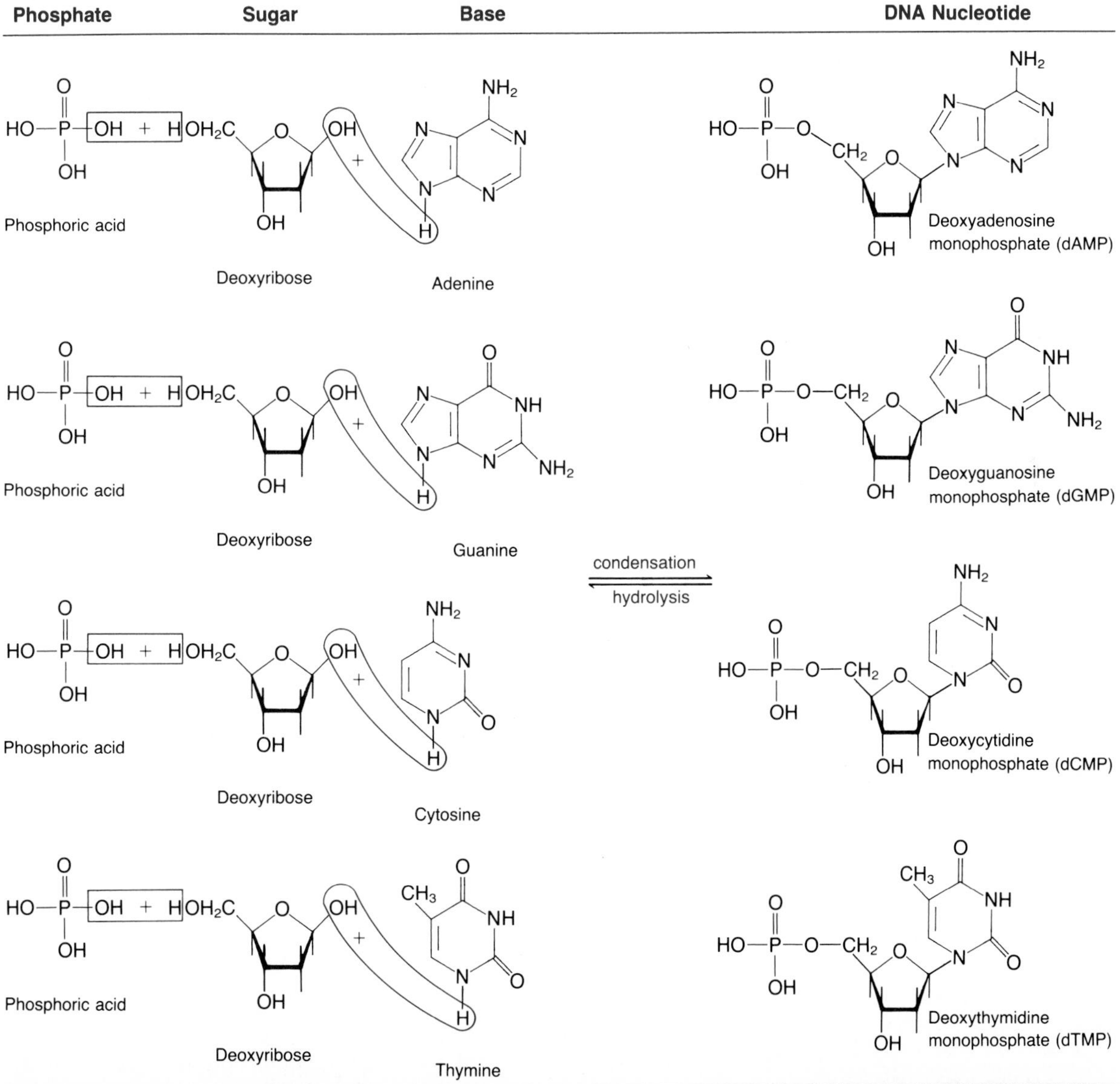
Phosphate
Sugar
Base
DNA Nucleotide
Phosphoric acid
Deoxyribose
Adenine
Deoxyadenosine monophosphate (dAMP)
Phosphoric acid
Deoxyribose
Guanine
Deoxyguanosine monophosphate (dGMP)
condensation
hydrolysis
Phosphoric acid
Deoxyribose
Cytosine
Deoxycytidine monophosphate (dCMP)
Phosphoric acid
Deoxyribose
Thymine
Deoxythymidine monophosphate (dTMP)

Table 20.1 shows the assembly, structures, and names of the four characteristic nucleotides found in DNA. The condensation reactions are the same in every assembly; only the identity of the base varies.

Problem 20.1 Without looking at Table 20.1, combine phosphoric acid, deoxyribose, and guanine to form the nucleotide deoxyguanosine monophosphate.

In addition to nucleotides present as monophosphates, cells contain nucleotides with two or three phosphoric acid residues, diphosphates and triphosphates. We can think of the nucleotide di- and triphosphates forming from the nucleotide monophosphate by acid anhydride formation between a phosphoric acid molecule and a phosphate residue, as you have seen in Section 17.8. For example,

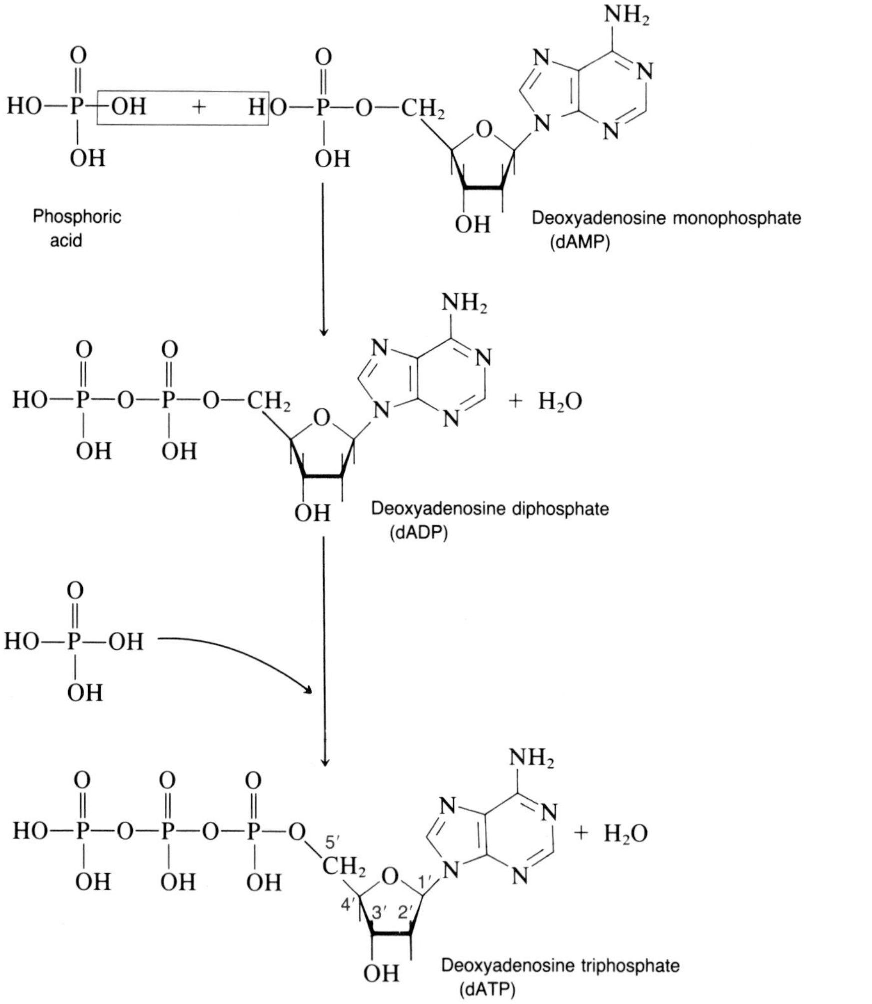
Phosphoric acid
Deoxyadenosine monophosphate (dAMP)
Deoxyadenosine diphosphate (dADP)
Deoxyadenosine triphosphate (dATP)
+ H₂O

Experiments indicate that nucleotide triphosphates are the actual monomers which polymerize together by condensation to form a DNA molecule.

20.3 THE PRIMARY STRUCTURE OF DNA—A POLYMER OF NUCLEOTIDES

A DNA molecule is a polymer of nucleotides, but how are the monomer units connected together? The complexity of the nucleotide unit presents several possibilities. It turns out that nucleotides are bound to one another by way of phosphodiester linkages connecting the 5′ carbon of the sugar in one nucleotide to the 3′ carbon of the sugar in the next nucleotide in the sequence.

Using the joining of deoxyadenosine triphosphate (dATP) and deoxythymidine triphosphate (dTTP) as an example, you can see how the phosphodiester linkage forms by condensation wherein a diphosphate group is split out.

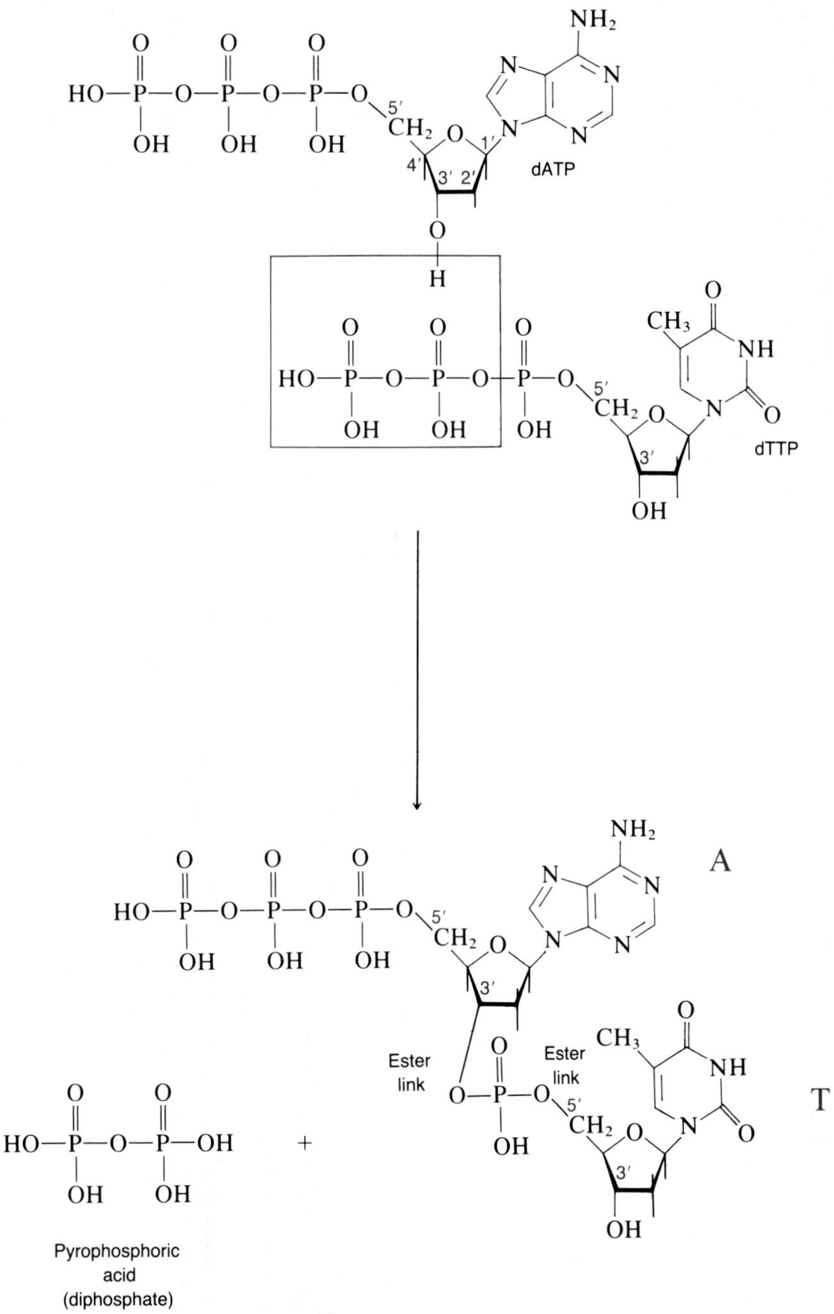
dATP
dTTP
A
T
Ester link
Ester link
Pyrophosphoric acid (diphosphate)

The chain can be continued by linking a third nucleotide, for example, guanosine triphosphate (dGTP), to the 3′ carbon of the deoxythymidine sugar. The resulting trinucleotide structure is

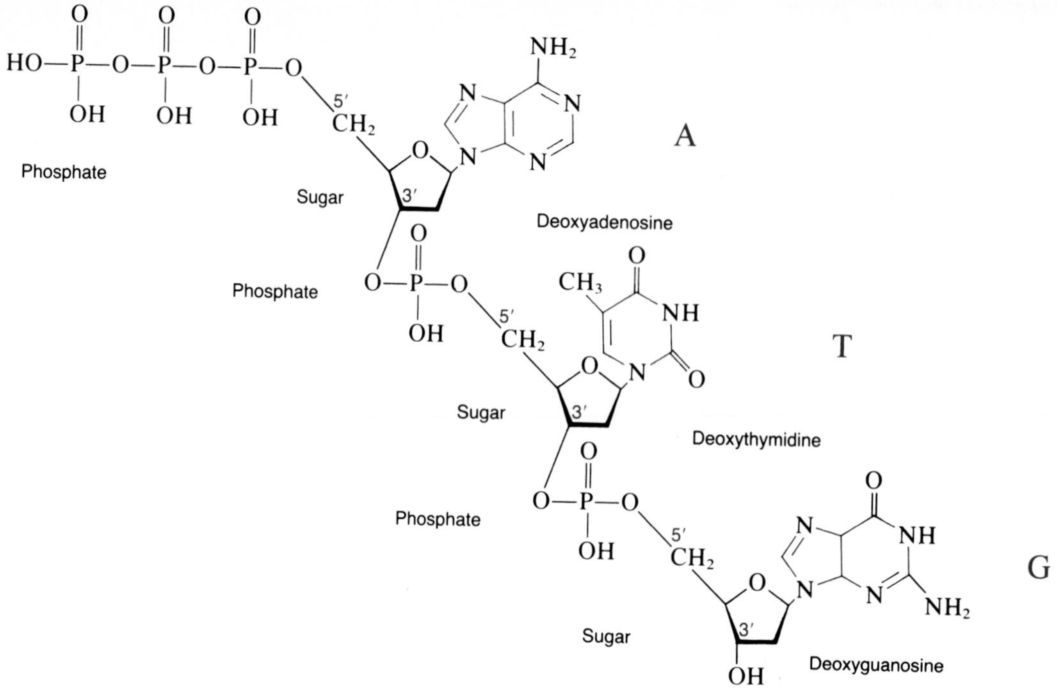

HO—P—O—P—O—P—O
OH OH OH
Phosphate
5′
CH_2
NH_2
A
Sugar
3′
Deoxyadenosine
Phosphate
O—P—O
OH
CH_3
NH
T
Sugar
Deoxythymidine
Phosphate
O—P—O
OH
NH
G
NH_2
Sugar
OH
Deoxyguanosine

The polynucleotide chain can be extended by continuing this process of forming phosphodiester linkages between the 3′ carbon of the sugar residue in the growing chain and the 5′ carbon of another nucleotide.

The growing chain develops a backbone of alternating phosphate and sugar residues, i.e., phosphate-sugar-phosphate-sugar-phosphate-sugar. . . . The nitrogen base bonded to each sugar extends from the backbone like rungs connected to *one* long sidepiece of a ladder.

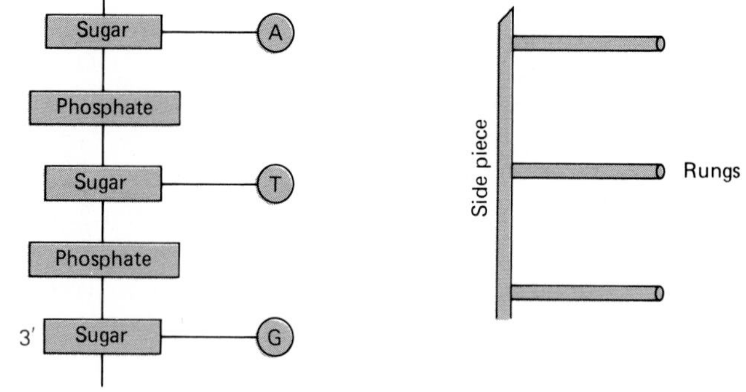

5′
Phosphate
Sugar
A
Phosphate
Sugar
T
Phosphate
3′
Sugar
G
Side piece
Rungs

However, unlike many ladders, there are two distinctive ends to the phosphate-sugar sidepiece, or backbone: One end has an intact phosphate group and is known as the 5′ end; the other end, known as the 3′ end, has a free —OH on the 3′ carbon of the sugar residue.

Because the nitrogen bases define the nucleotide sequence, an abbreviated form can be written for any polynucleotide which uses only the symbols for the nitrogen bases. For the trinucleotide shown above, this abbreviated form is A—T—G—. As with protein primary structure, the sequence A—T—G— is distinct from G—T—A— or T—A—G—. Therefore, also by analogy to protein primary structure, the sequence of nitrogen bases determines the nature of the polynucleotide, and is called the primary structure of the polynucleotide.

20.4 THE SECONDARY STRUCTURE OF DNA—THE DOUBLE HELIX

In the late 1940s, even before the nature of the linkage between nucleotides was known, it was discovered that, regardless of the DNA's origin, the molar amount of adenine present is always equal to the molar amount of thymine and there is a similar equality between cytosine and guanine. In short, in any DNA:

Number of A nucleotides = number of T nucleotides

Number of C nucleotides = number of G nucleotides

The primary structure does not explain this equality between A and T and between C and G. These base equalities pointed the way to the elucidation of the higher levels of DNA structure.

In 1953 James Watson and Francis Crick proposed a model structure for DNA that accounted for the equivalences between the A and T and between the C and G bases. The Watson-Crick hypothesis states that DNA is a **double helix** of two polynucleotide chains wrapped around each other in a right-handed helical fashion. (See Figure 20.3.) The sugar-phosphate backbones (or ladder sidepieces) are on the outside of the double helix, and the nitrogen bases (ladder rungs) are on the inside. Hydrogen bonding between the bases holds the helix together. Figure 20.3 also shows that the two sugar-phosphate backbones run *antiparallel,* that is, one runs $3' \longrightarrow 5'$ and the other $5' \longrightarrow 3'$. The coiling of two strands into a double helix is the secondary structure of DNA.

A key feature of the double helix is that the hydrogen bonding between bases is not random. Rather, an adenine base on one chain is always paired with and hydrogen-bonded to a thymine on the other chain and a cytosine is paired and hydrogen-bonded to a guanine. This unique pairing arises for two reasons:

1 The space between the two helically coiled phosphate-sugar chains allows effective hydrogen bonding only between a purine and a pyrimidine base. Two pyrimidines would be too far apart, and two purines would be squeezed too close together.
2 The strongest hydrogen bonding occurs between the A—T and G—C purine-pyrimidine pairs. The A—C and G—T pairs are less strongly attracted to one another.

The strength of the hydrogen-bonding attraction between them accounts for the molar equivalency between A and T and between C and G bases. Wherever there is an A on one polynucleotide chain, there must be a T on the other chain; similarly, for every C there must be a G. We call the relationship between adenine and thymine and between cytosine and guanine **complementary base pairing.** One polynucleotide chain of the DNA double helix is the complement of the other chain. Thus, if we know the sequence

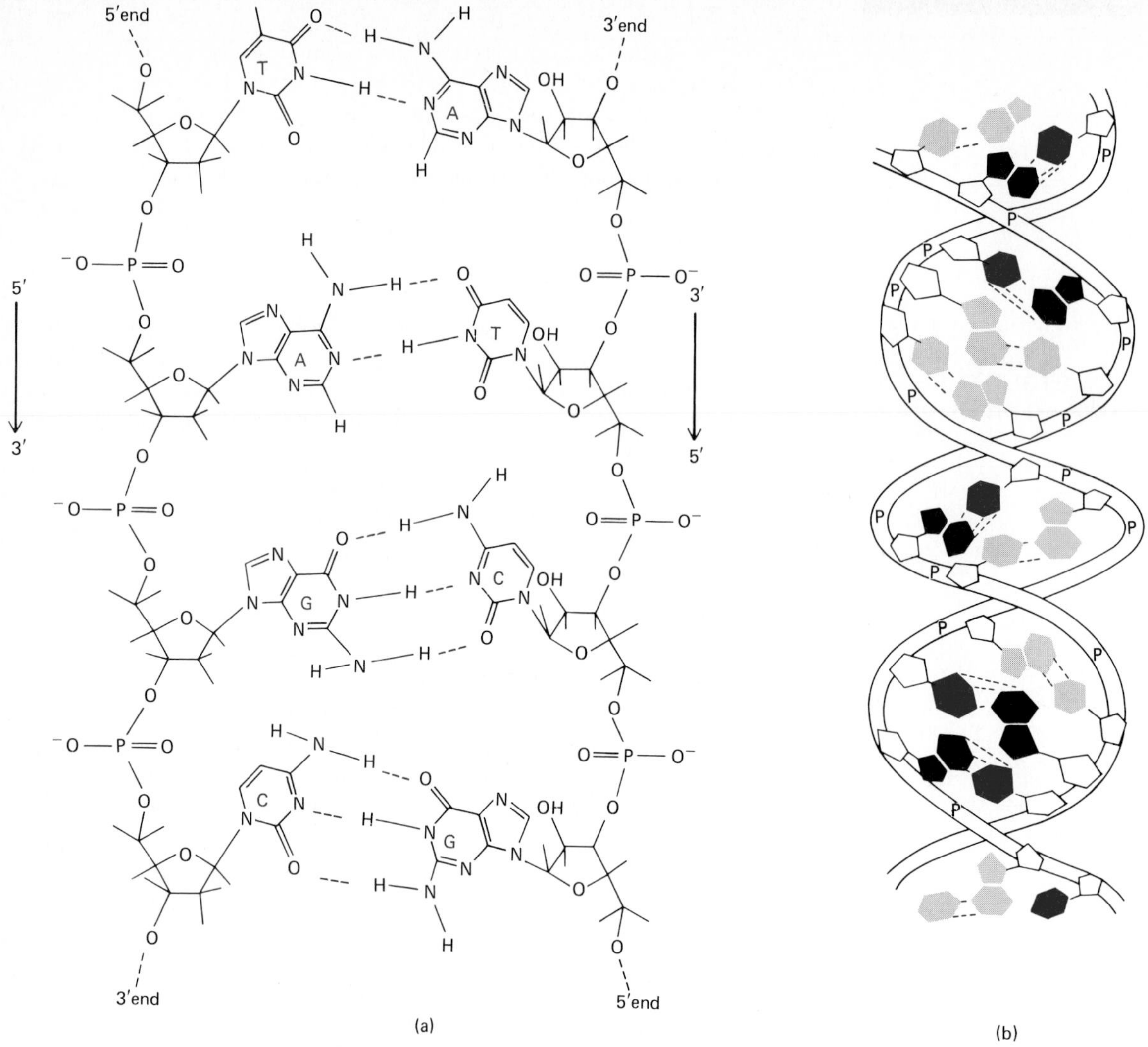
5′end
3′end
5′
3′
3′
5′
3′end
5′end
(a)
(b)

FIG. 20.3
The double helix. (*a*) All structural details of the double helix are shown. There are two sugar-phosphate backbones held together by hydrogen bonding between complementary bases (A and T, C and G). Notice that the sugar-phosphate backbones are antiparallel; that is, one runs 3′ → 5′ top to bottom, and the other runs in the opposite direction (5′ → 3′). (*b*) This picture has less structural detail, but it shows the helical coiling of the backbones and the bases in the center of the helix. Also, see the front cover.

of bases on one chain, we can immediately write the sequence of bases on the complementary chain by using the A—T, C—G relationship.

SAMPLE EXERCISE 20.1 Write the nucleotide sequence in the complementary chain of the following fragment found in a DNA molecule:

—T—C—A—A—G—C—

Solution For each base symbol write the symbol of the complementary base, for example, for T, write A. Thus:

Original chain	—T—C—A—A—G—C—	
	: : : : : :	Base pairing
Complementary chain	—A—G—T—T—C—G—	

Problem 20.2 Write the complementary chain nucleotide sequence for the following:

—C—G—G—T—A—C—G—A—

20.5 DNA REPLICATION

The Watson-Crick model of DNA was very satisfying because it accounted for all known experimental structural data, including the equal abundance of A and T bases and C and G bases. Perhaps of even greater importance was the fact that the double helix immediately suggested a hypothesis for a method of DNA **replication** that would conserve the chemical information in the DNA molecule. Replication is the formation of a *replica* or exact duplicate. We will see this structural model explain how cell division produces two cells each with the identical informational content of the original cell.

Watson and Crick were well aware of the biological significance of their double helix, and in their first publication on the structure of DNA, they ended the one-page manuscript with the following understatement: "It has not escaped our notice that the specific pairing we have postulated immediately suggests a possible copying mechanism for the genetic material."

In the Watson-Crick hypothesis for DNA replication each strand of an original double helix serves as a template for the formation of a complementary *daughter* strand. The details of how the overall process occurs are now well known. At some point along the double helix, the helix begins to unwind and the hydrogen bonds "unzip," as shown in Figure 20.4*a*. This leads to partial separation of the parental strands. Part of the DNA molecule maintains the two strands bound and coiled together, but there is a separated section which begins at a replication *fork*. Assembly of daughter strands is directed by the principle of complementary base pairing, and Figure 20.4*b* shows the daughter strands which form as complements to the unwound parent strands.

The assembly of daughter strands occurs by a series of hydrogen-bonding base pairings followed by condensation reactions between nucleotides, as shown in Figure 20.5. At the end of one separated parental strand an appropriate nucleotide triphosphate, that is, one that contains a nitrogen base complementary to the end base, fits into position by forming a C—G or A—T hydrogen bond. (See Figure 20.5*a*.) A second appropriate nucleotide triphosphate hydrogen-bonds to the next open base on the parental strand. These two nucleotides are linked by a phosphodiester bond between the 3′ position (OH) of the first nucleotide triphosphate and the 5′ position (triphosphate) of the second nucleotide. (See Figure 20.5*b*.) The second nucleotide of the growing daughter strand has a free 3′ OH, and the next (third) nucleotide triphosphate will bond to this 3′ OH. Reactions with nucleotide triphosphates continue, and the daughter strand grows from its 5′ end to the 3′ end toward the replication fork. The replication fork moves along the molecule as the parental double helix unwinds and separates. Eventually a complete daughter strand is synthesized. It is the complement of its parental template, and assuming there have been no errors, it is identical to the other separated parental strand. Several enzymes and proteins are required for this process; the DNA polymerases are an especially important group. These catalyze the linking together of the nucleotides to form the daughter strand.

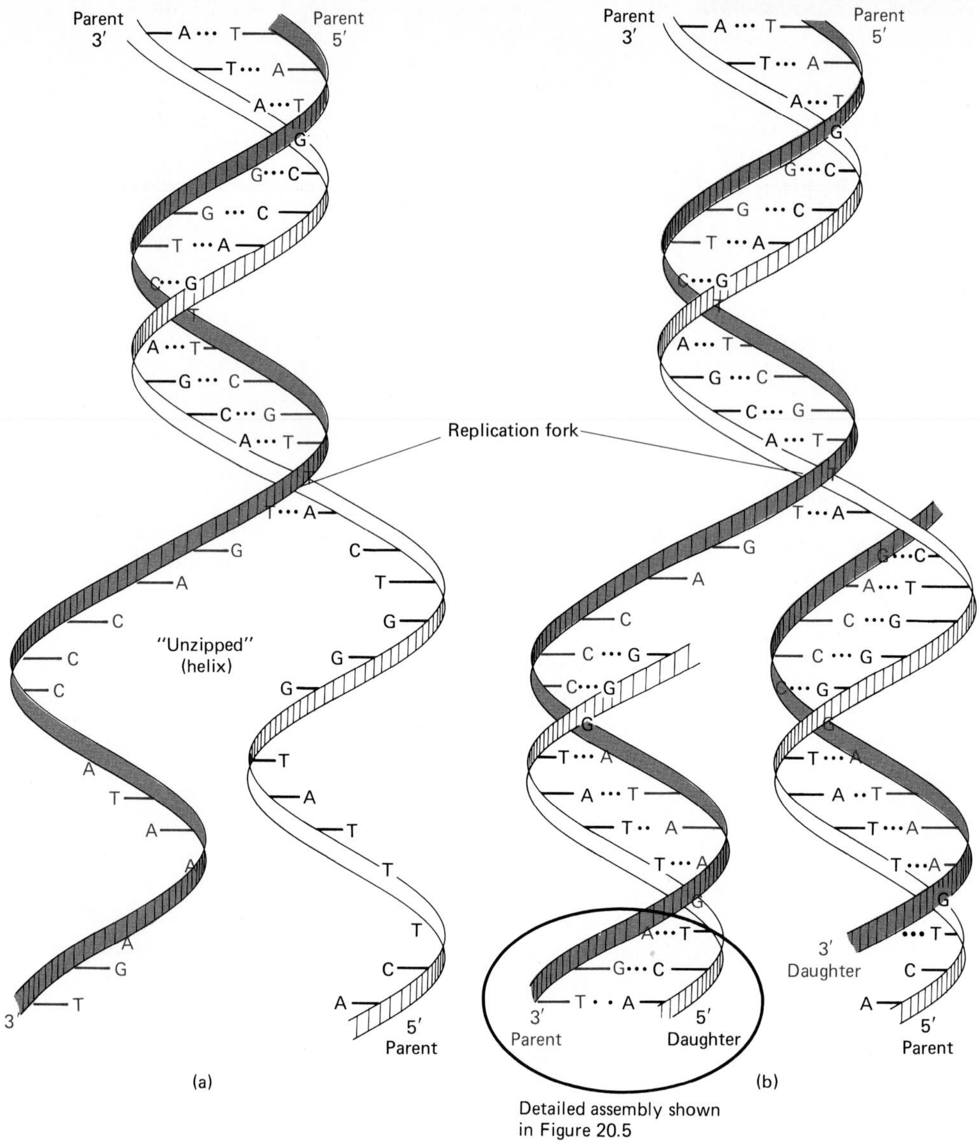
Parent
3′
Parent
5′
Replication fork
"Unzipped"
(helix)
3′
5′
Parent
(a)
3′
Parent
5′
Daughter
3′
Daughter
5′
Parent
(b)
Detailed assembly shown
in Figure 20.5

FIG. 20.4
(*a*) The first step in replication is the partial "unzipping" of the double helix; that is, base pairs separate. The replication fork is the point of opening of the double helix into two separate strands. (*b*) The daughter strands are assembled by the principle of base pairing in the 5′ → 3′ direction. Consequently, one daughter builds up toward the fork and the other away from it. Figure 20.5 shows the details of assembly of the daughter strand complementary to the left parent strand.

We have been focusing on the replication of one parental strand, for which the daughter strand is assembled from the separated or open end toward the fork. Of course, the other parent strand is also replicated. However, its daughter strand grows in the opposite direction, away from the replication fork. Apparently DNA polymerases can only catalyze the phosphodiester formation in a chain growing in the 5′⟶3′ direction. Recall from Section 20.4 that the original parental strands are antiparallel; that is, one goes up in a 5′⟶3′ direction and the other goes down in a 5′⟶3′ direction.

The cell is very precise in its assembly of nucleotides in new daughter strands. Experimental evidence indicates that the chance of the error of placing an incorrect (i.e. noncomplementary to parent) base in position on

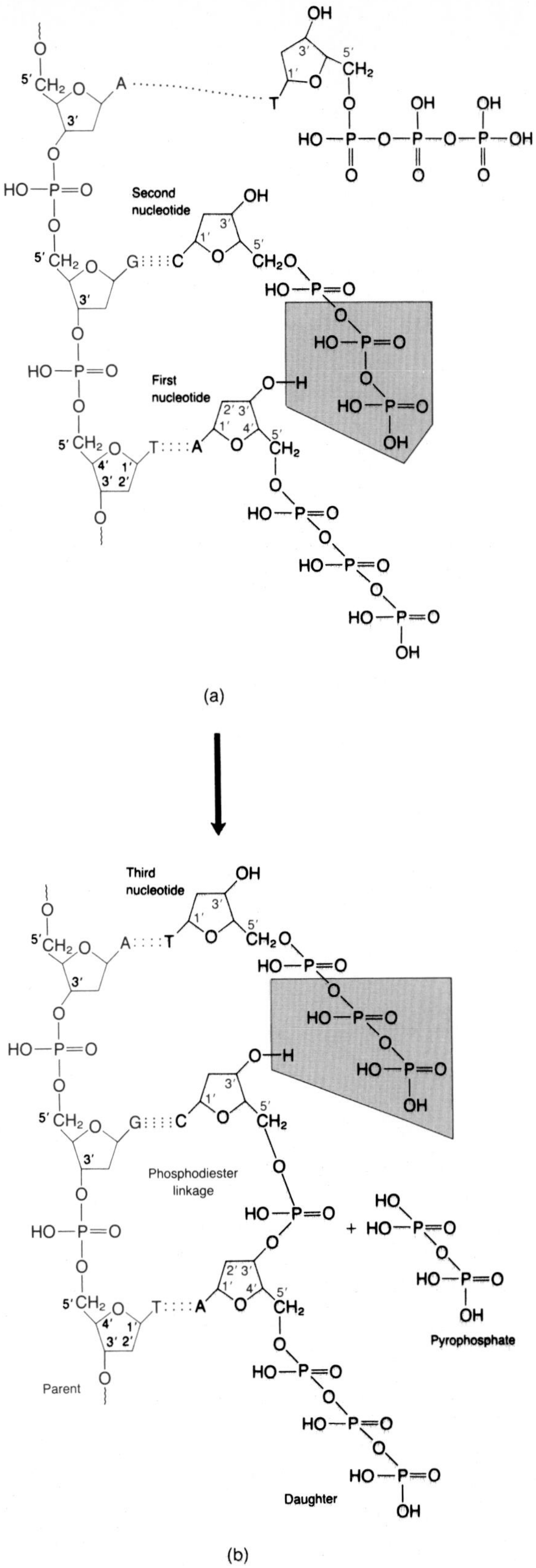
Second nucleotide
First nucleotide
(a)
Third nucleotide
Phosphodiester linkage
Pyrophosphate
Parent
Daughter
(b)

FIG. 20.5
Assembly of nucleotides during replication. This figure is a "close-up" of the action shown in Figure 20.4 and shows the structural details and actual reactions. (*a*) One nucleotide takes its place through base pairing. In this example, the first nucleotide fits in place through A-T pairing. The second nucleotide takes its place, and the two nucleotides are then lined up so that they can condense through the attack of the 3′ OH of the first nucleotide on the triphosphate group at the 5′ position of the second nucleotide. (*b*) The first and second nucleotides are linked by phosphodiester linkage as pyrophosphate is split off. A third nucleotide takes its place so that another condensation can occur.

a daughter strand is no more that one in a billion. What accounts for this high degree of accuracy in replication? One factor must be the precise fit requirement necessary for hydrogen bonding between the bases on the parent and the daughter strands. Another factor appears to be an editing or proofreading function of the DNA polymerases. Not only do these enzymes control the growth of the chain by catalyzing linkages between nucleotides, but also after putting a nucleotide in position they apparently "check" for possible errors. They excise at the 3′ position if a nucleotide is incorrectly paired. Accuracy in replication is crucial if the hereditary information of the species is to be maintained.

20.6 DNA, RNA, PROTEINS, AND THE CENTRAL DOGMA

You have now seen how DNA can direct replication, or synthesis of itself. Next we want to explore how DNA directs the synthesis of proteins. Through the synthesis of structural proteins (skin, hair, etc) and proteins with essential functions (enzymes, hormones, transporters, etc.), DNA controls the appearance and physiological processes that establish similarities among the members of a particular species and make each species distinct from other species.

When scientists first began to unravel the mystery of protein biosynthesis, they directed their efforts to answering two key questions:

1 How is information about the proper sequence of amino acids (primary structure) in a protein encoded in a DNA molecule? That is, what is the genetic code? This question will be answered in Section 20.8.
2 How is the coded message transmitted from DNA in the cell nucleus to structures in the cytoplasm called ribosomes where the amino acids are assembled? This transmission is accomplished by ribonucleic acids (RNA), which we will explore in Section 20.7.

The so-called **central dogma** of biochemistry states that information and thereby chemical product formation flow from DNA to RNA to protein. This flow can be represented:

$$\text{Replication} \circlearrowright \text{DNA} \xrightarrow{\text{transcription}} \text{RNA} \xrightarrow{\text{translation}} \text{protein}$$

The information or message in a DNA molecule about the order of assembly of amino acids is given to an RNA molecule in a process called **transcription.** Various types of RNA are involved in accomplishing the assembly of the amino acids into the protein according to the message. This assembly process is the **translation** of the message into the actual protein structure. The loop around DNA reflects DNA's ability to replicate itself.

An exception to this one-directional flow of information has been found with certain cancer-causing RNA viruses which can send messages from the cytoplasm to DNA in the nucleus.

It is necessary to explore the structure of RNA before the details of the transcription and translation processes can be elaborated. Throughout the discussion of the central dogma, or protein biosynthesis, you will repeatedly see the importance of the concept of base pairing and the significance of the ordering or primary structure of monomer units in nucleic acids (nucleotide monomers) and proteins (amino acid monomers).

20.7 RIBONUCLEIC ACIDS

RNA molecules, like DNA molecules, are polymers of nucleotides joined together by 5′⟶3′ phosphodiester linkages. The differences between RNA and DNA lie in the makeup of the nucleotides and in secondary structure.

Constituents of RNA Nucleotides

The sugar unit in a RNA nucleotide is **ribose** rather than the deoxyribose in DNA. The thymine base, present in DNA, does not occur in RNA. Rather, there are nucleotides with the base **uracil.** The four bases in RNA are adenine, cytosine, guanine, and uracil. Uracil can base-pair (hydrogen-bond) with adenine because it is structurally very similar to thymine.

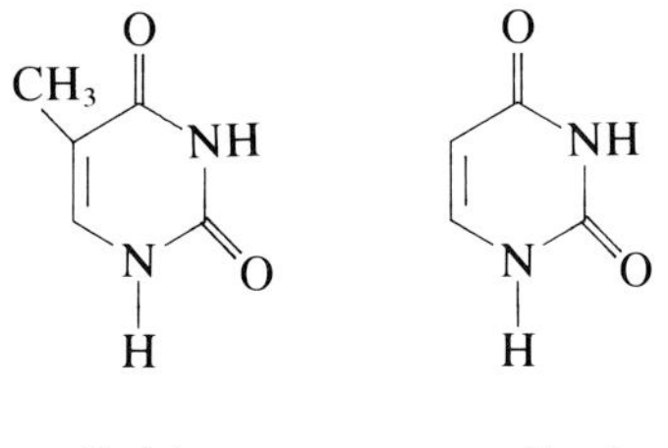
Thymine
Uracil

The following diagram shows three nucleotides in a RNA molecule. All the sugar units are ribose, and the first nucleotide at the 5′ end contains a uracil base.

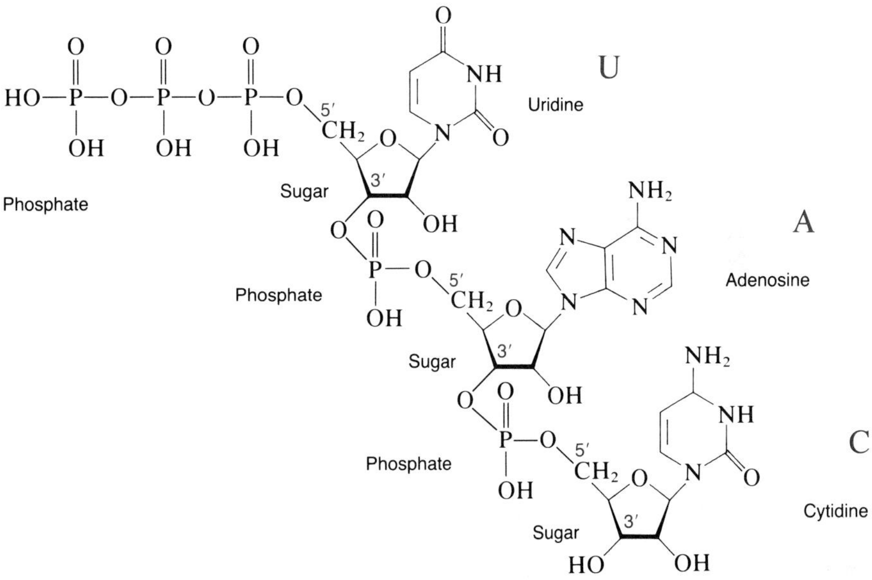
Phosphate
Sugar
U
Uridine
Phosphate
Sugar
A
Adenosine
Phosphate
Sugar
C
Cytidine

We can summarize the similarities and differences between RNA and DNA by using schematic diagrams of the sugar-phosphate backbones and attached bases:

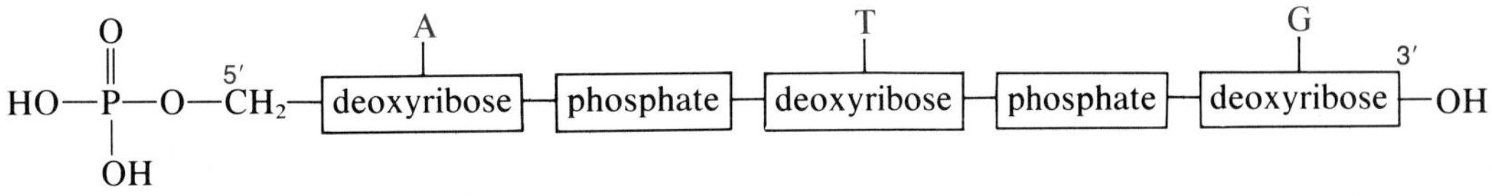
A
T
G
deoxyribose
phosphate
deoxyribose
phosphate
deoxyribose

DNA trinucleotide or simply ATG

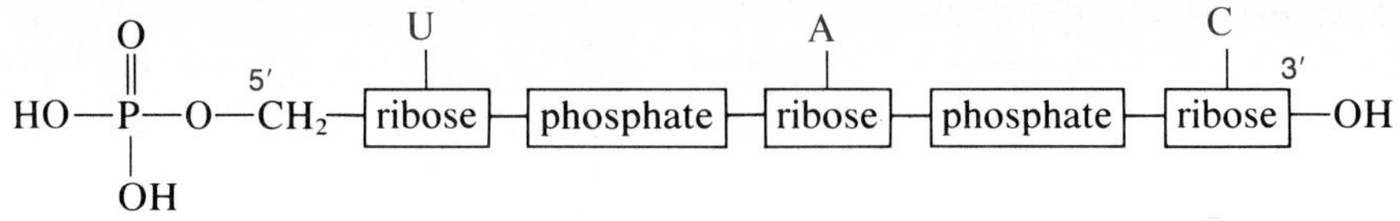

U
A
C
5′
3′
HO—P—O—CH₂
ribose
phosphate
ribose
phosphate
ribose
—OH
O
OH

RNA trinucleotide or simply UAC

RNA Secondary Structure

In general, RNA is single-stranded, rather than double-stranded like DNA. Consequently, there is no base pairing and no necessary equality between cytosine and guanine or adenine and uracil bases.

Three types of RNA have been found: *messenger* RNA (mRNA), *transfer* RNA (tRNA), and *ribosomal* RNA (rRNA). They vary in molecular weight and secondary structure, as discussed later in this section. The most important distinctions among the types are in their functions.

Messenger RNA (mRNA)

Messenger RNA is synthesized in the nucleus by the directions of a portion of a DNA molecule serving as a template. This process is known as transcription and serves to encode genetic information in mRNA. The synthesized mRNA is a single-stranded, open-chain molecule. Like DNA replication, transcription occurs according to the principles of base pairing. The DNA strand directs the assembly of ribonucleotides in RNA by base pairing (G—C, T—A, A—U), just as it can direct the assembly of deoxyribonucleotides into daughter strands. Figure 20.6 shows the synthesis of mRNA directed by DNA.

mRNA carries the genetic information from the nucleus to organelles in the cytoplasm known as ribosomes. Protein synthesis takes place at the ribosomes when the genetic message is received.

Many different mRNA molecules are synthesized; each one corresponds to a different protein in a way we will explore in Section 20.9. An individual mRNA molecule exists only long enough to carry the message and have its corresponding protein synthesized; then the mRNA is degraded. A new mRNA molecule of the same type will be prepared when more of its protein is needed.

Transfer RNA (tRNA)

Transfer RNA brings amino acids to the site of protein synthesis, the ribosome, for assembly into a protein. Each of the 20 amino acids is transported by a different tRNA; some amino acids are carried by two or more different tRNAs. The structure of tRNA (Figure 20.7) resembles a cloverleaf. One end of the cloverleaf has an acceptor site for an amino acid, and the opposite loop has an affinity site, called an anticodon, which can base-pair to a group of complementary bases, called a codon on mRNA. The codon and anticodon concepts will be discussed in Section 20.8, and the function of tRNA will be further explored in Section 20.10.

Ribosomal RNA (rRNA)

Ribosomal RNA complexed with protein forms the cellular structures known as ribosomes and provides the physical site for protein synthesis; ribosomes are often called "protein factories." mRNA molecules bring the genetic

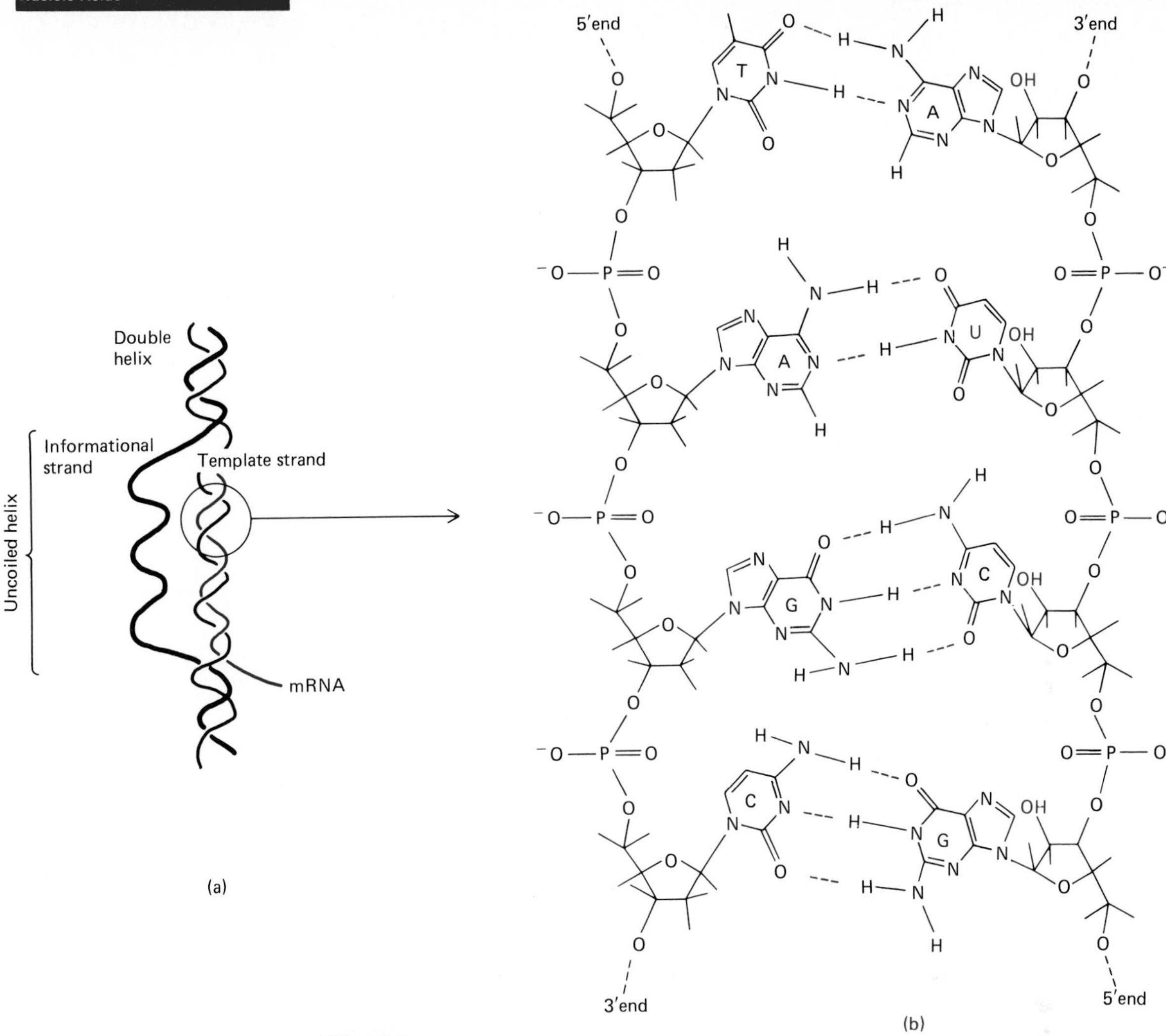
Double helix
Informational strand
Template strand
Uncoiled helix
mRNA
(a)
5′end
3′end
T
A
U
C
G
OH
(b)

FIG. 20.6
An mRNA strand is synthesized as a complement to a DNA strand. Ribonucleotides are the raw materials. They line up as directed by base pairing; C and G bases pair, T bases on DNA pair with A bases on ribonucleotides, A bases on DNA pair with U bases on ribonucleotides. The lined-up ribonucleotides condense, as was shown in Figure 20.5. (*a*) The DNA helix must unwind to expose bases for pairing. (*b*) This blow-up of the DNA template strand and mRNA strand shows the base pairing and the ribose sugars in the mRNA backbone.

message for protein synthesis from the nucleus to the ribosomes. The ribosome is a general protein-synthesizing site, the specificity for synthesis of a particular protein comes from the mRNA, not the ribosome.

These descriptions of the functions of the RNAs provide a superficial answer to the second question posed in Section 20.6, that is, how are messages about protein synthesis transmitted from DNA? Before a more complete description of transcription and translation can be appreciated we must return and answer the first question posed earlier. That is, how is information about the primary structure of proteins encoded in DNA?

20.8 THE GENETIC CODE

As we have seen, all DNA molecules have identical sugar-phosphate backbones. Distinctions between DNA molecules are made by the sequence of heterocyclic bases. Because different DNAs specify directions for synthesis

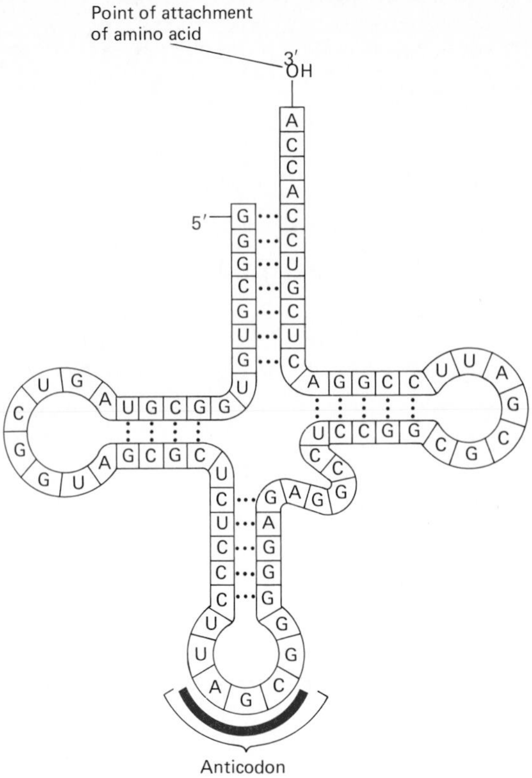
Point of attachment
of amino acid
3′
OH
5′
Anticodon

FIG. 20.7
Transfer RNA is a single strand folded into the peculiar shape shown. An amino acid becomes attached at the site marked by a condensation reaction. The three bases at the opposite loop are the *anticodon* corresponding to the attached amino acid. The anticodon guides the amino acid into its proper place in a protein chain, as you will see in Section 20.10.

of different proteins, the code for protein synthesis must lie in the sequence of bases. However, since there are 20 amino acids and only four bases, a single base cannot code for an amino acid. Four single bases could specify only four different amino acids. We might then ask, "Could two bases in sequence code for all the amino acids?" To answer this question, we consider how many different "words" of two letters can be composed from an "alphabet" of four letters: A, C, G, and T. There are 16 possibilities: AA, AC, AG, AT, CA, CC, CG, CT, GA, GC, GG, GT, TA, TC, TG, TT—too few to code for all 20 amino acids.

A general relationship which tells us the total number of possible combinations from four letters taken some number (n) at a time is 4^n. Thus as we saw, if $n=2$, there are 16 possibilities. If we consider three-letter "words," or *triplets* of bases, then there are 64 possibilities ($4^3=64$). Three sequential bases offer more than enough "words" to code for all 20 amino acids. Indeed, it is now known that almost all the amino acids can be specified by more than one triplet. In addition, three triplets do not designate amino acids but rather provide punctuation for genetic messages. **Triplets of bases specifying amino acids or punctuation are called codons.**

The following sequence of bases on a DNA template strand, for example, codes for the tetrapeptide Ser-Ala-Thr-Val:

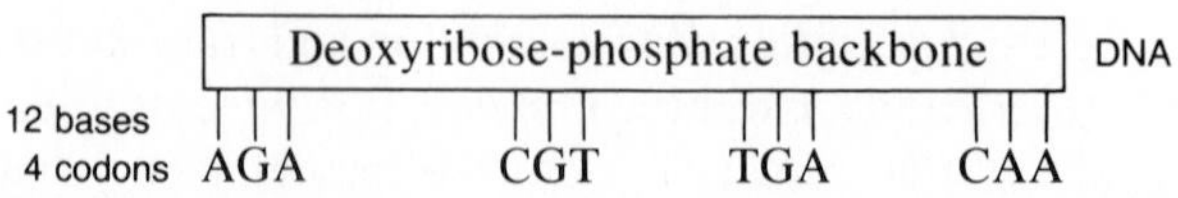
Deoxyribose-phosphate backbone
DNA
12 bases
4 codons
AGA
CGT
TGA
CAA

This message would be encoded in mRNA as the base complements with the use of U as the complement to A. Thus the message would leave the nucleus as follows:

TABLE 20.2 The Genetic Code

First Base	Second Base: U		Second Base: C		Second Base: A		Second Base: G		Third Base
U	UUU	Phe	UCU	Ser	UAU	Tyr	UGU	Cys	U
	UUC		UCC		UAC		UGC		C
	UUA	Leu	UCA		UAA	Stop	UGA	Stop	A
	UUG		UCG		UAG	Stop	UGG	Trp	G
C	CUU	Leu	CCU	Pro	CAU	His	CGU	Arg	U
	CUC		CCC		CAC		CGC		C
	CUA		CCA		CAA	Gln	CGA		A
	CUG		CCG		CAG		CGG		G
A	AUU	Ile	ACU	Thr	AAU	Asn	AGU	Ser	U
	AUC		ACC		AAC		AGC		C
	AUA		ACA		AAA	Lys	AGA	Arg	A
	AUG	Met	ACG		AAG		AGG		G
G	GUU	Val	GCU	Ala	GAU	Asp	GGU	Gly	U
	GUC		GCC		GAC		GGC		C
	GUA		GCA		GAA	Glu	GGA		A
	GUG		GCG		GAG		GGG		G

12 bases
4 codons
UCU
GCA
ACU
GUU
Ribose-phosphate backbone
mRNA

Thus we see there are DNA codons which specify an amino acid and complementary RNA codons which specify the same amino acid. Because mRNA actually directs protein synthesis, it is customary to express the code in terms of the RNA codon–amino acid relationships, as is done in Table 20.2.

Table 20.2 shows that most amino acids have at least two corresponding codons and many have four or more. Notice, for example, that there are six codons which specify leucine. If you wish to find a codon for a specified amino acid, scan the table for that amino acid and find its codon(s).

Problem 20.3 Consult Table 20.2 to find a codon specifying lysine.

If you are given a codon and are asked to tell what amino acid is specified, you can find it most efficiently by recognizing how Table 20.2 is constructed. To find a codon, let us say AGC, in this table, locate the first base (A) along the left side; this then specifies the horizontal row in which the codon will appear. Next locate the column marked by the middle base (G in our example). The intersection of row and column puts you in the box in which the codon will appear. Locate the third base (C) along the right side and this completes the sequence AGC, which is one of the codons for serine.

As you might imagine, unraveling the genetic code was an awesome experimental undertaking. However, the basic procedure followed was conceptually quite simple and can be explained by recounting the first experiment done. A synthetic mRNA which contained only uracil bases was made. This construction assured the presence of only one kind of codon, UUU. The synthetic mRNA was introduced into a working cell which followed the mRNA message and synthesized a polypeptide made up exclusively of phenylalanine. This experiment established that the triplet UUU tells the cell to put phenylalanine into a protein chain. The synthesis of a homogeneous mRNA was relatively simple. More sophisticated methods of synthesis had to be developed in order to establish all the amino acid–codon relationships. Marshall Nirenberg and H. Khorana received the Nobel Prize in 1968 for their work in unraveling this essential code.

The genetic code as shown in Table 20.2 appears to be the same for all species, whether *Escherichia coli,* an oak tree, or a human being, the code is universal. This tends to support the theory that all plants and animals had a common evolutionary origin.

Now that you know the structure of the "actors" in the play, that is, DNA and RNA, and know the "script" of the play, i.e., the genetic code, the complete drama of protein synthesis can unfold.

20.9 PROTEIN BIOSYNTHESIS I: TRANSCRIPTION

Transcription, the first step in protein biosynthesis, involves the formation of mRNA from a portion of a DNA molecule acting as an information template. The name given to a portion of a DNA molecule which directs the synthesis of a particular mRNA and thereby codes for a particular protein is **gene.** A single DNA molecule, i.e., a chromosome, contains many genes. For example, the one chromosome in *E. coli* contains at least 3000 and perhaps as many as 5000 genes.

The strand of DNA containing the gene for a particular protein is called the **informational strand.** The mRNA is formed under the direction of the complement of the informational strand. This complement is the **template strand.** Because the mRNA is the complement of the complement of the informational strand, the codons in the mRNA are the *same* as those on the informational strand of DNA (except that U replaces T).

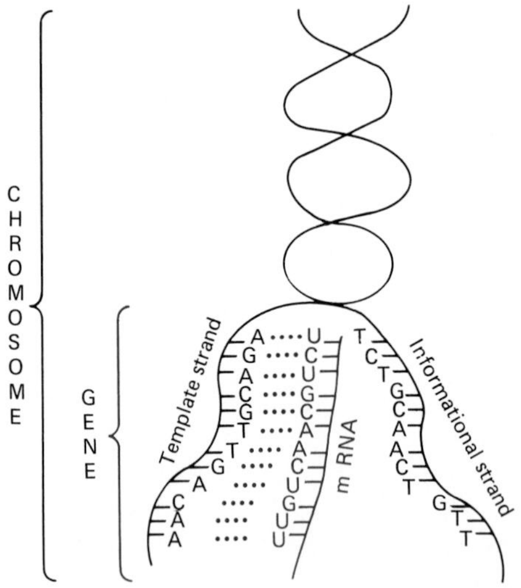
CHROMOSOME
GENE
Template strand
m RNA
Informational strand

SAMPLE EXERCISE 20.2

A portion of the informational strand of the gene for protein X has the nucleotide sequence A-A-C-T-G-G-C-C-C. Write the sequence of bases that will appear in mRNA for this portion of protein X.

Solution The template strand which directs mRNA synthesis is the complement to the informational strand. mRNA is the complement to the template, but uses uracil bases. The mRNA is always the same as the informational strand except for the substitution of U's for T's:

A-A-C-T-G-G-C-C-C	Informational strand
T-T-G-A-C-C-G-G-G	Template strand
A-A-C-U-G-G-C-C-C	mRNA

Problem 20.4 Use Table 20.2 to discover the amino acid sequence encoded in the mRNA in Sample Exercise 20.2.

What indicates the start of a gene along a DNA molecule? Apparently there are locations on DNA template strands known as **promoter sites.** RNA polymerase, the enzyme required for mRNA synthesis, "recognizes" and binds to the promoter site. The enzyme is thus activated and can begin the synthesis of an mRNA. In contrast to DNA polymerase, RNA polymerase does not appear to "edit" the developing RNA chain. The error rate in transcription is about one mistake per 10^5 bases. A cell is better able to tolerate mistakes in transcription than in replication because the turnover of mRNA molecules is rapid, whereas an incorrect DNA molecule remains in a cell.

Termination of mRNA synthesis is indicated by "stop" codons on the DNA template strand. The synthesis of mRNA occurs in the nucleus of cells of higher organisms (eucaryotic cells) or the nucleoid of procaryotes (review Section 18.11 for a discussion of cell types). After synthesis, mRNA moves into the cytoplasm to the ribosomes of cells.

Some poisons exert their toxic effect by inhibiting mRNA synthesis. For example, α-amanitin, the culprit of the poisonous mushroom *Amanita phalloides,* deactivates the mRNA polymerase required for mRNA synthesis in eucaryotic (animal) cells, but does not affect mRNA polymerases found in procaryotic mushroom cells. Thus, *A. phalloides* does not inhibit its own RNA synthesis.

20.10 PROTEIN BIOSYNTHESIS II: TRANSLATION

Translation of the genetic message encoded in mRNA into a protein molecule can conveniently be discussed in terms of four steps:

STEP 1: **Message Positioning**
mRNA lines up on the ribosome so that it can accept the incoming amino acids for orderly assembly.

STEP 2: **Amino Acid Transport by tRNA**
tRNA brings amino acids to their proper position alongside the mRNA message. This assures the proper placement of amino acids in the protein chain.

STEP 3: **Protein Chain Growth**
The protein chain grows as amino acids take their proper positions and are linked together.

STEP 4: **Protein Chain Termination**
The protein chain is terminated and released from RNA and "finishing touches" to construction are applied.

We will discuss each of these steps.

Message Positioning

Ribosomes have two distinct sites for protein synthesis, the *peptidyl* or *P site* and the *aminoacyl* or *A site*. The overall shape of a ribosome with labeled

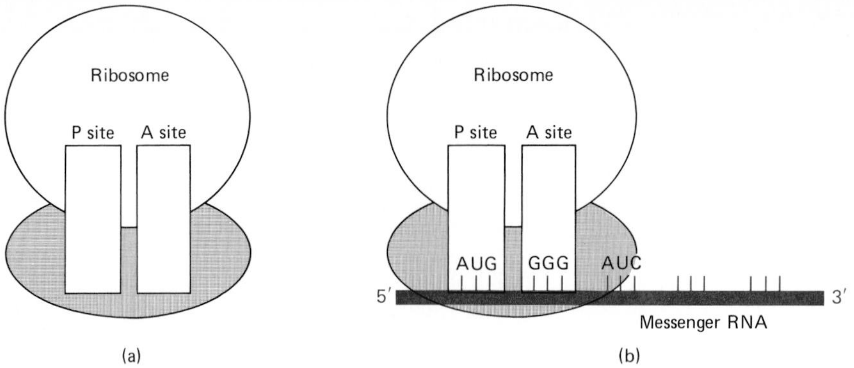
Ribosome
P site
A site
(a)
Ribosome
P site
A site
AUG
GGG
AUC
5′
3′
Messenger RNA
(b)

FIG. 20.8 (*a*) Ribosomes are ribonucleoproteins, i.e., complexes of RNA and protein material. Assembly of amino acids into protein chains occurs at the so-called P sites and A sites of ribosomes. (*b*) Ribosome with aligned mRNA.

P and A sites is shown in Figure 20.8*a*. mRNA attaches to a ribosome so that the first codon (reading in a 5′⟶3′ direction) is aligned in the P site. Figure 20.8*b* shows mRNA lined up on the ribosome. The first codon of mRNA to line up in the P site is always AUG. This codon codes for the amino acid methionine and dictates beginning the protein chain with methionine. However, the significance of AUG is as a start signal, or initiator. In many cases the initial methionine is clipped off after the protein is fully formed. Once AUG is at the P site, amino acid assembly can begin.

Amino Acid Transport by tRNA

As we have seen, tRNA is the smallest form of RNA and has a unique shape, shown in Figure 20.7. Specific tRNAs bind specific amino acids. Cells contain at least 32 different tRNAs; each tRNA recognizes and bonds to only one amino acid. Some amino acids bond to more than one kind of tRNA, but the critical point is that there must be a match between the amino acid attached and the anticodon, i.e., the base complement to the codon for the amino acid. For example, phenylalanine could only be attached to a tRNA with an anticodon AAA or AAG because the codons for phenylalanine are UUU and UUC.

Problem 20.5 Specify the anticodon on tRNA that could carry the amino acid lysine.

Amino acids are attached to tRNA by ester linkages between the acyl group of the amino acid and 3′ alcohol group on tRNA (See Figure 20.7), and the ester is called an aminoacyl-tRNA. Aminoacyl-tRNA synthetase is the general name given to the enzyme required for this reaction. There are 20 different aminoacyl-tRNA synthetases; each one is specific for one of the 20 different amino acids.

Because AUG is always the first codon on mRNA, the first amino acid to be brought into position in the protein chain is methionine attached to a tRNA with the anticodon, UAC. As Figure 20.9*a* shows, the anticodon bases, UAC, hydrogen-bond to their complementary bases, AUG, on the mRNA codon. **All amino acids are brought into position by this principle of base pairing between codons on mRNA and anticodons on tRNA.**

Protein Chain Growth

After the first codon at the P site is base-paired to Met-tRNA, the second codon is aligned on the A site. In Figures 20.8 and 20.9 the codon shown at the A site is GGG. Because GGG specifies glycine, glycyl-transfer RNA

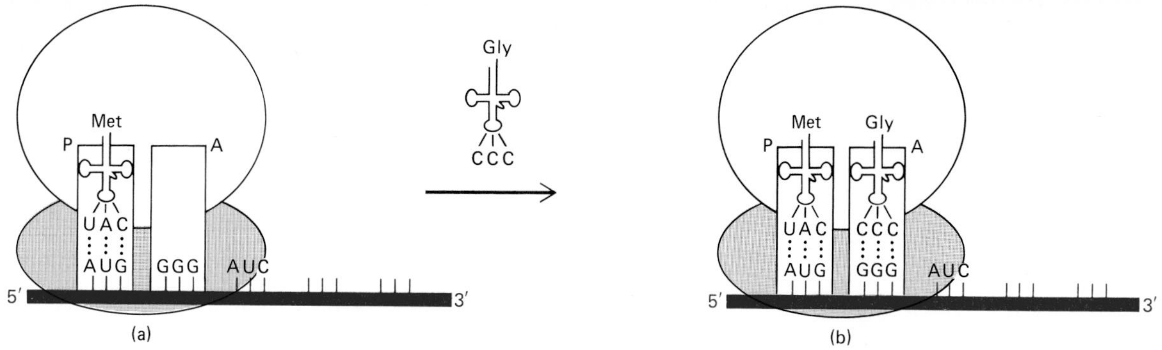
Met
P
A
UAC
AUG
GGG
AUC
5′
3′
(a)
Gly
CCC
Met
Gly
P
A
UAC
CCC
AUG
GGG
AUC
5′
3′
(b)

FIG. 20.9
(*a*) Initiation of the protein chain synthesis begins by the tRNA for methionine bringing that amino acid to the P site. (*b*) The next amino acid specified by the message takes its position at the A site. In this example, the codon GGG indicates glycine.

bearing the anticodon, CCC, lines up at the A site, as shown in Figure 20.9*b*.

The two amino acids, methionine and glycine, are now adjacent, and a peptide bond can form between them. This occurs by the transfer of the acyl group of methionine at site P to the amino group of the amino acid at site A, in this case glycine. The attack of methionine on glycine is shown by the arrow in Figure 20.10*a*. The A site now has a tRNA bound to a dipeptide, as shown in Figure 20.10*b*. The enzyme required for this peptide bond formation is peptidyl transferase.

The tRNA at the P site no longer is attached to an amino acid; it is released from the ribosome and can move back into the cytoplasm, where it can regain an amino acid and be available for further use. Simultaneously, upon this release, the ribosome moves in the 3′ direction of the mRNA. This shifts the dipeptide-bound tRNA to the P site, and the third codon of mRNA to the A site, as shown in Figure 20.10*c*. The aminoacyl-tRNA which complements the third codon moves into the A site, as in Figure 20.10*d*. The entire dipeptide in the P site is transferred and bonded to the amino group of the A site amino acid, forming a tripeptide bound to a tRNA in this A site (Figure 20.10*e*). Again the empty tRNA in the P site is released, the ribosome shifts, and the process repeats with the entrance of another aminoacyl-tRNA to the A site.

With the exception of the first, or initiating, tRNA, all other incoming aminoacyl-tRNAs base-pair to mRNA at the A site; hence the designation A for aminoacyl. Protein growth continues by a transfer of the formed peptide chain present in the P (peptidyl site) to the amino group of the next amino acid found on the A site.

Protein Chain Termination

Termination of the polypeptide chain is signaled by the appearance at the A site of one of three stop codons, UAA, UAG, or UGA. These termination codons are recognized by release factors, proteins that then aid in the hydrolytic cleavage of the polypeptide chain from its tRNA on the P site. Although we have focused on one ribosome, in actuality a single mRNA is translated by many ribosomes moving along the chain simultaneously (Figure 20.11). The formation of one mRNA molecule leads to the synthesis of many molecules of a particular protein; nature is highly efficient.

After the construction of the amino acid sequence, or primary structure, the polypeptide or protein chain is normally modified in a number of ways.

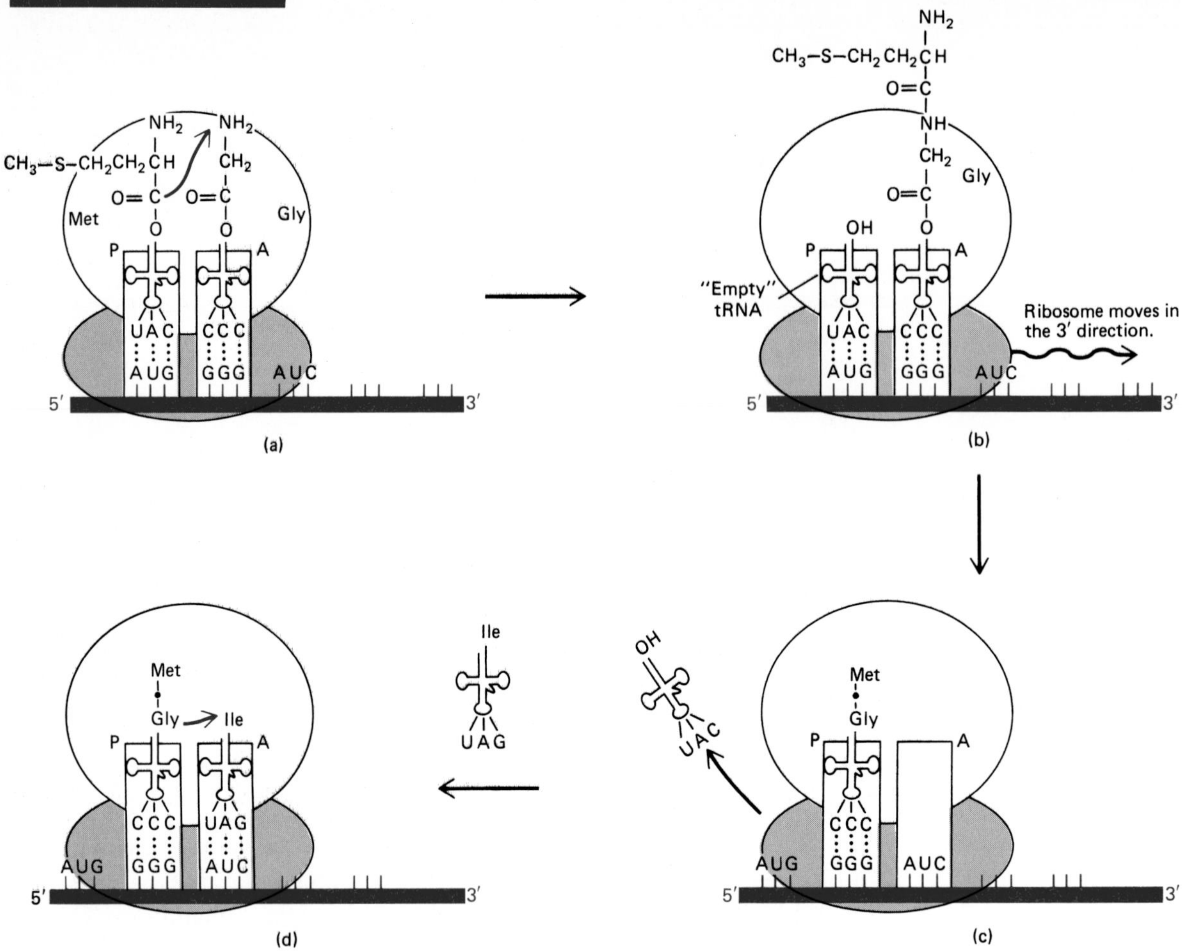

NH2
NH2
CH3—S—CH2CH2CH
CH2
O=C
O=C
O
O
Met
Gly
P
A
UAC
CCC
AUG
GGG
AUC
5′
3′
(a)
CH3—S—CH2CH2CH
O=C
NH
CH2
Gly
OH
"Empty" tRNA
Ribosome moves in the 3′ direction.
(b)
Met
Gly
OH
UAC
AUC
(c)
Ile
UAG
Met
Gly
Ile
CCC
UAG
GGG
AUC
AUG
(d)

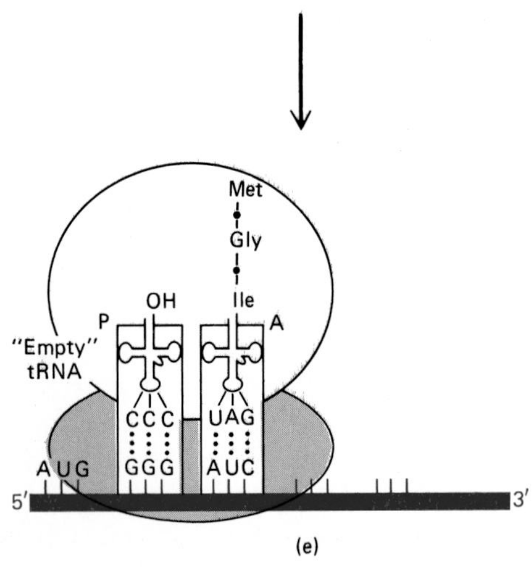

Met
Gly
OH
Ile
P
A
"Empty" tRNA
CCC
UAG
GGG
AUC
AUG
5′
3′
(e)

FIG. 20.10

Protein synthesis occurs by condensation of the growing chain at the P site with the amino acid at the A site as the ribosome moves along mRNA. (*a*) Methionine and glycine are lined up for condensation. The arrow shows the joining of the methionine carboxyl group to the glycine amino group. (*b*) The condensation results in a dipeptide at site A and an "empty" tRNA at site P. (*c*) The "empty" tRNA is released into the cytoplasm, the ribosome shifts, and the dipeptide is now at the P site. (*d*) Another amino acid (isoleucine) takes its place at the A site through its attachment to a tRNA. The dipeptide can link to isoleucine as shown by the arrow. (*e*) When the dipeptide attacks isoleucine, the result is a tripeptide at the A site and "empty" tRNA at the P site. Once again the "empty" tRNA can leave, and the ribosome can shift toward the 3′ end.

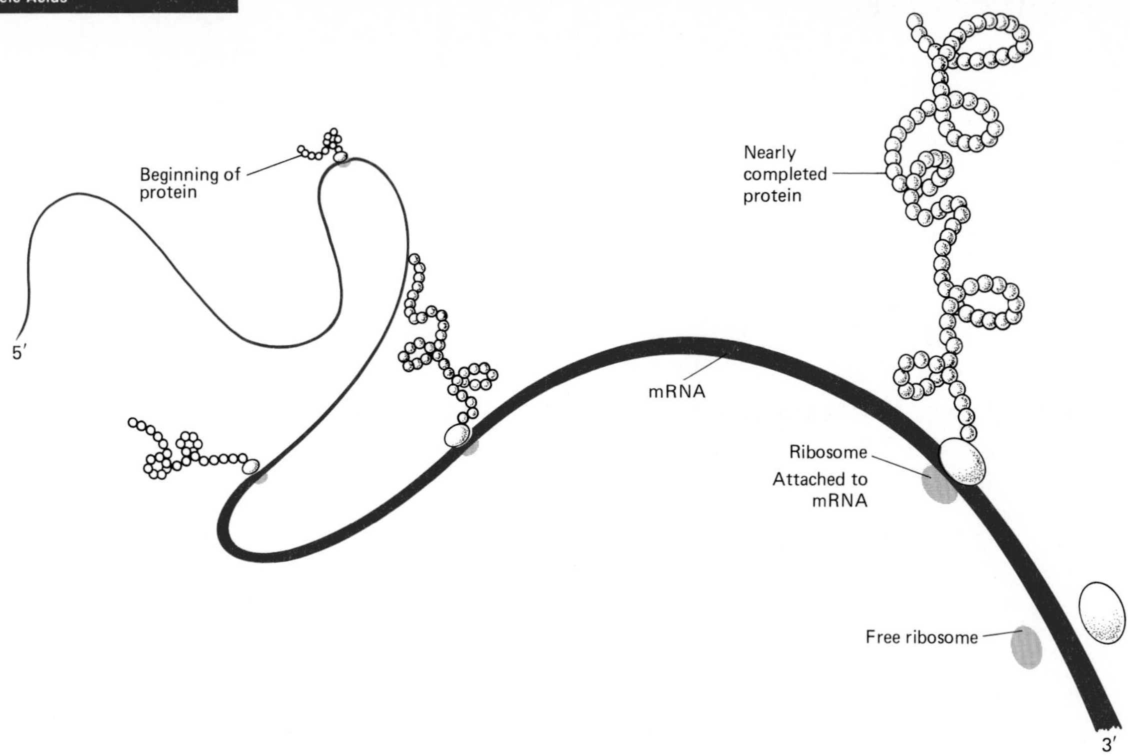
Beginning of protein
Nearly completed protein
5′
mRNA
Ribosome
Attached to mRNA
Free ribosome
3′

FIG. 20.11
Translation. Several ribosomes simultaneously move along a single mRNA from the 5′ to the 3′ end. Each ribosome assembles a complete protein by the procedure shown in Figure 20.10. The protein chain grows as the ribosome moves. When the protein is completed, the ribosome detaches from mRNA. *(From The DNA Story by J. D. Watson and J. Tooze. © 1968 by W. H. Freeman and Company. All rights reserved.)*

These include cleavage of the initial methionine group, oxidation of cysteine —SH groups to disulfide linkages, and attachment of prosthetic groups such as the heme unit of hemoglobin. Finally the chain coils and folds to yield the characteristic three-dimensional shape of the protein, i.e., the secondary, tertiary, and quaternary structures.

20.11 REGULATION OF PROTEIN SYNTHESIS

The DNA content in all cells of a particular organism is identical. That is, every cell has a set of instructions for the synthesis of all body proteins. But every cell does not need every type of protein. For example, liver cells do not need to synthesize the polypeptide neurotransmitters found in brain cells. Also, cells don't need the same amounts of various proteins at all times. So the question arises, how is protein synthesis regulated? How does a cell know when to make a particular protein and when to suppress synthesis?

Although these questions are still unanswered for eucaryotic cells, a hypothesis for protein regulation in the procaryotic cells of bacteria has been developed and put to the acid test of experimental verification. This theory was proposed by Francois Jacob and Jacques Monod.

In the Jacob-Monod theory the gene which codes for a particular protein, i.e., specifies its primary structure, is called a **structural gene.** The model proposes that there are two locations, called a **regulatory gene** and a **control site,** adjacent to the structural gene (Figure 20.12*a*). The entire combination of structural gene, regulatory gene, and control site is called an **operon.** The regulatory gene codes for a regulatory protein molecule called a **repressor.** When the repressor molecule binds to a segment of the control site known

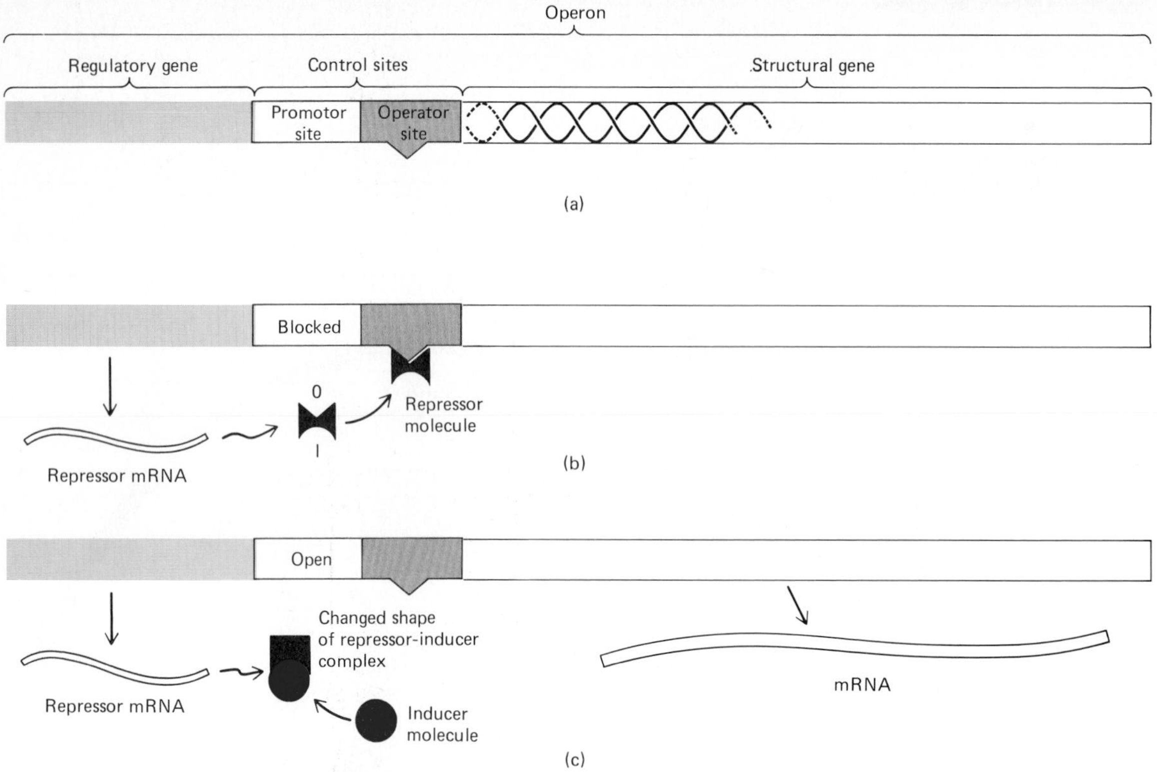
Operon
Regulatory gene
Control sites
Structural gene
Promotor site
Operator site
(a)
Blocked
0
Repressor molecule
I
Repressor mRNA
(b)
Open
Changed shape of repressor-inducer complex
Repressor mRNA
Inducer molecule
mRNA
(c)

FIG. 20.12
(*a*) An operon consists of a regulatory gene, control sites, and a structural gene. RNA polymerase, the necessary enzyme for transcription, initiates transcription of the structural gene by binding to the promoter site if that site is open. (*b*) Regulatory genes produce repressor mRNA, which directs the synthesis of repressor protein molecules. Repressor molecules have a site (marked O) which can bind to the operator and block the promoter site, thus suppressing transcription of the structural gene. The site marked I can bind so-called inducer molecules. (*c*) Formation of a repressor-induced complex changes the shape of the repressor so that it cannot bind to the operator site. Thus the promoter site is open, RNA polymerase can bind to it, and transcription of the structural gene can occur.

as the operator, transcription is prevented because attachment of RNA polymerase to the promoter site is blocked (Figure 20.12*b*). However, the repressor molecule has a second binding region to which another molecule called an **inducer** can bind and form a repressor-inducer complex. This repressor-inducer complex cannot fit the operator control site; thus transcription can take place unhindered on the original structural gene (Figure 20.12*c*). Varying concentrations of repressor and/or inducer molecules can thus regulate protein synthesis.

The Jacob-Monod theory was proposed to account for the variable concentration of the enzyme protein β-galactosidase in *E. coli*. β-Galactosidase catalyzes the hydrolysis of lactose to galactose and glucose in *E. coli*. When lactose is absent from the medium in which *E. coli* are growing, the concentration of β-galactosidase is vanishingly small. However, when *E. coli* are grown in the presence of lactose, the cell produces a high level of β-galactosidase and thus hydrolyzes and uses the lactose as a fuel and a source of carbon. According to the Jacob-Monod theory, the lactose acts as an inducer molecule. Lactose binds to the repressor molecule from the β-galactosidase operon and frees the operator site. This permits the structural gene to be in a turned-on state in which the mRNA for β-galactosidase can be produced. When lactose is not present, the repressor binds to the operator of the β-galactosidase operon and locks the structural gene in a turned-off state. Thus the binding of the repressor molecule to either the inducer or the operator allows the structural gene to be in a switched-on or switched-off position. A great deal of current research is being carried out to develop similar theories for eucaryotic cells.

20.12 MUTATIONS

Alterations in base sequences in DNA produce alterations in synthesized proteins. When these changes are permanent, they are called **mutations.** Mutations can be beneficial or they can be disastrous for an organism.

Chemical modifications in DNA bases can occur from the impact of high-energy radiation. The radiation may raise an electron in the molecule to a higher-energy level or completely remove the electron and produce an unstable intermediate which then proceeds to react further. Ultraviolet radiation from sunlight can lead to base changes in the DNA of skin cells. X-rays, radioactivity, and cosmic rays are all forms of ionizing radiation. DNA bases can also be altered by chemical agents, especially substances reacting with amine groups. Nitrous acid (see Section 16.11) can convert cytosine into uracil. Nitrogen mustards are amine alkylating agents (Section 16.6). Dimethylsulfate converts the —OH group of guanine into a —OCH_3 group.

Nature has developed "repair" enzymes which can remove and replace altered bases. However, in some cases these "repair" enzymes are missing. For example, in the rare skin disease xerodema pigmentosum individuals lack the ultraviolet radiation "repair" enzyme, uv-endonuclease. Their skin is extremely sensitive to sunlight and skin cancer often develops because bases altered by the sun's ultraviolet rays are not removed and replaced.

Mutations are said to occur when a base alteration in DNA is not repaired and replication continues, thereby propagating the error. When the damage occurs in a somatic cell (a body cell other than a sperm or egg cell), uncontrolled growth of that cell may result. This is what we call cancer. Substances that cause mutations are called **mutagens.** Those chemicals which cause a cancerous condition are called **carcinogens.**

If damage to DNA bases occurs in a germ cell (sperm or egg cell), the offspring of the damaged organism will inherit the defect. Most often the informational defect produces a defect in the child. In rare cases a mutation may be a positive change which better equips the offspring, in the Darwinian sense, for the struggle of the survival of the fittest. In some of these rare cases a new species may eventually evolve. Radiations from natural sources and cosmic rays have always been present on the earth, and the mutations they produced are one source of the raw material for the evolutionary process, on the molecular level.

Mutations occur most often by changes in just one base: a substitution of one base for another, the insertion of a new base, or the complete deletion of a base. Examples of each of these possibilities are shown in Figure 20.13.

Substitution alters only one codon; therefore no more than one amino acid is changed in the finished polypeptide or protein. Some substitutions simply produce other codons which correspond to the same amino acid. A change of one amino acid in a protein chain does not necessarily have an effect on the biological activity. However, if the change is at the active site of an enzyme (Section 21.5), or if the change affects the hydrophilic character of the exterior of a globular protein, then there may be a dramatic change in biological functioning. Sickle-cell anemia, which we examined in Section 19.13, results from the latter type of change; a hydrophobic valine replaces a hydrophilic glutamic acid in the hemoglobin molecule.

Normal gene	DNA template strand	AAA	TAG	CGG	TCC
	mRNA	UUU	AUC	GCC	AGG
	Normal protein chain	Phe ·	Ile ·	Ala ·	Arg ·
Substitution of G by A on DNA	Template	AAA	TAG	CAG	TCC
	mRNA	UUU	AUC	GUC	AGG
	Mutant protein	Phe ·	Ile ·	Val ·	Arg ·
Insertion of a C base in DNA	Template	AAA	TAG	CCG	GTCC
	mRNA	UUU	AUC	GGC	CAGG
	Mutant protein	Phe ·	Ile	Gly ·	Gln ·
Deletion of an A base from DNA	Template	AAT (A deleted)	AGC	GGT	CCG
	mRNA	UUA	UCG	CCA	GGC
	Mutant protein	Leu ·	Ser ·	Pro ·	Gly ·

FIG. 20.13
Mutations are changes in DNA base sequences which result in altered protein primary structure (mutant protein). (*a*) Normal, that is, unmutated DNA, mRNA and protein chain sequences. (*b*) The substitution of one base for another changes one codon and causes the substitution of one amino acid in the chain. (*c*) Insertion of one base in the sequence shifts the entire sequence so that there is a more dramatic alteration in amino acid sequence. In this case two amino acids are different from those in the normal sequence. (*d*) Deletion of a base also has a dramatic effect. In this case all four amino acids are different from those in the original.

Problem 20.6 If the mutation UUC ⟶ UUU occurs, what change is produced in the amino acid sequence? Explain.

Insertion or deletion of DNA bases leads to more extensive genetic damage because *all* the codons and thus all the amino acids beginning with and continuing past the point of insertion or deletion are affected. Often such an alteration generates a termination codon and the protein chain is prematurely cut short.

Defects in DNA which lead to the production of malformed and malfunctioning proteins produce medical conditions known as *molecular diseases*. Most often the faulty protein is an enzyme. We will be examining enzymes in more detail in the next chapter. Table 20.3 lists a few of the known molecular diseases and the affected proteins. Several of these diseases have already been discussed; others will be covered in later chapters. The sections in which these discussions appear are given in the table.

20.13 GENETIC REARRANGEMENTS

Children receive their genetic makeup from their parents. At conception they receive 23 chromosomes from the egg cell of their mother and 23 chromosomes from the sperm cell of their father. This gives them 23 unique pairs of chromosomes distinct from the makeup of either parent. Figure 20.14 illustrates this process and shows the variations that can occur by recombinations of genes between chromosomes. That is, chromosomes may be passed on with genes combined along the strand exactly as in the parent or there may be a recombination of genes in the egg or sperm cell chromosomes.

TABLE 20.3 Molecular Diseases and the Affected Protein

Molecular Disease	Affected Enzyme or Other Protein	Section Where Discussed
Acatalasia	Catalase	
Albinism	Tyrosinase	
Ateliotic dwarfism	Growth hormone	
Cystic fibrosis	α_2-Macroglobulin-protease complex	
Diabetes, childhood	Insulin	21.14, 23.9, 24.10
Galactosemia	Galactose 1-phosphate uridyl transferase	
Gaucher's disease	Glucocerebrosidase	
Glycogen storage disease I	Glucose 6-phosphatase	
Goiter	Iodotyrosine dehalogenase	26.10
Gout	Hypoxanthine-guanine phosphoribosyl-transferase	25.11
Hemolytic anemia (one form)	Glucose 6-phosphate dehydrogenase	
Hemophilia	Antihemophilic globulin	
Lesch-Nyhan disease	Hypoxanthine-guanine phosphoribosyl-transferase	20.13
Niemann-Pick disease	Sphingomyelin-hydrolyzing enzyme	18.12
Phenylketonuria	Phenylalanine hydroxylase	
Sickle-cell anemia	Hemoglobin	19.13
Tay-Sachs disease	Hexosaminidase A	18.12

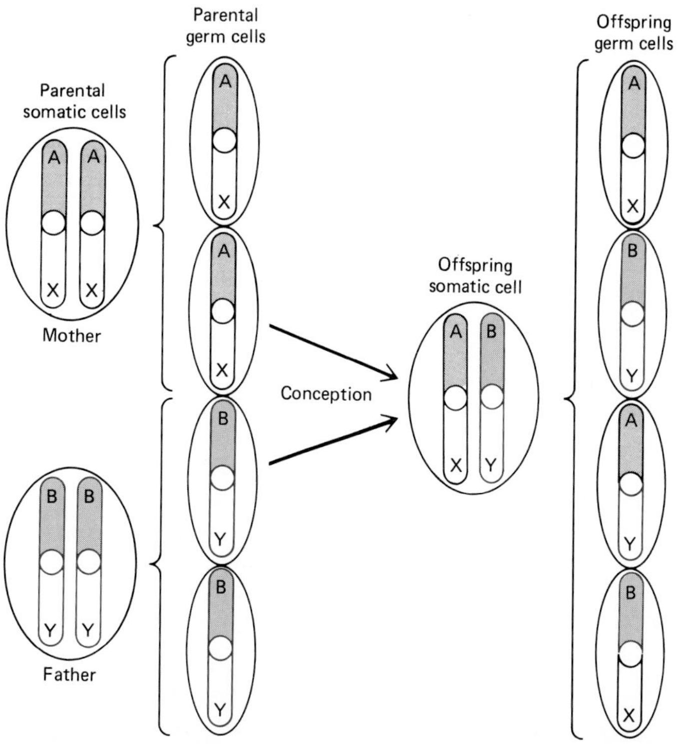
Parental germ cells
Parental somatic cells
Mother
Father
Conception
Offspring somatic cell
Offspring germ cells
A
B
X
Y

FIG. 20.14
Somatic (body) cells contain chromosomes in pairs. In humans each cell contains 23 pairs. In this simple example, which shows only one pair of chromosomes, the parents each have pairs with the same set of two genes on each member of the pair (
A
X
is a chromosome; A is a gene and X is a gene). Germ cells in eggs or sperm have only one chromosome, rather than pairs. The germ cells for our hypothetical mother and father are shown. The somatic cells of the offspring of the mother and father have the chromosome pairs shown, 
A
X
from the mother and 
B
Y
from the father. This more heterogeneous combination leads to a greater variety of possible germ cells through recombination of genes between the members of the chromosome pair.

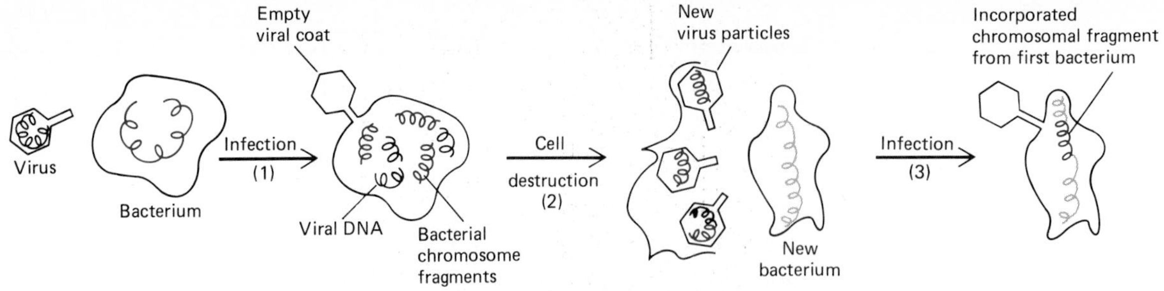
Empty
viral coat
New
virus particles
Incorporated
chromosomal fragment
from first bacterium
Virus
Bacterium
Infection
(1)
Viral DNA
Bacterial
chromosome
fragments
Cell
destruction
(2)
New
bacterium
Infection
(3)

FIG. 20.15
(1) When a virus infects a bacterium, the bacterial DNA is broken into fragments. (2) When the bacterial cell is destroyed as a result of this viral infection, some of the bacterial DNA fragments (color) are incorporated into the newly formed viruses and are mixed with the viral DNA (black). (3) When these new viruses infect a second bacterium, some of the bacterial DNA from the first bacterium (color) is incorporated into the DNA of the second bacterium (gray). Thus recombination of genes occurs.

This recombination in higher organisms is a natural process made possible by the mechanism of reproduction which involves two parents. In single-celled organisms, such as bacteria, recombination of genes is not possible because bacteria have only one chromosome which is simply replicated during reproduction of the cell during cell division. However, genetic recombination can be caused to occur within a bacterium by some agent working on the bacterial chromosome. For example, one such process occurs by the action of viruses which infect and kill one bacterium and then incorporate pieces of this bacterium's DNA into the DNA of another infected bacterium (Figure 20.15).

Genetic Engineering Techniques

Recently scientists have learned how to combine genes from different species synthetically. This exciting area of biochemistry is sometimes called genetic engineering or recombinant DNA technology. In this process a structural gene, for example, the one coding for the synthesis of human insulin, is "cut" from its natural chromosome and then spliced into a **plasmid,** a small fragment of DNA found in bacterial cells. The plasmid, now containing the foreign gene, is reinserted into a bacterium. The bacterium divides and redivides very rapidly, producing a large colony of bacteria, all of which are now producing the protein insulin from directions of the inserted gene. The formation of many gene copies from a single gene is known as gene cloning. This insertion process has converted the bacteria into a protein factory. Let us briefly look at some of the research developments that have made this process possible.

Restriction endonucleases are enzymes that can recognize specific base sequences in a DNA and then cleave the phosphodiester linkages between bases at a specific site within the sequence. For example, the restriction enzyme EcoRI cleaves between G and A bases in the ($5' \longrightarrow 3'$) direction in the sequence, G↓A-A-T-T-C. The arrow marks the cleavage spot. When a DNA molecule is cleaved with EcoRI the double helix is not cut straight across; rather the individual strands are cut at different points (Figure 20.16), leaving unpaired bases. These single-stranded ends are often called "sticky ends" because they tend to associate with strands containing complementary bases. When two different DNAs are cleaved with a restriction endonuclease such as EcoRI, the cleaved pieces can be recombined in new ways to produce new recombinant DNAs (Figure 20.16).

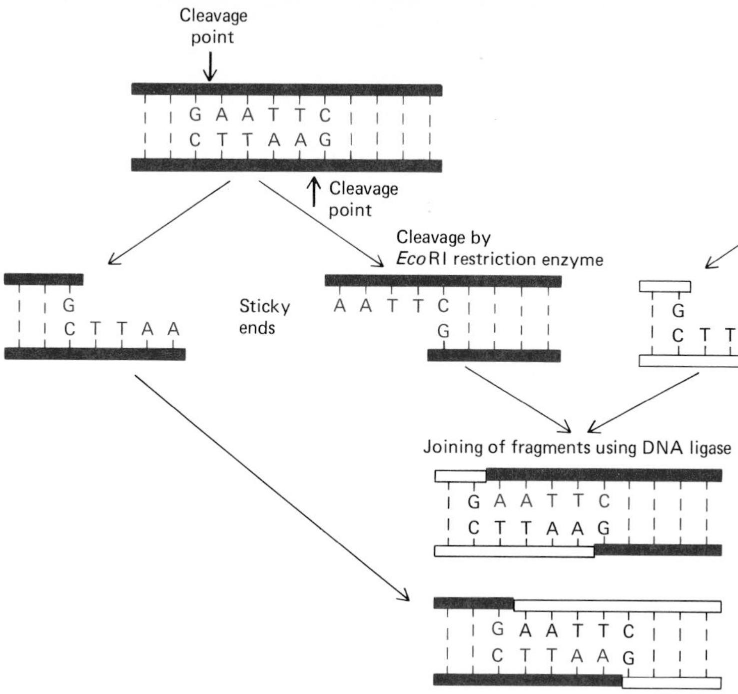
Cleavage point
G A A T T C
C T T A A G
Cleavage point
Cleavage point
G A A T T C
C T T A A G
Cleavage point
Cleavage by
EcoRI restriction enzyme
G
C T T A A
Sticky ends
A A T T C
G
G
C T T A A
Sticky ends
A A T T C
G
Joining of fragments using DNA ligase
G A A T T C
C T T A A G
G A A T T C
C T T A A G

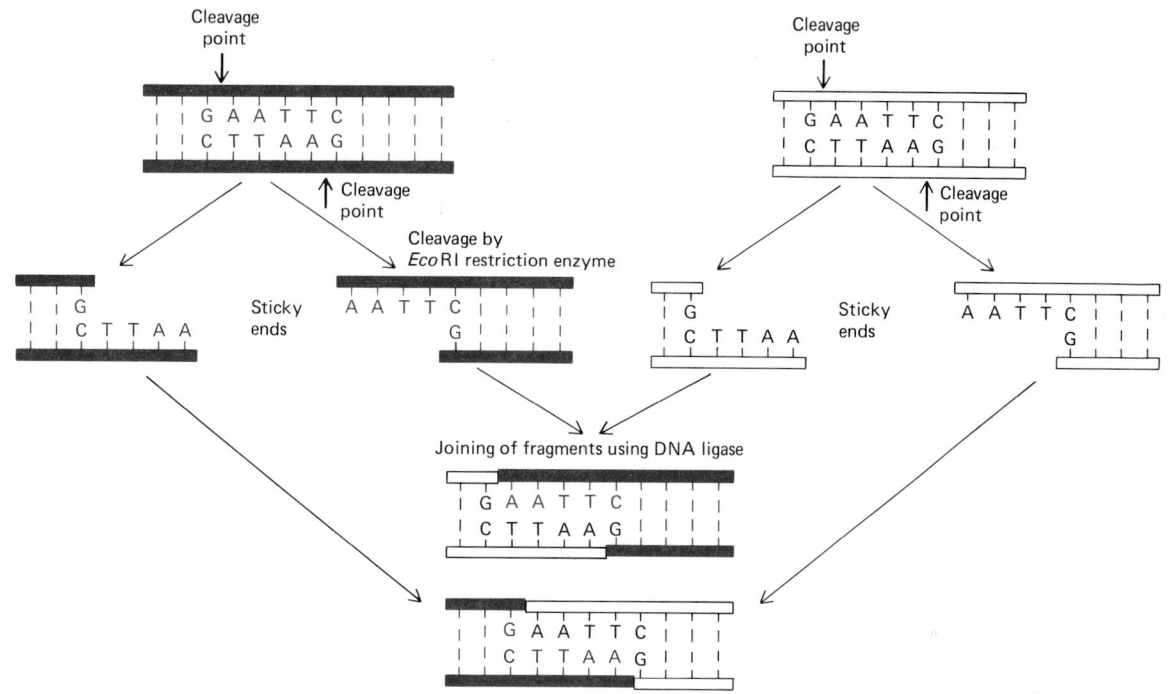

FIG. 20.16
The different color representations indicate two different DNA molecules; the base sequences on the unmarked portions of the strands differ. If these two different DNAs are cleaved simultaneously, recombination of fragments produces new DNA molecules. Because the enzyme EcoRI cleaves only at the specific site between G and A shown, the helix is not cleaved straight across. Rather, the cleavage produces long and short fragments with "sticky ends," so called because these ends can be joined to other fragments.

After genes are cut, a carrier is needed to insert the foreign DNA into the bacterial host cell. In most bacteria the genetic material consists of a single large chromosome in the nuclear zone and a plasmid, a small circular portion of DNA found in the cytoplasm. Plasmids have only a few genes and can be extracted from and reinserted into the bacterial cell. This makes the plasmids excellent carriers.

Figure 20.17 outlines the insertion method using plasmids. Both the donor chromosome, which contains the gene to be inserted, and a plasmid are cleaved with an enzyme such as EcoRI. The complementary fragments with sticky ends base-pair, and the phosphodiester linkages are rejoined using DNA ligase. The plasmid now contains the foreign gene. It is reinserted into the bacterium, where it will reproduce and also synthesize the desired protein according to the instructions for synthesis from the foreign gene.

Applications of Genetic Engineering

Many practical applications of recombinant DNA technology can be made in the efficient synthesis of proteins needed in medicine or scientific research. For example, insulin, required for the treatment of diabetic patients, is now obtained from slaughterhouse animals. However, this animal insulin is not effective for all patients. The synthesis of human insulin by recombining the human insulin gene into a plasmid of *E. coli* cells has been achieved and the human insulin product is being marketed commercially.

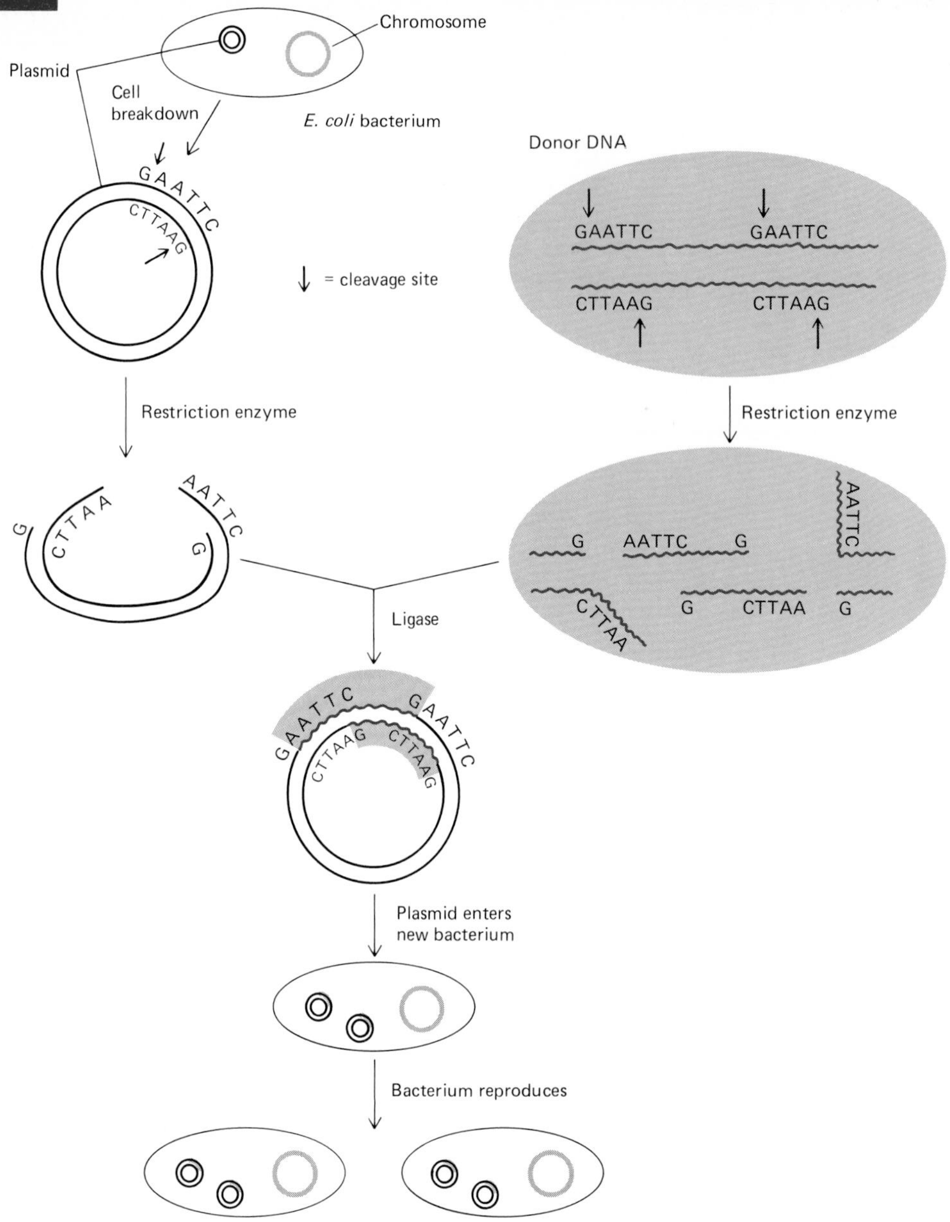
Chromosome
Plasmid
Cell
breakdown
E. coli bacterium
Donor DNA
GAATTC
CTTAAG
GAATTC
CTTAAG
GAATTC
CTTAAG
↓ = cleavage site
Restriction enzyme
Restriction enzyme
G
AATTC
G
AATTC
CTTAA
G
CTTAA
Ligase
Plasmid enters
new bacterium
Bacterium reproduces

FIG. 20.17
Recombinant DNA technology. Both a plasmid and a donor DNA are cleaved by a restriction enzyme. "Sticky ends" are joined so that the donor gene is inserted into the plasmid. The plasmid enters a new bacterium and is reproduced as the bacterium reproduces.

Interferon is a protein which appears to have antiviral activity, and may be of benefit in treatment of certain types of cancer. However, its scientific study has been limited because only very small amounts are naturally available. In 1980 the gene for white blood cell interferon was isolated and recombined into *E. coli* plasmids. It is now possible to produce much more interferon for scientific research than was previously possible. Human growth hormone is another protein where former problems of supply and purity have been overcome by recombinant genetic synthesis.

Another major goal of genetic engineers is the cure of molecular diseases through the insertion of a missing gene into a mammalian cell. In principle this could lead to a real cure of the molecular diseases listed in Table 20.3. Some progress has been made in this area. For example, the *E. coli* gene for the enzyme hypoxanthine-guanine phosphoribosyltransferase, which is defective in Lesch-Nyhan disease, has been incorporated into cells of human connective tissue. The enzyme was formed from the incorporated gene. However, difficulties have developed in the incorporation of eucaryotic genes into mammalian cells. Techniques for inserting such genes have been developed, but normal transcription and translation do not always happen. Much still needs to be learned about the regulation of eucaryotic genes.

CHAPTER ACCOMPLISHMENTS

After completing this chapter you should be able to define **key words** and do the following:

20.1 INTRODUCTION

1. Name the substance which is the bearer of hereditary information.

20.2 CHEMICAL COMPOSITION OF DNA

2. State the specific monomer units that make up DNA.
3. Give the components of each of the four DNA nucleotides.
4. Write the structure of a DNA nucleotide that could form from a given base.
5. Write the structure of a nucleotide triphosphate that could form from a given nucleotide monophosphate.
6. State what is meant by the 5′ and 3′ positions in a nucleotide.

20.3 THE PRIMARY STRUCTURE OF DNA—A POLYMER OF NUCLEOTIDES

7. State what is meant by the primary structure of DNA.
8. Convert a DNA structural formula into the abbreviated form.

20.4 THE SECONDARY STRUCTURE OF DNA—THE DOUBLE HELIX

9. Describe the mathematical relationship between the various kinds of DNA nucleotides.
10. Give a description of the Watson-Crick DNA structure.
11. Write the complementary nucleotide sequence to a given nucleotide sequence in a DNA molecule.

20.5 DNA REPLICATION

12. Outline the process of DNA replication.
13. Give two functions of the DNA polymerases.

20.6 DNA, RNA, PROTEINS, AND THE CENTRAL DOGMA

14. State the central dogma of biochemistry.

20.7 RIBONUCLEIC ACIDS

15. Describe the differences in component composition between DNA and RNA.
16. Name the three types of RNA and their function in the cell.

20.8 THE GENETIC CODE

17. Given Table 20.2, state the amino acid or punctuation conveyed by a given codon.

20.9 PROTEIN BIOSYNTHESIS I: TRANSCRIPTION

18. Given the sequence of bases on a gene's informational strand of DNA, write the sequence that would be found on the template strand and on an mRNA strand synthesized from the gene.
19. Given Table 20.2, tell the sequence of amino acids which is encoded on a given mRNA strand.
20. State the function of RNA-polymerase.

20.10 PROTEIN BIOSYNTHESIS II: TRANSLATION

21. State the four general steps in the translation process.
22. Describe the process of protein chain growth starting at the point when a given mRNA is bound to a ribosome.

23. State the signal which terminates protein chain growth.

20.11 REGULATION OF PROTEIN SYNTHESIS

24. List the components of an operon and state the function of each component.
25. Describe the basic ideas of the Jacob-Monod theory of protein synthesis regulation in procaryotic cells.

20.12 MUTATIONS

26. List two ways by which mutations can occur.
27. Distinguish between mutagens and carcinogens.
28. Given a specific mutation in the DNA bases, describe the corresponding change in amino acid sequence.
29. Describe the general cause of a molecular disease.

20.13 GENETIC REARRANGEMENTS

30. Describe the function of restriction endonucleases and plasmids in genetic recombination.
31. Explain how genetic engineering could lead to a cure for a molecular disease.

PROBLEMS

20.7 In what chemical form is hereditary information transferred from parent to child?

20.8 What chemical compounds are formed from the mild hydrolysis of a chromosome?

20.9 Give the origin for each of the following pieces of the chemical name *deoxyribonucleic acid:*
a *Deoxyribo*
b *Nucleic*
c *Acid*

20.10 What type of chemical functional group links together the nucleotides in a DNA strand? Explain the name given to this functional group.

20.11 Give the names of the hydrolysis products that would be formed in each of the following two reactions:

a A DNA molecule $\xrightarrow[\text{hydrolysis}]{\text{mild}}$

b Products from part *a* $\xrightarrow[\text{hydrolysis}]{\text{vigorous}}$

20.12 *a* What similarities in composition are there in the four DNA nucleotides?
b What compositional difference distinguishes the four DNA nucleotides?

20.13 *a* Without examining Table 20.1, combine phosphoric acid, deoxyribose, and thymine to form deoxythymidine monophosphate.

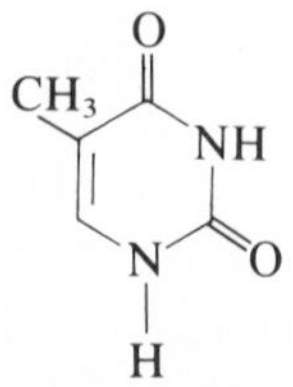
CH3
O
NH
N
O
H

Thymine

b Combine deoxythymidine monophosphate with two molecules of phosphoric acid to form the structure deoxythymidine triphosphate.
c What type of reaction is being carried out in part *b?*

20.14 *a* Write the structures of the two products that could form by bonding together the two nucleotides deoxythymidine triphosphate and deoxyadenosine triphosphate in a $5' \longrightarrow 3'$ direction.
b Bond a third nucleotide deoxycytidine triphosphate to each of the products you formed in part *a*.
c Express each chain of three nucleotides in an abbreviated format, stating only the nitrogen bases.

20.15 Define the primary structure of a DNA molecule.

20.16 A particular DNA molecule is 27 percent adenine.
a What is the percent of thymine in this DNA?
b Determine the percent of guanine and cystosine in this DNA.

20.17 Explain how the Watson-Crick DNA model accounts for the equality between A and T bases and between C and G bases.

20.18 *a* What attractive interaction holds together the two chains in a DNA molecule?
b What changes in function might occur if the two DNA chains were held together by covalent bonds.

20.19 Write the complementary chain nucleotide sequence for the following:

-C-A-T-G-A-C-T-

20.20 *a* Discuss in your own words the process of DNA replication.
b Why does a G rather than any of the other

three nucleotides come into position next to a C nucleotide on the parent chain?

c How many molecules are formed when a DNA molecule undergoes replication?

20.21 According to the discussion within the chapter, how are the original parental strands of a DNA molecule proportioned among the daughter molecules?

20.22 Describe two functions of the DNA polymerases.

20.23 If an error is made in the replication of a single DNA molecule, what happens to that error as the daughter strands undergo further replication?

20.24 DNA directs the synthesis of what component of the biomass?

20.25 *a* If a new protein is synthesized in the laboratory and then inserted within a cell, can it direct the synthesis of DNA to produce more of this protein?

b What principle relates to this question?

20.26 *a* List the differences in component composition between a DNA and an RNA molecule.

b Discuss the difference in secondary structure between a DNA and an mRNA molecule. What does this imply about the equality between A-U and C-G bases in an mRNA molecule?

20.27 *a* In what way is the process of transcription similar to that of replication?

b What result is obtained from each process?

20.28 *a* Compare the lifetimes of a DNA and an RNA molecule.

b In which molecule would an error in nucleotide sequence be more critical?

20.29 *a* What is the functional role of tRNA?

b What is the minimum number of tRNAs in a cell?

c Where in the cell is tRNA found?

20.30 *a* What is the functional role of rRNA?

b Where in the cell is rRNA found?

20.31 In what way does one DNA molecule differ from another DNA molecule?

20.32 Discuss why sequences of two bases in DNA cannot form an appropriate code for amino acids.

20.33 *a* What is meant by a codon?

b Give an example of a codon.

c What information is encoded in your example?

20.34 Which amino acid is coded for in each of the following codons:

a CCC

b GGG

c AGU

d ACG

e UUA

20.35 Why is the base T missing from the codon list in Table 20.2?

20.36 *a* Is the genetic code of Table 20.2 restricted to human beings? Animals? Vertebrates? Eucaryotic cells?

b How general is Table 20.2?

20.37 *a* Explain in your own words the meaning of the word *gene*.

b Is a chromosome made up of one gene? Give an example to support your answer.

c Does the mRNA formed in transcription resemble more the informational or the template strand of DNA?

20.38 Determine the amino acid sequences that would be formed from each of the following mRNA fragments:

a -A-U-U-C-A-C-U-G-U-C-G-A-C-C-C-

b -G-U-A-G-U-G-A-U-A-A-C-U-U-G-G-

c -A-U-G-U-U-U-A-A-A-G-A-U-U-G-A-

20.39 The template strand of a gene has the following sequence of nucleotides.

-T-A-C-G-G-A-G-C-T-C-C-T-C-T-A-

a Write the sequence of nucleotides that will appear in a mRNA formed on this template.

b What sequence of amino acids will be coded for by the mRNA of part *a?*

20.40 *a* The informational strand of a gene on a DNA molecule contains the following sequence of nucleotides:

-A-T-G-T-T-T-T-A-C-A-C-T-G-A-T-

a What will be the sequence of nucleotides on the mRNA formed in the transcription process from this DNA?

b What sequence of amino acids will appear in the synthesized polypeptide?

20.41 *a* What codon(s) typically initiate protein synthesis?

b What codon(s) signify the termination of protein synthesis?

20.42 List the four general steps in the translation process of protein biosynthesis.

20.43 *a* What is the first codon at the 5′ end of an mRNA molecule?

b What is the functional role of this codon in protein biosynthesis?

c What amino acid does this codon code for?

d Onto which site in the ribosome is this first codon aligned?

e With the first codon aligned in the site of part *d*, what is aligned on the other site of the ribosome?

20.44 Explain why only the amino acid proline will be aligned in the ribosome site in which the mRNA codon is CCC. Be sure to include the role of tRNA in your answer.

20.45 *a* Specify an anticodon in tRNA that could carry the amino acid valine.

b Could there be another tRNA that could carry valine?

20.46 *a* In the growth step of protein biosynthesis, at which site in the ribosome does new peptide bond formation actually take place?
b What two changes occur immediately after peptide bond formation?
c What occurs after the ribosome has shifted in the 3′ direction?
d When is termination signaled?

20.47 After the primary structure of a protein has been formed in the translation process, what steps remain before the protein is physiologically active?

20.48 Describe how protein biosynthesis is regulated according to the Jacod-Monod theory.

20.49 Describe the role played by each of the following in the Jacob-Monod theory.
a Regulatory gene
b Repressor molecule
c Control site
d Operator site
e Inducer molecule
f Promoter site
g Structural gene

20.50 What is a mutation with respect to a DNA molecule?

20.51 *a* Describe how x-rays can lead to a mutation of a DNA molecule.
b What chemical functional group on the DNA bases is especially susceptible to alteration by chemical agents?

20.52 Which alteration do you think would have more serious consequences, one on the DNA backbone (phosphate-sugar) or one on the nitrogen bases? Explain your answer.

20.53 A mutation in a DNA molecule leads to the following examples of changes in a mRNA molecule. In each example state the change in amino acid sequence between the nonmutated and mutated protein:
a -C-U-U-A-A-U-C-G-A-U-G-U- to -C-U-U-A-A-G-G-G-A-U-G-U-
b -C-U-U-A-A-U-C-G-A-U-G-U- to -C-U-C-A-A-U-C-G-A-U-G-U-
c -C-U-U-A-A-U-C-G-A-U-G-U- to -C-U-U-A-A-U-C-C-G-A-U-G-U-
d -C-U-U-A-A-U-C-G-A-U-G-U- to -C-U-U-A-A-U-C-G-U-G-U-

20.54 Sickle-cell anemia involves a change in one of the amino acids of hemoglobin from glutamic acid in normal hemoglobin to valine in sickle-cell hemoglobin. What base substitution could account for this amino acid change?

20.55 What is the functional role of each of the following in gene recombination?
a Donor structural gene
b Plasmid
c Restriction endonuclease
d Bacterium

20.56 The restriction endonuclease Hin dIII cleaves within the sequence (5′ ⟶ 3′) $^{5'}$-A↓A-G-C-T-T-$^{3'}$ between A and A, as shown by the arrow. The two DNA fragments shown below are cleaved by Hin dIII. Write the structures of the DNA products that could result after the cleaved pieces are recombined in new ways using DNA ligase.

$^{5'}$-C-C-A-A-G-C-T-T-G-$^{3'}$
$_{3'}$-G-G-T-T-C-G-A-A-C-$_{5'}$ Fragment 1

$^{5'}$-G-G-A-A-G-C-T-T-A-$^{3'}$
$_{3'}$-C-C-T-T-C-G-A-A-T-$_{5'}$ Fragment 2

20.57 *a* Genetic engineering synthesis is restricted to what general type of product?
b Could a fat-soluble vitamin be synthesized by genetic recombination techniques?

20.58 The protein insulin consists of 51 amino acids. What is the minimum number of nucleotides in the insulin gene?

CHAPTER TWENTY-ONE

ENZYMES

21.1 INTRODUCTION

Many bodily reactions can be carried out in a test tube. However, unless drastic conditions of pH and/or temperature are used, the reactions take place at very slow rates. For example, in a test tube at the typical pH of the small intestine (5–8) and body temperature (37°C), it would take years for the sucrose (table sugar) you used in your morning coffee to be digested. But this digestive reaction takes place very rapidly in the body because of the presence of the enzyme sucrase. **Enzymes** are biological catalysts; they increase the rate of a reaction, often as much as a billion times, so that even at a mild temperature and neutral pH, the physiological chemistry of the cell continues at an active pace. In Chapter 10 you learned that the rate of a reaction is the change in concentration of product (or reactant) per unit of time.

Because of their presence in all physiological reactions, we have had occasion to mention enzymes earlier when discussing such reactions as protein hydrolysis (digestion). Up until now we have told you only that enzymes are catalysts and have provided no other description. In this chapter you will see how enzymes interact with a reactant, how their effects are inhibited, the role vitamins play in enzyme activity, and the medical and food science uses of enzymes.

Unlike inorganic catalysts, which affect reaction rate for a variety of reactants and reaction types, enzymes have the additional property of **specificity.** That is, a given enzyme catalyzes only certain specific reactants or substrates. The reactant acted on by the enzyme is called the **substrate.** Often an enzyme will be effective toward only a single reactant or substrate.

A strong inorganic acid, such as HCl, can catalyze the hydrolysis of carbohydrates. It will also catalyze the hydrolysis of proteins and lipids. In contrast, the enzyme sucrase catalyzes *only* the hydrolysis of the disaccharide sucrose. Indeed, the specificity of sucrase is absolute. It will not even hydrolyze other disaccharides, such as maltose or lactose. These two properties, dramatic rate acceleration and high substrate specificity, are the hallmarks of an enzyme. These properties can be explained in terms of the structure and molecular operation of enzymes. Let us begin the explanation by looking at the composition of enzymes.

21.2 ENZYME COMPOSITION

Enzymes are proteins. Consequently, they have primary, secondary, tertiary, and quarternary structures and are subject to alteration by denaturing agents. In addition to their protein structure, some enzymes contain a nonprotein component called a **cofactor.** The cofactor may be an inorganic metal such as Fe^{2+}, Cu^{2+}, or Zn^{2+}, or it may be a nonprotein organic molecule called a **coenzyme.** The protein portion of an enzyme is given the name **apoenzyme.** The whole enzyme, that is, the apoenzyme plus the metal ion or coenzyme, is called a **holoenzyme.** The component parts of an enzyme are related as follows

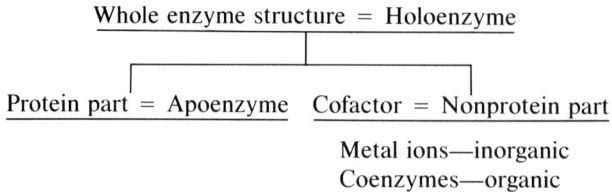

Whole enzyme structure = Holoenzyme

Protein part = Apoenzyme   Cofactor = Nonprotein part

Metal ions—inorganic
Coenzymes—organic

Many coenzymes have, as part of their structure, fragments of compounds we call vitamins. **Vitamins** are defined as substances which human beings cannot internally synthesize, but which are essential to our diet for good health. Their presence as necessary parts of coenzymes, and therefore of active holoenzymes, at least partly explains their requirement in our diet. Our dietary need for certain metal ions or minerals can also be understood in view of their presence in active enzymes. Section 21.11 will further discuss the relationship between coenzymes and vitamins, and Chapter 26 includes a discussion of our nutritional requirements for vitamins and inorganic elements.

21.3 ENZYME CLASSIFICATION AND NOMENCLATURE

To facilitate the discussion of different enzymes, some knowledge of their classification and nomenclature is useful. The suffix *-ase* is used to name an enzyme. The suffix is usually appended to the root name of the substrate being acted on or the type of reaction catalyzed. As examples of enzymes whose names are based on substrates, sucrase is the enzyme that acts on sucrose, and arginase catalyses a reaction of arginine. An example of an enzyme whose name is formed from reaction type is malate dehydrogenase, which catalyzes a reaction in which hydrogen is removed from malate. This is a dehydrogenation reaction; *de-* indicates "to take away." As we have seen, DNA polymerase is an enzyme required for the polymerization reaction of DNA nucleotides.

Unfortunately, some enzymes such as trypsin, pepsin, and chymotrypsin are identified by names that are not based on substrate or reaction type and do not even end in *-ase*. These enzymes were identified before the systematic nomenclature was devised, and their historical names persist.

Enzymes can be classified into one of six major categories, depending on the type of reaction they catalyze. The class names describe the kind of reaction, as you can see by reading the first two columns of Table 21.1. The reaction type is obvious for oxidoreductase, transferase, hydrolase, and isomerase. The term *lyase* comes from a root meaning "to break," and *ligase* from a root meaning "to link together." It is not necessary to memorize any of the subclasses or examples in Table 21.1. But when you study the metabolic reactions in Chapters 22 to 25, you will probably want to refer to the table often.

21.4 MECHANISM OF ENZYME ACTIVITY

Enzymes show all of the same properties as inorganic catalysts. That is, they speed up reactions without being themselves chemically changed. Like other catalysts, enzymes increase reaction rate by lowering activation energy. Energy barriers and activation energy were first introduced in Section

TABLE 21.1 Classification of Enzymes Based on Reaction Catalyzed

Major Classes of Enzymes	General Type of Reaction Catalyzed	Some Subclasses	Typical Example
Oxidoreductases	Oxidation-reduction	Dehydrogenase Oxidase Reductase	$CH_3—CH_2—OH \underset{\text{dehydrogenase}}{\overset{\text{alcohol}}{\rightleftharpoons}} CH_3—C(=O)—H$
Transferases	Transfer of atoms or groups of atoms from one compound to another	Transaminase—transfers an amino group Transmethylase—transfers a methyl group Kinase—transfers phosphate groups	Glucose + ATP $\overset{\text{hexokinase}}{\rightleftharpoons}$ glucose 6-phosphate + ATP
Hydrolases	Hydrolysis reactions	Lipase Carbohydrate hydrolase Protease Nuclease	Sucrose + $H_2O \overset{\text{sucrase}}{\rightleftharpoons}$ glucose + fructose
Lyases	Addition or removal of a molecule from a substrate without hydrolysis or oxidation-reduction	Hydrase Decarboxylase	$HOOC—CH=CH—COOH + H_2O \overset{\text{fumarase}}{\rightleftharpoons}$ $HOOC—CH_2—CH(OH)—COOH$ Fumaric acid → Malic acid
Isomerases	Rearrangements of groups within a molecule to form isomers	Epimerase Racemase	Glucose 6-phosphate $\overset{\text{phosphoglucoisomerase}}{\rightleftharpoons}$ fructose 6-phosphate
Ligases	New bond formation from condensation of two substrates coupled with energy provided by ATP	Synthetase Carboxylase	Fatty acid + ATP $\underset{\text{synthetase}}{\overset{\text{acyl coenzyme A}}{\rightleftharpoons}}$ fatty acyl coenzyme A + PP_i + AMP

10.7. Figure 21.1 reviews the concept of the lower energy barrier in a catalyzed reaction.

The enzyme provides an alternate low-energy reaction pathway for conversion from reactant to product. We would like to know the nature of the interaction between enzyme and substrate as conversion into product occurs; information about this interaction can be gleaned by noting how varying the substrate concentration affects the rate of reaction. Figure 21.2 shows what happens to the rate of an enzyme-catalyzed reaction as the substrate concentration is increased and the enzyme concentration is held constant. The sharp initial rise of the plot shows that the rate, at least in the beginning of the reaction, increases as the substrate concentration is increased. This is

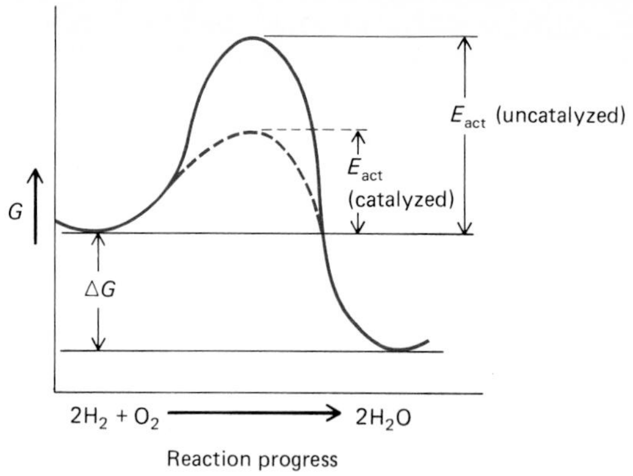

E_{act} (uncatalyzed)
E_{act} (catalyzed)
G
ΔG
$2H_2 + O_2 \longrightarrow 2H_2O$
Reaction progress

FIG. 21.1
This reaction progress diagram demonstrates the difference in energetics of the pathways for an uncatalyzed (solid line) versus a catalyzed (dotted line) reaction. In the catalyzed case, the activation energy (E_{act}) is lower and hence the reaction proceeds more quickly. One catalyst for this decomposition of hydrogen peroxide (H_2O_2) is the enzyme blood catalase. When H_2O_2 is applied to a cut as an antiseptic, the intense bubbling observed is the oxygen released when blood catalase speeds the decomposition of H_2O_2.

the usual behavior for any reaction; an increase in reactant concentration increases rate, as was discussed in Section 10.8. However, at some higher substrate concentration the rate levels off to a maximum value. This behavior, which is observed for almost all enzymes, led biochemists to postulate the idea of formation of an intermediate **enzyme-substrate complex** on the route to product.

According to this hypothesis, the pathway begins with an initial fast reversible step between substrate and enzyme to form the complex. In a second step, which is slow and thereby rate determining for the overall reaction, the enzyme-substrate complex breaks apart to yield product and the return of the free enzyme. This hypothesis can be represented in equation form using E for enzyme, S for substrate, P for product and [ES] for the enzyme-substrate complex:

$$\begin{aligned} E + S &\xrightleftharpoons{\text{fast}} [ES] \\ [ES] &\xrightleftharpoons{\text{slow}} P + E \end{aligned} \qquad (21.1)$$

FIG. 21.2
Rate versus substrate concentration when enzyme concentration is held constant. At first, rate increases rapidly with increased substrate concentration; increasing reactant increases reaction rate. Because substrate must complex with enzyme to go on to product, a maximum rate is reached when all of the fixed amount of enzyme is tied up. Increasing substrate will not increase rate because there is no available enzyme.

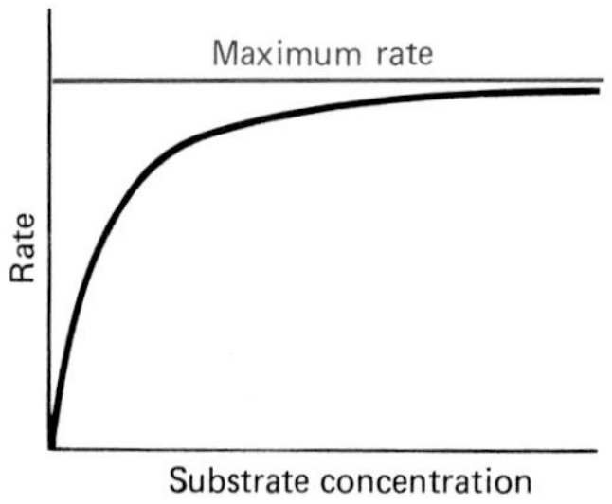

Maximum rate
Rate
Substrate concentration

Because [ES] is the reactant in the slow, rate-determining step, the rate of overall reaction depends on the concentration of [ES].

Returning to a consideration of the plot in Figure 21.2, at first, as the S concentration increases, E combines with S and increases the [ES] concentration. Thus the reaction proceeds faster. At some point, however, because the E concentration is fixed, all of the E will be tied up in the [ES] complex; that is, the maximum [ES] concentration will be reached. Adding additional S does not lead to a rate increase because there is no more E to form [ES]. When the [ES] concentration reaches a maximum, so does the rate. This maximum is marked on Figure 21.2.

The major message to derive from these data is that **enzyme activity depends on the formation of an enzyme-substrate complex.**

21.5 SUBSTRATE SPECIFICITY AND THE ENZYME-SUBSTRATE COMPLEX

All enzymes display some type of substrate specificity; that is, they catalyze the reaction of only one specific reactant or group of structurally related reactants. Sucrase is an example of an enzyme which exhibits **absolute specificity;** it catalyzes a single reaction (hydrolysis) of a single substrate (sucrose). Other enzymes show **group specificity;** they act only on certain functional groups in similar chemical environments. For example, chymotrypsin catalyzes the hydrolysis of ester and peptide linkages having a large nonpolar group, such as an aromatic ring, as a neighbor to the functional group. Enzymes displaying **linkage specificity** catalyze the reaction of a particular functional group regardless of the surrounding structural environment. Lipases are examples of linkage-specific enzymes; they will catalyze the hydrolysis of an ester linkage in waxes, triglycerides, or other lipids. Pancreatic lipase catalyzes the hydrolysis of the ester linkages at the 1 and 3 positions of all triglycerides.

Most enzymes show **stereochemical specificity** in that they will only catalyze one enantiomeric form of a substrate. So, for example, enzymes typically act on L-amino acids but are ineffective toward D-amino acids. Protein-hydrolyzing enzymes in human beings are active only on peptide bonds of L-amino acids.

These substrate specificities lead us to conclude that enzyme substrate interactions involve a spatial ''fit'' between enzyme and substrate. This concept of ''fit'' is often referred to as the **lock-and-key theory** of enzyme action. The enzyme and substrate fit together because of complementary shapes, just as a key fits in a lock.

The enzyme structure is very large, but there is a relatively small area to which the substrate binds. This area is called the **active site** of the enzyme. The active site has some three-dimensional shape to which only a substrate with a complementary shape can fit to form the enzyme-substrate complex. Figure 21.3 shows this diagrammatically. The attractive forces that hold the substrate to the enzyme arise from interactions between the bound substrate and the R groups of the amino acids in the active site region of the enzyme. They are the hydrophobic, hydrogen-bonding, and ionic interactions which you saw earlier holding together the tertiary structure of proteins (Section 19.10).

In some cases, as shown in Figure 21.4, the initial contact of substrate with enzyme induces the enzyme to undergo a change in shape so that effective binding can occur. The substrate may need to change shape slightly also. This **induced-fit theory** of enzyme-substrate interaction explains how similar, but not identical, substrates can be acted on by the same enzyme.

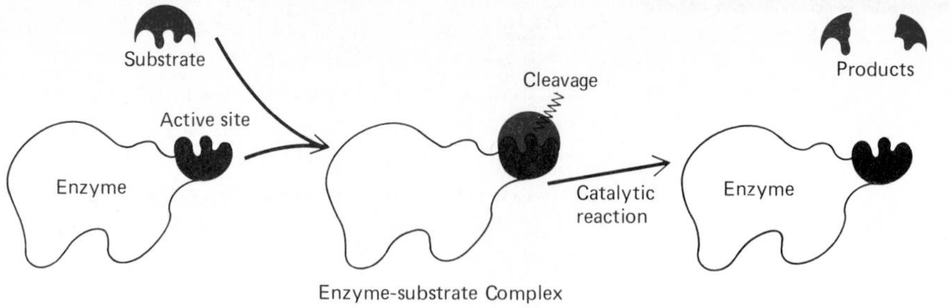
Substrate
Active site
Enzyme
Cleavage
Catalytic reaction
Enzyme-substrate Complex
Products
Enzyme

FIG. 21.3
The interaction of an enzyme and substrate. The active site on the enzyme and the substrate have complementary structures so that they fit together as a key fits a lock. The catalytic reaction (in this case, cleavage) occurs while the substrate is bonded to the enzyme. The product(s) of the reaction leave(s) the surface of the enzyme. This frees the enzyme to combine with another molecule of substrate.

21.6 HOW THE ENZYME-SUBSTRATE STRUCTURE ENHANCES RATE

The induced fit that occurs during binding of some substrates and enzymes may be a contributing factor to the rate enhancement produced by enzyme catalysis. When the substrate induces the enzyme to change shape and bind it, strain may occur in the substrate (Figure 21.5). The reacting bonds in the substrate may be stretched and weakened, and thereby more easily broken.

Enzymes may also enhance the rate of reactions by assuring the proximity and aligment of reactants. Two substrates may be brought together on the active site and properly oriented so that their reacting groups are close to one another. Figure 21.6 shows an example of this.

Finally, the active site of an enzyme may contain proton donors or acceptors. The enzyme can thus catalyze the reaction by providing a medium of acidity or basicity for the substrate without altering the pH of the surrounding solution (Figure 21.7). Some combination of these three structural possibilities probably plays a role in most enzyme reactions.

21.7 FACTORS AFFECTING ENZYME CATALYSIS

The activity of an enzyme often depends critically on pH and temperature.

FIG. 21.4
The induced-fit model of enzyme-substrate binding. As the substrate contacts the enzyme, the enzyme changes shape to accommodate and bind the substrate.

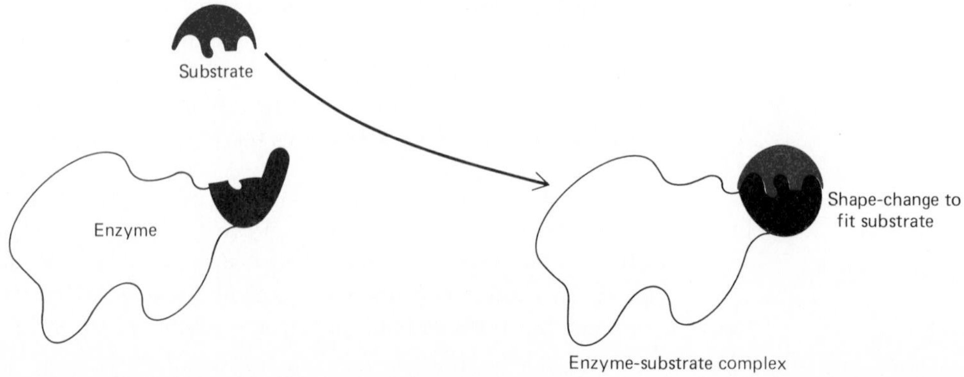
Substrate
Enzyme
Shape-change to fit substrate
Enzyme-substrate complex

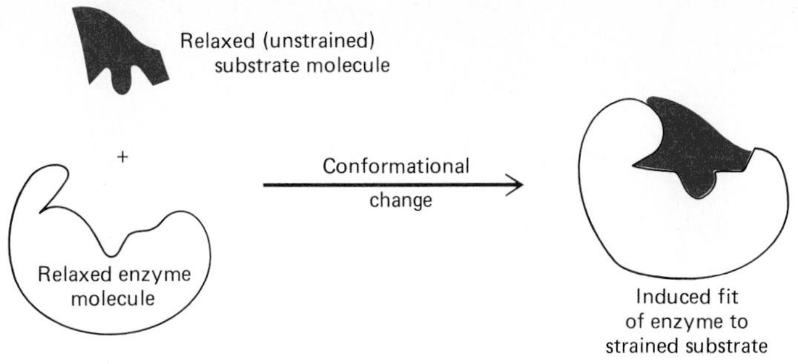
Relaxed (unstrained)
substrate molecule
+
Conformational
change
Relaxed enzyme
molecule
Induced fit
of enzyme to
strained substrate

FIG. 21.5
When the enzyme is induced to change shape, the substrate shape may also be altered. This alteration in substrate shape may be part of the catalytic process that leads to conversion to product.

pH

The catalytic activity of an enzyme varies dramatically as the pH of the surrounding medium changes. Figure 21.8*a* shows the variation of rate with pH graphically. This behavior is typical of enzyme-catalyzed reactions; there is a single optimum pH value at which the rate is at a maximum.

Changes in pH alter reaction rate by altering structure. pH values determine the extent of ionization and the degree of hydrogen bonding of some functional groups on the side chains of amino acids making up the enzyme. When these groups are at the active site of the enzyme, the ability of the enzyme to either bind the substrate or catalyze the chemical change is affected and the reaction rate is influenced. As is the case for any protein, extreme changes in pH can lead to denaturation of the enzyme; the consequence of this is complete loss of catalytic activity.

FIG. 21.6
Enzymes can facilitate reactions by binding substrates so that reacting functional groups are in close proximity. (*a*) For this acid and alcohol to react to form an ester, their functional groups must approach each other. This is a random event as they move around freely. (*b*) The enzyme holds the substrates close to each other, allowing their functional groups to react.

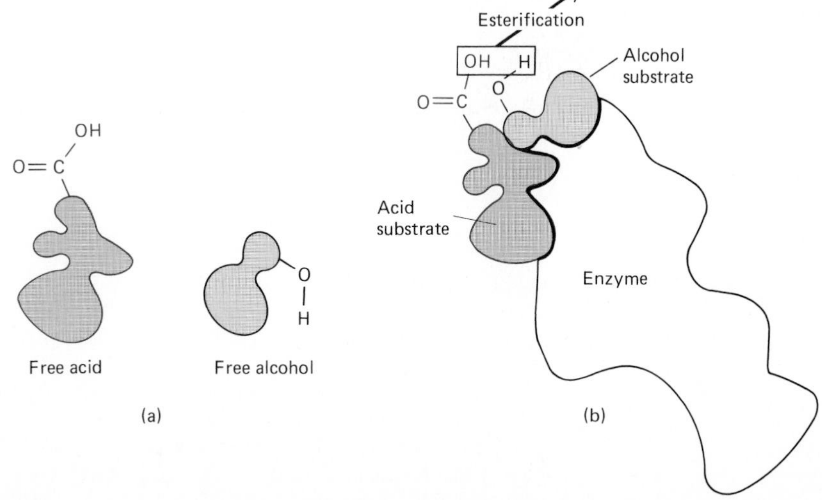
Esterification
OH H
O
O=C
Alcohol
substrate
OH
O=C
Acid
substrate
O
H
Enzyme
Free acid
Free alcohol
(a)
(b)

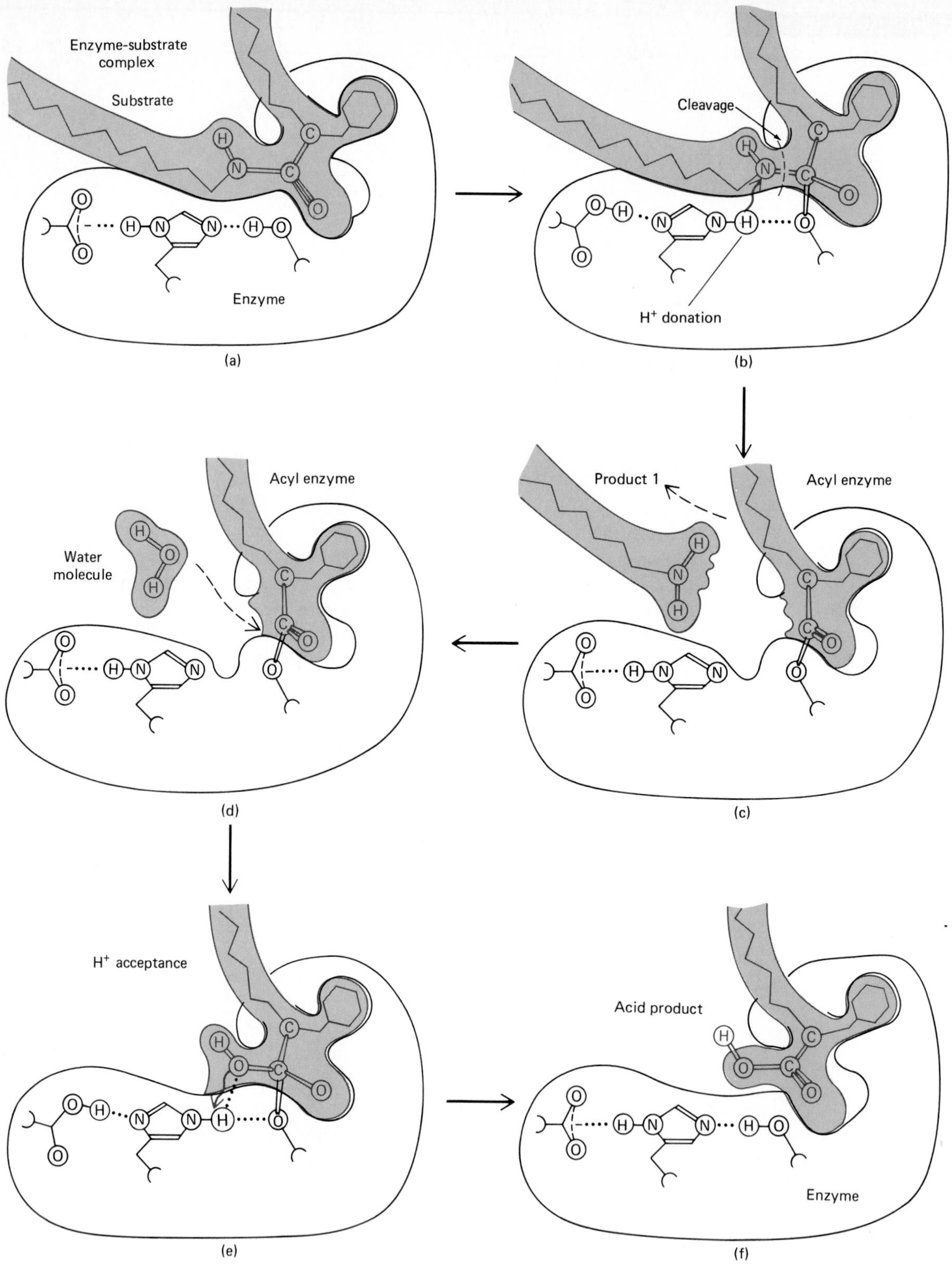
Enzyme-substrate complex
Substrate
Enzyme
(a)
Cleavage
H⁺ donation
(b)
Product 1
Acyl enzyme
(c)
Acyl enzyme
Water molecule
(d)
H⁺ acceptance
(e)
Acid product
Enzyme
(f)

FIG. 21.7

Enzymes can catalyze by acting as proton donors or acceptors. The overall reaction here is the hydrolysis of an amide to an amine and an acid. (*a*) An amide substrate is bound to the enzyme. (*b*) Cleavage of amide bond is accomplished, facilitated by proton donation by the enzyme. (*c*) Amine product of this cleavage departs, leaving the acyl group of the amide still attached to the enzyme. (*d*) Water attacks. (*e*) The elements of water complete the acid structure; simultaneously a proton is returned to the enzyme. (*f*) The acid product can now leave the enzyme.

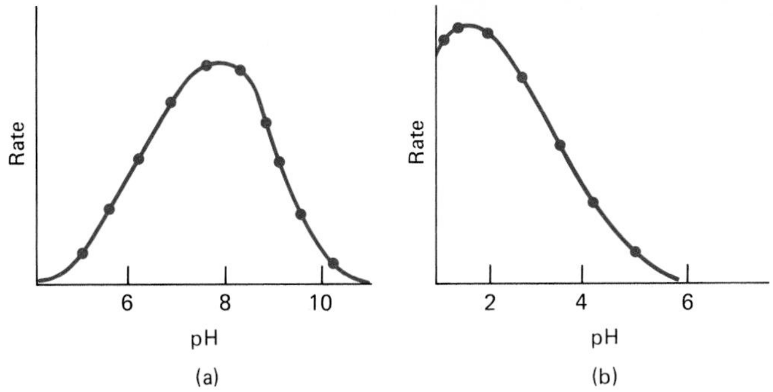
Rate
6
8
10
pH
(a)
Rate
2
4
6
pH
(b)

FIG. 21.8
Plots of the rates of enzyme-catalyzed reactions versus pH show that the rate is maximized at a specific value. (*a*) This plot of the activity of glucose 6-phosphatase shows a maximum rate at pH = 7.8. This enzyme catalyzes the release of glucose in liver cells, where the environment is typically at pH = 7.2. (*b*) The optimum pH of pepsin is 1.5, according to this plot. Gastric juice provides that medium.

The optimum pH of most enzymes lies in the range 7–8. The body's buffer systems work to maintain the pH of bodily fluids within the range required by enzymes. Nature provides a match between the optimum pH of an enzyme and the pH range of the bodily fluid within which the enzyme is acting. Figure 21.8*b* shows that the optimum pH for the protein digestive enzyme, pepsin, is 1.5. Pepsin is found in gastric fluid, which typically has a pH of 1–2 provided by hydrochloric acid.

Temperature

As was discussed in Section 10.8, the rate of a chemical reaction usually increases with an increase in temperature. However, enzyme-catalyzed reactions have an optimum temperature at which the reaction rate is at a maximum. Figure 21.9 shows that the curve for temperature dependence has the same general shape as the one for pH. The optimum temperature for most enzymes is normally very close to the body temperature of the species (in human beings, 37°C).

Above the optimum temperature the secondary and tertiary structure of the enzyme begins to break down. Many enzymes are denatured above 60°C. This fact finds practical application in the sterilization of medical instruments by heating them in an autoclave or boiling water. These procedures denature the enzymes of bacteria present on the instrument. The pasteurization of milk and the canning of food involve similar heat treatment to denature bacterial enzymes.

FIG. 21.9
Plots of the rates of enzyme-catalyzed reactions versus temperature show that the rate is maximized at a temperature close to body temperature, 37°C.

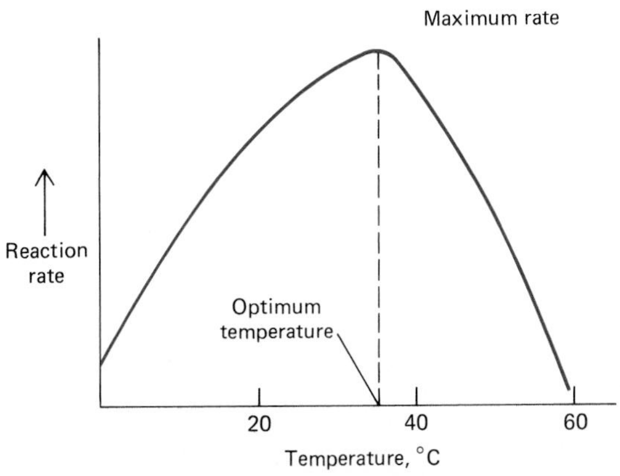
Maximum rate
Reaction rate
Optimum temperature
20
40
60
Temperature, °C

Enzymes which are stored at temperatures much below the optimum, for example those in tissue or sperm samples, return to full catalytic activity at the optimum temperature. Apparently low temperatures do not denature the enzyme, but only decrease reaction rate by the same principles by which cooling affects other reactions (review Section 10.8). Some surgical operations are done on patients whose body temperature has been reduced so that their cell chemistry is slowed down. This reduces the cells' need for oxygen and consequently minimizes the risk of brain damage.

21.8 ENZYME INHIBITION

pH and temperature are factors which affect all enzymes and generally can increase or decrease the rate of any enzyme-catalyzed reaction. Other factors are more specific in their effects. Some chemical substances can decrease the activity of specific enzymes. These substances are known as **inhibitors.** Inhibitors typically function by binding to the enzyme and thereby decreasing the concentration of free enzyme available to the substrate. Many drugs function as enzyme inhibitors in microorganisms. Many poisons are inhibitors of enzymes in the human organism.

Irreversible Inhibition

An irreversible inhibitor forms a strong bond either to a functional group on an amino acid side chain of the apoenzyme or to a metal ion cofactor. In either case the bonding permanently alters the catalytic site so that the substrate can no longer bind to the enzyme. Diisopropylfluorophosphate (DFP), a nerve gas, functions as an irreversible inhibitor of the enzyme acetylcholinesterase. This enzyme normally catalyzes a hydrolysis reaction of acetylcholine, a neurotransmitter (Section 16.9) which provides nerve impulse communication between the nervous and muscular systems.

DFP reacts with the alcohol group on a serine side chain in acetylcholinesterase, as shown in Figure 21.10. The product, DFP-acetylcholinesterase, is inactive toward acetylcholine. Acetylcholine remains attached to the receptor sites of certain neurons, causing a breakdown in the transmission of nerve signals and inducing the muscular system to experience uncontrolled tremors and convulsions. Death is the eventual result.

Fluoride (F^-) and cyanide (CN^-) ions are toxic because they form strong bonds to metal ions such as Fe^{2+}, Cu^{2+}, and Mg^{2+}. A very small amount (milligrams) of cyanide can tie up all the body's Fe^{2+}, a metal ion component of the enzyme cytochrome oxidase. This enzyme, as you will see in Section 22.6, is required in the respiratory chain, a key set of metabolic reactions. Its malfunction means that cell respiration stops and death follows in a very brief time.

Competitive Inhibition

A competitive inhibitor, which we'll represent as I, structurally resembles the substrate and can bind to the active site of the enzyme, thus depriving the substrate of the opportunity to do so. The substrate and the competitive inhibitor compete for the same site on the enzyme. Both cannot be bound to the enzyme at the same time. The cell contains a fixed amount of enzyme. In the presence of a competitive inhibitor, a smaller amount of that enzyme will be bound to substrate and the reaction rate will thereby be diminished.

Like that of the substrate, the binding of competitive inhibitor to the enzyme is reversible. This distinguishes its action from that of an irreversible inhibitor.

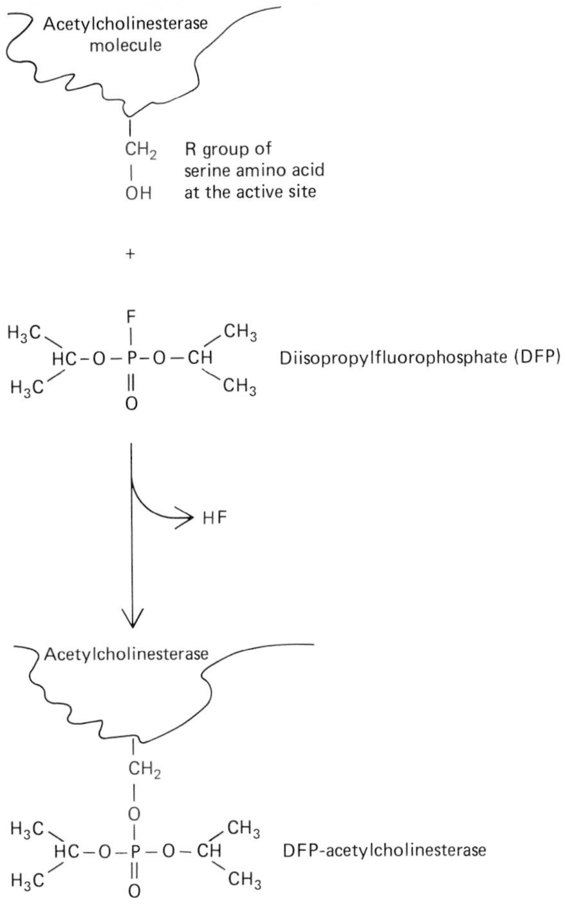
Acetylcholinesterase molecule
CH2
OH
R group of serine amino acid at the active site
+
F
H3C
CH3
HC-O-P-O-CH
Diisopropylfluorophosphate (DFP)
H3C
CH3
O
HF
Acetylcholinesterase
CH2
O
H3C
CH3
HC-O-P-O-CH
DFP-acetylcholinesterase
H3C
CH3
O

FIG. 21.10
Irreversible inhibition of acetylcholinesterase. DFP reacts with a serine side chain and thereby renders the enzyme inactive toward the substrate.

Enzyme-substrate equilibrium: $E + S \rightleftharpoons ES$ (21.2)

Enzyme-inhibitor equilibrium: $E + I \rightleftharpoons EI$ (21.3)

The decreased rate effect of competitive inhibition can be overcome by increasing the concentration of substrate. Increasing S in Equation (21.2) provides a stress (LeChatelier's principle—Section 10.12) which shifts the position of equilibrium toward the product, ES. This shift results in a decreased concentration of E. The effect of lowered E in Equation (21.3) is to shift that equilibrium to the left; the ultimate result is a lowered EI concentration. The net effect of increased substrate concentration is a lower concentration of EI and raised concentration of ES, resulting in an increased rate of reaction. Figure 21.11*a* shows competitive inhibition pictorially.

Competitive inhibition can be taken advantage of in the treatment of the toxic condition resulting from the accidental ingestion of ethylene glycol, a component of automobile antifreeze. Ethylene glycol is itself nontoxic, but the enzyme alcohol dehydrogenase catalyzes its oxidation to oxalic acid, a lethal toxin. The conversion to oxalic acid can be blocked by administering a large dose of ethanol. Ethanol is a competitive inhibitor and competes with the substrate ethylene glycol for the enzyme alcohol dehydrogenase. Unreacted ethylene glycol can then be excreted from the body. Ethanol functions similarly as an antidote for methanol poisoning. As will be seen in Section 21.9, competitive inhibition is also used in the design of drugs.

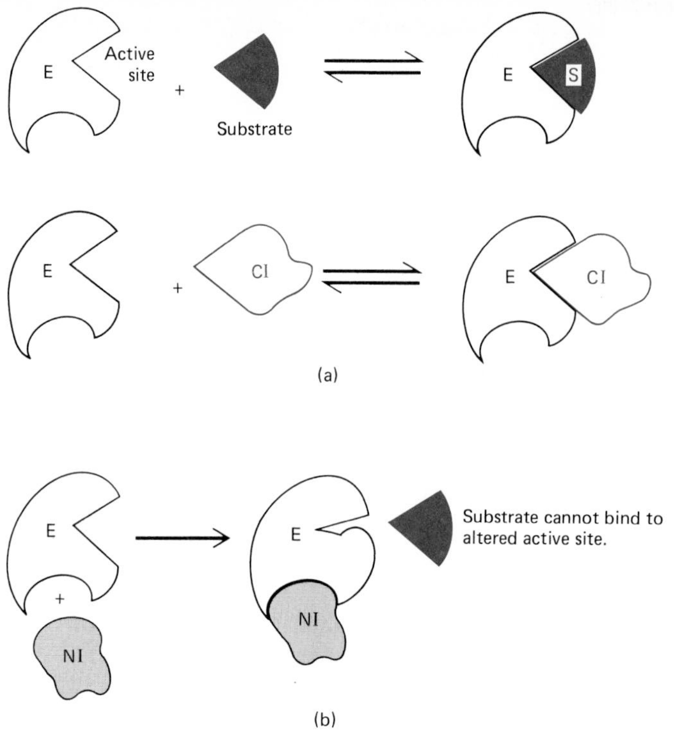
Active site
E
+
Substrate
E
S
E
+
CI
E
CI
(a)
E
+
NI
E
NI
Substrate cannot bind to altered active site.
(b)

FIG. 21.11
Comparison of competitive and noncompetitive inhibition. (*a*) A competitive inhibitor (CI) reversibly binds to the active site on the enzyme (E) and inhibits formation of the enzyme-substrate complex, thereby lowering the rate of formation of the product. (*b*) A noncompetitive inhibitor (NI) binds to the enzyme at a location other than the active site and changes the active site so that the substrate can no longer bind to the enzyme.

Noncompetitive Inhibition

Noncompetitive inhibitors do not resemble the substrate and bind to a site on the enzyme other than the active site. Substrate and inhibitor are not in competition for the enzyme. Indeed, the substrate and the noncompetitive inhibitor may bind simultaneously to the enzyme. Typically, the noncompetitive inhibitor, once bound to the enzyme, either alters the enzyme's shape to prevent further binding of the substrate or alters the catalytic action so that a bound substrate is not converted into product. Figure 21.11*b* shows this pictorially. The final result is a decrease in rate of formation of product. In contrast to competitive inhibition, noncompetitive inhibition *cannot* be overcome by an increase in substrate concentration because substrate and inhibitor are not competing for the same site on the enzyme.

The poisonous character of heavy metal ions such as Hg^{2+} and Pb^{2+} is a consequence of their action as noncompetitive inhibitors. The heavy metal ions combine with thiol (—SH) groups disrupting the formation of disulfide linkages and denaturing the enzyme.

21.9 ANTIMICROBIALS

Antimicrobial is the general name given to a chemical substance that inhibits the growth or destroys disease-producing microorganisms. There are several classes of antimicrobials, each of which has certain characteristics and modes of action. For example, **antimetabolites** are synthetic substances which act by competitively inhibiting a microorganism's enzymes. **Antibiotics** are natural substances secreted by one microorganism, but toxic to another. Let us now explore these types of antimicrobials more closely.

Antimetabolites

Many drugs function as inhibitors of enzymes. Drug design often involves finding or synthesizing a substance which will inhibit an enzyme required by a microorganism but not affect a human patient's enzymes. One approach is to find or make a substance which structurally resembles an essential substrate of the microorganism. The sulfonamide drugs are so designed.

Bacteria utilize the substrate *p*-aminobenzoic acid in the synthesis of folic acid, a necessary coenzyme for bacteria. Sulfanilamide, the first of the sulfonamides, or sulfa drugs, competes with *p*-aminobenzoic acid for the enzyme dihydrofolate synthetase which is required for the formation of folic acid.

ENZYME + *p*-aminobenzoic acid + other bacterial metabolite ⟶ folic acid

ENZYME + sulfanilamide ⟶ [enzyme-sulfanilamide complex]

The bacterial enzymes are "fooled" into accepting sulfanilamide because of its close structural similarity to *p*-aminobenzoic acid.

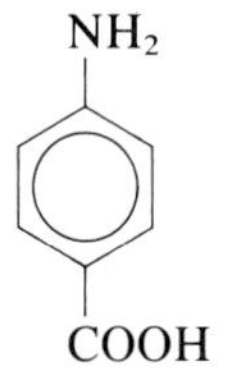

NH_2
COOH

p-Aminobenzoic acid

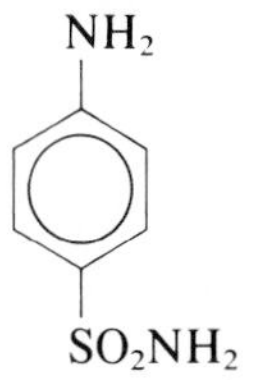

NH_2
SO_2NH_2

Sulfanilamide

This is fatal to the bacteria because folic acid synthesis is then suspended. Human beings do not possess enzymes for the synthesis of folic acid. Rather, the vitamin folacin provides a dietary source of folic acid. Hence, sulfanilamide has no effect on human enzymes and meets the necessary selective toxicity requirements of a good drug. Substances such as sulfanilamide which competitively inhibit an enzyme because they resemble a substrate normally used by a microorganism in a metabolic reaction (metabolite) are known as **antimetabolites.**

During World War II soldiers in tropical regions routinely spread sulfathiazole, a derivative of sulfanilamide, on wounds to prevent bacterial infections.

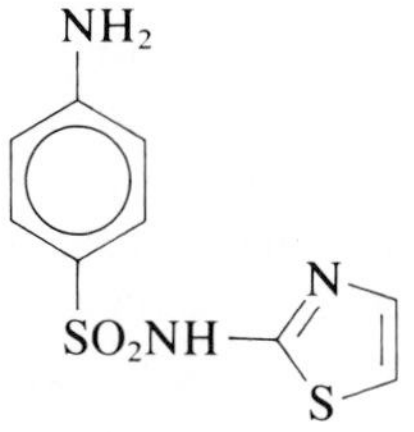

NH_2
SO_2NH
N
S

Sulfathiazole

Before the development of sulfa drugs soldiers in the tropics died from infection more often than from their wounds. Thus sulfathiazole has been credited with saving many lives during World War II.

Because of their undesirable side effects sulfa drugs have now been largely replaced by antibiotics; however, they are still used in the treatment of urinary tract infections.

Many structural derivatives of metabolites have been synthesized in the laboratory in efforts to find effective new antimetabolites. The antimetabolite 6-mercaptopurine has been useful in effecting the remission of childhood leukemia. Compare its structure with that of the metabolite adenine:

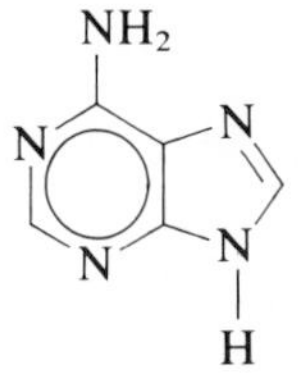

Adenine

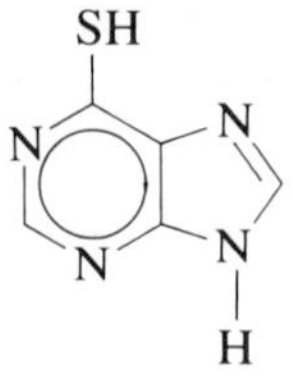

6-Mercaptopurine

6-mercaptopurine is thought to act by inhibiting adenine and guanine synthesis. This prevents DNA formation and checks the growth of the leukemia condition by inhibiting cell replication.

Antibiotics

Antibiotics are substances secreted by certain microorganisms and toxic to other species. The antibiotic action of penicillin was first observed by Sir Alexander Fleming in 1929. Fleming noticed that a strain of the mold *Penicillium notatium* secreted a substance which was toxic to some forms of bacteria but showed very low toxicity toward mammals. Later (1940), Florey and Chain isolated the active substance from the mold and demonstrated its wide usefulness for controlling bacterial infections.

Penicillin kills bacteria by irreversibly inhibiting an enzyme required for cell wall synthesis. Bacteria have cell walls but mammals do not; therefore, penicillin is selectively toxic. Without a structurally sound cell wall, the bacteria swell and burst. Unfortunately, many bacteria have "waged war" on penicillin by manufacturing an enzyme, generally called a penicillinase, which breaks down the ring structure of penicillin. Those bacteria that produced penicillinase survived and reproduced, and by 1960, the resistance of bacteria to penicillin G, the original penicillin, presented a major medical problem. A major effort was mounted to prepare penicillinase-resistant derivatives of penicillin. Methicillin was the first of such drugs to be synthesized. In addition, the original penicillins decomposed in acid solutions, such as gastric juice, and so were ineffective if taken orally. Derivatives such as ampicillin and penicillin V have improved acid stability and can be administered orally. Table 21.2 shows the structures of several penicillin derivatives.

More recently developed antibiotics, such as the cephalosporins, tetracyclines, and erythromycins also function by enzyme inhibition. It appears that the cephalosporins also inhibit an enzyme involved with bacterial cell wall synthesis. Table 21.2 also depicts some cephalosporin structures.

21.10 REGULATION OF ENZYME ACTIVITY

Enzyme activity and the products of the reactions enzymes catalyze are essential to healthy bodily functions. However, overactivity and overproduction must be avoided.

The activity of most enzymes is regulated so that it is greatly diminished or absent when the organism "senses" little demand for, or a sufficient supply of, the product of the enzyme-catalyzed reaction. There are several ways this regulation can be accomplished.

TABLE 21.2 Some Clinically Useful Penicillin and Cephalosporin Structures

GENERAL PENICILLIN STRUCTURE

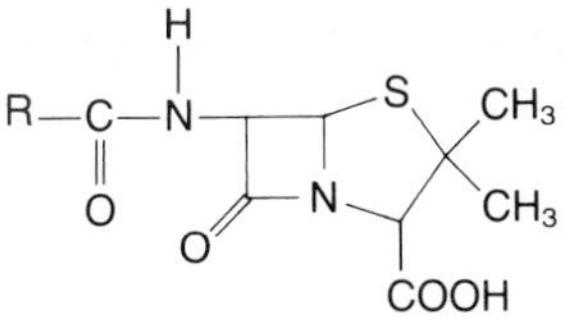

Penicillin Name	Structure of R	Comments
Penicillin G (benzylpenicillin potassium salt)	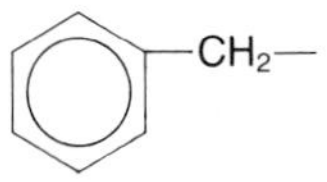	Administered by injection; decomposed by pencillinase
Penicillin V (phenoxymethylpenicillin)	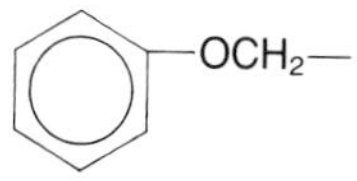	Stable to acid and therefore effective orally; decomposed by penicillinase
Methicillin	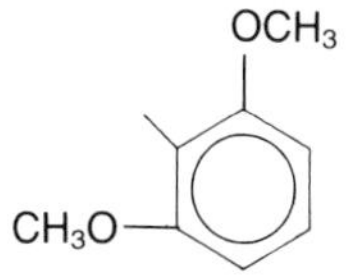	Administered by injection; resists penicillinase
Ampicillin	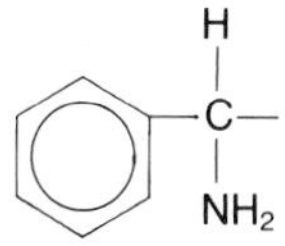	Stable to acid and therefore effective orally; decomposed by penicillinase

GENERAL CEPHALOSPORIN STRUCTURE

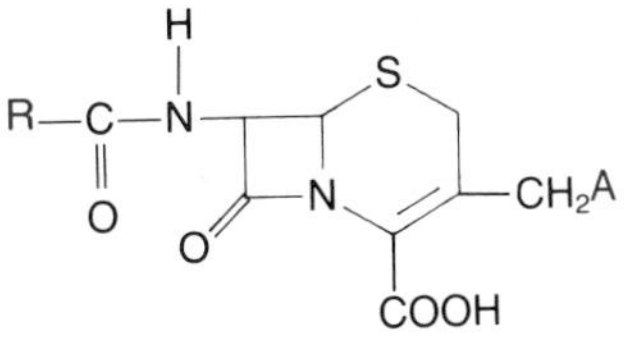

Cephalosporin Name	Structure of R	Structure of A	Comments
Cephalothin	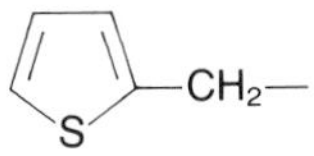	$—O—C(=O)—CH_3$	Administered by injection; resists penicillinase
Cephalexin	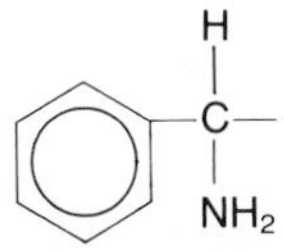	—H	Administered orally; effective against a wide variety of bacteria

Control of Protein Synthesis in DNA

You encountered the regulation of protein synthesis in Section 20.11. The repressor-inducer theory is one method of regulating enzyme concentrations in cells. In the example of the Jacob-Monod model described in Section 20.11, the inducer molecule was the substrate on which the enzyme acts. Thus the presence of substrate triggers the synthesis of enzymes by switching the DNA operon into the on position for transcription of the enzyme's mRNA. In the absence of substrate the repressor turns the operon off. Enzymes controlled by this on-off mechanism are known as **induced enzymes.**

Activation of an Inactive Enzyme: Zymogens

In another mode of enzyme control, an inactive precursor of the enzyme molecule, called a **zymogen**, is formed. Chemical modification of the zymogen produces the enzyme at the appropriate time and site in the body. The protein-digesting enzymes pepsin, trypsin, and chymotrypsin are examples of enzymes first synthesized in an inactive form.

The conversion of a zymogen to an active enzyme involves modification of the primary structure of the zymogen by cleavage of one or more peptide bonds. Figure 21.12 shows an example of such a conversion. Removal of a peptide fragment from the zymogen alters the three-dimensional shape of the molecule and an active enzyme is produced.

The zymogen pepsinogen is synthesized in the pancreas, where it shows no catalytic activity. When it enters the stomach, the high acidity of gastric juice promotes the hydrolysis of a 42–amino acid peptide fragment from the N-terminal end of pepsinogen. The remaining protein is the active enzyme pepsin. Interestingly, once formed, pepsin itself further catalyzes the conversion of pepsinogen to pepsin.

In the disease pancreatitis, digesting enzymes, rather than the zymogens, are formed in the pancreas. The active enzymes then catalyze the hydrolysis of proteins in the pancreas and its blood vessels. In other words, the pancreas is digested. This is a painful and frequently fatal occurrence.

Regulation of the body's blood clotting mechanism employs activation of zymogens. The formation of a blood clot requires a series of steps, the last of which is the conversion of the soluble protein fibrinogen to the insoluble fibrin. This final step requires the enzyme thrombin, which is not normally present in blood. Thrombin is formed from the zymogen prothrombin. Prothrombin must bind Ca^{2+} ions before being activated; hence the role of Ca^{2+} ions in blood clotting. The conversion of prothrombin to thrombin occurs as the body's response to events such as blood loss or tissue injury. In summary,

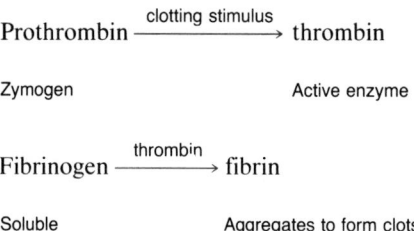

$$\text{Prothrombin} \xrightarrow{\text{clotting stimulus}} \text{thrombin}$$

Zymogen — Active enzyme

$$\text{Fibrinogen} \xrightarrow{\text{thrombin}} \text{fibrin}$$

Soluble — Aggregates to form clots

FIG. 21.12
An active enzyme is formed by hydrolytic cleavage of a peptide fragment out of the protein chain of the zymogen. This is the typical way that activation occurs; the shape of the molecule is changed by cleavage.

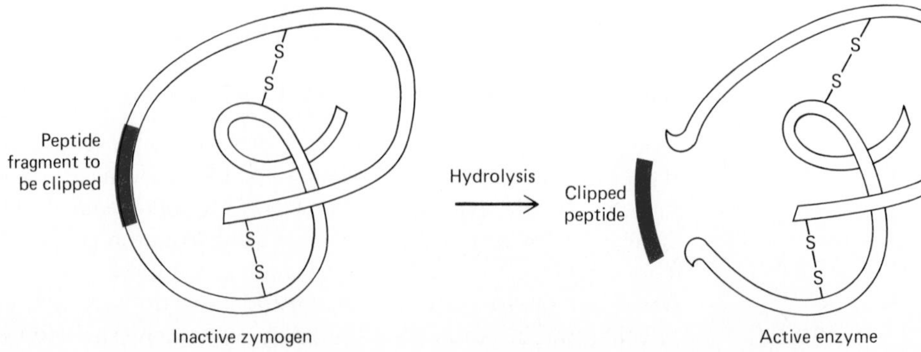
Peptide fragment to be clipped
S
S
S
S
Hydrolysis
Clipped peptide
S
S
S
S
Inactive zymogen
Active enzyme

Allosteric Regulation

Enzyme regulation by control of mRNA and protein synthesis and by zymogen formation are one-way processes because once the active enzyme is formed, it does not revert back to an inactive form. In contrast, allosteric regulation provides a reversible mechanism for control of active enzyme concentration.

In many metabolic reactions the product of the reaction can act as a noncompetitive inhibitor (Section 21.8) and thereby alter the activity of an enzyme necessary for formation of that product. This phenomenon has been observed in cases where a final product is formed after a sequence of reactions from an initial substrate. The final product inhibits an enzyme in the sequence, usually in the first step, in a process called **feedback inhibition.**

The sequence by which conversion of the amino acid threonine to isoleucine occurs is an example of a system regulated by feedback inhibition. The reaction sequence has five steps. The product, isoleucine, noncompetitively inhibits threonine dehydratase, the enzyme of the first step in the sequence. In the following representation, E stands for enzyme and S for intermediate substrates in the sequence.

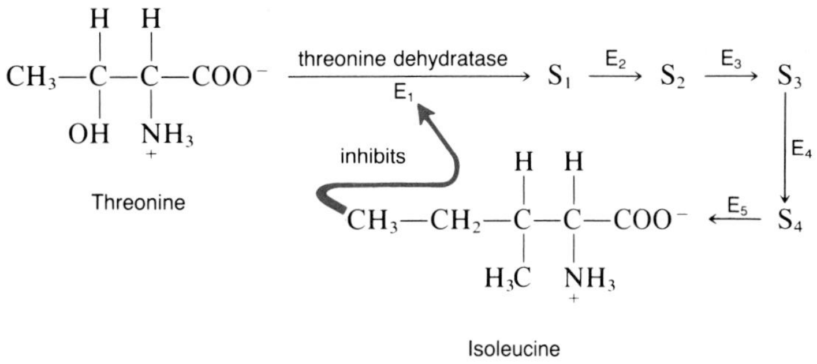
Threonine
threonine dehydratase
E_1
inhibits
S_1
E_2
S_2
E_3
S_3
E_4
S_4
E_5
Isoleucine

Like other noncompetitive inhibitors, isoleucine does not bind to the active site on the enzyme, but rather to another site called the regulatory site. For this reason, this kind of enzyme activity control is referred to as **allosteric,** from the Greek words for "other site." The enzyme regulated is called an allosteric enzyme.

Allosteric regulation is reversible. When the concentration of isoleucine in the cell decreases, its binding to threonine dehydratase must also decrease and the activity of the enzyme thereby increases. Increased enzyme activity increases the concentration of isoleucine. When the cell's needs are exceeded, excess isoleucine binds to the enzyme, leading to a resultant decrease in the formation of isoleucine.

Feedback inhibition is an efficient biological process by which a product can regulate its own formation by allosterically regulating an enzyme necessary for the synthesis of the product. If the product, is scarce, formation is allowed. If there is an abundance, the product inhibits formation of itself until a scarcity develops.

Hormone Regulation

One function of hormones is to influence enzyme activity by activating inactive enzymes. Because this topic requires some background knowledge

TABLE 21.3 Water-Soluble Vitamins and the Physiological Role of Their Coenzyme Derivatives

Water-Soluble Vitamin	Coenzyme Where Found	Coenzyme Function
Thiamine (B_1)	Thiamine pyrophosphate	Carrier of aldehyde group in decarboxylation reactions of α-keto acids
Riboflavin (B_2)	Flavin mononucleotide (FMN) Flavin adenine dinucleotide (FAD)	Carrier of pair of H· atoms in oxidation-reduction reactions
Nicotinic acid, nicotinamide (niacin, B_3)	Nicotinamide adenine dinucleotide (NAD^+) Nicotinamide adenine dinucleotide phosphate ($NADP^+$)	Carrier of hydride ion ($H:^-$) in oxidation-reduction reactions
Pyridoxine (B_6)	Pyridoxal phosphate	Carrier of amino group in transamination reactions
Cobalamin (B_{12})	Deoxyadenosyl cobalamin	Participates in the exchange of a hydrogen atom with an alkyl group between two adjacent carbons
Biotin	Biocytin	Carrier of $-\underset{O}{\underset{\|}{C}}-O^-$ group in carboxylation reactions
Folacin (folic acid)	Tetrahydrofolic acid	Carrier of 1-carbon groups ($-CH_3$, $-CH_2-$, $-\underset{H}{\underset{\mid}{C}}=O$, $-\underset{H}{\underset{\mid}{C}}=N-H$) from one molecule to another
Pantothenic acid	Coenzyme A	Carrier of acyl groups
Ascorbic acid	None known	Involved in the formation and maintenance of connective tissue

about hormones, we will defer a discussion of this type of regulation to Section 21.14.

21.11 COENZYMES AND VITAMINS

To be catalytically active, many enzymes require a nonprotein organic fragment called a coenzyme. Coenzymes may be bound to apoenzymes by noncovalent interactions such as hydrogen bonding. Alternatively, strong covalent bonds may bind the protein and nonprotein structures together. In this case the coenzyme is called a **prosthetic group** (Section 19.14). Many coenzymes incorporate a vitamin fragment as part of their structure. Indeed, all water-soluble vitamins, with the exception of vitamin C, have been identified with a coenzyme. Table 21.3 lists these vitamins and associated coenzymes.

Coenzymes play their role at the catalytic active site of the enzyme. Their function is to accept and transfer small organic or inorganic fragments, such as an aldehyde group ($-\overset{H}{\overset{\mid}{C}}=O$), an acetyl group ($CH_3-\overset{\mid}{C}=O$), a hydride ion ($H:^-$), or a hydrogen cation (H^+). For example, the coenzyme nicotinamide adenine dinucleotide (NAD^+), which we will see in Section 23.4 playing a critical role in metabolism, accepts a hydride ion (a hydrogen nucleus with two electrons) from a metabolic substrate molecule. The substrate thereby undergoes oxidation (or dehydrogenation). The actual hydride acceptor in NAD^+ is the vitamin component nicotinamide, the amide of niacin. NADH later gives up its hydride ion in the electron transport system (Section 22.6) thus regenerating NAD^+.

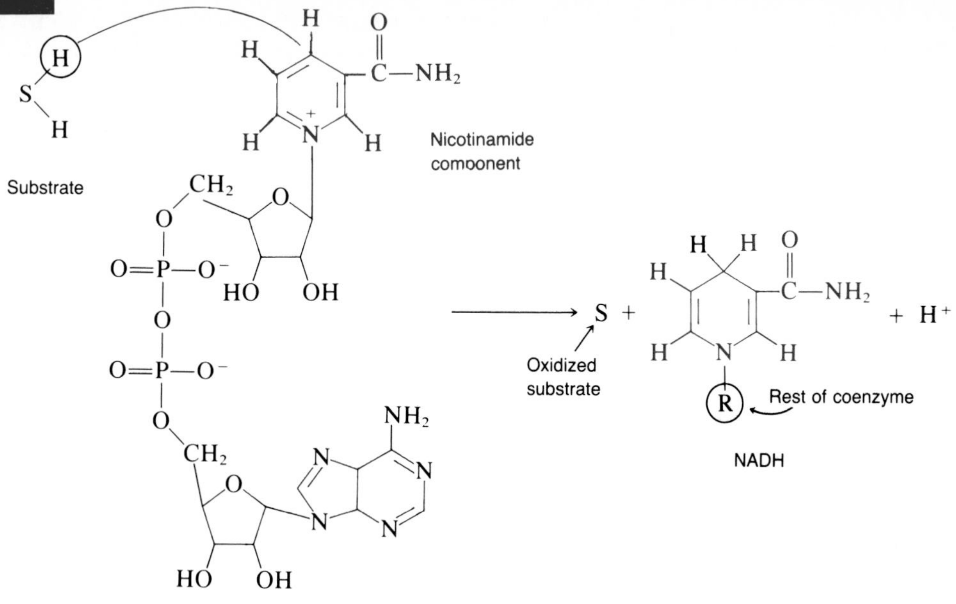

Substrate
Nicotinamide component
Oxidized substrate
Rest of coenzyme
NADH

NAD$^+$

Coenzyme A, which contains the vitamin pantothenic acid, provides another example of the acceptor and carrier role of a coenzyme. Coenzyme A accepts and transfers acyl groups in general ($R-\overset{\overset{O}{\|}}{C}-$), but most commonly acetyl groups ($CH_3-\overset{\overset{O}{\|}}{C}-$), in many enzyme-catalyzed metabolic reactions.

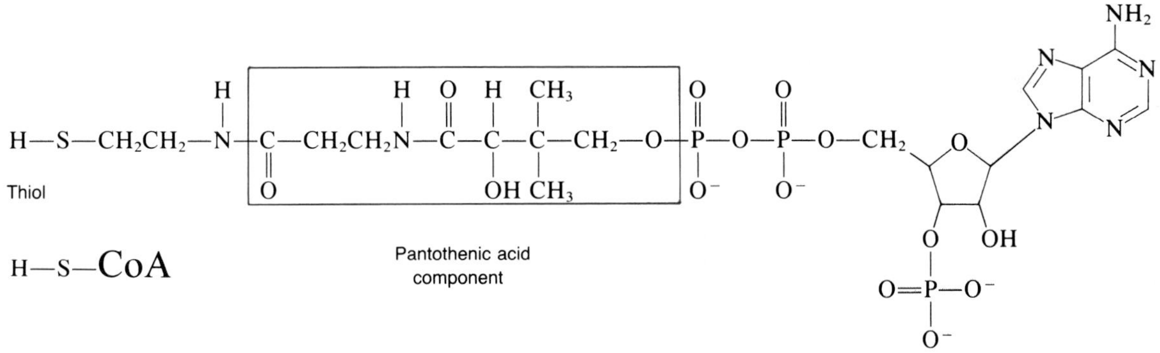

Thiol
H—S—CoA
Pantothenic acid component

The thiol group (—SH) of coenzyme A is the reactive site that binds to the carbonyl carbon of the acyl group and forms a thioester. At the end of glycolysis, an essential metabolic sequence that you will study in Section 23.2, coenzyme A is involved in the conversion of pyruvate to acetyl CoA. (Coenzyme A is represented as CoA—SH.)

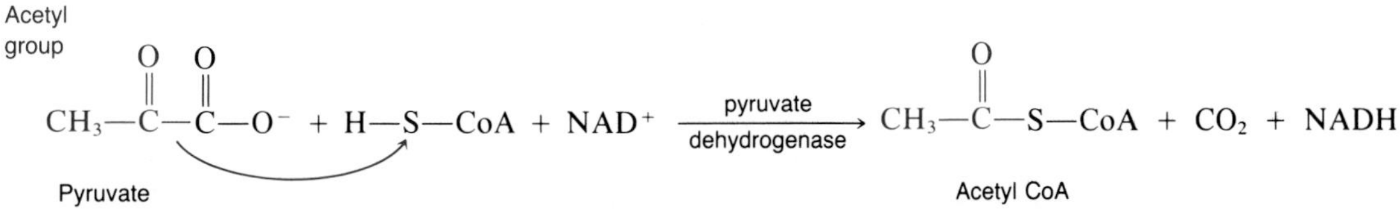

Acetyl group
pyruvate dehydrogenase
Pyruvate
Acetyl CoA

(21.4)

In the next metabolic step, the first reaction of the citric acid cycle (Section 23.4), acetyl coenzyme A transfers its acetyl group to oxaloacetic acid. This regenerates coenzyme A, which is then ready for the transfer of another acyl group.

As these two coenzyme examples (NAD^+ and coenzyme A) illustrate, although the coenzyme is chemically modified in the course of reaction with the substrate, it is eventually regenerated so that the total enzyme (apoenzyme and coenzyme) is left chemically unchanged.

21.12 MEDICAL USES OF ENZYMES

Enzymes are normally confined to the interior of cells, and therefore their concentration in blood serum is usually very low. However, when a cell dies either through disease or injury, its contents, including the particular enzymes within, are released into the blood serum. For example, the enzyme pancreatic amylase appears in elevated concentrations in the blood of patients suffering from diseases of the pancreas.

Clinical chemists can readily measure the concentration and activity of an enzyme by determining the rate at which it catalyzes the conversion of a known amount of substrate into product. The enzyme-substrate relationship is highly specific so that the choice of a particular substrate assures that the activity of only one specific enzyme will be measured. The experiment is called an **enzyme assay.** Clinical laboratories have worked out testing procedures for determining the concentration and activity of many enzymes in blood serum.

Enzyme activity is reported in international units (IU). The reference standard, one international unit, is the amount of enzyme that will catalyze the conversion of one micromole (1×10^{-6} mol) of substrate to product in one minute at a standard pH and temperature. As part of a blood chemistry profile, physicians obtain numerical values for a patient's blood serum enzyme activity and compare them with a chart of normal reference values. Deviations from the norms are indicative of particular medical conditions and can thus be used in diagnosis. Table 21.4 offers some examples.

Diagnosis

As can be seen from Table 21.4, elevated levels of some enzymes, such as LDH, can indicate more than one condition. However, diagnosis can be

TABLE 21.4 Blood Serum Enzyme Levels in Medical Diagnosis

Enzyme Assayed	Condition Indicated by Elevated Level
Acid phosphatase	Prostate cancer
Alkaline phosphatase	Bone disease; hepatitis or cirrhosis of the liver
Pancreatic amylase	Pancreatic disorders
Creatine phosphokinase (CPK)	Heart disease; skeletal muscle disorders; brain disease
Gamma glutamyl transpeptidase (GGT)	Liver disease
Lactate dehydrogenase (LDH)	Heart disease; liver metastases; pernicious anemia
Serum glutamate oxaloacetate transaminase (SGOT)	Viral hepatitis; heart disease
Serum glutamate pyruvate transaminase (SGPT)	Viral hepatitis; heart disease

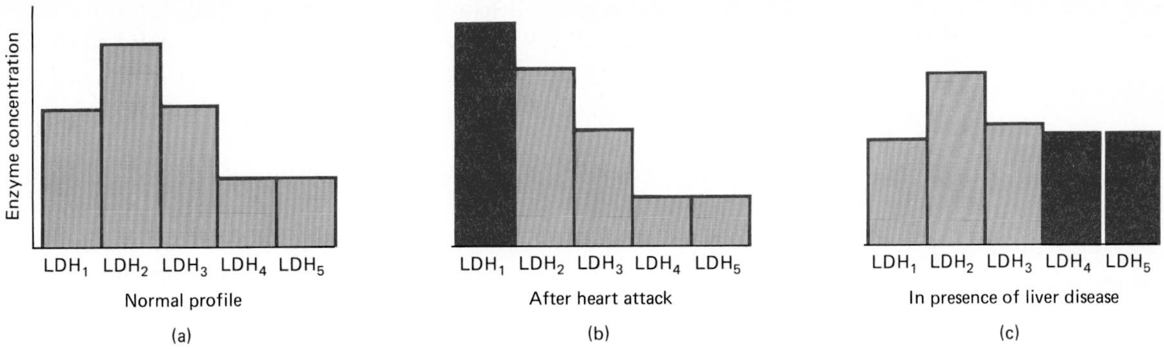
Enzyme concentration
LDH$_1$ LDH$_2$ LDH$_3$ LDH$_4$ LDH$_5$
Normal profile
(a)
LDH$_1$ LDH$_2$ LDH$_3$ LDH$_4$ LDH$_5$
After heart attack
(b)
LDH$_1$ LDH$_2$ LDH$_3$ LDH$_4$ LDH$_5$
In presence of liver disease
(c)

FIG. 21.13
The enzyme lactate dehydrogenase (LDH) is actually a mixture of at least five isoenzymes (designated here by subscripts). The composition of the mixtures varies among (*a*) normal individuals, (*b*) heart attack patients, and (*c*) patients with liver disease. The bar graphs show the relative concentrations of isoenzymes for the three cases. Heart attack victims have LDH_1 concentrations 45 to 65 percent above normal. The concentrations of LDH_4 and LDH_5 are elevated 33 to 80 percent in liver patients.

narrowed by taking advantage of the fact that LDH is actually a mixture of several enzymes with similar enzymatic functions but slightly different structures. Such groups of similarly acting enzymes are called **isoenzymes.** The overall composition of the LDH isoenzyme mixture depends on the tissue in which it is found. The LDH mixture is different in the liver than in the heart muscle; thus the composition of LDH isoenzymes in a heart attack victim is distinguishable from the composition of those of a patient suffering from a liver disease. Figure 21.13 shows a normal LDH profile and two abnormal profiles for two different conditions.

The enzyme activities of CPK, SGOT, and LDH are monitored over a period of days after it is believed that a patient has suffered a myocardial infarction (heart attack). The variation in the activity of these three enzymes over a period of time is highly indicative of a myocardial infarction.

The enzyme glucose oxidase is used in some diagnostic tests for the presence of glucose in the urine, a condition which may indicate hyperglycemia or diabetes mellitus. Glucose oxidase is impregnated on a dipstick and the stick is placed in a urine sample. If glucose is present, it will be oxidized and the product of the oxidation will cause a change in dye color to green or brown, depending upon the concentration of glucose in the urine. Glucose oxidase will act only on glucose; thus, the false results sometimes obtained using Benedict's or Tollens' tests, which were discussed in Section 15.14, are avoided.

Treatment

Enzymes can be used in treatment as well as in diagnosis. Again, heart attacks provide us with an example. The most common cause of a myocardial infarction is the development of a clot in a coronary artery; the clot prevents the flow of oxygenated blood to a portion of the heart muscle. The enzyme streptokinase is found in human blood and is involved in the dissolution of blood clots. Streptokinase has recently been used clinically to dissolve clots in the coronary arteries of patients exhibiting the initial symptoms of a heart attack. The streptokinase is injected directly onto the clot using an inserted catheter. Urokinase is another clot-dissolving enzyme that may prove effective in treating heart attacks. However, the costs of extracting urokinase from human urine and kidneys have been high. Synthesis using recombinant DNA techniques (Section 20.13) may remedy the cost problem.

21.13 ENZYMES AND FOOD SCIENCE

Enzymes are widely used in the food and other commercial industries. Because of their high substrate specificity, enzymes can be employed to alter just one substance in a complex food mixture chemically.

Fat Hydrolysis

For example, pancreatin, an enzyme extract of animal pancreas, functions as a lipase and is used in the selective hydrolysis and removal of yolk fat from egg whites. The presence of fat inhibits successful whipping of egg whites into a fluffy, voluminous airy mass. Egg white is mostly protein, and a general hydrolysis agent would break down the protein as well as the lipid material.

Other lipases preferentially hydrolyze ester linkages of short-chain fatty acids such as butyric (4C) and caproic (6C), but leave untouched the long-chain fatty acid ester links. These lipases find commercial use in the preparation of partially hydrolyzed butter and to speed the development of "natural" flavors in certain cheeses. When the cheese is treated with a lipase that cleaves some butyric or caproic acid fragments from the fat in the cheese, these acids give some of the characteristic odor and taste to cheese. Natural aging accomplishes the same process. Not everyone finds the odors and taste pleasant. They are distinctly unpleasant in high concentrations.

Fruit Juice Consistency

Pectin is a mixture of high-molecular-weight (30,000 to 300,000) carbohydrate derivatives; it is found principally in the fruit of plants. The pectin substances in fruit juices aggregate in the presence of sugar and acid molecules, forming a highly viscous liquid. In such juices as tomato and apricot, this thickness is considered desirable. However, in juices such as apple and grape a clear flowing liquid is desired, and pectin enzymes are added to the juice to achieve this. Pectinases catalyze the cleavage of 1,4-glycosidic linkages (Section 15.15) and convert the pectin polymer into soluble oligomers and monomers. The result is a clear, low-viscosity liquid. If pectinases are naturally present in "thick" juices, they are denatured by the pasteurization of the juice. Pectin enzymes are also added to wine must to ensure a clear wine product. Wine must is the crushed grape liquid before fermentation.

Regular vs. Light Beer

The chemical conversion of polymeric starch molecules to dextrins and maltose is a key part of the brewing step in the production of beer. The conversion is catalyzed by amylase enzymes (carbohydrases) which are added to the grain mixture in the form of brewer's malt. Approximately one-third of the calories in regular beer are contributed by carbohydrate remaining in the beer after fermentation. The carbohydrates remain because amylase is unable to break down the 1,6-glycosidic linkages in the amylopectin portion of the starch mixture. To make a lighter beer, less starch is used in the initial brewing step and an enzyme, amyloglucosidase, which can cleave 1,6-glycosidic linkages in amylopectin, is added in the fermentation step. The final light beer has a lower carbohydrate composition because of the added amyloglucosidase and a lower alcohol content because of the reduced starch in the initial fermentation mixture.

Protein Hydrolysis

Proteases are enzymes which catalyze the hydrolysis of peptide bonds in proteins (Section 19.16). The protease papain is often injected into slaughtered animals to tenderize the meat by partially breaking down muscle protein. The same enzyme can also be applied to the surface of the meat and is the active ingredient of meat tenderizers for consumer use. Papain is also used in beer to digest protein material which might precipitate and give a cloudy appearance when the beer is chilled. In the brewing industry this is called "chillproofing" the beer.

Rennin, a protease enzyme taken from the stomach of calves, is added to milk to separate the so-called curds and whey. The curd is precipitated protein; rennin hydrolyzes κ-casein, one of the four proteins making up the major portion of milk proteins. The cleavage of κ-casein breaks up the solubilizing interaction among the milk proteins and causes the casein to precipitate and form curds. The whey is a water solution of all truly soluble components. Precipitated protein curd is used to make cheese.

21.14 HORMONES

Hormones are chemical messengers secreted directly into the bloodstream by organs known as the endocrine glands, which are shown in Figure 21.14. There is a hierarchy among the endocrine glands in human beings such that ultimate control comes from the hypothalamus, which is located in the brain and receives its major input information from the central nervous system. The hypothalamus secretes hormonal messages to the anterior and posterior pituitary glands located directly below the hypothalamus, which in turn release hormones which stimulate specific endocrine glands throughout the body to secrete their hormones. The pituitary glands also secrete hormones

FIG. 21.14
Human endocrine glands manufacture hormones.

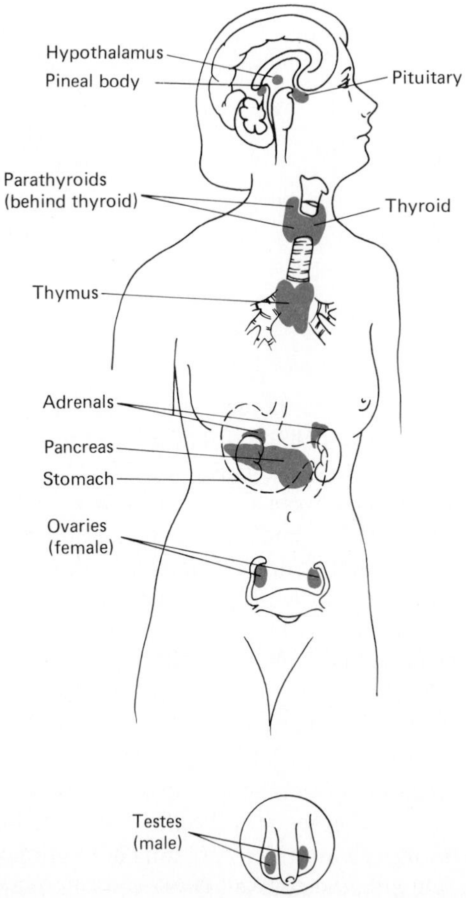
Hypothalamus
Pineal body
Pituitary
Parathyroids
(behind thyroid)
Thyroid
Thymus
Adrenals
Pancreas
Stomach
Ovaries
(female)
Testes
(male)

which directly affect tissues. The tissues affected by hormones are designated **target tissues.**

Hormones travel in the blood until they meet a target tissue, where they interact by a complementary fit with a receptor site either at the surface of a cell membrane or, for lipid-soluble hormones, within the cell membrane. As with enzymes, three-dimensional shape and fit play a significant role in the physiological function of hormone molecules.

Hormones can be classified into three categories based on their composition.

1 Polypeptide or protein hormones—insulin, vasopressin, glucagon, etc.
2 Nonpeptide hormones containing the amine group—epinephrine (adrenaline), thyroxine, etc.
3 Lipid-soluble or steroid hormones (See Table 18.4.)

Table 21.5 lists and classifies some hormones by the major bodily endocrine glands and tells their target tissues and functions.

The roles of hormones and enzymes are intertwined because a major function of hormones is the activation of an inactive enzyme. Often hormones set off a series of reactions beginning at the cell membrane and continuing inside the cell. Hormone regulation of cellular reaction proceeds in this way. For example, the binding of the hormone epinephrine to a receptor site on the cell membrane leads to activation of the enzyme adenylate cyclase which is bound at the inner surface of the cell membrane. Active adenylate cyclase is the enzyme required for the conversion of ATP into cyclic AMP (cAMP).

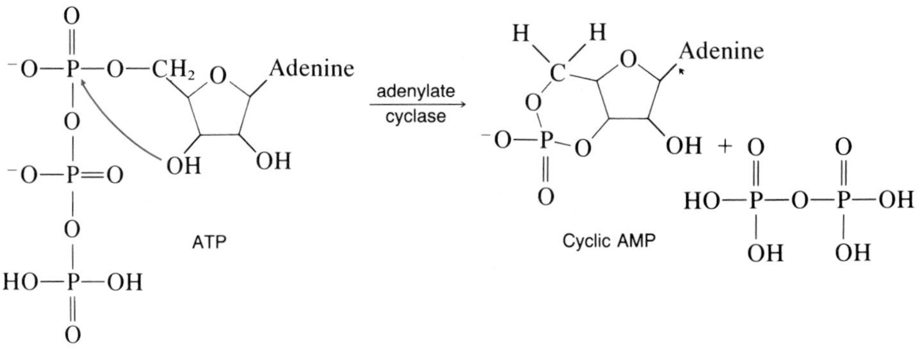
Adenine
OH
OH
adenylate
cyclase
Adenine
OH
ATP
Cyclic AMP

Cyclic AMP carries the activation message from the hormone at the cell membrane into the cell's cytoplasm, where it in turn activates a particular enzyme depending on the nature of the target cell. Because it picks up and carries the message from the initial primary hormone messenger to the enzyme, cyclic AMP is known as a *secondary messenger*. Figure 21.15 uses the hormone epinephrine to summarize this series of activation steps. Besides epinephrine, corticotropin, glucagon, thyroid-stimulating hormone, and vasopressin also exert their effects through the formation of the intracellular secondary messenger cyclic AMP. In Chapter 23 you will see a specific example wherein cyclic AMP activates the enzyme protein kinase, which goes on to stimulate enzymes involved in carbohydrate metabolism.

Varying hormone concentrations regulate cell reactions. A drop in the concentration of a stimulating hormone, such as epinephrine, results in a cutback in the synthesis of cyclic AMP. Ordinarily, the cyclic AMP remaining in the cell after delivering its activation message is destroyed by an enzyme, phosphodiesterase. However, alkaloids such as caffeine inhibit

TABLE 21.5 A Few Endocrine Glands of Human Beings and Their Hormones

Site of Secretion	Hormone	Chemical Classification of Hormone	Target Tissue	Major Functions
Adrenal cortex	Aldosterone	Steroid	Many tissues	Regulation of Na^+ and K^+
	Cortisone	Steroid	Liver	Regulation of carbohydrate and protein metabolism
Adrenal medulla	Epinephrine	Amine	Liver, skeletal muscle, heart tissue	Glucose metabolism, regulation of cardiovascular system
Ovary	Estradiol	Steroid	Many tissues	Secondary sex characteristics in female
	Progesterone	Steroid	Uterus, mammary glands	Preparation of uterus for implantation of ovum
Pancreas	Insulin	Protein	All tissues	Decrease of blood glucose level
Parathyroid	Parathyroid hormone	Polypeptide	Bones, kidney	Regulation of calcium and phophorus levels in blood
Pituitary, anterior	Corticotropin (ACTH)	Polypeptide	Adrenal cortex	Controls synthesis and secretion of adrenal cortex steroids
	Follicle-stimulating hormone (FSH)	Polypeptide	Ovary	Development of follicles, secretion of estrogen and ovulation
			Testis	Spermatogenesis
	Somatotropin, growth hormone (GH)	Polypeptide	All tissues	Growth of bone and muscle; metabolism of carbohydrates and lipids
	Thyrotropin, thyroid-stimulating hormone (TSH)	Polypeptide	Thyroid	Formation and secretion of thyroid hormones
Pituitary, posterior	Oxytocin	Polypeptide	Uterus; other smooth muscle	Contraction
	Vasopressin	Polypeptide	Kidneys, arteries	Blood pressure, water reabsorbtion
Testis	Testosterone	Steroid	Many tissues	Development of secondary sex characteristics
Thyroid	Thyroxine and triiodothyronine	Amino acid	Most tissues	Metabolic rate and oxygen consumption

phosphodiesterase. Thus, caffeine exerts its stimulant action by maintaining elevated levels of cAMP and thereby prolonging the effect of a hormone such as epinephrine.

The hormone insulin binds to specific receptor sites on muscle and fat cells. The molecular consequences of this binding are not understood, but it is known that the binding of insulin leads to a change in the permeability of the cell membrane to glucose molecules and small ions. A counterpart of cyclic AMP, that is, an agent that takes a message from insulin to the cell interior, may be involved, but none has as yet been identified. Greater understanding of the sequence of molecular events that enhance membrane permeability could have vast beneficial consequences in the medical treat-

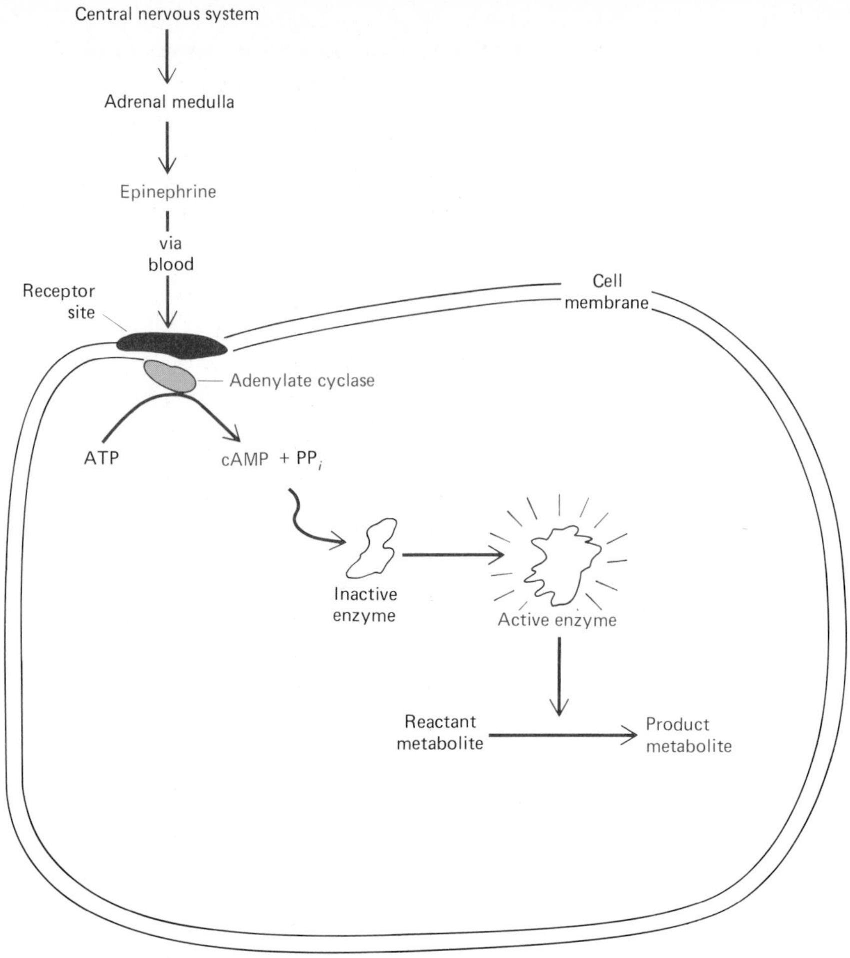
Central nervous system
Adrenal medulla
Epinephrine
via
blood
Receptor
site
Cell
membrane
Adenylate cyclase
ATP
cAMP + PP$_i$
Inactive
enzyme
Active enzyme
Reactant
metabolite
Product
metabolite

FIG. 21.15
Upon an appropriate stimulus, a hormone such as epinephrine travels from its endocrine gland via the blood to the cell of a target tissue. It does not enter the cell. Rather it fits into a receptor site and thereby causes an enzyme to produce a secondary chemical messenger inside the cell. In this case the secondary messenger is cyclic AMP. The secondary messenger activates an enzyme inside the cell and thereby promotes the occurrence of a metabolic reaction. Frequently there are several intermediate activations of enzymes before the actual desired reaction occurs.

ment of diabetes. The role of insulin in glucose and fat metabolism will be discussed more fully in Section 24.10.

In contrast to the extracellular mechanisms described so far, the lipid-soluble hormones are soluble in the cell membrane and so are able to enter the cell, where they bind to protein-type receptors in the cytoplasm. The hormone-receptor complexes travel to the cell nucleus, where in some way they activate a gene so that the synthesis of a specific protein is stimulated. Estradiol, for example, activates the synthesis of proteins that thicken the uterine lining in preparation for the deposition of a fertilized ovum.

21.15 ANTIBODIES

Enzyme-substrate complexes and hormone-receptor site interactions are but two illustrations of the significance of lock-and-key fit in physiological function. The mechanism of antibody action provides us with another example of the dependence of physiological function on a complementary fit between two molecules.

Antibodies are large protein molecules which are soluble in blood serum. They function in the defense (immune) system of human beings and other vertebrates by binding to an invading foreign molecule or group of molecules, such as a bacterium or virus. The foreign molecule that provokes the antibody is called an **antigen,** and when they bind together, the combination is termed an **antibody-antigen complex.** The antibody-antigen complex is generally insoluble in the blood plasma and is engulfed and destroyed by other types of cells in the immune system.

Antibodies are not synthesized in the body until an antigen is introduced. Each antigen causes the formation of a specific antibody. The molecular mechanism by which the synthesis takes place is not completely understood but it probably involves binding of the antigen to a specific β-lymphocyte, the cell type which produces antibodies. The antigen binding causes the cell to divide and increase the production of the specific antibody of that β-lymphocyte.

Antibodies are composed of two distinguishable polypeptide chains: one of high molecular weight, dubbed an H (heavy) chain, and one of lower molecular weight, called an L (light) chain. Each antibody has two H and two L chains arranged in a forked structure, as shown in Figure 21.16. The amino acid sequences within L or H chains are the same for different antibodies except for small regions at the ends of the fork. These regions are marked in Figure 21.16. The regions of varying amino acid sequence are believed to be the antigen binding sites and provide the high degree of specificity of a particular antibody toward a specific antigen.

The attractive forces that bind antibody to antigen are the relatively weak, noncovalent interactions: hydrogen bonding, electrostatic attractions, and hydrophobic interactions. As you have seen, they are also responsible for protein tertiary structure and for enzyme-substrate and hormone-receptor site interactions.

It is possible to manipulate the antibody defense system and provide immunity against a bacterial or viral disease to which a person is not presently immune. A bacterial causative agent of a disease is killed and then injected into a human being. The dead bacterium acts as an antigen and elicits the formation of antibodies specific for that bacterium. The antibodies remain

FIG. 21.16
Antibodies are composed of four polypeptide chains: two identical larger chains, designated H (heavy) chains, and the two identical L (light) chains. The amino acid sequences are essentially the same for all antibodies, except for the ends designated in color. The variations in amino acids in the variable regions impart specificity for an antigen to an antibody.

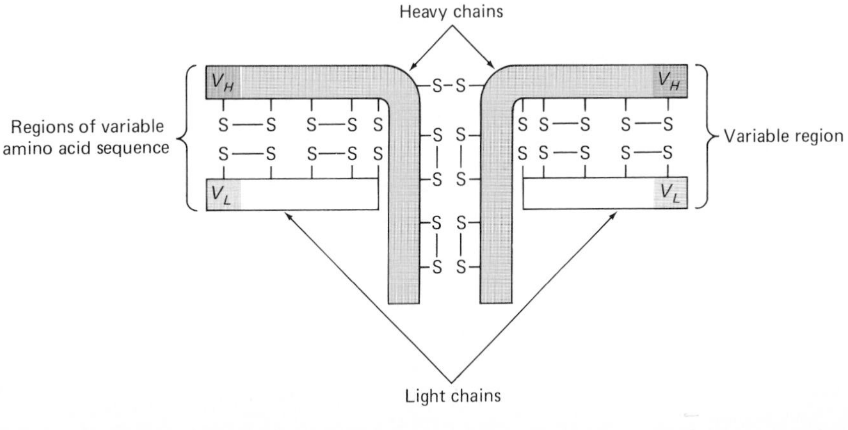
Heavy chains
Regions of variable amino acid sequence
Variable region
V_H
V_H
V_L
V_L
Light chains

TABLE 21.6 Blood Types

Blood Type	Antigen on Red Blood Cells	Antibody in Serum	Incompatible Donor Type*
A	A	Anti-B	B, AB
B	B	Anti-A	A, AB
AB	A,B	None	None
O	H	Anti-A, anti-B	A, B, AB

* Patients with the blood type given in Column 1 cannot receive these types. AB persons are universal acceptors.

in the blood serum, and the individual will be resistant to any future encounter with a live bacterium. For example, the injection of dead *Hemophilus pertussis,* the microorganism responsible for whooping cough, leads to immunity against this disease, because the body develops antibodies to fight the whooping cough microorganisms. Similarly, antibodies to microbiological toxins can be developed by injection of deactivated toxins. The toxins of the diphtheria or tetanus bacilli are made nontoxic by treatment with formaldehyde, and innoculation with the treated toxin brings about the production of antibodies which protect the individual against subsequent contact with untreated toxin.

Antibody-antigen reactions can be a problem in tissue transplants between organisms. Tissue cells contain antigens on their surfaces which the receiving organism may treat as a foreign invader and therefore reject the transplanted tissue. A similar problem can occur in whole blood transfusions. Human blood can be classified into four groups: A, B, AB, and O. Each blood type is distinguished by the antigens present on the red blood cells and the antibodies present in the serum. Let us examine how blood groups interact.

Type A blood contains antibodies which will combine with and precipitate B antigens. Thus, a patient with type A blood must not be transfused with whole blood of type B or AB because both B and AB blood contain B antigens. These antigens would be precipitated by the antibodies in the patient's type A blood, and furthermore, the B antigens would stimulate the patient's immune system to produce more anti-B antibodies. Patients can always receive their own type of blood, and type O blood can be given to all patients with type A, B, or AB blood because O blood does not contain antigens of either the A or B variety. For this reason, people with type O blood are sometimes called universal donors. Serum, which is devoid of red blood cells, and thereby of antigens, can be used in any transfusion providing that it does not contain an abnormally high concentration of soluble antibodies. Table 21.6 summarizes the compatibility of blood types.

CHAPTER ACCOMPLISHMENTS

After completing this chapter you should be able to define **key words** and do the following:

21.1 INTRODUCTION

1. State the function of enzymes in physiological chemistry.

21.2 ENZYME COMPOSITION

2. Describe the chemical nature of each of the potential component parts of a holoenzyme.

21.3 ENZYME CLASSIFICATION AND NOMENCLATURE

3. Relate a systematic enzyme name to the substrate acted upon.
4. State the class of enzyme required for a given reaction type.

21.4 MECHANISM OF ENZYME ACTIVITY

5. Describe the experimental evidence for the existence of enzyme-substrate complexes.
6. List the two steps postulated for the enzyme-catalyzed conversion of a substrate into product.

21.5 SUBSTRATE SPECIFICITY AND THE ENZYME-SUBSTRATE COMPLEX

7. Distinguish among absolute specificity, group specificity, linkage specificity, and stereochemical specificity.
8. Describe the lock-and-key theory of enzyme action.
9. Contrast the induced-fit theory with the lock-and-key theory.

21.6 HOW THE ENZYME-SUBSTRATE STRUCTURE ENHANCES RATE

10. State three possible ways by which an enzyme-substrate structure may enhance the rate of a reaction.

21.7 FACTORS AFFECTING ENZYME CATALYSIS

11. Explain how pH and temperature affect enzyme catalysis.

21.8 ENZYME INHIBITION

12. Describe and distinguish among irreversible, competitive, and noncompetitive inhibition.

21.9 ANTIMICROBIALS

13. Describe the role of antimetabolite drugs as competitive inhibitors.
14. Describe the action of penicillin as an antibiotic.

21.10 REGULATION OF ENZYME ACTIVITY

15. Describe the process of enzyme regulation through control of protein synthesis, zymogen formation, and allosterism.

21.11 COENZYMES AND VITAMINS

16. State the relationship between vitamins and enzyme activity.
17. Describe the role of a coenzyme in an enzyme-catalyzed reaction.

21.12 MEDICAL USES OF ENZYMES

18. Explain the usefulness of enzymes in medical diagnosis.
19. State a use of enzymes in medical treatment.

21.13 ENZYMES AND FOOD SCIENCE

20. Describe several uses of enzymes in food processing.

21.14 HORMONES

21. Describe the general role of hormones in the body.
22. Explain the relationship among hormones, cyclic AMP, and enzyme regulation.
23. Describe the function of the lipid-soluble hormones.

21.15 ANTIBODIES

24. Describe the role of antibodies in the body.
25. State the similarity in physiological function among enzymes, hormones, and antibodies.
26. Describe how a vaccine containing inactive bacteria can provide immunity against an active bacterial disease.
27. Explain why precautions are necessary in the selection of the type of blood used in a transfusion.

PROBLEMS

21.1 How do enzymes differ from inorganic catalysts in (*a*) composition and (*b*) reactions catalyzed?

21.2 What is meant by the specificity of an enzyme?

21.3 What problem might result if high temperatures or strong acidic or basic conditions had to be used to increase the rate of a reaction in a cell of the body?

21.4 List the two key chemical properties of an enzyme.

21.5 Describe the relationship between a holoenzyme and an apoenzyme, a cofactor, a metal ion, and a coenzyme.

21.6 *a* What is the compositional difference between an apoenzyme and a coenzyme?
b Which one is physically larger?
c What function does each serve?

21.7 What component of a holoenzyme is affected in the denaturation of an enzyme?

21.8 Give the name of the substrate which each of the following enzymes most likely acts on:

a Lactase   d Carbonic
b Glucosidase   anhydrase
c Acetylcholin-   e Ribonuclease
   esterase

21.9 Give the class of enzyme which would catalyze each of the following reactions:

a $CH_3CH_2-\overset{\displaystyle O}{\overset{\|}{C}}-OCH_3 + H_2O \xrightarrow{\text{enzyme}}$ 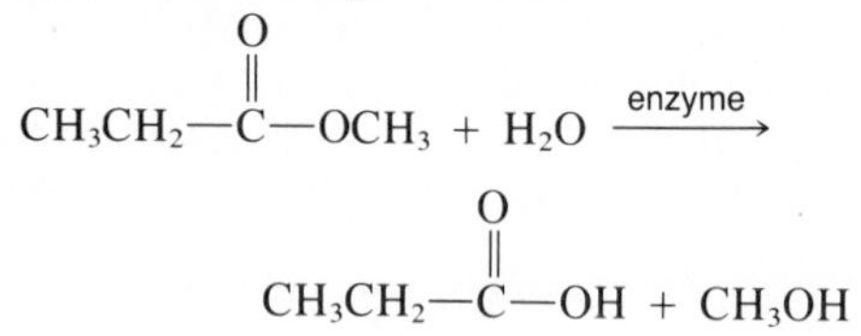

$CH_3CH_2-\overset{\displaystyle O}{\overset{\|}{C}}-OH + CH_3OH$

b $HO-\overset{\displaystyle O}{\overset{\|}{C}}-\overset{\displaystyle H}{\overset{|}{\underset{|}{\underset{\displaystyle H}{C}}}}-\overset{\displaystyle OH}{\overset{|}{\underset{|}{\underset{\displaystyle COOH}{C}}}}-CH_2-\overset{\displaystyle O}{\overset{\|}{C}}-OH \xrightarrow{\text{enzyme}}$ 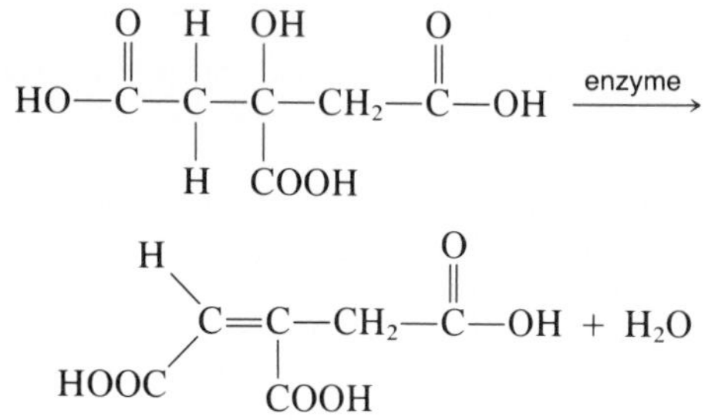

$\overset{\displaystyle H}{\underset{\displaystyle HOOC}{C}}=\overset{}{\underset{\displaystyle COOH}{C}}-CH_2-\overset{\displaystyle O}{\overset{\|}{C}}-OH + H_2O$

c $HOOC-\overset{\displaystyle OH}{\overset{|}{\underset{|}{\underset{\displaystyle H}{C}}}}-CH_2-COOH \xrightarrow{\text{enzyme}}$ 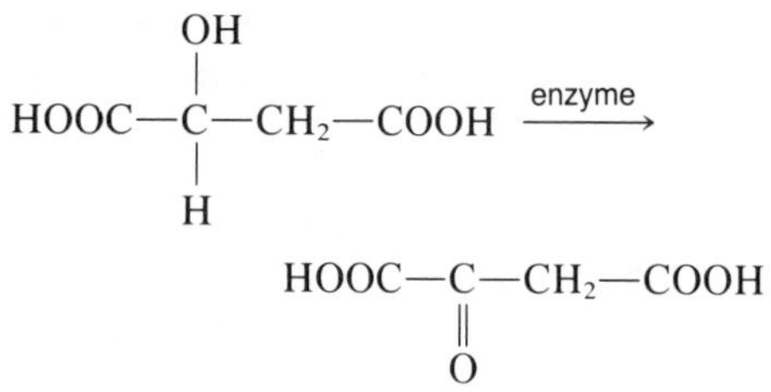

$HOOC-\overset{}{\underset{\|}{\underset{\displaystyle O}{C}}}-CH_2-COOH$

d 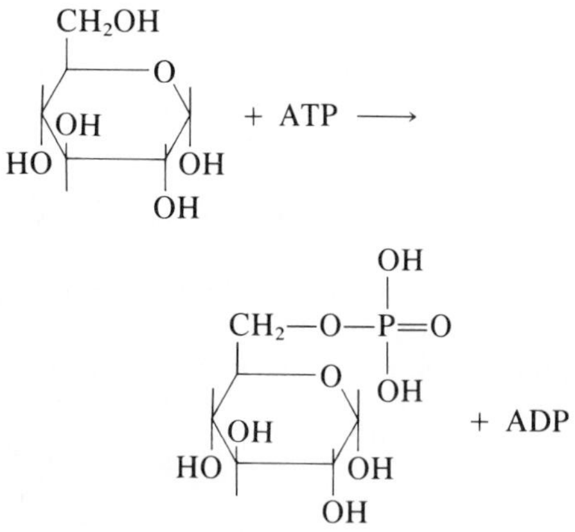

+ ATP $\longrightarrow$

+ ADP

21.10 a What chemical change is catalyzed by a carbohydrate hydrolase?
b What chemical change is catalyzed by a hydrase?

21.11 a Using the symbols E for enzyme, S for substrate, and P for product, list the two steps in the postulated general pathway for enzyme catalysis.
b Which is the slow step in this pathway?
c The rate of the overall reaction depends on the concentration of _____.

d Why does the rate of an enzyme-catalyzed reaction level off to a constant value when the enzyme concentration is held constant and the substrate is continuously increased?

21.12 An enzyme in which class of specificity—absolute, group, or linkage—would be most limited in its catalytic scope?

21.13 Distinguish between absolute specificity and linkage specificity. Give an example of each to illustrate your answer.

21.14 Distinguish between group specificity and linkage specificity. Give an example of each to illustrate your answer.

21.15 a An enzyme catalyzing the hydrolysis of all phosphodiester linkages would fit into which specificity class?
b An enzyme which only catalyzed the rearrangement of a substance X into a substance Y would fit into which specificity class?

21.16 In what way(s) does the catalytic action of an enzyme resemble the action of a lock and key?

21.17 What role do the hydrophobic, hydrogen-bonding, and ionic attractive forces play in the lock-and-key fit between an enzyme and substrate?

21.18 How does the lock-and-key theory of enzyme action account for enzyme specificity?

21.19 a Describe how the induced-fit theory differs from the lock-and-key theory of enzyme action.
b Which examples of enzyme specificity are best accounted for by the lock-and-key theory?
c Which example of enzyme specificity might best model the induced-fit theory?

21.20 Describe the three ways by which an enzyme may enhance the rate of conversion of a substrate into product.

21.21 What general statement can be made about reactions in which enzyme enhancement of the rate is at least partly achieved by a proximity effect?

21.22 Explain how pH changes, which *do not* denature the protein component of the enzyme, may still alter the rate of an enzyme-catalyzed reaction.

21.23 Although reaction rate usually increases with increasing temperature, the rate of an enzyme-controlled reaction typically decreases above 37°C. Explain.

21.24 The enzyme pepsin functions in the stomach. Predict a possible optimum pH for this enzyme.

21.25 a Define what is meant by an enzyme inhibitor.
b Compare the generality of inhibitor effects to pH and temperature effects on the rate of an enzyme-catalyzed reaction.
c Give two examples of enzyme inhibitors.

21.26 Distinguish among irreversible, competitive, and noncompetitive inhibition.

21.27 *a* Describe how a competitive inhibitor slows the rate of an enzyme-catalyzed reaction.
b Explain how increasing the concentration of substrate reverses the effect of a competitive inhibitor.

21.28 Discuss how ethanol functions as an antidote for ethylene glycol poisoning.

21.29 *a* Explain on a molecular level how heavy metal ions, such as Pb^{2+}, may function as poisons.
b By what type of enzyme inhibition process do the heavy metal ions function?
c How may an antidote for heavy metal poisoning function? What chemical principle is being taken advantage of here?
d Which type of enzyme inhibition poisoning do you think would be *most* difficult to treat medically? Why?

21.30 Why is the binding of a noncompetitive inhibitor to an enzyme site *not* overcome by an increase in substrate concentration?

21.31 What properties (structural and chemical) would you seek in designing a new drug which will function as an antimetabolite?

21.32 *a* How does penicillin function as an antibiotic?
b Why are human beings not also affected by this action?
c How do bacteria become resistant to penicillin?

21.33 What chemical property of the original penicillins prevented them from being taken orally?

21.34 Although drugs functioning as bacterial antimetabolites often structurally resemble a bacterial substrate, this is not true of antibiotics. Explain, based on the functional role of antimetabolites and antibiotics.

21.35 Describe how you would experimentally distinguish between competitive and noncompetitive inhibition of an enzyme-catalyzed reaction.

21.36 List four general ways by which enzyme activity is regulated.

21.37 What is the relationship between a zymogen and an active enzyme?

21.38 What is the relationship between the concentration of a substrate and the regulation of an induced enzyme?

21.39 Illustrate the role of zymogen $\longrightarrow$ active enzyme conversion in the blood-clotting process.

21.40 Allosteric regulation is said to provide a "finer" control of enzyme activity than either protein synthesis or zymogen formation. Discuss why this is the case.

21.41 Can increasing the substrate concentration overcome the feedback inhibition present in allosteric regulation? Explain your answer.

21.42 Why can't allosteric regulation be an irreversible process?

21.43 *a* What is the general function of coenzymes?
b Give two specific examples of this type of function.

21.44 *a* What vitamin is incorporated in coenzyme A?
b In NAD^+?

21.45 In the reaction

$$\underset{\text{Substrate}}{HSH} + NAD^+ \longrightarrow S + NADH + H^+$$

a What type of chemical change has NAD^+ undergone?
b What chemical change has occurred to the substrate HSH?
c What is the original source of the substrate HSH?

21.46 Describe how the concentration of enzymes in blood serum can provide information about particular disease conditions.

21.47 Why might the blood serum concentration of an enzyme such as LDH increase after a myocardial infarction (heart attack)?

21.48 Table 21.4 shows that the serum LDH concentration is increased in heart disease or liver disease. What enzyme assay can be done to distinguish between these two conditions?

21.49 *a* What peaks in the concentration of the LDH isoenzymes distinguish a myocardial infarction from a normal result?
b What peaks in the concentration of the LSH isoenzymes distinguish heart from liver disease?

21.50 Tests for sugar in the urine such as "Multistix" reagent strips contain glucose oxidase. What functional role does glucose oxidase play in the test?

21.51 What general physiological role do hormones play in the body?

21.52 *a* Discuss the compositional differences between hormones and enzymes.
b In what way are the roles of hormones and enzymes connected?

21.53 *a* Describe the classification scheme for hormones.
b Classify the hormones progesterone, insulin, cortisone, thyroxine, and tyrosine into the categories you listed in part *a*. Use the index of this text to help you find the necessary structures.

21.54 Describe the structural differences between cyclic AMP and ATP.

21.55 Caffeine functions as a stimulant. Provide an explanation on a molecular level for caffeine's stimulant action.

21.56 Give a hypothesis relating the necessity of insulin in glucose metabolism.

21.57 Describe a structural similarity in the mode of action for enzymes, hormones, and antibodies.

21.58 Describe on a molecular level how a vaccine may provide immunity against a bacterial or viral disease.

CHAPTER TWENTY-TWO

BIOCHEMICAL ENERGY TRANSFER

22.1 INTRODUCTION

We commonly talk about food supplying us with energy. We also describe ourselves as using our energy in work or play. We are able to move about and exhibit kinetic energy because our bodies can convert the stored chemical energy into kinetic energy.

The principles by which energy is converted from one form to another or transferred within the body are the same thermodynamic principles discussed in Chapter 10. Now that you are familiar with the structures of biological molecules, especially the foodstuffs, we can explore the details of important bodily reactions that keep us "running."

Digestion of food produces molecules which are small enough to cross cell membranes. Once inside the cell these food molecules are subject to cellular reactions which break them down further and produce energy. The series of reactions which transform food molecules and produce energy changes and maintain body tissues is called **metabolism**. There are two categories of metabolic activity. **Catabolism** is the breakdown of larger molecules into smaller ones. **Anabolism** is the putting together of larger molecules from smaller ones.

Metabolism

$$\text{Larger molecules} \underset{\textbf{anabolism}}{\overset{\textbf{catabolism}}{\rightleftharpoons}} \text{smaller molecules} \quad + \quad \text{energy}$$

22.2 THERMODYNAMIC PRINCIPLES

Let us begin the study of cellular reactions by reviewing and expanding upon some essential ideas about the energetics of chemical reactions (see also Sections 7.17 and 10.4). All chemical reactions involve energy changes. Energy may be released; that is, it may be a "product" of a chemical reaction:

$$\text{Reactants} \longrightarrow \text{products} \quad + \quad \text{energy}$$

This is the exergonic case: energy is released because the chemical products have a lower energy content than the reactants. Because the change in free energy is defined as $G_{\text{products}} - G_{\text{reactants}}$, ΔG for an exergonic reaction is always negative.

Alternatively, energy may be absorbed; that is, it may be a "reactant" in a chemical reaction:

Reactants + energy ⟶ products

This is the endergonic case wherein energy must be supplied in order to form the chemical products which have a higher energy content than the reactants. ΔG for an endergonic reaction is always positive.

Problem 22.1 Identify the following reactions as exergonic or endergonic:

a $CH_3-C(=O)-H + H_2 \rightarrow CH_3CH_2OH + 10\ \text{kcal}$

b $2CH_3OH(l) + 3O_2(g) \rightarrow 2CO_2(g) + 4H_2O(l) \quad \Delta G = -336\ \text{kcal}$

c $Fe^{2+} \rightarrow Fe^{3+} + e^- \quad \Delta G = +17.8\ \text{kcal}$

d $H_2N-C(=O)-NH_2(aq) + H_2O(l) + 3\ \text{kcal} \rightarrow CO_2(g) + 2NH_3(g)$

If a reaction is exergonic in one direction, then the reverse of that reaction is endergonic. A reaction and its exact reverse have ΔG of the same magnitude, but opposite sign. For example, consider the equilibrium reaction by which ammonia is formed or broken down.

$$N_2(g) + 3H_2(g) \rightleftharpoons 2NH_3(g) + 8.0\ \text{kcal}$$

The formation of ammonia (the reaction from left to right →) is exergonic; $\Delta G = -8.0$ kcal for the reaction as it is written (-4.0 kcal/mol of NH_3). The reverse reaction is endergonic and $\Delta G = +8.0$ kcal for the decomposition of 2 mol of ammonia ($+4.0$ kcal/mol of NH_3).

Problem 22.2 Indicate the value (sign and magnitude) of ΔG for both the forward (→) and reverse (←) of the reactions given in Problem 22.1.

Exergonic reactions are described as spontaneous, whereas energy must be supplied for an endergonic reaction to occur. Energy to promote an endergonic reaction may be supplied by the energy released in an exergonic reaction. When this occurs, we say the two reactions are *coupled*. The exergonic member of the couple drives the endergonic member.

For example, the formation of water from its elements is highly exergonic ($\Delta G = -56.7$ kcal/mol), whereas the formation of acetylene (C_2H_2) is quite endergonic ($\Delta G = +50.0$ kcal/mol).

$$H_2(g) + \tfrac{1}{2}O_2(g) \longrightarrow H_2O(l) + 56.7\ \text{kcal}$$

$$2C(s) + H_2(g) + 50.0\ \text{kcal} \longrightarrow C_2H_2(g)$$

These reactions can be coupled; that is, the energy released as water forms can be used to fuel the formation of acetylene. The coupling can be represented

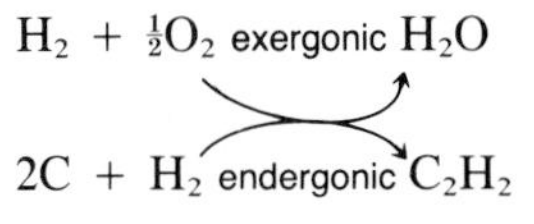
H2 + ½O2 exergonic H2O
2C + H2 endergonic C2H2

As the exergonic reaction occurs, energy is transferred to the endergonic system. This transfer is shown by the intersection of arrows. Catabolic

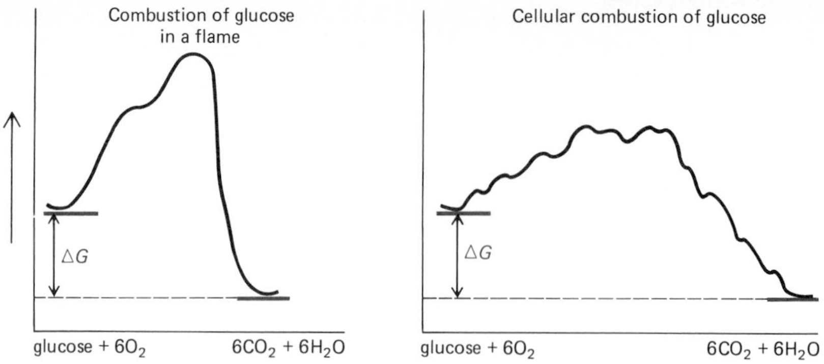
Combustion of glucose
in a flame
ΔG
glucose + 6O2
6CO2 + 6H2O
Cellular combustion of glucose
ΔG
glucose + 6O2
6CO2 + 6H2O

FIG. 22.1
The diagrams show schematically that combustion of glucose leads to the same liberation of free energy ($\Delta G = -686$ kcal/mol) whether glucose is burned in a flame or by a more complex process in the body. ΔG depends only on the differences in free energy between initial and final states. Different pathways will be reflected in different rates.

processes are usually exergonic and anabolic processes are generally endergonic. We will see, therefore, that these processes are coupled in the cell.

One last thermodynamic principle to review is the idea that the value of ΔG depends only on the initial and final states, that is, the difference in free energy between products and reactants. ΔG is independent of the pathway between reactants and products. As Figure 22.1 shows, ΔG is the same for a one-step transition or a multistep transition between the same reactants and products. The particular pathway, or the number of steps by which a reaction proceeds, affects only the rate of the reaction.

22.3 ENERGY FROM THE SUN

The ultimate source of all energy (other than nuclear energy) is the sun. The sun's energy is converted into "food energy" through the process of photosynthesis. In this process carbon dioxide from air reacts with water in air and the soil through the catalysis of chlorophyll complexes in plants to yield the carbohydrate glucose.

Photosynthesis

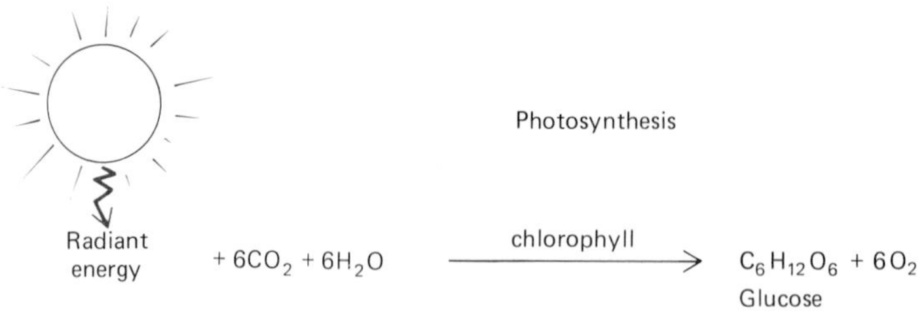
Photosynthesis
Radiant
energy
+ 6CO2 + 6H2O
chlorophyll
C6H12O6 + 6O2
Glucose

Glucose can then polymerize into plant polysaccharides.

Photosynthesis is a complex stepwise endergonic process. Because this is so, the reverse reaction, the oxidation of glucose is exergonic; in fact, $\Delta G = -686$ kcal/mol.

$$C_6H_{12}O_6 + 6O_2 \longrightarrow 6CO_2 + 6H_2O + 686 \text{ kcal}$$

Problem 22.3 How much energy must the sun provide to form 1 mol of glucose during photosynthesis?

When we say that carbohydrates give us energy, we mean that the body can oxidize glucose to produce 686 kcal/mol. We use this energy for body heat, motion, growth, and repair.

In Chapter 23 you shall see that in the body oxidation of glucose is a complex stepwise process. However, as Figure 22.1 demonstrated, the thermodynamics remain the same; whether simple combustion (burning) of glucose or complex biochemical oxidation of glucose to CO_2 and H_2O occurs, the total energy that may be released is 686 kcal/mol.

22.4 THE ROLE OF ATP

Now you know that glucose is a major source of bodily energy, but how is energy transferred (and momentarily stored) in the body? We cannot just "burn" glucose in the body as a direct energy source. This would be as inefficient and dangerous as using bonfires to keep our houses warm; too much heat would be generated at one time, rather than a steady, moderate amount of energy. We need an efficient system of energy "storage" and distribution just as we need controlled home heating systems. Molecules of *a*denosine *t*ri*p*hosphate (ATP) are the "storage" and distribution vehicles for energy in the body. ATP links together essentially all energy-consuming and energy-producing cellular reactions.

We have qualified with quotation marks the word *storage* as applied to ATP because ATP serves as the main *immediate* donor of free energy in the body, rather than as a static energy depot. Typically an ATP molecule is consumed as an energy source within a minute following its formation. That is, the turnover rate of ATP is very high; it is constantly being consumed and remade. On the other hand, you will see that we will frequently talk about replacing ATP reserves. This is because ATP is the *dynamic* reservoir of energy in the body. You will see this dynamic nature of the ATP reservoir in Figure 22.2.

Examination of the complete structure of ATP shows that it is a nucleotide composed of the nitrogenous base adenine, the sugar ribose, and three phosphate groups.

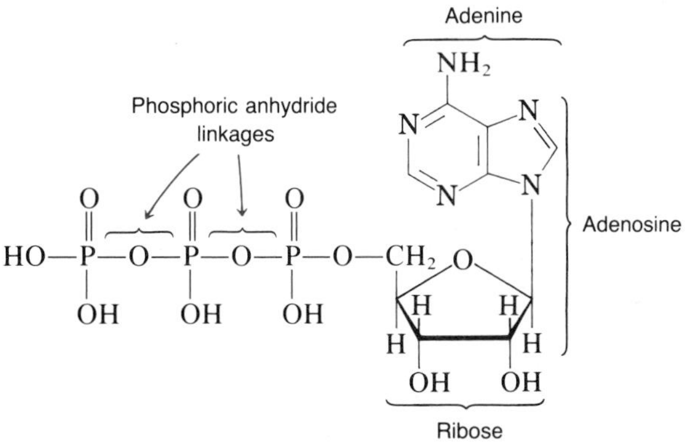
Adenine
Phosphoric anhydride linkages
Adenosine
Ribose

ATP is often abbreviated:

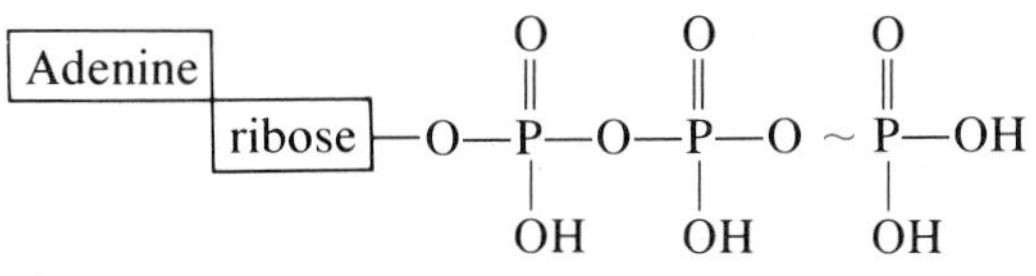
Adenine
ribose

This abbreviated structure is sufficient for use in explaining the role of ATP in energy transfer in the body, because the significant reactions of ATP involve cleavage or synthesis of a phosphoric anhydride bond, most often the one designated ~ in the abbreviated structure.

Hydrolysis of a phosphoric anhydride linkage is exergonic and proceeds much the same way as ester linkage hydrolysis (Section 17.13). For example, cleavage of the terminal phosphate group of ATP yields *a*denosine *di*phosphate (ADP) and 7.3 kcal/mol of energy; cleavage of the other phosphoric anhydride linkage (to form *a*denosine *m*ono*p*hosphate, AMP) releases less energy.

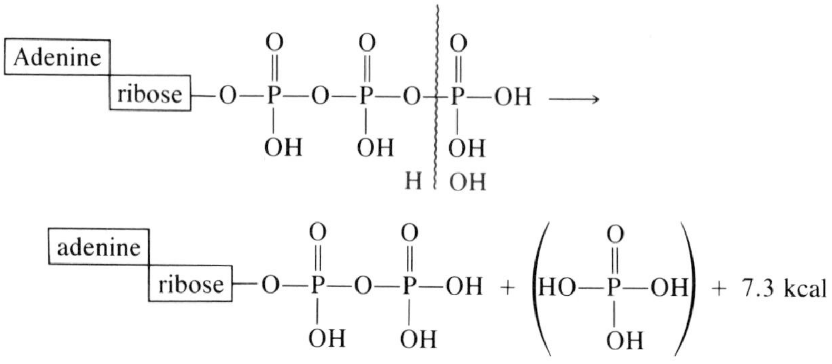
Adenine
ribose
adenine
ribose
+ 7.3 kcal

(22.1)

$$\text{ATP} + \text{HOH} \longrightarrow \text{ADP} + \text{P}_i + 7.3\ \text{kcal} \qquad (22.1)$$

The cleaved phosphate group is shown in parentheses or represented P_i (i for inorganic) because the exact species in solution depends on the pH. This is because the position of the multiple equilibria for the triprotic acid, H_3PO_4, depends on pH.

$$H_3PO_4 \rightleftharpoons H_2PO_4^- + H^+ \rightleftharpoons HPO_4^{2-} + 2H^+ \rightleftharpoons PO_4^{3-} + 3H^+$$

The equilibria shift further to the right as pH increases.

Whenever the body needs immediate energy, it breaks down ATP to ADP. Move your arm. You just cleaved some ATP molecules. The body must have a way to put ATP back together again or else we would quickly run out of these energy-rich and usable molecules. Note that the reverse of the reaction in Equation (22.1) must be endergonic; that is, 7.3 kcal of energy is required to combine ADP and P_i to make ATP. Because a phosphorus-oxygen bond is made when ADP + P_i form ATP, the process is termed *phosphorylation*. Some of the energy for making ATP is supplied by other phosphate compounds higher in energy than ATP in a process called substrate-level phosphorylation, but most of the energy is supplied by *oxidative* phosphorylation in the electron transport system. We will explore both these topics in this chapter. Figure 22.2 summarizes our energy uses and sources.

22.5 THE ROLES OF EATING AND BREATHING

The complex series of reactions which provide most of the energy for making ATP essentially involve the combination of hydrogen ions, electrons, and oxygen into water. The formation of water from its elements is highly exergonic.

Food is the source of hydrogen and the air we breathe is the source of oxygen for the exergonic formation of water in the electron transport system.

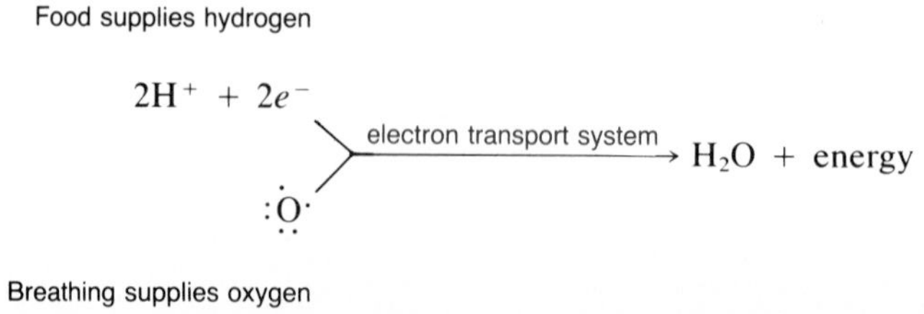
Food supplies hydrogen
2H⁺ + 2e⁻
electron transport system
H₂O + energy
Breathing supplies oxygen

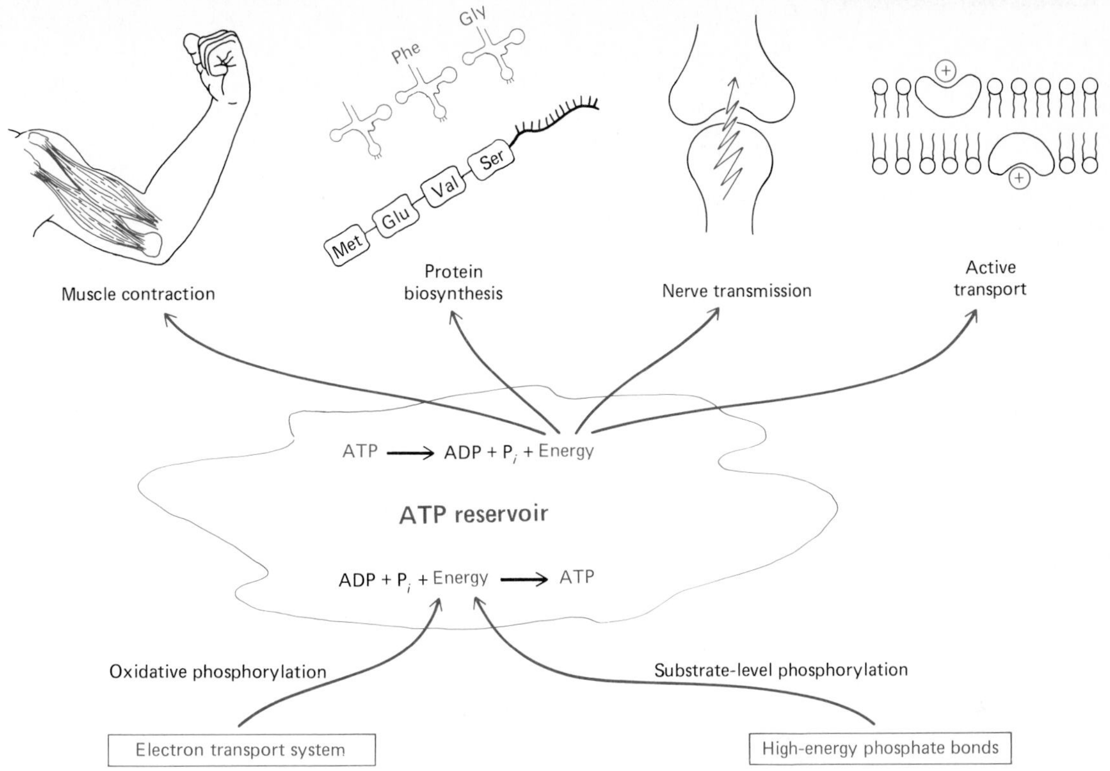
Phe
Gly
Met
Glu
Val
Ser
Muscle contraction
Protein biosynthesis
Nerve transmission
Active transport
ATP ⟶ ADP + P_i + Energy
ATP reservoir
ADP + P_i + Energy ⟶ ATP
Oxidative phosphorylation
Substrate-level phosphorylation
Electron transport system
High-energy phosphate bonds

FIG. 22.2
Many essential bodily activities require energy. Hydrolysis of ATP is the immediate source of that energy. The replacement of ATP is accomplished largely by oxidative phosphorylation. ATP molecules are short-lived because the two opposing processes, hydrolysis and phosphorylation, occur continuously.

Subsequent chapters will explore the ways in which hydrogen is removed from food. Now in Section 22.6 we turn to the details of the electron transport system, the series of reactions that produce water and energy to make ATP.

22.6 THE ELECTRON TRANSPORT SYSTEM

The **electron transport system** is a series of reactions involving food, coenzymes, and enzymes. During the course of these reactions, the overall result of which is the exothermic formation of water, energy is provided for making ATP. Figure 22.3 shows the full details of the electron transport system. It begins on the left with food, represented MH_2 or M, and ends on the right with the production of water. Along the way, you can see three places where energy is released to make ATP. The electron transport system is also called the **respiratory chain.** For convenience, we will use the abbreviation ETS for the electron transport system. Let us now examine each step in the ETS one at a time.

$MH_2 \longrightarrow M$

The first step at the far left of Figure 22.3 is removal of hydrogen from food. Food is represented as MH_2; after the H's are removed, the food molecule is simpiy represented M for metabolite, a participant in metabolism. ''Whole''

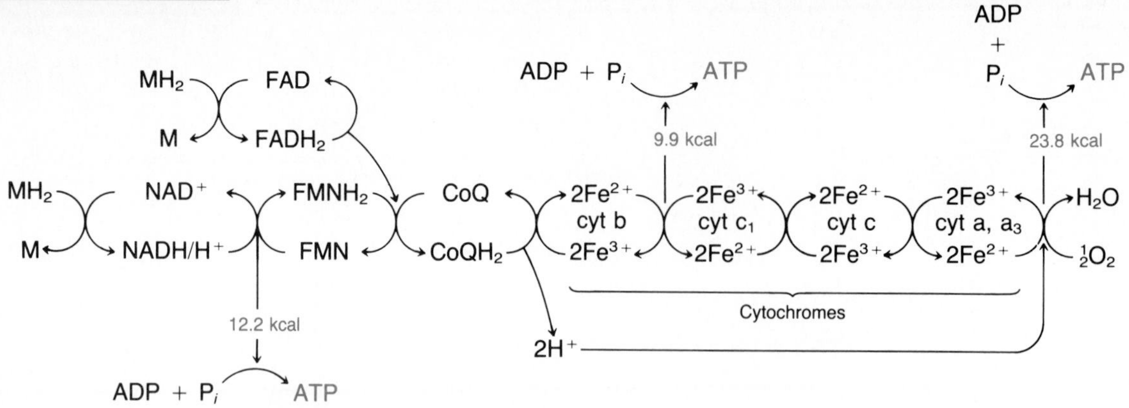

ADP
+
P_i
ATP
ADP + P_i
ATP
MH_2
FAD
M
$FADH_2$
9.9 kcal
23.8 kcal
MH_2
NAD^+
$FMNH_2$
CoQ
$2Fe^{2+}$
cyt b
$2Fe^{3+}$
$2Fe^{3+}$
cyt c_1
$2Fe^{2+}$
$2Fe^{2+}$
cyt c
$2Fe^{3+}$
$2Fe^{3+}$
cyt a, a_3
$2Fe^{2+}$
H_2O
M
$NADH/H^+$
FMN
$CoQH_2$
$\frac{1}{2}O_2$
Cytochromes
12.2 kcal
$2H^+$
ADP + P_i
ATP

FIG. 22.3
The electron transport system, also known as the respiratory chain. Note that ATP is produced at three sites. At these sites more than the minimum amount of energy (7.3 kcal/mol) necessary to synthesize ATP is released.

hydrogen atoms are being carried into the chain, that is, 2H·. Recall that in Lewis terms H· represents a hydrogen nucleus (a bare proton) surrounded by one electron. Recall also that stripping off electrons produces ions:

$$H\cdot + H\cdot \longrightarrow 2H^+ + 2e^-$$

2H· and $2\,H^+ + 2e^-$ represent the same numbers of protons and electrons.

Once hydrogen enters the chain, the H nuclei and/or their electrons are "passed along" the chain by successive reactions with the various coenzymes and enzymes in the chain. Each reaction is an oxidation-reduction reaction. Review Sections 5.6 and 7.12, which define oxidation and reduction in terms of the loss or gain of both electrons and hydrogen.

Hydrogen is brought into the ETS by one of two coenzymes, either nicotinamide adenine dinucleotide (NAD^+, Figure 22.4) or flavin adenine dinucleotide (FAD, Figure 22.5). Let us first explore the interactions of NAD^+ with metabolites and the rest of the ETS.

FIG. 22.4
Nicotinamide adenine dinucleotide (NAD^+). Oxidation-reduction reactions occur in the nicotinamide portion of the molecule. Nicotinamide is a derivative of the vitamin niacin. Nicotinamide adenine dinucleotide phosphate ($NADP^+$) has the same structure shown here, except that the starred OH group is phosphorylated.

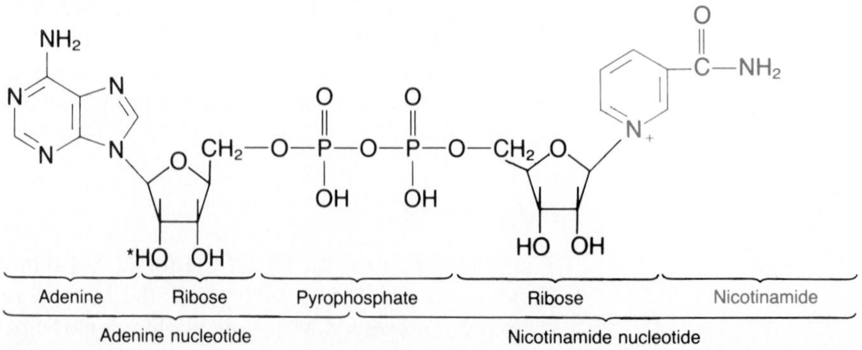

Adenine
Ribose
Pyrophosphate
Ribose
Nicotinamide
Adenine nucleotide
Nicotinamide nucleotide

NAD$^+$ $\longrightarrow$ NADH/H$^+$

NAD$^+$ is reduced, i.e., it gains hydrogen, and the metabolite is oxidized, i.e., it loses hydrogen, when the coenzyme and metabolite interact. We can represent this most simply as

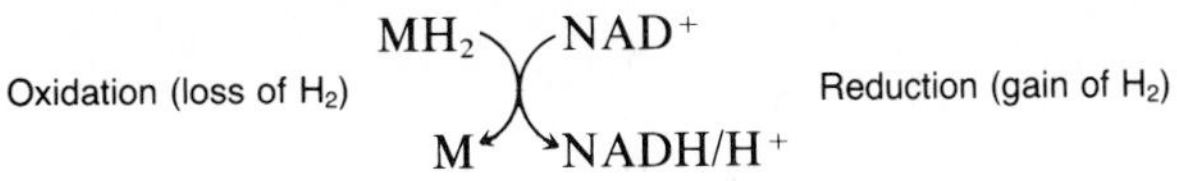

$$\text{Oxidation (loss of H}_2) \quad \underset{M}{\overset{MH_2}{\bigg\rangle}}\underset{NADH/H^+}{\overset{NAD^+}{\bigg\langle}} \quad \text{Reduction (gain of H}_2)$$

Oxidations of carbon-oxygen bonds, such as the oxidation of an alcohol to a ketone, are the typical metabolic oxidations for which NAD$^+$ is the required coenzyme. (Alcohol oxidation was discussed in Section 14.7.) For example, in the body, the oxidation of the —OH group of lactic acid to the keto group of pyruvic acid is catalyzed by lactic acid dehydrogenase and NAD$^+$ is required as a coenzyme. This can be represented as

(MH$_2$)
$$\underset{\substack{\text{H}}}{\overset{\substack{OH}}{CH_3{-}C{-}COOH}}$$

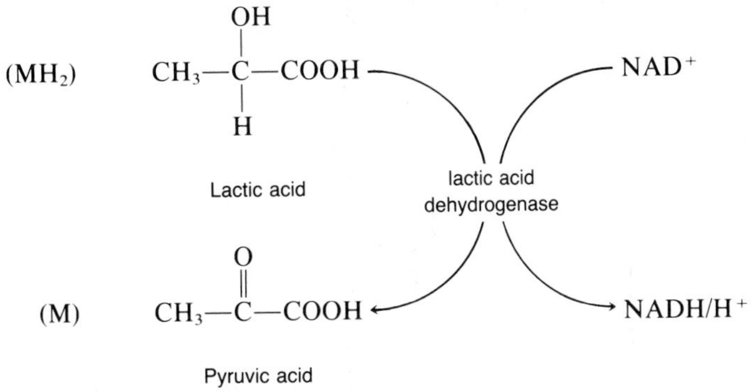

Lactic acid

lactic acid dehydrogenase

NAD$^+$

(M)
$$\overset{\substack{O}}{CH_3{-}\overset{\|}{C}{-}COOH}$$

Pyruvic acid

NADH/H$^+$

Problem 22.4 Write an equation for a coupled reaction which shows the reduction of pyruvic acid by the reduced form (NADH/H$^+$) of NAD$^+$.

If you examine the structural details of the reduction of NAD$^+$, you will find that it is the nicotinamide portion (see Figure 22.4) of the molecule that is involved. Therefore, it is convenient to represent the rest of the NAD$^+$ molecule, i.e., the nonnicotinamide part, simply as R:

$$\overset{\substack{O}}{\underset{\substack{N^+ \\ | \\ R}}{\overset{\|}{C}{-}NH_2}}$$

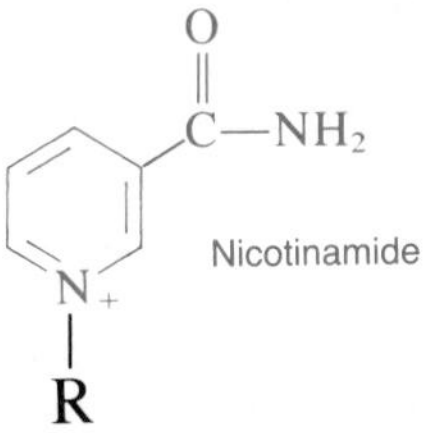

Nicotinamide

You can also see why we use the representations NAD$^+$ and NADH/H$^+$. Because there are four bonds to N, there is a positive charge on N (see Section 16.2), thus the representation NAD$^+$. The reduction of NAD$^+$ actually involves the transfer of just one H nucleus, but both electrons, from the 2H· removed from the metabolite, MH$_2$.

$$\underset{\substack{N^+ \\ | \\ R}}{\overset{\substack{H \quad O \\ \overset{\|}{C}{-}NH_2}}{}} \quad + \; 2H^{\cdot} \longrightarrow \underset{\substack{N \\ | \\ R}}{\overset{\substack{H \quad H \quad O \\ \overset{\|}{C}{-}NH_2}}{}} \quad + \; H^+$$

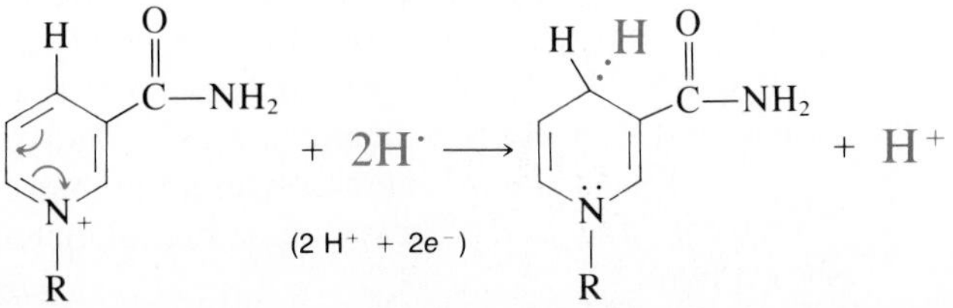

$$(2\,H^+ \, + \, 2e^-)$$

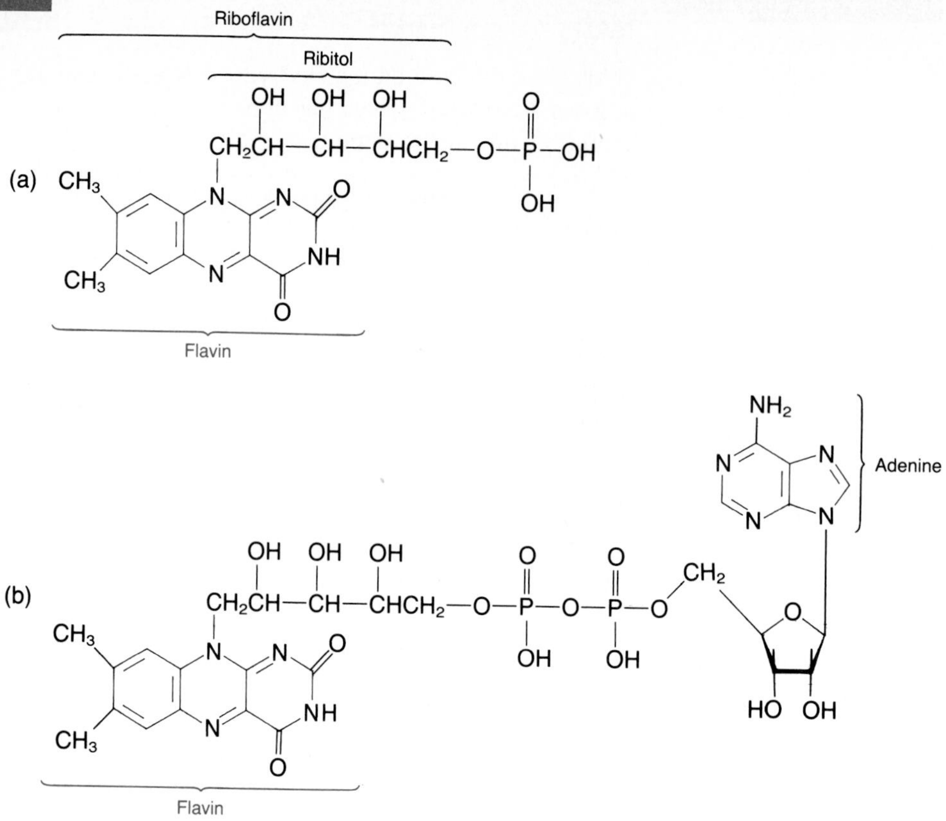
Riboflavin
Ribitol
Flavin
Adenine
(a)
(b)

FIG. 22.5
Flavin nucleotides. (*a*) Flavin mononucleotide (FMN) and (*b*) Flavin adenine dinucleotide (FAD). Oxidation-reduction reactions occur in the flavin portion of the molecule.

or

$$NAD^+ \quad + \quad 2H\cdot \longrightarrow NADH \quad + \quad H^+$$

Recall that the dash used to show a bond represents two electrons. The curved arrows show the rearrangement of electrons (traveling in pairs) as H^+ and $2e^-$ are added.

The reduced form of NAD^+ is represented $NADH/H^+$ even though both H's do not attach to NAD^+, because in the subsequent step both H nuclei and both electrons ($2H\cdot$) are transferred to *flavin mononucleotide* (FMN, Figure 22.5).

FMN ⟶ $FMNH_2$

The oxidation-reduction reaction wherein FMN "picks up" the two hydrogens from $NADH/H^+$ and carries them along is significant in two ways. First, the reaction oxidizes $NADH/H^+$ to NAD^+ and therefore restores the NAD^+ coenzyme to the condition necessary for it to interact with other metabolites (MH_2). Thus we see the chain is self-sustaining because it regenerates the necessary coenzymes. Secondly, this reaction is exergonic ($\Delta G = -12.2$ kcal/mol) and produces more than enough energy to synthesize ATP from ADP + P_i. These events are all included in Figure 22.3 and also shown here:

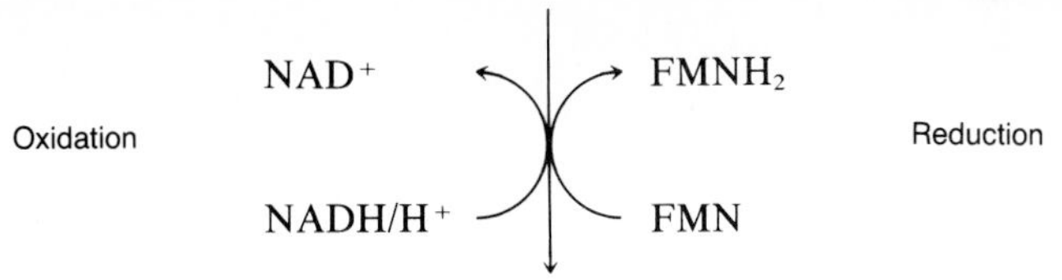
NAD+
FMNH2
Oxidation
Reduction
NADH/H+
FMN

$$ADP + P_i + 12.2\ kcal \longrightarrow ATP$$

The reduction of FMN involves the flavin portion of the molecule (see Figure 22.5):

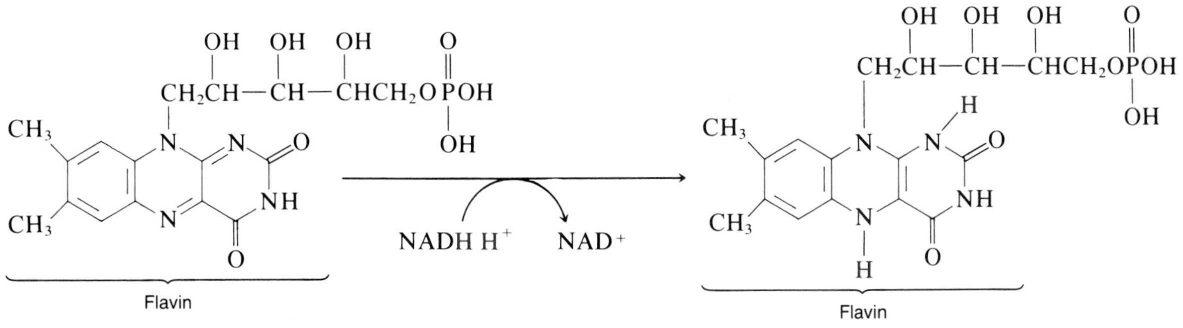
NADH H+
NAD+
Flavin
Flavin

FAD ⟶ $FADH_2$

Before we continue along the chain, let us return to a consideration of FAD, the other coenzyme which can interact directly with metabolites. FAD is a required coenzyme in reactions involving the oxidation of carbon-carbon single bonds to carbon-carbon double bonds. For example, during carbohydrate metabolism (Section 23.4) succinic acid is converted to fumaric acid:

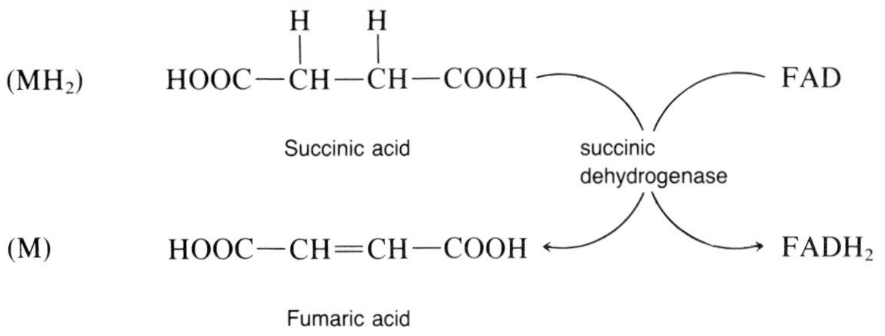
(MH2)
HOOC—CH—CH—COOH
Succinic acid
FAD
succinic
dehydrogenase
(M)
HOOC—CH=CH—COOH
Fumaric acid
FADH2

Like FMN, FAD is reduced through the flavin portion of the molecule.

SAMPLE EXERCISE 22.1 Which of the following reactions require NAD^+ as a coenzyme and which require FAD?

a

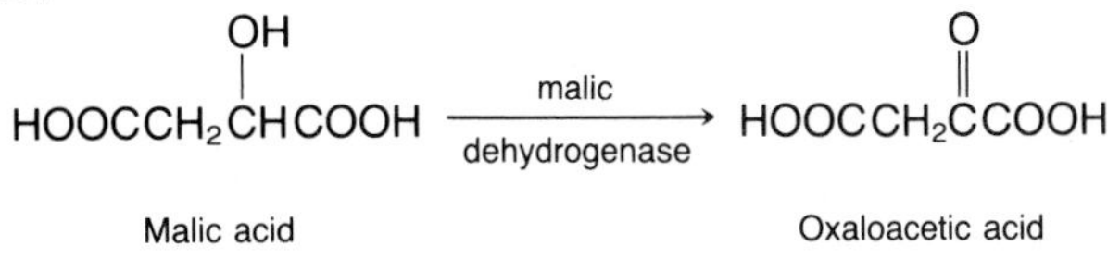
HOOCCH2CHCOOH
malic
dehydrogenase
HOOCCH2CCOOH
Malic acid
Oxaloacetic acid

b

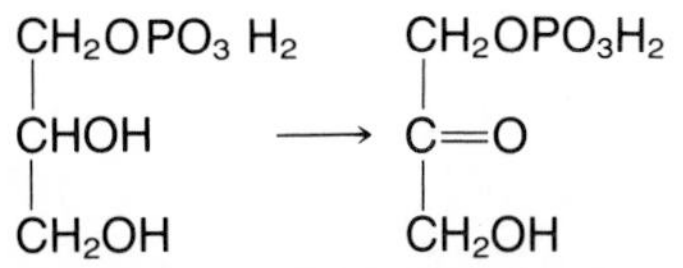
CH2OPO3H2
CHOH
CH2OH
CH2OPO3H2
C=O
CH2OH

c $CH_3CH_2-CH_2-\overset{O}{\overset{\|}{C}}-S-CoA \rightarrow CH_3CH=CH-\overset{O}{\overset{\|}{C}}-S-CoA$

Solution NAD$^+$ oxidizes (removes hydrogen from) carbon-oxygen bonds, whereas FAD oxidizes carbon-carbon bonds. Examining the reactants and products given, we find:

a and *b* $H-\overset{|}{\underset{|}{C}}-O-H \rightarrow \;\rangle C=O$; therefore, NAD$^+$ is required.

c $-CH_2-CH_2- \longrightarrow -CH=CH-$; therefore, FAD is needed.

CoQ ⟶ $CoQH_2$

In the next step of the ETS, the reduced forms of both flavin nucleotides pass hydrogen nuclei and their electrons to coenzyme Q. By so doing, FAD and FMN are regenerated for further reactions with MH_2 and NADH/H$^+$, respectively. Figure 22.6 shows the oxidized (CoQ) and reduced ($CoQH_2$) forms of coenzyme Q, and their interconversion through the agency of $FADH_2$/FAD.

SAMPLE EXERCISE 22.2

Complete the diagram of sequential oxidation-reduction reactions that follows by filling in the blanks:

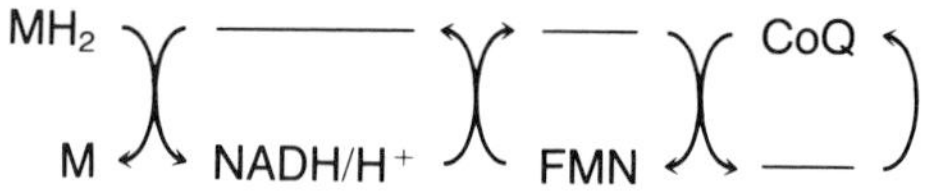

MH2
M
NADH/H+
FMN
CoQ

Solution The intersection of the arrows corresponds to the transfer of hydrogen (and electrons); therefore the two connected reactions are always one oxidation and one reduction. MH_2 is oxidized to M; at the same time NAD$^+$ is reduced to NADH/H$^+$. When NADH/H$^+$ is oxidized to NAD$^+$, FMN must be reduced to $FMNH_2$. Check your answer in Figure 22.3.

FIG. 22.6

Hydrogens from $FMNH_2$ or $FADH_2$ add across the quinone structure (1,4-diketone) to yield a 1,4-diphenol. The quinone structural feature is what gives the Q to the name coenzyme Q. Section 14.17 introduced quinone-hydroquinone interconversions.

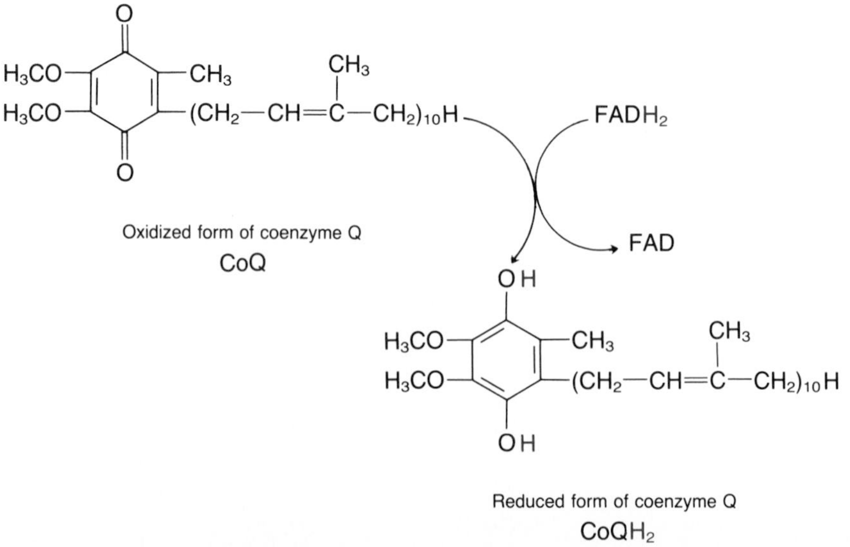

FADH2
FAD
Oxidized form of coenzyme Q
CoQ
Reduced form of coenzyme Q
CoQH2

Cytochrome Reactions: $2Fe^{3+} \longrightarrow 2Fe^{2+}$

The coenzymes NAD^+, FAD, FMN, and Q are both electron carriers and hydrogen ion carriers. The cytochromes, which are enzymes and the next carriers we encounter, carry only electrons. Hydrogen ions are released into the surrounding medium when $2e^-$ are passed from $CoQH_2$ to cytochrome b. The electrons reduce two Fe^{3+} ions to Fe^{2+} ions in the hemelike prosthetic group of cytochrome b. Schematically,

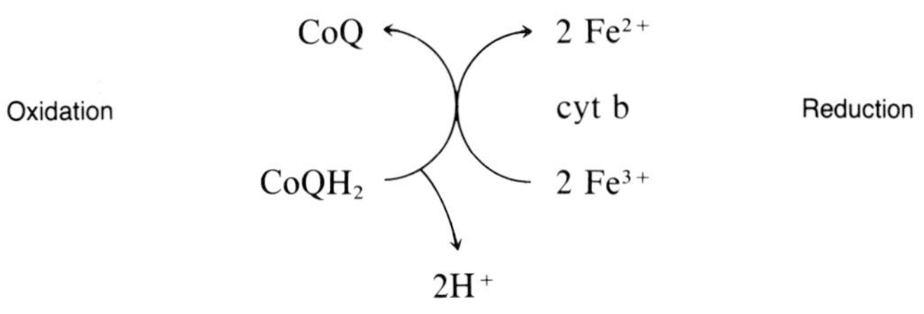
CoQ
2 Fe2+
Oxidation
cyt b
Reduction
CoQH2
2 Fe3+
2H+

Figure 22.7 shows a typical cytochrome structure. All cytochromes are enzymes with a hemelike coenzyme portion. The differences between the cytochromes are in:

1 Apoenzyme structure
2 Bonds between apoenzyme and coenzyme
3 Side chains attached to the heme coenzyme

The iron ions of the cytochromes are all capable of reversible one-electron transfers or redox reactions:

FIG. 22.7
Typical cytochrome structure. Cytochromes are similar in that they all have a protein (apoenzyme) part and heme coenzyme part. Differences are in (1) protein structure, (2) side chains in the heme portion, and (3) the nature of the association between heme and protein. In cytochromes c and c_1, heme and protein are covalently bonded as shown. In cytochrome b, there is no covalent linkage. For all cytochromes, ETS activity involves the central Fe ion of the heme coenzyme.

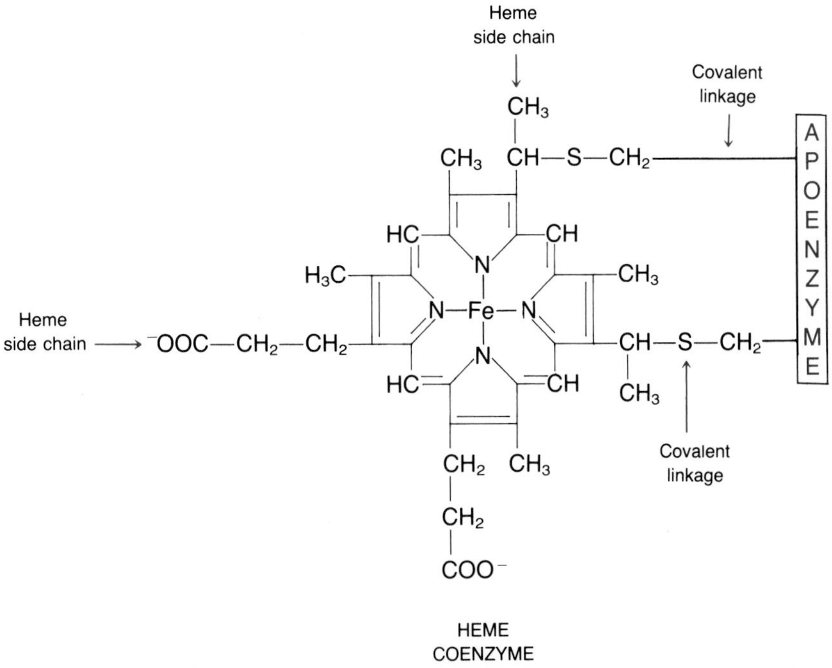
Heme
side chain
Covalent
linkage
APOENZYME
Heme
side chain
Covalent
linkage
HEME
COENZYME

Reduction $Fe^{3+} + e^- \longrightarrow Fe^{2+}$

Oxidation $Fe^{2+} \longrightarrow Fe^{3+} + e^-$

Because $CoQH_2$ transfers $2e^-$ from metabolites and $2e^-$ are needed in the final production of water from $2H^+$ and $\frac{1}{2}O_2$, but each cytochrome iron ion can handle only one electron, two cytochrome molecules must work together here and at each of the other steps in the cytochrome chain.

SAMPLE EXERCISE 22.3 Identify each of the following reactions as oxidation or reduction:

a Fe^{3+} (cyt c) $\longrightarrow$ Fe^{2+} (cyt c)
b Fe^{2+} (cyt b) Fe^{3+} (cyt c_1)

Fe^{3+} (cyt b) Fe^{2+} (cyt c_1)

Solution *a* A negative electron must be added if Fe^{3+} is to become Fe^{2+}. Gain of electrons is reduction.

$Fe^{3+} + e^- \longrightarrow Fe^{2+}$

b The left-hand reaction is oxidation; Fe^{2+} must lose an electron to become Fe^{3+}. The right-hand reaction must then necessarily be reduction. We see that it is reduction, as in part *a*, where $Fe^{3+} + e^- \longrightarrow Fe^{2+}$.

Problem 22.5 Identify each reaction in the cytochrome portion of the ETS as an oxidation or reduction.

Sufficient energy ($\Delta G = -9.9$ kcal/mol) is released during the electron transfer reaction between cytochrome b and cytochrome c to allow the synthesis of ATP from ADP and P_i. Another mole of ATP is produced as energy ($\Delta G = -23.8$ kcal/mol) is released during electron transfer from cytochromes a and a_3 to oxygen.

$$2H^+ + 2e^- + :\ddot{\underset{\cdot\cdot}{O}}\cdot \longrightarrow H_2O$$

Cytochromes a and a_3 work together, and because they transfer electrons directly to oxygen, they are sometimes called cytochrome oxidase. In the last step of the ETS two electrons are transferred from cytochrome oxidase in order to bond $2H^+$ to an oxygen atom, thereby forming water.

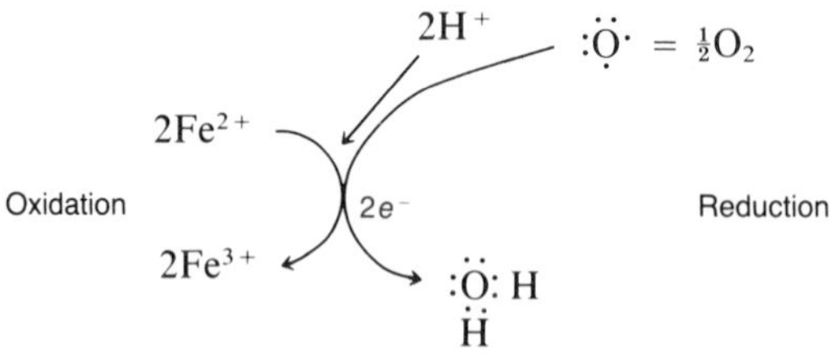
2H+
:Ö· = ½O2
2Fe2+
Oxidation
2e−
Reduction
2Fe3+
:O: H
H

This last step makes it clear that the ETS requires oxygen to operate. Most of the oxygen we breathe is used in the ETS and is thereby indirectly oxidizing the food we eat. **Because oxygen is required in the ETS and because energy released during electron transport is used to drive the synthesis of ATP, the coupling of the exergonic ETS and endergonic ATP synthesis is called oxidative phosphorylation.**

Now that we have discussed the individual steps of the ETS, you should survey Figure 22.3 once again. The most essential features are the overall exergonic synthesis of water and the endergonic synthesis of ATP at three points along the chain.

22.7 OXIDATIVE PHOSPHORYLATION

Oxidative phosphorylation is the coupling of the exergonic ETS and endergonic ATP synthesis. Let us count up the amount of ATP produced by oxidative phosphorylation as the coenzymes NAD^+ and FAD bring hydrogen from food to the electron transport system.

As Figure 22.3 shows, each mole of $NADH/H^+$ oxidized in the ETS results in the production of 3 mol of ATP. As $NADH/H^+$ is oxidized to NAD^+ by FMN, 1 mol of ATP is generated, and then 2 mol of ATP is produced as the electrons from $NADH/H^+$ are carried through the cytochrome system. This net production of 3 mol of ATP from 1 mol of $NADH/H^+$ depends on the $NADH/H^+$ being formed in the cell parts known as mitochondria. If the $NADH/H^+$ is formed elsewhere in the cell, some ATP may need to be used to transport the reduced coenzyme across the mitochondrial membrane.

Figure 22.3 also shows that each mole of $FADH_2$ oxidized in the ETS results in the production of only 2 mol of ATP. The oxidation of $FADH_2$ to FAD does not produce sufficient energy to make ATP. Therefore, the net yield of ATP from $FADH_2$ is the 2 mol of ATP produced as the electrons from it are carried through the cytochrome system. There will be frequent references in forthcoming chapters to the facts that in the ETS

$NADH/H^+$ produces 3 mol of ATP

$FADH_2$ produces 2 mol of ATP

SAMPLE EXERCISE 22.4

The following reactions occur in the Krebs cycle, an important metabolic cycle you will explore in Chapter 23. How much ATP can be synthesized as a subsequent result of each reaction?

a

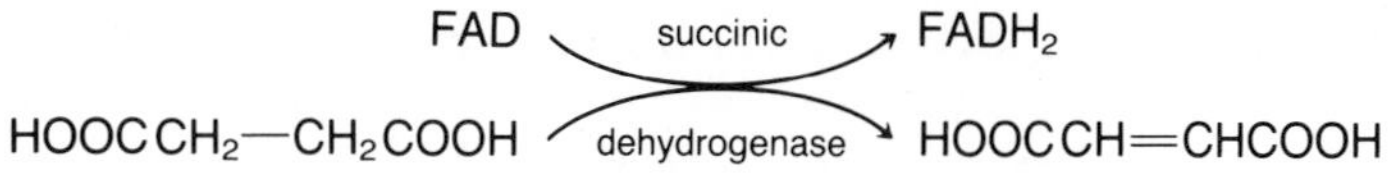
FAD
succinic
FADH2
HOOCCH2—CH2COOH
dehydrogenase
HOOCCH=CHCOOH

b

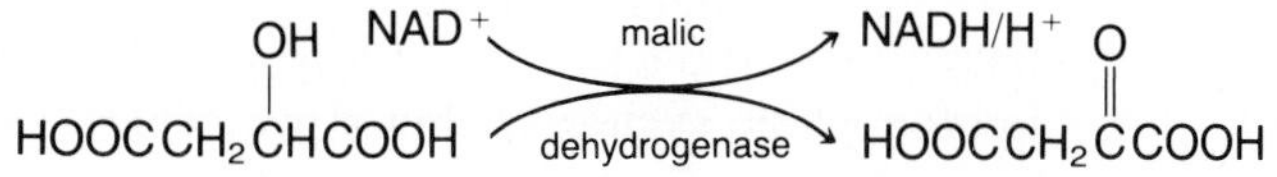
OH
NAD+
malic
NADH/H+
O
HOOCCH2CHCOOH
dehydrogenase
HOOCCH2CCOOH

Solution ATP synthesis will occur because $FADH_2$ and $NADH/H^+$ will react in the ETS.

a Every mole of $FADH_2$ will produce 2 mol of ATP.

b Every mole of $NADH/H^+$ will produce 3 mol of ATP.

Problem 22.6 How much ATP is produced by the oxidation of 1 mol of glyceryl phosphate?

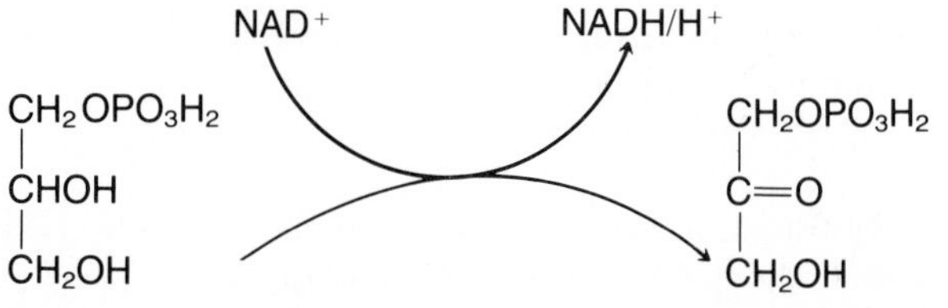
NAD+
NADH/H+
CH2OPO3H2
CHOH
CH2OH
CH2OPO3H2
C=O
CH2OH

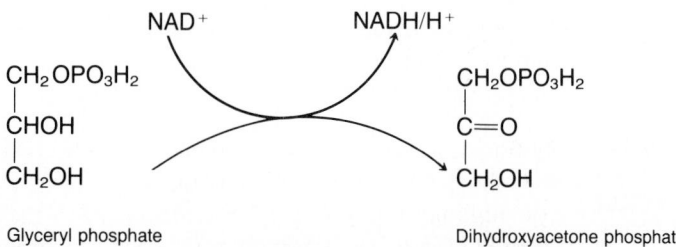

Glyceryl phosphate — Dihydroxyacetone phosphate

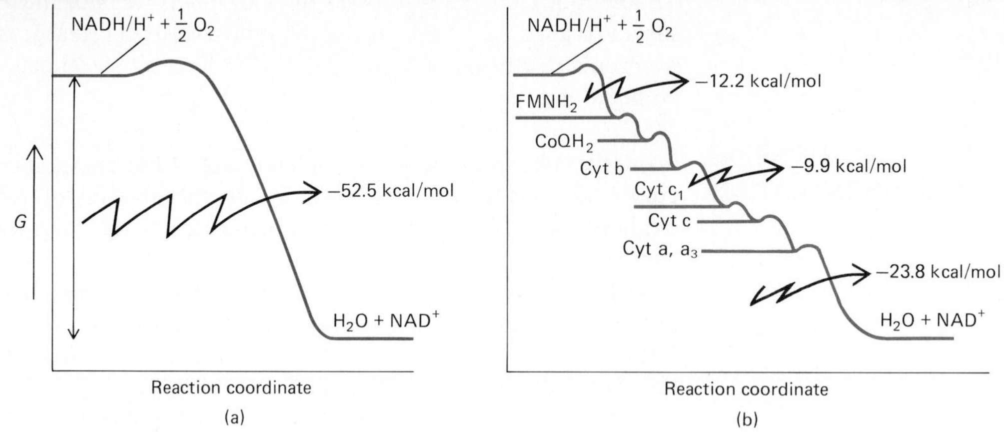
NADH/H⁺ + ½ O₂
G
−52.5 kcal/mol
H₂O + NAD⁺
Reaction coordinate
(a)
NADH/H⁺ + ½ O₂
−12.2 kcal/mol
FMNH₂
CoQH₂
Cyt b
−9.9 kcal/mol
Cyt c₁
Cyt c
Cyt a, a₃
−23.8 kcal/mol
H₂O + NAD⁺
Reaction coordinate
(b)

FIG. 22.8
These reaction coordinate diagrams correspond to the ETS. (*a*) The overall energy release. (*b*) The energy release as it actually happens in a more efficient stepwise manner. Steps that release more energy than 7.3 kcal/mol are sufficiently exergonic to be coupled with endergonic ATP synthesis.

Figure 22.8 shows an energy diagram for the electron transport system. The total energy output as water is made in this process is −52.6 kcal/mol. This energy is used more efficiently for ATP synthesis by coupling the individual steps in the chain which produce enough energy to combine ADP and P_i into ATP. Thus, steps which release more than 7.3 kcal/mol are coupled to endergonic ATP synthesis which requires 7.3 kcal/mol of energy input.

Problem 22.7 The reaction between $FMNH_2$ and CoQ does not result in ATP synthesis. What can we conclude about the magnitude of ΔG for this exergonic reaction?

The coupling of ATP synthesis to the ETS also requires a pH difference and consequent electrochemical potential (energy difference) at the site of the coupling, as will be discussed more fully in Section 22.9. This requirement is the reason that coupling occurs only at certain junctures along the respiratory chain, or ETS.

Though coupling occurs only at certain steps, all steps in the chain are equally important because if any step is inhibited, electron transfer stops and the whole system is shut down. The ETS is like a relay team in track. In a relay race, if the baton is not transferred at some point, the race is lost. In the ETS if electrons (and protons) are not transferred, the process stops.

Certain poisons operate by inhibiting a step in the ETS. For example, the insecticide rotenone and the barbiturate amytal prevent the transfer of electrons and protons from $FMNH_2$ to coenzyme Q. Excess ingestion of these materials can be fatal because of the breakdown in the ETS they produce. Carbon monoxide gas and cyanide ion are poisons because they can bind to the iron ions in cytochrome oxidase and inhibit the iron ions' ability to transfer electrons. CO also affects the oxygen-carrying capacity of hemoglobin through interaction with iron ions.

22.8 SUBSTRATE-LEVEL PHOSPHORYLATION

By far, most of the ATP produced in the body is synthesized by oxidative phosphorylation. But especially during extreme exertion, some ATP is made by coupling endergonic ATP synthesis with the exergonic hydrolysis of a phosphate bond in some other phosphate compound. This hydrolysis must be more exergonic than the hydrolysis of ATP for which $\Delta G = -7.3$ kcal. An example of such a phosphate compound is creatine phosphate: hydrolysis of the P—N bond in this compound liberates 10.3 kcal/mol.

$$^-OOC-CH_2-\underset{\underset{CH_3}{|}}{N}-\underset{\underset{H}{|}}{\overset{\overset{+NH_2}{\|}}{C}}-\underset{\underset{O^-}{|}}{N}\sim\overset{\overset{O}{\|}}{P}-O^- \xrightarrow{\ H_2O\ }$$

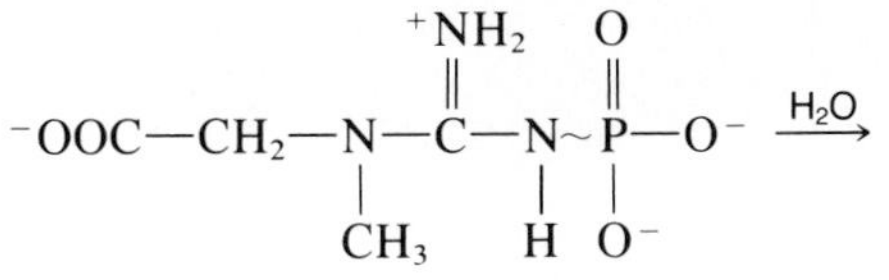

Creatine phosphate

$$^-OOC-CH_2-\underset{\underset{CH_3}{|}}{N}-\overset{\overset{+NH_2}{\|}}{C}-NH_2 + P_i + 10.3\ \text{kcal}$$

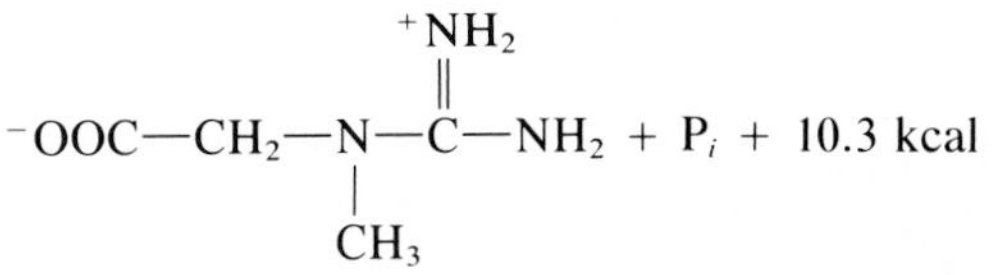

Creatine

This exergonic hydrolysis can be coupled to endergonic ATP synthesis.

$$\text{ADP} + P_i + 7.3\ \text{kcal} \longrightarrow \text{ATP}$$

The sum of these two equations is

$$\text{Creatine phosphate} + \text{ADP} \xrightarrow[\text{kinase}]{\text{creatine}} \text{creatine} + \text{ATP} + 3.0\ \text{kcal} \qquad (22.2)$$

This coupling can also be expressed in the usual manner:

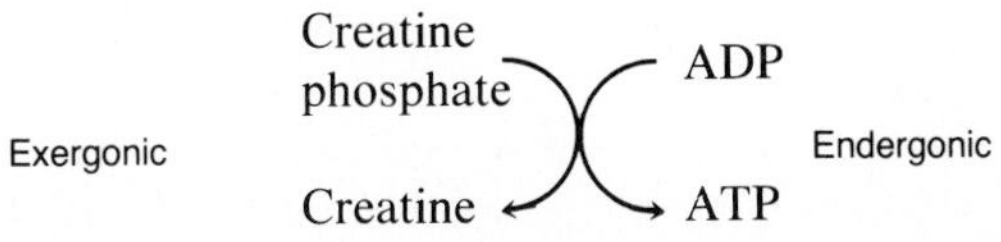

Exergonic   Creatine phosphate ⟩⟨ ADP   Endergonic

Creatine ⟵ ATP

Coupling of this kind is called **substrate-level phosphorylation.**

During vigorous exercise the body uses its reserves of creatine phosphate in muscle tissue to make ATP. During rest, creatine phosphate reserves are restored by a reversal of Equation (22.2) or by phosphorylation of creatine by even higher-energy phosphate compounds.

22.9 MITOCHONDRIA   The reactions of the ETS take place in cell structures called mitochondria (singular = mitochondrion), shown in Figure 22.9. The mitochondria are uniquely designed to accommodate the sequential reactions of the ETS and coupling to ATP synthesis. Note that mitochondria have outer and inner membranes. The space between these two membranes is called the *intermembrane space*. The inner membrane is extensively folded, and the folds are given the name *cristae*. The spaces between the cristae are called the *matrix*. The enzymes of the ETS are an integral part of the cristae, and the reactions of the ETS occur at or on these inner folded membranes. ATP synthesis occurs at "knobs" on these inner folds. Other metabolic cycles that we will examine in later chapters occur in the matrix space. Because of all this energy-producing metabolic activity, the mitochondria are called the "powerhouses" of cells.

Successful coupling of the ETS and ATP synthesis (oxidative phosphor-

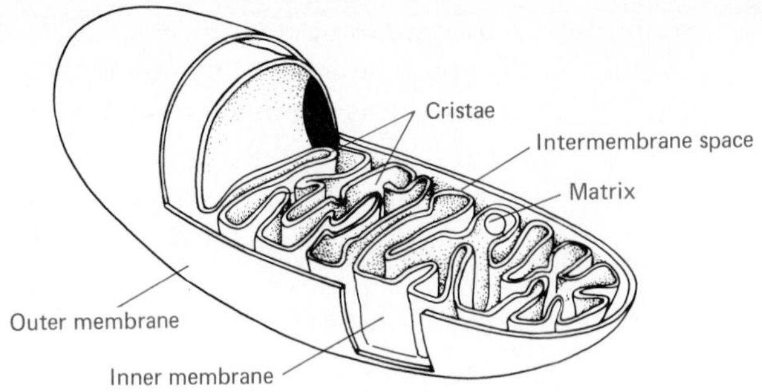
Cristae
Intermembrane space
Matrix
Outer membrane
Inner membrane

FIG. 22.9
A mitochondrion.

ylation) requires a pH difference between the matrix (inner compartment) and the intermembrane space (outer compartment) of the mitochondria. This is the **chemiosmotic hypothesis** of Peter Mitchell (Nobel Prize, 1978). According to the chemiosmotic hypothesis, the energy released as two electrons are transferred from NADH/H^+ through the ETS to oxygen pumps $6H^+$ out of the matrix across the inner membrane of the mitochondria. This movement of H^+ sets up a H^+ concentration gradient. Because of the charge difference thus created, an electric potential also develops. As Figure 22.10 shows, the $6H^+$ (per $2e^-$) are pumped across the membrane $2H^+$ at a time at three steps of the ETS; we have previously indicated that there are three steps where there is release of energy. Along most of the inner membrane H^+ ions cannot flow back into the matrix. However, there are knoblike shapes

FIG. 22.10
Protons (H^+) must move for ATP synthesis to occur. The ETS pumps H^+ from the matrix into the intermembrane space. Protons diffuse back across the membrane at knobs where ATP synthesis occurs.

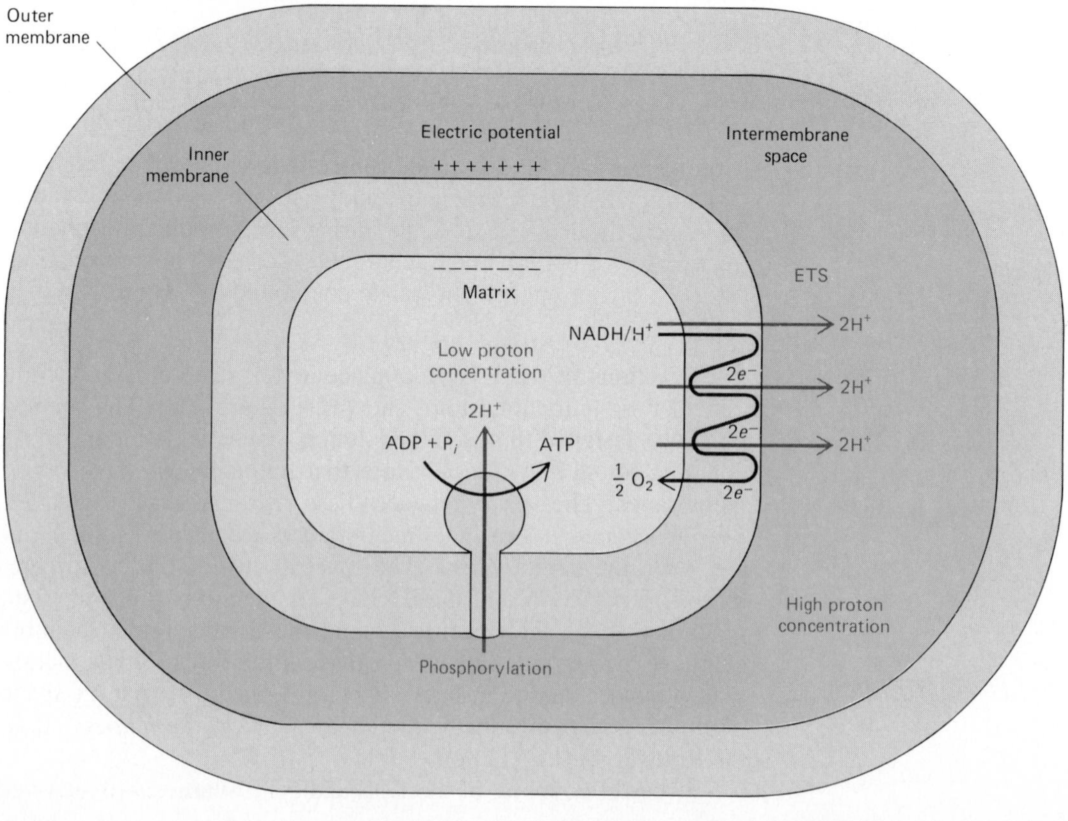
Outer membrane
Electric potential
+ + + + + + +
Intermembrane space
Inner membrane
– – – – – – –
Matrix
ETS
NADH/H⁺
2H⁺
Low proton concentration
2e⁻
2H⁺
2H⁺
ADP + Pᵢ
ATP
2e⁻
2H⁺
½ O₂
2e⁻
High proton concentration
Phosphorylation

on the inner membrane which extend into the matrix and to which are attached the enzyme complex, ATP synthetase.

The chemiosmotic theory proposes that these knobs act as channels through which the H^+ ions diffuse back into the matrix. The diffusion of $2H^+$ ions from the region of higher concentration and positive charge in the intermembrane space into the matrix releases sufficient energy to cause the synthesis of one ATP from ADP + P_i at the ATP synthetase site. Thus the $6H^+$ ions pumped out of the matrix two at a time flow back into the matrix two at a time and can result in the synthesis of three ATP molecules. Experimental evidence that the structural integrity of the mitochondria must be maintained for ATP synthesis to occur tends to confirm the chemiosmotic hypothesis. Indeed, it has been observed that if the membrane is not intact the ETS continues, but ATP is not produced. In order for oxidative phosphorylation to occur, there must be compartments separated by a membrane to create a pH gradient and diffusion.

A complete understanding of the chemiosmotic theory requires a background in chemistry and physics beyond the scope of this text. A brief discussion is included here for the purpose of pointing out nature's provision of a proper receptacle for the ETS reactions in the mitochondria. The importance of the provision is demonstrated by the necessity to maintain the structural integrity of the mitochondria in order for the ETS to make ATP.

22.10 AN OVERVIEW OF METABOLISM

Now that you have seen how the ETS works, we can explore the other major metabolic reactions. Through metabolism the body extracts energy from food by combining hydrogen from food with oxygen in the ETS. Most other metabolic reactions that you will study are catabolic ones which essentially involve the removal of hydrogen from digested food molecules. The waste products of metabolism are H_2O, which we excrete, and CO_2, which we exhale. Figure 22.11 summarizes the major catabolic processes of metabolism.

Digestion breaks large molecules into smaller molecules that can enter cells. Within the cell these molecules are broken down further and the fragments are combined with coenzyme A to form acetyl CoA. Some of these catabolic reactions involve removal of hydrogen atoms which are taken to the ETS by NAD^+ and FAD.

Acetyl coenzyme A then enters the citric acid cycle, a series of reactions in which hydrogen atoms are removed and carried to the ETS by NAD^+ and FAD. CO_2 is a waste product in some of these reactions. Chapters 23 to 25 discuss the reactions of the citric acid cycle and the details of the formation of acetyl coenzyme A from glucose, fatty acids, and amino acids.

Figure 22.11 represents only the major metabolic pathways in broadest outline. Other events occur; for example, amino acids can enter the citric acid cycle without going through acetyl coenzyme A, and some ATP is made directly during the citric acid cycle by substrate-level phosphorylation.

22.11 ACETYL COENZYME A

Figure 22.11 shows that acetyl coenzyme A plays a central role in metabolism. The structure of coenzyme A appears in Figure 22.12. It is a derivative of a nucleotide: a molecule of pantothenic acid is attached to the phosphate-sugar-base portion through a phosphate linkage. Pantothenic acid is an essential vitamin because it is an integral part of the coenzyme A structure.

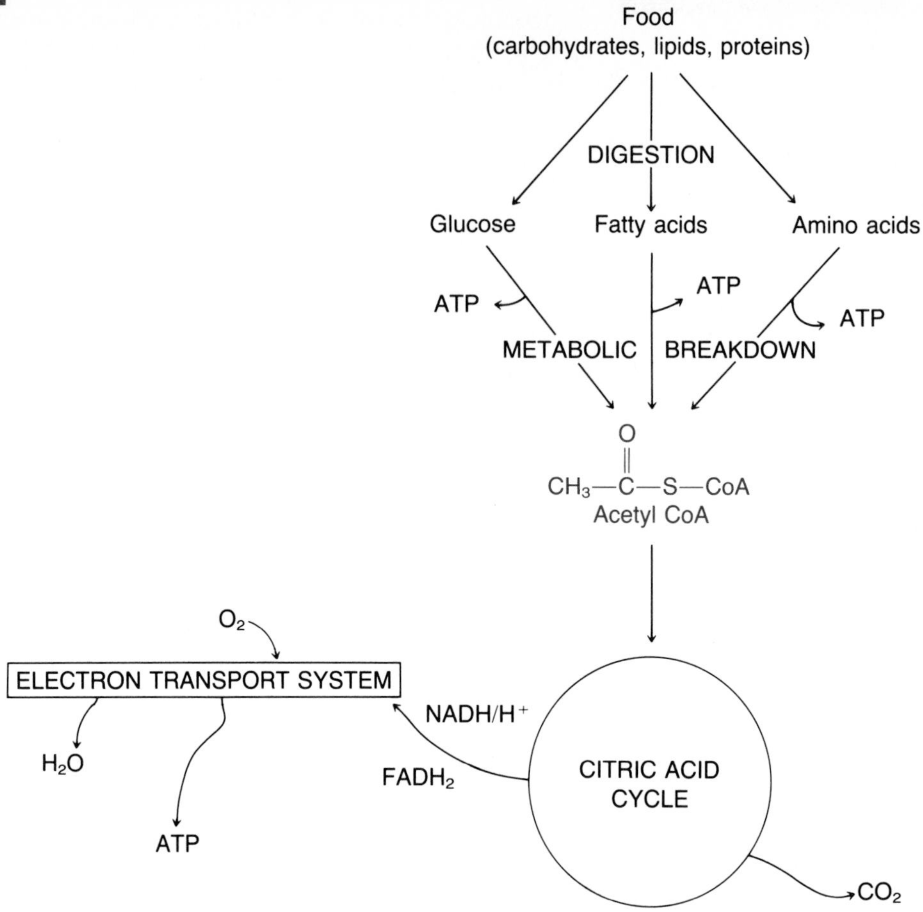
Food
(carbohydrates, lipids, proteins)
DIGESTION
Glucose
Fatty acids
Amino acids
ATP
ATP
ATP
METABOLIC
BREAKDOWN
CH3—C—S—CoA
Acetyl CoA
O2
ELECTRON TRANSPORT SYSTEM
H2O
ATP
NADH/H+
FADH2
CITRIC ACID
CYCLE
CO2

FIG. 22.11
A broad outline of the major metabolic pathways. The central role of acetyl CoA is evident. Most of our ATP needs are met by channeling degraded food through acetyl CoA and the citric acid cycle coupled to the ETS.

FIG. 22.12
Coenzyme A and acetyl coenzyme A. The nucleotide portion (base-sugar-phosphate) and the structural portion derived from the vitamin pantothenic acid are labeled. The thiol functional group is the site of the significant reactions of coenzyme A. Thus the abbreviations CoA-SH for coenzyme A and $CH_3—C(=O)—S—CoA$.

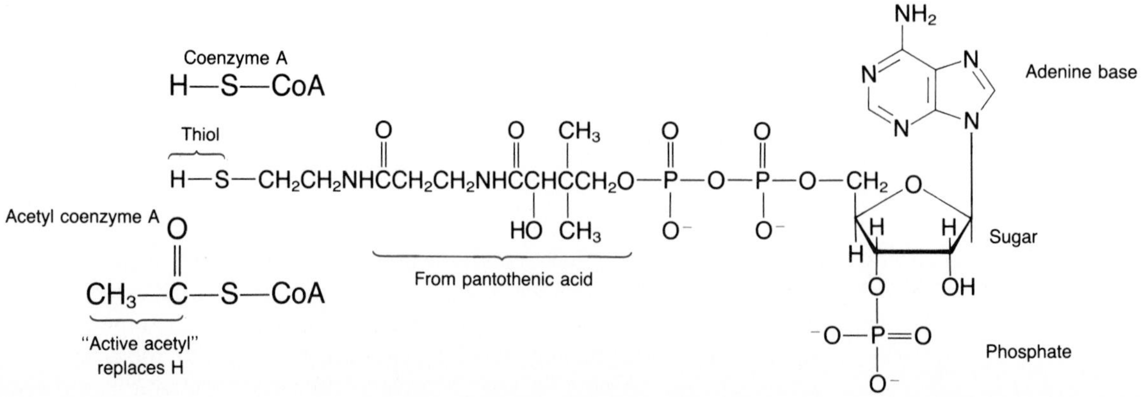
Coenzyme A
H—S—CoA
Thiol
Acetyl coenzyme A
CH3—C—S—CoA
"Active acetyl"
replaces H
From pantothenic acid
NH2
Adenine base
Sugar
Phosphate

The most significant reactions of coenzyme A involve the thiol (—S—H) functional group of the molecule. For this reason coenzyme A is often abbreviated CoA—SH. Acetyl coenzyme A is a thioester of the thiol and hence is represented $CH_3-\overset{\overset{\displaystyle O}{\|}}{C}-S-CoA$. We can envision formation of this compound through a typical esterification reaction:

$$\underset{\text{Acetic acid}}{CH_3-\overset{\overset{\displaystyle O}{\|}}{C}}-\boxed{OH + H}-\underset{\text{Coenzyme A}}{S-CoA} \longrightarrow \underset{\text{Acetyl coenzyme A}}{CH_3-\overset{\overset{\displaystyle O}{\|}}{C}-S-CoA} + HOH$$

The catabolism of glucose, fatty acids, and amino acids shown in Figure 22.11 effectively breaks these molecules down into acetyl ($CH_3-\overset{\overset{\displaystyle O}{\|}}{C}-$) fragments which become attached to HS—CoA. Chapter 23 will discuss the reactions that convert glucose to $CH_3-\overset{\overset{\displaystyle O}{\|}}{C}-SCoA$ and then explore the citric acid cycle.

CHAPTER ACCOMPLISHMENTS

After completing this chapter you should be able to define **key words** and do the following:

22.1 INTRODUCTION

1. Distinguish between anabolism and catabolism.

22.2 THERMODYNAMIC PRINCIPLES

2. Given a chemical equation which includes free energy information, classify a reaction as exergonic or endergonic.
3. Interpret equations which use intersecting curved arrows to represent coupled reactions.

22.3 ENERGY FROM THE SUN

4. State the role of photosynthesis in supplying energy to plants and animals.

22.4 THE ROLE OF ATP

5. Recognize ATP as a nucleotide.
6. State the role of ATP in cellular energy transfer.
7. Complete equations for the hydrolysis of ATP.

22.5 THE ROLES OF EATING AND BREATHING

8. Write an equation for the overall reaction that occurs in the electron transport system.
9. State the contributions which eating and breathing make to the overall reaction that occurs in the electron transport system.

22.6 THE ELECTRON TRANSPORT SYSTEM

10. Identify the reactions of the ETS as oxidations or reductions.
11. State the type of metabolic oxidations for which NAD^+ is the coenzyme.
12. State the type of metabolic oxidations for which FAD is the coenzyme.
13. Complete equations for coupled oxidation-reduction reactions.

22.7 OXIDATIVE PHOSPHORYLATION

14. State the yield of ATP per mole of $NADH/H^+$ oxidized in the ETS.
15. State the yield of ATP per mole of $FADH_2$ oxidized in the ETS.
16. Give examples of poisons which function by inhibiting steps in the ETS.

22.8 SUBSTRATE-LEVEL PHOSPHORYLATION

17. Distinguish substrate-level phosphorylation from oxidative phosphorylation.

22.9 MITOCHONDRIA

18. Explain the description of mitochondria as the cell's "powerhouses."

22.10 AN OVERVIEW OF METABOLISM

22.11 ACETYL COENZYME A

19. Recognize the central role of acetyl CoA in metabolism.

PROBLEMS

22.8 Why is digestion of food into small molecules a necessary prerequisite to metabolism?

22.9 Name the two general categories of metabolism.

22.10 Classify anabolism and catabolism as energy-producing or energy-requiring processes.

22.11 Classify anabolism and catabolism as synthetic or degradative processes.

22.12 Identify the following reactions as exergonic or endergonic:

a $ATP + H_2O \rightarrow ADP + P_i$ $\Delta G = -7.3$ kcal/mol

b Adenosine $+ P_i + 3.4$ kcal $\rightarrow AMP + H_2O$

c Glucose $+ P_i \rightarrow$ glucose 1-phosphate $+ H_2O$ $\Delta G = +5$ kcal/mol

d $ATP + H_2O \rightarrow AMP + PP_i + 10$ kcal

22.13 Which reactions in Problem 22.12 could be coupled to yield a resultant negative ΔG?

22.14 Classify anabolism and catabolism as exergonic or endergonic.

22.15 Write the following in the form of a traditional equation:

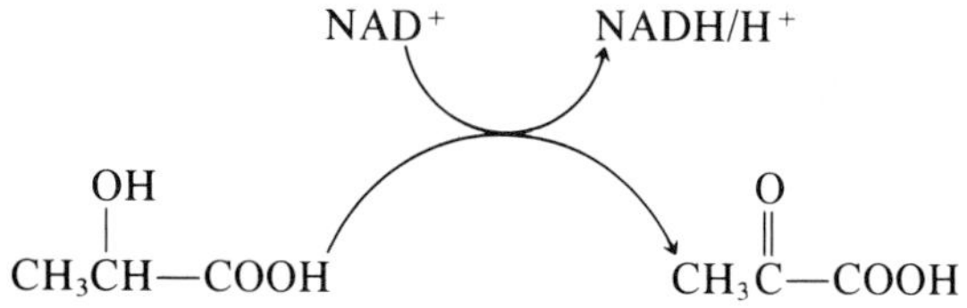
NAD+
NADH/H+
OH
CH3CH—COOH
O
CH3C—COOH

22.16 Write the following equation in the coupled form used in Problem 22.15:

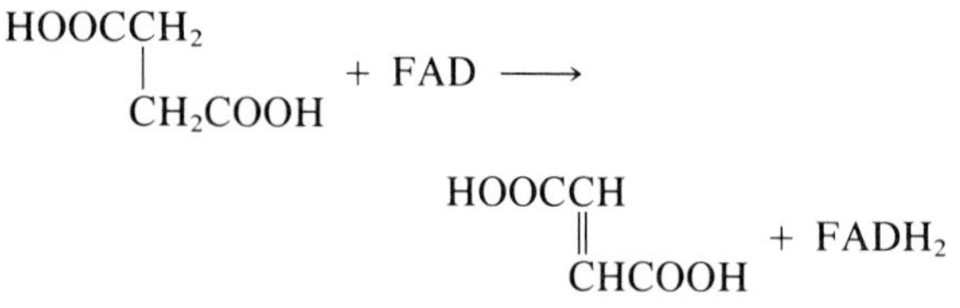
HOOCCH2
CH2COOH
+ FAD
HOOCCH
CHCOOH
+ FADH2

22.17 What is the ultimate source of most energy on earth?

22.18 Write a balanced equation for the overall process of photosynthesis.

22.19 Write a balanced equation for the complete combustion of glucose.

22.20 Into what class of compounds does ATP fall? Name its three component parts.

22.21 Complete this abbreviated structure for ATP:

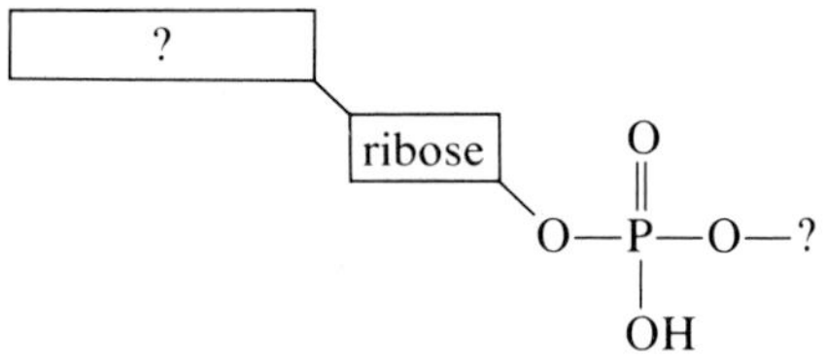
?
ribose
O
O—P—O—?
OH

22.22 What does the symbol P_i represent?

22.23 What bond(s) in ATP are cleaved or made during cellular energy transfer?

22.24 Classify ATP hydrolysis and ATP synthesis as exergonic or endergonic.

22.25 Explain the statement: "Eating and breathing provide energy by forming water."

22.26 Food energy is used to make ______.

22.27 Atoms of what element are supplied to the ETS by metabolites?

22.28 What other element is essential to the operation of the ETS?

22.29 What ETS coenzymes interact directly with metabolites?

22.30 Explain in your own words why the electron transport system is so named.

22.31 What kinds of reactions occur in the ETS?

22.32 Write the abbreviations for the reduced and oxidized forms of nicotinamide adenine dinucleotide and flavin adenine dinucleotide.

22.33 Classify each transformation as an oxidation or reduction:

a $FADH_2 \rightarrow FAD$

b $Fe^{3+} + e^- \rightarrow Fe^{2+}$

c $NAD^+ \rightarrow NADH/H^+$

d $CH_3\overset{O}{\overset{\|}{C}}CH_2\overset{O}{\overset{\|}{C}}—SCoA \rightarrow CH_3\overset{OH}{\overset{|}{C}}HCH_2\overset{O}{\overset{\|}{C}}—SCoA$

e $MH_2 \rightarrow M$

22.34 Which of the following reactions require NAD^+ as the coenzyme and which require FAD?

a $HC(=O)CHOHCH_2—O—Ⓟ \rightarrow HOC(=O)CHOHCH_2O—Ⓟ$

b $CH_3(CH_2)_{10}CH(OH)CH_2C(=O)SCoA \rightarrow CH_3(CH_2)_{10}C(=O)CH_2C(=O)SCoA$

c $CH_3(CH_2)_6CH_2CH_2C(=O)SCoA \rightarrow CH_3(CH_2)_6CH{=}CHC(=O)SCoA$

22.35 What is the class name for the enzymes containing Fe^{2+} or Fe^{3+} in the ETS?

22.36 How does the reduction of the cytochromes differ from the reduction of coenzyme Q?

22.37 Complete the diagram of sequential oxidation-reduction reactions that follows by filling in the blanks:

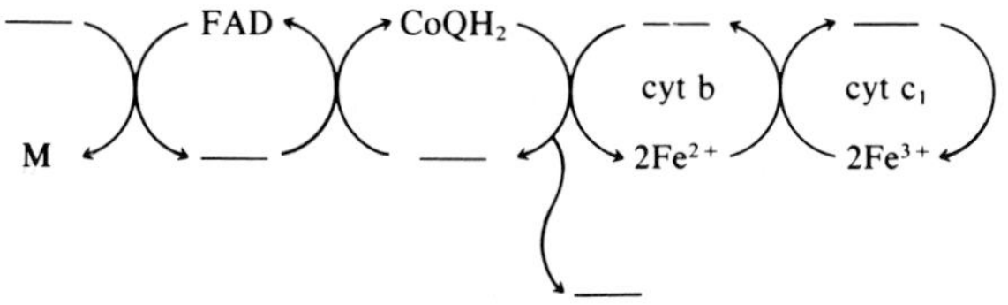
FAD
CoQH2
cyt b
cyt c1
M
2Fe2+
2Fe3+

22.38 What happens in the last step of the ETS?

22.39 How many steps in the ETS are sufficiently exergonic to run endergonic ATP synthesis?

22.40 How many moles of ATP are produced as 1 mol of electrons flows from cyt b to cyt c to cyt a?

22.41 How much energy (in terms of ATP) is produced by the subsequent oxidation of the reduced coenzymes produced in the following reactions:

a

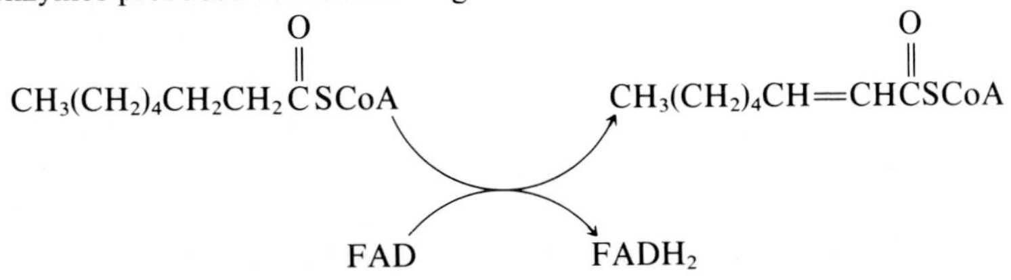
CH3(CH2)4CH2CH2CSCoA
CH3(CH2)4CH=CHCSCoA
FAD
FADH2

b

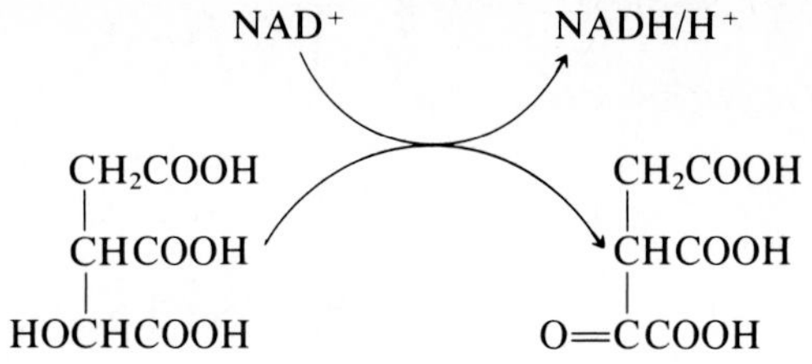
NAD+
NADH/H+
CH2COOH
CHCOOH
HOCHCOOH
CH2COOH
CHCOOH
O=CCOOH

c $MH_4 + 2FAD \rightarrow M + 2FADH_2$

22.42 Explain the effect of oxygen deficiency on the ETS.

22.43 Why is cyanide poisonous?

22.44 What is the difference between substrate-level and oxidative phosphorylation?

22.45 1,3-Diphosphoglyceric acid (1,3-DPGA) is found in cells. Cleavage of its phosphate bond designated ~ releases 11.8 kcal.

a Can the following coupled reactions proceed together? Explain.

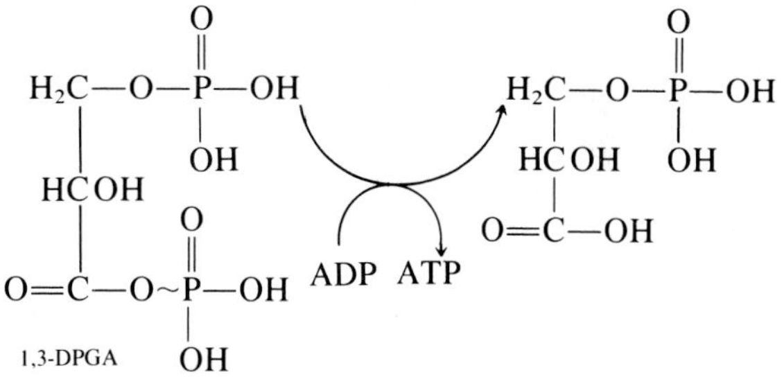
ADP
ATP
1,3-DPGA

b What name would you give to this coupling process?

22.46 Where in the cell do the reactions of the ETS take place?

22.47 Why is pantothenic acid a vitamin?

22.48 What functional group in coenzyme A accounts for most of its reactions in the body?

22.49 Into what structural fragments are the carbon skeletons of food molecules broken down?

CHAPTER TWENTY-THREE

CARBOHYDRATE METABOLISM

23.1 INTRODUCTION

As you saw in Chapter 15, most carbohydrates are composed of glucose monosaccharide units, and the principal product of carbohydrate digestion is glucose. Thus, after one eats carbohydrates, glucose enters the cell for metabolism. Glucose is a principal fuel (energy source) in the body.

The most efficient, largest energy-producing metabolic route for glucose to undergo is its conversion to acetyl CoA through the intermediacy of pyruvic acid, followed by the entry of acetyl CoA into the citric acid cycle.

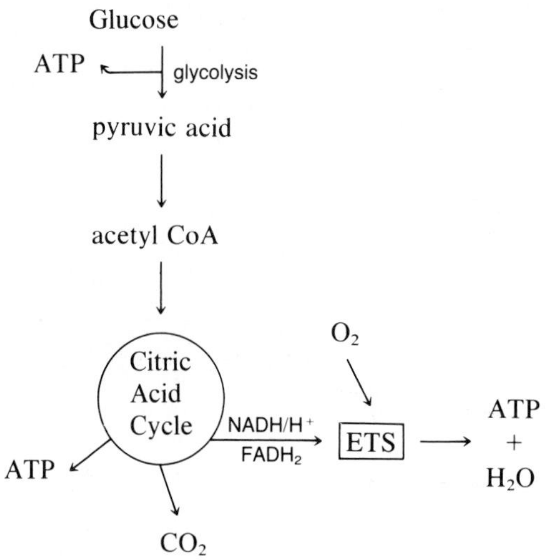

Glucose
ATP
glycolysis
pyruvic acid
acetyl CoA
Citric Acid Cycle
ATP
CO2
NADH/H+
FADH2
O2
ETS
ATP + H2O

Glycolysis, the conversion of glucose to pyruvic acid, will be discussed in Section 23.2, and the citric acid cycle in Section 23.4. These two metabolic pathways are multistep processes just as is the electron transport system, the first set of metabolic reactions which you studied.

Most of the ATP made from glucose is made when the reduced coenzymes, $NADH/H^+$ and $FADH_2$, take hydrogen to the ETS, although some ATP is made in prior events, as shown in the diagram. Because the ETS requires oxygen to run, the entire sequence shown is dependent on sufficient oxygen being available in cells. During rest or mild exertion this condition is met. If insufficient oxygen is present (as is the case during vigorous exercise), the body can function for short periods by converting pyruvic acid to lactic acid (Section 23.3). Under these anaerobic (without O_2) conditions one gets ATP only from the glycolysis pathway. When oxygen is available, the lactic acid is reoxidized to pyruvic acid and the aerobic process diagramed operates.

The details of these processes and some other tangential events will now unfold.

Problem 23.1 What major metabolic sequence of reactions requires oxygen to operate?

23.2 GLYCOLYSIS

Carbohydrate metabolism begins with **glycolysis,** a sequence of reactions that converts glucose into pyruvic acid. Glycolysis is also known as the Embden-Meyerhof pathway for the scientists who elucidated the sequence. The glucose–to–pyruvic acid transformation occurs in almost all biological systems. However, the fate of the pyruvic acid after it is formed varies among organisms and under different conditions.

Figure 23.1 shows the glycolysis pathway. Note that from every mole of glucose, which has six carbon atoms, the yield is 2 mol of pyruvic acid, which has three carbon atoms.

Let us now begin a consideration of the individual steps of the glycolysis sequence.

STEP 1 The first step is formation of a phosphate ester from the alcohol functional group at carbon atom 6 of glucose. Recall that the glucopyranose ring is numbered

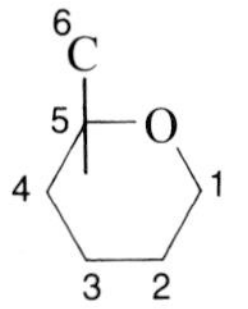
6
C
5
O
4
1
3 2

The source of the phosphate group is ATP; that is, ATP is converted to ADP as it transfers its terminal phosphate grouping to glucose. Hexokinase catalyzes the reaction.

STEP 2 The next step is an isomerization reaction. The "open-chain" forms (see Section 15.2) of the glucose and fructose phosphate esters show the essence of the reaction more clearly:

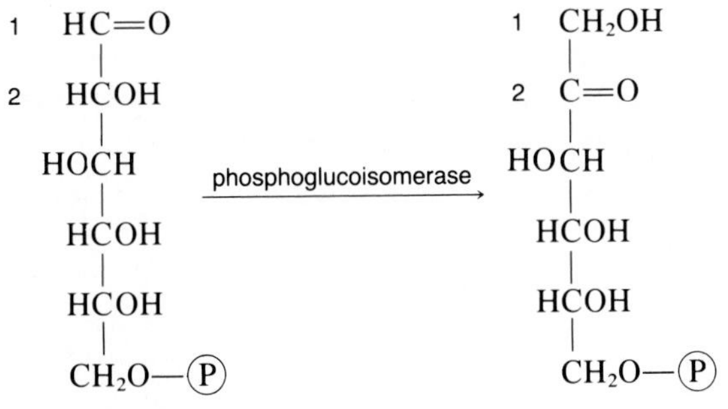
1 HC=O
2 HCOH
HOCH
HCOH
HCOH
CH₂O—(P)
phosphoglucoisomerase
1 CH₂OH
2 C=O
HOCH
HCOH
HCOH
CH₂O—(P)

Glucose 6-phosphate — Fructose 6-phosphate

The isomerization reaction involves an interchange of hydroxyl and carbonyl groups between carbon atoms 1 and 2 in the hexose chain. Fructose itself is easily metabolized by the body because fructose can readily enter the glycolysis pathway through direct phosphorylation,

$$\text{Fructose} \xrightarrow[\text{ATP} \quad \text{ADP}]{} \text{fructose 6-phosphate}$$

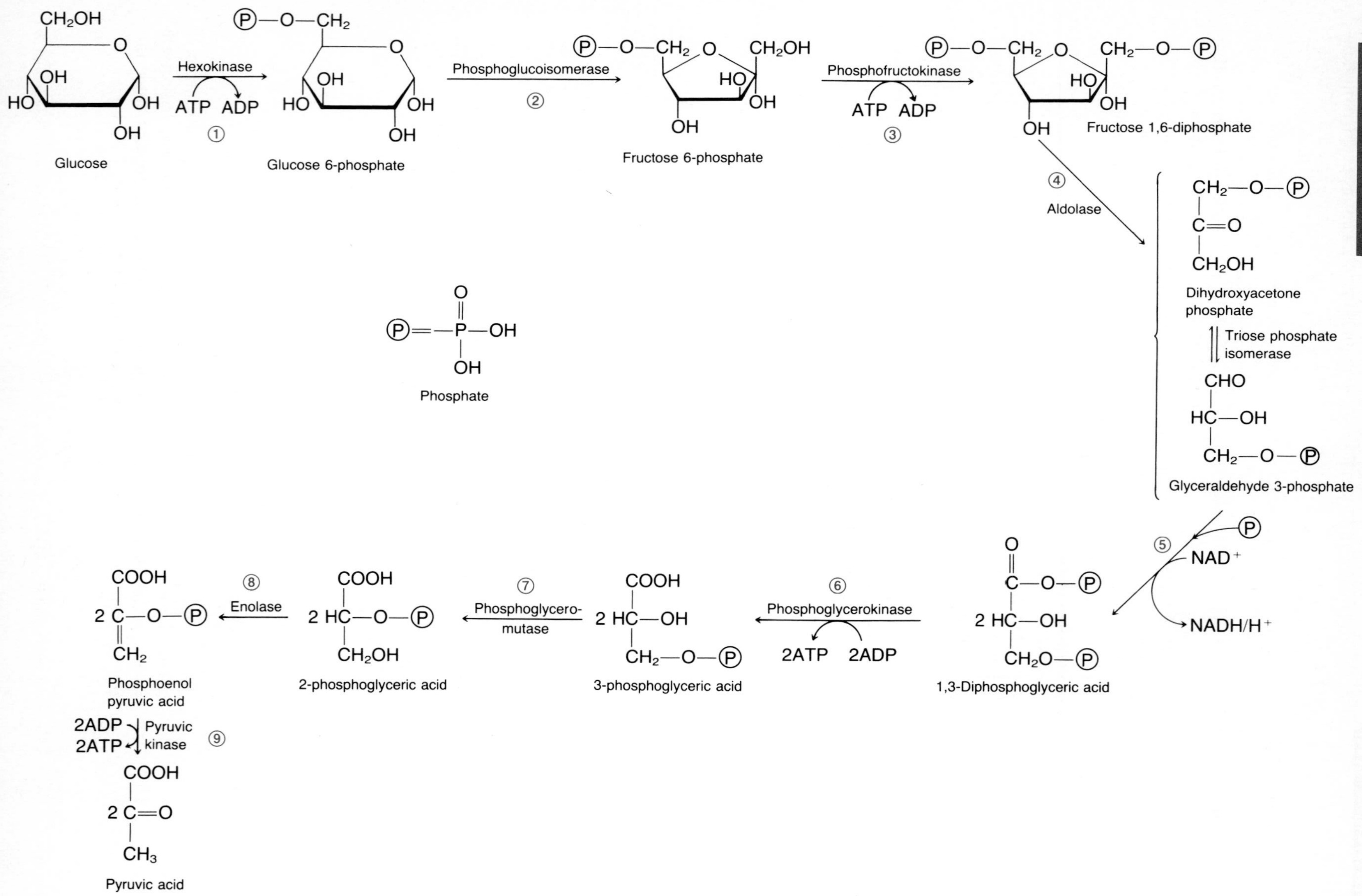

Glucose
Hexokinase
ATP ADP
①
Glucose 6-phosphate
Phosphoglucoisomerase
②
Fructose 6-phosphate
Phosphofructokinase
ATP ADP
③
Fructose 1,6-diphosphate
④
Aldolase
Dihydroxyacetone phosphate
Triose phosphate isomerase
Glyceraldehyde 3-phosphate
Phosphate
⑤
NAD+
NADH/H+
1,3-Diphosphoglyceric acid
⑥
Phosphoglycerokinase
2ATP 2ADP
3-phosphoglyceric acid
⑦
Phosphoglycero-mutase
2-phosphoglyceric acid
⑧
Enolase
Phosphoenol pyruvic acid
2ADP
2ATP
Pyruvic kinase
⑨
Pyruvic acid

FIG. 23.1
The glycolysis pathway. The overall result is the degradation of 1 mol of glucose into 2 mol of pyruvic acid. Notice the coefficient of 2 before the three-carbon compounds that arise from cleavage of the original six-carbon sugar.

STEP 3 Next is another phosphorylation by ATP. This step, which is catalyzed by phosphofructokinase, produces fructose 1,6-diphosphate, which is shown in Figure 23.1 in its furanose ring form and appears here as the open-chain form:

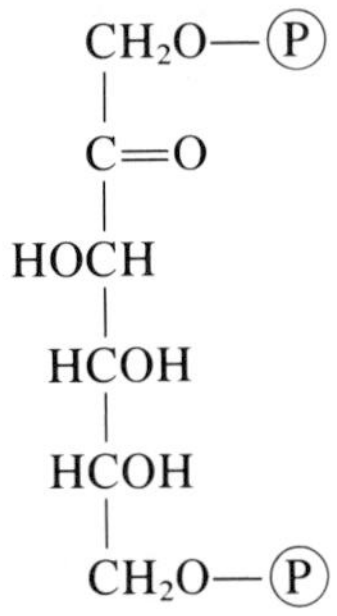

Fructose 1,6-diphosphate

STEP 4 Next is the cleavage of the six-carbon chain into two three-carbon fragments, each of which will eventually become pyruvic acid. This cleavage reaction is catalyzed by the enzyme aldolase, so named because the reverse of this reaction wherein the two three-carbon units would be joined together belongs to a class of organic reactions known as *aldol* condensations (see Section 14.14). Once again the structural details of the reaction are more readily appreciated through an examination of the open-chain form of fructose:

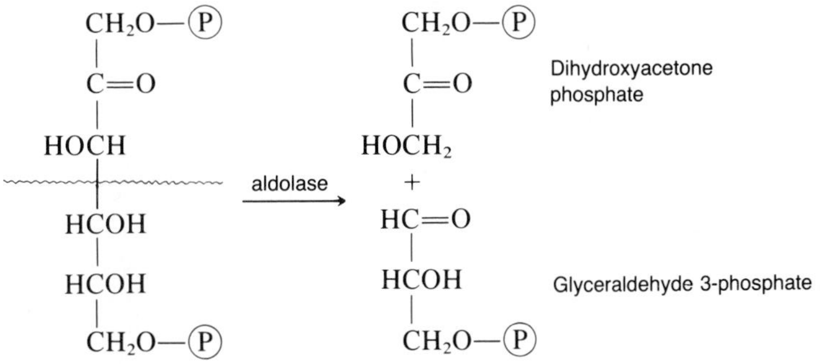
aldolase
Dihydroxyacetone phosphate
Glyceraldehyde 3-phosphate

Fructose 1,6-diphosphate

Only glyceraldehyde 3-phosphate can continue along the glycolysis pathway. However, dihydroxyacetone phosphate is readily isomerized to glyceraldehyde 3-phosphate by the action of the enzyme triose phosphate isomerase:

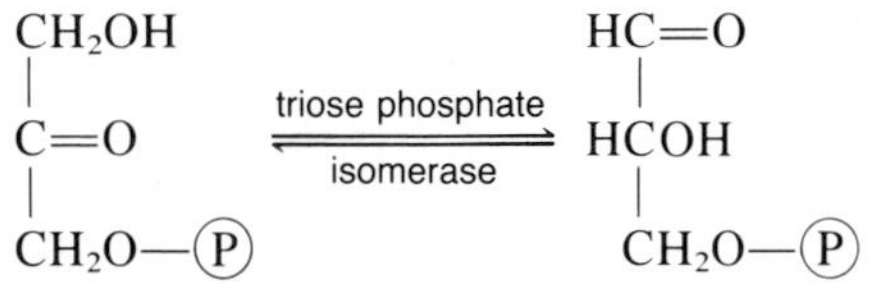
triose phosphate
isomerase

Dihydroxyacetone phosphate (written "upside down" compared with the previous structure)

Glyceraldehyde 3-phosphate

Although this equilibrium lies far to the left, it is shifted toward glyceraldehyde 3-phosphate as that molecule reacts in step 5 of the glycolysis pathway. Thus, effectively, step 4 is

$$\text{1 molecule of fructose 1,6-diphosphate} \xrightarrow{(4)} \text{2 molecules of glyceraldehyde 3-phosphate}$$

STEP 5 Step 5 is an oxidation and phosphorylation. The oxidation is of a type previously encountered (Section 14.12), that is, the oxidation of an aldehyde (H—C=O) to a carboxylic acid (HO—C=O) functional group. In this case the carboxylic acid is being esterified (phosphate ester) at the same time, so that the reaction may not seem so familiar. Because a carbon-oxygen bond is being oxidized, NAD^+ is the coenzyme involved. One of the two hydrogens that interact with NAD^+ comes from the aldehyde carbon and one from "P_i." Effectively,

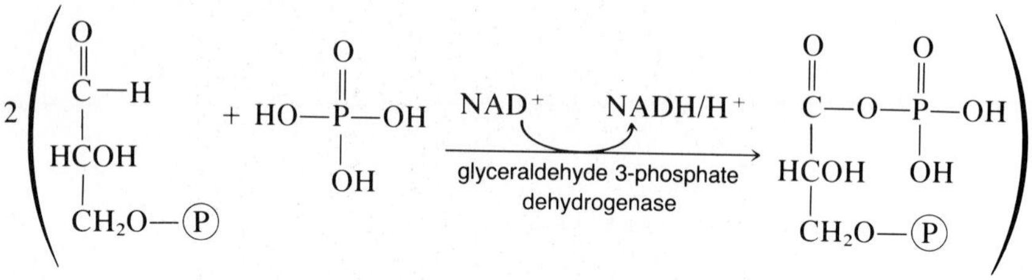

2
C—H
HCOH
CH₂O—Ⓟ
+ HO—P—OH
OH
NAD⁺
NADH/H⁺
glyceraldehyde 3-phosphate
dehydrogenase
C—O—P—OH
HCOH OH
CH₂O—Ⓟ

We multiply this reaction by 2 because 2 mol of three-carbon compounds is produced for every 1 mol of hexose which enters the glycolysis pathway.

STEP 6 Step 6 is an example of *substrate-level phosphorylation* (Section 22.8). ATP is synthesized from ADP as a phosphate group is removed from 1,3-diphosphoglyceric acid. ΔG for the hydrolysis of the 1,3-diphosphoglyceric acid–phosphate bond is 11.8 kcal/mol. In this step of the glycolysis pathway 2 mol of ATP is produced for every mole of hexose which enters the pathway because each mole of hexose produces 2 mol of 1,3-diphosphoglyceric acid.

STEP 7 Step 7 is another isomerization reaction. In this case there is an interchange of alcohol and phosphate ester functional groups between positions 3 and 2. The catalyzing enzyme is phosphoglyceromutase.

STEP 8 Step 8 is a dehydration reaction catalyzed by the enzyme enolase. The enzyme is so named because the product is the phosphate ester of an *enol,* i.e., a compound which has two functional groups: a carbon-carbon double bond (*-ene*) and an alcohol group (*-ol*).

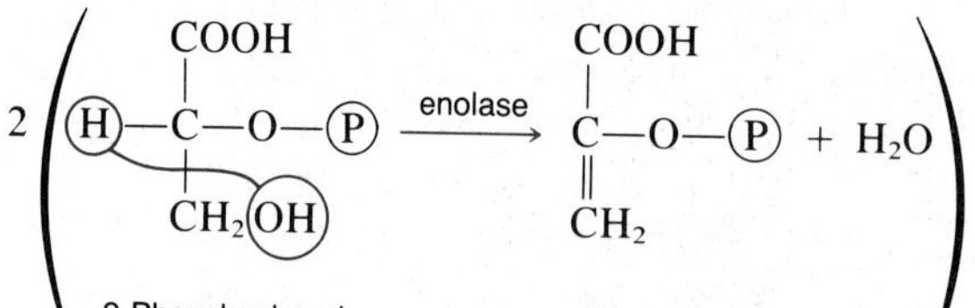

2
COOH
Ⓗ—C—O—Ⓟ
CH₂Ⓞ̲Ḥ
enolase
COOH
C—O—Ⓟ + H₂O
CH₂
2-Phosphoglyceric
acid
Phosphoenol-
pyruvic acid

This step sets the stage for another substrate-level phosphorylation because the phosphate bond in phosphoenolpyruvic acid is a very high-energy P—O bond. ΔG for its hydrolysis is -14.8 kcal/mol.

STEP 9 In step 9, substrate-level phosphorylation occurs as the phosphate group is transferred from phosphoenolpyruvic acid to ADP to make ATP. The initially produced enol immediately isomerizes to the keto form of pyruvic acid:

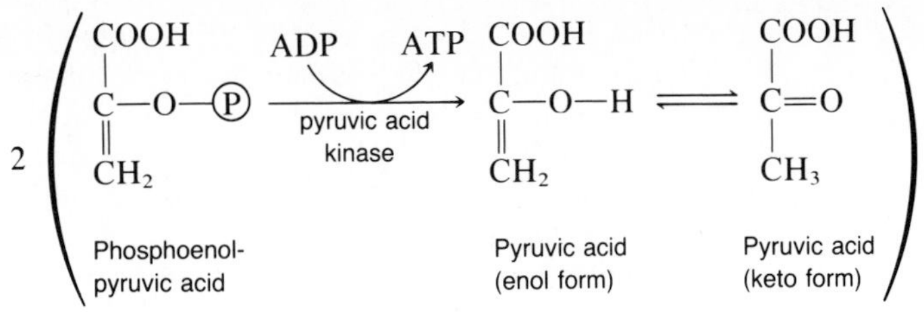
COOH
ADP
ATP
pyruvic acid kinase
Phosphoenol-pyruvic acid
Pyruvic acid (enol form)
Pyruvic acid (keto form)

Enol-keto conversions are commonplace events in aldehydes and ketones.

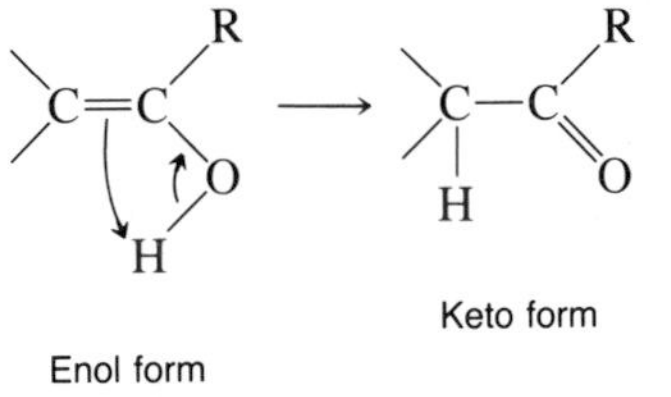
Keto form
Enol form

The glycolysis pathway provides energy for making ATP. The maximum energy yield in terms of ATP is 6 mol of ATP per mole of glucose (or fructose) converted to pyruvic acid (2 mol). Look back at the individual steps and see the ATP generated or consumed. In summary:

		Change in Moles of ATP per Mole of Glucose
Step 1:	Glucose ⟶ glucose 6-phosphate	−1
Step 3:	Fructose 6-phosphate ⟶ fructose 1,6-diphosphate	−1
Step 5:	2 NAD^+ ⟶ 2 $NADH/H^+$	+4
Step 6:	2 (1,3-diphosphoglyceric acid) ⟶ 2(3-phosphoglyceric acid)	+2
Step 9:	2 (phosphoenolpyruvic acid) ⟶ 2 (pyruvic acid)	+2
		+6

Step 5 needs a little more explanation. ATP is consumed or made directly in steps 1, 3, 6, and 9, but in step 5, ATP is generated because the 2 mol of $NADH/H^+$ formed produce ATP in the ETS. As you learned in Section 22.7, 1 mol of $NADH/H^+$ produces 3 mol of ATP in the ETS if the $NADH/H^+$ is in the mitochondria where the ETS operates. However, glycolysis occurs in the cytosol and transport of 1 mol of $NADH/H^+$ from the cytosol to the mitochondria requires 1 mol of ATP. Therefore, the net yield of ATP from $NADH/H^+$ produced in the cytosol is only 2 mol of ATP per mole of reduced NAD^+. This applies to skeletal muscle cells which we are discussing here and to which all problems refer. In heart and liver cells $NADH/H^+$ is brought into the mitochondria through a concentration gradient without consumption of ATP.

Because the ETS requires oxygen, these 4 mol of ATP in step 5 will not be produced if no oxygen is available to allow oxidation of $NADH/H^+$ in the ETS. Thus under **anaerobic** conditions (absence of oxygen) a total of only 2 mol of ATP will be produced by glycolysis. You will see the implications of this later in the chapter.

SAMPLE EXERCISE 23.1

One mole of glucose 6-phosphate enters the glycolysis pathway and begins to undergo the sequence of reactions shown in Figure 23.1. How many moles of ATP are generated or consumed as glucose 6-phosphate undergoes conversions through step 6?

Solution Examining Figure 23.1, we see that glucose 6-phosphate will begin reacting at step 2. Step 3 uses up an ATP as the conversions take place. We would count this as a loss of 1 mol of ATP (-1) per mole of glucose 6-phosphate. Then in step 5, 2 mol of NAD^+ is reduced to $NADH/H^+$ because 2 mol of 3-phosphoglyceraldehyde is produced from each mole of glucose 6-phosphate. If 2 mol of $NADH/H^+$ enter the mitochondria and react in the ETS, 4 mol of ATP ($+4$) will be produced. Step 6 leads to the synthesis of 2 mol of ATP ($+2$) per mole of glucose 6-phosphate. If oxygen is available to run the ETS, the net yield of ATP is $-1+4+2 = +5$ mol of ATP. If oxygen is not available to run the ETS, then the net yield is $-1+2 = +1$ mole of ATP.

23.3 THE FATES OF PYRUVIC ACID

Glycolysis is nearly a universal process in biological systems, but the pyruvic acid it produces undergoes different transformations in different organisms under different conditions. Figure 23.2 summarizes the three fates of pyruvic acid in living systems.

Fermentation is the conversion of glucose to ethyl alcohol which yeast organisms can accomplish. After glucose is converted into pyruvic acid, the pyruvic acid is converted to ethyl alcohol through the following decarboxylation ($-CO_2$) and reduction ($+2H\cdot$) reactions:

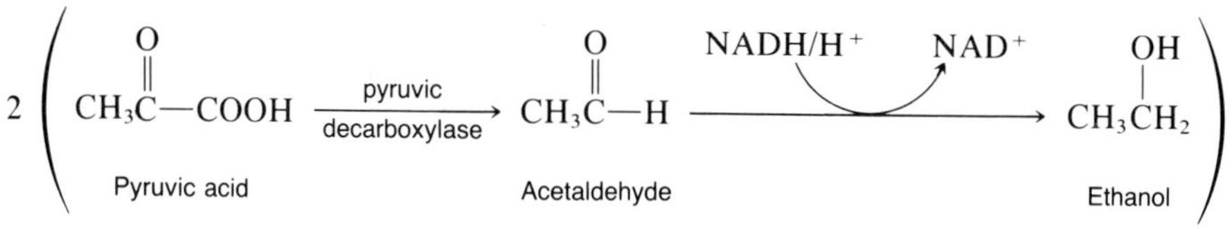

2
CH₃C—COOH
pyruvic decarboxylase
CH₃C—H
NADH/H⁺
NAD⁺
CH₃CH₂
OH
Pyruvic acid
Acetaldehyde
Ethanol

FIG. 23.2
The three fates of pyruvic acid in living systems. In yeast and some other microorganisms fermentation occurs and the final product of metabolism is ethanol. In human beings the fate of pyruvic acid depends on the extent of the availability of oxygen.

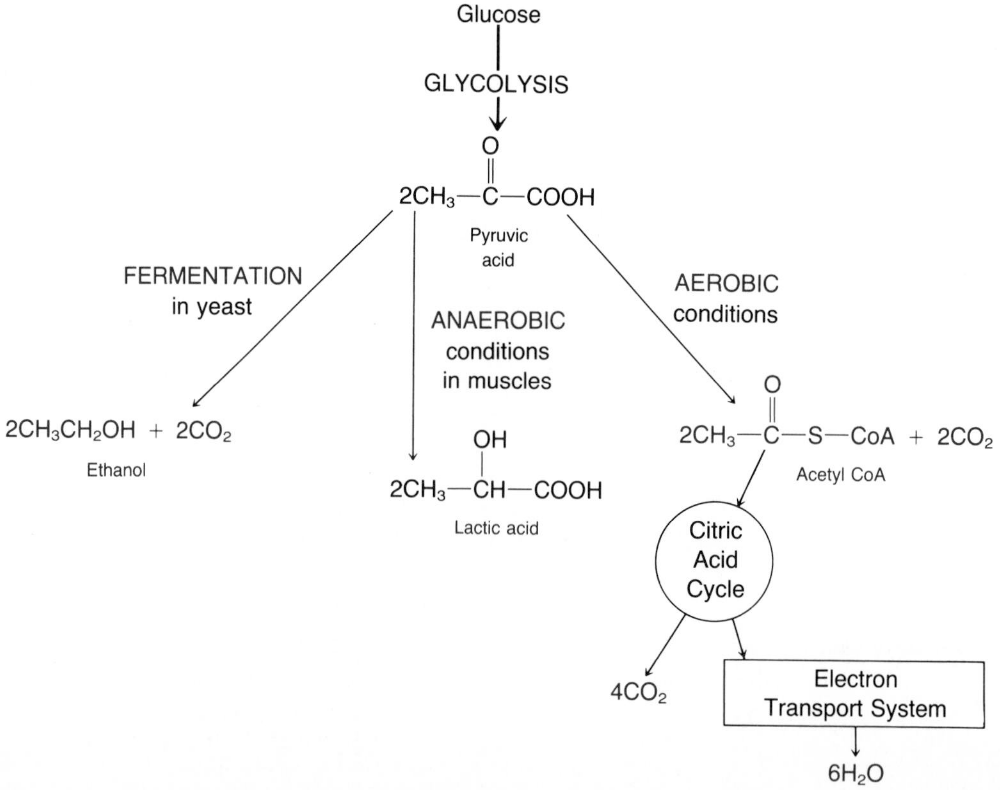

Glucose
GLYCOLYSIS
2CH₃—C—COOH
Pyruvic acid
FERMENTATION in yeast
ANAEROBIC conditions in muscles
AEROBIC conditions
2CH₃CH₂OH + 2CO₂
Ethanol
2CH₃—CH—COOH
OH
Lactic acid
2CH₃—C—S—CoA + 2CO₂
Acetyl CoA
Citric Acid Cycle
4CO₂
Electron Transport System
6H₂O

The overall energy yield is 2 mol of ATP per mole of glucose. In fermentation the 2 mol of NADH/H^+ produced in step 5 of glycolysis does not enter an ETS. Rather, it is used to reduce acetaldehyde to ethanol. Therefore, the yield of ATP is simply the sum of steps 1, 3, 6, and 9, as identified previously. Notice that the reduction of acetaldehyde to ethanol provides for the oxidation of NADH/H^+ to NAD^+ and the fermentation is thus self-sustaining. Fermentation is an *anerobic* process, i.e., it occurs in the absence of oxygen.

Glycolysis is also an anaerobic process, since no oxygen is required for the sequence of reactions shown in Figure 23.1. However, in order for glycolysis to continue in cells, there must be a renewed supply of NAD^+. If ample oxygen is available, NAD^+ is regenerated from NADH/H^+ in the ETS. This is an **aerobic** process.

In muscle cells during vigorous exercise there is insufficient oxygen to sustain glycolysis aerobically. Under these conditions the body regenerates NAD^+ anaerobically through the reduction of pyruvic acid to lactic acid:

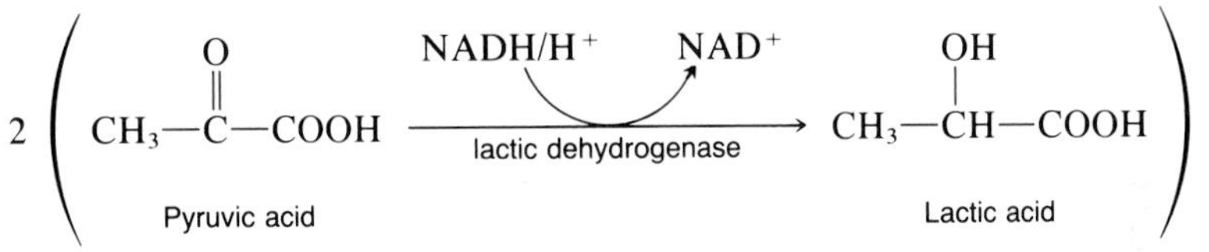

2
NADH/H+
NAD+
lactic dehydrogenase
Pyruvic acid
Lactic acid

This allows glycolysis to continue. The overall energy yield for anaerobic glycolysis, glucose ⟶ (2) lactic acid, is 2 mol of ATP per mole of glucose. As in the case of fermentation, the 2 mol of NADH/H^+ produced in step 5 of glycolysis is not oxidized in the ETS.

Problem 23.2 How many moles of ATP are generated from the metabolism of 2 mol of fructose under anaerobic conditions?

Under aerobic conditions the full energy potential (6 mol of ATP per mole of glucose) of the glycolysis pathway is realized and an additional 6 mol of ATP is generated as pyruvic acid is converted into acetyl CoA by reacting with CoASH:

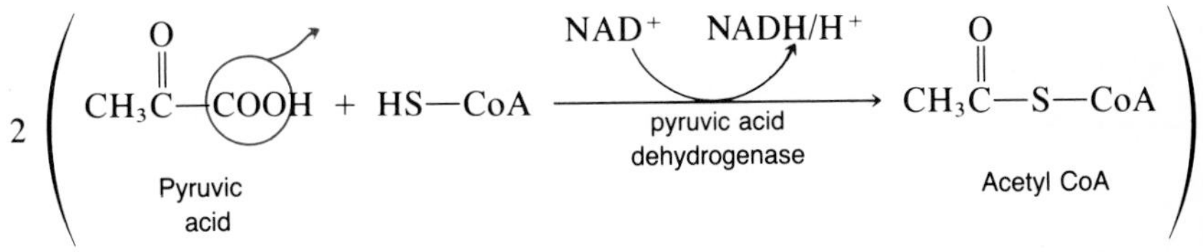

2
NAD+
NADH/H+
pyruvic acid
dehydrogenase
Pyruvic
acid
Acetyl CoA

Notice that what has happened to glucose up to this point is that it has been converted to two acetyl groups which become attached to CoASH. Two CO_2 molecules are also produced as waste products of the metabolism:

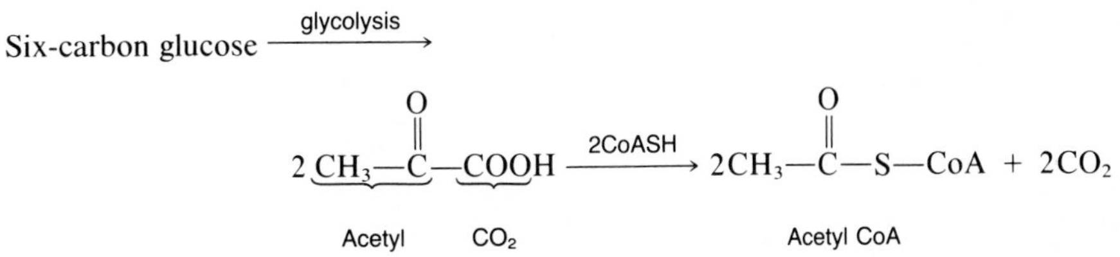

Six-carbon glucose
glycolysis
2CoASH
Acetyl
CO2
Acetyl CoA

The oxidative decarboxylation of pyruvic acid and conversion to acetyl CoA sends 2 mol of NADH/H^+ (per mole of glucose) to the ETS. Because this process happens in the mitochondria, the full potential of $NADH/H^+$ (3 mol of ATP per mole of $NADH/H^+$) is realized, and 6 mol of ATP is produced. Thus, under aerobic conditions the conversion of each mole of glucose to acetyl CoA results in the production of 12 mol of ATP (6+6) prior to the entry of the acetyl CoA into the citric acid cycle.

Problem 23.3 How many moles of ATP are generated if 1 mol of fructose is metabolized into acetyl CoA?

23.4 THE CITRIC ACID CYCLE

The sequence of metabolic reactions shown in Figure 23.3, operating in conjunction with the ETS, is the largest energy producer (in the form of ATP) in the body. This sequence is called the citric acid cycle because citric acid is the first compound formed in step 1. It is also known as the Krebs cycle in honor of Hans Krebs, who elucidated the sequence in 1937 and who won the Nobel Prize in 1953 for this work. Another commonly used name is the *tri*carboxylic *a*cid (TCA) cycle; this name applies because four compounds in the sequence are carboxylic acids with three —COOH groups.

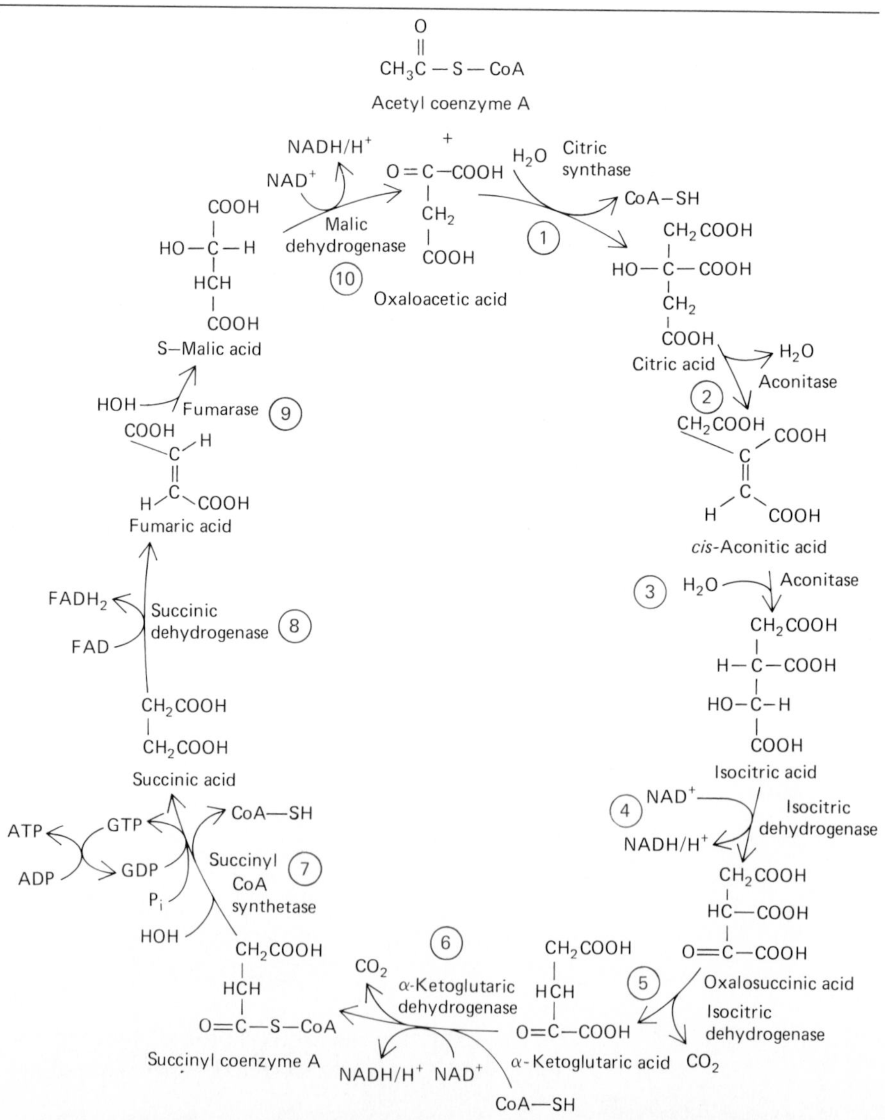
Acetyl coenzyme A
Citric synthase
CoA–SH
Citric acid
Aconitase
cis-Aconitic acid
Aconitase
Isocitric acid
Isocitric dehydrogenase
Oxalosuccinic acid
Isocitric dehydrogenase
α-Ketoglutaric acid
α-Ketoglutaric dehydrogenase
Succinyl coenzyme A
Succinyl CoA synthetase
Succinic acid
Succinic dehydrogenase
Fumaric acid
Fumarase
S–Malic acid
Malic dehydrogenase
Oxaloacetic acid

FIG. 23.3
The citric acid cycle. The ten sequential steps regenerate oxaloacetic acid, a necessary reactant in step 1. Four steps (4, 6, 8, and 10) produce reduced coenzymes which go to the ETS to generate ATP. Step 7 results in ATP production directly. As explained in Section 23.4, at physiological pH citric acid and other acids of the cycle are found predominantly in the anionic form (all carboxylic protons are ionized).

The sequence is cyclic because the product of the last step (step 10), oxaloacetic acid, is a necessary reactant in step 1.

Recall from Chapter 11 that acids in solution are equilibrium mixtures:

$$R{-}COOH \rightleftharpoons H^+ + RCOO^-$$

The position of the equilibrium depends on the pH ($[H^+]$ concentration) of the solution. At physiological pH values, the equilibrium lies to the right; that is, in the body the acid exists predominantly as the anion. For example, pyruvic acid ($CH_3{-}\overset{\overset{\displaystyle O}{\|}}{C}{-}COOH$) is present in cells mostly in the form of the pyruvate anion ($CH_3{-}\overset{\overset{\displaystyle O}{\|}}{C}{-}COO^-$). Although the anionic form predominates, all the components of the equilibrium mixture are present and the equilibrium can shift under conditions of stress. Despite the fact that "RCOOH is really mostly $RCOO^-$" in the body, we have elected to portray the acid structures in figures such as Figure 23.3 and to call compounds by their acid names rather than the corresponding anionic names. This practice is in keeping with what is found in most anatomy and physiology texts and other books which deal with medically related topics.

Some of the reactions of the citric acid cycle are straightforward examples of simple organic reactions you have seen before. For example, step 2 is a simple dehydration. Other reactions are more complicated and less familiar, but it is possible to see what has transpired by examining the structures of the starting materials and the products into which they are converted. Let us begin a stepwise consideration of the citric acid cycle.

STEP 1 In step 1 there are actually two reactions taking place. First, acetyl CoA adds to oxaloacetic acid through an aldol addition. Then the thioester is hydrolyzed to the acid and thiol:

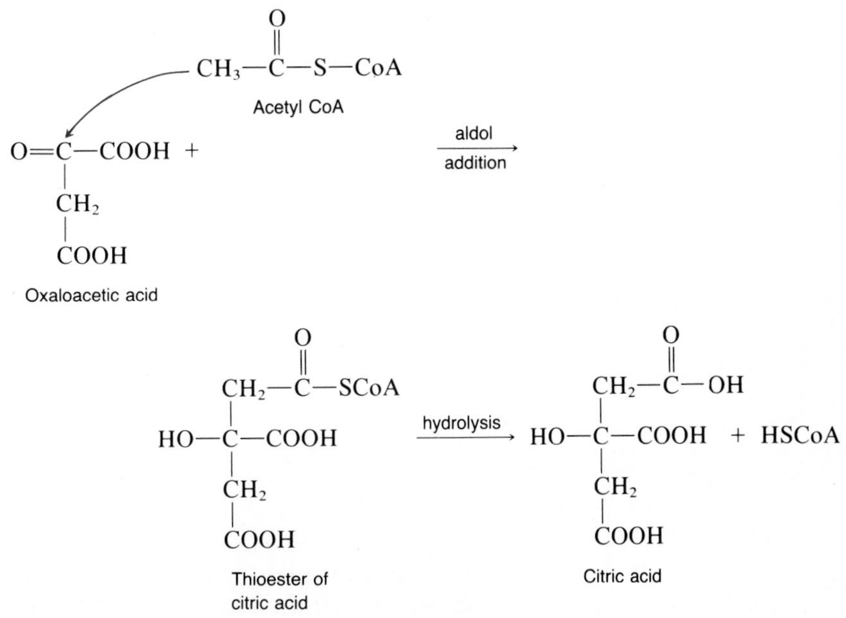
Acetyl CoA
aldol addition
Oxaloacetic acid
hydrolysis
+ HSCoA
Thioester of citric acid
Citric acid

Notice that coenzyme A is regenerated in this step, enabling it to go back and react with pyruvic acid or some other molecule.

STEPS 2 AND 3 Step 2 is a dehydration (−HOH), and step 3 is the addition of water (+HOH) to a carbon-carbon double bond. The overall result of steps 2 and 3 is the isomerization of citric acid into isocitric acid through the intermediacy of *cis*-aconitic acid:

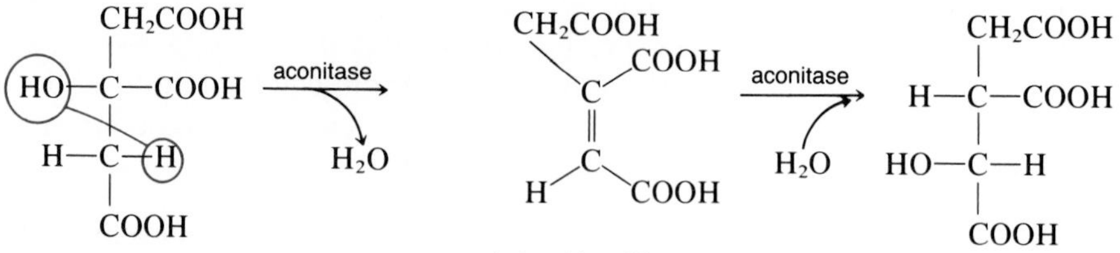
aconitase
aconitase

Citric acid
cis-Aconitic acid
Isocitric acid

STEPS 4 AND 5 Step 4 is a straightforward oxidation of a secondary alcohol to a ketone. NAD^+ is the coenzyme for the reaction. The same enzyme, isocitric dehydrogenase, catalyzes step 4 and step 5, which is a decarboxylation.

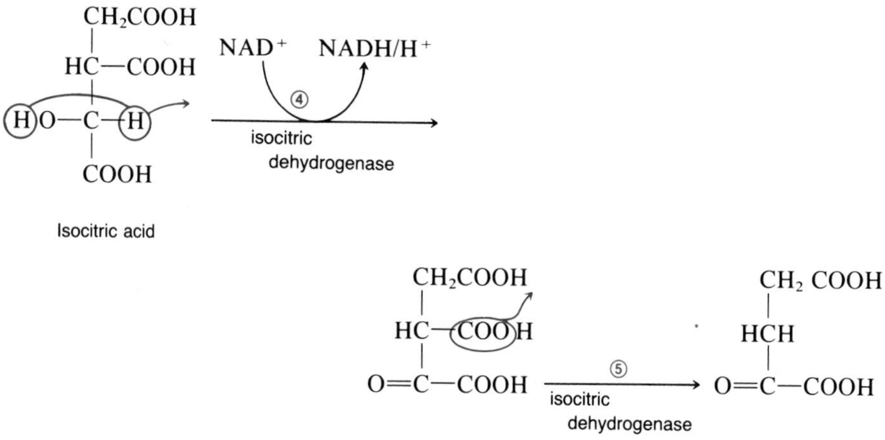
isocitric dehydrogenase
isocitric dehydrogenase

Isocitric acid
Oxalosuccinic acid
α-Ketoglutaric acid

STEP 6 Step 6 can be described as an oxidative decarboxylation. It is very similar to the reaction whereby pyruvic acid (also an α-keto acid) is converted to acetyl CoA.

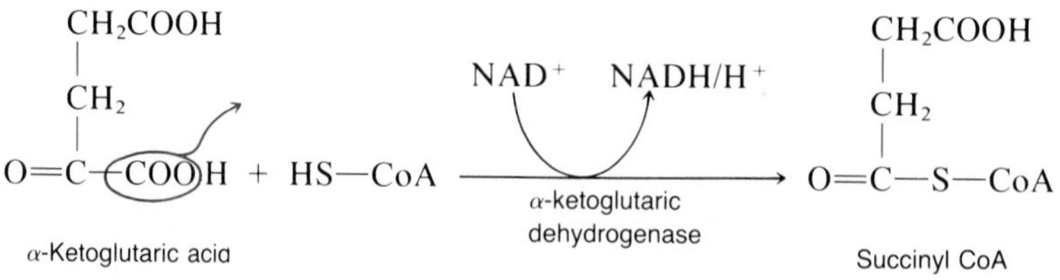
α-ketoglutaric dehydrogenase

α-Ketoglutaric acid
Succinyl CoA

STEP 7 Step 7 involves hydrolysis of a thioester to its corresponding acid (succinic acid). The hydrolysis is sufficiently exergonic to be coupled to endergonic phosphate bond formation. The overall result is the production of ATP through substrate-level phosphorylation, the only example of this process in the citric acid cycle. In this case, another nucleotide, GTP, guanosine triphosphate, is an intermediate which passes the energy from the hydrolysis of succinyl CoA for use in making ATP. The sequence that occurs in step 7 is the hydrolysis of succinyl CoA coupled to GTP formation.

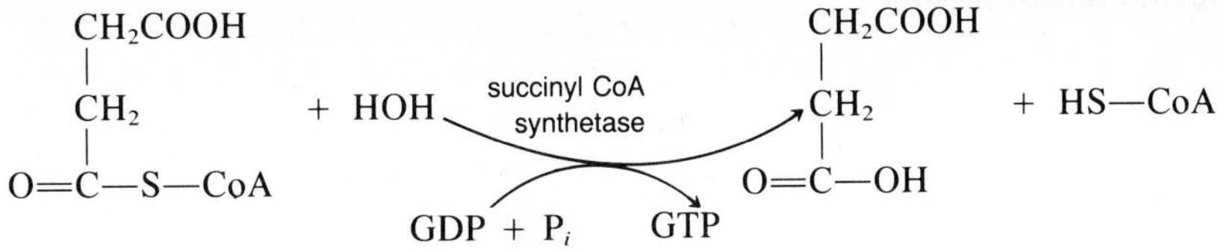
+ HOH
succinyl CoA synthetase
GDP + P_i
GTP
+ HS—CoA

Succinyl CoA   Succinic acid

GTP then phosphorylates ADP to make ATP.

$$\text{GTP} + \text{ADP} \xrightarrow[\text{diphosphokinase}]{\text{nucleoside}} \text{ATP} + \text{GDP}$$

STEP 8 Step 8 is the oxidation of a carbon-carbon bond, and hence FAD is the coenzyme. This is the only step in the Krebs cycle which requires FAD. Notice that only the trans geometric isomer is formed.

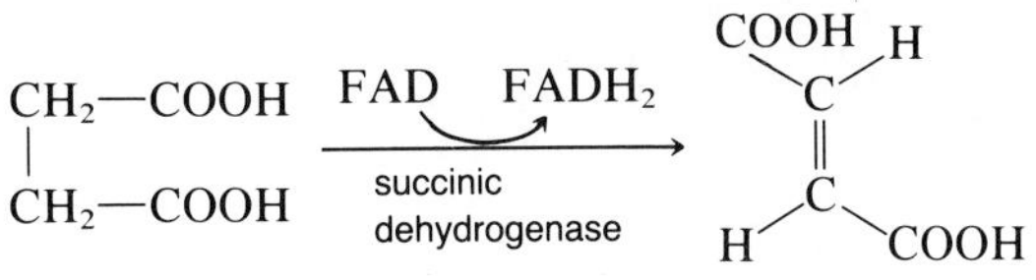
FAD
$FADH_2$
succinic dehydrogenase

Succinic acid   Fumaric acid

Problem 23.4 The cis isomer of fumaric acid is maleic acid, HC—COOH ‖ HC—COOH. Maleic acid cannot enter the citric acid cycle at step 9 even though it is functionally identical to fumaric acid. Based on your knowledge of enzymes, explain why this is so.

STEP 9 Step 9 is the fumarase-catalyzed addition of water to fumaric acid. The addition occurs in such a way that only the one enantiomer of malic acid shown, and designated *S*, is obtained.

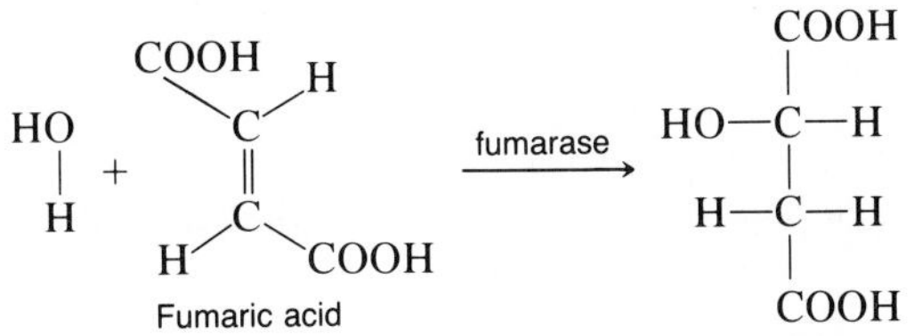
fumarase

Fumaric acid   *S*-Malic acid

Enzyme catalysis accounts for this "stereospecific" result. Hydration reactions (Section 13.8) catalyzed by acids result in mixtures of isomers.

STEP 10 Step 10 is the oxidation of a C—O bond, a secondary alcohol to a ketone. NAD^+ is therefore the coenzyme. This oxidation produces oxaloacetic acid and thus completes the cycle.

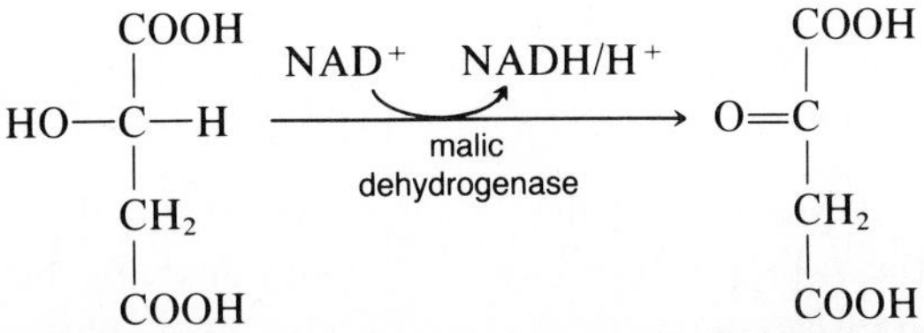
NAD^+
$NADH/H^+$
malic dehydrogenase

S-Malic acid   Oxaloacetic acid

Let us now count up the energy yield in terms of ATP per mole of acetyl CoA processed in the citric acid cycle by looking back at the individual steps.

Because the reactions of the citric acid cycle occur in the matrix of the mitochondria, each step involving NAD^+ (steps 4, 6, and 10) produces 3 mol of ATP when $NADH/H^+$ is oxidized in the ETS. The oxidation of $FADH_2$ (from step 8) in the ETS produces 2 mol of ATP. Step 7 produces 1 mol of ATP through substrate-level phosphorylation.

In summary:

	Moles of ATP Produced per Mole of Acetyl CoA
Step 4: $NAD^+ \longrightarrow NADH/H^+$	3
Step 6: $NAD^+ \longrightarrow NADH/H^+$	3
Step 7: $ADP \longrightarrow ATP$	1
Step 8: $FAD \longrightarrow FADH_2$	2
Step 10: $NAD^+ \longrightarrow NADH/H^+$	3
	12

That is, 12 mol of ATP is produced per mole of acetyl CoA entering the citric acid cycle.

23.5 ENERGY YIELD FROM GLUCOSE

Because 1 mol of glucose leads to the production of 2 mol of acetyl CoA, the energy yield of the citric acid cycle from 1 mol of glucose is 24 mol of ATP (2 × 12 mol of ATP per mole of acetyl CoA). Under aerobic conditions the total energy yield from 1 mol of glucose is 36 mol of ATP, as outlined here:

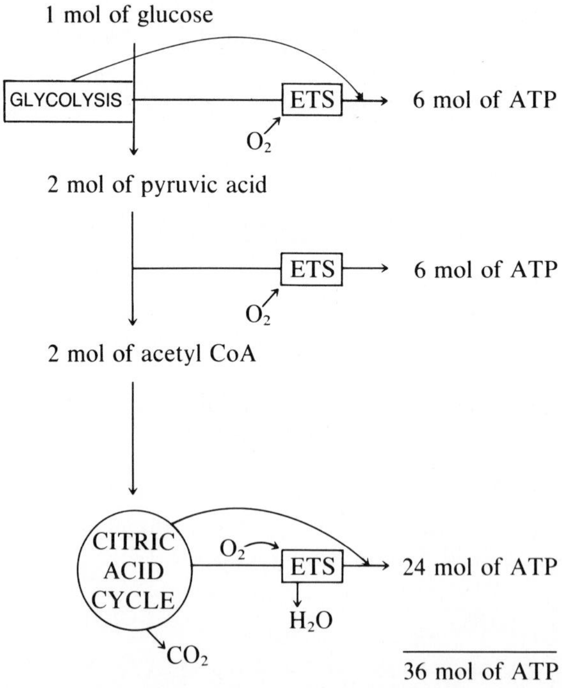
1 mol of glucose
GLYCOLYSIS
ETS
O_2
6 mol of ATP
2 mol of pyruvic acid
ETS
O_2
6 mol of ATP
2 mol of acetyl CoA
CITRIC ACID CYCLE
O_2
ETS
H_2O
CO_2
24 mol of ATP
36 mol of ATP

This process requires oxygen because it involves the ETS; that is, most of the ATP produced is not formed directly, but is derived from oxidation of coenzymes in the ETS. The glycolysis pathway produces only 2 mol of

ATP directly, but yields an additional 4 mol of ATP if NADH/H^+ can be oxidized in the ETS. Conversion of 2 mol of pyruvic acid to 2 mol of acetyl CoA produces 6 mol of ATP because 2 NADH/H^+ are oxidized in the ETS. Most of the energy from the citric acid cycle is derived through oxidation of coenzymes in the ETS.

If glucose is metabolized anaerobically (without oxygen), that is, without utilizing the ETS, the total energy yield is only 2 mol of ATP per mole of glucose. NADH/H^+ reduces pyruvic acid to lactic acid, rather than being oxidized in the ETS.

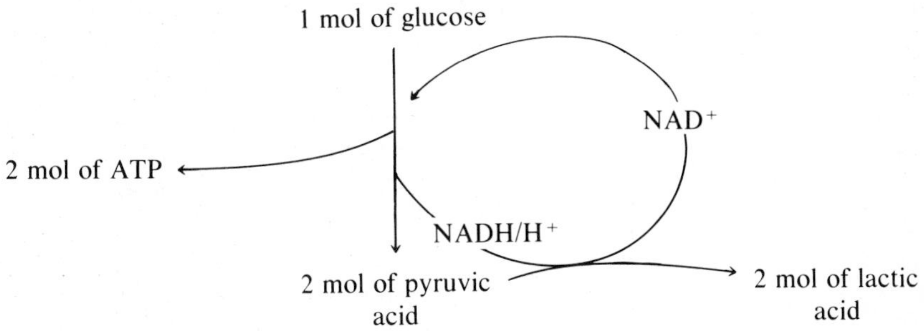
1 mol of glucose
NAD+
2 mol of ATP
NADH/H+
2 mol of pyruvic acid
2 mol of lactic acid

23.6 INTERPLAY OF AEROBIC AND ANAEROBIC METABOLISM

As mentioned in Sections 23.1 and 23.3, during vigorous exercise an *oxygen debt* is created in muscle cells. That is, oxygen cannot be brought to the cells fast enough, and the ETS and processes dependent on it shut down. The body must then depend on anaerobic metabolism of glucose to lactic acid to meet its immediate energy (ATP) needs.

SAMPLE EXERCISE 23.2

Two people consume 2 g of glucose apiece (2 g of glucose is approximately 0.01 mol of glucose). One person then sits down to read a book and the other sets off to run a race. How much ATP is likely to be made from the sugar each person has consumed?

Solution The reader will probably get the full energy yield from the 2 g because sufficient oxygen will be available for aerobic metabolism. Thus:

$$0.01 \text{ \sout{mol of glucose}} \times \frac{36 \text{ mol of ATP}}{1 \text{ \sout{mol of glucose}}} = 0.36 \text{ mol of ATP}$$

The runner is more likely to metabolize the glucose anaerobically and generate only 0.02 mol of ATP.

$$0.01 \text{ \sout{mol of glucose}} \times \frac{2 \text{ mol of ATP}}{1 \text{ \sout{mol of glucose}}} = 0.02 \text{ mol of ATP}$$

(This solution ignores the possibility of the body storing glucose rather than using it for energy production.)

The reduction to lactic acid provides for the regeneration of NAD^+ for subsequent use in glycolysis and also shifts some of the metabolic burden during strenuous exercise from muscle cells to liver cells. This happens in the following way. Lactic acid (lactate) can diffuse across cell membranes even more readily than pyruvic acid (pyruvate). (At physiological pH values, it is the anions that cross the membrane.) Lactic acid leaves muscle cells and is transported by blood to liver cells. In the liver it is oxidized back to

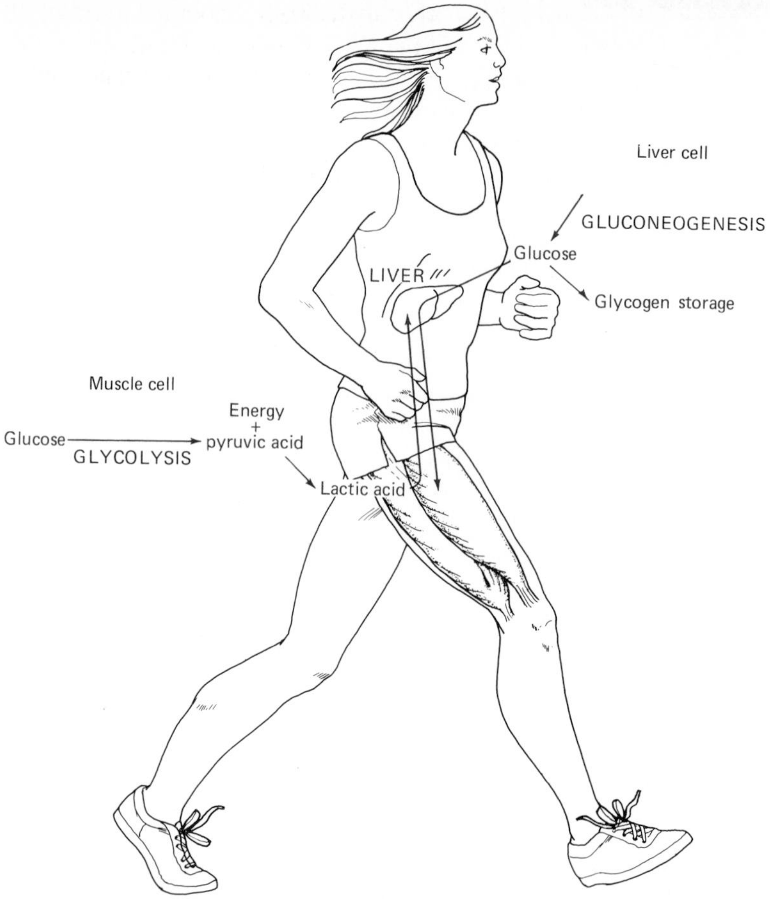
Liver cell
GLUCONEOGENESIS
Glucose
LIVER
Glycogen storage
Muscle cell
Energy
+
pyruvic acid
Glucose
GLYCOLYSIS
Lactic acid

FIG. 23.4
The Cori cycle. Lactic acid in muscles travels in the bloodstream to the liver, where it is converted into glucose. The glucose can return to active muscle for additional glycolysis or it may be stored as glycogen.

pyruvic acid, which is then converted into *glucose* through the process of gluconeogenesis, which we will examine in the next section. Glucose then returns via the bloodstream to muscle cells, where it undergoes glycolysis. This interrelationship between muscle cell and liver cell metabolic processes is known as the Cori cycle, and the details of it are summarized in Figure 23.4.

The body can only buy time, it cannot go on indefinitely employing this less efficient anaerobic metabolism. The accumulation of lactic acid in muscles during exercise leads to metabolic acidosis (Section 11.14) as the lactic acid diffuses into the blood. Acidosis would eventually cause us to collapse if we didn't stop to rest. The initial bodily response to acidosis is heavy breathing, an attempt to replenish oxygen in cells. The soreness in muscles after exercising is produced largely by excess lactic acid deposits.

When we rest, the body can oxidize lactic acid back to pyruvic acid and then convert this to acetyl CoA for entry into the citric acid cycle and aerobic metabolism. Alternatively, the pyruvic acid can be converted to glucose, and the glucose units assembled into glycogen for storage, in processes we will now examine. Figure 23.5 summarizes how the body fulfills its energy needs during and after exercise.

23.7 GLUCONEOGENESIS

The body has a great need for glucose as metabolic fuel. This is especially true in brain tissue, where it is essentially the only fuel source. Glucose enters the blood through digestion of carbohydrates, and the body stores carbohydrates as glycogen, which can be hydrolyzed to glucose as it is

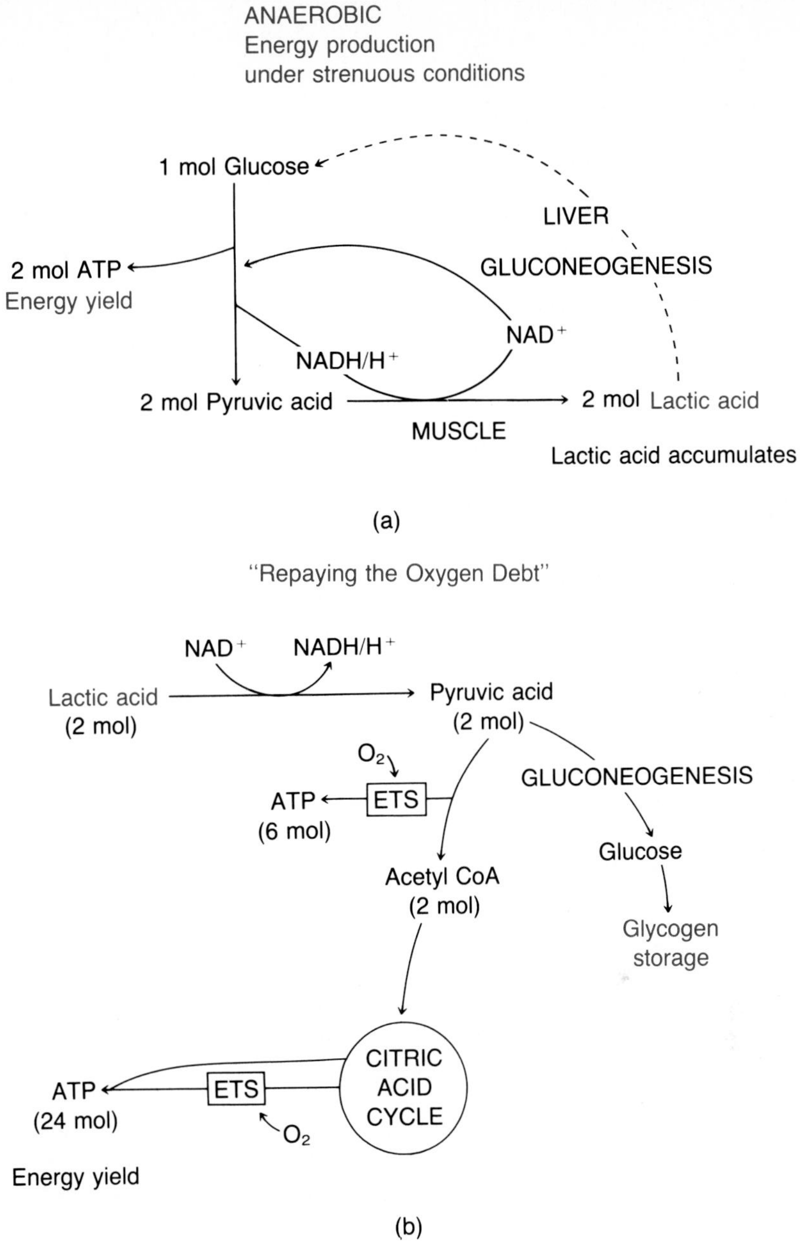
ANAEROBIC
Energy production
under strenuous conditions
1 mol Glucose
LIVER
GLUCONEOGENESIS
2 mol ATP
Energy yield
NAD+
NADH/H+
2 mol Pyruvic acid
2 mol Lactic acid
MUSCLE
Lactic acid accumulates
(a)
"Repaying the Oxygen Debt"
NAD+
NADH/H+
Lactic acid
(2 mol)
Pyruvic acid
(2 mol)
O2
GLUCONEOGENESIS
ATP
ETS
(6 mol)
Glucose
Acetyl CoA
(2 mol)
Glycogen
storage
CITRIC
ACID
CYCLE
ATP
ETS
(24 mol)
O2
Energy yield
(b)

FIG. 23.5
Interrelationships of anaerobic and aerobic metabolism of glucose. (*a*) Under anaerobic conditions glycolysis is the principal energy source yielding 2 ATP per mole of glucose. Glycolysis can continue because NAD^+ is replenished by the reduction of pyruvic to lactic acid. Lactic acid accumulates; only some is reconverted in the liver (Cori cycle). This cannot go on indefinitely. (*b*) When sufficient oxygen is available to run the ETS, energy needs are met by the more efficient citric acid cycle. Lactic acid is oxidized to pyruvic acid which enters the citric acid cycle, or if energy needs are met, is converted to glucose and glycogen for storage.

needed. In addition, the body has the ability to make glucose from noncarbohydrate sources.

This synthesis of glucose from noncarbohydrate precursors is called *gluconeogenesis*. Knowledge of the parts of this word helps us understand it: *gluco-* stands for "glucose," *neo-* means "new," and *genesis* means "a beginning or making of." Thus, gluconeogenesis is the making of "new"

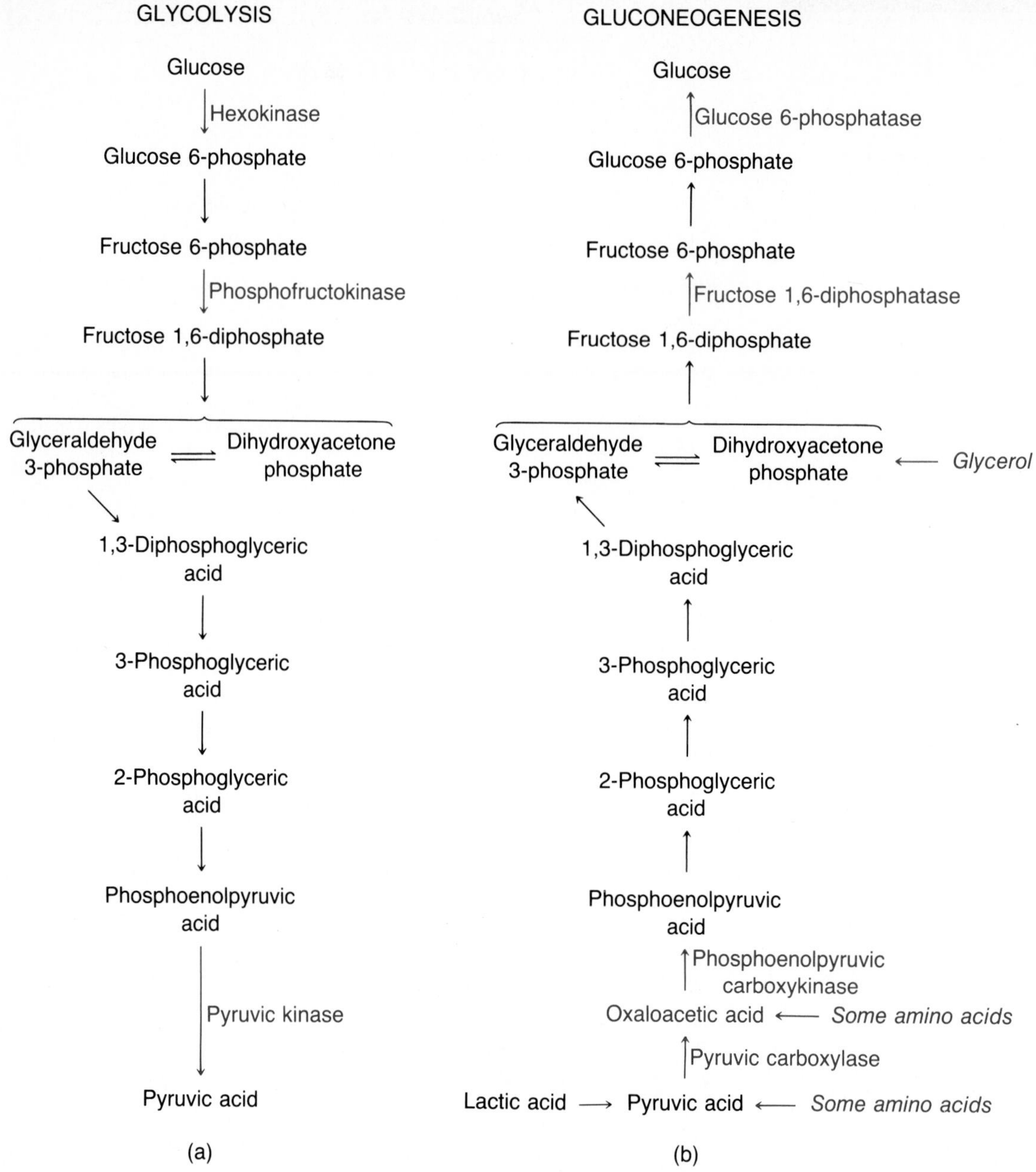
GLYCOLYSIS
Glucose
Hexokinase
Glucose 6-phosphate
Fructose 6-phosphate
Phosphofructokinase
Fructose 1,6-diphosphate
Glyceraldehyde 3-phosphate
Dihydroxyacetone phosphate
1,3-Diphosphoglyceric acid
3-Phosphoglyceric acid
2-Phosphoglyceric acid
Phosphoenolpyruvic acid
Pyruvic kinase
Pyruvic acid
(a)
GLUCONEOGENESIS
Glucose
Glucose 6-phosphatase
Glucose 6-phosphate
Fructose 6-phosphate
Fructose 1,6-diphosphatase
Fructose 1,6-diphosphate
Glyceraldehyde 3-phosphate
Dihydroxyacetone phosphate
Glycerol
1,3-Diphosphoglyceric acid
3-Phosphoglyceric acid
2-Phosphoglyceric acid
Phosphoenolpyruvic acid
Phosphoenolpyruvic carboxykinase
Oxaloacetic acid
Some amino acids
Pyruvic carboxylase
Lactic acid
Pyruvic acid
Some amino acids
(b)

FIG. 23.6
The glycolysis pathway (*a*) in which glucose is converted to pyruvic acid and the pathway of gluconeogenesis (*b*) in which pyruvic acid is converted to glucose are not the exact reverse of one another. Note the intermediacy of oxaloacetic acid in gluconeogenesis and its absence in glycolysis. Enzymes are specified only in steps in which they differ and are shown in color, as are the arrows of steps that differ. The entry points of noncarbohydrate sources of glucose are also shown in color.

glucose. The major noncarbohydrate precursors are lactic acid, amino acids from proteins, and glycerol from fat digestion. Gluconeogenesis occurs principally in the liver.

In Section 23.6 we alluded to gluconeogenesis and stated that the route was lactic acid ⟶ pyruvic acid ⟶ glucose. This might imply to you that gluconeogenesis is just the reverse of glycolysis (glucose ⟶ pyruvic acid), but this is not the case. Figure 23.6 compares and contrasts the two pathways.

It also shows the entry points for the three principal precursors, lactic acid, amino acids, and glycerol.

The most noticeable difference between glycolysis and gluconeogenesis involves pyruvic acid. In glycolysis, phosphoenolpyruvic acid is directly converted to pyruvic acid:

$$\text{Phosphoenolpyruvic acid} \xrightarrow[\text{pyruvic kinase}]{\text{ADP} \curvearrowright \text{ATP}} \text{pyruvic acid}$$

In gluconeogenesis, pyruvic acid is first converted to oxaloacetic acid before conversion to phosphoenolpyruvic acid:

$$\text{Pyruvic acid} \xrightarrow[\text{pyruvic carboxylase}]{CO_2 \searrow} \text{oxaloacetic acid} \xrightarrow[\text{phosphoenolpyruvic carboxykinase}]{\nearrow CO_2} \text{phosphoenolpyruvic acid}$$

The other differences are in the enzymes employed in the interconversions between fructose 1,6-diphosphate and fructose 6-phosphate and between glucose and glucose 6-phosphate.

Problem 23.5 What metabolic intermediate appears in the gluconeogenesis sequence, but not in the glycolysis pathway? What other major metabolic sequence does this intermediate appear in?

23.8 GLUCOSE AS A BIOSYNTHETIC PRECURSOR

So far we have considered glucose only as a fuel source, i.e., as a generator of ATP. Glucose is also needed in order to provide certain biosynthetic raw materials for the body's use in the construction of larger molecules. For example, cleavage of the hexose glucose provides pentose sugars for nucleotide synthesis. The process by which glucose is converted into pentose sugars is called the **pentose phosphate pathway**.

During the course of the pentose phosphate pathway, the coenzyme $NADP^+$ is reduced to $NADPH/H^+$. $NADP^+$ is a derivative of NAD^+, and the relationship of the two structures was shown in Figure 22.5. $NADPH/H^+$ is a more readily available source of reducing power than $NADH/H^+$. The production of $NADPH/H^+$ in the course of the pentose phosphate pathway is as important as the production of ribose sugars. Indeed, the body's need for $NADPH/H^+$ is usually greater than its need for ribose. Consequently, the pentose phosphate pathway can interface with the glycolysis pathway in the cytosol in such a way that both $NADPH/H^+$ and ribose are produced but the ribose is then "disposed of." The ribose produced is often converted to fructose 6-phosphate and glyceraldehyde 3-phosphate, intermediates in glycolysis. Figure 23.7 summarizes the pentose phosphate pathway.

These metabolic events occur mainly in adipose (fat) tissue where, as you will see in the next chapter, the body needs $NADPH/H^+$ for fatty acid synthesis.

23.9 BLOOD SUGAR LEVEL

How does the body regulate and balance these various metabolic processes to avoid excesses or deficiencies of ATP, $NADPH/H^+$, and metabolic intermediates? The regulatory mechanisms are not so well known as the details of the metabolic cycles. Regulation often involves some type of *feedback* mechanism (Section 21.10). For example, if excess ATP is produced, the accumulated ATP may inhibit the action of enzymes necessary for its synthesis and hence inhibit its production.

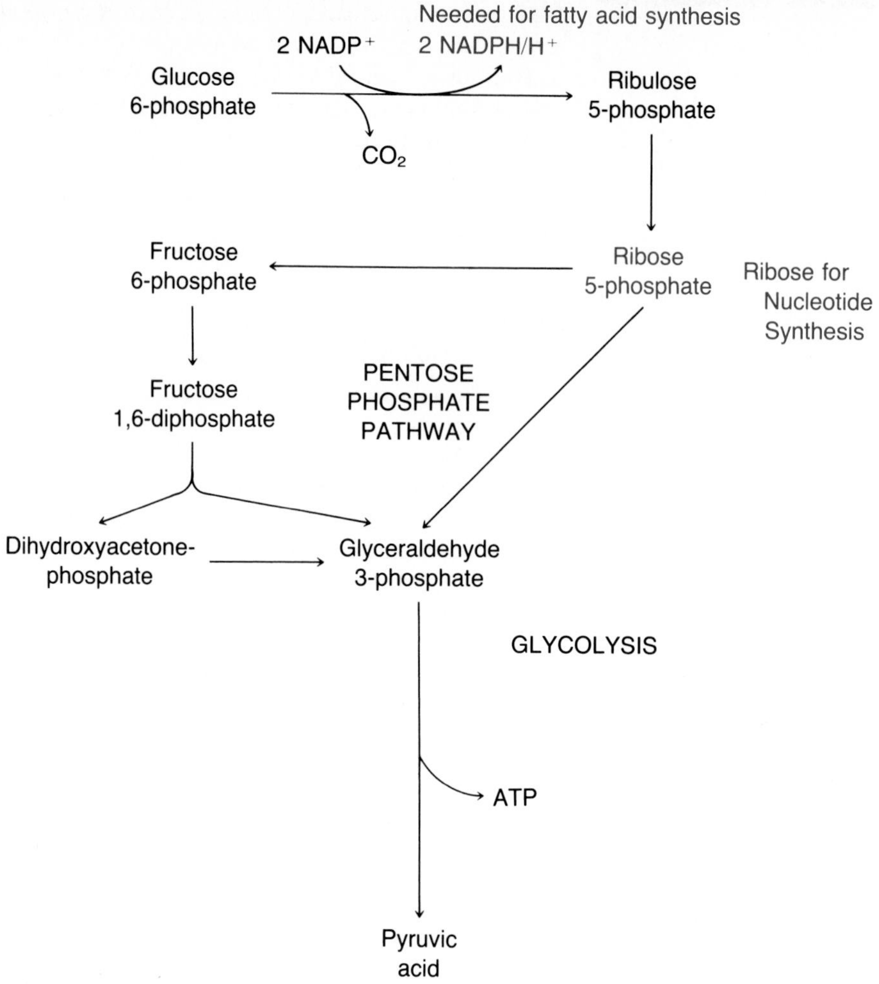
Needed for fatty acid synthesis
2 NADP+
2 NADPH/H+
Glucose 6-phosphate
Ribulose 5-phosphate
CO2
Fructose 6-phosphate
Ribose 5-phosphate
Ribose for Nucleotide Synthesis
Fructose 1,6-diphosphate
PENTOSE PHOSPHATE PATHWAY
Dihydroxyacetone-phosphate
Glyceraldehyde 3-phosphate
GLYCOLYSIS
ATP
Pyruvic acid

FIG. 23.7
The pentose phosphate pathway converts glucose into 5-carbon sugars and through subsequent steps also provides alternative entry points into the glycolysis pathway. Many intermediate steps have not been shown. The pentose phosphate pathway is also called the pentose shunt, the hexose monophosphate pathway, or the phosphogluconate oxidative pathway.

Because the principal fuel source in metabolism is glucose, one of the most important regulatory systems the body must have is the ability to maintain the amount of glucose in the blood at an appropriate level. The bloodstream is the most useful location for a constant supply of glucose, because blood's mobility makes glucose available to all bodily cells. Glucose can readily cross cell membranes in healthy individuals. Blood sugar level is generally defined as the concentration of glucose in the blood. There are other metabolically useful monosaccharides in the blood, but the amount of glucose in the blood sugar mixture is so overwhelmingly large that "blood sugar" and "blood glucose" are essentially synonymous.

The so-called **normal fasting level** (after 8 to 12 h of not eating) of blood sugar is 70 to 110 mg of glucose/100 mL of blood. This is about one teaspoon of glucose distributed throughout the entire body. Soon after a meal is eaten, the level may rise to 120 to 140 mg/100 mL, but it levels off after about 2 h.

Hypoglycemia is the condition in which blood sugar level is *below* the normal fasting level (*hypo-* means "below"; *gly-* stands for "monosaccharide''; *-emia* refers to ''blood''). Extreme hypoglycemia can cause unconsciousness and lowered blood pressure, and could result in death. Loss of consciousness occurs because brain tissue has no capacity for carbohydrate storage and depends almost totally on the availability of glucose in blood for its energy requirements.

Hyperglycemia is the condition of abnormally *high* blood sugar concentration. If hyperglycemia is extreme, the kidneys begin to transfer some glucose from the blood into urine. The blood sugar level at which this occurs is called the **renal threshold** for glucose. **Glucosuria** is the condition of having detectable glucose in the urine.

Diabetes mellitus is the major cause of hyperglycemia. This disease is characterized by faulty glucose metabolism. Because fat metabolism is also affected, we will delay further discussion of the metabolic details of diabetes until Section 24.10.

Regulation of blood sugar level is accomplished principally in the liver by the synthesis (glycogenesis) or breakdown (glycogenolysis) of glycogen.

$$\text{Glycogen} \underset{\text{glycogenesis}}{\overset{\text{glycogenolysis}}{\rightleftarrows}} \text{glucose}$$

You will shortly see that these two processes are not simply the reverse of one another. However, clearly glycogenolysis tends to increase the blood glucose level, while glycogenesis tends to decrease it.

Several hormones control blood sugar level mainly by stimulating either glycogenolysis or glycogenesis. The pancreas produces two hormones, insulin and glucagon, with opposing effects:

Insulin is secreted in response to a high blood sugar level and lowers this level by stimulating glycogenesis in muscles and in the liver. It also suppresses gluconeogenesis and promotes the entry of glucose into cells, where it is subject to catabolism.

Glucagon is secreted in response to a low blood sugar level and raises that level by stimulating glycogenolysis, especially in the liver. It also enhances the rate of gluconeogenesis.

SAMPLE EXERCISE 23.3

Write simple equations for the processes by which insulin decreases blood sugar levels and glucagon increases the level.

Solution

Insulin stimulates glycogenesis:

$$\text{Glucose} \longrightarrow \text{glycogen}$$

Insulin suppresses gluconeogenesis:

$$\text{Pyruvic acid} \not\longrightarrow \text{glucose}$$

Insulin promotes entry of glucose into cells where it is catabolized or degraded:

$$\text{Glucose} \longrightarrow \text{glucose 6-phosphate} \xrightarrow{\text{glycolysis}} \text{pyruvic acid}$$

In these cases, glucose is consumed or not produced.

Glucagon stimulates glycogenolysis:

Glycogen ⟶ glucose

Glucagon stimulates gluconeogenesis:

Pyruvic acid ⟶ glucose

Glucose is produced in both cases.

Typically, after a meal when the blood sugar level rises, secretion of insulin from the pancreas increases and secretion of glucagon decreases. The increased insulin level in the "fed" state promotes the entry of glucose into muscle and fat cells through interaction with receptors in the cell membrane. Glycogenesis is stimulated both in muscle tissue and in the liver. As you will see in Chapter 24, entry of glucose into fat cells provides an intermediate needed in the synthesis of fats.

When the blood glucose level begins to drop several hours after a meal, glucagon secretion increases and insulin secretion decreases. Glucagon stimulates glycogenolysis in the liver, which leads to release of glucose into the blood. Glucose cannot enter muscle and fat cells in the absence of insulin and the cells switch to fatty acids as a fuel source.

Epinephrine is a hormone secreted by the adrenal medulla in times of sudden danger or stress. It stimulates glycogenolysis in muscles, thus providing glucose as a mobilized energy source to meet the emergency. Epinephrine also stimulates the secretion of glucagon and suppresses that of insulin.

23.10 GLYCOGENOLYSIS

The construction of the word *glycogenolysis* reveals its meaning:

glycogen—olysis
glycogen *a breaking or cleavage*

Glycogen is a branched polysaccharide in which glucose monomer units are hooked together by α-1,4- and α-1,6-glycosidic linkages (review Section 15.16 and see Figure 23.8). Breakdown of the glycogen polymer is accomplished by phosphorolysis, i.e., cleavage by phosphate rather than water. Cleavage catalyzed by glycogen phosphorylase occurs at the 1,4-linkages only. One monomer unit at a time is removed. Schematically,

FIG. 23.8
A schematic representation of glycogen: solid black hexagons depict the nonreducing chain ends, that is, the end of the polysaccharide chain in which C-1 is involved in an acetal linkage. Red hexagons mark the 1,6 branching points. Review glycogen structure in Section 15.16.

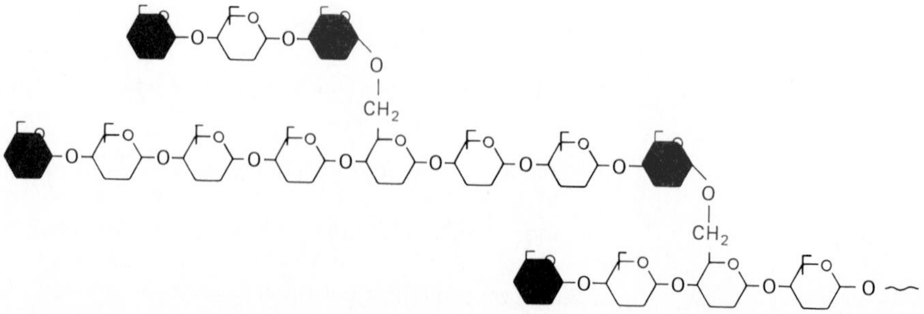
O
CH2

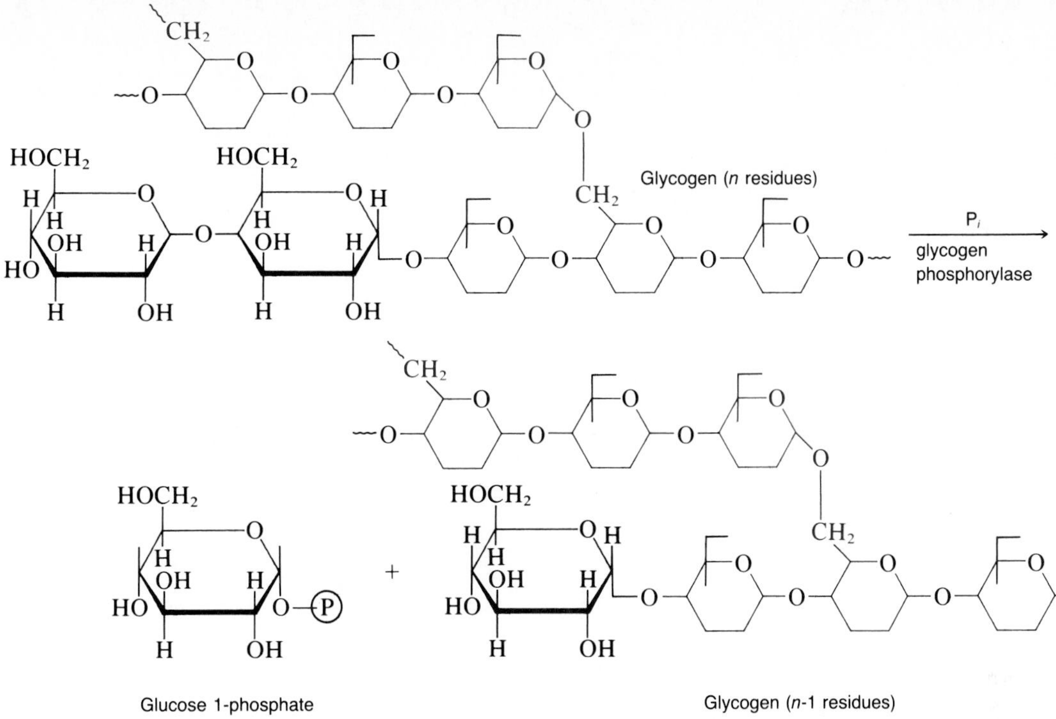
Glycogen (n residues)
P_i
glycogen phosphorylase
Glucose 1-phosphate
Glycogen (n-1 residues)

In terms of energy yield, the production of glucose 1-phosphate is very efficient because this product can be easily isomerized to glucose 6-phosphate and enter the glycolysis pathway without using any ATP. In contrast, converting glucose to glucose 6-phosphate requires the expenditure of ATP.

Problem 23.6 How much ATP is generated if 1 mol of glucose 1-phosphate is isomerized to glucose 6-phosphate and then undergoes glycolysis under anaerobic conditions?

Other enzymes are necessary to cleave at branched points (1,6-linkages) in the chain. Breakdown at the branches involves a transfer of the branch to "straighten out" the chain and allow phosphorolysis to continue at 1,4-linkages (see Figure 23.9). Thus, it is the action of the phosphorylase enzyme that is most important in glycogenolysis. Blood-sugar-regulating hormones either activate or deactivate this enzyme.

23.11 GLYCOGENESIS

Once again the construction of the word gives the meaning: *glyco-* refers to "glycogen," and *genesis* means "the making of." Thus the term *glycogenesis* means "the making of glycogen." Glycogen is not synthesized in the body by directly linking glucose units. Indeed, once glucose enters the cell, it is phosphorylated to glucose 6-phosphate in order to keep it there. That is, glucose can cross cell membranes, but glucose phosphate esters cannot. Nor does the buildup of glycogen happen through the reverse of the glycogenolysis process, the linkage of glucose 1-phosphate units. Rather glucose is converted to uridine diphosphate glucose (UDP-glucose) which then condenses with an alcohol group at C-4 of a monomer unit in the growing polysaccharide chain.

The process can be outlined briefly beginning with the isomerism of glucose 6-phosphate to glucose 1-phosphate and the condensation of glucose 1-phosphate with uridine triphosphate to form UDP-glucose:

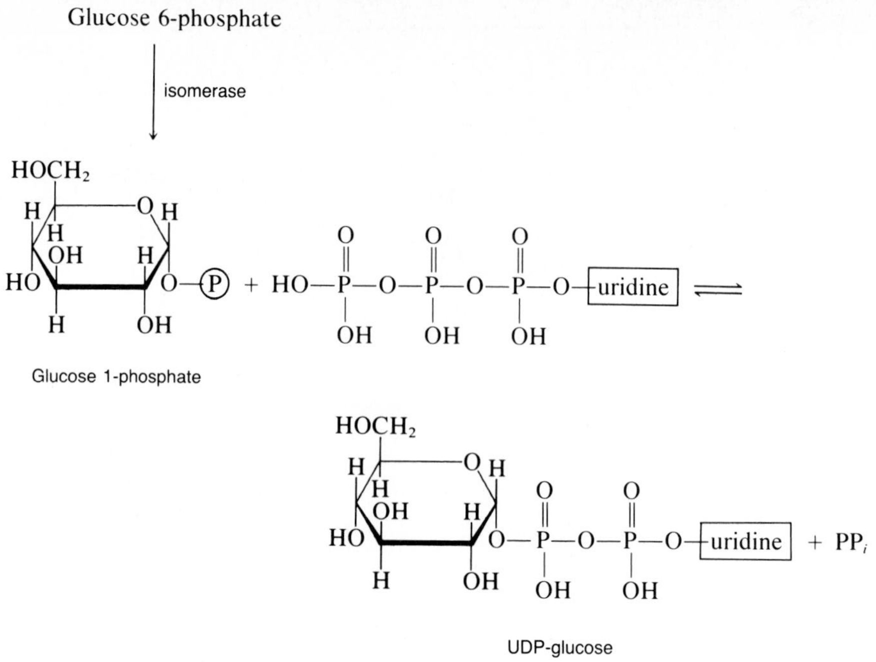
Glucose 6-phosphate
isomerase
Glucose 1-phosphate
uridine
+ PP_i
UDP-glucose

FIG. 23.9
Phosphorylase catalyzes the cleavage of 1,4-glycosidic linkages, that is, the disassembly of straight portions of glycogen chains, one glucose unit at a time. Branched portions undergo rearrangement as shown to achieve straight chains susceptible to phosphorylase action.

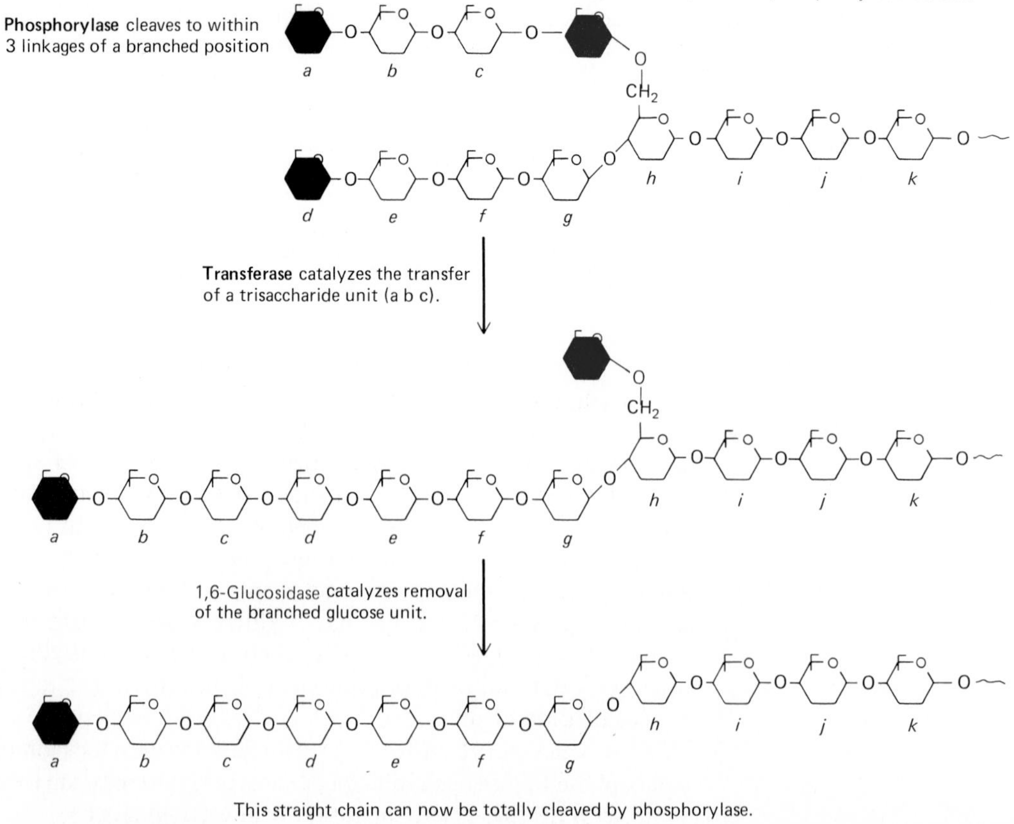
Phosphorylase cleaves to within 3 linkages of a branched position
Transferase catalyzes the transfer of a trisaccharide unit (a b c).
1,6-Glucosidase catalyzes removal of the branched glucose unit.
This straight chain can now be totally cleaved by phosphorylase.

Then the condensation of UDP-glucose with a growing polysaccharide chain occurs:

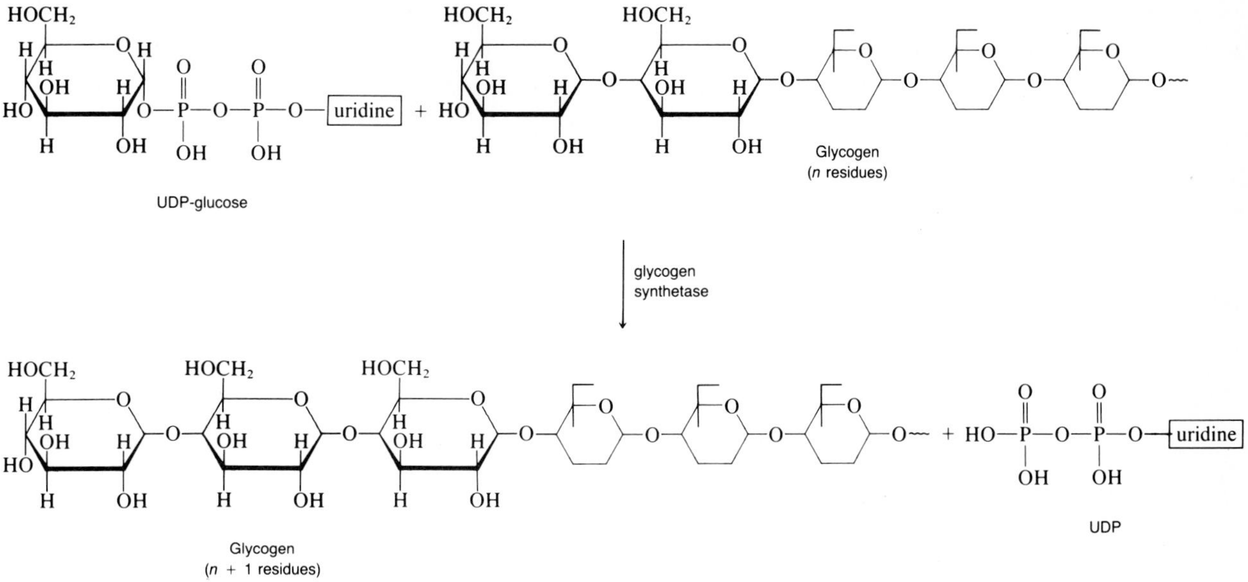
UDP-glucose
uridine
Glycogen
(n residues)
glycogen
synthetase
Glycogen
(n + 1 residues)
uridine
UDP

Glycogen synthetase catalyzes the formation of the α-1,4-linkages. Other enzymes are necessary to form α-1,6-linkages and make the chain branches.

It is an important principle that biosynthetic and degradative processes almost always occur by different pathways. Glycogen metabolism was one of the first examples of this principle scientists discovered. Separate pathways provide the opportunity for independent control of the two often opposing processes. The relationship between glycolysis and gluconeogenesis you saw in Section 23.7 is another example of this principle.

23.12 REGULATION OF GLYCOGEN METABOLISM

Blood-sugar-regulating hormones control glycogenesis by either activating or deactivating the enzyme glycogen synthetase, but they do not do so directly. Rather, the secondary messenger, cyclic AMP (cAMP), which you first saw in Section 21.14, is involved. Figure 23.10 shows the chain of events by which epinephrine enhances glycogenolysis and suppresses glycogenesis. Such a chain is called a *cascade* of reactions.

Epinephrine does not actually enter the cell. Rather, it binds to the membrane of the muscle cell and activates the enzyme adenylate cyclase. Recent research indicates that prostaglandins are involved in this process. Adenylate cyclase then catalyzes the formation of cyclic AMP in the cell. cAMP then directs the subsequent activation of enzymes which ultimately produce active phosphorylase and enhanced glycogenolysis, raising the blood glucose level. At the same time, glycogen synthetase is deactivated so that glycogenesis essentially stops.

The action of the hormone epinephrine is amplified by this process because activation of one enzyme molecule stimulates the reaction of many substrate molecules. Thus the production of intermediates in the diagram in Figure 23.10 is not 1:1. Rather activation of one adenylate cyclase leads to the formation of many cAMP molecules. This effect is then further amplified as cAMP activates protein kinase, which then catalyzes the phosphorylation reaction of many molecules of phosphorylase kinase or glycogen synthetase. Phosphorylation activates phosphorylase and deactivates the synthetase.

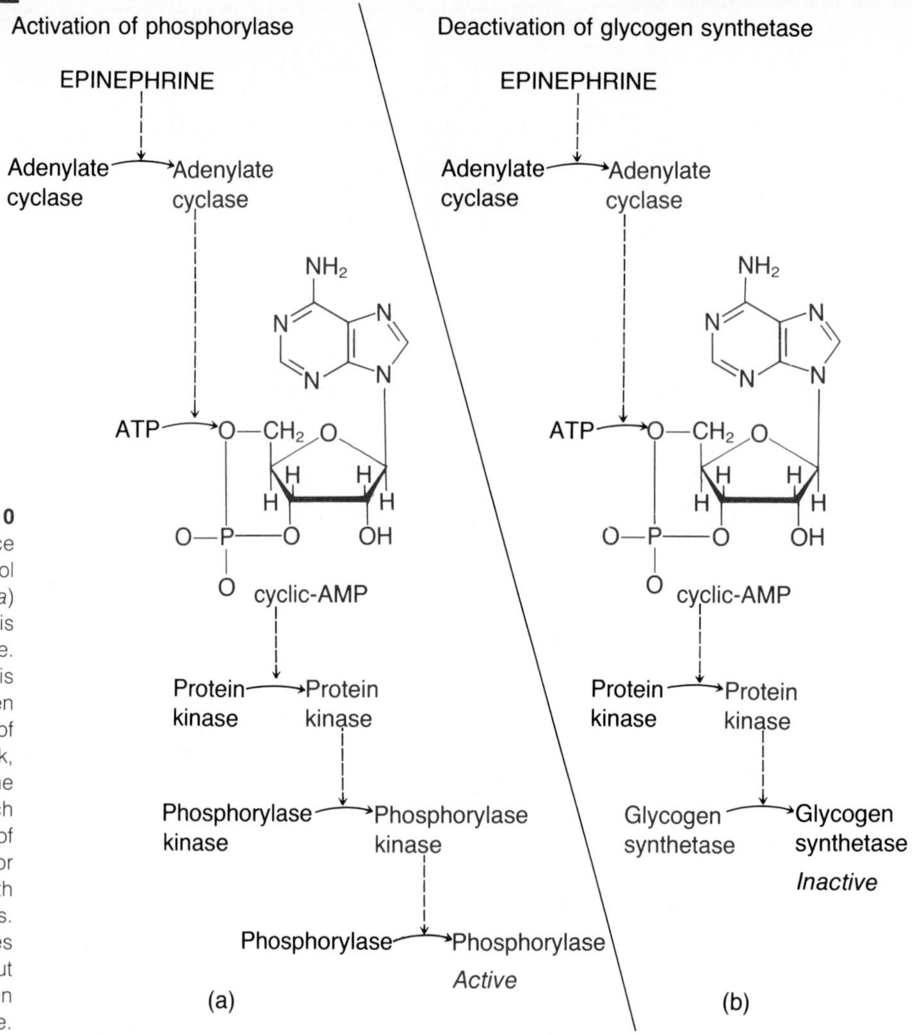
Activation of phosphorylase
EPINEPHRINE
Adenylate cyclase
Adenylate cyclase
NH2
N
N
N
N
ATP
O—CH2
O
H
H
H
H
O—P—O
OH
O
cyclic-AMP
Protein kinase
Protein kinase
Phosphorylase kinase
Phosphorylase kinase
Phosphorylase
Phosphorylase
Active
(a)
Deactivation of glycogen synthetase
EPINEPHRINE
Adenylate cyclase
Adenylate cyclase
cyclic-AMP
Protein kinase
Protein kinase
Glycogen synthetase
Glycogen synthetase
Inactive
(b)

FIG. 23.10
The "cascade" or sequence of reactions that control glycogen metabolism. (*a*) Glycogen degradation is catalyzed by phosphorylase. (*b*) Glycogen synthesis is catalyzed by glycogen synthetase. Inactive forms of enzymes are shown in black, and active ones in color. The sequence of reactions which leads to the activation of protein kinase is the same for the regulation of both degradation and synthesis. Protein kinase activates phosphorylase, but deactivates glycogen synthetase.

Problem 23.7 Predict the effect of insulin on the concentration of cAMP in cells.

23.13 FUEL STORAGE

Glycogen serves as a form of storage for glucose, our carbohydrate fuel. The body has a limited capacity for carbohydrate storage. Our glycogen reserves could keep us going for only about a day if we did not eat. To accomplish long-term storage, the body converts carbohydrates into fats. Our fat reserves could supply us with energy for about a month.

You saw in Section 18.3 that fats are triacylglycerols, that is, they are constructed of glycerol and fatty acids. Look back at the glycolysis pathway in Figure 23.1 and you can see that dihydroxyacetone phosphate could easily be a precursor for glycerol,

$$\begin{array}{l} CH_2OH \\ | \\ CHOH \\ | \\ CH_2OH \end{array}$$

In Chapter 24 we'll see that the body synthesizes fatty acids from acetyl CoA, to which glucose is metabolized. Hence, glucose can supply the raw materials for fat construction. That is how excess carbohydrate consumption can cause us to gain weight. We'll pursue the interactions between carbohydrate and fat metabolism further in the next chapter.

CHAPTER ACCOMPLISHMENTS

After completing this chapter you should be able to define **key words** and do the following:

23.1 INTRODUCTION

1. State the effect of oxygen availability on glucose metabolism.

23.2 GLYCOLYSIS

2. Given equations for the steps in glycolysis, indicate the type of reaction that has occurred in a specified step.
3. Given equations for the steps in glycolysis, indicate the energy yield in terms of ATP for each step under aerobic and anaerobic conditions.

23.3 THE FATES OF PYRUVIC ACID

4. State the fate of pyruvic acid under anaerobic conditions.
5. State the fate of pyruvic acid under aerobic conditions.
6. Compute energy yields from specified amounts of monosaccharides as they pass through the glycolysis pathway under aerobic or anaerobic conditions.

23.4 THE CITRIC ACID CYCLE

7. State the reactants and product of the first step of the citric acid cycle.
8. State the product of the last step of the citric acid cycle.
9. Given equations for the steps in the citric acid cycle, indicate the type of reaction that has occurred in a specified step.
10. Given equations for the steps in the citric acid cycle, indicate the energy yield in terms of ATP for each step.

23.5 ENERGY YIELD FROM GLUCOSE

11. State the total energy yield in ATP from glucose under aerobic and anaerobic conditions.

23.6 INTERPLAY OF AEROBIC AND ANAEROBIC METABOLISM

12. Explain what is meant by an *oxygen debt* in cells.
13. Explain the role of the Cori cycle.

23.7 GLUCONEOGENESIS

14. Cite two reasons that gluconeogenesis is not simply the reverse of glycolysis.

23.8 GLUCOSE AS A BIOSYNTHETIC PRECURSOR

15. Name the major products of the *pentose phosphate pathway*.

23.9 BLOOD SUGAR LEVEL

16. Explain the roles of glycogenesis and glycogenolysis in controlling blood sugar level.
17. State the effects of the hormones insulin, glucagon, and epinephrine on glycogenesis, glycogenolysis, and blood sugar level.

23.10 GLYCOGENOLYSIS

18. Name the enzyme which catalyzes glycogenolysis.

23.11 GLYCOGENESIS

19. State the important principle regarding the relationship between biosynthetic and degradative processes.

23.12 REGULATION OF GLYCOGEN METABOLISM

20. Describe the role of cAMP in the control of blood sugar level.

23.13 FUEL STORAGE

21. Explain how excess carbohydrate consumption leads to fat production.

PROBLEMS

23.8 What is the major product of carbohydrate digestion?

23.9 Give another name for the glycolysis pathway.

23.10 During glycolysis glucose is converted into ______.

23.11 What molar relationship exists between glucose degraded and pyruvic acid obtained?

23.12 Refer to Figure 23.1 in order to answer the following questions:

a Which steps in the glycolysis pathway involve phosphorylation?

b What is being phosphorylated in each case?

c What is the energy yield in ATP for each step involving phosphorylation?

23.13 Refer to Figure 23.1.

a Which step in glycolysis is an oxidation?

b Which step is a dehydration?

23.14 After the cleavage step in the glycolysis pathway, why does each compound have a coefficient of 2?

23.15 Where in the cell does glycolysis occur?

23.16 ATP energy is required for transport of coenzymes across some cell membranes. How does this fact relate to energy yields from $NADH/H^+$ during glycolysis?

23.17 What element must be available in order to realize ATP yields from reduced coenzymes?

23.18 How is NAD^+ regenerated from $NADH/H^+$ under aerobic conditions?

23.19 How is NAD^+ regenerated from $NADH/H^+$ under anaerobic conditions?

23.20 What is the end product of glycolysis under anaerobic conditions?

23.21 What is the energy yield in terms of ATP for the following amounts of monosaccharides under anaerobic conditions?

a 1 mol of glucose

b 1 mol of fructose

c 0.2 mol of glucose

d 0.5 mol of glucose

e 0.1 mol of fructose

23.22 Into what is pyruvic acid converted under aerobic conditions?

23.23 Each mole of glucose which enters the glycolysis pathway results in the production of ____ moles of acetyl CoA under aerobic conditions.

23.24 Under aerobic conditions the following overall transformations occur:

$$\text{Glucose or fructose} \xrightarrow{\text{A}} \text{pyruvic acid} \xrightarrow{\text{B}} \text{acetyl CoA}$$

What is the energy yield in terms of ATP from 1 mol of monosaccharide

a In transformation A?

b In transformation B?

23.25 Look back at Problem 23.21 and compute the ATP yield from conversion of the monosaccharides to acetyl CoA.

23.26 Is glycolysis itself an aerobic or anaerobic process?

23.27 Give two other names commonly applied to the citric acid cycle.

23.28 Explain in your own words why the term *cycle* is applied to the series of reactions called the citric acid cycle.

23.29 Name the reactants and products in the first step of the citric acid cycle.

23.30 Refer to Figure 23.3 to answer the following questions:

a Which steps in the citric acid cycle are oxidations?

b Which steps are hydrations?

c In which step of the citric acid cycle does substrate-level phosphorylation occur?

23.31 Where in the cell does the citric acid cycle occur?

23.32 What happens to the hydrogen atoms removed in oxidation reactions during glycolysis and the citric acid cycle?

23.33 Refer to Figure 23.3 and identify the steps which yield ATP. How much ATP is generated at each step (per mole of acetyl CoA)?

23.34 Under aerobic conditions the following overall transformations occur:

$$\text{Glucose or fructose} \xrightarrow[\text{(ETS)}]{\text{glycolysis}} \text{acetyl CoA} \xrightarrow[\text{(ETS)}]{\text{citric acid cycle}} CO_2 + H_2O$$

What is the energy yield in terms of ATP from 1 mol of monosaccharide

a From glycolysis?

b From the citric acid cycle?

23.35 Look back at Problem 23.21 and compute the ATP yield from conversion of the monosaccharides to CO_2 and H_2O.

23.36 What is an oxygen debt?

23.37 How does the liver help muscles cope with an oxygen debt?

23.38 Why do we breathe heavily during vigorous exercise?

23.39 What causes soreness in muscles after strenuous exercise?

23.40 In what two ways does the body dispose of lactic acid during rest periods after vigorous exercise?

23.41 What name is given to the process whereby the body can make glucose from noncarbohydrate sources?

23.42 What major bodily organ depends almost exclusively on glucose as its fuel source?

23.43 What is the significance of oxaloactic acid, a citric acid cycle intermediate, in gluconeogenesis?

23.44 Is oxaloacetic acid an intermediate in glycolysis?

23.45 Cite several similarities and differences between glycolysis and gluconeogenesis.

23.46 What important substances are made in the pentose phosphate pathway?

23.47 How are NAD^+ and $NADP^+$ related?

23.48 What is the chemical name for blood sugar?

23.49 What is the *normal fasting level* of blood sugar in blood?

23.50 What is hypoglycemia?

23.51 How might a normal person develop a mild hypoglycemic condition?

23.52 What is hyperglycemia?

23.53 What is the major cause of hyperglycemia?

23.54 Insulin and glucagon are both secreted by the pancreas. Compare their opposing effects on glycogenesis, gluconeogenesis, glycogenolysis, and blood sugar level.

23.55 How does the hormone epinephrine help us counter stress?

23.56 Is glycogenolysis degradative or synthetic?

23.57 Describe the monomer units that are hooked together during glycogenesis.

23.58 Explain in your own words what is meant by a "cascade of reactions" which transmit hormonal messages.

23.59 In what form does the body store carbohydrates? How long will these reserves last if we do not eat?

23.60 What is the long-term fuel storage material of the body?

CHAPTER TWENTY-FOUR

LIPID METABOLISM

24.1 INTRODUCTION Carbohydrates provide us with the most readily available source of energy. Marathon runners indulge in an abundance of pasta or other carbohydrates before a race to take advantage of this readily mobilized fuel. Lipids, on the other hand, have a greater energy density, which makes them more suitable for energy storage. Higher energy density means that lipids provide more energy per gram than carbohydrates, as you will see in Section 24.6. In Chapter 18 you saw that the most commonly ingested lipids are the fats and oils, or triacylglycerols. It is their metabolism that will concern us in this chapter.

Digestion and Catabolism

Triacylglycerols are digested into smaller component parts, diacylglycerols, monacylglycerols, and some fatty acids and glycerol, in the small intestine, but these parts are recombined before they enter the lymph and blood. The water-insoluble triacylglycerols attach to proteins to form *lipoproteins,* which can then be transported in the blood. Some lipoproteins are carried to the liver, the organ responsible for regulating lipid concentrations in the blood. Most lipid material is deposited in fat storage cells in adipose tissue under the skin and in the abdominal area. Some lipoprotein remains circulating in the bloodstream. Figure 24.1 shows the absorption and distribution of lipids to the body. Figure 24.2 is a picture of an actual fat cell; the oily appearance is due to the presence of triacylglycerols in the cytoplasm of the cell.

In order for the body cells to use triacylglycerols for energy, the fat must first be hydrolyzed into its component parts. This process is catalyzed by the enzyme lipase:

$$
\begin{array}{c}
CH_2\!-\!O\!-\!\overset{\displaystyle O}{\overset{\|}{C}}\!-\!R_1 \\[2mm]
CH\!-\!O\!-\!\overset{\displaystyle O}{\overset{\|}{C}}\!-\!R_2 \;+\; 3H_2O \;\xrightarrow{\text{lipase}}\; \\[2mm]
CH_2\!-\!O\!-\!\overset{\displaystyle O}{\overset{\|}{C}}\!-\!R_3
\end{array}
\qquad
\begin{array}{c}
CH_2\!-\!OH \\[2mm]
CH\!-\!OH \\[2mm]
CH_2\!-\!OH
\end{array}
\qquad
\begin{array}{c}
HO\!-\!\overset{\displaystyle O}{\overset{\|}{C}}\!-\!R_1 \\[2mm]
+\;HO\!-\!\overset{\displaystyle O}{\overset{\|}{C}}\!-\!R_2 \\[2mm]
HO\!-\!\overset{\displaystyle O}{\overset{\|}{C}}\!-\!R_3
\end{array}
$$

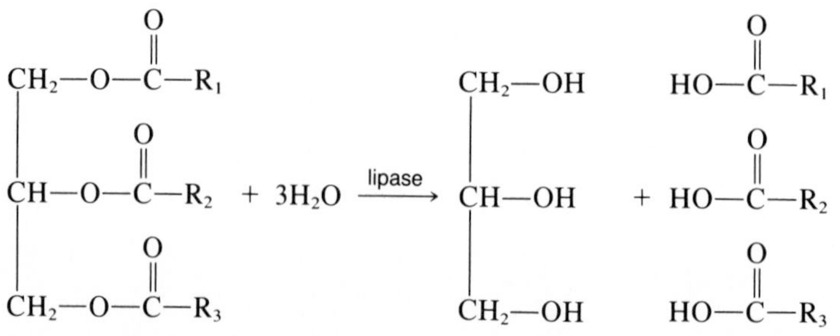

Triacylglycerol   Glycerol   Fatty acids

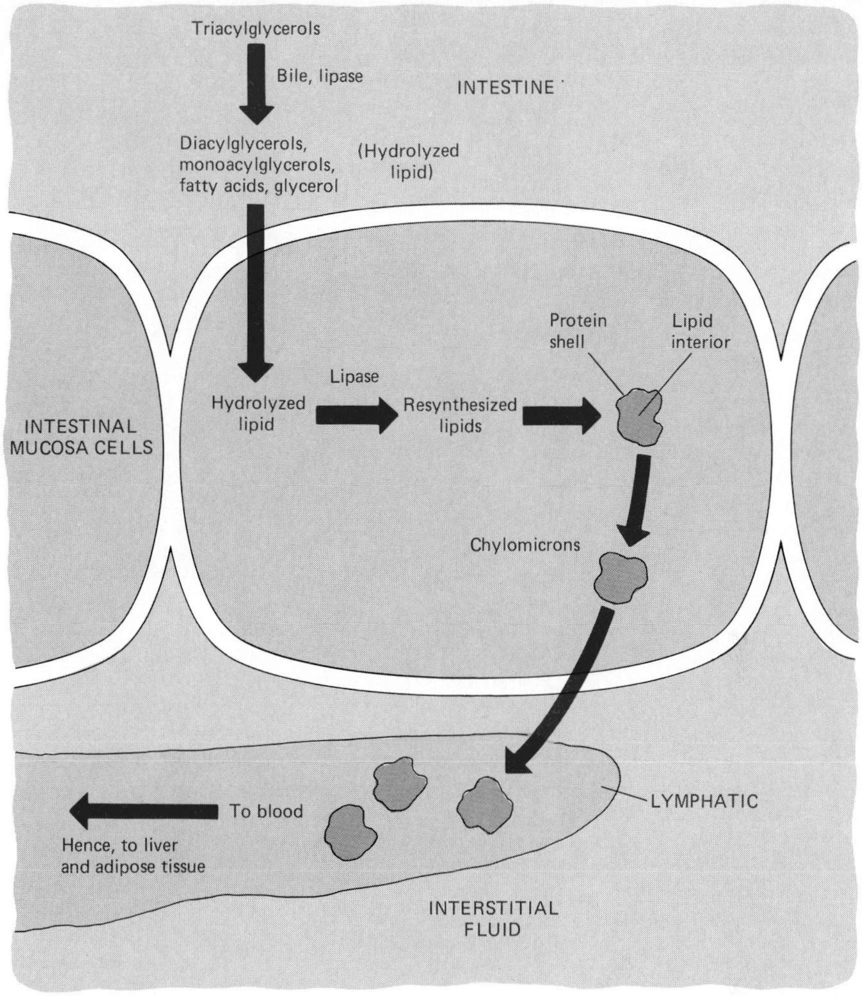
Triacylglycerols
Bile, lipase
INTESTINE
Diacylglycerols, monoacylglycerols, fatty acids, glycerol
(Hydrolyzed lipid)
Protein shell
Lipid interior
Lipase
Hydrolyzed lipid
Resynthesized lipids
INTESTINAL MUCOSA CELLS
Chylomicrons
LYMPHATIC
To blood
Hence, to liver and adipose tissue
INTERSTITIAL FLUID

FIG. 24.1
Dietary triacylglycerols are hydrolyzed (digested) at least partially in the small intestine. The hydrolysis products cross the cell membranes and enter the intestinal mucosa cells, where they are recombined into triacylglycerols. In order to solubilize these lipids, they are complexed with proteins to form lipoproteins called *chylomicrons* (Section 18.15). The chylomicrons carry lipids into the lymph, then into the bloodstream and to the liver and adipose tissue, where the lipids are again hydrolyzed into their component parts. Phospholipids follow a similar course. Cholesterol can enter intestinal cells without digestion and be incorporated into lipoproteins.

Hormonal control of fat utilization is exercised at this point through activation or deactivation of lipase. Upon hydrolysis, glycerol is metabolized by entering the glycolysis pathway. You studied glycolysis in general in Chapter 23. The specific details for glycerol will unfold in the next section. Fatty acids are broken down in the fatty acid spiral, which you will see in Section 24.5.

Anabolism

The body can also make fat (triacylglycerols) from nonlipid sources. Glycerol is available from carbohydrate metabolism. Both carbohydrates and proteins are catabolized to acetyl CoA from which the body can make fatty acids (Section 24.7). When excess nutrients are consumed, the body makes fat as its long-term storage facility. In fact, fat is constantly being synthesized for

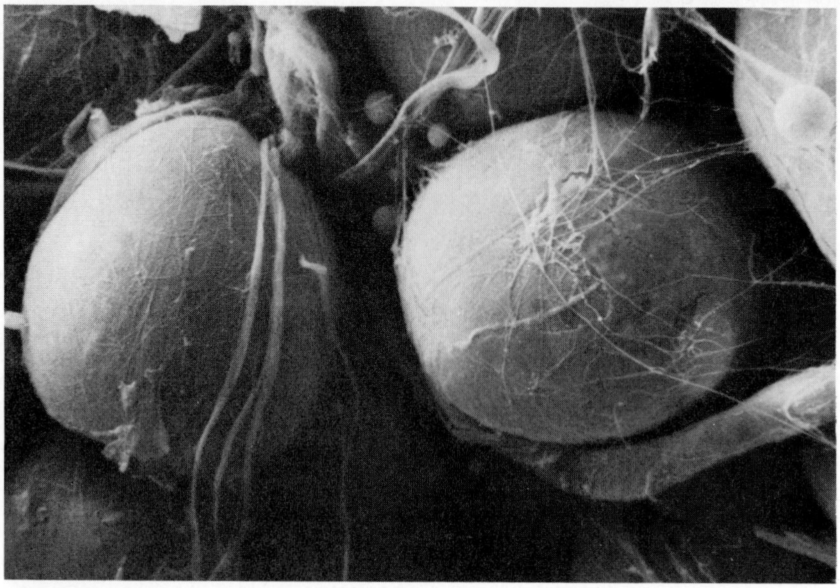

FIG. 24.2
Scanning electron micrograph of fat cells × 1450. (*From Tissues and Organs: Text-Atlas of Scanning Electron Microscopy by Richard G. Kessel and Randy H. Kardon. W. H. Freeman and Company. Copyright © 1979.)*

storage, but if we avoid overeating, it is broken down at least as quickly as it is made. During fasting or dieting, fat breakdown is an important energy source. Let us now explore the mechanisms by which fat metabolism occurs.

24.2 GLYCEROL METABOLISM

The glycerol released by hydrolysis of triacylglycerols is readily converted to dihydroxyacetone phosphate, which in turn can be isomerized to glyceraldehyde 3-phosphate:

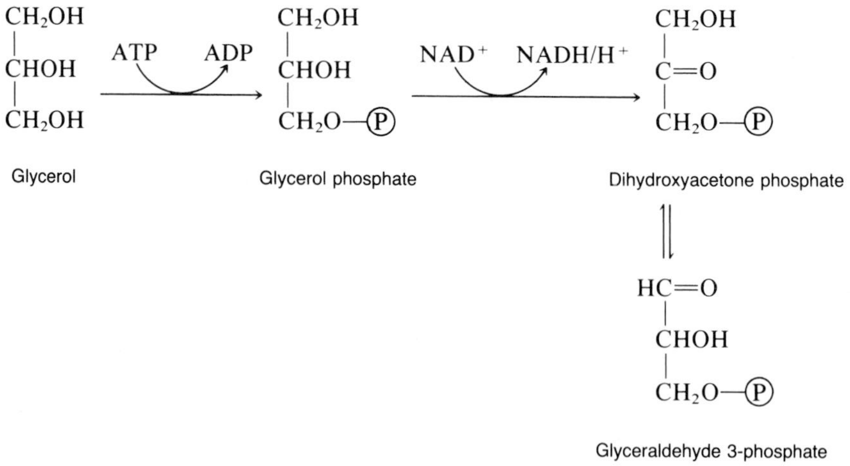

CH₂OH
CHOH
CH₂OH
ATP
ADP
CH₂OH
CHOH
CH₂O—(P)
NAD⁺
NADH/H⁺
CH₂OH
C=O
CH₂O—(P)
Glycerol
Glycerol phosphate
Dihydroxyacetone phosphate
HC=O
CHOH
CH₂O—(P)
Glyceraldehyde 3-phosphate

Glyceraldehyde 3-phosphate is an intermediate in both the glycolysis and gluconeogenesis pathways that you encountered in the previous chapter. Therefore, glycerol can be converted into either pyruvic acid or into glucose.

Problem 24.1 The conversion of glycerol to glyceraldehyde 3-phosphate occurs in the cytosol. How much energy (in ATP) is generated by the catabolism of 1 mol of glycerol to CO_2 and H_2O? (See Figure 23.1.)

24.3 OVERVIEW OF FATTY ACID METABOLISM

If you look back at Table 18.2, you will see that most naturally occurring fatty acids have an *even* number of carbon atoms in their chain. This design permits the most efficient use of all the carbon atoms in the fatty acid chain. The end product of fatty acid catabolism is acetyl CoA, $CH_3-\overset{O}{\overset{\|}{C}}-S-CoA$. Just as glucose is degraded during glycolysis into acetyl groups, so too are fatty acids broken down into these two-carbon fragments ($CH_3-\overset{O}{\overset{\|}{C}}-$). A fatty acid chain with an even number of carbons can be efficiently cleaved two carbon atoms at a time to acetyl groups with no carbons "left over."

The successive cleavages begin at the carboxyl end of the fatty acid molecule:

$$\overset{\omega}{CH_3}\ CH_2\ CH_2\ (CH_2)_n-\overset{\delta}{CH_2}-\overset{\gamma}{CH_2}-\overset{\beta}{CH_2}-\overset{\alpha}{CH_2}-\overset{O}{\overset{\|}{C}}-OH$$

This means the first cleavage occurs between what are labeled here as the α and β carbon atoms. These labels represent an alternative scheme for numbering the carbon chain of a carboxylic acid, as you first saw in Section 17.2. In the IUPAC system numbers are used and the carbon of the —COOH group is designated number 1. In this alternative system the lowercase letters of the Greek alphabet (α, β, γ, δ, . . . , ω) are used and the first position (α) is the carbon attached to $-\overset{O}{\overset{\|}{C}}-OH$; that is, α begins the counting of the hydrocarbon chain. We will use this lettering system throughout the discussion of fatty acid metabolism. The β-position is particularly significant.

SAMPLE EXERCISE 24.1

Label the β-position in the following acids:

a $CH_3CH_2CH_2COOH$

b $HOOCCH_2CH_2CH_2CH_2CH_3$

c $CH_3(CH_2)_{16}-\overset{O}{\overset{\|}{C}}-OH$

Solution

a The α-position is the carbon next to $-\overset{O}{\overset{\|}{C}}-$ and the β-position is one beyond that, i.e.,

$$CH_3-\overset{\beta}{CH_2}-\overset{\alpha}{CH_2}-\overset{O}{\overset{\|}{C}}-OH$$

b The —COOH group can appear at either end of the chain, but the rule remains as stated in *a*, thus:

$$HO-\overset{O}{\overset{\|}{C}}-\overset{\alpha}{CH_2}-\overset{\beta}{CH_2}-CH_2CH_2CH_3$$

c In this case the structure must be rewritten to show some of the individual carbons in the chain rather than having them all "contained" in the parentheses. Take out from the parentheses the two carbons at the α- and β-positions. This leaves only 14 $-CH_2-$ groups in the parentheses.

$$CH_3(CH_2)_{16}{-}\overset{\overset{\displaystyle O}{\|}}{C}{-}OH \equiv CH_3(CH_2)_{14}\overset{\beta}{C}H_2{-}\overset{\alpha}{C}H_2{-}\overset{\overset{\displaystyle O}{\|}}{C}{-}OH$$

In general,

$$CH_3(CH_2)_n{-}\overset{\overset{\displaystyle O}{\|}}{C}{-}OH \equiv CH_3(CH_2)_{n-2}\overset{\beta}{C}H_2{-}\overset{\alpha}{C}H_2{-}\overset{\overset{\displaystyle O}{\|}}{C}{-}OH$$

Even without a detailed knowledge of how fatty acid metabolism occurs, you can predict how many acetyl CoA molecules will be obtained from a specified fatty acid. Because the chain is cleaved into two-carbon fragments, the number of $CH_3{-}\overset{\overset{\displaystyle O}{\|}}{C}{-}S{-}CoA$ molecules obtained is half the number of C atoms in the fatty acid chain, e.g.,

$$CH_3{-}CH_2 \,\big|\, CH_2{-}CH_2 \,\big|\, CH_2{-}\overset{\overset{\displaystyle O}{\|}}{C}{-}OH \qquad 3(CH_3{-}\overset{\overset{\displaystyle O}{\|}}{C}{-})$$

Six-carbon chain — Three acetyl units

Problem 24.2 Look up the formulas of the following fatty acids in Table 18.2 and predict how many moles of acetyl CoA will be obtained from 1 mol of each:

a Palmitic acid
b Stearic acid
c Lauric acid

24.4 ACTIVATION OF FATTY ACIDS

Before cleavage of a fatty acid chain begins, the fatty acid must be activated and transported into the mitochondrial matrix from the cytosol. Activation involves reaction with both ATP and coenzyme A (CoA—SH) and prepares the fatty acid for the subsequent cleavage reactions it will undergo. The overall activation reaction is

$$R{-}\overset{\overset{\displaystyle O}{\|}}{C}{-}OH + ATP + HS{-}CoA \longrightarrow R{-}\overset{\overset{\displaystyle O}{\|}}{C}{-}S{-}CoA + AMP + 2P_i$$

The reaction is a three-step process:

$$R{-}\overset{\overset{\displaystyle O}{\|}}{C}{-}OH + ATP \longrightarrow R{-}\overset{\overset{\displaystyle O}{\|}}{C}{-}AMP + PP_i \qquad (24.1)$$

$$R{-}\overset{\overset{\displaystyle O}{\|}}{C}{-}AMP + HS{-}CoA \longrightarrow R{-}\overset{\overset{\displaystyle O}{\|}}{C}{-}S{-}CoA + AMP \qquad (24.2)$$

$$PP_i \longrightarrow 2P_i \qquad (24.3)$$

The energetics of the conversion is such that two phosphate bonds must be cleaved in order for the overall process to be exergonic. Notice that phosphate bonds are cleaved in Equations (24.1) and (24.3).

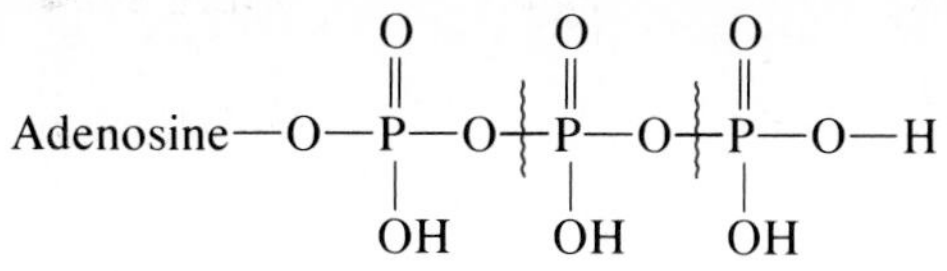

$$
\text{Adenosine}\!-\!\text{O}\!-\!\overset{\overset{\text{O}}{\|}}{\underset{\underset{\text{OH}}{|}}{\text{P}}}\!-\!\text{O}\!+\!\overset{\overset{\text{O}}{\|}}{\underset{\underset{\text{OH}}{|}}{\text{P}}}\!-\!\text{O}\!+\!\overset{\overset{\text{O}}{\|}}{\underset{\underset{\text{OH}}{|}}{\text{P}}}\!-\!\text{O}\!-\!\text{H}
$$

Cleaved in   Cleaved in
Equation (24.1)   Equation (24.3)

This is approximately equivalent to the energy expenditure of 2 ATPs. When we count up energy yield from fatty acid metabolism in Section 24.6, we will count the activation step as -2 ATP.

Fatty acids are activated, that is, converted to thioesters on the outer surface of the mitochondrial membrane. The thioesters cannot cross the membrane. Rather, they are carried across the membrane by carnitine:

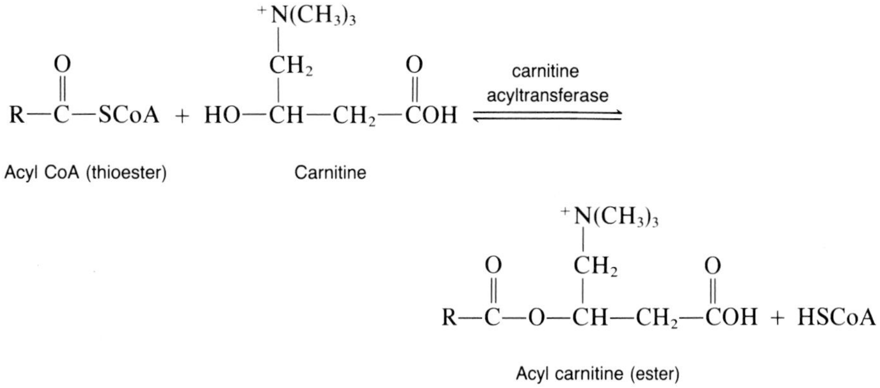

$$
\underset{\text{Acyl CoA (thioester)}}{\text{R}\!-\!\overset{\overset{\text{O}}{\|}}{\text{C}}\!-\!\text{SCoA}} + \underset{\text{Carnitine}}{\text{HO}\!-\!\overset{\overset{\overset{+\text{N(CH}_3)_3}{|}}{\overset{\text{CH}_2}{|}}}{\text{CH}}\!-\!\text{CH}_2\!-\!\overset{\overset{\text{O}}{\|}}{\text{C}}\text{OH}} \underset{\xrightleftharpoons{}}{\overset{\text{carnitine}}{\overset{\text{acyltransferase}}{}}}
$$

$$
\underset{\text{Acyl carnitine (ester)}}{\text{R}\!-\!\overset{\overset{\text{O}}{\|}}{\text{C}}\!-\!\text{O}\!-\!\overset{\overset{\overset{+\text{N(CH}_3)_3}{|}}{\overset{\text{CH}_2}{|}}}{\text{CH}}\!-\!\text{CH}_2\!-\!\overset{\overset{\text{O}}{\|}}{\text{C}}\text{OH}} + \text{HSCoA}
$$

Notice that the reaction is reversible. Acyl carnitine forms ($\longrightarrow$) on the outside and crosses the membrane, and then the reaction reverses ($\longleftarrow$) and the thioester re-forms inside the mitochondrial matrix.

Activation essentially converts a carboxyl group to a thioester group, or we might say that the *acyl* ($\text{R}\!-\!\overset{\overset{\text{O}}{\|}}{\text{C}}\!-$) portion of the acid becomes attached to the S of coenzyme A. The thioester is called an *acyl* CoA. For example,

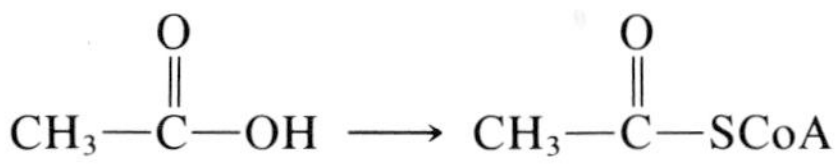

$$
\underset{\text{Acetic acid}}{\text{CH}_3\!-\!\overset{\overset{\text{O}}{\|}}{\text{C}}\!-\!\text{OH}} \longrightarrow \underset{\text{Acetyl CoA}}{\text{CH}_3\!-\!\overset{\overset{\text{O}}{\|}}{\text{C}}\!-\!\text{SCoA}}
$$

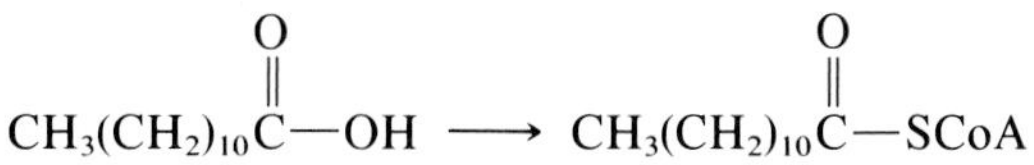

$$
\underset{\text{Lauric acid}}{\text{CH}_3(\text{CH}_2)_{10}\overset{\overset{\text{O}}{\|}}{\text{C}}\!-\!\text{OH}} \longrightarrow \underset{\text{Lauryl CoA}}{\text{CH}_3(\text{CH}_2)_{10}\overset{\overset{\text{O}}{\|}}{\text{C}}\!-\!\text{SCoA}}
$$

This is the essence of activation, the conversion of $\text{R}\!-\!\overset{\overset{\text{O}}{\|}}{\text{C}}\!-\!\text{OH}$ to $\text{R}\!-\!\overset{\overset{\text{O}}{\|}}{\text{C}}\!-\!\text{SCoA}$.

Problem 24.3   Write structures for the thioesters formed from stearic acid ($\text{CH}_3(\text{CH}_2)_{16}\text{COOH}$) and palmitic acid ($\text{CH}_3(\text{CH}_2)_{14}\text{COOH}$).

24.5 THE FATTY ACID SPIRAL

Inside the mitochondrion the thioester (acyl CoA) is broken down by a recurring sequence of four reactions:

1 Oxidation to form an α,β-alkene
2 Addition of water to the alkene to form a β-alcohol
3 Oxidation of the β-alcohol to a β-ketone
4 Cleavage between the α- and β-positions to form acetyl CoA and an acyl CoA two carbons shorter than the original.

The continuous repetition of these four steps is called the fatty acid spiral and is shown in Figure 24.3 breaking down lauryl CoA,

$$CH_3(CH_2)_{10}-\overset{\overset{\displaystyle O}{\|}}{C}-S-CoA, \quad \text{which can also be written} \quad CH_3(CH_2)_8CH_2CH_2-\overset{\overset{\displaystyle O}{\|}}{C}-S-CoA.$$

Each individual repetition of the four steps is called a turn of the spiral and results in two carbons being cleaved from the acyl thioester chain. Let us now look at each step in more detail.

STEP 1 Step 1 of the four repeating steps is the oxidation, that is, the loss of hydrogen from the α- and β-positions. The result is a trans carbon-carbon double bond between the α- and β-positions. FAD is the coenzyme involved.

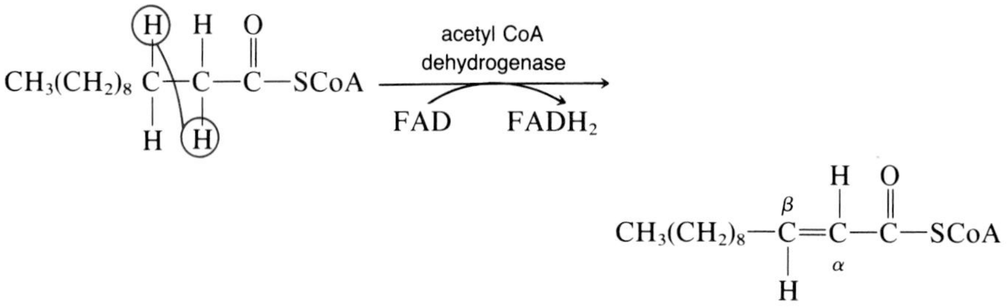
acetyl CoA
dehydrogenase
FAD
FADH2

The subsequent step in the spiral can produce an alcohol with the proper stereochemistry only if the geometry of the double bond is trans. Notice that the β-carbon in the alcohol is a chiral center. Many naturally occurring unsaturated fatty acids have double bonds with the cis geometry; the double bond must be isomerized before entering the fatty acid spiral. The body has the enzymes to accomplish this.

STEP 2 Step 2 is the addition of water to the α-, β-double bond. The —OH group of water always becomes attached to the β-position with a unique configuration:

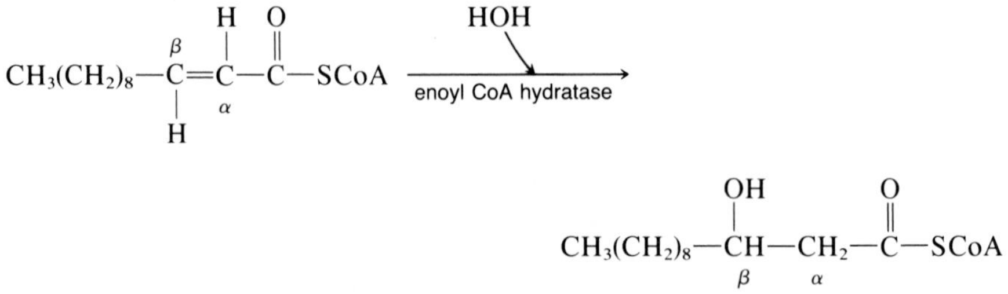
HOH
enoyl CoA hydratase

The proper stereoisomer must be formed in this step or the enzyme for step 3 will not function because the enzyme displays absolute specificity (Section 21.5).

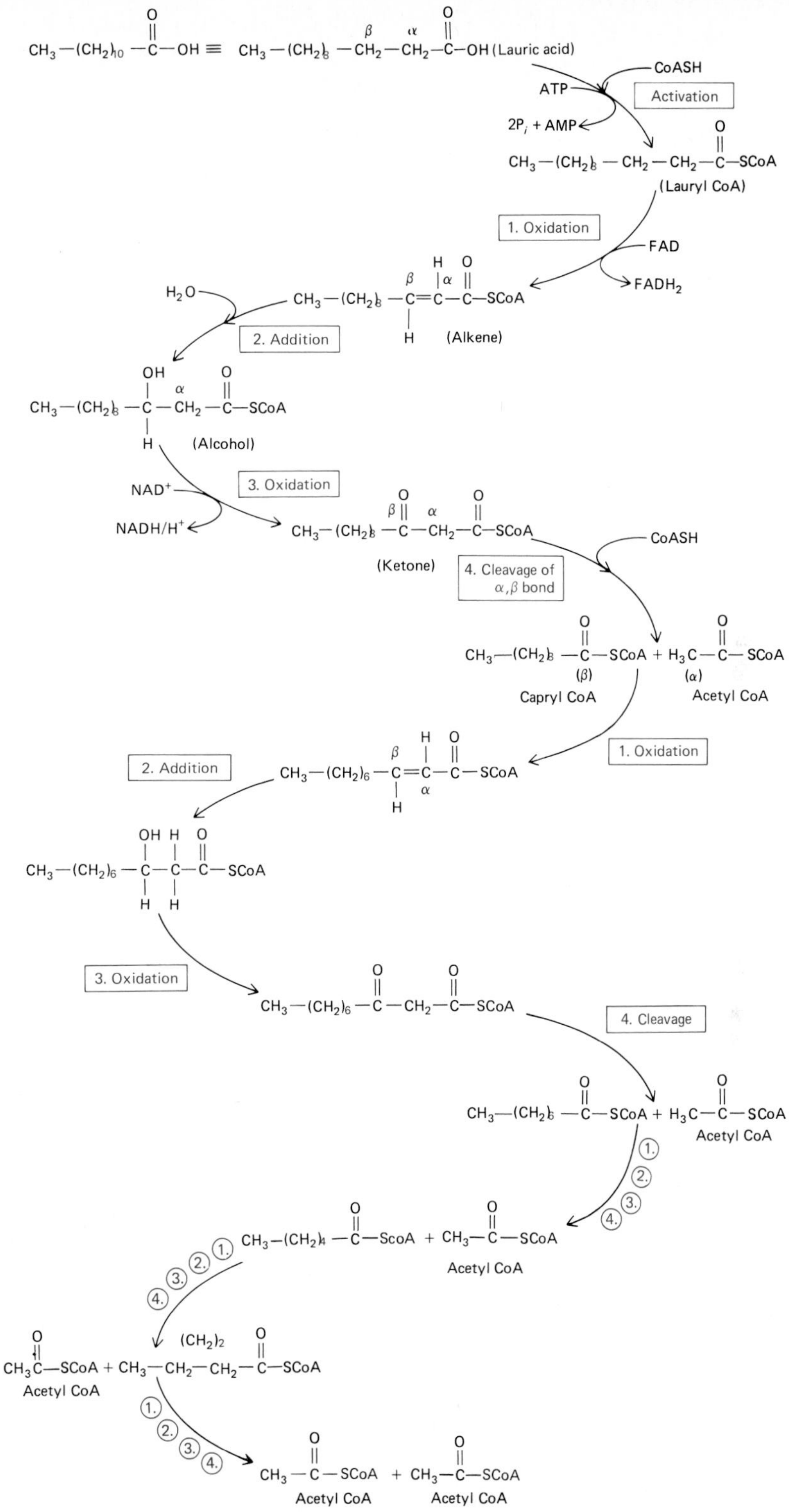
(Lauric acid)
CoASH
ATP
Activation
2P_i + AMP
(Lauryl CoA)
1. Oxidation
FAD
FADH2
H2O
2. Addition
(Alkene)
(Alcohol)
NAD+
NADH/H+
3. Oxidation
(Ketone)
4. Cleavage of α,β bond
CoASH
Capryl CoA
Acetyl CoA
1. Oxidation
2. Addition
3. Oxidation
4. Cleavage
Acetyl CoA
Acetyl CoA
Acetyl CoA
Acetyl CoA
Acetyl CoA

FIG. 24.3

The fatty acid spiral completely breaking down lauric acid into acetyl CoA. The colored subscripts $(CH_2)_8$ emphasize the loss of two carbon atoms at each turn of the spiral. Count up the acetyl CoA's shown and you will find six produced from this 12-carbon acid.

STEP 3 Step 3 is the oxidation of a secondary alcohol to a ketone. Because this oxidation occurs at the β-position, the fatty acid spiral is also known as the *β-oxidation pathway*. NAD^+ is the required coenzyme.

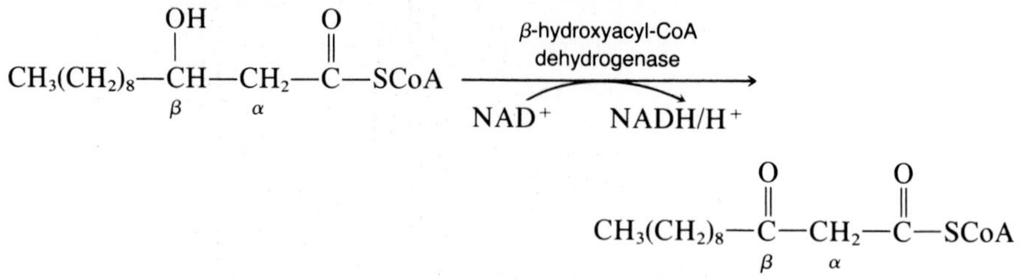
β-hydroxyacyl-CoA
dehydrogenase
NAD+
NADH/H+

STEP 4 Step 4 is the cleavage of the β-ketothioester between the α- and β-positions. The cleavage involves CoAS—H; the H— of CoASH attaches to the α-side after cleavage and CoAS— attaches to the β-side:

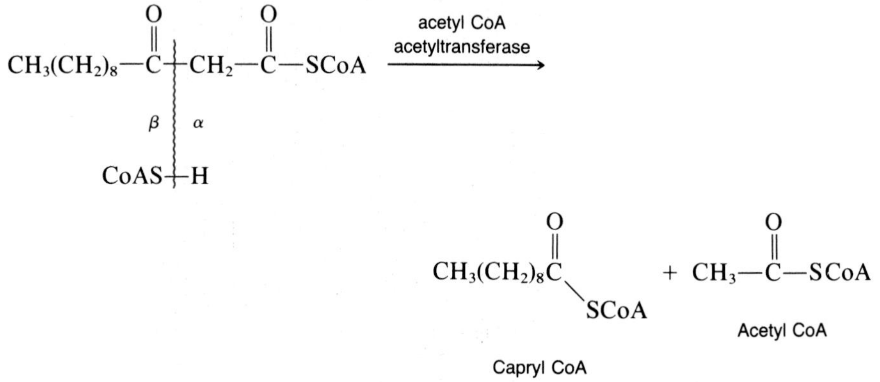
acetyl CoA
acetyltransferase
Capryl CoA
Acetyl CoA

The products are therefore acetyl CoA (which can enter the citric acid cycle) and capryl CoA, a thioester two carbons shorter than lauryl CoA. Capryl CoA then undergoes the four reactions of the fatty acid spiral and is thereby shortened by two carbons. $CH_3(CH_2)_8\text{—}\overset{O}{\overset{\|}{C}}\text{—}SCoA$, which can also be written $CH_3(CH_2)_6\overset{\beta}{C}H_2\overset{\alpha}{C}H_2\text{—}\overset{O}{\overset{\|}{C}}\text{—}SCoA$, undergoes

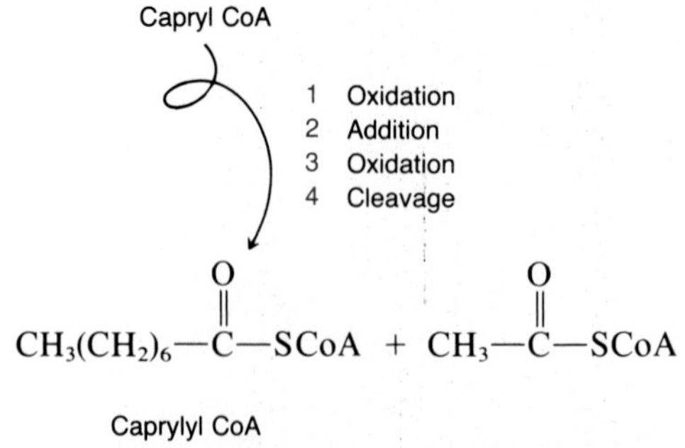
Capryl CoA
1 Oxidation
2 Addition
3 Oxidation
4 Cleavage
Caprylyl CoA

Look at Figure 24.3 to see the subsequent turns of the fatty acid spiral as applied to caprylyl CoA.

SAMPLE EXERCISE 24.2 Begin with capryl CoA, $CH_3(CH_2)_8-\overset{O}{\overset{\|}{C}}-SCoA$, and show the intermediates that form as this compound goes through the four reactions of the fatty acid spiral.

Solution "Expose" the α- and β-positions because all the reactions occur there:

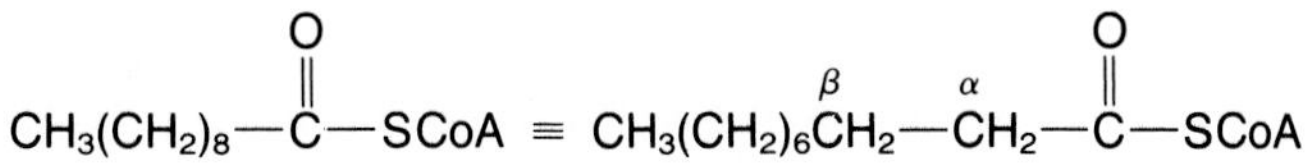

Carry out the reactions step by step as described above:

STEP 1 Oxidation at the α-, β-positions:

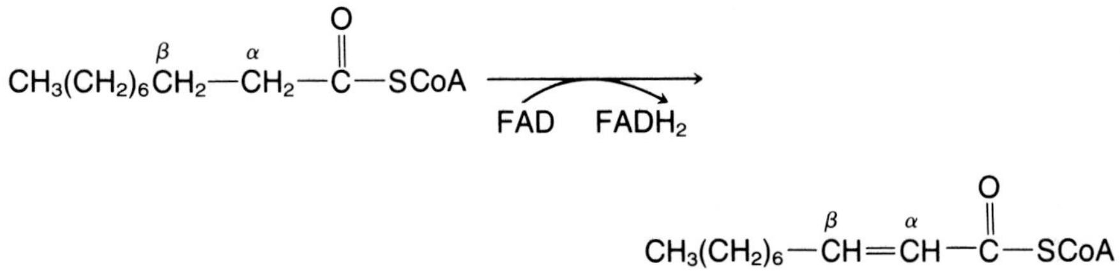

STEP 2 Addition of water to give a β-alcohol:

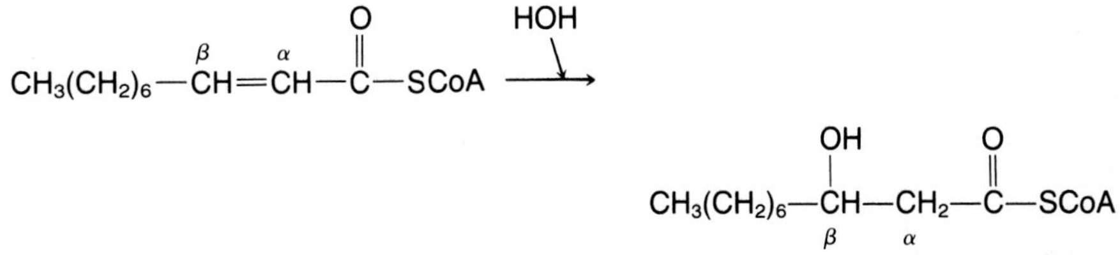

STEP 3 β-Oxidation:

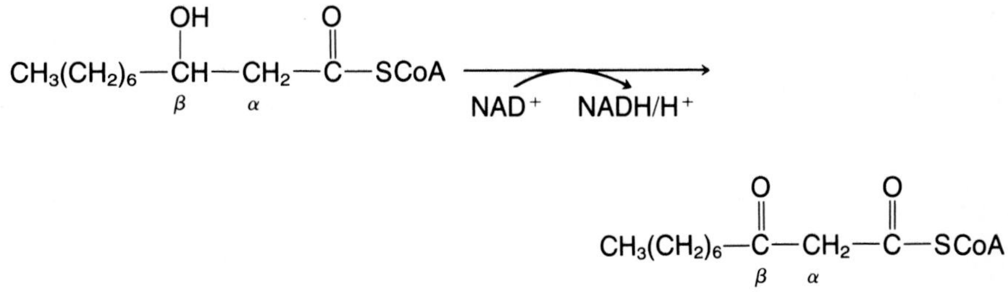

STEP 4 Cleavage between the α- and β-positions:

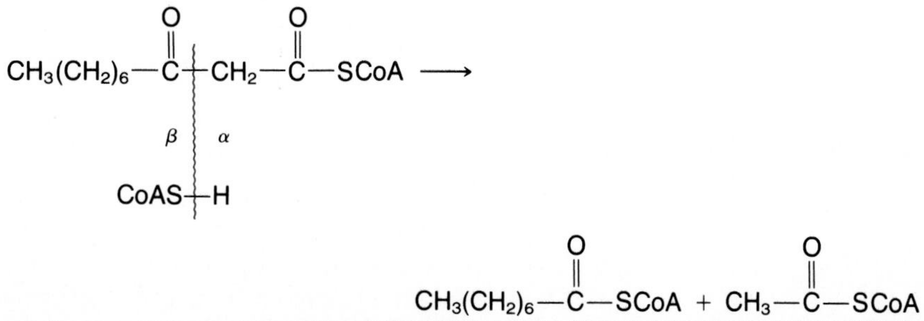

Problem 24.4 Take caprylyl CoA, $CH_3(CH_2)_6\text{—}\overset{O}{\overset{\|}{C}}\text{—S—CoA}$, through the four reactions of the fatty acid spiral.

It requires five complete turns of the fatty acid spiral to cleave lauryl CoA into 6 mol of acetyl CoA. For fatty acids with even numbers of C atoms, it is always true that the number of turns required for total cleavage is equal to one less than the number of moles of acetyl CoA generated. This is so because whereas most turns of the spiral produce just 1 mol of acetyl CoA, the last turn of the spiral for all fatty acids produces 2 mol of acetyl CoA. The last turn cleaves 1 mol of butyryl CoA into 2 mol of acetyl CoA.

Last turn in the cleavage of fatty acids with even numbers of carbons

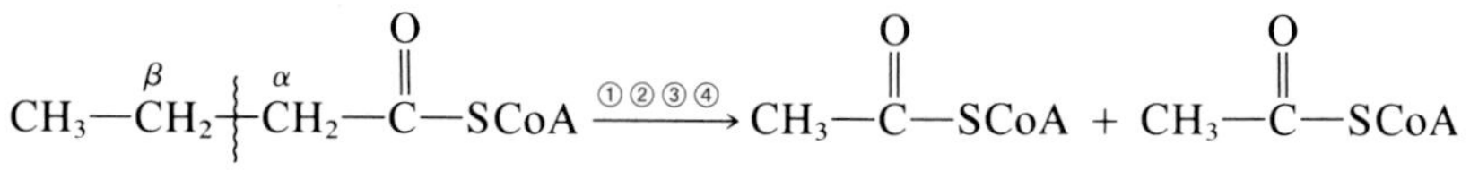

Butyryl CoA — Acetyl CoA — Acetyl CoA

To ascertain the number of turns of the spiral required to cleave 1 mol of a fatty acid totally and the number of moles of acetyl CoA obtained thereby, follow the procedure:

STEP 1 Count the number of carbons in the fatty acid. This is *n*.

STEP 2 Divide *n* by 2; $\frac{n}{2}$ = number of moles of acetyl CoA generated.

STEP 3 Subtract 1 from $\frac{n}{2}$; $\frac{n}{2} - 1$ = number of spiral turns.

SAMPLE EXERCISE 24.3 Stearic acid has the formula $CH_3(CH_2)_{16}COOH$. From 1 mol of stearic acid how many moles of acetyl CoA are produced after the acid is activated by conversion to stearyl CoA? How many turns of the fatty acid spiral are needed to generate this amount?

Solution Follow the stepwise procedure just given:

STEP 1 Count the number of carbons in stearic acid; $n = 18$.

STEP 2 Divide *n* by 2; $18/2 = 9$; 9 mol of acetyl CoA is produced.

STEP 3 Subtract 1 from $n/2$; $9 - 1 = 8$; thus eight turns of the fatty acid spiral are needed.

24.6 ENERGY YIELD FROM FATTY ACID METABOLISM

The energy yield in terms of ATP synthesis can be calculated easily for any fatty acid from the number of moles of acetyl CoA produced from that acid and the number of turns of the fatty acid spiral required to break it down completely. Each mole of acetyl CoA produced goes to the citric acid cycle and produces 12 mol of ATP, as you saw in Section 23.4. Thus for a fatty acid with some number of carbon atoms (*n*C), one-half that number (*n*/2) of moles of acetyl CoA is produced and $12 \times n/2$ mol of ATP is synthesized.

In addition, during each turn of the spiral, $FADH_2$ and $NADH/H^+$ are sent to the ETS, where they generate 2 mol of ATP and 3 mol of ATP, respectively, for a total of 5 mol of ATP per turn. Thus, we get a yield of 5 mol of ATP × number of turns of spiral, that is, one less than $n/2$ turns, as developed previously.

However, 2 mol of ATP was required to activate the fatty acid, so we must remember to subtract 2 mol of ATP from the total we calculate. In summary, the ATP yield for any fatty acid can be calculated from the sum:

$$\text{Number of moles of acetyl CoA produced} \times 12 \text{ mol of ATP}$$
$$+ \text{ number of turns of spiral} \times 5 \text{ mol of ATP}$$
$$+ \text{ 1 activation step} \times -2 \text{ mol of ATP}$$

SAMPLE EXERCISE 24.4 Calculate the ATP yield from 1 mol of stearic acid.

Solution In Sample Exercise 24.3 we calculated the number of moles of acetyl CoA produced and the number of turns of the spiral for $CH_3(CH_2)_{16}COOH$ and found 9 mol of acetyl CoA and eight turns. Thus,

$$9 \cancel{\text{mol}} \text{ of acetyl CoA} \times \frac{12 \text{ mol of ATP}}{1 \cancel{\text{mol}} \text{ of acetyl CoA}} = 108 \text{ mol of ATP}$$

$$8 \cancel{\text{turns}} \text{ of spiral} \times \frac{5 \text{ mol of ATP}}{1 \cancel{\text{turn}}} = 40 \text{ mol of ATP}$$

$$\text{Activation} = \underline{-2 \text{ mol of ATP}}$$

$$146 \text{ mol of ATP}$$

Problem 24.5 Calculate the ATP yield from 1 mol of lauric acid, $CH_3(CH_2)_{10}COOH$.

The body stores energy in fat reserves rather than as glycogen because fat has a higher energy density than carbohydrates. That is, there is a significantly higher yield of ATP from a gram of fat than from the same mass of carbohydrate. Recall from Section 23.5 that 1 mol of glucose (molecular weight = 180) yields 36 mol of ATP. In Sample Exercise 24.4 we saw that 1 mol of stearic acid (molecular weight = 284) yields 146 mol of ATP. From these data we can readily calculate the number of moles of ATP generated per gram of nutrient:

$$\text{Glucose:} \quad \frac{36 \text{ mol of ATP}}{180 \text{ g of glucose}} = \frac{0.20 \text{ mol of ATP}}{\text{gram of glucose}}$$

$$\text{Stearic acid:} \quad \frac{146 \text{ mol of ATP}}{284 \text{ g of stearic acid}} = \frac{0.51 \text{ mol of ATP}}{\text{gram of stearic acid}}$$

Fats typically yield more than twice the amount of energy per gram as do carbohydrates. A 150-lb person with adequate energy reserves stored in fat tissue would have to weigh about 375 lb if glycogen were our energy storage facility; that is, more than twice the mass of glycogen is required to produce the same amount of energy as a mass of fat.

24.7 LIPOGENESIS

The body degrades fatty acids into acetyl CoA and also makes fatty acids from acetyl CoA. However, the two processes are not the exact reverse of one another. Once again we encounter the principle that synthetic and degradative pathways in biological systems are usually different. Previously we have seen the similarities and differences between the reverse processes of gluconeogenesis and glycolysis and between glycogenesis and glycogenolysis (Section 23.11).

The synthesis of fatty acids by the body is called **lipogenesis.** Lipogenesis occurs in the cytosol in contrast to the fatty acid spiral, which occurs in the mitochondrial matrix. Other distinctions between the synthetic and degradative pathways will emerge as we discuss the individual steps in lipogenesis.

Figure 24.4 illustrates the overall process of lipogenesis. Steps 3 to 5 of lipogenesis closely resemble the fatty acid spiral in structural detail, although the enzymes or coenzymes involved may vary. But the first two steps of lipogenesis are very distinctive.

STEP 1 Step 1 is the combination of acetyl CoA with CO_2. Energy is required for this reaction; that is, it is endergonic and so the reaction is coupled to ATP hydrolysis:

$$\underset{\text{Acetyl CoA}}{CH_3{-}\overset{\displaystyle O}{\overset{\|}{C}}{-}S{-}CoA} + CO_2 \xrightarrow{\text{ATP} \quad \text{ADP} + P_i} \underset{\text{Malonyl CoA}}{HOOC{-}CH_2{-}\overset{\displaystyle O}{\overset{\|}{C}}{-}S{-}CoA}$$

The enzyme is acetyl CoA carboxylase; the coenzyme is the vitamin biotin.

Whereas fatty acids are "carried through" the fatty acid spiral as thioesters of coenzyme A, that is, they are activated by the conversion

$$\underset{\text{Fatty acid}}{R{-}\overset{\displaystyle O}{\overset{\|}{C}}{-}OH} + HSCoA \longrightarrow \underset{\text{Thioester}}{R{-}\overset{\displaystyle O}{\overset{\|}{C}}{-}S{-}CoA}$$

a different carrier operates in lipogenesis. Before lipogenesis continues, the thioesters acetyl CoA and malonyl CoA are converted to the corresponding thioesters of the so-called *a*cyl *c*arrier *p*rotein (ACP), a partial structure of which is shown in Figure 24.5. Like coenzyme A, acyl carrier protein is a thiol and so in its reactions may be represented HS—ACP. The conversion is represented schematically:

$$\underset{\text{Acetyl CoA}}{CH_3{-}\overset{\displaystyle O}{\overset{\|}{C}}{-}S{-}CoA} + HS{-}ACP \xrightarrow[\text{transacylase}]{\text{acetyl}} \underset{\text{Acetyl ACP}}{CH_3{-}\overset{\displaystyle O}{\overset{\|}{C}}{-}S{-}ACP} + HSCoA$$

$$\underset{\text{Malonyl CoA}}{HOOC{-}CH_2{-}\overset{\displaystyle O}{\overset{\|}{C}}{-}S{-}CoA} + HS{-}ACP \xrightarrow[\text{transacylase}]{\text{malonyl}} \underset{\text{Malonyl ACP}}{HOOC{-}CH_2{-}\overset{\displaystyle O}{\overset{\|}{C}}{-}S{-}ACP} + HSCoA$$

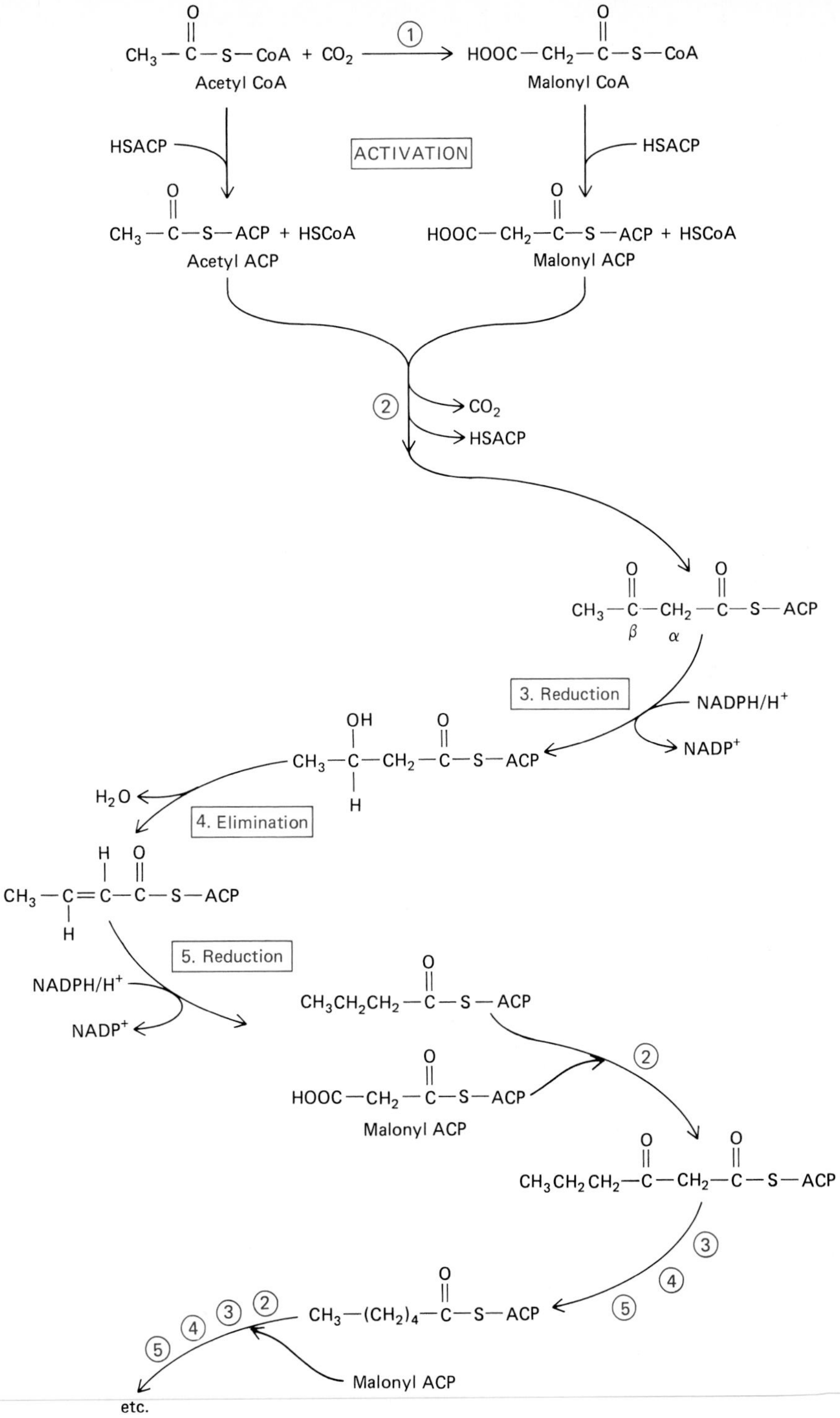

$CH_3-C(=O)-S-CoA + CO_2$
①
$HOOC-CH_2-C(=O)-S-CoA$
Acetyl CoA
Malonyl CoA
HSACP
ACTIVATION
HSACP
$CH_3-C(=O)-S-ACP + HSCoA$
$HOOC-CH_2-C(=O)-S-ACP + HSCoA$
Acetyl ACP
Malonyl ACP
②
CO_2
HSACP
$CH_3-C(=O)-CH_2-C(=O)-S-ACP$
β α
3. Reduction
$NADPH/H^+$
$NADP^+$
$CH_3-CH(OH)-CH_2-C(=O)-S-ACP$
H_2O
4. Elimination
$CH_3-CH=CH-C(=O)-S-ACP$
5. Reduction
$NADPH/H^+$
$NADP^+$
$CH_3CH_2CH_2-C(=O)-S-ACP$
$HOOC-CH_2-C(=O)-S-ACP$
Malonyl ACP
②
$CH_3CH_2CH_2-C(=O)-CH_2-C(=O)-S-ACP$
③ ④ ⑤
$CH_3-(CH_2)_4-C(=O)-S-ACP$
② ③ ④ ⑤
Malonyl ACP
etc.

FIG. 24.4

Lipogenesis. Fatty acids are made from acetyl CoA. First, acetyl CoA combines with CO_2 to yield malonyl CoA. These two molecules are then activated and prepared for condensation (combination) by conversion to acyl carrier proteins (ACP). Steps 3 through 5, reduction, elimination, and reduction, are the reverse of steps 1 through 3 (oxidation, addition, oxidation) of the fatty acid spiral, in which fatty acids are broken down to acetyl CoA. Compare with Figure 24.3.

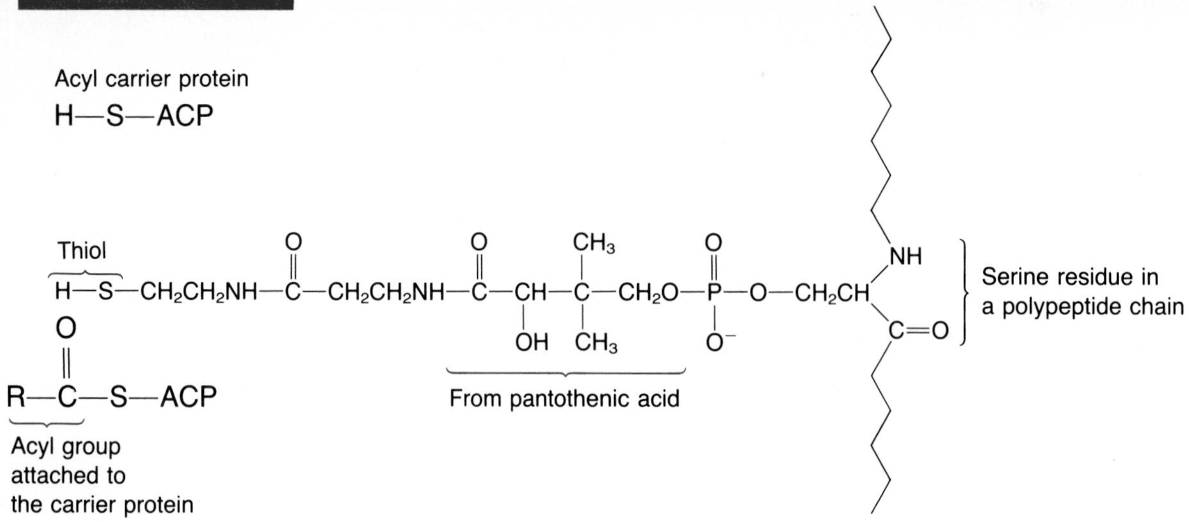

Acyl carrier protein
H—S—ACP
Thiol
H—S—CH₂CH₂NH—C—CH₂CH₂NH—C—CH—C—CH₂O—P—O—CH₂CH
OH CH₃
O⁻
NH
C=O
Serine residue in a polypeptide chain
From pantothenic acid
R—C—S—ACP
Acyl group attached to the carrier protein

FIG. 24.5
Acyl carrier protein. Like coenzyme A, the thiol functional group is the site of the significant reactions of acyl carrier protein. Thus the abbreviations HS—ACP and R—C(=O)—ACP arise. Also like CoASH, the vitamin pantothenic acid is incorporated in the ACP structure. Whereas a nucleotide completes the structure of CoASH, in ACP pantothenic acid is attached through a phosphate group to a serine residue of a polypeptide chain. Compare this figure to Figure 22.12, which shows CoASH and acetyl CoA.

STEP 2 Step 2 of lipogenesis is the combination of acetyl ACP and malonyl ACP. The combination reaction is coupled with the exergonic loss of CO_2.

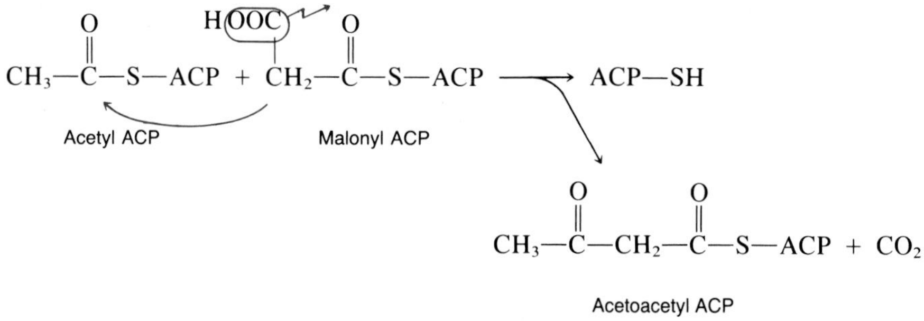

Acetyl ACP
Malonyl ACP
ACP—SH
CO₂
Acetoacetyl ACP

The step 2 reaction increases a growing fatty acid chain by two carbon atoms. In this case, the two-carbon acetyl ACP becomes the four-carbon acetoacetyl ACP.

All of the subsequent reactions (steps 3, 4, and 5) occur at the α- and β-positions and resemble the reverse of the fatty acid spiral (steps 3, 2, and 1). A significant difference in the synthesis is that the coenzyme reducer is NADPH/H^+, not NADH/H^+ or $FADH_2$.

STEP 3 Step 3 is reduction of a β-ketone to a β-alcohol:

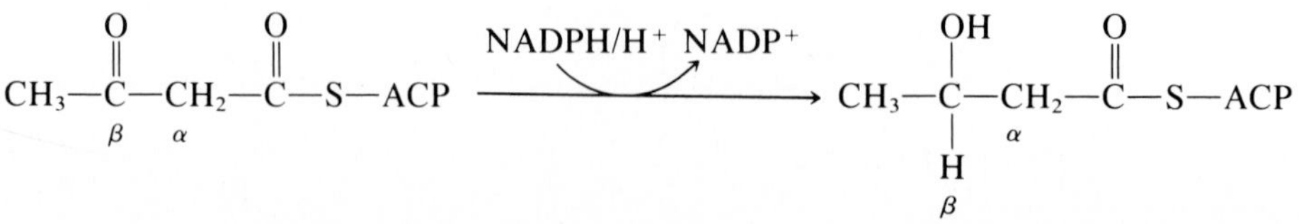

NADPH/H⁺ NADP⁺

STEP 4 Step 4 is loss of water from the α- and β-positions to yield an α, β-alkene:

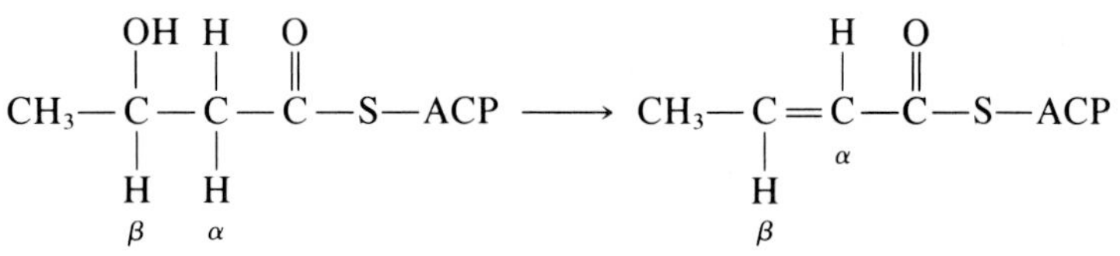

STEP 5 Step 5 is addition of hydrogen (reduction) which saturates the α-, and β-positions.

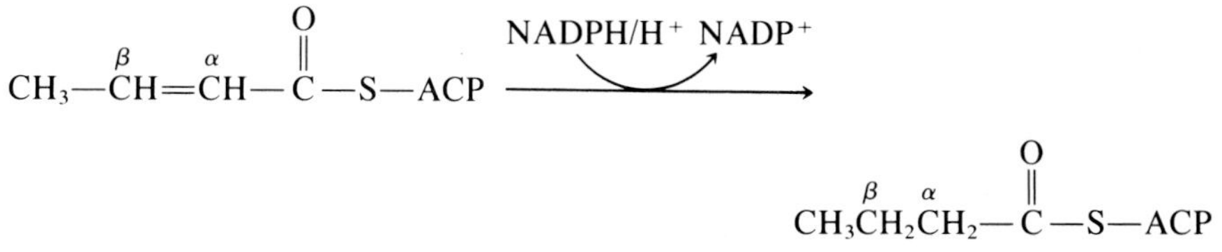

Butyryl ACP

This overall process has elongated the carbon chain by two carbon atoms, acetyl ACP (2C) $\longrightarrow$ butyryl ACP (4C). The actual elongation occurs in step 2. Steps 2 through 5 are repeated on butyryl ACP with the result that two more carbons are added:

Step 2 repeated produces

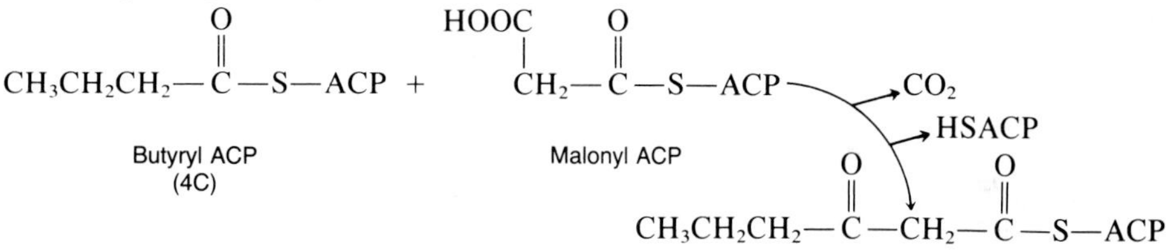

Butyryl ACP (4C)

Malonyl ACP

β-Ketohexanoyl ACP (6C)

Steps 3 to 5 produce $CH_3CH_2CH_2CH_2CH_2{-}\overset{\overset{O}{\|}}{C}{-}S{-}ACP$, which is then ready for another turn of the lipogenesis cycle. The buildup stops at palmityl ACP, $CH_3(CH_2)_{14}{-}\overset{\overset{O}{\|}}{C}{-}SACP$. Other enzyme systems and processes are required to extend chains beyond 16 carbons and to introduce unsaturation into the chain.

Problem 24.6 Hexanoyl ACP, $CH_3(CH_2)_4{-}\overset{\overset{O}{\|}}{C}{-}S{-}ACP$, reacts with malonyl ACP and goes through steps 3 to 5. Write the structure that results from these reactions.

To complete the synthesis of a typical triacylglycerol, the fatty acids made in lipogenesis must react with glycerol. In addition to dietary sources, another source of glycerol in the body is glycolysis. The reduction of dihydroxyacetone phosphate yields glycerol phosphate, which can readily condense with fatty acids.

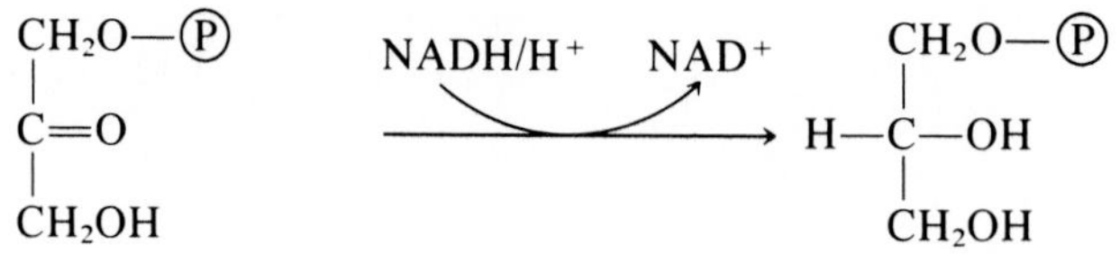
CH₂O—Ⓟ
C=O
CH₂OH
NADH/H⁺
NAD⁺
CH₂O—Ⓟ
H—C—OH
CH₂OH

Dihydroxyacetone phosphate

Glycerol phosphate

24.8 INTERRELATIONSHIPS OF CARBOHYDRATE AND LIPID METABOLISM

Figure 24.6 shows the interfacing and interactions of the carbohydrate and lipid metabolic cycles. It is important to observe which processes are reversible, at least in the sense that a pathway exists in each direction, and which processes go in only one direction. The most notable one-directional

FIG. 24.6
The big picture of carbohydrate and fat metabolism. Most of our energy needs are met by acetyl CoA from carbohydrates or fats entering the citric acid cycle. Note that fat metabolism in the absence of carbohydrate metabolism is not likely to lead to entry of acetyl CoA into the citric acid cycle because oxaloacetic acid is required in the citric acid cycle. Only carbohydrate metabolism produces significant amounts of oxaloacetic acid. Fat metabolism cannot produce it because acetyl CoA cannot be converted to pyruvic acid; i.e., the process pyruvic acid → acetyl CoA is irreversible. Also, lipogenesis cannot run in the absence of glucose metabolism to produce NADPH/H⁺.

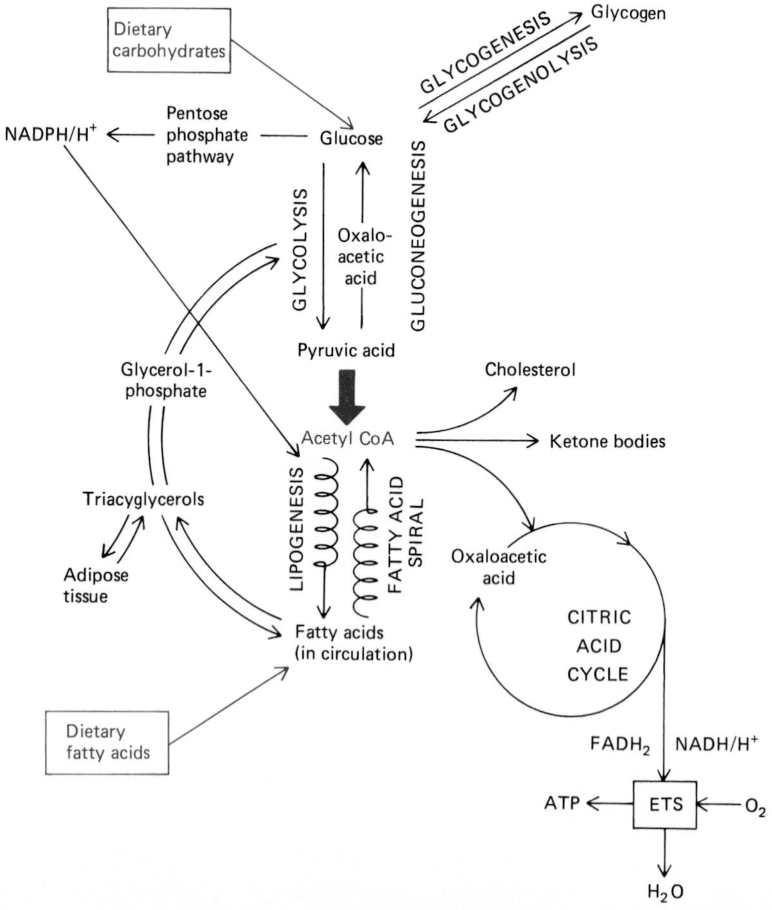
Dietary carbohydrates
Glycogen
GLYCOGENESIS
GLYCOGENOLYSIS
NADPH/H⁺
Pentose phosphate pathway
Glucose
GLYCOLYSIS
Oxalo-acetic acid
GLUCONEOGENESIS
Pyruvic acid
Glycerol-1-phosphate
Cholesterol
Acetyl CoA
Ketone bodies
Triacyglycerols
LIPOGENESIS
FATTY ACID SPIRAL
Oxaloacetic acid
Adipose tissue
CITRIC ACID CYCLE
Fatty acids (in circulation)
Dietary fatty acids
FADH₂
NADH/H⁺
ATP
ETS
O₂
H₂O

process is the conversion of pyruvic acid to acetyl CoA. Because this process is irreversible, the body can make fats from carbohydrates,

$$\text{Glucose} \rightleftharpoons \text{pyruvic acid} \longrightarrow \text{acetyl CoA} \rightleftharpoons \text{fatty acids}$$

but fatty acids cannot be converted into carbohydrates. It is true that the fat component, glycerol, can enter the gluconeogenesis pathway, but glycerol constitutes only a very small portion of the lipid biomass.

Another observation to make about fat-carbohydrate metabolic relations is that oxaloacetic acid must be available for acetyl CoA to enter the citric acid cycle. Carbohydrate metabolism provides a source of both acetyl CoA and oxaloacetic acid through the intermediacy of pyruvic acid. Fat breakdown provides only acetyl CoA. Therefore, fatty acid metabolism in the absence of sufficient glucose metabolism leads to an accumulation of acetyl CoA unable to enter the citric acid cycle. A small amount of oxaloacetic acid can be derived from amino acids, that is, from protein digestion and metabolism, as you will see in Chapter 25.

24.9 KETONE BODIES

The accumulation of acetyl CoA can lead to fatty acid synthesis and fat storage. However, another bodily response to excess acetyl CoA is the production of three structurally related compounds called *ketone bodies*. The first of these formed is acetoacetic acid. Without considering the details of the metabolic route, the overall process is

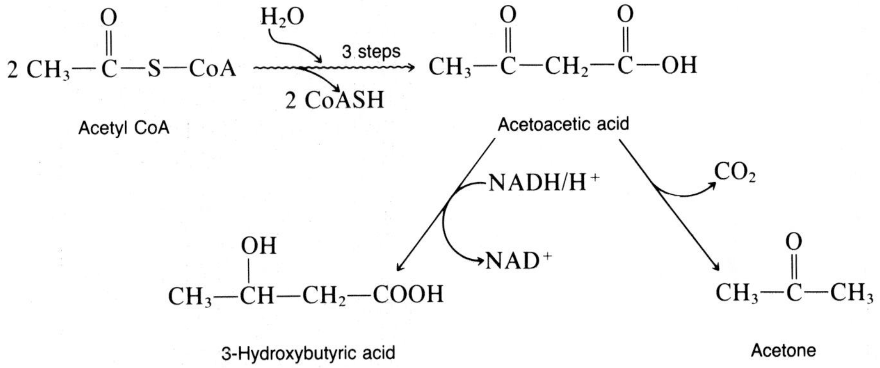
2 CH3—C—S—CoA
H2O
3 steps
2 CoASH
CH3—C—CH2—C—OH
Acetyl CoA
Acetoacetic acid
NADH/H+
NAD+
CO2
CH3—CH—CH2—COOH
OH
CH3—C—CH3
3-Hydroxybutyric acid
Acetone

Two acetyl CoA molecules condense into acetoacetic acid (a β-keto acid), which may be reduced to 3-hydroxybutyric acid (a β-hydroxy acid) or decarboxylated to form acetone. These three compounds are collectively known as **ketone bodies.** The lipid cholesterol is also formed from acetyl CoA starting with condensation of two molecules to form an acetoacetic acid derivative. The carbon chain is lengthened further by more acetyl CoA and eventually twisted into the tetracyclic steroid ring structure of cholesterol (Section 18.14). The production of both ketone bodies and cholesterol from acetyl CoA is shown in Figure 24.6.

The production of acetoacetic acid is a normal metabolic event, and in fact, acetoacetic acid is a normal and important energy source. The heart muscle actually uses acetoacetic acid as a fuel in preference to glucose. Acetoacetic acid can also be thought of as a transportable form of acetyl units since it can readily be reconverted to acetyl CoA. However, *excess* production of acetoacetic acid and the other ketone bodies can have serious medical consequences. Such excess production occurs as a result of the disease diabetes mellitus.

24.10 THE DIABETIC CONDITION

Diabetes mellitus is a disease often caused by an insufficient supply of the hormone insulin in the body and is characterized by an abnormally high blood sugar level (review Section 23.9). Profiles of the blood sugar levels for normal and diabetic patients appear in Figure 24.7. These curves are obtained as part of glucose tolerance tests. After a period of fasting, patients are given 100 g of glucose. Initially, blood sugar levels rise from this ingestion. In a normal patient the level comes back down to the *normal fasting level* (70 to 110 mg/100 mL) within 2 h. A diabetic patient's blood sugar level is consistently above normal; that is, moderate to severe hyperglycemia is displayed.

In a normal person, insulin regulates (lowers) blood sugar level by signaling to the body that it is in the "fed" state. That is, it signals that there is sufficient glucose and no more need be produced. This signal is given by stimulating anabolic processes and inhibiting catabolic processes. Insulin is required for glucose to cross cell membranes and enter cells. Thus, insulin stimulates glucose to enter cells where it undergoes glycolysis. This removes glucose from blood, provides energy, and produces the raw materials for fat synthesis (glycerol phosphate and acetyl CoA). Insulin also inhibits the breakdown of fat. Insulin stimulates glycogenesis and inhibits glycogenolysis and gluconeogenesis. All these effects lower the blood sugar level. Because insulin permits the body to make full use of glucose and the body cannot use glucose without insulin, an absence or deficiency of insulin (the diabetic condition) greatly disrupts metabolism. Figure 24.8 summarizes these effects.

What is the body's response to an absence of insulin? Because glucose is not entering cells, the body "thinks" there is a lack of glucose even in the face of extremely high blood sugar levels (hyperglycemia), i.e., with the exception of brain cells, the membranes of which can be transversed by glucose without insulin, other bodily cells crave glucose even though the blood sugar level is very high. Therefore, glycogenolysis takes place in the liver. This releases more glucose into the blood, but, of course, it still cannot

FIG. 24.7
Typical blood glucose levels for normal and diabetic patients during a 5-h glucose tolerance test. The normal response to ingestion of glucose is a sharp rise in blood sugar level which then returns to the normal fasting level (70 to 110 mg/100 mL). Blood sugar levels of diabetics are consistently elevated.

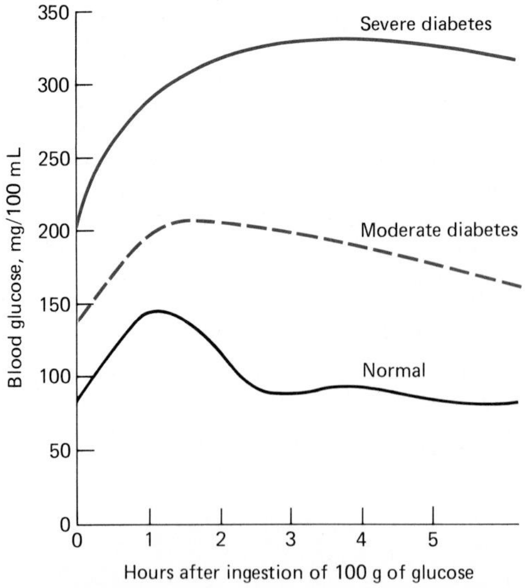
Severe diabetes
Moderate diabetes
Normal
Blood glucose, mg/100 mL
350
300
250
200
150
100
50
0
0
1
2
3
4
5
Hours after ingestion of 100 g of glucose

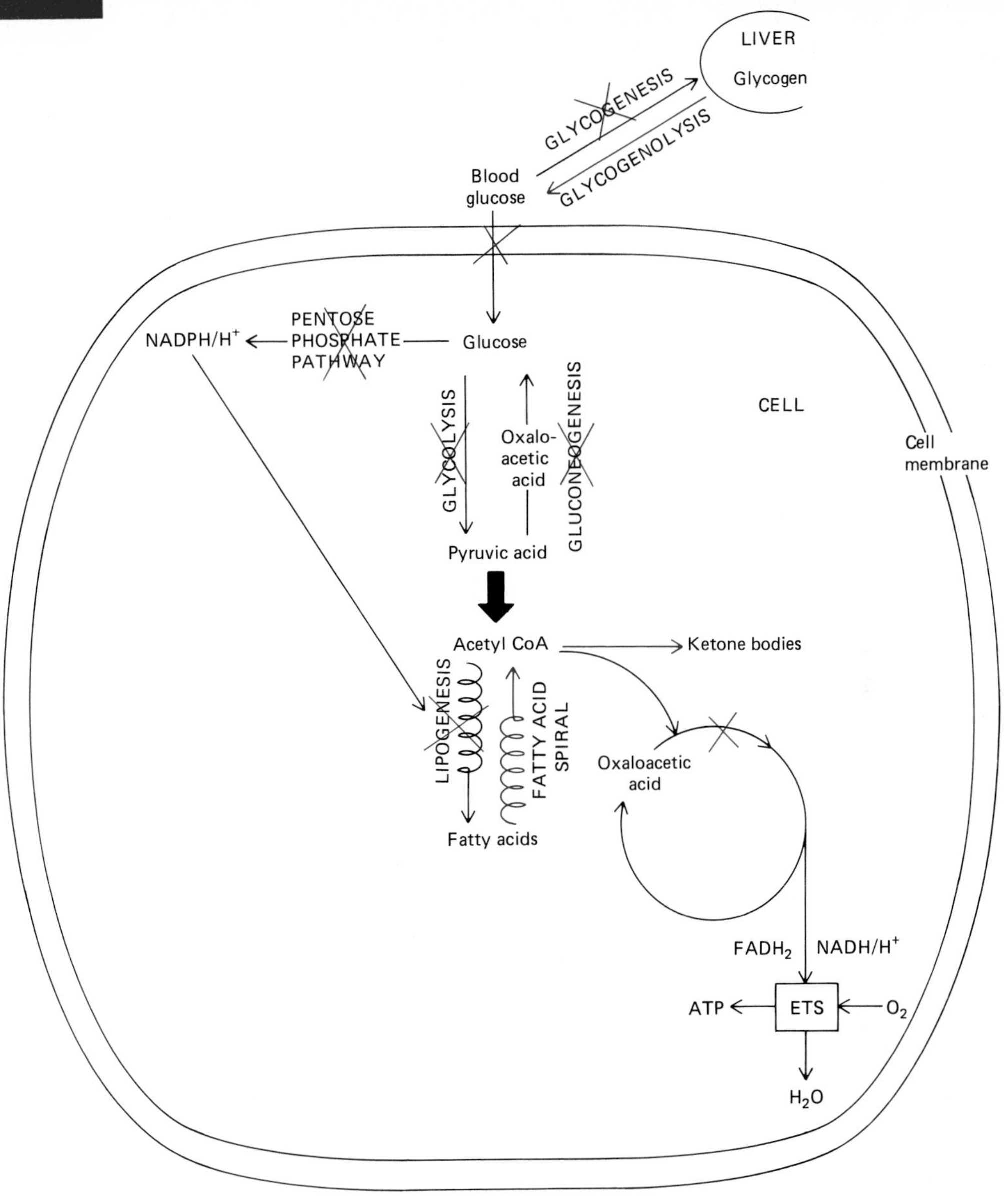
LIVER
Glycogen
GLYCOGENESIS
GLYCOGENOLYSIS
Blood glucose
NADPH/H+
PENTOSE PHOSPHATE PATHWAY
Glucose
GLYCOLYSIS
Oxalo-acetic acid
GLUCONEOGENESIS
CELL
Cell membrane
Pyruvic acid
Acetyl CoA
Ketone bodies
LIPOGENESIS
FATTY ACID SPIRAL
Oxaloacetic acid
Fatty acids
FADH2
NADH/H+
ATP
ETS
O2
H2O

FIG. 24.8
The disruption to metabolism that occurs in the absence of insulin. The processes crossed out do not operate; that is, there is no glucose metabolism because glucose cannot enter the cell. The body responds by stimulating the processes marked with colored arrows. The body "thinks" there is a lack of glucose because it is not being metabolized; thus glycogenolysis is tried as a correction. Blood glucose level rises as glycogenolysis produces ever more glucose which cannot enter the cell. Fatty acids are broken down to acetyl CoA for fuel, but produce mostly ketone bodies as the citric acid cycle fails in the absence of oxaloacetic acid from glucose metabolism.

enter the cells. Thus the hyperglycemic condition worsens. The cells begin to break down fats into acetyl CoA for fuel in the absence of glucose. But as we saw in Section 24.8, acetyl CoA cannot be used fully and efficiently in the citric acid cycle in the absence of glucose metabolism. Thus excess acetyl CoA accumulates and the body forms acetoacetic acid and other ketone bodies which enter the blood. It is often possible to detect the odor of acetone in a diabetic patient's breath because this ketone body is highly volatile.

The diabetic condition is thus characterized by:

1 Hyperglycemia—high blood sugar level
2 Glucosuria—sugar spilling over into the urine
3 Ketosis—excess ketone bodies in the blood

24.11 KETOSIS, ACIDOSIS, AND DEHYDRATION

What are the consequences of the ketosis caused by diabetes? Because two of the ketone bodies are carboxylic acids, $CH_3C(=O)CH_2COOH$ and $CH_3CH(OH)CH_2COOH$, their excess concentration in the blood leads to acidosis. The severe consequences of acidosis were discussed in Section 11.13. Most notably, lowered blood pH interferes with oxygen transport by hemoglobin. Rapid breathing is a first result; prolonged cellular deprivation of oxygen leads to coma and death.

Dehydration is another severe consequence of ketosis. The body's buffer systems try to neutralize the acidic ketone bodies. The bicarbonate ion, HCO_3^-, is consumed in this process. The depletion of this anion leads to a series of events which disrupt electrolyte balance and ultimately dehydrate cells. This presents another life-threatening situation.

Initial clinical treatment of severe diabetes involves alleviation of dehydration through the intravenous administration of physiological saline. Intraveneous supplies of bicarbonate may also be necessary. Of course, these are only temporary, stopgap measures; insulin must be supplied in order to reestablish the correct balance of metabolic processes in the body.

Some diabetic patients do produce insulin in the pancreas, but its primary protein structure is incorrect and disables its function. Alternatively, the patient may produce insulin but lack the receptor sites in cell membranes that interact with the insulin. Thus not all diabetes is the result of an absence of insulin.

If insulin is in short supply, but not absent, mild cases of diabetes can sometimes be treated without insulin injections. Careful regulation of diet, especially limiting sugar intake, allows the body to make its own adjustments and maintain suitable balances between carbohydrate and fat metabolism.

CHAPTER ACCOMPLISHMENTS

After completing this chapter you should be able to define **key words** and do the following:

24.1 INTRODUCTION

1. Describe the absorption and distribution of lipids in the body.

24.2 GLYCEROL METABOLISM

2. Name the metabolic pathway by which glycerol is catabolized.

24.3 OVERVIEW OF FATTY ACID METABOLISM

3. Name the end product of fatty acid catabolism.
4. Explain the efficiency associated with the fact that most naturally occurring fatty acids have even numbers of carbon.
5. Use the α, β lettering system for the hydrocarbon chain of a fatty acid.
6. Given the formula for a fatty acid, predict the number of moles of acetyl CoA obtained from 1 mol of that acid.

24.4 ACTIVATION OF FATTY ACIDS

7. Given the formula for a fatty acid, write the structure for the corresponding acyl CoA thioester.
8. State the energy requirement for activation of a fatty acid in terms of ATP.

24.5 THE FATTY ACID SPIRAL

9. Describe the recurring four reactions of the fatty acid spiral.
10. Given the formula for a fatty acid or its acyl thioester, write chemical equations to show the given compound reacting in the four reactions of the fatty acid spiral.
11. Name the coenzymes which are essential to the operation of the fatty acid spiral and indicate the steps in which they are involved.
12. Given the formula for a fatty acid, predict the number of turns of the fatty acid spiral required to cleave the acid totally to acetyl CoA.

24.6 ENERGY YIELD FROM FATTY ACID METABOLISM

13. State the ATP yield per mole of acetyl CoA processed in the citric acid cycle.
14. State the ATP yield per turn of the fatty acid spiral from the reduced coenzymes $FADH_2$ and $NADH/H^+$.
15. Given the formula for a fatty acid, calculate the total ATP yield for its catabolism.
16. Recognize fats as being more energy-rich than carbohydrates.

24.7 LIPOGENESIS

17. State three differences between the processes of lipogenesis and the fatty acid spiral.

24.8 INTERRELATIONSHIPS OF CARBOHYDRATE AND LIPID METABOLISM

18. Explain the significance of the fact that the conversion pyruvic acid $\longrightarrow$ acetyl CoA is irreversible.
19. Explain why acetyl CoA cannot be metabolized effectively in the absence of carbohydrate metabolism.

24.9 KETONE BODIES

20. Write structures for and name the three ketone bodies.

24.10 THE DIABETIC CONDITION

21. State the primary cause and primary symptom of diabetes mellitus.
22. State three ways in which insulin lowers blood sugar level.
23. State the body's response to an absence of insulin.

24.11 KETOSIS, ACIDOSIS, AND DEHYDRATION

24. Explain why ketosis leads to acidosis.
25. Explain why dehydration is a consequence of ketosis.

PROBLEMS

24.7 What are the products of digestion of triacylglycerols?

24.8 Describe the process of absorption of the products of fat digestion from the small intestine into the blood.

24.9 What is another name for fat tissue?

24.10 Describe the metabolism of glycerol.

24.11 What is the end product of fatty acid catabolism?

24.12 How many carbon atoms are in each of the following acids?

a Lauric, $CH_3(CH_2)_{10}COOH$
b Palmitic, $CH_3(CH_2)_{14}COOH$
c Stearic, $CH_3(CH_2)_{16}COOH$

24.13 What similarity do you note about the numbers of carbons in fatty acids such as those in Problem 24.12? What is the significance of this similarity?

24.14 What is the significance of the fragment $CH_3-\overset{\overset{\displaystyle O}{\|}}{C}-$ in both glucose and fatty acid metabolism?

24.15 Label the α- and β-positions in the following acids:

a $CH_3CH=CHCOOH$
b $HOOCCH_2\underset{\underset{\displaystyle O}{\|}}{C}CH_2CH_3$
c $CH_3(CH_2)_8CHOHCH_2\overset{\overset{\displaystyle O}{\|}}{C}-OH$
d $CH_3(CH_2)_6COOH$

24.16 Predict the numbers of moles of acetyl CoA obtained from 1 mol of each of the following:
 a $CH_3CH_2CH_2COOH$
 b $CH_3(CH_2)_{12}COOH$
 c $CH_3(CH_2)_6CH_2CH_2COOH$

24.17 Write structures for the thioesters formed from capric acid, $CH_3(CH_2)_8COOH$, and myristic acid, $CH_3(CH_2)_{12}COOH$, by reaction with ATP and HSCoA.

24.18 How much energy (in ATP) is required to activate 1 mol of a fatty acid?

24.19 Where in the cell does fatty acid activation occur?

24.20 Where in the cell does fatty acid catabolism occur?

24.21 How is a fatty acid prepared for entry into a mitochondrion?

24.22 Briefly describe the four reactions that recur in the fatty acid spiral.

24.23 Why is catabolism via the fatty acid spiral also known as β-oxidation?

24.24 The reactions of the fatty acid spiral affect just two carbon atoms of an activated fatty acid. Which two?

24.25 Fill in the blanks:

a _____

$$\overset{FAD\ FADH_2}{\curvearrowright}$$

$$CH_3(CH_2)_6CH{=}CH\overset{\displaystyle O}{\overset{\|}{C}}{-}SCoA$$

b $CH_3(CH_2)_6CH{=}CH\overset{\displaystyle O}{\overset{\|}{C}}{-}SCoA \xrightarrow{\overset{HOH}{\curvearrowright}}$

_____

c _____

$$\overset{NAD^+\ NADH/H^+}{\curvearrowright}$$

$$CH_3(CH_2)_6\overset{\displaystyle O}{\overset{\|}{C}}{-}CH_2\overset{\displaystyle O}{\overset{\|}{C}}{-}SCoA$$

d $CH_3(CH_2)_6\overset{\displaystyle O}{\overset{\|}{C}}{-}CH_2\overset{\displaystyle O}{\overset{\|}{C}}{-}SCoA \xrightarrow{\overset{CoASH}{\curvearrowright}}$

_____ + _____

$$\overset{\displaystyle O}{\overset{\|}{}}$$

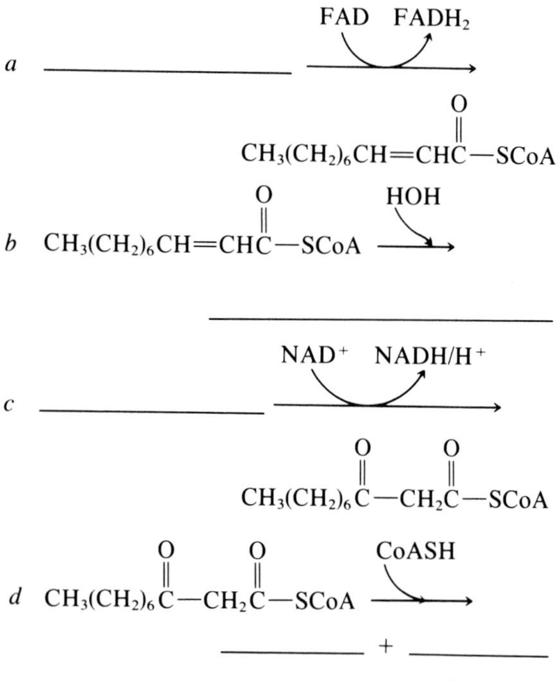

24.26 Take myristyl CoA, $CH_3(CH_2)_{12}{-}\overset{\displaystyle O}{\overset{\|}{C}}{-}SCoA$, through one turn of the fatty acid spiral; i.e., write four equations for the four recurring steps in the catabolic sequence.

24.27 Why are different coenzymes required in the two oxidation steps of the fatty acid spiral?

24.28 One product of the four recurring steps of the fatty acid spiral is always _____.

24.29 Write an equation for the *last* turn of the fatty acid spiral for a fatty acid with an even number of carbons.

24.30 Why is catabolism of fatty acids called the fatty acid *spiral* rather than the fatty acid *cycle?*

24.31 How many moles of acetyl CoA are produced from 1 mol of myristyl CoA (see Problem 24.26)? How many turns of the fatty acid spiral are required to produce this amount?

24.32 For each fatty acid in Problem 24.12 how many turns of the fatty acid spiral are required for total conversion to $CH_3{-}\overset{\displaystyle O}{\overset{\|}{C}}{-}SCoA$?

24.33 What is the ATP yield per mole of acetyl CoA which enters the citric acid cycle?

24.34 How much ATP is generated per turn of the fatty acid spiral from reduced coenzymes being oxidized in the ETS?

24.35 Calculate the total ATP yield from 1 mol of myristic acid, $CH_3(CH_2)_{12}COOH$.

24.36 Calculate the ATP yield from 0.5 mol of each acid in Problem 24.12.

24.37 Explain what is meant by saying that fats have a higher energy density than carbohydrates.

24.38 Recall that calories are a unit of energy. Which foods are higher in calories (gram for gram), fats or carbohydrates?

24.39 Lipogenesis is not simply the reverse of the fatty acid spiral. Explain the five differences noted below:
 a The two processes occur in different cell locations.
 b A small molecule is gained and lost in lipogenesis.
 c The thioester carriers vary.
 d Different coenzymes are involved.
 e An odd-numbered carbon chain is involved in two steps of lipogenesis.

24.40 Fill in the blanks:

$$CH_3(CH_2)_6\overset{\displaystyle O}{\overset{\|}{C}}{-}SACP + \overset{\displaystyle HOOC}{\underset{\displaystyle }{\overset{\displaystyle |}{CH_2}}}\overset{\displaystyle O}{\overset{\|}{C}}{-}SACP$$

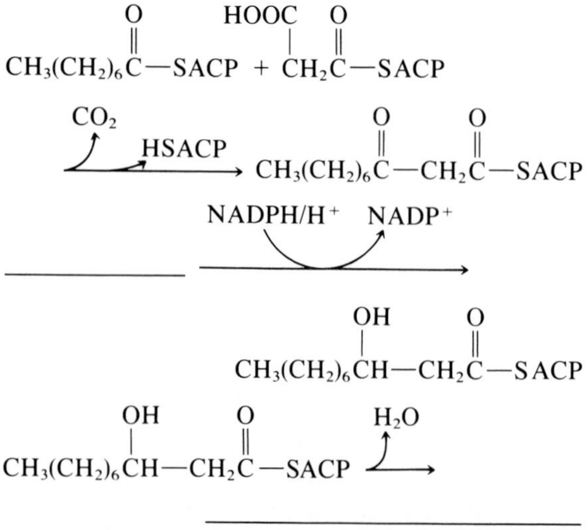

$$\xrightarrow[\overset{\curvearrowright}{HSACP}]{CO_2} CH_3(CH_2)_6\overset{\displaystyle O}{\overset{\|}{C}}{-}CH_2\overset{\displaystyle O}{\overset{\|}{C}}{-}SACP$$

$$\overset{NADPH/H^+\ NADP^+}{\curvearrowright}$$

_____

$$CH_3(CH_2)_6\overset{OH}{\overset{|}{CH}}{-}CH_2\overset{\displaystyle O}{\overset{\|}{C}}{-}SACP$$

$$CH_3(CH_2)_6\overset{OH}{\overset{|}{CH}}{-}CH_2\overset{\displaystyle O}{\overset{\|}{C}}{-}SACP \xrightarrow{\overset{H_2O}{\curvearrowright}}$$

_____

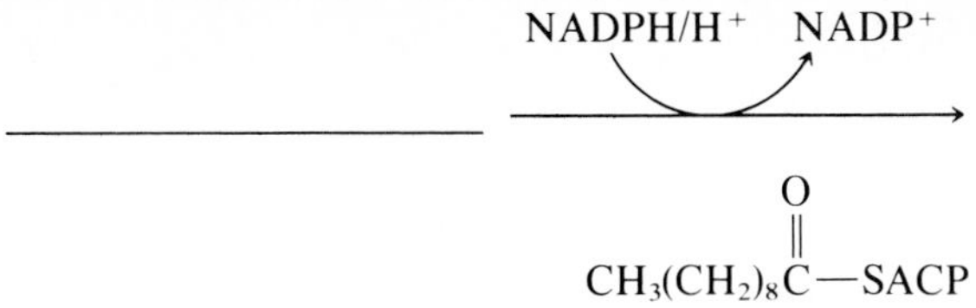

24.41 How is glycerol phosphate involved in fat synthesis?

24.42 Discuss the mobilization of chemical energy (ATP production) from fat tissue. How is the citric acid cycle necessary to it?

24.43 Write structures for acetoacetic acid, β-hydroxybutyric acid, and acetone. What are these three known as collectively?

24.44 How does acetone differ chemically from the other ketone bodies?

24.45 Why is "ketone body" an inappropriate name for β-hydroxybutyric acid?

24.46 What bodily organ uses acetoacetate as a fuel in preference to glucose?

24.47 Is the production of acetoacetic acid an abnormal metabolic event? Explain.

24.48 Name the reaction type in each case for the conversion of acetoacetic acid to the other ketone bodies:

a 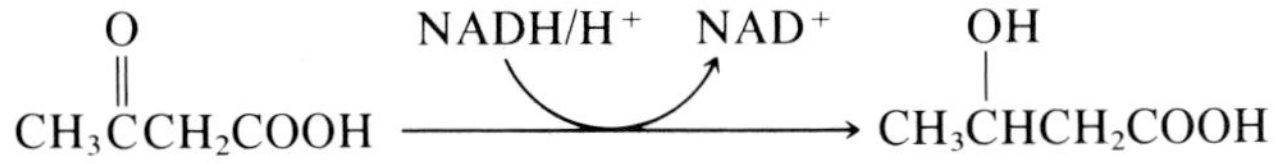

b 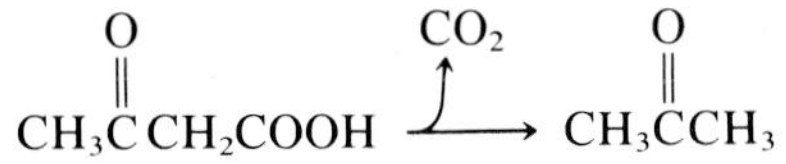

24.49 What hormone signals the body that it is in the "fed" state?

24.50 What is the cause of diabetes mellitus?

24.51 What is the first symptom of diabetes?

24.52 Why does the liver produce glucose even in the presence of the hyperglycemia of diabetes?

24.53 Name three ways by which insulin lowers blood sugar levels.

24.54 Why can't a diabetic patient metabolize acetyl CoA in the citric acid cycle?

24.55 How does a diabetic patient typically metabolize acetyl CoA?

24.56 What are the pathological consequences of a prolonged increase in ketone body concentration in the blood?

24.57 Describe how untreated diabetes mellitus leads to coma and death.

24.58 Explain how excess insulin (from injection or internal secretion) causes insulin shock, i.e., extreme hypoglycemia.

24.59 What is the raw material for the synthesis of cholesterol in the body?

CHAPTER TWENTY-FIVE

PROTEIN METABOLISM

25.1 INTRODUCTION

You have seen that the most significant metabolic reactions of carbohydrates and fats are *catabolic*. That is, the body relies on the breakdown of these materials as its source of energy. In contrast, the most significant aspect of protein metabolism is *anabolic*. The main event of protein metabolism is the biosynthesis of proteins from amino acids, which you have already seen in Chapter 20.

Structurally we are made up largely of protein material. Skin, muscles, and brain are all proteins. These tissues must constantly be repaired and replenished. Also, the body's metabolic reactions do not occur at reasonable rates without enzymes which are largely proteinaceous. Thus the importance of the body's ability to synthesize proteins is apparent.

Ingested proteins are digested (Section 19.16) into their component amino acids in the small intestine and enter the so-called amino acid pool. This raw material depot is also called the body's "nitrogen pool" because it is also the source of the nitrogen needed for important N-containing compounds other than proteins, such as nucleic acids and heme. Note that of the three foodstuffs, only proteins contain nitrogen. (The amount of nitrogen in complex lipids is insignificant.)

Catabolism of amino acids can provide energy to the body. The route to this energy is the conversion of amino acids to intermediates in the citric acid cycle. In the course of the conversions, nitrogen is removed from amino acids as ammonia (NH_3), which is quite toxic to the body. Thus these conversions present a waste disposal problem to the body. The urea cycle detoxifies the NH_3 by converting it into urea, which can be excreted from the body.

Because we have already considered protein anabolism, when we discussed protein biosynthesis in Chapter 20, this chapter will largely be concerned with protein catabolism.

25.2 NITROGEN FIXATION

Before continuing, it is useful to note the means of entry of nitrogen from the atmosphere into the foodstuffs and metabolic cycles of plants and animals. That is, we have seen the entry of the elements C, H, and O into the food chain through plant photosynthesis:

$$6CO_2 + 6H_2O \xrightarrow[\text{light}]{\text{chlorophyll}} \underset{\text{Glucose}}{C_6H_{12}O_6} + 6O_2$$

Plants cannot use nitrogen (N_2) in the air directly. They rely on *nitrogen-fixing* bacteria in soil to convert elemental nitrogen into usable forms such as ammonia (NH_3) or nitrite (NO_2^-) and nitrate (NO_3^-) ions.

Plants use NH_3 to make amino acids and proteins. Animals eat plants (and other animals) as their source of amino acids. Dead plants and waste products from animals restore nitrogen compounds to the soil, and the decay process, aided by *denitrifying* bacteria, returns nitrogen to the air. You are familiar with the return of CO_2 and H_2O to the atmosphere through exhalation and excretion. Thus there is a cycle of capture of elements from the atmosphere and return of these elements as end products of metabolism.

25.3 AMINO ACID POOL

Figure 25.1 summarizes the fates of the amino acids which enter the amino acid pool from digestion of proteins. The amino acid pool is a concept, not some location in the body. All cells contain a supply of amino acids for synthesis or the other processes shown in Figure 25.1. This supply is not static, nor is protein tissue static. Both are *dynamic;* that is, there is a constant breakdown of bodily proteins into amino acids and a constant buildup of new protein material to replace the old.

Different proteins break down and build up at different rates. This is known as the *turnover* rate of a protein. For example, liver and blood protein have a fast turnover rate, or short half-life. Half the protein molecules in your blood today will be broken down and replaced by new molecules in about six days. Muscle tissue is replaced more slowly; the turnover rate is about 180 days. Connective tissue or collagen is the longest-lived. Nearly three years is required for the replacement of half of your collagen molecules.

Section 25.11 discusses the use of amino acids as a nitrogen source for nonprotein N-containing compounds. Syntheses for these compounds usually require the degradation of the amino acids. Degradation of amino acids involves removal of the amino group by *transamination* or *deamination,* as

FIG. 25.1
The amino acid pool is supplied by ingested protein, by breakdown of tissue protein, and by synthesis of some amino acids from carbohydrate sources. Amino acids are used to make proteins and nitrogen-containing compounds that are not proteins. Amino acids are degraded by removal of the amino group which is excreted; the remaining carbon skeleton is metabolized in carbohydrate or lipid cycles.

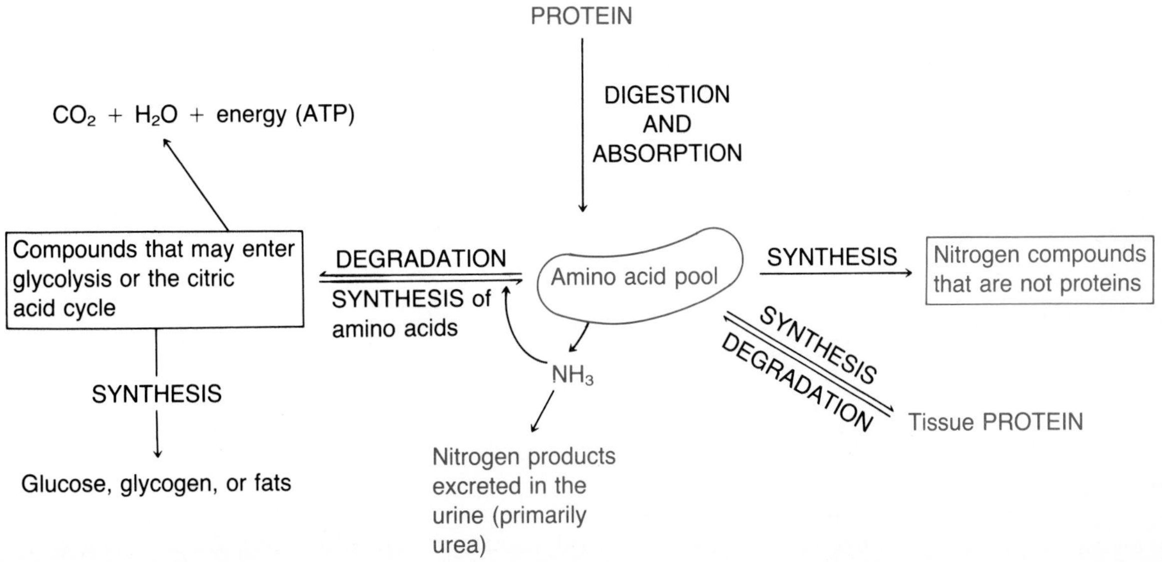
PROTEIN
DIGESTION AND ABSORPTION
CO2 + H2O + energy (ATP)
Compounds that may enter glycolysis or the citric acid cycle
DEGRADATION
SYNTHESIS of amino acids
Amino acid pool
SYNTHESIS
Nitrogen compounds that are not proteins
SYNTHESIS
DEGRADATION
Tissue PROTEIN
NH3
SYNTHESIS
Glucose, glycogen, or fats
Nitrogen products excreted in the urine (primarily urea)

you will shortly see in the next sections. The removed amino group is then used for synthesis or transformed into urea for excretion. The carbon skeleton of the degraded amino acid enters a carbohydrate or lipid metabolic pathway, as Section 25.8 will show.

Because there is no real means for the body to store nitrogen—it is either used for synthesis or excreted—we must constantly replenish our nitrogen supply by eating proteins. A healthy, well-nourished adult exhibits *nitrogen* balance, that is, he or she excretes as much nitrogen each day as dietary protein supplies. Growing children should be in a state of positive nitrogen balance, taking in more nitrogen to build new tissues than is excreted. During starvation or certain wasting diseases, negative nitrogen balance occurs—more nitrogen is excreted from tissue breakdown than is ingested.

Let us now pursue the course of amino acid degradation in more detail.

25.4 TRANSAMINATION

The typical degradative route which most amino acids follow begins with a reaction called **transamination.** In this reaction there is a transfer of amino and keto groups between an α-amino acid and an α-keto acid. Most often the reactant α-keto acid is α-ketoglutaric acid, which is an intermediate in the citric acid cycle. For example, the transamination of alanine occurs as follows:

$$\underset{\text{Alanine}}{CH_3-\overset{\alpha}{C}H(NH_2)-COOH} + \underset{\alpha\text{-Ketoglutaric acid}}{HOOC-\overset{\alpha}{C}(=O)-CH_2CH_2COOH} \xrightarrow{\text{transaminase}}$$

$$\underset{\alpha\text{-Ketopropanoic (pyruvic acid)}}{CH_3-\overset{\alpha}{C}(=O)-COOH} + \underset{\text{Glutamic acid}}{HOOC-\overset{\alpha}{C}H(NH_2)-CH_2CH_2COOH} \quad (25.1)$$

Notice that only the α-positions are involved; the rest of the reacting molecules remain the same. Glutamic acid is always the product amino acid when α-ketoglutaric acid is the reactant keto acid.

The α-keto acid produced in transamination is frequently an intermediate metabolite encountered in carbohydrate metabolism. Notice in this case that pyruvic acid is the product of transaminating alanine. Transaminases catalyze these reactions. The specific enzyme is often named for the products of the reaction. In this example, the enzyme is named *g*lutamic-*p*yruvic *t*ransaminase (GPT).

Transaminases require pyridoxal phosphate (PLP-CHO), a derivative of pyridoxine (vitamin B_6), as a coenzyme.

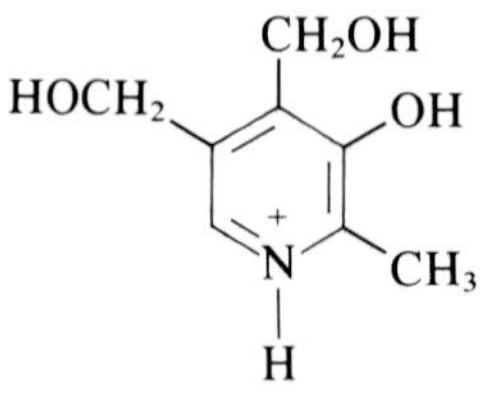
CH₂OH
HOCH₂
OH
N
CH₃
H

Pyridoxine (vitamin B_6)

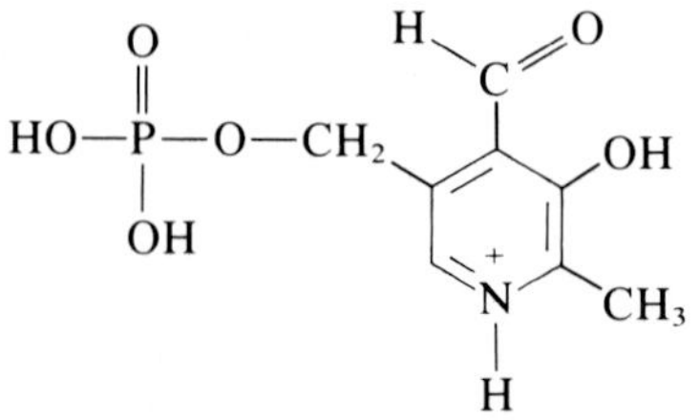
O
H
C
O
HO—P—O—CH₂
OH
OH
N
CH₃
H

Pyridoxal phosphate (PLP-CHO)

Problem 25.1 What two reactions convert pyridoxine into pyridoxal phosphate?

Pyridoxal phosphate is abbreviated PLP-CHO because it is the aldehyde group of that molecule which actually removes the amino group from the reacting amino acid:

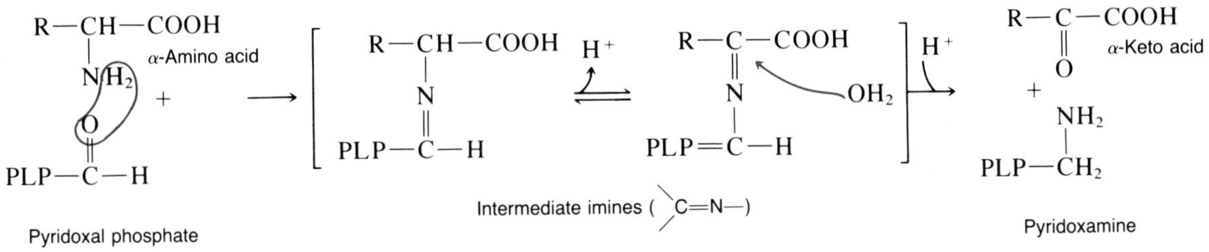
α-Amino acid
Pyridoxal phosphate
Intermediate imines
α-Keto acid
Pyridoxamine

The coenzyme PLP-CHO is then restored by reaction of pyridoxamine with α-ketoglutaric acid:

Pyridoxamine ($PLP\text{-}CH_2NH_2$) + α-ketoglutaric acid ⟶ PLP-CHO + glutamic acid

Thus the transfer reaction is actually accomplished in two steps rather than just one, as implied by the overall Equation (25.1), and a vitamin is involved.

Although most transaminations employ α-ketoglutaric acid, the other α-keto acids can react as well. For example, pyruvic acid could be a *reactant* in transamination:

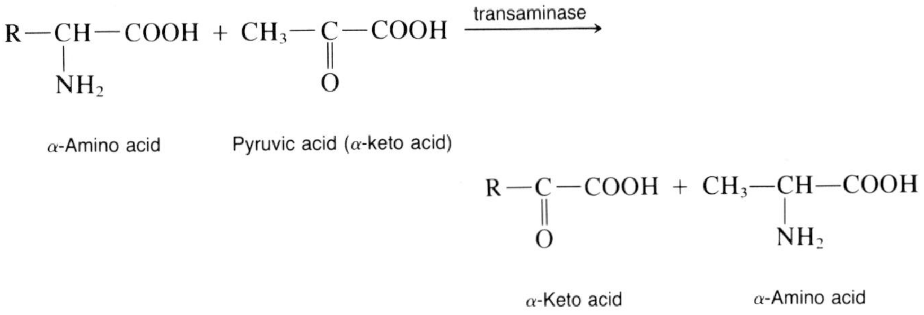
transaminase
α-Amino acid
Pyruvic acid (α-keto acid)
α-Keto acid
α-Amino acid

SAMPLE EXERCISE 25.1 What α-keto acid is produced as aspartic acid undergoes transamination with α-ketoglutaric acid?

Solution Look up the structure of aspartic acid in Table 19.1. Write down this structure and that of α-ketoglutaric acid:

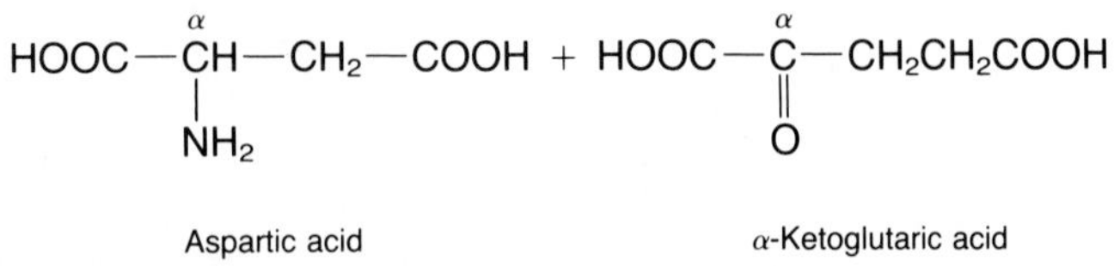
Aspartic acid
α-Ketoglutaric acid

Realize that the transamination reaction involves only the α-position of each acid. The functional groups at the α-position are "swapped." Thus the products are

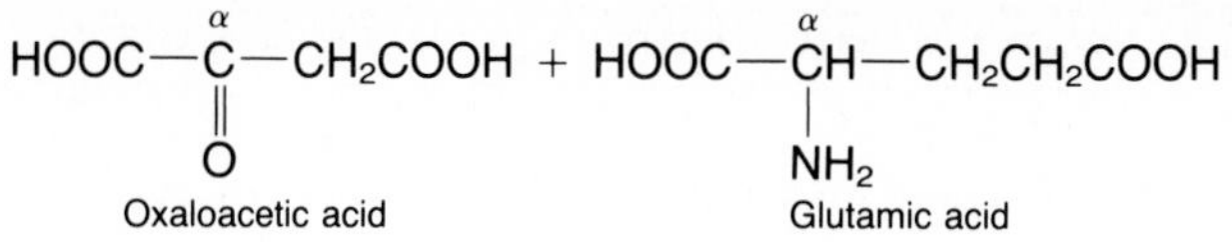
HOOC—C—CH2COOH + HOOC—CH—CH2CH2COOH
O
NH2
Oxaloacetic acid
Glutamic acid

Once again the keto acid produced is an important citric acid cycle intermediate, oxaloacetic acid.

Problem 25.2 Write structures for the α-keto acids produced by transamination of (*a*) valine, (*b*) serine, and (*c*) phenylalanine.

25.5 DEAMINATION

Glutamic acid produced in transamination can be *deaminated;* that is, its α-amino group can be removed. The result of this removal is the production of NH_3 and the regeneration of α-ketoglutaric acid for reaction in the citric acid cycle or for subsequent transamination reactions.

The deamination reaction proceeds through an imine (>C=N—) intermediate, as does transamination. Because hydrogen is being lost (and oxygen gained) in the reaction, it is an oxidation reaction, i.e., *oxidative* deamination. Either NAD^+ or $NADP^+$ can function as the hydrogen carrier.

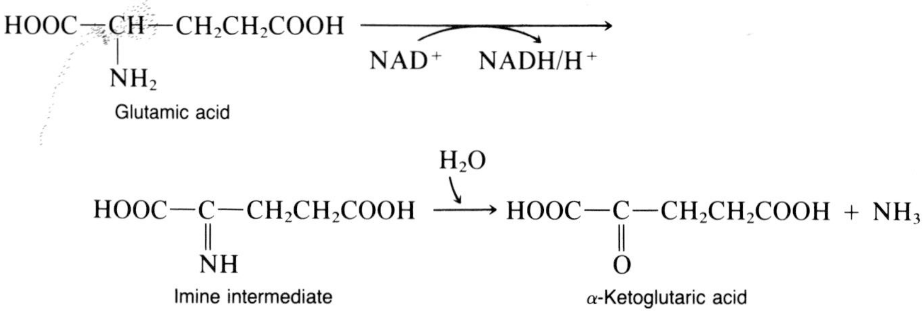
HOOC—CH—CH2CH2COOH
NH2
Glutamic acid
NAD+
NADH/H+
H2O
HOOC—C—CH2CH2COOH
NH
Imine intermediate
HOOC—C—CH2CH2COOH + NH3
O
α-Ketoglutaric acid

The reverse process is also possible in the body; that is, a keto acid can be aminated and become an amino acid. Hydrogen would be added (and oxygen lost) in this case, and thus it is *reductive* amination.

As in previous chapters, the structures written do not necessarily reflect the actual form of the molecules at physiological pH. In the body, carboxylic acids are generally in their anionic forms, and when ammonia is released, it is most often in the form of the ammonium (NH_4^+) ion. Biochemists use the word *ammonia* to mean either NH_3 or NH_4^+ in the body.

Serine and threonine can be deaminated directly; that is, their amino group need not be transferred in a transamination step to make glutamic acid. It is the presence of the β-hydroxy group that allows for direct deamination of these amino acids. Dehydration of these β-hydroxy acids leads to an isomer of the imine intermediate of the deamination pathway. For example, threonine can react:

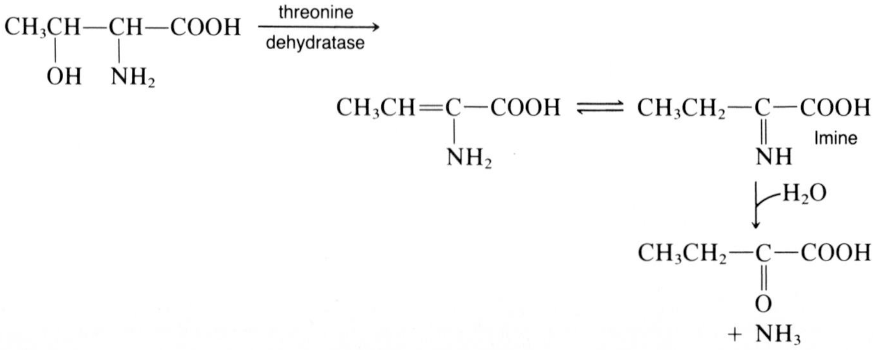
CH3CH—CH—COOH
OH NH2
threonine
dehydratase
CH3CH=C—COOH
NH2
CH3CH2—C—COOH
NH
Imine
H2O
CH3CH2—C—COOH
O
+ NH3

Problem 25.3 Show the reaction sequence for the deamination of serine.

This behavior of serine and threonine is exceptional. The usual route for removal of nitrogen from an amino acid is transamination to produce glutamic acid and then deamination of glutamic acid to regenerate α-ketoglutaric acid and produce NH_3:

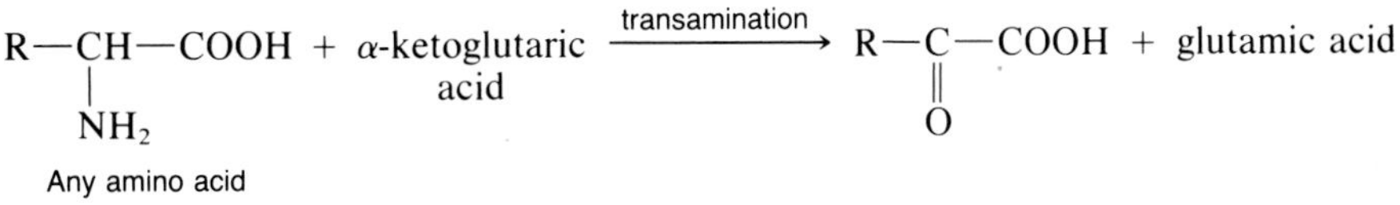
R—CH—COOH + α-ketoglutaric acid
NH2
Any amino acid
transamination
R—C—COOH + glutamic acid
O

$$\text{Glutamic acid} \xrightarrow[\text{H}_2\text{O} \quad \text{NADP}^+ \quad \text{NADPH/H}^+]{\text{deamination}} \text{NH}_3 + \alpha\text{-ketoglutaric acid}$$

Figure 25.2 summarizes the degradation routes of amino acids and the fates of NH_3 in the body.

25.6 AMMONIA DISPOSAL

The ammonia produced by deamination must be disposed of because NH_3 is extremely toxic. The exact reasons for this toxicity are not clear, but are undoubtedly related to the disruption of the citric acid cycle which excess ammonia causes. Accumulation of NH_3 reverses the deamination equilibria, thus depriving the Krebs cycle of the α-ketoglutaric acid which reacts with excess NH_3 to form glutamic acid. In addition, ammonia's basicity can overtax the blood's buffer systems.

As Figure 25.2 shows, the immediate fate of generated NH_3 depends

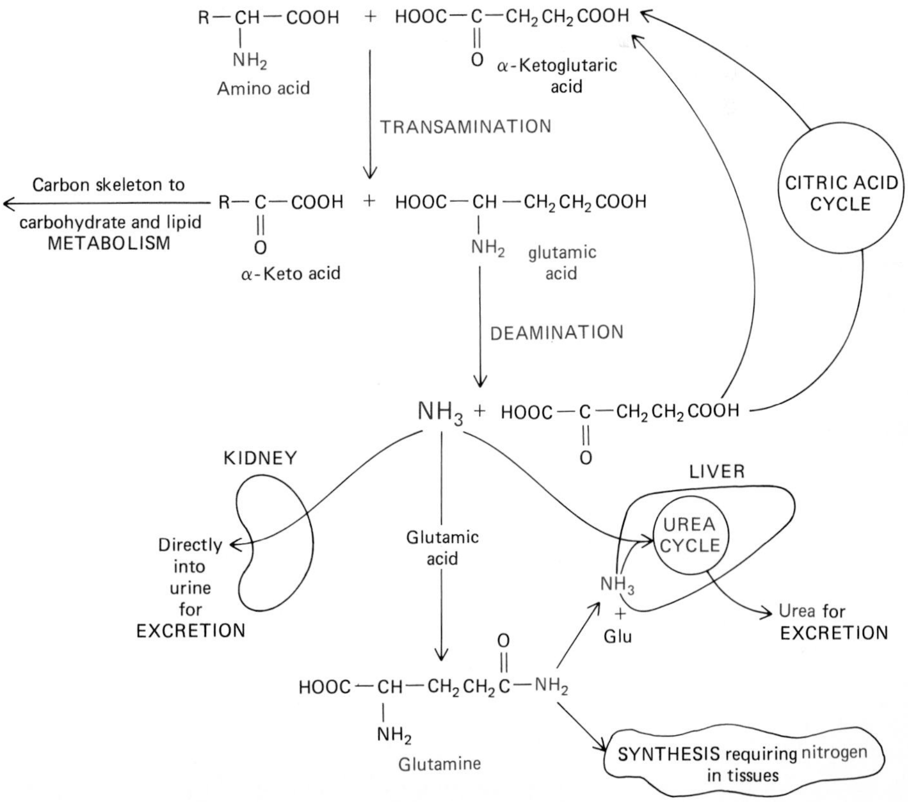
R—CH—COOH + HOOC—C—CH2CH2COOH
NH2
Amino acid
O α-Ketoglutaric acid
TRANSAMINATION
Carbon skeleton to carbohydrate and lipid METABOLISM
R—C—COOH + HOOC—CH—CH2CH2COOH
O
α-Keto acid
NH2 glutamic acid
CITRIC ACID CYCLE
DEAMINATION
NH3 + HOOC—C—CH2CH2COOH
O
KIDNEY
Directly into urine for EXCRETION
Glutamic acid
LIVER
UREA CYCLE
NH3
+
Glu
Urea for EXCRETION
HOOC—CH—CH2CH2C—NH2
O
NH2
Glutamine
SYNTHESIS requiring nitrogen in tissues

FIG. 25.2
The overall degradative route followed by most amino acids and the fates of NH_3. A very small amount of NH_3 produced in kidney cells is excreted directly into the urine. Ammonia produced in the liver is converted by the urea cycle to urea for excretion. NH_3 produced elsewhere combines with glutamic acid to form glutamine, which can transport ammonia to tissues for use in synthesis or to the liver for disposal.

somewhat on where it is formed. Ammonia produced in kidney cells may be released directly into the urine. However, because of ammonia's caustic nature, the NH_3 concentration must be kept quite dilute and thus a large volume of water would need to be excreted if substantial amounts of NH_3 were to be disposed of in this way. This could lead to dehydration. Fish need not fear dehydration in their environment and therefore can excrete dilute NH_3 solutions directly. Land-dwelling animals have all evolved alternative disposal systems that conserve water. Human beings convert ammonia to nontoxic, very soluble urea for excretion.

Conversion of NH_3 to urea ($H_2N{-}\overset{\overset{\displaystyle O}{\|}}{C}{-}NH_2$) occurs in the liver, where most deamination takes place. This conversion happens through a series of reactions called the urea cycle, which we will explore in detail in the next section.

Ammonia generated elsewhere in the body reacts with glutamic acid to form glutamine:

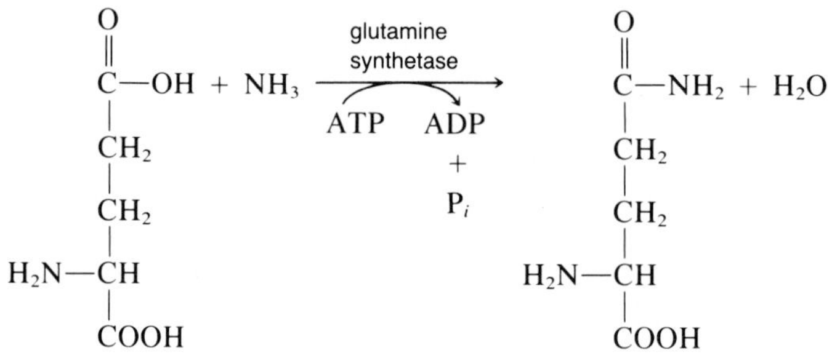
glutamine
synthetase
ATP ADP
+
P$_i$

Glutamic acid   Glutamine   (25.2)

This reaction is the formation of an amide from the reaction of an acid and ammonia. Glutamine is nontoxic and serves as a temporary "storage" and transport vehicle for NH_3. Glutamine can safely travel through the bloodstream to a site where NH_3 is needed for synthesis of nitrogen-containing compounds. There glutamine is hydrolyzed [the reverse of Equation (25.2), catalyzed by glutaminase] and the released NH_3 is thereby made available for synthesis. Alternatively, glutamine can travel to the liver, where it is hydrolyzed and the NH_3 enters the urea cycle for ultimate disposal.

Problem 25.4 Write an equation for the hydrolysis of glutamine.

25.7 THE UREA CYCLE

This important metabolic cycle was elucidated by Hans Krebs, the same biochemist who proposed the citric acid cycle. The urea cycle is also called the ornithine cycle for the compound ornithine, which is regenerated at the last step in the cycle. See Figure 25.3 for an overview of the reactions in the urea cycle. The overall result of the cycle is the combination of ammonia with carbon dioxide to make urea:

$$2NH_3 + CO_2 \longrightarrow H_2N{-}\overset{\overset{\displaystyle O}{\|}}{C}{-}NH_2 + H_2O$$

Urea

Let us now look at the individual steps in the reaction cycle.

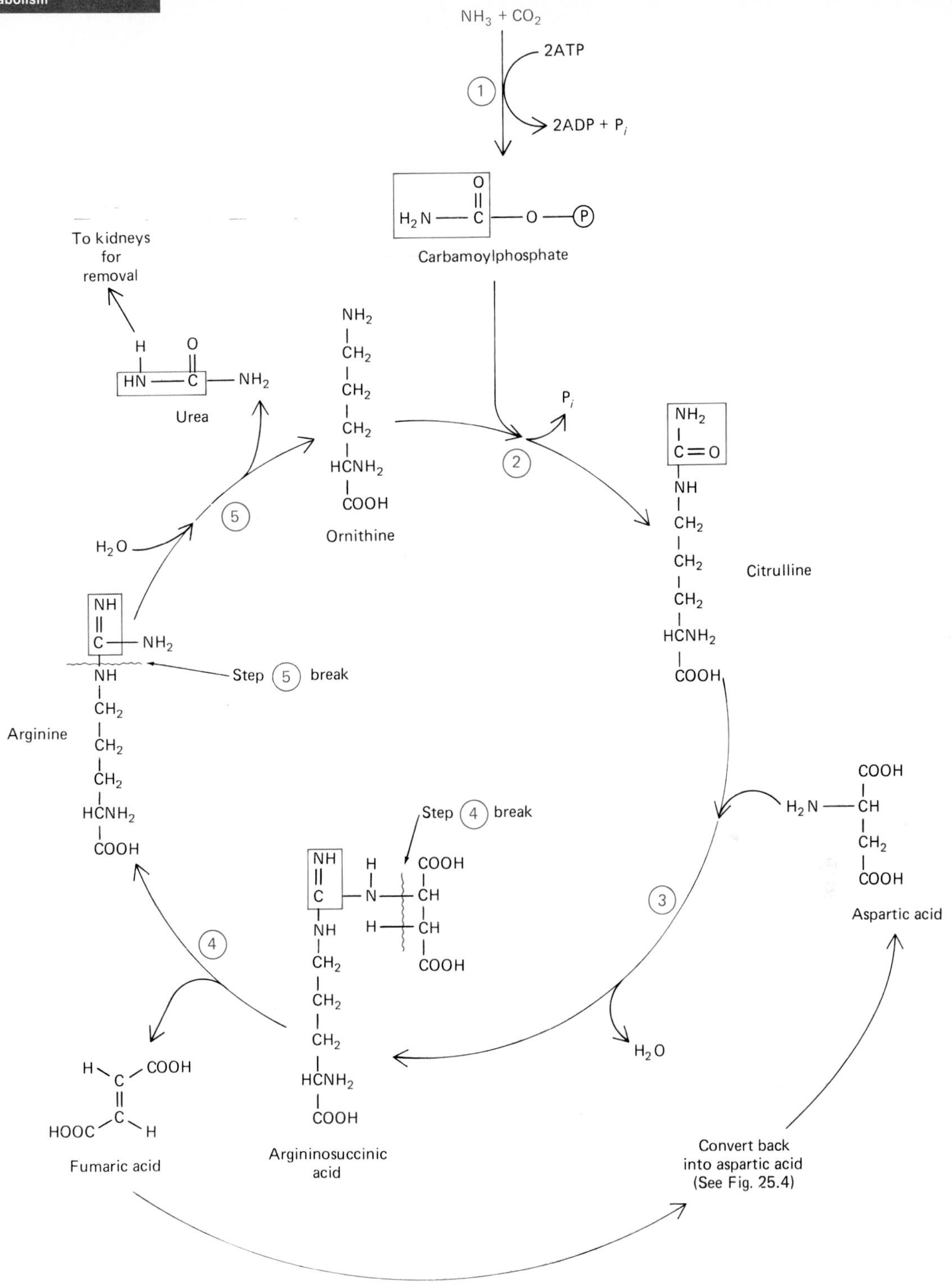
NH3 + CO2
2ATP
1
2ADP + Pi
H2N — C — O — P
Carbamoylphosphate
To kidneys
for
removal
H O
HN — C — NH2
Urea
NH2
CH2
CH2
CH2
HCNH2
COOH
Ornithine
Pi
2
NH2
C=O
NH
CH2
CH2
CH2
HCNH2
COOH
Citrulline
5
H2O
NH
C — NH2
NH
CH2
CH2
CH2
HCNH2
COOH
Arginine
Step 5 break
Step 4 break
COOH
H2N — CH
CH2
COOH
Aspartic acid
3
4
NH
C — N — CH
NH
H — CH
COOH
CH2
CH2
CH2
HCNH2
COOH
Argininosuccinic
acid
H2O
H COOH
C
C
HOOC H
Fumaric acid
Convert back
into aspartic acid
(See Fig. 25.4)

FIG. 25.3
The urea cycle. Two amino groups, one from NH_3 and one from aspartic acid, are excreted in every urea molecule synthesized in this cycle. Boxes in color trace the course of NH_3 from a deaminated amino acid and CO_2 in the cell through the cycle.

STEP 1 Step 1 is the combination of NH_3 and CO_2 in the presence of ATP to form the highly energetic molecule carbamoyl phosphate. A quantity of 2 mol of ATP is required per mol of ammonia.

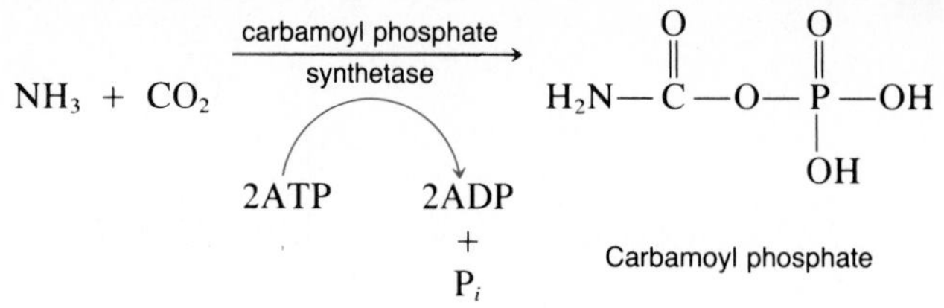
carbamoyl phosphate synthetase
2ATP
2ADP + P_i
Carbamoyl phosphate

(As previously noted, the structures shown do not necessarily reflect those present at physiological pH values.)

STEP 2 Step 2 is the reaction of carbamoyl phosphate with ornithine. This reaction combines the carbamoyl group, the remnants of NH_3 and CO_2 entering the cycle, with ornithine. The amine group of ornithine is bonded to the carbamoyl group of carbamoyl phosphate, replacing the phosphate group and forming citrulline. A high-energy phosphate bond is cleaved in this process. The enzyme for this step is ornithine transcarbamoylase. Henceforth, a colored outline in the structures traces the path of the ammonia N and carbon dioxide C.

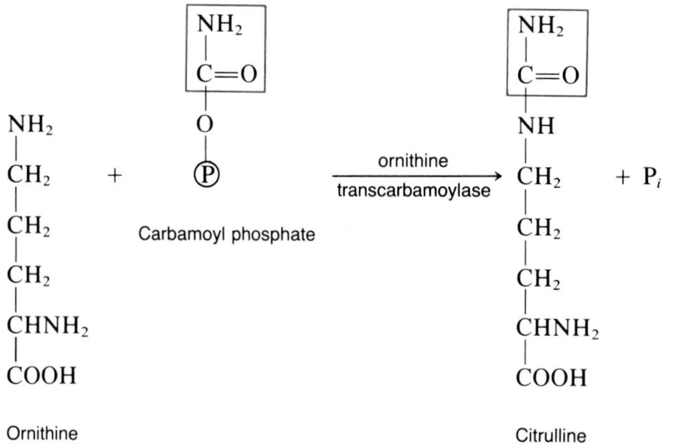
ornithine transcarbamoylase
+ P_i
Carbamoyl phosphate
Ornithine
Citrulline

STEP 3 Step 3 is the addition of the amino nitrogen of aspartic acid to the amide carbonyl of citrulline. The product, argininosuccinic acid, has the imine group and the N—C—N skeleton necessary for urea formation. Two phosphate bonds are cleaved as an ATP becomes AMP.

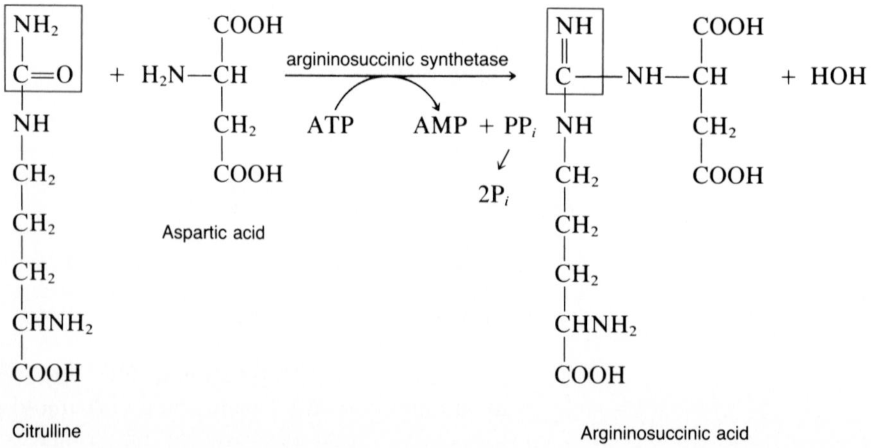
argininosuccinic synthetase
ATP
AMP + PP_i
$2P_i$
+ HOH
Aspartic acid
Citrulline
Argininosuccinic acid

STEP 4 Step 4 is a cleavage of argininosuccinic acid to form the amino acid arginine and fumaric acid; it is catalyzed by the enzyme argininosuccinase.

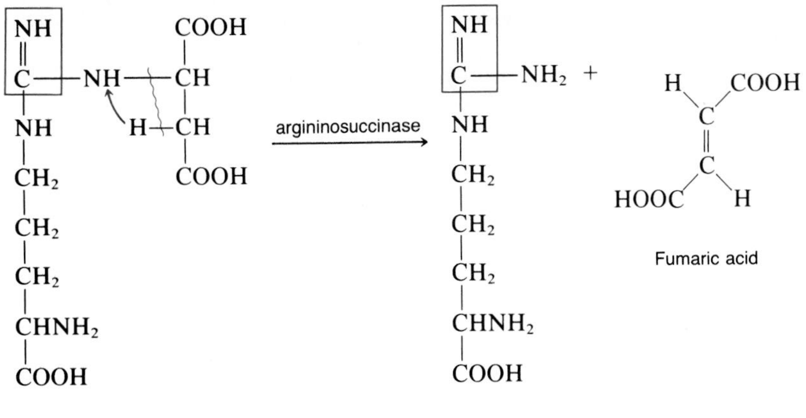

NH
C
NH
COOH
CH
H
CH
COOH
NH
CH2
CH2
CH2
CHNH2
COOH
Argininosuccinic acid
argininosuccinase
NH
C
NH2 +
NH
CH2
CH2
CH2
CHNH2
COOH
Arginine
H
COOH
C
C
HOOC
H
Fumaric acid

Figure 25.4 shows how fumaric acid, a citric acid cycle intermediate, is reconverted to aspartic acid for reuse in the urea cycle. The N—C—N urea skeleton in arginine is now ready for removal of urea.

FIG. 25.4
Fumaric acid formed in the urea cycle is converted to aspartic acid, a needed urea cycle intermediate, by citric acid cycle reactions and transaminations with glutamic acid. The glutamic acid is formed by reductive amination of α-ketoglutaric acid that is formed in the citric acid cycle, and is also regenerated during transamination reactions.

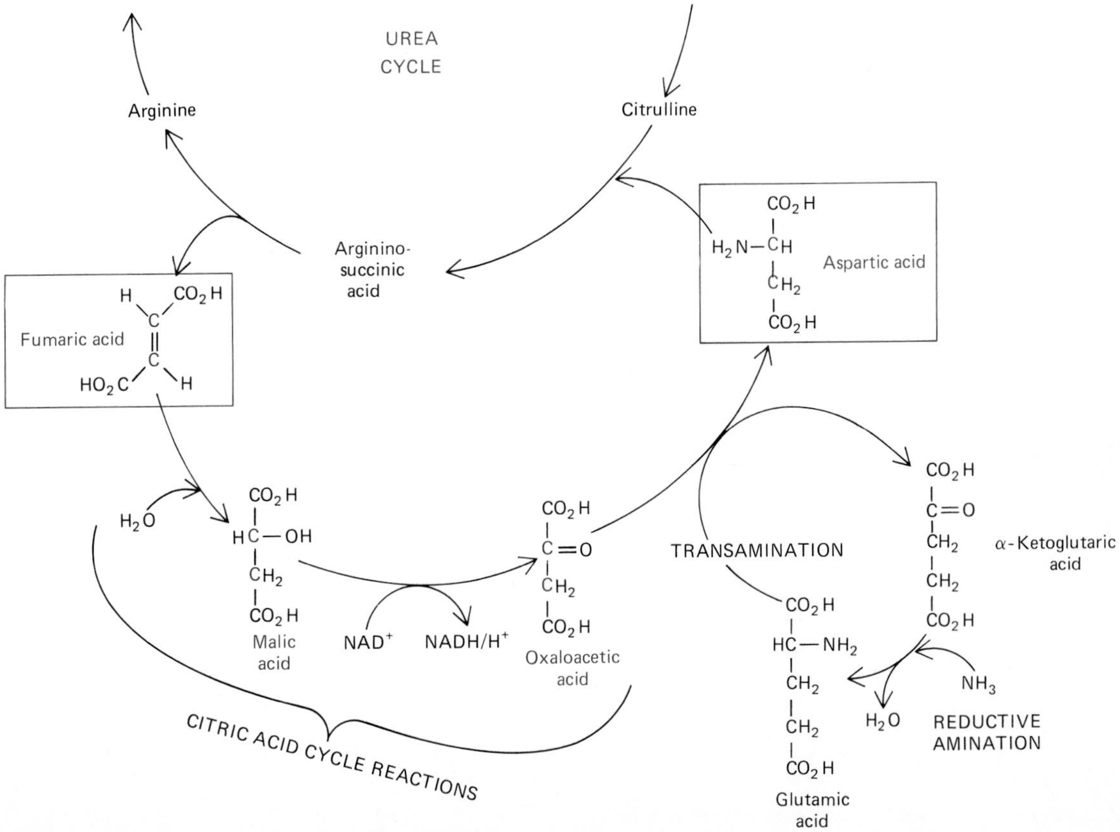

UREA
CYCLE
Arginine
Citrulline
Argininosuccinic acid
Aspartic acid
Fumaric acid
H2O
Malic acid
NAD+
NADH/H+
Oxaloacetic acid
TRANSAMINATION
α-Ketoglutaric acid
Glutamic acid
NH3
H2O
REDUCTIVE AMINATION
CITRIC ACID CYCLE REACTIONS

STEP 5   Step 5 is the last step. Water hydrolyzes a urea molecule away from the arginine structure. The other product is ornithine, which is then ready to go around the cycle again by reacting with another carbamoyl phosphate. Arginase catalyzes the hydrolysis:

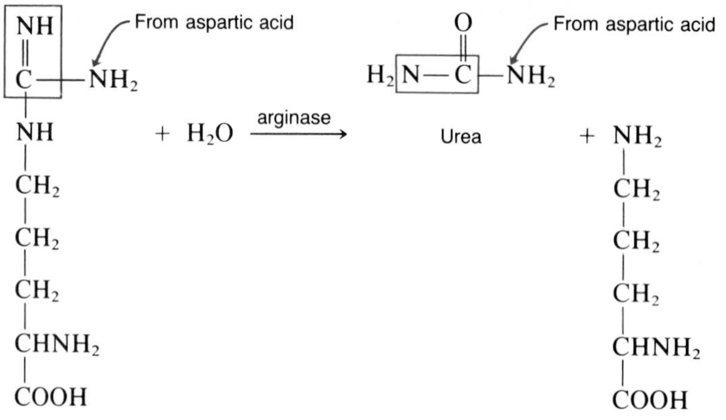

$$\begin{array}{ccc}
\begin{array}{l}
\text{NH} \\
\| \\
\text{C}-\!-\text{NH}_2 \\
| \\
\text{NH} \\
| \\
\text{CH}_2 \\
| \\
\text{CH}_2 \\
| \\
\text{CH}_2 \\
| \\
\text{CHNH}_2 \\
| \\
\text{COOH}
\end{array}
& + \text{H}_2\text{O} \xrightarrow{\text{arginase}} &
\end{array}$$

From aspartic acid

Arginine

$$\text{H}_2\text{N}-\overset{\overset{\text{O}}{\|}}{\text{C}}-\text{NH}_2$$

From aspartic acid

Urea

$$+\ \begin{array}{l}
\text{NH}_2 \\
| \\
\text{CH}_2 \\
| \\
\text{CH}_2 \\
| \\
\text{CH}_2 \\
| \\
\text{CHNH}_2 \\
| \\
\text{COOH}
\end{array}$$

Ornithine

Urea can then enter the bloodstream and go to the kidneys for disposal into the urine.

25.8 FATE OF AMINO ACID CARBON SKELETONS

Thus far we have been following the destiny of the amino group from an amino acid during catabolism and have found that it is either used in synthesis or converted to urea. It was mentioned in Section 25.4 that the α-keto acid formed when an amino acid is subjected to transamination is frequently an intermediate of the carbohydrate or lipid metabolic pathways. That is, the carbon skeleton of an amino acid undergoes carbohydrate or lipid metabolism through conversion to an α-keto acid. Briefly,

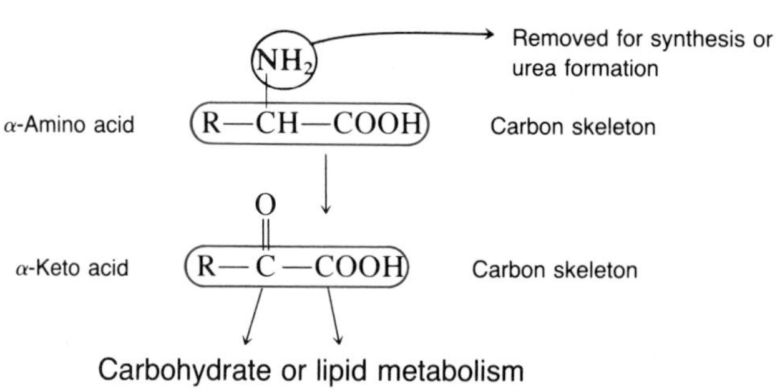

α-Amino acid   (R—CH—COOH)   Carbon skeleton
(NH₂) Removed for synthesis or urea formation

α-Keto acid   (R—C—COOH)   Carbon skeleton

Carbohydrate or lipid metabolism

Problem 25.5   Name three α-keto acids which occur in the glycolysis or gluconeogenesis pathways or in the citric acid cycle.

Figure 25.5 shows the entry points of many amino acids into the carbohydrate metabolic sequences. In some cases transamination alone is sufficient to transform an amino acid into a carbohydrate cycle metabolite. For example,

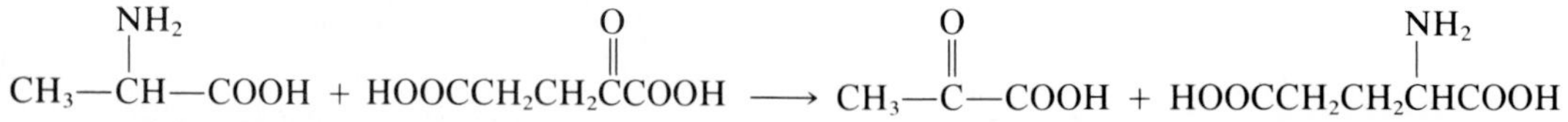

$$\underset{\text{Alanine}}{\text{CH}_3-\overset{\overset{\text{NH}_2}{|}}{\text{CH}}-\text{COOH}} + \underset{\text{α-Ketoglutaric acid}}{\text{HOOCCH}_2\text{CH}_2\overset{\overset{\text{O}}{\|}}{\text{C}}\text{COOH}} \longrightarrow \underset{\text{Pyruvic acid}}{\text{CH}_3-\overset{\overset{\text{O}}{\|}}{\text{C}}-\text{COOH}} + \underset{\text{Glutamic acid}}{\text{HOOCCH}_2\text{CH}_2\overset{\overset{\text{NH}_2}{|}}{\text{CH}}\text{COOH}}$$

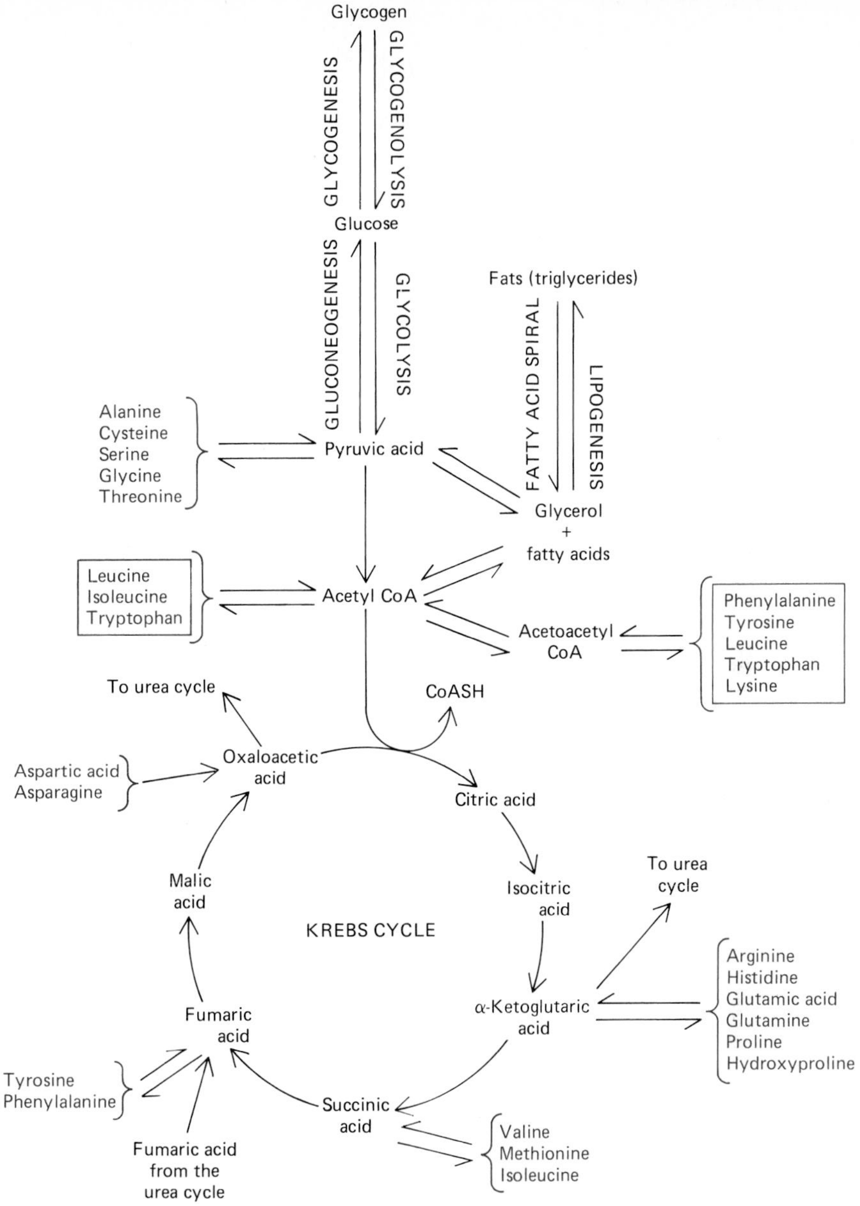
Glycogen
GLYCOGENESIS
GLYCOGENOLYSIS
Glucose
GLUCONEOGENESIS
GLYCOLYSIS
Fats (triglycerides)
FATTY ACID SPIRAL
LIPOGENESIS
Alanine
Cysteine
Serine
Glycine
Threonine
Pyruvic acid
Glycerol
+
fatty acids
Leucine
Isoleucine
Tryptophan
Acetyl CoA
Acetoacetyl CoA
Phenylalanine
Tyrosine
Leucine
Tryptophan
Lysine
To urea cycle
CoASH
Aspartic acid
Asparagine
Oxaloacetic acid
Citric acid
Malic acid
Isocitric acid
To urea cycle
KREBS CYCLE
Arginine
Histidine
Glutamic acid
Glutamine
Proline
Hydroxyproline
α-Ketoglutaric acid
Fumaric acid
Tyrosine
Phenylalanine
Succinic acid
Valine
Methionine
Isoleucine
Fumaric acid from the urea cycle

FIG. 25.5
Points of entry for the amino acids into the carbohydrate and lipid metabolic cycles. Ketogenic amino acids are enclosed in a colored box.

$$\mathrm{HOOCCH_2{-}\underset{\textstyle Aspartic\ acid}{\overset{\displaystyle NH_2\atop|}{CH}}{-}COOH} + \alpha\text{-ketoglutaric acid} \longrightarrow$$

$$\mathrm{HOOCCH_2{-}\overset{\displaystyle O\atop\|}{C}{-}COOH} + \text{glutamic acid}$$

Aspartic acid

Oxaloacetic acid

Other amino acids require additional reactions to modify their skeleton into a Krebs cycle intermediate. For example, proline can enter the citric acid cycle as α-ketoglutaric acid through the following sequence:

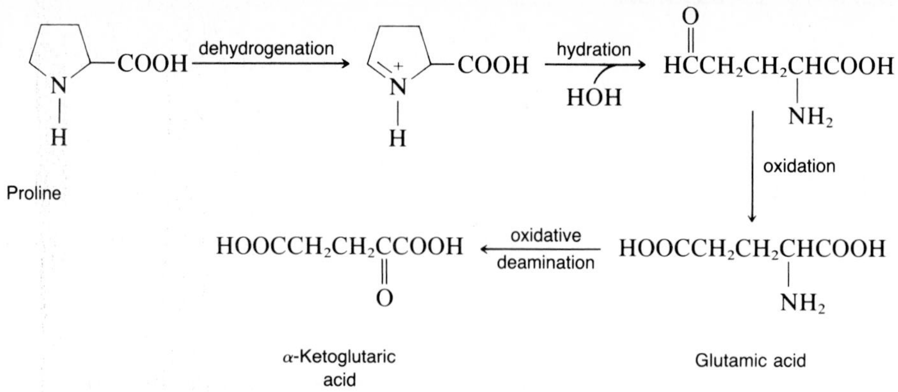

dehydrogenation
hydration
HOH
oxidation
oxidative
deamination
Proline
α-Ketoglutaric acid
Glutamic acid

Amino acids are classified as being **glucogenic** or **ketogenic,** depending on their entry points into the energy-producing metabolic cycles, as shown in Figure 25.5, and listed in Table 25.1. **Glucogenic** amino acids are degraded to pyruvic, α-ketoglutaric, oxaloacetic, succinic, or fumaric acids and can thereby be made into glucose. **Ketogenic** amino acids are degraded to acetyl CoA or acetoacetyl CoA and can thereby be made into *ketone bodies*. Ketogenic amino acids could also be termed **lipogenic** because their conversion to acetyl CoA puts them in the lipogenesis pathway. All ketogenic amino acids except leucine are also glucogenic.

25.9 ESSENTIAL AND NONESSENTIAL AMINO ACIDS

The interplay between amino acid catabolism and carbohydrate and lipid metabolism is a two-way street. That is, not only can skeletons from amino acids end up in the citric acid cycle or lipogenesis, but metabolic intermediates from carbohydrate pathways can be transformed into amino acids. Amino acids that the body can synthesize are termed **nonessential** because it is not essential that they be in our diets. If the body cannot synthesize an amino acid, then it is essential to our diet, and is called an **essential** amino acid. Table 25.2 categorizes amino acids, and Figure 25.6 gives some examples of how the body synthesizes nonessential amino acids.

Problem 25.6 Explain the statement, "Alanine is a nonessential amino acid because the body has large supplies of pyruvic and glutamic acids."

TABLE 25.1 Amino Acids Classified as Glucogenic or Ketogenic

Glucogenic	
Alanine	Histidine
Arginine	Hydroxyproline
Aspartic acid	Methionine
Asparagine	Proline
Cysteine	Serine
Glutamic acid	Threonine
Glutamine	Valine
Glycine	
Ketogenic*	
Isoleucine	Phenylalanine
Leucine	Tryptophan
Lysine	Tyrosine

* All but leucine are also glucogenic.

TABLE 25.2 Essential and Nonessential Amino Acids for Human Beings

Essential	Nonessential
Histidine	Alanine
Isoleucine	Asparagine
Leucine	Cysteine
Lysine	Glutamine
Methionine	Glutamic acid
Phenylalanine	Glycine
Threonine	Proline
Tryptophan	Serine
Valine	Tyrosine

Dietary proteins are described as *complete* proteins if they contain adequate amounts of essential amino acids and *incomplete* if they lack one or more. Animal proteins are generally complete, but proteins from grains are usually incomplete. The amino acid lysine is typically the missing essential amino acid in grains.

FIG. 25.6 Transamination between glutamic acid and some carbohydrate metabolic intermediates produces many nonessential amino acids, e.g., aspartic acid, alanine, and the phosphate ester of serine. In other cases more complicated multistep syntheses occur. The nonessential amino acid tyrosine is made from the essential amino acid phenylalanine by oxidation.

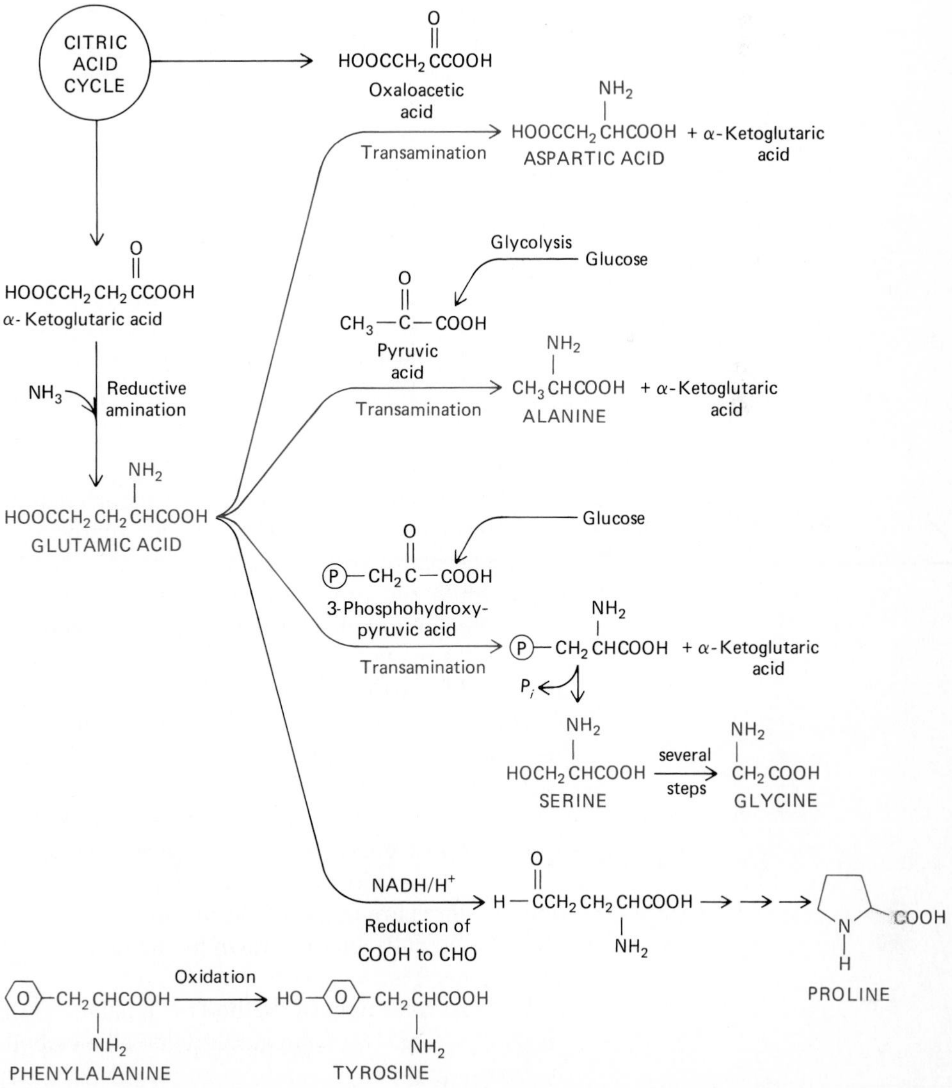
CITRIC ACID CYCLE
HOOCCH2CCOOH
Oxaloacetic acid
Transamination
HOOCCH2CHCOOH + α-Ketoglutaric acid
ASPARTIC ACID
HOOCCH2CH2CCOOH
α-Ketoglutaric acid
NH3
Reductive amination
HOOCCH2CH2CHCOOH
GLUTAMIC ACID
Glycolysis
Glucose
CH3—C—COOH
Pyruvic acid
Transamination
CH3CHCOOH + α-Ketoglutaric acid
ALANINE
Glucose
P—CH2C—COOH
3-Phosphohydroxy-pyruvic acid
Transamination
P—CH2CHCOOH + α-Ketoglutaric acid
Pi
HOCH2CHCOOH
SERINE
several steps
CH2COOH
GLYCINE
NADH/H+
Reduction of COOH to CHO
H—CCH2CH2CHCOOH
COOH
PROLINE
Oxidation
CH2CHCOOH
PHENYLALANINE
HO—CH2CHCOOH
TYROSINE

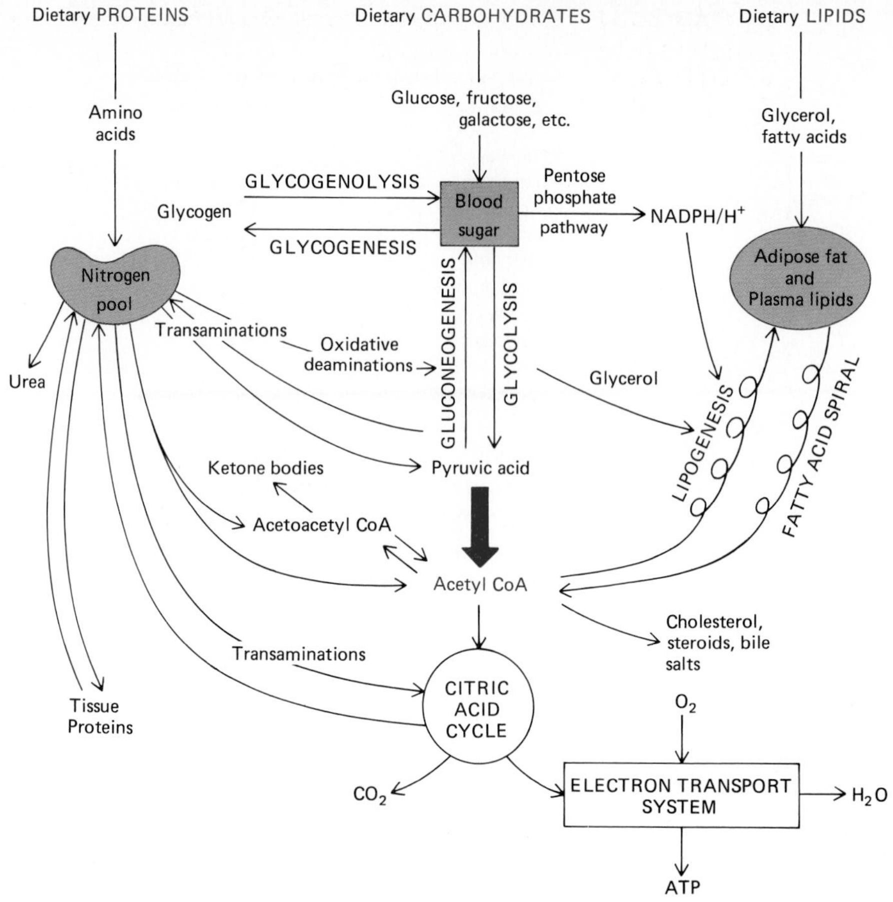
Dietary PROTEINS
Dietary CARBOHYDRATES
Dietary LIPIDS
Amino acids
Glucose, fructose, galactose, etc.
Glycerol, fatty acids
GLYCOGENOLYSIS
Glycogen
GLYCOGENESIS
Blood sugar
Pentose phosphate pathway
NADPH/H+
Adipose fat and Plasma lipids
Nitrogen pool
Urea
Transaminations
Oxidative deaminations
GLUCONEOGENESIS
GLYCOLYSIS
Glycerol
LIPOGENESIS
FATTY ACID SPIRAL
Ketone bodies
Pyruvic acid
Acetoacetyl CoA
Acetyl CoA
Cholesterol, steroids, bile salts
Transaminations
Tissue Proteins
CITRIC ACID CYCLE
O2
CO2
ELECTRON TRANSPORT SYSTEM
H2O
ATP

FIG. 25.7
The big picture of metabolism. Most of our energy needs are met by carbohydrates and lipids being broken down to acetyl CoA and entering the citric acid cycle. We can also store carbohydrate and lipid resources. Amino acids are used primarily for tissue building and repair. Proteins can be energy sources through conversion of amino acids into carbohydrate or lipid metabolic intermediates. Nonessential amino acids can be made from carbohydrate sources.

25.10 THE BIG PICTURE OF FOOD METABOLISM

Now that we have considered the major anabolic and catabolic sequences of all three major foodstuffs, we can appreciate the interworkings of these metabolic pathways. Figure 25.7 summarizes the interweaving of carbohydrate, lipid, and protein metabolism.

The body catabolizes food to produce energy. If immediate energy needs are met, then "energy" is stored in glycogen or fat. A great deal of energy is needed for protein biosynthesis and also for nitrogen waste disposal. Thus usually our energy needs are met by carbohydrates and lipids and our structural needs by proteins.

However, as we have seen in this chapter, proteins can supply energy by conversion of amino acids into appropriate metabolites. We could even gain weight from excessive ingestion of protein since amino acids can be converted to acetyl CoA and hence to lipids. On the other hand, if our glycogen and fat stores are depleted during starvation, body structures (such as muscles) are actually broken down as the body tries to meet energy needs from protein sources.

The best treatment of the body is a balanced diet which provides all three foodstuffs in moderate amounts so that all metabolic cycles can work smoothly and harmoniously.

25.11 METABOLISM OF NONPROTEIN NITROGEN COMPOUNDS

In the previous discussion we frequently alluded to the use of amino acids as sources of nonprotein nitrogen compounds in the body. We will now look briefly at the synthesis and degradation of such compounds.

Some of these syntheses are quite simple if the compound to be made bears a skeleton similar to some amino acid. For example, the stress hormone, epinephrine, is made from tyrosine through simple modification of the tyrosine skeleton:

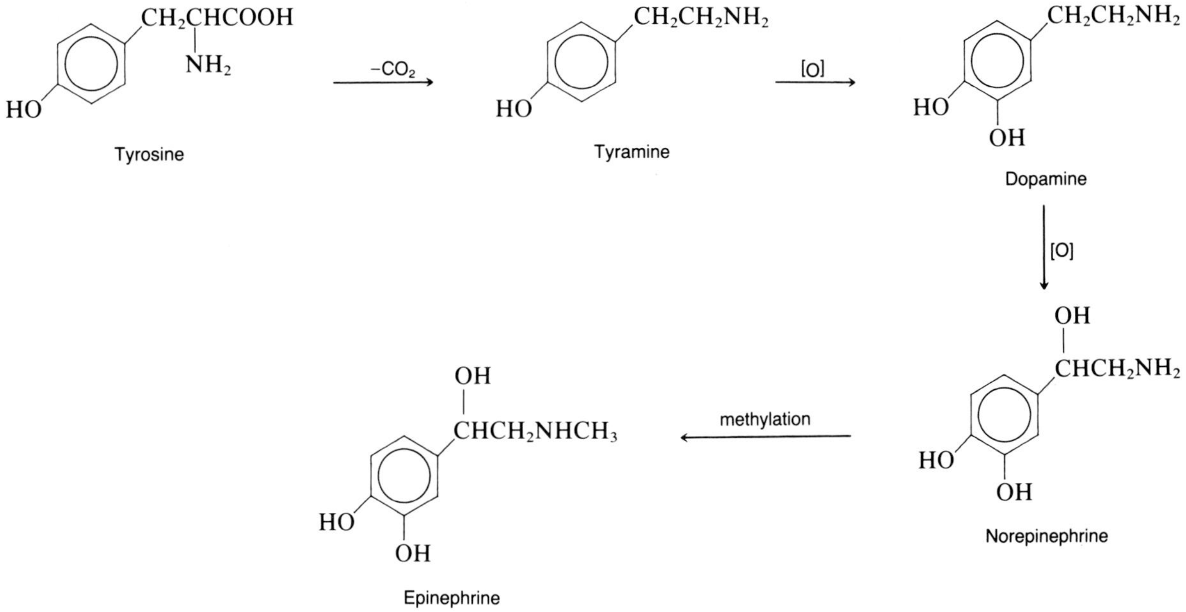

−CO₂
[O]
[O]
methylation
Tyrosine
Tyramine
Dopamine
Norepinephrine
Epinephrine

Biosynthesis of pyrimidine rings needed for nucleotide and nucleic acid synthesis is more complex; the initial reactant is indicated in color to clarify the course of the construction reactions.

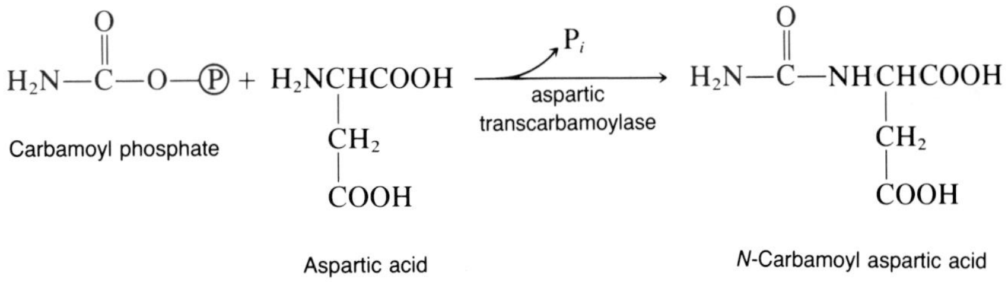

aspartic transcarbamoylase
Carbamoyl phosphate
Aspartic acid
N-Carbamoyl aspartic acid

Rewriting the N-carbamoyl aspartic acid structure makes the subsequent steps clearer:

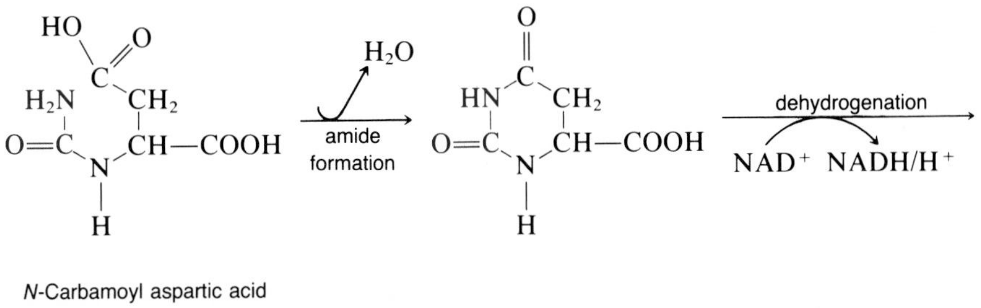

amide formation
dehydrogenation
NAD⁺ NADH/H⁺
N-Carbamoyl aspartic acid

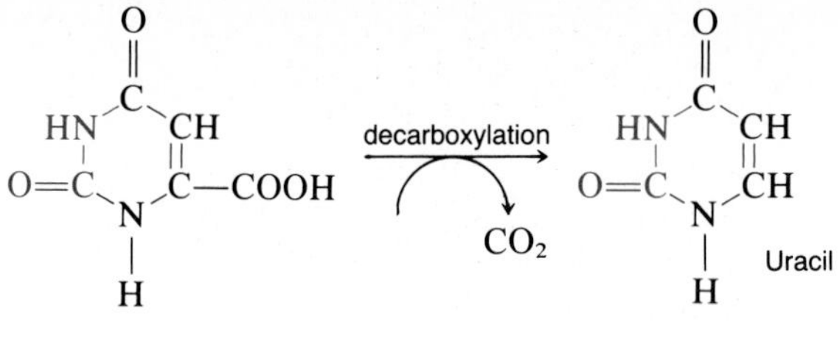
decarboxylation
CO_2
Uracil

Pyrimidine biosynthesis contrasted with the urea cycle offers another example of the principle of the separation of synthetic and degradative processes in the cell. We saw in Section 25.7 that any amino acid can provide the amino group to make carbamoyl phosphate, which then can enter the urea cycle by reaction with ornithine. The urea cycle occurs in the mitochondria, and the formation of carbamoyl phosphate is from NH_3 obtained by transamination and deamination. Pyrimidine biosynthesis occurs in the cytosol, and the nitrogen for carbamoyl phosphate is provided by the hydrolysis of glutamine. A different carbamoyl phosphate synthetase is required for this biosynthesis. The degradative and synthetic processes occur in different places in the cell (a physical separation or difference) and require different enzymes (a chemical difference).

The biosynthesis of the purine ring system requires amino acids, CO_2, and derivatives of the vitamin folic acid. Figure 25.8 shows the origin of each atom in the purine ring skeleton. Rather than make purine itself, the body builds purine rings onto ribophosphates so that a nucleotide is the actual final product of the synthesis of this nitrogenous base. That is, the body begins with ribose-5-phosphate and constructs the purine ring at C-1 from NH_3 (from amino acids), CO_2, and folic acid:

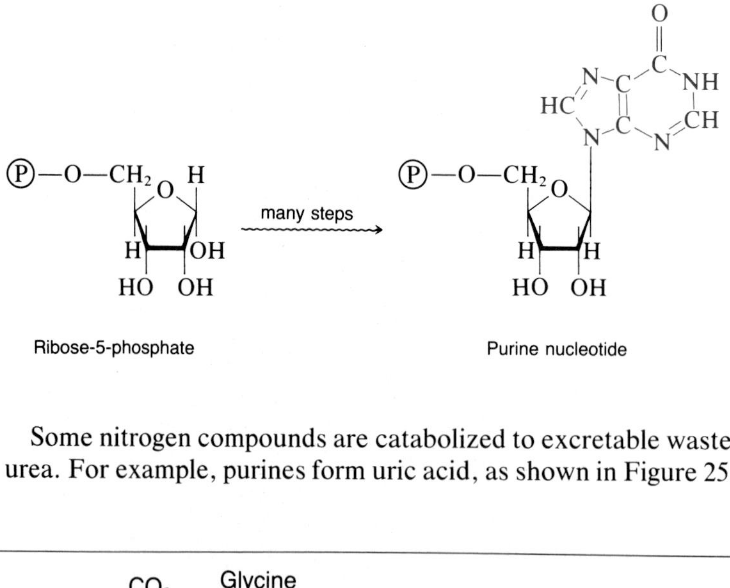
many steps
Ribose-5-phosphate
Purine nucleotide

Some nitrogen compounds are catabolized to excretable wastes other than urea. For example, purines form uric acid, as shown in Figure 25.9. Whereas

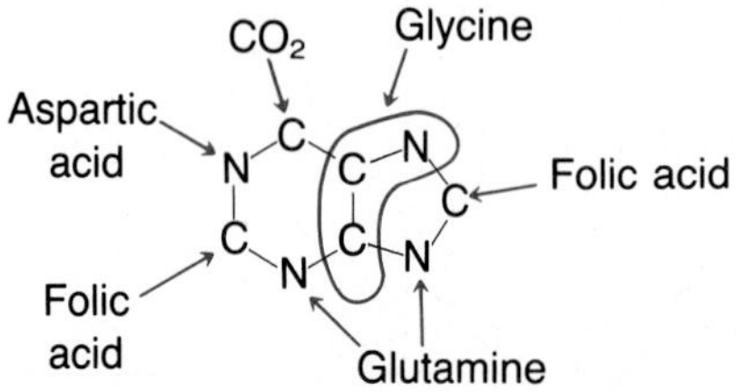
CO_2
Glycine
Aspartic acid
Folic acid
Folic acid
Glutamine

FIG. 25.8
Origins of the atoms in the purine ring in biosynthesis. Folic acid is a vitamin.

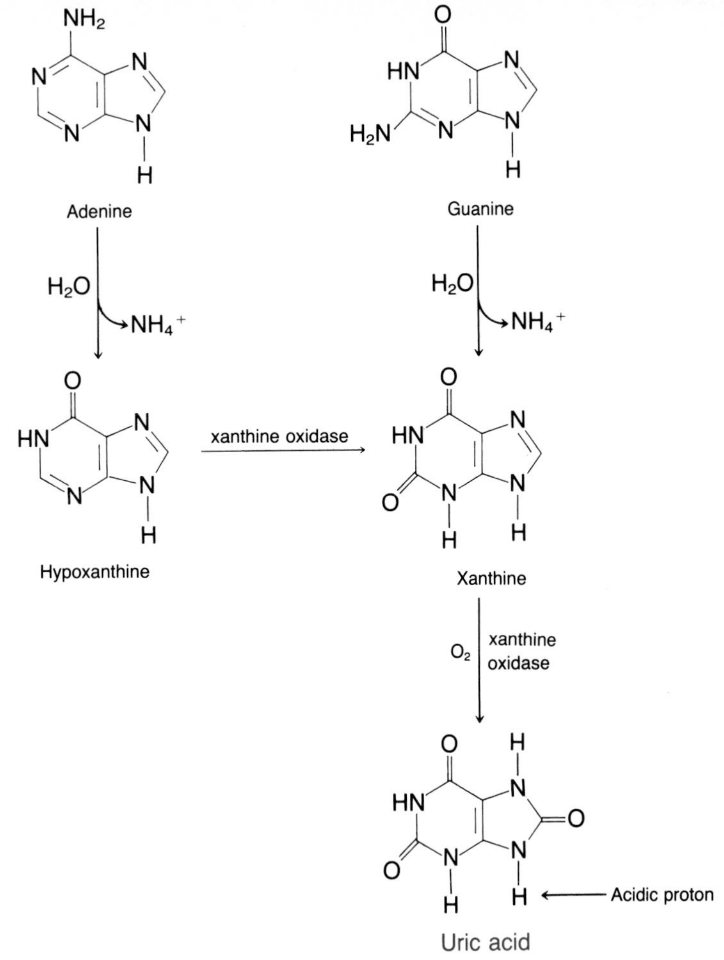
NH2
Adenine
O
Guanine
H2O
NH4+
Hypoxanthine
xanthine oxidase
Xanthine
O2
xanthine oxidase
Acidic proton
Uric acid

FIG. 25.9
Purines are degraded to uric acid, which is therefore a nitrogenous waste product of human beings. It is a minor product compared with urea. Excess production of uric acid leads to the condition known as gout.

uric acid is a minor nitrogenous waste product in human beings, it is the major nitrogen metabolic waste product of birds and reptiles.

Some people, almost always males, make excessive amounts of uric acid, leading to the condition known as gout. Uric acid and its salts are fairly insoluble and so may precipitate out of blood plasma. In gout, deposits of needlelike crystals of uric acid occur in joints, most notably those in the big toe. The resultant pain and inflammation are described by victims as agonizing. Eighteenth-century literature often attributes gout to a life of debauchery, but there is no scientific evidence to support this common misconception. Gout is treated by restricting purine sources in the diet such as shellfish and turkey. The drug allopurinol may be prescribed.

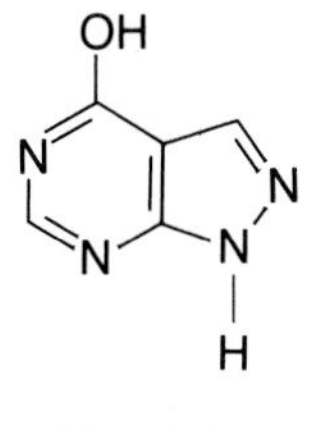

Allopurinol

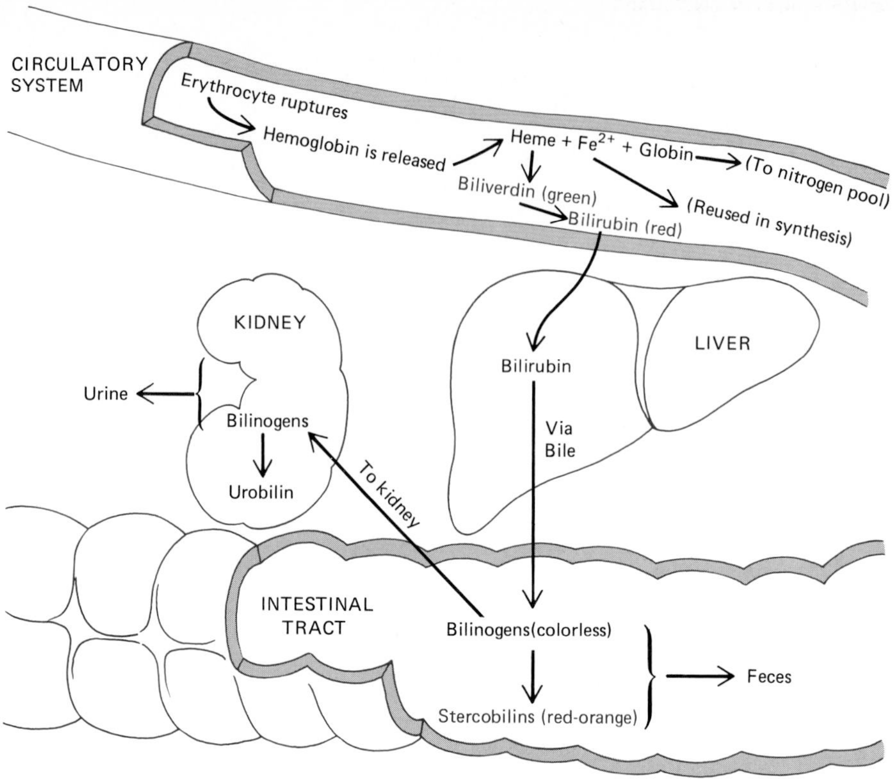
CIRCULATORY SYSTEM
Erythrocyte ruptures
Hemoglobin is released
Heme + Fe2+ + Globin
(To nitrogen pool)
(Reused in synthesis)
Biliverdin (green)
Bilirubin (red)
KIDNEY
LIVER
Bilirubin
Via Bile
Urine
Bilinogens
Urobilin
To kidney
INTESTINAL TRACT
Bilinogens(colorless)
Stercobilins (red-orange)
Feces

FIG. 25.10
Heme from hemoglobin is converted enzymatically into so-called bile pigments, first green biliverdin, and then red bilirubin. Bilirubin is the principal bile pigment in human beings. In the intestine, bacterial flora convert bilirubin into colorless bilinogens. Oxidation of bilinogens produces deep red-orange bilins. Bilinogens and bilins with the prefix *sterco-* are found in the feces; those with the prefix *uro-* go to the kidneys and urine. Stercobilin and urobilin give feces and urine their characteristic color.

Because the structure of allopurinol is very similar to that of xanthine, it inhibits the enzyme xanthine oxidase and thus disrupts the synthetic route to uric acid (see Figure 25.9).

The nitrogen compound heme that is the nonprotein part of hemoglobin is degraded into several colorful compounds known as bile pigments which give urine and feces their characteristic colors. A buildup of bile pigments in blood leads to the condition known as *jaundice*, wherein the skin and the whites of the eyes take on a yellow coloration from these pigments. *Hemolytic jaundice* is caused by excessive breakdown of red blood cells and hemoglobin, while *obstructive jaundice* is a result of an obstruction to the elimination of bile pigments in feces or urine. Figure 25.10 shows the degradation of hemoglobin and the usual disposition of its component parts.

CHAPTER ACCOMPLISHMENTS

After completing this chapter you should be able to define **key words** and do the following:

25.1 INTRODUCTION

1. Explain the greater relative importance of anabolic processes for proteins than for carbohydrates or lipids.

25.2 NITROGEN FIXATION

2. Describe the nitrogen cycle.

25.3 AMINO ACID POOL

3. State at least three fates of amino acids which enter the amino acid pool from digestion of proteins.

25.4 TRANSAMINATION

4. Describe transamination.
5. Given an α-amino acid and α-keto acid as reactants, write the products of their transamination reaction.
6. Explain the relationship between α-ketoglutaric acid and glutamic acid.

25.5 DEAMINATION

7. Describe deamination.
8. Write an equation for the deamination of glutamic acid.
9. Describe the usual route by which NH_3 is removed from an amino acid.

25.6 AMMONIA DISPOSAL

10. State the name and write the structure of the body's principal nitrogenous waste product.
11. State the purpose of glutamine synthesis.

25.7 THE UREA CYCLE

12. State the overall result of the urea cycle.

25.8 FATE OF AMINO ACID CARBON SKELETONS

13. Distinguish between glucogenic and ketogenic amino acids.

25.9 ESSENTIAL AND NONESSENTIAL AMINO ACIDS

14. Distinguish between essential and nonessential amino acids.
15. Distinguish between complete and incomplete proteins.

25.10 THE BIG PICTURE OF FOOD METABOLISM

16. State the usual source of our energy needs and the usual source of our structural needs.

25.11 METABOLISM OF NONPROTEIN NITROGEN COMPOUNDS

17. Name some nonprotein nitrogen compounds which the body can synthesize from amino acids.
18. Name the nitrogenous waste product of purine catabolism.
19. Explain how heme catabolism can lead to jaundice.
20. Distinguish between hemolytic and obstructive jaundice.

PROBLEMS

25.7 Why is protein anabolism described as the "main event" of protein metabolism?

25.8 How does atmospheric nitrogen (N_2) become incorporated into plant material?

25.9 Describe the nitrogen cycle in your own words.

25.10 Name three uses for amino acids derived from protein digestion.

25.11 Why is "nitrogen pool" probably a better name than "amino acid pool"?

25.12 Where is the amino acid pool located?

25.13 What is meant by saying that protein tissue is dynamic?

25.14 What is the turnover rate of a protein?

25.15 Why do growing infants exhibit positive nitrogen balance?

25.16 What two kinds of compounds react in a transamination reaction?

25.17 Why is vitamin B_6 a vitamin?

25.18 Write equations for transamination reactions between α-ketoglutaric acid and each of the following:

a Glycine, H_2NCH_2COOH

b Lysine, $H_2N(CH_2)_4CH(NH_2)COOH$ (the NH_2 is drawn above the CH)

c Methionine, $CH_3SCH_2CH_2CH(NH_2)COOH$ (the NH_2 is drawn below the CH)

25.19 Which amino acid is most often deaminated in the body? Write an equation for this reaction.

25.20 Glycine cannot be directly deaminated in the body. Use equations to show how a molecule of NH_3 is produced from glycine.

25.21 Why is NH_3 toxic?

25.22 In what form is most of the nitrogen in amino acids excreted from the body? Name the compound and show its structure.

25.23 How does the excretion of nitrogen as urea rather than NH_3 conserve water in land animals?

25.24 In which organ of the body do the reactions of the urea cycle occur?

25.25 How are glutamic acid and glutamine related?

25.26 How is ammonia transported to the liver from other production sites?

25.27 The overall result of the urea cycle is the combination of NH_3 with ______ to give ______.

25.28 Why is the urea cycle also known as the ornithine cycle?

25.29 Name the reaction type for each step in the sequence by which fumaric acid is converted to aspartic acid for further use in the urea cycle:

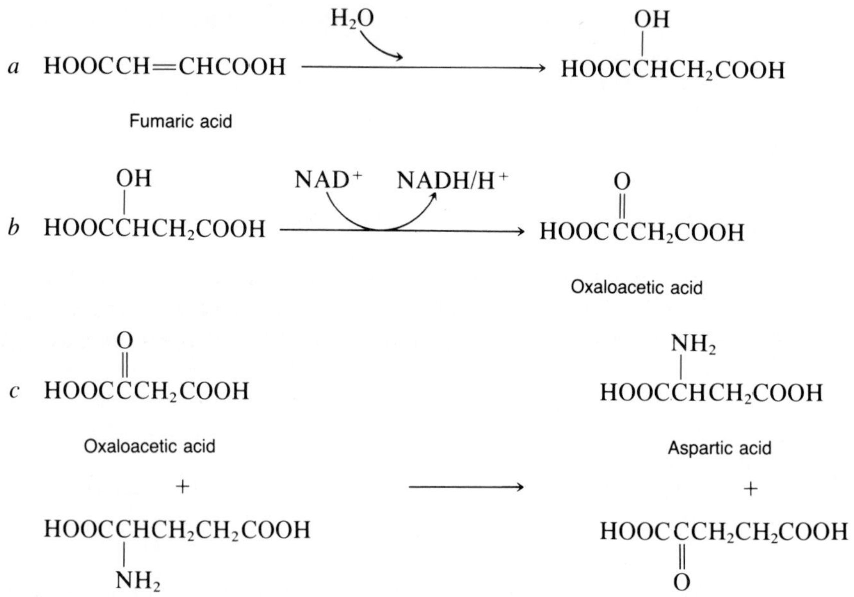

a
H2O
HOOCCH=CHCOOH
OH
HOOCCHCH2COOH
Fumaric acid
b
OH
HOOCCHCH2COOH
NAD+
NADH/H+
O
HOOCCCH2COOH
Oxaloacetic acid
c
O
HOOCCCH2COOH
Oxaloacetic acid
+
HOOCCHCH2CH2COOH
NH2
NH2
HOOCCHCH2COOH
Aspartic acid
+
HOOCCCH2CH2COOH
O

25.30 In the body's synthesis of urea, $H_2N{-}\overset{\overset{\displaystyle O}{\|}}{C}{-}NH_2$, what is the source of the carbonyl group and of each amino group?

25.31 How does transamination link protein catabolism and carbohydrate metabolism?

25.32 Given the conversion information below, classify each amino acid as either glucogenic or ketogenic.

Amino Acid	Metabolic Intermediate	Classification
Histidine ⟶	α-Ketoglutaric acid	______
Leucine ⟶	Acetyl CoA	______
Tyrosine ⟶	Acetoacetyl CoA	______
Aspartic acid ⟶	Oxaloacetic acid	______

25.33 What does the body get by processing amino acid carbon skeletons in carbohydrate and lipid pathways?

25.34 Nitrogen disposal is an endergonic process. How do amino acids provide at least some energy for this process?

25.35 Look back at Problem 25.29*c*. Is aspartic acid an essential amino acid?

25.36 How may amino acids be used in diabetes mellitus?

25.37 Why might an untreated diabetic display negative nitrogen balance?

25.38 Which of the following conversions are possible?
a Lipids ⟶ amino acids
b Lipids ⟶ glucose
c Glucose ⟶ lipids
d Amino acids ⟶ lipids
e Glucose ⟶ amino acids
f Amino acids ⟶ glucose

25.39 What is the excretable waste product of purine catabolism?

25.40 What causes gout?

25.41 Heme is degraded into colored compounds called ______.

25.42 Describe the appearance of a jaundiced patient.

25.43 Name two causes of jaundice.

CHAPTER TWENTY-SIX

NUTRITION

26.1 INTRODUCTION

We consume food for a variety of reasons: as nutrients for growing and maintaining our health, as the supply of energy for work and play, for the enjoyment of eating and the relaxation of dining with family and friends. Food and eating are interesting to most people because they are not only necessary but pleasurable. In this chapter we shall address several questions about **nutrition,** the study of the food we eat and the body's utilization of that food.

What types and how much food do we need to take in to supply our energy, growth, and health needs? What substances within foods do human beings require? What symptoms are associated with their deficiency? Can drugs affect the absorption of minerals? These questions are particularly relevant to health professionals who are concerned not only with their own well-being, but also with that of their patients. In this brief introduction we shall try to show how chemical principles you have encountered in previous chapters can help in answering questions of the type we have posed.

26.2 NUTRITIONAL STANDARDS

Let us begin by considering the nutritional information supplied to us on food packages. Figure 26.1 shows us the information on a typical cereal box. A $1\frac{1}{3}$-oz serving of the cereal contains 3 g of proteins, 31 g of carbohydrate, and 1 g of fat. Note that the sum of the components is 35 g. Since $1\frac{1}{3}$ oz is 37.8 g, the bulk of the cereal, like our biomass, consists of the three nutrients: proteins, carbohydrates, and fat. **Nutrients** are substances which are required in the diet for energy, growth, maintenance, or reproduction. You already know from the discussion of metabolism that carbohydrates and fats provide most of our energy, whereas proteins are used for such other functions as tissue growth and hemoglobin, hormone, enzyme, and nucleic acid synthesis. Because proteins, carbohydrates, and fats, along with water, make up the bulk of our food needs, they are known as **macronutrients.**

The amounts of other substances found in $1\frac{1}{3}$ oz of the cereal are quoted as percentages of United States Recommended Daily Allowances (USRDA). The USRDAs are established by the Food and Drug Administration (FDA) based on the Recommended Dietary Allowances (RDAs) developed by the Food and Nutrition Board of the National Academy of Sciences. Table 26.1 summarizes the National Academy's RDAs. The recommended nutrient levels are given in the table according to sex and for different age groups; the specific requirements of pregnant and lactating women are also included.

EACH SERVING CONTAINS 2 g DIETARY FIBER, INCLUDING 0.6 g (2% BY WEIGHT) NON-NUTRITIVE CRUDE FIBER.

NUTRITION INFORMATION PER SERVING

SERVING SIZE: ¾ CUP WHEAT & RAISIN CHEX (1⅓ OUNCE CEREAL AND RAISINS) AND IN COMBINATION WITH ½ CUP VITAMIN D FORTIFIED WHOLE MILK.

SERVINGS PER CONTAINER 15

	WHEAT & RAISIN CHEX 1⅓ OZ. CEREAL & RAISINS	WITH ½ CUP WHOLE MILK
CALORIES	130	210
PROTEIN	3 g	7 g
CARBOHYDRATE	31 g	37 g
FAT	1 g	5 g
SODIUM	215 mg	275 mg
POTASSIUM	120 mg	295 mg

PERCENTAGE OF U.S. RECOMMENDED DAILY ALLOWANCES (U.S. RDA)

PROTEIN	4	15
VITAMIN A	*	4
VITAMIN C	2	4
THIAMINE (VITAMIN B_1)	25	25
RIBOFLAVIN (VITAMIN B_2)	4	15
NIACIN	25	25
CALCIUM	*	15
IRON	30	30
VITAMIN D	*	10
VITAMIN B_6	25	25
FOLIC ACID	25	25
VITAMIN B_{12}	25	30
PHOSPHORUS	10	20
MAGNESIUM	8	10
ZINC	4	4
COPPER	8	8

*CONTAINS LESS THAN 2 PERCENT OF THE U.S. RDA OF THESE NUTRIENTS.

INGREDIENTS: WHOLE WHEAT, RAISINS, SUGAR, SALT, PARTIALLY HYDROGENATED COTTONSEED AND SOYBEAN OILS, SODIUM ASCORBATE (VITAMIN C), INVERT SUGAR, FERROUS FUMARATE, NIACINAMIDE, BHT (A PRESERVATIVE), THIAMINE MONONITRATE (VITAMIN B_1), PYRIDOXINE HYDROCHLORIDE (VITAMIN B_6), FOLIC ACID AND VITAMIN B_{12}.

MADE BY
RALSTON PURINA COMPANY
GENERAL OFFICES
CHECKERBOARD SQUARE
ST. LOUIS, MISSOURI 63164

CARBOHYDRATE INFORMATION

	WHEAT & RAISIN CHEX 1⅓ OZ. CEREAL & RAISINS	WITH ½ CUP WHOLE MILK
STARCH AND RELATED CARBOHYDRATES	19 g	19 g
SUCROSE AND OTHER SUGARS	10 g**	16 g
DIETARY FIBER	2 g	2 g
TOTAL CARBOHYDRATES	31 g	37 g

VALUES BY FORMULATION AND ANALYSIS

**SIX OF THE TEN GRAMS OF SUGAR OCCUR NATURALLY IN RAISINS.

On any correspondence regarding this product, please include the production number printed on the top of this carton and send to the Office of Consumer Affairs.

FIG. 26.1 Nutritional information from a box of Wheat & Raisin Chex cereal. To determine the milligram amounts of micronutrient present in a serving, multiply the percentage given (in decimal form) by the USRDA value in Table 26.3. (*Courtesy of Ralston Purina Company.*)

RDAs

In Table 26.1 you can see that with the exception of protein, the nutrients are needed in very small amounts, either milligram or microgram quantities. For this reason the vitamin and mineral nutrients are referred to as **micronutrients.** According to the Food and Nutrition Board, RDAs are defined as "the levels of intake of essential nutrients considered to be adequate to meet the known nutritional needs of practically all healthy persons." Some guidelines may be helpful in the interpretation of RDAs.

TABLE 26.1 Food and Nutrition Board, National Academy of Sciences—National Research Council: Recommmended Daily Dietary Allowances[a]

(Revised 1980. Designed for the maintenance of good nutrition of practically all healthy people in the United States)

Age, years	Weight kg	Weight lb	Height cm	Height in	Energy,[b] kcal	Protein, g	Fat-Soluble Vitamins: Vitamin A, μg[c]	Vitamin D, μg[d]	Vitamin E, mg[e]	Water-Soluble Vitamins: Vitamin C, mg	Thiamin, mg	Riboflavin, mg	Niacin, mg	Vitamin B_6, mg	Folacin, μg	Vitamin B_{12}, μg	Minerals: Calcium, mg	Phosphorus, mg	Magnesium, mg	Iron, mg	Zinc, mg	Iodine, μg
INFANTS																						
0.0–0.5	6	13	60	24	kg × 115	kg × 2.2	420	10	3	35	0.3	0.4	6	0.3	30	0.5[f]	360	240	50	10	3	40
0.5–1.0	9	20	71	28	kg × 105	kg × 2.0	400	10	4	35	0.5	0.6	8	0.6	45	1.5	540	360	70	15	5	50
CHILDREN																						
1–3	13	29	90	35	1300	23	400	10	5	45	0.7	0.8	9	0.9	100	2.0	800	800	150	15	10	70
4–6	20	44	112	44	1700	30	500	10	6	45	0.9	1.0	11	1.3	200	2.5	800	800	200	10	10	90
7–10	28	62	132	52	2400	34	700	10	7	45	1.2	1.4	16	1.6	300	3.0	800	800	250	10	10	120
MALES																						
11–14	45	99	157	62	2700	45	1000	10	8	50	1.4	1.6	18	1.8	400	3.0	1200	1200	350	18	15	150
15–18	66	145	176	69	2800	56	1000	10	10	60	1.4	1.7	18	2.0	400	3.0	1200	1200	400	18	15	150
19–22	70	154	177	70	2900	56	1000	7.5	10	60	1.5	1.7	19	2.2	400	3.0	800	800	350	10	15	150
23–50	70	154	178	70	2700	56	1000	5	10	60	1.4	1.6	18	2.2	400	3.0	800	800	350	10	15	150
51 +	70	154	178	70	2400	56	1000	5	10	60	1.2	1.4	16	2.2	400	3.0	800	800	350	10	15	150
FEMALES																						
11–14	46	101	157	62	2200	46	800	10	8	50	1.1	1.3	15	1.8	400	3.0	1200	1200	300	18	15	150
15–18	55	120	163	64	2100	46	800	10	8	60	1.1	1.3	14	2.0	400	3.0	1200	1200	300	18	15	150
19–22	55	120	163	64	2100	44	800	7.5	8	60	1.1	1.3	14	2.0	400	3.0	800	800	300	18	15	150
23–50	55	120	163	64	2000	44	800	5	8	60	1.0	1.2	13	2.0	400	3.0	800	800	300	18	15	150
51 +	55	120	163	64	1800	44	800	5	8	60	1.0	1.2	13	2.0	400	3.0	800	800	300	10	15	150
Pregnant, supplemental needs					+300	+30	+200	+5	+2	+20	+0.4	+0.3	+2	+0.6	+400	+1.0	+400	+400	+150	[g]	+5	+25
Lacatating, supplemental needs					+500	+20	+400	+5	+3	+40	+0.5	+0.5	+5	+0.5	+100	+1.0	+400	+400	+150	[g]	+10	+50

[a] The allowances are intended to provide for individual variations among most normal persons as they live in the United States under usual environmental stresses. Diets should be based on a variety of common foods in order to provide other nutrients for which human requirements have been less well defined.

[b] The energy allowances for the young adults are for men and women doing light work. The allowances for the older age group represent mean energy needs, allowing for a 2 percent decrease in basal (resting) metabolic rate per decade and a reduction in activity of 200 kcal/day for men and women between 51 and 75 years, 500 kcal for men over 75 years, and 400 kcal for women over 75 years. Energy allowances for children through age 18 are based on median energy intakes of children of these ages followed in longitudinal growth studies.

[c] As retinol.

[d] As cholecalciferol.

[e] As D-α-Tocopherol.

[f] The recommended dietary allowance for vitamin B_{12} in infants is based on average concentration of the vitamin in human milk. The allowances after weaning are based on energy intake (as recommended by the American Academy of Pediatrics) and consideration of other factors, such as intestinal absorption.

[g] The increased requirement during pregnancy cannot be met by the iron content of habitual American diets nor by the existing iron stores of many women; therefore the use of 30 to 60 mg of supplemental iron is recommended. Iron needs during lactation are not substantially different from those of nonpregnant women, but continued supplementation of the mother for 2 to 3 months after parturition is advisable in order to replenish stores depleted by pregnancy.

Guidelines and Limitations to RDAs

1 The RDAs are given as those for healthy individuals and may not meet the nutritional needs of individuals suffering from a particular illness or recovering from surgery.
2 The RDAs are not minimum requirements but are allowances to meet the needs of almost all persons within a particular category. Taking in less than is recommended by one of the allowances may not be a deficiency in a particular case.
3 The RDAs are *daily* allowances and, except for the fat-soluble vitamins, must be provided on a regular daily-intake basis.
4 The RDAs specify only recommended nutrient intake; they do not tell us a diet that will supply these nutrients. However, by varying the diet among a classified set of food groups shown in Figure 26.2, it is likely that all the required allowances would be met. Nutrition labeling, such as Figure 26.1 shows, can aid in the selection of foods for their nutrient value. The Department of Agriculture publishes an extensive survey of nutrient values of different foods in its booklet *Nutritive Value of Foods*. It is available through the United States Government Printing Office.
5 Table 26.2 shows RDA values for some other vitamins and minerals. These nutrients appear in a separate table because there is less firm quantitative knowledge about our need for them. They are nonetheless essential to our diet.

USRDAs

The FDA establishes the USRDAs for a given nutrient from the highest RDA level of that nutrient for any age group (excluding pregnant and lactating women) in Table 26.1. USRDAs are given in Table 26.3. As an example of the relationship of RDA and USRDA, note that for vitamin C the USRDA is 60 mg, which is the highest RDA level for vitamin C in Table 26.1.

FIG. 26.2
Nutritionists traditionally classify food into four major categories; milk, meat, fruits and vegetables, and grains. A diet which includes selections from all four groups should ensure attainment of all necessary nutrients.

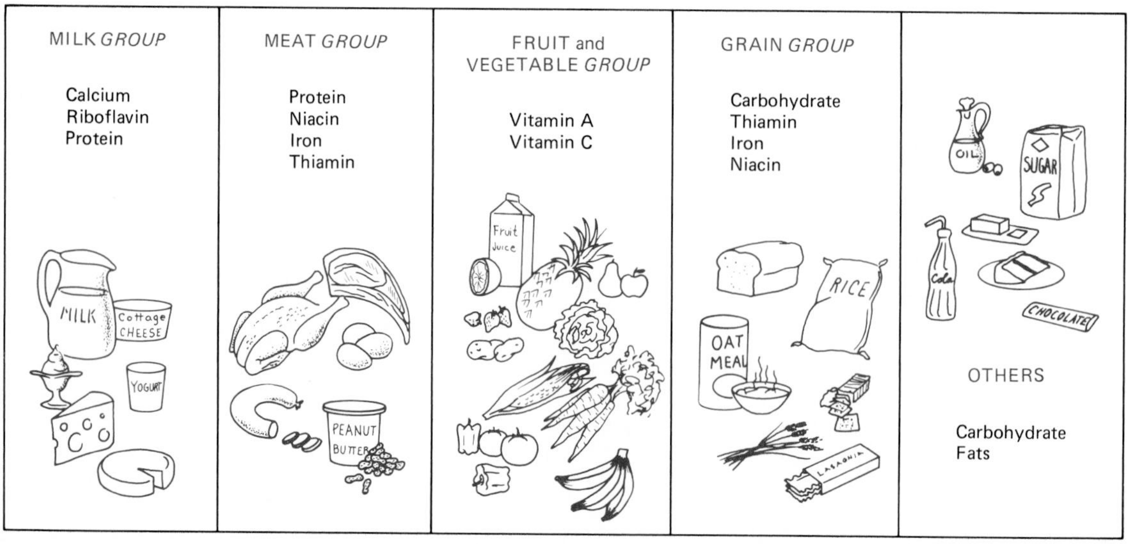
MILK *GROUP*
Calcium
Riboflavin
Protein
MILK
Cottage CHEESE
YOGURT
MEAT *GROUP*
Protein
Niacin
Iron
Thiamin
PEANUT BUTTER
FRUIT and VEGETABLE *GROUP*
Vitamin A
Vitamin C
Fruit Juice
GRAIN *GROUP*
Carbohydrate
Thiamin
Iron
Niacin
RICE
OAT MEAL
OIL
SUGAR
Cola
CHOCOLATE
OTHERS
Carbohydrate
Fats

TABLE 26.2 Estimated Safe and Adequate Daily Dietary Intakes of Selected Vitamins and Minerals (United States)*

	Age, years	Vitamins: Vitamin K, μg	Biotin, μg	Pantothenic Acid, mg
Infants	0–0.5	12	35	2
	0.5–1	10–20	50	3
Children and adolescents	1–3	15–30	65	3
	4–6	20–40	85	3–4
	7–10	30–60	120	4–5
	11+	50–100	100–200	4–7
Adults		70–140	100–200	4–7

	Age, years	Trace Elements†: Copper, mg	Manganese, mg	Fluoride, mg	Chromium, mg	Selenium, mg	Molybdenum, mg
Infants	0–0.5	0.5–0.7	0.5–0.7	0.1–0.5	0.01–0.04	0.01–0.04	0.03–0.06
	0.5–1	0.7–1.0	0.7–1.0	0.2–1.0	0.02–0.06	0.02–0.06	0.04–0.08
Children and adolescents	1–3	1.0–1.5	1.0–1.5	0.5–1.5	0.02–0.08	0.02–0.08	0.05–0.1
	4–6	1.5–2.0	1.5–2.0	1.0–2.5	0.03–0.12	0.03–0.12	0.06–0.15
	7–10	2.0–2.5	2.0–3.0	1.5–2.5	0.05–0.2	0.05–0.2	0.10–0.3
	11+	2.0–3.0	2.5–5.0	1.5–2.5	0.05–0.2	0.05–0.2	0.15–0.5
Adults		2.0–3.0	2.5–5.0	1.5–4.0	0.05–0.2	0.05–0.2	0.15–0.5

	Age, years	Electrolytes: Sodium, mg	Potassium, mg	Chloride, mg
Infants	0–0.5	115–350	350–925	275–700
	0.5–1	250–750	425–1275	400–1200
Children and adolescents	1–3	325–975	550–1650	500–1500
	4–6	450–1350	775–2325	700–2100
	7–10	600–1800	1000–3000	925–2775
	11+	900–2700	1525–4575	1400–4200
Adults		1100–3300	1875–5625	1700–5100

* Because there is less information on which to base allowances, these figures are not given in the main table of RDA and are provided here in the form of ranges of recommended intakes.

† Since the toxic levels for many trace elements may be only several times usual intakes, the upper levels for the trace elements given in this table should not be habitually exceeded.

Problem 26.1 Calculate the number of milligrams of vitamin C in 1⅓ oz of the cereal for which data are given in Figure 26.1.

26.3 ENERGY

Perhaps the most important function of nutrients is the production of energy. The energy content for each of the three macronutrients is as follows:

Macronutrient	Calories/Gram
Carbohydrates	4
Fats	9
Proteins	4

TABLE 26.3 The USRDAs

Nutrient	USRDA	Nutrient	USRDA
Protein	65 g	Biotin	0.3 mg
Vitamin A	1.5 mg	Pantothenic Acid	10.0 mg
Vitamin D	10 μg	Vitamin B_{12}	6.0 μg
Vitamin E	30 mg	Calcium	1.0 g
Vitamin C	60 mg	Iron	18.0 mg
Thiamin	1.5 mg	Phosphorus	1.0 g
Riboflavin	1.7 mg	Magnesium	400 mg
Niacin	20.0 mg	Iodine	150 μg
Folacin	0.4 mg	Zinc	15 mg
Vitamin B_6	2.0 mg	Copper	2.0 mg

Source: Code of Federal Regulations, Title 21, Section 101.9 (1981).

The nutritionists' Calorie is the chemists' kilocalorie, as you saw in Section 2.9. To release this amount of energy, the macronutrient must be "burned," that is, it must react with oxygen so that it is completely converted to CO_2, H_2O, and in the case of a protein, urea. To determine caloric value in the laboratory, burning, that is, direct combustion with oxygen, is done. However, in the body, as you saw in the metabolism chapters, "burning" proceeds through a stepwise set of reactions eventually ending with the combination of hydrogen atoms with oxygen in the respiratory chain. The energy released is used in three ways: to maintain the basal metabolic rate, for physical activity, and for specific dynamic action.

Basal Metabolic Rate

The basal metabolic rate (BMR) measured in Calories per day, is the energy consumed when an individual is at rest and fasting. It includes the energy required to maintain basic functions such as heartbeat, brain and nerve activity, kidney action, and body temperature. The BMR can be determined indirectly by measuring the oxygen consumption of a resting, fasting individual. Each liter of consumed and metabolically reacted oxygen translates into an average body expenditure of 4.8 Cal. Typically the BMR is about 10.9 Cal/lb of body weight or about 1750 Cal per day for a 160-lb person.

Problem 26.2 Calculate the BMR for an individual who consumed 3.0 L of oxygen in a 12-min test period.

Physical Activity

When a person is not completely at rest, energy is consumed to support physical activity. Table 2.6 lists energy expenditures for a variety of physical activities.

Specific Dynamic Action

Energy is also consumed in the digestion, absorption, transport, and metabolism of ingested food. This is known as the specific dynamic action (SDA) of food. On the average, SDA consumes an amount of energy equal to 10 percent of the sum of BMR and physical activity. For example, if a person is expending 2800 Cal for BMR and physical activity, an additional

280 Cal must be added to the diet intake for SDA, yielding 3080 Cal as the total dietary energy requirement.

Energy and Weight Maintenance

It is the balance of energy or Calorie intake and expenditure that determines body weight. If intake and output are exactly equal, then there is no energy storage. However, if, for example, food with a potential energy of 3500 Cal is ingested and 3100 Cal is expended for BMR, physical activity, and SDA, then the remaining 400 Cal must be stored as potential energy. Because fat has the highest Calorie count per gram, the most efficient way for the body to store excess potential energy is in the form of fat. Since 1 lb of body weight represents 3500 Cal, whenever your consumption exceeds your expenditure by 3500 Cal, you gain 1 lb of weight. Despite what some fad diet books might say, there are only two ways of losing weight:

1 Reduce Calorie intake below Calorie output.
2 Increase Calorie output above Calorie intake.

Actually, these are just two ways of saying the same thing: Intake must be less than output to lose weight. If you don't eat enough Calories to meet your energy expenditure, your body will use stored fat for energy. You must have a caloric deficiency of 3500 Cal to lose 1 lb of weight.

Of course, food is a source of other necessities besides energy.

26.4 PROTEINS AS MACRONUTRIENTS

Protein is a required macronutrient in the body for two reasons:

1 Proteins are the main source of elemental nitrogen for maintaining nitrogen balance (Section 25.3) in the body.
2 Proteins provide the nine essential amino acids (Section 25.9) which the body cannot synthesize.

Nitrogen Balance

Nitrogen is lost from the body in three ways: through the urine, through the feces, and through the skin. The major output is in the form of urea in the urine. As described in Chapter 25, urea is produced by the catabolism of amino acids from the digestion of dietary proteins and the breakdown of bodily protein. Further nitrogen loss results from undigested dietary protein which appears in the feces along with some digestive tract enzymes. Also, we shed skin protein, hair, and finger- and toenails.

The nitrogen loss must be balanced by an intake of dietary protein, or the body will begin to waste away. When the nitrogen intake exactly equals the nitrogen loss, the person is said to be in a state of **nitrogen balance.** This is the normal situation for a healthy, nonpregnant, nonlactating adult.

A positive nitrogen balance, wherein the intake of nitrogen exceeds the loss, is indicative of a condition in which new proteins are synthesized in addition to those formed for replacement of old protein. Growing children, pregnant or lactating women, and individuals recovering from a malnourished condition exhibit positive nitrogen balance. Negative nitrogen balance exists when nitrogen output is greater than intake. It can occur when there is a deficit in either quantity or quality of dietary protein, or through inactivity leading to muscle waste, or from a disease condition such as cancer. Table 26.4 lists pathological conditions which alter protein intake requirements.

TABLE 26.4 Pathological Conditions Altering Protein Requirements

CONDITIONS REQUIRING INCREASED PROTEIN CONSUMPTION
Sepsis
Fever
Trauma (injury)
Fractures
Burns
Gastrointestinal disorders (ileostomy, colostomy, diarrhea and other malabsorption states, ulcerative colitis)
Respiratory infections
Parasitic infections
Bacterial infections
Viral infections
Liver disease
Proteinuria
Renal disease (glomerulonephrosis)
Cancer
Marasmus (protein-energy malnutrition)
Kwashiorkor (protein malnutrition)
Pain
Anxiety or other psychological stress
Profuse sweating
CONDITIONS REQUIRING DECREASED PROTEIN CONSUMPTION
Hepatic coma
Massive hepatic necrosis
Renal failure, acute and chronic

Quantitative Statements of Protein Needs

Measurements of nitrogen output show that an average adult requires 0.45 g (450 mg) of complete protein per kilogram of body weight. A **complete protein** is one that contains all the essential amino acids in the required proportions for human beings, as shown in Table 26.5. That is, for example, within the total requirement for 450 mg/kg there should be twice as much leucine (16 mg) as threonine (8 mg). Because egg protein most closely matches the amino acid requirements for human beings, it is assigned a rating of 100 for quality of amino acid. Other foods are assigned quality ratings based on their amino acid composition. However, even a complete protein such as egg is not completely digested, so that the full benefit of the protein in the egg is not received. To take into account the lack of completeness in most food proteins and the fact that even complete proteins, such as egg protein, are not fully retained by the body, the adult RDA of protein is set at 0.8 g/kg of body weight. This value is nearly twice the actual measured value (0.45 g/kg) based on nitrogen output. Note that in Table 26.1 the protein RDA for an average adult male (70 kg) is given as 56 g (56 g/70 kg = 0.8 g/kg).

Perhaps the best measurement of protein quality in food is the net protein utilization (NPU). The NPU measures how well the body is utilizing the protein ingested by comparing the grams of nitrogen retained in the body with the grams of nitrogen ingested.

$$\text{NPU} = \frac{\text{grams of nitrogen retained}}{\text{grams of nitrogen ingested}} \times 100$$

Table 26.6 gives the NPU values and the usable protein per typical serving of several common foods. In general, proteins derived from animal sources such as meat, fish, milk, and cheese have a high NPU percent because they

TABLE 26.5 Estimated Requirements for Essential Amino Acids

Amino Acid	Requirement, mg/kg of Body Weight per Day Infant (3–6 Months)	Child (10–12 Years)	Adult
Isoleucine	80	28	12
Leucine	128	42	16
Lysine	97	44	12
Total sulfur-containing (methionine + cysteine)	45	22	10
Total aromatic (phenylalanine + tyrosine)	132	22	16
Threonine	63	28	8
Tryptophan	19	4	3
Valine	89	25	14
Histidine*	33	?	?

* Originally thought to be essential only for infants, histidine is now thought to be essential for adults as well. According to the 1980 edition of the RDAs, arginine is not essential for healthy human beings.

Source: Food and Nutrition Board, *Recommended Dietary Allowances,* 9th ed., National Academy of Sciences, Washington, 1980, p. 43.

TABLE 26.6 The Quantity and Quality of Protein in Selected Foods

Food	Percent Protein by Weight, g of Protein per 100 g of Food	NPU, %	Serving Size	Usable Protein per Serving, g
Eggs	13.8	94	1 egg (46 g)	5.9
Whole milk	3.5	82	1 cup (244 g)	7.0
Cheddar cheese	25.0	70	1 oz (28 g)	4.9
Sirloin steak (choice grade, separable fat removed, cooked)	32.2	67	3 oz (85 g)	18.4
Hamburger (regular, cooked)	24.2	67	3 oz (85 g)	13.8
Haddock (breaded, fried)	19.6	83	4-oz fillet (110 g)	17.9
Chicken (flesh only, fried)	31.2	73	3 oz (85 g)	19.4
Soybeans (mature seeds, cooked)	11.0	61	1 cup (180 g)	12.1
White beans (average of several types of mature seeds, cooked)	7.8	38	1 cup (180 g)	5.3
Pecans (halves)	3.2	42	½ cup (59 g)	0.7
Cashews (whole)	17.2	58	½ cup (70 g)	4.8
Peanuts (shelled, roasted)	26.2	43	½ cup (72 g)	8.1
Whole wheat bread	10.0	60	2 slices (50 g)	3.0
White rice (cooked and cooled)	2.0	57	1 cup (145 g)	2.2
Rolled oats (oatmeal, cooked)	2.0	66	1 cup (240 g)	3.2
Sesame seeds	18.6	53	½ cup (75 g)	7.4
Sunflower seeds (kernels)	24.0	58	½ cup (72 g)	10.0
Cabbage (coarsely shredded)	1.3	35	1 cup (70 g)	0.3
Potato (baked in skin)	2.6	60	1 potato (202 g)	3.1
Mushrooms (raw)	2.7	72	1 cup (70 g)	0.9
Corn (cooked)	3.5	51	1 cup (165 g)	2.8

Sources: C. F. Adams, *Nutritive Value of Foods in Common Units,* Agriculture Handbook No. 456, USDA, Washington, 1975; Food and Agriculture Organization, *Amino Acid Content of Foods and Biological Data on Proteins,* FAO, Rome, 1970. (NPU data used with permission.)

contain all the essential amino acids in approximately the required proportions for human beings. For this reason, animal protein is referred to as complete protein.

Although they also contain all the essential amino acids, vegetable, grain, and bean products do not contain them in the necessary proportions. In these foods one or more of the essential amino acids is lacking in relative proportion to the other required amino acids. For example, Table 26.5 tells you that an adult requires 12 mg of lysine per kilogram per day. One slice of whole wheat bread supplies 1500 mg of total protein, but hardly any lysine. Table 26.7 lists the limiting amino acids, i.e., amino acids which are missing or deficient in common foods and thereby limit the completeness of the protein in the food. Thus, the usable human protein that can be formed from such a food is limited. A diet based solely on one of these products, for example, rice, would be inadequate in protein unless an absurdly large amount of the food was eaten. For example, one would need to eat 4.6 lb of rice to supply the daily need of approximately 31 g of usable protein for a 70-kg adult. Even if one could consume this much rice in a day, it would not be a desirable diet. The rice would meet or exceed one's Calorie needs and leave little room for variety in food consumption so that other necessary nutrients would not be eaten. Consequently, vegetable, grain, and bean protein are labeled **incomplete.** A diet composed predominantly of one such food product, without animal protein and without variety, could lead to protein malnourishment.

Protein Complementarity

Individuals who do not consume any animal food, including egg and dairy items, are called *vegans*. Vegans must give careful thought to their proper intake of dietary protein. They should make use of **protein complementarity** to assure themselves of an adequate intake of the essential amino acids. Protein complementarity means combining foods individually deficient in at least one essential amino acid, so that the combined foods contain all the essential amino acids in the proper proportions.

Table 26.7 shows that rice is deficient in lysine and isoleucine, whereas beans are deficient in tryptophan and cysteine (or methionine). If one eats a meal containing rice and beans, the amino acid deficiencies of rice are compensated for by the beans and vice versa. The NPU of beans and rice eaten alone are about 40 percent and 60 percent, respectively. The NPU of the combination increases to about 80 percent. In protein complementarity the whole can be greater than the sum of the parts. The increase in NPU is particularly striking in the case of beans, which have large amounts of protein in an average serving which go unutilized if the amino acid deficiencies are

TABLE 26.7 The Limiting Amino Acid in Some General Food Types

Food Type	Limiting Amino Acids
Legumes (e.g., peas, lentils, soybeans, kidney beans)	Cysteine or methionine, tryptophan
Nuts and Seeds (e.g., cashews, peanuts)	Lysine, isoleucine
Grains, Cereals (e.g., wheat, rice, spaghetti)	Lysine, isoleucine
Flour (e.g., oatmeal, cornmeal, whole wheat flour)	Lysine, isoleucine
Vegetables (e.g., broccoli, mushrooms, potatoes, spinach)	Cysteine or methionine, isoleucine

not complemented. Table 26.7 suggests some general complementary protein combinations:

Legumes + grains

Legumes + nuts and seeds

Vegetables + seeds and nuts

Vegetables + grains

Because amino acids are not stored in the body, complementary foods should be eaten at the same meal or within a few hours of each other.

Protein-Energy Malnutrition

Although people in developed countries generally have access to sufficient protein in their diet, this is not always true for people in certain underdeveloped regions of Africa, Southeast Asia, and Latin America. Many children under 5 years of age suffer from protein-energy malnutrition (PEM); that is, their intake of both protein for growth and food in general is deficient.

One form of PEM known as **marasmus** occurs when there is a severe deficiency of Calories, including protein, in the diet. Marasmus may develop when a child under 1 year of age is removed from breast-milk feeding and put on a soupy diet which is insufficient in total Calories and protein. The outward symptoms of marasmus are growth retardation, severe weight loss, and muscle wasting. In this condition, the blood levels of proteins and organ functions remain about normal, but the body breaks down its muscle protein to supply caloric energy.

Kwashiorkor is another form of PEM which occurs in children 2 to 4 years of age. This is a severe protein deficiency disease. The typical starchy diet of a kwashiorkor victim consists of cassava (a root), plantain, maize, or rice. The diet may be minimally sufficient in Calories, but protein is deficient in both quantity and quality. In the kwashiorkor patient, serum protein levels are low, the skin tends to be covered with discolorations, and edema (fluid retention) is present so that the child is bloated.

The name *kwashiorkor* originated in the language of Ghana where it means "the disease the first child gets when the second is on the way." During the first-year nursing period the baby receives sufficient protein, but the adequate diet is lost when the child is displaced from the breast and given a starchy diet to make way for the next baby.

PEM can be treated by supplying the missing Calories and protein, usually in the form of milk. However, there is evidence that PEM, if prolonged, can lead to permanently stunted growth and mental retardation.

26.5 CARBOHYDRATES AS MACRONUTRIENTS

Carbohydrates do not appear to be absolutely essential in the diet, and indeed, there is no RDA for them in Table 26.1. However, there are several reasons for eating at least 100 g of carbohydrate a day:

1 Glucose is the main fuel source for metabolic reactions, and can be easily obtained from carbohydrates.
2 Carbohydrates are a less expensive source of Calories than fats or proteins.
3 Very low carbohydrate intake can lead to excessive breakdown of adipose tissue and the eventual production of ketosis.

4 Low carbohydrate intake may also lead the body to break down proteins for use as an energy source, which can result in muscle wasting.
5 Sufficient dietary carbohydrate intake is necessary to maintain glycogen reserves which are needed for prolonged physical activity.
6 Carbohydrates are the source of dietary fiber or roughage. Polysaccharides such as cellulose, hemicellulose (a polysaccharide composed of xylose, glucose, and mannose), and pectin cannot be hydrolyzed to simpler monosaccharides by human digestive enzymes and therefore pass through the digestive tract and contribute bulk to the stools. Low intake of dietary fiber in peoples of developed countries has been implicated in the increased incidence of disorders of the large intestine and in heart disease.

Increased consumption (as a percentage of total dietary intake) of complex (polysaccharide) carbohydrate was recommended by a committee of the U.S. Senate in 1977. One of the rationales for this is that increased carbohydrate consumption will be matched by decreased fat consumption. High fat consumption is generally accepted as a major risk factor in the development of heart disease. Although the Senate committee recommends an increase in *complex* carbohydrates, i.e., polysaccharides, it also recommends a significant decrease in mono- and disaccharides from about 18 to 10 percent of the total caloric intake. These sugars are the ones we find added to processed or fast foods; the usual sweetener is sucrose or table sugar. Simple sugars are said to provide "empty" Calories because the foods they are commonly found in, such as candy, cakes, pies, and soft drinks, are relatively low in nutrients. Furthermore, the simple sugars lead to dental caries if they are not properly cleaned from the teeth. There is no clear evidence that links sucrose intake to heart disease, diabetes, or hypoglycemia. Table sugar can be found in a number of forms including white sugar, brown sugar, honey, and molasses. However, aside from molasses, which contains significant amounts of calcium, iron, potassium, and B vitamins, none of the other three sugars is nutritionally superior to another.

Maintaining a relatively constant level of blood glucose is critically important to diabetics and may be of importance to the general population. High blood glucose concentrations tend to make people sleepy, and food substances which give rapid increases in blood glucose concentration may be a factor in the development of adult-onset diabetes. Nutritionists and physicians have taught that simple sugars such as glucose, sucrose, lactose, and fructose lead to a rapid rise in glucose concentration compared with complex carbohydrates found in starches. However, these old ideas on how simple sugars and complex sugars affect glucose levels had apparently never been experimentally tested, and recent research has given results contrary to the old beliefs.

For example, although diabetics are told to avoid ice cream in their diet, the experimental tests show that ice cream does not affect the blood glucose level. Intuitively, it might seem that the opposite should be true, but a white potato, not a sweet potato, greatly alters the blood glucose level. Indeed, a white potato affects glucose level as if it were candy containing pure glucose. Foods such as pasta, beans, and sweet potatoes have much smaller effects on the glucose level than bread, cereals, and white potatoes.

It has not been possible to predict the effect on glucose levels that a given food source will produce. In fact, the level can even depend on the physical state (e.g., rice slurry or rice grains) of the food. However incomplete, the recent research should warn us about the danger of accepting experimentally untested assumptions even when (or especially when) the expected answers appear obvious.

26.6 LIPIDS AS MACRONUTRIENTS

No RDA has been set for lipids. One reason for this is that almost all necessary lipids can be synthesized in the body from carbohydrate origins. Two polyunsaturated acids, linoleic and arachidonic, cannot be synthesized by human beings, apparently because we lack the enzymes needed to form the double bonds; these fatty acids must therefore be supplied in the dietary intake.

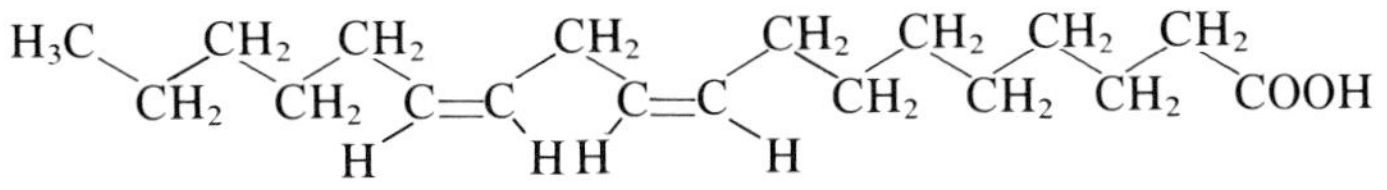

Linoleic acid

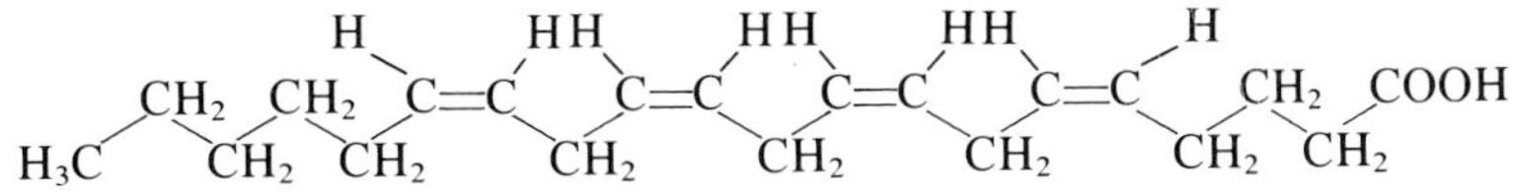

Arachidonic acid

Linoleic acid is required for growth and healthy skin, and arachidonic acid is necessary for the synthesis of prostaglandins. However, since arachidonic acid can be formed in the body from linoleic acid, only linoleic acid is really essential in the diet. Some fat in the digestive system is necessary for proper absorption of the fat-soluble vitamins, but this can be met by a daily intake of only 15 to 25 g of fat. This amount is far exceeded in the diet of most people in developed countries.

The same Senate committee mentioned in the carbohydrate section has proposed that fats should contribute no more than 30 percent of the daily caloric intake and that these fats should be divided equally among polyunsaturated, monounsaturated, and saturated fatty acids. High dietary intake of saturated, and to some extent, monounsaturated, fatty acids increases the level of blood lipids and may predispose an individual to atherosclerosis. Furthermore, saturated fats tend to increase the level of low-density lipoproteins (LDL) and decrease the level of high-density lipoproteins (HDL). These lipoproteins carry cholesterol in the blood. High LDL levels have been associated with increased deposition of cholesterol in the arteries, whereas increased HDL levels favor transfer of excess cholesterol to the liver, where it is degraded and thus removed from the body. Review Section 18.15, which discussed the relationships among HDL, LDL, and coronary disease. When polyunsaturated oils are substituted for animal fat in the diet, the level of HDLs increases with a corresponding decrease in LDL. However, increased polyunsaturate consumption beyond the 10 percent recommended in the Senate committee's dietary guidelines may not be a good idea because there may be a relationship between cancer and polyunsaturate intake.

Animal fats which adversely affect LDL and HDL levels in the blood also generally contain cholesterol. Many experiments with human beings have shown that lowering animal fat intake in the diet can lower blood serum levels of cholesterol; furthermore, increased intake of polyunsaturated fatty acids also lowers blood serum levels of cholesterol. Recent studies indicate that individuals with lower blood cholesterol do show a reduced incidence of heart disease. It may be that only a small segment of the population can benefit by way of reduced heart disease from dietary changes. However, since most of us do not know whether we are a member of that segment, the best advice for avoiding a heart attack, or surviving one, may be to follow the cautionary advice of reducing total fat intake and particularly animal fat intake.

26.7 VITAMINS AS MICRONUTRIENTS

Vitamins are organic substances required in the diet in small amounts. They are necessary for proper growth, maintenance, and reproduction. If a vitamin is missing, or lacking in quantity, a deficiency symptom, characteristic for that vitamin, will develop. The exact bodily function of a vitamin is not always known, but the symptoms of their deficiency or malfunction are well-characterized. Most vitamins cannot be synthesized by the body. However, niacin, vitamin A, and vitamin D can be formed in the body if the appropriate starting materials, called **provitamins,** are present. When neither the vitamin or provitamin is available, the deficiency symptom develops.

A substance which is a vitamin for one species may not be a vitamin for another. For example, folic acid is a vitamin for human beings but not for bacteria, which are capable of synthesizing folic acid. The distinction allows the use of drugs such as sulfanilamide (Section 21.9), which structurally resembles the raw material bacteria use for making folic acid and interferes with the synthesis. The drug is not harmful to human beings who ingest already formed folic acid. Although most animals internally synthesize vitamin C, human beings require it as a vitamin in their diet.

Solubility characteristics divide the vitamins into two major classes: the fat-soluble and the water-soluble vitamins. Fat-soluble vitamins are lipids themselves or are found in nature dissolved in plant and animal lipids. Water-soluble vitamins, of course, are so named because they dissolve in water. The water-soluble vitamins must be constantly replenished in the diet because they are rapidly eliminated from the body. There is little danger of their accumulation; hence they are generally nontoxic even in large dosages.

The fat-soluble vitamins are vitamins A, D, E, and K. They fit into recognizable lipid categories. Vitamin D is a steroid, and the others are terpenes, compounds largely hydrocarbon in nature and characterized by a pattern of alternating single and double bonds. The dietary source of these vitamins is fat, and they are stored in bodily fatty tissue, mostly around the liver. Persons who have problems with fat absorption are susceptible to experiencing deficiencies of the fat-soluble vitamins. On the other hand, excess ingestion of these vitamins can be toxic because of their solubility properties. They accumulate in fatty tissue and are not readily "washed away" by our water-based excretion system. A dangerous accumulation of vitamins in tissues is called **hypervitaminosis.**

All of the water-soluble B vitamins function as coenzymes (Section 21.11) in various physiological reactions. In the forthcoming sections of this chapter we will examine other functions of vitamins, the food sources of vitamins, and the clinical symptoms of vitamin deficiencies.

The known bodily requirements for vitamins are the amounts necessary to meet the RDAs listed in Table 26.1. The large doses, called megadoses or megavitamin therapy, which are claimed to be of benefit for some vitamins, particularly niacin, vitamin C, and vitamin E, really exemplify the use of vitamins as drugs in a nonnutritionally related disease. That is, ingestion of only small amounts of vitamins meets our nutritional needs. Excesses are essentially drugs taken in the hope of preventing or treating conditions such as the common cold, cancer, or the deterioration of aging. Table 26.8 summarizes vitamin functions, deficiency diseases, food sources, and toxicity from megadoses.

TABLE 26.8 Summary of Vitamin Functions, Deficiencies and Food Sources

Vitamin	Function	Deficiency Condition	Some Food Sources	Toxicity from Megadoses
		WATER-SOLUBLE VITAMINS		
B vitamins:				
Thiamine	Coenzyme particularly in carbohydrate metabolism	Beriberi	Pork, legumes, whole grain cereals and bread	Not known
Niacin	Coenzyme in oxidation-reduction reactions, NAD^+	Pellagra	Meat, milk	Skin flush, heartbeat arrhythmia
Riboflavin	Coenzyme in oxidation-reduction reactions, FAD	Dermatitis, scaly skin, itchy eyes	Liver, dairy products, broccoli, enriched bread	Not known
Vitamin B_6	Coenzyme in transamination ana decarboxylation reactions	Anemia, nervous system disorders	Liver, legumes, potatoes, bananas	Dependency on B_6, sleepiness
Folic Acid	Coenzyme in transfer of single-carbon groups	Anemia	Organ meats, green leafy vegetables, legumes	Not known
Vitamin B_{12}	Coenzyme in transfer of single-carbon groups in nucleic acid synthesis	Pernicious anemia	Animal products	Not known
Biotin	Coenzyme in CO_2 transfer reactions	Dermatitis, loss of appetite, depression	Liver, chicken, eggs, nuts	Not known
Pantothenic acid	Part of Coenzyme A	Rare; insomnia, fatigue	Eggs, liver, salmon; widespread in most foods	Diarrhea
Vitamin C (ascorbic acid)	Collagen synthesis, iron metabolism	Scurvy	Citrus fruit, strawberries, melons, broccoli	Diarrhea, kidney stones, possible dependency on vitamin C
		FAT-SOLUBLE VITAMINS		
Vitamin A	Vision in dim light, maintenance of epithelial tissue	Night blindness, xerophthalmia	Liver; eggs; yellow, orange, and dark green leafy vegetables	Fatigue, loss of hair, pain in bones and joints
Vitamin D	Promotes absorption of dietary calcium and phosphorus	Rickets in children, osteomalacia in adults	Fortified milk, fish liver oil	Loss of appetite, nausea, elevated blood calcium levels
Vitamin E	Antioxidant, protects polyunsaturated fatty acids from oxidation	Rare; hemolysis of red blood cells	Vegetable oils, wheat germ, legumes	Prolonged blood clotting time, headache, extreme fatigue
Vitamin K	Blood clotting	Prolonged clotting time	Dark green leafy vegetables, liver	Anemia and jaundice in infants

26.8 THE WATER-SOLUBLE VITAMINS AS MICRONUTRIENTS

The group of B vitamins (thiamine, niacin, B_6, folacin, pantothenic acid, biotin, and B_{12}) and vitamin C make up the water-soluble vitamins.

Thiamine

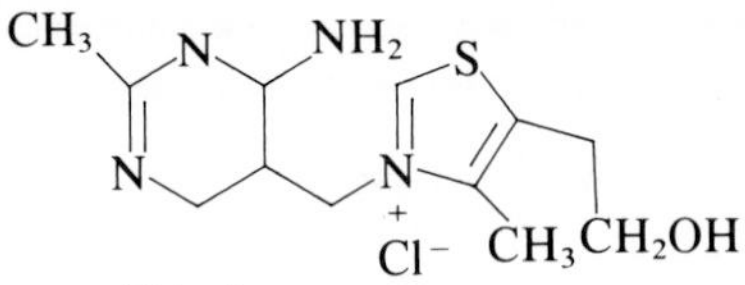
CH3
N
NH2
S
N
N
+
Cl−
CH3CH2OH

Thiamine

Beriberi is a disease of thiamine deficiency; it affects the nervous system and, in severe cases, causes heart arrhythmias, heart enlargement, and eventually heart failure. Beriberi is a particular problem in countries in which rice is a major staple in the diet. The rice bran and germ, which contain most of the thiamine, are removed in the polishing of the rice. Replacing polished rice by brown rice alleviates the condition. Chronic alcoholics, among their other problems, may also suffer from beriberi because of low thiamine intake and poor intestinal absorption. For the most part, beriberi has been eradicated in the United States by thiamine enrichment of grain products.

Niacin

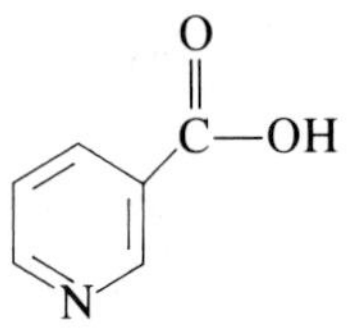
O
C—OH
N

Nicotinic acid (niacin)

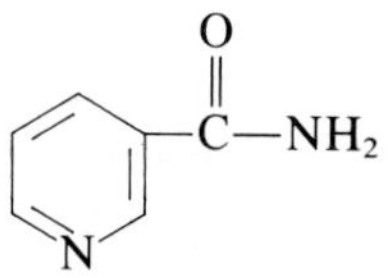
O
C—NH2
N

Nicotinamide (niacinamide)

Nicotinic acid and nicotinamide, the actual metabolic form in the body, make up the vitamin niacin. A deficiency of niacin results in pellagra, a disease which is characterized by diarrhea, dermatitis, dementia, and in untreated cases, death. Pellagra occurs in people whose diet consists mainly of corn or corn products and in which meat is lacking. Corn contains considerable amounts of niacin (an ear of corn contains about 15 percent of an adult's RDA). However, the niacin is bound to a protein and is not released for use in the body by normal enzyme activity in the digestive tract. Although their diet is corn-based, Indians in Mexico and Central America who soak their corn in lime water (lime is the common name for calcium oxide, which dissolved in water is calcium hydroxide) before using it to make tortillas are found to be resistant to pellagra. Apparently the base releases the niacin for use. Human beings can synthesize niacin from the amino acid tryptophan; however, corn contains only low concentrations of tryptophan.

Megadoses of niacin have been used in some experimental treatments of schizophrenia with varying success. However, large doses of niacin can have such adverse effects as elevated blood glucose and uric acid levels, liver injury, and arrhythmic heartbeats.

Riboflavin

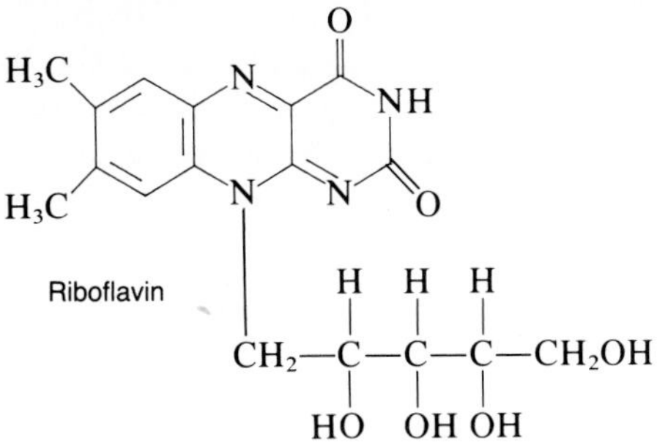
H3C
H3C
N
NH
O
N
N
O
H H H
CH2—C—C—C—CH2OH
HO OH OH

Riboflavin

Riboflavin deficiency often accompanies thiamine or niacin deficiency, and its symptoms are masked by those of beriberi or pellagra. Riboflavin deficiency leads to growth retardation and abnormal skin tissue around the eyes and cracks at the corners of the mouth. Milk and milk products are good sources of riboflavin, as are green leafy vegetables and organ meats. Strict vegetarians who do not consume milk products may need to be especially careful about their riboflavin intake.

Vitamin B_6

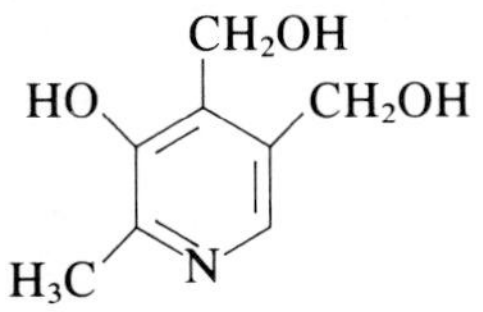
CH2OH
HO
CH2OH
H3C
N

Pyridoxine

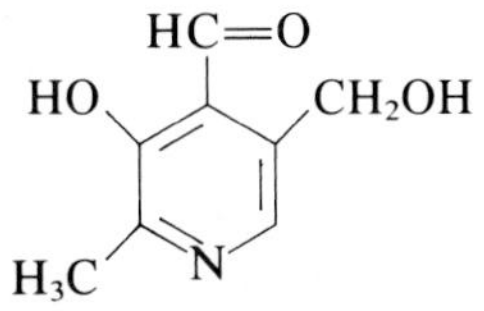
HC=O
HO
CH2OH
H3C
N

Pyridoxal

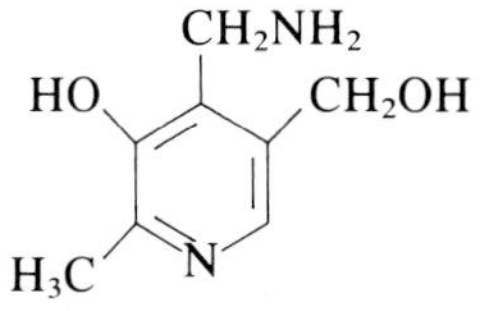
CH2NH2
HO
CH2OH
H3C
N

Pyridoxamine

Vitamin B_6 is actually a group of three compounds—pyridoxine, pyridoxal, and pyridoxamine—each of which is physiologically active. B_6 appears to play an important role in protein metabolism. It functions in the conversion of tryptophan to niacin, and so a deficiency in B_6 can result in a niacin deficiency with the consequent symptoms of pellagra. Vitamin B_6 deficiency alone leads to symptoms involving the central nervous system; the incidence of such deficiencies is particularly high among alcoholics. Good sources of B_6 are whole grain cereals, pork, organ meats, and bananas. There is some evidence that excessive intake of vitamin B_6 can lead to a dependency condition with subsequent withdrawal symptoms.

Folic Acid (Folacin)

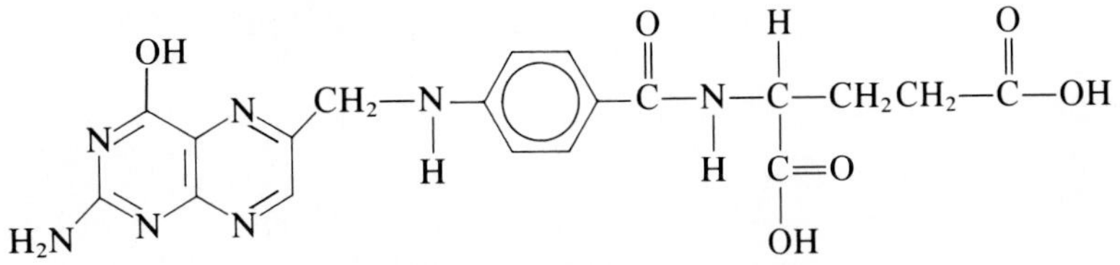
OH
N
N
CH2—N
H
O
C—N—C—CH2CH2—C—OH
H
H
C=O
OH
O
H2N
N
N

Folic acid

Folic acid is the parent component of the folacin group of vitamins. Folic acid deficiency is thought to be the most common vitamin deficiency in the

United States. This may be due to the increased amount of processed food in our diet and the instability of this vitamin in food processing. Typically 50 to 95 percent of folic acid is destroyed by heating or other food processing. Folic acid is involved in the synthesis of the heme unit (hemoglobin) and the formation of red blood cells. Folic acid deficiency is characterized by anemia, weight loss, and gastrointestinal disorders. Good sources of folic acid are green leafy vegetables, organ meats, legumes, and whole grains.

Vitamin B_{12}

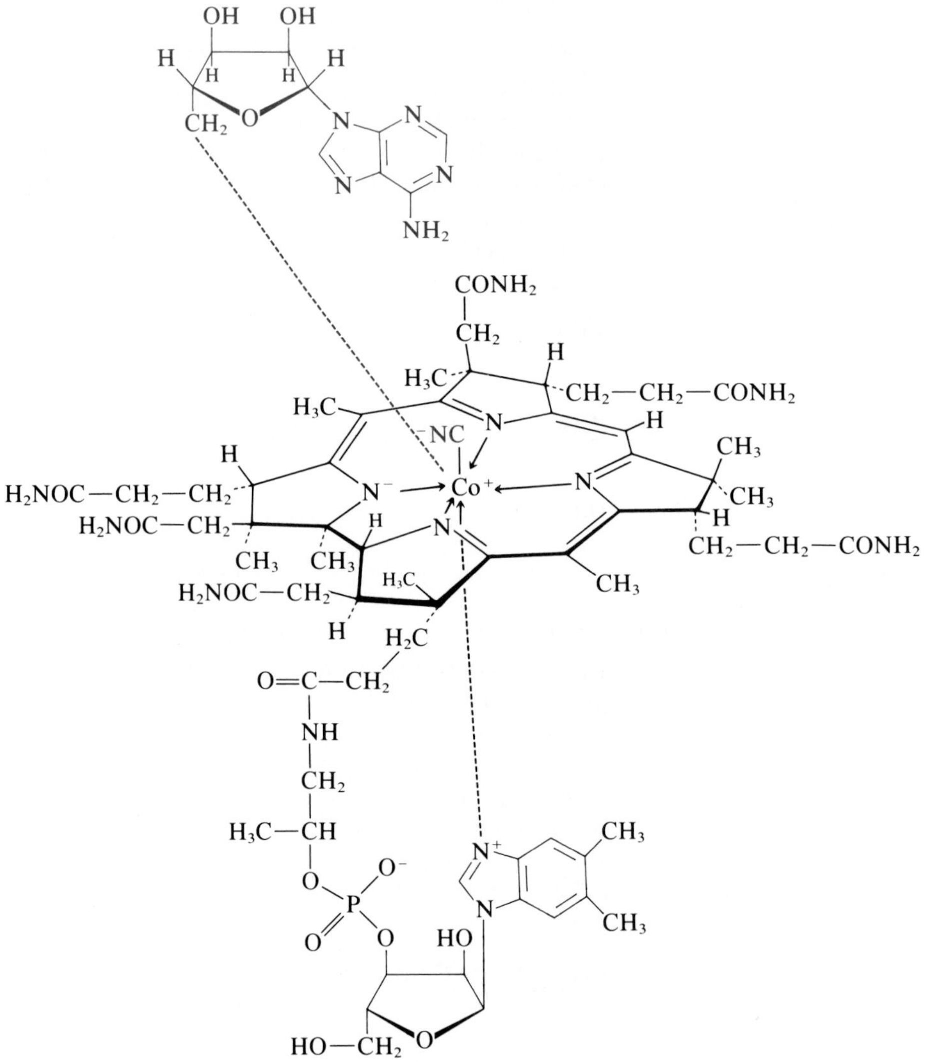
CONNH₂
CH₂—CH₂—CONH₂
H₂NOC—CH₂—CH₂
H₂NOC—CH₂
H₂NOC—CH₂
NC
Co⁺
CH₂—CH₂—CONH₂
NH₂
HO—CH₂

Vitamin B_{12}

As you can see, vitamin B_{12} has the most complex structure of the vitamins. It is generally isolated in natural sources with the cyano group (shown in color) bound to cobalt. In the coenzyme form the other group shown in color replaces —CN. Both vitamin B_{12} and folic acid are required for the proper formation of cells. Thus vitamin B_{12} deficiency also has as its major symptom anemia, a low red blood cell count. Dietary deficiency of B_{12} is rare because the needed intake is very small (3 μg) and sufficient amounts are normally stored in the liver and can last several years. However, vitamin B_{12} cannot be made by either animals or plants; it is found in foods of animal origin, where it was originally formed by microorganisms. Individuals who have been strict vegetarians for a number of years may show deficiency symptoms. However, even when meat is rigorously excluded from the diet, adequate B_{12} may be obtained from human intestinal bacteria.

A serious medical problem involving vitamin B_{12} is pernicious anemia, an anemic condition accompanied by neurological damage. This condition is not due to a dietary insufficiency of B_{12}, but a malfunction in the absorption of B_{12} from the intestine. Victims of pernicious anemia lack a glycoprotein required in the B_{12} absorption process. Pernicious anemia is treated by direct injection of vitamin B_{12}.

Biotin

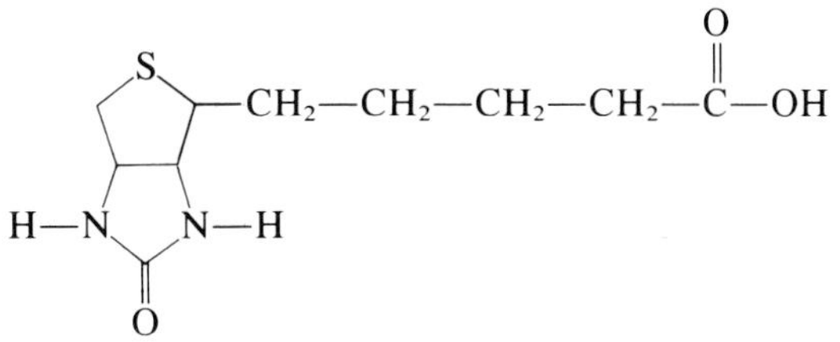
S
CH₂—CH₂—CH₂—CH₂—C—OH
O
H—N
N—H
O

Biotin

Biotin is present in a wide variety of foods, and a deficiency of this vitamin is very rare. However, a deficiency can occur if the diet includes a very large number of raw egg whites (more than 20 per day). Raw egg white contains a protein which binds to biotin and prevents it from being absorbed from the intestine. Cooked egg white does not cause the deficiency. (What do you think happened to the protein upon cooking?) Symptoms of biotin deficiency include dermatitis, anorexia, and muscle pain. Biotin is found in milk, liver, egg yolk, mushrooms, and legumes.

Pantothenic Acid

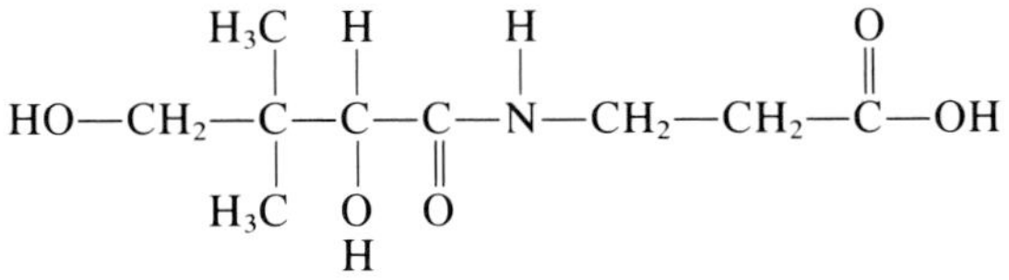
H₃C H H O
HO—CH₂—C—C—C—N—CH₂—CH₂—C—OH
H₃C O O
H

Pantothenic acid

No specific deficiency symptoms of pantothenic acid are known in human beings. Perhaps this is a consequence of the fact that pantothenic acid is found in all plants and animals. Indeed the vitamin gets its name from the Greek word *pantos,* meaning "everywhere." Contrary to some claims, there is no evidence that pantothenic acid can prevent the appearance of gray hair.

Ascorbic Acid (Vitamin C)

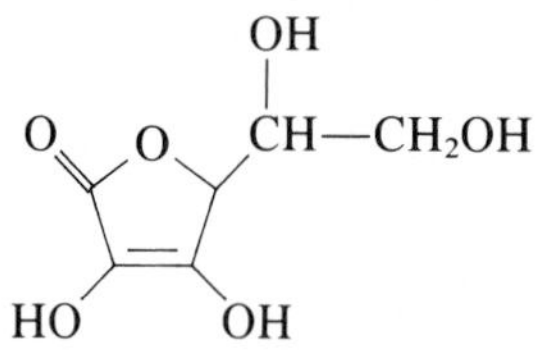
OH
O
O
CH—CH₂OH
HO
OH

Ascorbic acid

A deficiency of vitamin C in the diet results in scurvy, a disease characterized by weakness, swollen joints, bleeding gums and loose teeth, and delayed wound healing. In 1735 James Lind was the first to connect the symptoms of scurvy with a dietary deficiency. Lind found that sailors provided with two oranges and lemons daily were cured of the disease. After citrus fruits were added to their diet, British sailors received the nickname "limeys."

Vitamin C, in contrast to all the other water-soluble vitamins, does not function as a coenzyme. Its full role in the body is not understood, but it does appear to be required in the synthesis of collagen. Ascorbic acid plays a role in wound healing, and physicians often recommend doses of vitamin C in postsurgical and burn patients. Vitamin C also promotes the absorption of iron from dietary foods.

Since the original publication of *Vitamin C and the Common Cold* by Linus Pauling in 1970, there has been much interest in the relationship, if any, between megadoses of ascorbic acid and the prevention of colds. Pauling claims that 1 to 2 g per day of vitamin C (RDA = 60 mg) will aid in the prevention of common colds, and 4 to 10 g per day will help relieve an already present cold. Although the final verdict has not been pronounced, the experimental results to date do not appear to support Pauling's hypothesis. Other claims have been made for megadoses of vitamin C as a preventive measure against heart disease and cancer. Here again experimental evidence supporting these claims is lacking. Some contraindications to megadoses of ascorbic acid include diarrhea, possible induction of kidney stones, and possible vitamin C dependency in infants whose mothers took large doses of ascorbic acid during pregnancy.

26.9 THE FAT-SOLUBLE VITAMINS AS MICRONUTRIENTS

Vitamins A, D, E, and K constitute the fat-soluble vitamins.

Vitamin A

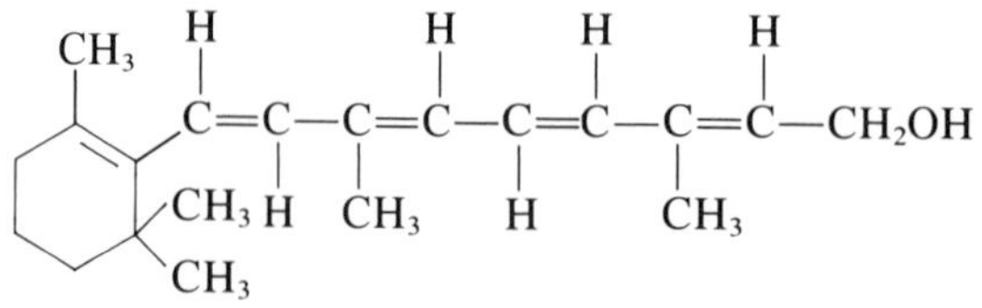
CH₃
H
H
H
H
C=C—C=C—C=C—C=C—CH₂OH
CH₃ H
CH₃
H
CH₃
CH₃

Vitamin A

Vitamin A is found in animal tissue; it can also be formed in the body from the carotenoid pigments which give the yellow-orange color to many fruits and vegetables. Vitamin A is one of the two chemical substances making up the rods, the structure in the eye responsible for vision in dim light (see Section 13.7). Deficiency of vitamin A results in a condition known as night blindness, wherein normal vision in dim light returns very slowly after a bright light is observed. The chemical name for the structure of Vitamin A shown is retinol.

Vitamin A is also necessary for the maintenance of epithelial tissue, which includes cells on the body surface as well as of the gastrointestinal, respiratory, and urogenital tracts of the body. Deficiency of this vitamin can result in a condition called xerophthalmia in which excess keratin, a structural protein, forms in the epithelial cells; the cells become excessively dry and hard. When this occurs in the eye, the cornea becomes opaque and permanent blindness can result if the condition is not immediately treated with vitamin A. Night blindness is an early sign of vitamin A deficiency.

Good sources of vitamin A are liver and yellow, orange, and dark green vegetables and fruits. Because it is fat-soluble, vitamin A is stored in the liver. A supplemental vitamin pill containing 30 to 60 mg of vitamin A would provide sufficient vitamin A for a child for about 6 months. Daily consumption of such pills clearly constitutes an overdose. Persons who daily consume large doses of vitamin A (7.5 mg) may show such toxicity symptoms as fatigue, headache, joint pains, loss of hair, and irregular menstruation.

Vitamin D

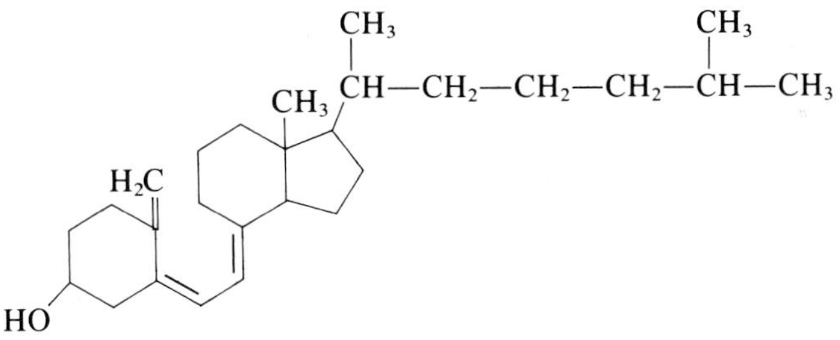
CH3
CH3
CH3
CH—CH2—CH2—CH2—CH—CH3
H2C
HO

Vitamin D_3 (cholecalciferol)

Two forms of vitamin D are active in the body: ergocalciferol (vitamin D_2) and cholecalciferol (vitamin D_3). Cholecalciferol can be formed from the reaction of the ultraviolet rays of sunlight on 7-dehydrocholesterol, a substance present in the skin.

$$\text{7-Dehydrocholesterol} \xrightarrow[\text{in sunlight}]{\text{uv radiation}} \text{cholecalciferol (vitamin } D_3\text{)}$$

Aside from growing children and pregnant or lactating women, individuals who are regularly exposed to sunlight do not require vitamin D as part of their dietary intake.

The body converts cholecalciferol to dihydroxycholecalciferol, a hormone secreted by the kidneys; it regulates the level of calcium in the blood and in bones. When vitamin D is deficient, bones do not retain calcium properly and become soft and misshapen. In children this condition is called rickets and is characterized by bowed legs because the weight of the body deforms inadequately calcified bone. Osteomalacia is the adult counterpart of rickets; decalcified bone fractures more readily. Neither rickets nor osteomalacia is a result of an inadequate calcium intake, but rather is caused by a defect in the body's calcium regulation.

Cases of rickets are rare in the United States because cow's milk is enriched with vitamin D. Pregnant women are perhaps most susceptible to osteomalacia. Natural sources of vitamin D are cod liver oil, egg yolk, tuna, and salmon.

Excess vitamin D can be toxic. Continuous ingestion of high doses (25 to 75 μg per day) can lead to high blood calcium levels and deposition of this calcium in blood vessels and the kidneys. Cases of irreversible kidney damage have been reported.

Vitamin E

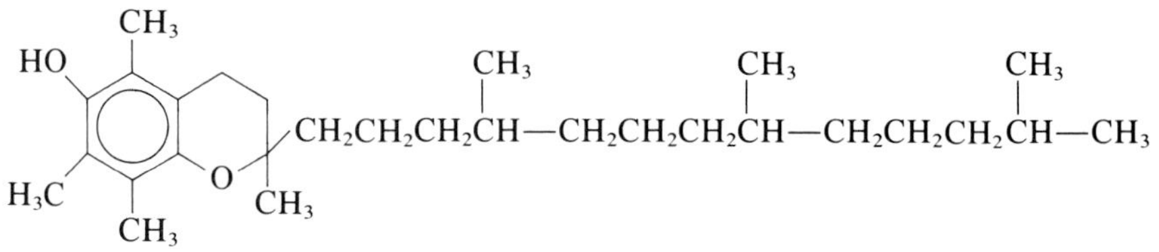
CH3
HO
CH3
CH3
CH3
CH2CH2CH2CH—CH2CH2CH2CH—CH2CH2CH2CH—CH3
O
H3C
CH3
CH3

Vitamin E (α-tocopherol)

Vitamin E is a group of substances, the most potent of which is α-tocopherol. The exact function of vitamin E in the body is not known. However, in test tubes with oxidizing agents, α-tocopherol acts as an antioxidant. That is, since vitamin E is readily oxidized at the phenol-type hydroxyl group, other substances in a mixture are protected from being oxidized. It is believed that vitamin E in the body may prevent oxidation of polyunsaturated fatty acids, sulfur-containing enzymes, and lipids making up cell membranes. Some researchers have theorized that the aging of a cell may be accelerated by intermediate products known as free radicals, which tend to form in these oxidation reactions. These researchers conclude that vitamin E, acting as an antioxidant, may be able to prevent or at least slow down the aging process. However, the evidence to support the theory is still inconclusive.

Some of the popular benefits that have been attributed to vitamin E include increased sexual potency, cure of male infertility, disappearance of wrinkles, and prevention of cancer and heart disease. There are no known controlled experiments that support any of these claims.

Premature infants sometimes show a vitamin E deficiency because of poor placental transport and storage. The deficiency results in anemia due to an excess breakage of red blood cells. However, vitamin E deficiency in adults is rare. Indeed, in test subjects even after 3 years of a diet designed to be low in vitamin E, no deficiency symptoms were observed. Vitamin E is found in many foods and is particularly concentrated in vegetable oils.

Large doses of vitamin E are not believed to present the same toxicity problems as vitamins A and D. However, it is fat-soluble, and the long-term accumulation effects are simply not known.

Vitamin K

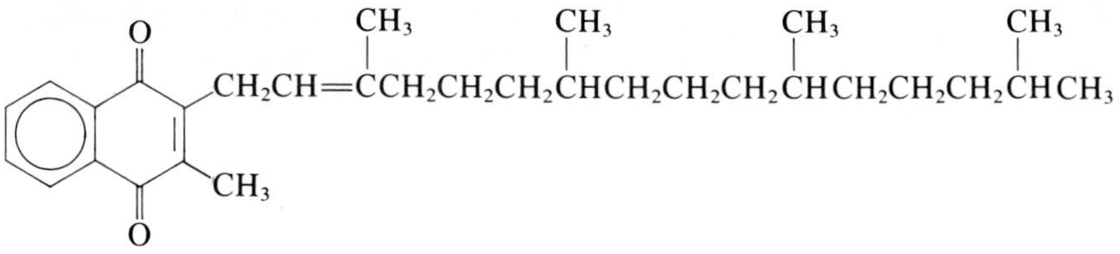

Vitamin K

Vitamin K is required for proper blood clotting; it has been found to serve as a coenzyme in the formation of prothrombin. Properly formed prothrombin can bind calcium ion and be converted to thrombin, the enzyme required for the conversion of fibrinogen into fibrin, the actual protein material of blood clots. Vitamin K is sometimes given to newborn infants, or gallbladder or liver surgical patients to assure adequate prothrombin levels and prevent hemorrhaging problems.

Deficiencies of vitamin K are rare, possibly because it is formed by bacteria in the human intestine. The only known symptom is improper blood coagulation.

Food sources of vitamin K include leafy vegetables, egg yolk, and liver. Jaundice and anemia in infants are the only known toxic symptoms of large doses of vitamin K.

26.10 MINERALS AS NUTRIENTS

Good nutrition requires the dietary intake of a number of elements in addition to vitamins and the elements C, H, N, O, and S, which make up the essential amino acids. These are often referred to as the nutritional **mineral** requirements and are listed in Table 26.9. These elements are ingested in their simple ionic form (e.g., Na^+) or as part of a polyatomic ion. The required minerals can be classified as macronutrients or micronutrients depending on whether the necessary intake exceeds 100 mg per day.

TABLE 26.9 Elements Required as Macronutrients or Micronutrients

Macronutrients	Micronutrients
Calcium	Arsenic*
Chloride	Cadmium*
Magnesium	Chromium
Phosphorus	Cobalt*
Potassium	Copper
Sodium	Fluoride
	Iodine
	Iron
	Manganese
	Molybdenum
	Nickel*
	Selenium
	Silicon*
	Tin
	Vanadium*
	Zinc

* These elements are known to be essential in animals and are thought to be necessary for human beings.

Minerals serve in a variety of functions: they may be cofactors to complete an active enzyme, they are an integral part of the principal compounds making up bone structure, and they are regulators of the fluid-electrolyte balance within the body. In most cases, a clear deficiency symptom results from the absence or insufficient intake of a necessary mineral.

Sodium, Potassium, Chloride

Sodium is the principal cation in the extracellular fluid. Along with the balancing anion, chloride, sodium plays the major role in the maintenance of osmotic pressure between intra- and extracellular fluids. To maintain a normal osmotic pressure, a complex system is employed by the body. During periods of low Na^+ and Cl^- intake, the kidneys decrease the amount of Na^+ and Cl^- passing into the urine. High Na^+ retention in body fluids is accompanied by water retention, which can lead to edema.

Potassium is the principal intracellular cation. The maintenance of Na^+ outside the cell and K^+ ions inside the cell is necessary for proper transmission of nerve impulses. Although most K^+ is found in the cell, a small amount (156 mg/L) resides in the serum, and this serum K^+ is required for cardiac function. Diuretic pills, used by individuals who suffer from hypertension, prevent reabsorption of Na^+ in the kidneys, resulting in an increased urinary excretion of Na^+ and water. However, K^+ is also excreted, and individuals consuming diuretics must have their serum K^+ levels monitored to avoid cardiac problems.

Na^+ and Cl^- ion deficiencies are very rare because sodium chloride is so ubiquitous. Indeed, the modern problem is probably an excess (greater than 3 g per day) of Na^+ in the diet. Excess Na^+ may be a significant risk factor in the development of hypertension and its associated cardiovascular and kidney ailments.

Calcium and Phosphorus

The insoluble mineral calcium hydroxyapatite contains both calcium and phosphorus and is the major structural component of bones and teeth. Serum calcium functions in a number of roles, including muscle contraction, blood coagulation, enzyme activation, and cardiac function. A few of the critical roles of phosphorus are in the formation of RNA, DNA, ATP, and cellular membranes.

A dietary deficiency of calcium, particularly among older people, may be a cause of osteoporosis, a condition in which bones decrease in size and become increasingly fragile. Recent research indicates that low intake of Ca^{2+} may be correlated with a higher incidence of hypertension. Milk and milk products are the best natural sources of calcium.

Dietary deficiencies of phosphorus are unknown. However, as with calcium, there are sometimes absorption problems which may reflect a vitamin D deficiency. Milk and milk products, legumes, cereals, meat, fish, and poultry are all rich in phosphorus.

Magnesium

Magnesium is an essential cofactor in enzymes involved in phosphate transfer reactions and in many of the glycolysis reactions. It is also a component of bone structure. Magnesium deficiencies can occur through malabsorption of the ion or through fluid loss such as occurs in chronic diarrhea or in urine excretion in individuals taking diuretics. Tremors, mental disorientation, and rapid heartbeat are some symptoms of magnesium deficiency.

Micronutrient Minerals

The physiological role, known deficiency symptoms, and food sources for the minerals required in amounts less than 100 mg per day are given in Table 26.10.

TABLE 26.10 Micronutrient Minerals in the Diet

Mineral	Physiological Role	Deficiency Symptom	Food Source
Chromium	Necessary for insulin action in metabolism of glucose	Insulin less effective, diabetes-like condition	Meat, milk, and whole grains
Cobalt	Formation of vitamin B_{12}	Same as vitamin B_{12}	Same as vitamin B_{12}
Copper	Cofactor in many enzymes, such as cytochrome C oxidase; also required in synthesis of hemoglobin	Similar to iron; anemia, low white cell count	Liver, shellfish, dried fruits, nuts
Fluoride	Protection against dental caries	Increased incidence of dental caries	Fluoridated drinking water, seafood, tea
Iodine	Component of thyroid hormones thyroxine and triiodothyronine	Goiter	Seafood, iodized salt
Iron	Constituent of hemoglobin and myoglobin; also a cofactor in the cytochrome enzymes	Anemia (red cells small and pale)	Meat, poultry, fish, grains, fruit, eggs
Manganese	Cofactor in enzymes such as arginase, which hydrolyzes arginine to urea	Decrease in glucose tolerance, linked with some forms of epilepsy	Nuts, seeds, and whole grains
Molybdenum	Cofactor in xanthine oxidase, enzyme responsible for conversion of xanthine to uric acid	Unknown in human beings; for animals, loss of weight	Meats, grains, legumes
Nickel	Involved in fixation of CO_2 in bacteria, cofactor in urease	Unknown in human beings; various pathological conditions in animals; may stabilize RNA and DNA against thermal decomposition	Grains, vegetables, oysters
Selenium	Cofactor for enzyme glutathione peroxidase, which protects red blood cells against peroxidation	Unknown in human beings, although some believe deficiency may cause fibrosis; animals show degeneration of pancreas	Meat, seafood
Silicon	Bone calcification	Unknown in human beings, but essential for proper growth in rats and chicks	Grains, brown sugar, organ meats
Tin	Proper development of skeletal system	Growth and reproduction in animals; unknown in human beings	Grains, foods preserved in tin cans
Vanadium	Cofactor in flavin dehydrogenases	Unknown in human beings; animals show impaired growth, reproduction, and lipid metabolism	Seafoods, grains, seeds
Zinc	Cofactor in a number of enzymes such as DNA polymerase and carbonic anhydrase; also a component of insulin required in protein synthesis	Diminished taste sensation, anorexia, growth retardation and sexual immaturity, poor wound healing	Red meat, seafood

CHAPTER ACCOMPLISHMENTS

After completing this chapter, you should be able to define **key words** and do the following:

26.1 INTRODUCTION

26.2 NUTRITIONAL STANDARDS

1. Use Tables 26.1 and 26.2 to list the RDA value for protein or a particular vitamin or mineral in a given individual.
2. Distinguish between RDA and USRDA.

26.3 ENERGY

3. State and distinguish among the three forms of energy consumption in the body.
4. Calculate BMR for an individual, given the oxygen consumption in a particular time period.
5. State the relationship between caloric intake and weight gain or loss.

26.4 PROTEINS AS MACRONUTRIENTS

6. Give two reasons for proteins being considered a macronutrient.
7. Give examples of positive and negative nitrogen balance conditions.
8. State why the recommended quantity of protein per day is greater than the experimentally measured necessary amount.
9. Calculate the NPU of a protein, given the grams of nitrogen in a fixed quantity of that protein and the grams of nitrogen retained from eating a fixed quantity of the protein.
10. Describe the concept of protein complementarity.
11. State and distinguish between the two forms of protein-energy malnutrition.

26.5 CARBOHYDRATES AS MACRONUTRIENTS

12. Give general reasons for the maintenance of carbohydrate in the diet.

26.6 LIPIDS AS MACRONUTRIENTS

13. Name the two essential polyunsaturated acids.
14. State the hypothesis for the relationship between lipoprotein levels and heart disease.
15. Describe the relationship between polyunsaturate consumption and lipoprotein levels.

26.7 VITAMINS AS MICRONUTRIENTS

16. List the properties a substance must have to be considered a vitamin.
17. State the two major subdivisions of vitamins.
18. Explain why fat-soluble vitamins have a greater potential of toxicity in large doses than water-soluble vitamins.
19. State the general function common to all water-soluble vitamins except C.

26.8 THE WATER-SOLUBLE VITAMINS AS MICRONUTRIENTS

20. List all the water-soluble vitamins.
21. State the deficiency condition associated with a given water-soluble vitamin.
22. Indicate the vitamin thought to be associated with the largest number of cases of vitamin deficiency.
23. Describe the relationship between pernicious anemia and vitamin B_{12}.

26.9 THE FAT-SOLUBLE VITAMINS AS MICRONUTRIENTS

24. List the fat-soluble vitamins.
25. State the deficiency condition associated with a given fat-soluble vitamin.

26.10 MINERALS AS NUTRIENTS

26. State three possible roles for minerals in the body.
27. List the minerals which serve as macronutrients.
28. Distinguish between the principal locations of Na^+ and K^+ in the body.
29. State the medical problem that may be associated with excessive Na^+ intake.
30. State a medical condition which may be associated with Ca^{2+} deficiency.

PROBLEMS

26.3 What general concerns should a nutritionist consider in recommending a diet for a healthy 10-year-old child?

26.4 Give a definition and an example for each of the following:
a Nutrient
b Macronutrient
c Micronutrient

26.5 *a* How has the Food and Nutrition Board designed the RDAs so that they will provide good nutrition for practically all healthy people in the United States?
b Do the RDAs give nutritional requirements for any particular individual? Explain.

26.6 *a* What is the RDA of riboflavin in a 17-year-old male?
b What is the RDA of iodine in a 25-year-old female?
c What is the RDA of iodine in a 25-year-old pregnant female?

26.7 *a* What nutritional information is given by the table of RDAs?
b What necessary nutritional information is *not* given by the table of RDAs?

26.8 *a* How are the USRDAs obtained?
b What is the USRDA for vitamin D?
c What is the USRDA for magnesium?

26.9 The nutrition information on a cereal package indicates that 1 oz supplies 25 percent of the USRDA for thiamine. How many milligrams of thiamine are present in 1 oz of this cereal?

26.10 List the three categories of energy consumption in the body.

26.11 State at least five physiological activities that contribute to the BMR.

26.12 Calculate the BMR of an individual who under basal test conditions consumed 1.2 L of oxygen in 6 min.

26.13 *a* How long would a 150-lb person have to run at 10 mi/h to lose 1 lb of body weight? Use Table 2.6.
b How much wood chopping would this individual have to do to lose the number of calories gained in eating a hamburger? Use Tables 2.5 and 2.6.

26.14 What would be the result if the body stored its excess calories in the form of carbohydrate, not fat?

26.15 After returning from a half hour of jogging, one of the authors found he had lost 2 lb according to scale weight. Had he really lost 2 lb of "body" weight? What was really lost? Will it stay off?

26.16 What physiological needs are satisfied by the dietary intake of proteins?

26.17 *a* What condition is satisfied when an individual is in a state of nitrogen balance?
b In what types of individuals would you expect to observe a positive nitrogen balance?
c What conclusions about a person's health may you come to upon observing a negative nitrogen balance?

26.18 An individual living on a diet which is adequate in quantity of protein but deficient in one of the essential amino acids shows a negative nitrogen balance. Explain.

26.19 In which of the following legumes is the protein used most efficiently by the body?
a Soybeans, NPU = 61
b Peas, NPU = 48
c Lima beans, NPU = 52
d Navy beans, NPU = 38

26.20 *a* What is meant by an incomplete protein?
b What is meant by the limiting amino acid in a protein?

26.21 What food(s) would you recommend to a person whose diet consisted predominantly of nuts and seeds?

26.22 *a* Distinguish between the causes of marasmus and kwashiorkor.
b Distinguish between the symptoms of marasmus and kwashiorkor.

26.23 Why is there no RDA for carbohydrates?

26.24 Give at least two reasons for recommending an increased percent consumption of complex carbohydrates in the diet.

26.25 Give at least two reasons for recommending *against* a no-carbohydrate diet.

2.26 What disease condition has been definitely linked to the high consumption of simple sugars?

26.27 How can we determine the effect a food will have on the blood glucose level?

26.28 *a* What food class contributes fiber to the diet?
b Give examples of compounds which make up dietary fiber.

26.29 Why has no RDA been set for lipids?

26.30 *a* What two polyunsaturated acids cannot be synthesized by human beings?
b Are both necessary in the human diet?

26.31 *a* What desirable function is served by the high-density lipoproteins?
b State one way of increasing the percent of HDLs relative to LDLs.

26.32 *a* What changes in the diet are recommended to reduce blood cholesterol levels?
b If the changes in part *a* are made, will there necessarily be a reduction in heart disease?

26.33 *a* Give the definition of a vitamin.
b If a substance is a vitamin for one species,

is it necessarily a vitamin for all species? Explain.

c What are the two general classes of vitamins?

d Which vitamin has the highest RDA in Table 26.1? What is the value of that RDA?

26.34 Give the names of all the water-soluble vitamins.

26.35 Of the water-soluble vitamins, state which ones match the following descriptions:

a Contains a pyridine ring

b Is a carboxylic acid

c Contains an amide linkage

26.36 Give the name of the disease condition associated with a deficiency of each of the following vitamins:

a Vitamin B_{12} *c* Vitamin C

b Niacin *d* Thiamine

26.37 A student purchased a multivitamin supplement. Each tablet contained the following quantities of vitamins:

Thiamine, 10 mg	Vitamin B_{12}, 10 μg
Riboflavin, 10 mg	Vitamin C, 250 mg
Niacin, 100 mg	Vitamin A, 10,000 μg
Vitamin B_6, 10 mg	Vitamin E, 200 mg

Has the student made a wise choice for a vitamin supplement? Why or why not?

26.38 *a* Which vitamin deficiency is thought to be the most common in the United States?

b Which vitamin in corn is unreleased in the human digestive tract?

c Which vitamin is found in all plants and animals?

26.39 Consumption of a large number of raw eggs can lead to biotin deficiency, but consumption of the same number of cooked eggs does not lead to this result. Explain.

26.40 Give the names of the fat-soluble vitamins.

26.41 *a* In general, would you describe the fat-soluble vitamins as polar or nonpolar?

b What general differences in structure can you observe between water-soluble and fat-soluble vitamins?

26.42 Give the name of the disease condition associated with a deficiency of each of the following vitamins:

a Vitamin A *b* Vitamin D

26.43 What chemical property of vitamin E has provoked interest in its action as an antiaging pill? Explain.

26.44 Why are megadoses of the fat-soluble vitamins considered more dangerous than megadoses of the water-soluble vitamins?

26.45 Potassium is one of the required minerals. Does this mean that we should be eating potassium metal every day? Explain.

26.46 State three functions that minerals may serve in the body.

26.47 Explain why people who take diuretic pills need to have their K^+ levels carefully monitored.

26.48 Are deficiencies of Na^+ or excesses of Na^+ more of a medical problem? Why?

26.49 What is the difference between a dietary deficiency of a certain nutrient and an absorption problem of that nutrient?

26.50 What difference is there between a mineral classified as a macronutrient and one listed as a micronutrient?

26.51 In what ways are the needs for calcium and phosphorus interlocked?

26.52 What function do each of the following minerals serve?

a Iodine *c* Manganese

b Cobalt *d* Magnesium

26.53 Name the mineral that may be deficient in each of the following conditions:

a Osteoporosis *d* Loss of taste, anorexia

b Anemia

c Goiter

26.54 An individual placed on a high-protein diet has a higher than normal requirement for vitamin B_6. Why?

APPENDIX

ANSWERS TO SELECTED PROBLEMS

CHAPTER 1

1.1 *a.* Mixture, *b.* mixture, *c.* pure substance, *d.* mixture
1.2 *a.* Chemical, *b.* physical, *c.* physical, *d.* chemical
1.3 Table 1.2 lists the boiling point of carbon tetrachloride as 76.5°C. Since the boiling point of chloroform is given as 61°C, an experimental measurement of the boiling points of the two unlabeled liquids would distinguish between them.
1.4 *a.* Mixture, *b.* compound, *c.* element, *d.* compound
1.5 *b, c, d*
1.7 Matter is everything that occupies space and has mass.
1.9 Classified into pure substances (constant composition) and mixtures (variable composition)
1.11 *a.* Mixture, *b.* pure substance, *c.* mixture, *d.* pure substance, *e.* pure substance, *f.* mixture, *g.* mixture, *h.* pure substance, *i.* mixture
1.13 An element
1.15 *a.* Chemical, *b.* physical, *c.* chemical, *d.* chemical, *e.* physical, *f.* chemical, *g.* chemical, *h.* chemical
1.17 Two: elements and compounds
1.19 Compounds; compounds can be formed from various combinations of elements.
1.21 Melting point, boiling point, burns in air?, etc.
1.23 Energy
1.25 *a.* Potential, *b.* kinetic, *c.* kinetic, *d.* potential, *e.* potential
1.27 K, Na, Cl, Fe, H, Ag, Au
1.29 *a.* Correct, *b.* Al (the ell must be small), *c.* numbers in formulas are subscripts; R is not an element symbol, *d.* correct, *e.* H (must be capital)
1.31 *a.* 4, *b.* 1, *c.* 2, *d.* 2, *e.* 5

CHAPTER 2

2.1 *a.* $\frac{1\text{ L}}{1000\text{ mL}}, \frac{1000\text{ mL}}{1\text{ L}}$; *b.* $\frac{1\text{ kg}}{2.20\text{ lb}}, \frac{2.20\text{ lb}}{1\text{ kg}}$
2.2 34 mL
2.3 0.601 lb
2.4 1.6 g/mL
2.5 30. mL
2.6 20°C
2.7 3 cal
2.8 *a.* 4, *b.* 4, *c.* 3, *d.* 1
2.9 25
2.10 1032.91
2.11 *a.* 1.89×10, *b.* 1.3×10^{-2}, *c.* 5.342×10^{6}, *d.* 5.24×10^{-6}, *e.* $7.73 \times 10^{+2}$, *f.* 7.2
2.12 *a.* 11,900; *b.* 0.0873; *c.* 593,000,000; *d.* 0.00000219

2.13 kilo: one thousand times; deci: one-tenth of; centi: one-hundredth of; milli: one-thousandth of; micro: one-millionth of

2.15 *a.* $\frac{1000\ g}{1\ kg}, \frac{1\ kg}{1000\ g}$; *b.* $\frac{1\ L}{1000\ mL}, \frac{1000\ mL}{1\ L}$; *c.* $\frac{1\ m}{100\ cm}, \frac{100\ cm}{1\ m}$; *d.* $\frac{1{,}000{,}000\ \mu m}{1\ m}, \frac{1\ m}{1{,}000{,}000\ \mu m}$; *e.* $\frac{1\ mL}{1\ cm^3}, \frac{1\ cm^3}{1\ mL}$

2.17 *a.* 6000 (6×10^3) cm, *b.* 7.5 cm, *c.* 0.004330 (4.330×10^{-3}) kg, *d.* 0.610 m, *e.* 0.00148 (1.48×10^{-3}) lb, *f.* 410 (4.1×10^2) mL

2.19 *a.* 35 mL, *b.* 908 g, *c.* 1890 (1.89×10^3) mL, *d.* 114 g, *e.* 2740 (2.74×10^3) mm, *f.* 56 ft

2.21 91.4 m

2.23 *a.* 1800 (1.8×10^3) m, *b.* 1.1 mi

2.25 89 km/h

2.27 *a.* 1.5×10^{13} cm, *b.* 5.0×10^2 s

2.29 0.7 g

2.31 1.039

2.33 *a.* 13 g, *b.* 475 g, *c.* 5.8 g

2.35 500 (5×10^2) g

2.37 *a.* 2.14 lb, *b.* 28.3 lb

2.39 −15°C

2.41 −90°C

2.43 2.8×10^3 cal

2.45 0.12 cal/(g·°C)

2.47 *a.* 103°F, *b.* 86 kg, *c.* 4.3 g, *d.* 170 mL

2.49 *a.* 10., *b.* 0.02, *c.* 0.002, *d.* 14.3, *e.* 0.446

2.51 *a.* 6.554 lb, *b.* 6.55 lb

2.53 28.9 mL

2.55 *a.* 9.0×10^{14}, *b.* 1.46×10^{-5}, *c.* 3.0×10^3, *d.* 5.0×10^{-12}

CHAPTER 3

3.1 Compounds are pure substances with a constant composition because atoms are combined in simple, whole-number ratios. These ratios lead to a fixed proportion by weight. Chemical changes can disrupt the combination of atoms.

3.2 *a.* 13, 30, 33; *b.* 13, 30, 33

3.3 *a.* Br, *b.* Mg

3.4 3, 15, 80

3.5 *a.* 11, 5, 26; *b.* sodium, boron, iron

3.6 17

3.7

	Proton	Neutron	Electron
a.	17	20	17
b.	12	12	12
c.	19	20	19

3.8 *a.*

	$^{24}_{11}Q$	$^{24}_{12}Q$	$^{75}_{35}Q$	$^{25}_{12}Q$	$^{81}_{35}Q$
Protons	11	12	35	12	35
Neutrons	13	12	44	13	46
Electrons	11	12	35	12	35

b. Sodium-24, magnesium-24, bromine-79, magnesium-25, bromine-81; *c.* $^{24}_{12}Q$ and $^{25}_{12}Q$ isotopes of Mg, $^{79}_{35}Q$ and $^{81}_{35}Q$ isotopes of Br

3.9 2×12 amu = 24 amu

3.10 $^{239}_{94}Pu \rightarrow {}^{4}_{2}He + {}^{235}_{92}U$
3.11 $^{32}_{15}P \rightarrow {}^{0}_{-1}e + {}^{32}_{16}S$
3.12 0.13 g
3.13 $^{242}_{96}Cm + {}^{4}_{2}He \rightarrow {}^{1}_{0}n + {}^{245}_{98}Cf$
3.15 All atoms have protons and neutrons in the nucleus and electrons outside the nucleus.

3.17

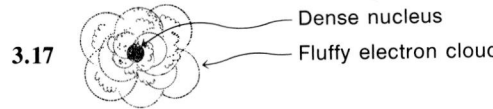

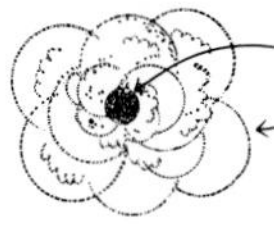
Dense nucleus
Fluffy electron cloud

3.19 Electron
3.21 Proton ≈ neutron ≈ 1 amu; electron ≈ 0
3.23 Protons

3.25

	p^+	n^0	e^-
a.	26	30	26
b.	15	16	15
c.	5	6	5

3.27

Symbol	Protons	Neutrons	Electrons
$^{27}_{13}Al$	13	14	13
$^{79}_{35}Br$	35	44	35
$^{40}_{20}Ca$	20	20	20
$^{84}_{36}Kr$	36	48	36

3.29 $^{24}_{11}X$ and $^{23}_{11}X$ are isotopes of sodium.
3.31 *a.* $^{15}_{7}N$, *b.* $^{14}_{7}N$

3.33

	Ion	Protons	Electrons	Charge
a.	Mg	12	10	+2
b.	F	9	10	−1
c.	S	16	18	−2
d.	Au	79	76	+3

3.35 *a.* Protium $^{1}_{1}H$, deuterium $^{2}_{1}H$, tritium $^{3}_{1}H$; *b.* tritium $^{3}_{1}H$
3.37 Spontaneous emission of energy

3.39

	Mass	Charge
α	4	+2
β	0	−1
γ	0	0

3.41 γ radiation is more penetrating. The curtain would stop the α radiation.
3.43 Radiation ionizes the argon gas, allowing Ar^+ ions to move to the cathode and electrons to the anode. This movement of charged particles produces an electric current in the connected wires and can be used to yield an audible click.
3.45 Drop in white blood cell count
3.47 $^{235}_{92}U \rightarrow {}^{4}_{2}He + {}^{231}_{90}Th$ or any α or β emission

3.49 Since the numbers of protons in the nucleus change, a new element is formed.

3.51 ${}^{42}_{19}\mathrm{K} \rightarrow {}^{0}_{-1}e + {}^{42}_{20}\mathrm{Ca}$ ${}^{131}_{53}\mathrm{I} \rightarrow {}^{0}_{-1}e + {}^{131}_{54}\mathrm{Xe}$ ${}^{14}_{6}\mathrm{C} \rightarrow {}^{0}_{-1}e + {}^{14}_{7}\mathrm{N}$

3.53 *a.* Beta, *b.* alpha, *c.* alpha, *d.* beta

3.55 Because it has a much shorter half-life

3.57 No

3.59 Because it has a shorter half-life

3.61 *a.* Fission, *b.* fusion, *c.* fission

CHAPTER 4

4.1 Less stable

4.2 Case A

4.3 Si: $1s^2\ 2s^2\ 2p^6\ 3s^2\ 3p^2$

4.4 Cs: $1s^2 2s^2 2p^6 3s^2 3p^6 4s^2 3d^{10} 4p^6 5s^2 4d^{10} 5p^6 6s^1$

4.5 Three

4.6 Eight, except helium, which has two

4.7 Na· :N· :Br· ·C· :Ne: Mg·

4.8 *a.* Group IA, *b.* alkali metals

4.9 Metals: Atomic size decreases across a period from left to right, and metals are on the left of each period.

4.10 *a.* Helium, *b.* francium

4.11 *a.* The apple is in a lower potential energy state on the ground. *b.* Water seeks a lower energy state. *c.* Close approach of negatively charged e^- to the positively charged nucleus is a low energy condition.

4.13 *s* orbitals are spherically symmetric; i.e., all orientations of a sphere in space are identical. *p* orbitals can be oriented in three different ways along the three cartesian axes.

4.15 *a.* H: $1s^2$, *b.* C: $1s^2\ 2s^2\ 2p^2$, *c.* O: $1s^2\ 2s^2\ 2p^4$, *d.* Na: $1s^2\ 2s^2\ 2p^6\ 3s^1$, *e.* Si: $1s^2\ 2s^2\ 2p^6\ 3s^2\ 3p^2$

4.17 *a.* B, *b.* K, *c.* Co, *d.* Ga, *e.* Kr

4.19

	MAIN LEVEL			
	1	2	3	4
$_1$H	1	0	0	0
$_2$He	2	0	0	0
$_3$Li	2	1	0	0
$_4$Be	2	2	0	0
$_5$B	2	3	0	0
$_6$C	2	4	0	0
$_7$N	2	5	0	0
$_8$O	2	6	0	0
$_9$F	2	7	0	0
$_{10}$Ne	2	8	0	0
$_{11}$Na	2	8	1	0
$_{12}$Mg	2	8	2	0
$_{13}$Al	2	8	3	0
$_{14}$Si	2	8	4	0
$_{15}$P	2	8	5	0
$_{16}$S	2	8	6	0
$_{17}$Cl	2	8	7	0
$_{18}$Ar	2	8	8	0
$_{19}$K	2	8	8	1
$_{20}$Ca	2	8	8	2

4.21

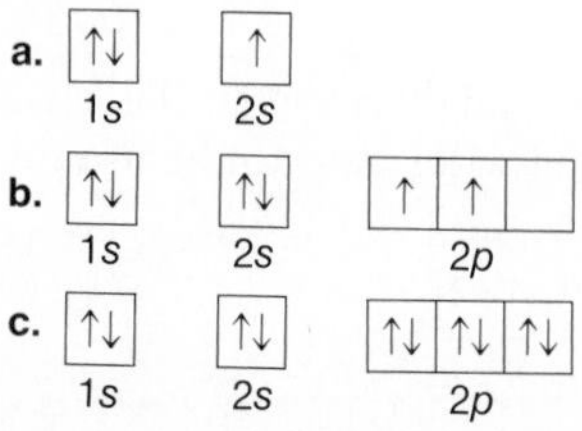
a.
1s
2s
b.
1s
2s
2p
c.
1s
2s
2p

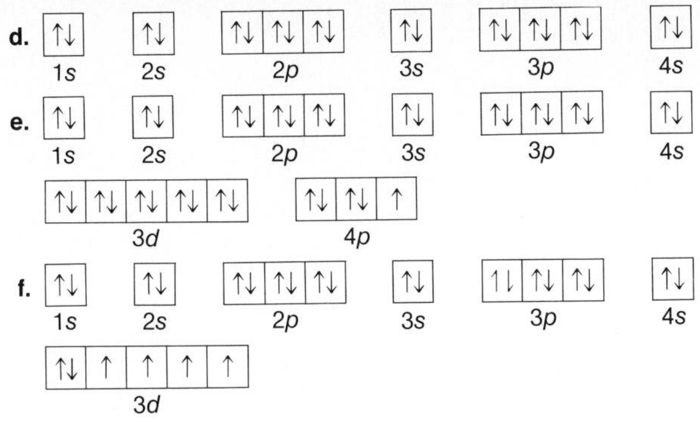
d.
1s
2s
2p
3s
3p
4s
e.
1s
2s
2p
3s
3p
4s
3d
4p
f.
1s
2s
2p
3s
3p
4s
3d

4.23 Absorbed, emitted
4.25 Light energy is emitted as excited electrons fall back to the ground state. The wavelength of light energy corresponds to that of yellow light.
4.27 Group VIA elements have the outer-level configuration $ns^2\ np^4$; Group VIIA elements have the outer-level configuration $ns^2\ np^5$.
4.29 Elements with atomic numbers 4, 20, and 38 are all in Group IIA.
4.31 *a.* 2, *b.* 8, *c.* 18
4.33 *a.* 7, *b.* 1, *c.* 4, *d.* 6
4.35 Group VIIA: fluorine, chlorine, bromine, iodine
4.37 F_2, Cl_2, Br_2, I_2, N_2, O_2, H_2
4.39 A Group B element
4.41 K· ·P· :Cl· ·Si· :Kr: Ca·
4.43 *a.* To the left; *b.* to the right, nonmetals; *c.* inert gases, highly stable
4.45 *a.* C, *b.* F, *c.* Cs, *d.* Cs
4.47 Atomic number, atomic weight, relative size, relative ionization energy, electron configuration, nature of element (metal or nonmetal), and physical and chemical properties

CHAPTER 5

5.1 *a.* 3 electrons, neon; *b.* 2 electrons, argon; *c.* Al^{3+} cation, S^{2-} anion
5.2 *a.*

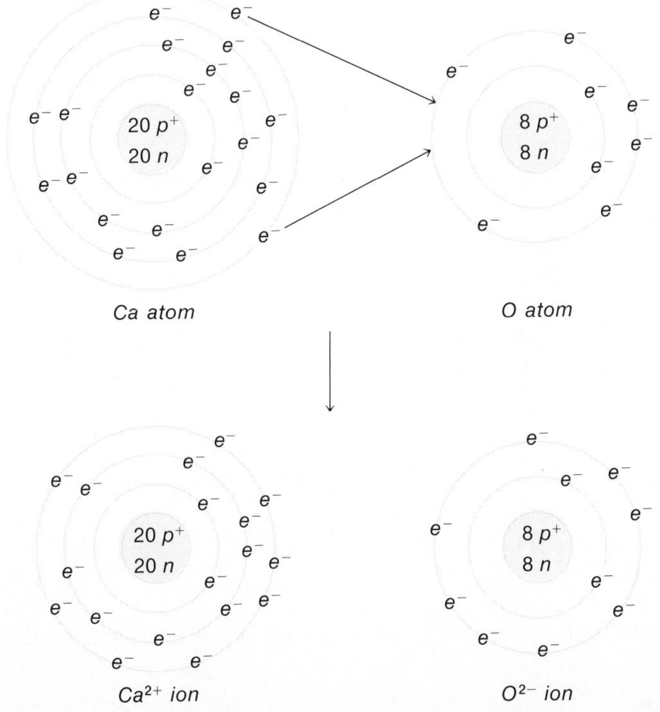
20 p+
20 n
Ca atom
8 p+
8 n
O atom
20 p+
20 n
Ca2+ ion
8 p+
8 n
O2− ion

b.

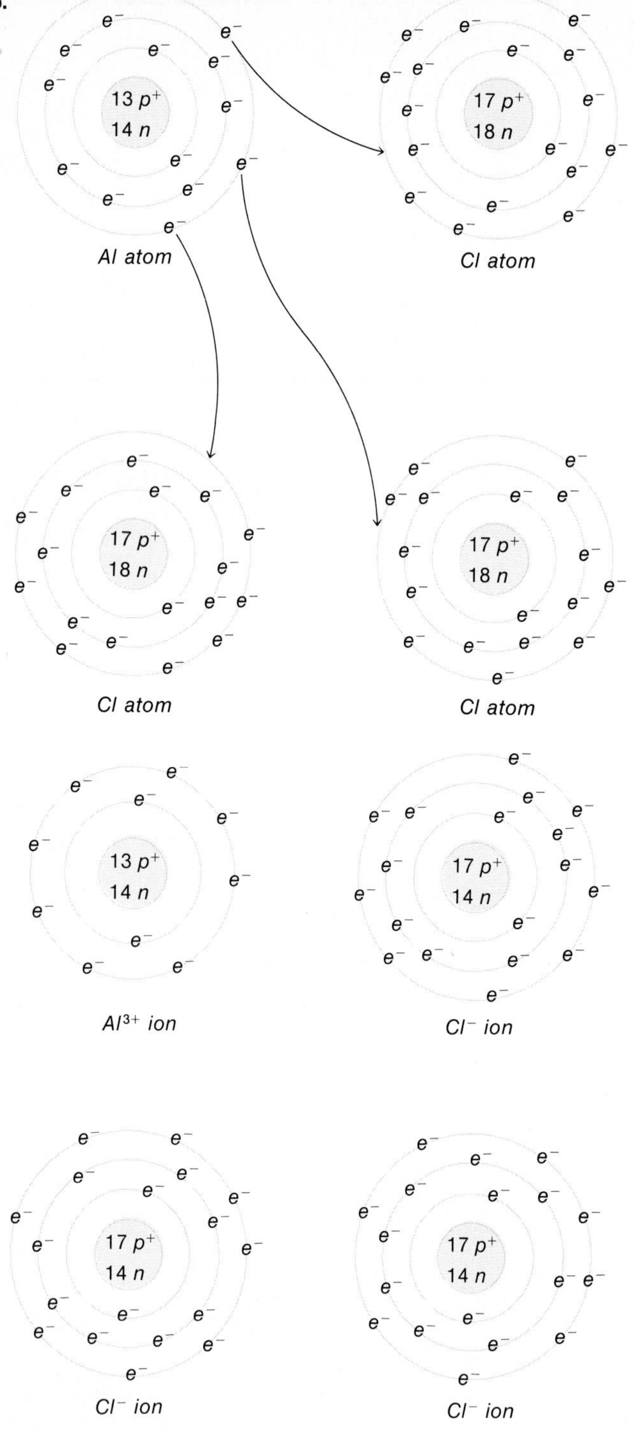
13 p+
14 n
Al atom
17 p+
18 n
Cl atom
Cl atom
Cl atom
Al3+ ion
Cl− ion
Cl− ion
Cl− ion

5.3 -2

5.4 *a.* NaI, *b.* $(NH_4)_2S$, *c.* $Al(NO_3)_3$, *d.* $Al_2(CO_3)_3$, *e.* $AlPO_4$

5.5 *a.* $1Ca^{2+}$, $1O^{2-}$; *b.* $1Mg^{2+}$, $2F^-$; *c.* $3Na^+$, $1PO_4^{3-}$; *d.* $1K^+$, $1C_2H_3O_2^-$; *e.* $3Ca^{2+}$, $2PO_4^{3-}$

5.6 *a.* MgS, *b.* $CrCl_3$, *c.* Ca_3N_2, *d.* PbO_2

5.7 See Problem 5.4

5.8 Sodium chloride, ammonium nitrate, calcium bromide, potassium carbonate, iron(II) oxide, iron(III) oxide, lead(II) phosphate, copper(II) oxide

5.9 40.0; 111; 234

5.11 Ionic and covalent
5.13 By losing one electron they attain a noble-gas electron configuration.
5.15 *a*. $1s^2$, *b*. helium
5.17 Rb^+ and Sr^{2+}
5.19 They gain two electrons to form a noble-gas electron configuration.
5.21 As^{3-}, Se^{2-}, Br^-
5.23 *a*. Br^-, *b*. Sr^{2+}, *c*. Cs^+, *d*. P^{3-}, *e*. Al^{3+}, *f*. Li^+
5.25 *a*.

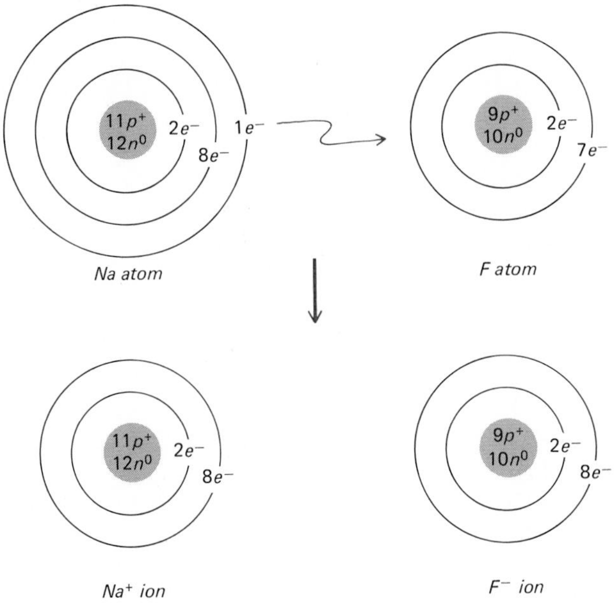
11p⁺ 12n⁰
2e⁻
8e⁻
1e⁻
9p⁺ 10n⁰
2e⁻
7e⁻
Na atom
F atom
11p⁺ 12n⁰
2e⁻
8e⁻
9p⁺ 10n⁰
2e⁻
8e⁻
Na⁺ ion
F⁻ ion

b. Na needs to lose only $1e^-$ and F needs to gain only $1e^-$ in order for them both to attain the Ne noble-gas configuration. Then the +1 charge on Na^+ is exactly neutralized by F^-.

5.27 *a*.

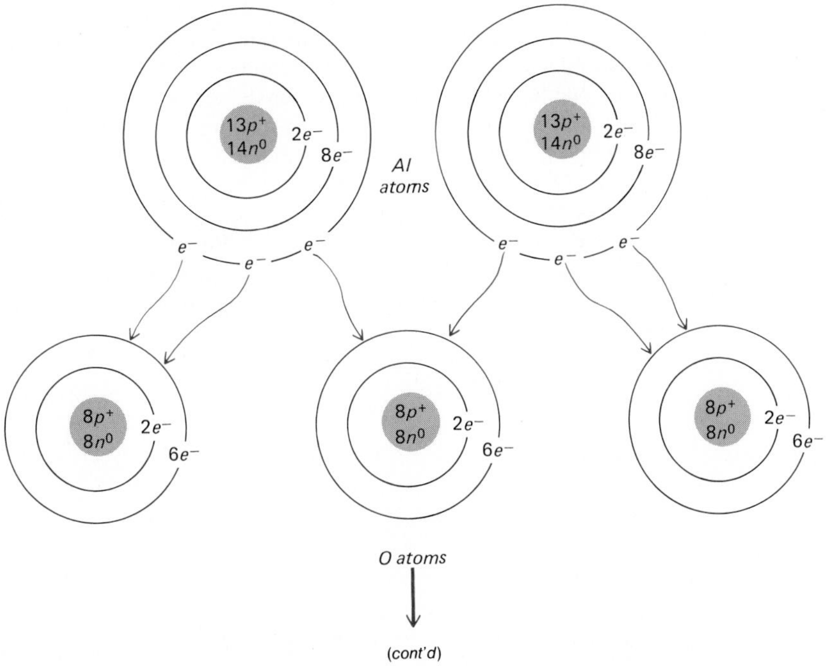
13p⁺ 14n⁰
2e⁻
8e⁻
Al atoms
13p⁺ 14n⁰
2e⁻
8e⁻
e⁻
e⁻
e⁻
e⁻
e⁻
e⁻
8p⁺ 8n⁰
2e⁻
6e⁻
8p⁺ 8n⁰
2e⁻
6e⁻
8p⁺ 8n⁰
2e⁻
6e⁻
O atoms
(cont'd)

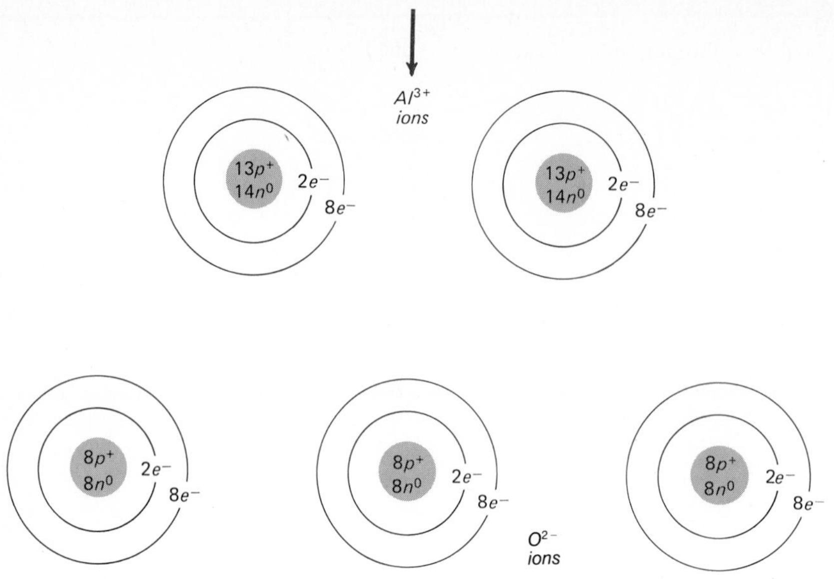

Al3+
ions
13p+
14n0
2e−
8e−
8p+
8n0
O2−
ions

b. Al atoms need to lose $3e^-$ to attain a Ne noble-gas configuration. Each O atom needs only $2e^-$ to achieve the Ne configuration. For equal numbers of e^- to be lost and gained, 2Al atoms lose $3e^-$ each ($6e^-$), and 3 O atoms gain $2e^-$ each ($6e^-$). The total positive charge of +6 ($2Al^{3+}$) is exactly neutralized by the total negative charge of −6 ($3O^{2-}$).

5.29 *a*. Oxidation, *b*. reduction

5.31 *a*. Reduction, *b*. oxidation, *c*. oxidation, *d*. reduction

5.33 *a*. CsF, *b*. BaI_2, *c*. BaO, *d*. $(NH_4)_2CO_3$, *e*. $Al_2(SO_4)_3$, *f*. $NaMnO_4$, *g*. Li_3PO_4, *h*. $KC_2H_3O_2$, *i*. Mg_3N_2, *j*. $ZnHPO_4$, *k*. $Pb(SO_4)_2$, *l*. Fe_2O_3, *m*. $Ca(HCO_3)_2$

5.35 *a*. Li_2O, *b*. $BaCl_2$, *c*. Na_2S, *d*. Ni_3N_2 *e*. AgBr, *f*. FeO, *g*. Fe_2O_3, *h*. $SnCl_2$, *i*. $SnCl_4$, *j*. Ga_2S_3, *k*. CaI_2

5.37 *a*. One potassium ion, one bromide ion; *b*. one aluminum ion, one phosphate ion; *c*. one magnesium ion, two bromide ions; *d*. two lithium ions, one carbonate ion; *e*. two ammonium ions, one sulfate ion; *f*. one calcium ion, two hydroxide ions; *g*. one barium ion, two nitrate ions; *h*. one sodium ion, one hydrogen carbonate (bicarbonate) ion; *i*. one copper(II) ion, one oxide ion; *j*. one copper(I) ion, one bromide ion; *k*. one iron(III) ion, three chloride ions; *l*. one lead(II) ion, one oxide ion

5.39 *a*. 3, *b*. 13, *c*. 16, *d*. 15

5.41 This compound also contains sodium ion.

5.43 *a*. Ag_2O, *b*. KCl, *c*. $Mg(HCO_3)_2$, *d*. $Al(H_2PO_4)_3$, *e*. Na_3PO_4, *f*. $BaSO_4$, *g*. CuI_2, *h*. PbS_2, *i*. Cu_2O, *j*. Sn_3N_2

5.45 The cations and anions are held together strongly by electrostatic attraction. Much (heat) energy is required to disrupt the attraction and allow flow into the liquid state.

CHAPTER 6

6.1 H:Br: H:F: H:I:

6.2

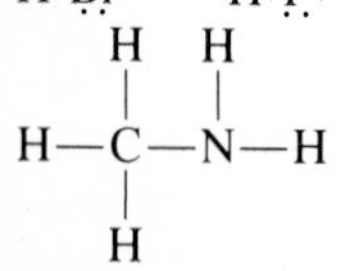

6.3

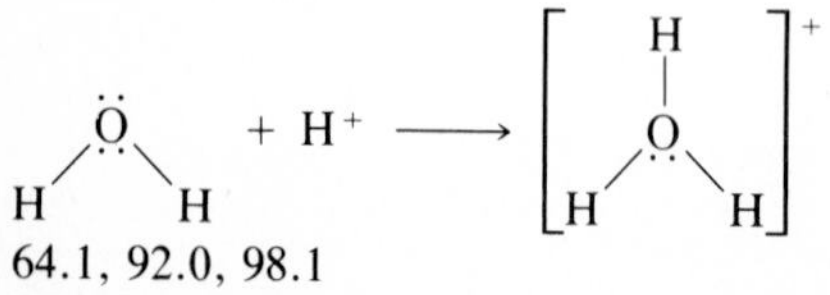

6.4 64.1, 92.0, 98.1

6.5 $C_2:H_2$ = 24.0:2.0 = 12:1

6.7 Ionic, covalent

6.9 *a*. :I:Cl: *b*. :Br:Cl: *c*. :Cl:F:

6.11 *a.* H· ·F: *b.* Attractions between H nucleus and F electrons and F nucleus and H electrons; repulsions between electrons and between nuclei
6.13 *a.* H one, S two; *b.* C four, H one, Br one
6.15 *a.* :F:N:F: (with :F: below N) *b.* :O::C::O: *c.* :C:::O: *d.* H:P:H (with H below P) *e.* :F:O:F: *f.* H:O:O:H
6.17 :O—S=O: :O—O=O:
6.19 Anions
6.21 Extra electrons
6.23 Electronegativity increases across a period (left to right) and up a group (bottom to top).
6.25 *a.* polar, *b.* polar, *c.* nonpolar, *d.* polar
6.27 P↔Br, N↔H, H↔O
6.29 *a.* F↔N↔F (with ↕ F below N) *b.* O↔C↔O, *c.* C↔O, *e.* F↔O↔F, *f.* H↔O—O↔H
6.31 (Answer based solely on metallic or nonmetallic nature of combined elements): *a.* Molecular, molecule; *b.* ionic, formula unit; *c.* ionic, formula unit; *d.* molecular, molecule; *e.* ionic, formula unit; *f.* molecular, molecule; *g.* ionic, formula unit; *h.* ionic, formula unit; *i.* ionic, formula unit; *j.* molecular, molecule
6.33 *a.* Tetrahedral, *b.* linear, *c.* tetrahedral
6.35 *a.* Reduces repulsive interactions. *b.* The electron pairs would actually be closer together (within 160°) in this case than in the linear arrangement (180°).
6.37 *a.* Tetrahedral about both

b.

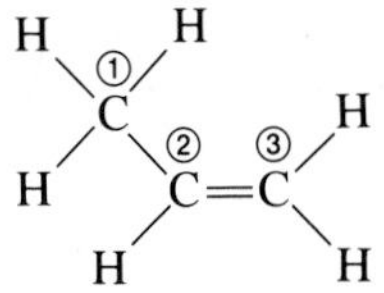

1 tetrahedral; 2, 3 planar triangular

c. Tetrahedral about all three carbons

d.

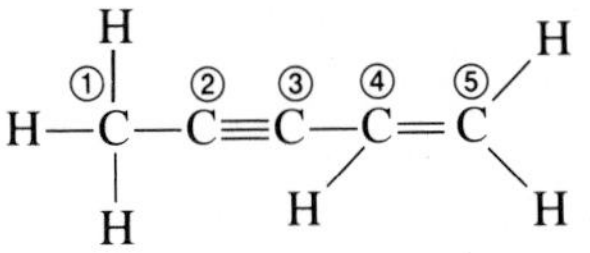

1 tetrahedral; 2, 3 linear; 4, 5 planar triangular

e.

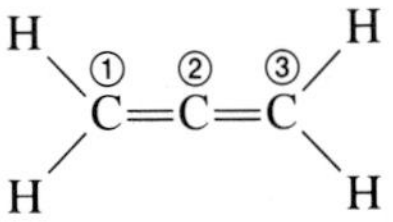

1, 3 planar triangular; 2 linear

6.39 HI < HBr < HCl < HF
6.41 *a.* IF, *c.* NF_3, *d.* HCN
6.43 C:H = 36.0:8.08 = 4.49:1.00
6.45 *a.* 1C;3H, *b.* 1S:4F, *c.* 1C:1H:3Cl, *d.* 2H:1S:4O

CHAPTER 7

7.1 27.0 g, 19.0 g, 79.9 g
7.2 3 × 27.0 g = 81.0 g
7.3 111 g
7.4 1.65 mol
7.5 *a.* Not balanced, *b.* not balanced, *c.* balanced
7.6 $H_2 + Br_2 \rightarrow 2HBr$
7.7 *a.* 2 mol $Al(OH)_3$ + 3 mol $H_2SO_4 \rightarrow$ 1 mol $Al_2(SO_4)_3$ + 6 mol H_2O
b. 4 mol Li + 1 mol $O_2 \rightarrow$ 2 mol Li_2O
7.8 $\dfrac{1 \text{ mol Sn}}{1 \text{ mol } H_2}$
7.9 $2K + Br_2 \rightarrow 2KBr$; 15.0 mol KBr
7.10 70.9 g DDT
7.11 *a.* Endothermic, *b.* exothermic, *c.* exothermic
7.12 58.5 kcal
7.13 A pair (2); a gross (144); a case (some specified number of items); a six-pack

7.15 *a*. 39.95 g, *b*. 32.06 g, *c*. 126.9 g, *d*. 107.9 g, *e*. 39.10 g, *f*. 28.09 g
7.17 *a*. 13.9 g, *b*. 184. g, *c*. 2010 g (2.01×10^3 g)
7.19 *a*. 0.00286 (2.86×10^{-3}) mol K; *b*. 2.86 mol K
7.21 8.3 mol K
7.23 *a*. 12.9 g Cu, *b*. 25.8 g Cu
7.25 *a*. 2.016 g, *b*. 70.90 g, *c*. 73.89 g, *d*. 601.8 g, *e*. 39.10 g, *f*. 96.06 g
7.27 *a*. 2.00 mol H_2, *b*. 1.08 mol Br, *c*. 0.0284 mol H_2O, *d*. 3.84×10^{-1} mol $Ca_3(PO_4)_2$, *e*. 0.530 mol $(NH_4)_2CO_3$, *f*. 1.00 mol K^+
7.29 *a*. 3.00 mol Li^+, 3.00 mol Br^-; *b*. 4.25 mol Ca^{2+}, 8.50 mol Cl^-; *c*. 3.30×10^3 mol Mg^{2+}, 2.20×10^3 mol PO_4^{3-}; *d*. 0.750 mol Ba^{2+}, 0.750 mol CO_3^{2-}
7.31 97.94 g
7.33 *a*. $Cu + S \rightarrow CuS$; *b*. $6K + N_2 \rightarrow 2K_3N$; *c*. $Ca(OH)_2 \rightarrow CaO + H_2O$
7.35 A chemical equation is balanced to satisfy the Law of Conservation of matter.
7.37 *a*. Unbalanced, *b*. balanced, *c*. unbalanced, *d*. balanced, *e*. unbalanced
7.39 *a*. $N_2 + O_2 \rightarrow 2NO$
b. $2NaBr + Cl_2 \rightarrow 2NaCl + Br_2$
c. $2P + 5Cl_2 \rightarrow 2PCl_5$
d. $BaCl_2 + (NH_4)_2SO_4 \rightarrow BaSO_4 + 2NH_4Cl$
e. $K_2O + H_2O \rightarrow 2KOH$
f. $4Fe + 3O_2 \rightarrow 2Fe_2O_3$
g. $CaC_2 + 2H_2O \rightarrow C_2H_2 + Ca(OH)_2$
h. $Zn + 2HNO_3 \rightarrow Zn(NO_3)_2 + H_2$
i. $NH_4NO_2 \rightarrow N_2 + 2H_2O$
j. $2PbO + O_2 \rightarrow 2PbO_2$
7.41 *a*. $2Fe(OH)_3 + 3H_2SO_4 \rightarrow Fe_2(SO_4)_3 + 6H_2O$
b. $2C_2H_6 + 7O_2 \rightarrow 4CO_2 + 6H_2O$
c. $NaOH + Al(OH)_3 \rightarrow NaAlO_2 + 2H_2O$
d. $P_4O_{10} + 6H_2O \rightarrow 4H_3PO_4$
e. $6K_2O + P_4O_{10} \rightarrow 4K_3PO_4$
f. $MgI_2 + H_2SO_4 \rightarrow 2HI + MgSO_4$
g. $PCl_5 + 4H_2O \rightarrow 5HCl + H_3PO_4$
h. $2Al + 3Sn(NO_3)_2 \rightarrow 2Al(NO_3)_3 + 3Sn$
7.43 *Combination:* Problems 7.33 *a, b;* 7.34 *b, c;* 7.37 *b, d;* 7.39 *a, c, e, f, j;* 7.40 *a, d;* 7.41 *d, e;* *Decomposition:* Problems 7.33 *c;* 7.37 *e;* 7.39 *i;* 7.40 *b;* *Single Replacement:* Problems 7.37 *a, c;* 7.39 *b, h;* 7.41 *h;* *Double Replacement:* Problems 7.34 *a;* 7.39 *d, g;* 7.40 *c;* 7.41 *a, c, f, g*
7.45 *a*. $Mg + I_2 \rightarrow MgI_2$
b. $4Li + O_2 \rightarrow 2Li_2O$
c. $2Al + 3S \rightarrow Al_2S_3$
d. $3K + P \rightarrow K_3P$
e. $3Ca + N_2 \rightarrow Ca_3N_2$
7.47 Mg; Li; Al; K; Ca; Mg; Al
7.49 Like all the halogens, it is an oxidizing agent and kills microorganisms by oxidizing vital bacterial components.
7.51 Reduced: hydrogen atoms are gained.
7.53 $4\ Fe + 6H_2O + 3O_2 \rightarrow 4\ Fe(OH)_3$
7.55 $6\ CO_2 + 6\ H_2O \rightarrow C_6H_{12}O_6 + 6\ O_2$
7.57 *a*. 1.0 mol I_2, *b*. 3.0 mol HI, *c*. 16 mol H_2O, *d*. 0.12 mol HI
7.59 *a*. 0.50 mol H_2SO_4, *b*. 0.400 mol H_2O, *c*. 5.55 mol KOH
7.61 *a*. 0.16 mol Cl_2, *b*. 13 g NaOH
7.63 $2Na + Cl_2 \rightarrow 2NaCl$, 62.5 g Cl_2
7.65 *a*. 1.3 g $PbSO_4$, *b*. 1.1 g PbS
7.67 *a*. 536 kcal, *b*. 115 kcal
7.69 0.300 mol CO_2

CHAPTER 8

8.1 762 torr (760 to 2 significant figures)
8.2 3 atm

8.3 *a.* 273 K, *b.* 298 K, *c.* 373 K

8.4 *a.* $\frac{9.6}{263} = 0.37$ $\frac{10.0}{273} = 0.37$ $\frac{10.5}{283} = 0.37,$ etc.

b. $\frac{9.6}{-10.0} = -0.96$ $\frac{10.0}{0} = ?!?$ $\frac{10.5}{10} = 1.05,$ etc.

8.5 2.6 mL

8.6 0.05 mol

8.7 83.3 L

8.8 *b, c*

8.9 *a.* NH_3, *b.* H_2O, *c.* Br_2

8.10 12.2 kcal

8.11 Easily compressed; low density; exert pressure; expand when heated; diffuse (see Section 8.2)

8.13 Original = 1.9 g/L; final = 0.080 g/L

8.15 1.24 atm or 946 mmHg

8.17 *a.* $V \propto \frac{1}{P}$, *b.* $V = \frac{K}{P}$ or $PV = K$

8.19 *Boyle's:* Bubbles of gas in a liquid expand as they rise to the top of the liquid because the pressure is lower at the top than the bottom of the liquid. *Charles:* An inflated balloon expands when heated and contracts when cooled.

8.21 *a.* No, *b.* 323°C

8.23 2.15 L

8.25 17.1 mL

8.27 *a.* Fewer air molecules per unit volume at the mountain top indicate a lower density of air and consequently a lower p_{O_2} than at sea level. *b.* The lower p_{O_2} at the mountain top will decrease the amount of oxygen taken into the alveoli and subsequently into the blood and ultimately bodily tissues. A person must breath more rapidly to compensate for the lower p_{O_2}.

8.29 1.1×10^2 L

8.31 84.8 L

8.33 *a.* 10.2 L, *b.* 4.14 L, *c.* 0.431 L, *d.* 15.5 L

8.35 *a.* A gas with no attractive forces between its particles; *b.* high *T* and low *P;* *c.* low *T* and high *P*

8.37 Increased temperature causes gas molecules to move faster and collide harder with the walls of the container. The consequent increased pressure may lead to an explosion.

8.39 *a.* No, *b.* only at STP

8.41 *a.* Dispersion, dipole-dipole, hydrogen bonding *b.* All molecules show dispersion forces; polar molecules undergo dipole-dipole interactions; only molecules with —OH or —NH bonds (and H—F) exhibit hydrogen bonding.

8.43 *a.* Dispersion, *b.* hydrogen bonding, *c.* dipole-dipole

8.45 The numbers of electrons increase, resulting in larger dispersion forces.

8.47 *a, d,* and *e*

8.49 More heat energy is needed to overcome increased attractive forces.

8.51 Since there is more space between gas molecules, other molecules can move between (diffuse) the gas molecules faster than between the more closely arranged molecules in liquids.

8.53 Heat energy, needed for evaporation, is taken from the skin, giving a cooling effect.

8.55 Increasing *T* increases the average kinetic energy of molecules so that more can escape into the gas phase.

8.57 *Evaporation:* Occurs only at the liquid surface and is independent of the external pressure. *Boiling:* Bubbles of gas form throughout the liquid. Occurs only when the vapor pressure equals the external pressure.

8.59 *a.* Vapor pressure of water = 760 mmHg at 100°C; *b.* yes, by altering external *P* to coincide with the vapor pressure of water at *any T*

8.61 Increased temperature would decrease surface tension.

8.63 To prevent the escape of volatile components of the cheese, which give it its aroma and flavor. Also, retards loss of water vapor, which loss would dry out the cheese.

8.65 48 kcal

8.67 *a.* Potassium bromide, solid methane, sand, and copper; *b.* ions, molecules, atoms, and cations, respectively

8.69 912 cal

8.71 17,222 cal

CHAPTER 9

9.1 *a*. Unsaturated, *b*. unsaturated, *c*. saturated, *d*. unsaturated
9.2 Dilute
9.3 700 mL
9.4 0.25 *M* NaOH
9.5 *a*. 3 osM, *b*. 2 osM, *c*. 6 osM
9.7 A solution is a mixture. Although the components of a solution are uniformly mixed, their proportion varies and they are separable by physical means.
9.9 *a*. Yes, *b*. no, only homogeneous mixtures are solutions

9.11

Solute	Solvent
He	N_2
Alcohol, caramel	Water
CO_2	Water
Tin	Copper

9.13 *a*. No, if solubility is low, solution can be saturated but dilute; *b*. silver acetate has a solubility limit of 1.04 g per 100 g H_2O; this solution is saturated but dilute.
9.15 *a*. Saturated, *b*. unsaturated, *c*. unsaturated, *d*. saturated
9.17 *a*. Mixtures of liquids with liquids, *b*. ethylene glycol and water (antifreeze mixture), *c*. turpentine and water
9.19 *a*. Electrostatic forces (the ionic bond), *b*. ion-dipole attractions
9.21 *a*. Solutes tend to dissolve in solvents having similar intermolecular forces. Polar solutes dissolve in polar solvents; nonpolar solutes in nonpolar solvents. *b*. polar solutes; *c*. nonpolar solutes
9.23 *a*. Opening the bottle relieves the pressure on the solution. The solubility of CO_2 in water decreases as pressure decreases *b*. No, solubility of a gas in water decreases with increasing temperature.
9.25 *a*.

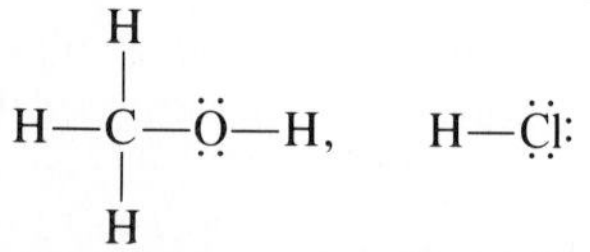

b. HCl will dissolve in CH_3OH because both substances are polar.
9.27 Solution conducts electricity.
9.29 Some covalent materials react with H_2O to form many ions in solution (e.g., HBr), other covalent materials form only a few ions (e.g., $HC_2H_3O_2$).
9.31 HCl reacts with H_2O to form ions. In benzene no ions are produced.
9.33 *a*. K^+ and Br^- ions; *b*. Ca^{2+} and OH^- ions; *c*. H^+ and NO_3^- ions; *d*. CH_3COOH molecules predominantly, a few H^+ and CH_3COO^- ions; *e*. Mostly H_2SO_3 molecules, some H^+ and HSO_3^- ions; *f*. $C_6H_{12}O_6$ molecules; *g*. CH_3OH molecules
9.35 Solubility refers to the maximum amount of solute that can dissolve in a given amount of solvent at a given temperature. For example, a maximum of 36 g NaCl can dissolve in 100 g H_2O at 20°C. Concentration is a statement of the amount of solute dissolved in a given amount of solvent or solution. 10 g NaCl in 90 g H_2O produces a 10% solution. Concentration is variable.
9.37 38.7%
9.39 31 mL of 4.0% (w/v) $Ba(NO_3)_2$ solution
9.41 56 mg
9.43 26.3%
9.45 1.00 *M* KNO_3
9.47 135 g glucose
9.49 0.056 *M* NaOH
9.51 66.7 mL
9.53 33 mL
9.55 *a*. 1 meq, *b*. 3 meq, *c*. 1 meq, *d*. 0.50 meq
9.57 Vapor pressure, boiling-point elevation, freezing-point lowering, osmotic pressure
9.59 *a*. Pure water > glucose solution > NaCl solution *b*. NaCl solution > glucose > pure water *c*. Pure water > glucose solution > NaCl solution *d*. NaCl solution > glucose solution > pure water

9.61 *a.* A→B, *b.* A→B, *c.* B→A, *d.* no net flow
9.63 *a.* Hypertonic, *b.* hypotonic, *c.* hypertonic, *d.* hypertonic
9.65 Solutions, colloids, and suspensions differ in particle size and visibility, whether particles settle on standing, whether they can be separated on filter paper, how they respond to light, and whether the particles can pass through a dialyzing membrane (see Table 9.4).
9.67 The Tyndall effect results when colloidal particles reflect light; as a consequence a beam of light can be seen in a colloidal dispersion. Solutions do not produce this effect because their particles are too small to reflect light.
9.69 *a.* Water X→Y, Na^+ and Cl^- ions flow X→Y, glucose passes Y→X, starch remains in compartment Y *b.* no net flow of water or Cl^- ions, Na^+ ions flow X→Y, K^+ ions flow Y→X *c.* Water Y→X, K^+ and Cl^- ions flow X→Y, protein remains in compartment X *d.* Water Y→X, starch remains in compartment X
9.71 Active transport
9.73 Edema, improper removal of metabolic waste, and/or improper transport of nutrients and fluids into tissues

CHAPTER 10

10.1 The waste products have less energy than the starting foodstuffs. The chemical energy in the food we eat is released as the food is metabolized and enables us to function.

10.2 Negative; energy is released.

10.3 The enzyme-catalyzed reaction is the fastest, as shown by the fact that this reaction has the lowest activation energy.

10.4 A decrease in reactant concentration leads to more "room" in the reaction vessel, lowered collision frequency, and decreased probability of product formation.

10.5 A decrease in temperature causes a decrease in the average kinetic energy of the molecules and hence in the average molecular speed. Molecules will collide less often, and each collision that does occur will be less energetic, leaving it with a lessened chance of overcoming the activation barrier. Therefore, a decrease in temperature yields a decrease in the rate of product formation.

10.6 *a.* $K = \dfrac{[\text{glucose-6-phosphate}]}{[\text{glucose-1-phosphate}]}$ *b.* $K = \dfrac{[\text{glucose}]^2}{[\text{maltose}]}$

10.7 *a.* Significant amounts of products and reactants at equilibrium; *b.* there is a much greater amount of product than reactant at equilibrium.

10.9 A spontaneous reaction is one in which the energy of the product(s) is lower than that of the reactant(s).

10.11 When gasoline burns, to yield CO_2 and H_2O, a large amount of heat energy is released, which is why it is used as a fuel. Water does not burn. This everyday observation leads to the conclusion that gasoline has a greater amount of chemical energy than water.

10.13 *a.* The O═O double bond, *b.* two C═O double bond

10.15 There is a tendency for a system to go to increasing disorder; e.g., when gas is allowed to escape from a closed container, it will randomly fill the entire room. Any ordered array of objects thrown into the air will fall randomly in disarray.

10.17 A negative change, i.e., a lowering in enthalpy, and a positive change, or increase in entropy

10.19 *a, b, e*

10.21 Activation energy is the difference between the energy of the reactants and the highest point, or value of energy, on the reaction pathway; i.e., it is the amount of energy input needed to surmount the energy barrier between reactants and products.

10.23 *a.*

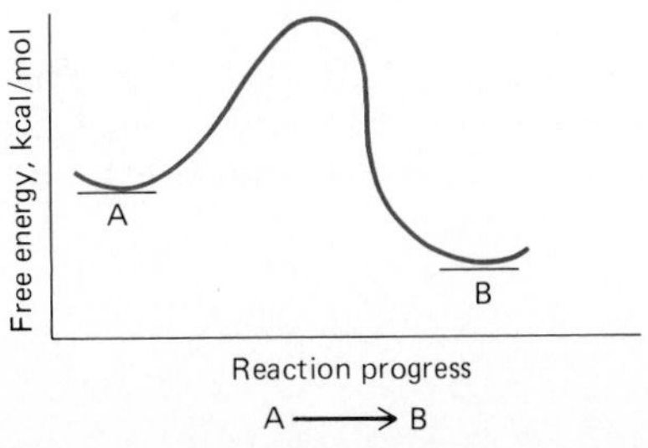
Free energy, kcal/mol
A
B
Reaction progress
A ⟶ B

b.

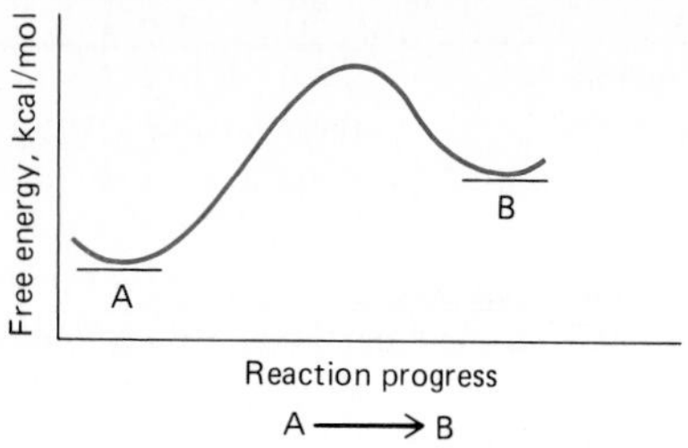
Free energy, kcal/mol
B
A
Reaction progress
A ⟶ B

10.25 By heating the reactants or adding some other form of energy such as light energy
10.27 By lowering E_{act}
10.29 The magnitude of E_{act} is inversely related to the rate at which the reaction proceeds; i.e., the larger E_{act} the slower the reaction rate.
10.31 Temperature, reactant concentration
10.33 No; the effectiveness of the collision depends on its energy.
10.35 *a.* The yellow color gradually becomes greener and greener as *D* forms (green is the color that arises from a yellow-blue mix). *b.* The blue color becomes green as yellow *A* forms. *c.* Remains some intermediate green color depending on the equilibrium concentrations of yellow *A* and blue *D*.
10.37 Rate forward = rate reverse
10.39 *a.* $K_{eq} = \frac{[HCl]^2}{[H_2][Cl_2]}$ *b.* $K_{eq} = \frac{[PCl_3][Cl_2]}{[PCl_5]}$ *c.* $K_{eq} = \frac{[CO_2][H_2]}{[CO][H_2O]}$ *d.* $K_{eq} = \frac{[NO_4]^2[H_2O]^6}{[NH_3]^4[O_2]^5}$
10.41 *a.* More products than reactants at equilibrium; *b.* more reactants than products at equilibrium; *c.* slightly more reactant than product at equilibrium; *d.* significant amounts of both products and reactants at equilibrium
10.43 Le Chatelier's principle states that when a stress is applied to a system at equilibrium, the equilibrium will shift so as to relieve that stress.
10.45 $[HCO_3^-]$ is lowered.
10.47 *a.* Shifts right (⟶), *b.* shifts left (⟵), *c.* shifts right (⟶), *d.* shifts left (⟵), *e.* shifts right (⟶)
10.49 *a.* Shifts right (⟶), *b.* shifts left (⟵), *c.* shifts right (⟶), *d.* shifts right (⟶), *e.* shifts right (⟶)
10.51 *a.* Lowers activation energy; *b.* no effect

CHAPTER 11

11.1 $KOH(s) \rightleftharpoons K^+(aq) + OH^-(aq)$ $Ca(OH)_2(s) \rightleftharpoons Ca^{2+}(aq) + 2OH^-(aq)$
11.2 *a.* $HCl + NH_3 \rightleftharpoons NH_4^+ + Cl^-$
Acid, Base, Conjugate acid, Conjugate base

b. $OH^- + CH_3C(=O){-}OH \rightleftharpoons CH_3C(=O){-}O^- + H_2O$
Base, Acid, Conjugate base, Conjugate acid

11.3 $HNO_3 > CH_3CH(OH)COOH > H_2CO_3 > H_3BO_3 > H_2O$
11.4 $OH^- > H_2BO_3^- > HCO_3^- > CH_3CH(OH)COO^- > NO_3^-$
11.5 $[OH^-] = 3.1 \times 10^{-6}$ *M;* basic
11.6 *a.* 5.0, *b.* 5.0
11.7 1.5
11.8 1.5 *M* H^+; 0.4 *M* H^+; 0.45 *M* H^+
11.9 pH = 7.21 + 0.2
11.11 $HNO_3 + H_2O \rightleftharpoons H_3O^+ + NO_3^-$
11.13 $HSO_4^- + H^+ \rightleftharpoons H_2SO_4$ (Base) $HSO_4^- + OH^- \rightleftharpoons H_2O^+ SO_4^{2-}$ (Acid)
11.15 *a.* HBr acid, Br^- conjugate base, H_2O base, H_3O^+ conjugate acid *b.* H_3O, acid, H_2O conjugate base, NH_3 base, NH_4^+ conjugate acid *c.* HSO_4^- acid, SO_4^{2-} conjugate base, OH^- base, H_2O conjugate acid *d.* H_3O^+ acid, H_2O conjugate base, HCO_3^- base, H_2CO_3 conjugate acid *e.* H_3PO_4 acid, $H_2PO_4^-$ conjugate base, NH_3 base, NH_4^+ conjugate acid
11.17 *a.* H_3O^+, *b.* HF, *c.* H_3PO_4, *d.* NH_4^+
11.19 Litmus test; neutralization reaction, pH meter

11.21 $NO_3^- < SO_4^{2-} < CH_3CH_3CH(OH)COO^- < H_2BO_3^- < OH^-$

11.23 **Acid** **Base**

a. $NH_4^+ + NH_3 \rightleftharpoons NH_3 + NH_4^+$

b. $H_2SO_4 + H_2O \rightleftharpoons HSO_4^- + H_3O^+$

c. $HCl + CN^- \rightleftharpoons HCN + Cl^-$

d. $H_2CO_3 + NO_3^- \rightleftharpoons HCO_3^- + HNO_3$

e. $HSO_4^- + OH^- \rightleftharpoons SO_4^{2-} + H_2O$

f. $HSO_4^- + NO_3^- \rightleftharpoons SO_4^{2-} + HNO_3$

11.25 Pure water is a nonelectrolyte; it is neither sour nor bitter, and it is neutral to litmus.

11.27 Crops grow and produce best when the soil is within a certain pH range. Farmers must be able to measure soil pH so that they can make necessary adjustments to assure maximum yields.

11.29 *a.* pH = 7.0, neutral; *b.* pH = 2.0, acidic; *c.* pH = 10, basic; *d.* pH = 4, acidic

11.31 *a.* Basic, *b.* Acidic, *c.* Acidic, *d.* Acidic, *e.* Basic, *f.* Acidic, *g.* Acidic, *h.* Basic

11.33 *a.* $KOH + HCl \rightarrow KCl + H_2O$

b. $Ba(OH)_2 + 2HNO_3 \rightarrow Ba(NO_3)_2 + 2H_2O$

c. $3NaOH + H_3PO_4 \rightarrow Na_3PO_4 + 3H_2O$

d. $Ca(OH)_2 + H_2SO_4 \rightarrow CaSO_4 + 2H_2O$

e. $3Ca(OH)_2 + 2H_3PO_4 \rightarrow Ca_3(PO_4)_2 + 6H_2O$

11.35 *a.* $HNO_3 + KHCO_3 \rightarrow KNO_3 + H_2CO_3\ (\rightarrow H_2O + CO_2)$

b. $2HCl + Li_2CO_3 \rightarrow 2LiCl + H_2CO_3\ (\rightarrow H_2O + CO_2)$

11.37 0.9009 *M*

11.39 There are 2 mol of H^+ in each mole of H_2SO_4.

11.41 0.06000 *M* HCl

11.43 *a.* 0.00981 mol CH_3COOH, *b.* 0.589 g, *c.* 5.89%

11.45 A buffer is a weak acid–weak base pair which by reacting with added base or acid can resist large changes in the pH of a solution and so maintain pH levels within certain desirable ranges.

11.47 Benzoic acid, a weak acid, will ionize to a small extent in solution

$$C_6H_5COOH(aq) \rightleftharpoons C_6H_5COO^-(aq) + H^+(aq) \qquad \text{(i)}$$

The benzoate salt will completely dissociate to produce greater concentrations of the conjugate base, $C_6H_5COO^-$

$$C_6H_5COO^-Na^+ \longrightarrow C_6H_5COO^-(aq) + Na^+(aq) \qquad \text{(ii)}$$

If acid (H^+) is added to this buffer, the equilibrium of Equation (i) will shift to the left, consuming the added H^+ and relieving the stress on the equilibrium (Le Chatelier's principle). If base (OH^-) is added, it will react with H^+ to yield H_2O, thus decreasing the H^+ concentration. The equilibrium will therefore shift to the right and form more H^+. The new ratio of acid to anion after equilibrium has been established is essentially the same as before acid or base was added, assuming the initial concentrations of weak acid and conjugate base in the buffer are sufficiently large and the amount of added acid or base relatively small.

11.49 $[H^+]$ will increase; $[H_2CO_3]$ will increase; $[HCO_3^-]$ will decrease.

11.51 pH = 6.4

11.53 pH = 6.1 (= pK_a; log 1 = 0)

11.55 The coupling of the equilibria between H_2CO_3 and the unlimited supply of CO_2 in the lungs provides for the high buffering capacity toward base through appropriate shifts in equilibrium.

11.57 In respiratory acidosis $[HCO_3^-]$ and $[H_2CO_3]$ are higher than normal, whereas metabolic acidosis shows lowered $[HCO_3^-]$ and $[H_2CO_3]$, providing criteria by which the two conditions can be distinguished.

CHAPTER 12

12.1 The differences between the properties of table salt and gasoline arise from different types of bonding. The atoms of gasoline molecules are covalently bonded, whereas table salt is an ionic compound. The intermolecular forces are relatively weak in organic compounds (such as gasoline) compared to the very strong electrostatic attraction between ions in ionic compounds (like table salt).

12.2 The five —OH (hydroxyl) groups present in glucose represent a substantial fraction of the molecule, giving glucose good potential for hydrogen bonding with water and hence water solubility.

12.3 *a.*

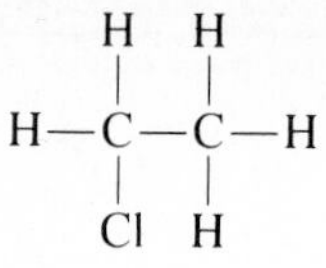

b.

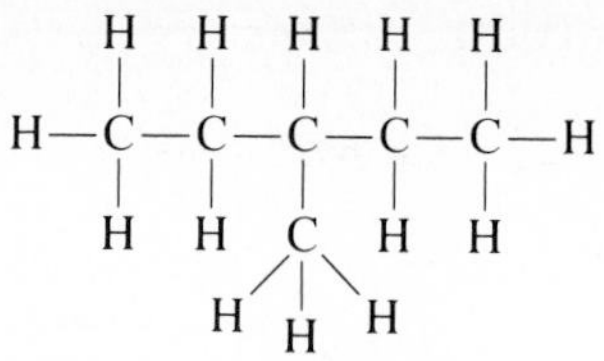

c.

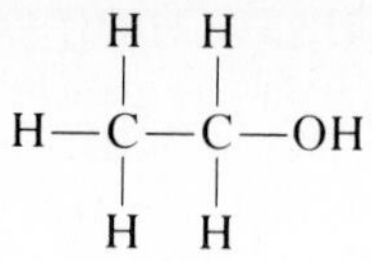

12.4 *a.* Same, *b.* structural isomers, *c.* unrelated, *d.* same, *e.* structural isomers
12.5 *a.* 2-Methylbutane, *b.* 2,3-dimethylpentane
12.6 *a.* 2-Iodobutane, *b.* 3-chloro-2-methylpentane
12.7

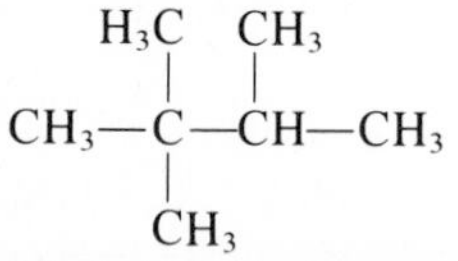

12.9 Methyl alcohol is a molecular compound; it dissolves in water due to the presence of an —OH (hydroxyl) group in the molecule which hydrogen-bonds with water. However, since no ions are present in an aqueous solution of methyl alcohol, it cannot conduct electricity.
12.11 Methane and O_2 have a higher free energy than the combination of CO_2 and H_2O. When methane burns, it combines with oxygen in a highly exothermic (energy-*releasing*) reaction, and we can see that the energy is released as great amounts of heat and light.
12.13 *a.* Organics contain C; most inorganics do not; *b.* hydrogen
12.15 Si
12.17 *a.*

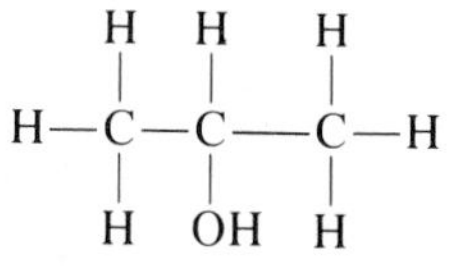

b.

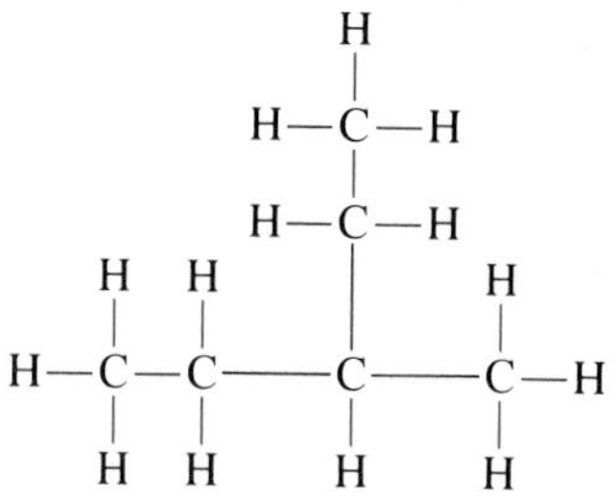

c.

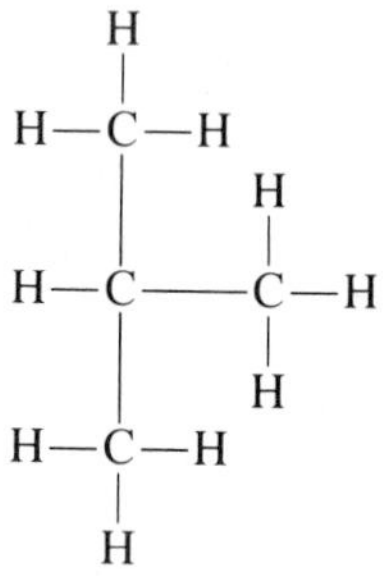

d.

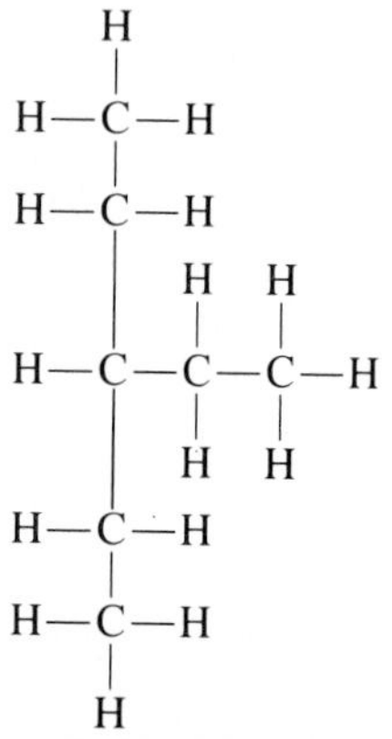

12.19 *a.* Alkane, *b.* ketone, *c.* alkyl halide, *d.* alkyne, *e.* aromatic, *f.* carboxylic acid, *g.* ether, *h.* ester, *i.* amine, *j.* alkene, *k.* amide, *l.* aldehyde
12.21 $CH_3CH_2CH_2Cl$, $CH_3(CH_2)_2CH_2Cl$, $CH_3(CH_2)_3CH_2Cl$, $CH_3(CH_2)_4CH_2Cl$, $CH_3(CH_2)_5CH_2Cl$, $CH_3(CH_2)_6CH_2Cl$
12.23 R represents any carbon chain (with attached hydrogens) which is part of this carboxylic acid. For example, R might be CH_3—.
12.25 *a.* Isomeric, *b.* identical, *c.* isomeric, *d.* isomeric, *e.* unrelated, *f.* isomeric, *g.* isomeric
12.27

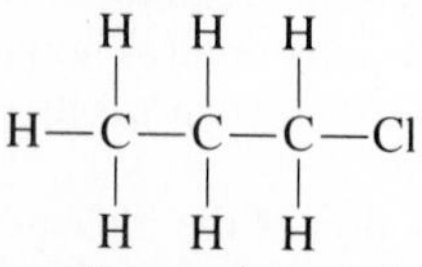

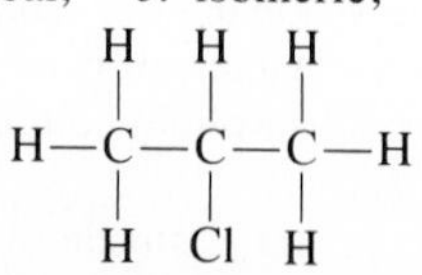

12.29 *a.* Tetrahedral, *b.* planar triangular, *c.* planar triangular, *d.* linear
12.31 From left to right: *a.* Both tetrahedral; *b.* Tetrahedral, planar triangular, planar triangular; *c.* All

tetrahedral; *d.* All tetrahedral; *e.* Tetrahedral, linear, linear, planar triangular, planar triangular; *f.* Planar triangular, linear, planar triangular

12.33 No; since rotation of 180° about one of the C—C bonds will convert one into the other, they are identical.

12.35 *a.* Same *b.* isomers *c.* unrelated *d.* identical *e.* unrelated *f.* isomers *g.* identical

12.37 $CH_3CH_2CH_2$— CH_3CHCH_3 (bond from central C)

n-Propyl Isopropyl

12.39 *a.* 2-Iodobutane, *b.* 2-iodo-2-methylpropane, *c.* 2-bromo-3-methylbutane, *d.* 2-chloropropane, *e.* 1-chloro-2,2-dimethylpropane, *f.* 2-bromo-4,4-dimethylhexane

CHAPTER 13

13.1 $2C_4H_{10} + 13O_2 \rightarrow 8CO_2 + 10H_2O$

13.2 Two products:

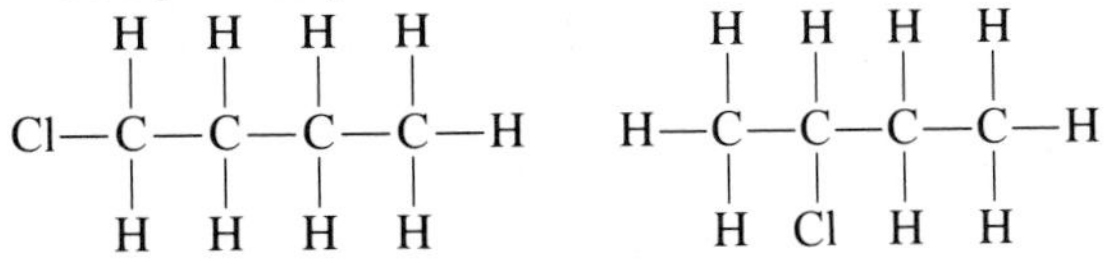

13.3

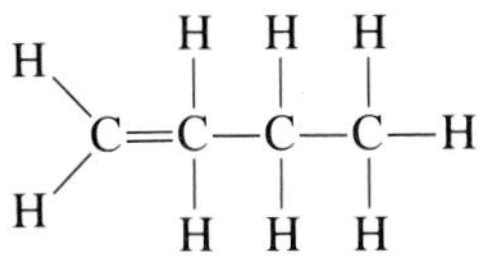

1-Butene

13.4 *a.* 2-Butene, *b.* 2,4-dimethyl-1-pentene

13.5 *b*

13.6

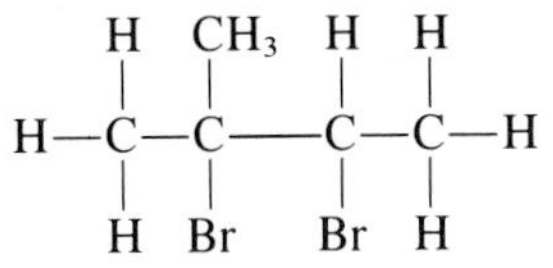

13.7

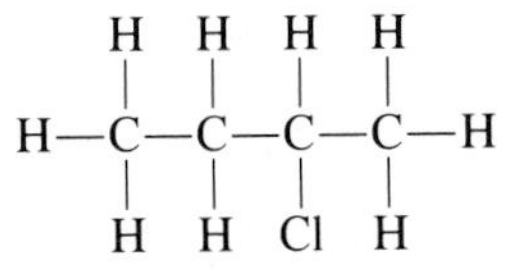

13.8

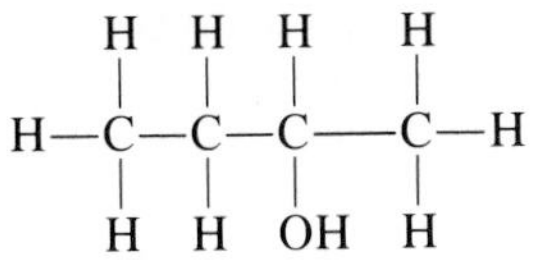

13.9 *a.*

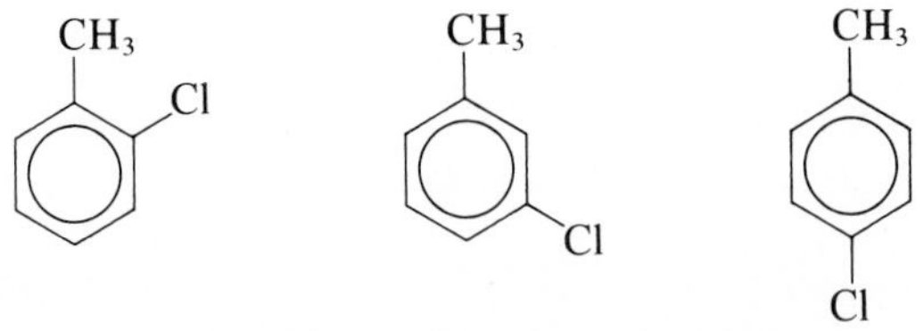

b. *o*-Chlorotoluene *m*-Chlorotoluene *p*-Chlorotoluene

13.11 *a.* Alkane, *b.* alkyne, *c.* aromatic, *d.* alkene, *e.* alkane, *f.* cycloalkane

13.13 CO_2 and H_2O

13.15 They are the most volatile (low boiling).

13.17 $a < c < b$

13.19 Since hydrocarbons are less dense than water and insoluble in it, oil floats on water and does not diffuse, allowing it to burn on the ocean.

13.21 It will be insoluble because of the nature of its intermolecular forces. Sodium chloride is an ionic compound which dissolves best in polar solvents. Hydrocarbons such as pentane are nonpolar.

13.23 Alkanes have a great deal more chemical energy than their combustion products and are therefore less stable. We can see this from the large amounts of energy released when they burn.

13.25 $2C_8H_{18} + 25O_2 \rightarrow 16CO_2 + 18H_2O$

13.27 *a.* Ethene, *b.* 1-butene, *c.* 4,4-dimethyl-2-pentene, *d.* 2-ethyl-1-butene

13.29 *b.*

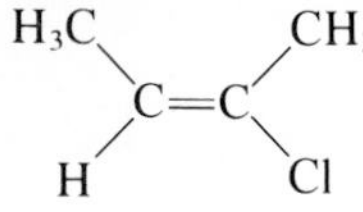

cis

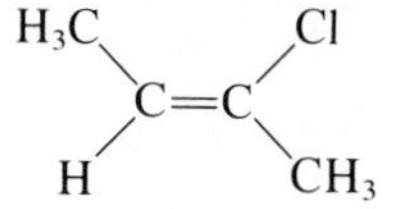

trans

c.

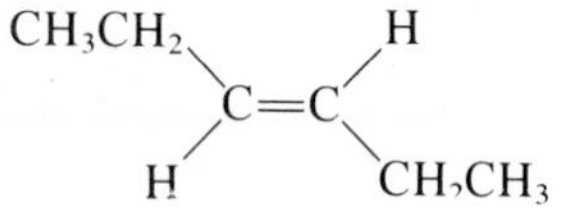

trans

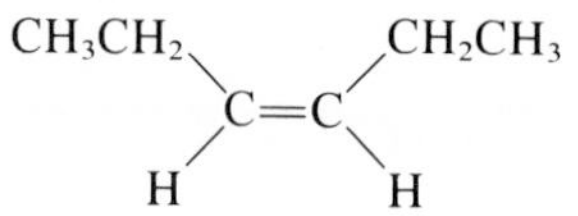

cis

13.31 *a.*

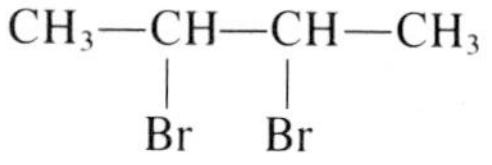

b.

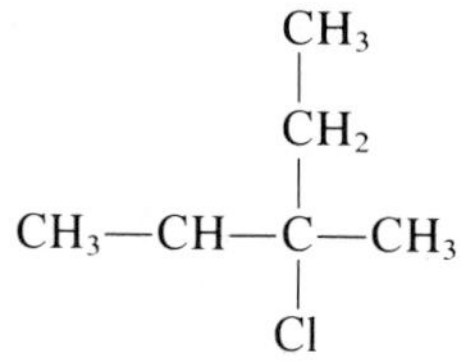

c.

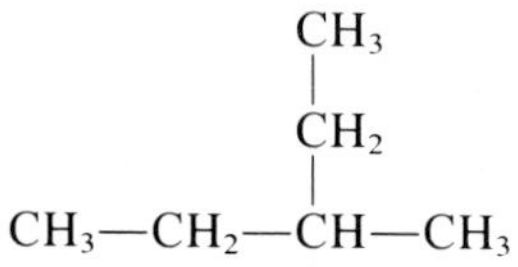

d.

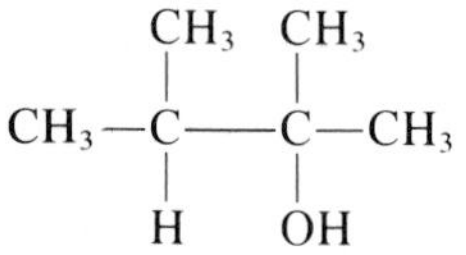

e.

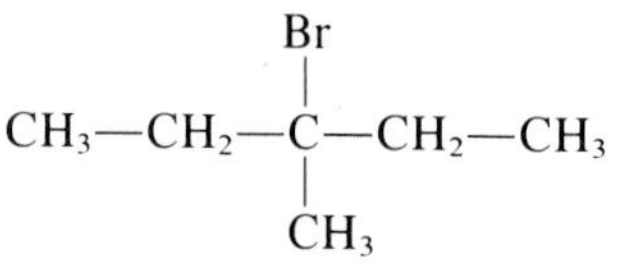

13.33 *a.*

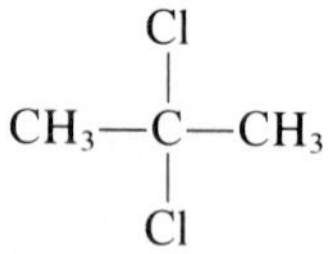

b. $CH_3—CH=CH—CH_3$

13.35 *a.* $H—C\equiv C—H$ *b.* $CH_3—C\equiv C—CH_2—CH_3$ *c.*

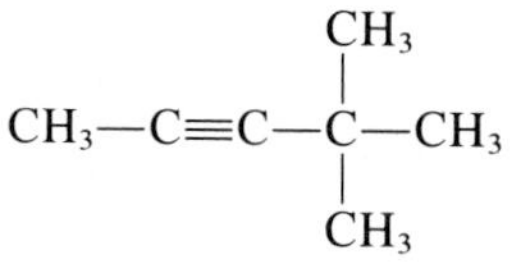

13.37 *a.*

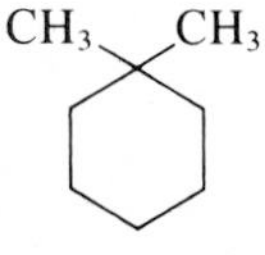

b.

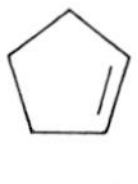

c.

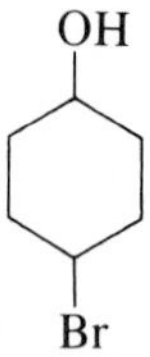

13.39 Angle strain

13.41 *a.* Addition, *b.* substitution

13.43 *a.*

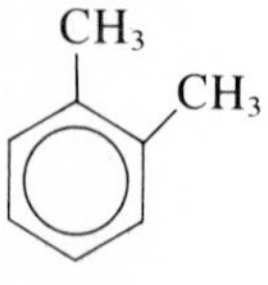

b.

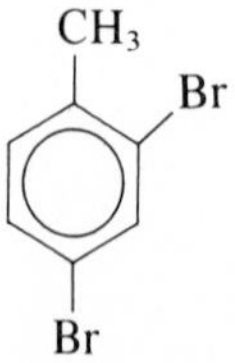

c.

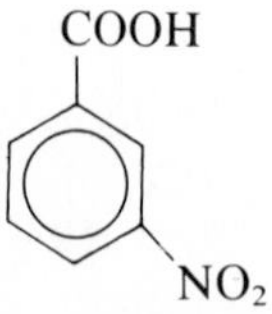

13.45 *a.* addition
b. substitution
c. elimination
d. substitution

13.47 Cyclopropane, ethyl chloride, and haloethane are used as anesthetics; mineral oil is used for skin lubrication, and fluorinated hydrocarbons are used as blood substitutes.

13.49 *a.* 1-Bromobutane, *b.* 3-chlorotoluene, *c.* 1-pentene, *d.* *cis*-2-hexene, *e.* 3,4-dimethyl-1-pentyne, *f.* *cis*-1,2-dichlorocyclohexane, *g.* *p*-ethyltoluene, *h.* 2-methyl-2-butene

13.51 *a.* 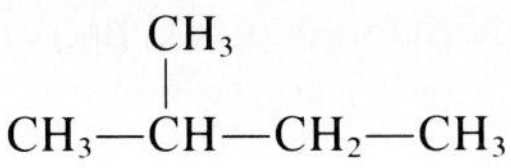 *b.* 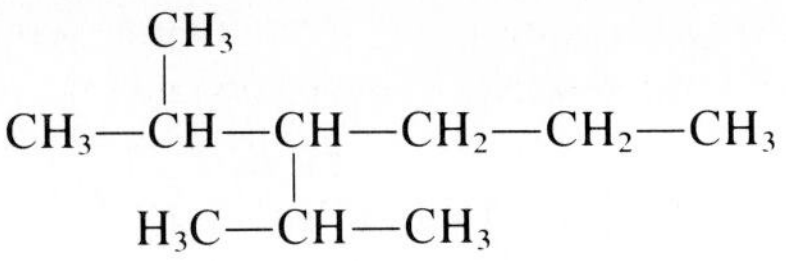 *c.* 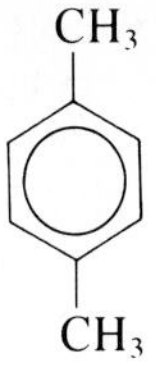

d. 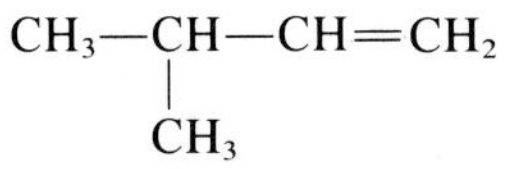 *e.*  *f.* 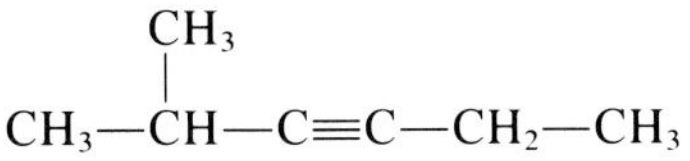

13.53 *a.* A bromine solution, which is red-brown in color, loses its color upon addition to an alkene but retains its color when added to water.
b. Alkynes
13.55 $CH_3—CH=CH—CH_3$

CHAPTER 14

14.1 *a.* 2-Propanol, *b.* 2,4-dimethyl-3-pentanol

14.2 *a.* $CH_3—CH=CH_2$, *b.* $CH_3—CH_2—CH=CH—CH_3$, *c.* 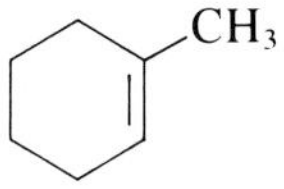

14.3 *a.* 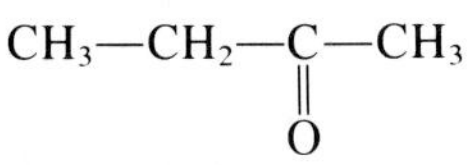 *b.* 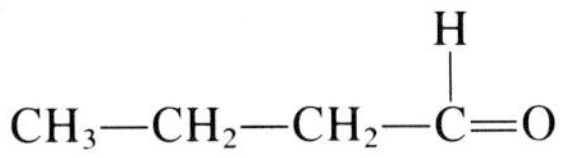 *c.* N.R.

14.4 3-Ethyl-2-methylhexanal
14.5 2-Methyl-3-pentanone
14.6 *b.* Hemiacetal, *d.* hemiacetal

14.7 *a.* 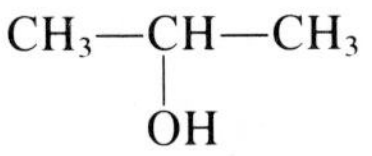 *b.* 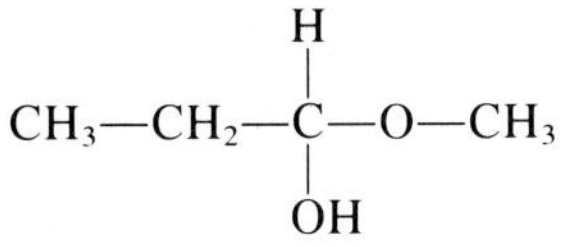 *c.* $CH_3—CH=N—CH_3$

d. 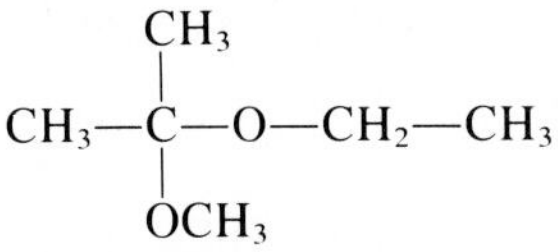

14.9 Ethyl alcohol molecules can hydrogen-bond to each other but dimethyl ether molecules cannot. The presence of this intermolecular force allows ethanol to exist as a liquid at room temperature. Because the molecules of dimethyl ether are not as strongly attracted to each other, it exists as a gas.
14.11 CH_3OH is more soluble in water than $CH_3CH_2CH_2OH$, which has a larger nonpolar portion, the carbon chain.
14.13 *a.* 1-Propanol, *b.* 2-butanol, *c.* 1,3-propanediol, *d.* cyclohexanol, *e.* 1,2-ethanediol, *f.* 2,2-dimethyl-1-propanol
14.15 Liquor, rubbing alcohol, moisturizing lotion, suppositories, tinctures, perfumes and colognes, hairspray, mouthwash, antiseptics, disinfectants
14.17 Denatured alcohol is alcohol to which 2-propanol has been added, rendering it unfit for internal consumption.
14.19 Sugar

14.21 *a.* $CH_3—O—CH_3$ *b.* 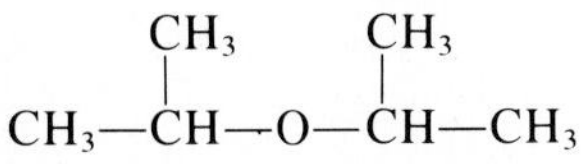 *c.* 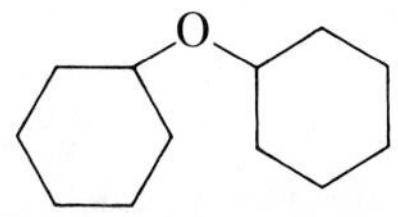

14.23 Ethanol competes with toxic methanol for the oxidizing enzyme. Since ethanol's oxidation is more effective than that of methanol, ethanol essentially ties up the enzyme, allowing unreacted methanol to be excreted in the urine.

14.25 *a*. Ethyl methyl ether, *b*. diisopropyl ether, *c*. ethyl propyl ether

14.27 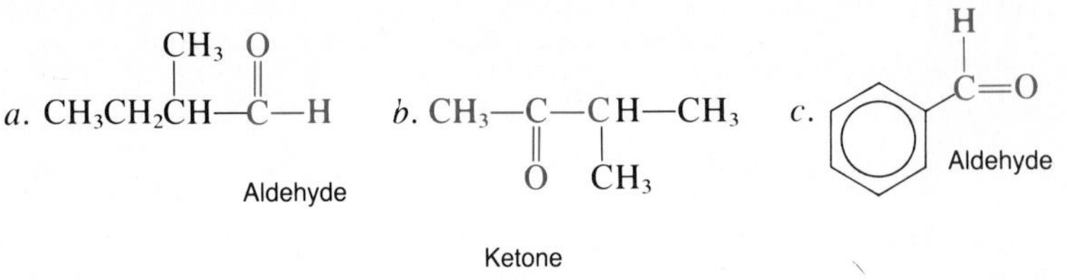
Aldehyde
Ketone
Aldehyde

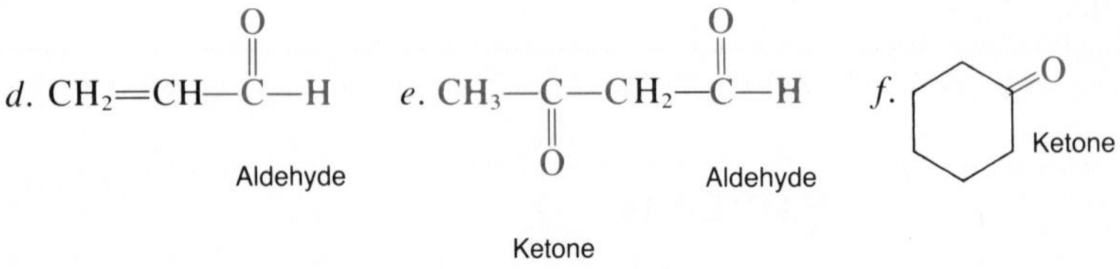
Aldehyde
Aldehyde
Ketone
Ketone

14.29 *a*. No, there can be no hydrogen bonding in aldehydes or ketones because no hydrogen is bound to oxygen.

b. The polarity of the carbonyl group is responsible for the added intermolecular force which causes these compounds to boil at higher temperatures.

14.31 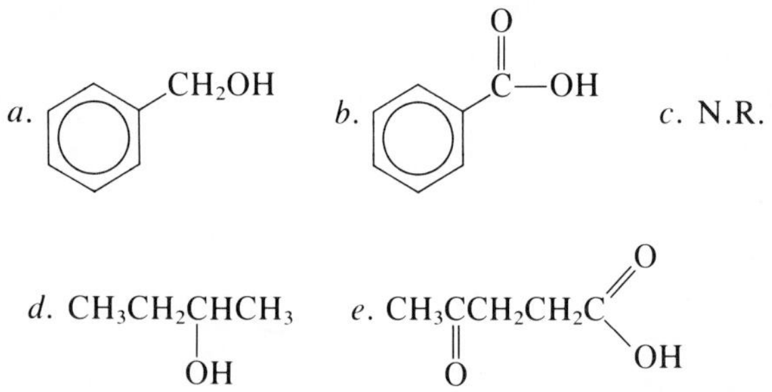

14.33 *b*, *c*

14.35 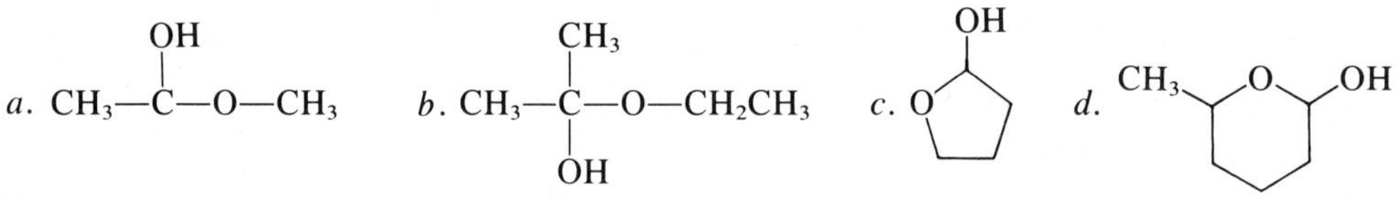

14.37 *a*. Gain of electrons, *b*. oxidizing agent

14.39 *b*, *c*

14.41 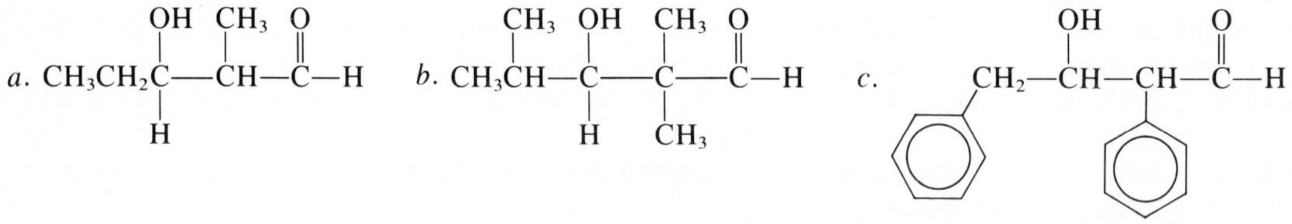

14.43 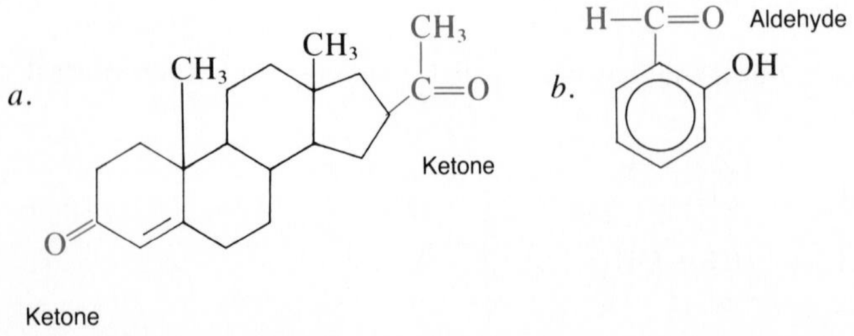
Ketone
Aldehyde
Ketone

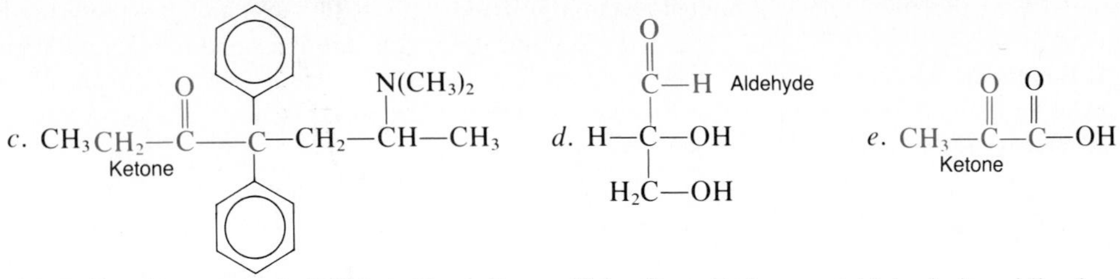
c.
Ketone
d.
Aldehyde
e.
Ketone

14.45 With Tollens' reagent, $Ag(NH_3)_2^+$. Free silver will be deposited as acetaldehyde is oxidized and Tollens' reagent reduced.
14.47 In a phenol, the hydroxyl group is bonded directly to an aromatic ring.
14.49 Phenol; H is more electropositive and O is more electronegative.
14.51 Ethyl alcohol hydrogen-bonds; ethyl thiol does not.
14.53 Aldehydes give positive Benedict's tests. On oxidation a primary alcohol yields an aldehyde; therefore the original alcohol was primary.
14.55 Ten-carbon phenol. Phenols are more acidic than alcohols and react more readily with base.

14.57 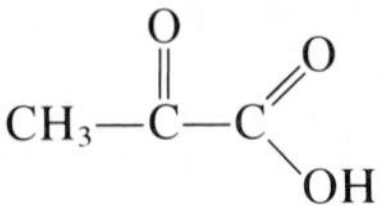

14.59 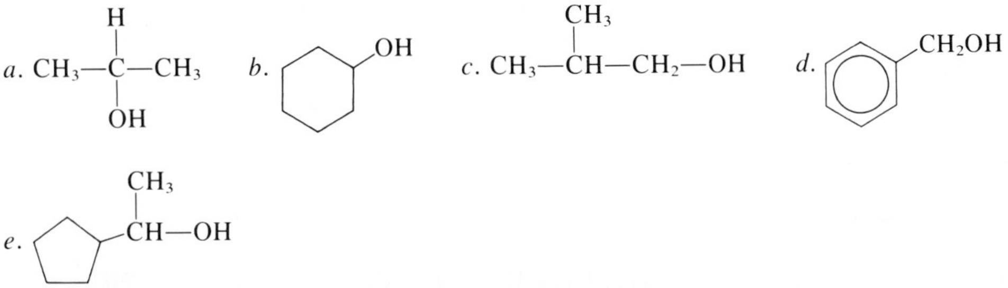

14.61 *a.* Reduction, *b.* oxidation, *c.* oxidation, *d.* oxidation, *e.* oxidation

14.63 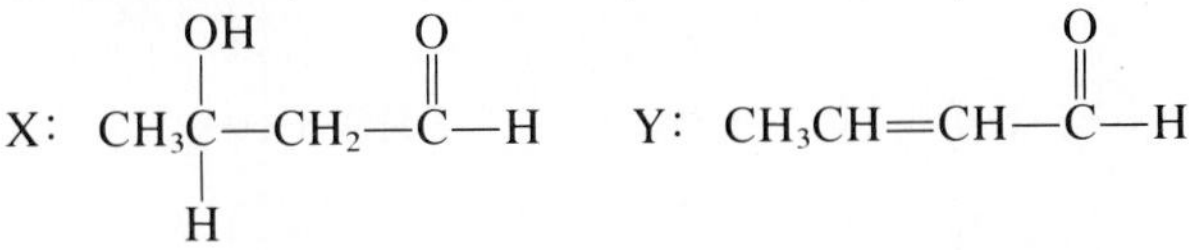

CHAPTER 15

15.1 *a.* Aldose, *b.* ketose, *c.* ketose, *d.* aldose
15.2 *a.* Aldoheptose, *b.* ketopentose, *c.* ketotriose, *d.* aldotetrose
15.3 +41°
15.4 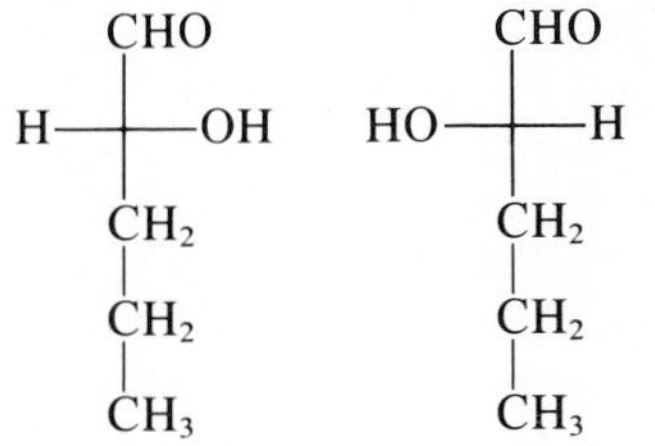

15.5 *a.* L, *b.* L, *c.* D
15.6 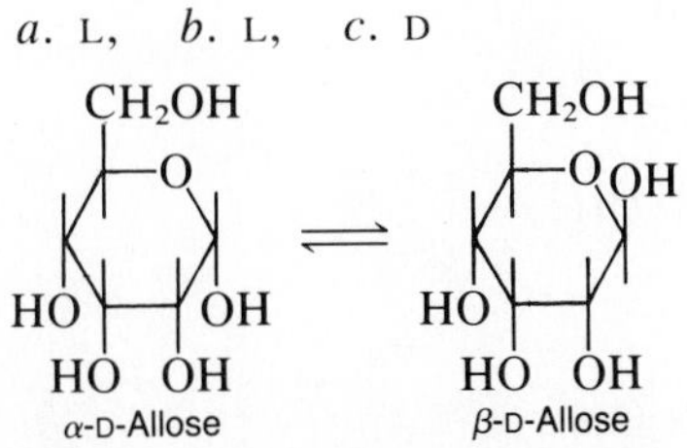
α-D-Allose
β-D-Allose

15.7 One should *not* expect the values of optical rotation of α and β forms to be related. The only predictable values of optical rotation are those of enantiomers, whose optical rotations are the same in magnitude but different in sign. Enantiomers are mirror images whose stereochemistry is different at *every single chiral carbon atom*. The cyclic hemiacetal form of glucose has *five* chiral carbons. The α and β forms of glucose are *anomers* whose stereochemistry differs only at the *first* carbon.

15.8 *a*. Trisaccharide, *b*. polysaccharide, *c*. monosaccharide, *d*. monosaccharide

15.9 *a*. β-1,4 *b*. α-1,6

15.10 The hemiacetal group (written at the far right in the structure) enables maltose to be a reducing sugar and undergo mutarotation.

15.11 Invert sugar is a reducing sugar. Its components, glucose and fructose, are both monosaccharides with hemiacetal groups and both are reducing sugars. Invert sugar is a 50-50 *mixture* of glucose and fructose.

15.13 Sugars, saccharides

15.15 *a*. Aldotriose, *b*. aldotetrose, *c*. aldohexose, *d*. ketohexose, *e*. ketotetrose, *f*. ketopentose, *g*. aldotetrose, *h*. aldohexose, *i*. ketotriose

15.17 *a*.

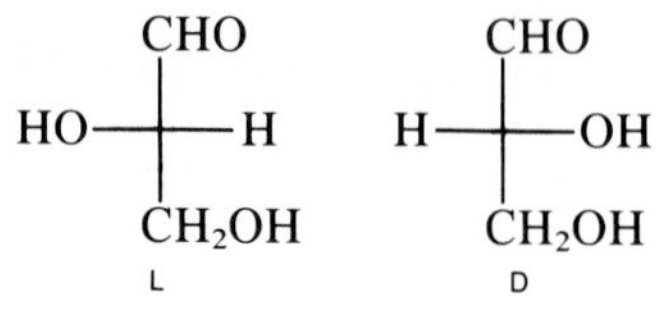
CHO
HO
H
CH₂OH
L
CHO
H
OH
CH₂OH
D

g.

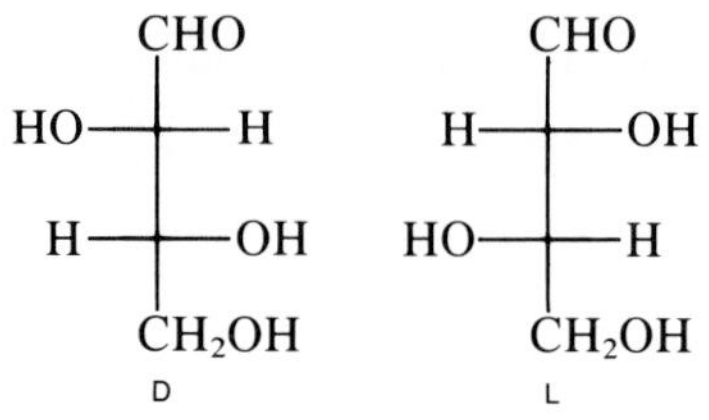
CHO
HO
H
H
OH
CH₂OH
D
CHO
H
OH
HO
H
CH₂OH
L

e.

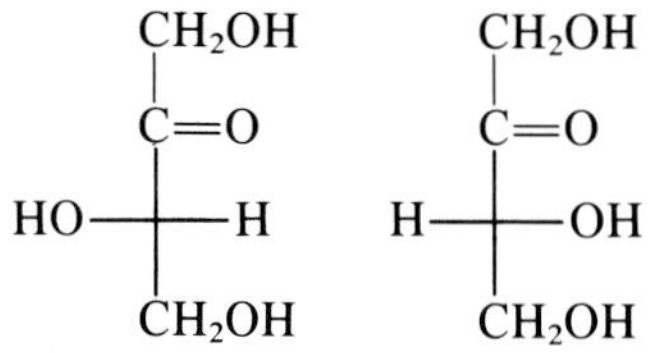
CH₂OH
C=O
HO
H
CH₂OH
CH₂OH
C=O
H
OH
CH₂OH

15.19 *a*.

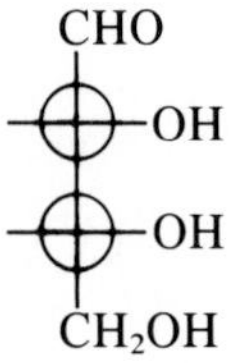
CHO
OH
OH
CH₂OH

b.

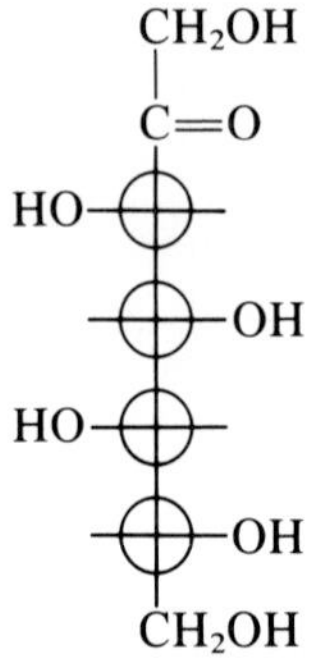
CH₂OH
C=O
HO
OH
HO
OH
CH₂OH

c.

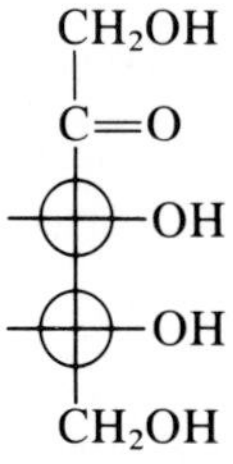
CH₂OH
C=O
OH
OH
CH₂OH

d.

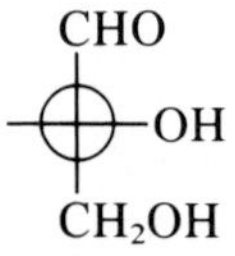
CHO
OH
CH₂OH

e.

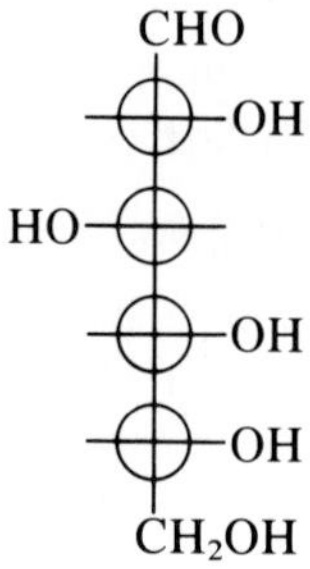
CHO
OH
HO
OH
OH
CH₂OH

15.21 There are six diasteromers of ribose:

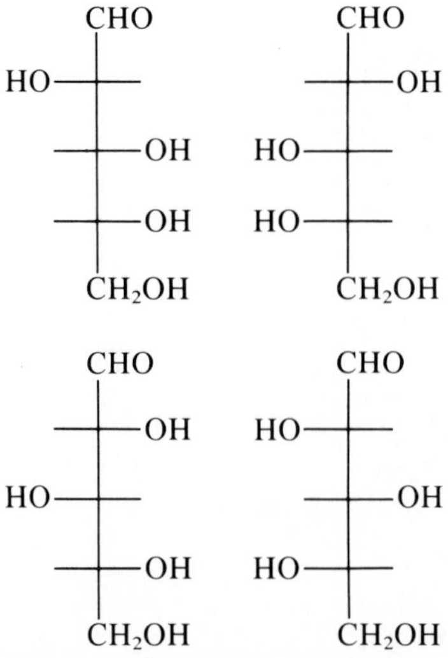
CHO
HO
OH
OH
CH₂OH
CHO
OH
HO
HO
CH₂OH
CHO
OH
HO
OH
CH₂OH
CHO
HO
OH
HO
CH₂OH

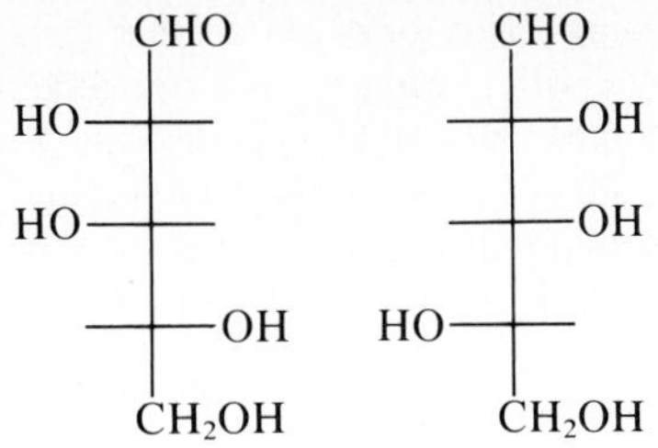

15.23

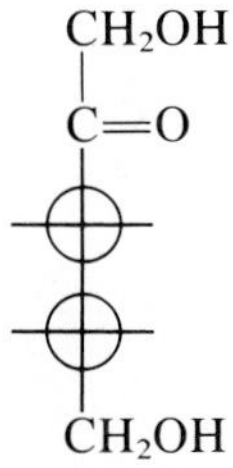

Ketopentoses have two chiral carbons. Maximum number of stereoisomers = $2^n = 2^2 = 4$.

15.25 Yes, b

15.27

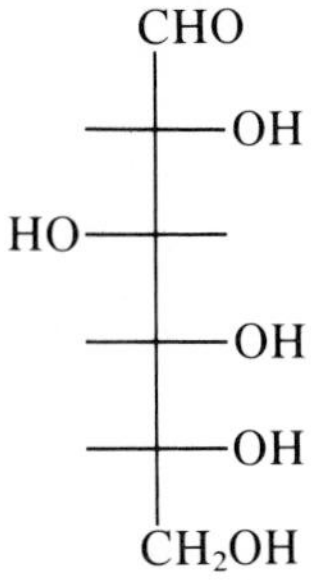

D-Glucose

15.29 Monosaccharides form cyclic structures because of the ability of the hydroxyl group and the aldehyde or ketone group to react intramolecularly to form a hemiacetal or a hemiketal.

15.31 *a.*

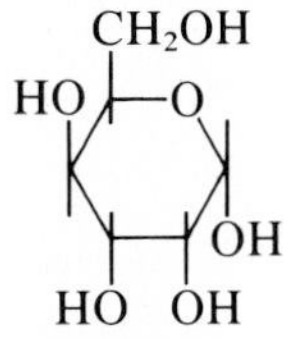

b.

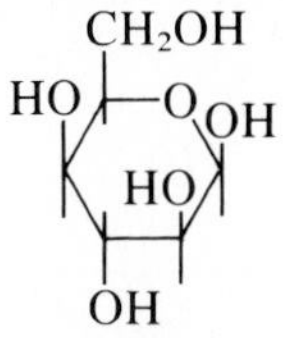

15.33

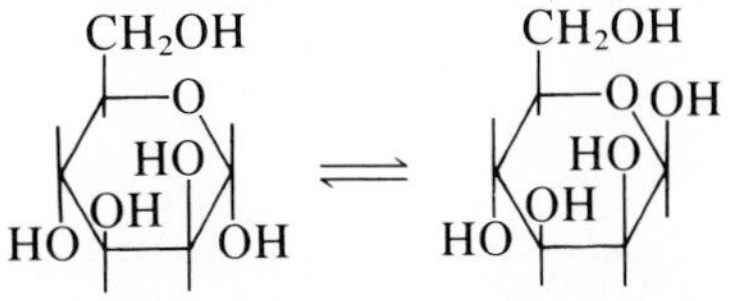

α-D-Mannose β-D-Mannose

15.35 *a.*

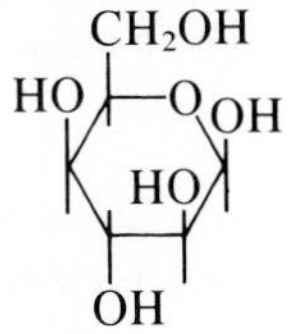

b.

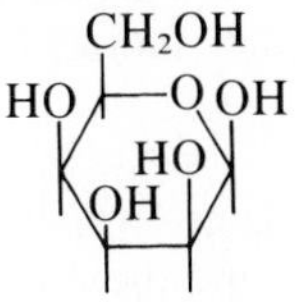

15.37 Fructose

15.39 *a.*

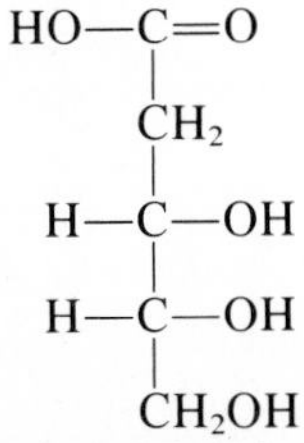

b.

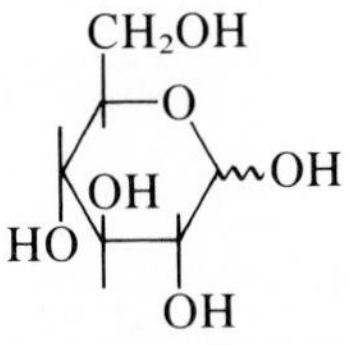

α or β

c.

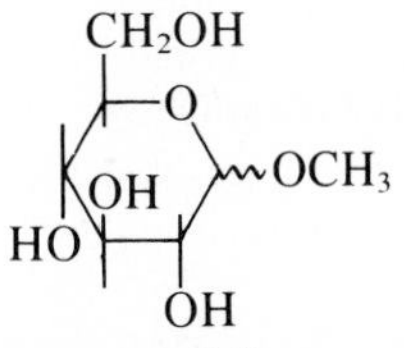

α or β

15.41 α-1,6 and β-1,4

15.43 Mutarotation and the reducing ability of monosaccharides are related by the fact that the cyclic hemiacetal or hemiketal is in equilibrium with the open-chain free aldehyde or free α-hydroxy ketone. It is the open-chain form (present in a small amount) that reacts with Benedict's or Tollens' reagent; when it is depleted, more is formed as the equilibrium shifts to compensate for the loss of the open-chain form. It is also the equilibrium between the open-chain form and the cyclic α or β form which allows the interconversion between the α and β forms: $\alpha \rightarrow$ open chain $\rightarrow \beta$; $\beta \rightarrow$ open chain $\rightarrow \alpha$. Gradually α reaches an equilibrium with β; this process is called mutarotation.

15.45 *a.* Trehalose (left) and gentiobiose (right); *b.* glucose

15.47 Isotrehalose (left); and neotrehalose (right)

15.49 *a.*

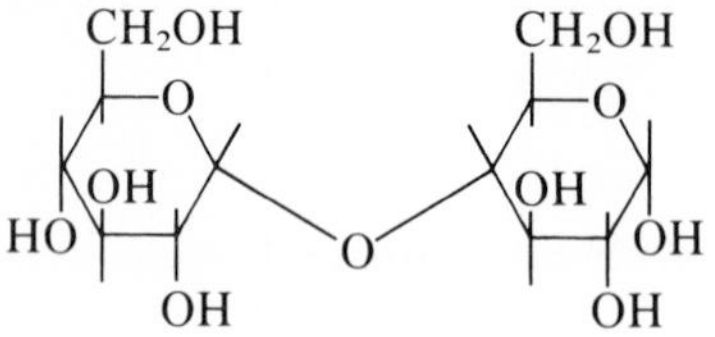
CH₂OH
O
OH
HO
OH
CH₂OH
O
OH
OH
OH
O

α-1,4

b.

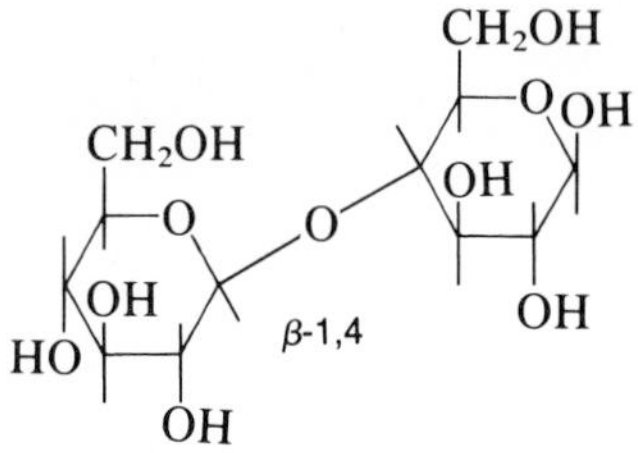
CH₂OH
O
OH
HO
OH
O
CH₂OH
O
OH
OH
OH

β-1,4

c.

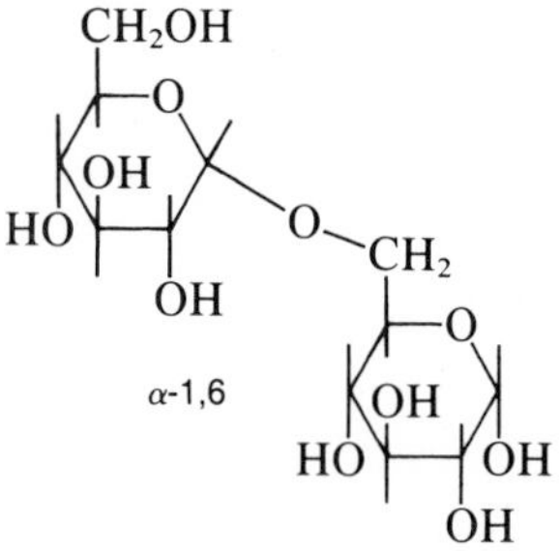
CH₂OH
O
OH
HO
OH
O
CH₂
O
OH
HO
OH
OH

α-1,6

d.

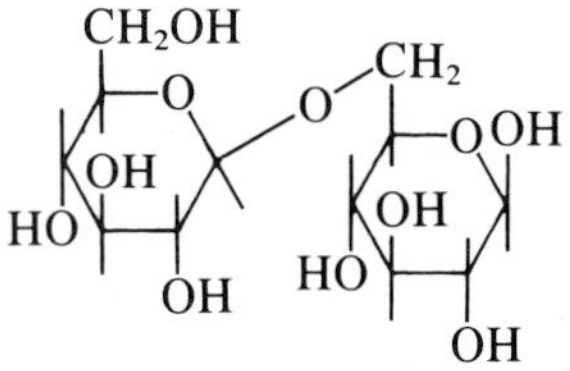
CH₂OH
O
OH
HO
OH
O
CH₂
O
OH
OH
HO
OH

β-1,6

15.51 Amylose has α-1,4 glycosidic linkages; cellulose has β-1,4 glycosidic linkages.

15.53 Glycogen is most like amylopectin because it is a polymer of glucose with a branched structure and all α glycosidic linkages (both 1,4 and 1,6). Glycogen has more branching and shorter branching chains than amylopectin.

15.55 *(1) c, (2) d, (3) a, (4) b, (5) g, (6) h, (7) f, (8) e*

15.57 Cellulose is a polymer of glucose but contains β glycosidic linkages which are *not* hydrolyzed by human enzymes. Therefore people cannot use cellulose as an energy source, and there are no calories in cellulose because the glucose units are never hydrolyzed from the cellulose polymer. If it were tasty, cellulose would be filling but would not supply *any* calories and simply be excreted in the feces.

15.59 Glucose (it need not be "digested") is already the chemical form required (right size, shape, etc.) to cross cell membranes and enter a metabolic pathway.

15.61 Enantiomers must be both nonsuperimposable *and* mirror images.

15.63 *a, b, e, f, g*

CHAPTER 16

16.1 *a.* Diethylamine, secondary; *b. n*-propylamine, primary; *c. N*-ethylaniline, secondary; *d.* 4-chloroaniline, primary

16.2 $\left[\begin{array}{c} \mathrm{CH_3} \\ | \\ \mathrm{CH_3CH_2\overset{\displaystyle}{N}-H} \\ | \\ \mathrm{H} \end{array} \right]^+ \mathrm{Br^-}$ 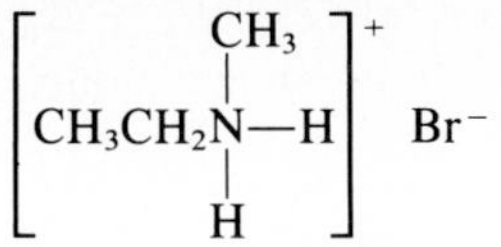

16.3 *a.* $\left[\begin{array}{c} \mathrm{H} \\ | \\ \mathrm{CH_3CH_2CH_2\overset{\displaystyle}{N}-H} \\ | \\ \mathrm{H} \end{array} \right]^+ \mathrm{Br^-}$ *b.* $\underset{\text{Amine}}{\mathrm{CH_3CH_2CH_2NH_2}} + \underset{\text{Water}}{\mathrm{H_2O}} + \underset{\text{Salt}}{\mathrm{NaBr}}$

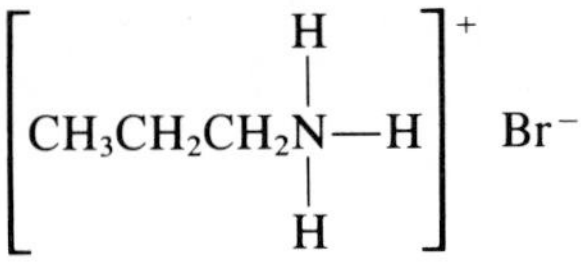

16.4 *a.* $\mathrm{CH_3Br}$

b. 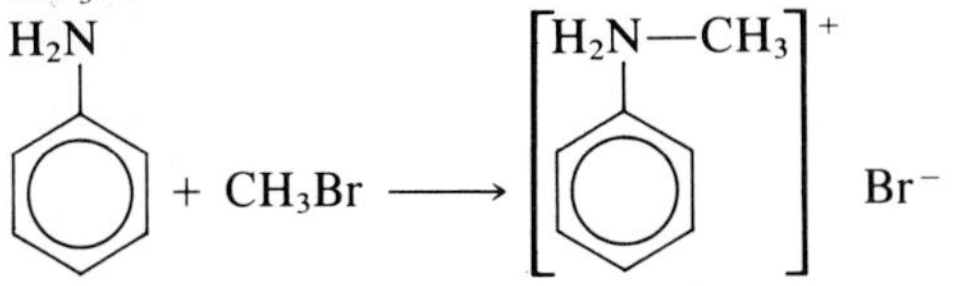

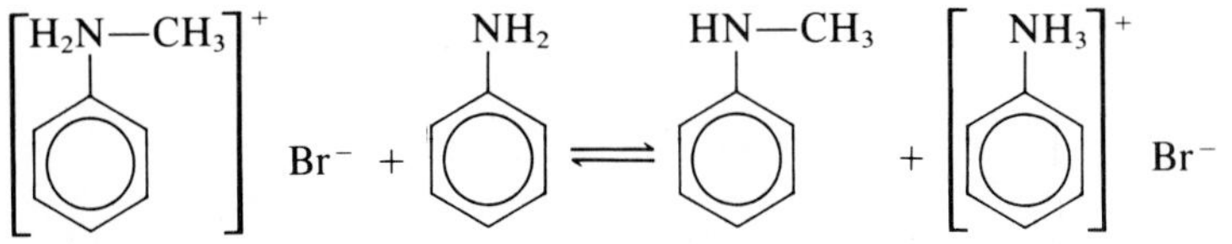

16.5 *a.* Proteins, nucleic acids;   *b.* choline, acetylcholine, seratonin, histamine, dopamine, norepinephrine

16.7 *a.* Methylamine,   *b.* isopropylamine,   *c.* triethylamine,   *d.* aniline,   *e.* N-ethylaniline,   *f.* 2-bromoaniline,   *g.* ethylisopropylmethylamine,   *h.* N-ethyl-N-methylaniline

16.9 *n*-Butylamine can exhibit intermolecular hydrogen bonding; since pentane does not have hydrogen bonding, the boiling point is lower.

16.11 $\mathrm{R-\overset{\displaystyle R}{\overset{|}{N}}-R}$ 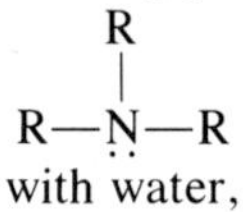 contains no H to hydrogen-bond to another $\mathrm{NR_3}$ molecule. When a tertiary amine hydrogen-bonds with water, the water molecule donates the H for the hydrogen bond between $\mathrm{H_2O}$ and the amine nitrogen.

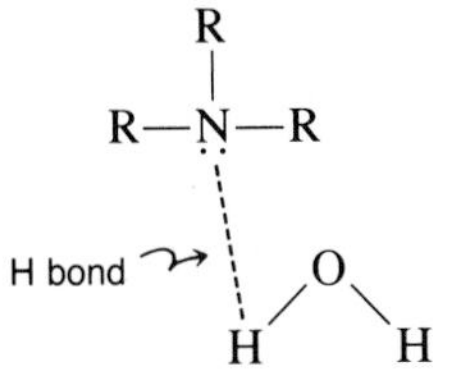

16.13 *a.* $\mathrm{CH_3CH_2NH_2} + \mathrm{H_2O} \rightleftharpoons \mathrm{CH_3CH_2NH_3^+} + \mathrm{OH^-}$ *b.* Reactants;   *c.* both will be blue.

16.15 *a.* $(\mathrm{CH_3CH_2NH_3^+})\,\mathrm{Br^-}$
ethylammonium bromide

b. $\left(\mathrm{CH_3CHCH_3} \atop | \atop \underset{+}{\mathrm{NH_3}} \right) \mathrm{NO_3^-}$ 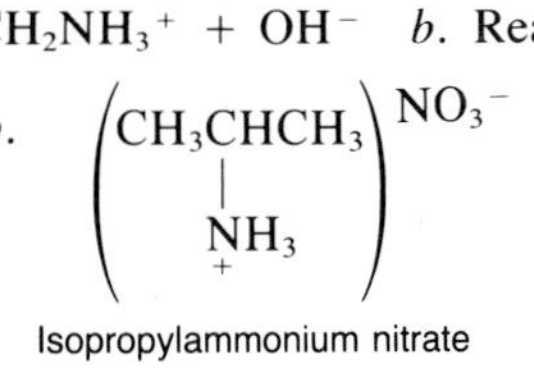
Isopropylammonium nitrate

c. 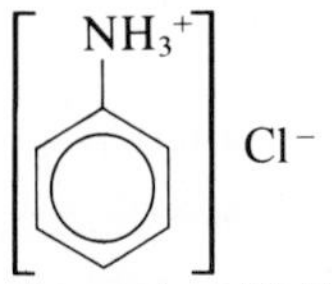
Phenylammonium chloride
(anilium chloride)

16.17 Brønsted-Lowry definition of a base: a base is a proton acceptor.

16.19 *a.* $\left[\begin{array}{c} \mathrm{H} \\ | \\ \mathrm{CH_3-\overset{\displaystyle}{N}-CH_2CH_3} \\ | \\ \mathrm{H} \end{array} \right]^+ \mathrm{Br^-}$ 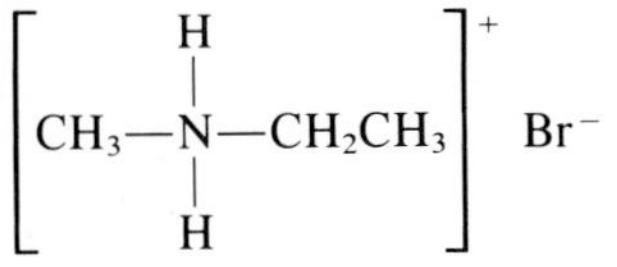   *b.* 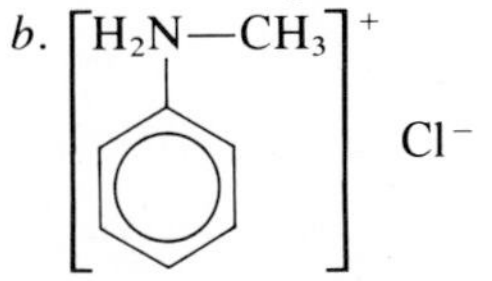

16.21 *a.* Pyridine,   *b.* imidazole,   *c.* pyrimidine,   *d.* indole,   *e.* pyrolidine

16.23 Ethylammonium chloride will be more soluble in water. It is an ionic compound, whereas methylamine is a covalent compound.

16.25 An alkyl halide could disrupt cellular processes by alkylating an amine nitrogen of DNA, thus disrupting cell-division processes.

16.27 Antihistamines *block* the receptor sites on which histamines would bind. They are similar in structure to histamine. By binding themselves to the receptor sites and preventing histamine from binding they relieve the *symptoms* of the allergic reaction.

16.29 Amines

16.31 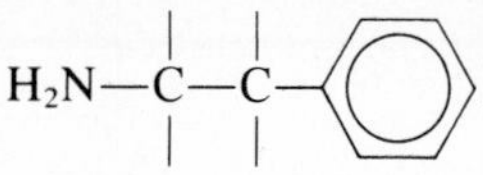 All have β-phenylethylamine and either ethers or phenols on aromatic ring.
16.33 Indole fused rings and ethylamine group
16.35 The N in the indole ring (secondary amine)

16.37 *a.* 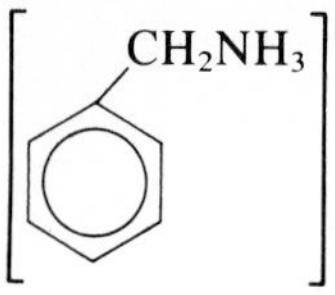 Br^- *b.* 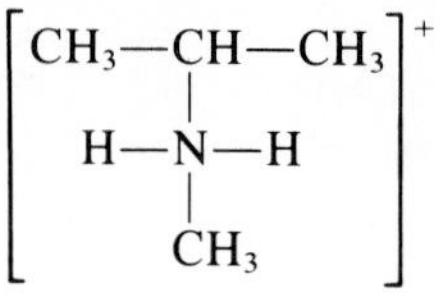 Br^- *c.* N.R.

d. 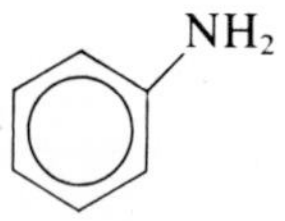 *e.* 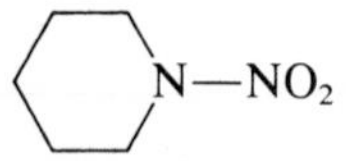 *f.* N.R.

16.39 *a.* Heroin; *b.* methylamine; *c.* acetylcholine; *d.* pyrrole; *e.* histamine; *f.* butorphanol, nalbuphine; *g.* cocaine; *h.* caffeine; *i.* LSD; *j.* codeine; *k.* seratonin; *l.* cimetidine

CHAPTER 17

17.1 $CH_3CH_2CH_2\overset{O}{\overset{\|}{C}}O{-}CH_3$

17.2 *a.* $CH_3\overset{O}{\overset{\|}{C}}OH$ and CH_3CH_2OH *b.* $CH_3\overset{O}{\overset{\|}{C}}OH$ and 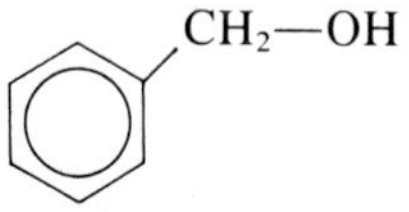

17.3 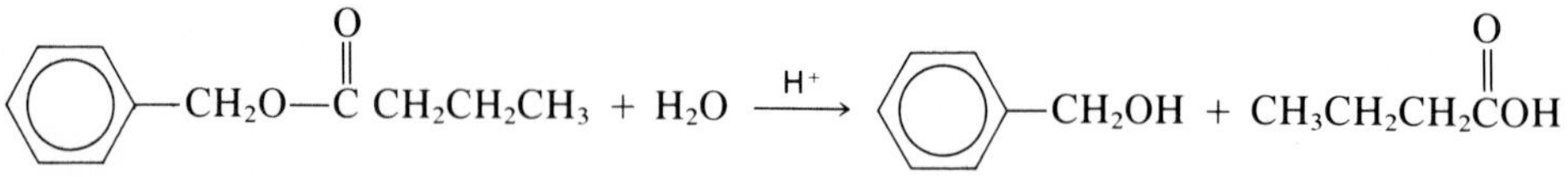

17.4 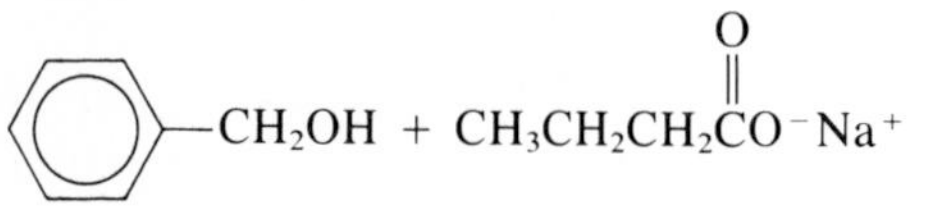

17.5 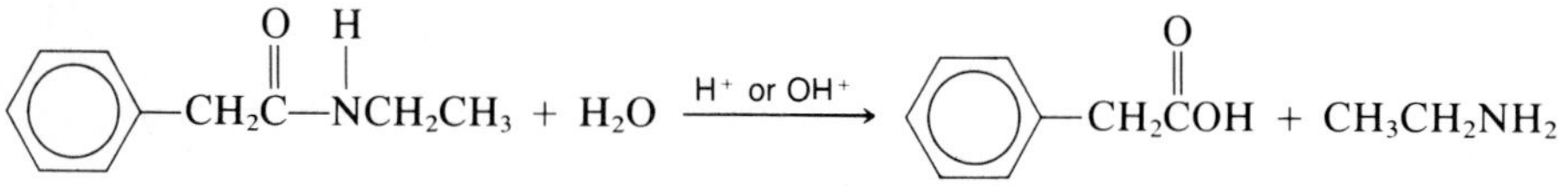

17.7 $CH_3CH_2\overset{O}{\overset{\|}{C}}OH$ *a.* $CH_3CH_2{-}\overset{O}{\overset{\|}{C}}{-}Cl$ *b.* $CH_3CH_2{-}\overset{O}{\overset{\|}{C}}{-}O{-}CH_3$ *c.* $CH_3CH_2{-}\overset{O}{\overset{\|}{C}}{-}N\langle$

d. $CH_3CH_2{-}\overset{O}{\overset{\|}{C}}{-}S{-}\underset{|}{\overset{|}{C}}{-}$

17.9 *a.* Lactic acid, *b.* salicylic acid, *c.* butyric acid
17.11 The 2 position, α; the 3 position, β
17.13 $b > c > a > d$
17.15 Tartness of fruits and vegetables is caused by the carboxylic acids they contain.

17.17 *a.* $CH_3\overset{O}{\overset{\|}{C}}O^-\ Na^+$ *b.* $CH_3\overset{O}{\overset{\|}{C}}O^-\left[CH_3\overset{CH_3}{\underset{CH_3}{N}}{-}H\right]^+$ *c.* $CH_3\overset{O}{\overset{\|}{C}}OCH_2CH_3$ *d.* $CH_3\overset{O}{\overset{\|}{C}}{-}O^-K^+$

17.19 *a.* $CH_3(CH_2)_{22}\overset{O}{\overset{\|}{C}}O^-\ Na^+$ *b.* soap
17.21 Phosphoglycerides also contain a hydrophobic, nonpolar hydrocarbon portion and a hydrophilic, polar, ionic portion. They emulsify by allowing the nonpolar portion to attract and "dissolve" other nonpolar sub-

stances; again the nonpolar portions stay together, shielded from the water, and the ionic phosphate ester remains on the outside, where it can interact with the water.

17.23 A detergent does *not* form an insoluble Ca^{2+} or Mg^{2+} salt as a soap does. These insoluble salts are soap scum.

17.25 *a* and *b* are structural isomers; *c* and *d* are structural isomers.

17.27 Condensation, also known as phosphorylation

17.29 *a.* $CH_3CH_2CH_2CH_2-\overset{\displaystyle O}{\overset{\|}{C}}-S-CH_2CH_3$ *b.* $CH_3-\overset{\displaystyle O}{\overset{\|}{C}}-S-CH_2CH_2CH_2CH_2CH_3$ *c.* $H-\overset{\displaystyle O}{\overset{\|}{C}}-S-CH\overset{\displaystyle CH_3}{\underset{\displaystyle CH_3}{<}}$ 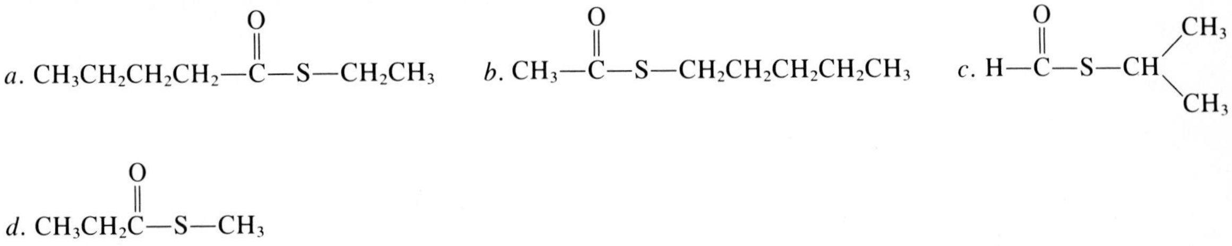

d. $CH_3CH_2\overset{\displaystyle O}{\overset{\|}{C}}-S-CH_3$

17.31 *a.* $\underset{\text{Acid}}{CH_3-\overset{\displaystyle O}{\overset{\|}{C}}}\!\!+\!\!\underset{\text{Alcohol}}{OCH_3}$ *b.* $\underset{\text{Acid}}{H-\overset{\displaystyle O}{\overset{\|}{C}}}\!\!+\!\!\underset{\text{Alcohol}}{OCH_3}$ *c.* $\underset{\text{Acid}}{CH_3CH_2-\overset{\displaystyle O}{\overset{\|}{C}}}\!\!+\!\!\underset{\text{Alcohol}}{O-\bigcirc}$ *d.* $\underset{\text{Acid}}{CH_3-\overset{\displaystyle O}{\overset{\|}{C}}}\!\!+\!\!\underset{\text{Alcohol}}{O-CH_2CH_2CH_3}$ 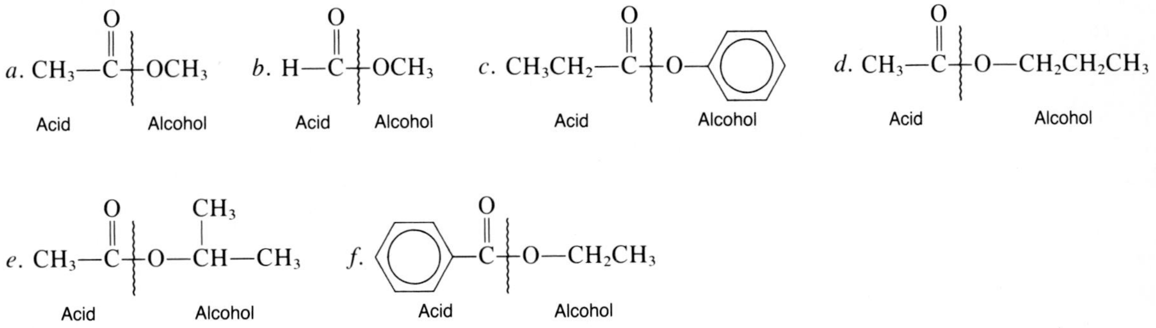

e. $\underset{\text{Acid}}{CH_3-\overset{\displaystyle O}{\overset{\|}{C}}}\!\!+\!\!\underset{\text{Alcohol}}{O-\overset{\displaystyle CH_3}{\overset{\displaystyle |}{CH}}-CH_3}$ *f.* $\underset{\text{Acid}}{\bigcirc-\overset{\displaystyle O}{\overset{\|}{C}}}\!\!+\!\!\underset{\text{Alcohol}}{O-CH_2CH_3}$

17.33 $c < e < a < d < b$

17.35 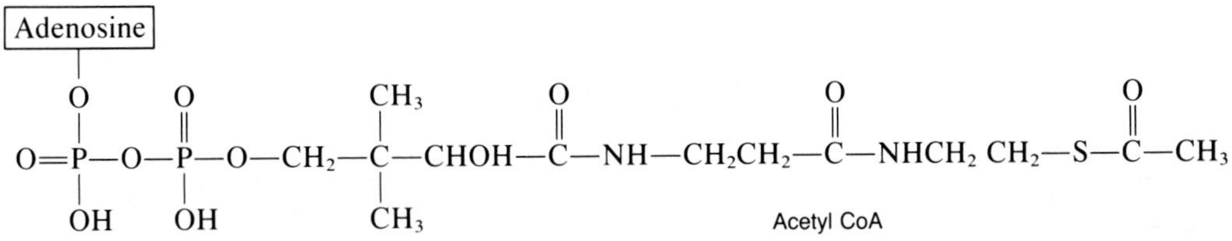 $\boxed{\text{Adenosine}}$

$O=\overset{\displaystyle O}{\underset{\displaystyle OH}{\overset{\|}{P}}}-O-\overset{\displaystyle O}{\underset{\displaystyle OH}{\overset{\|}{P}}}-O-CH_2-\overset{\displaystyle CH_3}{\underset{\displaystyle CH_3}{\overset{\displaystyle |}{\underset{\displaystyle |}{C}}}}-CHOH-\overset{\displaystyle O}{\overset{\|}{C}}-NH-CH_2CH_2-\overset{\displaystyle O}{\overset{\|}{C}}-NHCH_2\ CH_2-S-\overset{\displaystyle O}{\overset{\|}{C}}-CH_3$

                                                              Acetyl CoA

17.37 *a.* $\bigcirc-CH_2-\overset{\displaystyle O}{\overset{\|}{C}}OH + CH_3OH$ 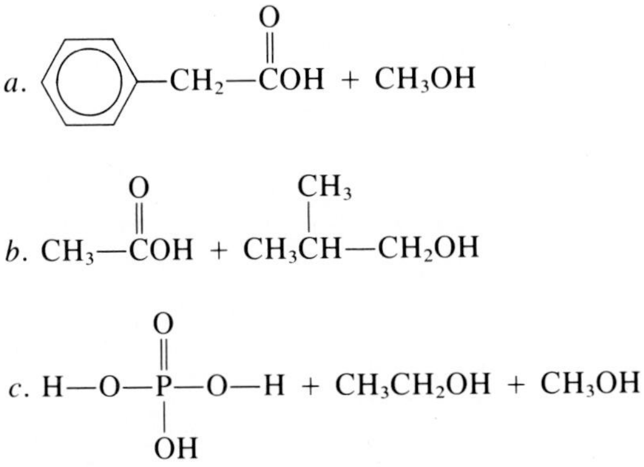

b. $CH_3-\overset{\displaystyle O}{\overset{\|}{C}}OH + CH_3\overset{\displaystyle CH_3}{\overset{\displaystyle |}{CH}}-CH_2OH$

c. $H-O-\overset{\displaystyle O}{\underset{\displaystyle OH}{\overset{\|}{P}}}-O-H + CH_3CH_2OH + CH_3OH$

17.39 Hydrolysis (ester cleavage by water): $CH_3CH_2-\overset{\displaystyle O}{\overset{\|}{C}}-O-CH_3 + HOH \xrightarrow{\ H^+\ } CH_3CH_2-\overset{\displaystyle O}{\overset{\|}{C}}OH + CH_3OH$ 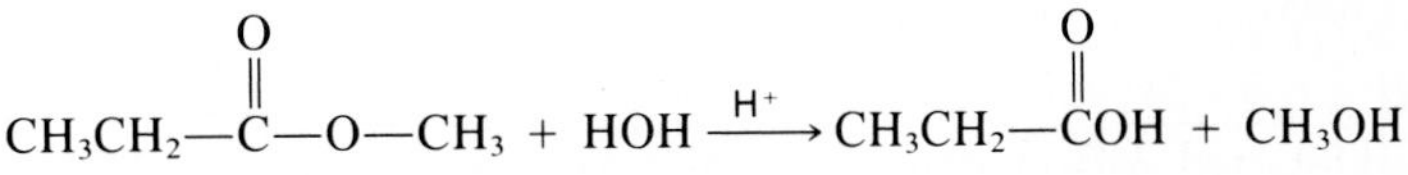

Saponification (ester cleavage in the presence of base): $CH_3CH_2\overset{\displaystyle O}{\overset{\|}{C}}OCH_3 + NaOH \longrightarrow CH_3CH_2\overset{\displaystyle O}{\overset{\|}{C}}O^-Na^+ + CH_3OH$ 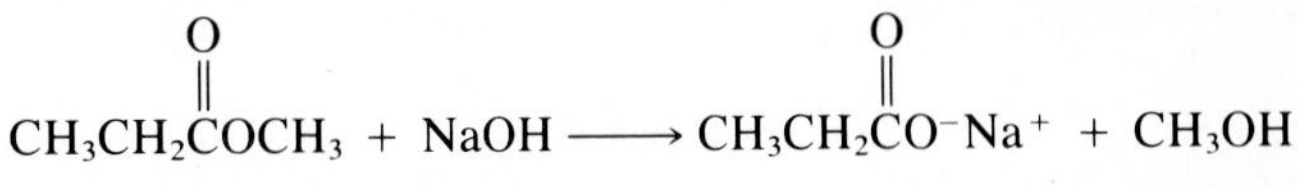

17.41 $3\ CH_3(CH_2)_{16}\overset{\displaystyle O}{\overset{\|}{C}}O^-\ Na^+ + \overset{\displaystyle CH_2-OH}{\underset{\displaystyle CH_2-OH}{\overset{\displaystyle |}{\underset{\displaystyle |}{CH-OH}}}}$ 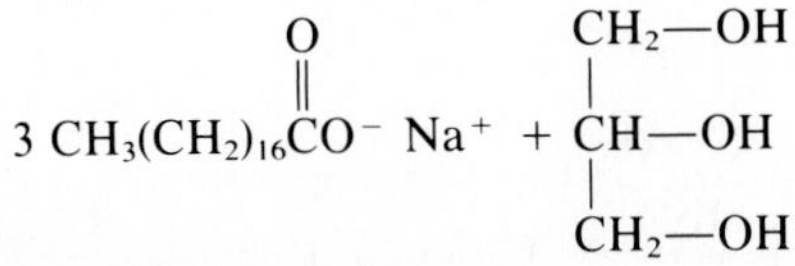

17.43 *a*. acid-amine salt or $CH_3\overset{\overset{O}{\|}}{C}-\overset{\overset{H}{|}}{N}-CH_2CH_3$,   *b*. acid-amine salt or $CH_3\overset{\overset{O}{\|}}{C}-\underset{\underset{CH_3}{|}}{N}-CH_2CH_3$

c. $CH_3\overset{\overset{O}{\|}}{C}O^- \left[H-\underset{\underset{CH_3}{|}}{\overset{\overset{CH_3}{|}}{N}}CH_2CH_3 \right]^+$

17.45 Isomers: *a, b, c*

17.47 $HOC\overset{\overset{O}{\|}}{}-\bigcirc-\overset{\overset{O}{\|}}{C}OH + H_2N-\bigcirc-NH_2$

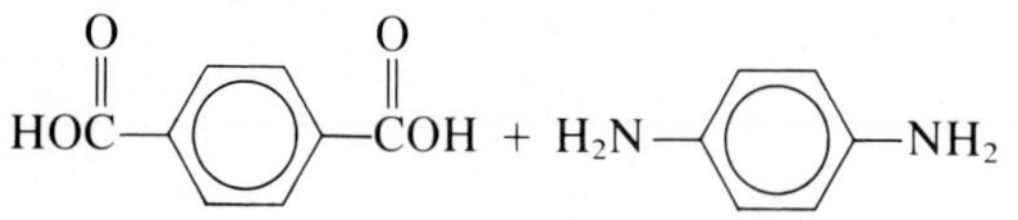

17.49 $NH_2\overset{\overset{O}{\|}}{C}NH_2 + HO\overset{\overset{O}{\|}}{C}-\underset{\underset{\bigcirc}{|}}{\overset{\overset{CH_2CH_3}{|}}{C}}-\overset{\overset{O}{\|}}{C}OH$

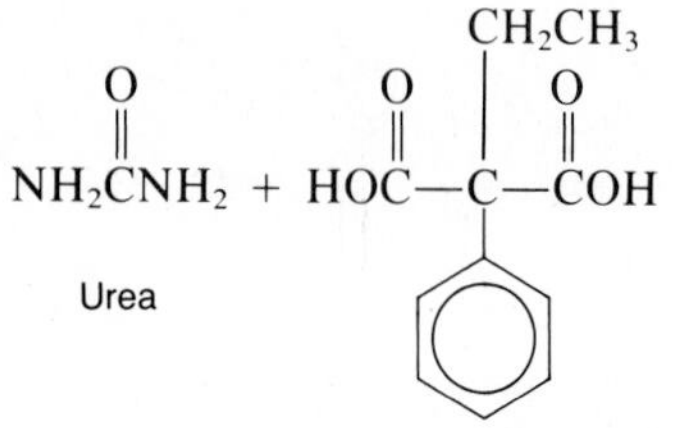

Urea

α-Ethyl-α-phenylmalonic acid
(2-ethyl-2-phenylmalonic acid)

17.51 Nylon is a polyamide. Acid solutions cause acid-catalyzed hydrolysis of the amide linkages; the nylon filament comes apart, and holes are formed.

CHAPTER 18

18.1 $CH_3(CH_2)_{12}\overset{\overset{O}{\|}}{C}O-CH_2$

$CH_3(CH_2)_{12}\overset{\overset{O}{\|}}{C}O-CH$

$CH_3(CH_2)_{12}\overset{\overset{O}{\|}}{C}O-CH_2$

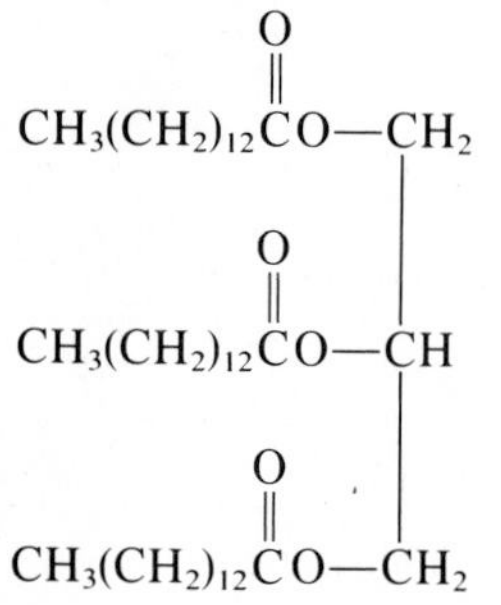

18.2 $CH_3(CH_2)_7CH=CH(CH_2)_7\overset{\overset{O}{\|}}{C}OH$ $CH_3(CH_2)_{14}\overset{\overset{O}{\|}}{C}OH$ $CH_3(CH_2)_{16}\overset{\overset{O}{\|}}{C}OH$ + $\underset{\underset{CH_2OH}{|}}{\overset{\overset{CH_2OH}{|}}{CHOH}}$

Oleic acid   Palmitic acid   Stearic acid   Glycerol

18.3 $CH_3(CH_2)_7CH=CH(CH_2)_7\overset{\overset{O}{\|}}{C}-O-CH_2$

$CH_3(CH_2)_{14}\overset{\overset{O}{\|}}{C}-O-CH$

$CH_2-O-\overset{\overset{O}{\|}}{\underset{\underset{O^-}{|}}{P}}-CH_2-CH_2-\overset{+}{N}\overset{H}{\underset{H}{-H}}$

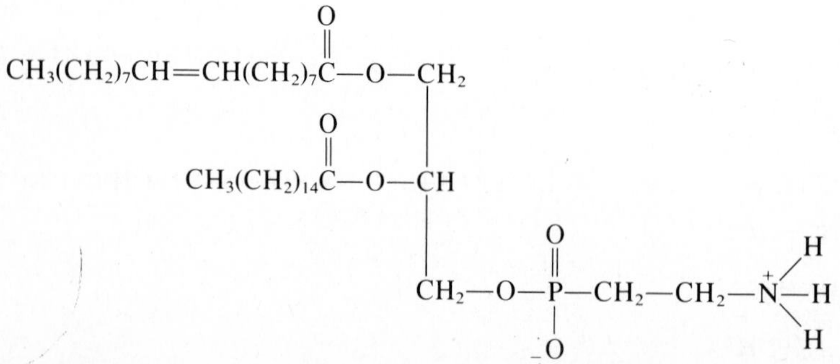

18.4 Sphingosine, a fatty acid, phosphoric acid, and an amine-containing alcohol

18.5 Estrone contains an *aromatic* ring and the —OH on that ring is a *phenolic* —OH; androsterone contains completely saturated rings and an *alcoholic*—OH. The remaining functional groups are identical.

18.7 Not *all* lipids contain an ester functional group, e.g., the steroids.

18.9 Waxes (carnauba wax), fats and oils (butter, corn oil, etc), vitamins, steroidal drugs, cell membranes

18.11 $CH_3(CH_2)_{14}\overset{O}{\overset{\|}{C}}OH + CH_3(CH_2)_{15}OH$

18.13 Waxes are protective water-resistant coatings on leaves.

18.15

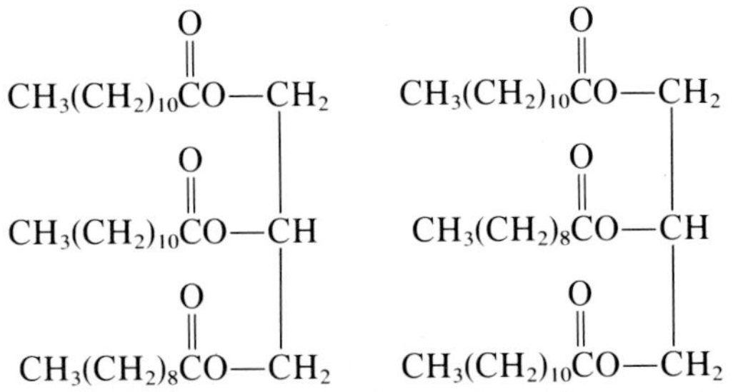

18.17 Oleic acid is unsaturated; stearic acid is completely saturated. Unsaturated fatty acids have much lower melting points than their saturated analogs of similar molecular weight. Therefore the mp of oleic acid (14°C) is much lower than that of stearic acid (70°C).

18.19 *a.* Oleic acid, *b.* oleic acid, *c.* lauric acid

18.21 Hydrogenation: addition of hydrogen to the double bonds

18.23 The iodine number is the number of *grams* of iodine absorbed per 100 g of fat or oil.

18.25 *a.* 50, *b.* 137

18.27 Bile salts have both a hydrophobic end and hydrophilic end and disperse the lipid by breaking it up into tiny droplets.

18.29

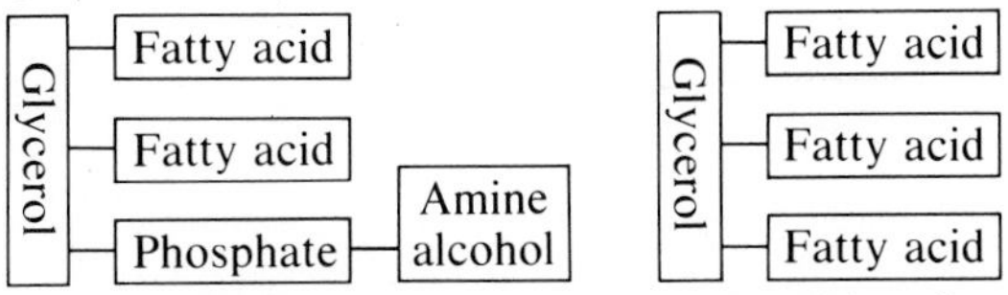
Glycerol
Fatty acid
Fatty acid
Phosphate
Amine alcohol
Glycerol
Fatty acid
Fatty acid
Fatty acid

Phospholipid Triacylglycerol

Both phospholipids and triacylglycerols have a glycerol backbone. In a triacylglycerol, the glycerol is esterified by three fatty acids. In a phospholipid, glycerol is esterified by two fatty acids and one phosphoric acid. In addition, the phosphoric acid is also esterified by an amine-containing alcohol.

18.31

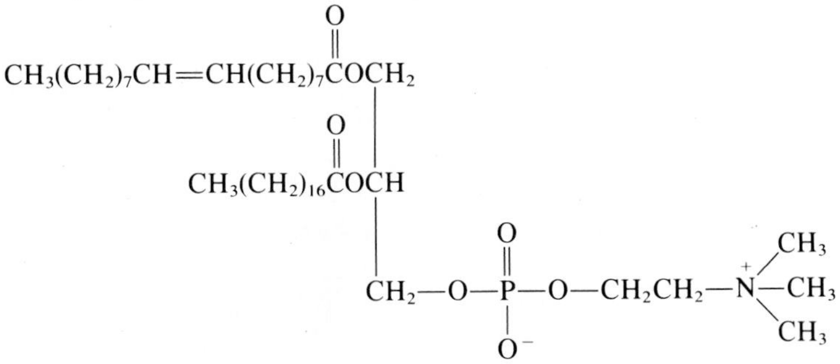

18.33 Since triacylglycerols *do not* contain an ionic hydrophilic portion, they are insoluble in H_2O, whereas cephalins, with a hydrophilic head, are partially soluble in H_2O.

18.35 The nonpolar region (hydrophobic) of the phospholipid bilayer contains alkyl chains which provide a jellylike consistency, especially when the double bonds are cis.

18.37 *a.* K^+ would have to move from higher to lower concentration, which is *not* a spontaneous process; it *requires energy* to move across a concentration gradient. *Active transport* requiring ATP energy would be needed. *b.* To move *into* the cell, Na^+ would move from higher to lower concentration; no energy is required; *simple diffusion* would be utilized. To move Na^+ from the cell into the blood plasma requires transportation across a concentration gradient. *Active transport* requiring ATP energy would be utilized.

18.39 No nuclear membrane in procaryotics (or any other membrane-enclosing organelles)

18.41 $CH_3(CH_2)_{12}CH=CH-CH-OH$ 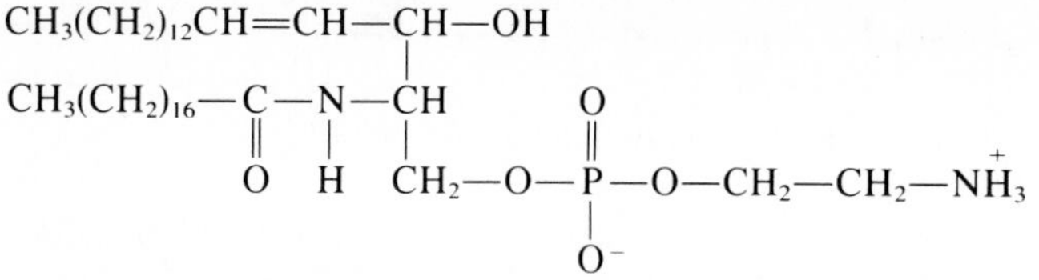

$CH_3(CH_2)_{16}-\underset{\underset{O}{\|}}{C}-\underset{\underset{H}{|}}{N}-\underset{\underset{CH_2-O-\underset{\underset{O^-}{|}}{\overset{\overset{O}{\|}}{P}}-O-CH_2-CH_2-\overset{+}{N}H_3}{|}}{CH}$

18.43 *a.* Sphingosine, a fatty acid, phosphate, and a nitrogen-containing alcohol   *b.* sphingosine, a fatty acid, and a carbohydrate   *c.* sphingomyelin has a phosphate and nitrogen-containing alcohol; glycolipids contain a carbohydrate   *d.* A sphingomyelin has an ionic hydrophilic head; a glycolipid does not have an ionic portion but is polar from all the —OH groups on the carbohydrate. The sphingomyelin is *more* polar.

18.45 In Niemann-Pick disease the enzyme necessary to degrade sphingomyelins is lacking.

18.47 *a.* Prostaglandins are formed from phospholipids, i.e., phosphoglycerides that have arachidonic acid at position 2.   *b.* Arachidonic acid

18.49 *a.* Prostaglandins (*1*) cause contraction or relaxation of smooth muscle, (*2*) lower blood pressure, (*3*) stop flow of gastric juice, (*4*) induce labor, (*5*) induce fever, (*6*) are involved in inflammation mechanism, and (*7*) influence blood coagulation.   *b.* Treatment of (*1*) hypertension, (*2*) arthritis, (*3*) asthma, and (*4*) ulcers, (*5*) prevention of conception and (*6*) blood clots, (*7*) relief of nasal congestion.

18.51 Cholesterol is *synthesized* by the body in the liver from acetyl CoA. If cholesterol is eliminated from the diet, it can still be made by the body and the blood level may not change.

18.53 Lipids are transported through the body by complexing them with proteins to form lipoproteins.

18.55 *a, e, g*

18.57 Ketone, alkene, secondary alcohol, aldehyde, ketone, primary alcohol, steroid

CHAPTER 19

19.1 $\overset{+}{N}H_3\underset{\underset{\underset{CH_3\quad CH_3}{|}}{CH}}{\overset{\overset{O}{\|}}{C}}HCO^-$ 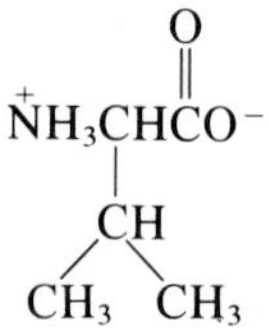

19.2 *a.* $NH_2-\underset{\underset{\underset{CH_3\quad CH_3}{|}}{CH}}{CH}-\overset{\overset{O}{\|}}{C}-\overset{\overset{H}{|}}{N}-CH_2-\overset{\overset{O}{\|}}{C}-\overset{\overset{H}{|}}{N}-\underset{\underset{CH_3}{|}}{CH}-\overset{\overset{O}{\|}}{C}OH$ 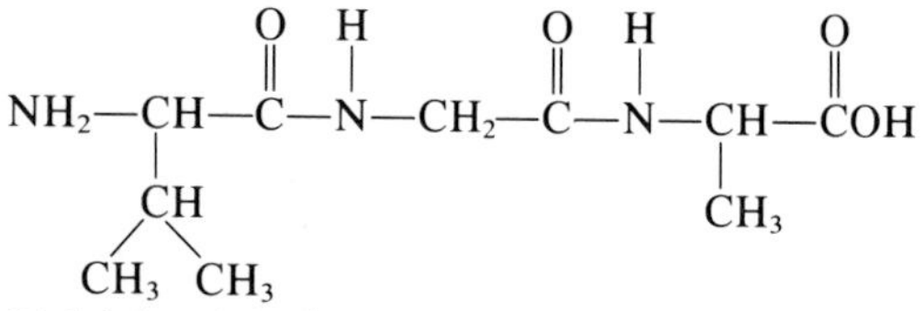

b. Valylglycylalanine

19.3 Tetrapeptide $+ 3H_2O \longrightarrow$ $\overset{+}{N}H_3\underset{\underset{\underset{CH_3\ CH_3}{|}}{CH}}{\overset{\overset{H\ O}{|\ \|}}{C}}-CO^-$ $+$ $\overset{+}{N}H_3\underset{\underset{CH_2}{|}}{C}HCO^-$ $+$ $\overset{+}{N}H_3\underset{\underset{\underset{+NH_3}{|}}{(CH_2)_4}}{CH}-\overset{\overset{O}{\|}}{C}O^-$ $+$ $\overset{+}{N}H_3CH_2\overset{\overset{O}{\|}}{C}O^-$ 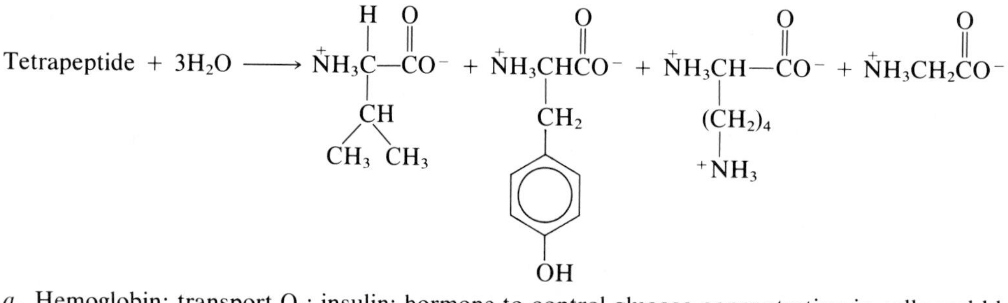

19.5 *a.* Hemoglobin: transport O_2; insulin: hormone to control glucose concentration in cells and blood; all enzymes, many structural materials like collagen and keratin.

19.7 *a.* $\overset{+}{N}H_3\underset{\underset{R}{|}}{C}H\overset{\overset{O}{\|}}{C}O^-$ *b.* $NH_2-\underset{\underset{R}{|}}{CH}-\overset{\overset{O}{\|}}{C}-O^-$ *c.* $\overset{+}{N}H_3-\underset{\underset{R}{|}}{CH}-\overset{\overset{O}{\|}}{C}-OH$ 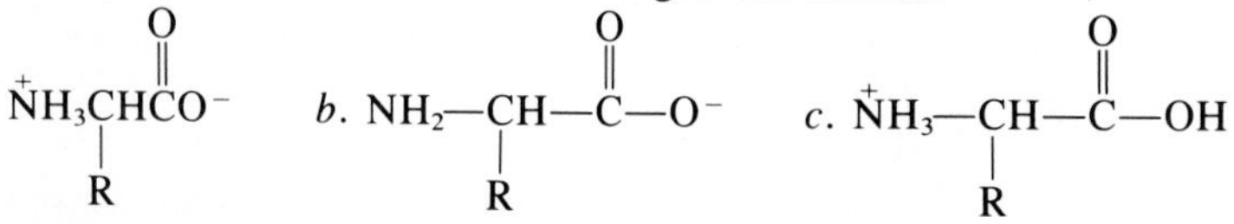

19.9 *a.* Hydrophobic, *b.* hydrophobic, *c.* hydrophilic, *d.* hydrophilic, *e.* hydrophobic, *f.* hydrophilic, *g.* hydrophilic

19.11 *a.* L, *b.* L

19.13

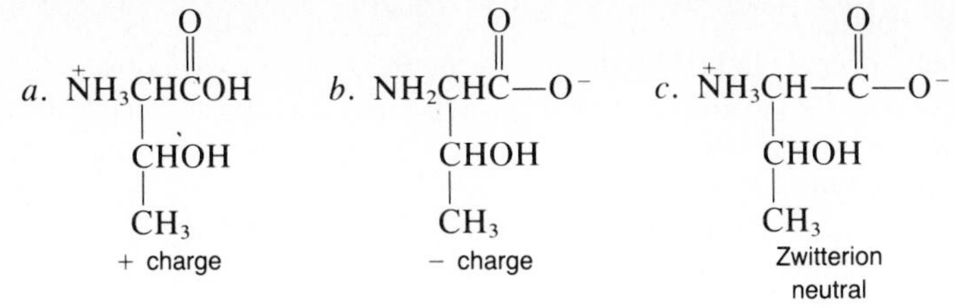
a.
+ charge
b.
− charge
c.
Zwitterion
neutral

19.15 No, at 7.4, the cation predominates, zwitterion at 9.74 (pI)

19.17 At a pH < pI, amino acid particles are + charged, repel each other, and do not aggregate. At a pH > pI, amino acid particles are − charged, repel each other, and do not aggregate. At the isoelectric point the zwitterion has a + charged end and a − charged end. The + end of one amino acid molecule is attracted to the − end of another amino acid molecule. There is attraction between particles; they aggregate and precipitate out of solution.

19.19 Glycine is a neutral zwitterion at pH 6 and would not be strongly attracted to the anode. Glutamic acid would be negatively charged at pH 6 and would migrate to the anode.

19.21 (*1*) Val-Leu-Ile, (*2*) Val-Ile-Leu, (*3*) Leu-Val-Ile, (*4*) Leu-Ile-Val, (*5*) Ile-Val-Leu, (*6*) Ile-Leu-Val

19.23

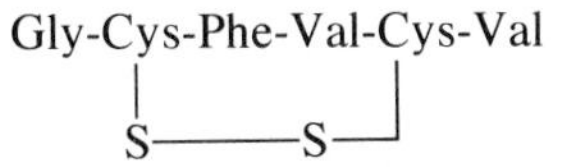
Gly-Cys-Phe-Val-Cys-Val
S——S

19.25

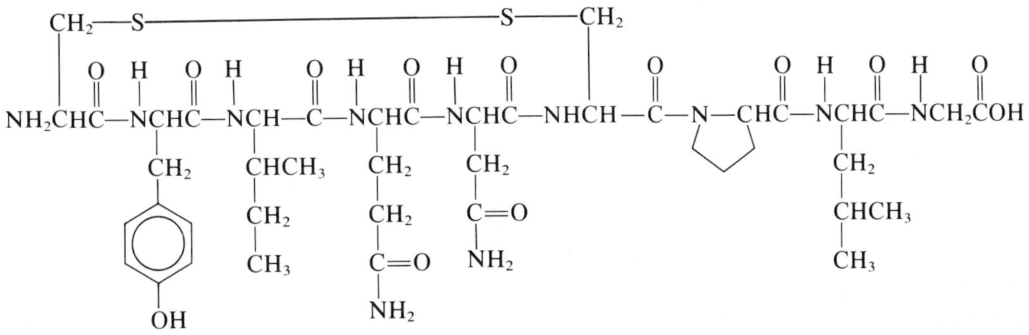

19.27 (*1*) Substance P, a neurotransmitter, responds to pain stimuli; (*2*) enkaphalins (Met and Leu) suppress release of substance P and regulate input to the brain of pain stimuli.

19.29 Primary structure: the *sequence* of amino acids (number, kind, and order)

19.31

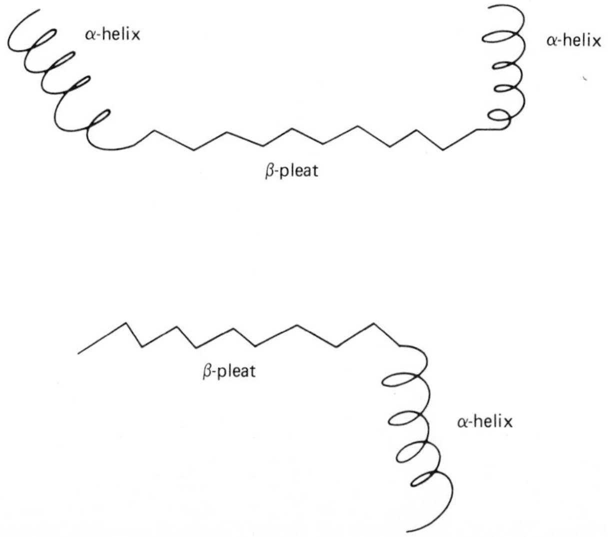
α-helix
α-helix
β-pleat
β-pleat
α-helix

19.33 Hydrogen bonding between peptide bonds.
19.35 Each chain in the collagen triple helix is cross-linked to another of the three chains in the triple helix by hydrogen bonds; each triple helix is cross-linked by covalent bonds to an adjacent triple helix. This tight coiling and extensive cross-linking give collagen its high tensile strength.
19.37 Vitamin C is required for the proper formation of collagen.
19.39 *a.* Ionic attractions (salt bridges), hydrogen bonding between side chains, disulfide bonds, and hydrophobic attractions *b. Tertiary*: attractive interactions between R *side chains*; *Secondary*: attraction is hydrogen-bonding between *peptide bonds*
19.41 *a.* Salt bridge

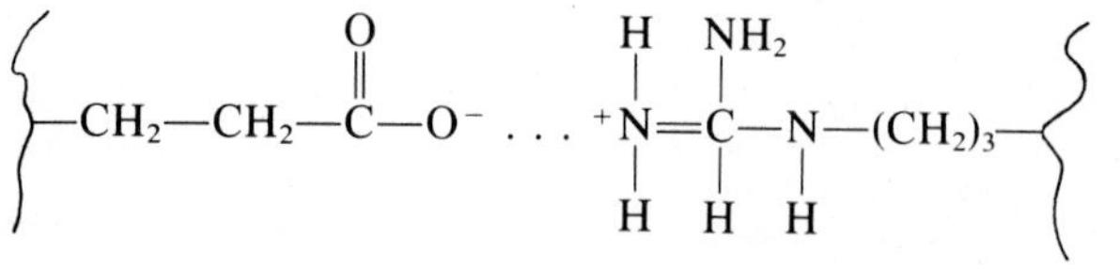

b. Hydrogen bonding

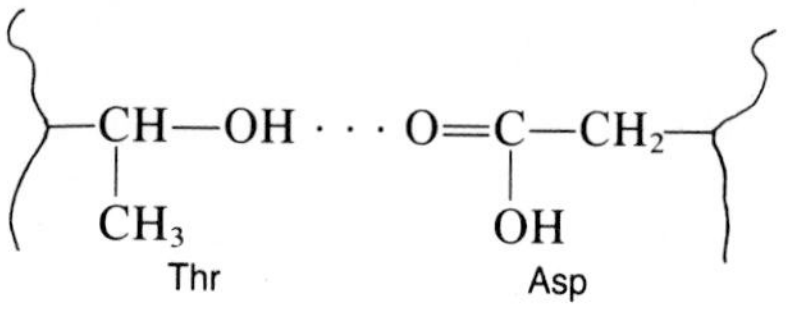

c. Hydrophobic

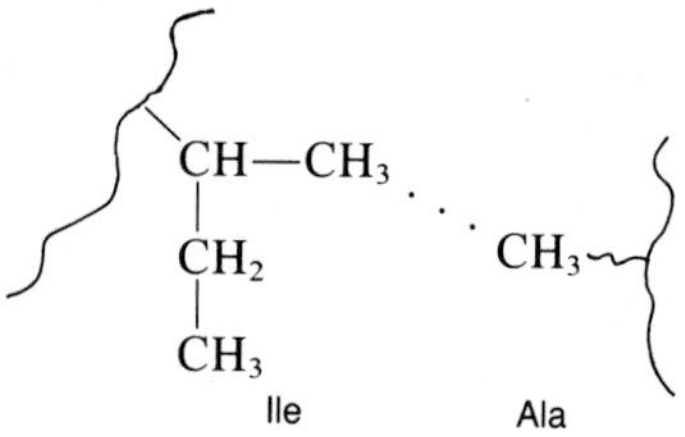

19.43 To lower the energy of the protein by maximizing the attractive forces
19.45 Pyrrole
19.47 *a.* Hemoglobin is a tetramer containing four heme units and four polypeptide chains. It has quaternary structure. Myoglobin consists of a single polypeptide chain and one heme unit; it has no quaternary structure. *b.* Myoglobin (Mb) picks up O_2 rapidly and becomes completely saturated. Hemoglobin (Hb) takes up O_2 more slowly and releases it at a higher partial pressure than Mb. This difference is due to the cooperativity of Hb (preference for all four O_2 on or all four O_2 off). When one O_2 binds to one chain of deoxyhemoglobin, a structural shift occurs between the other four chains so that they have a higher affinity for oxygen.
19.49 pH is lower; Hb binds less O_2 and is less efficient in delivering O_2 to the cells.
19.51 Bound to hemoglobin or as bicarbonate ion
19.53 *a.* In sickle-cell anemia the amino acid valine replaces glutamic acid on position 6 of each β chain. *b.* Val is nonpolar (Glu is polar), so that a hydrophobic interaction *between* hemoglobin molecules causes the Hb molecules to associate. This aggregation of Hb molecules tends to block small blood vessels.
19.55 Simple protein: composed solely of amino acids, e.g., human serum albumin; conjugated protein: composed of amino acids *plus* another component (prosthetic group), e.g., hemoglobin, in which the prosthetic group is heme
19.57 The protein of microrganisms is being denatured; the microorganisms are killed.
19.59 Pb^{2+} ions (not Pb^0 metal) are electrostatically attracted to COO^- or —SH groups.
19.61

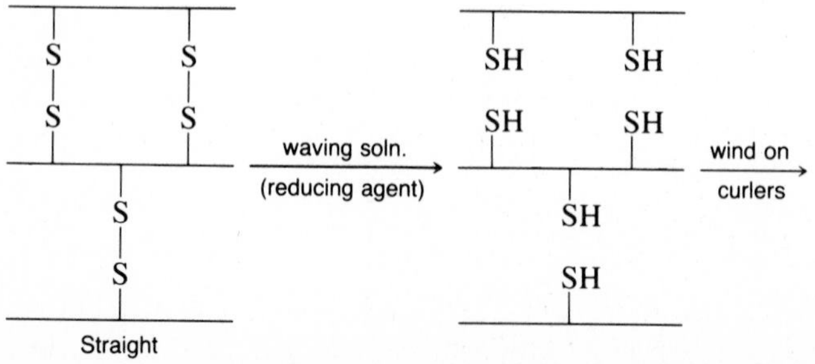
waving soln.
(reducing agent)
wind on
curlers
Straight

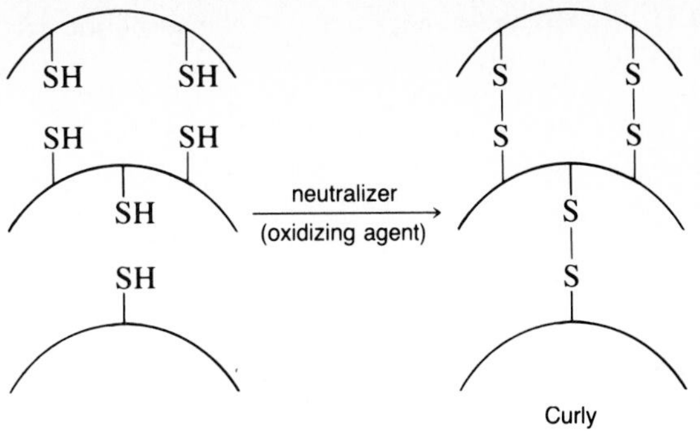
SH
SH
SH
SH
SH
SH
neutralizer
(oxidizing agent)
S
S
S
S
S
S
Curly

19.63 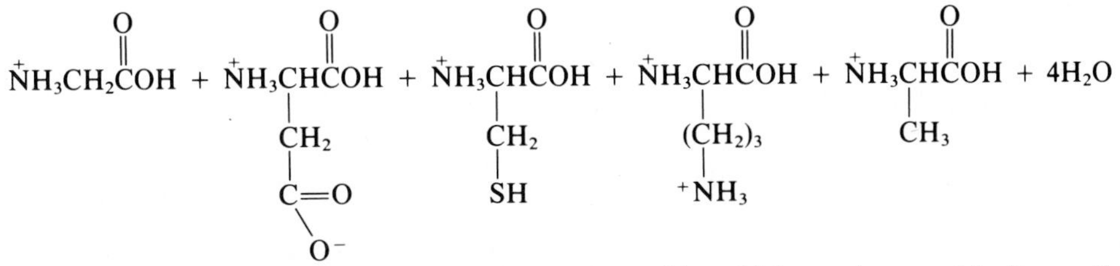

19.65 No; they are all hydrolyzed into constituent amino acids, which are then used in the synthesis of our body's proteins.

19.67 If insulin were taken by mouth, it would be digested (hydrolyzed) into constituent amino acids, just like any dietary protein, and would therefore be useless.

19.69 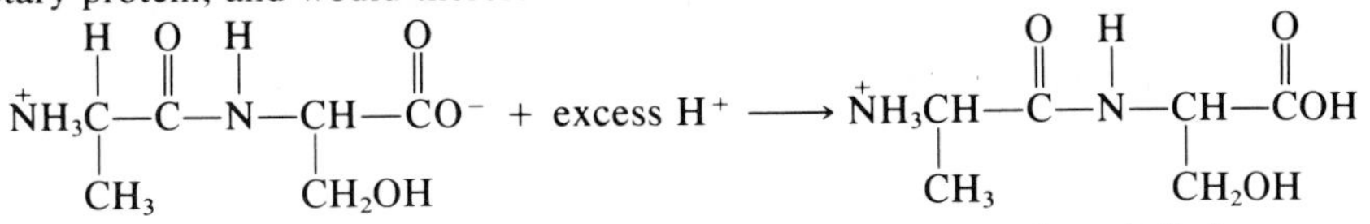
+ charge (cation)

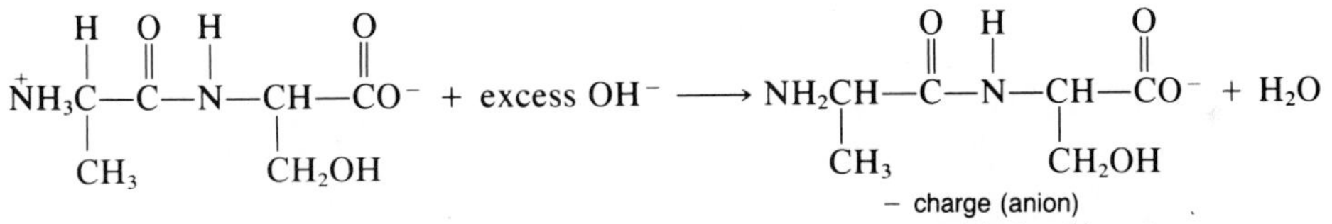
− charge (anion)

Proteins function as a buffer by being able to neutalize excess added acid or base.

CHAPTER 20

20.1 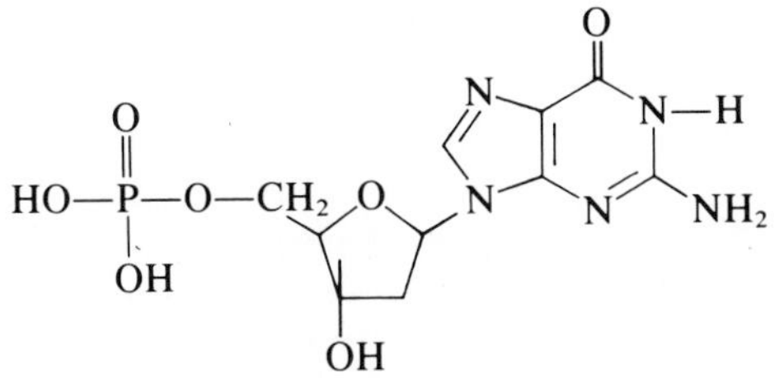

20.2 -G-C-C-A-T-G-C-T-

20.3 AAA, AAG

20.4 Asn-Trp-Pro

20.5 UUU, UUC

20.6 No change in sequence; UUC and UUU both code for Phe

20.7 DNA

20.9 *a.* DNA has as a component 2-deoxyribose sugar; *b.* DNA is found in the nucleus; *c.* DNA is acidic by virtue of phosphoric acid groups.

20.11 *a.* Mixture of nucleotides; *b.* phosphoric acid, deoxyribose, and nitrogenous heterocyclic bases

20.13 *a.*

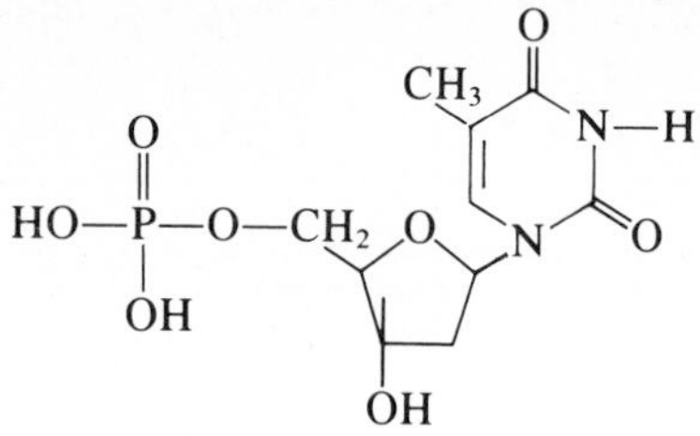

b.

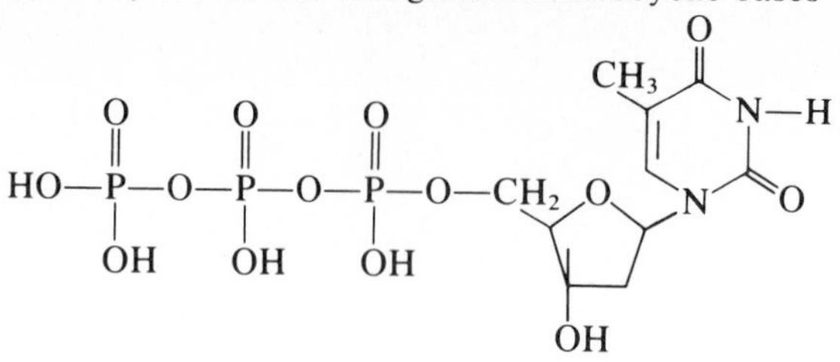

c. Formation of an acid anhydride

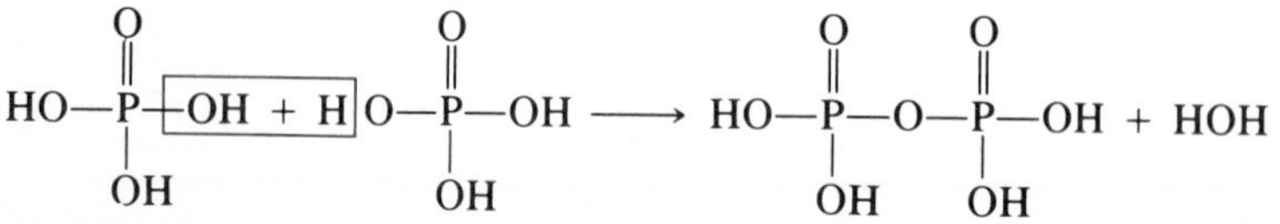

20.15 The *primary structure* of DNA is the sequence of nitrogen-containing heterocyclic bases, specifying the number, kind, and order of nucleotides.

20.17 Hydrogen bonding between bases holds the helix together. The spacing between the two chains allows for a perfect fit of one purine and one pyrimidine. The strongest hydrogen bonding is between an A and T and a C and G. Wherever there is an A on one chain, there is a T on the other; wherever there is a C on one chain, there is a G on the other.

20.19 -G-T-A-C-T-G-A

20.21 Each daughter molecule contains one strand from the parent and one newly synthesized strand.

20.23 All the strands replicated from the daughter strand with the error will contain the same error.

20.25 *a.* No; *b.* central dogma of biochemistry: DNA $\rightarrow$ RNA $\rightarrow$ protein

20.27 *a.* In both transcription and replication, DNA unwinds, and a strand of nucleic acid is synthesized by complementary base pairing. *b.* replication $\Rightarrow$ 2 daughter DNAs; transcription $\Rightarrow$ mRNA

20.29 *a.* tRNA delivers amino acids to the site of protein synthesis; *b.* 20, one for each amino acid; *c.* cytoplasm

20.31 In the sequence of heterocyclic bases (number, kind, and order)

20.33 *a.* A *codon* is a sequence of three bases (triplet) which specifies a specific amino acid; *b.* UUA; *c.* codes for the amino acid leucine

20.35 T is only in DNA; RNA contains U instead of T. mRNA carries codons.

20.37 *a.* A gene is a section of a chromosome (DNA molecule) which codes for specific protein. *b.* No; many genes. One chromosome in *E. coli* contains 3000 genes. *c.* mRNA is the same as the informational DNA strand except that U replaces T.

20.39 *a.* mRNA AUG CCU CGA GGA GAU
b. Met — Pro — Arg — Gly — Asp

20.41 *a.* AUG; *b.* UAA, UAG, UGA

20.43 *a.* AUG; *b.* initiation of the synthesis of the protein-start signal; *c.* methionine; d. P site; *e.* the second codon

20.45 *a.* Val, codons GUU, GUC, GUA, GUG tRNA anticodons CAA, CAG, CAU, CAC
b. Four different tRNAs can carry Val because there are four codons for Val.

20.47 The protein can be modified by cleavage of Met, addition of prosthetic group, etc. The protein chain then coils and folds to yield secondary, tertiary, and quaternary structure.

20.49 *a.* Regulatory gene: codes for the synthesis of the repressor molecule; *b.* Repressor molecule: binds to the control site (operator) and prevents transcription by preventing (blocking) binding of RNA polymerase; *c.* Control site: contains the promoter-operator sites (site of binding of the repressor molecule or site of binding of RNA polymerase) and controls transcription; *d.* Operator site: site where the repressor molecule binds; *e.* Inducer molecule: binds to the repressor molecule, forms an inducer-repressor complex, and removes the repressor bound to operator site, thus turning transcription on; *f.* Promoter site: site where RNA polymerase is bound; if it is bound, transcription occurs; otherwise it does not; *g.* Structural gene: codes for the mRNA which synthesizes a particular protein

20.51 *a.* X-rays can lead to mutations by raising an electron in the molecule to a higher energy level or completely ionizing the molecule and producing an unstable intermediate, which reacts further to alter a base. *b.* The amine group

20.53 Unmutated = Leu-Asn-Arg-Cys; *a.* mutant = Leu-Lys-Gly-Cys, *b.* mutant = Leu-Asn-Arg-Cys, *c.* mutant = Leu-Asn-Pro-Met, *d.* mutant = Leu-Asn-Arg

20.55 *a.* Donor structural gene: gene which codes for the desired substance such as insulin; it is spliced into a

foreign DNA (plasmid). *b.* Plasmid: small circular portion of DNA used to carry the foreign donor gene back into the bacterial cell. *c.* Restriction endonuclease: enzyme which cleaves DNA phosphodiester linkages at a specific base sequence. It is used to cut the genes. *d.* Bacterium: foreign substance into whose DNA a donor structural gene is incorporated. The bacterium DNA now produces the substance coded for by the foreign gene.
20.57 *a.* Proteins; *b.* no, only proteins

CHAPTER 21

21.1 *a.* Enzymes are largely proteins; inorganic catalysts show a variety of composition. *b.* Enzymes are biological catalysts; enzymes catalyze reactions of the body and are specific. Inorganic catalysts are more general in their action.
21.3 Tissue damage. In general the heat or pH necessary to catalyze cell reactions in the absence of enzymes would destroy the cell.
21.5

$$\underbrace{\text{Holoenzyme}}_{\text{Whole enzyme}} = \underbrace{\text{Apoenzyme}}_{\text{Protein portion}} + \underbrace{\text{Cofactor}}_{\text{Nonprotein portion}}$$

The cofactor may be an inorganic metal ion or an organic molecule, called a coenzyme, which is usually a vitamin.
21.7 The apoenzyme, the *protein* portion
21.9 *a.* Hydrolase, *b.* lyase, *c.* oxidoreductase, *d.* transferase
21.11 *a.* $E + S \overset{(1)}{\rightleftarrows} [ES] \overset{(2)}{\rightarrow} E + P$; *b.* second step is slow; *c.* [ES]; *d.* when all of the enzyme is tied up in the [ES]
21.13 *Absolute specificity*: single reactions of a single substrate

$$\text{Sucrose} \xrightarrow[H_2O]{\text{sucrase}} \text{glucose} + \text{fructose}$$

Linkage specificity: reaction of a functional group regardless of surrounding structure. Lipase will hydrolyze ester linkage in wax or other lipids.
21.15 *a.* Linkage specificity, *b.* absolute specificity
21.17 These forces hold the substrate and the enzyme together at the active site.
21.19 *a.* In the induced-fit theory, contact of the substrate with enzyme induces the enzyme to undergo a change in shape so that effective binding can occur. In the lock-and-key theory, the enzyme never changes shape; it only accepts a substrate of a certain shape. *b.* Absolute specificity, group specificity, and stereochemical specificity *c.* Linkage
21.21 These reactions usually involve two separate reacting groups or two separate substrates.
21.23 Above 37°C, an enzyme (which is a *protein*) experiences a loss of secondary and tertiary structure. Hence the rate of the reaction decreases.
21.25 *a.* An inhibitor is a substance which *decreases* the activity of a specific enzyme. *b.* pH effects and temperature effects are observed for *all* enzymes. Inhibitors are specific for a certain enzyme or type of enzyme. *c.* DFP (diisopropylfluorophosphate) inhibits acetylcholinesterase; CN^- inhibits cytochromes and hemoglobin.
21.27 *a.* A competitive inhibitor has a structure similar to that of the substrate. It binds to the active site, ties it up, and prevents the substrate from binding there. If the substrate does not bind to the active site, the enzyme-catalyzed reaction is prevented. *b.* The substrate and the inhibitor are both competing for the active site. If the substrate concentration is increased, there is a greater chance that the substrate will bind to the enzyme. Therefore increasing the substrate concentration increases the amount of substrate bound to enzyme and increases the reaction rate.
21.29 *a.* Pb^{2+} denatures the enzyme by disrupting disulfide linkages. *b.* Noncompetitive *c.* If a substance with a high concentration of protein is given (like milk or egg white), the heavy metal will denature that protein before destroying the body's proteins. Vomiting can then be induced to rid the patient of the heavy-metal-protein complex. The chemical principle is Le Chatelier's principle. *d.* Irreversible: the bonding of the inhibitor permanently and irreversibly alters the catalytic site, and the substrate can no longer bind.
21.31 An antimetabolite *competitively* inhibits a microorganism's enzymes. The antimetabolite must have a structure similar to that of the microorganism substrates but must *not* be chemically toxic or affect human enzymes.
21.33 They decompose in acid solutions, such as gastric juice.
21.35 Experimentally you would increase the concentration of substrate; in competitive equilibrium, the reac-

tion rate would *increase* with increasing substrate concentration; in noncompetitive equilibrium it would remain unchanged.

21.37 A zymogen is an inactive precursor of an active enzyme. The zymogen is chemically modified to produce the active enzyme at the right time and the right place in the body.

21.39

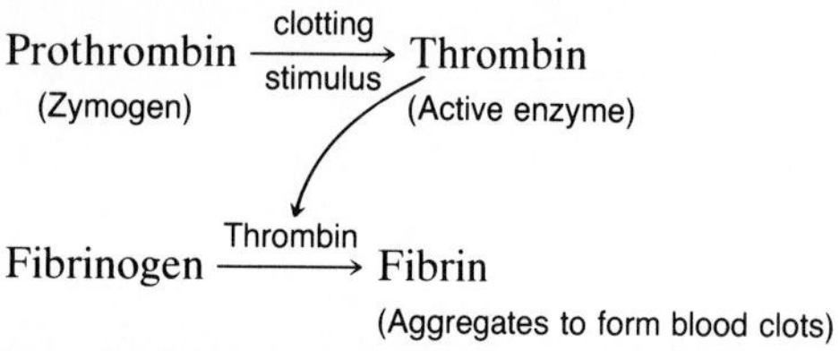
Prothrombin
clotting
stimulus
Thrombin
(Zymogen)
(Active enzyme)
Fibrinogen
Thrombin
Fibrin
(Aggregates to form blood clots)

21.41 No; the inhibition by the product of the metabolic pathway is *noncompetitive,* and increasing the substrate concentration would have *no* effect.

21.43 *a.* Coenzymes allow an enzyme to be catalytically active. They accept and transfer small organic or inorganic fragments. *b.* Coenzyme A: carrier of acyl groups; biocytin: carrier of CO_2 group

21.45 *a.* NAD^+ has been reduced; *b.* HSH has been oxidized; *c.* food

21.47 LDH is present in the cells of the heart. When some of these cells die as a result of the myocardial infarction, the LDH is released into the blood.

21.49 *a.* LDH_1 becomes the largest peak; *b.* LDH_4 and LDH_5 are elevated.

21.51 Hormones are chemical messengers secreted directly into the bloodstream by an endocrine gland. They travel directly to a target cell, where their major function is to activate inactive enzymes. They regulate cell reactions.

21.53 *a.* (*1*) Polypeptide or protein hormones, (*2*) nonpeptide hormones containing amino group, (*3*) lipid soluble or steroid; *b.* (*1*) progesterone: steroid, (*2*) insulin: protein, (*3*) cortisone: steroid, (*4*) thyroxine: nonpeptide hormones containing amino group, (*5*) tyrosine: nonpeptide hormones containing amino group

21.55 Caffeine maintains elevated levels of cAMP and prolongs the effect of epinephrine, a stimulating hormone.

21.57 In each case the physiological function depends on a complementary fit between two molecules: enzyme and substrate; hormone and receptor; antigen and antibody.

CHAPTER 22

22.1 *a.* Exergonic, *b.* exergonic, *c.* endergonic, *d.* endergonic

22.2

Forward (→)	Reverse (←)
a. −10	+10
b. −336	+336
c. +17.8	−17.8
d. +3	−3

22.3 686 kcal

22.4

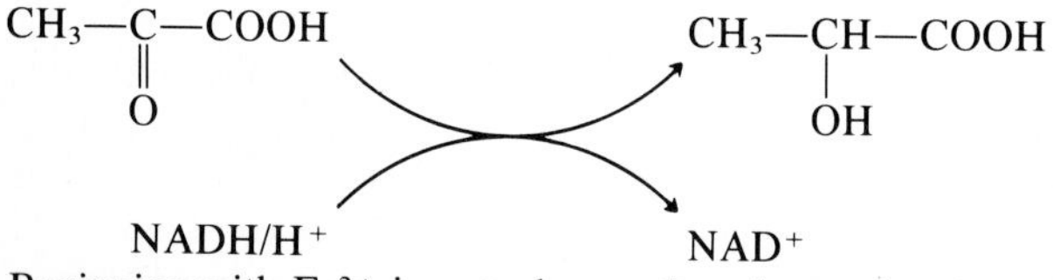
CH₃—C—COOH
O
CH₃—CH—COOH
OH
NADH/H⁺
NAD⁺

22.5 Beginning with Fe^{2+} in cytochrome *b* each step involves the oxidation $Fe^{2+} \rightarrow Fe^{3+}$ in the cytochrome (on the left) giving up electrons and the reduction $Fe^{3+} \rightarrow Fe^{2+}$ in the cytochrome that receives electrons. In the last step iron is oxidized and H^+ is reduced.

22.6 3 mol ATP

22.7 <7.3 kcal/mol is released

22.9 Anabolism and catabolism

22.11 Anabolism: synthetic; catabolism: degradative

22.13 *a* and *b; a* and *c; b* and *d; c* and *d*

22.15

$$CH_3\underset{\displaystyle|}{\overset{OH}{CH}}\text{—}COOH + NAD^+ \longrightarrow CH_3\text{—}\overset{O}{\overset{\|}{C}}\text{—}COOH + NADH/H^+$$

22.17 The sun

22.19 $C_6H_{12}O_6 + 6O_2 \rightarrow 6CO_2 + 6H_2O + 686$ kcal
22.21 Adenine; diphosphate
22.23 Phosphorus-oxygen bonds
22.25 Food molecules provide hydrogen; breathing provides oxygen. The combination of H_2 and O_2 to form H_2O is highly exergonic.
22.27 Hydrogen
22.29 FAD, NAD^+
22.31 Oxidations and reductions
22.33 *a*. oxidation, *b*. reduction, *c*. reduction, *d*. reduction, *e*. oxidation
22.35 Cytochromes
22.37

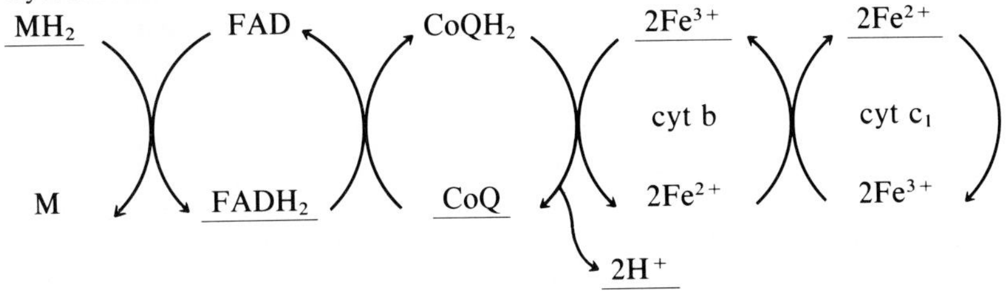

MH2
FAD
CoQH2
2Fe3+
2Fe2+
cyt b
cyt c1
M
FADH2
CoQ
2Fe2+
2Fe3+
2H+

22.39 Three steps
22.41 *a*. 2 mol ATP, *b*. 3 mol ATP, *c*. 4 mol ATP
22.43 Cyanide ion binds to iron ions in cytochromes and inhibits electron transfer in the ETS.
22.45 *a*. Yes. Cleavage of the 1,3-DPGA bond is more exergonic than cleavage of ATP.
b. Substrate-level phosphorylation.
22.47 It is a structural part of coenzyme A.
22.49 Acetyl groups

CHAPTER 23

23.1 Electron transport system
23.2 4 mol ATP
23.3 12 mol ATP
23.4 The cis geometric isomer has a different shape from the trans, which apparently does not fit the enzyme operating at step 9.
23.5 Oxaloacetic acid; citric acid cycle
23.6 3 mol ATP
23.7 Lowers the concentration
23.9 Embden-Meyerhof pathway
23.11 1 mol glucose yields 2 mol pyruvic acid
23.13 *a*. Step 5, *b*. none
23.15 Cytosol
23.17 Oxygen
23.19 By reduction of pyruvic acid to lactic acid. $NADH/H^+$ is the reducing agent and is thereby oxidized to NAD^+.
23.21 *a*. 2 mol ATP, *b*. 2 mol ATP, *c*. 0.4 mol ATP, *d*. 1.0 mol ATP, *e*. 0.2 mol ATP
23.23 2
23.25 *a*. 12 mol ATP, *b*. 12 mol ATP, *c*. 2.4 mol ATP, *d*. 6 mol ATP, *e*. 1.2 mol ATP
23.27 Krebs cycle; tricarboxylic acid cycle
23.29 Acetyl CoA + oxaloacetic acid $\rightarrow$ citric acid + coenzyme A
23.31 In the mitochondria
23.33 Step 4: 3 mol; step 6: 3 mol; step 7: 1 mol; step 8: 2 mol; step 10: 3 mol
23.35 *a*. 36 mol ATP, *b*. 36 mol ATP, *c*. 7.2 mol ATP, *d*. 18 mol ATP, *e*. 3.6 mol ATP
23.37 The liver oxidizes lactic acid to pyruvic acid and converts it to glucose, which then goes to muscle cells to enter the glycolysis pathway.
23.39 Accumulation of lactic acid
23.41 Gluconeogenesis
23.43 It is an intermediate in the gluconeogenesis pathway, pyruvic $\rightarrow$ glucose, but not in the reverse process, glucose $\rightarrow$ pyruvic.

23.45 Similar terminal points

$$\text{Glucose} \underset{\text{gluconeogenesis}}{\overset{\text{glycolysis}}{\rightleftarrows}} \text{pyruvic acid}$$

Differences: Intermediacy of oxaloacetic acid in gluconeogenesis and not in glycolysis. Different enzymes in several steps.

23.47 See Figure 22.5. NAD^+ is converted to $NADP^+$ by phosphorylation of an —OH group

23.49 70 to 110 mg glucose per 100 mL of blood

23.51 Fasting, especially omission of carbohydrates from the diet

23.53 Diabetes mellitus

23.55 It stimulates glycogenolysis providing glucose as an energy source to meet an emergency.

23.57 Uridine phosphate glucose

23.59 Glycogen; approximately 1 day

CHAPTER 24

24.1 20 mol ATP

24.2 *a*. 8, *b*. 9, *c*. 6

24.3 $CH_3(CH_2)_{16}—C(=O)—S—CoA$ $\quad$ $CH_3(CH_2)_{14}—C(=O)—S—CoA$

24.4

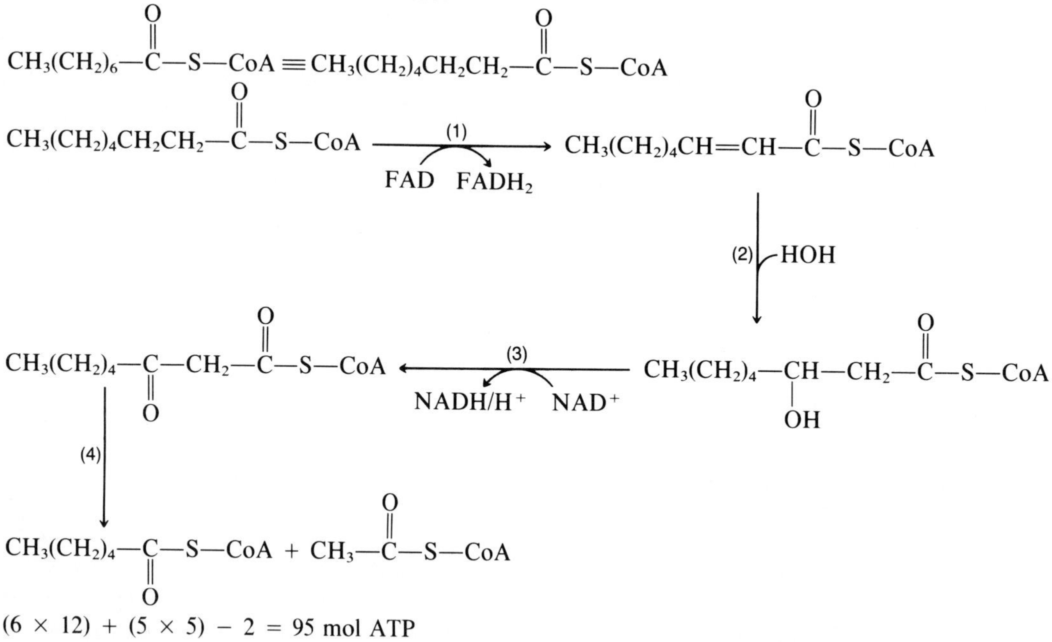

24.5 $(6 \times 12) + (5 \times 5) - 2 = 95$ mol ATP

24.6

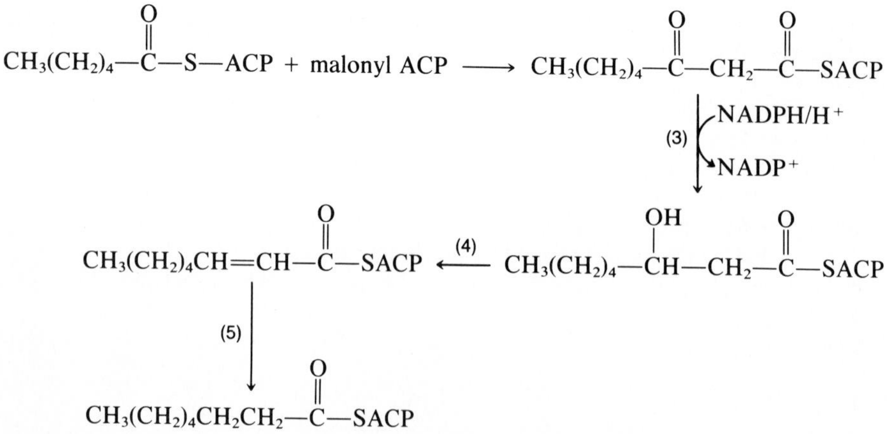

24.7 Diacylglycerols, monoacylglycerols, fatty acids and glycerol
24.9 Adipose tissue
24.11 Acetyl CoA
24.13 Fatty acids have even numbers of carbon atoms so that they are metabolized to 2-carbon acetyl units without leftover carbons.

24.15 *a.* $CH_3\overset{\beta}{C}H{=}\overset{\alpha}{C}H{-}COOH$ *b.* $HOOC{-}\overset{\alpha}{C}H_2{-}\underset{\underset{O}{\|}}{\overset{\beta}{C}}{-}CH_2CH_3$

c. $CH_3(CH_2)_8\overset{\beta}{C}HOH{-}\overset{\alpha}{C}H_2{-}\overset{\overset{O}{\|}}{C}{-}OH$ *d.* $CH_3(CH_2)_4\overset{\beta}{C}H_2\overset{\alpha}{C}H_2{-}COOH$

24.17 $CH_3(CH_2)_8{-}\overset{\overset{O}{\|}}{C}{-}S{-}CoA$ $CH_3(CH_2)_{12}{-}\overset{\overset{O}{\|}}{C}{-}S{-}CoA$
Capryl CoA Myristyl CoA

24.19 Cytosol
24.21 After activation, it undergoes reversible attachment to carnitine which carries it across the membrane
24.23 Because of the oxidation of an alcohol group to a ketone group at the β position.

24.25 *a.* $CH_3(CH_2)_6CH_2CH_2{-}\overset{\overset{O}{\|}}{C}{-}S{-}CoA$ *b.* $CH_3(CH_2)_6CHOHCH_2{-}\overset{\overset{O}{\|}}{C}{-}S{-}CoA$

c. $CH_3(CH_2)_6CHOHCH_2{-}\overset{\overset{O}{\|}}{C}{-}S{-}CoA$ *d.* $CH_3(CH_2)_6{-}\overset{\overset{O}{\|}}{C}{-}S{-}CoA + CH_3{-}\overset{\overset{O}{\|}}{C}{-}S{-}CoA$

24.27 FAD is the coenzyme specific for the oxidation of C—C bonds. NADH/H^+ is specific for C—O bonds.

24.29
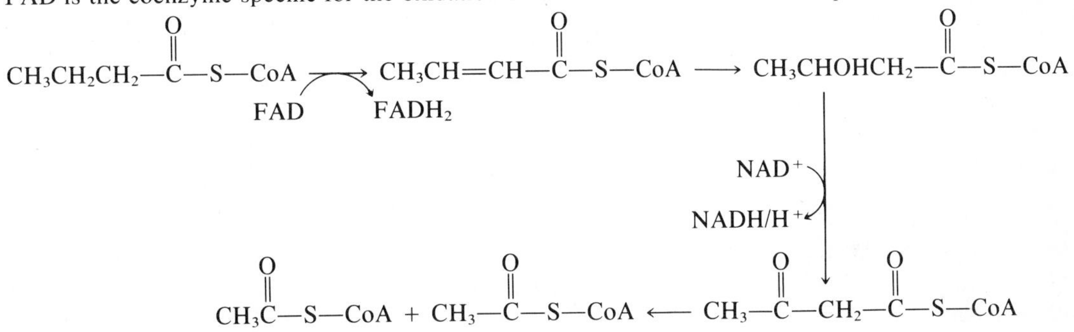

CH3CH2CH2—C(=O)—S—CoA
FAD
FADH2
CH3CH=CH—C(=O)—S—CoA
CH3CHOHCH2—C(=O)—S—CoA
NAD+
NADH/H+
CH3C(=O)—S—CoA + CH3—C(=O)—S—CoA
CH3—C(=O)—CH2—C(=O)—S—CoA

24.31 7 mol acetyl CoA through six turns
24.33 12 mol ATP
24.35 $(7 \times 12) + (6 \times 5) - 2 = 112$ mol ATP
24.37 More energy (ATP) is obtained per gram of fat than per gram of carbohydrate.
24.39 *a.* Fatty acid spiral in mitochondria; lipogenesis in the cytosol *b.* CO_2 is added in step 1 and lost in step 2. *c.* The carriers are thioesters of coenzyme A for the fatty acid spiral and thioesters of acyl carrier protein for lipogenesis. *d.* As stated in part *c*, the thioester carriers are derived from different coenzymes. *e.* The addition of CO_2 produces an odd-numbered carbon chain, malonyl CoA in step 1. In step 2 malonyl CoA reacts with acetyl CoA with loss of CO_2.
24.41 It combines with fatty acids to form triacyglycerols.

24.43 $CH_3{-}\overset{\overset{O}{\|}}{C}{-}CH_2COOH$ $CH_3{-}\overset{\overset{OH}{|}}{C}H{-}CH_2{-}COOH$ $CH_3{-}\overset{\overset{O}{\|}}{C}{-}CH_3$ ketone bodies

24.45 There is no ketone functional group.
24.47 No, it normally occurs to a certain extent. Only excess production is abnormal.
24.49 Insulin
24.51 Elevated blood sugar level
24.53 Allows glucose to enter cells, stimulates glycogenesis, and suppresses gluconeogenesis.
24.55 Converts it to ketone bodies
24.57 Through ketosis leading to acidosis which interfers with oxygen transport by hemoglobin
24.59 Acetyl CoA

CHAPTER 25

25.1 Phosphate esterification and oxidation of an alcohol to an aldehyde.

25.2

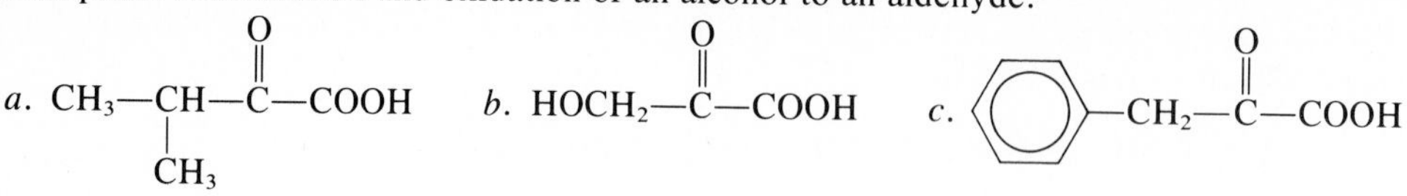

25.3

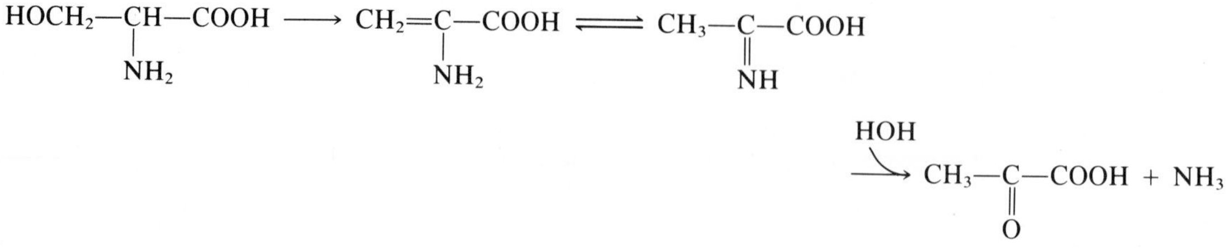

25.4

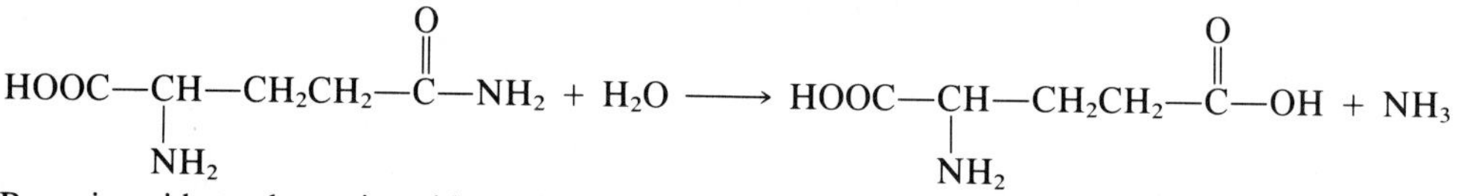

25.5 Pyruvic acid, oxaloacetic acid, oxalosuccinic acid, α-ketoglutaric acid

25.6 Alanine is formed by transamination between pyruvic acid and glutamic acids.

25.7 Structurally the body is largely protein material, as are enzymes. These materials must continually be made by anabolic processes. Catabolism to yield energy is a minor event for proteins. Most bodily energy is derived from carbohydrates and fats.

25.9 Bacteria in soil convert N_2 from the air into nitrogen compounds, which plants use to make amino acids and proteins. Animals eat plants. Waste products of animals and decayed plants, aided by denitrifying bacteria, release nitrogen back into the air.

25.11 Because it is the source of all nitrogen needed by the body to synthesize both protein and nonprotein compounds

25.13 Protein is constantly being broken down and resynthesized.

25.15 Their nitrogen intake is greater than nitrogen output because they are building new tissue as they grow.

25.17 Pyridoxal phosphate, a derivative of vitamin B_6, is a coenzyme for transamination.

25.19 Glutamic acid

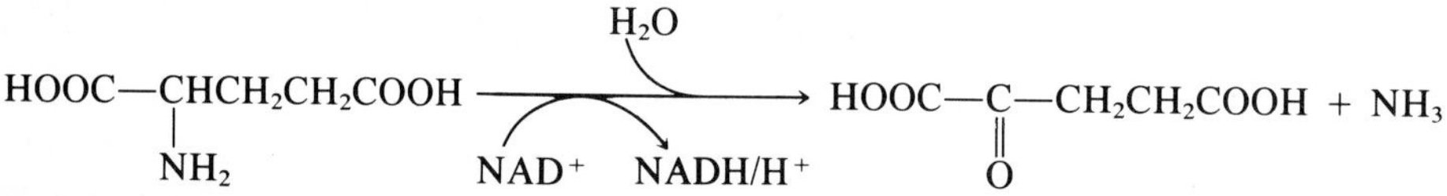

25.21 It disrupts the citric acid cycle, and it is quite basic.

25.23 Urea is nontoxic and very soluble in water so that only a minimum amount of water is lost in its excretion. NH_3 requires a large volume of water to dilute its toxic effects.

25.25 Glutamine is the amide of glutamic acid.

25.27 CO_2, urea

25.29 *a*. Addition (hydration), *b*. oxidation, *c*. transamination

25.31 Amino acids are converted by transamination to α-ketoacids, many of which are intermediates in carbohydrate metabolism.

25.33 Energy

25.35 No

25.37 If excessive amounts of amino acids are deaminated for use of their carbon skeletons for energy, excretion of nitrogen will exceed intake, i.e., negative nitrogen balance will occur.

25.39 Uric acid

25.41 Bile pigments

25.43 Excessive breakdown of red blood cells; obstruction of the elimination of bile pigments in urine and feces

CHAPTER 26

26.1 1.2 mg
26.2 1728 (1.7×10^3) cal/day
26.3 The RDAs necessary for a 10-year-old; a cross section of food from the groups in Figure 26.2 that will contain these RDAs; the child's particular likes and dislikes among foods
26.5 *a.* By using values greater than the minimum requirements and classifying individuals in different age brackets and by sex *b.* No, they are allowances which are sufficient to meet the needs of almost all persons in a given category.
26.7 *a.* Daily allowance of nutrients sufficient for most individuals in a category, *b.* dietary sources of the nutrients
26.9 0.38 mg
26.11 Maintenance of heart beat, breathing, brain and nerve activity, kidney action, maintenance of body temperature.
26.13 *a.* 3.9 h, *b.* 1 h just for the meat, not counting the bun
26.15 Most of the weight loss is water through perspiration, not a loss of adipose tissue. This weight is quickly regained.
26.17 *a.* Input and output of nitrogen are equal; *b.* growing children, pregnant and lactating women; *c.* body protein is being broken down more rapidly than it is being made.
26.19 *a*
26.21 Legumes
26.23 They do not appear to be absolutely essential to the diet.
26.25 (*1*) Excess breakdown of fat in the body can lead to ketosis and acidosis. (*2*) Carbohydrates are a source of dietary fiber. (*3*) Excess protein breakdown can occur to replace the missing carbohydrate energy source.
26.27 By experiment; food is eaten under controlled conditions and blood sugar level is monitored.
26.29 Almost all lipids can be made from carbohydrate sources.
26.31 *a.* HDL transfers cholesterol from arteries to the liver for disposal. *b.* Increase polyunsaturated fat in diet and decrease saturated fats
26.33 *a.* Organic substances required in the diet in small amounts; *b.* no, different species have different abilities to synthesize nutrients; *c.* fat-soluble and water-soluble; *d.* vitamin C, 60 mg
26.35 *a.* Niacin, B_6; *b.* niacin, folacin, biotin, pantothenic acid; *c.* niacinamide, folacin, B_{12}, biotin, pantothenic acid, riboflavin
26.37 No; the water-soluble vitamins are present in amounts 2 to 10 times that required, but the excess does not matter since they are water-soluble. However, the tenfold excess of vitamin A and twentyfold excess of vitamin E could lead to accumulation of these fat-soluble vitamins and possible toxicity.
26.39 A protein in raw eggs binds biotin and makes it unusable. This protein is denatured by cooking.
26.41 *a.* Nonpolar; *b.* Fat-soluble structures are almost all hydrocarbon in nature with only a small oxygen content. Water-soluble vitamins have more functional groups with O and N atoms.
26.43 Vitamin E is an antioxidant. It may slow down cell aging by stopping oxidation reactions and the formation of free radicals.
26.45 No, potassium ions are needed in small amounts.
26.47 To avoid cardiac problems arising from depletion of K^+ along with Na^+ ions.
26.49 Dietary deficiency means that not enough of the mineral is ingested. In some people when even sufficient mineral is ingested, it is not absorbed through the intestines into the bloodstream.
26.51 Both are needed for the compound, calcium hydroxapatite, which is the major structural component of bones and teeth.
26.53 *a.* Calcium, *b.* Iron, *c.* Iodine, *d.* Zinc

Page numbers in **boldface** indicate definitions; page numbers in *italic* indicate tables or illustrations.